HANDBOOK OF BRAIN TUMOR
CHEMOTHERAPY

HANDBOOK OF BRAIN TUMOR CHEMOTHERAPY

Edited by

HERBERT B. NEWTON, M.D., FAAN

Professor of Neurology & Oncology
Director, Division of Neuro-Oncology
Esther Dardinger Chair in Neuro-Oncology
Co-Director, Dardinger Neuro-Oncology Center
Ohio State University Medical Center
James Cancer Hospital & Solove Research Institute

ELSEVIER

AMSTERDAM • BOSTON • HEIDELBERG • LONDON • NEW YORK • OXFORD • PARIS
SAN DIEGO • SAN FRANCISCO • SINGAPORE • SYDNEY • TOKYO
2006

ACADEMIC
PRESS

ELSEVIER B.V.
Radarweg 29
P.O. Box 211, 1000 AE Amsterdam
The Netherlands

ELSEVIER Inc.
525 B Street, Suite 1900
San Diego, CA 92101-4495
USA

ELSEVIER Ltd
The Boulevard, Langford Lane
Kidlington, Oxford OX5 1GB
UK

ELSEVIER Ltd
84 Theobalds Road
London WC1X 8RR
UK

First edition 2006

Library of Congress Cataloging in Publication Data
A catalog record is available from the Library of Congress.

British Library Cataloguing in Publication Data
A catalogue record is available from the British Library.

ISBN: 0-12-088410-0/978-0-12-088410-0

⊚ The paper used in this publication meets the requirements of ANSI/NISO Z39.48-1992 (Permanence of Paper).
Printed in the United States of America.

Dedication

I would like to thank my wife, Cheryl, and my two children, Alex and Ashley, for their unconditional love, patience, and support while I worked on this book.

I would also like to thank Dr. Harry Greenberg, for his mentoring and friendship over the years, and for initially pointing me in the right direction.

Finally, I would like to thank my Neuro-Oncology patients and their families, for their courage and strength during adversity. They are a constant inspiration in the battle against this dreaded disease.

Herbert B. Newton

Preface

The treatment of patients with brain tumors remains a challenge that is unique in modern oncology. No other target organ is as delicate and fragile as the brain, where even a small tumor, when localized within a region of eloquent neural tissue (e.g., brain stem, primary motor, or speech cortex), can have devastating consequences on a patient's neurological function and quality of life. Because of this fragility, attempts at surgical resection are often limited to minimize further neurological injury, thereby leaving behind millions of viable tumor cells. In addition, the infiltrative and diffuse nature of many of these tumors also reduces the ability of the neurosurgeon to perform an extensive resection. These limitations place a heavy burden on other forms of treatment, such as irradiation and chemotherapy, to control residual disease. The efficacy of chemotherapy against most brain tumors has been modest. However, recent advances in Neuro-Oncology suggest that the role of chemotherapy is expanding and that it will become a more important aspect of multi-modality treatment.

The inspiration for this book was slowly realized over the past few years as I followed the explosive growth in the field of Neuro-Oncology and the broadening application of conventional and molecular forms of chemotherapy to brain tumor patients. I felt there was a growing need for a single source, comprehensive, reference handbook that would encompass the most up to date information regarding conventional forms of cytotoxic chemotherapy, as well as the basic science and clinical application of molecular therapeutics, in this group of patients. With the publication of this book, an important void has now been filled.

This book should have broad appeal to anyone interested in the field of Neuro-Oncology and, in particular, to the treatment of patients with brain tumors. It will satisfy clinicians interested in a thorough overview of the use of chemotherapy in patients with a broad range of brain tumors and, in addition, can serve as a source of background information to basic scientists and pharmaceutical researchers with an interest in the molecular therapeutics of brain tumors. I sincerely hope that the world-class group of authors assembled herein can assist you in providing the best oncologic care for your brain tumor patients.

Herbert B. Newton, M. D., FAAN
Columbus, Ohio 2005

Contents

Contributors

Lauren E. Abrey (395) Department of Neurology, Memorial Sloan-Kettering Cancer Center, New York, NY, USA

Till Acker (219) Karolinska Institute, CMB, Stockholm, Sweden

Eric Amundson (274) Department of Neurological Surgery, The Johns Hopkins University School of Medicine, Baltimore, MD, USA

Kaveh Asadi-Moghaddam (332) Department of Neurological Surgery, The Ohio State University Medical Center, Columbus, OH, USA

Michael E. Berens (115) TGen, The Translational Genomics Research Institute, Phoenix, AZ, USA

Henry Brem (274) Departments of Neurological Surgery, Oncology, and Opthalmology, The Johns Hopkins University School of Medicine, Baltimore, MD, USA

William C. Broaddus (295) Department of Neurosurgery, Medical College of Virginia Hospitals, Richmond, VA, USA

Anna Butturini (305) Children's Hospital Los Angeles and the Norris Comprehensive Cancer Center, Keck School of Medicine, University of Southern California, Los Angeles, CA, USA

Nicholas Butowski (105) Department of Neuro-Oncology, University of California, San Francisco, San Francisco, CA, USA

Matthew Carabasi (305) Children's Hospital Los Angeles and the Norris Comprehensive Cancer Center, Keck School of Medicine, University of Southern California, Los Angeles, CA, USA

Marc C. Chamberlain (316) Department of Interdisciplinary Oncology, University of South Florida, Tampa, FL, USA

Susan Chang (105) Department of Neuro-Oncology, University of California, San Francisco, San Francisco, CA, USA

Mike Yue Chen (295) Department of Neurosurgery, Medical College of Virginia Hospitals, Richmond, VA, USA

Zhi-jian Chen (295) Department of Neurosurgery, Medical College of Virginia Hospitals, Richmond, VA, USA

Antonio E. Chiocca (332) Department of Neurological Surgery, The Ohio State University Medical Center, Columbus, OH, USA

Tim Demuth (115) TGen, The Translational Genomics Research Institute, Phoenix, AZ, USA

Nancy D. Doolittle (262) Department of Neurology, Oregon Health & Science University, Portland, OR, USA

Michael Dorsi (274) Department of Neurological Surgery, The Johns Hopkins University School of Medicine, Baltimore, MD, USA

Patricia K. Duffner (490) Departments of Pediatrics and Neurology, State University of New York at Buffalo School of Medicine and The Women and Children's Hospital of Buffalo, Buffalo, NY, USA

Francois G. El Kamar (395) Department of Neurology, Memorial Sloan-Kettering Cancer Center, New York, NY, USA

Herbert H. Engelhard (236) Departments of Neurosurgery and Pathology, The University of Illinois at Chicago, Chicago, IL, USA

Christopher Fahey (371) Department of Neurology, Northwestern University, Evanston, IL, USA

Jonathan L. Finlay (305) Children's Hospital Los Angeles and the Norris Comprehensive Cancer

Center, Keck School of Medicine, University of Southern California, Los Angeles, CA, USA

Karen L. Fink (44) Baylor University Medical Center, Dallas, TX, USA

Lorna K. Fitzpatrick (490) Departments of Pediatrics and Neurology, State University of New York at Buffalo School of Medicine and The Women and Children's Hospital of Buffalo, Buffalo, NY, USA

George T. Gillies (295) Department of Neurosurgery, Medical College of Virginia Hospitals, Richmond, VA, USA

Abhijit Guha (173) Division of Neurosurgery and the Arthur and Sonia Labatts Brain Tumor Research Center, Hospital for Sick Children, University of Toronto, Toronto, ON, Canada

Peter J. Haar (295) Department of Neurosurgery, Medical College of Virginia Hospitals, Richmond, VA, USA

Daphne A. Haas-Kogan (185) Department of Radiation Oncology, University of California, San Francisco, CA, USA

Raqeeb M. Haque (274) Department of Neurological Surgery, The Johns Hopkins University School of Medicine, Baltimore, MD, USA

Stacey M. Ivanchuk (123) The Division of Neurosurgery and The Arthur and Sonia Labatt Brain Tumor Research Centre, The Hospital for Sick Children, Toronto, ON, Canada

Mark T. Jennings (448) Department of Pediatrics and Neurology, University of Illinois College of Medicine, Peoria, IL, USA

Mark G. Malkin (364, 426) Department of Neurology, Medical College of Wisconsin, Milwaukee, WI, USA

Tom Mikkelsen (193) Department of Neurosurgery, Henry Ford Hospital, Detroit, MI, USA

Nimish Mohile (432) Department of Neurology, Northwestern Univesity, Chicago, IL, USA

Joydeep Mukherjee (173) Arthur and Sonia Labatts Brain Tumor Center, Hospital for Sick Children, University of Toronto, Toronto, ON, Canada

Tulio P. Murillo (262) Departments of Neurology and Neurosurgery, Oregon Health & Science University, Portland, OR, USA

Jean L. Nakamura (185) Department of Radiation Oncology, University of California, San Francisco, CA, USA

Edward A. Neuwelt (262) Departments of Neurology and Neurosurgery, Oregon Health & Science University, Portland, OR, USA

Herbert B. Newton (3, 21, 247, 347, 407, 439, 463, 475) Division of Neuro-Oncology, Dardinger Neuro-Oncology Center, The Ohio State University Medical Center and James Cancer Hospital and Solove Research Institute, Columbus, OH, USA

Nina A. Paleologos (371, 382) Evanston Northwestern Healthcare and Feinberg School of Medicine, Northwestern University, Evanston, IL, USA

Karl H. Plate (219) Institute of Neurology (Edinger Institute), Johann-Wolfgang Goethe University, Frankfurt, Germany

Ian F. Pollack (155) Department of Neurosurgery, Children's Hospital of Pittsburgh, University of Pittsburgh School of Medicine, Pittsburgh, PA, USA

Scott L. Pomeroy (74) Department of Neurology, Children's Hospital Boston, Boston, MA, USA

Jeffrey J. Raizer (432) Department of Neurology, Northwestern University, Chicago, IL, USA

Abhik Ray-Chaudhury (3) Department of Pathology, The Ohio State University Medical Center and James Cancer Hospital and Solove Research Institute, Columbus, OH, USA

Sandra A. Rempel (193) Department of Neurosurgery, Henry Ford Hospital, Detroit, MI, USA

James T. Rutka (123) The Division of Neurosurgery and The Arthur and Sonia Labatt Brain Tumor Research Centre, The Hospital for Sick Children, Toronto, ON, Canada

Adrienne C. Scheck (89) Ina Levine Brain Tumor Center, Neuro-Oncology and Neurosurgery Research, Barrow Neurological Institute of SJHMC, Phoenix, AZ, USA

Joachim P. Steinbach (141) Department of General Neurology, Hertie Institute for Clinical Brain Research, University of Tübingen, Tübingen, Germany

Beverly A. Teicher (58) Genzyme Corporation, Framingham, MA, USA

Nicole J. Ullrich (74) Department of Neurology, Children's Hospital Boston, Boston, MA, USA

Tibor Valyi-Nagy (236) Departments of Neurosurgery and Pathology, The University of Illinois at Chicago, Chicago, IL, USA

Allison L. Weathers (382) Department of Neurological Sciences, Rush University, Chicago, IL, USA

Michael Weller (141) Department of General Neurology, Hertie Institute for Clinical Brain Research, University of Tübingen, Tübingen, Germany

PHARMACOLOGY AND CLINICAL APPLICATIONS

1

Overview of Brain Tumor Epidemiology and Histopathology

Herbert B. Newton and Abhik Ray-Chaudhury

ABSTRACT: Primary brain tumors (PBT) comprise a diverse group of neoplasms that are often malignant and refractory to treatment. Between 30 000 and 35 000 new PBT are diagnosed each year in the USA (approximately 14 per 100 000). Tumors of neuroepithelial origin are the largest histological class of PBT and include the glioma sub-group (e.g., glioblastoma multiforme, anaplastic astrocytoma, oligodendroglioma), which represent the most frequently diagnosed tumors in adults. Other important tumors in adults include meningiomas, primary brain lymphoma, and oligoastrocytoma. Commonly diagnosed tumors in children include medulloblastoma, cerebellar astrocytoma, and optic pathway glioma. This chapter will review the microscopic and molecular pathology of PBT that are most likely to require treatment with chemotherapy, utilizing the classification system of the World Health Organization.

EPIDEMIOLOGY OF BRAIN TUMORS

Brain tumors remain a significant health problem in the USA and worldwide. Overall, they comprise some of the most malignant tumors known to affect humans and are generally refractory to all modalities of treatment. It is estimated that between 30 000 and 35 000 new cases of primary brain tumors (PBT) will be diagnosed in the upcoming year in the USA (1–2 per cent of newly diagnosed cancers overall) [1–6]. Metastatic brain tumors (MBT) are even more common and affect between 100 000 and 150 000 new patients each year in this country [7]. Most studies suggest that approximately 14 per 100 000 people in

the USA will be diagnosed with a PBT each year. Among this cohort with newly diagnosed tumors, 6 to 8 per 100 000 will have a high-grade neoplasm. Recent epidemiological studies suggest an increasing incidence rate for development of PBT in children less than 14 years of age and in patients 70 years or older [8]. For people in the 15- to 44-year-old age group, the overall incidence rates have remained fairly stable. The cause of the increased incidence of PBT in some age groups remains unclear, but may be due to improvements in diagnostic neuro-imaging such as magnetic resonance imaging (MRI), greater availability of neurosurgeons, improved patterns of access to medical care for children and elderly patients, and more aggressive approaches to health care for elderly patients [5,8].

The prognosis and survival of patients with brain tumors remains poor [1–7]. Although an uncommon neoplasm, PBT are among the top 10 causes of cancer-related deaths in the USA and account for 2.4 per cent of all yearly cancer-related deaths [9]. The median survival for a patient with glioblastoma multiforme (GBM) is approximately 12–14 months, and has not improved substantially over the past 30 years. For patients with a low-grade astrocytoma or oligodendroglioma, the median survival is still significantly curtailed and is about 6–10 years. For PBT patients in the USA as a whole, across all age groups and tumor types, the 5-year survival rate is 20 per cent [3]. If a patient with a PBT survives for an initial 2 years, the probability of surviving another 3 years is 76.2 per cent. In general, for any given tumor type, survival is better for younger patients than for older patients. The only exception to this generalization is for children with medulloblastoma and embryonal tumors, in which patients under three years of age have poorer survival rates than children between 3 and 14 years of age [10]. The 5-year survival rate for all children less than 14 years of age with a malignant PBT is 72 per cent.

The median age for diagnosis of PBT is between 54 and 58 years [1–6]. Among different histological varieties of PBT, there is significant variability in the age of onset. A small secondary peak is also present in the pediatric age group, in children between the ages of 4 and 9. Overall, PBT are more common in males than females, with the exception of meningiomas, which are almost twice as common in females. Tumors of the sellar region, and of the cranial and spinal nerves, are almost equally represented among males and females. In the USA, gliomas are more commonly diagnosed in Whites than Blacks, while the incidence of meningiomas is relatively equal between the two groups.

Numerous epidemiological studies have been performed in an attempt to define risk factors involved in the development of brain tumors (see Table 1.1) [2–6]. The vast majority of these potential risk factors have not been associated with any

TABLE 1.1 Risk Factors that have been Investigated in Epidemiological Studies of Primary Brain Tumors

Hereditary syndromes (proven): tuberous sclerosis, neurofibromatosis types 1 and 2, nevoid basal cell carcinoma syndrome, Turcot's syndrome, and Li-Fraumeni syndrome

Family History of brain tumors

Constitutive polymorphisms: glutathione transferases, cytochrome P-450 2D6 and 1A1, N-acetyltransferase, and other carcinogen metabolizing, DNA repair, and immune function genes

History of prior cancer

Exposure to infectious agents

Allergies (possible reduced risk)

Head trauma

Drugs and medications

Dietary history: N-nitroso compounds, oxidants, antioxidants

Tobacco usage

Alcohol consumption

Ionizing radiation exposure (proven)

Occupational and industrial chemical exposures: pesticides, vinyl chloride, synthetic rubber manufacturing, petroleum refining and production, agricultural workers, lubricating oils, organic solvents, formaldehyde, acrylonitrile, phenols, polycyclic aromatic hydrocarbons

Cellular telephones

Power frequency electromagnetic field exposure

Data adapted from references [2–6,11–23]

significant predisposition to brain tumors. One risk factor that has proven to be important is the presence of a hereditary syndrome with a genetic predisposition for developing tumors, some of which can affect the nervous system [4,5,11]. Several hereditary syndromes are associated with PBT, including tuberous sclerosis, neurofibromatosis types 1 and 2, nevoid basal cell carcinoma syndrome, Li-Fraumeni syndrome, and Turcot's syndrome. However, it is estimated that hereditary genetic predisposition may be involved in only 2–8 per cent of all cases of PBT. Familial aggregation of brain tumors has also been studied, with conflicting results [5,11]. The relative risk for developing a tumor among family members of a patient with a PBT are quite variable and range from 1 to 10. One study that performed a segregation analysis of families of more than 600 adult glioma patients showed that a polygenic model most accurately explained the inheritance pattern [12]. A similar analysis of 2141 first-degree relatives of 297 glioma families did not reject a multifactorial model, but concluded that an autosomal recessive model fit the inheritance pattern more accurately [13]. Critics of these studies suggest that the common exposure of a family to a similar pattern of environmental agents could lead to a similar clustering of tumors. Other investigators have focused on genetic polymorphisms that might influence genetic and environmental factors to increase the risk for a brain tumor [4,5]. Alterations in genes involved in oxidative metabolism, detoxification of carcinogens, DNA stability and repair, and immune responses might confer a genetic predisposition to tumors. For example, Elexpuru-Camiruaga and colleagues demonstrated that cytochrome P-4502D6 and glutathione transferase theta were associated with an increased risk for brain tumors [14]. Other studies have not supported these results, but have found an increased risk for rapid N-acetyltransferase acetylation and intermediate acetylation [15]. In general, further studies with larger cohorts of patients will be necessary to determine if genetic polymorphisms of key metabolic enzyme systems play a significant role in the risk for developing a brain tumor.

Cranial exposure to therapeutic ionizing radiation is a potent risk factor for subsequent development of a brain tumor, and is known to occur after a wide range of exposures [1–6]. Application of low doses of irradiation (1000 to 2000 cGy), such as for children with tinea capitis or skin hemangiomas, have been associated with relative risks of 18 for nerve sheath tumors, 10 for meningiomas, and 3 for gliomas [5,16]. Gliomas and other PBT are also known to occur after radiotherapy for diseases such as leukemia, lymphoma, and head and neck cancers [5,17,18]. In addition, alternative methods of radiation exposure, such as nuclear bomb blasts and employment at nuclear production facilities, have also been implicated as significant risk factors for development of brain tumors [19,20].

Many other risk factors have been evaluated for their potential role in the genesis of brain tumors [1–6]. The majority of these factors have been proven to have little, if any, relationship to brain tumor development or have an indeterminate association due to a mixture of positive and negative studies. Factors in this category include the history of a prior primary systemic malignancy, head injury, prenatal or premorbid ingestion of various types of medications, exposure to viruses and other types of infection (except for the human immunodeficiency virus, which is known to be associated with brain lymphoma), dietary history (i.e., ingestion of N-nitroso compounds, oxidants, and antioxidants), alcohol ingestion, smoking tobacco, cellular telephone use, residential chemical exposures, and proximity to electromagnetic fields. The relationship between industrial and occupational chemical exposures and brain tumors is very complex and remains unclear [2,4,5]. Numerous chemicals that workers are potentially exposed to are carcinogenic or neurotoxic, or both, and include lubricating oils, organic solvents, formaldehyde, acrylonitrile, phenols and phenolic-based compounds, vinyl chloride, and polycyclic aromatic hydrocarbons. Pre-clinical studies have proven the ability of vinyl chloride to induce brain tumors in rat models and some studies suggest an increased risk for chemical workers that handle this compound [21]. However, more recent and extensive analyses suggest that the relationship between vinyl chloride exposure and brain tumors remains inconclusive [22]. Similar inconclusive results for other chemicals are common in the epidemiology literature and demonstrate the difficulty of proving an association between workplace exposures and an uncommon form of cancer. At this time, no definitive associations have been proven between brain tumors and any specific chemicals found in the occupational or industrial setting, including those that are known to be definite or putative carcinogens. Of all the potential risk factors studied, the only one that might be associated with a protective effect for developing a brain tumor is the presence of an allergy [23]. The presence of any form of allergy was inversely associated with the development of a glioma (OR = 0.7), but not with meningiomas or acoustic neuromas. Similar inverse associations were noted for the presence of autoimmune diseases and the presence of both

gliomas and meningioma. The authors suggested that allergy-related immunological factors might play a protective role in the genesis of certain brain tumors.

PATHOLOGY OF CENTRAL NERVOUS SYSTEM TUMORS—INTRODUCTION

Clinical neuro-oncology and the application of appropriate therapeutic strategies are dependent on knowing the type of tumor affecting a given patient. Therefore, it is critical that the neuro-oncology treatment team have a close working relationship with the clinical neuropathologist, who can perform a histological review of biopsied or resected tissues and make a pathological diagnosis, consisting of a tumor classification and grade. In addition to assisting with treatment decisions, the tumor classification and grade provide important information regarding prognosis. This chapter will follow the World Health Organization (WHO) classification, as outlined in the recent monograph by Kleihues and Cavanee, that separates nervous system tumors into different nosological entities and assigns a grade of I to IV to each lesion (see Table 1.2), with grade I being biologically benign and grade IV being biologically most malignant and having the worst prognosis [24,25].

The most common method of histopathological examination remains light microscopic analysis of brain tumor tissue after tinctorial staining with hematoxylin and eosin (H&E) [26]. Analysis of specimens after H&E staining is an excellent method for delineating the general histological features of a given tumor. Other tinctorial stains that may be helpful for distinguishing specific aspects of different tumor types include the reticulin stain (defines the presence of reticulin frameworks around blood vessels and also groups or individual tumor cells), periodic acid-Schiff stain (defines the presence of glycogen or mucins),

TABLE 1.2   WHO Classification: Tumors of the Central Nervous System

Tumors of Neuroepithelial Tissue

Tumors of Cranial Nerves and Spinal Nerves

Tumors of the Meninges

Lymphomas and Hemopoietic Neoplasms

Germ Cell Tumors

Tumors of the Sellar Region

Cysts and Tumor-like Lesions

Metastatic Tumors

Singh and Masson-Fontana stains (define the presence of melanin), and the luxol fast blue stain (defines the presence of myelin). Other techniques that are frequently applied to brain tumor diagnosis include immunohistochemical staining and cell proliferation indices. Immunohistochemical staining is used to investigate cellular differentiation in tumors and often recognizes different classes of intermediate filaments, including glial fibrillary acidic protein (GFAP), neurofilament protein, cytokeratins, and desmin [27]. GFAP is helpful to define tumors of glial origin, including astrocytomas, ependymomas, oligodendrogliomas, medulloblastomas and, to a lesser degree, choroid plexus tumors. Staining with GFAP is typically more prominent in the lower grade tumors (i.e., grade I/II). Cytokeratin staining (e.g., AE1/AE3, CK7, CK20) is useful for discriminating poorly differentiated MBT and primary CNS tumors. Among PBT, cytokeratin staining can also be noted in meningiomas and chordomas and, rarely, in focal regions of GBM and gliosarcoma with squamous-cell morphology. Other helpful stains include synaptophysin (tumors of neuronal or neuroendocrine origin), neuron-specific enolase (tumors of neuronal or neuroendocrine origin), S-100 protein (melanoma, nerve sheath tumors), and proteins specific for germ cell tumors, such as placental alkaline phosphatase, alpha-fetoprotein, beta-human chorionic gonadotropin, and carcinoembryonic antigen. For a more detailed listing of brain tumors and their associated immunohistochemical staining patterns, see Table 1.3 In addition to antibodies employed to define cellular differentiation, other antibodies can be applied to detect the percentage of proliferating cells of brain tumors, such as Ki-67 and proliferating cell nuclear antigen (PCNA) [28,29].

PATHOLOGY OF TUMORS OF NEUROEPITHELIAL ORIGIN

Tumors of neuroepithelial origin comprise a large and diverse group of neoplasms, with a mixture of benign and malignant tumor types (see Table 1.4) [24,25,30]. Gliomas (e.g., GBM, AA, oligodendrogliomas, medulloblastoma) are the largest sub-group within the neuroepithelial class of neoplasms, and are also the most common type of PBT. Tumors of neuroepithelial origin, and gliomas in particular, can grow diffusely within the brain or be more circumscribed. Diffusely growing tumors are most common and include the astrocytomas, oligodendrogliomas, and mixed oligoastrocytomas. Any of these subtypes can undergo malignant transformation and

TABLE 1.3   Immunohistochemical Staining Patterns of CNS Tumors

Tumor Type	Antibodies Used
Astrocytic tumors	GFAP
Oligodendrogliomas	GFAP
	Leu7
Ependymomas	S-100 protein
	GFAP
	Epithelial membrane antigen
Choroid plexus tumors	GFAP
	Cytokeratins
	Transthyretin
	Cathepsin D
	Epithelial membrane antigen
Medulloblastoma	GFAP
	Synaptophysin
	Neuron-specific enolase
	Neurofilament protein
Neuronal tumors	Synaptophysin
	Neuron-specific enolase
	Neurofilament protein
	MAP-2
Germ cell tumors	Placental alkaline phosphatase
	Alpha-fetoprotein
	Beta-human chorionic
	Gonadotropin
	Carcinoembryonic antigen
Lymphomas	CD3, CD20, CD45, CD68, CD79a
	Immunoglobulins
	Epstein-Barr virus latent proteins
Meningiomas	Epithelial membrane antigen
	Vimentin
	Cytokeratins
	Progesterone receptors
Melanoma	S-100 protein
	Neuron-specific enolase
	HMB45
	MART-1
Nerve Sheath Tumors	S-100 protein
	Neurofilament protein
Pituitary Tumors	GH, PRL, ACTH, FSH, LH, TSH
	Alpha-glyprotein subunit
Metastatic Tumors	Cytokeratins
	Epithelial membrane antigen
	Chromogranin
	Neuron-specific enolase
	Cell-specific markers

Data derived from references [24–26]

Abbreviations: GFAP — glial fibrillary acidic protein; GH — growth hormone; PRL — prolactin; ACTH — adrenocorticotrophic hormone; FSH — follicle stimulating hormone; LH — leuteinizing hormone; TSH — thyroid stimulating hormone.

TABLE 1.4 WHO Classification: Tumors of Neuroepithelial Tissue

Astrocytic tumors

Oligodendroglial tumors

Ependymal tumors

Mixed gliomas

Choroid plexus tumors

Neuronal and mixed neuronal-glial tumors

Pineal parenchymal tumors

Neuroepithelial tumors of uncertain origin

Embryonal tumors

TABLE 1.5 WHO Classification: Astrocytic Tumors

Diffuse Astrocytomas

　Astrocytoma (WHO grade II)

　　Fibrillary

　　Protoplasmic

　　Gemistocytic

　Anaplastic astrocytoma (WHO grade III)

　Glioblastoma multiforme (WHO grade IV)

　　Giant cell glioblastoma

　　Gliosarcoma

Localized Astrocytomas (WHO grade I)

　Pilocytic astrocytoma

　Pleomorphic xanthoastrocytoma

　Subependymal giant cell astrocytoma

degenerate into the most aggressive form of glioma, the GBM.

Diffuse Astrocytomas. The current WHO classification divides astrocytomas into diffuse and localized varieties (see Table 1.5) [24,25,30]. The diffuse astrocytomas are intrinsically invasive and often travel along white matter tracts deep into normal brain. There are three groups of diffuse astrocytic neoplasms: astrocytoma (WHO grade II; peak age of 30–39 years), AA (WHO grade III; peak age of 40–49 years), and GBM (WHO grade IV; peak age of 50–69 years), including the GBM variants giant cell glioblastoma and gliosarcoma (both WHO grade IV). Diffuse astrocytic tumors can be divided into fibrillary, protoplasmic, and gemistocytic forms, with the fibrillary form being most common. The presence of gemistocytic and protoplasmic cellular variations are most often visualized in WHO grade II tumors. WHO grade II astrocytomas are considered low-grade tumors and usually occur in the cerebral white matter. These tumors are characterized by a relatively uniform population of proliferating neoplastic astrocytes in a fibrillary matrix, with minimal cellular and nuclear pleomorphism or atypia (see Fig. 1.1). Grade II tumors may display diffuse and intense staining with GFAP. Tumor margins are poorly delineated and suggest significant infiltration into surrounding brain. Within infiltrated regions of brain, "secondary structures of Scherer" may be noted, with the presence of tumor cells around cortical neurons (i.e., perineuronal satellitosis), along white matter tracts, and around blood vessels. Mitotic figures are absent and there is

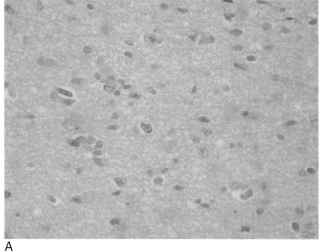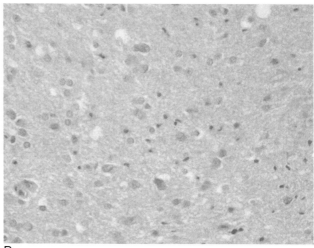

A　　　　　　　　　　　　　　　　　　　　　　　B

FIGURE 1.1 WHO grade II fibrillary astrocytoma (1A and 1B). Note the neoplastic astrocytes in a fibrillary matrix, with mildly increased cellularity and pleomorphism. No mitoses or hypervascularity is present. H&E @ 200x. See Plate 1.1 in Color Plate Section.

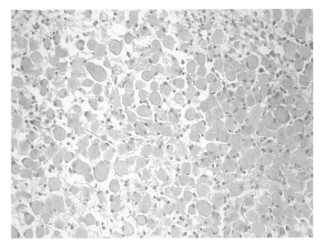

FIGURE 1.2   WHO grade II gemistocytic astrocytoma. Low-grade astrocytic tumor demonstrating numerous, plump gemistocytic astrocytes, with eosinophilic cytoplasm and short cellular processes. H&E @ 200x. See Plate 1.2 in Color Plate Section.

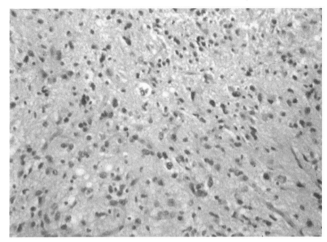

FIGURE 1.3   WHO grade III fibrillary astrocytoma (AA). The tumor is more densely cellular than grade II, with significant cellular and nuclear pleomorphism and atypia. Mitotic figures are evident. H&E @ 200x. See Plate 1.3 in Color Plate Section.

no evidence for vascular hyperplasia. Microcystic change is commonly noted in all variants of grade II astrocytoma. The fibrillary variant is characterized by tumor cells with mild to moderately pleomorphic astrocyte-appearing nuclei and cellular processes that form a fibrillary background. Gemistocytic cells (i.e., plump astrocytes with eosinophilic cytoplasm and short cellular processes) may be present in fibrillary tumors, but as a minor cell population. If the tumor has gemistocytes representing greater than 20 per cent of neoplastic cells, then the histological designation is changed to a gemistocytic astrocytoma (see Fig. 1.2) [31]. Gemistocytic astrocytomas tend to behave in a more aggressive fashion than other grade II tumors, with a high propensity (approximately 80 per cent) for degeneration into AA and GBM. The gemistocytic cells appear to be resistant to apoptosis and express high levels of mutant p53 and bcl-2 [32]. Fibrillary tumors with a high percentage of gemistocytes (i.e., 5–19 per cent) appear to be biologically affected by the presence of these cells, since there is a significant negative correlation with progression to higher grades of astrocytoma [33]. Protoplasmic astrocytomas are rare and have features of inconspicuous cytoplasm and cellular processes. The Ki-67 labeling index of WHO grade II astrocytomas is typically less than 4 per cent, with a mean of approximately 2.0–2.5 per cent.

Higher grade diffuse astrocytomas include AA (WHO grade III) and GBM (WHO grade IV), as well as the GBM variants giant cell glioblastoma and gliosarcoma (WHO grade IV) (see Table 1.5) [24,25, 30,34]. Anaplastic astrocytomas are similar to grade II tumors, except for the presence of more prominent cellular and nuclear pleomorphism and atypia, and

mitotic activity (see Fig. 1.3). In addition, grade III and IV tumors usually do not stain as intensely or as homogeneously with GFAP. According to WHO criterion, the critical feature that upgrades a grade II tumor to an AA is the presence of mitotic activity, with anaplastic tumors having Ki-67 indices in the range of 5–10 per cent in most cases. Other features of anaplasia can be present, such as multinucleated tumor cells and abnormal mitotic figures. Regions of vascular proliferation and necrosis are absent in grade III astrocytomas.

Glioblastoma multiforme is classified as a WHO grade IV tumor and has similar histological features to AA, but with more pronounced anaplasia (see Fig. 1.4) [24,25,30,34]. The presence of microvascular proliferation and/or necrosis in an otherwise malignant astrocytoma upgrades the tumor to a GBM. Vascular proliferation is defined as blood vessels with "piling up" of endothelial cells, including the formation of glomeruloid vessels. The glomeruloid vessels can form undulating garlands that surround necrotic zones in some cases. Necrosis can be noted in large amorphous areas, which appear ischemic in nature, or can appear as more serpiginous regions with surrounding palisading tumor cells (i.e., perinecrotic pseudopalisading). Necrosis with nuclear pseudopalisading is essentially pathognomonic for GBM. Other features of GBM that are typically prominent include marked cellular and nuclear pleomorphism and atypia, mitotic figures and multinucleated giant cells, and pronounced infiltrative capacity into surrounding brain, with extensive production of "secondary structures" within the cerebral cortex. In addition to giant cells, other cell types that may be

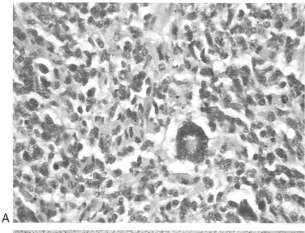

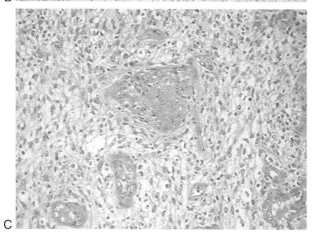

FIGURE 1.4 WHO grade IV fibrillary astrocytoma (GBM). A highly cellular tumor with marked cellular and nuclear pleomorphism, numerous mitoses, giant cells (4A; H&E @ 400x), regions of necrosis with pseudopallisading tumor nuclei (4B; H&E @ 100x), and dense vascular proliferation (4C; H&E @ 200x). See Plate 1.4 in Color Plate Section.

noted include "small cells", gemistocytes, granular cells, lipidized cells, perivascular lymphocytes, and regions of metaplasia. Labeling indices with Ki-67 are usually in the range of 15–20 per cent, but can be much higher in some tumors.

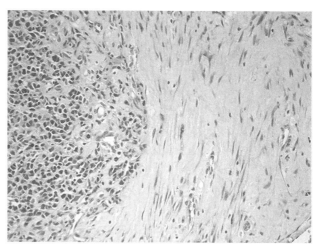

FIGURE 1.5 WHO grade IV gliosarcoma. The biphasic nature of the tumor is demonstrated, with high-grade astrocytoma consistent with GBM in the left hemifield and sarcomatous, spindle-shaped cells with pleomorphic nuclei in the right hemifield. H&E @ 200x. See Plate 1.5 in Color Plate Section.

Recent molecular biological studies would suggest that GBM can be further classified based on the spectrum of amplified oncogenes and mutations or deletions of tumor suppressor genes [34–37]. GBM can be designated as a secondary tumor when derived from a lower grade neoplasm (e.g., WHO grade II or III), or as a primary GBM when it arises *de novo* from normal astrocytes without any intervening stages. Secondary GBM (mean age of onset 38–42 years) typically have mutations of p53 (60–70 per cent), activation of the ras pathways, mutation or inactivation of the retinoblastoma gene (43 per cent), inactivation of $p14^{ARF}$ and p16 (35–60 per cent), and amplification of the platelet-derived growth factor (PDGF) gene. In contrast, primary GBM (mean age of onset 54 to 56 years) have amplification of the epidermal growth factor receptor (EGFR) gene (40 per cent), mutation of the PTEN tumor suppressor gene (32 per cent), and other genetic abnormalities such as loss of the p16 tumor suppressor gene (35 per cent).

Variants of GBM that are also classified as WHO grade IV tumors include the giant cell glioblastoma and gliosarcoma [24,25,30]. Giant cell glioblastoma (also known as monstrocellular sarcoma) is a rare histological variant of GBM that has a predominance of multinucleated giant cells within a more prominent reticulin network [38]. The giant cells may measure more than 500 microns in diameter and have up to 20 nuclei. Giant cell glioblastomas consistently stain with GFAP and are usually more circumscribed than GBM. Molecular analysis demonstrates a high rate (75–90 per cent) of p53 mutations. Gliosarcomas are a GBM variant that display a biphasic histological pattern, with alternating areas of glial and mesenchymal differentiation (see Fig. 1.5) [39]. The gliomatous regions are similar in appearance to typical GBM, with dense cellularity and nuclear atypia, necrosis, and a variable degree of GFAP expression. The sarcomatous regions are often well demarcated from the glial regions by a reticulin network, which may be very dense in some tumors. Spindle cells within the sarcomatous portion of the tumor also have anaplastic features, such as nuclear atypia and mitotic activity. Regions of necrosis may also be noted in sarcomatous regions of tumor. Gliosarcomas have a similar molecular phenotype to GBM, with frequent mutation and/or deletion of p53, PTEN, and p16, as well as amplification of mdm2. However, in contrast, the frequency of amplification of the EGFR is quite low.

Localized Astrocytomas. In the WHO classification, the localized astrocytomas include the pilocytic astrocytoma (WHO grade I), pleomorphic xanthoastrocytoma (PXA; WHO grade II), and the subependymal giant cell astrocytoma (WHO grade I) [24,25,30]. Pilocytic astrocytomas are slow growing, relatively circumscribed tumors that usually occur in children (peak age 10–12 years) and young adults (peak age 20–24 years). These tumors can arise anywhere in the brain, but have a predilection for the cerebellum, optic nerves and optic pathways, and hypothalamus. Pilocytic astrocytomas can present as a homogeneous, well-demarcated mass, with a gliotic surrounding margin, or as an enhancing mural nodule associated with a large cyst [40]. The distinctive histological feature is the presence of cells

with slender, elongated nuclei and thin, hair-like (i.e., piloid), GFAP-positive, bipolar processes (see Fig. 1.6). These cells are found in a biphasic background, which consists of dense fibrillary regions alternating with loose, microcystic areas. Also frequently noted are Rosenthal fibers, which are eosinophilic, refractile, corkscrew-shaped deposits. The presence of Rosenthal fibers is highly suggestive of a pilocytic astrocytoma. Another characteristic feature is the eosinophilic granular body, which is a circular profile of granular, amorphous material. In some cases, pilocytic astrocytomas can demonstrate mild degrees of cellular and nuclear pleomorphism, mitotic activity, vascular proliferation, regions of necrosis, and invasion of the subarachnoid space. However, in these tumors, it does not denote a poor prognosis, as in higher grade, diffuse astrocytomas. Labeling index studies with Ki-67 report values of 0.5–1.5 per cent in most tumors.

The PXA is a supratentorial tumor with a predilection for the superficial temporal lobes that usually occurs in younger patients (mean age 15–18 years) with a longstanding history of seizure activity [24,25,30]. On histological examination, PXA demonstrates significant pleomorphism, with numerous atypical giant cells and astrocytes with prominent nucleoli (see Fig. 1.7) [41]. Also present are large foamy (xanthomatous) cells with lipidized cytoplasm that express GFAP. Regions of focal lymphocytic and plasma cell infiltration can occur, along with a reticulin-positive stroma that delineates fascicles of cells. Typical PXA do not have mitotic activity, necrosis, or vascular pleomorphism and are classified

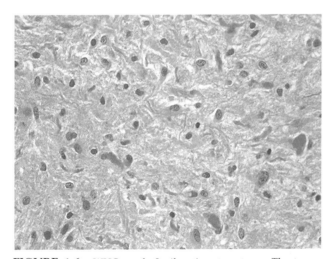

FIGURE 1.6 WHO grade I pilocytic astrocytoma. The tumor demonstrates "piloid" neoplastic astrocytes, with minimal cellularity, pleomorphism, cells with delicate hair-like processes, and the presence of numerous Rosenthal fibers. H&E @ 200x. See Plate 1.6 in Color Plate Section.

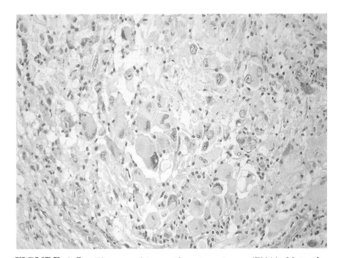

FIGURE 1.7 Pleomorphic xanthoastrocytoma (PXA). Note the presence of mild cellular and nuclear pleomorphism, with numerous atypical giant cells and large astrocytes with foamy (xanthomatous) cytoplasm. H&E @ 200x. See Plate 1.7 in Color Plate Section.

as WHO grade II. However, a sub-group of PXA can have anaplastic features, including increased mitotic activity (>5 mitoses per high power field), necrosis, and endothelial proliferation. By WHO criteria, these tumors should be classified as "PXA with anaplastic features". The term "anaplastic PXA" (grade III) is not recommended at this time.

Subependymal giant cell astrocytoma is a benign, slow-growing tumor that typically arises in the walls of the lateral ventricles and is almost invariably associated with tuberous sclerosis [24,25,30]. The tumor is composed of cells with extensive glassy, eosinophilic cytoplasm. Some cells may have a neuronal appearance, with vesicular nuclei and large nucleoli. Subependymal giant cell astrocytomas are classified as WHO grade I.

Oligodendrogliomas and Oligoastrocytomas. Oligodendrogliomas are a form of diffuse glioma that can be of pure or mixed histology and are classified as WHO grade II or III [24,25,30]. They typically occur in young to middle-aged adults (peak age 35–45 years) with a history of seizures, within the white matter of the frontal and temporal lobes. Pure low-grade oligodendroglial tumors (WHO grade II) are characterized histologically by a moderately cellular, monotonous pattern of cells with round nuclei and perinuclear halos (the classic "fried egg" appearance; see Fig. 1.8) [42]. The perinuclear halos are an artifact of the formalin fixation process of the tumor tissue. Foci of calcification are frequent, and can be quite dense in some cases. Delicately branching blood vessels are prominent (i.e., "chicken-wire" vasculature), but do not display endothelial proliferation. GFAP staining is limited in the majority of oligodendrogliomas, except for the variant that has a significant component of mini-gemistocytes (small cells with round nuclei and eosinophilic cytoplasm), which strongly express GFAP. Oligodendrogliomas have a pronounced invasive capacity and are known to invade the gray and white matter diffusely, with a strong tendency to form secondary structures of Scherer, in particular perineuronal satellitosis. Mitoses are absent or rare and necrosis is not present. Labeling studies with Ki-67 usually demonstrate indices less than 5 per cent, with a mean of approximately 2 per cent. The diagnosis of an anaplastic oligodendroglioma (WHO grade III) requires the presence of additional histologic features, including a higher degree of cellularity and mitotic activity, vascular endothelial hyperplasia, nuclear pleomorphism, and regions of necrosis (see Fig. 1.9). These tumors behave in a more aggressive fashion, with a higher proliferative rate (Ki-67 labeling index >5 per cent) and capacity for invasion of

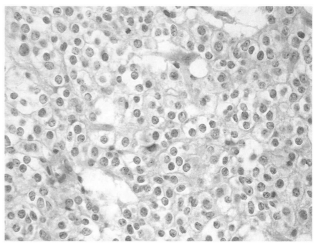

FIGURE 1.8 WHO grade II oligodendroglioma. Demonstrates the classic features of typical oligodendroglioma, with moderate cellularity and numerous round cells with the "fried egg" pattern of perinuclear halos, and delicate "chicken wire" vasculature. H&E @ 400x. See Plate 1.8 in Color Plate Section.

surrounding brain. In some cases, anaplastic oligodendrogliomas can develop further malignant features and degenerate into a tumor similar to GBM. By WHO criteria, an oligodendroglioma with predominant astrocytic areas that look like GBM should be classified as an oligoastrocytoma grade III, rather than a GBM.

Advances in molecular neuropathology have begun to clarify the biological underpinnings of variability in response to treatment of oligodendrogliomas [42–44]. The majority of tumors demonstrate genetic losses on chromosome 1p (40 to 92 per cent) and/or 19q (50–80 per cent). There is a strong predilection for deletions of 1p and 19q to occur together, but in some tumors they can be singular events. Loss of 1p and/or 19q are mutually exclusive to certain other molecular abnormalities, such as mutation of p53, amplification of EGFR, or mutations of the PTEN tumor suppressor gene. Patients with oligodendrogliomas that contain deletions of 1p and 19q are consistently more responsive to irradiation and chemotherapy, and have an overall median survival of 8–10 years. In contrast, patients with tumors that do not have deletion of 1p and 19q are more resistant to all forms of therapy, and have an overall median survival of only 3–4 years.

Mixed oligoastrocytomas can be classified as WHO grade II or III tumors [24,25,30]. Distinct populations of neoplastic oligodendroglial cells and astrocytes can be identified within the mass that have similar features to pure versions of the tumor (see Fig. 1.10) [45]. The percentage of each cell population can be quite variable, with an even mixture of cell types or

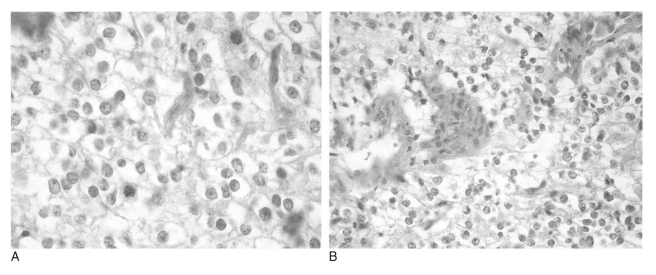

A   B

FIGURE 1.9 WHO grade III oligodendroglioma. A more densely cellular tumor with prominent cellular and nuclear pleomorphism, mitotic activity (9A), and increased vascularity (9B). H&E @ 400x. See Plate 1.9 in Color Plate Section.

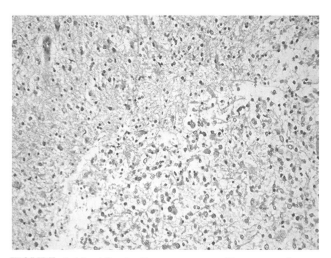

FIGURE 1.10 Mixed oligoastrocytoma. The tumor demonstrates cellular regions of neoplastic astrocytic cells and adjacent, less cellular areas more consistent with low-grade oligodendroglioma, with the presence of some "fried egg" cells. H&E @ 200x. See Plate 1.10 in Color Plate Section.

TABLE 1.6   WHO Classification: Ependymal Tumors

Ependymoma (WHO grade II)

   Cellular

   Papillary

   Clear cell

   Tanycytic

Anaplastic ependymoma (WHO grade III)

Myxopapillary ependymoma (WHO grade I)

Subependymoma (WHO grade I)

with one cell type predominating. The astrocytic and oligodendroglial populations can be totally separate within the tumor (biphasic pattern) or more intermingled (diffuse). Tumors with WHO grade II features will have mild to moderate cellularity, absent or low mitotic activity, minimal cellular and nuclear pleomorphism, and lack vascular proliferation and necrosis. Grade III tumors demonstrate more pronounced cellularity and anaplasia of cells and nuclei, a high mitotic rate, vascular endothelial proliferation, and necrosis. At the molecular level, oligoastrocytomas are often heterogeneous, with the astrocytic portion having the typical genotype of low-grade diffuse astrocytomas, with p53 mutations and loss of chromosome 17p. Allelic loss of 1p and/or 19q is commonly present in the oligodendroglial component (30–70 per cent), but may also occur in astrocytic tumor cells.

Ependymomas. Ependymomas are slow-growing tumors derived from the ependymal lining cells of the ventricular system and are classified by the WHO as grades I, II, and III (see Table 1.6) [24,25,30]. These tumors are most common in children and young adults and typically grow as intraventricular masses, often from the floor of the fourth ventricle. They are commonly divided into typical ependymoma (grade II) and anaplastic ependymoma (grade III). In addition, the WHO recognizes four histological variants of ependymoma, including the cellular, papillary, clear cell, and tanycytic subtypes, as well as two low-grade (grade I) forms, myxopapillary ependymoma and subependymoma. On histological examination, WHO grade II ependymomas are densely cellular and

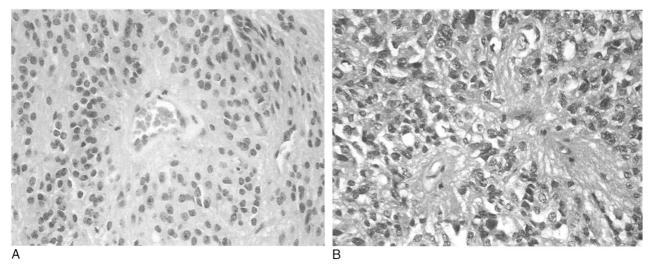

A   B

FIGURE 1.11   WHO grades II and III ependymoma. The typical low-grade tumor is densely cellular and composed of oval to carrot-shaped cells, with dense speckled nuclei and tapering eosinophilic cytoplasm (11A; H&E @ 200x). Note the ependymal perivascular psuedorosette in the center of the field. The anaplastic ependymoma is more pleomorphic and demonstrates mitotic activity (11B; H&E @ 400x). See Plate 1.11 in Color Plate Section.

composed of oval to carrot-shaped cells, with a dense speckled nucleus and tapering eosinophilic cytoplasm (see Fig. 1.11) [46]. Some tumors have cells with a more glial appearance and more background fibrillarity or more epithelioid cells and architecture. Perivascular pseudorosettes, which are commonly observed, are circular arrangements of tumor cells that send processes towards vessel walls, creating a perivascular "nuclear-free zone" that can be noted at low-power. Less commonly, true ependymal rosettes, surrounding a true lumen, can be observed. Ependymomas are usually GFAP positive. Anaplastic ependymomas (WHO grade III) have additional features such as increased cellularity, mitotic activity, pleomorphic nuclei, vascular hyperplasia, nuclear atypia, and necrosis. Myxopapillary ependymomas are WHO grade I tumors that arise in the lower spine within the cauda equina. They display an admixture of fibrillated and epithelioid cells with an exuberant connective tissue stroma. Papillary arrangements of elongated fibrillary cells that extend delicate processes to hyalinized vessels are prominent. In addition, prominent myxoid degeneration of the cell cytoplasm and blood vessel walls is a constant feature. Subependymomas are WHO grade I tumors that develop in the walls of the ventricular system (65–70 per cent fourth ventricle) and may present with hydrocephalus. They are composed of tumor cells exhibiting features of ependymal and astrocytic differentiation, with the tendency to form distinct clusters ("glomerate" structures) separated by cell-free areas.

Choroid Plexus Tumors.   Neoplasms of the choroid plexus are classified by the WHO as choroid plexus papillomas (WHO grade I) or choroid plexus carcinomas (WHO grade III) [24,25,30]. These tumors arise from the specialized epithelium constituting the choroid plexus and therefore develop within the lateral and fourth ventricles. Choroid plexus papillomas are usually diagnosed in children and present with increased intracranial pressure from hydrocephalus secondary to obstruction of cerebrospinal fluid (CSF) pathways or overproduction of CSF. Histological examination reveals delicate fibrovascular connective tissue fronds covered by a single layer of uniform cuboidal to columnar epithelial cells [47]. The fronds are similar in appearance to normal choroid plexus, but with slightly more cellularity and pleomorphism. Choroid plexus carcinoma has additional features of anaplasia including increased cellularity, frequent mitoses, cellular and nuclear pleomorphism, loss of papillary structure, regional brain invasion, and areas of necrosis.

Neuronal and Mixed Neuronal-Glial Tumors. Neuronal and mixed neuronal-glial tumors are a heterogeneous group of neoplasms that are mainly classified by the WHO as grades I and II (see Table 1.7) [24,25,30]. Gangliogliomas are well differentiated, slow-growing tumors with a mixture of neoplastic glial and neuronal elements. They are composed of irregular groups of large, multipolar neurons with dysplastic features, and intermixed with regions of neoplastic glial cells, usually of astrocytic origin (WHO grade I or II; see Fig. 1.12) [48]. The astrocytic

TABLE 1.7   WHO Classification: Neuronal and Mixed
Neuronal-Glial Tumors

Gangliocytoma

Dysplastic gangliocytoma of cerebellum (Lhermitte-Duclos)

Desmoplastic infantile

Astrocytoma/ganglioglioma

Dysembryoplastic neuroepithelial tumor

Ganglioglioma

Anaplastic ganglioglioma

Central neurocytoma

Cerebellar liponeurocytoma

Paraganglioma of the filum terminale

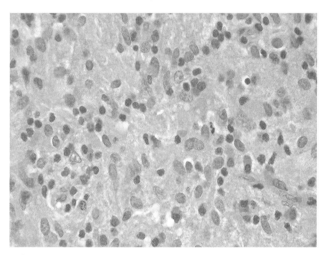

FIGURE 1.12   WHO grade II ganglioglioma. The tumor demon-strates a mixture of large neoplastic ganglion cells within a back-ground of diffuse, low-grade fibrillary astrocyoma. H&E @ 400x. See Plate 1.12 in Color Plate Section.

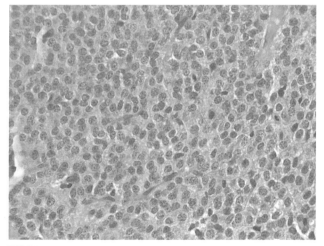

FIGURE 1.13   Central neurocytoma. Note the presence of dense sheets of uniform round cells with neuronal features and the delicate vasculature. H&E @ 200x. See Plate 1.13 in Color Plate Section.

portion is fibrillary and similar to low-grade astro-cytomas as described earlier. Eosinophilic granular bodies are often noted, along with frequent peri-vascular lymphocytic infiltration. Occasional mitoses may be present, but significant pleomorphism, necro-sis, and vascular hyperplasia are absent. Anaplastic ganglioglioma (WHO grade III or IV) are similar to low-grade gangliogliomas, except for additional malignant features within the glial component of the tumor [30,48]. These malignant features include increased cellularity, frequent mitoses, nuclear pleo-morphism and atypia, regional brain invasion, and areas of necrosis. Gangliocytomas are well differen-tiated, slow-growing tumors composed of neoplastic, mature ganglion cells, without a glial component. The ganglion cells are large and multipolar, and often show dysplastic features. There is a non-neoplastic glial stroma and network of reticulin fibers that surround the ganglion cells.

Central neurocytomas are low-grade tumors (WHO grade II) composed of uniform cells with neuronal differentiation that arise in the lateral ventricles near the foramen of Monro [24,25,30]. Histological examination reveals sheets of uniform round cells with neuronal features, along with fibrillary areas that mimick neuropil (see Fig. 1.13) [30,49]. In some tumors, regions with oligodendroglioma-like features may be noted, with a honeycomb appearance and large fibrillary areas. The tumors stain diffusely with synaptophysin, neuron specific enolase, and calcineurin.

Pineal Parenchymal Tumors. The WHO recognizes three forms of pineal parenchymal tumors, including the pineocytoma (WHO grade II), pineoblastoma (WHO grade IV), and tumors of intermediate differ-entiation [24,25,30]. The pineocytoma is a slow-growing pineal tumor that usually occurs in young adults. On histological examination, the tumor is a well-differentiated neoplasm composed of sheets of small, uniform, mature cells resembling pineocytes. A lobular pattern may be present, with groups of tumor cells separated by mesenchymal septae. The cells are small in size, with inconspicuous nucleoli, a fine chromatin pattern, and moderate quantities of eosinophilic cytoplasm. Another characteristic feature is the presence of pineocytomatous rosettes, which can vary in size and number. Mitoses and regions of necrosis are either absent or rare. Pineoblastomas are malignant tumors with a more aggressive growth rate, tendency to invade surrounding brain, and disseminate along CSF pathways. They are composed of patternless sheets of densely packed small blue cells with round-to-irregular nuclei and scant cytoplasm, similar to other small cell tumors

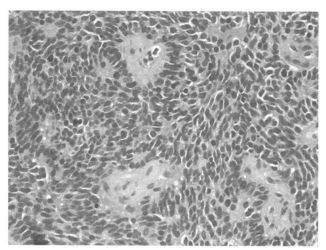

FIGURE 1.14 WHO grade IV pineoblastoma. Densely cellular neoplasm with large, pleomorphic and hyperchromatic nuclei. Cells forming Homer-Wright rosettes are present. H&E @ 400x. See Plate 1.14 in Color Plate Section.

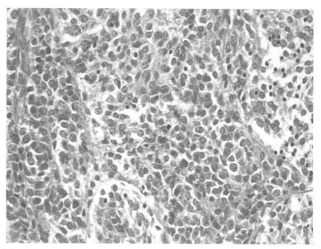

FIGURE 1.15 WHO grade IV medulloblastoma. Note the dense cellularity and presence of undifferentiated cells with hyperchromatic, oval to carrot-shaped nuclei with scant cytoplasm. The nuclei have a tendency to mold against one another. H&E @ 200x. See Plate 1.15 in Color Plate Section.

TABLE 1.8   WHO Classification: Embryonal Tumors

Medulloepithelioma

Ependymoblastoma

Medulloblastoma

    Desmoplastic medulloblastoma

    Large-cell medulloblastoma

    Medullomyoblastoma

    Melanotic medulloblastoma

Supratentorial PNET

    Neuroblastoma

    Ganglioneuroblastoma

Atypical teratoid/rhabdoid tumor

(i.e., primitive neuroectodermal tumors, PNET; see Fig. 1.14). The tumor cells have a high nuclear cytoplasmic ratio, hyperchromatic nuclei, and indistinct cell borders. Typical pineocytomatous rosettes of lower-grade pineal tumors are absent. However, Homer-Wright pseudo-rosettes may be noted. Mitotic activity is variable and can be high in some tumors. Regions of necrosis are commonly noted.

Medulloblastoma and other Embryonal Tumors. Embryonal tumors are a group of aggressive, malignant neoplasms that usually affect children that are classified by the WHO as grade IV in all cases (see Table 1.8) [24,25,30]. All embryonal tumors share the common features of high cellularity, frequent mitoses, regions of necrosis, and a propensity for metastases along CSF pathways. Medulloblastoma is the most common of the embryonal tumors and is considered a PNET of the cerebellum. It usually arises in the

midline in children, within the cerebellar vermis, while in adults it is more likely to have an off-center location within the cerebellar hemispheres. The typical medulloblastoma is densely cellular and composed of undifferentiated cells with hyperchromatic, oval to carrot-shaped nuclei with scant cytoplasm (see Fig. 1.15) [30,50]. The nuclei have a tendency to mold against one another. Mitoses and single-cell necrosis are frequently present. Evidence of anaplasia is variable and may include increased nuclear size, abundant mitoses, and the presence of large-cell or similar aggressive cellular morphology. Some tumors may display immunohistochemical and morphological evidence for differentiation along neuronal, glial, or mesenchymal lines. Tumors with neuronal differentiation may display mature ganglion cells, immunoreactivity for neuronal markers (e.g., synaptophysin), neuroblastic Homer Wright rosettes, and ultrastructural findings such as dense core granules and synaptic vesicles. Several histologic variants of medulloblastoma are recognized by the WHO including the desmoplastic, large cell, and melanotic sub-types, as well as medullomyoblastoma. The desmoplastic variant shows nodular, reticulin-free zones ("pale islands") surrounded by densely packed, highly proliferative cells within a dense reticulin fiber network. Large-cell medulloblastoma is an uncommon variant with large cells that contain large, rounded nuclei with prominent nucleoli and more abundant cytoplasm. Nuclear molding is less frequent; however, mitotic activity, pleomorphism, and necrosis are more prominent. Melanotic medulloblastoma has similar histological features to more typical cases, except for the presence of melanotic tumor cells that often form tubular epithelial structures. Medulloblastomas are highly proliferative tumors, with Ki-67 labeling indices ranging from 15 to 50 per cent.

Atypical teratoid/rhabdoid tumors (WHO grade IV) are rare neoplasms that usually develop in the posterior fossa and affect children less than 3 years of age [24,25,30]. Histological examination reveals a cellular tumor with large and polygonal rhabdoid cells with conspicuous eosinophilic or pink cytoplasm, that contain spherical fibrillary intracytoplasmic inclusions composed of bundles of intermediate filaments [51]. Other tissue elements may be present, such as primitive neuroepithelium, mesenchyme, and mature epithelium. Mitoses and regions of necrosis are common. The rhabdoid cells lack true muscle differentiation and do not react positively with desmin. They may show variable reactivity with epithelial-membrane antigen, vimentin, smooth muscle actin, and GFAP. These tumors have marked proliferative capacity as shown by Ki-67 studies, with labeling indices ranging from 50 to 80 per cent. Molecular studies consistently reveal a loss of chromosome 22 or specific loss of 22q11, with mutations in the chromatin remodeling tumor suppressor gene hSNF5/INI1 (human sucrose nonfermentor/integrase interactor 1). This molecular signature is unique to teratoid/rhabdoid tumors and is not present in medulloblastomas or other embryonal tumors.

PATHOLOGY OF TUMORS OF THE MENINGES

Tumors of the meninges comprise a large and diverse group of neoplasms that mostly have meningothelial or mesenchymal, non-meningothelial origins (see Table 1.9) [24,25]. The most common primary tumor of this group is the meningioma (18–20 per cent of intracranial tumors), which has meningothelial cell origins and is composed of neoplastic arachnoidal cap cells of the arachnoidal villi and granulations. Meningiomas can occur anywhere within the intracranial cavity, but favor the sagittal area along the superior longitudinal sinus, over the lateral cerebral convexities, at the tuberculum sellae and parasellar region, the sphenoidal ridge, and along the olfactory grooves. Numerous histologic variants of meningioma are described and recognized by the WHO (see Table 1.9). However, the biological behavior of most of these variants does not impact on the clinical behavior of the tumor. Meningioma sub-types that have a more indolent nature and low risk for

TABLE 1.9   WHO Classification: Tumors of the Meninges

Tumors of meningothelial cells

Meningioma

   Meningothelial

   Fibrous (fibroblastic)

   Transitional (mixed)

   Psammomatous

   Angiomatous

   Microcystic

   Secretory

   Lymphoplasmacyte-rich

   Metaplastic

   Clear cell

   Chordoid

   Atypical

   Papillary

   Rhabdoid

   Anaplastic meningioma

Mesenchymal, non-meningothelial tumors

Lipoma

Angiolipoma

Hibernoma

Liposarcoma (intracranial)

Solitary fibrous tumor

Fibrosarcoma

Malignant fibrous histiocytoma

Leiomyoma

Leiomyosarcoma

Rhabdomyoma

Rhabdomyosarcoma

Chondroma

Chondrosarcoma

Osteoma

Osteosarcoma

Osteochondroma

Hemangioma

Epithelioid haemangioendothelioma

Hemangiopericytoma

Angiosarcoma

Kaposi sarcoma

Primary melanocytic lesions

Diffuse melanocytosis

Melanocytoma

Malignant melanoma

Meningeal melanomatosis

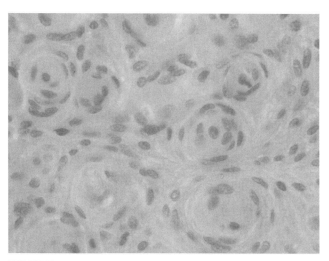

FIGURE 1.16 WHO grade II meningioma. The tumor demonstrates a moderately dense, uniform pattern of cells with oval shaped nuclei and the presence of many cellular whorl patterns. H&E @ 200x. See Plate 1.16 in Color Plate Section.

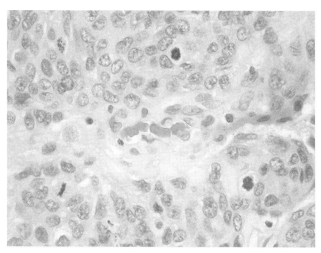

FIGURE 1.17 WHO grade III anaplastic meningioma. This higher power view demonstrates increased nuclear pleomorphism and frequent mitoses. H&E @ 400x. See Plate 1.17 in Color Plate Section.

aggressive growth or recurrence are classified as WHO grade I, and include the meningothelial, fibrous/fibroblastic, transitional (mixed), secretory, psammomatous, angiomatous, microcystic, lymphoplasmocyte-rich, and metaplastic variants [52,53]. Of this group, the meningothelial, fibrous, and transitional variants are most frequently diagnosed. The histological features common to most low-grade meningiomas are the presence of whorls (tightly wound, rounded collections of cells), psammoma bodies (concentrically laminated mineral deposits that often begin in the center of whorls), intranuclear pseudoinclusions (areas in which pink cytoplasm protrudes into a nucleus to produce a hollowed-out appearance), and occasional pleomorphic nuclei and mitoses (see Fig. 1.16) [52,53]. Meningothelial meningiomas are composed of lobules of typical meningioma cells, with minimal whorl formation. The tumor cells are uniform in shape, with oval nuclei that may show central clearing. Fibrous variants have spindle-shaped cells resembling fibroblasts that form parallel and interlacing bundles within a matrix of collagen and reticulin. More typical meningothelial type tumor cells may be intermixed with the spindle cells. Whorl formation and psammoma bodies are uncommon. The transitional meningioma have features between those of the meningothelial and fibrous variants. Regions of typical meningothelial cells are present, and may display lobular and fascicular patterns. In other areas, spindle cells may be more prominent. Whorls and psammoma bodies are very common in transitional tumors.

Meningioma sub-types that are more likely to display aggressive clinical behavior and to recur

are classified by the WHO as grade II (atypical, clear cell, chordoid) and grade III (rhabdoid, papillary, anaplastic) [24,25,52,53]. On histological examination, all of the grade II tumors are likely to demonstrate increased cellularity, more frequent mitoses, diffuse or sheetlike growth, nuclear pleomorphism and atypia, and evidence for micronecrosis. The atypical meningiomas are similar to low-grade meningothelial tumors, with additional features including increased mitotic activity, small cells with high nuclear/cytoplasmic ratio, high cellularity, and prominent nucleoli. Clear cell tumors are composed of polygonal cells with a clear, glycogen-rich cytoplasm that is PAS positive. The chordoid meningioma contains regions that are similar to chordoma, with trabeculae of eosinophilic, vacuolated cells in a myxoid background. Chordoid regions are interspersed with more typical areas of meningioma. Grade III tumors such as anaplastic meningioma, show features consistent with frank malignancy, including a high mitotic rate, advanced cytological atypia, nuclear pleomorphism, and necrosis (see Fig. 1.17). Invasion of underlying brain is frequently noted in grade III meningiomas, but can also occur in lower grade variants. Proliferation studies using Ki-67 demonstrate labeling indices ranging from 8 to 15 per cent.

PATHOLOGY OF PRIMARY CNS LYMPHOMA

Primary CNS lymphomas (PCNSL) are malignant tumors classified as WHO grade IV, that affect adults in the sixth and seventh decade of life [24,25].

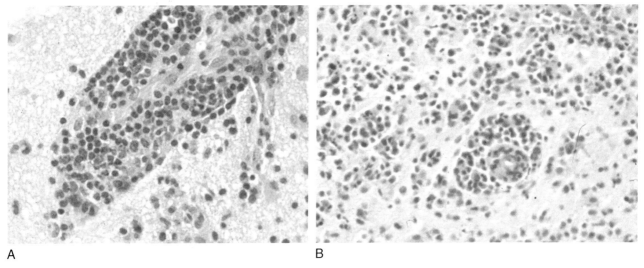

A   B

FIGURE 1.18   WHO grade IV primary CNS lymphoma (PCNSL). In panels A and B, note the presence of neoplastic lymphocytes in an angiocentric growth pattern, with nuclear pleomorphism and mitoses. H&E @ 100x. See Plate 1.18 in Color Plate Section.

They are often multifocal and usually arise in the deep supratentorial white matter, with a predilection for the periventricular region and basal ganglia. PCNSL are composed of a clonal expansion of neoplastic lymphocytes, typically of the diffuse, large cell or immunoblastic variety. In 95 per cent of the tumors, the cells have a B-cell lineage, often with monoclonal IgM kappa production. On histological examination, PCNSL display a perivascular cellular orientation, with expansion of vessel walls and reticulin deposition (see Fig. 1.18) [54]. Regions of necrosis are common, especially if steroids have been administered prior to the biopsy. The lymphomatous cells are noncohesive and usually have large, irregular nuclei, prominent nucleoli, and scant cytoplasm. From the perivascular region, tumor cells are noted to invade the surrounding brain parenchyma, either in compact cellular aggregates or as singly infiltrating tumor cells. PCNSL are highly proliferative tumors, with Ki-67 labeling indices ranging from 20 to 50 per cent in most studies. The diagnosis can be confirmed by immunohistochemical positivity for leukocyte common antigen (CD45) and specific B cell markers (CD19, CD20, and CD79a).

PATHOLOGY OF INTRACRANIAL GERM CELL TUMORS

Primary intracranial germ cell tumors are rare neoplasms that typically affect children and teenagers.

TABLE 1.10   WHO Classification: Intracranial Germ Cell Tumors

Germinoma
Embryonal carcinoma
Yolk sac tumor
Choriocarcinoma
Teratoma
Mature
Immature
Teratoma with malignant transformation
Mixed germ cell tumors

They are analogous to their gonadal and extragonadal systemic counterparts and arise almost exclusively in the midline, affecting the pineal and sellar regions, third ventricle, and hypothalamus [24,25]. Several histologic varieties of germ cell tumor are recognized by the WHO and are listed in Table 1.10, including germinomas, embryonal carcinomas, yolk sac tumors, choriocarcinomas, and teratomas. All of these tumors are considered malignant (i.e., WHO grade IV) except for teratoma, which has a more benign histology and clinical course. Germinomas are the most common of all germ cell tumors and are composed of uniform cells resembling primitive germ cells, with large vesicular nuclei, prominent nucleoli and clear, glycogen-rich cytoplasm (see Fig. 1.19) [55]. The tumor cells are present in monomorphous sheets or lobules.

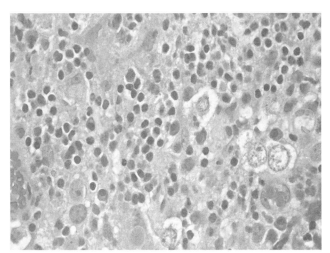

FIGURE 1.19 WHO grade IV germinoma. High power view demonstrating cells with vesicular nuclei, prominent nucleoli, and scattered infiltrating mature lymphocytes. H&E @ 400x. See Plate 1.19 in Color Plate Section.

Mitoses are plentiful but necrosis is uncommon. Also, delicate fibrovascular septa infiltrated by mature lymphocytes are a common feature. Analysis by immunohistochemistry typically demonstrates positivity for placental alkaline phosphatase. However, patchy positivity for beta-HCG and human placental lactogen can be observed.

ACKNOWLEDGMENTS

The author would like to thank Ryan Smith for research assistance. Dr. Newton was supported in part by National Cancer Institute grant, CA 16058 and the Dardinger Neuro-Oncology Center Endowment Fund.

References

1. Newton, H. B. (1994). Primary brain tumors: Review of etiology, diagnosis, and treatment. *Am Fam Physician* **49**, 787–797.
2. Preston-Martin, S. (1996). Epidemiology of primary CNS neoplasms. *Neurol Clin* **14**, 273–290.
3. Davis, F. G., and McCarthy, B. J., (2001). Current epidemiological trends and surveillance issues in brain tumors. *Expert Rev Anticancer Ther* **1**, 395–401.
4. Osborne, R. H., Houben, M. P. W. A., Tijssen, C. C., Coebergh, J. W. W., and van Duijn, C. M. (2001). The genetic epidemiology of glioma. *Neurol* **57**, 1751–1755.
5. Wrensch, M., Minn, Y., Chew, T., Bondy, M., and Berger, M. S. (2002). Epidemiology of primary brain tumors: Current concepts and review of the literature. *Neuro-Oncol* **4**, 278–299.
6. Hess, K. R., Broglio, D. R., and Bondy, M. L. (2004). Adult glioma incidence trends in the United States, 1977–2000. *Cancer* **101**, 2293–2299.
7. Newton, H. B. (1999). Neurological complications of systemic cancer. *Am Fam Phys* **59**, 878–886.
8. Legler, J. M., Ries, L. A., Smith, M. A. *et al.* (1999). Cancer surveillance series [corrected]: Brain and other central nervous system cancers: recent trends in incidence and mortality. *J Natl Cancer Inst* **91**, 1382–1390.
9. ACS (American Cancer Society) (2002). Cancer Facts and Figures 2002. American Cancer Society, Atlanta.
10. Grovas, A., Fremgen, A., Rauck, A. *et al.* (1997). The National Cancer Data Base report on patterns of childhood cancers in the United States. *Cancer* **80**, 2321–2332.
11. Bondy, M., Wiencke, J., Wrensch, M., and Kryitsis, A. P. (1994). Genetics of primary brain tumors: A review. *J Neuro-Oncol* **18**, 69–81.
12. de Andrade, M., Barnholtz, J. S., Amos, C. I., Adatto, P., Spencer, C., and Bondy, M. L. (2001). Segregation analysis of cancer in families of glioma patients. *Genet Epidemiol* **20**, 258–270.
13. Malmer, B., Gronberg, H., Bergenheim, A. T., and Lenner, P., and Henriksson, R. (1999). Familial aggregation of astrocytoma in northern Sweden: An epidemiological cohort study. *In J Cancer* **81**, 366–370.
14. Elexpuru-Camiruaga, J., Buxton, N., Kandula, V. *et al.* (1995). Susceptibility to astrocytoma and meningioma: Influence of allelism at glutathione S-transferase (GSTT1 and GSTM1) and cytochrome P-450 (CYP2D6) loci *Cancer Res* **55**, 4237–4239.
15. Trizna, Z., de Andrade, M., Kyritsis, A. P. *et al.* (1998). Genetic polymorphisms in glutathione S-transferase mu and theta, N-acetyltransferase, and CYP1A1 and risk of gliomas. *Cancer Epidemiol Biomarkers Prev* **7**, 553–555.
16. Karlsson, P., Holmberg, E., Lundell, M., Mattsson, A., Holm, L. E., and Wallgren, A. (1998). Intracranial tumors after exposure to ionizing radiation during infancy: A pooled analysis of two Swedish cohorts of 28,008 infants with skin hemangioma. *Radiat Res* **150**, 357–364.
17. Salvatai, M., Aratico, M., Caruso, R., Rocchi, G., Orlando, E. E. R., and Nucci, F. (1991). A report on radiation-induced gliomas. *Cancer* **67**, 392–397.
18. Yeh, H., Matanoski, G. M., Wang, N., Sandler, D. P., and Comstock, G. W. (2001). Cancer incidence after childhood nasopharyngeal radium irradiation: A follow-up study in Washington County, Maryland. *Am J Epidemiol* **153**, 749–756.
19. Shintani, T., Hayakawa, N., Hoshi, M. *et al.* (1999). High incidence of meningioma among Hiroshima atomic bomb survivors. *J Radiat Res* **40**, 49–57.
20. Loomis, D. P., and Wolf, S. H. (1996). Mortality of workers at a nuclear materials production plant at Oak Ridge, Tennessee, 1947–1990. *Am J Ind Med* **29**, 131–141.
21. Wong, O., Whorton, M. D., Follart, D. E., and Ragland, D. (1991). An industry-wide epidemiologic study of vinyl choride workers, 1942–1982. *Am J Ind Med* **20**, 317–334.
22. McLaughlin, J. K., and Lipworth, L. (1999). A critical review of the epidemiologic literature on health effects of occupational exposure to vinyl chloride. *J Epidemiol Biostat* **4**, 253–275.
23. Brenner, A. V., Linet, M. S., Fine, H. A. *et al.* (2002). History of allergies and autoimmune diseases and risk of brain tumors in adults. *Int J Cancer* **99**, 252–256.

24. Kleihues, P., and Cavanee, W. K. (2000). *Pathology and Genetics of Tumours of the Nervous System*. International Agency for Research on Cancer, Lyon.

25. Kleihues, P., Louis, D. N., Scheithauer, B. W. *et al.* (2002). The WHO classification of tumors of the nervous system. *J Neuropathol Exp Neurol* **61**, 215–225.

26. Okazaki, H. (Ed.) (1989). General methodology and pathologic cellular reactions. *In Fundamentals of Neuropathology. Morphologic Basis of Neurologic Disorders*, 2nd ed., pp. 1–26. Igaku-Shoin, New York.

27. Morrison, C. D., and Prayson, R. A. (2000). Immunohistochemistry in the diagnosis of neoplasms of the central nervous system. *Sem Diagn Pathol* **17**, 204–215.

28. Raghavan, R., Steart, P., and Weller, R. O. (1990). Cell proliferation patterns in the diagnosis of astrocytomas, anaplastic astrocytomas, and glioblastoma multiforme: a Ki-67 study. *Neuropathol Appl Neruobiol* **16**, 123–133.

29. Revesz, T., Alsanjari, N., Darling, J. L. *et al.* (1993). Proliferating cell nuclear antigen (PCNA): expression in samples of human astrocytic gliomas. *Neuropathol. Appl Neurobiol* **19**, 152–158.

30. McLendon, R. E., Enterline, D. S., Tien, R. D., Thorstad, W. L., and Bruner, J. M. (1998). Tumors of central neuroepithelial origin. *In Russell & Rubinstein's Pathology of Tumors of the Nervous System* (D. D. Bigner, R. E. McLendon, and J. M. Bruner, Eds.), 6th ed., Vol.1, pp.307–571. Arnold, London.

31. Krouwer, H. G. J., Davis, R. L., Silver, R., and Prados, M. (1991). Gemistocytic astrocytomas: a reappraisal. *J Neurosurg* **74**, 399–406.

32. Watanabe, K., Peraud, A., Gratas, C. *et al.* (1998). p53 and PTEN gene mutations in gemistocytic astrocytomas. *Acta Neuropathol (Berl)* **95**, 559–564.

33. Peraud, A., Ansari, H., Bise, K., and Reulen, H. J. (1998). Clinical outcome of supratentorial astrocytoma WHO grade II. *Acta Neurochir (Wien)* **140**, 1213–1222.

34. von Deimling, A., Louis, D. N., and Wiestler, O. D. (1995). Molecular pathways in the formation of gliomas. *Glia* **15**, 128–138.

35. Louis, D. N., Holland, E. C., and Cairncross, J. G. (2001). Glioma classification. A molecular reappraisal. *Am J Pathol* **159**, 779–786.

36. Maher, E. A., Furnari, F. B., Bachoo, R. M. *et al.* (2001). Malignant glioma: genetics and biology of a grave matter. *Genes & Develop.* **15**, 1311–1333.

37. Ichimura, K., Ohgaki, H., Kleihues, P., and Collins, V. P. (2004). Molecular pathogenesis of astrocytic tumours. *J Neuro-Oncol* **70**, 137–160.

38. Palma, L., Celli, P., Maleci, A. Di Lorenzo, N., and Cantore, G. (1989). Malignant monstrocellular brain tumours. A study of 42 surgically treated cases. *Acta Neurochir (Wien)* **97**, 17–25.

39. Meis, J. M., Ho, K. L., and Nelson, J. S. (1990). Gliosarcoma: a histologic and immunohistochemical reaffirmation. *Mod Pathol* **3**, 19–24.

40. Forsyth, P. A., Shaw, E. G., Scheithauer, B. W. *et al.* (1993). Supratentorial pilocytic astrocytomas: a clinicopathologic, prognostic, and flow cytometric study of 51 patients. *Cancer* **72**, 1335–42.

41. Kepes, J. J. (1993). Pleomorphic xanthoastrocytoma: the birth of a diagnosis and a concept. *Brain Pathol* **3**, 269–274.

42. Engelhard, H. H., Stelea, A., and Cochran, E. J. (2002). Oligodendrglioma: Pathology and molecular biology. *Surg Neurol* **58**, 111–117.

43. Reifenberger, G., and Louis, D. N. (2003). Oligodendroglioma: Toward molecular definitions in diagnostic neuro-oncology. *J Neuropathol Exp Neurol* **62**, 111–126.

44. Jeuken, J. W. M., von Deimling, A., and Wesseling, P. (2004). Molecular pathogenesis of oligodendroglial tumors. *J Neuro-Oncol* **70**, 161–181.

45. Beckmann, J. M., and Prayson, R. A. (1997). A clinicopathologic study of 30 cases of oligoastrocytoma including p53 immunohistochemistry. *Pathol* **29**, 159–164.

46. Rosenblum, M. K. (1998). Ependymal tumors: A review of their diagnostic surgical pathology. *Pediatr Neurosurg* **28**, 160–165.

47. Greene, K. A., Dickman, C. A., Marciano, F. F., Coons, S. W., and Rekate, H. L. (1994). Pathology and management of choroid plexus tumors. *BNI Quarterly* **10**, 13–21.

48. Luyken, C., Blümcke, I., Fimmers, R., Urbach, H., Wiestler, O. D., and Schramm, J. (2004). Supratentorial ganglioglioams: Histopathologic grading and tumor recurrence in 184 patients with a median follow-up of 8 years. *Cancer* **101**, 146–155.

49. Massoun, J., Soylemezoglu, F., Gambarelli, D., Figarella-Branger, D., von Ammon, K., and Kleihues, P. (1993). Central neurocytoma: A synopsis of clinical and histological features. *Brain Pathol* **3**, 297–306.

50. Eberhart, C. G., and Burger, P. C. (2003). Anaplasia and grading in medulloblastomas. *Brain Pathol* **13**, 376–385.

51. Rorke, L. B., Packer, R. J., and Biegel, J. A. (1996). Central nervous system atypical teratoid/rhabdoid tumors of infancy and childhood: definition of an entity. *J Neurosurg* **85**, 56–65.

52. Lamszus, K. (2004). Meningioma pathology, genetics, and biology. *J Neuropathol Exp Neurol* **63**, 275–286.

53. Perry, A., Gutmann, D. H., and Reifenberger, G. (2004). Molecular pathogenesis of meningiomas. *J Neuro-Oncol* **70**, 183–202.

54. Jellinger, K. A., and Paulus, W. (1992). Primary central nervous system lymphomas – an update. *J Cancer Res Clin Oncol* **119**, 7–27.

55. Matsutani, M., Sano, K., and Takakura, K., *et al.* (1997). Primary intracranial germ cell tumors: a clinical analysis of 153 histologically verified cases. *J Neurosurg* **86**, 446–455.

2

Clinical Pharmacology of Brain Tumor Chemotherapy

Herbert B. Newton

ABSTRACT: Chemotherapy has become a more common adjunctive treatment for most patients with malignant brain tumors and for selected patients with benign neoplasms. In general, the most effective chemotherapeutic drugs are alkylating agents, such as the nitrosoureas, procarbazine, temozolomide, cisplatin, carboplatin, and cyclophosphamide. Other agents that have been applied to brain tumors include methotrexate, vincristine, etoposide, irinotecan, hydroxyurea, and paclitaxel. Molecular therapeutic strategies are also under development, such as tyrosine kinase inhibitors and angiogenesis inhibitors. Overall, the efficacy of chemotherapy for brain tumors remains modest in the majority of patients. This chapter will review the detailed pharmacology and clinical application of chemotherapy in brain tumor patients.

INTRODUCTION

Although surgical resection and irradiation continue to be the focus of initial treatment for most patients with primary and metastatic brain tumors (PBT, MBT), chemotherapy has begun to play a more prominent role in the last fifteen years [1–4]. Chemotherapy can be broadly defined as the administration of one or more pharmacological agents with the intention of either killing the tumor cells or controlling their biological behavior such that further tumor growth is impaired. The expanding role of chemotherapy is partially due to the realization that, despite numerous improvements in the fields of neurosurgery and radiation oncology, malignant

brain tumors cannot be cured or well controlled by the use of these modalities alone. In addition, the recent data have demonstrated a modest yet significant survival advantage for those patients receiving chemotherapy, and suggest great potential for further improvements in the field [5–7]. In this chapter, the basic pharmacology and biology of cancer chemotherapy will be reviewed, along with an in-depth overview of chemotherapy agents and supportive drugs that have been applied to the treatment of brain tumors.

PHARMACOLOGY AND DRUG DISTRIBUTION

For a chemotherapy agent to be effective, it must be exposed to the brain tumor cells at concentrations that can overcome intrinsic resistance mechanisms [8,9]. This is often difficult for many drugs, due to the privileged location of central nervous system tumors, which are protected by the blood–brain and blood–tumor barriers (BBB, BTB). The biology of the BBB and BTB are discussed in Chapter 16. In general, drugs with the best penetration into the nervous system and brain tumors tend to be non-ionized, small molecules that are very lipid soluble. Lipid solubility (i.e., lipophilicity; equivalent to lipid/aqueous partition coefficient) is an essential factor for allowing the drugs to have a high brain capillary permeability (P_c) and be able to transfer across the intact BBB or to the regions of relatively intact BTB [10–12]. For small drugs, increasing lipophilicity has a positive impact on P_c and drug transfer into the brain. However, for large drugs above 400 Da, or for those with strong polarity, the P_c will remain low regardless of the degree of lipophilicity. Other factors of importance in the final distribution pattern of a given drug are the regional blood flow, tumor capillary permeability, tumor vascular surface area, drug stability, and intracerebral drug metabolism [10–12]. Once the drug has transferred across the brain or tumor capillaries, further distribution will be determined by diffusion (i.e., down concentration gradients) and convection (i.e., *via* bulk flow). The interstitial fluid pressure (IFP) of the tumor will partially determine the ease of ingress of drugs from the interstitial spaces [13]. Larger tumors tend to have elevated IFP, which generate a pressure gradient and an outward convection of interstitial fluid towards the periphery of the mass. This outward flow of interstitial fluid works against the gradient of drug diffusion and impedes the delivery of drugs to neoplastic cells.

SUPPORTIVE THERAPEUTICS

Supportive treatment of brain tumor patients is primarily concerned with controlling the symptoms caused by elevated intracranial pressure and minimizing seizure activity [1,2]. The use of other medications to ameliorate symptoms, such as nausea and emesis (i.e., secondary to edema or chemotherapy), constipation, pain, anxiety, and depresssion are beyond the scope of this chapter but are summarized in recent review articles [14–20].

Corticosteroids. The use of corticosteroids is often necessary in PBT and MBT patients to control the symptoms caused by increased intracranial pressure (e.g., headache, nausea and emesis, confusion, and weakness) [1,2,21]. Peritumoral edema is the principal cause of elevated intracranial pressure and is mediated through numerous mechanisms, including the leaky neovasculature associated with tumor angiogenesis, as well as increased permeability induced by factors secreted by the tumor and surrounding tissues, such as oxygen-free radicals, arachidonic acid, glutumate, histamine, bradykinin, atrial natriuretic peptide, and vascular endothelial growth factor (VEGF) [22–24]. Dexamethasone is the high-potency steroid used most often to treat the edema associated with brain tumors [1,21]. It has several advantages over other synthetic glucocorticoids, including a longer half-life, reduced mineralocorticoid effect, lower incidence of cognitive and behavioral complications, and diminished inhibition of leukocyte migration [25]. The mechanisms by which dexamethasone and other glucocorticoids reduce peritumoral edema remain unclear. It is

known that both PBT and MBT have high concentrations of glucocorticoid receptors. The effects of these drugs on tumor-induced edema are most likely mediated through binding to these receptors, with subsequent transfer to the nucleus and the expression of novel genes [24]. In a recent MRI study, dexamethasone was able to induce a dramatic reduction in blood–tumor barrier permeability and regional cerebral blood volume, without significant alteration in cerebral blood flow or the degree of edema [26]. The inhibition of production and/or release of vasoactive factors secreted by tumor cells and endothelial cells, such as VEGF and prostacyclin, appears to be involved in this process [23,24]. In addition, glucocorticoids appear to inhibit the reactivity of endothelial cells to several substances that induce capillary permeability.

The exact dose of steroids necessary for each patient will vary depending on the histology (i.e., benign or malignant), size and location of the tumor, and amount of peritumoral edema. In general, most patients with malignant tumors will require between 8 and 16 mg of dexamethasone per day to remain clinically stable. The lowest dose of steroid that can control the patient's pressure-related symptoms should be used [1,21]. This approach will minimize some of the toxicity and complications that can arise from long-term corticosteroid usage, which includes hyperglycemia, peripheral edema, proximal myopathy, gastritis, infection, osteopenia, weight gain, bowel perforation, and psychiatric or behavioral changes (e.g., euphoria, hypomania, depression, psychosis, and sleep disturbance) [1,27–31]. Patients with dexamethasone-induced proximal myopathy will often improve when the dosage is reduced [30,31]. In addition, the proximal leg muscles can usually be strengthened if the patient is placed on a lower extremity exercise regimen. Some authors have also reported an improvement in the myopathy when dexamethasone is replaced by an equivalent dosage of prednisone or hydrocortisone [30,31]. The neuropsychiatric complications of steroids can often be improved by dosage reduction or discontinuation of the drug [28]. For those patients in whom continued steroid usage is necessary, symptomatic pharmacological intervention is appropriate. For example, patients experiencing steroid-induced delerium or psychosis will often improve with low-dose haloperidol (0.5–1.0 mg PO, IM, or IV), titrated to control the symptoms. Steroid-induced sleep disturbances often respond to dosage reduction or by eliminating any doses after dinner. In refractory cases, the use of a hypnotic medication at bedtime (e.g., triazolam, 0.25 mg) will often be of benefit.

TABLE 2.1 Antiepileptic Drugs Commonly Used for the Treatment of Seizures in Brain
Tumor Patients

Drug	Dose (mg/day)	Metabolism	Enzyme inducing	Mechanism	Bound fraction (%)
Traditional AEDs					
Phenytoin	300–400	Hepatic +++	+++	Sodium channel	90–95
Carbamazepine	800–1600	Hepatic +++	+++	Sodium channel	75
Valproic acid	1000–3000	Hepatic +++	No; inhibitory	Sodium channel; enhanced GABA	80–90
Phenobarbital	90–180	Hepatic +++	+++	EAA antagonist; enhanced GABA	45
Newer AEDs					
Felbamate	2400–3600	Hepatic ++	+	EAA antagonist; enhanced GABA	25
Lamotrigine	100–500	Hepatic +++	None	Sodium channel	55
Gabapentin	1800–3600	Renal +++	None	Enhanced GABA	<5
Topiramate	200–400	Hepatic +	None	Sodium channel; EAA antagonist; enhanced GABA	9–17
Tiagabine	32–56	Hepatic +++	None	Enhanced GABA	95
Oxcarbazine	600–1800	Hepatic +++	+	Sodium channel	40
Levetiracetam	1000–3000	Renal ++	None	N-type calcium channels	<10
Zonisamide	100–400	Hepatic ++	None	Sodium & calcium channels; enhanced GABA	40

Adapted from references [1, 36, 40, 41, 43]

Abbreviations: AED – antiepileptic drug; + – mild; ++ – moderate; +++ – severe; EAA – excitatory amino acids; GABA – gamma amino butyric acid

Corticosteroid-induced osteoporosis is a common problem, affecting 30–50 per cent of patients receiving treatment for a year or more [27,32,33]. Patients on long-term dexamethosone require a preventive program to minimize osteoporosis, including calcium and vitamin D supplements, and weight-bearing exercises. These measures should be started early, since bone loss is the greatest in the first 2–4 months of chronic steroid treatment. For patients on long-term steroid therapy (i.e., ≥3 months), or in those with established osteoporosis or evidence of an osteoporotic fracture, bisphosphonate therapy (e.g., risedronate, 2.5–5.0 mg/day; alendronate, 5–10 mg/day) should be added to the regimen of calcium and vitamin D supplements [33].

Gastric acid inhibitors. Patients on long-term corticosteroids are at increased risk for gastrointestinal complications (i.e., gastritis, ulceration, and bowel perforation); an incidence of 14 per cent was noted in one study [28]. Patients at risk should be treated prophylactically with a gastric acid inhibitor, such as ranitidine hydrochloride (150 mg po bid), famotidine (20 mg po bid), or omeprazole (20–40 mg po qd) [34,35].

Anticonvulsants. Seizures occur at presentation in 20–50 per cent of patients with PBT and MBT [1,2,36]. It is important to note that more than 25 per cent of the adults between 25 and 64 years of age with newly diagnosed seizures will have an underlying brain tumor [36]. At the time of tumor progression, seizure activity often becomes more frequent and severe, affecting another 10–20 per cent of patients. The overall incidence of seizures is the highest in patients with PBT of low histological grade and slow-growth potential, and becomes less frequent in those with high-grade PBT and MBT.

There is a general consensus that any brain tumor patient with a well documented, unequivocal seizure (generalized or focal) should be placed on an antiepileptic drug (AED; see Table 2.1) [1,2,36–39]. For adult patients with generalized seizures, phenytoin, carbamazepine, and valproate have relatively equivalent efficacy for reducing the seizure activity [40,41]. Similarly, all three drugs are effective for partial motor, partial sensory, and partial complex seizures. However, a comparative trial of carbamazepine and valproate has demonstrated better control of complex partial seizure activity with carbamazepine [42]. Monotherapy with phenytoin, carbamazepine, or valproate should be the initial management approach in most of the patients [1,40,41]. In some of the patients, a second drug must be added if high therapeutic concentrations of several of the first-line drugs are unable to control the seizure activity. Phenytoin or carbamazepine in combination with valproate is a common strategy. Alternatively, one of the new anticonvulsants (e.g., levetiracetam, gabapentin, topiramate, and zonisamide) could be added to one of the first-line agents [43–45]. Levetiracetam may be an excellent choice, since initial experience suggests it is effective and well tolerated in brain tumor patients, and has minimal potential to interact with other drugs, such as corticosteroids or chemotherapy agents [45]. Ongoing studies will determine if levetiracetam and other new agents might be appropriate for first-line use or as secondary, stand-alone agents. Serum drug concentrations must be monitored and optimized in all patients whenever clinically indicated (e.g., phenytoin, carbamazepine, and valproate).

Anticonvulsant drugs are frequently administered to brain tumor patients at the time of diagnosis or after craniotomy, as prophylaxis for potential seizure activity [36]. This practice was given early support in the literature, despite the fact that the data in these reports was modest at best [46,47]. All subsequent reports on the use of AED prophylaxis do not support this practice, including a randomized, blinded, placebo-controlled trial of valproate in patients with newly diagnosed PBT and MBT [48–52]. In this study, the odds ratio for a seizure in the valproate arm relative to the placebo arm was 1.7 ($p = 0.3$) [52]. Most of the authors would now recommend withholding implementation of AED in newly diagnosed patients until a seizure has been documented. This approach is supported by a recent meta-analysis by Glantz and associates for the American Academy of Neurology [53]. In addition, for patients that have not had a seizure and have received AED for craniotomy, tapering and discontinuing the AED after the first postoperative week is recommended [53].

DELIVERY OF CHEMOTHERAPEUTIC AGENTS

Chemotherapeutic drugs are commonly given orally or by bolus intravenous infusion. Several agents used for PBT are administered orally, including temozolomide, lomustine, procarbazine (PCB), and tamoxifen. The main benefit of oral chemotherapy is the ease of administration [54]. However, the oral bioavailability of many drugs is variable and depends on factors, such as drug stability in gastric acid, ease of absorption through the intestinal mucosa, inactivation by intestinal enzymes (e.g., CYP3A4, cytidine deaminase), hepatic metabolism, biliary excretion, and treatment-induced emesis [54,55]. Administration by bolus intravenous infusion provides maximal peak drug concentrations in plasma, followed by a rapid decline as drug is metabolized and excreted. Agents that are cell-cycle-nonspecific and interact chemically with DNA (e.g., BCNU, cisplatin) are most appropriate for this route of administration [55]. The dosing and schedule of treatment will depend on the timing of chemotherapy induced nadirs and the toxicity profiles of the individual drugs used.

Continuous infusion. Continuous intravenous infusion is defined as parenteral administration of a drug for 24 h or longer [56]. The rationale for treatment by continuous infusion is based on the fact that most of the antineoplastic agents kill tumor cells when they are in an active phase of the cell cycle (i.e., non-G_0) [56]. In most of the brain tumors, the "active" population of cells is small, often 5–10 per cent or less [57]. Treatment by continuous infusion maximizes the number of actively cycling cells exposed to a sustained, therapeutic plasma concentration of the drug. This approach is especially important for cell-cycle-specific drugs, such as methotrexate, fluorouracil, and cytarabine. Recent data suggest that continuous infusion may enhance the beneficial effects of alkylating agents and topoisomerase inhibitors by reducing the toxicity (as seen with ifosfamide and cisplatin) or increasing the cell kill (as seen with etoposide) [56].

Intra-arterial. Intra-arterial (IA) administration of drugs into the carotid or vertebral arteries is another method to improve the efficacy of chemotherapy for brain tumors. The rationale for IA treatment is that it augments the local plasma "peak" drug concentration and local area under the concentration–time curve as compared to an equivalent intravenous dose [8,58]. This method can potentially increase the drug concentrations by twofold to fourfold while reducing the risk of systemic toxicity. The benefits of IA administration are maximized with drugs that have a high rate of biotransformation, metabolism, and excretion during the first pass (i.e., total body drug clearance; CL_{tb}). BCNU, cisplatin, etoposide, methotrexate, and carboplatin are only some of the drugs that have been used intra-arterially to treat PBT and MBT [3,8]. The risks of IA treatment are significant and include stroke, retinal toxicity, leukoencephalopathy, vascular dissection, and seizures [58]. For more information regarding the use of IA chemotherapy for brain tumor treatment, see Chapter 17.

Interstitial. Interstitial chemotherapy involves administration of drugs by various methods directly into, or adjacent to, the brain tumor. Numerous strategies have been explored, including topical application, indwelling catheters, biodegradable polymers, ommaya reservoirs, and implantable drug delivery systems [8,59,60]. The most promising form of interstitial treatment uses a biodegradable polyanhydride polymer wafer containing 3.8 per cent

BCNU (i.e., Gliadel® wafer) [60,61]. The wafer allows controlled, sustained high concentrations of BCNU in tumor tissues with minimal systemic exposure to drug. This has been verified in animal models (e.g., rodents, primates) and patients with PBT [60–65]. For more information regarding the use of interstitial chemotherapy for brain tumor treatment, see Chapter 19.

Blood–brain barrier disruption. The blood–brain barrier (BBB) results from specific structural alterations of endothelial cells of the central nervous system vasculature. When compared to vessels in other organs, central nervous system endothelial cells lack fenestrations, contain few pinocytotic vesicles, and have tight junctions between the adjacent cells [8,66,67]. Although the integrity of the BBB is compromised within many tumors (i.e., contrast enhancement), passage of chemotherapeutic agents into the tumor tissues is often prohibited. Only drugs with low molecular weight, high lipid solubility, low ionization, and minimal protein binding can readily pass the blood–brain barrier [10–12]. Unfortunately, many beneficial chemotherapeutic agents do not fit into this profile. In order to circumvent the restrictions of the BBB and increase the intratumoral concentrations of drug, the barrier can be transiently disrupted with osmotic agents (i.e., 25 per cent mannitol) before chemotherapy administration [8,66–69]. The mannitol is administered intra-arterially at a rate of 8–10 ml/s for over 30 s. Following the mannitol infusion, chemotherapy is then given by the IA (e.g., methotrexate or carboplatin) and intravenous (e.g., cyclophosphamide) routes. Limitations of BBB disruption treatment include the invasive nature of the procedure (i.e., general anesthesia), the inherent risks of IA chemotherapy (e.g., stroke, seizure, vascular dissection, and vasospasm), and the potential for toxicity to normal brain within the disrupted vascular territory [70]. For more information regarding the use of BBB disruption chemotherapy for brain tumor treatment, see Chapter 18.

CHEMOTHERAPEUTIC AGENTS

Chemotherapy is considered as an adjunctive treatment after surgical resection and irradiation for patients with malignant brain tumors and selected recurrent or progressive benign neoplasms [1,2]. As mentioned above, several meta-analyses have suggested that chemotherapy does confer a survival advantage to the adult patients with malignant gliomas [5,7]. Numerous chemotherapeutic agents have been tested, as single agents and in various

combinations, for efficacy against brain tumors [3,8,21,71–73]. In general, these agents can be grouped into cell-cycle-specific and cell-cycle-nonspecific classes. Cell-cycle-specific chemotherapy agents attack the tumor cells during certain phases of the cell cycle (e.g., S-phase, mitosis) and are more efficacious when tumors have a high mitotic rate [8]. Examples of cell-cycle-specific agents are methotrexate, vincristine, and etoposide. Cell-cycle-nonspecific agents attack the tumor cells during all the phases of the cell cycle, and are not limited by the size of the growth fraction. Examples of cell-cycle-nonspecific agents are temozolomide, BCNU, and cisplatin. Chemotherapeutic agents are further classified by pharmacological mechanism (e.g., alkylating agents, antimetabolites, and topoisomerase inhibitors). The following section will group chemotherapeutic agents by mechanism and discuss their pharmacology, toxicity, complications, and applications to brain tumor therapy. The emphasis will be on drugs that have demonstrated at least minimal activity against brain tumors in clinical trials, either as single agents or in combination regimens. See Table 2.2 for a review of various chemotherapy regimens and their results, listed according to brain tumor type.

ALKYLATING AGENTS

Nitrogen mustards. In this category, there are several agents that have been used in the treatment of brain tumors, including mechlorethamine, melphalan, chlorambucil, cyclophosphamide, and ifosfamide (See Figs. 2.1A–2.1D) [8,21]. The general mechanism of action of drugs in this class involves the formation of reactive carbonium ions that subsequently interact with electrophilic regions of susceptible molecules (i.e., DNA). Mechlorethamine and melphalan have both been used in small phase II trials of IA administration with minimal efficacy and significant neurologic toxicity (e.g., encephalopathy, cerebellar syndrome, hemiparesis, and neuropathy) [8,21]. When used in combination with vincristine, PCB, and prednisone (i.e., MOPP), mechlorethamine demonstrated mild activity against recurrent high-grade glioma [74]. Chlorambucil is an oral congener of melphalan, which has demonstrated minimal activity against gliomas. Cyclophosphamide and ifosfamide are related compounds that require hepatic activation and must be administered orally or by intravenous infusion [8,21]. Cyclophosphamide is initially hydroxylated to form 4-hydroxycyclophosphamide. Further activation generates an intermediate compound, aldophosphamide, which then breaks down

TABLE 2.2 Selected Chemotherapy Regimens and their Results Listed According to Tumor Type

Regimen	Trial type	No. Cases	Results (weeks)	Reference #
Glioblastoma multiforme				
RT + IV BCNU	Phase III	290	MST 50–56	[85,86,142]
RT + IA BCNU	Phase III	56	MST 44.8	[142]
RT + Oral PCB	Phase III	114	MST 47	[85]
RT + PCV	Phase III	31	MST 50.4	[103]
Oral PCB	Phase II	10	MTP 24	[94,95]
IV CTX/VCR	Phase II	4	MNTP 19	[77]
IV DDP	Phase II	10	MNTP 9.5	[110]
IA DDP	Phase II	19	MNTP 14	[112]
IV Carbo	Phase II	15	MST 32	[118]
IV Carbo + VP-16	Phase II	30	MST 43.5	[119]
IA MTX, IV CTX, oral PCB, BBB	Phase II	28	MST 70	[68]
Oral TZM	Phase II	225	MPFS 11	[138]
Oral VP-16	Phase II	21	MNTP 7.5	[159]
Oral DBD	Phase III	64	MST 41	[170]
Anaplastic astrocytoma				
RT + PCV	Phase III	36	MTP 125.6	[103]
Oral PCB	Phase II	15	MTP 42	[94]
IA DDP	Phase II	11	MNTP 18	[112]
IV Carbo	Phase II	14	MST 32	[118]
IA MTX, IV CTX, oral PCB, BBB	Phase II	10	MST 70	[68]
Oral VP-16	Phase II	15	MNTP 9.1	[159]
Oral DBD	Phase III	44	MST 41	[170]
RT + IV Carbo, oral tamoxifen	Phase II	32	MST 52	[187]
Pediatric recurrent gliomas				
IV CTX	Phase II	7	MNTP 29.6	[75]
Oral VP-16	Phase II	14	MTP 32	[160]
Oligodendroglioma				
PCV	Phase II	23	MNTP 75	[100]
Paclitaxel	Phase II	20	MTP 40	[192]
Oligoastrocytoma				
PCV	Phase II	14	MNTP 37.3	[101]
PCV	Phase II	25	MNTP 52.4	[102]
Anaplastic oligodendroglioma				
PCV	Phase II	5	MNTP 53.2	[101]
PCV	Phase II	7	MTP 254	[102]
PCV	Phase II	39	MST 324	[99]
Medulloblastoma				
RT + CCNU/VCR (high risk)	Phase III	233	5-year DFS 59%	[87]
RT + CCNU/VCR, DDP	Phase II	51	5-year DFS 88%	[108]
IV CTX	Phase II	14	MTP 24	[75]
IV CTX/VCR	Phase II	12	MNTP 22.4	[76]
IV DDP	Phase II	14	MTP 32	[106]
IV Carbo	Phase I/II	14	MTP 40	[116]
Oral DBD	Phase II	20	MTP 10	[170]
Primary CNS lymphoma				
RT + IV/IT MTX, Ara-C	Phase II	31	MST 170	[145]
IA MTX, IV CTX, oral PCB, BBB	Phase II	17	MST 178	[150]
IV BCNU, MTX, VCR, VP-16	Phase II	19	MNTP 71	[151]

Abbreviations: RT – radiation therapy; IV – intravenous; MST – median survival time; IA – intra-arterial; PCB – procarbazine; PCV – procarbazine, CCNU, vincristine; MTP – median time to progression; CTX – cyclophosphamide; VCR – vincristine; MNTP – mean time to progression; MPFS – median progression-free survival; TZM – temozolomide; DDP – cisplatin; carbo – carboplatin; MTX – methotrexate; BBB – blood–brain barrier; DBD – dibromodulcitol; DFS – disease-free survival; IT – intrathecal; ara-C – cytosine arabinoside

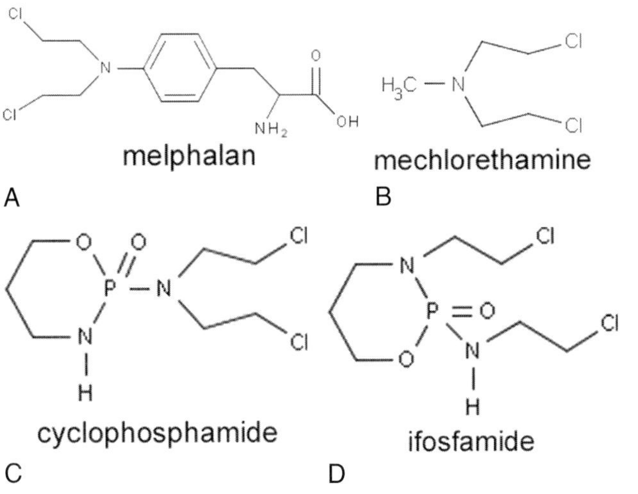

FIGURE 2.1

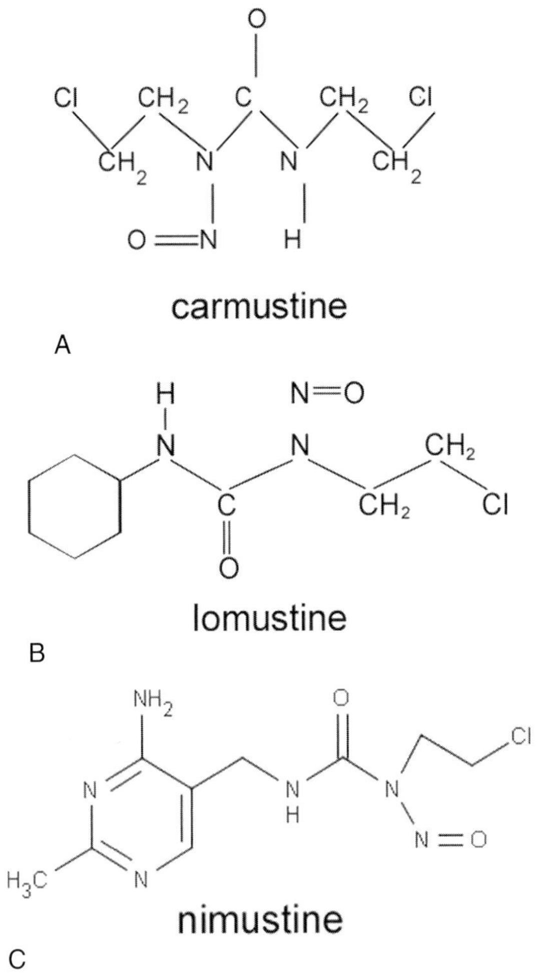

FIGURE 2.2

into phosphoramide mustard and acrolein. Ifosfamide undergoes similar metabolic activation by hepatic P450 mixed-function oxidases. Cyclophosphamide has shown activity as salvage treatment against recurrent PBT in adults and children, used as a single agent or in combination regimens [8,21,75–78]. In adult patients, the typical doses have ranged from 500 to 1000 mg/m^2; in children doses have been between 80 and 165 mg/kg. Dose-intensified regimens using cyclophosphamide (2000 mg/m^2) have failed to improve response rates or survival and have significant toxicity [79]. When administered to adults with recurrent gliomas at a dose of 2500 mg/m^2 per day, ifosfamide failed to demonstrate significant activity [80]. Ifosfamide appears to be more active against recurrent pediatric PBT when used in combination regimens. The primary toxicities of cyclophosphamide and ifosfamide are myelosuppression and hemorrhagic cystitis. The risk of hemorrhagic cystitis can be minimized by concomitant hydration and administration of mesna. Mesna binds to the metabolic breakdown product, acrolein, inhibiting its irritation of the bladder mucosa. Those patients treated with high-dose cyclophosphamide, and all patients receiving ifosfamide, require mesna (500–600 mg/m^2 IV × 3 doses at 0, 4, and 8 h after chemotherapy) in addition to hydration [79–81]. Recent research also suggests that oral mesna may have a similar uroprotective effect, when used alone or in combination with intravenous mesna [82]. Ifosfamide is associated with significant neurotoxicity (e.g., encephalopathy, seizures, hallucinations, and coma) in 5–30 per cent of the patients [83].

Nitrosoureas. Nitrosoureas are the most commonly used chemotherapeutic agents used in the treatment of malignant PBT, and have also been applied to MBT [3,8,21,71–73]. Included in this class are CCNU (lomustine), BCNU (carmustine), ACNU (nimustine), PCNU, fotemustine, streptozotocin, and MeCCNU (see Figs. 2.2A–2.2C). As a group, nitrosoureas are highly lipid soluble, non-ionized, cell-cycle-nonspecific agents that readily cross the blood–brain barrier [8,21]. They spontaneously decompose into two active intermediates, a chloroethyldiazohydroxide and an isocyanate group. DNA alkylation leads to the formation of DNA–DNA and DNA–protein cross-links; mediated by the chloroethyl-diazohydroxide intermediate. The isocyanate intermediate produces carbamoylation of amino groups, which depletes glutathione, inhibits DNA repair, and interferes with RNA synthesis. After systemic administration, BCNU concentrations within brain reach 15–70 per cent of plasma concentrations, while CCNU concentrations are equivalent to that of plasma. The nitrosoureas have been used either as single agents (BCNU – most effective) or in combination with other cytotoxic drugs (e.g., PCV: procarbazine, CCNU, and vincristine). In addition, several of the nitrosoureas (BCNU, ACNU) have been applied to PBT and MBT by IA delivery techniques (see Chapters 17 and 35).

Seminal studies by the Brain Tumor Study Group demonstrated that the addition of nitrosoureas could improve survival in patients with malignant PBT [84–86]. Walker and colleagues analyzed 358 patients with malignant glioma and demonstrated that survival at 18 months was superior for patients receiving irradiation plus BCNU in comparison to those treated with irradiation alone [84]. Green and associates studied 527 patients with malignant glioma after surgery and irradiation, randomizing them into four groups (BCNU, methylprednisolone, procarbazine, or BCNU plus methylprednisolone) [85]. Median survival for patients treated with single-agent BCNU (50 weeks) and procarbazine (47 weeks) were superior to survival in the groups receiving other regimens or irradiation alone (36 weeks).

CCNU (110 mg/m^2 every 6–8 weeks) is an integral component of the PCV regimen that is commonly used for anaplastic oligodendroglioma, recurrent oligodendroglioma, anaplastic astrocytoma, and mixed gliomas [8,21,71–73]. In addition, CCNU has proved beneficial in multi-agent regimens for treatment of high-risk medulloblastoma patients [87,88]. In a Children's Cancer Study Group prospective, randomized trial of radiation therapy *versus* radiation plus chemotherapy in 233 patients with medulloblastoma, the addition of CCNU, vincristine, and prednisone extended event-free and overall survival in the high-risk group [87].

The major acute toxicity of the nitrosoureas is gastrointestinal (i.e., nausea and emesis). The dose-limiting side effect is myelosuppression, which develops at 3–5 weeks after the treatment. In up to 20 per cent of the patients receiving nitrosoureas (mainly BCNU), pulmonary fibrosis can occur. Risk factors for the development of pulmonary fibrosis include the total cumulative dose (>1100 mg/m^2) and number of cycles of drug (>5 cycles), co-morbid lung disease, age, smoking history, and platelet count nadir after the first course of treatment [89].

Procarbazine. Procarbazine is a drug that requires hepatic activation to intermediate forms before developing potent activity as an alkylating agent (see Fig. 2.3) [90,91]. PCB is taken orally and is rapidly absorbed by the gastrointestinal tract. Following absorption, it is first metabolized into an azo-PCB derivative, which has similar potency to PCB. Further metabolism by the cytochrome P-450 system converts

procarbazine

FIGURE 2.3

azo-PCB into two separate azoxy-PCB derivatives, which have significantly greater antitumor activity than PCB or azo-PCB [90,91]. Once activated, PCB alkylates DNA at the O^6 position of guanine [92]. In addition, PCB can induce DNA strand breakage and inhibit DNA, RNA, and protein synthesis. PCB has further pharmacological properties, including activity as a monoamine oxidase-inhibitor and a disulfiram-like effect [8,21]. Potential interactions (i.e., acute hypertension) can occur if PCB is taken concomitantly with sympathomimetic drugs, antihistamines, tricyclic antidepressants, and food, high in tyramine content (e.g., wine, beer, cheese, chocolate, bananas, and yogurt) [93]. Due to the disulfiram-like effect, alcohol should be avoided while taking PCB, or severe gastrointestinal distress will develop. Although water soluble, PCB and its metabolites readily cross the blood–brain barrier, with rapid equilibration between plasma and cerebrospinal fluid.

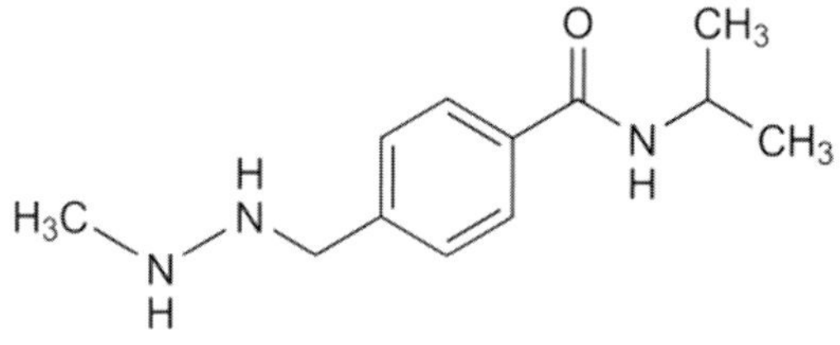

Procarbazine has been used to treat malignant PBT as a single-agent or in multi-agent regimens [8,21,71–73]. As mentioned above, single-agent PCB was compared to intravenous BCNU, methylprednisolone, and intravenous BCNU plus methylprednisolone in a randomized trial of malignant glioma patients by Green and colleagues [85]. Overall median survival was similar for the PCB (47 weeks) and BCNU (50 weeks) groups. Although the survival percentage was similar for PCB and BCNU at 12 months (44.0 *versus* 48.5 per cent), long-term survival at 18 and 24 months was superior in the PCB group (28.8 and 22.8 per cent *versus* 23.8 and 15.6 per cent, respectively). Newton and colleagues found PCB beneficial for glioma patients after failure of irradiation and nitrosourea chemotherapy [94]. In a series of 35 patients they noted 2 complete responses, 7 partial responses, and 11 patients with stable disease. Furthermore, when responses were compared in a cohort of malignant glioma patients who were initially treated with BCNU and then received PCB (after BCNU failure), there was a significant difference in time to progression between the groups [95]. The PCB group had a greater percentage of patients without disease progression at 6 and 12 months as compared to BCNU (48 and 35 per cent *versus* 26 and 3 per cent, respectively). Other investigators have found single-agent PCB to be less efficacious, with fewer complete and partial responses [96].

The most commonly used multi-agent regimen that incorporates PCB is PCV (PCB 60 mg/m^2 per days, 8–21, CCNU 110 mg/m^2 day 1, vincristine 1.4 mg/m^2 days 8 and 28; every 8 weeks) [8,21,71–73]. Numerous reports document the efficacy of PCV against anaplastic oligodendrogliomas, anaplastic astrocytomas, recurrent oligodendrogliomas, and mixed gliomas. The most dramatic results have been in patients with anaplastic or recurrent oligodendrogliomas [97–99]. Cairncross and colleagues have determined that anaplastic oligodendrogliomas are relatively chemosensitive, especially to PCV. Several series of patients have reported durable response rates in excess of 50 per cent [73,97–99]. A recent report suggests that anaplastic oligodendrogliomas with a specific molecular genetic profile are more sensitive to PCV [99]. Those patients with tumors that had allelic deletion or loss of heterozygosity of chromosomes 1p and 19q had significantly longer recurrence-free and overall survival than patients that retained 1p and 19q. The use of PCV appears to be beneficial for recurrent typical (i.e., WHO grade II) oligodendroglioma as well, especially after failure of non-PCV regimens [100]. Mixed gliomas with an oligodendroglial component also appear sensitive to PCV [101,102]. PCV has also been used for the treatment of anaplastic and mixed astrocytoma by Levin and colleagues [103]. They found PCV superior to BCNU when comparing time to progression and overall survival. Other multi-agent regimens for malignant gliomas combine PCB with vincristine and either mechlorethamine or etoposide (VP-16) [74,104]. Results with these regimens are similar to that reported for intravenous BCNU.

The most frequent acute toxicity of PCB is nausea and emesis (80–85 per cent) [94,95]. Other less common side effects include fatigue (17 per cent), rash, and neurotoxicity (usually when administered intravenously). For most patients, the dose-limiting toxicity is a combination of neutropenia and thrombocytopenia.

Platinum compounds. Cisplatin and carboplatin are divalent platinum coordination compounds that act as bifunctional alkylating agents in a cell-cycle-nonspecific manner (see Figs. 2.4A,B) [8,21,71–73]. Alkylation occurs at the N^7 position of guanine, producing intra- and inter-strand DNA cross-links [105]. Both the agents are water soluble, excreted primarily in the urine, and demonstrate poor penetration

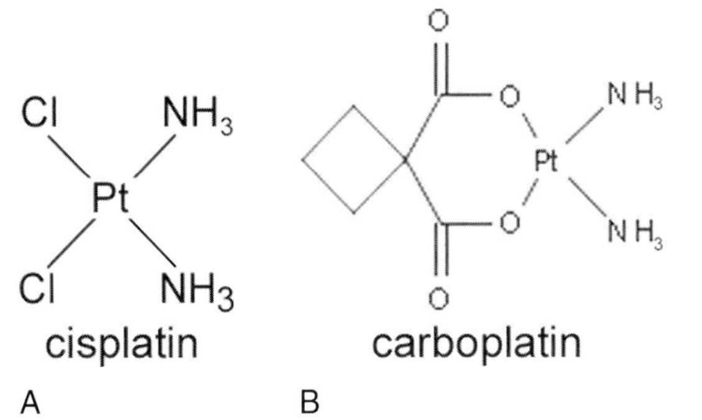

FIGURE 2.4

of the blood–brain barrier. Cisplatin and carboplatin can be administered intravenously or intra-arterially. They have both demonstrated efficacy for the treatment of adult and pediatric PBT [71–73].

Cisplatin has variable activity against a wide range of tumors including anaplastic astrocytoma, glioblastoma multiforme, medulloblastoma, central nervous system lymphoma, germ-cell tumors, recurrent gliomas, and primitive neuroectodermal tumors [8,21,71–73]. It can be administered as a single agent or in combination with other drugs, such as BCNU, etoposide, cyclophosphamide, or vincristine. In the pediatric population, cisplatin as a single agent (60–120 mg/m^2 per cycle IV) has proved beneficial for patients with recurrent medulloblastoma, ependymoma, and primitive neuroectodermal tumors [106,107]. In combination with CCNU or cyclophosphamide plus vincristine, cisplatin has efficacy against high risk and recurrent medulloblastoma (see Nitrosourea section) [88,108,109]. In adult patients with malignant gliomas after nitrosourea failure, single-agent intravenous cisplatin showed modest activity, although the duration of responses was brief [110]. When used in combination with etoposide or BCNU, intravenous cisplatin has efficacy against a variety of recurrent malignant PBT [71–73].

Numerous investigators have attempted to administer cisplatin intra-arterially to patients with newly diagnosed and recurrent gliomas [8,21,71–73, 111–114]. Early reports suggested significant activity of the drug (60–100 mg/m^2 per cycle) when used by this route [111,112]. Response rates of 40–60 per cent (i.e., partial responses plus stable disease) were noted, with overall time to progression ranging from 13 to 22 weeks [111–113]. However, reports by other authors suggest minimal efficacy of IA cisplatin against these tumors, as well as unacceptable toxicity [114]. In general, IA cisplatin is associated with increased risk for encephalopathy, cerebral edema, seizure activity, stroke, and retinal damage.

The major acute toxicity of cisplatin is severe nausea and emesis, which can be acute or delayed (24–120 h after treatment) in onset. Patients will require a potent prophylactic antiemetic regimen, such as a $5\text{-}HT_3$-receptor antagonist, plus or minus dexamethasone, for acute emesis and prochlorperazine for delayed emesis [14,15]. Nephrotoxicity is another common side effect that involves several mechanisms, including coagulative necrosis, drug–protein interactions, and inactivation of specific renal brush border enzymes [8]. The areas of the kidney most severely affected on histological examination include the loops of Henle, distal tubules, and collecting ducts. To reduce the risk of renal damage, aggressive hydration (often in conjunction with mannitol) is necessary. Other potential side effects include ototoxicity, myelosuppression, and peripheral neuropathy.

Carboplatin is an analog of cisplatin that has a similar profile of activity against low-grade and malignant PBT [8,21,71–73]. *In vitro* drug testing of cisplatin and carboplatin has shown comparable cytotoxicity against glioma cell lines at clinically relevant concentrations [115]. Although it is more myelosuppressive than cisplatin (i.e., thrombocytopenia), carboplatin causes less ototoxicity, nephrotoxicity, nausea and emesis, and peripheral neuropathy. Single-agent intravenous carboplatin (175 mg/m^2 weekly or 560 mg/m^2 monthly) has demonstrated efficacy in pediatric patients for the treatment of low-grade gliomas and recurrent malignant PBT [116,117]. In adult patients, single-agent carboplatin (400–450 mg/m^2 IV every 4 weeks) was moderately effective for the treatment of recurrent malignant gliomas [118]. The overall response rate was 48 per cent, with a median time to progression of 26 weeks in responders. In multi-agent regimens, carboplatin has been combined with etoposide and teniposide, with results similar to that achieved by cisplatin or BCNU [119,120]. Carboplatin has also been administered intra-arterially, either alone or in combination with etoposide and BBB disruption, with a similar level of activity to IA cisplatin [121,122]. However, significantly less neurological and retinal toxicity is noted with IA carboplatin.

Temozolomide. Temozolomide is an imidazotetrazine derivative of the alkylating agent dacarbazine with activity against systemic and CNS malignancies [123–125]. The drug undergoes chemical conversion at physiological pH to the active species 5-(3-methyl-1-triazeno)imidazole-4-carboxamide (MTIC) (see Fig. 2.5). Temozolomide exhibits schedule-dependent

temozolomide

FIGURE 2.5

antineoplastic activity by interfering with DNA replication through the process of methylation. The methylation of DNA is dependent upon the formation of a reactive methyldiazonium cation, which interacts with DNA at the following sites: N^7-guanine (70 per cent), N^3-adenine (9.2 per cent), and O^6-guanine (5 per cent). The cytotoxicity of temozolomide can be modulated by the degree of activity of three DNA-repair enzyme systems: DNA-mismatch repair, O^6-alkylguanine-DNA alkyltransferase (AGT), and poly (ADP-ribose) polymerase [124,126–128]. The DNA-mismatch repair pathways must be at a normal functional capacity to confer temozolomide cytotoxicity. The mechanism remains unclear, but may involve initiation of apoptosis in those cells that cannot repair the methylated sites. Tumor cells with mutations in, or phenotypically low expression of, DNA-mismatch repair genes are more resistant to temozolomide and other methylating agents. Conversely, high expression of AGT confers resistance to temozolomide by removing methyl groups from DNA before cell injury and death can occur. This relationship has been noted *in vitro* and is also clinically relevant, as shown by Friedman and colleagues in a series of 33 newly diagnosed patients with malignant gliomas [128]. Patients with tumors that stained strongly for AGT did not respond well to temozolomide. Depletion of AGT levels with O^6-benzylguanine can restore sensitivity to temozolomide and other alkylating agents [129].

The antitumor activity of temozolomide is schedule dependent as shown by *in vitro* and *in vivo* experiments [124,125]. A five-day administration schedule is superior to single-day dosing for numerous malignancies, including lymphoma, leukemia, and CNS tumor xenografts. As this drug is stable at acid pH, it can be taken orally in capsules. Oral bioavailability is approximately 100 per cent, with rapid absorption of the drug. Mean peak plasma concentrations (C_{max}) are reached 60 min after oral dosing. Temozolomide pharmacokinetics appear to be linear, so that C_{max} and the area under the concentration–time curve (AUC) increase proportionally with single oral doses over a range of 200–1200 mg/m^2. Administration of temozolomide with food results in a 33 per cent decrease in C_{max} and a 9 per cent decrease in AUC. Therefore, it is recommended that the drug be given without food. The drug is eliminated by pH-dependent degradation to MTIC, followed by further degradation to 4-amino-5-imidazole-carboxamide. Temozolomide has excellent penetration of the blood–brain barrier and brain tumor tissue. In glioma patients, tumor uptake and concentration of ^{11}C-temozolomide, as measured by positron emission tomography, correlated with response duration ($p < 0.01$) [130].

Pre-clinical testing of temozolomide using glioma cell lines and animal models has consistently demonstrated significant activity [124,125]. Temozolomide was found effective against U251 and SF-295 human xenografts implanted subcutaneously or intracerebrally into athymic mice [131]. A synergistic response was noted when temozolomide was used in conjunction with BCNU. Similarly, temozolomide was effective against subcutaneously or intracerebrally implanted tumor xenografts derived from adult glioma, pediatric glioma, and ependymoma cell lines [132]. The cytotoxicity of temozolomide was increased after pre-treatment with O^6-benzylguanine.

Temozolomide has shown significant activity and excellent tolerability in clinical trials against both adult and pediatric malignant gliomas [124,125, 128,133–141]. Initial clinical reports from Europe by Newlands and colleagues suggested that temozolomide was very active against newly diagnosed, as well as recurrent and progressive, high-grade gliomas [124,133,134]. Over 100 patients with glioblastama multiforme (GBM) and AA were treated with temozolomide (150–200 mg/m^2 per day × 5 days, every 28 days); some newly diagnosed patients received the drug prior to irradiation, while others were treated after recurrence or progression. The objective response rate was 30 per cent for newly diagnosed patients and 11–25 per cent for patients with progressive disease. The response rates were similar between AA and GBM patients, with a median duration of response of 4.6 months in patients with progressive disease. Toxicity was mainly hematological, but also included nausea and emesis, headache, fatigue, and constipation. Several phase I studies using temozolomide in pediatric patients with advanced solid tumors have demonstrated objective responses in PNET, medulloblastoma, high-grade astrocytoma, and brainstem glioma [135,136]. The drug was generally well tolerated, although nausea and emesis occurred frequently. A phase II study of AA and anaplastic mixed glioma at first relapse

also found temozolomide (150–200 mg/m^2 per day × 5 days) to have significant activity [137]. The objective response rate was 25 per cent (8 per cent CR, 27 per cent PR), with an additional 26 per cent having stable disease. The median progression-free survival was 5.4 months, with a median overall survival of 13.6 months. In 225 patients with GBM at first relapse, a randomized phase II trial by Yung and coworkers compared temozolomide to PCB (125–150 mg/m^2 per day × 28 days, every 56 days) [138]. Progression-free survival at 6 months was significantly better for temozolomide than for PCB (21 *versus* 8 per cent; $p < 0.008$). Median progression-free survival showed a similar advantage for temozolomide over PCB (2.89 *versus* 1.88 months; $p < 0.0063$). The six-month overall survival was also significantly different between the two drugs: 60 per cent for temozolomide and 44 per cent for PCB ($p < 0.019$). In a further comparison of temozolomide and PCB, Osoba and colleagues reviewed health-related quality of life in a cohort of patients with recurrent GBM [140]. Treatment with temozolomide was well tolerated and associated with overall improvement in quality of life scores. PCB therapy was more toxic, and patients generally had deterioration in quality of life scores, independent of disease progression. Temozolomide is now FDA (Food and Drug Adminstration) approved for adult patients with recurrent anaplastic astrocytoma and is undergoing extensive clinical testing in other clinical trials (see Chapter 24).

NATURAL PRODUCTS

Vinca alkaloids.   The vinca alkaloids, vincristine, and vinblastine, are derived from the periwinkle plant and function as cell-cycle-specific spindle poisons (see Figs. 2.6A,B) [8,21]. They enter cells *via* an energy dependent carrier-mediated transport system and bind to tubulin during S-phase, preventing polymerization and inducing metaphase arrest. Both the agents are water soluble, penetrate the blood–brain barrier poorly, undergo extensive metabolism in the liver, and are excreted primarily in the bile. Vincristine has demonstrated activity (1.4 mg/m^2 IV; maximum 2 mg) against low-grade and malignant PBT (e.g., medulloblastoma, oligodendroglioma, anaplastic astrocytoma) as part of multi-agent regimens, such as PCV or CCNU and vincristine [71–73, 87,101,103]. The dose-limiting toxicity is usually peripheral neuropathy.

Antibiotics.   The antitumor antibiotics consist of dactinomycin, adriamycin, bleomycin, mithramycin,

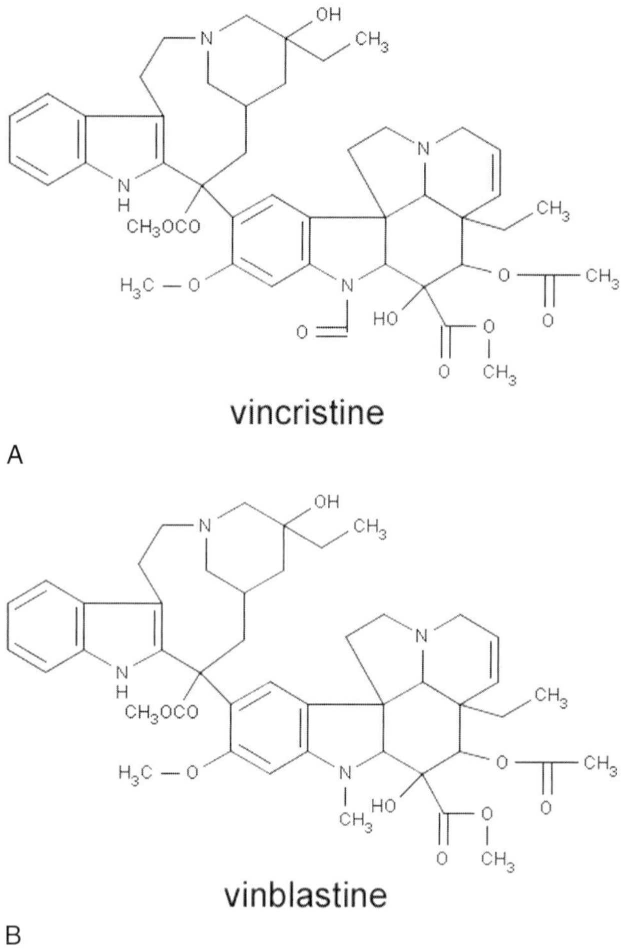

vincristine

A

adriamycin

FIGURE 2.7

vinblastine

B

FIGURE 2.6

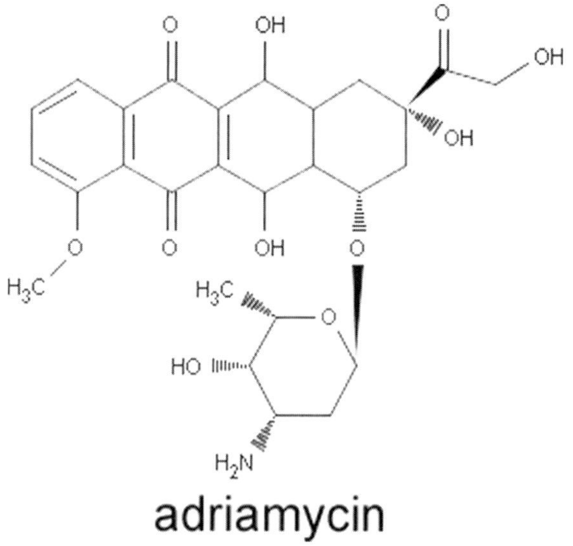

5 - flourouracil

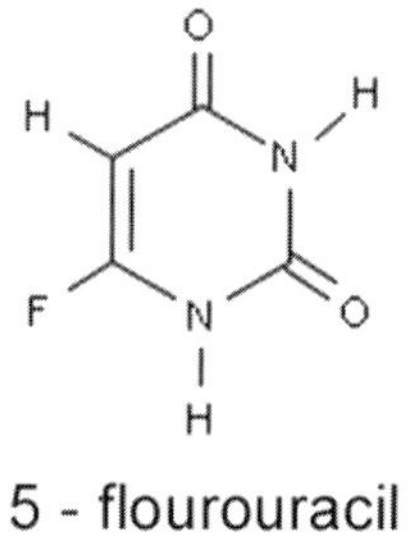

FIGURE 2.8

and mitomycin-C. They function as cell-cycle-nonspecific DNA intercalating agents, interfering with DNA and RNA synthesis [8,21]. Penetration of the blood–brain barrier is poor. In general, the antibiotic drugs have minimal activity against brain tumors [8,71–73]. In combination with vinblastine and cisplatin, bleomycin may have efficacy in patients with primary intracranial germ-cell tumors. Adriamycin has demonstrated some activity in combination with dacarbazine for treatment of malignant meningiomas (see Fig. 2.7). The primary side effects of these drugs include bone marrow suppression, mucositis, extravasation injury, and cardiotoxicity.

ANTIMETABOLITES

Pyrimidine analogs. The pyrimidine analogs are S-phase-specific drugs that include fluorouracil and cytarabine. Fluorouracil is a fluorinated pyrimidine that becomes phosphorylated intracellularly and then binds covalently with thymidylate synthetase, thereby inhibiting its action and interfering with DNA and RNA synthesis (see Fig. 2.8) [8,21]. There is little evidence that fluorouracil has activity against PBT as a single agent or in combination with other drugs. When administered as part of a phase III trial investigating intravenous *versus* intra-arterial BCNU, fluorouracil did not improve survival in either of the BCNU groups [142]. In a phase II trial of 30 patients with various PBT, Cascino and coworkers administered fluorouracil (370 mg/m^2) with leucovorin modulation for five days every five weeks [143]. Only three objective responses were noted, with a median time to progression of 6.7 weeks and an overall median survival of 16.1 weeks. The most frequent toxicities of fluorouracil are myelosuppression, stomatitis, and diarrhea. Rarely, an acute cerebellar syndrome can develop that appears to be correlated with peak plasma concentrations of drug.

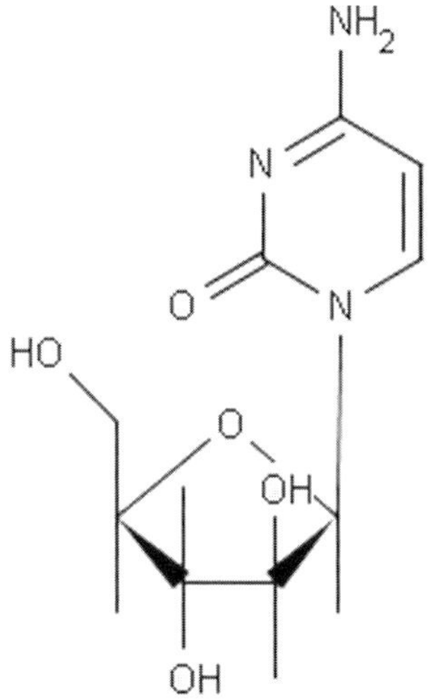

cytarabine (AraC)

FIGURE 2.9

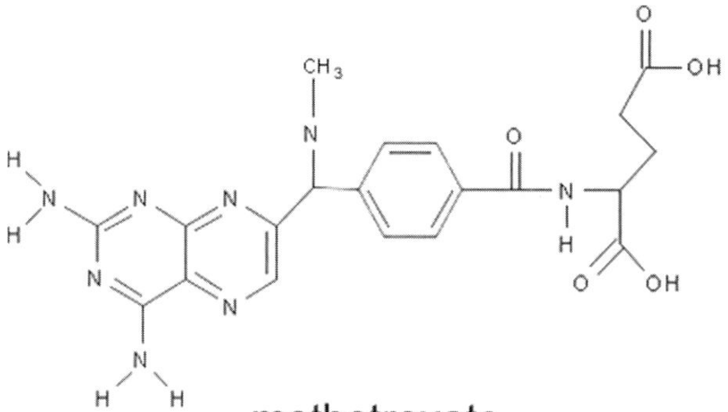

methotrexate

FIGURE 2.10

Cytarabine is structurally similar to cytidine except for the sugar moiety (arabinose replaces ribose; see Fig. 2.9) [8,21]. After intracellular phosphorylation to ara-C triphosphate, the drug is incorporated into DNA and inhibits DNA polymerase. In addition, cytarabine can slow the elongation of DNA and induce premature chain termination [144]. Cytarabine can be administered intravenously or intrathecally and does penetrate the blood–brain barrier. It is usually a component of methotrexate-based regimens for the treatment of primary central nervous system lymphoma [71–73,145,146]. DeAngelis and colleagues recommend using cytarabine as consolidative therapy ($3 \ g/m^2$ per dose IV for two days) after methotrexate or irradiation for two cycles. For patients with leptomeningeal metastasis from systemic malignancies or PBT, cytarabine ($30 \ mg/m^2$ per dose) is often efficacious. Common toxicities include myelosuppression, nausea, emesis, and neurological dysfunction, such as cerebellar syndrome, encephalopathy, seizures, and myelopathy [144].

Purine analogs. This class consists of 6-mercaptopurine and 6-thioguanine, which have a thiol moiety substituted for a hydroxyl group on the purine ring [8,21]. Both the agents are orally administered and require activation by the enzyme hypoxanthineguanine phosphoribosyl-transferase before incorporation into DNA. Antineoplastic activity involves inhibition of purine synthesis, DNA strand breaks, and interference with DNA synthesis. Neither 6-mercaptopurine or 6-thioguanine are very effective against PBT as single agents or in combination regimens [71–73]. Levin and Prados have used

6-thioguanine in combination with other agents to minimize nitrosourea resistance in recurrent gliomas [147]. Toxicity consists of moderately severe delayed myelosuppression and mild gastrointestinal upset.

Folic acid analogs. The most important antifolate drug is methotrexate, an S-phase-specific 4-amino-N_{10}-methyl analog of folic acid that binds reversibly to dihydrofolate reductase, effectively blocking the production of tetrahydrofolate and reducing the intracellular pool of reduced folates (see Fig. 2.10) [8,21,145,148]. Cells require tetrahydrofolate for transfer of one-carbon groups during synthesis of purines and thymidylate. In addition, polyglutamation of methotrexate produces metabolites that inhibit thymidylate and purine biosynthesis. Methotrexate can be administered intravenously, intra-arterially, or intrathecally. It is used most often as a single-agent or in multi-agent regimens for CNS lymphoma [145,146,148–151]. DeAngelis and colleagues recommend high-dose intravenous methotrexate ($1 \ g/m^2$ per day for two doses) before the administration of external beam radiation therapy [145]. For elderly patients, in whom irradiation can cause severe cognitive deficits, similar doses can be used as the sole mode of therapy [146]. Neuwelt and others recommend IA methotrexate (2.5 g; plus IV cyclophosphamide and oral PCB) in combination with BBB disruption as the method of treatment [8,150]. This approach precludes the use of radiation therapy, with reduction of post-treatment cognitive deficits and may extend survival over conventional treatment [150]. Cheng and colleagues report the use of intravenous methotrexate ($1.5 \ g/m^2$) in combination with BCNU, vincristine, etoposide, and methylprednisolone for central nervous system lymphoma patients [151]. Responses were noted in 7 of 8 previously untreated patients and 5 of 6 patients who had failed prior therapy. Methotrexate has also been used in multi-agent regimens for the treatment of medulloblastoma and other high-grade pediatric PBT [71–73]. In PBT patients with leptomeningeal spread, methotrexate can be administered (12 mg per dose in patients older than three years of age) *via* the intrathecal route (lumbar puncture or ommaya reservoir) [152]. High-dose systemic methotrexate is associated with numerous side effects including myelosuppression, nephrotoxicity, nausea, emesis, diarrhea, mucositis, interstitial pneumonitis, and neurotoxicity [8,21]. The neurological sequelae can manifest as an acute encephalopathy syndrome, myelopathy, arachnoiditis, and progressive leukoencephalopathy [153]. Systemic toxicity of methotrexate can be ameliorated by leucovorin rescue.

TOPOISOMERASE INHIBITORS

Etoposide (VP-16). Etoposide is a natural product that derives from podophyllotoxin (see Fig. 2.11). Its mechanism of action is to cause single-strand and double-strand breaks in DNA through interaction with DNA topoisomerase II, inducing arrest in the G_2-phase of the cell cycle [8,21,154]. This activity is mediated through the formation of a stable complex with DNA and topoisomerase II. In addition, etoposide binds to tubulin and inhibits microtubular assembly. Although etoposide is highly lipophilic, it does not readily pass the blood–brain barrier due to its large size; concentration in the cerebrospinal fluid after an intravenous bolus is less than 10 per cent of plasma. Etoposide can be administered orally, by bolus intravenous infusion, or by continuous intravenous infusion over several days. Some authors report improved efficacy when etoposide is given by

etoposide

FIGURE 2.11

continuous infusion [56,155]. This method may increase the dose intensity and efficacy of an etoposide-based regimen without significant augmentation of toxicity. Etoposide has been evaluated as a single-agent and as part of multi-agent regimens for the treatment of pediatric and adult PBT [79,104,119, 156–160]. Several researchers have administered etoposide as a long-term oral regimen [159,160]. Oral etoposide was well tolerated and demonstrated only mild myelosuppression. Fulton and colleagues noted modest efficacy of oral etoposide (50 mg/day) in patients with malignant gliomas [159]. The overall response rate was 42 per cent, with a median time to progression of 8.8 weeks. In a similar pediatric trial, Chamberlain treated 14 children at a dose of 50 mg/m^2 per day [160]. The overall response rate was 50 per cent, with a median duration of response of 8 months. Tirelli and associates administered etoposide as a single-agent in a phase II trial (50–100 mg/m^2 per day IV continuously for five days every three weeks) to 18 patients with malignant PBT [156]. Although the overall response rate was 50 per cent, the duration of the responses and the overall median survival were brief. The majority of PBT patients receive etoposide as part of multi-agent regimens that include cisplatin, carboplatin, cyclophosphamide, vincristine, or PCB [79,104,119,157,158]. The most important toxicities are myelosuppression, nausea, and emesis.

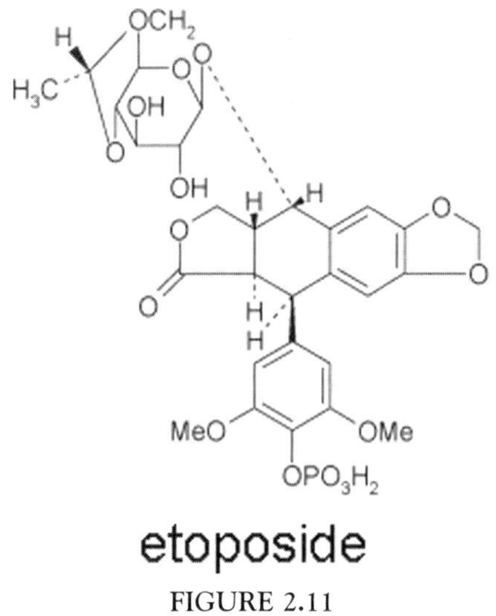

Camptothecin derivatives. Camptothecin is the active agent derived from the bark extract of the *Camptotheca acuminata* tree. Two analogs of camptothecin have been developed that are clinically active and less toxic than the parent compound: irinotecan (CPT-11) and topotecan (see Figs. 2.12A,B) [161]. Both the agents are under investigation for efficacy against solid tumors, including PBT and MBT. The mechanism of action appears to be stabilization of the covalent adduct between topoisomerase I and the 3'-phosphate group of the DNA backbone (i.e., the cleavable complex) [161]. The re-ligation reaction catalyzed by topoisomerase I is inhibited, ultimately leading to multiple DNA single-strand breaks. In addition, the camptothecins also inhibit DNA and RNA synthesis, causing arrest of cells in the G_2-phase of the cell cycle. Irinotecan is a prodrug that requires de-esterification by carboxylesterases to yield SN-38, a metabolite that is 1000 times more active at inhibiting topoisomerase I [161,162]. Topotecan is not a prodrug and binds directly to topoisomerase I without activation. In studies of glioma xenografts in athymic nude mice, topotecan demonstrated activity against ependymomas, adult and pediatric high-grade gliomas, and medulloblastomas [163].

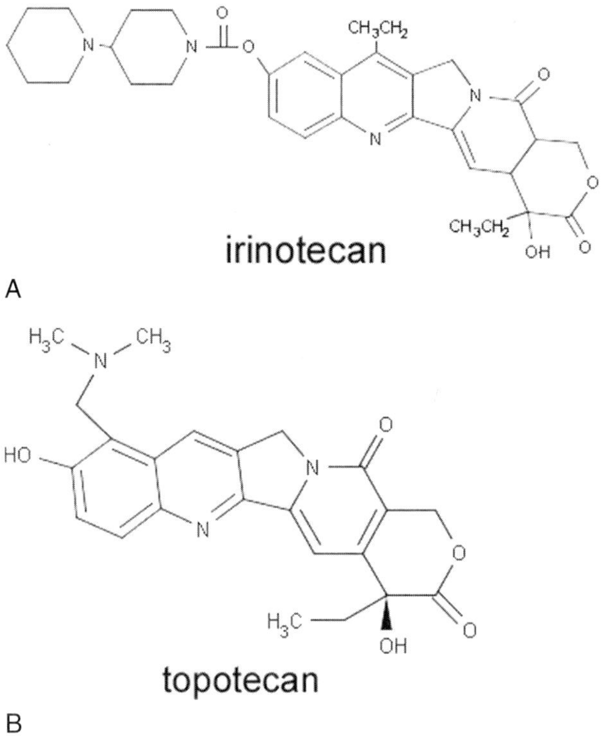

irinotecan

A

topotecan

B

FIGURE 2.12

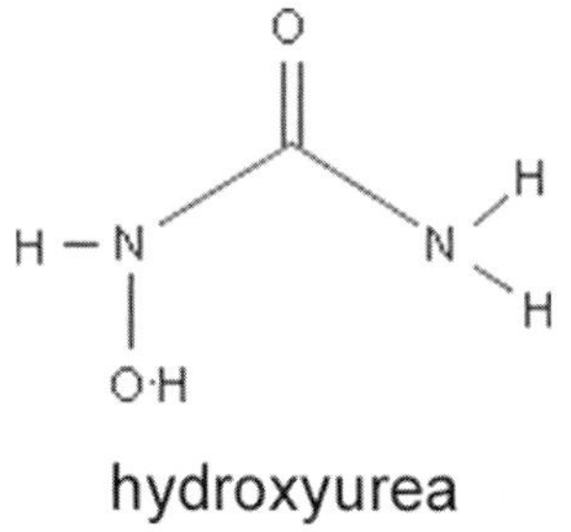

hydroxyurea

FIGURE 2.13

Frequent tumor regressions and improved median survival were noted in treated animals *versus* controls. These preclinical results were the impetus for a phase II study of topotecan (5.5–7.5 mg/m^2 per day, continuous IV infusion every 21 days) in 44 children with low- and high-grade gliomas, medulloblastomas, and brainstem gliomas [164]. One pateint had a partial response, and 5 patients were stabilized. The authors concluded that topotecan as used in this study was inactive against pediatric PBT. Wong and others are evaluating the efficacy of topotecan against MBT, especially those from lung and breast primaries [165]. Other phase II studies are currently investigating the efficacy of irinotecan (125 mg/m^2 per week × 4, every 6 weeks) against adult and pediatric PBT. Toxicity consists of myelosuppression, mild nausea and emesis, hypotension, and, in the case of irinotecan, severe acute and delayed diarrhea [161,166].

MISCELLANEOUS AGENTS

Dibromodulcitol. Dibromodulcitol is a hexitol compound that requires activation to mono- and diepoxides to acquire antineoplastic activity [167]. The most active metabolite is the diepoxide 1,2:5,6-dianhydrogalactitol, which functions as a potent alkylating agent, cross-linking DNA. Dibromodulcitol is administered orally and undergoes rapid and complete absorption from the gastrointestinal tract [167]. The first use of dibromodulcitol was as a sensitizer during radiation therapy of malignant gliomas [168]. Studies of single-agent dibromodulcitol have demonstrated activity against a variety of PBT [169,170]. Levin and colleagues administered dibromodulcitol (100–150 mg/m^2 per day) to patients with recurrent PBT and noted moderate efficacy in patients with medulloblastoma and ependymoma [169]. In a recent randomized trial of 238 patients with high-grade astrocytoma, Elliot and associates found that dibromodulcitol (200 mg/m^2 days 1–10 every 5 weeks) demonstrated efficacy similar to that of intravenous BCNU [170]. The median survival was 41 weeks in both arms of the study, with identical median times to progression (22 weeks). Other investigators have used dibromodulcitol in multi-drug regimens that usually include BCNU and PCB [147,171,172]. Although these regimens have activity against malignant gliomas, it remains unclear if there is a significant survival advantage over single-agent temozolomide or intravenous BCNU. The major toxicity of dibromodulcitol is myelosuppression; nausea, emesis, nephrotoxicity, and hepatotoxicity are mild.

Hydroxyurea. Hydroxyurea is a cell-cycle-specific urea analog that inhibits the enzyme ribonucleotide diphosphate reductase, thereby interfering with DNA synthesis by reducing the available pool of deoxyribonucleotides (see Fig. 2.13) [8,21]. It is administered orally and is well absorbed by the gastrointestinal tract. Hydroxyurea is most often used as a radiation sensitizer or in combination with other agents for the treatment of malignant PBT [8,147]. Recently, other investigators have evaluated hydroxyurea as a therapy for unresectable and recurrent meningiomas [173–177]. Meningioma cell cultures were inhibited after the treatment with hydroxyurea and exhibited arrest in S-phase of the cell cycle [173]. Ultrastructural evaluation of the affected cells was consistent with the induction of the apoptotic pathway. Similar findings were noted in tumors transplanted into nude mice. The same authors administered hydroxyurea (20 mg/kg per day) to four patients with unresectable meningiomas [174]. Objective shrinkage occurred in 3 of the 4 patients, ranging from 15 to 75 per cent, while the fourth patient remained stable. Toxicity was mild and consisted mainly of myelosuppression. Several larger phase II studies have demonstrated activity against meningioma, but had infrequent objective responses [175–177]. For more information regarding the chemotherapy of meningiomas, see Chapter 34.

Mifepristone (RU 486). Mifepristone is an antiprogestin derived from norethindrone that has recently been evaluated as a therapy for patients with unresectable meningiomas [21,178]. Its mechanism of action is to block the progesterone receptors, which occur at higher concentrations (60–75 per cent) in meningiomas and are involved in mitogenic pathways [179]. When used on meningioma cultures, mifepristone demonstrated 18–36 per cent growth inhibition [180]. Grunberg and colleagues performed a phase II trial of mifepristone (200 mg/day orally) in 14 patients with unresectable meningiomas [181]. Four patients had minor regression of tumor by computed tomograpy or MRI; one patient had stable disease with improvement of visual fields. The most common side effect of mifepristone was fatigue; less frequent side effects included hot flashes, gynecomastia, alopecia, and cessation of menses. A large, multicenter phase III randomized, placebo-controlled, trial of mifepristone has recently been reported and did not reveal a significant difference in survival between the treatment arm and the placebo [182].

Tamoxifen. Tamoxifen is a nonsteroidal antiestrogen that is used most often for the treatment of breast cancer patients. However, recent research suggests that tamoxifen also inhibits the activity of protein kinase C, a cytoplasmic enzyme involved in intracellular signaling pathways that can induce cell proliferation [183,184]. Glioma cells strongly express protein kinase C and known mitogens of glioma cells, such as fibroblast growth factor and epidermal growth factor are able to enhance protein kinase C activity. Cultured glioma cells are inhibited by clinically relevant concentrations of tamoxifen, with concomitant decrease in ^{3}H-thymidine incorporation [184]. Couldwell and colleagues administered tamoxifen (80–100 mg bid orally) to 11 patients with recurrent malignant gliomas that had failed irradiation [185]. Three patients had objective responses that lasted longer than 12 months; one patient had

paclitaxel

FIGURE 2.14

stable disease. Other investigators have used tamoxifen in multi-agent regimens for glioma patients [186,187]. Chang and colleagues combined tamoxifen (120–240 mg/m^2 per day) with interferon alpha-2a in 18 patients with recurrent gliomas [186]. There was only one patient with a minor response and three with stable disease; the remaining twelve patients progressed within six weeks. Toxicity was significant and mostly affected the nervous system (e.g., dizziness and unsteady gait). A regimen consisting of tamoxifen (40–120 mg/day) and carboplatin was used by Mastronardi and coworkers to treat 40 patients with newly diagnosed gliomas [187]. The overall median survival was 13 months, with a median time to progression of 7 months. Patients receiving higher doses of tamoxifen (i.e., 80–120 mg/day) tended to survive longer.

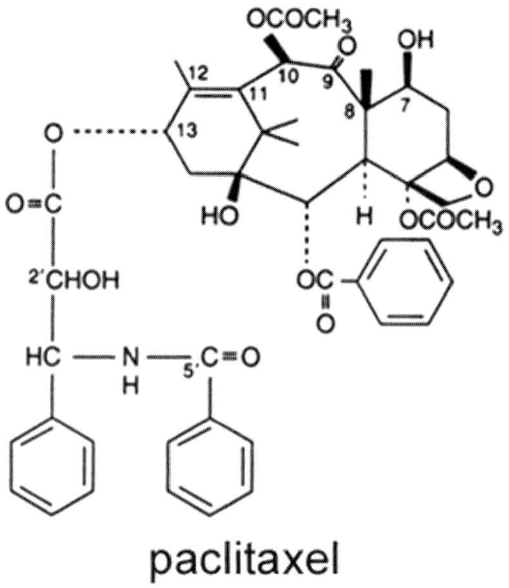

Paclitaxel. Paclitaxel is derived from the bark of the Pacific yew, *Taxus brevifolia*, and has a wide range of clinical activity (see Fig. 2.14) [188]. The mechanism of action is to bind to the N-terminal amino acids of the beta-tubulin subunit of microtubules and induce polymerization. This reaction inhibits the disassembly of microtubules, killing the cells by disrupting normal microtubular function during cell division [188]. Paclitaxel penetrates the blood–brain and blood–tumor barriers (BBB, BTB) poorly, but is measurable in cerebrospinal fluid and tumors with a disrupted barrier [189]. Phase I and II studies of single-agent paclitaxel (210–240 mg/m^2 every three weeks) have demonstrated an overall response rate of 35 per cent in patients with recurrent gliomas

[190,191]. These studies also found that the pharmacokinetics of paclitaxel were altered by concomitant usage of anticonvulsants [190]. Furthermore, patients on anticonvulsants that received paclitaxel were more likely to develop neurotoxicity (i.e., encephalopathy and seizures). In a trial of 20 patients with recurrent oligodendrogliomas, Chamberlain and Kormanik administered single-agent paclitaxel (175 mg/m^2 every 3–4 weeks) [192]. Three patients had a partial response, and seven patients had stable disease. The median time to progression was 10 months. The most common side effects of paclitaxel include myelosuppression, alopecia, nausea, peripheral neuropathy, and arthralgias.

MOLECULAR THERAPEUTIC AGENTS

Signal transduction modulators. Advances in the molecular biology of brain tumors have led to the development of "targeted therapeutics", designed to exploit the signal transduction pathways mediating the malignant phenotype [193–197]. The pathways that have been most intensely studied include the epidermal growth factor and receptor (EGF, EGFR), platelet-derived growth factor and receptor (PDGF, PDGFR), Ras, phosphoinositide 3′ kinase (PI3K), Akt, mammalian target of rapamycin (mTOR), and p53, as well as the processes of tumor invasion, apoptosis, and the cell cycle [193–200]. The drugs that have been advanced the furthest in clinical testing include tyrosine kinase inhibitors of PDGFR (e.g., imatinib) and EGFR (e.g., geftinib, erlotinib), farnesyltransferase inhibitors designed to reduce Ras activity (e.g., R115777), and mTOR inhibitors (e.g., CCI-779, RAD-001) [193,194]. Numerous other "targeted agents" are in development or early phases of clinical evaluation. Thus far, single-agent molecular therapeutic drugs have demonstrated minimal, if any, activity against malignant gliomas in clinical trials. Ongoing and new clinical trials will investigate the use of these drugs in combination with cytotoxic agents (e.g., temozolomide) and other molecular agents that target different pathways.

Angiogenesis inhibitors. For tumors to enlarge beyond 2–3 mm^3 size (approximately one million cells), they must acquire the angiogenic phenotype and induce neovascularization [194,201]. The switch to the angiogenic phenotype involves up-regulation of the angiogenic factors (e.g., basic fibroblast growth factor, vascular endothelial growth factor (VEGF)) and down-regulation of angiogenesis inhibitors (e.g., thrombospondin-1) in the local environment. This process is important for the growth of PBT as

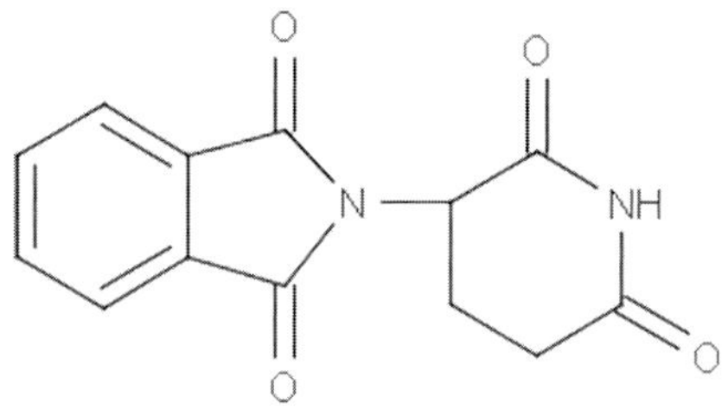

thalidomide
FIGURE 2.15

well as systemic neoplasms [202]. For example, there is a significant inverse correlation between the density of vessels in astrocytic gliomas and postoperative survival [203]. Patients with the highest microvessel density had the most malignant tumors and the shortest survivals. Several investigators have demonstrated extensive expression of VEGF and its receptors in gliomas and other PBT [204,205]. Bernsen and colleagues treated glioma xenografts in nude mice with TNP-470, a semisynthetic analog of fumagillin with antiangiogenic properties, but were unable to consistently demonstrate growth inhibition or alteration of vascular parameters [206]. In a phase II trial of the antiangiogenic agent thalidomide, Fine and coworkers treated 35 patients with recurrent high-grade gliomas (see Fig. 2.15) [207]. There were two minor responses noted among the first 10 evaluable patients. Other angiogenesis inhibitors, such as angiostatin, endostatin, and SU11248 are under consideration for clinical trials in brain tumor patients.

CONCLUSIONS

Chemotherapy has demonstrated modest efficacy for the treatment of selected patients with brain tumors. It is an effective adjunctive therapy for patients with sensitive anaplastic oligodendrogliomas (i.e., deletions of chromosomes 1p and 19q) and certain pediatric tumors (e.g., high-risk medulloblastoma). However, the majority of high-grade astrocytomas and other malignant tumors remain insensitive to currently available chemotherapeutic agents. Further work is needed to characterize low-grade oligodendrogliomas and other brain neoplasms according to molecular phenotype and to correlate this information with chemosensitivity and patient survival. New chemotherapeutic agents must be discovered that are more specific and can attack tumor cells at the molecular level of tumorigenesis and tumor angiogenesis. In addition, it is important to further our understanding of brain tumor chemoresistance and to design improved strategies for counteracting this pervasive problem.

ACKNOWLEDGMENTS

The author would like to thank Ryan Smith for research assistance. Dr. Newton was supported in part by National Cancer Institute grant, CA 16058 and the Dardinger Neuro-Oncology Center Endowment Fund.

References

1. Newton, H. B. (1994). Primary brain tumors: Review of etiology, diagnosis, and treatment. *Am Fam Phys* **49**, 787–797.
2. Newton, H. B. (1999). Neurological complications of systemic cancer. *Am Fam Phys* **59**, 878–886.
3. Newton, H. B. (2002). Chemotherapy for the treatment of metastatic brain tumors. *Expert Rev Anticancer Ther* **2**, 495–506.
4. Newton, H. B. (2000). Novel chemotherapeutic agents for the treatment of brain cancer. *Expert Opin Investig Drugs* **12**, 2815–2829.
5. Fine, H. A., Dear, K. B. G., Loeffler, J. S., Black, P, M., and Canellos, G. P. (1993). Meta-analysis of radiation therapy with and without adjuvant chemotherapy for malignant gliomas in adults. *Cancer* **71**, 2585–2597.
6. Reni, M., Cozzarini, C., Ferreri, A. J. M. *et al.* (2000). A retrospective analysis of postradiation chemotherapy in 133 patients with glioblastoma multiforme. *Cancer Investig* **18**, 510–515.
7. Stewart, L. A., Burdett, S., Parmar, M. K. B. *et al.* (2002). Chemotherapy in adult high-grade glioma: a systematic review and meta-analysis of individual patient data from 12 randomised trials. *Lancet* **359**, 1011–1018.
8. Gumerlock, M. K., and Neuwelt, E. A. (1987). Principles of chemotherapy in brain neoplasia. *In Therapy of Malignant Brain Tumors* (K. Jellinger Ed.), pp. 277–348. Springer Verlag, New York.
9. Parney, I. F., and Prados, M. D. (2005). Chemotherapy principles. *In Textbook of Neuro-Oncology* (M. S. Berger, and M. D. Prados, Eds.), pp. 75–79. Elsevier Saunders, Philadelphia.
10. Rapoport, S. I., Ohno, K., and Pettigrew, K. D. (1979). Drug entry into the brain. *Brain Res* **172**, 354–359.
11. Levin, V. A. (1980). Relationship of octanol/water partition coefficient and molecular weight to rat brain capillary permeability. *J Med Chem* **23**, 682–684.
12. Blasberg, R. G., and Groothuis, D. R. (1986). Chemotherapy of brain tumors: physiological and pharmacokinetic considerations. *Semin Oncol* **13**, 70–82.
13. Stohrer, M., Boucher, Y., Stangassinger, M., and Jain, R. K. (2000). Oncotic pressure in solid tumors is elevated. *Cancer Res* **60**, 4251–4255.
14. Gralla, R. J., Osoba, D., Kris, M. G. *et al.* (1999). Recommendations for the use of antiemetics: Evidence-based, clinical practice guidelines. *J Clin Oncol* **17**, 2971–2994.
15. Hesketh, P. J. (2000). Comparative review of 5-HT$_3$ receptor antagonists in the treatment of acute chemotherapy-induced nausea and vomiting. *Cancer Investig* **18**, 163–173.
16. Portenoy, R. K. (1987). Constipation in the cancer patient: Causes and management. *Med Clin North Am* **71**, 303–311.
17. Weinstein, S. M. (1998). New pharmacological strategies in the management of cancer pain. *Cancer Investig* **16**, 94–101.
18. Cherny, N. I. (2004). The pharmacologic management of cancer pain. *Oncology* **18**, 1499–1515.
19. Breitbart, W. (1994). Psycho-oncology: Depression, anxiety, delerium. *Semin Oncol* **21**, 754–769.
20. Winnell, J., and Roth, A. J. (2004). Depression in cancer patients. *Oncology* **18**, 1554–1560.
21. Newton, H. B., Turowski, R. C., Stroup, T. J., and McCoy, L. K. (1999). Clinical presentation, diagnosis, and pharmacotherapy of patients with primary brain tumors. *Ann Pharmacother* **33**, 816–832.
22. Ohnishi, T., Sher, P. B., Posner, J. B., and Shapiro, W. B. (1990). Capillary permeability factor secreted by malignant brain tumor. Role in peritumoral edema and possible mechanism for anti-edema effect of glucocorticoids. *J Neurosurg* **72**, 245–251.
23. Del Maestro, R. F., Megyesi, J. F., and Farrell, C. L. (1990). Mechanisms of tumor-associated edema: A review. *Can J Neurol Sci* **17**, 177–183.
24. Samdani, A. F., Tamargo, R. J., and Long, D. M. (1997). Brain tumor edema and the role of the blood-brain barrier. *In Handbook of Clinical Neurology, Vol. 23 (67), Neuro-Oncology, Part I* (C. J. Vecht, Ed.), pp. 71–102. Elsevier Science, Amsterdam.
25. Mukwaya, G. (1988). Immunosuppressive effects and infections associated with corticosteroid therapy. *Pediatr Infect Dis J* **7**, 499–504.
26. Østergaard, L., Hochberg, F. H., Rabinov, J. D. *et al.* (1999). Early changes measured by magnetic resonance imaging in cerebral blood flow, blood volume, and blood-brain barrier permeability following dexamethasone treatment in patients with brain tumors. *J Neurosurg* **90**, 300–305.
27. Lester, R. S., Knowles, S. R., and Shear, N. H. (1998). The risks of systemic corticosteroid use. *Dermatol Clin* **16**, 277–286.
28. Weissman, D. E., Dufer, D., Vogel, V., and Abeloff, D. D. (1987). Corticosteroid toxicity in neuro-oncology patients. *J Neuro-Oncol* **5**, 125–128.
29. Fadul, C. E., Lemann, W., Thaler, H. T., and Posner, J. B. (1988). Perforation of the gastrointestinal tract in patients receiving steroids for neurologic disease. *Neurol* **38**, 348–352.
30. Stiefel, F. C., Breitbart, W. S., and Holland, J. C. (1989). Corticosteroids in cancer: Neuropsychiatric complications. *Cancer Investig* **7**, 479–491.
31. Dropcho, E. J., and Soong, S. J. (1991). Steroid-induced weakness in patients with primary brain tumors. *Neurol* **41**, 1235–1239.
32. Joseph, J. C. (1994). Corticosteroid-induced osteoporosis. *Am J Hosp Pharm* **51**, 188–197.
33. McIlwain, H. H. (2003). Glucocorticoid-induced osteoporosis: pathogenesis, diagnosis, and management. *Preventive Med* **36**, 243–249.
34. Garnett, W. R., and Garabedian-Ruffalo, S. M. (1997). Identification, diagnosis, and treatment of acid-related diseases in the elderly: Implications for long-term care. *Pharmacother* **17**, 938–958.
35. Sachs, G. (1997). Proton pump inhibitors and acid-related diseases. *Pharmacother* **17**, 22–37.
36. Glantz, M., and Recht, L. D. (1997). Epilepsy in the cancer patient. *In Handbook of Clinical Neurology, Vol. 25 (69), Neuro-Oncology, Part III* (C. J. Vecht, Ed.), pp. 9–18. Elsevier Science, Amsterdam.

37. Moots, P. L., Maciunas, R. J., Eisert, D. R., Parker, R. A., Laporte, K., and Abou-Khalil, B. (1995). The course of seizure disorders in patients with malignant gliomas. *Arch Neurol* **52**, 717–724.

38. Elisevich, K. (1999). Epilepsy and low-grade gliomas. *In The Practical Management of Low-Grade Primary Brain Tumors* (J. P. Rock, M. L. Rosenblum, E. G. Shaw, and J. G. Cairncross, Eds.), pp. 149–169. Lippincott Williams & Wilkins, Philadelphia.

39. Schaller, B., and Rüegg, S. J. (2003). Brain tumor and seizures: Pathophysiology and its implications for treatment revisited. *Epilepsia* **44**, 1223–1232.

40. Brodie, M. J., and Dichter, M. A. (1996). Antiepileptic drugs. *New Engl J Med* **334**, 168–175.

41. Britton, J. W., and So, E. L. (1996). Selection of antiepileptic drugs: A practical approach. *Mayo Clin Proc* **71**, 778–786.

42. Mattson, R. H., Cramer, B. S., Collins, J. F., and the Department of Veterans Affairs Epilepsy Cooperative Study Group (1992). A comparison of valproate with carbamazepine for the treatment of complex partial seizures and secondarily generalized tonic-clonic seizures in adults. *New Engl J Med* **327**, 765–771.

43. Dichter, M. A., and Brodie, M. J. (1996). New antiepileptic drugs. *New Eng J Med* **334**, 1583–1590.

44. Rosenfeld, W. E. (1997). Topiramate: A review of preclinical, pharmacokinetic, and clinical data. *Clin Therap* **19**, 1294–1308.

45. Wagner, G. L., Wilms, E. B., Van Donselaar, C. A., and Vecht, C. J. (2003). Levetiracetam: preliminary experience in patients with primary brain tumours. *Seizure* **12**, 585–586.

46. North, J. B., Penhall, R. K., Hanieh, A., Frewin, D. B., and Taylor, W. B. (1983). Phenytoin and postoperative epilepsy. A double-blind study. *J Neurosurg* **58**, 672–677.

47. Boarini, D. J., Beck, D. W., and VanGilder, J. C. (1985). Postoperative prophylactic anticonvulsant therapy in cerebral gliomas. *Neurosurg* **16**, 290–292.

48. Cohen, N., Strauss, G., Lew, R., Silver, D., and Recht, L. (1988). Should prophylactic anticonvulsants be administered to patients with newly-diagnosed cerebral metastases? A retrospective analysis. *J Clin Oncol* **6**, 1621–1624.

49. Franceschetti, S., Binelli, S., Casazza, M. *et al.* (1990). Influence of surgery and antiepileptic drugs on seizures symptomatic of cerebral tumours. *Acta Neurochir* **103**, 47–51.

50. Shaw, M. D. M. (1990). Post-operative epilepsy and the efficacy of anticonvulsant therapy. *Acta Neurochir Suppl* **50**, 55–57.

51. Foy, P. M., Chadwick, D. W., Rajgopalan, N., Johnson, A. L., and Shaw, M. D. M. (1992). Do prophylactic anticonvulsant drugs alter the pattern of seizures after craniotomy? *J Neurol Neurosurg Psych* **55**, 753–757.

52. Glantz, M. J., Cole, B. F., Friedberg, M. H. *et al.* (1996). A randomized, blinded, placebo-controlled trial of divalproex sodium prophylaxis in adults with newly diagnosed brain tumors. *Neurol* **46**, 985–991.

53. Glantz, M. J., Cole, B. F., Forsyth, P. A. *et al.* (2000). Practice parameter: Anticonvulsant prophylaxis in patients with newly diagnosed brain tumors. Report of the Quality Standards Subcommittee of the American Academy of Neurology. *Neurol* **54**, 1886–1893.

54. DeMario, M. D., and Ratain, M. J. (1998). Oral chemotherapy: Rationale and future directions. *J Clin Oncol* **16**, 2557–2567.

55. Wilkinson, G. R. (2001). Pharmacokinetics: The dynamics of drug absorption, distribution, and elimination. *In Goodman & Gilman's The Pharmacological Basis of Therapeutics.* (L. G. Hardman, L. E. Limbird, and A. G. Gilman Eds.), 10th ed., pp. 3–30. McGraw-Hill, New York.

56. Anderson, N., and Lokich, J. J. (1994). Cancer chemotherapy and infusional scheduling. *Oncology* **8**, 99–111.

57. Shibuya, M., Ito, S., Miwa, T., Davis, R. L., Wilson, C. B., and Hoshino, T. (1993). Proliferative potential of brain tumors. Analyses with Ki-67 and anti-DNA polymerase alpha monoclonal antibodies, bromodeoxyuridine labeling, and nucleolar organizer region counts. *Cancer* **71**, 199–206.

58. Stewart, D. J. (1989). Pros and cons of intra-arterial chemotherapy. *Oncology* **3**, 20–26.

59. Tomita, T. (1991). Interstitial chemotherapy for brain tumors: review. *J Neurooncol* **10**, 57–74.

60. Sipos, E. P., and Brem, H. (1995). New delivery systems for brain tumor therapy. *Neurol Clin* **13**, 813–825.

61. Grossman, S. A., Reinhard, C., Colvin, O. M. *et al.* (1992). The intracerebral distribution of BCNU delivered by surgically implanted biodegradable polymers. *J Neurosurg* **76**, 640–647.

62. Tamargo, R. J., Myeseros, J. S., Epstein, J. I., Yang, M. B., Chasin, M., Brem, H. (1993). Interstitial chemotherapy of the 9L gliosarcoma: Controlled release polymers for drug delivery in the brain. *Cancer Res* **53**, 329–333.

63. Brem, H., Tamargo, R. J., Olivi, A., Pinn, M., Wharam, M., and Epstein, J. (1994). Biodegradable polymers for controlled delivery of chemotherapy with and without radiation therapy in the monkey brain. *J Neurosurg* **80**, 283–290.

64. Brem, H., Piantadosi, S., Burger, P. C. *et al.* (1995). Placebo-controlled trial of safety and efficacy of intraoperative controlled delivery by biodegradable polymers of chemotherapy for recurrent gliomas. *Lancet* **354**, 1008–1012.

65. Valtonen, S., Timonen, U., Toivanen, P. *et al.* (1997). Interstitial chemotherapy with carmustine-loaded polymers for high-grade gliomas: A randomized double-blind study. *Neurosurg* **41**, 44–49.

66. Neuwelt, E. A., Hill, S. A., Frenkel, E. P. *et al.* (1981). Osmotic blood-brain barrier disruption: Pharmacodynamic studies in dogs and a clinical phase I trial in patients with malignant brain tumors. *Cancer Treat Rep* **65**, 39–43.

67. Neuwelt, E. A., and Kroll, R. A. (1994). Osmotic blood-brain barrier modification: Increasing delivery of diagnostic and therapeutic agents to the brain. *Meth Neurosci* **21**, 52–67.

68. Neuwelt, E. A., Howieson, J., Frenkel, E. P. *et al.* (1986). Therapeutic efficacy of multiagent chemotherapy with drug delivery enhancement by blood-brain barrier modification in glioblastoma. *Neurosurg* **19**, 573–582.

69. Gumerlock, M. K., Belshe, B. D., Madsen, R., and Watts, C. (1992). Osmotic blood-brain barrier disruption and chemotherapy in the treatment of high grade malignant glioma: patient series and literature review. *J Neurooncol* **12**, 33–46.

70. Zünkeler, B., Carson, R. E., Olson, J. *et al.* (1996). Quantification and pharmacokinetics of blood-brain barrier disruption in humans. *J Neurosurg* **85**, 1056–1065.

71. Kornblith, P. L., and Walker, M. (1988). Chemotherapy for malignant gliomas. *J Neurosurg* **68**, 1–17.

72. Kyritsis, A. P. (1993). Chemotherapy for malignant gliomas. *Oncology* **7**, 93–100.

73. Pech, I.V., Peterson, K., and Cairncross, J. G. (1998). Chemotherapy for brain tumors. *Oncology* **12**, 537–547.

74. Coyle, T., Baptista, J., Winfield, J. *et al.* (1990). Mechlorethamine, vincristine, and procarbazine chemotherapy for recurrent high-grade glioma in adults: A phase II study. *J Clin Oncol* **8**, 2014–2018.

75. Allen, J. C., and Helson, L. (1981). High-dose cyclophosphamide chemotherapy for recurrent CNS tumors in children. *J Neurosurg* **55**, 749–756.

76. Friedman, H. S., Mahaley, M. S., Schold, S. C. *et al.* (1986). Efficacy of vincristine and cyclophosphamide in the therapy of recurrent medulloblastoma. *Neurosurg* **18**, 335–340.

77. Longee, D. C., Friedman, H. S., Albright, R. E. *et al.* (1990). Treatment of patients with recurrent gliomas with cyclophosphamide and vincristine. *J Neurosurg* **72**, 583–588.

78. Carpenter, P. A., White, L., McCowage, G. B. *et al.* (1997). A dose-intensive, cyclosphosphamide-based regimen for the treatment of recurrent/progressive or advanced solid tumors of childhood. A report from the Australia and New Zealand Children's Cancer Study Group. *Cancer* **80**, 489–496.

79. Newton, H. B., and Newton, C. L. (1995). Attempted dose intensified cyclophosphamide, etoposide, and granulocyte colony-stimulating factor for treatment of malignant astrocytoma. *J Neurooncol* **24**, 285–292.

80. Elliot, T. E., Buckner, J. C., Cascino, T. L., Levitt, R., O'Fallon, J. R., and Scheithauer, B. W. (1991). Phase II study of ifosfamide with mesna in adult patients with recurrent diffuse astrocytoma. *J Neurooncol* **10**, 27–31.

81. Siu, L. L., and Moore, M. J. (1998). Use of mesna to prevent ifosfamide-induced urotoxicity. *Support Care Cancer* **6**, 144–154.

82. Goren, M. P. (1996). Oral administration of mesna with ifosfamide. *Sem Oncol* **23**, 91–96.

83. Pratt, C. B., Green, A. A., Horowitz, M. E. *et al.* (1986). Central nervous system toxicity following the treatment of pediatric patients with ifosfamide/mesna. *J Clin Oncol* **4**, 1253–1261.

84. Walker, M. D., Green, S. B., Byar, D. P. *et al.* (1980). Randomized comparisons of radiotherapy and nitrosoureas for the treatment of malignant glioma after surgery. *N Engl J Med* **303**, 1323–1329.

85. Green, S. B., Byar, D. P., Walker, M. D. *et al.* (1983). Comparisons of carmustine, procarbazine, and high-dose methylprednisolone as additions to surgery and radiotherapy for the treatment of malignant glioma. *Cancer Treat Rep* **67**, 121–132.

86. Shapiro, W. R., Green, S. B., Burger, P. C. *et al.* (1989). Randomized trial of three chemotherapy regimens and two radiotherapy regimens in postoperative treatment of malignant glioma. Brain tumor cooperative group trial 8001. *J Neurosurg* **71**, 1–9.

87. Evans, A. E., Jenkin, R. D. T., Sposto, B. S. R. *et al.* (1990). The treatment of medulloblastoma. Results of a prospective randomized trial of radiation therapy with and without CCNU, vincristine, and prednisone. *J Neurosurg* **72**, 572–582.

88. Packer, R. J. (1990). Chemotherapy for medulloblastoma/primitive neuroectodermal tumors of the posterior fossa. *Ann Neurol* **28**, 823–828.

89. Aronin, P. A., Mahaley, M. S., Rudnick, S. A. *et al.* (1980). Prediction of BCNU pulmonary toxicity in patients with malignant gliomas. An assessment of risk factors. *N Engl J Med* **303**, 183–188.

90. Shiba, D. A., and Weinkam, R. J. (1983). The *in vivo* cytotoxic activity of procarbazine and procarbazine metabolites against L1210 ascites leukemia cells in CDF$_1$ mice and the effects of pre-treatment with procarbazine, phenobarbital, diphenylhydantoin, and methylprednisolone upon in vivo procarbazine activity. *Cancer Chemother Pharmacol* **11**, 124–129.

91. Erickson, J. M., Tweedie, D. J., Ducore, J. M., and Prough, R. A. (1989). Cytotoxicity and DNA damage caused by the azoxy metabolites of procarbazine in L1210 tumor cells. *Cancer Res* **49**, 127–133.

92. Schold, S. C., Brent, T. P., von Hofe, E. *et al.* (1989). O^6-alkylguanine-DNA alkyltransferase and sensitivity to procarbazine in human brain-tumor xenografts. *J Neurosurg* **70**, 573–577.

93. Maxwell, M. B. (1980). Reexamining the dietary restrictions with procarbazine (an MAOI). *Cancer Nurs* **3**, 451–457.

94. Newton, H. B., Junck, L., Bromberg, J., Page, M. A., and Greenberg, H. S. (1990). Procarbazine chemotherapy in the treatment of recurrent malignant astrocytomas after radiation and nitrosourea failure. *Neurol* **40**, 1743–1746.

95. Newton, H. B., Bromberg, J., Junck, L., Page, M. A., and Greenberg, H. S. (1993). Comparison between BCNU and procarbazine chemotherapy for treatment of gliomas. *J Neurooncol* **15**, 257–263.

96. Rodriguez, L. A., Prados, M., Silver, P., and Levin, V. A. (1989). Reevaluation of procarbazine for the treatment of recurrent malignant central nervous system tumors. *Cancer* **64**, 2420–2423.

97. Cairncross, J. G., and Macdonald, D. R. (1988). Successful chemotherapy for recurrent malignant oligodendroglioma. *Ann Neurol* **23**, 360–364.

98. Cairncross, J. G., Macdonald, D. R., and Ramsay, D. A. (1992). Aggressive oligodendroglioma: A chemosensitive tumor. *Neurosurg* **31**, 78–82.

99. Cairncross, J. G., Ueki, K., Zlatescu, M. C. *et al.* (1998). Specific genetic predictors of chemotherapeutic response and survival in patients with anaplastic oligodendrogliomas. *J Natl Cancer Inst* **90**, 1473–1479.

100. Peterson, K., Paleologos, N., Forsyth, P., Macdonald, D. R., Cairncross, J. G. (1996). Salvage chemotherapy for oligodendroglioma. *J Neurosurg* **85**, 597–601.

101. Glass, J., Hochberg, F. H., Gruber, M. L., Louis, D. N., Smith, D., and Rattner, B. (1992). The treatment of oligodendrogliomas and mixed oligodendroglioma-astrocytomas with PCV chemotherapy. *J Neurosurg* **76**, 741–745.

102. Kim, L., Hochberg, F. H., Thornton, A. F. *et al.* (1996). Procarbazine, lomustine, and vincristine (PCV) chemotherapy for grade III and grade IV oligoastrocytomas. *J Neurosurg* **85**, 602–607.

103. Levin, V. A., Silver, P., Hannigan, J. *et al.* (1990). Superiority of post-radiotherapy adjuvant chemotherapy with CCNU, procarbazine, and vincristine (PCV) over BCNU for anaplastic gliomas: NCOG 6G61 final report. *Int J Radiat Oncol Biol Phys* **18**, 321–324.

104. Hellman, R. M., Calogero, J. A., and Kaplan, B. M. (1990). VP-16, vincristine and procarbazine with radiation therapy for treatment of malignant brain tumors. *J Neurooncol* **8**, 163–166.

105. Huang, H., Zhu, L., Reid, B. R., Drobny, G. P., and Hopkins, P. B. (1995). Solution structure of a cisplatin-induced DNA interstrand cross-link. *Science* **270**, 1842–1845.

106. Walker, R. W., and Allen, J. C. (1988). Cisplatin in the treatment of recurrent childhood primary brain tumors. *J Clin Oncol* **6**, 62–66.

107. Bertolone, S. J., Baum, E. S., Krivit, W., and Hammond, G. D. (1989). A phase II study of cisplatin therapy in recurrent childhood brain tumors. A report from the Childrens Cancer Study Group. *J Neurooncol* **7**, 5–11.

108. Packer, R. J., Sutton, L. N., Goldwein, J. W. *et al.* (1991). Improved survival with the use of adjuvant chemotherapy in the treatment of medulloblastoma. *J Neurosurg* **74**, 433–440.

109. Mosijczuk, A. D., Nigro, M. A., Thomas, P. R. M. *et al.* (1993). Preradiation chemotherapy in advanced medulloblastoma. A Pediatric Oncology Group pilot study. *Cancer* **72**, 2755–2762.

110. Spence, A. M., Berger, M. S., Livinston, R. B., Ali-Osman, F., and Griffin, B. (1992). Phase II evaluation of high-dose cisplatin for treatment of adult malignant gliomas recurrent after chloroethylnitrosourea failure. *J Neurooncol* **12**, 187–191.

111. Feun, L. G., Wallace, S., Stewart, D. J. *et al.* (1984). Intracarotid infusion of cis-diamminedichloro-platinum in the treatment of recurrent malignant brain tumors. *Cancer* **54**, 794–799.

112. Mahaley, M. S., Hipp, S. W., Dropcho, E. J. *et al.* (1989). Intracarotid cisplatin chemotherapy for recurrent gliomas. *J Neurosurg* **70**, 371–378.

113. Dropcho, E. J., Rosenfeld, S. S., Morawetz, R. B. *et al.* (1992). Preradiation intracarotid cisplatin treatment of newly diagnosed anaplastic gliomas. *J Clin Oncol* **10**, 452–458.

114. Newton, H. B., Page, M. A., Junck, L., and Greenberg, H. S. (1989). Intra-arterial cisplatin for the treatment of malignant gliomas. *J Neuro-Oncol* **7**, 39–45.

115. Doz, F., Berens, M. E., Dougherty, D. V., and Rosenblum, M. L. (1991). Comparison of the cytotoxic activities of cisplatin and carboplatin against glioma cell lines at pharmacologically relevant drug exposures. *J Neurooncol* **11**, 27–35.

116. Allen, J. C., Walker, R., Luks, E., Jennings, M., Barfoot, S., and Tan, C. (1987). Carboplatin and recurrent childhood brain tumors. *J Clin Oncol* **5**, 459–463.

117. Friedman, H. S., Krischer, J. P., Burger, P. *et al.* (1992). Treatment of children with progressive or recurrent brain tumors with carboplatin or iproplatin: A Pediatric Oncology Group randomized phase II study. *J Clin Oncol* **10**, 249–256.

118. Yung, W. K. A., Mechtler, L., and Gleason, M. J. (1991). Intravenous carboplatin for recurrent malignant glioma: A phase II study. *J Clin Oncol* **9**, 860–864.

119. Jeremic, B., Grujicic, D., Jevremovic, S. *et al.* (1992). Carboplatin and etoposide chemotherapy regimen for recurrent malignant glioma: A phase II study. *J Clin Oncol* **10**, 1074–1077.

120. Brandes, A. A., Rigon, A., Zampieri, P. *et al.* (1998). Carboplatin and teniposide concurrent with radiotherapy in patients with glioblastoma multiforme. A phase II study. *Cancer* **82**, 355–361.

121. Follézou, J. Y., Fauchon, J., and Chiras, J. (1989). Intra-arterial infusion of carboplatin in the treatment of malignant gliomas: A phase II study. *Neoplasm* **36**, 349–352.

122. Williams, P. C., Henner, W. D., Roman-Goldstein, S. *et al.* (1995). Toxicity and efficacy of carboplatin and etoposide in conjunction with disruption of the blood-brain tumor barrier in the treatment of intracranial neoplasms. *Neurosurg* **37**, 17–28.

123. Stevens, M. F. G., and Newlands, E. S. (1993). From triazines and triazenes to temozolomide. *Eur J Cancer* **20A**, 1045–1047.

124. Newlands, E. S., Stevens, M. F. G., Wedge, S. R., Wheelhouse, R. T., and Brock, C. (1997). Temozolomide: a review of its discovery, chemical properties, pre-clinical development and clinical trials. *Cancer Treat Rev* **23**, 35–61.

125. Hvizdos, K. M., and Goa, K. L. (1999). Temozolomide. *CNS Drugs* **12**, 237–243.

126. Liu, L., Markowitz, S., and Gerson, S. L. (1996). Mismatch repair mutations override alkyltransferase in conferring resistance to temozolomide but not to 1,3-Bis(2-chloroethyl) nitrosourea. *Cancer Res* **56**, 5375–5379.

127. Friedman, H. S., Johnson, S. P., Dong, Q. *et al.* (1997). Methylator resistance mediated by mismatch repair deficiency in a glioblastoma multiforme xenograft. *Cancer Res* **57**, 2933–2936.

128. Friedman, H. S., McLendon, R. E., Kerby, T. *et al.* (1998). DNA mismatch repair and O^6-alkylguanine-DNA alkyltransferase analysis and response to Temodal in newly diagnosed malignant glioma. *J Clin Oncol* **16**, 3851–3857.

129. Baer, J. C., Freeman, A. A., Newlands, E. S. *et al.* (1993). Depletion of O^6-alkylguanine-DNA alkyltransferase correlates with potentiation of temozolomide and CCNU toxicity in human tumour cells. *Br J Cancer* **67**, 1299–1302.

130. Brock, C. S., Matthews, J. C., Brown, G. *et al.* (1998). Response to temozolomide in recurrent high grade gliomas is related to tumour drug concentration (abstract 667). *Ann Oncol* **2**, 174.

131. Plowman, J., Waud, W. R., Koutsoukos, A. D., Rubinstein, L. V., Moore, T. D., Grever, M. R. (1994). Preclinical antitumor activity of temozolomide in mice: Efficacy against human brain tumor xenografts and synergism with 1,3-Bis (2-chloroethyl)-1-nitrosurea. *Cancer Res* **54**, 3793–3799.

132. Friedman, H. S., Dolan, M. E., Pegg, A. E. *et al.* (1995). Activity of temozolomide in the treatment of central nervous system tumor xenografts. *Cancer Res* **55**, 2853–2857.

133. Newlands, E. S., O'Reilly, S. M., Glaser, M. G. *et al.* (1996). The Charing Cross Hospital experience with temozolomide in patients with gliomas. *Eur J Cancer* **32A**, 2236–2241.

134. Bower, M., Newlands, E. S., Bleehen, N. M. *et al.* (1997). Multicentre CRC phase II trial of temozolomide in recurrent or progressive high-grade glioma. *Cancer Chemother Pharmacol* **40**, 484–488.

135. Estlin, E. J., Lashford, L., Ablett, S. *et al.* (1998). Phase I study of temozolomide in paediatric patients with advanced cancer. *Br J Cancer* **78**, 652–661.

136. Nicholson, H. S., Krailo, M., Ames, M. M. *et al.* (1998). Phase I study of temozolomide in children and adolescents with recurrent solid tumors: A report from the Children's Cancer Group. *J Clin Oncol* **16**, 3037–3043.

137. Yung, W. K. A., Prados, M. D., Yaya-Tur, R. *et al.* (1999). Multicenter phase II trial of temozolomide in patients with anaplastic astrocytoma or anaplastic oligoastrocytoma at first relapse. *J Clin Oncol* **17**, 2762–2771.

138. Yung, W. K. A., Levin, V. A., Albright, R. *et al.* (1997). Randomized trial of temodal (TEM) vs. procarbazine (PCB) in glioblastoma multiforme (GBM) at first relapse (abstract 532). *Proc ASCO* **18**, 139a.

139. Yung, W. K. A. (2000). Temozolomide in malignant glioma. *Semin Oncol* **27** *(Suppl 6)*, 27–34.

140. Osoba, D., Brada, M., Yung, W. K. A., and Prados, M. (2000). Health-related quality of life in patients treated with temozolomide versus procarbazine for recurrent glioblastoma multiforme. *J Clin Oncol* **18**, 1481–1491.

141. Prados, M. D. (2000). Future directions in the treatment of malignant gliomas with temozolomide. *Semin Oncol* **27** *Suppl* **6**, 41–46.

142. Shapiro, W. R., Green, S. B., Burger, P. C. *et al.* (1992). A randomized comparison of intra-arterial versus intravenous BCNU, with or without intravenous 5-fluorouracil, for newly diagnosed patients with malignant glioma. *J Neurosurg* **76**, 772–781.

143. Cascino, T. L., Veeder, M. H., Buckner, J. C. *et al.* (1996). Phase II study of 5-fluorouracil and leucovorin in recurrent primary brain tumor. *J Neurooncol* **30**, 243–246.

144. Baker, W. J., Royer, G. L., and Weiss, R. B. (1991). Cytarabine and neurologic toxicity. *J Clin Oncol* **9**, 679–693.

145. DeAngelis, L. M., Yahalom, J., Thaler, H. T., and Kher, U. (1992). Combined modality therapy for primary CNS lymphoma. *J Clin Oncol* **10**, 635–643.

146. Freilich, R. J., Delattre, J. Y., Monjour, A., and DeAngelis, L. M. (1996). Chemotherapy without radiation therapy as initial treatment for primary CNS lymphoma in older patients. *Neurol* **46**, 435–439.

147. Levin, V. A., and Prados, M. D. (1992). Treatment of recurrent gliomas and metastatic brain tumors with a polydrug protocol designed to combat nitrosourea resistance. *J Clin Oncol* **10**, 766–771.

148. Chu, E., and Allegra, C. J. (1996). Antifolates. *In Cancer Chemotherapy and Biotherapy. Principles and Practice* (B. A. Chabner, and D. L. Longo, Eds.), 2nd ed., pp. 109–148. Lippincott-Raven, Philadelphia.

149. Gabbai, A. A., Hochberg, F. H., Linggood, R. M., Bashir, R., and Hotleman, K. (1989). High-dose methotrexate for non-AIDS primary central nervous system lymphoma. Report of 13 cases. *J Neurosurg* **70**, 190–194.

150. Neuwelt, E. A., Goldman, D. L., Dahlborg, S. A. *et al.* (1991). Primary CNS lymphoma treated with osmotic blood-brain barrier disruption: Prolonged survival and preservation of cognitive function. *J Clin Oncol* **9**, 1580–1590.

151. Cheng, A. L., Yeh, K. H., Uen, W. C. *et al.* (1998). Systemic chemotherapy alone for patients with non-acquired immuno-deficiency syndrome-related central nervous system lymphoma. A pilot study of the BOMES protocol. *Cancer* **82**, 1946–1951.

152. Grant, R., Naylor, B., Junck, L., and Greenberg, H. S. (1992). Clinical outcome in aggressively treated meningeal gliomatosis. *Neurol* **42**, 252–254.

153. Macdonald, D. R. (1991). Neurologic complications of chemotherapy. *Neurol Clin* **9**, 955–967.

154. van Maanen, J. M. S., Retèl, J., de Vries, J., and Pinedo, H. M. (1988). Mechanism of action of antitumor drug etoposide: A review. *J Natl Cancer Inst* **80**, 1526–1533.

155. Goldberg, S. L., and Ahlgren, J. D. (1991). Recent clinical experience with infusional etoposide. *J Infusion Chemo* **1**, 37–43.

156. Tirelli, U., D'Incalci, M., Canetta, R. *et al.* (1984). Etoposide (VP-16-213) in malignant brain tumors: A phase II study. *J Clin Oncol* **2**, 432–437.

157. Buckner, J. C., Brown, L. D., Cascino, T. L. *et al.* (1990). Phase II evaluation of infusional etoposide and cisplatin in patients with recurrent astrocytoma. *J Neurooncol* **9**, 249–254.

158. Corden, B. J., Strauss, L. C., Killmond, T. *et al.* (1991). Cisplatin, ara-C and etoposide (PAE) in the treatment of recurrent childhood brain tumors. *J Neurooncol* **11**, 57–63.

159. Fulton, D., Urtasun, R., and Forsyth, P. (1996). Phase II study of prolonged oral therapy with etoposide (VP16) for patients with recurrent malignant gliomas. *J Neurooncol* **27**, 149–155.

160. Chamberlain, M. C. (1997). Recurrent supratentorial malignant gliomas in children. Long-term salvage therapy with oral etoposide. *Arch Neurol* **54**, 554–558.

161. Slichenmyer, W. J., Rowinsky, E. K., Donehower, R. C., and Kaufmann, S. H. (1993). The current status of camptothecin analogues as antitumor agents. *J Natl Cancer Inst* **85**, 271–291.

162. Kuhn, J. G. (1998). Pharmacology of irinotecan. *Oncology* **12**, 39–42.

163. Friedman, H. S., Houghton, P. J., Schold, S. C., Keir, S., and Bigner, D. D. (1994). Activity of 9-dimethylaminomethyl-10-hydroxycamptothecin against pediatric and adult central nervous system tumor xenografts. *Cancer Chemother Pharmacol* **34**, 171–174.

164. Blaney, S. M., Phillips, P. C., Packer, R. *et al.* (1996). Phase II evaluation of topotecan for pediatric central nervous system tumors. *Cancer* **78**, 527–531.

165. Wong, E. T., and Berkenblit, A. (2004). The role of topotecan in the treatment of brain metastases. *Oncologist* **9**, 68–79.

166. Hecht, J. R. (1998). Gastrointestinal toxicity of irinotecan. *Oncology* **12**, 72–78.

167. Horváth, I. P., Csetényi, J., Hindy, I. *et al.* (1982). Metabolism and pharmacokinetics of dibromodulcitol (DBD, NSC-104800) in man – II. Pharmacokinetics of DBD. *Eur J Cancer Clin Oncol* **18**, 1211–1219.

168. Áfra, D., Kocsis, B., Dobay, J. *et al.* (1983). Combined radiotherapy and chemotherapy with dibromodulcitol and CCNU in the postoperative treatment of malignant gliomas. *J Neurosurg* **59**, 106–110.

169. Levin, V. A., Edwards, M. S. B., Gutin, P. H. *et al.* (1984). Phase II evaluation of dibromodulcitol in the treatment of recurrent medulloblastoma, ependymoma, and malignant astrocytoma. *J Neurosurg* **61**, 1063–1068.

170. Elliot, T. E., Dinapoli, R. P., O'Fallon, J. R. *et al.* (1997). Randomized trial of radiation therapy (RT) plus dibromodulcitol (DBD) versus RT plus BCNU in high grade astrocytoma. *J Neurooncol* **33**, 239–250.

171. Hildebrand, J., Sahmoud, T., Mignolet, F., Brucher, J. M., and Afra, D. (1994). Adjuvant therapy with dibromodulcitol and BCNU increases survival of adults with malignant gliomas. *Neurol* **44**, 1479–1483.

172. Hildebrand, J., de Witte, O., and Sahmoud, T. (1998). Response of recurrent glioblastoma and anaplastic astrocytoma to dibromodulcitol, BCNU, and procarbazine. *J Neurooncol* **37**, 155–160.

173. Schrell, U. M. H., Rittig, M. G., Anders, M. *et al.* (1997). Hydroxyurea for treatment of unresectable and recurrent meningiomas. I. Inhibition of primary human meningioma cells in culture and in meningioma transplants by induction of the apoptotic pathway. *J Neurosurg* **86**, 845–852.

174. Schrell, U. M. H., Rittig, M. G., Anders, M. *et al.* (1997). Hydroxyurea for treatment of unresectable and recurrent meningiomas. II. Decrease in the size of meningiomas in patients treated with hydroxyurea. *J Neurosurg* **86**, 840–844.

175. Newton, H. B., Slivka, M. A., and Stevens, C. (2000). Hydroxyrea chemotherapy for unresectable or residual meningioma. *J Neurooncol* **49**, 165–170.

176. Mason, W. P., Gentill, F., Macdonald, D. R., Hariharan, S., Cruz, C. R., and Abrey, L. E. (2002). Stabilization of disease progression by hydroxyurea in patients with recurrent or unresectable meningioma. *J Neurosurg* **97**, 341–346.

177. Newton, H. B., Scott, S. R., and Volpi, C. (2004). Hydroxyurea chemotherapy for meningiomas: enlarged cohort with extended follow-up. *Br J Neurosurg* **18**, 495–499.

178. Spitz, I. M., and Bardin, C. W. (1993). Mefepristone (RU 486) – A modulator of progestin and glucocorticoid action. *New Engl J Med* **329**, 404–412.

179. Carroll, R. S., Glowacka, D., Dashner, K., and Black, P. M. (1993). Progesterone receptor expression in meningiomas. *Cancer Res* **53**, 1312–1316.

180. Olson, J. J., Beck, D. W., Schlechte, J., and Loh, P. M. (1986). Hormonal manipulation of meningiomas in vitro. *J Neurosurg* **65**, 99–107.

181. Grunberg, S. M., Weiss, M. H., Spitz, I. M. *et al.* (1991). Treatment of unresectable meningiomas with the antiproges-terone agent mifepristone. *J Neurosurg* **74**, 861–866.

182. Grunberg, S. M., Rankin, C., Townsend, J. *et al.* (2001). Phase III double-blind randomized placebo-controlled study of mifepristone (RU) for the treatment of unresectable meningioma. *Proc ASCO* **20**, 56a.

183. Friedman, Z. Y. (1998). Recent advances in understanding the molecular mechanisms of tamoxifen action. *Cancer Investig* **16**, 391–396.

184. Pollack, I. F., Randall, M. S., Kristofik, M. P., Kelly, R. H., Selker, R. G., Vertosick, F.T., (1990). Effect of tamoxifen on DNA synthesis and proliferation of human malignant glioma lines in vitro. *Cancer Res* **50**, 7134–7138.

185. Couldwell, W. T., Weiss, M. H., DeGiorgio, C. M. *et al.* (1993). Clinical and radiographic response in a minority of patients with recurrent malignant gliomas treated with high-dose tamoxifen. *Neurosurg* **32**, 485–490.

186. Chang, S. M., Barker, F. G., Huhn, S. L. *et al.* (1998). High dose oral tamoxifen and subcutaneous interferon alpha-2a for recurrent glioma. *J Neurooncol* **37**, 169–176.

187. Mastronardi, L., Puzzilli, F., Couldwell, W. T., Farah, J. O., and Lunardi, P. (1998). Tamoxifen and carboplatin combinational treatment of high-grade gliomas. Results of a clinical trial on newly diagnosed patients. *J Neurooncol* **38**, 59–68.

188. Rowinsky, E. K., and Donehower, R. C. (1995). Paclitaxel (taxol). *New Engl J Med* **332**, 1004–1014.

189. Glantz, M. J., Choy, H., Kearns, C. M. *et al.* (1995). Paclitaxel distribution in plasma and central nervous systems of humans and rats with brain tumors. *J Natl Cancer Inst* **87**, 1077–1081.

190. Chang, S. M., Kuhn, J. G., Rizzo, J. *et al.* (1998). Phase I study of paclitaxel in patients with recurrent malignant glioma: A North American Brain Tumor Consortium report. *J Clin Oncol* **16**, 2188–2194.

191. Prados, M. D., Schold, S. C., Spence, A. M. *et al.* (1996). Phase II study of paclitaxel in patients with recurrent malignant glioma. *J Clin Oncol* **14**, 2316–2321.

192. Chamberlain, M. C., and Kormanik, P. A. (1997). Salvage chemotherapy with paclitaxel for recurrent oligodendrogliomas. *J Clin Oncol* **15**, 3427–3432.

193. Newton, H. B. (2003). Molecular neuro-oncology and the development of "targeted" therapeutic strategies for brain tumors. Part 1 – growth factor and ras signaling pathways. *Expert Rev Anticancer Ther* **3**, 595–614.

194. Newton, H. B. (2004). Molecular neuro-oncology and the development of "targeted" therapeutic strategies for brain tumors. Part 2 – PI3K/Akt/PTEN, mTOR, SHH/PTCH, and angiogenesis. *Expert Rev Anticancer Ther* **4**, 105–128.

195. Newton, H. B. (2004). Molecular neuro-oncology and the development of "targeted" therapeutic strategies for brain tumors. Part 3 – brain tumor invasiveness. *Expert Rev Anticancer Ther* **4**, 803–821.

196. Newton, H. B. (2005). Molecular neuro-oncology and the development of "targeted" therapeutic strategies for brain tumors. Part 4 – p53 signaling pathways. *Expert Rev Anticancer Ther* **5**, 177–191.

197. Newton, H. B. (2005). Molecular neuro-oncology and the development of "targeted" therapeutic strategies for brain tumors. Part 5 – apoptosis and cell cycle. *Expert Rev Anticancer Ther* (In Press)

198. von Deimling, A., Louis, D. N., and Wiestler, O. D. (1995). Molecular pathways in the formation of gliomas. *Glia* **15**, 328–338.

199. Shapiro, J. R., and Coons, S. W. (1998). Genetics of adult malignant gliomas. *BNI Q* **14**, 27–42.

200. Maher, E. A., Furnari, F. B., Bachoo, R. M. *et al.* (2001). Malignant glioma: genetics and biology of a grave matter. *Genes & Dev* **15**, 1311–1333.

201. Folkman, J. (1995). Clinical applications of research on angiogenesis. *New Engl J Med* **333**, 1757–1763.

202. Wesseling, P., Ruiter, D. J., and Burger, P. C. (1997). Angiogenesis in brain tumors; pathobiological and clinical aspects. *J Neurooncol* **32**, 253–265.

203. Leon, S. P., Folkerth, R. D., and Black, P. M. (1996). Microvessel density is a prognostic indicator for patients with astroglial brain tumors. *Cancer* **77**, 362–372.

204. Plate, K. H., Breier, G., Weich, H. A., and Risau, W. (1994). Vascular endothelial growth factor and glioma angiogenesis: Coordinate induction of VEGF receptors, distribution of VEGF protein and possible in vivo regulatory mechanisms. *Int J Cancer* **59**, 520–529.

205. Pietsch, T., Valter, M. M., Wolf, H. K. *et al.* (1997). Expression and distribution of vascular endothelial growth factor protein in human brain tumors. *Acta Neuropathol* **93**, 109–117.

206. Bernsen, H. J. J. A., Rijken, P. F. J. W., Peters, H., Bakker, H., and van der Kogel, A. J. (1998). The effect of the anti-angiogenic agent TNP-470 on tumor growth and vascularity in low passaged xenografts of human gliomas in nude mice. *J Neurooncol* **38**, 51–57.

207. Fine, H. A., Figg, W. D., Jaeckle, K. *et al.* (2000). A phase II trial of the antiangiogenic agent thalidomide in patients with recurrent high-grade gliomas. *J Clin Oncol* **18**, 708–715.

3

Chemotherapy and Anti-Epileptic Drug Interactions

Karen L. Fink

ABSTRACT: Patients with primary or metastatic brain tumors are often on medications that alter the pharmacokinetics of chemotherapeutic agents (CTAs). The most common drugs administered to neuro-oncology patients that can affect their chemotherapy are anti-epileptic drugs (AEDs). Oncology patients who require chemotherapy may also be on AEDs for pre-existing epilepsy, neuropathic pain or headaches. These enzyme-inducing anti-epileptic drugs (EIAEDs) induce the production and activity of hepatic enzymes that are responsible for the metabolism of many drugs. The most common mechanism for this effect is induction of the cytochrome p450 (CYP) system. The induction of CYP by EIAEDs can affect the metabolism and effectiveness of CTAs. When patients require chemotherapy, it is important to know, how their concomitant medications affect the pharamacokinetics of the chosen CTA, because significant under or overdosing can occur. It is also important that these interactions be taken into account when clinical trials are designed for neuro-oncology patients. Concerns about the potential for AED interactions with CTAs have led to a change in prescribing practices for some AEDs among neuro-oncologists. While there are no firm guidelines to cover every situation, the oncologist and neuro-oncologist must be familiar with the medications that can alter the dosing and effectiveness of CTAs, and be prepared to make adjustments.

INTRODUCTION

Neuro-oncology patients present a special challenge to the physician who wants to offer effective chemotherapy. In patients with primary brain tumors, the blood–brain barrier, tumor heterogeneity, and the intrinsic drug resistance of these tumors can hamper the effectiveness of chemotherapy. Add to this the potential for drug–drug interactions, and the problem is compounded. Neuro-oncology patients are often on multiple medications in addition to their chemotherapy, and it is increasingly recognized that co-administration of medications that change the activity of the systems that metabolize drugs can lead to changes in the pharmacokinetics and effectiveness of chemotherapeutic agents (CTAs). Co-administration of drugs that affect each other's metabolism can lead to increased plasma levels, causing toxicity, but more commonly, causes diminished plasma concentrations and creates the potential for ineffective chemotherapy. The most common class of drugs that produce this effect in neuro-oncology patients are anti-epileptic drugs (AEDs).

ANTI-EPILEPTIC DRUGS IN NEURO-ONCOLOGY PATIENTS

Approximately 25–30 per cent of the adults and 10–15 per cent of the children with primary or metastatic brain tumors will experience a seizure during the course of their illness [1–4]. Seizure incidence is even higher in patients with primary glioma, particularly low-grade glioma and in patients with metastases from melanoma [3,5–10]. Although routine prophylaxis with AEDs is not recommended for patients with a brain tumor who have not had a seizure, patients in the community are often on an AED, usually phenytoin [2]. This practice can lead to ineffective chemotherapy, since some AEDs are well-known inducers of the hepatic systems responsible for drug metabolism. These inducers include phenytoin, and carbamazepine, the most frequently prescribed AEDs, and phenobarbital.

Central nervous system (CNS) metastases are more common than primary gliomas and represent an even more common cause of seizures in patients who will require chemotherapy. While there are approximately 17 000 primary brain tumors diagnosed in the U.S. each year, the figure for CNS metastases is 100–150 000 [11]. The incidence of seizures in patients with CNS metastases ranges from 10 to 30 per cent [3,11]. This allows a rough estimate of the number of patients with CNS metastases who will be on AEDs of 10 000–50 000 patients per year. Chemotherapy can be effective for some patients with CNS metastases, so the potential for interaction with co-administered medications can be important in these patients [12].

Even patients with systemic cancer who do not have central nervous system metastases or seizures may be on AEDs for other reasons, and may also require adjustments to their chemotherapy. The extent of AED use among patients with cancer is unknown; but is expected to be at least as common as in the general population, in whom the cumulative incidence of epilepsy through age 74 years is 3.1 per cent [13]. AED use is undoubtedly even more widespread than this, since AEDs are also used to treat some non-epileptic conditions, including neuropathic pain [14–16], migraines [17,18], post-herpetic neuralgia [19,20], movement disorders [21–23], and some psychiatric conditions [24–26].

DRUG METABOLISM AND RESISTANCE

Drug metabolism is generally classified in two phases, the initial oxidation or reduction reactions termed Phase I and Phase II reactions that create more water soluble and easily excreted forms of the drugs.

Phase I reactions include oxidation or reduction reactions, usually through the actions of cytochrome P450 oxidative enzymes or reductases. These enzymes prepare fat-soluble molecules for further metabolism *via* Phase II reactions by creating a reactive group suitable for conjugation. Drug interactions are usually the result of interactions with Phase I drug metabolism mechanisms. Conjugation reactions performed during Phase II metabolism of drugs usually involve the addition of a peptide (glutathione), sugar, or sulfur group to the drug molecule. These groups are usually common and easily accessible in well-nourished cells, so these Phase II reactions are rarely rate-limiting, and therefore rarely involved in drug interactions.

Phase I reactions carried out by cytochrome P450 enzymes, flavin monooxygenases, and reductases are more frequently rate-limiting, and are more often the target of clinically significant drug interactions. Phase I oxidative enzymes are mostly found in the endoplasmic reticulum within liver cells. The main enzymes responsible for Phase I reactions involve the microsomal mixed function oxidation system. This system requires the presence of NADPH and NADPH-cytochrome P450 reductase (see Fig. 3.1). "Cytochrome P450" refers to a superfamily of enzymes that comprise the terminal oxidase of this system. They form part of the cascade that shuttles electrons from molecular oxygen to oxidize drugs. Six cytochrome P450 (CYP) isoforms have been recognized as important in human drug metabolism: CYP1A2, CYP3A, CYP2C9, CYP2C10, and CYP2D6. Three of these isoforms—CYP2C9, CYP2C19, and CYP2D6—can be genetically absent. The enzyme-inducing AEDs (EIAEDs) are typically metabolized by CYP3A4, CYP2C9, and CYP2C19 [27]. Table 3.1 lists the specific CYP enzymes induced or inhibited by AEDs.

Advances in the understanding of the basic biology of the induction of drug metabolism provide a scientific scaffold for understanding drug interactions between EIAEDs and chemotherapeutics. The molecular mechanisms underlying induction and increased elimination of drugs from the body involve drug interactions with the nuclear receptor superfamily of ligand-activated transcription factors (ligands include drugs, hormones, nutrients, and metabolites) [28]. Several classes of nuclear receptors that mediate drug enzyme induction have been identified. In particular these include

(1) CAR or constitutive androstane receptor,
(2) PXR or pregnane X receptor,
(3) PPAR or peroxisome proliferators-activated receptor,
(4) AhR or aryl hydrocarbon receptor [29].

EIAEDs like phenobarbital and phenytoin actually led to the discovery of CAR, and this remains one of the best experimental models for studying drug induction [30]. CAR mediates the induction of CYP enzymes 2B6, 2C9, and 3A4, and in addition induces some drug conjugation enzymes responsible for Phase II reactions. Both the CAR and PXR receptors act as drug sensors that regulate an overlapping set of genes that increase oxidative drug metabolism,

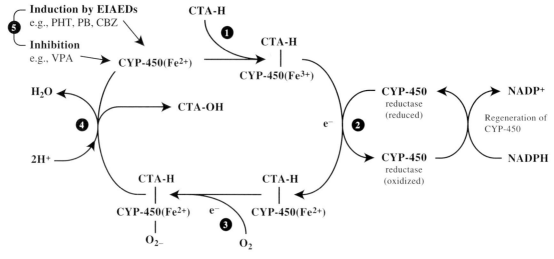

FIGURE 3.1   Explaining the EIAED–CTA interaction.
Graphic depiction of the mechanism by which the EIAED-CTA interactions occur via the cytochrome P-450 enzyme system. P-450 enzymes metabolize CTAs via a variety of reactions, including dealkylation, hydroxylation, and oxidation (pictured).

Sequence:
1. Oxidized CYP-450 combines with CTA substrate to form a complex.
2. and 3. NADPH releases an electron to CYP-450-reductase. A second electron (from NADPH) reduces molecular oxygen to form "activated oxygen" CYP-450-CTA complex.
4. This complex transfers oxygen to the CTA molecule, rendering it water soluble for renal elimination, and (in most cases) inactive as a chemotherapeutic agent.
5. Coadministration of an EIAED induces increased P-450 enzyme production by activation of gene transcription resulting in increased P-450 enzyme production and decreased plasma concentrations of the CTA substrate. Valproic acid exerts the opposite effect, inhibiting P-450 metabolism, probably through competition.

conjugation, and transport. These genes include several CYPs, Phase II conjugation enzymes, multiple drug resistance-associated proteins (MRPs), and organic anion transporters (OATs) [31].

Phase II Metabolism

Phase II reactions conjugate a water soluble group to a drug in order to allow biliary or renal excretion. These Phase II enzymes include UDP-glucuronyl transferase (UGTs), sulfotransferases (SULTs), and glutathione S-transferases (GSTs), and also some drug transporter enzymes [32–34]. UGTs, SULTs, and GSTs prepare polar drugs for biliary or urinary excretion. Glucuronide conjugation is performed by the two main families, UGT1 and UGT2, each of which have several isoforms [35]. The activity of UGT enzymes varies within the human population, and low UGT enzyme activity can lead to severe toxicity of some CTAs such as irinotecan, which requires glucuronidation for inactivation [36]. SULT-mediated sulfation also plays an essential role in drug removal,

and is performed by two major families, SULT1 and SULT2, which can be induced by phenobarbital *via* the CAR receptor [34]. GSTs are a family of enzymes that are also inducible by phenobarbital *via* the CAR receptor. They conjugate glutathione to a wide variety of drugs to inactivate them. Potential upregulation of the conjugation phase II enzymes for those CTAs affected by conjugation processes is a consideration with EIAED treatment, although this type of interaction is less frequent than interaction *via* the induction of CYP. A molecular understanding of EIAED effects can be useful in choosing specific EIAEDs and NEIAEDs in the clinical setting.

AED EFFECTS ON HEPATIC METABOLISM

As described above, when drugs interact pharmacokinetically, the reaction is usually as a result of the induction of hepatic enzymes that are responsible for drug degradation. Many drugs are metabolized in the liver *via* oxidation, hydroxylation, or dealkylation (see Fig. 3.1). These degradative pathways are mediated by the cytochrome P-450 (CYP) enzyme system, specifically the CYP 1, 2, and 3 families [27,37,38]. The CYP system employs a homeostatic mechanism whereby a regulator gene adjusts enzyme production levels to a steady state level. CYP-inducing drugs interact with the regulator gene, increasing the rate of CYP enzyme production [39,40]. This inducing effect is negligible for most drugs, but some AEDs such as carbamazepine, phenobarbital, and phenytoin, and to a lesser extent felbamate and oxcarbazepine, are potent inducers of CYP enzymes. Table 3.1 lists these enzyme-inducing AEDs (EIAEDs), along with the CYP enzymes they are known to induce. EIAED administration increases the rate of clearance of any drug that is a substrate for the P-450 enzymes it induces. For example, in a patient taking phenytoin, hepatic production of CYP enzymes 1A2, 2C9, and 3A4 is increased [37]. If the patient is also taking paclitaxel, which is metabolized by these enzymes [41–43], the rate of clearance of the paclitaxel is increased, and its effectiveness may be compromised.

Not all AEDs interact significantly with the CYP enzymes; AEDs classified as non-EIAEDs include gabapentin, lamotrigine, levetiracetam, tigabine, topiramate, vigabatrin, and zonisamide (see Table 3.1).

TABLE 3.1 AED–Cytochrome P-450 Enzyme System Interactions

	CYP Enzyme(s) Induced	CYP Enzyme Inhibited
EIAEDs		
Carbamazepine	1A2, 2C9, 3A4	2C19
Felbamate	2C9, 3A4 [27]	—*
Phenobarbital	1A2, 2B, 2C9 [27], 3A4	—*
Phenytoin	1A2, 2C9, 3A4	—*
Primidone	1A2, 2B, 3A4	—*
Oxcarbazepine	3A4	2C19 [27]
Non-EIAEDs		
Valproic acid[†]	—*	2C9, 3A4? [27]
Gabapentin, lamotrigine, levetiracetam, tiagabine, topiramate, vigabatrin, zonisamide	—*	—*

Adapted with permission from Elsevier [37].
* No significant interaction currently recognized.
† Valproic acid is listed separately from other non-EIAEDs because, unlike the others, it is a CYP enzyme inhibitor.

Gabapentin, levetiracetam, and vigabatrin are the non-EIAEDs with the lowest drug-interaction profiles [27,44,45]. Note that while valproic acid is classified as a non-EIAED, it is an *inhibitor* of CYP enzyme 2C9, and possibly 3A4 [27]. The consequent increase in CTA levels in patients taking valproic acid may be responsible for the increased incidence of thrombocytopenia that has been reported during co-administration of valproic acid and CTAs, compared to co-administration of CTAs and carbamazepine or phenytoin [46].

EIAEDs EFFECTS ON CHEMOTHERAPY

EIAEDs have been found to affect the metabolism of drugs in many categories of CTAs. They can affect vinca alkaloids, nitrogen mustards, taxanes, epipodophyllotxins, and other alkylating agents. Table 3.2 lists some of the interactions that are known to occur between CTAs and AEDs. CTA doses must often be substantially but variably increased in order to compensate for the hepatic induction of CYP enzymes caused by EIAEDs. Although the literature on CTA dose increases required in patients on concurrent EIAEDs is incomplete, clinical trials of CTAs in patients with malignant gliomas have reported maximum tolerated doses 50–400 per cent higher in patients taking paclitaxel [42,47–49], irinotecan [50–55], and newer experimental CTAs [56–59] in patients who are also taking EIAEDs.

Irinotecan

Irinotecan is a striking example of the potential for ineffective chemotherapy in patients taking EIAEDs, and of the need for dose adjustments in these patients. When irinotecan was given every three weeks in a clinical trial for patients with recurrent malignant gliomas, the maximum tolerated dose (MTD) in patients who were not taking EIAEDs was 350 mg/m^2. In the patients who were taking EIAEDs, the MTD was more than twice as high—750mg/m^2 [51]. When irinotecan was given on the more frequently prescribed regimen of four weekly infusions followed by a two week break, the maximum tolerated dose for patients who were not taking EIAEDs was 117 mg/m^2, and the dose for patients on such drugs was 344 mg/m^2 [54]. The MTD of irinotecan was thus 3.5 times greater in patients with malignant glioma who were concurrently receiving EIAEDs than in those who were not. This study also confirmed

TABLE 3.2   Known AED–CTA Drug Interactions

	Chemotherapeutic Agent	Known AED Interaction(s)
Alkyl sulfonates	Busulfan	PHT [98]
Antimetabolites	Methotrexate	CBZ, PB, PHT, VPA [74]
Camptothecin analogs	9-aminocamptothecin	CBZ, PB, PHT [50,99]
	Irinotecan (CPT-11)	CBZ, PB, PHT [52–54,100]
	Tamoxifen	PHT [101]
Epipodophyllotoxins	Etoposide, teniposide	CBZ, PB, PHT [74]
Taxanes	Docetaxel	CBZ, PB, PHT [47]
	Paclitaxel	CBZ, PB, PHT [47–49]
Vinca alkaloids	Vinblastine, vincristine, vindesine	CBZ, PB, PHT [102]
Novel agents	Erlotinib (OSI-774)	CBZ, PB, PHT [56,60]
	Gefitinib (ZD1839)	CBZ, PB, PHT [57,59,61]
	GW572016	CBZ [103]
	Topotecan	PHT [104]
Combination therapy	Cisplatin/carmustine	PHT [69]
	Doxorubicin/cisplatin	CBZ, VPA [70]
	5-fluorouracil/leucovorin	PHT [71]
	Fotemustine/cisplatin/etoposide	VPA [46]

With the exception of VPA, these interactions increase the dose requirement for the CTA.
Abbreviations: CBZ – carbamazepine; PHT – phenytoin; PB – phenobarbital; VPA – valproic acid.

that the concomitant administration of EIAEDs markedly increased the clearance of irinotecan [54].

Erlotinib

Newer agents that are in trials for treating patients with brain tumors are also affected by the use of EIAEDs. Much of this data is in abstract form at the time of this writing. Erlotinib HCL (Tarceva) is an orally active, highly potent and selective inhibitor of the epidermal growth factor receptor. It is metabolized by CYP3A4, so in a recent clinical trial led by Yung and colleagues, patients receiving EIEADs were given a higher starting dose (300 mg/day), while patients who were not on EIEADs were started at 150 mg/day [60]. In another study of erlotinib for patients with recurrent malignant glioma, pharmacokinetics showed that exposure to erlotinib and its active metabolite was markedly reduced (~50–75 per cent) when it was given in combination with EIAEDs [56].

Gefitinib

In a trial of gefitinib (Iressa) for patients with recurring malignant glioma or progressive unresectable meningioma, gefitinib was well tolerated at 500 mg/day in patients who were not on EIAEDs. The MTD for patients who were taking EIAEDs was 1500 mg/day [59] Similarly, in a phase I study of

gefitinib plus temozolomide (Temodar) for malignant glioma, the maximum tolerated dose of gefitinib was 1000 mg/day in patients who were taking an EIAED, and 250 mg/day in patients who were not [57]. In a phase I/II study of gefitinib in combination with radiation for patients with newly diagnosed glioblastoma, there were no dose-limiting toxicities up to a dose of 750 mg/day in the patients receiving EIAEDs, while 23 per cent (3 of 13) of those who received non-EIAEDs experienced dose-limiting toxicities at or below a dose of 500 mg/day [61].

Methotrexate

While most EIAEDs affect the serum concentrations of CTAs by interacting with hepatic drug metabolism systems, special mention must be made regarding methotrexate (MTX) and its interaction with EIAEDs. Even though methotrexate is primarily renally excreted, it can be affected by the co-administration of EIAEDs. It is theorized that when MTX is given as a long infusion (24 h), the liver metabolizes approximately 40 per cent of the drug, and the usual EIAED effect is seen. When MTX is given by shorter infusions (4–6 h), the excretion is nearly entirely renal, but the effect on serum levels of MTX by EIAEDs persists. There has, therefore, been some speculation that EIAEDs might alter other aspects of drug metabolism, perhaps even by increasing the renal excretion of other drugs [62].

Temozolomide

No discussion of the treatment of patients with brain tumors would be complete without mentioning temozolomide. Most patients with primary malignant gliomas will receive temozolomide at some point during the course of their disease, and many will be on AEDs. Fortunately for neuro-oncologists, temozolomide does not require metabolism by hepatic mechanisms. It spontaneously hydrolyzes to active intermediates, and its serum levels and effectiveness should not be affected by the co-administration of EIAEDs [63].

AED–Corticosteroid Interactions

In addition to their interactions with CTAs, EIAEDs are known to alter the action of corticosteroids, which are often required in the management of patients with brain metastases or glioma. Phenytoin has been shown to increase the rate of clearance of steroids [64], and phenytoin and phenobarbital both shorten the half-life and increase total body clearance of both dexamethasone and prednisone [65,66]. Conversely, dexamethasone induces several of the CYP isoenzymes [66–68], and may decrease the serum levels, and therefore the effectiveness of concomitant EIAEDs [68]. Thus, when EIAEDs are used with corticosteroids, adjustments to the AED and/or corticosteroid doses must often be made.

CHEMOTHERAPY EFFECTS ON AEDs

In addition to the potential for inadequate dosing of chemotherapy in patients with cancer who are taking EIAEDs, there is the potential for interaction in the opposite direction, with the CTA affecting the pharmacokinetics of the AED. In the cases where the CTA inhibits hepatic enzymes, this can lead to toxic levels of the AED. If on the other hand, the CTA causes induction of the hepatic microsomal system and thereby increases the clearance of the AED, there is a risk for breakthrough seizures in patients who are taking the AED to prevent seizures. For instance, Grossman and co-workers showed in a retrospective review that patients who were receiving phenytoin during chemotherapy with carmustine (BCNU) and cisplatin often required an increase in their dose of phenytoin to maintain a therapeutic level. All the patients who received three or more cycles of BCNU/cisplatin required an increase in their maintenance phenytoin dose. The average required increase in phenytoin dose to maintain therapeutic levels was 41 per cent [69]. The change in phenytoin levels occurred as early as two days after chemotherapy, and was related to cisplatin rather than to BCNU [69].

Similarly, Neef and associates reported a patient with epilepsy who required treatment with adriamycin and cisplatin, and had breakthrough seizures during treatment. Peak and trough plasma levels of phenytoin, carbamazepine, and valproate sodium showed lower plasma levels of carbamazepine and valproate sodium that normalized 2–3 days after cisplatin. Phenytoin levels were reduced by more than 60 per cent, even though phenytoin was given intravenously, hence gastrointestinal absorption of phenytoin was not a factor [70].

CTAs can cause elevation as well as reductions in plasma concentrations of AEDs. Gilbar et al. reported a case of phenytoin toxicity during chemotherapy with weekly fluorouracil that required a reduction in the dose of phenytoin during treatment [71]. Phenytoin is primarily metabolized by CYP2C9, and inhibition of this microsomal isoenzyme by fluorouracil was felt to be the most likely explanation. When CTAs increase EIAED levels, patients may experience toxicity, and might even affect compliance with the AED regimen, putting the patient at risk of seizures.

It is important to know the directions of potential interactions between drugs that are co-administered during chemotherapy. Unfortunately, induction or inhibition of the CYP system is difficult to assess; there is no simple test that allows the determination of the extent of hepatic enzyme induction to help guide chemotherapy dose adjustments. In addition, the changes in CTA metabolism induced by EIAEDs vary substantially between patients, making it difficult to predict how doses should be adjusted for an individual. Dose adjustments of AEDs that are required in order to control seizures may lead to further inconsistencies in serum levels of CTAs. In practice, it may be easier to switch to a non-EIAED than to determine the appropriate dose of a given CTA for a patient who requires both chemotherapy and an AED.

PRACTICAL EFFECTS OF AED–CHEMOTHERAPY INTERACTIONS

Effects on Clinical Trial Design in Neuro-Oncology

The potential for subtherapeutic chemotherapy dosing in patients on EIAEDs has affected clinical trial design for patients with malignant gliomas, and should also be taken into account in trials for patients with central nervous system metastases as well. Clinical trials for patients with brain tumors must stratify patients by the use or non-use of EIAEDs, and seek separate maximum tolerated doses for any CTA that might be metabolized by the CYP system [49,51,54]. In effect, there has to be a separate Phase I clinical trial in patients on EIAEDs, because the dose determined in Phase II trials done in patients who are not taking such agents is an inadequate estimate of the dose that will produce toxicity or effectiveness in patients on EIAEDs. This leads to a complicated accrual process for clinical trials for brain tumor patients. To find the proper dose in patients on EIAEDs, there must be one arm that dose escalates to find the MTD in patients on EIAEDs. At the same time, another arm accrues patients who are not on EIAEDs at the dose of the CTA that was determined to be tolerated and/or effective in clinical trials in patients with other types of cancer. Sometimes, when a patient and/or their physician wants to participate in a clinical trial, the decision has been made to change the patients AED based on which arm of the clinical trial is open. Other trials allow intra-patient dose escalation in patients who are taking EIAEDs [60,72]. The need for CTA dose adjustment and dose finding in patients who are taking EIAEDs adds to the complexity and expense of clinical trials in neuro-oncology, such that some trials now exclude patients on EIAEDs [73]. Fortunately, there are now several non-EIAEDs, and many neuro-oncologists have come to prefer these agents for their patients with a brain tumor who may require chemotherapy.

CTA Metabolism Induction Effects on the Cost of Cancer Care

Obviously, the requirement for increased doses of chemotherapy in patients who are also taking EIAEDs, increases the cost of treating these patients. In practice, it may be difficult to get payment for these higher doses of CTAs. The higher doses of CTAs required for patients on AEDs are unlikely to be studied sufficiently to receive official FDA approval, and some insurance companies restrict payment for drugs to FDA-approved indications and doses. In addition to cost considerations, prolonged administration of the higher required doses of CTAs in patients taking EIAEDs may increase the potential for side effects. If on the other hand CTA doses are not increased, inadequate treatment doses may be administered, potentially lessening the CTA efficacy, response rates, and even survival in patients who require AEDs. Indeed effects of AEDs on survival have been shown for patients receiving chemotherapy for lymphoblastic leukemia [74].

PRACTICAL MANAGEMENT

Choosing an AED

For cancer patients who require an AED, the selection of an appropriate AED must be based on effectiveness, side effect profile, cost, and available routes of administration. To add to this complicated decision-making, we must consider the potential for drug–drug interactions, particularly with CTAs. The ideal AED for neuro-oncology patients would have a long half-life to allow once daily dosing to improve compliance, would have a low incidence of tolerable or transient side effects, would be highly effective as monotherapy for both partial and generalized seizures, would allow a rapid titration to a therapeutic dose, would have minimal interaction with other medications, and would be low cost. No currently available agent meets all of these requirements. Table 3.3 summarizes some of the considerations when choosing an AED.

The advantages and disadvantages of specific AEDs in treating cancer patients (see Table 3.3) vary with the clinical situation. For surgical candidates, the potential of valproic acid to produce abnormal hematologic parameters such as thrombocytopenia [75–78] argues against its use perioperatively— although there does not appear to be an increased risk of perioperative blood loss [79,80]. Phenytoin is favored by many neurosurgeons because of its history of effectiveness in preventing perioperative seizures, and its availability in intravenous formulation. However, phenytoin's potential to decrease the effectiveness of CTAs may make it a less than ideal choice for long-term use in patients with cancer. For patients who do not require intravenous AED administration, the lack of AED–CTA drug interactions makes choosing a non-EIAED attractive. At this point however, the newer non-EIAEDs have not been studied extensively as monotherapy agents, so there is some hesitance for clinicians to use the newer agents alone [81].

For patients who need radiation therapy, the skin rash that can be associated with lamotrigine [82], oxcarbazepine [83], carbamazepine [84], and phenytoin [85–88], may become important, because local skin reactions due to radiation may be exacerbated

TABLE 3.3 Considerations in AED Selection for Patients with Cancer

	EIAEDs				Non-EIAEDs						
General	PHT	CBZ	OXC	PB	GBP	LEV	LTG	TGB	TPM	VPA	ZNS
AED–CTA drug interactions	Yes	Yes	?	Yes	No[a]	No[a]	No[a]	No[a]	No[a]	Yes	No[a]
Serious side effects [81]	Yes[b]	Yes[b]	Some[c]	Yes[b]	Fewer	Fewer	Some[c]	Some[c]	Some[c]	Yes[b]	Some[c]
Drug cost	Low	Low	High	Low	High	High	High	High	High	Low	High
Efficacy as monotherapy	Yes	Yes	Yes[d] [105,106]	Yes	Yes[d,e] [107]	?[81]	Yes[e] [108–110]	?[e] [81]	Yes[d,e] [111,112]	Yes	?[e] [81]
IV formulation	Yes	No	Yes	Yes	No	No	No	No	No	Yes	No
Children											
Neurological side effects [81]	Yes	Yes	Fewer	Yes	Some	Some	Some	Some	Some	Some	Some
Efficacy in children	Yes	Yes [113]	Yes [113]	Yes	? [81]	? § [81]	Yes [114]	? [81]	Yes [112,115]	Yes [115]	? § [81]

This is not meant to be a comprehensive description of all considerations involved in selecting an AED for cancer patients.

[a] Suggested in Patsalos, et al. [27].

[b] Includes serious systemic side effects such as aplastic anemia, agranulocytosis, pancreatitis, and hepatic failure.

[c] Includes serious side effects such as rash (Stevens Johnson, topical epidermal necrolysis), hypersensitivity reactions (hepatic and renal failure, disseminated intravascular coagulation, arthritis), hyponatremia, stupor, nephrolithiasis, open angle glaucoma, hypohidrosis, renal calculi.

[d] Efficacy as monotherapy has been demonstrated for newly diagnosed partial seizures but not for newly diagnosed absence seizures.

[e] Not approved by the Food and Drug Administration for this indication.

Abbreviations: CBZ – carbamazepine; EIAED – enzyme-inducing antiepileptic drug; GBP – gabapentin; LEV – levetiracetam; LTG – lamotrigine; OXC – oxcarbazepine; PB – phenobarbital; PHT – phenytoin; TGB – tiagibine; TPM – topiramate; VPA – valproic acid; ZNS – zonisamide.

by a generalized rash caused by the AED. There was an initial report in 1988 of multiple cases of toxic epidermolysis in patients receiving phenytoin and concurrent radiation therapy [89]. Subsequent reports have not found a similar high incidence, but rash is still a significant side effect to consider when choosing an AED for a patient who has cancer. Mamon and colleagues reviewed the records of patients with brain tumors treated with cranial radiation and AEDs to assess the frequency of both severe and mild skin reactions. Only one of 289 patients developed erythema multiforme, but milder rashes, occurred in 18 per cent of exposures to AEDs and in 22 per cent of patients taking phenytoin. This was higher than the expected rate of 5–10 per cent. There appears to be an increased frequency of mild drug rashes among patients with brain tumors who take phenytoin that does not appear related to radiation [86].

Rash can be a side effect of some CTAs, and procarbazine is a frequent offender. The likelihood of rash can be influenced by the use of EIAEDs. Lehman *et al.* reported that patients taking EIAEDs had an increased risk of developing hypersensitivity when given procarbazine [90]. The increased incidence of procarbazine hypersensitivity reactions was felt to be related to a reactive intermediate of procarbazine that was generated by CYP3A induction, since phenobarbital is known to cause the generation of such an intermediate *via* CYP3A induction. The authors advocated trials of newer non-EIADs for the prophylaxis of seizures in patients with brain tumors who required procarbazine.

Neuro-oncologists are also influenced by the desire to achieve a rapid therapeutic response in their brain tumor patients. Since many patients with a brain tumor have a reduced life expectancy, the need for a prolonged titration to avoid side effects is undesirable. This makes lamotrigine less attractive, since a slow titration is preferred to prevent the development of rash. Neuro-oncologists also prefer to avoid AEDs that impair cognition, so have been hesitant to use topiramate because of its reputation for cognitive impairment at higher doses.

Data regarding the effectiveness of monotherapy with the newer AEDs in patients with brain tumors is limited, but expanding. Gabapentin is one non-EIAED with a favorable side effect profile, and it has been used to treat seizures in children undergoing chemotherapy with good results [91]. Khan reports that seizures were controlled in 74 per cent of the 50 children given gabapentin monotherapy as initial treatment and in 49 per cent of the 59 children when gabapentin was added to other AEDs [91].

Only 8 children (7 per cent) reported adverse effects. Gabapentin was thus found to be effective and well tolerated. Chemotherapy effectiveness in this population was not reported.

Levitiracetam has been used as monotherapy in several small retrospective and prospective trials in patients with brain tumors [92,93]. The largest retrospective series thus far has been reported in abstract form and included 278 patients with brain tumors who received levetiracetam. The institution had adopted a policy of using levetiracetam in patients who required an AED. Levetiracetam was well tolerated, and only 3.5 per cent of the patients discontinued the drug, usually due to behavioral issues. The authors reported that 60 per cent of the patients receiving levetiracetam had a 50 per cent or greater reduction in their seizure frequency, although these patients were not necessarily patients with medically refractory seizures. Monotherapy was achieved in 70 per cent of the patients [93].

When patients have neurologic impairment that prevents swallowing, the availability of alternate routes of administration of AEDs becomes important. As patients who have a brain tumor are not able to swallow pills, and the ability to administer an AED *via* sublingual or rectal routes, is convenient. Carbamazepine and valproic acid have not traditionally been given rectally, because of concerns about irritation of the rectal mucosa or lack of efficacy. Phenytoin can be administered rectally, but absorption is significantly reduced compared to oral or intravenous administration, and the dose must be substantially increased [94,95]. There are no data regarding rectal administration of the newer AEDs.

Chemotherapy Dose Adjustment in Patients on EIAEDs

There is no straightforward recommendation about how chemotherapeutic agents should be adjusted in patients on EIAEDs because the effects vary depending on the chemotherapeutic agent and its mode of metabolism. The issue is further complicated by the fact that many cancer patients are on drugs other than anticonvulsants that induce the CYP system. We have listed in Table 3.4, examples of some of the drugs that induce CYP enzymes other than the EIAEDs. This is a varied list, and emphasizes the importance of knowing all the drugs and even alternative remedies that a patient is taking.

TABLE 3.4   Examples of Inducers of Cytochromes P-450
(Other than EIAEDs)

HIV Protease- and RT Inhibitors

Ritonavir

Saquinavir

Efavirenz

Nevirapine

Antidepressants

Hyperforin, component of St. John's Wort

Antibiotics

Rifampicin

Rifabutin

Others

Bosentan

Tamoxifen

Troglitazone

Omeprazole

Smoking

Ethanol

Isoniazid

Dexamethasone

Prednisone

Prednisolone

Methylprednisolone

Modified from Figure 1, Handschin C, Meyer UA. Induction of drug metabolism: the role of nuclear receptors. *Pharmacol Rev.* 2003;55:649–673, and Table 1, Tanaka E. Clinically important pharmacokinetic drug–drug interactions: role of cytochrome P450 enzymes. *J Clin Pharm Ther.* 1998;23:403–416 [116,117]

Changing AEDs

Patients may need to switch from one AED to another for a variety of reasons:

(1) to improve seizure control,
(2) to avoid side effects or improve tolerability,
(3) to avoid drug interactions,
(4) to avoid teratogenicity during pregnancy, and
(5) for cost considerations.

Often, the decision to switch AEDs takes more than one of these considerations into account. There is very little data in the literature to guide the clinician who wants to advise a patient in these transitions. The following is a compilation of information from experienced clinicians and epileptologists.

There are two main approaches to changing AEDs. In one approach, the new AED is titrated to a target blood level or to a specific dose, and then the second AED is tapered slowly or rapidly, depending on the potential for withdrawal effects. The second approach is to titrate the new AED while simultaneously reducing the old AED. The first approach is probably the safest, whereas the second approach has less potential for AED-induced toxicity. Of the non-EIAEDs, rapid titration to a standard therapeutic dose (or at least a potentially effective dose), can be accomplished with valproic acid, gabapentin, topiramate, levetiracetam, and oxcarbazepine, although these rates of titration are "off label" when compared to the rates used during pivotal, double blind trials in medically refractory epilepsy. There is limited literature describing the process of switching AEDs and the published data usually involves the older EIAEDs such as carbamazepine [96]. However, anecdotal experience at epilepsy centers where rapid inpatient medication changes are accomplished safely can provide guidance for rapid AED changes in oncology patients. Initial daily doses of gabapentin 1200–1800 mg/day, topiramate 100–200 mg/day, levetiracetam 2000 mg/day, and oxcarbazepine 600–1200 mg/day are generally well tolerated, and produce anticonvulsant effects within 24 h. If there are side effects, they are usually transient, and can be managed by rapidly tapering the EIAED.

If a patient requires chemotherapy that can be altered by P-450 enzyme induction, ideally, changing to a non-EIAED should be accomplished before starting chemotherapy, so that deinduction of cytochrome P-450 can occur. This process can take up to several weeks. Schaffler calculated a $T_{1/2}$ for deinduction after discontinuing carbamazepine of 3.84 days [97]. This time frame is not always practical for patients who need rapid treatment of their underlying cancer. In addition, patients who are taking an EIAED that has a long half-life or the potential for withdrawal seizures during rapid withdrawal cannot use a rapid taper. Phenobarbital or primidone are both problematic for these reasons. Patients requiring chemotherapy who are on EIAEDs with a long half-life should probably stay on them, and have their chemotherapy dose adjusted appropriately.

There is no standard protocol for tapering non-barbiturate AEDs, but when time allows, a 5–6 week taper is suggested. Barbiturates such as phenobarbital or primidone may require a longer taper of 2 months or more, particularly if phenobarbital levels are initially higher than 10 µg/ml. When the AED to be tapered has a low level or dose, discontinuation is generally low risk. For example, if a patient taking phenytoin 200 mg/day has a blood level of 5 µg/ml, abrupt discontinuation or a one-week taper carries little risk for seizures, particularly if a non-EIAED has been introduced at a potentially effective dose at the same time. If the patient is taking several EIAEDs, most epileptologists would allow a rapid taper of each AED, one at a time, as long as there is at least one drug at therapeutic doses or levels. If seizures recur, during the taper, the EIAED may be restored at its previous dose, or the remaining AED dose may be increased.

CONCLUSIONS

Drug interactions between AEDs and CTAs have become increasingly recognized. Induction of the hepatic cytochrome P-450 enzyme system by EIAEDs increases the rate of clearance of some CTAs, resulting in decreased and potentially subtherapeutic CTA plasma concentrations unless the CTA dose is increased. Increasing the CTA dose to overcome this effect substantially increases the cost of treatment, and also increases the potential for toxicities and adverse side effects, particularly if the EIAED administration is interrupted because of compliance issues or adverse events. The use of a non-EIAED in cancer patients who require an AED seems preferable. In our experience, the switch from an EIAED to a non-EIAED can usually be made safely. Further clinical trials are necessary to determine which non-EIAEDs are most effective as monotherapy in patients with cancer or a brain tumor.

At this time, the decision about which AED to use in any given patient cannot be guided by clinical trial evidence. Although no firm recommendations can be made to cover all situations, physicians should know the likely status of the hepatic microsomal system in their patients, by considering all the medications they are taking. Reasonable adjustments should be made to their chemotherapy and/or AED as needed, and patients should be followed closely for toxicity. It is critical, whenever chemotherapy is initiated, that the treating physician be aware of whether the patient is taking an EIAED, both to guide the dosing of the CTA, and to consider whether a switch to a non-EIAED is desirable.

ACKNOWLEDGMENT

I wish to thank B. Alex Merrick, PhD, for his assistance in writing the sections of this manuscript concerning drug metabolism and resistance. Dr. Merrick is Head, Proteomics Group, National Center for Toxicogenomics, NIEHS. I also acknowledge the valuable input from Paul Van Ness, M.D at the University of Texas Southwestern Medical Center in Dallas for his comments regarding practical management of AED changes.

References

1. Cohen, N., Strauss, G., Lew, R., Silver, D., and Recht, L. (1988). Should prophylactic anticonvulsants be administered to patients with newly-diagnosed cerebral metastases? A retrospective analysis. *J Clin Oncol* **6**, 1621–1624.
2. Forsyth, P. A., Weaver, S., Fulton, D. *et al.* (2003). Prophylactic anticonvulsants in patients with brain tumour. *Can J Neurol Sci* **30**, 106–112.
3. Liigant, A., Haldre, S., Oun, A. *et al.* (2001). Seizure disorders in patients with brain tumors. *Eur Neurol* **45**, 46–51.
4. Maytal, J., Grossman, R., Yusuf, F. H. *et al.* (1995). Prognosis and treatment of seizures in children with acute lymphoblastic leukemia. *Epilepsia* **36**, 831–8.
5. Pace, A., Bove, L., Innocenti, P. *et al.* (1998). Epilepsy and gliomas: incidence and treatment in 119 patients. *J Exp Clin Cancer Res* **17**, 479–4.
6. Im, S. H., Chung, C. K., Cho, B. K. *et al.* (2002). Intracranial ganglioglioma: preoperative characteristics and oncologic outcome after surgery. *J Neurooncol* **59**, 173–1.
7. Razack, N., Baumgartner, J., and Bruner, J. (1998). Pediatric oligodendrogliomas. *Pediatr Neurosurg* **28**, 121–1.
8. Moots, P. L., Maciunas, R. J., Eisert, D. R., Parker, R. A., Laporte, K., and Abou-Khalil, B. (1995). The course of seizure disorders in patients with malignant gliomas. *Arch Neurol* **52**, 717–7.
9. Hirsch, J. F., Sainte Rose, C., Pierre-Kahn, A., Pfister, A., and Hoppe-Hirsch, E. (1989). Benign astrocytic and oligodendrocytic tumors of the cerebral hemispheres in children. *J Neurosurg* **70**, 568–5.
10. Oberndorfer, S., Schmal, T., Lahrmann, H., Urbanits, S., Lindner, K., and Grisold, W. (2002). The frequency of seizures in patients with primary brain tumors or cerebral metastases. An evaluation from the Ludwig Boltzmann Institute of Neuro-oncology and the Department of Neurology, Kaiser Franz Josef Hospital, Vienna. *Wien Klin Wochenschr* **114**, 911–9.
11. Barnholtz-Sloan, J. S., Sloan, A. E., Davis, F. G., Vigneau, F. D., Lai, P., and Sawaya, R. E. (2004). Incidence proportions of brain metastases in patients diagnosed (1973 to 2001) in the Metropolitan Detroit Cancer Surveillance System. *J Clin Oncol* **22**(14), 2865–2872.
12. van den Bent, M. J. (2003). The role of chemotherapy in brain metastases. *Eur J Cancer* **39**, 2114–20.
13. Hauser, W. A., Annegers, J. F., and Kurland, L. T. (1993). Incidence of epilepsy and unprovoked seizures in Rochester, Minnesota: 1935–1984. *Epilepsia* **34**, 453–4.
14. Chandramouli, J. (2002). Newer anticonvulsant drugs in neuropathic pain and bipolar disorder. *J Pain Palliat Care Pharmacother* **16**, 19–37.
15. Levendoglu, F., Ogun, C. O., Ozerbil, O., Ogun, T. C., and Ugurlu, H. (2004). Gabapentin is a first line drug for the treatment of neuropathic pain in spinal cord injury. *Spine* **29**, 743–7.
16. McCleane, G. (2003). Pharmacological management of neuropathic pain. *CNS Drugs* **17**, 1031–1043.
17. Silberstein, S. D., Neto, W., Schmitt, J., and Jacobs, D. (2004). Topiramate in migraine prevention: results of a large controlled trial. *Arch Neurol* **61**, 490–495.

18. Freitag, F. G., Collins, S. D., Carlson, H. A. *et al.* (2002). A randomized trial of divalproex sodium extended-release tablets in migraine prophylaxis. *Neurology* **58**, 1652–1659.

19. Sabatowski, R., Galvez, R., Cherry, D. A. *et al.* (2004). Pregabalin reduces pain and improves sleep and mood disturbances in patients with post-herpetic neuralgia: results of a randomised, placebo-controlled clinical trial. *Pain* **109**, 26–35.

20. Modi, S., Pereira, J., and Mackey, J. R. (2000). The cancer patient with chronic pain due to herpes zoster. *Curr Rev Pain* **4**, 429–436.

21. Hardoy, M. C., Hardoy, M. J., Carta, M. G., and Cabras, P. L. (1999). Gabapentin as a promising treatment for antipsychotic-induced movement disorders in schizoaffective and bipolar patients. *J Affect Disord* **54**, 315–317.

22. Pedley, T. A., and Guilleminault, C. (1977). Episodic nocturnal wanderings responsive to anticonvulsant drug therapy. *Ann Neurol* **2**, 30–35.

23. Shimoda, M., Nakayasu, H., Nakashima, K., Shimomura, T., and Takahashi, K. (1995). Effective anticonvulsant therapy in a patient with limb shaking: a case report. *Psychiatry Clin Neurosci* **49**, 71–72.

24. Goldsmith, D. R., Wagstaff, A. J., Ibbotson, T., and Perry, C. M. (2003). Lamotrigine: A review of its use in bipolar disorder. *Drugs.* **63**, 2029.

25. Kim, E. (2002). The use of newer anticonvulsants in neuropsychiatric disorders. *Curr Psychiatry Rep* **4**, 331–337.

26. DelBello, M. P., Kowatch, R. A., Warner, J. *et al.* (2002). Adjunctive topiramate treatment for pediatric bipolar disorder: a retrospective chart review. *J Child Adolesc Psychopharmacol* **12**, 323–330.

27. Patsalos, P. N., Froscher, W., Pisani, F., and van Rijn, C. M, (2002). The importance of drug interactions in epilepsy therapy. *Epilepsia* **43**, 365–385.

28. Gronemeyer, H., Gustafsson, J. A., and Laudet, V. (2004). Principles for modulation of the nuclear receptor superfamily. *Nat Rev Drug Discov* **3**, 950–964.

29. Johnson, D. R., and Klaassen, C. D. (2002). Regulation of rat multidrug resistance protein 2 by classes of prototypical microsomal enzyme inducers that activate distinct transcription pathways. *Toxicol Sci* **67**, 182–189.

30. Swales, K., and Negishi, M. (2004). CAR, driving into the future. *Mol Endocrinol* **18**, 1589–1598.

31. Goodwin, B., and Moore, J. T. (2004). CAR: detailing new models. *Trends Pharmacol Sci* **25**, 437–441.

32. Sueyoshi, T., and Negishi, M., (2001). Phenobarbital response elements of cytochrome P450 genes and nuclear receptors. *Annu Rev Pharmacol Toxicol* **41**, 123–143.

33. Huang, W., Zhang, J., Chua, S. S. *et al.* (2003). Induction of bilirubin clearance by the constitutive androstane receptor (CAR). *PNAS* **100**, 4156–4161.

34. Saini, S. P., Sonoda, J., Xu, L. *et al.* (2004). A novel constitutive androstane receptor-mediated and CYP3A-independent pathway of bile acid detoxification. *Mol Pharmacol* **65**, 292–300.

35. de Wildt, S. N., Kearns, G. L., Leeder, J. S., and van den Anker, J. N. (1999). Glucuronidation in humans. Pharmacogenetic and developmental aspects. *Clin Pharmacokinet* **36**, 439–452.

36. Toffoli, G., Cecchin, E., Corona, G., and Boiocchi, M., (2003). Pharmacogenetics of irinotecan. *Curr Med Chem Anti-Canc Agents* **3**, 225–237.

37. Vecht, C. J., Wagner, G. L., and Wilms, E. B. (2003). Interactions between antiepileptic and chemotherapeutic drugs. *Lancet Neurol* **2**, 404–409.

38. Kivisto, K. T., Kroemer, H. K., and Eichelbaum, M. (1995). The role of human cytochrome P450 enzymes in the metabolism of anticancer agents: implications for drug interactions. *Br J Clin Pharmacol* **40**, 523–530.

39. Nallani, S. C., Glauser, T. A., Hariparsad, N. *et al.* (2003). Dose-dependent induction of cytochrome P450 (CYP) 3A4 and activation of pregnane X receptor by topiramate. *Epilepsia* **44**, 1521–1528.

40. Ueda, A., Hamadeh, H. K., Webb, H. K. *et al.* (2002). Diverse roles of the nuclear orphan receptor car in regulating hepatic genes in response to phenobarbital. *Mol Pharmacol* **61**:1–6.

41. Harris, J., Rahman, A., Kim, B., Guengerich, F., and Collins, J. (1994). Metabolism of taxol by human hepatic microsomes and liver slices: participation of cytochrome P450 3A4 and an unknown P450 enzyme. *Cancer Res* **54**, 4026–4035.

42. Rahman, A., Korzekwa, K., Grogan, J., Gonzalez, F., and Harris, J. (1994). Selective biotransformation of taxol to 6 alpha-hydroxytaxol by human cytochrome P450 2C8. *Cancer Res* **54**, 5543–5546.

43. Desai, P. B., Duan, J. Z., Zhu, Y. W., and Kouzi, S. (1998). Human liver microsomal metabolism of paclitaxel and drug interactions. *Eur J Drug Metab Pharmacokinet* **23**, 417–424.

44. Vecht, C. J., Wagner, G. L., and Wilms, E. B. (2003). Treating seizures in patients with brain tumors: Drug interactions between antiepileptic and chemotherapeutic agents. *Semin Oncol* **30**, 49–52.

45. Wagner, G. L., Wilms, E. B., and Vecht Ch, J. (2002). Levetiracetam: an anti-epileptic drug with interesting pharmacokinetic properties. *Ned Tijdschr Geneeskd* **146**, 1218–1221.

46. Bourg, V., Lebrun, C., Chichmanian, R. M., Thomas, P., and Frenay, M. (2001). Nitroso-urea-cisplatin-based chemotherapy associated with valproate: increase of haematologic toxicity. *Ann Oncol* **12**, 217–219.

47. Baker, A. F., and Dorr, R. T. (2001). Drug interactions with the taxanes: clinical implications. *Cancer Treat Rev* **27**, 221–233.

48. Chang, S., Kuhn, J., Rizzo, J. *et al.* (1998). Phase I study of paclitaxel in patients with recurrent malignant glioma: a North American Brain Tumor Consortium report. *J Clin Oncol* **16**, 2188–2194.

49. Chang, S. M., Kuhn, J. G., Robins, H. I. *et al.* (2001). A Phase II study of paclitaxel in patients with recurrent malignant glioma using different doses depending upon the concomitant use of anticonvulsants: a North American Brain Tumor Consortium report. *Cancer* **91**, 417–422.

50. Grossman, S. A., Hochberg, F., Fisher, J. *et al.* (1998). Increased 9-aminocamptothecin dose requirements in patients on anticonvulsants. NABTT CNS Consortium. The New Approaches to Brain Tumor Therapy. *Cancer Chemother Pharmacol* **42**, 118–126.

51. Prados, M. D., Yung, W. K., Jaeckle, K. A. *et al.* (2004). Phase 1 trial of irinotecan (CPT-11) in patients with recurrent malignant glioma: a North American Brain Tumor Consortium study. *Neuro-Oncol* **6**, 44–54.

52. Cloughesy, T. F., Filka, E., Kuhn, J. *et al.* (2003). Two studies evaluating irinotecan treatment for recurrent malignant glioma using an every-3-week regimen. *Cancer* **97**, 2381–2386.

53. Gilbert, M., Supko, J., Grossman, S., Mikkelsen, T., Priet, R., and Carson, K. (2000). Dose requirements, pharmacology and activity of CPT-11 in patients with recurrent high-grade glioma. A NABTT CNC Consortium trial. *Proc Am Soc Clin Oncol* **19**:A622.

54. Gilbert, M. R., Supko, J. G., Batchelor, T. *et al.* (2003). Phase I clinical and pharmacokinetic study of irinotecan in

adults with recurrent malignant glioma. *Clin Cancer Res* **9**, 2940–2949.

55. Kuhn, J. G. (2002). Influence of anticonvulsants on the metabolism and elimination of irinotecan. A North American Brain Tumor Consortium preliminary report. *Oncology (Huntington)* **16**, 33–40.

56. Prados, M., Chang, S., Burton, E. *et al.* (2003). Phase I study of OSI-774 alone or with temozolomide in patients with malignant glioma. *Proc Am Soc Clin Oncol* **22**, 99.

57. Prados, M. D., Yung, W. A., Wen, P. Y. *et al.* (2004). Phase I study of ZD1839 plus temozolomide in patients with malignant glioma. A study of the North American Brain Tumor Consortium. *J Clin Oncol* **22**, 108.

58. Lieberman, F. S., Cloughesy, T., Malkin, M. *et al.* (2003). Phase I-II study of ZD-1839 for recurrent malignant gliomas and meningiomas progressing after radiation therapy. *Proc Am Soc Clin Oncol* **22**, 103.

59. Lieberman, F. S., Cloughesy, T. F., Fine, H. A. *et al.* (2004). NABTC phase I/II trial of ZD-1839 for recurrent malignant gliomas and unresectable meningiomas. *J Clin Oncol* **22**, 109.

60. Yung, A., Vredenburgh, J., Cloughesy, T. *et al.* (2004). Erlotinib HCL for glioblastoma multiforme in first relapse, a phase II trial. *J Clin Oncol* **22**, 120.

61. Chakravarti, A., Seiferheld, W., Robins, H. I. *et al.* (2004). An update of phase I data from RTOG 0211: A phase I/II clinical study of gefitinib + radiation for newly-diagnosed glioblastoma (GBM) patients. *J Clin Oncol* **22**, 124.

62. Schroder, H., and Ostergaard, J. R. (1994). Interference of high-dose methotrexate in the metabolism of valproate? *Pediatr Hematol Oncol* **11**, 445–449.

63. Schering-Plough, (2004). Temozolomide [package insert]. Schering-Plough, Kenilworth, NJ.

64. Chalk, J., Ridgeway, K., Brophy, T., Yelland, J., and Eadie, M. (1984). Phenytoin impairs the bioavailability of dexamethasone in neurological and neurosurgical patients. *J Neurol Neurosurg Psychiatry* **47**, 1087–1090.

65. Gambertoglio, J. G., Holford, N. H., Kapusnik, J. E. *et al.* (1984). Disposition of total and unbound prednisolone in renal transplant patients receiving anticonvulsants. *Kidney Int* **25**, 119–123.

66. Brooks, S. M., Werk, E. E., Ackerman, S. J., Sullivan, I., and Thrasher, K. (1972). Adverse effects of phenobarbital on corticosteroid metabolism in patients with bronchial asthma. *N Engl J Med* **286**, 1125–1128.

67. Liddle, C., Goodwin, B. J., George, J., Tapner, M., and Farrell, G. C. (1998). Separate and interactive regulation of cytochrome P450 3A4 by triiodothyronine, dexamethasone, and growth hormone in cultured hepatocytes. *J Clin Endocrinol Metab* **83**, 2411–2416.

68. Pascussi, J. M., Gerbal-Chaloin, S., Fabre, J. M., Maurel, P., and Vilarem, M. J. (2000). Dexamethasone enhances constitutive androstane receptor expression in human hepatocytes: consequences on cytochrome P450 gene regulation. *Mol Pharmacol* **58**, 1441–1450.

69. Grossman, S. A., Sheidler, V. R., and Gilbert, M. R. (1989). Decreased phenytoin levels in patients receiving chemotherapy. *Am J Med* **87**, 505–510.

70. Neef, C., and de Voogd-van der Straaten, I. (1988). An interaction between cytostatic and anticonvulsant drugs. *Clin Pharmacol Ther* **43**, 372–375.

71. Gilbar, P., and Brodribb, T. (2001). Phenytoin and fluorouracil interaction. *Ann Pharmacother* **35**, 1367–1370.

72. Rich, J. N., Reardon, D. A., Peery, T. *et al.* (2004). Phase II Trial of Gefitinib in recurrent glioblastoma. *J Clin Oncol* **22**, 133–142.

73. Raizer, J. J., Abrey, L. E., Wen, P. *et al.* (2004). A phase II trial of erlotinib (OSI-774) in patients (pts) with recurrent malignant gliomas (MG) not on EIAEDs. *J Clin Oncol* 22:107.

74. Relling, M. V., Pui, C. H., Sandlund, J. T. *et al.* (2000). Adverse effect of anticonvulsants on efficacy of chemotherapy for acute lymphoblastic leukaemia. *Lancet* **356**, 285–290.

75. Acharya, S., and Bussel, J. B. (2000). Hematologic toxicity of sodium valproate. *J Pediatr Hematol Oncol* **22**, 62–65.

76. Chambers, H. G., Weinstein, C. H., Mubarak, S. J., Wenger, D. R., and Silva, P. D. (1999). The effect of valproic acid on blood loss in patients with cerebral palsy. *J Pediatr Orthop* **19**, 792–795.

77. Serdaroglu, G., Tutuncuoglu, S., Kavakli, K., fand Tekgul, H. (2002). Coagulation abnormalities and acquired von Willebrand's disease type 1 in children receiving valproic acid. *J Child Neurol* **17**, 41–43.

78. Zeller, J. A., Schlesinger, S., Runge, U., and Kessler, C. (1999). Influence of valproate monotherapy on platelet activation and hematologic values. *Epilepsia* **40**, 186–189.

79. Ward, M. M., Barbaro, N. M., Laxer, K. D., and Rampil, I. J. (1996). Preoperative valproate administration does not increase blood loss during temporal lobectomy. *Epilepsia* **37**, 98–101.

80. Winter, S. L., Kriel, R. L., Novacheck, T. F., Luxenberg, M. G., Leutgeb, V. J., and Erickson, P. A. (1996). Perioperative blood loss: the effect of valproate. *Pediatr Neurol* **15**, 19–22.

81. French, J. A., Kanner, A. M., Bautista, J. *et al.* (2004). Efficacy and tolerability of the new antiepileptic drugs I: Treatment of new onset epilepsy: Report of the Therapeutics and Technology Assessment Subcommittee and Quality Standards Subcommittee of the American Academy of Neurology and the American Epilepsy Society. *Neurology* **62**, 1252–1260.

82. GlaxoSmithKline, (2004). Lamictal [package insert]. GlaxoSmithKline, Research Triangle Park, NC.

83. FDA. Trileptal [FDA-approved package insert]. FDA, Washington, DC.

84. Novartis (2003). Tegretol [package insert]. Novartis, East Hanover, NJ.

85. Delattre, J. Y., Safai, B., and Posner, J. B. (1988). Erythema multiforme and Stevens-Johnson syndrome in patients receiving cranial irradiation and phenytoin. *Neurology* **38**, 194–198.

86. Mamon, H. J., Wen, P. Y., Burns, A. C., and Loeffler, J. S. (1999). Allergic skin reactions to anticonvulsant medications in patients receiving cranial radiation therapy. *Epilepsia* **40**, 341–344.

87. Micali, G., Linthicum, K., Han, N., and West, D. P. (1999). Increased risk of erythema multiforme major with combination anticonvulsant and radiation therapies. *Pharmacotherapy* **19**, 223–227.

88. Aguiar, D., Pazo, R., Duran, I. *et al.* (2004). Toxic epidermal necrolysis in patients receiving anticonvulsants and cranial irradiation: a risk to consider. *J Neurooncol* **66**, 345–350.

89. Delattre, J., Safai, B., and Posner, J. (1988). Erythema multiforme and Stevens-Johnson syndrome in patients receiving cranial irradiation and phenytoin. *Neurology* **38**, 194–198.

90. Lehmann, D. F., Hurteau, T. E., Newman, N., and Coyle, T. E. (1997). Anticonvulsant usage is associated with an increased risk of procarbazine hypersensitivity reactions in patients with brain tumors. *Clin Pharmacol Ther* **62**, 225–229.

91. Khan, R. B., Hunt, D. L., and Thompson, S. J. (2004). Gabapentin to control seizures in children undergoing cancer treatment. *J Child Neurol* **19**, 97–101.

92. Wagner, G. L., Wilms, E. B., Van Donselaar, C. A., and Vecht Ch, J. (2003). Levetiracetam: preliminary experience in patients with primary brain tumours. *Seizure* **12**, 585–586.

93. Stevens, G. H. J, Vogelbaum, M. A., Suh, J. A., Peereboom, D. M., and Barnett, G. H. (2005). Levetiracetam use in brain tumor patients. *Neurology* **64**, A48 P01.066.

94. Burstein, A. H., Fisher, K. M., McPherson, M. L., and Roby, C. A. (2000). Absorption of phenytoin from rectal suppositories formulated with a polyethylene glycol base. *Pharmacotherapy* **20**, 562–567.

95. Chang, S. W., da Silva, J. H., and Kuhl, D. R. (1999). Absorption of rectally administered phenytoin: a pilot study. *Ann Pharmacother* **33**, 781–786.

96. Kanner, A. M., Bourgeois, B. F., Hasegawa, H., and Hutson, P. (1998). Rapid switchover to carbamazepine using pharmacokinetic parameters. *Epilepsia* **39**, 194–200.

97. Schaffler, L., Bourgeois, B. F., and Luders, H. O. (1994). Rapid reversibility of autoinduction of carbamazepine metabolism after temporary discontinuation. *Epilepsia* **35**, 195–198.

98. Hassan, M., Oberg, G., Bjorkholm, M., Wallin, I., and Lindgren, M., (1993). Influence of prophylactic anticonvulsant therapy on high-dose busulphan kinetics. *Cancer Chemother Pharmacol* **33**, 181–186.

99. Minami, H., Lad, T. E., Nicholas, M. K., Vokes, E. E., and Ratain, M. J. (1999). Pharmacokinetics and Pharmacodynamics of 9-Aminocamptothecin Infused Over 72 Hours in Phase II Studies. *Clin Cancer Res* **5**, 1325–1330.

100. Murry, D. J., Cherrick, I., Salama, V. *et al.* (2002). Influence of phenytoin on the disposition of irinotecan: a case report. *J Pediatr Hematol Oncol* **24**, 130–133.

101. Rabinowicz, A. L., Hinton, D. R., Dyck, P., and Couldwell, W. T. (1995). High-dose tamoxifen in treatment of brain tumors: interaction with antiepileptic drugs. *Epilepsia* **36**, 513–515.

102. Villikka, K., Kivisto, K. T., Maenpaa, H., Joensuu, H., and Neuvonen, P. J. (1999). Cytochrome P450-inducing antiepileptics increase the clearance of vincristine in patients with brain tumors. *Clin Pharmacol Ther* **66**, 589–593.

103. Herendeen, J. M., Smith, D. A., Stead, A., Bowen, C., Koch, K. M., and Beelen, A. P. (2004). An open-label, fixed sequence, two period study to evaluate the potential induction of GW572016 metabolism by carbamazepine. *J Clin Oncol* 22:215.

104. Zamboni, W., Gajjar, A., Heideman, R. *et al.* (1998). Phenytoin alters the disposition of topotecan and N-desmethyl topotecan in a patient with medulloblastoma. *Clin Cancer Res* **4**, 783–789.

105. Bill, P. A., Vigonius, U., Pohlmann, H. *et al.* (1997). A double-blind controlled clinical trial of oxcarbazepine versus phenytoin in adults with previously untreated epilepsy. *Epilepsy Res* **27**, 195–204.

106. Dam, M., Ekberg, R., Loyning, Y., Waltimo, O., and Jakobsen, K. (1989). A double-blind study comparing oxcarbazepine and carbamazepine in patients with newly diagnosed, previously untreated epilepsy. *Epilepsy Res* **3**, 70–76.

107. Chadwick, D., Anhut, H., Greiner, M. *et al.* (1998). A double-blind trial of gabapentin monotherapy for newly diagnosed partial seizures. International Gabapentin Monotherapy Study Group 945-77. *Neurology* **51**, 1282–1288.

108. Brodie, M. J., Overstall, P. W., and Giorgi, L. (1999). Multicentre, double-blind, randomised comparison between lamotrigine and carbamazepine in elderly patients with newly diagnosed epilepsy. The UK Lamotrigine Elderly Study Group. *Epilepsy Res* **37**, 81–87.

109. Brodie, M. J., Richens, A., and Yuen, A. W. (1995). Double-blind comparison of lamotrigine and carbamazepine in newly diagnosed epilepsy. UK Lamotrigine/Carbamazepine Monotherapy Trial Group. *Lancet* **345**, 476–479.

110. Steiner, T. J., Dellaportas, C. I., Findley, L. J. *et al.* (1999). Lamotrigine monotherapy in newly diagnosed untreated epilepsy: a double-blind comparison with phenytoin. *Epilepsia* **40**, 601–607.

111. Gilliam, F. G., Veloso, F., Bomhof, M. A. M. *et al.* (2003). A dose-comparison trial of topiramate as monotherapy in recently diagnosed partial epilepsy. *Neurology* **60**, 196–202.

112. Privitera, M. D., Brodie, M. J., Mattson, R. H., Chadwick, D. W., Neto, W., and Wang, S., (2003). Topiramate, carbamazepine and valproate monotherapy: double-blind comparison in newly diagnosed epilepsy. *Acta Neurol Scand* **107**, 165–175.

113. Guerreiro, M. M., Vigonius, U., Pohlmann, H. *et al.* (1997). A double-blind controlled clinical trial of oxcarbazepine versus phenytoin in children and adolescents with epilepsy. *Epilepsy Res* **27**, 205–213.

114. Frank, L. M., Enlow, T., Holmes, G. L. *et al.* (1999). Lamictal (lamotrigine) monotherapy for typical absence seizures in children. *Epilepsia* **40**, 973–979.

115. Gilliam, F. G., Veloso, F., Bomhof, M. A., *et al.* (2003). A dose-comparison trial of topiramate as monotherapy in recently diagnosed partial epilepsy. *Neurology* **60**, 196–202.

116. Handschin, C., and Meyer, U. A. (2003). Induction of Drug Metabolism: The Role of Nuclear Receptors. *Pharmacol Rev* **55**, 649–673.

117. Tanaka, E. (1998). Clinically important pharmacokinetic drug-drug interactions: role of cytochrome P450 enzymes. *J Clin Pharm Ther* **23**, 403–416.

4

Brain Tumor Models for Cancer Therapy

Beverly A. Teicher

There are at least 18 000 new cases of primary brain and CNS-malignant neoplasms diagnosed in the United States per year. The most common malignant CNS tumor is glioblastoma multiforme. The mean age of patients with glioblastoma multiforme is 52 years. These tumors cause approximately 12 000 deaths per year. Despite the relatively small numbers of CNS tumors, the morbidity and mortality they cause are significant [1,2]. Surgery remains the primary treatment for CNS tumors. After surgery, patients with glioblastoma multiforme or anaplastic astrocytoma are treated with radiation therapy. Chemotherapy is used as an adjuvant with surgery and radiation therapy. The most commonly used chemotherapeutic agents are the nitrosoureas, including 1,3-bis(2-chloroethyl)-1-nitrosourea (BCNU), 1-(2-chloroethyl)-3-cyclohexyl-1-nitrosourea (CCNU), and methyl-1-(2-chloroethyl)-3-cyclohexyl-1-nitrosourea (methyl-CCNU). The platinum complexes cisplatin and carboplatin are also used. Most recently, temozolomide has been approved in the United States for the treatment of astrocytoma and is in clinical trial for other CNS tumor indications [3–5].

The most frequently used preclinical *in vivo* models of brain tumors include the rat carcinogen-induced syngeneic models and human tumor xenografts. Both types of tumors are routinely implanted subcutaneously or intracranially in host animals including conventional and nude rats or nude or SCID mice. These models have been used to explore the efficacy of diverse therapeutic strategies.

RAT BRAIN TUMOR MODELS

Rat brain tumor models have been widely used in experimental neuro-oncology studies for more than 30 years [6]. The 9L gliosarcoma of the Fischer 344 rat is a carcinogen-induced cell line that results from nitrosourea exposure [7, 8]. The C6 glioma is derived from a nitrosourea-induced rat glial tumor [9–11]. The F98 and D74-RG2 gliomas were developed by injection of a single dose of *N*-ethyl-*N*-nitrosourea (ENU) (50 mg/kg) into pregnant CD Fischer rat on gestation day 20 [12]. Tumors were harvested and cloned. The F98 and D74-RG2 malignant gliomas produce an infiltrative pattern of growth in the brain resembling human glioblastoma [13]. The F98 tumor is weakly immunogenic in the syngeneic host and an intracranial implant of as few as 100 cells is lethal [14]. The RG2 tumor is non-immunogenic in the syngeneic host.

BLOOD–BRAIN BARRIER IN RAT BRAIN TUMOR MODELS

The blood–brain barrier remains a very active area of investigation even though the blood–brain tumor barrier is more permeable than the blood–brain barrier. There is still significantly decreased delivery of anticancer drugs to brain tumors. The endothelial cells that form the vasculature of the tumor express protein markers that are absent or expressed at very low levels in the normal brain endothelium. Ningaraj *et al.*, [15] found the calcium-dependent potassium (K(Ca)) channel is increased in expression in brain tumors. They used a specific K(Ca) channel agonist, NS-1619, to obtain sustained enhancement of drug delivery to syngeneic rat brain tumors. NS-1619 administration was able to facilitate increased delivery of carboplatin to brain tumors resulting in increased survival of tumor-bearing rats. Erucylphosphocholine represents a potential anticancer agent for brain tumors but has limited access to the brain. Intra-arterial administration of alkylglycerols has been described as a method of opening the blood–brain barrier [16]. Erucylphosphocholine was given to C6 glioma-bearing rats as a single intracarotid bolus injection along with or without 300 mM 1-O-pentylglycerol or as an intracarotid infusion along with bradykinin. The combination of erucylphosphocholine administration with pentylglycerol results in a 17-fold increased concentration of erucylphospholine in the tumor compared to administration of erucylphosphocholine alone and a 2.5-fold increase in the tumor compared with surrounding normal brain.

Although liposomes have been used as a vehicle for delivery of therapeutic agents in oncology, the penetration of liposomes through the blood–brain barrier is very poor. Saito et al., [17] explored the use of convection-enhanced delivery (CED) of liposomes to improve targeting to brain tumors. Liposomes were labeled with gadolinium (Gd) and a fluorescent marker and were administrated into the striatal region in rats. In vivo MRI was used to monitor liposome distribution in C6 and 9L bearing rats. The distribution of liposomes carrying Gd and fluorescence to the tumor was monitored by MRI. Finally, Gd-labeled liposomes with optimum convection-enhanced delivery carrying doxorubicin were administered to rats bearing 9L gliosarcoma. This method allowed in vivo MRI monitoring of therapeutic distribution to the tumor and optimal local drug delivery. The feasibility of imaging pentose cycle glucose metabolism in gliomas with positron emission tomography (PET) was explored in rats bearing C6 tumors [18]. In vivo, rats bearing C6 gliomas at six subcutaneous sites received simultaneous intravenous injections of either [1–11C]glucose and [6–14C] glucose or [1–14C]glucose and [6–11C]glucose. The results were analyzed with a model of glucose metabolism that simultaneously optimized parameters for C-1 and C-6 glucose kinetics by simulating the C-1 and C-6 tumor time–activity curves. The rate constant for loss of radiolabeled carbon from the tumors was higher for C-1 than for C-6 indicating that PET may be useful in monitoring brain tumor response to therapy. The rate constant for loss of radiolabeled carbon from the tumors was higher for C-1 than for C-6 in all groups of rats (19 per cent higher for T-36B-10 unirradiated and 32 per cent higher for T-36B-10 irradiated and for T-C6 unirradiated). Mathematical modeling, Monte Carlo simulations and construction of receiver-operator-characteristic curves show that if human gliomas have a similar fractional use of the pentose cycle, it should be measurable with PET using sequential studies with [1–11C]glucose and [6–11C] glucose.

LOCAL DELIVERY IN RAT BRAIN TUMOR MODELS

Gene therapy remains very attractive as a treatment strategy for brain tumors since local delivery is feasible and likely to have a major impact on the disease [19, 20]. Several gene therapy approaches are based upon prodrug-activating enzymes, inhibition of tumor neovascularization, and stimulation of antitumor immune responses. The most widely explored paradigm is based on the activation of ganciclovir to a cytotoxic compound by a viral enzyme, thymidine kinase, which is expressed by tumor cells, after infection by a retroviral vector containing the gene. Thus, the transfer of the gene coding for the thymidine kinase of the herpes simplex virus (HSV-tk), followed by ganciclovir administration, has been described as a treatment for a variety of cancers but especially for brain tumors. When Cool et al., [21] stereotactically injected cells producing up to 3×10^5 HSV-tk retroviral particles per ml into 9L brain tumors and administered ganciclovir to the animals, there was no increase in survival. However, when the 9L tumor cells were transduced in vitro with the HSV-tk, then implanted into animals, 26 per cent of intracranial and 67 per cent of subcutaneous tumors responded to treatment with ganciclovir. In another study, the HSV-tk gene was inserted into a retroviral vector (pMFG) which was produced using the amphotropic packaging cell line CRIP-MFG-S-HSV-tk. The packaging cells were injected intra-tumorally into established intracranial 9L tumors to affect gene delivery to the tumor cells.

Approximately 45 per cent of the 9L cells in the CRIP-MFG-S-HSV-tk-injected tumors showed immunohistochemical staining for HSV-tk, demonstrating efficient retrovirus-mediated gene transfer. Treatment of animals with tumors injected with the CRIP-MFG-S-HSV-tk cells with ganciclovir resulted in positive response to the therapy [22]. The expression of major histocompatibility complex (MHC) class I antigens by 9L tumor cells can influence response to therapy [23]. High and low MHC class I expressing clones of 9L tumor cells transduced by the HSV-tk gene were selected and grown as intracranial tumors. The response of tumors developed from both the high and low MHC class I expressing tumor cells had the same level of response to ganciclovir. Histological examination of the tumors developed from the high MHC class I expressing cells had lymphocyte infiltration while tumors developed from the low MHC class I expressing clone did not have infiltrating cells.

A second generation replication-conditional herpes simplex virus type 1 (HSV) vector defective for both ribonucleotide reductase (RR) and the neurovirulence factor gamma 34.5 was developed and tested for therapeutic efficacy in rats bearing intracranial 9L tumors [24]. This modified viral vector was designated MGH-1. The modified viruses were injected intratumorally or intrathecally with or without administration of ganciclovir. There was no toxicity due to administration of the MGH-1 virus, however, there was a decreased therapeutic effect with the MGH-1 virus compared with the parental vector. To investigate the potential of the thymidine kinase gene from Varicella zoster virus (VZVtk) to act as a suicide gene, VZVtk was transferred *via* a dicistronic retroviral construct in 9L rat gliosarcoma cells [25, 26]. The 9L gliosarcoma cells infected with the VZVtk-carrying vector were implanted *in vivo*. When the tumor-bearing animals were treated with (E)-5-(2-bromovinyl)-2′-deoxyuridine (BVDU) a significant bystander effect was observed. A survival benefit was not seen.

Substrates for monitoring HSV1-tk gene expression include uracil and acycloguanosine derivatives [27, 28]. To assess the value of ^{123}I-1-(2-fluoro-2-deoxy-d-ribofuranosyl)-5-iodouracil (FIRU) as an imaging agent, rat 9L gliosarcoma cells stably transfected with the HSV-tk gene and parental 9L cells were subcutaneously implanted into both flanks of NIH-bg-nu-xid mice. Biodistribution studies for ^{123}I-FIRU were performed using gamma camera imaging. The tumor/muscle, tumor/blood, and tumor/brain ratios of the ^{123}I reached maximal values of 32, 12.5, and 171 at 4 h post administration of the tracer. Subsequently, NIH-bg-nu-xid mice were implanted subcutaneously with human U87MG glioblastoma cells. When the tumors reached 1000 mm^3, 108 and 109 Infectious Units of a vector carrying the HSV-tk gene were injected intratumorally. After 48 h, the ^{123}I-FIRU was injected and followed by gamma camera imaging. The images and direct measurement were able to identify greater uptake of the tracer in the tumors with the higher I.U. injections indicating that ^{123}I-FIRU could be a useful tracer for monitoring adenoviral gene transfer.

The observation that impaired p53 expression is present in a proportion of malignant gliomas suggests that restoration of p53 function may provide tumor control [29, 30]. Badie *et al.*, [29] studied the effect of two replication-defective adenovectors bearing human wild type tumor suppressor gene p53 (Adp53) and *Escherichia coli* beta-galactosidase gene (AdLacZ) on 9L glioma cells. Transduction of 9L cells with the Adp53 inhibited cell proliferation and induced phenotypic changes consistent with cell death at low titers, while AdLacZ caused little cell death and only at high titers. Stereotactic injection of AdLacZ (10^7 plaque forming units) intratumorally stained 25 to 30 per cent of tumor cells at the site of vector delivery. Injection of Adp53 (10^7 plaque forming units), but not AdLacZ controls, into established 4-day-old 9L glioma brain tumors decreased tumor volume by 40 per cent after 14 days. The 9L and RG2 intracranial rodent tumor models are utilized to assess SCH58500, an adenoviral p53 delivery system. The RG2 tumors demonstrate a greater propensity for transfection with this vector *in vitro* than the 9L tumors. *In vivo*, little tumor transfection beyond the immediate area of the needle tract used for direct SCH58500 injection was observed in either tumor type. Intracarotid injection resulted in no tumor transfection. Intratumoral administration of SCH58500 enhances the survival of animals with established 9L tumors. Both SCH58500 and its control viral construct not containing the p53 gene enhance survival in animals with RG2 tumors. Driessens *et al.*, [31] compared apoptosis induction and micronuclei formation to assess the DNA damage produced *in vivo* by cytotoxic agents in established 9L rat gliosarcoma tumors expressing a mutated p53 gene. Results from TUNEL assays revealed the efficiency of local gamma-irradiation at the tumor site to induce apoptosis within the 9L tumor mass.

Antifolate anticancer agents are selectively cytotoxic to cells in S-phase and therefore are less effective against slow-growing tumors and are cytotoxic toward actively proliferating normal tissues. Intracellularly, most antifolates are converted to polyglutamate derivatives by the enzyme folylpolyglutamyl synthetase (FPGS). Since tumors with high expression of FPGS are more responsive to antifolate anticancer agents, Aghi *et al.*, [32] investigated the effects of transfecting the gene for human FPGS into rat 9L gliosarcoma cells on response to treatment of tumor-bearing animals with methotrexate or edatrexate. Rat 9L gliosarcoma cells were stably transfected with a human FPGS complementary DNA (cDNA) to produce 9L/FPGS cells. Subcutaneous 9L/FPGS tumors responded as well to methotrexate administered every third day as did parental 9L tumors treated daily. A modest bystander effect was observed with edatrexate treatment in culture and *in vivo*. The observed bystander effect appeared to result from release of active drug species by the transfected cells. One approach to improving the specificity of gene therapy involves using radiosensitive promoters to activate gene expression selectively in the radiation field. Manome et al., [33] evaluated the ability of irradiation to regulate the transcription of a recombinant replication-defective adenovirus vector, Ad.Egr-1/lacZ, containing the radiation-inducible Egr-promoter driving the beta-galactosidase reporter gene in glioma cells. Irradiation or intracerebral 9L tumors infected with the Ad.Egr-1/lacZ virus, using either external beam radiotherapy (2Gy) or the thymidine analog 5-iodo-2′-deoxyuridine radiolabeled with the Auger electron emitter iodine-125 ([125]IdUrd), also resulted in increased beta-galactosidase activity in the tumor cells. Rainov et al., [34] investigated intraarterial delivery of genetically engineered replication-deficient adenovirus vectors versus cationic liposome-plasmid DNA complexes to rats bearing intracranially implanted 9L gliosarcoma. Adenovirus or liposome-DNA was injected into the internal carotid artery of tumor-bearing rats using bradykinin to selectively permeabilize the blood–tumor barrier. The liposome-DNA-mediated gene transfer showed increased efficacy compared with the adenovirus-mediated gene transfer, but had less specificity since a larger number of endothelial and glial cells were seen to express the transgene. Both the adenovirus and liposome-DNA injections, in the absence of bradykinin, resulted in transduction of peripheral organs including liver, lung, testes, lymph nodes, and spleen.

IMMUNOLOGICAL APPROACHES IN RAT BRAIN TUMOR MODELS

The tolerance of the central nervous system (CNS) for activated host immune reactions indicates that the effects of immunotherapy might be decreased for brain tumors. Iwadate et al., [35] developed an interleukin-2 producing 9L rat gliosarcoma and assessed the effect of secretion of interleukin-2 on the growth and immunological responses to the tumor implanted subcutaneously and intracranially. The Fischer 344 rats rejected the 9L/IL-2 tumors when implanted subcutaneously; however, the same tumors cells implanted intracranially grew tumors albeit with a slower growth rate than with the parental 9L cells. Lefranc et al., [36] carried out similar studies with 9L gliosarcoma cells transfected to produce GM-CSF and demonstrated that implantation of these cells prevented tumor growth. Local delivery of carmustine (BCNU) from biodegradable polymers prolongs survival in rats bearing intracranial 9L gliosarcoma [37]. Interleukin-2 can also produce a potent antitumor immune response and improve the survival of animals bearing brain tumors. Several methods have been developed and studied for encapsulation of interleukin-2 into polymeric vehicles that can be implanted into brain tumor models [38]. Fischer 344 rats bearing intracranial 9L gliosarcoma received an intracranial implant of empty microspheres or microspheres containing interleukin-2. The combination treatment of microspheres containing interleukin-2 along with 10 per cent BCNU polymer resulted in a median survival of 45.5 days while treatment with the interleukin-2-containing microspheres alone or the 10 per cent BCNU polymer alone resulted in median survival times of 24 and 32.5 days, respectively. The untreated controls survived a median of 18 days [37].

Transforming growth factor-beta is well known to be an immunosupressive cytokine and inactivating it may prolong survival of tumor-bearing animals. The cDNA for simian TGF-beta 2 was ligated in antisense orientation into the episomal plasmid mammalian vector pCEP-4 and transfected into rat C6 glioma cells [39]. Adult female Wistar rats bearing intracranial C6 glioma were injected subcutaneously with saline, C6 glioma cells or TGF-beta-antisense-modified C6 glioma cells. The survival of tumor-bearing rats injected with the TGF-beta-antisense-modified C6 glioma cells was significantly longer than that of animals injected with saline or with parental C6 glioma cells.

Malignant astrocytic tumors are characterized by aggressive, diffuse migration of tumor astrocytes into the brain parenchyma. Gastrin is a brain neuropeptide that is able to significantly modulate astrocytic tumor migration controlling both invasion and motility [40–42]. Gastrin belongs to the cholecystokinin (CCK) peptide family and is widely distributed in the brain. The rat C6 glioma expresses the CCKB receptor while the 9L gliosarcoma expresses the CCKA and CCKC receptors. When animals bearing intracranial C6 glioma and the 9L gliosarcoma tumors were treated with gastrin, animals with the C6 glioma but not animals with the 9L gliosarcoma showed a survival benefit.

CHEMOTHERAPY AND BORON NEUTRON CAPTURE IN RAT BRAIN TUMOR MODELS

Chloroethyl-nitrosourea (CENU) is a useful chemotherapeutic agent for treatment of brain tumors. Acquired resistance to CENU is a serious problem. The main mechanism of resistance to CENU is upregulation of O^6-methylguamine-DNA-methyl-transferase (MGMT) in the tumor cells. Manome et al., [43] examined the response to CENU therapy of animals bearing 9L gliosarcoma retrovirally transduced with MGMT cDNA that was sterotactically implanted into the brain. These animals had significantly decreased response to CENU treatment. The spatial distribution of the photosensitizer hematoporphyrin derivative was studied after intratumoral injection into intracerebral implant of 9L gliosarcoma [44]. The fluorescence volume was measured in histological sections from 10 min up to 5 days after injection. Whether the administration was performed by slow stereotactic injection or increased velocity higher volume injection, complete sensitization of the tumor could not be achieved.

Boron neutron capture therapy (BNCT) depends upon the accumulation and retention of boron-10 in brain tumor cells and not in normal brain cells. BNCT is based on the nuclear reaction that occurs when boron-10 is irradiated with neutrons of the appropriate energy to produce high-energy alpha particles and lithium-7 nuclei [45, 46]. Numerous boron-containing compounds have been developed. Hydroxyl forms of boronophenylalanine (BPA) were prepared and tested for tumor:brain ratio in rats bearing 9L gliosarcoma. The tumor:brain ratio of the singly hydroxylated form was 1.2 and that of the dihydroxylated derivative was 1.4 compared with dl-boronophenylalanine [47]. Ion microscopy allowed evaluation of the microdistribution of boron-10 from p-boronophenylalanine (BPA) in the rat 9L gliosarcoma and the rat F98 glioma brain tumors [48]. The p-boronophenylalanine was administered by intraperitoneal injection, intracarotid artery injection with or without blood-brain barrier disruption and by continuous intravenous infusion. Intraperitoneal injection of p-boronophenylalanine resulted in a tumor:brain ratio of 3.7:1 for the 9L gliosarcoma, while intracarotid injection of p-boronophenylalanine resulted in a tumor:brain ratio of 2.9:1 in animals bearing the F98 glioma. Continuous 3- and 6-h intravenous infusions of p-boronophenylalanine in the 9L gliosarcoma resulted in similar high boron-10 concentrations in the main tumor mass. The boron-10 concentration in infiltrating tumor cells was 2-times lower than in the main tumor mass after a 3 h infusion. After a 6 h infusion, the boron-10 concentration in the infiltrating tumor cells increased nearly 90 per cent relative to the 3 h infusion concentrations. Both the F98 and 9L rat glioma models have been used to study boron neutron capture therapy [49, 46]. Treatment with 1200 mg/kg administered by intraperitoneal injection and followed by exposure to

neutron flux at the Brookhaven National Laboratory Medical reactor appeared curative for rats bearing 9L gliosarcoma. On the other hand, the F98 tumor was refractory to therapy. Barth et al., [49] went on to determine whether the efficacy of boron neutron capture therapy could be increased by combination with standard radiation therapy. F98 glioma-bearing rats treated with intracarotid or intravenous sodium borocaptate (30 mg/kg) plus p-boronophenylalanine (250 mg/kg followed 2.5 h later by exposure to a neutron flux and 7–10 days later by standard radiation therapy to a total dose of 15 Gray had survival times (61–53 days) similar to those of rats that did not receive the standard radiation therapy (52–40 days). However, animals in the same study bearing MRA 27 melanoma showed a significantly enhanced survival when boron neutron capture therapy was followed with conventional radiation therapy.

HUMAN BRAIN TUMOR XENOGRAFT MODELS

Xenografts of human brain tumors implanted intracranially or subcutaneously have proven useful as models for the study of human brain tumors. Nuclear magnetic resonance imaging (MRI) methods have been developed that can achieve volumetric comparison with histological methods and submillimeter resolution, improved by contrast enhancement with intravenous administration of a Gd agent [50–52]. Direct injection or convection-enhanced delivery of liposomes containing attached or encapsulated fluorochromes and/or encapsulated gold particles were able to distribute throughout intracranially implanted U-87 glioma xenografts and into surrounding normal brain tissue by MRI scanning, thus demonstrating that convection-enhanced delivery of liposomes is feasible and may allow targeting of therapeutic agents to brain tumors [53]. MRI of murine brain tumors in different locations has been carried out with a 1.5 T MR system and a surface coil along with Gd injected via a tail vein to better delineate the tumors [54]. Mice harboring nascent brain tumors were followed sequentially by serial MR imaging. Cha et al., [55] used MRI to evaluate the growth and vascularity of implanted GL261 mouse gliomas. Both conventional T(1)- and T(2)-weighted imaging and dynamic, contrast-enhanced T(2)*-weighted imaging was performed with varied-sized tumors. Post contrast enhancement on T(1)-weighted images was observed at all stages of Gl261 glioma progression, even before evidence of angiogenesis, indicating that the mechanism of conventional contrast enhancement in MRI does not require neovascularization.

BLOOD–BRAIN BARRIER IN HUMAN BRAIN TUMOR XENOGRAFT MODELS

Even though the blood–brain tumor barrier (BTB) is more permeable than the blood–brain barrier (BBB), the blood–brain tumor barrier still significantly restricts the delivery of anticancer drugs to brain tumors. Ningaraj et al., [15] were able to modulate the calcium-dependent potassium (K(Ca)) channel using a specific K(Ca) channel agonist, NS-1619, to obtain sustained enhancement of selective drug delivery in xenograft brain tumor models. NS-1619 administration was used to increase delivery of carboplatin, Her-2 monoclonal antibody and green fluorescence protein-adenoviral vectors to human brain tumor xenografts. P-glycoprotein (Pgp) in the blood–brain barrier limits the uptake of drugs such as paclitaxel into the brain. Kemper et al., [56] assessed the ability of several putative inhibitors of Pgp including cyclosporin A, PSC833, GF120918, and Cremophor EL, to improve the penetration of paclitaxel into mouse brain. Of the Pgp inhibitors tested, GF120918 was most effective in increasing paclitaxel concentrations in the brain. The influence of anesthetic, $PaCO_2$ and infusion rate on osmotic blood–brain barrier disruption in rats bearing brain tumor xenografts was explored [57]. Propofol/N_2O anesthesia was found to be better than isoflurane/O_2 for optimizing osmotic blood–brain barrier disruption for delivery of chemotherapeutic drugs to brain tumor and normal brain.

The blood vessel structure in xenograft brain tumor models has been examined by several investigators [58–60]. Schlageter et al., [58] studied microvessel organization in five xenograft brain tumor models by determination of microvessel diameter, intermicrovessel distance, microvessel density vessel surface area, and vessel orientation. Microvessel diameter and inter-microvessel distance were larger and microvessel density was lower in tumors than normal brain. Vessel surface area in tumors overlapped normal brain values and orientation was random both in tumors and normal brain. Three microvessel populations were identified in brain tumors: (1) continuous nonfenestrated, (2) continuous fenestrated, and (3) discontinuous (with and without fenestrations). While continuous nonfenestrated vessels may be unique to brain tumors, continuous fenestrated and discontinuous vessels are generally found in malignant tumors. Vascular endothelial growth factor (VEGF) is over expressed in most glioblastoma multiforme and the level of expression is correlated with tumor grade. VEGF isoforms were overexpressed in genetically modified, mutant H-Ras-transformed human astrocytes that when implanted intracranially for astrocytoma-like tumors [59,61]. Expression of modest levels of VEGF were insufficient to drive oxygenation and glioblastoma formation by these cells, expression of very high levels of VEGF resulted in cells which, after intracranial implantation, formed tumors that were larger, more vascular, and better oxygenated than those formed by the mutant H-Ras-transformed parental cells. Chen et al., [60] radiolabeled small peptide antagonists of alpha-v-integrins, RGD peptides and found that the overall molecular charge and characteristics of the radiolabels had a profound effect on the tumor accumulation and in vivo kinetics of the RGD peptides and therefore further work was needed to optimize these agents as markers.

Green fluorescent protein can be used for labeling of glioma cells in experimental brain tumor models and neural progenitor cells by retrovirus vectors and for efficient, non-toxic delivery of genes to postmitotic cells of the nervous system using helpervirus free HSV-1 amplicon vectors [62]. Pyles et al., [63] generated a HSV-1 double-mutant, designated 3616UB by interrupting the uracil DNA glycosylase gene in the R3616 mutant. The uracil glycosylase gene is required for efficient replication of HSV-1 replication in non-dividing cells but is dispensable for virus replication in rapidly dividing cells. The 3616UB strain did not replicate in primary neuronal cultures in vitro or in mouse brain but efficiently killed six of six human tumor cell lines and replicated within brain human xenografts. Intratumoral injection of 3616UB into human medulloblastoma or angiosarcoma xenografts established in SCID mice resulted in a slowing of tumor growth and some tumor regressions.

LOCAL DELIVERY IN HUMAN BRAIN TUMOR XENOGRAFT MODELS

The Escherichia coli gpt gene sensitizes cell expressing the gene to the prodrug 6-thioxanthine (6TX) [64]. Rat C6 glioma cells were infected with a retrovirus vector that transduces the E. coli gpt gene and a clonal cell line designed C6GPT-7 was derived. In vivo both 6-thioxanthine and 6-thioguanine inhibited the growth of subcutaneously transplanted C6GPT-7 cells but not C6 parental cells in nude mice. In an intracerebral model, both 6-thioxanthine and 6-thioguanine exhibited antiproliferative effects against C6GPT-7 tumors.

Dendritic cell (DC)-based immunotherapy is being assessed for treatment of malignant brain tumors. Dendritic cells were isolated from C57BL mouse bone marrow and pulsed with phosphate-buffered saline, Semliki Forest virus (SFV)-LacZ, retrovirus vector GCsap-interleukin (IL)-12, respectively, to treat mice bearing intracranial B16 melanoma [65]. Intratumorally injected dendritic cells that have been transiently transduced with interleukin-12 do not require pulsing of a source of tumor antigens to induce tumor regression. Adoptive therapy with TALL-104 cells, an IL-2-dependent, major histocompatibility complex nonrestricted, cytotoxic T-cell line, has demonstrated significant antitumor activity against a broad range of implanted or spontaneously arising tumors. Geoerger et al., [66] investigated distribution of systemically and locally administered TALL-104 cells and their efficacy in effecting survival of a nude rat bearing xenografts of human glioblastomas U-87 MG, U-251 MG, or A1690; medulloblastomas DAOY, D238 Med, or D341 Med; or the epidermoid cancer line A431. Overall, local therapy with TALL-104 cells may be a novel and effective treatment approach for malignant brain tumors. Varied antibody-based approaches to immunotherapy for malignant brain tumors also continue to be explored [67–70].

O^6-Methylguanine-DNA methyltransferase (MGMT), a constitutively expressed DNA repair protein, removes alkyl groups from the O^6-position of guanine in DNA. Tumor cells with high MGMT activity are resistant to nitrosoureas and other agents that form toxic O^6-alkyl adducts. O^6-Benzylguanine (BG) inactivates the MGMT protein and thereby enhances the sensitivity of tumor cells to methylating drugs [71]. The therapeutic potential of O^6-benzylguanine is limited by its poor solubility and nonspecific inactivation of MGMT in normal tissues as well as in tumor tissues. Schold et al., [71] evaluated O^6-benzyl-2'-deoxyguanosine (dBG), the 2'-deoxyribonucleoside analog of BG for inhibition of MGMT and potentiation of 1,3-bis (2-chloroethyl)-1-nitrosourea (BCNU) in a MGMT-positive human DAOY brain tumor xenograft. O^6-benzyl-2'-deoxyguanosine potentated the efficacy of BCNU against MGMT-positive, BCNU-resistant DAOY human medulloblastoma tumor xenograft in nude mice. Kokkinakis et al., [72] optimized the therapeutic effect of the O^6-benzyl-2'-deoxyguanosine and BCNU combination against brain tumor xenografts without inducing substantial toxicity in the host. They were able to achieve tumor suppression of >90 days without dose-limiting host toxicity. The effect of pretreatment with O^6-benzylguanine on the activity of BCNU against Mer– human central nervous tumor xenografts D-54 MG and D-245 MG has been evaluated

in nude mice [73]. The results indicated that the combination of BCNU and O^6-benzylguanine could be a promising treatment for both Mer+ and Mer– tumors. Ewesuedo et al., [74] compared the usefulness of various 8-substituted O^6-benzylguanine analogs as modulators of the DNA repair protein O^6-alkylguanine-DNA methyltransferase (AGT). Each of the 8 analogs had AGT inhibitory activity that was comparable to or greater than O^6-benzylguanine especially in brain and kidney tissues. The level of inhibition of AGT activity in brain and D456 brain tumor xenografts after administration of O^6-benzylguanine and the various analogs could be explained in large part by the relative distribution of the compounds into these tissues.

CHEMOTHERAPY IN HUMAN BRAIN TUMOR XENOGRAFT MODELS

The efficacy of protracted schedules of therapy of the topoisomerase I inhibitors 9-dimethyl-aminomethyl-10-hydroxycamptothecin (topotecan) and 7-ethyl-10-[4-(1-piperidino)-1-piperidino]carbonyloxycamptothecin (irinotecan; CPT-11) were evaluated against a panel of human tumor xenografts [75]. Both agents demonstrated good activity against brain tumor xenografts with irinotecan causing complete regressions in 2 of 3 lines and topotecan inducing complete regressions in 1 of 3 lines. The results indicate that low-dose protracted schedules of daily administration of these topoisomerase I inhibitors is either equi-effective or more efficacious than more intense shorter schedules. Keir et al., [76] assessed the efficacy of karenitecin, a highly lipophilic camptothecin derivative in nude mice bearing xenografts of childhood high-grade gliomas (D212 MG, D-456 MG), adult high-grade gliomas (D-54 MG, D-245 MG), medulloblastomas (D-341 MED, D-487 MED) and ependymomas (D-528 EP, D-612 EP) as well as sublines resistant to procarbazine (D-245 MG(PR)) and busulfan (D-456 (BR)). Karenitecin was active against all of these central nervous system xenografts. The novel toposiomerase I, J-107088, was evaluated against human pediatric and adult malignant central nervous system tumors implanted subcutaneously or intracranially in nude mice [77]. The results indicate that J-107088 could be promising for the treatment of childhood and adult malignant brain tumors.

Temozolomide, an imidazole tetrazinone, and irinotecan have been shown to have antitumor activity against CNS tumors. Patel et al., [78] evaluated the activity of temozolomide plus irinotecan against a malignant glioma-derived xenograft D-54 MG growing subcutaneously in nude mice.

The combination of these two agents produced greater than additive antitumor activity in a schedule-dependent manner. Heimberger *et al.*, [79] evaluated the efficacy and toxicity of intracerebral microinfusion (ICM) with temozolomide in a nude rat bearing human D-54 intracerebral xenografts. Treatment of rats by intracerebral microinfusion with temozolomide 3 days after tumor implant increased median survival by 128 per cent compared with rats treated with saline.

Several new drugs have been assessed in pre-clinical brain tumor models. S16020 is a cytotoxic olovacine derivative and an inhibitor of topo-isomerase II. S16020 was examined in several subcutaneously implanted medulloblastomas and glioblastomas grown in nude mice and compared with doxorubicin [80]. S16020 demonstrated anti-tumor activity in two out of three medulloblastomas and three out of three glioblastoma xenografts. Some ellipticine derivatives, including 9-chloro-2-methylel-lipticinium (CME), were selectivity against eight brain tumor cell lines of the NCI disease-oriented *in vitro* screen [81]. When 9-chloro-2-methylellipticinium was tested in animals bearing orthotopically implanted U251 gliomas in the brain of nude mice, the survival of the treated mice was not better than vehicle controls either as a single agent or in combinations. The lack of benefit may be due to poor penetration into the brain. Epidermal growth factor receptor (EGFR) is upregulated in primary malignant tumors of the central nervous system (CNS) and in many systemic tumors that metastasize to brain. Heimberger *et al.*, [82] evaluated the efficacy and toxicity of orally administered Iressa (ZD 1839) for treatment of established intracerebral tumors expressing EGFR or the tumorigenic mutated variant EGFRvIII which is constitutively phosphorylated. ZD1839 is active in a brain tumor model expressing EGFR and clinical trials of ZD1839 against tumors expressing EGFR could be warranted. The efficacy of irofulven (6-hydroxy-methylacylfulvene) was assessed in a series of glioblastoma multiforme-derived xenografts growing subcutaneously and intracranially in nude mice [83]. Irofuloven was active against all tumor lines tested with growth delays ranging from 6 to 81 days. The combination of RSU 1069 and radiation was evaluated in U251 MG and U87 MG tumors grown as intra-cerebral xenografts in nude rats [84]. Using the comet assay, there appeared to be a relationship between tumor size and degree of hypoxia for the U251 tumor but not for the U87 tumor model.

Brem and Gabikian [85] demonstrated the feasibility of polymer-mediated drug delivery using the cytotoxic agent 1,3-bis(2-chloroethyl)-1-nitrosourea (BCNU) and showed that local treatment of gliomas by this treatment was effective in animal tumor models. This research led to clinical trials and approval by the FDA for Gliadel. This research has been expanded in many directions looking at other polymers, other chemotherapeutic agents and multi-modality therapeutic regimens [85, 86].

Depletion of plasma methionine might be expected to inhibit or reverse growth of methionine-dependent tumors [87, 88]. Plasma methionine can be lowered to a steady state of <5 micromolar in mice with a combination of dietary restriction of methionine, homocystine, and choline and synchronous treatments with intraperitoneal injections of 1000 U/kg L-methioninase and 25–50 mg/kg homocystine administered twice per day. When this regimen was used for 10 days to treat nude mice bearing subcutaneously implanted human medulloblastoma (DAOY) tumor, tumor growth was inhibited. The combination of recombinant L-methionine-alpha-deamino-gamma-lyase and BCNU was also studied. Animals bearing DAOY, SWB77, and D-54 MG xenografts were treated with depletion of plasma methionine with a combination of a methionine and choline-free diet and recombinant L-methionine-alpha-deamino-gamma-lyase along with rescue with daily intraperitoneal homocystine for 10–12 days followed by BCNU [88]. Tumor growth delays of 80 days were observed in animals bearing DAOY and D-54 MG tumors and a tumor growth delay of 20 days was observed in animals bearing SWB77 tumors.

ANGIOGENESIS IN BRAIN TUMOR MODELS

Marked neovascularization is a hallmark of many neoplasms in the CNS. Vascular pathology is a key feature of glioblastoma multiforme characterized by hypervascularity, vascular permeability and hyper-coagulability. Vascular endothelial growth factor (VEGF) has been investigated as a potent mediator of brain tumor angiogenesis, vascular permeability, and glioma growth and is known to be upregulated in most cases of glioblastoma multiforme [89–101]. Microvessel density and VEGF levels have been shown to be independent prognostic markers of survival in fibrillary low-grade astrocytoma. Tumors with a larger number of microvessels also had a greater probability of undergoing malignant transformation [90]. Another study examined the activated phosphorylated form of the KDR receptor in astrocytic neoplasms and found the phosphorylated form of KDR in fresh surgical specimens of glioblastomas (71 per cent) and anaplastic gliomas (15 per cent), but not in low grade gliomas indicating that onset of angiogenesis is an important event during the disease progression of gliomas [96]. Chan et al., [100] found the VEGF receptors, KDR and Flt-1 to be upregulated in the tumor vasculature of glioblastoma multiforme, anaplastic oligodendrogliomas, and ependymomas with necrosis but not in astrocytomas grade II, anaplastic astrocytomas, or oligodendroglioma tumors. In meningiomas, VEGF was associated both with tumor vascularity and peritumoral edema [91]. In looking for a correlation between angiographic neovascularization, peritumoral brain edema and the expression of vascular endothelial growth factor, Bitzer et al., [95] found that tumors with high VEGF staining had a significantly higher edema index and a higher edema incidence. In addition, all of the meningiomas with very high VEGF expression were associated with vascular tumor supply from cerebral arteries. Takano et al., [93] found that VEGF concentrations of glioblastoma cyst fluid were 200- to 300-fold higher than those of serum in the patients. VEGF concentration in the tumors was significantly correlated with the vascularity measured by counting vessels stained with von Willebrand factor antibody. VEGF is expressed in a wide spectrum of brain tumors and is associated with neovascularization.

However, other angiogenic factors also appear to contribute to the vascularization of CNS neoplasms [92]. The expression of angiopoietin-1 and angiopoietin-2 in human astrocytomas was investigated by in situ hybridization [99]. Angiopoietin-1 mRNA was localized in tumor cells and angiopoietin-2 mRNA was detected in endothelial cells. The results suggested that angiopoietins are involved in the early stage of vascular activation and in advanced angiogenesis and indicate that angiopoietin-2 may be an early marker of glioma-induced neovascularization. Takano et al., [102] investigated the expression of the angiogenic factor thymidine phosphorylase in human astrocytic tumors and found that thymidine phosphorylase was expressed in the tumor cells, macrophages, and endothelial cells. The influence of antiangiogenic treatment on 9L gliosarcoma oxygenation and response to cytotoxic therapy was investigated [103]. In another study, the mean concentrations of VEGF were found to be 11-fold higher in high grade gliomas and the mean concentrations of hepatocyte growth factor/scatter factor (HGF/SF) were found to be 7-fold higher in high grade gliomas than in low grade tumors [104]. In addition, VEGF and HGF/SF appeared to be independent predictive parameters for glioma

microvessel density. The findings of this study also suggested that basic fibroblast growth factor (bFGF) is an essential cofactor for angiogenesis in gliomas.

The signal transduction from extracellular protein growth factors occurs by a variety of mechanisms that share many common features. Activation of specific receptor kinases do not activate unique intracellular kinases which then result in a linear signaling pathway; rather multiple signaling cascades can be activated producing combinatorial effects that allow more refined regulation of the biological outcome [105]. The intracellular signal transduction pathways for VEGF and bFGF in endothelial cells have not been fully elucidated; however, it is likely that protein kinase C is an important pathway component for both mitogens [106–114]. Neoangiogenesis in the eyes of rats bearing corneal micropocket implants of either VEGF or bFGF was inhibited by treatment of the animals with LY317615 orally twice per day [115]. Treatment of human SW-2 small cell lung carcinoma-bearing mice with LY317615 orally twice per day resulted in a dose-dependent decrease in the number of countable intratumoral vessels in the tumors. The number of intratumoral vessels stained by Factor VIII was decreased to one-half of the controls in animals treated with LY317615 and the number of vessels stained by CD31 was decreased to one-quarter of the controls in animals treated with LY317615 [115]. The effects of LY317615 have been explored in a variety of tumor models [116–120].

The human T98G glioblastoma multiforme line was used as a brain tumor model [121]. The protein kinase Cβ inhibitor 317615 was not very cytotoxic toward T98G cells in culture and was additive in cytotoxicity with BCNU. When nude mice bearing subcutaneous T98G tumors were treated with LY317615 orally twice daily on days 14 through 30 post-implant, the number of intratumoral vessel stained by CD31 was decreased to 37 per cent of control and vessels stained by CD105 was decreased to 50 per cent of control. The compound LY317615 was an active antitumor agent against subcutaneously growing T98G xenografts (Fig. 4.1) [122]. A treatment regimen administering LY317615 prior to, during and after BCNU was compared with a treatment regimen administering LY317615 sequentially after BCNU. In the tumor growth delay determination of the subcutaneous tumor, the sequential treatment regimen was more effect than the simultaneous treatment regimen (Figs. 4.1 and 4.2). However, when the same treatments were administered to animals bearing intracranial T98G tumors, the survival of animals receiving the simultaneous treatment regimen increased from 41 days for BCNU alone to 102 days for the

RESPONSE OF HUMAN T98G GLIOBLASTOMA MULTIFORME TO TREATMENT WITH A PKCβ INHIBITOR or SU5416 PRIOR TO, DURING AND AFTER BCNU

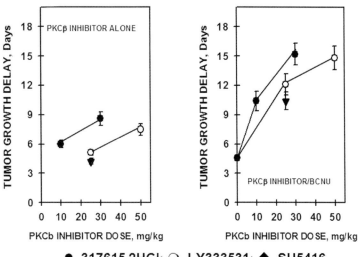

●, 317615·2HCl; ○, LY333531; ◆, SU5416
PKCβ INHIBITOR or SU5416, d4-18/ BCNU, d7-11

FIGURE 4.1 Growth delay of subcutaneously implanted human T98G glioblastoma multiforme after treatment with LY317615 (10 or 30 mg/kg) orally twice per day on days 4 through 18 alone or along with BCNU (15 mg/kg, ip) on days 7 through 11, or with SU5416 (25 mg/kg, ip). Points are the means of 5 animals; bars are SEM.

RESPONSE OF HUMAN T98G GLIOBLASTOMA MULTIFORME TO SEQUENTIAL TREATMENT WITH A PKCβ INHIBITOR AFTER BCNU

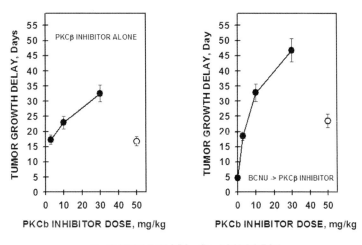

●, 317615·2HCl; ○, LY333531
BCNU, d7-11 -> PKCβ INHIBITOR, 2x d12-30

FIGURE 4.2 Growth delay of subcutaneously implanted human T98G glioblastoma multiforme after treatment with LY317615 (3, 10, or 30 mg/kg) orally twice per day on days 12 through 30 alone or after administration of BCNU (15 mg/kg, ip) on days 7 through 11, or with LY333531 (50 mg/kg, ip). Points are the means of 5 animals; bars are SEM.

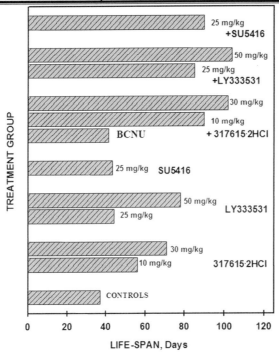

SURVIVAL OF MICE BEARING IC T98G GLIOBLASTOMA MUTLIFORME TREATED WITH A PKCβ INHIBITOR or SU5416 ALONG WITH BCNU

FIGURE 4.3 Survival of animals bearing intracranial human T98G glioblastoma multiforme after treatment with LY317615 (10 or 30 mg/kg) orally twice per day on days 4 through 18 alone or along with BCNU (15 mg/kg, ip) on days 7 through 11, or with SU5416 (25 mg/kg, ip). Data are the means of 5 animals.

combination while animals receiving the sequential treatment regimen survived 74 days (Figs. 4.3 and 4.4). Treatment with the protein kinase Cβ inhibitor decreased T98G glioblastoma multiforme angiogenesis and improved treatment outcome with BCNU [122].

The potential of antiangiogenic agents to augment the antitumor activity of standard cytotoxic chemotherapeutic agents is becoming well established [123, 124]. Among the antiangiogenic agents under investigation, TNP-470, an inhibitor of endothelial cell proliferation, has been shown to delay the growth of gliomas and other brain tumors in several studies [124, 125]. The antiangiogenic combination of TNP-470 and minocycline increased the response of both intracranial and subcutaneous rat 9L gliosarcoma to BCNU or adriamycin [103]. Lund et al., [125] found that TNP-470 treatment increased the response of subcutaneous human U87 glioblastoma xenografts to radiation therapy but did not increase the response of intracranial U87 to radiation therapy. Angiostatin, an antiangiogenic internal fragment of plasminogen, has been shown to suppress the growth of rat C6 and rat 9L gliomas as well as human U87 glioma whether implanted subcutaneously or intracranially [126]. Angiostatin used in combination with fractionated

radiation therapy had a greater-than-additive effect on the growth of a human glioma in nude mice [127]. The PKCβ inhibitor LY317615 is an orally administered small molecule without toxicity in rodents at the antiangiogenic doses [115]. The T98G glioblastoma multiforme tumor model allowed the comparison of combination treatment regimens that examined the efficacy of the LY317615 given simultaneously or sequentially with BCNU and the examination of two experimental endpoints, tumor growth delay and survival [122]. The cell culture studies indicate there can be an interaction between PKC inhibition and BCNU to enhance cytotoxicity in the malignant cells. Interestingly, the sequential treatment regimen, giving the protein kinase Cβ inhibitor after the cytotoxic agent resulted in a greater effect on tumor growth delay but a lesser effect on survival although these differences did not reach statistical significance. There are several possible reasons for the difference. The first possibility is that the sequential regimen involved treatment of a larger tumor burden allowing the impact on subcutaneous tumor growth to be manifest, but the larger tumor burden in the cranium impacted negatively on survival. A second possibility is that the angiogenic factors operative in the subcutaneous

CONCLUSIONS   69

SURVIVAL OF MICE BEARING IC T98G GLIOBLASTOMA MULTIFORME
TREATED SEQUENTIALLY WITH BCNU -> PKCβ INHIBITOR

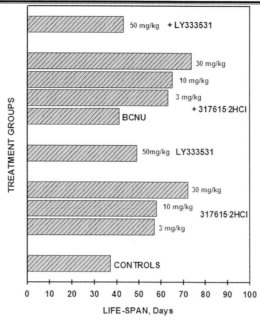

FIGURE 4.4 Survival of animals bearing intracranial human T98G glioblastoma multiforme after treatment with LY317615 (3, 10, or 30 mg/kg) orally twice per day on days 12 through 30 alone or after administration of BCNU (15 mg/kg, ip) on days 7 through 11, or LY333531 (50 mg/kg, ip). Data are the means of 5 animals.

tumor are different than the angiogenic factors expressed in the intracranial tumor and that the protein kinase Cβ inhibitor is more effective against the subcutaneous tumor. It has often been noted that chemotherapeutic agents have very different levels of efficacy depending upon the organ environment of the tumor [128–130].

A normal volunteer and a phase I trial in solid tumor patients demonstrated the drug was very well tolerated at doses that achieve a biologically active serum concentration [131]. Based on the dependence of glioma growth on VEGF-mediated angiogenesis, and the promising preclinical and clinical data, Fine et al., [132] initiated a phase II trial of LY317615 in patients with recurrent and progressive high grade gliomas following standard therapy. Treatment consisted of oral LY317615 administered daily on an every 6-week cycle after which patients underwent a complete physical/neurological, biochemical, and radiographic reevaluation. Patients were stratified based on those taking enzyme-inducing antiepileptic drugs (EIAED; Group B) and those not taking EIAED (Group A) and conducted pharmacokinetic studies. To date, 32 patients (17 patients in Group A and 15 patients in Group B) have been accrued to the trial and 28 patients were evaluable for response. Treatment was well tolerated with only one possible case of

drug-related toxicity > grade 1 (Grade 2 thrombocytopenia). Eleven patients have received more than one cycle of treatment (6 patients in Group A and 5 patients in Group B) and several patients have been stable on treatment for greater then 3 months and a number of other patients continue treatment with LY317615. Objective radiographic responses have been seen in 5 patients. LY317615 appears to have antitumor activity against recurrent malignant gliomas [132].

CONCLUSIONS

New genetic models continue to be developed to provide better biological mimics of human tumors [133,134].

References

1. American Cancer Society (2004). *Cancer Facts and Figures.* Atlanta, GA.
2. Levin, V. A., Leibel, S. A., and Gutin, P. H. (2001). Neoplasms of the central nervous system. *In Cancer Principles and practice of Oncology.* (V.T. Devita, S. Hellman, and S.A. Rosenberg, Eds.), 6th ed., pp. 2100–2160. Lippincott Williams & Wilkins, Philadelphia, PA.
3. Middlemas, D. S., Stewart, C. F., Kirstein, M. N. *et al.* (2000). Biochemical correlates of temozolomide sensitivity in pediatric solid tumor xenograft models. *Clin Cancer Res* **6**, 998–1007.

4. Batchelor, T. (2000). Temozolomide for malignant brain tumors. *Lancet* **355**, 1115–1116.

5. Osoba, D., Brada, M., Yung, W. K., and Prados, M. (2000). Health-related quality of life in patients treated with temozolomide versus procarbazine for recurrent glioblastoma multiforme. *J Clin Oncol* **18**, 1481–1491.

6. Barth, R. F. (1998). Rat brain tumor models in experimental neuro-oncology: the 9L, C6, T9, F98, RG2 (D74), RT-2 and CNS-1 gliomas. *J Neurooncol* **36**, 91–102.

7. Schmidek, H. H., Nielsen, S. L., Schiller, A. L., and Messer, J. (1971). Morphological studies of rat brain tumors induced by N-nitrosomethylurea. *J Neurosurg* **34**, 335–340.

8. Coderre, J. A., Glass, J. D., Fairchild, R. G., Micca, P. L., Fand, I., and Joel, D. D. (1990). Selective delivery of born by the melanin precursor analogue p-boronphenylalanine to tumors other than melanoma. *Cancer Res* **50**, 138–141.

9. Benda, P., Lightbody, J., Sato, G., Levine, L., and Sweet, W. (1968). Differentiated rat glial cell strain in tissue culture. *Science* **26**, 370–371.

10. Benda, P., Someda, K., Messer, J., and Sweet, W. H. (1971). Morphological and immunochemical studies of rat glial tumors and clonal strains propagated in culture. *J Neurosurg* **34**, 310–323.

11. Lightbody, J. J. *et al.* (1968). Establishment of differentiated clonal strains of glial brain cells in culture. *Fed Proc* **27**, 720.

12. Ko L Koestner, A., and Wechsler, W. (1980). Morphological characterization of nitrosourea-induced glioma cell lines and clones. *Acta Neuropathol* **51**, 23–31.

13. Clendenon, N. R., Barth, R. F., Gordon, W. A. *et al.* (1990). Boron neutron-capture therapy of a rat glioma. *Neurosurgery* **26**, 47–55.

14. Tzeng, J. J., Barth, R. F., Orosz, C. G., and James, S. M. (1991). Phenotype and functional activity of tumor-infiltrating lymphocytes isolated from immunogenic and nonimmunogenic rat brain tumors. *Cancer Res* **51**, 2373–2378.

15. Ningaraj, N. S., Rao, M., and Black, K. L. (2003). Calcium-dependent potassium channels as a target protein for modulation of the blood-brain tumor barrier. *Drug News & perspectives* **16**, 291–298.

16. Erdlebruch, B., Jendrossek, V., Kugler, W., Eibl, H., and Lakomek, M. (2002). Increased delivery of erucylphosphocholine to C6 gliomas by chemical opening of the blood-brain barrier using intracarotid pentylglycerol in rats. *Cancer Chemotherap Pharmacol* **50**, 299–304.

17. Saito, R., Bringas, J. R., McKnight, T. R. *et al.* (2004). Distribution of liposomes into brain and rat brain tumor models by convection-enhanced delivery monitored with magnetic resonce imaging. *Cancer Res* **64**, 2572–2579.

18. Spence, A. M., Graham, M. M., Muzi, M. *et al.* (1997). Feasibility of imaging pentose cycle glucose metabolism in gliomas with PET: studies in rat brain tumor models. *J Nuclear Med* **38**, 617–624.

19. Kramm, C. M., Sena-Esteves, M., Barnett, F. H. *et al.* (1995). Gene therapy for brain tumors. *Brian Pathol* **5**, 345–381.

20. Rainov, N. G., and Ren, H. (2003). Clinical trials with retrovirus mediated gene therapy- what have we learned? *J Neurooncol* **65**, 227–236.

21. Cool, V., Pirotte, B., Gerard, C. *et al.* (1996). Curative potential of herpes simplex virus thymidine kinase gene transfer in rats with 9L gliosarcoma. *Human Gene Therapy* **7**, 627–635.

22. Rainov, N. G., Kramm, C. M., Aboody-Guterman, K. *et al.* (1996). Retrovirus-mediated gene therapy of experimental brain neoplasms using the herpes simplex virus-thymidine kinase/ganciclovir paradigm. *Cancer Gene Therapy* **3**, 99–106.

23. Iwadate, Y., Namba, H., Tagawa, M., Takenaga, K., Sueyoshi, K., and Sakiyama, S. (1997). Induction of acquired immunity in rats that have eliminated intracranial gliosarcoma cells by the expression of herpes simplex virus-thymidine gene and ganciclovir administration. *Oncology* **54**, 329–334.

24. Kramm, C. M., Chase, M., Herrlinger, U. *et al.* (1997). Therapeutic efficiency and safety of a second-generation replication-conditional HSV1 vector for brain tumor therapy. *Human Gene Therapy* **8**, 2057–2068.

25. Grignet-Debrus, C., and Calberg-Bacq, C. M. (1997). Potential of Varicella zoster virus thymidine kinase as a suicide gene in breast cancer cells. *Gene Therapy* **4**, 560–569.

26. Grignet-Debrus, C., Cool, V., Baudson, N. *et al.* (2000). Comparitive in vitro and in vivo cytotoxic activity of (E)-5-(2-bromovinyl)-2′-deoxyuridine (BVDU) and its arabinosyl derivative, (E)-5-(2-bromovinyl)-1-beta-D-arabinofuranosyluracil (BVaraU), against tumor cells expressing either the Varicella zoster or the Herpes simplex virus thymidine kinase. *Cancer Gene Therapy* **7**, 215–223.

27. Sahu, S. K., Wen, P. Y., Foulon, C. F. *et al.* (1997). Intrathecal 5-[125I]iodo-2′-deoxyuridine in a rat model of leptomeningeal metastases. *J Nuc Med* **38**, 386–390.

28. Nanda, D., de Jong, M., Vogels, R. *et al.* (2002). Imaging expression of adenoviral HSV1-tk suicide gene transfer using the nucleoside analogue FIRU. *Europ J Nuclear Med Molec Imaging* **29**, 939–947.

29. Badie, B., Drazan, K. E., Kramar, M. H., Shaked, A., and Black, K. L. (1995). Adenovirus-mediated p53 gene delivery inhibits 9L glioma growth in rats. *Neuro Res* **17**, 209–216.

30. Bowers, G., He, J., Schulz, K., Maneval, D., and Loson, J. J. (2003). Efficacy of adenoviral p53 delivery with SCH58500 in the intracranial 9L and RG2 models. *Frontiers Bioscience* **8**, a54–a61.

31. Driessens, G., Harsan, L., Browaeys, P., Giannakopoulos, X., Velu, T., and Bruyns, C. (2003). Assessment of in vivo chemotherapy-induced DNA damage in a p53-mutated rat tumor by micronuclei assay. *Ann NY Acad Sci* **1010**, 775–779.

32. Aghi, M., Kramm, C. M., and Breakefield, X. O. (1999). Folylpolyglutamyl synthetase gene transfer and glioma antifolate sensitivity in culture and in vivo. *J Natl Cancer Inst* **91**, 1233–1241.

33. Manome, Y., Kunieda, T., Wen, P. Y., Koga, T., Kufe, D. W., and Ohno, T. (1998). Transgene expression in malignant glioma using a replication-defective adenoviral vector containing the Egr-1 promoter, activation by ioning radiation or uptake of radioactive iododeoxyuridine. *Human Gene Therapy* **9**, 1409–1417.

34. Rainov, N. G., Ikeda, K., Qureshi, N. H. *et al.* (1999). Intraarterial delivery of adenovirus vectors and liposome-DNA complexes to experimental brain neoplasms. *Human Gene Therapy* **10**, 311–318.

35. Iwadate, Y., Tagawa, M., Namba, H. *et al.* (2000). Immunological responsiveness to interleukin-2-producing brain tumors can be restored by concurrent subcutaneous transplantation of the same tumors. *Cancer Gene Therapy* **7**, 1263–1269.

36. Lefranc, F., Camby, I., Belot, N. *et al.* (2002). Gastrin significantly modifies the migratory abilities of experimental glioma cells. *Lab Invest* **82**, 1241–1252.

37. Rhines, L. D., Sampath, P., Dimeco, F. *et al.* (2003). Local immunotherapy with interleukin-2 delivered from biodegradable polymer microspheres combined with interstitial chemotherapy: a novel treatment for experimental malignant glioma. *Neurosurgery* **52**, 872–879.

38. Hanes, J., Sills, A., Zhao, Z. *et al.* (2001). Controlled local delivery of interleukin-2 by biodegradable polymers protects animals from experimental brain tumors and liver tumors. *Pharmaceutical Res* **18**, 899–906.

39. Liau, L. M., Fakhrai, H., and Black, K. L. (1998). Prolonged survival of rats with intracranial C6 gliomas by treatment with TGF-beta antisense gene. *Neurol Res* **20**, 742–747.

40. Lefranc, F., Sadeghi, N., Metens, T., Brotchi, J., Salmon, I., and Kiss, R. (2003). Characterization of gastrin-induced cytostatic effect on cell proliferation in experimental malignant gliomas. *Neurosurgery* **52**, 881–890.

41. Lefranc, F., Chaboteaux, C., Belot, N., Brotchi, J., Salmon, I., and Kiss, R. (2003). Determination of RNA expression for cholecystokinin/gastrin receptors (CCKA, CCKB and CCKC) in human tumors of the central and peripheral nervous system. *Int J Oncol* **22**, 213–219.

42. Lefranc, F., Cool, V., Velu, T., Brotchi, J., and De Witte, O. (2002). Granulocyte macrophage-colony stimulating factor gene transfer to induce a protective antitumoral immune response against the 9L rat gliosarcoma model. *Int J Oncol* **20**, 1077–1085.

43. Manome, Y., Watanabe, M., Futaki, K. *et al.* (1999). Development of a syngenic brain tumor model resistant to chloroethylnitrosourea using a methylguanine DNA methyltransferase cDNA. *Anticancer Res* **19**, 5313–5318.

44. Hebeda, K. M., Wolbers, J. G., Sterenborg, H. J., Kamphorst, W., van Gemert, M. J., and van Alphen, H. A. (1995). Fluorescence localization in tumor and normal brain after intratumoral injection of hematoporphyrin derivative into rat brain tumor. *J Phtochem Photobiol B* **27**, 85–92.

45. Barth, R. F., Yang, W., and Coderre, J. A. (2003). Rat brain tumor models to assess the efficacy of boron neutron capture therapy: a critical evaluation. *J Neurooncol* **62**, 61–74.

46. Barth, R. F. (2003). A critical assessment of boron neutron capture therapy: an overview. *J Neurooncol* **62**, 1–5.

47. Takagaki, M., Ono, K., Oda, Y. *et al.* (1996). Hydroxyforms of p-boronophenylalanine as potential boron carriers on boron neutron capture therapy for malignant brain tumors. *Cancer Res* **56**, 2017–2020.

48. Smith, D. R., Chandra, S., Barth, R. F., Yang, W., Joel, D. D., and Coderre, J. A. (2001). Quantitative imaging and microlocalization of boron-10 in brain tumors and infiltrating tumor cells by SIMS ion microscopy: relevance to neutron capture therapy. *Cancer Res* **61**, 8179–8187.

49. Barth, R. F., Grecula, J. C., Yang, W. *et al.* (2004). Combination of boron neutron-capture therapy and external beam radiotherapy for brain tumors. *Int J Radiat Oncol Biol Phys* **58**, 267–277.

50. Nelson, A. L., Algon, S. A., Munasinghe, J. *et al.* (2003). Magnetic resonance imaging of patched heterozygous and xenografted mouse brain tumors. *J Neurooncol* **62**, 259–267.

51. Engebraaten, O., Hjortland, G. O., Hirschberg, H., and Fodstad, O. (1999). Growth of precultured human glioma specimens in nude rat brain. *J Neurosurgery* **90**, 125–132.

52. Ozawa, T., Wand, J., Hu, L. J., Bollen, A. W., Lamborn, K. R., and Deen, D. F. (2002). Growth of human glioblastomas as xenografts in the brains of athymic rats. *In vivo* **16**, 55–60.

53. Mamot, C., Nguyen, J. B., Pourdehnad, M. *et al.* (2004). Extensive distribution of liposomes in rodent brains and brain tumors following convection-enhanced delivery. *J Neurooncol* **68**, 1–9.

54. van Furth, W. R., Laughlin, S., Taylor, M. D. *et al.* (2003). Imaging of murine brain tumors using a 1.5 Tesla clinical MRI system. *Canadian J Neuro Sci* **30**, 326–332.

55. Cha, S., Johnson, G., Wadghiri, Y. Z. *et al.* (2003). Dynamic, contrast-enhanced perfusion MRI in mouse gliomas: correlation with histopathology. *Magentic Resonance in Medicine* **49**, 848–855.

56. Kemper, E. M., van Zandbergen, A. E., Cleypool, C. *et al.* (2003). Increased penetration of paclitaxel into the brain by inhibition of P-glycoprotein. *Clin Cancer Res* **9**, 2849–2855.

57. Remsen, L. G., Pagel, M. A., McCormick, C. I. *et al.* (1999). The influence of anesthetic choice, PaCO2, and other factors on osmotic blood-brain barrier disruption in rats with brain tumor xenografts. *Anesth Analg* **88(3)**, 559–567.

58. Schlageter, K. E., Molnar, P., Lapin, G. D., and Groothuis, D. R. (1999). Microvessel organization and structure in experimental brain tumors: microvessel populations with distinctive structural and functional properties. *Microvascul Res* **58**, 312–328.

59. Sonoda, Y., Kanamori, M., Deen, D. F., Cheng, S. Y., Berger, M. S., and Pieper, R. O. (1999). Overexpression of vascular endothelial growth factor isoforms drives oxygenation and growth but not progression to glioblastoma multiforme in a human model of gliomagenesis. *Cancer Res* **63**, 1962–1968.

60. Chen, X., Park, R., Tohme, M., Shahinian, A. H., Bading, J. R., and Conti, P. S. (2004). MicroPET and autoradiographic imaging of breast cancer alpha v-integrin expression using 18F- and 64Cu-labeled RGD peptide. *Bioconjugate Chem* **15**, 41–49.

61. Sonoda, Y., Ozawa, T., Aldape, K. D., Deen, D. F., Berger, M. S., and Pieper, R. O. (2001). Akt pathway activation converts anaplastic astrocytoma to glioblastoma multiforme in a human astrocyte model of glioma. *Cancer Res* **61**, 6674–6678.

62. Aboody-Guterman, K. S., Pechan, P. A., Rainov, N. G. *et al.* (1997). Green fluorescent protein as a reporter for retrovirus and helper virus-free HSV-1 amplicon vector-mediated gene transfer into neural cells in culture and in vivo. *Neuroreport* **8**, 3801–3808.

63. Pyles, R. B., Warnick, R. E., Chalk, C. L., Szanti, B. E., and Parysek, L. M. (1997). A novel multiply-mutated HSV-1 strain for the treatment of human brain tumors. *Human Gene Therapy* **8**, 533–544.

64. Tamiya, T., Ono, Y., Wei, M. X., Mroz, P. J., Moolten, F. L., and Chiocca, E. A. (1996). Escherichia coli gpt gene sensitizes rat glioma cells to killing by 6-thioxanthine or 6-thioguanine. *Cancer Gene Therapy* **3**, 155–162.

65. Yamanaka, R., Zullo, S. A., Ramsey, J. *et al.* (2002). Marked enhancement of antitumor immune responses in mouse brain tumor models by genetically modified dendritic cells producing Semliki Forest virus-mediated interleukin-12. *J Neurosurgery* **97**, 611–618.

66. Geoerger, B., Tang, C. B., Cesano, A. *et al.* (2000). Antitumor activity of a human cytotoxic T-cell line (TALL-104) in brain tumor xenografts. *Neuro-Oncol* **2**, 103–113.

67. Wikstrand, C. J., Cokgor, I., Sampson, J. H. *et al.* (1999). Monoclonal antibody therapy of human gliomas: current status and future approaches. *Cancer Metastasis Rev* **18**, 451–464.

68. Muldoon, L. L., and Neuwelt, E. A. (2003). BR96-DOX immunoconjugate targeting of chemotherapy in brain tumor models. *J Neurooncol* **65**, 49–62.

69. Foulon, C. F., Reist, C. J., Bigner, D. D., and Zalutsky, M. R. (2000). Radioiodination via D-amino acid peptide enhances cellular retention and tumor xenograft targeting of an internalizing anti-epidermal growth factor receptor variant III monoclonal antibody. *Cancer Res* **60**, 4453–4460.

70. Vaidyanathan, G., Affleck, D. J., Bigner, D. D., and Zalutsky, M. R. (2002). Improved xenograft targeting of tumor-specific anti-epidermal growth factor receptor variant III antibody labeled using N-succinimidyl 4-guanidinomethyl-3-iodobenzoate. *Nuc Med Biol* **29**, 1–11.

71. Schold, S. C., Jr., Kokkinakis, D. M., Rudy, J. L., Moschel, R. C., and Pegg, A. E. (1996). Treatment of human brain tumor xenograft with O^6-benzyl-2'-deoxyguanosine and BCNU. *Cancer Res* **56**, 2076–2081.

72. Kokkinakis, D. M., Moschel, R. C., Pegg, A. E., and Schold, S. C. (1996). Eradication of human medulloblastoma tumor xenografts with a combination of O^6-benzyl-2'-deoxyguanosine and 1,3-bis(2-chloroethyl)1-nitrosourea. *Clin Cancer Res* **5**, 3676–3681.

73. Keir, S. T., Dolan, M. E., Pegg, A. E. *et al.* (2000). O^6-benzylguanine-mediated enhancement of nitrosourea activity in Mer-central nervous system tumor xenografts- implications for clinical trials. *Cancer Chemotherap Pharmacol* **45**, 437–440.

74. Ewesuedo, R. B., Wilson, L. R., Friedman, H. S., Moschel, R. C., and Dolan, M. E. (2001). Inactivation of O6-alkyloguanine-DNA alkyltransferase by 8-substituted O6-benzylguanine analogs in mice. *Cancer Chemotherap Pharmacol* **47**, 63–69.

75. Houghton, P. J., Cheshire, P. J., Hallman, J. D. *et al.* (1995). Efficacy of topoisomerse I inhibitors, topotecan and ironotecan, administered at low dose levels in protracted schedules to mice bearing xenografts of human tumors. *Cancer Chemotherap Pharmacol* **36**, 393–403.

76. Keir, S. T., Hausheer, F., Lawless, A. A., Bigner, D. D., and Friedman, H. S. (2001). Therapeuric activity of 7-[(2-trimethyl-silyl)ethyl]]-20(S)-camptothecin against central nervous system tumor-derived xenografts in athymic mice. *Cancer Chemotherap Pharmacol* **48**, 83–87.

77. Cavazos, C. M., Keir, S. T., Yoshinari, T., Bigner, D. D., and Friedman, H. S. (2001). Therapeutic activity of the topoisomerase I inhibitor J-107088 [6-N-(1-hydroxymethyl-2-hydroxyl)-ethylamino-12,13-dihydro-13-(beta-D-glucopyranosyl)-5H-indolo[2,3-a]-pyrrolo[3,4-c]-carbazole-5,7(6H)-dione]] against pediatric and adult central nervous system tumor xenografts. *Cancer Chemotherap Pharmacol* **48**, 250–254.

78. Patel, V. J., Elion, G. B., Houghton, P. J. *et al.* (2000). Schedule-dependent activity of temozolomide plus CPT-11 against a human central nervous system tumor-derived xenograft. *Clin Cancer Res* **6**, 4154–4157.

79. Heimberger, A. B., Archer, G. E., McLendon, R. E. *et al.* (2000). Temozolomide delivered by intracerebral microinfusion is safe and efficacious against malignant gliomas in rats. *Clin Cancer Res* **6**, 4148–4153.

80. Vassal, G., Merlin, J. L., Terrier-Lacombe, M. J. *et al.* (2003). In vivo antitumor activity of S16020, a topoisomerase II inhibitor and doxorubicin against human brain tumor xenografts. *Cancer Chemotherap Pharmacol* **51**, 385–394.

81. Arguello, F., Alexander, M. A., Greene, J. F., Jr. *et al.* (1998). Preclinical evaluation of 9-chloro-2-methylellipiticinium acetate and in combination with conventional anticancer drugs for the treatment of human brain tumor xenografts. *J Cancer Res Clin Oncol* **124**, 19–26.

82. Heimberger, A. B., Learn, C. A., Archer, G. E. *et al.* (2002). Brain tumors in mice are susceptible to blockade of epidermal growth factor receptor (EGFR) witgh the oral, specific, EGFR-tyrosine kinase inhibitor ZD 1839 (iressa). *Clin Cancer Res* **8**, 3496–3502.

83. Friedman, H. S., Keir, S. T., Houghton, P. J., Lawless, A. A., Bigner, D. D., and Waters, S. J. (2001). Activity of irofulven (6-hydroxymethylacylfulvene) in the treatment of glioblastoma multiforme-derived xenografts in athymic mice. *Cancer Chemotherap Pharmcol* **48**, 413–416.

84. Wang, J., Klem, J. B., Ozawa, T. *et al.* (2003). Detection of hypoxia in human brain tumor xenografts using a modified comet assay. *Neoplasia* **5**, 288–296.

85. Brem, H., and Gabbikian, P. (2001). Biodegradable polymer implants to treat brain tumors. *J Controlled Release* **74**, 63–67.

86. Ewend, M. G., Williams, J. A., Tabassi, K. *et al.* (1996). Local delivery o9f chemotherapy and concurrent external beam radiotherapy prolongs survival in metastatic brain tumor models. *Cancer Res* **56**, 5217–5223.

87. Kokkinakis, D. M., Schold, S. C., Jr., Hori, H. *et al.* (1997). Effect of long-term depletion of plasma methionine on the growth and survival of human brain tumor xenografts in athymic mice. *Nutrition & Cancer* **29**, 195–204.

88. Kokkinakis, D. M., Hoffman, R. M., Frenkel, E. P. *et al.* (2001). Synergy between methionine stress and chemotherapy in the treatment of brain tumor xenografts in athymic mice. *Cancer Res* **61**, 4017–4023.

89. Bernsen, H. J., and van der Kogel, A. J. (1999). Antiangiogenic therapy in brain tumor models. *J Neuro-Oncol* **45**, 247–255.

90. Adulrauf, S. I., Edvardsen, K., Ho, K. L., Yang, X. Y., Rock, J. P., and Rosenblum, M. L. (1998). Vascular endothelial growth factor expression and vascular density as prognostic markers of survival in patients with low-grade astrocytoma. *J Neurosurg* **88**, 513–520.

91. Provias, J., Claffey, K., delAguila, L., Lau, N., Feldkamp, M., and Guha, A. (1997). Meningiomas: role of vascular endothelial growth factor/vascular permeability factor in angiogenesis and peritumoral edema. *Neurosurgery* **40**, 1016–1026.

92. Pietsch, T., Valter, M. M., Wolf, H. K. *et al.* (1997). Expression and distribution of vascular endothelial growth factor protein in human brain tumors. *Acta Neuropathol* **93**, 109–117.

93. Takano, S., Yoshii, Y., Kondo, S. *et al.* (1996). Concentration of vascular endothelial growth factor in the serum and tumor tissue of brain tumor patients. *Cancer Res* **56**, 2185–2190.

94. Jensen, R. L., and Kornblith, P. L. (1998). Growth factor-mediated angiogenesis in the malignant progression of glial tumors: a review. Surgical *Neurology* **49**, 189–196.

95. Bitzer, M., Opitz, H., Popp, J. *et al.* (1998). Angiogenesis and brain edema in intracranial meningiomas: influence of vascular endothelial growth factor. *Acta Neurochirurgica* **140**, 333–340.

96. Carroll, R. S., Zhang, J., Bello, L., Melnick, M. B., Maruyama, T., and Black, P. (1999). KDR activation in astrocytic neoplasms. *Cancer* **86**, 1335–1341.

97. Avgeropoulos, N. G., and Batchelor, T. T. (1999). New treatment strategies for malignant gliomas. *Oncologist* **4**, 209–224.

98. Rubin, J. B., and Kieran, M. W. (1999). Innovative therapies for pediatric brain tumors. *Current Opinion in Pediatrics* **11**, 39–46.

99. Zagzag, D., Hooper, A., Friedlander, D. R. *et al.* (1999). In situ expression of angiopoietins in astrocytomas identifies angiop-poietin-2 as an early marker of tumor angiogenesis. *Exp Neurology* **159**, 391–400.

100. Chan, A. S., Leung, S. Y., Wong, M. P. *et al.* (1998). Expression of vascular endothelial growth factor and its receptors in the anaplastic progression of astrocytoma, oligodendroglioma and ependymoma. *Amer J Surg Pathol* **22**, 816–826.

101. Bodey, B., Bodey, B., Jr, Siegel, S. E., and Kaiser, H. E. (1998). Upregulation of endoglin (CD105) expression during childhood brain tumor-related angiogenesis. Antiangiogenic therapy. *Anticancer Res* **18**, 1485–1500.

102. Tarano, S., Tsuboi, K., Matsumura, A., Tomono, Y., Mitsui, Y., and Nose, T. (2000). Expression of the angiogenic factor thymidine phosphorylase in human astrocytic tumors. *J Cancer Res Clin Oncol* **126**, P 145–152.

103. Teicher, B. A., Holden, S. A., Ara, G. et al. (1995). Influence of an antiangiogenic treatment on 9L gliosarcoma: oxygenation and response to cytotoxic therapy. Int J Cancer 61, 732–737.

104. Schmidt, N. O., Westphal, M., Hagel, C. et al. (1999). Levels of vascular endothelial growth factor, hepatocyte growth factor/scatter factor and basic fibroblast growth factor in human gliomas and their relation to angiogenesis. Int J Cancer (Pred Oncol) 84, 10–18.

105. Larner, A. C., and Keightley, A. (2000). The Jak/Stat signaling cascade. In Signaling Networks and Cell Cycle Control (J. Silvio Gutkind, Ed.), pps. 393–409. Humana Press: Totowa, NJ.

106. Pluda, J. M. (1997). Tumor-associated angiogenesis: mechanisms, clinical implications, and therapeutic strategies. Sem Oncol 24, 203–218.

107. Norrby, K. (1997). Angiogenesis: new aspects relating to its initiation and control. APMIS 105, 417–437.

108. Fox, S. B., and Harris, A. L. (1997). Markers of tumor angiogenesis: clinical applications in prognosis and anti-angiogenic therapy. Invest New Drugs 15, 15–28.

109. Ellis, L. M., Takahashi, Y., Liu, W., and Shaheen, R. M. (2000). Vascular endothelial growth factor in human colon cancer: biology and therapeutic implications. Oncologist 5 (suppl. 1), 11–15.

110. Sawano, A., Takahashi, T., Yamaguchi, S., and Shibuya, M. (1997). The phosphorylated 1169-tyrosine containing region of flt-1 kinase (VEGFR-1) is a major binding site for, PLCgamma, Biochem Biophys Res Commun 238, 487–491.

111. McMahon, G. (2000). VEGF receptor signaling in tumor angiogenesis. Oncologist 5 (suppl.1), 3–11.

112. Buchner, K. (2000). The role of protein kinase C in the regulation of cell growth and in signalling to the cell nucleus. J Cancer Res Clin Oncol 126, 1–11.

113. Martelli, A. M., Sang, N., Borgatti, P., Capitani, S., and Neri, L. M. (1999). Multiple biological responses activated by nuclear protein kinase C. J Cellular Biochem 74, 499–521.

114. Teicher, B. A., Alvarez, E., Mendelsohn, L. G., Ara, G., Menon, K., and Ways, K. D. (1999). Enzymatic rationale and preclinical support for a potent protein kinase Cβ inhibitor in cancer therapy. Advs Enzyme Regul 39, 313–327.

115. Teicher, B. A., Alvarez, E., Menon, K., Considine, E., Shih, C., and Faul, M. M. (2002). Antiangiogenic effects of a protein kinase Cβ selective small molecule. Cancer Chemotherap Pharmacol 49, 69–77.

116. Teicher, B. A., Menon, K., Alvarez, E., Galbreath, E., Shih, C., Faul, M. M. (2001). Antiangiogenic and antitumor effects of a protein kinase Cβ inhibitor in human HT-29 colon carcinoma and human CaKi1 renal cell carcinoma xenografts. Anticancer Res 21, 3175–3184.

117. Teicher, B. A., Menon, K., Alvarez, E., Galbreath, E., Shih, C., and Faul, M. M. (2001). Antiangiogenic and antitumor effects of a protein kinase Cβ inhibitor in murine Lewis lung carcinoma and human Calu-6 non-small cell lung carcinoma xenografts. Cancer Chemotherapy and Pharmacology 48, 473–480.

118. Teicher, B. A., Menon, K., Alvarez, E., Liu, P., Shih, C., and Faul, M. M. (2001). Antiangiogenic and antitumor effects of a protein kinase Cβ inhibitor in human hepatocellular and gastric cancer xenografts. In vivo 15, 185–193.

119. Keyes, K., Cox, K., Treadway, P., et al. (2002). An in vitro tumor model: analysis of angiogenic factor expression. Cancer Res 62, 5597–5602.

120. Teicher, B. A., Menon, K., Alvarez, E., Shih, C., and Faul, M. M. (2002). Antiangiogenic and antitumor effects of a protein kinase Cβ inhibitor in human breast cancer and ovarian cancer xenografts. Invest New Drugs 20, 241–251.

121. Stein, G. H. (1979). T98G: an anchorage-independent human tumor cell line that exhibits stationary phase G1 arrest in vitro. J Cell Physiol 99, 43–54.

122. Teicher, B. A., Menon, K., Alvarez, E., Galbreath, E., Shih, C., and Faul, M. (2001). Antiangiogenic and antitumor effects of a protein kinase Cβ inhibitor in human T98G glioblastoma multiforme xenografts. Clin Cancer Res 7, 634–640.

123. Teicher, B. A. (1996). A systems approach to cancer therapy. Cancer Metastasis Rev 15, 247–272.

124. Teicher, B. A. (Ed.) (1999). Antiangiogenic Agents in Cancer Therapy. New Jersey, The Humana Press, Inc.

125. Lund, E. L., Bastholm, L., and Kristjansen, P. E. G. (2000). Therapeutic synergy of TNP-470 and ionizing radiation: effects on tumor growth, vessel morphology and angiogenesis in human glioblastoma multiforme xenografts. Clin Cancer Res 6, 971–978.

126. Kirsch, M., Strasses, J., Allende, R., Bello, L., Zhang, J., and Black, P. (1998). Angiostatin suppresses malignant glioma growth in vivo. Cancer Res 58, 4654–4659.

127. Mauceri, H. J., Hanna, N. N., Beckett, M. A. et al. (1998). Combined effects of angiostatin and ionizing radiation in antitumor therapy. Nature (London) 394, 287–291.

128. Fidler, I. J. (1999). Critical derterminants of cancer metastasis: rationale for therapy. Cancer Chemotherap Pharmacol 43 (suppl), 3–10.

129. Killion, J. J., Radinsky, R., Fidler, I. J. (1998–99). Orthotopic models are necessary to predict therapy of transplantable tumors in mice. Cancer Mets Rev 17, 279–284.

130. Fidler, I. J., Wilmanns, F. C., Staroselsky, A., Radinsky, R., Dong, Z., and Fan, D. (1994). Modulation of tumor cell response to chemotherapy by the organ environment. Cancer Mets Rev 13, 209–222.

131. Herbst, R. S., Thornton, D. E., Kies, M. S. et al. (2002). Phase 1 study of LY317615, a protein kinase Cβ inhibitor. Amer Soc Clin Oncol 2002, abstr 326.

132. Fine, H. A., Kim, L., Royce, C. et al. (2004). A phase II trial of LY317615 in patients with recurrent high grade gliomas. Proc Amer Soc Clin Oncol 40, abstr 1511.

133. Lampson, L. A. (2001). New animal models to probe brain tumor biology, therapy, and immunotherapy: advantages and remaining concerns. J Neurooncol 53, 275–287.

134. Hesselager, G., and Holland, E. C. (2003). Using mice to decipher the molecular genetics of brain tumors. Neurosurgery 53, 685–694.

Microarray Analysis and Proteomic Approaches to Drug Development

Nicole J. Ullrich and Scott L. Pomeroy

INTRODUCTION

The recent development of microarray and other genomic technologies has revolutionized research into the molecular mechanisms of disease. Genetic phenotyping of disease has advanced our understanding of several neurologic syndromes [1], such as multiple sclerosis [2–4], Alzheimer's disease [5], Huntington's disease [6] and complex disorders such as schizophrenia [7]. These advances have been enabled by the sequencing of the human genome and the development of novel approaches to assess gene expression of the entire genome.

The use of gene expression arrays is particularly relevant to investigations of cancer, where these approaches are beginning to have direct effects on tumor diagnosis, disease management, and the development of directed drug therapies. Gene expression data can be useful for identification of genetic factors that are involved in the etiology of disease, factors involved in the metabolism, transport, kinetics, and adverse effects of medications as well as determination of genetic variations at drug targets. In the arena of drug development and discovery, microarray technology has been incorporated at each phase of the process, from identification and validation of potential targets to correlation with clinical parameters for predicting patient responses. These advances are now revolutionizing approaches to drug development, as we move from relatively non-selective therapies towards a new generation of therapeutic agents that target specific molecular mechanisms of disease. This review covers approaches to microarray analysis and the application of microanalysis to development of novel therapeutics. Findings pertinent to primary brain t7umors are presented.

MICROARRAY ANALYSIS

There are several methods currently available for detection and quantification of the level of gene expression. These include conventional northern blot analysis [8], differential display [9], cDNA library sequencing [10,11] and serial analysis of gene expression (SAGE) [12]. Recently, DNA microarrays have become available that can simultaneously monitor the expression of thousands of genes. There are two primary technologies that have become commercially available, based either on the use of cDNA or oligonucleotide arrays. All DNA microarrays rely on the principle of complementary DNA base pairing (southern), which permits the analysis of relative levels of mRNA abundance as an indicator of relative levels of gene expression. One can compare levels of single genes among cohorts of individuals and also between different genes within an individual [13,14]. In addition, new genes are often detected, since modern microarrays contain thousands of DNA sequences from genes which have not been functionally characterized. Identification of potential gene targets within biological samples can then be validated and analyzed, typically by probing northern blots or other assays of RNA expression, by performing immunohistochemistry, or by performing functional assays in living cells.

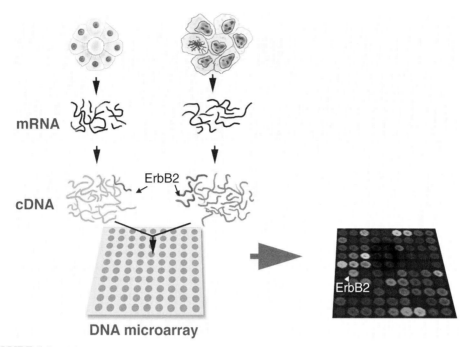

mRNA

cDNA ErbB2

DNA microarray

ErbB2

FIGURE 5.1 Schematic representation of a DNA microarray hybridization comparing gene expression of a malignant epithelial cancer with its normal tissue counterpart. (Journal of Pathology, Alizadeh *et al.* (2001). Copyright John Wiley and Sons Ltd. Reproduced with permission.) See Plate 5.1 in Color Plate Section.

Most microarray systems and methods share the same general principles, but may differ in the specific details. Basically, microarray probes tethered to an immobile surface are exposed to nucleic acid targets that are generated from RNA of the sample of interest (see Fig. 5.1). The signal generated by complementary binding of target to specific microarray probes is proportional to the level of RNA expression in the sample. Microarray probes often take the form of specific cDNA sequences, which are obtained and amplified by PCR, or they may be oligonucleotides, which are smaller than cDNAs. More recently, peptide nucleic acids (PNAs) have also been used. One advantage to the smaller oligonucleotide and PNA sequences is that they can be tiled at higher density allowing a larger number of probes per array. They also do not require PCR, which can introduce errors into the probe set. There is some data to suggest that PNA chips may also have a higher target affinity [15] and, therefore, provide more specific results.

The microarrays are produced by deposition and immobilization of selected probes onto a matrix. A number of matrix substrates have been utilized, including glass, silicon, nylon, and nitrocellulose. Glass is often selected, because it is nonporous and allows covalent attachment of probes. A computer-aided robot spots a sample of each gene product, with up to 10 000 cDNA probes/spots per cm^2 (Fig. 5.2).

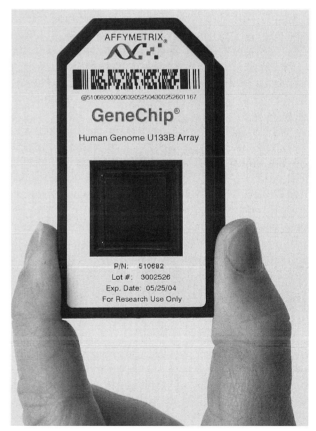

FIGURE 5.2 An example of an oligonucleotide-based DNA microarray is seen in this Affymetrix GeneChip array. See Plate 5.2 in Color Plate Section.

For cDNA arrays, fluorescently labeled target is generated from RNA derived from test and control samples by a single round of reverse transcription. These targets are then mixed and hybridized with the arrayed DNA spots. Image analysis of the microarray is utilized to assess the relative amount of the targets hybridized to each probe, indicative of either increased or decreased levels of gene expression in the test relative to the control sample [16,17].

Oligonucleotide probes may be synthesized *in situ* on the substrate, and may consist of up to 250 000 oligonucleotide probes/spots per cm^2. The arrays are hybridized with target from a single RNA sample, using differential hybridization of exact match oligonucleotide sequences with oligonucleotide probes that have a single base pair substitution to control for non-specific binding. The signal from exact match probes is then quantified as a measure of gene expression level.

A large number of data points are generated from each microarray experiment; all array methods require database management in order to compare the outputs of single and multiple array experiments using a number of data mining tools. These allow correlation of data into meaningful groups, which permit new hypotheses regarding possible cellular pathways and clues to disease pathophysiology. The correct interpretation of these immense datasets that result from multiple samples and from many patients in complex systems presents a major challenge. There is a clear need to prioritize the selection and investigation of candidate genes for further study.

Computational methods are used to help redefine relevant groups of tumors and genes; this hierarchical clustering can be used to identify groups of tumors with similar gene-expression profiles. These defined groups can subsequently be probed for other outcome parameters, such as survival and biologic or histologic activity [18]. Typically, interpretation of microarray data requires assistance from bioinformatics specialists familiar with the analysis of these very large datasets, and many resources are available on the internet to assist in this effort.

The expense of microarray experiments has decreased over time, due to the declining cost of commercial microarrays and the availability of core microarray facilities in many institutions. Moreover, studies can now be performed on a genome-wide basis for many different species as the completed genome sequences for a variety of organisms have permitted the development of DNA microarrays of model systems such as *Mycobacterium tuberculosis*, *Escherichia coli*, *Drosophila melanogaster*, *Caenorhabditis elegans*, and *Plasmodium falciparum*. As a result of the Human Genome Project, commercially prepared slides are now available to probe the human genome as well.

Limitations on these methods primarily rest on the availability of appropriate RNA samples that are derived from tissues of clinical interest. For brain tumors, handling of biopsy tissue currently is geared towards histopathologic examination to establish diagnosis and staging; methods of fixation and tissue embedding were not developed with microarray-based studies in mind. RNA is relatively unstable; therefore, tumor samples used for DNA microarrays must use RNA extracted from fresh or flash-frozen tissue, which is collected in liquid nitrogen or in isopentane on dry ice. RNA which has been extracted from paraffin-embedded tissue is, in general, too fragmented for use on arrays [19].

Finally, most tumor specimens are heterogeneous, in that they contain an intermixture of normal tissue, inflammatory cells, and connective tissue, in addition to malignant cells. Analysis of homogenized tissue from this sample will generate a mix of mRNA derived from malignant tissue in addition to these normal host cell types. Validation studies are needed to separate the gene expression signals of malignant from normal cell types.

DISEASE PROTEOMICS

The pattern of RNA expression in normal and developing tissue provides crucial, though sometimes inferential, information regarding gene function. The caveat is that RNA expression does not always correlate well with protein levels and expression. For example, in a comparison of the level of mRNA and protein levels for the same tissue from lung tumors, only a small subset of genes showed a correlation between the level of gene expression and corresponding proteins [20]. Moreover, higher levels of protein do not necessarily correlate with a higher activity or functional level, since proteins often require post-translational modification or may change function following relocalization within the cell. Post-translational modifications such as phosphorylation, sulfation, and glycosylation are not currently addressable by DNA microarray technology. In addition, the level and expression of proteins may not have clearly identifiable physiologic relevance. Despite these limitations, a proteomic approach may delineate altered protein expression, and help to develop new biomarkers for diagnosis and detection of disease. By profiling the full complement of proteins expressed within a tumor sample, proteomics may help identify new drug targets, thereby accelerating drug development [21–23].

The most widely used approach to the analysis of protein expression in disease relies on two-dimensional polyacrylamide gel electrophoresis for protein separation. Following separation, protein profiles from test and control samples are compared, and differentially expressed proteins are identified by mass spectrometry. Protein profiles have been successfully used to classify leukemia subtypes [21]. Other investigators have developed databases of protein expression in the myocardium [24], as well as in multiple cancer models including lung carcinoma [25], lymphoma [26], and ovarian cancer [27].

IMPLEMENTATION OF PROTEIN MICROARRAYS

Implementation of protein microarray strategies will help to address the different features of proteins that may be altered in disease. These include biochips that contain either peptides or proteins as well as recombinant proteins that can be synthesized directly onto the chip [28]. The development of protein microarrays relies on the use of capture agents, such as ribozymes, partial molecules, antibodies, and modified binding proteins (Fig. 5.3). Newer technologies have been developed to immobilize the entire protein repertoire of tissue using a reverse-phase

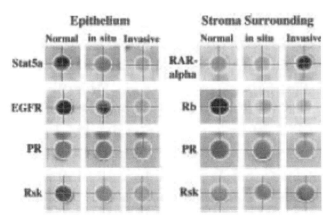

FIGURE 5.3 Antibody microarray example. Approximately 2500–3500 cells of each of six histologic types were microdissected and biotinylated protein from each dissection was used for a single incubation on a single antibody array. Examples of P-SCAN analysis indicated that Stat5a, EGFR-nonphosphoY1173, PR, and Rsk proteins decreased in expression in epithelium with advancing epithelial disease. Lines and circles result from the orientation function and data analysis by P-SCAN. (Proteomics, Knezevic *et al.* (2001). Copyright John Wiley and Sons Ltd. Reproduced with permission.) See Plate 5.3 in Color Plate Section.

protein array [29]. Using this approach, pro-survival checkpoints were determined showing the transition for histologically normal to neoplastic prostate tissue associated with a change in phosphorylation status of signal subpopulations. Similarly, protein microarrays can be incubated with patient serum and used to detect autoantibody binding to specific proteins of autoimmune disease, such as systemic lupus erythematosus and rheumatoid arthritis [30]. This new area of research is limited by the availability of the capture proteins and antibodies.

GENE EXPRESSION PATTERNS IN NORMAL AND DISEASED TISSUES

The gene expression profile of a cell gives insight into the mechanisms that lead to cell phenotype, function, and growth by helping to determine the levels of expression of regulatory genes, enzymes involved in biochemical pathways, and other clues to cellular function. One of the most attractive applications of microarrays is to study the differential pattern of gene expression in normal *versus* diseased tissues in order to identify potential targets for therapy. The first generation of microarray studies was focused on detection of these differences in gene expression and showed that subsets of tissue with different levels of aggressiveness also show distinctive differences in gene expression signatures [18,31–33]. Alteration in gene expression often accompanies the change from normal physiology, either as the cause of the disease process or as an end product of the disease and its treatments. Identification of specific genes that are implicated in the disease itself is desirable in an effort to halt or abort entirely the evolution of disease. Interfering with genes that result from abnormal physiology will also help to identify intolerable symptoms that accompany the disease process. Microarrays may also be used to screen for polymorphisms within a population that may either protect against disease or predispose to disease development [34].

The application of microarray technology to cancer is particularly relevant [35], as the tumorigenesis is likely to be accompanied by changes at the molecular level in a group of genes. Moreover, changes in gene expression during tumorigenesis have been linked to control mechanisms for tumor growth and survival [36]. One of the goals of the Cancer Genome Anatomy Project [37] was to produce and sequence cDNA libraries from five major human cancers (prostate, ovary, breast, lung, and gastrointestinal tract). Microarray technology has been used successfully to determine the genetic profile of a number of clinically relevant tumor subtypes, including B-cell lymphoma [38], leukemia [33], melanoma [39], and breast cancer [40]. In relation to cancer, DNA microarrays have also been used to distinguish between patients who respond to therapy *versus* those who do not, despite having identical histological appearance. In addition, it has been useful to identify molecular correlates and molecular signatures of tumors of the same and different lineages. Within the field of neuro-oncology, several examples now are emerging in the literature describing differential gene expression by microarrays. DNA microarrays have been used to profile tumor subtypes, most notably in astrocytomas [41,42] and medulloblastomas [41–43]. As each group of tumors had significantly different prognoses, treatment can then be optimized for tumor subtypes based on their molecular phenotypes.

Proteomic approaches towards identification of disease markers include comparison of protein expression in normal and diseased tissues, analysis of secreted proteins and direct patient serum profiling. In combination with mass spectrometry to identify the selected peptides and proteins, this becomes a powerful method for marker identification. One such approach has been used to detect tumor autoantibodies in the serum of patients with different cancer types [22,44]. This may ultimately be useful for cancer screening, diagnosis or in directed immunotherapy. Microarrays of proteins isolated from tumors have the potential to delineate the immune response to the tumor and to help discover tumor-specific antigens [45].

TUMOR EXPRESSION PROFILING: MEDULLOBLASTOMA

Genomic methods have been effective in providing new insights into the biology of brain tumors and other cancers [35,46–48]. These emerging technologies have been used within pediatric brain tumors to differentiate a group of tumors whose diagnosis based on conventional pathology remains controversial. Medulloblastomas are the most common malignant brain tumors of childhood, accounting for 20 per cent of pediatric brain tumors [49]. Because of the presence of dissemination at initial diagnosis and the long-term risks of metastases and recurrences, children are treated with high doses of radiation to the brain and spine, followed by multi-drug chemotherapy. Approximately 70 per cent of children survive 10 years after diagnosis [50]; most of these experience extensive neurologic and cognitive deficits largely due to aggressive treatment [51]. The most recent

standard risk medulloblastoma protocol of the Children's Oncology Group (COG) incorporates reduced-dose craniospinal irradiation with standard posterior fossa tumor boost [52]. The next generation of protocols will reduce radiation even further to children under eight years. For both of these trials, risk assignment uses traditional clinical criteria, which is based on the presence or absence of metastatic disease at diagnosis. Although widely used, these same clinical criteria have proven imprecise; many children with standard risk disease appear to have biologically aggressive disease.

Molecular Basis and Genomic Subclassification of Medulloblastomas

Medulloblastomas were once thought to be a subset of primitive neuroectodermal tumors, differing in their location within the cerebellum; however, the marker genes that supported the distinction also supported the hypothesis that these tumors are molecularly distinct and are derived from cerebellar granule cell precursors [42]. Gene expression profiles were found to be different for each histological type of tumor. Moreover, unsupervised analysis of medulloblastomas with self-organizing maps led to the hypothesis that there is a subclass of tumors that is distinct in having high levels of ribosomal gene expression and biogenesis. This hypothesis is now being tested in the laboratory in tumors that rely on the PI3K signaling pathway and ribosome biogenesis [53].

Differential gene expression does not necessarily imply causality and is, therefore, only the first of many steps towards identification of relevant therapeutic targets. Molecular prognostic markers have also been used to predict clinical outcome [42,47,54–56]. For example, in the early 1990s, northern blot analysis was used to show that TrkC can be highly expressed in medulloblastomas with favorable prognosis, serving as an independent marker of clinical outcome [54–56]. Moreover, TrkC activation induces apoptosis of medulloblastomas *in vitro* and slows the growth of medulloblastoma xenografts [54]. Amplification and overexpression of the proto-oncogene *MYCC* is associated with 17p loss and has been linked to poorer prognosis [57–59]. It is also thought that loss of 17p loss may be associated with loss of function of tumor suppressor genes, such as the hypermethylated in cancer-1 (*HIC-1*) gene [60,61]. In medulloblastomas associated with metastases, there is upregulation of the platelet-derived growth factor receptor and members of the RAS/mitogen-activated protein kinase signal transduction pathway [43].

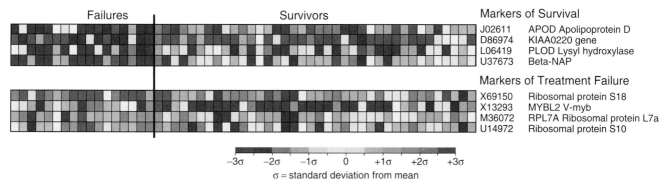

FIGURE 5.4 Signal to noise rankings of genes comparing tumors from surviving patients to those that experienced treatment failure. (J Clin Oncol, Fernandez-Teijeiro *et al.* (2004) Copyright American Society of Clinical Oncology. Reproduced with permission.) See Plate 5.4 in Color Plate Section.

There is also new evidence that overexpression of ERBB2, which encodes for class I receptor tyrosine kinases, predicts poor survival [62]. Many of these studies, however, were performed with small sample sizes and still need to be validated in large-scale institutional studies. For these reasons, it is now clear that all future diagnostic tools for risk stratification must incorporate clinical features and histologic subtype as well as the expression of specific genes and proteins.

Microarray analysis can also be used to identify gene expression profiles that indicate an activated regulatory pathway or multiple interacting processes that lead to a known cellular response. Medulloblastomas are known to express specific markers which indicate that they usually arise from granule cell progenitors [47,63–65]. For example, mutations of the Sonic hedgehog (Shh) receptor *patched* (PTCH) have been linked both to proliferation of granule cell progenitors in the developing cerebellum and to medulloblastoma oncogenesis [66–68]. Mice heterozygous for mutation of *Ptc* develop medulloblastomas in a multi-step process, first manifested as persistent proliferation of granule cell progenitors that then progress to overt tumor growth [54,69,70]. Children heterozygous for germline *PTCII* mutations (Gorlin syndrome) account for less than 1 per cent of medulloblastomas, but mutations of *PTCH* and related molecules are linked by gene expression profiling to the desmoplastic subclass of medulloblastomas (approximately 25 per cent of tumors) [42].

Mutations of *PTCH* and other molecular signaling pathways provide insight into the molecular pathogenesis of medulloblastomas as well as targets for the development of small molecule inhibitors. In a mouse model of spontaneous medulloblastoma, small molecule inhibitors of the Shh pathway eliminated tumor and reduced cell proliferation [71]. The mechanism of action is thought to be through downstream target inhibition and could be monitored by measuring dose-dependent gene downregulation resulting from inhibition of downstream effectors.

For several childhood cancers, such as neuroblastoma and leukemia, therapy is adapted according to a combined assessment of disease status and molecular profile [72,73]. In medulloblastomas, genomic methods that look beyond the expression and mutation of single genes have begun to reveal a biological subclassification based on gene-expression profiles that define invasive growth, response to therapy, and linkage to the Sonic hedgehog pathway [42,43]. Moreover, gene expression profiles were found to predict clinical outcome and survival with greater accuracy than current clinical risk criteria and outcome predictors or by single gene outcome predictors [42]. Gene expression differences not only were predictive of survival, but also successfully classified classic and desmoplastic medulloblastomas (Fig. 5.4) [65].

TUMOR EXPRESSION PROFILING: GLIOMA

Currently, patients with the most common form of glioma in adults, glioblastoma, have a median survival of twelve months after diagnosis. These dismal statistics remain remarkably stable, despite aggressive surgical resection, radiation therapy, and current forms of chemotherapy [74]. The present classification schema for gliomas is based on pathologic and microscopic criteria; this is based on the assumption that tumor cells share similarity to a presumed neural or glial precursor (Fig. 5.5). According to this schema, tumors that are considered less aggressive share more similarity with their cells of origin and normal tissue counterparts; by contrast, tumors that are more malignant, such as glioblastoma, share features of less differentiated precursor cells.

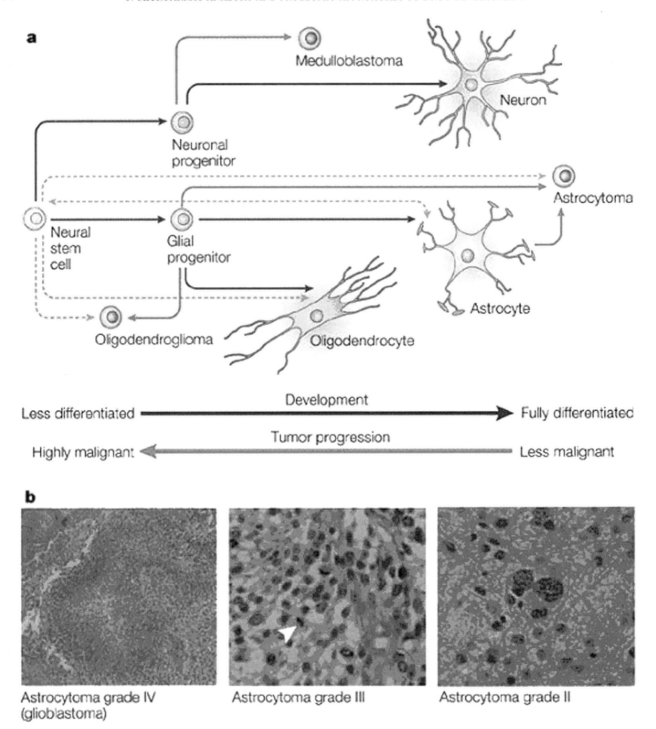

FIGURE 5.5   Classification scheme for brain tumors. (A) The present classification scheme for brain tumors. This classic model is based on the assumption that tumor cells of a specific lineage share microscopic similarity to a presumed neural or glial precursor. The black arrows indicate the hypothesized normal development and the red arrows indicate the hypothesized cell of origin of CNS tumors. (B) According to this scheme, less malignant tumors resemble their normal tissue counterparts; more malignant tumors resemble less differentiated precursor cells. Tumors are graded on the basis of the extent of anaplasia and other microscopic features that connote aggressive behavior — such as mitotic activity, tumor necrosis and angiogenesis. The white arrow points to a mitotic figure. (Nature Rev Neurosci, Mischel *et al.* (2004). Copyright Nature Publishing Group. Reproduced with permission). See Plate 5.5 in Color Plate Section.

Therefore, tumors are graded according to features such as the number of mitoses, presence or absence of neovascularization and necrosis, nuclear atypia, and mitotic activity. The intra- and inter-patient tumor heterogeneity has made this pathologic classification difficult. As with the prior example for medulloblastoma, this system is useful for stratifying patients according to histopathological diagnosis, but is less useful for identifying clinically relevant subsets of patients that might differ both in their clinical course and in their responsiveness to therapies.

Aberrations of signaling pathways have also been implicated in the pathogenesis and progression of glial tumors. Constitutive activation of the phosphatidylinositol 3-kinase and the Ras-MAPK (mitogen-activated protein kinase) signaling pathways have been shown to promote tumor formation and progression in mouse genetic models [75]. Patient-derived tumor samples have confirmed chronic activation of these pathways in correlative experiments [76,77]. As with medulloblastoma, it is hoped that subclassification of molecular subtypes of gliomas will then permit more specific targeting by biologic agents that target these abnormal pathways. The biggest challenge in the utility of such small molecule inhibitors is that brain tumors are both histologically and genetically heterogeneous. For example, the model for use of biologic agents as inhibitors of specific pathways is chronic myelogenous leukemia (CML), in which the target is a constitutively activated BCR-ABL breakpoint cluster region. The ABL-kinase inhibitor imatinib mesylate (Gleevac, STI-571, Novartis) promotes remission in up to 95 per cent of patients with CML [78,79]. For brain tumors, the molecular target may ultimately be present in only a small subset of pathologically identical tumors, which would complicate clinical trials substantially [80]. Moreover, if multiple molecular abnormalities are present, specific targeting of one step in the pathway may not take into account downstream modifiers or downstream mutations involving the same pathway [81].

Molecular Subsets of Brain Tumors

DNA microarray analyses have shown that different subtypes of gliomas have distinct gene-expression profiles, which can be distinguished from one another and from normal tissue [82–86]. Specifically, these differences involve pathways crucial to cell proliferation, energy metabolism, and signal transduction [86]. The genetic profile of low grade astrocytomas shows most frequently alterations in *p53* expression and in loss of heterozygosity on chromosome 17, which correlates with overexpression of PDGFR-α[87]. Overexpression of *c-myc* has also been detected both by immunohistochemistry [88] and by DNA microarrays [83]. In children, low grade astrocytomas show less elevation of components of the EGFR-FKB12-KIF2α (EGFR-FK506 binding protein 1A-hypoxia-inducible factor 2, α subunit) pathway when compared to high-grade tumors [89].

The distinction between high-grade glial and oligodendroglial tumors is important, because of the very different prognosis and responsiveness to chemotherapy. Gene-expression analyses of glial tumors has revealed molecular distinctions between low-grade astrocytomas, oligodendrogliomas, and glioblastomas (Fig. 5.6) [86,90]. Moreover, primary glioblastomas, which are thought to arise *de novo*, are distinct from secondary glioblastomas, which arise from lower grade gliomas. In adults, global expression profiling identified differences in the expression level of 360 genes between low grade tumors and glioblastoma; many of these genes encode proteins involved in cell proliferation or cell migration [84]. Subsets of glioblastoma multiforme have also been identified based on the expression of EGFR and on the overexpression of chromosome 12q13–15 [91].

Correlation with patient survival is difficult, however, because of the short survival time of patients with glioblastoma; however, gene-expression profiling was shown to be more accurate than pathologic grading in distinguishing anaplastic oligodendrogliomas and glioblastomas for prediction of survival [92]. This suggests that transcriptional information could be used to develop predictive molecular diagnostics [82]. These distinctions and prediction models were also cross-validated in comparison to normal tissue [86]. Again, as with medulloblastoma, the identification of clinically relevant subsets of patients based on gene-expression-based grouping of tumors is a more powerful predictor of survival than subtype of tumor pathology, tumor grade, or patient age (see Fig. 5.6) [90,91].

DRUG DISCOVERY

Rational methods of drug discovery in the past have typically used a biochemical approach. Target enzymes were identified — preferably the rate limiting step in a particular pathway — and then characterized and purified. Most often, these experiments were performed using animal tissue. Occasionally, enzymes could be targeted individually if enough information was known about mechanism of action and structure. Similar approaches were

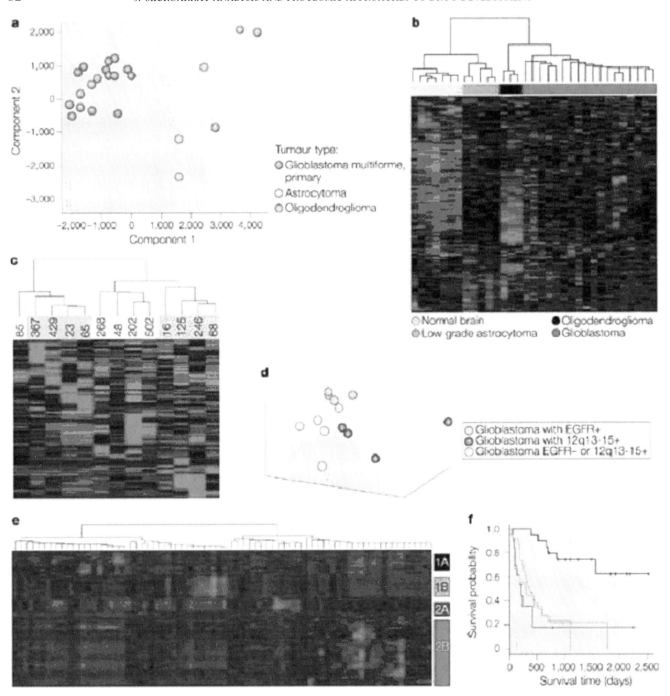

FIGURE 5.6 DNA-microarray analyses can identify relevant clinical subsets of gliomas. **A** and **B** show that different subtypes of gliomas have distinct gene-expression profiles. **C** and **D** show identification of molecular subsets of microscopically identical glioblastomas. **E** and **F** show the detection of clinically relevant, previously undetected subsets of patients with high-grade gliomas that have significantly different survival times; these genes can identify the subset of patients who are most likely to have prolonged survival. (Nature Rev Neurosci, Mischel *et al.* (2004). Copyright Nature Publishing Group. Reproduced with permission). See Plate 5.6 in Color Plate Section.

employed for identification and isolation of individual receptor subtypes as therapeutic targets. Biochemical techniques to identify drug-binding proteins rely on the affinity and specificity of small molecules to their protein targets; this affinity is quite low in some cases, leading to difficulty in specifying the primary binding partners. These often-laborious experimental approaches resulted in many effective therapies for a variety of diseases. Advances in molecular biology, however, have accelerated drug discovery, particularly in the field of oncology.

The new era of molecular biology and gene cloning techniques has enabled rapid advances in the field of drug discovery. Use of animal models and animal tissue has been helpful for initial studies; however, small variations in protein structure and amino acid sequences may render compounds identified in animal experiments ineffective for treating human disease [93]. The availability of human genes as targets through the efforts of the Human Genome Project has opened a new arena for high throughput testing of potential drug targets from human tissues. Gene expression and profiling can be used at all stages of discovery and development of therapeutic agents (Fig. 5.7). Gene sequencing and cloning can also permit production of targets that are difficult to isolate from natural sources. Site-directed mutagenesis can be utilized to test hypotheses regarding

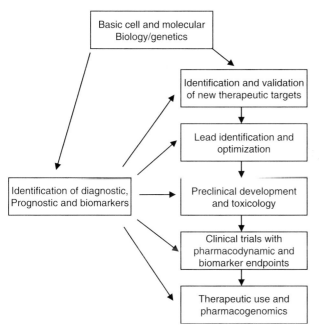

FIGURE 5.7 Various phases in the discovery and development of therapeutic agents and of diagnostic, prognostic and other biomarkers. Microarray expression profiling can be used to help advance all stages of this process (Adapted with permission from Biochem Pharmacol, Clarke *et al.*, 2001) [109].

target–drug interactions. Microarray analysis has now evolved to help both identify and validate potential novel therapeutic targets in a high throughput manner. This has greatly facilitated the testing of new biologic agents and the discovery of known and unpredicted drug targets.

New drug therapies identified with genomic methodologies have the potential to target molecular mechanisms of disease while avoiding side effects that arise from the activation or inhibition of normal molecular mechanisms within the body that do not have a role in the disease process. Selective tissue expression is attractive for use as drug targets, as this may reduce or eliminate unwanted and unexpected side effects.

Methodologies such as the DNA microarrays, which permit measurement of differential levels of gene expression of thousands of genes, may also provide an ideal framework for measuring how a compound effects metabolism, function and regulation on a much larger — e.g., genomic — scale [14,39].

For example, a combined approach of affinity purification, followed by array-based transcription profiling was used to identify candidate proteins of a novel class of anticancer agents [94]. This also permitted evaluation of the transcriptional effects of these agents on the tumor cells, specifically of sets of genes involved in cell metabolism, cell cycle control, and the immune response. This process highlighted a number of genes that were down-regulated in a time-dependent manner when treated with a pharmacologically relevant and clinically achievable drug concentration [94]. This integrated strategy led to the identification of drug-binding targets of biologically active small molecules.

Therapy for brain tumors based on small molecule inhibitors of growth factor receptors or their downstream targets can be individualized based on the results of gene expression patterns. This relies on microarray studies that demonstrate the presence of, and activity of, the intended target. Using yeast as a model system, drug inhibition of target genes can be compared to inactivation of the target by mutation [95]. Mutant strains with the most similar expression profile are then selected and treated to determine the 'ideal drug' with optimal specificity. These methods can be used to identify both co-regulated genes and unexpected target genes or 'off-target' effects. This approach was used to assess the calcineurin signaling pathway, which is inhibited by FK506 or cyclosporine A [95].

Microarrays are also useful in determination of the relationship between structure and activity by comparing active and inactive derivatives of novel drugs [96]. Using the induction of gene expression as a measure of drug function is useful for preclinical investigation of novel therapeutic agents to select newer drugs that have molecular selectivity.

GENE EXPRESSION AS A PREDICTOR OF CHEMOSENSITIVITY

As shown above, DNA microarrays have been quite useful both in subtyping of tumors as well as for outcome prediction, when correlated with information on patient survival and tumor biology. One of the next obvious applications is for prediction of therapeutic responsiveness. Several model systems are in place to address this important goal. The inhibitory activity of approximately 70 000 compounds has been assessed against a panel of 60 human tumor cell lines from the National Cancer Institute; the effects of several of these agents have been compared to the expression of 140 different proteins and this testing now has been extended to incorporate gene expression profiling [97–99]. Panels of cancer cell lines can be screened *in vitro* for chemosensitivity to a large number of anti-cancer agents [100,101] and further application for clinical use has been shown in preliminary studies of non-brain tumors, such as lymphoma and breast cancer [102,103]. Data related to cytotoxicity is now available by the clustering of drug sensitivity data and gene expression data following exposure to the various compounds [98]. These data were further utilized to evaluate the relationship between drug activities in response to gene expression.

GENE EXPRESSION IN RESPONSE TO DRUG THERAPY

Gene expression studies have also been utilized to investigate the mechanism of drug action and interactions. Molecular markers can be used as diagnostic tools to measure effects of drug exposure [104]. For example, novel agents can lead to changes in gene expression and protein expression patterns [96]. It is often difficult to determine the difference between effects on a drug target and the results of inhibition of target activity. In addition, there may be genetic variations in the metabolism, absorption, transport, kinetics, and adverse effects of drugs. For patients with acute lymphocytic leukemia, this application was used to compare pre- and post-treatment gene-expression profiles in patients who were treated with two different chemotherapeutic regimens. Though

combination therapy did not result in an additive response in terms of gene expression, there were profound differences in transcriptional response compared to each regimen when used alone [105]. One other example is shown in the alterations of the cytochrome p450 system, through which many chemotherapeutic agents are metabolized. Other agents induce enzyme activity through the cytochrome p450 system, either directly by enhanced enzymatic function or through competition for binding sites. This is particularly crucial for many patients with brain tumors who are treated with both chemotherapy and antiepileptic medications, which often compete for the same enzymes and can alter each other's effective serum concentrations. Microarray studies can be utilized not only to assess expression of different enzyme subtypes but also to help predict adverse side effects, such as the upregulation of hepatic enzymes [106,107]. These induced markers of gene products can be used as surrogate markers to follow the effect and dosages of drugs in the clinical setting. This is particularly useful when assessment of drug levels is difficult to achieve or unstable.

The development of drug resistance in response to antineoplastic chemotherapy remains an obstacle to successful treatment. Genome-wide studies with DNA/tissue microarrays and proteomic datasets have been used to identify candidate genes and proteins involved in the development of chemosensitivity or drug resistance [108]. These are also used to correlate drug response phenotypes in order to predict responsiveness to therapy. These validated markers will ideally facilitate not only the identification of novel therapies that bypass the development of drug resistance, but will also permit selection of optimal treatment regimens for individual patients. The merging of DNA microarray technology and proteomics will help to establish the links between gene expression profiles, mechanism of action of different drugs, and establishment of toxicity and toxic endpoints.

SUMMARY

Gene expression patterns of human tumor types have been studied using cDNA and protein microarray patterns; in many cases, new tumor subtypes have been identified based on this molecular classification. These initial findings still need to be validated using conventional methods of analysis and to search for clinical correlates. Microarray data is most useful when it can be correlated with informative clinical, imaging, and histologic data to create a genotype–phenotype connection.

This new categorization may lead to the development of improved and better-targeted therapeutics to distinct tumor types. One of the future challenges involved in targeted drug development is that there is a complex series of interactions in both the normal and diseased states. Both genes and gene products function in large complex systems; clearer understanding of the dysfunction in the pathways responsible for the disease will permit selection of the ideal drug target to compensate for the system malfunction.

DNA and protein microarrays continue to assume an increasingly larger role in all phases of drug discovery and development. Identification of potential targets by creation of molecular phenotypes, followed by profiling of drug effects and optimization will lead to clarification of the mechanisms of action and discovery of novel agents. These technologies will be crucial for pre-clinical drug development. Moreover, they will permit identification of the subset of patients in whom the targets are present and who make the best candidates. These technologies will likely have a substantial impact on the screening of chemical libraries and on the development of novel therapeutics. It will be important to take the microarray data back to the laboratory to determine targets that are important in disease and most suitable for drug development. This will take the approach to a disease involving small islands of genes to one that connects the genetic abnormalities to one cohesive story that explains the pathophysiology and permits identification of drug targets.

References

1. Sturla, L. M., Fernandez-Teijeiro, A., and Pomeroy, S. L. (2003). Application of microarrays to neurological disease. *Arch Neurol* **60**, 676–682.
2. Whitney, L. W., Ludwin, S. K., McFarland, H. F., and Biddison, W. E. (2001). Microarray analysis of gene expression in multiple sclerosis and EAE identifies 5-lipoxygenase as a component of inflammatory lesions. *J Neuroimmunol* **121**, 40–48.
3. Bitsch, A., and Bruck, W. (2002). Differentiation of multiple sclerosis subtypes: implications for treatment. *CNS Drugs* **16**, 405–418.
4. Goodin, D. S., Frohman, E. M., Garmany, G. P. Jr. *et al.* (2002). Disease modifying therapies in multiple sclerosis: report of the Therapeutics and Technology Assessment Subcommittee of the American Academy of Neurology and the MS Council for Clinical Practice Guidelines. *Neurology* **58**, 169–178.
5. Ginsberg, S. D., Hemby, S. E., Lee, V. M., Eberwine, J. H, and Trojanowski, J. Q. (2000). Expression profile of transcripts in Alzheimer's disease tangle-bearing CA1 neurons. *Ann Neurol* **48**, 77–87.
6. Luthi-Carter, R., Strand, A., Peters, N. L. *et al.* (2000). Decreased expression of striatal signaling genes in a mouse model of Huntington's disease. *Hum Mol Genet* **9**, 1259–1271.
7. Mirnics, K., Middleton, F. A., Lewis, D. A., and Levitt, P. (2001). Analysis of complex brain disorders with gene expression microarrays: schizophrenia as a disease of the synapse. *Trends Neurosci* **24**, 479–486.
8. Alwine, J. C., Kemp, D. J., and Stark, G. R. (1977). Method for detection of specific RNAs in agarose gels by transfer to diazobenzyloxymethyl-paper and hybridization with DNA probes. *Proc Natl Acad Sci USA* **74**, 5350–5354.
9. Liang, P., and Pardee, A. B. (1992). Differential display of eukaryotic messenger RNA by means of the polymerase chain reaction. *Science* **257**, 967–971.
10. Okubo, K., Hori, N., Matoba, R. *et al.* (1992). Large scale cDNA sequencing for analysis of quantitative and qualitative aspects of gene expression. *Nat Genet* **2**, 173–179.
11. Adams, M. D., Kelley, J. M., Gocayne, J. D. *et al.* (1991). Complementary DNA sequencing: expressed sequence tags and human genome project. *Science* **252**, 1651–1656.
12. Velculescu, V. E., Zhang, L., Vogelstein, B., and Kinzler, K. W. (1995). Serial analysis of gene expression. *Science* **270**, 484–487.
13. Schena, M., Shalon, D., Davis, R. W., and Brown, P. O. (1995). Quantitative monitoring of gene expression patterns with a complementary DNA microarray. *Science* **270**, 467–470.
14. Schena, M., Shalon, D., Heller, R., Chai, A., Brown, P. O., and Davis, R. W. (1996). Parallel human genome analysis: microarray-based expression monitoring of 1000 genes. *Proc Natl Acad Sci USA* **93**, 10614–10619.
15. Jacob, A., Brandt, O., Stephan, A. *et al.* (2004). Peptide nucleic acid microarrays. *Methods Mol Biol* **283**, 283–293.
16. Maughan, N. J., Lewis, F. A., and Smith, V. (2001). An introduction to arrays. *J Pathol* **195**, 3–6.
17. Duggan, D. J., Bittner, M., Chen, Y., Meltzer, P., and Trent, J. M. (1999). Expression profiling using cDNA microarrays. *Nat Genet* **21**, 10–14.
18. Ramaswamy, S., Tamayo, P., Rifkin, R. *et al.* (2001). Multiclass cancer diagnosis using tumor gene expression signatures. *Proc Natl Acad Sci USA* **98**, 15149–15154.
19. Lewis, F., Maughan, N. J., Smith, V., Hillan, K., and Quirke, P. (2001). Unlocking the archive–gene expression in paraffin-embedded tissue. *J Pathol* **195**, 66–71.
20. Chen, G., Gharib, T. G., Huang, C. C. *et al.* (2002). Proteomic analysis of lung adenocarcinoma: identification of a highly expressed set of proteins in tumors. *Clin Cancer Res* **8**, 2298–2305.
21. Hanash, S. M. (2001). Global profiling of gene expression in cancer using genomics and proteomics. *Curr Opin Mol Ther* **3**, 538–545.
22. Hanash, S. (2003). Disease proteomics. *Nature* **422**, 226–232.
23. Petricoin, E. F., Zoon, K. C., Kohn, E. C., Barrett, J. C., and Liotta, L. A. (2002). Clinical proteomics: translating benchside promise into bedside reality. *Nat Rev Drug Discov* **1**, 683–695.
24. Evans, G., Wheeler, C. H., Corbett, J. M., and Dunn, M. J. (1997). Construction of HSC-2DPAGE: a two-dimensional gel electrophoresis database of heart proteins. *Electrophoresis* **18**, 471–479.
25. Li, C., Chen, Z., Xiao, Z. *et al.* (2003). Comparative proteomics analysis of human lung squamous carcinoma. *Biochem Biophys Res Commun* **309**, 253–260.
26. Joubert-Caron, R., and Caron, M. (2005). Proteome analysis in the study of lymphoma cells. *Mass Spectrom Rev* **24**, 455–468.
27. Alexe, G., Alexe, S., Liotta, L. A., Petricoin, E., Reiss, M., and Hammer, P. L. (2004). Ovarian cancer detection by logical analysis of proteomic data. *Proteomics* **4**, 766–783.

28. Pellois, J. P., Zhou, X., Srivannavit, O., Zhou, T., Gulari, E., and Gao, X. (2002). Individually addressable parallel peptide synthesis on microchips. *Nat Biotechnol* **20**, 922–926.

29. Paweletz, C. P., Charboneau, L., Bichsel, V. E. *et al.* (2001). Reverse phase protein microarrays which capture disease progression show activation of pro-survival pathways at the cancer invasion front. *Oncogene* **20**, 1981–1989.

30. Robinson, W. H., DiGennaro, C., Hueber, W. *et al.* (2002). Autoantigen microarrays for multiplex characterization of autoantibody responses. *Nat Med* **8**, 295–301.

31. Sorlie, T., Perou, C. M., Tibshirani, R. *et al.* (2001). Gene expression patterns of breast carcinomas distinguish tumor subclasses with clinical implications. *Proc Natl Acad Sci USA* **98**, 10869–10874.

32. Perou, C. M., Sorlie, T., Eisen, M. B. *et al.* (2000). Molecular portraits of human breast tumours. *Nature* **406**, 747–752.

33. Golub, T. R., Slonim, D. K., Tamayo, P. *et al.* (1999). Molecular classification of cancer: class discovery and class prediction by gene expression monitoring. *Science* **286**, 531–537.

34. Hacia, J. G. (1999). Resequencing and mutational analysis using oligonucleotide microarrays. *Nat Genet* **21**, 42–47.

35. Ramaswamy, S., and Golub, T. R. (2002). DNA microarrays in clinical oncology. *J Clin Oncol* **20**, 1932–1941.

36. Hanahan, D., and Weinberg, R. A. (2000). The hallmarks of cancer. *Cell* **100**, 57–70.

37. Strausberg, R. L. (2001). The Cancer Genome Anatomy Project: new resources for reading the molecular signatures of cancer. *J Pathol* **195**, 31–40.

38. Alizadeh, A. A., Eisen, M. B., Davis, R. E. *et al.* (2000). Distinct types of diffuse large B-cell lymphoma identified by gene expression profiling. *Nature* **403**, 503–511.

39. DeRisi, J., Penland, L., Brown, P. O. *et al.* (1996). Use of a cDNA microarray to analyse gene expression patterns in human cancer. *Nat Genet* **14**, 457–460.

40. Marx, J. (2000). Medicine. DNA arrays reveal cancer in its many forms. *Science* **289**, 1670–1672.

41. Ho, I. A., Lam, P. Y., and Hui, K. M. (2004). Identification and characterization of novel human glioma-specific peptides to potentiate tumor-specific gene delivery. *Hum Gene Ther* **15**, 719–732.

42. Pomeroy, S. L., Tamayo, P., Gaasenbeek, M. *et al.* (2002). Prediction of central nervous system embryonal tumour outcome based on gene expression. *Nature* **415**, 436–442.

43. MacDonald, T. J., Brown, K. M., LaFleur, B. *et al.* (2001). Expression profiling of medulloblastoma: PDGFRA and the RAS/MAPK pathway as therapeutic targets for metastatic disease. *Nat Genet* **29**, 143–152.

44. Hanash, S. (2003). Harnessing immunity for cancer marker discovery. *Nat Biotechnol* **21**, 37–38.

45. Madoz-Gurpide, J., Wang, H., Misek, D. E., Brichory, F., and Hanash, S. M. (2001). Protein based microarrays: a tool for probing the proteome of cancer cells and tissues. *Proteomics* **1**, 1279–1287.

46. Sorlie, T., Tibshirani, R., Parker, J. *et al.* (2003). Repeated observation of breast tumor subtypes in independent gene expression data sets. *Proc Natl Acad Sci USA* **100**, 8418–8423.

47. Pomeroy, S. L., and Sturla, L. M. (2003). Molecular biology of medulloblastoma therapy. *Pediatr Neurosurg* **39**, 299–304.

48. Wright, G., Tan, B., Rosenwald, A., Hurt, E. H., Wiestner, A., and Staudt, L. M. (2003). A gene expression-based method to diagnose clinically distinct subgroups of diffuse large B cell lymphoma. *Proc Natl Acad Sci USA* **100**, 9991–9996.

49. Giangaspero, F., Chieco, P., Ceccarelli, C. *et al.* (1991). 'Desmoplastic' versus 'classic' medulloblastoma: comparison of DNA content, histopathology and differentiation. *Virchows Arch A Pathol Anat Histopathol* **418**, 207–214.

50. Packer, R. J., Sutton, L. N., Elterman, R. *et al.* (1994). Outcome for children with medulloblastoma treated with radiation and cisplatin, CCNU, and vincristine chemotherapy. *J Neurosurg* **81**, 690–698.

51. Glauser, T. A., and Packer. R. J. (1991). Cognitive deficits in long-term survivors of childhood brain tumors. *Childs Nerv Syst* **7**, 2–12.

52. Packer, R. J., Goldwein, J., Nicholson, H. S. *et al.* (1999). Treatment of children with medulloblastomas with reduced-dose craniospinal radiation therapy and adjuvant chemotherapy: A Children's Cancer Group Study. *J Clin Oncol* **17**, 2127–2136.

53. Hidalgo, M., and Rowinsky, E. K. (2000). The rapamycin-sensitive signal transduction pathway as a target for cancer therapy. *Oncogene* **19**, 6680–6686.

54. Kim, J. Y., Nelson, A. L., Algon, S. A. *et al.* (2003). Medulloblastoma tumorigenesis diverges from cerebellar granule cell differentiation in patched heterozygous mice. *Dev Biol* **263**, 50–66.

55. Kim, J. Y., Sutton, M. E., Lu, D. J. *et al.* (1999). Activation of neurotrophin-3 receptor TrkC induces apoptosis in medulloblastomas. *Cancer Res* **59**, 711–719.

56. Segal, R. A., Goumnerova, L. C., Kwon, Y. K., Stiles, C. D., and Pomeroy, S. L. (1994). Expression of the neurotrophin receptor TrkC is linked to a favorable outcome in medulloblastoma. *Proc Natl Acad Sci USA* **91**, 12867–12871.

57. Herms, J., Neidt, I., Luscher, B. *et al.* (2000). C-MYC expression in medulloblastoma and its prognostic value. *Int J Cancer* **89**, 395–402.

58. Grotzer, M. A., Hogarty, M. D., Janss, A. J. *et al.* (2001). MYC messenger RNA expression predicts survival outcome in childhood primitive neuroectodermal tumor/medulloblastoma. *Clin Cancer Res* **7**, 2425–2433.

59. Scheurlen, W. G., Schwabe, G. C., Joos, S., Mollenhauer, J., Sorensen, N., and Kuhl, J. (1998). Molecular analysis of childhood primitive neuroectodermal tumors defines markers associated with poor outcome. *J Clin Oncol* **16**, 2478–2485.

60. Rood, B. R., Zhang, H., Weitman. D. M., and Cogen, P. H. (2002). Hypermethylation of HIC-1 and 17p allelic loss in medulloblastoma. *Cancer Res* **62**, 3794–3797.

61. Waha, A., Koch, A., Meyer-Puttlitz, B. *et al.* (2003). Epigenetic silencing of the HIC-1 gene in human medulloblastomas. *J Neuropathol Exp Neurol* **62**, 1192–1201.

62. Gajjar, A., Hernan, R., Kocak, M. *et al.* (2004). Clinical, histopathologic, and molecular markers of prognosis: toward a new disease risk stratification system for medulloblastoma. *J Clin Oncol* **22**, 984–993.

63. Aruga, J., Yokota, N., Hashimoto, M., Furuichi, T., Fukuda, M., and Mikoshiba, K. (1994). A novel zinc finger protein, zic, is involved in neurogenesis, especially in the cell lineage of cerebellar granule cells. *J Neurochem* **63**, 1880–1890.

64. Yokota, N., Aruga, J., Takai, S. *et al.* (1996). Predominant expression of human zic in cerebellar granule cell lineage and medulloblastoma. *Cancer Res* **56**, 377–383.

65. Fernandez-Teijeiro, A., Betensky, R. A., Sturla, L. M., Kim, J. Y., Tamayo, P., and Pomeroy, S. L. (2004). Combining gene expression profiles and clinical parameters for risk stratification in medulloblastomas. *J Clin Oncol* **22**, 994–998.

66. Wechsler-Reya, R. J., and Scott, M. P. (1999). Control of neuronal precursor proliferation in the cerebellum by Sonic Hedgehog. *Neuron* **22**, 103–114.

67. Hahn, H., Wojnowski, L., Specht, K. et al. (2000). Patched target Igf2 is indispensable for the formation of medulloblastoma and rhabdomyosarcoma. J Biol Chem 275, 28341–28344.

68. Gailani, M. R., Bale, S. J., Leffell, D. J. et al. (1992). Developmental defects in Gorlin syndrome related to a putative tumor suppressor gene on chromosome 9. Cell 69, 111–117.

69. Johnson, R. L., Rothman, A. L., Xie, J. et al. (1996). Human homolog of patched, a candidate gene for the basal cell nevus syndrome. Science 272, 1668–1671.

70. Goodrich, L. V., Johnson, R. L., Milenkovic, L., McMahon, J. A., and Scott, M. P. (1996). Conservation of the hedgehog/patched signaling pathway from flies to mice: induction of a mouse patched gene by Hedgehog. Genes Dev 10, 301–312.

71. Romer, J. T., Kimura, H., Magdaleno, S. et al. (2004). Suppression of the Shh pathway using a small molecule inhibitor eliminates medulloblastoma in Ptc1(+/−)p53(−/−) mice. Cancer Cell 6, 229–240.

72. Brodeur, G. M. (2003). Neuroblastoma: biological insights into a clinical enigma. Nat Rev Cancer 3, 203–216.

73. Rubnitz, J. E., and Pui, C. H. (2003). Recent advances in the treatment and understanding of childhood acute lymphoblastic leukaemia. Cancer Treat Rev 29, 31–44.

74. Mischel, P. S., and Cloughesy, T. F. (2003). Targeted molecular therapy of GBM. Brain Pathol 13, 52–61.

75. Holland, E. C., Celestino, J., Dai, C., Schaefer, L., Sawaya, R. E., and Fuller, G. N. (2000). Combined activation of Ras and Akt in neural progenitors induces glioblastoma formation in mice. Nat Genet 25, 55–57.

76. Ermoian, R. P., Furniss, C. S., Lamborn, K. R., et al. (2002). Dysregulation of PTEN and protein kinase B is associated with glioma histology and patient survival. Clin Cancer Res 8, 1100–1106.

77. Choe, G., Horvath, S., and Cloughesy, T. F., et al. (2003). Analysis of the phosphatidylinositol 3′-kinase signaling pathway in glioblastoma patients in vivo. Cancer Res 63, 2742–2746.

78. Druker, B. J. (2002). Perspectives on the development of a molecularly targeted agent. Cancer Cell 1, 31–36.

79. Sawyers, C. L. (2002). Disabling Abl-perspectives on Abl kinase regulation and cancer therapeutics. Cancer Cell 1, 13–15.

80. Betensky, R. A., Louis, D. N., and Cairncross, J. G. (2002). Influence of unrecognized molecular heterogeneity on randomized clinical trials. J Clin Oncol 20, 2495–2499.

81. Bianco, R., Shin, I., Ritter, C. A. et al. (2003). Loss of PTEN/MMAC1/TEP in EGF receptor-expressing tumor cells counteracts the antitumor action of EGFR tyrosine kinase inhibitors. Oncogene 22, 2812–2822.

82. Collins, V. P. (2004). Brain tumours: classification and genes. J Neurol Neurosurg Psychiatry 75 Suppl 2, ii2–11.

83. Huang, H., Colella, S., Kurrer, M., Yonekawa, Y., Kleihues, P., and Ohgaki, H. (2000). Gene expression profiling of low-grade diffuse astrocytomas by cDNA arrays. Cancer Res 60, 6868–6874.

84. Rickman, D. S., Bobek, M. P., Misek, D. E. et al. (2001). Distinctive molecular profiles of high-grade and low-grade gliomas based on oligonucleotide microarray analysis. Cancer Res 61, 6885–6891.

85. Sallinen, S. L., Sallinen, P. K., Haapasalo, H. K. et al. (2000). Identification of differentially expressed genes in human gliomas by DNA microarray and tissue chip techniques. Cancer Res 60, 6617–6622.

86. Shai, R., Shi, T., Kremen, T. J. et al. (2003). Gene expression profiling identifies molecular subtypes of gliomas. Oncogene 22, 4918–4923.

87. Hermanson, M., Funa, K., Koopmann, J. et al. (1996). Association of loss of heterozygosity on chromosome 17p with high platelet-derived growth factor alpha receptor expression in human malignant gliomas. Cancer Res 56, 164–171.

88. Orian, J. M., Vasilopoulos, K., Yoshida, S., Kaye, A. H., Chow, C. W., and Gonzales, M. F. (1992). Overexpression of multiple oncogenes related to histological grade of astrocytic glioma. Br J Cancer 66, 106–112.

89. Khatua, S., Peterson, K. M., Brown, K. M. et al. (2003). Overexpression of the EGFR/FKBP12/HIF-2alpha pathway identified in childhood astrocytomas by angiogenesis gene profiling. Cancer Res 63, 1865–1870.

90. Mischel, P. S., Cloughesy, T. F., and Nelson, S. F. (2004). DNA-microarray analysis of brain cancer: molecular classification for therapy. Nat Rev Neurosci 5, 782–792.

91. Mischel, P. S., Shai, R., Shi, T. et al. (2003). Identification of molecular subtypes of glioblastoma by gene expression profiling. Oncogene 22, 2361–2373.

92. Nutt, C. L., Mani, D. R., Betensky, R. A. et al. (2003). Gene expression-based classification of malignant gliomas correlates better with survival than histological classification. Cancer Res 63, 1602–1607.

93. Oksenberg, D., Marsters, S. A., O'Dowd, B. F. et al. (1992). A single amino-acid difference confers major pharmacological variation between human and rodent 5-HT1B receptors. Nature 360, 161–163.

94. Oda, Y., Owa, T., Sato, T. et al. (2003). Quantitative chemical proteomics for identifying candidate drug targets. Anal Chem 75, 2159–2165.

95. Marton, M. J., DeRisi, J. L., Bennett, H. A. et al. (1998). Drug target validation and identification of secondary drug target effects using DNA microarrays. Nat Med 4, 1293–1301.

96. Clarke, P. A., Hostein, I., Banerji, U. et al. (2000). Gene expression profiling of human colon cancer cells following inhibition of signal transduction by 17-allylamino-17-demethoxygeldanamycin, an inhibitor of the hsp90 molecular chaperone. Oncogene 19, 4125–4133.

97. Ross, D. T., Scherf, U., Eisen, M. B. et al. (2000). Systematic variation in gene expression patterns in human cancer cell lines. Nat Genet 24, 227–235.

98. Scherf, U., Ross, D. T., Waltham, M. et al. (2000). A gene expression database for the molecular pharmacology of cancer. Nat Genet 24, 236–244.

99. Weinstein, J. N., Myers, T. G., O'Connor, P. M. et al. (1997). An information-intensive approach to the molecular pharmacology of cancer. Science 275, 343–349.

100. Staunton, J. E., Slonim, D. K., Coller, H. A. et al. (2001). Chemosensitivity prediction by transcriptional profiling. Proc Natl Acad Sci USA 98, 10787–10792.

101. Wallqvist, A., Monks, A., Rabow, A. A. et al. (2003). Mining the NCI screening database: explorations of agents involved in cell cycle regulation. Prog Cell Cycle Res 5, 173–179.

102. Sotiriou, C., Powles, T. J., Dowsett, M. et al. (2002). Gene expression profiles derived from fine needle aspiration correlate with response to systemic chemotherapy in breast cancer. Breast Cancer Res 4, R3.

103. Bohen, S. P., Troyanskaya, O. G., Alter, O. et al. (2003). Variation in gene expression patterns in follicular lymphoma and the response to rituximab. Proc Natl Acad Sci USA 100, 1926–1930.

104. Guerreiro, N., Staedtler, F., Grenet, O., Kehren, J., and Chibout, S. D. (2003). Toxicogenomics in drug development. Toxicol Pathol 31, 471–479.

105. Cheok, M. H., Yang, W., Pui, C. H. *et al.* (2003). Treatment-specific changes in gene expression discriminate in vivo drug response in human leukemia cells. *Nat Genet* **34**, 85–90.

106. Smith, A. G., Davies, R., Dalton, T. P. *et al.* (2003). Intrinsic hepatic phenotype associated with the Cyp1a2 gene as shown by cDNA expression microarray analysis of the knockout mouse. *EHP Toxicogenomics* **111**, 45–51.

107. Sahi, J., Milad, M. A., Zheng, X. *et al.* (2003). Avasimibe induces CYP3A4 and multiple drug resistance protein 1 gene expression through activation of the pregnane X receptor. *J Pharmacol Exp Ther* **306**, 1027–1034.

108. Huang, Y., and Sadee, W. (2003). Drug sensitivity and resistance genes in cancer chemotherapy: a chemogenomics approach. *Drug Discov Today* **8**, 356–363.

109. Clarke, P. A., te Poele, R., Wooster, R., and Workman, P. (2001). Gene expression microarray analysis in cancer biology, pharmacology, and drug development: progress and potential. *Biochem Pharmacol* **62**(10), 1311–1336.

6

Chemotherapy Resistance

Adrienne C. Scheck

As described in Chapter 8 of this book, the molecular changes that occur during the formation and progression of malignant gliomas vary, leading to the phenotypic and genotypic heterogeneity demonstrated by numerous investigators [1–7]. Chemotherapy resistance can come about in a number of ways. The process by which a tumor cell becomes resistant to a particular chemotherapeutic agent is a function of the type of therapy and the phenotype of the particular tumor cell. The fact that some tumors do not respond to chemotherapy, and others rapidly recur and are often refractory to further therapy with the same or similar agents suggests that within the heterogeneous primary tumor there are cells that are intrinsically resistant to therapy. This has been demonstrated by Shapiro and coworkers who showed that glioma cells selected for resistance to clinically achievable concentrations (blood plasma levels) of 1,3-bis(2-chloroethyl)-1-nitrosourea (BCNU) *in vitro*, or *in vivo* through the use of primary/recurrent glioma pairs are near-diploid, and carry a specific non-random karyotypic deviation that includes over-representation of chromosomes 7 and 22 [8,9]. These cells were present in the primary tumor as a minor subpopulation, but they became the dominant population following selection for BCNU resistance. Thus, therapy can select for cells that are genetically capable of resistance.

COMMON CHEMOTHERAPEUTIC AGENTS IN GLIOMAS

The most common class of chemotherapeutic agents used against malignant brain tumors is the alkylating agents. These include chloroethylating agents such as 1,3-bis(2-chloroethyl)-1-nitrosourea (BCNU, carmustine), CCNU, ACNU, and methylators such as procarbazine, and temozolomide (TMZ). These compounds alkylate DNA at multiple positions; however, the most important modification is thought to be alkylation of the O^6 position of guanine (Fig. 6.1). This lesion can lead to a complex range of effects and the major mechanism of resistance involves removal of these DNA adducts, particularly O^6-alkylguanine, by O^6-methylguanine methyltransferase (MGMT, also called O^6-alkylguanine-DNA alkyltransferase, AGT) [10–15]. If not removed, these adducts can lead to a number of secondary lesions that result in genotoxicity due to mispairing by the mismatch repair system, mutation when the second round of replication uses the strand containing "T" as template and the formation of DNA interstrand cross-links which require repair by nucleotide excision factors [10,16,17].

Perhaps, the second most common class of therapeutic agents used against brain tumors are the platinum compounds, especially cisplatin and related drugs such as carboplatin. Cisplatin binds to the N^7 atom of guanine, particularly in regions of 2 consecutive guanines (GG) or a guanine followed by an adenine (GA) where they cause intrastrand and interstrand diadducts. These are repaired primarily by nucleotide excision repair (NER) and the mismatch repair (MMR) pathways (described below), and cells deficient in either of these are often resistant to these drugs [18–23]. Recent work in other tumor systems has revealed several interactions that may point to new mechanisms of cisplatin resistance in brain tumors. Activation of the AKT pathway downregulates apoptosis and increases cisplatin resistance in ovarian cancer cells [24]. In addition, RNAi-mediated knockdown of PTEN increased resistance, and treatment with PI3K inhibitors LY294002 and wortmannin

89

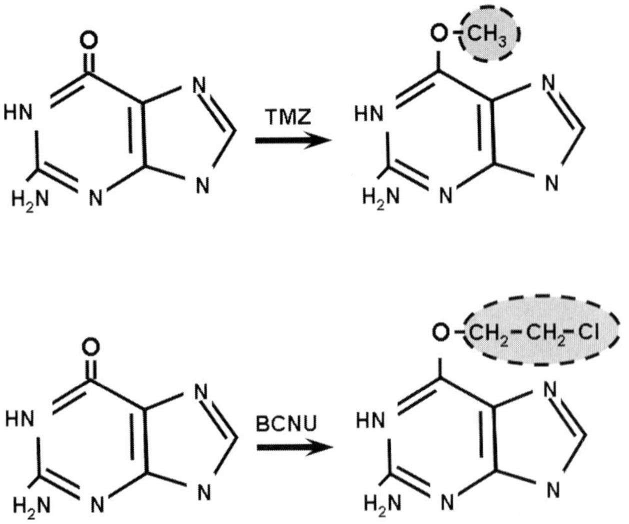

FIGURE 6.1 Modification of the O^6 position of guanine by TMZ and BCNU.

sensitized the cells to cisplatin-induced apoptosis. Furthermore, adenovirus E1a increases sensitivity to cisplatin by blocking AKT activation [18]. Finally, AKT can cause an increase in NF-κB which leads to decreased apoptosis and cisplatin resistance [25]. While studies of this type have not yet been done in gliomas, investigations of PTEN inactivation in these tumors and design of new therapies targeting AKT and other signal transduction pathways are ongoing and are likely to lead to new methodologies to reduce resistance to cisplatin in brain tumors. To date, most studies in brain tumors have centered on DNA repair.

In addition to the alkylating agents, the topoisomerase inhibitors are being used in the treatment of some gliomas. CPT-11 (irinotecan), an inhibitor of topoisomerase I has shown efficacy against gliomas, particularly in combination with BCNU or temozolomide [26–30]. This prodrug crosses the blood–brain barrier, is converted by carboxylesterases to its active metabolite SN-38 which then interacts directly with topoisomerase I, stabilizing covalent topo I-DNA complexes and forming single strand DNA breaks [31]. Resistance to this drug may come about through a point mutation in topoisomerase I resulting in an enzyme with reduced activity or reduced expression of the enzyme [32,33] or through increased efflux of the drug mediated by the ATP-binding cassette system [18].

Etoposide (VP-16), an inhibitor of topoisomerase II was originally thought to inhibit microtubules. It is now known that this drug acts by stabilizing the cleavable complex that is formed when topoisomerase II interacts with the DNA, cuts both strands of DNA and forms covalent bonds with the ends. This complex is normally short-lived and the DNA breaks are resealed afer the DNA goes through rotation or strand passage. In the presence of etoposide the DNA strands cannot rotate, resulting in a persistent cleavable complex and DNA strand breaks [34,35]. Clinical trials have included VP-16 with a variety of other compounds in the treatment of both pediatric and adult brain tumors [35–45], and the main mechanism of resistance is thought to be related to increased efflux *via* ATP-binding cassette proteins including the multidrug-resistant-associated protein (MRP) [46,47].

DNA REPAIR AND CHEMOTHERAPY RESISTANCE

Methylguanine Methyltransferase

One of the most important, and most studied mechanisms of resistance to alkylating agents is methylguanine methyltransferase (MGMT). The MGMT enzyme transfers the alkyl group from the O^6 position of guanine to a cysteine residue in its active site. This repairs the DNA, prevents the formation of most additional secondary lesions, and causes irreversible inactivation of the MGMT molecule. Thus, MGMT-mediated resistance to alkylating agents is correlated to MGMT levels [17,48–52].

MGMT inactivation occurs *in vivo* through the methylation of CpG islands in the gene promoter. Thirty to eighty eight percent [53–56] of malignant gliomas are thought to have reduced or absent expression of MGMT due to promoter methylation. These tumors are often referred to as Mer⁻ or Mex⁻, and numerous studies have shown that cells from these tumors are typically more sensitive to various alkylating agents [50,56–59]. Furthermore, a number of studies have been done to determine if methylation of the MGMT promoter can predict patient response to therapy with alkylating agents including BCNU, temozolomide, ACNU, ENU, and procarbazine [49,53, 55,60,61]. In particular, recent work has suggested that MGMT promoter methylation is highly correlated with response to temozolomide, suggesting that promoter methylation may be useful for the identification of patients whose tumors are more likely to respond to BCNU or temozolomide [53,55,60,62,63]. However, other investigators have suggested that patients whose tumors have methylated MGMT may have a better prognosis irrespective of alkylating agent therapy, and some have reported that there is no correlation with response to particular alkylating agents such as CENU [54]. This may be due to a number of things including patient age, tumor grade, and tumor heterogeneity. In addition, while it is reasonable to assume that methylation of the MGMT promoter should lead to the absence of the MGMT protein, immunohistochemical analyses in medulloblastoma did not demonstrate agreement between the promoter methylation and protein data [64]. In contrast, work done by Zuo et al., [65] in head and neck squamous cell carcinoma did find a correlation between MGMT promoter methylation, reduced protein expression and decreased patient survival.

The lack of strict correlation between MGMT promoter methylation and reduced protein expression may be due, at least in part, to the heterogeneity inherent in gliomas. The data obtained from PCR-based analyses are an average of all the cells in the sample. Some authors even note the presence of unmethylated sequences in samples that demonstrate MGMT promoter methylation; however, they often attribute this to the contribution of contaminating normal cells. It is equally likely that the tumor is comprised of cells with and without MGMT promoter methylation. For this reason, immunohistochemical analyses in which tumor heterogeneity can be taken into account may in fact be better than methylation-specific PCR. The impact of this on the use of MGMT promoter methylation for clinical studies such as those recently reported remains to be seen [53,55, 60,62,63].

Methylation of the MGMT promoter may have additional ramifications in gliomas. While there are conflicting results regarding whether p53 status is directly involved in alkylating agent resistance, numerous investigators have tried to correlate mutations in p53 with alkylating agent resistance and MGMT activity [52,59,66–69]. MGMT promoter methylation has been correlated to p53 mutations, particularly G:C → A:T, perhaps due to the cell's reduced capacity for repair [70–73]. In addition, MGMT promoter methylation correlates with younger age in patients with glioblastomas multiforme (GBMs). These patients typically have a better prognosis than older patients and this provides a potential link between GBMs that arise in these younger patients, the so-called secondary GBMs, and the p53 mutations which are a frequent finding in these tumors. In addition, the methylation status of this gene has been correlated to deletion of chromosomes 1p and 19q in oligodendrogliomas, suggesting that this may play a role in sensitivity to chemotherapy seen in these tumors; however, there is no known causal relationship between 1p/19q deletion and MGMT promoter methylation at this time [56].

The data demonstrating a correlation between MGMT promoter methylation and alkylating agent response suggests that reducing MGMT activity may sensitize tumor cells to therapy with alkylating agents such as BCNU and temozolomide. O^6-benzylguanine (O^6-BG) is an inhibitor of MGMT that sensitizes resistant cells to BCNU by binding to the active site of MGMT [74–77]. Inhibition of MGMT activity by O^6-BG potentiates the effects of BCNU in vitro and in tumor xenografts in a number of tumor systems including gliomas [78–89]. The results of dose escalation studies using O^6-BG to sensitize patients to treatment with alkylating agents have been published for adults and children, and O^6-BG in combination with BCNU or TMZ is currently in phase II clinical trials in adults [90–96].

The effectiveness of any drug regimen depends not only on the intrinsic resistance shown by the tumor cells, but also on the possibility of acquired resistance that can occur during therapy. To identify mechanisms of resistance that could arise during treatment, a number of studies have been conducted in vitro [97–101]. This work has demonstrated that MGMT resistance to the inhibitory effects of O^6-BG can arise due to mutations in the binding site of MGMT [97,98]. Furthermore, work by Ueda et al., [102] showed that dexamethasone induces MGMT, most likely through 2 glucocorticoid-response elements located in the gene promoter. Thus, studies designed to increase alkylating agent efficacy by depleting MGMT should consider the drug interactions that occur not only between chemotherapeutic agents, but also between chemotherapeutic agents and other medications. In addition, depletion of MGMT may require combinations of multiple inhibitors such as O^6-BG and/or other derivatives and streptozotocin [82,103].

DNA Mismatch Repair

Depletion of MGMT activity by promoter methylation or the use of inhibitors is not always sufficient to ensure alkylating agent sensitivity [52]. In the absence of functional MGMT, an important additional mechanism of resistance to alkylating agents is DNA mismatch repair (MMR) deficiency. At least 5 proteins organized as heterodimers are involved in MMR: MutSα (a heterodimer of hMSH2 and hMSH6), MutSβ (a heterodimer of hMSH2 and hMSH3) and MutLα (a heterodimer of hMLH1 and hPMS2). Mismatched heteroduplexes are recognized by MutSα or MutSβ which bind to the DNA. MutLα then binds to the complex and the segment of the DNA strand containing the mismatched base is excised. The gap is then filled in by DNA polymerase and ligation (Fig. 6.2) (for reviews see [19,104]).

MMR functions in resistance to alkylating agent chemotherapy when the O^6-methylguanine adduct is not repaired by MGMT prior to DNA replication, resulting in a O^6-meG:C base pair that is typically replicated as a O^6-meG:T. In the case of mismatch repair of O^6-meG:T the DNA gap is created but it cannot be correctly filled in, resulting in a "futile repair loop" that leads to additional DNA double strand breaks and apoptotic cell death [10,52]. Thus, a defect in mismatch repair can lead to alkylating agent resistance in a cell that lacks MGMT activity. It does not however, change the incidence of mutations that occur due to the insertion of a "T" in the daughter DNA strand. Thus, defective mismatch repair may also contribute to the genetic instability seen in these tumors by promoting additional mutational events.

Analysis of MMR in brain tumors and other cancers has included studies of the proteins, gene mutations, mRNA, and MMR activity. These studies have identified a link between MMR deficiency and resistance to a number of therapeutic agents. In fact, some studies suggest that intact MMR is required for cell death in MGMT deficient cells; however, the literature is not always consistent with respect to the contribution of MMR. For example, while a number of studies have established a relationship between resistance to alkylating agents such as BCNU and temozolomide with MMR, others have suggested that this relationship is true for some, but not all alkylating agents [10,57,66,70,87,104–112]. Similarly, some reports demonstrate a clear relationship between MMR deficiency and cisplatin resistance [20,22,106,113–115], others suggest that MMR may play a relatively small role in cisplatin resistance [104,116]. Houghton *et al.*, [58] used temozolomide in combination with irinotecan on mouse xenografts and found that while the 2 drugs did not appear to interact directly, their action was synergistic and was independent of *both* MGMT and mismatch repair.

Regulation of the genes involved in MMR occurs at a number of levels. Rellecke *et al.*, [108] demonstrated that all 5 MMR genes are transcribed in *de novo* GBMs prior to treatment. Gene transcription was approximately equal for all but hMSH2. The expression of this gene was high in resistant tumors. Srivastava *et al.*, [117] found higher expression of hMLH2 in high grade tumors when compared to low grade tumors, with a higher number of cells showing positive by immunohistochemistry. MSH2 and MSH6 were found to be regulated through post-translational

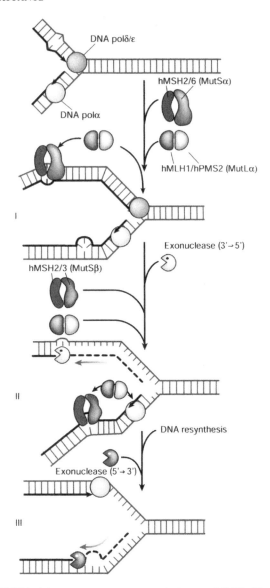

FIGURE 6.2 Mammalian Mismatch Repair (MMR). Heterodimers of hMSH2/6 (called hMutSα) focus on mismatches and single-base loops (stage I in the figure, upper strand), whereas hMSH2/3 dimers (hMutSβ) recognize insertion/deletion loops (II, lower strand). Heterodimeric complexes of the hMutL-like proteins hMLH1/hPMS2 (hMutLα) and hMLH1/hPMS1 (hMutLβ) interact with MSH complexes and replication factors. Strand discrimination may be based on contact with the nearby replication machinery. A number of proteins are implicated in the excision of the new strand past the mismatch and resynthesis steps, including polδ/ε, RPA, PCNA, RFC, exonuclease 1, and endonuclease FEN1 (II, III). MMR components also interact functionally with NER and recombination. Recent crystallographic studies have revealed that a MutS dimer detects the structural instability of a heteroduplex by kinking the DNA at the site of the mismatch, which is facilitated when base pairing is affected. However, DNA damage with similar characteristics, such as that caused by alkylating agents and intercalators, may fool MutS, triggering erroneous or futile MMR. (Reprinted from [122]). See Plate 6.2 in Color Plate Section.

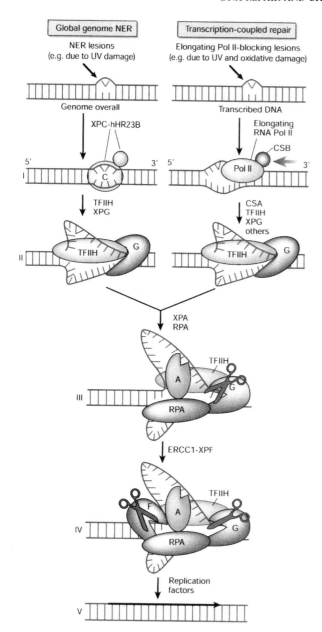

FIGURE 6.3 Nucleotide Excision Repair (NER). The global-genome (GG)-NER-specific complex XPC-hHR23B screens first on the basis of disrupted base pairing, instead of lesions per se. In transcription-coupled repair (TCR), the ability of a lesion (whether of the NER- or BER-type) to block RNA polymerase seems critical (stage I in this figure). The stalled polymerase must be displaced to make the injury accessible for repair, and this requires at least two TCR-specific factors: CSB and CSA. The subsequent stages of GG-NER and TCR may be identical. The XPB and XPD helicases of the multi-subunit transcription factor TFIIH open ~30 base pairs of DNA around the damage (II). XPA probably confirms the presence of damage by probing for abnormal backbone structure, and when absent aborts NER. The single-stranded-binding protein RPA (replication protein A) stabilizes the open intermediate by binding to the undamaged strand (III). The use of subsequent factors, each with limited capacity for lesion detection *in toto*, still allows very high damage specificity. The endonuclease duo of the NER team, XPG and ERCC1/XPF, respectively cleave 3' and 5' of the borders of

modification in that they are translocated from the cytoplasm to the nucleus following DNA damage, particularly in the absence of MGMT [107,118]. Finally, the gene encoding hMSH6 is thought to be regulated in part by methylation of CpG islands in the gene's promoter [110] in a manner similar to that seen for MGMT. In addition, hypermethylation of the hMLH1 gene promoter has recently been found to predict a patient's clinical response to nitrosoureas in much the same way as have been found for MGMT promoter methylation. Since cells with deficient MMR and MGMT tend to be drug resistant, the analysis of hMLH1 expression in combination with MGMT expression may provide a more robust clinical test to predict therapy resistance.

Nucleotide and Base Excision Repair

DNA damage that causes a distortion in the shape of the DNA helix such as those caused by intrastrand dimers can be repaired by nucleotide excision repair (NER) [120–122] (Fig. 6.3). There are many polypeptides involved in this repair system (for review see [18,122]). The DNA lesion is recognized by a polypeptide complex and a 27–29 nucleotide long region around the lesion is removed. The gap is then filled by a complex of DNA replication proteins and the repaired region is ligated to complete the patch [123,124]. NER is also called "transcription-coupled repair" because it preferentially repairs actively transcribed DNA strands. This pathway is thought to be responsible for the repair of DNA intrastrand diadducts resulting from cisplatin treatment [18,124,125].

Although alkylation of the O^6 position of guanine is thought to be the most important adduct that results from alkylating agents, N^7-methylguanine and N^3-methyladenine also result from TMZ. These adducts are repaired by the Base Excision Repair Pathway (BER; Reviewed in [122]; Fig. 6.4). Poly(ADP-ribose) polymerase-1 (PARP-1) is activated by DNA damage and participates in repair by interacting with the BER complex. Inhibitors of PARP-1 have been found to sensitize cells to some DNA methylators [87,126], and inhibitors of PARP have been found to

the opened stretch only in the damaged strand, generating a 24–32-base oligonucleotide containing the injury (IV). The regular DNA replication machinery then completes the repair by filling the gap (V). In total, 25 or more proteins participate in NER. *In vivo* studies indicate that the NER machinery is assembled in a step-wise fashion from individual components at the site of a lesion. After a single repair event (which takes several minutes), the entire complex is disassembled again. (Reprinted from [122].) See Plate 6.3 in Color Plate Section.

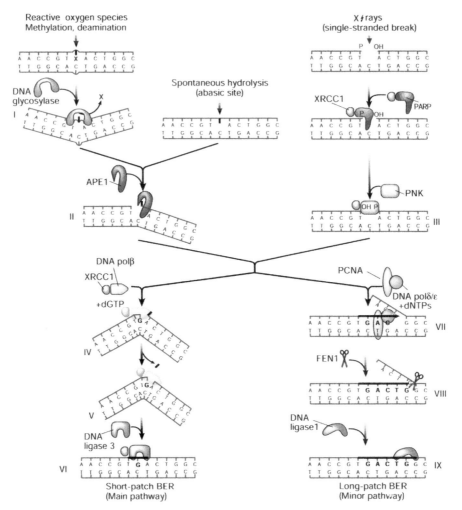

FIGURE 6.4   Base Excision Repair (BER). A battery of glycosylases, each dealing with a relatively narrow, partially overlapping spectrum of lesions, feeds into a core reaction. Glycosylases flip the suspected base out of the helix by DNA backbone compression to accommodate it in an internal cavity of the protein. Inside the protein, the damaged base is cleaved from the sugar-phosphate backbone (stage I in this figure). The resulting abasic site can also occur spontaneously by hydrolysis. The core BER reaction is initiated by strand incision at the abasic site by the APE1 endonuclease (II). Poly(ADP-ribose) polymerase (PARP), which binds to and is activated by DNA strand breaks, and polynucleotide kinase (PNK) may be important when BER is initiated from a SSB to protect and trim the ends for repair synthesis (III). In mammals, the so-called short-patch repair is the dominant mode for the remainder of the reaction. DNA pol β performs a one-nucleotide gap-filling reaction (IV) and removes the 58-terminal baseless sugar residue via its lyase activity (V); this is then followed by sealing of the remaining nick by the XRCC1–ligase3 complex (VI). The XRCC1 scaffold protein interacts with most of the above BER core components and may therefore be instrumental in protein exchange. The long-patch repair mode involves DNA pol β, polδ/ε and proliferating cell nuclear antigen (PCNA) for repair synthesis (2–10 bases) as well as the FEN1 endonuclease to remove the displaced DNA flap and DNA ligase 1 for sealing (VII–IX). The above BER reaction operates across the genome. However, some BER lesions block transcription, and in this case the problem is dealt with by the TCR pathway described in Fig. 6.3, including TFIIH, XPG (which also stimulates some of the glycosylases) and probably the remainder of the core NER apparatus. (Reprinted from [122]). See Plate 6.4 in Color Plate Section.

sensitize MMR deficient cells to TMZ [127]. The BER inhibitors AG14361, 3-aminobenzamide, PD128763 and NU1025 have all been shown to sensitize cells to TMZ [127].

DETOXIFICATION AND OTHER MECHANISMS OF RESISTANCE

Glutathione-S-transferase

In addition to DNA repair, resistance to alkylating agents and other therapies can come about through detoxification. The family of glutathione-S-transferase (GST) enzymes catalyze the conjugation of electrophilic compounds with glutathione. These include various carcinogens, chemotherapeutic agents, and their metabolites [97,128–134]. There are 5 families of GSTs: α(GSTA), μ(GSTM), π(GSTP), θ(GSTT) and ζ(GSTZ) [130,131]. Earlier work demonstrated the importance of the GSTP, GSTM, and GSTT isoforms in resistance to chemotherapeutic agents such as BCNU and cisplatin; however, there were conflicting reports in the literature as to the isoforms most involved in therapy resistance in various human and animal cancers [97,129,131–133,135,137–144]. Work in our laboratory agreed with data obtained in other laboratories in that cells from some tumors demonstrated over-expression of one or more of these genes, while others did not [145]. Furthermore, even when BCNU resistant cells were cloned from a single tumor that did not express MGMT, the expression of various GST isoforms varied and there was no clear correlation between the expression of specific GST isoforms and BCNU resistance (Scheck et al., unpublished results).

The conflicting data on the relative importance of specific GST isoforms to chemotherapy resistance may be due, in part, to the fact that the genes encoding GSTM1, GSTT1, and GSTP1 are polymorphic [146,147], and numerous studies have shown that polymorphisms in GST enzymes can alter their ability to metabolize various substrates including carcinogens and chemotherapeutic agents [147–149]. A number of studies have found relationships between specific genotypes and the incidence of various cancers [147], including brain tumors [150–153]; however, there is inconsistency in the literature. This may be due, at least in part, to the fact that correlations with specific GST genotypes vary with the histological and genetic subtype and grade of pediatric and adult brain tumors.

Similar studies have found correlations between GST polymorphisms and survival in a number of tumor systems [128,154–157], including brain tumors [158,159]. Recent additional work has suggested that the observed increase in survival may be due to the presence of specific GST genotypes that result in reduced GST activity [128,157]; however, some investigators have found that tumors that recur following treatment do not show increased expression of GST suggesting that in these tumors GST may not be correlated to therapy resistance [141]. Furthermore, patients with genotypes leading to reduced GST activity are also at increased risk for adverse effects as a result of chemotherapy [158,160]. It is likely that future studies will indeed identify a role(s) for specific GST genotypes in treatment decisions, therapy resistance and survival; however, these analyses will have to include not only the histological subtype of tumor, but also an understanding of the overall genetic profile of the tumor including MGMT and DNA repair activities.

Metallothioneins

The metallothionein (MT) family consists of 4 isoforms: MT-1 and MT-2 are expressed in all eukaryotes and MT-3 and MT-4 are expressed in mammals. These proteins are involved in metal detoxification. They can also act as scavengers of reactive oxygen species and are thought to protect cells against damage from radiation and some therapeutic agents, particularly cisplatin [161]. Finally, MT appears to protect cells from apoptosis induced by a variety of stimuli including etoposide, oxidative stress, and metals, although the mechanism of this protection is unclear [162].

Meningiomas have been found to express MT; however this expression does not appear to correlate to grade, nor is there a difference between expression in primary or recurrent meningiomas prior to or following irradiation [163,164]. Immunohistochemical analysis of MT expression in malignant gliomas have shown a correlation between expression and histologic grade [165]; however, no differences between primary and recurrent tumors have been seen. In addition, there are no differences in MT expression in primary versus secondary GBMs although there does appear to be an inverse relationship between extent of positivity and p53 expression [164]. Finally, in a study by Hiura et al., [165], patients with MT negative anaplastic astrocyomas had longer median survival than those whose tumors were MT positive (32.7 weeks vs 66.6 weeks, $p = 0.0275$). In contrast, a study of 76 pediatric and adult ependymomas showed a preponderance of the positive tumors were low grade. Furthermore, the median recurrence-free survival was significantly better for patients with MT-positive tumors [166]. These results were somewhat confusing in that tumors that were positive for MT by immunohistochemistry would be expected to be resistant to therapy, yet these tumors showed a more favorable biologic behavior and patient survival was better. Thus, while correlations exist between MT expression and therapy resistance in some tumors, the mechanisms of this resistance are not fully understood. Furthermore, most of these studies do not distinguish between the various isoforms of MT and the overall function of MT in tumor growth, progression, and therapy resistance is likely to be different depending on tumor type, MT isoform, and the background tumor genetics as suggested for many other mechanisms of resistance.

ATP-Binding Cassette or Multidrug Resistance Genes

The ATP-binding cassette (ABC) genes are a family of genes that encode proteins involved in multi-drug resistance in a number of cancers. These proteins reside in membranes and act as efflux pumps, removing toxic compounds from the cell. They are not thought to be involved in resistance to alkylating agents; however, they are involved in resistance to the topoisomerase inhibitors VP-16 (etoposide) and CPT-11 (irinotecan), as well as vincristine. Specifically, ABCB1 (or MDR1) encodes p-glycoprotein which is involved in resistance to VP-16, CPT-11, and vincristine. ABCC1 (which encodes the multidrug-resistant-associated protein or MRP1) has also been implicated in resistance to VP-16 and vincristine. ABCG2 is involved in resistance to VP-16 and SN-38, the active metabolite of CPT-11 [167].

Expression of both p-glycoprotein and MRP1 have been shown to be increased in some brain tumors, although not all reports demonstrate over-expression of both genes [46,47,163,168–172]. p-Glycoprotein was the first ABC transporter identified, and it has been shown to be involved in resistance to a number therapeutic agents commonly used in cancers [47,173,174]. It acts by transporting these compounds out of the cell in an energy-dependent manner. The multidrug-resistant-associated protein (MRP1, encoded by ABCC1) is the main transporter for compounds that are linked to glutathione [167,175,176]. A number of inhibitors have been developed for both p-glycoprotein and MDR1 including cyclosporin A, verapamil, indomethacin and others [167,177–180]. These inhibitors have been shown to increase sensitivity to various therapies *in vitro* [47,180], and a number of them are in clinical trials in a variety of tumor systems [167]. In addition to modulation by inhibitors, polymorphisms in this gene have recently been identified that may affect the efficacy of some drugs such as cardiac glycosides, tricyclic antidepressants, and others, but to date there is little in the literature to suggest that these polymorphisms have a major effect on therapy resistance in cancer [181]. These studies are still somewhat limited and more work is needed to definitively determine the effects, if any, of polymorphisms in these genes [167].

In vitro Analysis of Drug Resistance

In vitro chemosensitivity (or chemoresistance) assays are routinely used in the laboratory for preclinical assessments of therapeutic efficacy. While many different assays have been used, there are 2 main types of assays: viability assays and proliferation assays. Viability assays analyze whether cells survive drug treatment; however, they do not typically demonstrate whether the cell retains the ability to proliferate. The simplest way to quantitate cell survival is to use live/dead cell assays such as cell counts using trypan blue. The trypan blue dye is actively pumped out of live cells but remains in dead cells giving them a blue color. This is a measure of membrane integrity and requires that each sample is individually counted. More sophisticated live/dead cell assays that use spectrophotometers or flow cytometers for quantitation are commercially available; however, they still only provide information as to how many cells survived treatment.

Viability assays that rely on the cell's metabolic activity are commonly used for high-throughput analysis of therapy resistance. These colorimetric assays utilize a dye that altered by the cell's metabolic activity. An example of this is the tetrazolium-type indicator dye 3-(4,5-dimethylthiazolyl-2)-2,5-diphenyltetrazolium bromide (MTT). Cells are plated in a 96-well plate, treated with a range of concentrations of the test drug, and MTT is added after an appropriate length of time. MTT is reduced to an insoluble formazan product by the mitochondria which is then solubilized with a detergent and the color is quantitated using a spectrophotometer [182]. The color is directly proportional to the number of viable (metabolizing) cells. Many assays have been developed that rely on similar mechanisms such as Alamar Blue [183] which has the advantage of leaving the cells alive so the same cells can be assayed over a number of days. The results are graphed with the drug concentration on the X-axis and the spectrophotometer result on the Y-axis. While relatively fast and easy to perform, the main disadvantage of these types of assays is that they only measure cell survival, they do not measure a cell's ability to divide. This may not provide an accurate indication of the drug's efficacy, as some drugs such as alkylating agents may damage a cell such that it survives and continues to metabolize until it has to divide. DNA damage prevents the cell from successfully dividing more than a few times.

Colony forming assays (CFAs) are more stringent assays for therapy resistance in that they require a cell to be able to divide in order to be counted as resistant to therapy. Cells are treated with a range of concentrations of the test drug, the cells are counted and an equal number of cells are plated into petri dishes, typically in triplicate. Cells are allowed to grow to form colonies of at least 50 cells and after the cells are fixed and stained, the colonies are counted. The results of a CFA done to analyze the BCNU resistance of a set of clones grown from a single tumor can be seen in Fig. 6.5. While some clones have essentially horizontal lines indicating resistance to BCNU, others are quite sensitive to the same doses of drug [184], demonstrating the heterogeneity present

within a single tumor with regards to therapy resistance.

The ability to grow tumor cells as spheroids in the laboratory is allowing *in vitro* analyses of therapy resistance that takes into account the three-dimensional characteristics of tumors [111,185–188]. While these models have some obvious advantages, they also increase the number of variables inherent in the assay (reviewed in [189]), making it difficult to compare the results obtained in various laboratories.

Perhaps the most important question one must ask with regards to laboratory resistance assays is whether they should be used to identify patients that are likely to respond to a particular therapy. The American Society of Clinical Oncology (ASCO) organized a working group to compare various assays and determine if they should be used in a clinical setting [190]. They compared a subrenal capsule assay in which tumor cells are cultured and treated in the subrenal capsule of mice, CFAs, various viability assays and extreme drug resistance (EDR) assays in which cells are labeled with tritiated thymidine and exposed to concentrations of drugs equal to the maximum achievable concentrations *in vivo*. Resistance is indicated by continued uptake of tritiated thymidine. Their results demonstrated that none of these assays were currently suitable for routine incorporation into clinical trials. Similarly, Samson *et al.*, [190] compared the results obtained from studies in which patients were treated based on the results of resistance assays *versus* those treated empirically. They concluded that randomized trials were needed to conclusively demonstrate the relative effectiveness of assay guided *versus* empiric therapy decisions. Haroun and coworkers [191] specifically examined extreme drug resistance in 64 brain tumors, most of which were obtained from patients who had previously been treated. They compared their results to response rates reported in the literature; however, it was not clear that the results of the EDR assays matched the *in vivo* results reported for these same drugs. Nevertheless, they conclude that this may still be a reasonable strategy to identify patients who will not respond to a particular therapy. This same assay was used by Parker *et al.*, [31] in a prospective study of CPT-11 for the treatment of recurrent glioma. They correlated patient survival with the results of the EDR assay and found a significant correlation between the pre-treatment EDR results and both time to progression and overall survival, suggesting that this assay may have utility in clinical trials, although additional study is warranted.

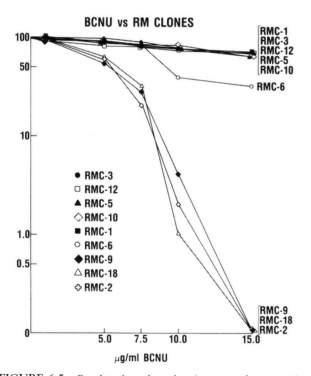

FIGURE 6.5 Results of a colony forming assay demonstrating differential BCNU resistance in clones isolated from a single tumor. Clones with an essentially horizontal dose response curve are resistant. (Reprinted from [184]).

Genetic Analyses and Selection for Resistant Cells Suggest the Existence of Additional Resistance Mechanisms

As described above, a great deal of work has been carried out to determine the cause(s) of intrinsic therapy resistance in gliomas, and a variety of genes have been implicated; however, reports on molecular and genetic aberrations that lead to therapy resistance can be conflicting. Tumor cells are inherently genetically unstable and as the tumor grows the genotype of the cells can change. The result is a heterogeneous tumor in which some cells may be resistant to a specific therapy by one mechanism, while others may be sensitive or resistant by other mechanisms. This can complicate analyses of therapy resistance, and suggests that long term effective therapy will probably require a combination of approaches. Despite this, there are a limited number of studies in which the genetic analysis of the tumor can be used to predict response to treatment. The most recent work is the analysis of MGMT promoter methylation to predict response to TMZ as described in this chapter. Although the results are very promising, it has been shown that the MGMT promoter methylation status in the primary tumor cannot be used to predict the response of the recurrent tumor to treatment with TMZ. This is likely to be due to the genetic changes and selective pressures that occur during treatment, resulting in selection for cells that are resistant to TMZ through a different mechanism [60].

In some cases, it may be possible to exploit the fact that a patient's therapy selects intrinsically resistant cells that share genetic changes which confer a selective survival advantage, enabling these cells to repopulate the tumor mass. Work in our laboratory has demonstrated that when cells from primary malignant gliomas are selected for resistance to BCNU *in vitro* or *in vivo* during patient treatment, the predominant cell is near-diploid, with overrepresentation of part or all of chromosomes 7 and 22. This is in contrast to cells from the primary tumor that are more heterogeneous and often have loss of chromosome 22 [9,184,192–194]. The identification of specific genes or regulatory sequences that are directly responsible for, or contribute to the growth of resistant cells would provide a novel target for the design of a new therapy. This could then be directed against the recurrent tumor—a tumor that is less heterogeneous than the original primary tumor.

Another example of genetic information that suggests a novel mechanism of resistance is the deletion of sequences on chromosomes 1p and 19q seen in oligodendroglioma. Patients whose tumors have LOH or deletions of 1p and 19q have a better prognosis and are more likely to respond to therapy. However, despite intensive efforts the specific genes responsible for this behavior have remained elusive [195–202]. Furthermore, it is not known what percentage of cells in the tumor must have this loss to affect the overall patient prognosis. Finally, we have not been able to demonstrate selection for cells with, or without this loss in samples from recurrent tumors. Nevertheless, oligodendrogliomas are now routinely screened for deletion of 1p/19q because of the diagnostic and prognostic information provided by this test, and it is likely that the identification of specific sequences on these chromosomes will provide additional targets for the design of new therapies.

SUMMARY

A great deal has been learned about the mechanisms of chemotherapy resistance in malignant gliomas. Despite this, there has been little impact in overall survival. Recent advances are suggesting adjuvant therapies to enhance the effectiveness of currently available therapeutic modalities by circumventing known mechanisms of resistance. Furthermore, the use of molecular analyses to predict response to specific therapies is coming into play. Microarray analyses are allowing us to study changes in global gene profiles that arise as a result of therapy resistance. This will provide information on pathways that lead to chemotherapy resistance and suggest additional ways to circumvent this resistance. Despite the enormous power of modern molecular biological analyses, overcoming therapy resistance in malignant brain tumors will still require an understanding of the genotypic and phenotypic heterogeneity inherent in these tumors.

References

1. Coons, S. W., Johnson, P. C., and Shapiro, J. R. (1995). Cytogenetic and flow cytometry, DNA analysis of regional heterogeneity in a low grade glioma. *Cancer Res* **55**, 1569–1577.
2. Scheck, A. C., Shapiro, J. R., Coons, S. W., Norman, S. A., and Johnson, P. C. (1996). Biological and molecular analysis of a low grade recurrence of a glioblastoma multiforme. *Clin Cancer Res* **2**, 187–199.
3. Scheck, A. C., Shapiro, J. R., Ballacer, C. *et al.* (1996). Clonal analysis of human malignant glioma cells selected for resistance to BCNU. *Proceedings of the American Association for Cancer Research* **37**, 338.
4. Maintz, D., Fiedler, K., Koopmann, J. *et al.* (1997). Molecular genetic evidence for subtypes of oligoastrocytomas. *Neuropathol Exp Neurol* **56**, 1098–1104.

5. von Deimling, A., Louis, D. N., von Ammon, K. *et al.* (1992) Association of epidermal growth factor receptor gene amplification with loss of chromosome 10 in human glioblastoma multiforme. *J Neurosurg* **77**, 295–301.

6. von Deimling, A., Louis, D. N., Schramm, J., and Wiestler, O. D. (1994). Astrocytic gliomas: characterization on a molecular genetic basis. *Recent Results in Cancer Res* **135**, 33–42.

7. von Deimling, A., Louis, D. N., and Wiestler, O. D. (1995). Molecular pathways in the formation of gliomas. *Glia* **15**, 328–338.

8. Shapiro, J. R. (1986). Cellular characterization and BCNU resistance of freshly resected and early passage human glioma cells. *Progress in Neuropathology* **6**, 133–143.

9. Scheck, A. C., Mehta, B. M., Beikman, M. K., and Shapiro, J. R. (1993). BCNU-resistant human glioma cells with over-representation of chromosomes 7 and 22 demonstrate increased copy number and expression of platelet-derived growth factor genes. *Genes, Chromosomes & Cancer* **8**, 137–148.

10. Drablos, F., Feyzi, E., Aas, P. A. *et al.* (2004). Alkylation damage in DNA and RNA–repair mechanisms and medical significance. *DNA Repair (Amst)* **3**, 1389–1407.

11. Margison, G. P., and Santibanez-Koref, M. F. (2002). O^6-alkylguanine-DNA alkyltransferase: role in carcinogenesis and chemotherapy. *BioEssays* **24**, 255–266.

12. Perez, R. P., Hamilton, T. C., and Ozols, R. F. (1990). Resistance to alkylating agents and cisplatin: Insights from ovarian carcinoma model systems. *Pharmac Ther* **48**, 19–27.

13. Schold, S. C., Jr., Brent, T. P., Von Hofe, E. *et al.* (1989). O^6-alkylguanine-DNA alkyltransferase and sensitivity to procarbazine in human brain-tumor xenografts. *J Neurosurg* **70**, 573–577.

14. Silber, J. R., Blank, A., Bobola, M. S., Ghatan, S., Kolstoe, D. D., and Berger, M. S. (1999). O^6-methylguanine-DNA methyltransferase-deficient phenotype in human gliomas: frequency and time to tumor progression after alkylating agent-based chemotherapy. *Clin Cancer Res* **5**, 807–814.

15. Friedman, H. S. (2000). Can O^6-alkylguanine-DNA alkyltransferase depletion enhance alkylator activity in the clinic? *Clin Cancer Res* **6**, 2967–2968.

16. Ludlum, D. B. (1990). DNA alkylation by the haloethylnitrosoureas: Nature of modifications produced and their enzymatic repair or removal. *Mutat Res Fundam Mol Mech Mutagen* **233**, 117–126.

17. Gerson, S. L. (2002). Clinical relevance of MGMT in the treatment of cancer. *J Clin Oncol* **20**, 2388–2399.

18. Wang, D., and Lippard, S. J. (2005). Cellular processing of platinum anticancer drugs. *Nat Rev Drug Discov* **4**, 307–320.

19. Fedier, A., and Fink, D. (2004). Mutations in DNA mismatch repair genes: Implications for DNA damage signaling and drug sensitivity (Review). *Int J Cancer* **24**, 1039–1047.

20. Aebi, S., Kurdi-Haidar, B., Gordon, R. *et al.* (1996). Loss of DNA mismatch repair in acquired resistance to cisplatin. *Cancer Res* **56**, 3087–3090.

21. Fink, D., Nebel, S., Aebi, S. *et al.* (1996). The role of DNA mismatch repair in platinum drug resistance. *Cancer Res* **56**, 4881–4886.

22. Fink, D., Zheng, H., Nebel, S. *et al.* (1997). In vitro and in vivo resistance to cisplatin in cells that have lost DNA mismatch repair. *Cancer Res* **57**, 1841–1845.

23. Fink, D., Aebi, S., and Howell, S. B. (2000). The role of DNA mismatch repair in drug resistance. *Clin Cancer Res* **4**, 1–6.

24. Asselin, E., Mills, G. B., and Tsang, B. K. (2001). XIAP regulates Akt activity and caspase-3-dependent cleavage during cisplatin-induced apoptosis in human ovarian epithelial cancer cells. *Cancer Res* **61**, 1862–1868.

25. Pommier, Y., Sordet, O., Antony, S., Hayward, R. L., and Kohn, K. W. (2004). Apoptosis defects and chemotherapy resistance: molecular interaction maps and networks. *Oncogene* **23**, 2934–2949.

26. Cloughesy, T. F., Filka, E., Kuhn, J. *et al.* (2003). Two studies evaluating irinotecan treatment for recurrent malignant glioma using an every-3-week regimen. *Cancer* **97**, 2381–2386.

27. Prados, M. D., Yung, W. K., Jaeckle, K. A. *et al.* (2004). Phase 1 trial of irinotecan (CPT-11) in patients with recurrent malignant glioma: a North American Brain Tumor Consortium study. *Neuro-oncol* **6**, 44–54.

28. Reardon, D. A., Friedman, H. S., Powell, J. B., Jr., Gilbert, M., and Yung, W. K. (2003). Irinotecan: promising activity in the treatment of malignant glioma. *Oncology (Huntingt)* **17**, 9–14.

29. Reardon, D. A., Quinn, J. A., Rich, J. N. *et al.* (2004). Phase 2 trial of BCNU plus irinotecan in adults with malignant glioma. *Neurooncol* **6**, 134–144.

30. Reardon, D. A., Quinn, J. A., Vredenburgh, J. *et al.* (2005). Phase II trial of irinotecan plus celecoxib in adults with recurrent malignant glioma. *Cancer* **103**, 329–338.

31. Parker, R. J., Fruehauf, J. P., Mehta, R., Filka, E., and Cloughesy, T. (2004). A prospective blinded study of the predictive value of an extreme drug resistance assay in patients receiving CPT-11 for recurrent glioma. *J Neurooncol* **66**, 365–375.

32. Kanzawa, F., Sugimoto, Y., Minato, K. *et al.* (1990). Establishment of a camptothecin analogue (CPT-11)-resistant cell line of human non-small cell lung cancer: characterization and mechanism of resistance. *Cancer Res* **50**, 5919–5924.

33. Kubota, N., Kanzawa, F., Nishio, K. *et al.* (1992). Detection of topoisomerase I gene point mutation in CPT-11 resistant lung cancer cell line. *Biochem Biophys Res Commun* **188**, 571–577.

34. Pratt, W. B., Ruddon, R. W., Ensminger, W. D., and Maybaum, J. (1994). Inhibitors of Chromatin Function. *In The Anticancer Drugs*. pp. 183–198. Oxford Universtiy Press, New York.

35. Parney, I. F., and Chang, S. M. (2003). Current chemotherapy for glioblastoma. *Cancer J* **9**, 149–156.

36. Beauchesne, P., Soler, C., Boniol, M., and Schmitt, T. (2003). Response to a phase II study of concomitant-to-sequential use of etoposide and radiation therapy in newly diagnosed malignant gliomas. *Am J Clin Oncol* **26**, e22–e27.

37. Fagioli, F., Biasin, E., Mastrodicasa, L. *et al.* (2004). High-dose thiotepa and etoposide in children with poor-prognosis brain tumors. *Cancer* **100**, 2215–2221.

38. Fiorillo, A., Maggi, G., Greco, N. *et al.* (2004). Second-line chemotherapy with the association of liposomal daunorubicin, carboplatin and etoposide in children with recurrent malignant brain tumors. *J Neurooncol.* **66**, 179–185.

39. Franceschi, E., Cavallo, G., Scopece, L. *et al.* (2004). Phase II trial of carboplatin and etoposide for patients with recurrent high-grade glioma. *Br J Cancer* **91**, 1038–1044.

40. Korones, D. N., Benita-Weiss, M., Coyle, T. E. *et al.* (2003). Phase I study of temozolomide and escalating doses of oral etoposide for adults with recurrent malignant glioma. *Cancer* **97**, 1963–1968.

41. Lopez-Aguilar, E., Sepulveda-Vildosola, A. C., Rivera-Marquez, H. *et al.* (2003). Preirradiation ifosfamide, carboplatin and etoposide (ICE) for the treatment of high-grade astrocytomas in children. *Childs Nerv Syst* **19**, 818–823.

42. Rao, R. D., Krishnan, S., Fitch, T. R. *et al.* (2005). Phase II trial of carmustine, cisplatin, and oral etoposide chemotherapy before radiotherapy for grade 3 astrocytoma (anaplastic astrocytoma): results of North Central Cancer Treatment Group trial 98-72-51. *Int J Radiat Oncol Biol Phys* **61**, 380–386.

43. Rutkowski, S., Bode, U., Deinlein, F. et al. (2005). Treatment of early childhood medulloblastoma by postoperative chemotherapy alone. N Engl J Med 352, 978–986.

44. Wolff, J. E., Westphal, S., Molenkamp, G. et al. (2002). Treatment of paediatric pontine glioma with oral trophosphamide and etoposide. Br J Cancer 87, 945–949.

45. Wolff, J. E., and Finlay, J. L. (2004). High-dose chemotherapy in childhood brain tumors. Onkologie 27, 239–245.

46. Matsumoto, Y., Tamiya, T., and Nagao, S. (2005). Resistance to topoisomerase II inhibitors in human glioma cell lines overexpressing multidrug resistant associated protein (MRP) 2. J Med Invest 52, 41–48.

47. Benyahia, B., Huguet, S., Decleves, X. et al. (2004). Multidrug resistance-associated protein MRP1 expression in human gliomas: chemosensitization to vincristine and etoposide by indomethacin in human glioma cell lines overexpressing MRP1. J Neurooncol 66, 65–70.

48. Fornace, A. J., Papathanasiou, M. A., Hollander, M. C., and Yarosh, D. B. (1990). Expression of the O6-methylguanine-DNA methyltransferasa gene MGMT in MER+ and MER− human tumor cells. Cancer Res 50, 7908–7911.

49. Gerson, S. L. (2004). MGMT: its role in cancer aetiology and cancer therapeutics. Nat Rev Cancer 4, 296–307.

50. Phillips, W. P., Jr., Willson, J. K. V., Markowitz, S. D. et al. (1997). O^6-methylguanine-DNA methyltransferase (MGMT) transfectants of a 1,3-bis(2-chloroethyl)-1-nitrosourea (BCNU)-sensitive colon cancer cell line selectively repopulate heterogeneous MGMT$^+$/MGMT$^-$ xenografts after BCNU and O^6-benzylguanine plus BCNU. Cancer Res 57, 4817–4823.

51. Ali-Osman, F., Srivenugopal, K., and Sawaya, R. (2001). The DNA-repair gene MGMT and the clinical response of gliomas to alkylating agents. N Engl J Med 344, 687–688.

52. Kaina, B., and Christmann, M. (2002). DNA repair in resistance to alkylating anticancer drugs. Int J Clin Pharmacol Ther 40, 354–367.

53. Esteller, M., Garcia-Foncillas, J., Andion, E. et al. (2000). Inactivation of the DNA-repair gene MGMT and the clinical response of gliomas to alkylating agents. N Engl J Med 343, 1350–1354.

54. Blancato, J. K., Wager, M., Guilhot, J. et al. (2004). Correlation of clinical features and methylation status of MGMT gene promoter in glioblastomas. J Neurooncol 68, 275–283.

55. Hegi, M. E., Diserens, A. C., Gorlia, T. et al. (2005) MGMT gene silencing and benefit from temozolomide in glioblastoma. N Engl J Med 352, 997–1003.

56. Mollemann, M., Wolter, M., Felsberg, J., Collins, V. P., and Reifenberger, G. (2005). Frequent promoter hypermethylation and low expression of the MGMT gene in oligodendroglial tumors. Int J Cancer 113, 379–385.

57. Hirose, Y., Kreklau, E. L., Erickson, L. C., Berger, M. S., and Pieper, R. O. (2003). Delayed repletion of O6-methylguanine-DNA methyltransferase resulting in failure to protect the human glioblastoma cell line SF767 from temozolomide-induced cytotoxicity. J Neurosurg 98, 591–598.

58. Houghton, P. J., Stewart, C. F., Cheshire, P. J. et al. (2000). Antitumor activity of temozolomide combined with irinotecan is partly independent of O^6-methylguanine-DNA methyltransferase and mismatch repair phenotypes in xenograft models. Clin Cancer Res 6, 4110–4118.

59. Srivenugopal, K. S., Shou, J., Mullapudi, S. R., Lang, F. F., Jr., Rao, J. S., and Ali-Osman, F. (2001). Enforced expression of wild-type p53 curtails the transcription of the O(6)-methylguanine-DNA methyltransferase gene in human tumor cells and enhances their sensitivity to alkylating agents. Clin Cancer Res 7, 1398–1409.

60. Paz, M. F., Yaya-Tur, R., Rojas-Marcos, I. et al. (2004). CpG island hypermethylation of the DNA repair enzyme methyltransferase predicts response to temozolomide in primary gliomas. Clin Cancer Res 10, 4933–4938.

61. Esteller, M., and Herman, J. G. (2004). Generating mutations but providing chemosensitivity: the role of O6-methylguanine DNA methyltransferase in human cancer. Oncogene 23, 1–8.

62. Hegi, M. E., Diserens, A. C., Godard, S. et al. (2004). Clinical trial substantiates the predictive value of O-6-methylguanine-DNA methyltransferase promoter methylation in glioblastoma patients treated with temozolomide. Clin Cancer Res 10, 1871–1874.

63. Stupp, R., Mason, W. P., van den Bent, M. J. et al. (2005). Radiotherapy plus concomitant and adjuvant temozolomide for glioblastoma. N Engl J Med 352, 987–996.

64. Rood, B. R., Zhang, H., and Cogen, P. H. (2004). Intercellular heterogeneity of expresison of the MGMT DNA repair gene in pediatric medulloblastoma. Neuro-oncol 6, 200–207.

65. Zuo, C., Ai, L., Ratliff, P. et al. (2004). O^6-methylguanine-DNA methyltransferase gene: epigenetic silencing and prognostic value in head and neck squamous cell carcinoma. Cancer Epidemiol Biomarkers Prev 13, 967–975.

66. Hickman, M. J., and Samson, L. D. (1999). Role of DNA mismatch repair and p53 in signaling induction of apoptosis by alkylating agents. Proc Natl Acad Sci USA 96, 10764–10769.

67. Biroccio, A., Bufalo, D. D., Ricca, A. et al. (1999). Increase of BCNU sensitivity by wt-p53 gene therapy in glioblastoma lines depends on the administration schedule. Gene Ther 6, 1064–1072.

68. Middlemas, D. S., Stewart, C. F., Kirstein, M. N. et al. (2000). Biochemical correlates of temozolomide sensitivity in pediatric solid tumor xenograft models. Clin Cancer Res 6, 998–1007.

69. Xu, G. W., Nutt, C. L., Zlatescu, M. C., Keeney, M., Chin-Yee, I., and Cairncross, J. G. (2001). Inactivation of p53 sensitizes U87MG glioma cells to 1,3-bis(2-chloroethyl)-1-nitrosourea. Cancer Res 61, 4155–4159.

70. Bocangel, D. B., Finkelstein, S., Schold, S. C., Bhakat, K. K., Mitra, S., and Kokkinakis, D. M. (2002). Multifaceted resistance of gliomas to temozolomide. Clin Cancer Res 8, 2725–2734.

71. Rolhion, C., Penault-Llorca, F., Kemeny, J. L. et al. (1999). O(6)-methylguanine-DNA methyltransferase gene (MGMT) expression in human glioblastomas in relation to patient characterisitics and p53 accumulation. Int J Cancer 84, 416–420.

72. Bello, M. J., Alonso, M. E., Aminoso, C. et al. (2004). Hypermethylation of the DNA repair gene MGMT: association with TP53 G:C to A:T transitions in a series of 469 nervous system tumors. Mutat Res 554, 23–32.

73. Nakamura, M., Watanabe, T., Yonekawa, Y., Kleihues, P., and Ohgaki, H. (2001). Promoter methylation of the DNA repair gene MGMT in astrocytomas is frequently associated with G:C → A:T mutations of the TP53 tumor suppressor gene. Carcinogenesis 22, 1715–1719.

74. Dolan, M. E., Stine, L., Mitchell, R. B., Moschel, R. C., Pegg, A. E. (1990). Modulation of mammalian O^6-alkylguanine-DNA alkyltransferase in vivo by O6-benzylguanine and its effect on the sensitivity of a human glioma tumor to 1-(2-chloroethyl)-3-(4-methylcyclohexyl)-1- nitrosourea. Cancer Communications 2, 371–377.

75. Dolan, M. E., Chae, M.-Y., Pegg, A. E., Mullen, J. H., Friedman, H. S., and Moschel, R. C. (1994). Metabolism of O^6-benzylguanine, an inactivator of O^6- alkylguanine-DNA alkyltransferase. Cancer Res 54, 5123–5130.

76. Dolan, M. E., and Pegg, A. E. (1997). O^6-Benzylguanine and its role in chemotherapy. Clinical Cancer Res 3, 837–847.

77. Pegg, A. E., Boosalis, M., Samson, L. *et al.* (1993). Mechanism of inactivation of human O^6-alkylguanine-DNA alkyltransferase by O^6-benzylguanine. *Biochemistry* **32**, 11998–12006.

78. Chen, J.-M., Zhang, Y.-P., Moschel, R. C., and Ikenaga, M. (1993). Depletion of O^6-methylguanine-DNA methyltransferase and potentiation of 1,3-bis(2-chloroethyl)-1-nitrosourea antitumor activity by O^6-benzylguanine *in vitro*. *Carcinogenesis* **14**, 1057–1060.

79. Dolan, M. E., Pegg, A. E., Moschel, R. C., and Grindey, G. B. (1993). Effect of O^6-benzylguanine on the sensitivity of human colon tumor xenografts to 1,3-bis(2-chloroethyl)-1-nitrosourea (BCNU). *Biochem Pharmacol* **46**, 285–290.

80. Felker, G. M., Friedman, H. S., Dolan, M. E., Moschel, R. C., and Schold, C. (1993). Treatment of subcutaneous and intracranial brain tumor xenografts with O^6-benzylguanine and 1,3-bis(2-chloroethyl)-1-nitrosourea. *Cancer Chemother Pharmacol* **32**, 471–476.

81. Kokkinakis, D. M., Bocangel, D. B., Schold, S. C., Moschel, R. C., and Pegg, A. E., (1907). Thresholds of O6-alkylguanine-DNA alkyltransferase which confer significant resistance of human glial tumor xenografts to treatment with 1,3-bis(2-chloroethyl)-1-nitrosourea or temozolomide. *Clin Cancer Res 2001* 421–428.

82. Kokkinakis, D. M., Ahmed, M. M., Chendil, D., Moschel, R. C., and Pegg, A. E. (2003). Sensitization of pancreatic tumor xenografts to carmustine and temozolomide by inactivation of their O6-Methylguanine-DNA methyltransferase with O6-benzylguanine or O6-benzyl-2′-deoxyguanosine. *Clin Cancer Res* **9**, 3801–3807.

83. Mitchell, R. B., Moschel, R. C., and Dolan, M. E. (1992). Effect of O^6-benzylguanine on the sensitivity of human tumor xenografts to 1,3-bis(2-chloroethyl)-1-nitrosourea and on DNA interstrand cross-link formation. *Cancer Res* **52**, 1171–1175.

84. Sarkar, A., Dolan, M. E., Gonzalez, G. G., Marton, L. J., Pegg, A. E., and Deen, D. F. (1993). The effects of O^6-benzylguanine and hypoxia on the cytotoxicity of 1,3-bis(2-chloroethyl)-1-nitrosourea in nitrosourea-resistant SF-763 cells. *Cancer Chemother Pharmacol* **32**, 477–481.

85. Srivenugopal, K. S., Yuan, X. H., Friedman, H. S., and Ali-Osman, F. (1996). Ubiquitination-dependent proteolysis of O^6-methylguanine-DNA methyltransferase in human and murine tumor cells following inactivation with O^6-benzylguanine or 1,3-bis(2-chloroethyl)-1-nitrosourea. *Biochemistry* **35**, 1328–1334.

86. Wedge, S. R., and Newlands, E. S. (1973). O6-benzylguanine enhances the sensitivity of a glioma xenograft with low O6-alkylguanine-DNA alkyltransferase activity to temozolomide and BCNU. *Br J Cancer 1996* 1049–1052.

87. Wedge, S. R., Porteous, J. K., and Newlands, E. S. (1974). 3-aminobenzamide and/or O6-benzylguanine evaluated as an adjuvant to temozolomide or BCNU treatment in cell lines of variable mismatch repair status and O6-alkylguanine-DNA alkyltransferase activity. *Br J Cancer 1996* 1030–1036.

88. Friedman, H. S., Keir, S., Pegg, A. E. *et al.* (2002). O6-benzylguanine-mediated enhancement of chemotherapy. *Mol Cancer Ther* **1**, 943–948.

89. Kanzawa, T., Bedwell, J., Kondo, Y., Kondo, S., and Germano, I. M. (2003). Inhibition of DNA repair for sensitizing resistant glioma cells to temozolomide. *J Neurosurg* **99**, 1047–1052.

90. Neville, K., Blaney, S., Bernstein, M. *et al.* (2004). Pharmacokinetics of O(6)-benzylguanine in pediatric patients with central nervous system tumors: a pediatric oncology group study. *Clin Cancer Res* **10**, 5072–5075.

91. Dolan, M. E., Posner, M., Karrison, T. *et al.* (2002). Determination of the optimal modulatory dose of O6-benzylguanine in patients with surgically resectable tumors. *Clin Cancer Res* **8**, 2519–2523.

92. Friedman, H. S., Kokkinakis, D. M., Pluda, J. *et al.* (1998). Phase I trial of O6-benzylguanine for patients undergoing surgery for malignant glioma. *J Clin Oncol* **16**, 3570–3575.

93. Friedman, H. S., Pluda, J., Quinn, J. A. *et al.* (2000). Phase I trial of carmustine plus O6-benzylguanine for patients with recurrent or progressive malignant glioma. *J Clin Oncol* **18**, 3522–3528.

94. Schilsky, R. L., Dolan, M. E., Bertucci, D. *et al.* (2000). Phase I clinical and pharmacological study of O6-benzylguanine followed by carmustine in patients with advanced cancer. *Clin Cancer Res* **6**, 3025–3031.

95. Schold, S. C., Jr., Kokkinakis, D. M., Chang, S. M. *et al.* (2004). O6-benzylguanine suppression of O6-alkylguanine-DNA alkyltransferase in anaplastic gliomas. *Neuro-oncol* **6**, 28–32.

96. Spiro, T. P., Gerson, S. L., Liu, L. *et al.* (1999). O^6-benzylguanine: a clinical trial establishing the biochemical modulatory dose in tumor tissue for alkyltransferase-directed DNA repair. *Cancer Res* **59**, 2402–2410.

97. Bacolod, M. D., Johnson, S. P., Ali-Osman, F. *et al.* (2002). Mechanisms of resistance to 1,3-bis(2-chloroethyl)-1-nitrosourea in human medulloblastoma and rhabdomyosarcoma. *Mol Cancer Ther* **1**, 727–736.

98. Bacolod, M. D., Johnson, S. P., Pegg, A. E. *et al.* (2004). Brain tumor cell lines resistant to O^6-benzylguanine/1,3-bis(2-chloroethyl)-1-nitrosourea chemotherapy have O^6-alkylguanine-DNA alkyltransferase mutations. *Mol Cancer Ther* **3**, 1127–1135.

99. Crone, T. M., and Pegg, A. E. (1993). A single amino acid change in human O6-alkylguanine-DNA alkyltransferase decreasing sensitivity to inactivation by O^6-benzylguanine. *Cancer Res* **53**, 4750–4753.

100. Crone, T. M., Kanugula, S., and Pegg, A. E. (1995). Mutations in the Ada O^6-alkylguanine-DNA alkyltransferase conferring sensitivity to inactivation by O6-benzylguanine and 2,4-diamino-6-benzyloxy-5-nitrosopyrimidine. *Carcinogenesis* **16**, 1687–1692.

101. Xu-Welliver, M., Leitao, J., Kanugula, S., Meehan, W. J., and Pegg, A. E. (1999). Role of codon 160 in the sensitivity of human O6-alkylguanine-DNA alkyltransferase to O6-benzylguanine. *Biochem Pharmacol* **58**, 1279–1285.

102. Ueda, S., Mineta, T., Nakahara, Y. *et al.* (2004). Induction of the DNA repair gene O6-methylguanine-DNA methyltransferase by dexamethasone in glioblastomas. *J Neurosurg* **101**, 659–663.

103. Marathi, U. K., Dolan, M. E., and Erickson, L. C. (1994). Anti-neoplastic activity of sequenced administration of O^6-benzylguanine, streptozotocin, and 1,3-bis(2-chloroethyl)-1-nitrosourea in vitro and in vivo. *Biochem Pharmacol* **48**, 2127–2134.

104. Karran, P. (2001). Mechanisms of tolerance to DNA damaging therapeutic agents. *Carcinogenesis* **22**, 1931–1937.

105. Liu, L., Markowitz, S., and Gerson, S. L. (1996). Mismatch repair mutations override alkyltransferase in conferring resistance to temozolomide but not to 1,3-bis(2-chloroethyl) nitrosourea. *Cancer Res* **56**, 5375–5379.

106. Pepponi, R., Marra, G., Fuggetta, M. P. *et al.* (2003). The effect of O6-alkylguanine-DNA alkyltransferase and mismatch repair activities on the sensitivity of human melanoma cells to temozolomide, 1,3-bis(2-chloroethyl)1-nitrosourea, and cisplatin. *J Pharmacol Exp Ther* **304**, 661–668.

107. Christmann, M., and Kaina, B. (2000). Nuclear translocation of mismatch repair proteins MSH2 and MSH6 as a response of cells to alkylating agents. *J Biol Chem* **275**, 36256–36262.

108. Rellecke, P., Kuchelmeister, K., Schachenmayr, W., and Schlegel, J. (2004). Mismatch repair protein hMSH2 in primary drug resistance in in vitro human malignant gliomas. *J Neurosurg* **101**, 653–658.

109. Luo, M., and Kelley, M. R. (2004). Inhibition of the human apurinic/apyrimidinic endonuclease (APE1) repair activity and sensitization of breast cancer cells to DNA alkylating agents with lucanthone. *Anticancer Res* **24**, 2127–2134.

110. Friedman, H. S., Johnson, S. P., Dong, Q. *et al.* (1997). Methylator resistance mediated by mismatch repair deficiency in a glioblastoma multiforme xenograft. *Cancer Res* **57**, 2933–2936.

111. Francia, G., Man, S., Teicher, B., Grasso, L., and Kerbel, R. S. (2004). Gene expression analysis of tumor spheroids reveals a role for suppressed DNA mismatch repair in multicellular resistance to alkylating agents. *Mol Cell Biol* **24**, 6837–6849.

112. Bignami, M., Casorelli, I., and Karran, P. (2003). Mismatch repair and response to DNA-damaging antitumor therapies. *Eur J Cancer* **39**, 2149.

113. Meyers M., Hwang A., Wagner, M. W., and Boothman, D. A. (2004). Role of DNA mismatch repair in apoptotic responses to therapeutic agents. *Environ Mol Mutagen* **44**, 249–264.

114. Mello, J. A., Acharya, S., Fishel, R., and Essigmann, J. M. (1996). The mismatch-repair protein hMSH2 binds selctively to DNA adducts of the anticancer drug cisplatin. *Chem Biol* **3**, 579–589.

115. Papouli, E., Cejka, P., and Jiricny, J. (2004). Dependence of the cytotoxicity of DNA-damaging agents on the mismatch repair status of human cells. *Cancer Res* **64**, 3391–3394.

116. Branch, P., Masson, M., Aquilina, G., Bignami, M., and Karran, P. (2000). Spontaneous development of drug resistance: mismatch repair and p53 defects in resistance to cisplatin in human tumor cells. *Oncogene* **19**, 3138–3145.

117. Srivastava, T., Chattopadhyay, P., Mahapatra, A. K., Sarkar, C., and Sinha, S. (2004). Increased hMSH2 protein expression in glioblastoma multiforme. *J Neurooncol* **66**, 51–57.

118. Lage, H., Christmann, M., Kern, M. A. *et al.* (1999). Expression of DNA repair proteins hMSH2, hMSH6, hMLH1, O6-methylguanine-DNA methyltransferase and N-methylpurine-DNA glycosylase in melanoma cells with acquired drug resistance. *Int J Cancer* **80**, 744–750.

119. Bearzatto, A., Szadkowski, M., Macpherson, P., Jiricny, J., and Karran, P. (2000). Epigenetic regulation of the MGMT and hMSH6 DNA repair genes in cells resistant to methylating agents. *Cancer Res* **60**, 3262–3270.

120. Kirkpatrick, D. T., and Petes, T. D. (1997). Repair of DNA loops involves DNA-mismatch and nucleotide-excision repair proteins. *Nature* **387**, 929–931.

121. Moggs, J. G., Szymkowski, D. E., Yamada, M., Karran, W., and Wood, R. D. (1997). Differential human nucleotide excision repair of paired and mispaired cisplatin-DNA adducts. *Nucleic Acids Res* **25**, 480–490.

122. Hoeijmakers, J. H. (2001). Genome maintenance mechanisms for preventing cancer. *Nature* **411**, 366–374.

123. Lodish, H., Berk, A., Matsudaira, P. *et al.* (2004). Cancer. In *Molecular Cell Biology.* pp. 935–973. W. H. Freeman and Company, New York.

124. Chaney, S. G., and Sancar, A. (1996). DNA repair: enzymatic mechanisms and relevance to drug response. *J Natl Cancer Inst* **88**, 1346–1360.

125. Reardon, J. T., Vaisman, A., Chaney, S. G., and Sancar, A. (1999). Efficient nucleotide excision repair of cisplatin, oxaliplatin, and Bis-aceto-ammine-dichloro-cyclohexylamine-platinum(IV) (JM216) platinum intrastrand DNA diadducts. *Cancer Res* **59**, 3968–3971.

126. Tentori, L., Portarena, I., Torino, F., Scerrati, M., Navarra, P., and Graziani, G. (2002). Poly (ADP-ribose) polymerase inhibitor increases growth inhibition and reduces G(2)/M cell accumulation induced by temozolomide in malignant glioma cells. *Glia* **40**, 44–54.

127. Curtin, N. J., Wang, L. Z., Yiakouvaki, A. *et al.* (2004). Novel poly(ADP-ribose) polymerase-1 inhibitor, AG14361, restores sensitivity to temozolomide in mismatch repair-deficient cells. *Clin Cancer Res* **10**, 881–889.

128. Stoehlmacher, J., Park, D. J., Zhang, W. *et al.* (2002). Association between glutathione S-transferase P1, T1, and M1 genetic polymorphism and survival of patients with metastatic colorectal cancer. *J Natl Cancer Inst* **94**, 936–942.

129. Britten, R. A., Green, J. A., and Warenius, H. M. (1992). Cellular glutathione (GSH) and glutathione S-transferase (GST) activity in human ovarian tumor biopsies following exposure to alkylating agents. *Int J Radiat Oncol, Biol, Phy* **24**, 527–531.

130. Evans, C. G., Bodell, W. J., Tokuda, K., Doane-Setzer, P., and Smith, M. T. (1987). Glutathione and related enzymes in rat brain tumor cell resistance to 1,3-bis(2-chloroethyl)-1-nitrosourea and nitrogen mustard. *Cancer Res* **47**, 2525–2530.

131. Hara, A., Yamada, H., Sakai, N., Hirayama, H., Tanaka, T., and Mori, H. (1990). Immunohistochemical demonstration of the placental form of glutathione S-transferase,a detoxifying enzyme in human gliomas. *Cancer* **66**, 2563–2568.

132. Smith, M. T., Evans, C. G., Doane-Setzer, P., Castro, V. M., Tahir, M. K., and Mannervik, B. (1989). Denitrosation of 1,3-bis(2-chloroethyl)-1-nitrosourea by class mu glutathione transferase and its role in cellular resistance in rat brain tumor cells. *Cancer Res* **49**, 2621–2625.

133. Waxman, D. J. (1990). Glutathione S-transferases: Role in alkylating agent resistance and possible target for modulation chemotherapy – a review. *Cancer Res* **50**, 6449–6454.

134. Yang, W. Z., Begleiter, A., Johnston, J. B., Israels, L. G., and Mowat, M. R. A. (1991). The role of glutathione (GSH) and glutathione S-transferase (GST) in Chlorambucil (CLB) resistance. *Proceedings of the American Association for Cancer Research* **32**, 360.

135. Hayes, J. D., and Pulford, D. J. (1995). The glutathione S-Transferase supergene family: Regulation of GST and the contribution of the isoenzymes to cancer chemoprotection and drug resistance. *Crit Rev Biochem Mol Biol* **30**, 445–600.

136. Yamayoshi, Y., Iida, E., and Tanigawara, Y. (2005). Cancer Pharmacogenomics: international trends. *The Japan Society of Clinical Oncology* **10**, 5–13.

137. Ali-Osman, F., Stein, D. E., and Renwick, A. (1990). Glutathione content and glutathione-S-transferase expression in 1,3-bis(2-chloroethyl)-1-nitrosourea-resistant human malignant astrocytoma cell lines. *Cancer Res* **50**, 6976–6980.

138. Brandt, T. Y., and Ali-Osman, F. (1997). Detection of DNA damage in transcriptionally active gencs by RT-PCR and assessment of repair of cisplatin-induced damage in the glutathione S-transferase-pi gene in human glioblastoma cells. *Toxicol Appl Pharmacol* 22–29.

139. Wang, Y., Teicher, B. A., Shea, T. C. *et al.* (1989). Cross-resistance and glutathione-S-transferase-π levels among four human melanoma cell lines selected for alkylating agent resistance. *Cancer Res* **49**, 6182–6192.

140. Lien, S., Larsson, A. K., and Mannervik, B. (2002). The polymorphic human glutathione transferase T1-1, the most

efficient glutathione transferase in the denitrosation and inactivation of the anticancer drug 1,3-bis(2-chloroethyl)-1-nitrosourea. *Biochem Pharmacol* **63**, 191–197.

141. Winter, S., Strik, H., Rieger, J., Beck, J., Meyermann, R., and Weller, M. (2000). Glutathione S-transferase and drug sensitivity in malignant glioma. *J Neurol Sci* **179**, 115–121.

142. Ali-Osman, F., Antoun, G., Wang, H., Rajagopal, S., and Gagucas, E. (1996). Buthionine sulfoximine induction of gamma-L-glutamyl-L-cysteine synthetase gene expression, kinetics of glutathione depletion and resynthesis, and modulation of carmustine-induced DNA-DNA cross-linking and cytotoxicity in human glioma cells. *Mole Pharmacol* **49**, 1012–1020.

143. Ali-Osman, F., Caughlan, J., and Gray, G. S. (1989). Decreased DNA interstrand cross-linking and cytotoxicity induced in human brain tumor cells by 1,3-bis(2-chloroethyl)-1-nitrosourea after in vitro reaction with glutathione. *Cancer Res* **49(21)**, 5954–5958.

144. Ali-Osman, F., Brunner, J. M., Kutluk, T. M., and Hess, K. (1997). Prognostic significance of glutathione S-transferase pi expression and subcellular localization in human gliomas. *Clin Cancer Res* **3**, 2253–2261.

145. Norman, S. A., Rhodes, S. N., Treasurywala, S., Hoelzinger, D. B., Shapiro, J. R., and Scheck, A. C. (2000). Identification of transforming growth factor-beta1 binding protein over-expression in BCNU-resistant glioma cells by differential mRNA display. *Cancer* **89**, 850–862.

146. Ali-Osman, F., Akande, O., Antoun, G., Mao, J. X., and Buolamwini, J. (1997). Molecular cloning, characterization, and expression in Escherichia coli of full-length cDNAs of three human glutathione S-transferase Pi gene variants. Evidence for differential catalytic activity of the encoded proteins. *J Biol Chem* **272**, 10004–10012.

147. Strange, R. C., Jones, P. W., and Fryer, A. A. (2000). Glutathione S-transferase: genetics and role in toxicology. *Toxicol Lett* **112–113**, 357–363.

148. Iyer, L., and Ratain, M. J. (1998). Pharmacogenetics and cancer chemotherapy. *Eur J Cancer* **34**, 1493–1499.

149. Pemble, S., Schroeder, K. R., Spencer, S. R. *et al.* (1994). Human glutathione S-transferase theta (GSTT1): cDNA cloning and the characterization of a genetic polymorphism. *Biochem J* **300**, 271–276.

150. Ezer, R., Alonso, M., Pereira, E. *et al.* (2002). Identification of glutathione S-transferase (GST) polymorphisms in brain tumors and assocation with susceptibility to pediatric astrocytomas. *J Neurooncol* **59**, 123–134.

151. Pinarbasi, H., Silig, Y., and Gurelik, M. (2005). Genetic polymorphisms of GSTs and their association with primary brain tumor incidence. *Cancer Genet Cytogenet* **156**, 144–149.

152. Wrensch, M., Kelsey, K. T., Liu, M. *et al.* (2004). Glutathione-S-transferase variants and adult glioma. *Cancer Epidemiol Biomarkers Prev* **13**, 461–467.

153. De Roos, A. J., Rothman, N., Inskip, P. D. *et al.* (2003). Genetic polymorphisms in GSTM1, -P1, -T1, adn CYP2E1 and the risk of adult brain tumors. *Cancer Epidemiol Biomarkers Prev* **12**, 14–22.

154. Yang, G., Shu, X. O., Ruan, Z. X. *et al.* (2005). Genetic polymorphisms in glutathione-S-transferase genes (GSTM1, GSTT1, GSTP1) and survival after chemotherapy for invasive breast carcinoma. *Cancer* **103**, 52–58.

155. Howells, R. E., Dhar, K. K., Hoban, P. R. *et al.* (2004). Association between glutathione-S-transferase GSTP1 genotypes, GSTP1 over-expression, and outcome in epithelial ovarian cancer. *Int J Gynecol Cancer* **14**, 242–250.

156. Sweeney, C., Nazar-Stewart, V., Stapleton, P. L., Eaton, D. L., and Vaughan, T. L. (2003). Glutathione S-transferase M1, T1, and P1 polymorphisms and survival among lung cancer patients. *Cancer Epidemiol Biomarkers Prev* **12**, 527–533.

157. Cabelguenne, A., Loriot, M. A., Stucker, I. *et al.* (2001). Glutathione-associated enzymes in head and neck squamous cell carcinoma and response to cisplatin-based neoadjuvant chemotherapy. *Int J Cancer* **93**, 725–730.

158. Okcu, M. F., Selvan, M., Wang, L. -E. *et al.* (2004). Glutathione S-transferase polymorphisms and survival in primary malignant glioma. *Clinical Cancer Res* **10**, 2618–2625.

159. Tanaka, S., Kobayashi, I., Oka, H. *et al.* (2001). Drug-resistance gene expression and progression of astrocytic tumors. *Brain Tumor Pathol* **18**, 131–137.

160. Peters, U., Preisler-Adams, S., Hebeisen, A. *et al.* (2000). Glutathione S-transferase genetic polymorphisms and individual sensitivity to the ototoxic effect of cisplatin. *Anticancer Drugs* **11**, 639–643.

161. Theocharis, S. E., Margeli, A. P., Klijanienko, J. T., and Kouraklis, G. P. (2004). Metallothionein expression in human neoplasia. *Histopathology* **45**, 103–118.

162. Shimoda, R., Achanzar, W. E., Qu, W. *et al.* (2003). Metallothionein is a potential negative regulator of apoptosis. *Toxicol Sci* **73**, 294–300.

163. Tews, D. S., Fleissner, C., Tiziani, B., and Gaumann, A. K. (2001). Intrinsic expression of drug resistance-associated factors in meningiomas. *Appl Immunohistochem Mol Morphol* **9**, 242–249.

164. Maier, H., Jones, C., Jasani, B. *et al.* (1997). Metallothionein overexpression in human brain tumours. *Acta Neuropathol (Berl)* **94**, 599–604.

165. Hiura, T., Khalid, H., Yamashita, H., Tokunaga, Y., Yasunaga, A., and Shibata, S. (1998). Immunohistochemical analysis of metallothionein in astrocytic tumors in relation to tumor grade, proliferative potential, and survival. *Cancer* **83**, 2361–2369.

166. Korshunov, A., Sycheva, R., Timirgaz, V., and Golanov, A. (1999). Prognostic value of immunoexpression of the chemoresistance-related proteins in ependymomas: an analysis of 76 cases. *J Neurooncol* **45**, 219–227.

167. Sparreboom, A., Danesi, R., Ando, Y., Chan, J., and Figg, W. D. (2003). Pharmacogenomics of ABC transporters and its role in cancer chemotherapy. *Drug Resist Updat* **6**, 71–84.

168. Abe, T., Mori, T., Wakabayashi, Y. *et al.* (1998). Expression of multidrug resistance protein gene in patients with glioma after chemotherapy. *J Neurooncol* **40**, 11–18.

169. Nagane, M., Asai, A., Shibui, S., Oyama, H., Nomura, K., and Kuchino, Y. (1999). Expression pattern of chemoresistance-related genes in human malignant brain tumors: a working knowledge for proper selection of anticancer drugs. *Jpn J Clin Oncol* **29**, 527–534.

170. Tanaka, S., Kamitani, H., Amin, M. R. *et al.* (2000). Preliminary individual adjuvant therapy for gliomas based on the results of molecular biological analyses for drug-resistance genes. *J Neurooncol* **46**, 157–171.

171. Tews, D. S., Nissen, A., Kulgen, C., and Gaumann, A. K. (2000). Drug resistance-associated factors in primary and secondary glioblastomas and their precursor tumors. *J Neurooncol* **50**, 227–237.

172. Tishler, D. M., Weinberg, K. I., Sender, L. S., Nolta, J. A., and Raffel, C. (1992). Multidrug resistance gene expression in pediatric primitive neuroectodermal tumors of the central nervous system. *J Neurosurg* **76**, 507–512.

173. Deng, L., Tatebe, S., Lin-Lee, Y. C., Ishikawa, T., and Kuo, M. T. (2002). MDR and MRP gene families as cellular determinant

factors for resistance to clinical anticancer agents. *Cancer Treat.Res* **112**, 49–66.

174. Tishler, D. M., and Raffel, C. (1992). Development of multidrug resistance in a primitive neuroectodermal tumor cell line. *J Neurosurg* **76**, 502–506.

175. Muller, M., Meijer, C., Zaman, G. J. R. *et al.* (1994). Overexpression of the gene encoding the multidrug resistance-associated protein results in increased ATP-dependent glutathione S-conjugate transport. *Proc Natl Acad Sci USA* **91**, 13033–13037.

176. Zaman, G. J. R., Flens, M. J., van Leusden, M. R. *et al.* (1994). Human multidrug resistance-associated protein MRP is a plasma membrane drug-efflux pump. *Proc Natl Acad Sci USA* **91**, 8822–8826.

177. Nito, S. (1989). Enhancement of cytogenetic and cytotoxic effects on multidrug-resistant (MDR) cells by a calcium antagonist (verapamil). *Mutat Res* **227**, 73–79.

178. Rothenberg, M., and Ling, V. (1989). Multidrug resistance: Molecular biology and clinical relevance. *JNCI* **81**, 907–910.

179. Mickley, L. A., Bates, S. E., Richert, N. D. *et al.* (1989). Modulation of the expression of a multidrug resistance gene (*mdr*-1/P-glycoprotein) by differentiating agents. *J Biol Chem* **264**, 18031–18040.

180. Abe, T., Koike, K., Ohga, T. *et al.* (1995). Chemosensitisation of spontaneous multidrug resistance by a 1,4-dihydropyridine analogue and verapamil in human glioma cell lines over-expressing *MRP* or *MDR*. *Br J Cancer* **72**, 418–423.

181. Eichelbaum, M., Fromm, M. F., and Schwab, M. (2004). Clinical aspects of the MDR1 (ABCB1) gene polymorphism. *Ther Drug Monit* **26**, 180–185.

182. Burton, J. D. (2005). The MTT assay to evaluate chemosensitivity. *Methods Mol Med* **110**, 69–78.

183. O'Brien, J., Wilson, I., Orton, T., and Pognan, F. (2000). Investigation of the Alamar Bluse (resazurin) fluorescent dye for the assessment of mammalian cell cytotoxicity. *Eur J Biochem* **267**, 5421–5426.

184. Shapiro, J. R., and Shapiro, W. R. (1985). The subpopulations and isolated cell types of freshly resected high grade human gliomas: Their influence on the tumor's evolution in vivo and behavior and therapy in vitro. *Cancer Metastasis Review* **4**, 107–124.

185. Santini, M. T., and Rainaldi, G. (1999). Three-dimensional spheroid model in tumor biology. *Pathobiology* **67**, 148–157.

186. Hamilton, G. (1998). Multicellular spheroids as an in vitro tumor model. *Cancer Lett* **131**, 29–34.

187. Kunz-Schughart, L. A., Kreutz, M., and Knuechel, R. (1998). Multicellular spheroids: a three-dimensional in vitro culture system to study tumour biology. *Int J Exp Pathol* **79**, 1–23.

188. Desoize, B., Gimonet, D., and Jardiller, J. C. (1998). Cell culture as spheroids: an approach to multicellular resistance. *Anticancer Res* **18**, 4147–4158.

189. Mueller-Klieser, W. (1997). Three-dimensional cell cultures: from molecular mechanisms to clinical applications. *Am J Physiol* **273**, C1109–C1123.

190. Schrag, D., Garewal, H. S., Burstein, H. J., Samson, D. J., Von Hoff, D. D., Somerfield, M. R., for the ASCO Working Group on Chemotherapy Sensitivity and Resistance Assays (2004). American Society of Clinical Oncology technology assessment: chemotherapy sensitivity and resistance assays. *J Clin Oncol* **22**, 3631–3638.

191. Haroun, R. I., Clatterbuck, R. E., Gibbons, M. C. *et al.* (2002). Extreme drug resistance in primary brain tumors: *in vitro* analysis of 64 resection specimens. *J Neurooncol* **58**, 115–123.

192. Shapiro, J. R., Yung, W.-K. A., and Shapiro, W. R. (1981). Isolation, karyotype, and clonal growth of heterogeneous subpopulations of human malignant gliomas. *Cancer Res* **41**, 2349–2359.

193. Shapiro, J. R., Pu, P.-Y., Mohamed, A. N., Galicich, J. H., Ebrahim, S. A. D., and Shapiro, W. R. (1993). Chromosome number and carmustine sensitivity in human gliomas. *Cancer* **71**, 4007–4021.

194. Yung, W.-K. A., Shapiro, J. R., and Shapiro, W. R. (1982). Heterogeneous chemosensitivities of subpopulations of human glioma cells in culture. *Cancer Res* **42**, 992–998.

195. Fuller, C. E., Schmidt, R. E., Roth, K. A. *et al.* (2003). Clinical utility of fluorescence in situ hybridization (FISH) in morphologically ambiguous gliomas with hybrid oligodendroglial/astrocytic features. *J Neuropathol Exp Neurol* **62**, 1118–1128.

196. van den Bent, M. J. (2004). Diagnosis and management of oligodendroglioma. *Semin Oncol* **31**, 645–652.

197. Perry, J. R. (2001). Oligodendrogliomas: clinical and genetic correlations. *Current Opinion in Neurology* **14**, 705–710.

198. Fallon, K. B., Palmer, C. A., Roth, K. A. *et al.* (2004). Prognostic value of 1p, 19q, 9p, 10q, and EGFR-FISH analyses in recurrent oligodendrogliomas. *J Neuropathol Exp Neurol* **63**, 314–322.

199. Buckner, J. C. (2003). Factors influencing survival in high-grade gliomas. *Semin Oncol* **30**, 10–14.

200. van den, B. M., Chinot, O. L., and Cairncross, J. G. (2003). Recent developments in the molecular characterization and treatment of oligodendroglial tumors. *Neuro-oncol* **5**, 128–138.

201. Stege, E. M., Kros, J. M., de Bruin, H. G. *et al.* (2005). Successful treatment of low-grade oligodendroglial tumors with a chemotherapy regimen of procarbazine, lomustine, and vincristine. *Cancer* **103**, 802–809.

202. van den Bent, M. J. (2004). Advances in the biology and treatment of oligodendrogliomas. *Curr Opin Neurol* **17**, 675–680.

7

Clinical Trial Design and Implementation

Nicholas Butowski and Susan Chang

ABSTRACT: This chapter is a succinct overview of clinical trial design and conduct as it pertains to oncology. The basic principles involved in phase I, II, and III cancer clinical trials for both the traditional cytotoxic agents and for the more recently emerging molecularly targeted agents and cytostatic agents are discussed. Additionally, special challenges that must be considered in the design and implementation of brain tumor clinical trials are given closer examination.

INTRODUCTION

The prognosis for patients with primary brain tumors remains poor. There is excitement in the oncology community that a shift away from the traditional cytotoxic agents toward molecularly targeted agents may provide more effective treatment options. However, before a new therapy can be tested in patients, pre-clinical research is performed to determine the new agent's mechanism of activity, its relevance to treatment, and its toxic effects. Only once the pre-clinical research is completed and thought promising is the new agent tested in clinical trials.

The general aim of therapeutic cancer clinical trials is to determine the effectiveness and safety of a new agent, whether used alone or in combination with another drug. These trials are carefully performed in an effort to avoid the use of ineffective or toxic therapies. Clinical trials are scientific experiments with four main requirements:

(1) a well-defined question within a specific patient population;

(2) a prospective plan with clear objectives and endpoints;

(3) conduct under controlled conditions; and

(4) statistical thoroughness.

Failure to conduct a study with precision and attention to these areas threatens the validity of the results.

No matter what the clinical trial phase, several components of a clinical trial require prospective planning. First, the scientific background of the experimental agent including the mechanism of action, pre-clinical data, and any clinical data needs careful review and documentation in the introduction of the protocol. Next, the objectives and endpoints should be thought out and defined clearly. Patient eligibility criteria should also be unambiguously defined with strict attention to the method of enrollment. All the patients should be included in the final report. The design of the study and the treatment plan including a detailed description of dose administration and schedule and frequency of evaluation should be constructed so that it can be reproduced. Expected toxicities and methods of toxicity assessment with dosage modifications or early stopping rules should be created. One must also determine which statistical considerations will be used to analyze the data. Finally, an informed consent document must be created and written in a manner that allows the patient to understand the objectives of the study and that he/she has the freedom to withdraw at any time.

A comprehensive review of cancer clinical trials is beyond the scope of this chapter. Instead we provide an overview of the basic principles of therapeutic cancer clinical trials and specifically address the challenges of clinical research in neuro-oncology.

PHASE I TRIALS

Cytotoxic Agents

Phase I trials are designed to identify the recommended phase II schedule and dose and describe the pharmacological characteristics of the drug and its toxic effects. Phase I studies of traditional cytotoxic drugs determine the recommended phase II dose based on the maximum tolerated dose (MTD) of a new agent. This toxicity based approach originates from the supposition that the therapeutic effect and toxic effects of the agent parallel each other as the dose is increased and that the same mechanism of action produces the therapeutic and toxic effects.

Customarily, the starting dose for a new drug is based on preclinical animal toxicologal studies unless other investigations warrant a modified dose (e.g., a study reveals that human bone marrow is more sensitive to the new drug than animal bone marrow). Previously, the starting dose was deduced on the basis of one-tenth of the dose at which 10 per cent of the most sensitive animal species died. However, a recent retrospective analysis of 21 phase I trials revealed that this starting dose was inappropriately low in most of the cases [1]. In an attempt to remedy this, the U.S. Food and Drug Administration released draft guidelines that may help to determine a better estimation of the recommended starting dose [2]. These guidelines incorporate factors, such as selection of the most appropriate species from which to extrapolate data and the human equivalent dose calculation. The dose schedule for the new agent is determined by the mechanism of action, clinical pharmacokinetics (half-life), and any previous clinical experience.

One well-accepted standard phase I design involves the accrual of cohorts of three to six patients, who are treated at an initial starting dose with pre-planned increases in dose levels after enough time has elapsed to observe acute toxic effects. Toxicity is standardly measured using the National Cancer Institute Common Terminology Criteria for Adverse Events Version 3.0 [3]. The recommended phase II dose is defined as one dose level below that which resulted in dose-limiting toxicity in two or more of the three to six patients treated.

Different dose-escalation designs exist. Historically, a modified Fibonacci escalation sequence (higher escalation steps have decreasing increments, e.g., 100,

65, 50, 40 per cent, etc.) is used. However, evidence suggests that the use of the Fibonacci method is not necessary to ensure safety or statistical rigor [1]. Other dose escalation designs, such as statistically based methods, continual reassessment methods, and accelerated titration designs can also be used. All the methods have the same goal of defining the MTD but differ in the amount of importance they assign to such factors as the number of patients treated at a specified dose level, rapidity of accrual, and pharmacokinetic modeling. In comparison to the Fibonacci method, these alternative dose-escalation designs may reduce the number of patients treated at potentially sub-therapeutic dose levels or at doses that exceed the MTD.

Bearing in mind that the goal of phase I trials is not therapeutic effect, patient selection is generally open to patients with a broad spectrum of cancers that are refractory to standard therapy or who have no other reasonable therapeutic option. Still, it is important to control patient characteristics that could confound toxicity evaluation. Factors such as general medical health, performance status of the patient, normal organ function, and concomitant medications can alter the new drug's toxic effect. Additionally, one must take into account that many cytotoxic agents possess similar toxicities (hematological, nausea, and vomiting) and as such, patients who have been pretreated with standard cytotoxic agents may be less tolerant of a new cytotoxic agent than a newly diagnosed patient would be.

Phase I studies also usually incorporate pharmacokinetic studies (absorption, distribution, metabolism, and excretion) of the new agent in relation to patient characteristics like age, organ function, and degree of toxic effect. Such considerations may aid in refinement of the recommended phase II dose. Also studied is the influence of agents that may alter metabolism of the study agent. For instance, several commonly used antiepileptic drugs (AED) induce the hepatic cytochrome P450 system and alter the metabolism of the treatment agent [4,5]. Corticosteroids may also produce this effect. Due to these effects, phase I neuro-oncology studies often stratify patients into two different dose-escalation cohorts of those on and off enzyme-inducing AED; they also document corticosteroid use.

Finally, phase I studies have a predefined schedule and method of reporting, monitoring, and documenting toxicities. Normally there are team meetings at pre-defined intervals to review the status of all the patients enrolled in the study. Patients are reviewed in the context of a data safety monitoring plan, which is set up to assure the timely reporting of adverse events, to monitor the progress of the study, and to assure data accuracy.

Novel Agents

The traditional paradigm of dose selection based on maximum tolerable toxicity may not apply to novel therapies that target cell-signaling pathways or the cellular environment. Unlike cytotoxic agents, which act on DNA, these novel therapies have different targets, such as membrane receptors, signaling pathways, and proteins or factors important in cell-cycle regulation or in angiogenesis. As such, in laboratory models these agents seem to inhibit tumor progression rather than cause tumor regression. Novel agents also seem to be more selective and less toxic to normal tissue. Considering these points, the dose of the targeted agent needed to achieve tumor inhibition may not be the one that produces significant organ toxicity. Therefore, while the goal of phase I trials of targeted agents remains the determination of the recommended phase II dose, this dose is likely to be determined by biological endpoints and not necessarily by the MTD. Biological endpoints are associated with a desired biological effect such as inhibition of an enzyme or immunological change, but not necessarily with a specific toxicity and may actually be reached below the MTD. Additionally, the toxic effects of molecular agents may be achieved through different mechanisms than the therapeutic effect and hence may not parallel one another at all.

An example of a biological endpoint as an alternative to toxicity is measurement of target inhibition. This approach requires a fixed understanding of the complexity of cellular pathways, and extensive laboratory data is needed to validate the mechanism of the new agent and its effect of inhibition on a specific target. After all, it is possible that the drug's interaction with another cell pathway produces a different effect at the cellular level that is responsible for the desired effect but has gone unmeasured. Another obstacle is that a reliable assay needs to be created to obtain the tissue for evaluation. Pre- and post-treatment results could be obtained by multiple tumor biopsies, but this is invasive to the patient and prone to sampling errors, not to mention the expenditure and that it is limited to those patients with accessible disease and to tumors that express the desired target. A possible solution is to use surrogate tissue like peripheral blood cells—provided that changes in the surrogate parallel those in the tumor. The chosen assay must also be performed in "real time" to allow decisions to be made regarding dose escalation.

Another possible alternative endpoint is functional imaging that quantifies the level of the target function *in vivo*. For example, enhanced magnetic resonance imaging has been used to assess changes in tumor blood flow after treatment with antiangiogenic agents [6]. However, prior to using such changes as primary endpoints in a clinical trial, these surrogate endpoints need extensive preclinical evaluation correlating the effect of the new agent on the target in the tumor.

Until such alternative endpoints are validated, phase I trials of novel agents incorporating multiple endpoints may be the most practical approach. For example, a useful design may be to define the recommended phase II dose on the basis of the MTD and the maximum target inhibition dose. Taking into account that many targeted agents may require longer-term treatment than the relatively short-term treatment of cytotoxic agents, the definition of tolerable toxicity may need to be adjusted. In order to limit the intensity of daily toxicity there may be more impetus to discover a therapeutic dose below the MTD. In this vein, concurrent pharmacokinetic studies with molecularly targeted agents may prove important in assessing the time that inhibitory concentrations are sustained for a given schedule. Clearly, further work is needed to optimize the strategies for dose selection for targeted therapies.

Limitations

The toxicity risk for patients on a phase I study is high. As the goal of phase I study is not patient benefit, practitioners may feel that patients are vulnerable to diminished quality of life due to toxic effects of new drugs. Additionally, phase I studies often only accrue small number of patients and therefore may not detect all the toxic effects of a particular drug. Also, phase I trials do not assess long-term toxicities (escalation measures are based on immediate toxicities), which may be especially problematic for cytotoxic agents given for repeated cycles or cytostatic agents that may require long-term administration.

PHASE II TRIALS

Cytotoxic Agents

The primary objectives of conventional phase II trials using a cytotoxic agent are to

(1) determine the efficacy of the agent at a certain dose and schedule as identified in phase I trials and

(2) determine whether this level of efficacy warrants further testing in a phase III trial.

Phase II trials are restricted to patients with a specific tumor type, based on the mechanism of the drug and its observed activity in phase I studies. To ensure detection of response and patient compliance, most patients also have a good performance status and little exposure to prior chemotherapy.

Most phase II trials are conducted in a multi-stage design. This design allows sequential treatment and evaluation of cohorts of patients with early stopping rules if the accumulated evidence for efficacy is poor or absent. The exact number of patients and number of responses required depends on the willingness to accept specified probabilities of false positive or negative results. Typically, phase II cancer studies are open label and single arm in design. However, if multiple agents are concurrently available for testing, or if various regimens of the same agent are under consideration, a randomized design may be used. The purpose of randomization is not to test which regimen is the "best" but to eliminate bias on the part of the investigator in assigning the treatment [7, 8]. As single-arm trials use historical data to assess efficacy, in some instances an appropriate control population may not be available and a concurrent control group is required for some novel strategies, e.g., autologous vaccine use. It must be noted, however, that these trials estimate and do not prove potential differences in determining the likely usefulness of therapy. Biostatistical input is crucial for the appropriate planning of the study as well as the analysis and interpretation of the results [9].

Efficacy is commonly assessed in terms of antitumor activity, traditionally measured as decrease in tumor size after treatment. These measurements are commonly made with standardized criteria such as those of RECIST or WHO [10]. It is important to note that tumor regression is the only endpoint that has been validated to be predictive of drug efficacy [11]. However, this method dictates that the patient has a measurable tumor.

Tumor response (regression) is not an appropriate endpoint for those patients who do not have measurable tumor. For example, those patients who undergo gross total resection do not have measurable tumor. Thus, surrogate endpoints such as time to tumor progression may be employed. This method requires prospective determination of the nature and frequency of evaluation methods used to assess tumor growth. Furthermore, such evaluation methods should be standardized in an effort to leave little to subjective interpretation. Another difficulty inherent in the use of time to progression as a surrogate endpoint is the need for a prognostically similar historical control group. This historical control serves as a comparison by which one can detect whether the new agent prolongs time to progression. If a historical control is not available, the study will require a prospectively planned, concurrent control group.

The use of a set time at which the status of the patient and the tumor is assessed has also been employed as a surrogate endpoint, e.g., six-month progression-free survival. This approach does not require that a certain event should have taken place (progression) and allows for more timely completion of studies. Obviously, though, this approach also requires a historical control group.

Novel Agents

Molecular agents may prevent tumor growth without shrinking the tumor. Thus, response measured as tumor regression is not an appropriate endpoint for these agents. Possible end points for molecularly targeted agents include time to tumor progression, change in tumor markers, measures of target inhibition, and metabolic imaging. Time to progression is a well-described endpoint in the literature, where benefit is measured by comparison with a historical cohort treated with the standard of care. Ideally, the only difference between the control and the treatment group is the treatment itself. Selection bias may unduly make such comparison invalid as groups often differ in comparability of such factors as response assessment, ancillary care, and patient characteristics. One way to minimize selection bias is to use groups evaluated at the same institution.

Change in tumor markers is an appealing endpoint but the technique is unproven and not widely employed in brain tumor therapy for lack of a marker to measure. If a marker did exist, physicians must be certain that the drug itself does not directly lower the level by protein degradation independent of tumor burden.

Measures of target inhibition require a definitive cause and effect relationship between the novel agent, the targeted molecule, and tumor growth. As a lot of the understanding of these relationships is still in its infancy, such targeting has yet to be shown clinically effective.

Metabolic imaging modalities like single photon emission computed tomography (SPECT) and positron emission tomography (PET) can reveal the biochemical changes in tumors as they are treated, or possibly measure the receptor density status—this usage is currently unproven and requires validation. These imaging techniques may allow an assessment of clinical efficacy and play a role in the management and understanding of tumors [12]. Currently, these techniques are costly and still require study and validation.

Secondary Evaluations

Another objective of phase II studies is a more thorough understanding of the toxicity associated with the agent being tested. Toxicity evaluation is an important secondary endpoint and can add to the knowledge gained in phase I studies. Such evaluation is even more important when considering that molecular agents may be given long term and phase II studies may require longer term toxicity evaluation. An additional secondary endpoint is quality of life assessment. In patients with recurrent malignant glioma who have a poor median survival, quality of life is an important goal and needs to be assessed during a clinical trial. It is a difficult endpoint to assess because of understandable subjectivity and as patients progress, their quality of life tends to decrease; nevertheless, techniques for evaluation exist and should be used [13].

Limitations

The relatively small number of patients in phase II trials raises the possibility of a false-positive or false-negative result. A false-negative result can eliminate further study of a potentially active agent while a false-positive study can commit further resources (patients, infrastructure, and finances) to evaluate an agent that may be of marginal effect. Appropriate statistical considerations with attention to sample size, relevant and valid endpoints, and comparable historical controls, and the determination *a priori* of what constitutes benefit can help minimize the error.

PHASE III CLINICAL TRIALS

The objective of a phase III trial is to establish whether the efficacy of an experimental therapy is better than the standard therapy [14–16]. These trials are usually large cooperative trials that randomize patients to new agents *versus* standard therapy. The goal of randomization is to prevent bias and usually leads to an equal distribution of the prognostic factors between the treatment arms.

Predefining risk groups (strata) and randomizing within strata further assures balance and can increase the power to detect differences in treatment [14,15]. Ideally patients and doctors would be blinded to the treatment to reduce bias and any subjective "placebo" effect. However, blinding is often not possible in oncology trials. Typical endpoints are time to progression or survival, though quality of life should also be assessed. Biostatistical input in the design and planned analysis of these studies are crucial to determine the appropriate sample size for the study based on the assumptions of what is expected with standard therapy and the prestated effect size that would warrant declaring "success" of the experimental arm [14,15].

Whether the agent is cytotoxic or molecularly targeted should not change the design in a substantial way. However, eligibility should be restricted to patients with a specific tumor type or, in the case of molecular-based therapy, only patients with the target should be enrolled.

SPECIAL CHALLENGES OF BRAIN TUMOR TRIALS

In addition to the principles of clinical trial design and conduct outlined above, there are special issues and challenges relevant to neuro-oncology clinical trials that necessitate closer examination. Several of these issues are described below.

Drug Selection

Many brain tumor patients take antiepileptic drugs (AED) for seizure prophylaxis. Some anticonvulsants may induce the hepatic cytochrome P450 system and alter the metabolism of a treatment agent being studied [17]. Such drug interactions may alter the type and severity of toxicity that patients experience. For example, it has been discovered that patients taking paclitaxel or CPT-11 while taking enzyme-inducing antiepileptic drugs (EIAED) may have lower-than-expected plasma levels and higher-than-expected tolerated doses [18,19]. Thus, phase I studies of agents known to be metabolized by this enzyme system should stratify patients into two different dose-escalation cohorts of those on and off EIAED. A more recent approach in early-phase studies in neuro-oncology is to initiate a phase II study in brain tumor patients using the established phase II dose from other systemic cancer patients not taking EIAEDs. Only if some measure of activity is demonstrated will a phase I study in patients taking EIAEDs be performed.

Additionally, study agents may cause neurological toxicity that may be difficult to distinguish from cerebral edema or effects of the tumor. Neurological toxicity assessments can be confounded by the disease process or concurrent medications. For instance, a seizure may be caused or influenced by the study drug, but could also be solely due to a preexisting seizure disorder, tumor progression, or lack of adequate treatment with anticonvulsants. Consequently, in phase I trials it is important to determine whether the patients may be experiencing neurological toxicity caused by the new agent or by pre-existing neurological status. If the toxicity is felt to be due to compromised neurological status and not the study agent, adverse events should be reported in the context of the patient's preexisting neurological condition. Of course, if the study agent is thought responsible, then this needs to be conveyed clearly. Taking such measures will assure that a study drug is not inappropriately blamed for adverse events and the study inappropriately halted.

Neuropathology

Histological confirmation of the diagnosis is necessary to predict clinical behavior and determine appropriate treatment. Unfortunately, multiple previously used grading systems exist for brain tumors. The WHO II classification is now accepted as the current system used to classify CNS tumors [20]. However, it remains difficult to compare historical controls that used older classification systems with studies using current histological classification. Additionally, vague terms are sometimes used to describe certain types of brain tumors. For instance, the term malignant glioma is commonly used but is not specific to one tumor type and includes anaplastic astrocytomas, anaplastic oligodendrogliomas, and glioblastomas. These histologies have different responses to therapy and corresponding prognoses. Thus, in an effort to allow for appropriate comparison between the groups, use of vague terms should be avoided in any clinical trial.

Another difficulty is inter-observer variability and subjectivity in classifying brain tumors, especially with grade II and III gliomas [21]. It is recommended that brain tumor clinical trials incorporate central pathology review to guarantee statistical power. Also, patients are often included in trials on the basis of their original histological diagnosis even though there has been obvious tumor progression and possible change in the grade of the tumor. In these cases, a second surgery should be performed to check if pathology has changed and whether the patient qualifies for the trial. Ideally, phase II studies should be performed with previously untreated patients, thereby avoiding the problems of acquired drug resistance and diminished tolerance associated with past treatment.

Endpoints

The most common endpoint chosen to evaluate efficacy in phase II brain tumor trials is tumor response as shown by the reduction in the size of tumor on imaging. Response is typically measured with either a computed tomography (CT) or magnetic resonance imaging (MRI) scan of the brain before therapy and at regular intervals after therapy has started. MRI is superior to CT in assessing the extent of tumor and response to treatment, and is the modality of choice, whenever possible [22,23]. A response is defined as reduction in tumor size and is usually assessed by standardized radiographic response criteria such as those described by MacDonald et al. [24]. Such response definitions are to some extent subjective and allow for degrees of variability. This variability is due to several factors, including that an MRI or CT scan does not directly measure tumor size but measures the extent of disruption of the blood–brain barrier that is indirectly related to tumor size. Also, brain tumors generally appear in intricate shapes that are prone to a wide range of interpretation. Imaging variations in timing, position, and technique also lead to significant variability. Even if left to central review, residual or progressive tumor may be difficult to distinguish from the effects of surgery or radiation. Corticosteroids may influence cerebral edema and contrast enhancement as assessed by CT or MRI, allowing even further variability [25]. For appropriate response assessment, patients should be on stable or decreasing doses of steroids. As treatment effects can mimic tumor progression on MRI, other studies such as MR spectroscopy may be used to confirm biological effect [26]. However, this technique is new and used in widely different manner from one institution to the next [27–29]. Nonetheless, every effort should be made to clearly and fully report which definition of measurable disease was used in a study as well as the criteria of response using clinical or radiographical means. The timing of the first image, imaging parameters, and when follow-up scans are to take place should also be clearly defined.

As discussed above, cytostatic agents are not expected to shrink a tumor and thus response as assessed by the reduction in the size of tumor on imaging is not appropriate. Instead, these agents are evaluated by different endpoints such as progression-free survival or time to progression. Both these endpoints require historical controls, which pose several challenges (see below). Additionally, a longer time to progression does not necessarily correlate with longer life.

Historical Controls

Most phase II neuro-oncology studies are open-label, single-arm studies that necessitate the use of a historical control. These historical control groups must be chosen with great care, applying the same inclusion and exclusion criteria to controls that are applied to patients receiving the study agent. If this same rigor is not used in choosing a control group, an experimental agent can appear better than the standard by virtue of the patient selection process. This type of error is evident in a recent intra-arterial chemotherapy study and in an interstitial brachytherapy study from the late 1980s. On review, the apparent benefit in both the studies was due to the patient selection process rather than the novel treatments [30]. To minimize this type of error, study populations should be compared to control groups that have been subjected to the same inclusion and exclusion criteria. Additionally, an appropriate historical control would have the same criteria for pathological assessment and the same endpoints used to assess the efficacy.

CONCLUSIONS

Clinical trials are evolving scientific experiments with constant effort being made on the part of clinicians to develop more efficient designs that quickly identify and discontinue ineffective agents without discarding beneficial ones. Neuro-oncologists must continue to use resources well and implement data and reporting standards to foster advances on par with most other vital scientific endeavors. Quality control, endpoint evaluation, and data reporting techniques should be standardized and followed with the same rigor inherent in any scientific experiment. Additionally, to deal with the complexity of long-term follow up, new methods of chronically evaluating patients need to be devised. This is especially true for cytostatic agents, but is also evident when clinicians prescribe cytotoxic agents for longer periods of time than traditionally used, with unknown long-term risk or benefit. Lastly, as the molecular pathogenesis of brain tumors has not been linked to a single genetic defect or target, molecular agents may be used in tandem with cytotoxic agents, which will require a combination of clinical trial designs to properly determine patient benefit.

References

1. Eisenhauer, E. A., O'Dwyer, P. J., Christian, M., and Humphrey, J. S. (2000). Phase I clinical trial design in cancer drug development. *J Clin Oncol* **18**, 684–692.
2. Estimating the safe starting dose in clinical trials for therapeutics in adult healthy volunteers (draft guidance for industry and reviewers). Washington, DC, USA.
3. National Cancer Institute. Common Terminology Criteria for Adverse Events v3.0. Available at http://ctep.cancer.gov/reporting/ctc.html. Accessed November 5, 2004.
4. Vecht, C. J., Wagner, G. L., and Wilms, E. B. (2003). Treating seizures in patients with brain tumors: Drug interactions between antiepileptic and chemotherapeutic agents. *Semin Oncol* **30**, 49–52.
5. Vecht, C. J., Wagner, G. L., and Wilms, E. B. (2003). Interactions between antiepileptic and chemotherapeutic drugs. *Lancet Neurol* **2**, 404–409.
6. Fox, E., Curt, G. A., and Balis, F. M. (2002). Clinical trial design for target-based therapy. *Oncologist* **7**, 401–409.
7. Fazzari, M., Heller, G., and Scher, H. I. (2000). The phase II/III transition. Toward the proof of efficacy in cancer clinical trials. *Control Clin Trials* **21**, 360–368.
8. Estey, E. H., and Thall, P. F. (2003). New designs for phase 2 clinical trials. *Blood* **102**, 442–448.
9. Liu, P. Y., LeBlanc, M., and Desai, M. (1999). False positive rates of randomized phase II designs. *Control Clin Trials* **20**, 343–352.
10. Therasse, P., Arbuck, S. G., Eisenhauer, E. A. *et al.* (2000). New guidelines to evaluate the response to treatment in solid tumors. European Organization for Research and Treatment of Cancer, National Cancer Institute of the United States, National Cancer Institute of Canada. *J Natl Cancer Inst* **92**, 205–216.
11. Parulekar, W. R., and Eisenhauer, E. A. (2002). Novel endpoints and design of early clinical trials. *Ann Oncol* **13**, (Suppl 4) 139–143.
12. Schiepers, C., and Hoh, C. K. (1998). Positron emission tomography as a diagnostic tool in oncology. *Eur Radiol* **8**, 1481–1494.
13. Recht, L., Glantz, M., Chamberlain, M., and Hsieh, C. C. (2003). Quantitative measurement of quality outcome in malignant glioma patients using an independent living score (ILS). Assessment of a retrospective cohort. *J Neurooncol* **61**, 127–136.
14. Pocock, S. J. (1997). Randomised clinical trials. *Br Med J* **1**, 1661.
15. Simon, R. (1994). Randomized clinical trials in oncology. Principles and obstacles. *Cancer* **74**, 2614–2619.
16. Perry, J. R., DeAngelis, L. M., Schold, S. C., Jr. *et al.* (1997). Challenges in the design and conduct of phase III brain tumor therapy trials. *Neurology* **49**, 912–917.
17. Kivisto, K. T., Kroemer, H. K., and Eichelbaum, M. (1995). The role of human cytochrome P450 enzymes in the

metabolism of anticancer agents: implications for drug inter-actions. *Br J Clin Pharmacol* **40**, 523–530.

18. Gilbert, M. R., Supko, J. G., Batchelor, T. *et al.* (2003). Phase I clinical and pharmacokinetic study of irinotecan in adults with recurrent malignant glioma. *Clin Cancer Res* **9**, 2940–2949.

19. Chang, S. M., Kuhn, J. G., Robins, H. I. *et al.* (2001). A Phase II study of paclitaxel in patients with recurrent malignant glioma using different doses depending upon the concomitant use of anticonvulsants: a North American Brain Tumor Consortium report. *Cancer* **91**, 417–422.

20. Radner, H., Blumcke, I., Reifenberger, G., and Wiestler, O. D. (2002). The new WHO classification of tumors of the nervous system 2000. Pathology and genetics. *Pathologe* **23**, 260–283.

21. Scott, C. B., Nelson, J. S., Farnan, N. C. *et al.* (1995). Central pathology review in clinical trials for patients with malignant glioma. A Report of Radiation Therapy Oncology Group 83–02. *Cancer* **76**, 307–313.

22. Schellinger, P. D., Meinck, H. M., and Thron, A. (1999). Diagnostic accuracy of MRI compared to CCT in patients with brain metastases. *J Neurooncol* **44**, 275–281.

23. Lee, B. C., Kneeland. J. B., Cahill, P. T., and Deck, M. D. (1985). MR recognition of supratentorial tumors. *AJNR Am J Neuroradiol* **6**, 871–878.

24. Macdonald, D. R., Cascino, T. L., Schold, S. C., Jr., and Cairncross, J. G. (1990). Response criteria for phase II studies of supratentorial malignant glioma. *J Clin Oncol* **8**, 1277–1280.

25. Cairncross, J. G., Macdonald, R., Pexman, J. H. *et al.* (1988). Steroid-induced CT changes in patients with recurrent malig-nant glioma. Neurology **38**, 724–726.

26. Graves, E. E., Nelson, S. J., Vigneron, D. B. *et al.* (2001). Serial proton MR spectroscopic imaging of recurrent malignant gliomas after gamma knife radiosurgery. *AJNR Am J Neuroradiol* **22**, 613–624.

27. Chan, A. A., Lau, A., Pirzkall, A. *et al.* (2004). Proton magnetic resonance spectroscopy imaging in the evaluation of patients undergoing gamma knife surgery for Grade IV glioma. *J Neurosurg* **101**, 467–475.

28. Nelson, S. J. (2003). Multivoxel magnetic resonance spectros-copy of brain tumors. *Mol Cancer Ther* **2**, 497–507.

29. Hsu, Y. Y., Chang, C. N., Wie, K. J., Lim, K. E., Hsu, W. C., and Jung, S. M. (2004). Proton magnetic resonance spectroscopic imaging of cerebral gliomas: correlation of metabolite ratios with histopathologic grading. *Chang Gung Med J* **27**, 399–407.

30. Haines, S. J. (2002). Moving targets and ghosts of the past: outcome measurement in brain tumour therapy. *J Clin Neurosci* **9**, 109–112.

MOLECULAR BIOLOGY AND BASIC SCIENCE

Molecular Genetics of Brain Tumors—An Overview

Tim Demuth and Michael E. Berens

ABSTRACT: Human brain tumors continue to benefit from ongoing investigation of the molecular genetic changes that explain gliomagenesis, malignant progression, patient outcome, and possibly therapeutic responsiveness. Optimal utilization of current therapies may be gained by using molecular genetic markers of glial tumors, but meaningful advances in improved patient treatment will emerge from new therapies matched to the specific chemo-vulnerabilities of these tumors as addressed by small-molecule agents specifically targeting the points of vulnerability.

INTRODUCTION

Classical molecular genetics as a descriptive science has afforded insight into the events that underlie tumorigenesis of intracranial neoplasms. From this orientation, clinical management of patients with brain tumors has gained more refined diagnostic and prognostic tools that enable more accurate assessment of the anticipated course of the disease process. The tools for such descriptions continue to expand in their pace, precision, and practice, warranting a review of the current understanding of the genetic and genomic basis for glioma behavior.

DESCRIPTIVE MODELING OF THE MOLECULAR PATHOLOGY OF BRAIN TUMORS

While classical cytogenetics paves the way to understand significant genetic changes that accompany malignant progression in cancer, comparative genomic hybridization (CGH) reveals even smaller portions of chromosomes that are gained or lost during malignant progression. From this closer look at the chromosomal aberrations in tumors, has emerged a description of brain tumors that includes foundational demarcations highlighting the roles of activated oncogenes and lost/dysfunctional tumor suppressor genes. From these studies, a coherent model for the role of different genes and chromosomal regions in gliomagenesis has been developed which informs changes based on the age of tumor onset, the pattern of progression of the disease, and the histological subtype at diagnosis [1]. The histological and molecular genetic descriptors of human brain tumors are maturing into routine applications in diagnostic and prognostic practices [2]. These cutting-edge developments are updated here.

Due to the distorted genetics of astrocytic tumors, it remains of interest to gain an appreciation for the underlying mechanisms of the genetic instability. Chromosomal instability in glioblastoma multiforme (GBM) is found in a large percentage of specimens (up to 70 per cent), resulting in a tumor process with a rich repertoire of permutations for both microenvironmental adaptation and survival under selection pressure such as chemotherapy [3]. Microsatellite instability among GBMs was found to be an infrequent process, and when it occurred, the degree was only low, which is in stark contrast to many other malignancies [4].

More commonly, aberrations in the epidermal growth factor receptor, EGFR, and the p53 tumor suppressor gene are associated with early events in the two predominant but disparate pathways for gliomagenesis in older and younger patients, respectively. Surprisingly, in those instances in older patients where a rigorous case is made for a critical role of EGFR in tumorigenesis, the autocatalytically active form of the receptor, EGFRvIII, is almost never found to be the predominantly expressed form of this receptor tyrosine kinase, but that it is co-expressed with wild-type EGFR as well [5]. The arena for one of the most extensively studied and thoroughly characterized signaling triggers for malignant behavior of GBMs still holds some surprises for the experimental oncologist.

Brain tumors join all other malignancies in their abundance of instances showing mutations in the p53 tumor suppressor gene. The biology of p53 in GBMs can not be entirely captured by catalogs of different mutations. Overexpression of TP53 protein, irrespective of the presence or absence of mutations in the conserved exons, inflicts a survival disadvantage to patients with GBM [6]. Paradoxically, gliomas arising in the thalamus and basal ganglia as primary GBMs (i.e., no history of tumor progression from lower grades), show a universal presentation of mutated p53 [7]; typically for all highly malignant gliomas, p53 mutations in primary GBMs are a very rare finding. This observation raises the possibility for an unanticipated or under-appreciated microenvironmental selection for genetic subtypes of GBM. The presence of p53 mutations in low-grade astrocytomas serves as a predictor of accelerated progression of disease [8] in keeping with its role as a tumor suppressor.

The interplay between various oncogenes and tumor suppressor genes serves as a model for the survival advantage conferred by adaptation or redundancy of signaling pathways in GBMs. Malignant brain tumors are demonstrated to avail themselves of either dysregulated behaviors driven by p53 mutations/deletions or through alternative but complimentary mediators affecting the same outcomes, such as phosphoinositide 3-kinase (PI3K) [9]. Similarly, PTEN aberrations (loss or inactivating mutations) characteristically and pathologically activate the PI3K/Akt signaling pathway in GBM; where the tumors show normal PTEN, some instances are reported where an activating mutation in PI3K is discerned [10]. Such demonstrations highlight a need to adopt a systems approach to tumor biology, seeking the broadest understanding of collateral or parallel signaling pathways that may foster the malignant behavior of cancers. The breadth of collateral routes for pathological signaling in tumor cells does show some limits. For example, despite a near-universal presentation of activated RAS-function in GBM cells and tissue, mutations in key RAS-family members in GBMs are exceptionally rare [11]. This being a point of interventional vulnerability in these tumors is an argument that awaits demonstration.

Among glial tumors, oligodendrogliomas have evidenced the most robust molecular genetic determinants predicting tumor behavior [12]. The pathogenesis of glial tumors with oligodendroglial elements of complete or mixed degrees demonstrates clear clinical consequences to the loss of heterozygosity (LOH) of chromosomes 1p and 19q, including remarkable chemosensitivity to standard interventions. The preponderance of 1p/19q LOH in oligodendroglial neoplasms support the use of molecular markers for these features as an adjunct in differential or confirmatory diagnosis. Fine detailed mapping for polymorphic markers at 1p among a large set of oligodendrogliomas and astrocytomas known to be informative for 1p LOH allowed narrowing of the minimal deleted region to 1p36.23, which maps to the calmodulin binding transcription activator 1 (CAMTA1) transcription factor [13]. Although currently unverified, the suggestion of a deleted transcription factor in the context of chemovulnerability offers a promising avenue to drive innovation for new treatments of brain tumors.

The expansion in genetic studies of glial tumors is anticipated to impact prognostic prowess, and these expectations are emerging in practice. Among astrocytomas analyzed for LOH of the 10q region, survival was found to be tightly linked to the magnitude of loss of this genetic material [14]. The grade of malignancy also correlated with the frequency of 10q LOH. The study suggests that unappreciated tumor suppressor genes may reside at 10q24 and 10q26. In a larger population-based study over 15 years, 10q LOH was the most frequently detected genetic aberration in GBM (69 per cent of the cases), as well as proved to be the strongest predictor of patient survival among the molecular genetic markers [15]. EGFR amplification, p53 mutation, and deletion of cyclin-dependent kinase inhibitor 2A (CDKN2A/p16) were each found in approximately one-third of the 715 cases of GBM. Identification of mutations in p53 in primary GBMs is an uncommon occurrence, but when detected, they cluster at CpG sites, in contrast to the majority of mutations in secondary GBMs that occur at hotspot codons 248 and 273. This increases the prospect that aberrant functioning of p53 may arise through disparate mechanisms when it occurs in primary *versus* secondary GBM. A separate report associating the mutation events of p53 with age of the patient at the time of GBM diagnosis corroborates this observation, and argues strongly that molecular markers in GBM are informed by chronology of the onset of disease [16]. Genetic markers such as amplification of EGFR as reported in primary GBMs, gain added prognostic strength when combined with analysis of levels of the translated protein [17].

Overexpression of the wild-type form of EGFR by GBMs is invariably associated with gene amplification, but those GBMs expressing the truncated EGFR (EGFRvIII) may do so in the absence of amplification. Unexpectedly, expression status of wild-type EGFR provides a stronger prognostic lead for GBM patients under 60 years than in older patients, but levels of expression of the mutated EGFRvIII predict overall survival of GBM patients. Comparisons of the genetic aberrations in GBMs from patients who succumbed to disease early (< 1.5 years) *versus* late (> 3.0 years) identified discriminating incidences of 10q del, 6q del, and amplified 19q in those tumors inflicting short-term survival; loss of 19q was associated with long-term survival [18]. These genetic descriptors of human GBMs offer prognostic guidance to the anticipated outcomes for patients.

Genetic markers provide insight to the underlying processes of gliomagenesis, and may inform the outcome for patients with various brain tumors. Therapeutically, there are some indications that genetic markers may indicate chemovulnerability. The specific response to topoisomerase I inhibitors can be predicted by mutated p53 in GBM xenografts [19], and also that GBM xenografts with mutated p53 show greater apoptosis rates and suppressed re-proliferation activity after chemotherapy. However, patient response rates to nitrosoureas were not related to mutation status of p53 [20]. Glioma cells showing LOH of 1p/19q are dramatically more responsive to chemotherapy than similar histological subtypes with normal copy numbers of these chromosomal regions; an observation of amplified EGFR was also associated with relative chemoresistance [21]. The incidence of hypermethylation of the MGMT promoter CpG-islands is also higher in gliomas with 1p/19q LOH [22]. As new therapies are developed against targets expressed in GBMs, the application of genetic markers for planning therapy is likely to greatly improve patient management.

PREDICTIVE MODELING OF THERAPEUTIC VULNERABILITY OF BRAIN TUMORS

Partly due to their heterogenic nature and early tendency to actively invade the surrounding brain parenchyma, glial tumors represent an intractable disease. All the current treatment strategies are derived from clinical trials comparing varieties of insufficient therapies prone to fail since tumor cells are not being specifically targeted. With the exception of LOH 1p/19q as predictor for increased sensitivity to chemotherapy in oligodendroglioma, no markers exist for allowing treatment stratification. It is this bleak situation for patients, as well as the disarmed neuro-oncologist, that mandates novel therapies directly targeting a tumor cell's point(s) of vulnerability. The implementation of high-throughput genomic and proteomic techniques to analyze an individual's tumor will bring personalized medicine within reach of brain tumor patients.

Histopathological evaluation of tumor samples and classification according to WHO criteria is the current standard for diagnosis and serves as a basis for therapeutic decisions. Since tissue morphology is in part dictated by gene expression, it is not surprising that clustering of expression data reveals classes recapitulating histopathological groupings of tumors. This tool becomes more significant as it allows identification of subgroups exhibiting significantly increased survival within histologically similar tumors rendering this a potential strategy suitable for patient stratification [23]. Nutt and coworkers used a 20-feature class prediction model to consistently classify histologically non-classic tumor samples. Gene-expression profiling provided more accurate prediction of prognosis than standard histology [24]. Out of the thousands of genes printed on standard microarrays, subsets with strong predictive value can be identified in such a way that this panel can be used as a prognostic tool. Freije and coworkers reported a set of 44 genes that accurately predicts patient survival [25]. Partial least squares were used to cluster samples into two prognosis-related groups revealing significantly different survival times (4.8 years *versus* 8 months). Each group consisted of different histological diagnoses, with the vast majority of glioblastoma multiforme falling into the poor prognosis group within which two subgroups were identified, one with an invasion and one with a proliferation signature suggesting alternate points of vulnerability.

Based on well-established gene amplifications and losses during glioma progression, expression profiling has been employed to discover further genes involved in glioma progression. Sallinen *et al.*, identified IGFBP2 (Insulin-like growth-factor-binding protein) to be correlated with progression from low-grade astrocytomas to glioblastoma multiforme [26] and Ljubimova and coworkers reported a variety of novel genes to be involved in glioma pathology [27].

Rickman and colleagues reported an expression signature characteristic for high-grade and low-grade gliomas [28] generating evidence for a molecular evolution driving progressive stages of astrocytoma malignancy.

Gene-expression profiles described so far are mostly signatures of the main tumor mass, while not necessarily depicting expression changes occurring as tumor cells egress and invade the surrounding brain parenchyma [29]. Proliferative cells from the tumor core and infiltrative cells show distinct differences in gene-expression profiles, implying differential vulnerabilities that may require separate therapies designed for each population. Microarray analysis of migration-induced glioma cells *in vitro* compared to stationary cells led to the discovery of death-associated protein (DAP3), which is up-regulated with glioma migration concomitantly increasing resistance to apoptosis induction [30]. Another gene candidate up-regulated in the context of invasion is Fn14 encoding cell-surface receptor for TNF-related weak inducer of apoptosis (TWEAK). Fn14 is correlated with glioma progression and acts as a pro-survival factor through activation of NFκB-pathway and expression of BCL-XL/BCL-W [31]. The receptor tyrosine kinase, EphB2, was also found to be up-regulated with glioma migration; its activation by phosphorylation is an important driver of glioma invasion [32]. Cell-adhesion molecule N-Cadherin, linking extracellular signals to the actin cytoskeleton through catenins, is up-regulated in glioblastoma samples while T-Cadherin was found to be up-regulated with glioma invasion; blocking of N-Cadherin results in decreased *in vitro* invasion [33].

Cell–cell communications are important in growth control and differentiation. This is in part accomplished by gap junction proteins, of which connexin 43 is the most abundant in the CNS. Increased malignancy of astrocytomas has been reported to be correlated with reduced connexin 43 expression [34] and increased expression of this gene leads to reduction of apoptosis inhibitor Bcl2, rendering it an interesting therapeutic target. Rho GTPases interact with the actin cytoskeleton and regulate many signal transduction pathways participating in regulation of cell polarity, gene transcription, and G1 cell-cycle progression. Rho is thought to act as a molecular switch controlling signal transduction pathways, linking membrane receptors to the cytoskeleton, and has been demonstrated to be inversely correlated with glioma progression. Rho is rapidly activated when glioma cells attach to a substrate, thereby arguing for its involvement in forming focal adhesions [35]. While RAC activation occurs much slower in that context [36], inhibition of RAC1 leads to apoptosis induction in glioma cells and is found to be necessary for the migratory phenotype [37]. The hypothesis that migratory glioma cells are more resistant to apoptosis induction is supported by the finding that the anti-apoptotic PI3K pathway is activated in migratory glioma cells [38]. Based on these findings, the invasive transcriptome of malignant gliomas were studied by comparing the gene-expression profile of stationary and invasive glioma cells isolated by laser capture microdissection (LCM) from GBM biopsy specimens [39]. Gene candidates up-regulated with active invasion included adhesion molecules, transducers of extracellular and intracellular signaling as well as apoptosis resistance genes; stationary glioma cells exhibited increased levels of cytoskeleton stabilizing elements as well as angiogenesis-related genes. Zagzag and coworkers in a similar approach reported down-regulation of MHC (major histocompatibility) antigens in invading glioma cells, facilitating escape from the patient's immune system [40].

Matrix metalloproteinases (MMPs) have been thought to be a promising target for anti-invasive therapies in many tumor types including glioblastoma multiforme. Conceptually, MMPs create room for invading tumor cells degrading extracellular matrix proteins. Several studies report matrix-degrading enzymes to be up-regulated compared to normal brain tissue as well as during glioma progression [41,42]. *In vitro*, several MMP-inhibitors have effectively down-regulated glioma invasion [43]. However, clinical trials with these agents failed to demonstrate significant improvements in the clinical course of glioma and other tumors [44,45] (reviewed in [46]). One of the reasons for this discordant finding might be the lack of adequate substrate (collagen-I, gelatin) for these enzymes in brain tissue or inefficient transition of drugs through an intact blood–brain barrier at the tumor's invasive edge. There seems to be some evidence for survival benefits from combination therapy of the MMP-inhibitor marimastat and temozolomide in brain tumors [47], which might make use of the more important cytostatic effect of marimastat which is potentiated by an alkylating agent.

Another interesting target developed from the studies on glioma biology is the extracellular matrix (ECM) protein Tenascin-C, which has been extensively studied over the past two decades. Its expression has been linked to malignancy in astrocytomas [48] coincident with higher concentration of Tenascin-C around hyperplastic vessels [49]. Tenascin-C is produced by cultured endothelial cells [50,51] and enhances their migration as a consequence of phosphorylation of focal adhesion kinase (FAK) [52]. Wenk and colleagues have shown that Tenascin-C induces cell motility in fibroblasts through inhibition of RhoA, which immediately leads to decreased formation of stress fibers and increased occurrence of actin-rich filopodia [53]. The significance of Tenascin-C in glioma pathobiology is complimented by findings of its overexpression in invasive glioma cells compared to their stationary cognates *in vivo* [30]. Kurpad and coworkers described a strong and homogenous distribution of Tenascin-C in invasive rat brain tumors, supporting the observation that Tenascin-C is produced by the tumor rather than by the invaded brain. Based on identification of tenascin as a "glioma-specific" antigen, a novel therapy for malignant gliomas was introduced by Bigner and coworkers [54,55]. Tenascin-specific antibodies were employed to deliver I[131] radio immunotherapy to glioma patients. Minimal increases in overall survival have been reported from a phase II study; interpretation of the outcome of the trial may have been confounded by vagaries of patient selection or deficiencies in control patient populations [56,57]. Due to its close association with angiogenesis, targeting tenascin may turn out to be more of an anti-angiogenic intervention, which in the end may increase tumor invasiveness [6].

Presently, there are few markers that correlate with chemosensitivity in brain tumors as are beginning to emerge for other cancers. Factors pointing towards increased sensitivity to chemotherapeutic agents apart from MGMT expression and promoter methylation have yet to be determined. Molecular analysis will likely be able to subcategorize brain tumors according to outcome in a manner similar to a study by Vijer and coworkers who described an expression signature that separates young women with breast cancer exhibiting good prognosis from those with poor prognosis [58], an approach that helps the physician to decide whether or not to treat the patient more aggressively.

Marker selection in glioma is rendered more intricate by recent findings from Singh and coworkers who presented evidence for stem cells as initiators and maintainers of glial tumors. Brain tumor stem cells were isolated from different human brain tumor specimens and expressed neural stem cell markers such as CD133. These cells exhibit clonogenic capacity and are capable of tumor propagation. In a xenograft model, isolated CD133 positive but not CD133 negative cells exhibited tumorigenicity and could be passaged through several animals while closely resembling the morphology of the original tumor [59]. The recognition of tumor stem cells may introduce an assortment of new therapeutic approaches aimed at converting these self-renewing cancer cells into differentiated ones.

It is important to keep in mind that differential gene expression does not necessarily translate into differentially expressed protein products. Post-translational modifications as well as protein–protein and protein–DNA interactions are important determinants for biological functions. Expression microarrays are powerful high-throughput tools capable of creating and testing hypotheses that need to be validated at the protein level as well as functionally. In recent years, proteomic techniques have become increasingly available and are employed to identify surrogate markers for early diagnosis of cancer to monitor response to therapy as well as to discover new therapeutic targets. To characterize the proteome of malignant gliomas, Iwadate and coworkers exploited a combination of two-dimensional (2D) gel electrophoresis and MALDI-TOF (Matrix Assisted Laser Desorption - Time Of Flight) identifying a set of 37 proteins differentially expressed between normal brain tissue and gliomas [60]. Subsets of more aggressive astrocytomas were identified and survival could be predicted based on a pattern of proteins. Zhang and coworkers used 2D gel-electrophoresis followed by LC-MS-MS (liquid chromatography–mass spectroscopy) to identify proteins differentially expressed in human glioma cell lines relative to cultured fetal human astrocytes [61]. Candidate proteins Hsp27, major vault protein (MVP) which has recently been related to chemoresistance, TTG (tissue transglutaminase), and cytostatin B were found to be overexpressed in GBM samples compared to normal brain. An easily accessible source for biomarker analysis in brain tumors is cerebrospinal fluid (CSF) comprised of fewer proteins than serum that can be analyzed by similar techniques as solid tumors. N-myc and l-CaD were detected in CSF of 12 patients with primary brain tumors and may serve as markers for intra-axial tumors. Low abundance proteins, however, escape detection through this technique, mandating use of alternate approaches such as reverse phase protein microarray, allowing for detection of the activation state of signaling pathways. The Akt survival pathway has been found to be activated with disease progression in prostate cancer employing this technique on laser-capture microdissected samples [62] and it can be expected that this method will help to discover pathways activated during glioma progression as well as in glioma invasion.

Epigenetic phenomena are increasingly appreciated modulators of transcription including DNA methylation and histone modification. The main epigenetic modification in humans is methylation of the nucleotide cytosine (C) when it precedes a guanine (G) forming the dinucleotide CpG. Hyper-methylation commonly occurs in CpG islands located in the promoters of tumor suppressor genes, such as p16INK4a, BRCA1, or hMLH1 [63–65]. With MGMT as a major factor determining sensitivity, Hegi demonstrated that methylation of MGMT promoter in glioma patients results in increased sensitivity to alkylating agents and radiation therapy [66] and it can be hypothesized that decreasing incidence of MGMT promoter methylation is correlated with malignant progression as observed in other tumors. Apoptosis regulator TMS1 was found to be down-regulated in 50 per cent of the glioma specimens coincident with gene methylation in more than 40 per cent of the cases [67]. Cancer antigen MAGEA1 which could be involved in immune response towards glioma was also found to be methylated [68]. Decreased expression of carboxyl-terminal modulator protein (CTMP), a negative regulator of protein kinase B/Akt, was found to correlate with hypermethylation of CTMP promoter in glioblastoma biopsies [69]. Presently methylation status is studied on a gene-by-gene basis using methylation-specific PCR. In an attempt to monitor CpG-island methylation in a high-throughput manner, microarray based approaches have been recently described. Yu and coworkers studied the methylation status of 100 genes in prostate cancer cell lines and correlated the results with gene-expression profiling allowing for identification of genes whose expression is regulated by methylation events [68]. Waha and coworkers identified protocadherin-gamma subfamily A11 (PCDH-gamma-A11) to be hypermethylated in low-grade astrocytomas through microarray based methylation analysis [70]. After demethylating treatment, transcription of PCDH-gamma-A11 was restored. These reports underline the importance of epigenetic modifications and their contribution to gene-expression changes in cancer; array based analysis of these events may help to shed light on expression changes observed during glioma progression as well as in phenotypically different cells from the same tumor (i.e., proliferative *versus* stationary) that might be translated into therapeutic targets. Known chemotherapeutic agents such as Azacytidine have been identified to act as hypomethylators in this manner explaining good response in high-risk myelodysplastic syndrome which shows a high prevalence of tumor suppressor gene hyper-methylation [71].

FUTURE DIRECTIONS

Presently different experimental approaches are taken to study individual features of a complex disease such as brain tumors. It is, however, not a single gene or protein that is the ultimate driver and origin of a malignant disease but rather a variety of complex networks of closely interacting proteins providing alternate interchangeable routes for maintenance and progression of autonomous growth. It is therefore imperative to integrate data sets (i.e., genetic, genomic, and proteomic) studying different facets of the disease in a systems biology approach that allows for better understanding of the complex nature of brain cancer. Information from promoter-methylation studies will be integrated with expression profiles to help explain observed signatures. Data from proteomics experiments and tissue micro-arrays will be related to this information enriched with clinical annotation regarding the patients from whom samples were derived.

Observations from *in vitro* studies perturbing pathways of interest will be fed back into the analysis process. Analysis of these complex data sets creates a new level of intricacy, requiring analytical techniques that go beyond the standard frequency based (i.e., t-test, ANOVA) statistical approaches, such as the implementation of machine-learning algorithms, pattern recognition [72], text-mining, and network modeling tools (i.e., Ingenuity's Pathway Analysis Tool). The cancer Biomedical Informatics Grid (caBIG - http://cabig.nci.nih.gov) is a collaborative project driven by the NCI, which aims at providing infrastructure to integrate the multitude of high-throughput "omics" information. Due to its cohesive nature, the brain cancer research community is piloting the application of caBIG in Rembrandt (REpository for Molecular BRAin Neoplasia DaTa), a database aimed at gathering gene-expression profiles, Comparative genomic hybridization (CGH) and Single nucleotide polymorphism (SNP) information, sequencing data, tissue array results, and clinical information from patients with brain tumors.

References

1. Kleihues, P. C. W. (2000). *Pathology & Genetics of Tumours of the Nervous System*. 2nd ed. IARC Press, Lyon.
2. Kleihues, P., Louis, D. N., Scheithauer, B. W. *et al.* (2002). The WHO classification of tumors of the nervous system. *J Neuropathol Exp Neurol* **61**, 215–225, (discussion 226–219).
3. Martinez, R., Schackert, H. K., Plaschke, J., Baretton, G., Appelt, H., and Schackert, G. (2004). Molecular mechanisms associated

with chromosomal and microsatellite instability in sporadic glioblastoma multiforme. *Oncology* 66, 395–403.

4. Martinez, R., Schackert, H. K., Appelt, H., Plaschke, J., Baretton, G., and Schackert, G. (2004). Low-level microsatellite instability phenotype in sporadic glioblastoma multiforme. *J Cancer Res Clin Oncol* 131, 87–93.

5. Biernat, W., Huang, H., Yokoo, H., Kleihues, P., and Ohgaki, H. (2004). Predominant expression of mutant EGFR (EGFRvIII) is rare in primary glioblastomas. *Brain Pathol* 14, 131–136.

6. Pardo, F. S., Hsu, D. W., Zeheb, R., Efird, J. T., Okunieff, P. G., and Malkin, D. M. (2004). Mutant, wild type, or overall p53 expression: freedom from clinical progression in tumours of astrocytic lineage. *Br J Cancer* 91, 1678–1686.

7. Hayashi, Y., Yamashita, J., and Watanabe, T. (2004). Molecular genetic analysis of deep-seated glioblastomas. *Cancer Genet Cytogenet* 153, 64–68.

8. Stander, M., Peraud, A., Leroch, B., and Kreth, F. W. (2004). Prognostic impact of TP53 mutation status for adult patients with supratentorial World Health Organization Grade II astrocytoma or oligoastrocytoma: a long-term analysis. *Cancer* 101, 1028–1035.

9. Mayo, L. D., Dixon, J. E., Durden, D. L., Tonks, N. K., and Donner, D.B. (2002). PTEN protects p53 from Mdm2 and sensitizes cancer cells to chemotherapy. *J Biol Chem* 277, 5484–5489.

10. Mizoguchi, M., Nutt, C. L., Mohapatra, G., and Louis, D. N. (2004). Genetic alterations of phosphoinositide 3-kinase subunit genes in human glioblastomas. *Brain Pathol* 14, 372–377.

11. Knobbe, C. B., Reifenberger, J., and Reifenberger, G. (2004). Mutation analysis of the Ras pathway genes NRAS, HRAS, KRAS and BRAF in glioblastomas. *Acta Neuropathol (Berl)* 108, 467–470.

12. Jeuken, J. W., von Deimling, A., and Wesseling, P. (2004). Molecular pathogenesis of oligodendroglial tumors. *J Neurooncol* 70, 161–181.

13. Barbashina, V., Salazar, P., Holland, E. C., Rosenblum, M. K., and Ladanyi, M. (2005). Allelic losses at 1p36 and 19q13 in gliomas: correlation with histologic classification, definition of a 150-kb minimal deleted region on 1p36, and evaluation of CAMTA1 as a candidate tumor suppressor gene. *Clin Cancer Res* 11, 1119–1128.

14. Daido, S., Takao, S., Tamiya, T. *et al.* (2004). Loss of heterozygosity on chromosome 10q associated with malignancy and prognosis in astrocytic tumors, and discovery of novel loss regions. *Oncol Rep* 12, 789–795.

15. Ohgaki, H., Dessen, P., Jourde, B. *et al.* (2004). Genetic pathways to glioblastoma: a population-based study. Cancer Res 64, 6892–6899.

16. Batchelor, T. T., Betensky, R. A., Esposito, J. M. *et al.* (2004). Age-dependent prognostic effects of genetic alterations in glioblastoma. *Clin Cancer Res* 10, 228–233.

17. Shinojima, N., Tada, K., Shiraishi, S. *et al.* (2003). Prognostic value of epidermal growth factor receptor in patients with glioblastoma multiforme. *Cancer Res* 63, 6962–6970.

18. Burton, E. C., Lamborn, K. R., Feuerstein, B. G. *et al.* (2002). Genetic aberrations defined by comparative genomic hybridization distinguish long-term from typical survivors of glioblastoma. *Cancer Res* 62, 6205–6210.

19. Wang, Y., Zhu, S., Cloughesy, T. F., Liau, L. M., and Mischel, P. S. (2004). p53 disruption profoundly alters the response of human glioblastoma cells to DNA topoisomerase I inhibition. *Oncogene* 23, 1283–1290.

20. Watanabe, T., Katayama, Y., Komine, C. *et al.* (2005). O6-methylguanine-DNA methyltransferase methylation and TP53 mutation in malignant astrocytomas and their relationships with clinical course. *Int J Cancer* 113, 581–587.

21. Leuraud, P., Taillandier, L., Medioni, J. *et al.* (2004). Distinct responses of xenografted gliomas to different alkylating agents are related to histology and genetic alterations. *Cancer Res* 64, 4648–4653.

22. Mollemann, M., Wolter, M., Felsberg, J., Collins, V. P., and Reifenberger, G. (2005). Frequent promoter hypermethylation and low expression of the MGMT gene in oligodendroglial tumors. *Int J Cancer* 113, 379–385.

23. Fuller, C.E., and Perry, A. (2001). Pathology of low- and intermediate-grade gliomas. *Semin Radiat Oncol* 11, 95–102.

24. Nutt, C. L., Mani, D. R., Betensky, R. A. *et al.* (2003). Gene expression-based classification of malignant gliomas correlates better with survival than histological classification. *Cancer Res* 63, 1602–1607.

25. Freije, W. A., Castro-Vargas, F. E., Fang, Z. *et al.* (2004). Gene expression profiling of gliomas strongly predicts survival. *Cancer Res* 64, 6503–6510.

26. Sallinen, S. L., Sallinen, P. K., Haapasalo, H. K. *et al.* (2000). Identification of differentially expressed genes in human gliomas by DNA microarray and tissue chip techniques. *Cancer Res* 60, 6617–6622.

27. Ljubimova, J. Y., Khazenzon, N. M., Chen, Z. *et al.* (2001). Gene expression abnormalities in human glial tumors identified by gene array. *Int J Oncol* 18, 287–295.

28. Rickman, D. S., Bobek, M. P., Misek, D. E. *et al.* (2001). Distinctive molecular profiles of high-grade and low-grade gliomas based on oligonucleotide microarray analysis. *Cancer Res* 61, 6885–6891.

29. Demuth, T., and Berens, M. E. (2004). Molecular mechanisms of glioma cell migration and invasion. *J Neurooncol* 70, 217–228.

30. Mariani, L., Beaudry, C., McDonough, W. S. *et al.* (2001). Glioma cell motility is associated with reduced transcription of proapoptotic and proliferation genes: a cDNA microarray analysis. *J Neurooncol* 53, 161–176.

31. Tran, N. L., McDonough, W. S., Savitch, B. A., Sawyer, T. F., Winkles, J. A., and Berens, M. E. (2005). The tumor necrosis factor-like weak inducer of apoptosis (TWEAK)-fibroblast growth factor-inducible 14 (Fn14) signaling system regulates glioma cell survival via NFkappaB pathway activation and BCL-XL/BCL-W expression. *J Biol Chem* 280, 3483–3492.

32. Nakada, M., Niska, J. A., Miyamori, H. *et al.* (2004). The phosphorylation of EphB2 receptor regulates migration and invasion of human glioma cells. *Cancer Res* 64, 3179–3185.

33. Asano, K., Duntsch, C. D., Zhou, Q. *et al.* (2004). Correlation of N-cadherin expression in high grade gliomas with tissue invasion. *J Neurooncol* 70, 3–15.

34. Soroceanu, L., Manning, T. J., Jr., and Sontheimer, H. (2001). Reduced expression of connexin-43 and functional gap junction coupling in human gliomas. *Glia* 33, 107–117.

35. Ding, Q., Stewart, J., Jr., Prince, C. W. *et al.* (2002). Promotion of malignant astrocytoma cell migration by osteopontin expressed in the normal brain: differences in integrin signaling during cell adhesion to osteopontin versus vitronectin. *Cancer Res* 62, 5336–5343.

36. Forget, M. A., Desrosiers, R. R., Del, M. *et al.* (2002). The expression of rho proteins decreases with human brain tumor progression: potential tumor markers. *Clin Exp Metastasis* 19, 9–15.

37. Ridley, A. J., and Hall, A. (1992). The small GTP-binding protein rho regulates the assembly of focal adhesions and actin stress fibers in response to growth factors. *Cell* 70, 389–399.

38. Joy, A. M., Beaudry, C. E., Tran, N. L. *et al.* (2003). Migrating glioma cells activate the PI3-K pathway and display decreased susceptibility to apoptosis. *J Cell Sci* **116**, 4409–4417.

39. Hoelzinger, D. B., Mariani, L., Weis, J. *et al.* (2005). Gene expression profile of glioblastoma multiforme invasive phenotype points to new therapeutic targets. *Neoplasia* **7**, 7–16.

40. Zagzag, D., Salnikow, K., Chiriboga, L. *et al.* (2005). Down-regulation of major histocompatibility complex antigens in invading glioma cells: stealth invasion of the brain. *Lab Invest* **85**, 328–341.

41. Forsyth, P. A., Wong, H., Laing, T. D. *et al.* (1999). Gelatinase-A (MMP-2), gelatinase-B (MMP-9) and membrane type matrix metalloproteinase-1 (MT1-MMP) are involved in different aspects of the pathophysiology of malignant gliomas. *Br J Cancer* **79**, 1828–1835.

42. Nakada, M., Nakamura, H., Ikeda, E. *et al.* (1999). Expression and tissue localization of membrane-type 1, 2, and 3 matrix metalloproteinases in human astrocytic tumors. *Am J Pathol* **154**, 417–428.

43. Tonn, J. C., Kerkau, S., Hanke, A. *et al.* (1999). Effect of synthetic matrix-metalloproteinase inhibitors on invasive capacity and proliferation of human malignant gliomas in vitro. *Int J Cancer* **80**, 764–772.

44. Tonn, J. C., and Goldbrunner R. (2003). Mechanisms of glioma cell invasion. *Acta Neurochir Suppl* **88**, 163–167.

45. Whittaker, M., Floyd, C. D., Brown, P., and Gearing, A. J. (1999). Design and therapeutic application of matrix metalloproteinase inhibitors. *Chem Rev* **99**, 2735–2776.

46. Coussens, L. M., Fingleton, B., and Matrisian, L. M. (2002). Matrix metalloproteinase inhibitors and cancer: trials and tribulations. *Science* **295**, 2387–2392.

47. Groves, M. D., Puduvalli, V. K., Hess, K. R. *et al.* (2002). Phase II trial of temozolomide plus the matrix metalloproteinase inhibitor, marimastat, in recurrent and progressive glioblastoma multiforme. *J Clin Oncol* **20**, 1383–1388.

48. Leins, A., Riva, P., Lindstedt, R., Davidoff, M. S., Mehraein, P., and Weis, S. (2003). Expression of Tenascin-C in various human brain tumors and its relevance for survival in patients with astrocytoma. *Cancer* **98**, 2430–2439.

49. Zagzag, D., Friedlander, D. R., Miller, D. C. *et al.* (1995). Tenascin expression in astrocytomas correlates with angiogenesis. *Cancer Res* **55**, 907–914.

50. Zagzag, D., Friedlander, D. R., Dosik, J. *et al.* (1996). Tenascin-C expression by angiogenic vessels in human astrocytomas and by human brain endothelial cells in vitro. *Cancer Res* **56**, 182–189.

51. Kostianovsky, M., Greco, M. A., Cangiarella, J., and Zagzag, D. (1997). Tenacin-C expression in ultrastructurally defined angiogenic and vasculogenic lesions. *Ultrastruct Pathol* **21**, 537–544.

52. Zagzag, D., Shiff, B., Jallo, G. I. *et al.* (2002). Tenascin-C promotes microvascular cell migration and phosphorylation of focal adhesion kinase. *Cancer Res* **62**, 2660–2668.

53. Wenk, M. B., Midwood, K. S., and Schwarzbauer, J. E. (2000). Tenascin-C suppresses Rho activation. *J Cell Biol* **150**, 913–920.

54. Bigner, D. D., Brown, M., Coleman, R. E. *et al.* (1995). Phase I studies of treatment of malignant gliomas and neoplastic meningitis with 131I-radiolabeled monoclonal antibodies anti-tenascin 81C6 and anti-chondroitin proteoglycan sulfate Me1-14 F (ab')2–a preliminary report. *J Neurooncol* **24**, 109–122.

55. Kurpad, S. N., Zhao, X. G., Wikstrand, C. J., Batra, S. K., McLendon, R. E., and Bigner, D. D. (1995). Tumor antigens in astrocytic gliomas. *Glia* **15**, 244–256.

56. Reardon, D. A., Akabani, G., Coleman, R. E. *et al.* (2002). Phase II trial of murine (131)I-labeled antitenascin monoclonal antibody 81C6 administered into surgically created resection cavities of patients with newly diagnosed malignant gliomas. *J Clin Oncol* **20**, 1389–1397.

57. Goetz, C., Riva, P., Poepperl, G. *et al.* (2003). Locoregional radioimmunotherapy in selected patients with malignant glioma: experiences, side effects and survival times. *J Neurooncol* **62**, 321–328.

58. van de Vijver, M. J., He, Y. D., van't Veer, L. J. *et al.* (2002). A gene-expression signature as a predictor of survival in breast cancer. *N Engl J Med* **347**, 1999–2009.

59. Singh, S. K., Hawkins, C., Clarke, I. D. *et al.* (2004). Identification of human brain tumour initiating cells. *Nature* **432**, 396–401.

60. Iwadate, Y., Sakaida, T., Hiwasa, T. *et al.* (2004). Molecular classification and survival prediction in human gliomas based on proteome analysis. *Cancer Res* **64**, 2496–2501.

61. Zhang, R., Tremblay, T. L., McDermid, A., Thibault, P., and Stanimirovic, D. (2003). Identification of differentially expressed proteins in human glioblastoma cell lines and tumors. *Glia* **42**, 194–208.

62. Grubb, R. L., Calvert, V. S., Wulkuhle, J. D. *et al.* (2003). Signal pathway profiling of prostate cancer using reverse phase protein arrays. *Proteomics* **3**, 2142–2146.

63. Esteller, M., Corn, P. G., Baylin, S. B., and Herman, J. G. (2001). A gene hypermethylation profile of human cancer. *Cancer Res* **61**, 3225–3229.

64. Esteller, M., Cordon-Cardo, C., Corn, P. G. *et al.* (2001). p14ARF silencing by promoter hypermethylation mediates abnormal intracellular localization of MDM2. *Cancer Res* **61**, 2816–2821.

65. Feinberg, A. P., and Tycko, B. (2004). The history of cancer epigenetics. *Nat Rev Cancer* **4**, 143–153.

66. Hegi, M. E., Diserens, A. C., Gorlia, T. *et al.* (2005). MGMT gene silencing and benefit from temozolomide in glioblastoma. *N Engl J Med* **352**, 997–1003.

67. Stone, A. R., Bobo, W., Brat, D. J., Devi, N. S., Van Meir, E. G., and Vertino, P. M. (2004). Aberrant methylation and downregulation of TMS1/ASC in human glioblastoma. *Am J Pathol* **165**, 1151–1161.

68. Yu, J., Zhang, H., Gu, J. *et al.* (2004). Methylation profiles of thirty four promoter-CpG islands and concordant methylation behaviours of sixteen genes that may contribute to carcinogenesis of astrocytoma. *BMC Cancer* **4**, 65.

69. Knobbe, C. B., Reifenberger, J., Blaschke, B., and Reifenberger, G. (2004). Hypermethylation and transcriptional downregulation of the carboxyl-terminal modulator protein gene in glioblastomas. *J Natl Cancer Inst* **96**, 483–486.

70. Waha, A., Guntner, S., Huang, T. H. *et al.* (2005). Epigenetic silencing of the protocadherin family member PCDH-gamma-A11 in astrocytomas. *Neoplasia* **7**, 193–199.

71. Leone, G., Voso, M. T., Teofili, L., and Lubbert, M. (2003). Inhibitors of DNA methylation in the treatment of hematological malignancies and MDS. *Clin Immunol* **109**, 89–102.

72. Pan, K. H., Lih, C. J., and Cohen, S. N. (2002). Analysis of DNA microarrays using algorithms that employ rule-based expert knowledge. *Proc Natl Acad Sci USA* **99**, 2118–2123.

Regulation of the Cell Cycle and Interventional Developmental Therapeutics

Stacey M. Ivanchuk and James T. Rutka

ABSTRACT: A complex series of molecular signaling events are responsible for cell division. Early in the cell cycle, cues determine whether a cell will initiate a round of replication or withdraw from cell division and enter a state of quiescence. Passage through the restriction point (RP), designated the "point of no return", marks cellular commitment to a new round of division. Genetic mutations affecting molecules that control proliferation predispose to tumorigenesis. Many of the identified mutations associated with cancer cells give rise to molecules that are no longer able to appropriately regulate the mammalian cell cycle and the end result is neoplasia. Here, we review the critical elements that enable cell cycle progression as well as the positive and negative regulators that facilitate the process. In addition, the development and implementation of therapeutic modalities that exploit cell cycle inhibition are examined.

OVERVIEW

Successful cellular division requires two alternating processes. The doubling of DNA content referred to as synthesis (S-phase) is followed by halving of the genome during mitosis (M-phase) to yield two identical cells. Gap (G) periods separate S- and M-phases of the cell cycle. G1 preceeds DNA synthesis and contains the restriction point (RP) which serves as a marker for a cell's decision to commit to

a new round of cell division or not (Fig. 9.1). During G1, growth-promoting signals in the microenvironment prompt cellular growth and prepare cells for chromosomal replication (DNA synthesis). In the absence of mitogenic (growth-stimulating) signals or in the presence of anti-mitogenic signals, cells cease to cycle and enter a state of quiescence (G0 phase). A second gap period (G2) follows DNA synthesis in order to prepare the cell for equal division of its DNA and cellular components between two identical daughter cells (M-phase).

"Checkpoints" exist at the transition between different phases of the cell cycle where integrity of the replication and division processes are carefully monitored prior to passage through the next phase. Cells have developed several mechanisms to interrupt the cell cycle in the event that errors occur. Progression through the cell cycle is inhibited in normal cells until errors have been corrected. Cancer cells almost invariably harbour defects in regulatory checkpoint molecules that enable them to escape arrest and repeat the cell cycle indefinitely regardless of DNA integrity.

There are two main classes of proteins that act in concert to form catalytic enzymatic complexes that maintain cell cycle coordination. The *cyclins* were initially identified in marine invertebrate embryos as proteins that were synthesized in interphase, but were rapidly destroyed with progression through the cell cycle [1]. The cyclical nature of their expression was the impetus for naming this class of proteins as "cyclins". The rapid and timely destruction of cyclins

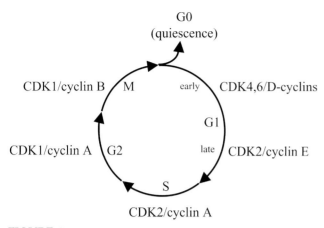

FIGURE 9.1   General schematic of the cell cycle. The cell cycle can be broken down into four distinct phases: Gap 1 (G1), DNA synthesis (S), Gap 2 (G2), and mitosis (M). The phase-specific CDK/cyclin complexes responsible for cell cycle regulation are indicated.

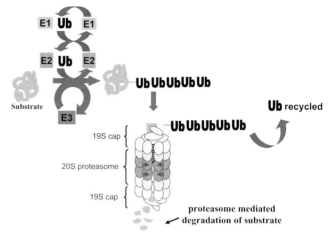

FIGURE 9.2   The ubiquitin/proteasome-mediated degradation pathway. The ubiquitin conjugating enzymes E1, E2, and E3 sequentially transfer a ubiquitin moiety (Ub) (or polyubiquitin chain) from one enzyme to the next followed by transfer of the Ub chain from E2 to the substrate. The polyubiquitinated product is targeted for destruction by the large, self-compartmentalized protease known as the 26S proteasome. The 19S subunit of the proteasome (regulatory subunit) is responsible for substrate recognition, removal of Ub and substrate unfolding. Free Ub is recycled. The 20S subunit (protease subunit) is responsible for substrate degradation. See Plate 9.2 in Color Plate Section.

pointed to protein degradation (proteolysis) as a key regulator of their expression [2]. It was discovered that the cyclins become covalently bound to ubiquitin moieties which target the proteins for degradation *via* the proteasome, a large, compartmentalized protease complex that destroys ubiquitinated substrates (Fig. 9.2) (reviewed in [3]). In mammalian cells, the cyclins can be divided into the G1 cyclins (D-type cyclins 1–3, cyclin E), the S-phase cyclins (cyclins A, E) and the mitotic cyclins (cyclins A, B) (Fig. 9.1). Cyclin-Dependent Kinases (CDKs) require cyclin binding in order to form enzymatically active heterodimeric complexes competent for substrate phosphorylation (addition of phosphate moieties). As cyclin levels are diminished, CDKs lose catalytic activity and are no longer capable of substrate phosphorylation. CDK substrates include the retinoblastoma protein (pRb), a key effector protein that mediates progression beyond the RP. Protein modification events such as phosphorylation (achieved by *kinases*) and de-phosphorylation (achieved by *phosphatases*) are critical regulatory mechanisms responsible for the activation or inactivation of proteins that control entry into and progression through the cell cycle, targeting proteins for degradation while activating others in order to maintain cell cycle integrity. As is the case with the cyclins, CDKs can be grouped according to their roles in the various phases of the cell cycle: the G1 CDKs (CDK4, CDK6, CDK2), the S-phase CDKs (CDK2) and the M-phase CDKs (CDK2, CDK1).

An additional level of cell cycle regulation is achieved by CDK Inhibitors (CDKIs). CDKIs are capable of abrogating catalytic kinase activity by binding to CDKs or CDK/cyclin complexes. There are two main families of CDKIs: the INhibitor of CDK4 (INK4) family of inhibitors and the CDK Inhibitor Protein/Kinase Inhibitor Protein (CIP/KIP) family of inhibitors. INK4 inhibitors specifically inhibit CDK4 and CDK6 activity during G1 to elicit a G1/S arrest whereas the CIP/KIP proteins are capable of inhibiting all CDK/cyclin complexes [4]. INK4 and CIP/KIP proteins are mechanistically different in that INK4 CDKIs only bind to CDKs while CIP/KIP proteins bind to both CDK and cyclin molecules simultaneously, stabilizing the proteins in a ternary complex.

Recently, a complex critical to M-phase progression was discovered. The Anaphase Promoting Complex (APC) is an E3 ubiquitin ligase complex that triggers the timed, sequential ubiquitination and proteolysis of mitotic regulators such as cyclin A, securin, and cyclin B (reviewed in [5]). In much the same way that proteasomes mediate degradation of cyclins in a timely manner, the APC coordinates the destruction of proteins involved in both entry into and exit from mitosis. Activation of the APC in late S-phase initiates a signaling cascade that enables sister chromatid separation. Subsequent binding of the APC to another regulatory molecule directs exit from mitosis.

The molecular associations amongst regulators of the cell cycle and the critical protein modification events required for cycling will be discussed below. Genetic analyses of primary human brain tumors have revealed common mutations in genes encoding proteins critical for cell cycle regulation. Such observations indicate the importance of tight control of the cell cycle in maintaining normal cellular cycling and appropriate quiescence. Dysregulation of the cell cycle in cancer has researchers looking for new ways to inhibit cell cycle progression. A number of small molecular inhibitors that target CDK inactivation have been investigated for use in clinical trials. Current research is also exploring the use of proteasome inhibitors to block the relentless cellular division associated with cancer cells. Some of the recent pharmacological interventions and their preclinical and clinical results are presented here.

G1 PHASE

Cells prepare themselves for DNA replication in G1 when cellular stimuli signal an upregulation in the expression levels of the D-type cyclins (growth factor sensors). Mitogenic stimuli activate *cyclin D* transcription as well as a cascade of phosphorylation events regulated by CDK/cyclin complexes [6–9]. Phosphorylation of critical amino acid residues results in conformational changes to CDKs that

contribute to both their catalytic activity and negative regulation [10,11] (Fig. 9.3). The opposing activities of Wee1/Myt1 kinases (phosphorylation) and cdc25 phosphatases (de-phosphorylation) are responsible for the maintenance of appropriate kinase activity [1,12].

CDK4,6/D cyclins

A requirement for the D-type cyclins to drive cells through G1 was identified in experimental cell culture studies where cyclin D1 overexpression was shown to be sufficient to enable G1 progression [13]. The D-type cyclins bind to CDKs 4 and 6 to form active kinase complexes that regulate progression through RP [14]. Whereas mice lacking CDK4 expression have a normal phenotype and growth profile, mice engineered to overexpress CDK4 are predisposed to tumorigenesis at an accelerated rate [15,16]. These results are in agreement with clinical findings which have found that approximately 10 per cent of high-grade astrocytic tumors contain amplification and consequent overexpression of *CDK4* and/or *CDK6* (reviewed in [17,18]). However, mutations or amplification in cyclin-encoding genes are not common in

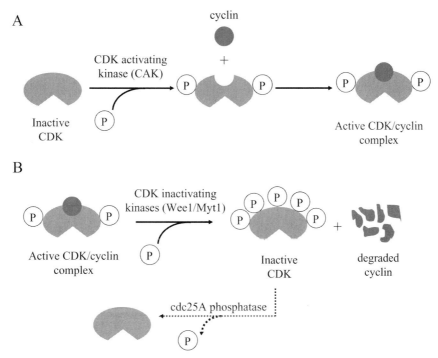

FIGURE 9.3 Activation and inactivation of CDK/cyclin complexes. (**A**) Inactive CDKs are initially phosphorylated by CDK-activating kinase (CAK) which promotes conformational alterations and permits cyclin binding activating the CDK/cyclin complex. (**B**) Wee1/Myt1 kinases additionally phosphorylate CDKs inducing conformational changes that release cyclins targeting them for degradation. cdc25 phosphatases de-phosphorylate CDKs returning them to an inactive state. See Plate 9.3 in Color Plate Section.

human brain tumors [12]. While these tumors preferentially contain alterations in genes encoding CDKs and CDKIs, the proliferative state of tumors and the invasiveness of gliomas have been found to be correlated with cyclin overexpression [19–21].

Assembly of CDK4/cyclin D complexes activates the kinase complex which partially phosphorylates pRb (hypophosphorylated pRb) (Fig. 9.4). The pRb residues that undergo phosphorylation are key docking sites for proteins that require tight regulation throughout the cell cycle such as members of the E2F family of transcription factors [22]. pRb binding to E2F proteins inactivates the E2F/DP1 transcription factor complex [23]. Initial pRb hypophosphorylation

liberates a small number of E2F transcription factor complexes permitting the transcription of genes encoding proteins required for passage through the G1 such as *cyclin E*. At this point, pRb still exerts an inhibitory effect on E2F-mediated transcription albeit incomplete. Mutations in *pRb* have been identified in 20 per cent of high-grade astrocytic tumors (grade III by WHO classification), however, mutations are more common in molecules directly involved in pRb signaling such as the cell cycle inhibitor p16^{INK4A} or CDK4 (reviewed in [18]). The fact that mice engineered to be deficient for *pRb* alone do not develop brain tumors suggests a role for *pRb* loss in glioma progression as opposed to early gliomagenesis.

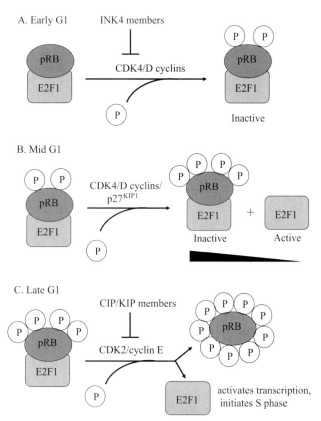

FIGURE 9.4   The G1/S transition. (**A**) In early G1, pRB is bound to the E2F transcription factor inhibiting E2F-mediated transcription. CDK4/cyclin D complexes are responsible for initial phosphorylation (indicated by encircled P) of pRB which remains bound to E2F. INK4 CDK inhibitors are capable of binding to CDK4 to prevent pRB phosphorylation. (**B**) During mid-G1, pRB is additionally phosphorylated by CDK4/cyclin D complexes (hypophosphorylated pRB). p27^{KIP1} binding to these kinase complexes does not affect their activity.While a portion of E2F is liberated from pRB at this stage and is transcriptionally active, the majority of E2F remains bound to pRB. (**C**) Late G1 signals completion of E2F activation by hyperphosphorylation of pRB achieved by CDK2/cyclin E complexes. CIP/KIP CDK inhibitors bound to CDK2/cyclin E complexes at this stage render the kinases inactive. See Plate 9.4 in Color Plate Section.

INK4 CDK Inhibitors

Each of the four members of the INK4 family, p16^{INK4A}, p15^{INK4B}, p18^{INK4C}, and p19^{INK4D}, is capable of binding to CDK4 and CDK6 (for reviews, see [11,24]). INK4 binding to CDK4 prevents the D-type cyclins from accessing CDKs, thus, inhibiting activation of the catalytic kinase complex. Mutations in CDK4 that confer resistance to INK4 inhibition are associated with both familial and sporadic melanoma in humans suggesting that INK4 proteins are a main regulator of CDK4 activity. When examined *in vitro*, all four INK4 proteins were found to have similar activity in inducing cell cycle arrest, however, there is experimental evidence to suggest unique roles for these proteins as well.

p16^{INK4A} and *p15^{INK4B}* have both been mapped to chromosome *9p21*, a region found frequently deleted or altered in a variety of human cancers including high grade astrocytomas [25]. Studies have shown that 60–80 per cent of high-grade astrocytic tumors and 25 per cent of anaplastic oligodendrogliomas contain homozygous deletion, mutation, or promoter hypermethylation of the *p16^{INK4A}* (*INK4A/ARF*) locus [26]. The close proximity of *p15^{INK4B}* to the *INK4A/ARF* locus is predicted to contribute to the homozygous deletion of both loci in anaplastic oligodendrogliomas [27]. While the *INK4/ARF* locus is the principal target of homozygous deletion, studies have shown that hypermethylation of the *p15^{INK4B}* promoter occurs frequently in gliomas without alterations in *INK4A/ARF* [28]. These clinical data help explain *in vitro* data that demonstrated expression of *p15^{INK4B}* alone effectively inhibited glioma cell growth [29]. The *p16$^{INK4A/ARF}$* locus (*INK4A/ARF*) additionally encodes an alternative reading frame product named p14^{ARF} (ARF) and tumor-associated chromosomal deletions frequently affects both genes [30–32].

Whereas p16^{INK4A} mediates G1 cell cycle arrest *via* binding to CDK4(6)/cyclin D complexes, ARF interacts with the ubiquitin ligase Human Double Minute 2 (HDM2) to stabilize and activate p53, a key checkpoint molecule in response to DNA damage, oncogenic stimuli and other cellular stresses. Both p16 and ARF are capable of eliciting a G1/S cell cycle arrest.

INK4A/ARF double knockout mice are viable, however, 90 per cent of knockout animals develop tumors upon carcinogen exposure [33,34]. In order to evaluate the individual contributions of p16^{INK4A} and ARF to tumorigenesis, knockout mice were engineered to be deficient for either p16^{INK4A} or ARF [35–37]. Deletion of either gene alone resulted in mice that were highly prone to tumors suggesting that both gene products play significant roles in tumor suppression [38]. Additionally, a small percentage of ARF-null mice (~10 per cent) develop brain tumors [37]. Elimination of this important tumor surveillance molecule (ARF) and cell cycle brakes (p16, ARF) is believed to directly contribute to progression from grade II to grade III astrocytoma and there are many mouse models to support this theory. Whereas *INK4A/ARF*-null mice do not develop brain tumors, *INK4A/ARF*-null background mice engineered to express an oncogenic variant of the epidermal growth factor receptor (EGFR) or K-Ras, a mitogenic factor, are predisposed to gliomagenesis [39,40].

Neither *p18^{INK4C}*- nor *p19^{INK4D}*-null mice develop tumors and neither are susceptible to tumor formation upon carcinogen treatment [41,42]. However, *p19^{INK4D}*-null mice exhibit progressive hearing loss while widespread hyperplasia and organomegaly are observed in mice lacking *p18^{INK4C}* [42,43]. Closer examination of the *p18^{INK4C}*-null mice identified slow-growing intermediate lobe pituitary tumors later in life. The hyperproliferative effects of *p18^{INK4C}* loss could be canceled in *p18^{INK4C}* /CDK4 mutant mice and suggests that *p18^{INK4C}* is functionally dependent on CDK4 [43].

CIP/KIP CDK Inhibitors

It has been shown that CIP/KIP family members (p21^{CIP1}, p27^{KIP1}, and p57^{KIP2}) bind to CDK4(6)/cyclin D complexes in G1 as well as to CDK2/cyclin E complexes. p21^{CIP1} was identified as a p53 target gene that becomes upregulated upon p53 activation with a key role in p53-mediated cell cycle arrest [44,45], reviewed in[46]). It exerts its effects on CDK2/cyclin E complexes rendering them inactive. p21^{CIP1} is also required for the assembly of CDK4/cyclin D complexes early in G1 [47,48]. p27^{KIP1}, on the other

hand, is an important mediator of the G1/S transition. During proliferation, most p27^{KIP1} is bound to CDK4/cyclin D complexes which are sensitive to INK4 inhibition but resistant to inhibition by CIP/KIP protein binding [49]. Distinct roles for p21^{CIP1}and p27^{KIP1} during tumor development have been suggested by studies of knockout mice. While p21^{CIP1} animals do not develop tumors spontaneously and show only a marginal increase in tumor predisposition upon exposure to ENU, *p27^{KIP1}*-/- mice spontaneously develop both pituitary and lung adenomas [50–53]. These data suggest that despite functional similarities between p21^{CIP1} and p27^{KIP1}, p27^{KIP1} is more critical for tumor suppression.

The functional importance of the CIP/KIP proteins was determined upon examination of *p21^{CIP1}/p27^{KIP1}* double null mouse embryonic fibroblasts (MEFs) cultured *in vitro* [54]. *p21^{CIP1}/p27^{KIP1}* double null MEFs exhibited a 10-fold reduction in CDK4/cyclin D complex formation resulting in inhibition of Rb phosphorylation. Re-introduction of *p21^{CIP1}* or *p27^{KIP1}* corrected this defect. The absence of phosphorylated Rb species in *p21^{CIP1}/p27^{KIP1}* -null MEFs suggested inactivation of CDK4/cyclin D complexes. In support of a requirement for p21^{CIP1} and p27^{KIP1} to induce active kinase complex formation, it has been demonstrated that restoration of p27^{KIP1} promotes CDK4/cyclin D assembly [55,56].

The preferential binding partner of p27^{KIP1} is the CDK2/cyclin E complex. In mid-G1, *cyclin E* transcription is activated by E2F transcription factors that retain partial activity despite being bound to hypophosphorylated pRb [57]. G0 and G1 pools of CDK2/cyclin E complexes are rendered inactive by p27 binding until targeted destruction of p27 at the G1/S boundardy [58–60]. In order to inhibit CDK kinase activity, p27^{KIP1} must be localized to the nucleus. Phosphorylation of p27^{KIP1} early in G1 has been shown to stabilize the protein [61]. The CDK inhibitory activity of CIP/KIP proteins is contained within the N-terminus while the C-terminal residues undergo key phosphorylations that either stabilize the protein or trigger ubiquitin-mediated proteolysis at the G1/S boundary. A phosphorylation-regulated nuclear localization signal has been identified in the C-terminal portion of p27^{KIP1} [62] that regulates its transport between the nucleus (where it is active) and cytoplasm (where it is degraded). Complete inactivation of p27^{KIP1} requires multiple phosphorylation events achieved by both CDK2 and other kinases which target it for degradation [63,64].

As is the case with *INK4* inactivation, mice engineered to be deficient for *p27^{KIP1}* or *p21^{CIP1}* do not develop tumors [50]. However, CIP/KIP tumor suppressor function was suggested by observations that mice lacking p21^{CIP1} were unable to undergo DNA-damage-induced p53-mediated cell cycle arrest [50,65].

In summary, tight regulation of p27^{KIP1} is critical for progression through the cell cycle. Despite the importance of CIP/KIP family members to appropriate cell cycle regulation, the genes encoding these proteins are not frequently altered in primary brain tumors. p27^{KIP1} expression in tumors, however, is a good prognostic indicator correlated with increased survival [66,67].

G1/S Checkpoints

The ARF tumor suppressor is not expressed at detectable levels either during development or in adult tissues. However, its expression becomes dramatically upregulated upon cellular exposure to oncogenic stress such as activated RAS or increased levels of c-MYC or E2F1 (reviewed in [68]). Upon upregulation of ARF, cells are found to arrest at both the G1/S and G2/M transition points in a p53-dependent manner (reviewed in [69–71]). p53 is a

known tumor suppressor with multiple cellular functions. Activation of p53 is a common response to cellular stress ranging from DNA damage to oncogenic stress. Normally, p53 expression is kept low by HDM2-mediated degradation, however, ARF expression inhibits HDM2's ability to target p53 for proteolysis. The consequent stabilization and activation of p53 results in upregulation of p21^{CIP1} inducing G1/S arrest as well as upregulation of CDK1 resulting in G2/M arrest.

S-PHASE

CDK2/Cyclin E Complexes

Upregulation of cyclin E expression begins late in G1 and is maintained into S-phase (reviewed in [24]). The initial CDK2/cyclin E complexes that form are maintained in an inactive state by CIP/KIP inhibitor binding. Complete activation of CDK2/cyclin E complexes requires titration of CIP/KIP proteins from CDK2/cyclin E complexes in mid-G1 inactivating CDK4/cyclin D associations (reviewed in [72]) (Fig. 9.5).

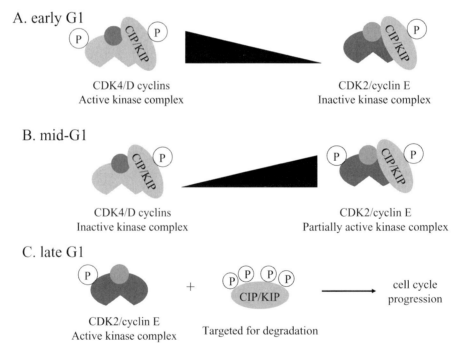

FIGURE 9.5 The balance of CIP/KIP inhibitor binding to CDK/cyclin complexes in G1. (**A**) During early G1, the majority of CIP/KIP inhibitors are bound to CDK4/cyclin D complexes. Binding of CIP/KIP inhibitors to these complexes does not affect their activity, however, CDK2/cyclin E/CIP/KIP complexes are inactive. (**B**) Mid-G1 signals a shift of CIP/KIP molecules from CDK4/cyclin D complexes to CDK2/cyclin E complexes. Phosphorylation of CDK2 at this stage partially activates the kinase complex which retains the ability to bind to CIP/KIP. (**C**) Late in G1 phosphorylation of CIP/KIP molecules targets the proteins for degradation leading to complete activation of CDK2/cyclin E kinases and, hence, progression through S-phase. See Plate 9.5 in Color Plate Section.

This event renders the CDK2/cyclin E kinase complex "active" to phosphorylate additional residues on pRB (hyperphosphorylated pRB) and permit progression through the G1/S transition [73–75]. These additional phosphorylation events (hyperphosphorylated pRB) result in complete inactivation of pRB's repressive function towards E2F transcription factors in late G1 facilitating the E2F-mediated transcription of genes required for DNA synthesis (refer to Fig. 9.3). In the absence of a functional interaction between cyclin E and CDK2, a cell is not able to pass beyond the restriction point [76].

The importance of CDK2 to cell cycle progression is underscored in cell culture studies that examined the function of dominant negative CDK2 (CDK2^{DN}) protein variants (isoforms). CDK2^{DN} remains constitutively inactive despite the presence of negative regulatory factors such as CDK inhibitors and association with cyclins A and E, however, cells are unable to progress beyond G1 [77]. CDK4^{DN} isoforms that assemble into kinase-dead complexes with D-type cyclins are not capable of eliciting similar growth arrest. This suggests that while CDK4 functions to sequester CIP/KIP proteins enabling CDK2/cyclin complex activation, CDK2 activation is critical to cell cycle progression [78]. The decline of CDK2 activity observed at the G1/S transition is controlled mainly by ubiquitin-mediated degradation of cyclin E.

M-PHASE

CDK1/cyclin A and CDK1/cyclin B

Mitosis is characterized by nuclear envelope breakdown and chromosome alignment on the mitotic spindle. Sister chromatid separation follows as they move towards opposite poles of the spindle permitting the formation of daughter nuclei. Central to mitotic entry and orchestration of mitotic events is the activation of CDK1. As mentioned above, CDK kinase activity is controlled by phosphorylation and de-phosphorylation events in addition to cyclin binding (reviewed in [79]). Critical de-phosphorylation events occur at the G2/M transition when de-phosphorylation events outnumber phosphorylation events (cdc25 phosphatase activity exceeds Wee1/Myt1 kinase activity). Regulation of CDK1 kinase activity must be strictly coordinated as constitutive CDK1 activation was found to prevent chromosomal condensation, nuclear envelope re-assembly, and cell division [80].

The formation of CDK1/cyclin A and CDK1/cyclin B complexes is important for M-phase initiation, progression, and exit. Associations between cyclins A and B and CDK1 are initiated during G2 and are required for activation of chromosome condensation, nuclear envelope breakdown, and spindle assembly (reviewed in [81]).

The periodic activation of maturation promoting factor (MPF), identified as CDK1/cyclin B complexes, results in the phosphorylation of critical mitotic regulators such as cyclin B and securin [82]. The upregulation of cyclin B promotes CDK1/cyclin B complex formation responsible for activation of the APC E3 ubiquitin ligase, which in turn controls mitotic cyclin degradation [83,84]. APC functions as a multi-subunit ubiquitin ligase complex and in conjunction with APC activator proteins (e.g., cdc20, cdh1), contributes to cell cycle regulation *via* the timed, sequential ubiquitination, and proteolysis of cyclins A and B as well as securin (reviewed in [85]). The APC itself is regulated by the spindle or kinetochore (point of chromosome attachment to mitotic spindle fibers) checkpoint which prevents sister chromatid separation until all chromosomes are correctly aligned on the mitotic spindle (reviewed in [86]). Early in S-phase, the Early Mitotic Inhibitor 1 (emi1) protein inhibits the APC by binding to cdc20 [87–89] (Fig. 9.6). Proteolysis of emi1 in prophase renders the APC active and capable of association with cdc20 and destruction of cyclin A occurs [90]. Upon exit from mitosis, the APC retains activity and binds to cdh1 which is held inactive by CDK1 [91,92]. It is not yet known to what extent the APC is phosphorylated in order to remain active.

Mitotic Checkpoints

APC, bound to the cdc20 subunit controls the activity of the Mad2 and BubR1, regulatory components of the spindle assembly checkpoint (reviewed in [91]). It has been demonstrated that the spindle checkpoint machinery is required to recruit APC to the kinetochores [93]. Recently, the RASSF1A tumor suppressor, the expression of which is frequently silenced in cancers, was found to act as an inhibitor of cdc20 preventing mitotic progression due to premature APC activation [94].

A critical target of mitotic control is *separase*, responsible for anaphase progression [95]. Separase is inactive when bound to *securin* which is in turn inhibited by the APC (Fig. 9.6). When kinetochores are properly attached to the spindle, APC binds to cdc20 and targets securin for degradation, liberating separase. Relieved from securin binding, separase is free to cleave cohesin molecules between sister

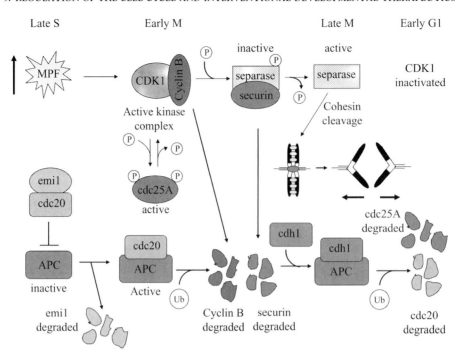

FIGURE 9.6   The Anaphase-Promoting Complex (APC). Maturation-promoting factor (MPF) consisting of CDK1/cyclin B complexes initiates entry into mitosis late in S-phase. cdc25A phosphatase contributes to maintenance of CDK1/cyclin B activity. At the same time, the APC co-factor cdc20, rendered inactive by binding to emi-1, is liberated by targeted degradation of emi-1. The active APC/cdc20 ubiquitin ligase targets mitotic molecules such as cyclin B and securin for degradation by ligation of ubiquitin moieties (encircled Ub). Securin normally functions to maintain separase in an inactive state. Once activated, separase promotes sister chromatid separation. Late in M-phase, APC associates with cdh1 targeting cdc20 as well as cdc25A for degradation. See Plate 9.6 in Color Plate Section.

chromatids facilitating the metaphase to anaphase transition. Another APC complex activator, cdh1, is phosphorylated by CDK1 until anaphase at which point cdh1 is subjected to de-phosphorylation events facilitating its binding to APC [96]. APC–cdh1 is responsible for the degradation of cdc20, inactivating the APC–cdc20 complex and orchestrating the cell's exit from M-phase [97]. Expression of APC–cdh1 in quiescent cells suggests its importance in G1 regulation. In fact, APC–cdh1 complexes are readily detectable and abundant in tissues composed of post-mitotic cells (e.g., brain). APC–cdh1 directly affects cell cycle during G0/G1 by preventing accumulation of S- and M-phase cyclins [98]. Identification of the APC and its role in cell cycle regulation is a relatively new field of study. Mutations in APC components have not yet been identified in human brain tumors.

THERAPEUTIC SMALL-MOLECULE INHIBITORS OF THE CELL CYCLE

Viral-based therapeutic strategies for the treatment of cancer have been designed and are discussed elsewhere in this book (see Chapter 23, Gene and Viral Therapy Approaches). More recently, research has focused on identifying small-molecule inhibitors of the cell cycle that act either by: (1) direct blockage of the catalytic activity of CDKs or (2) indirect inhibition of CDK activity by targeting the major effectors including factors that affect the expression and synthesis of CDKs, cyclins and their inhibitors, proteins that regulate phosphorylation (e.g., cdc25 phosphatases, Wee1/Myt1 kinases), and the machinery involved in the proteolytic degradation of subunits (e.g., proteasomes) [99]. There is both *in vitro* and *in vivo* data to suggest that CDK/cyclin complexes can be effectively targeted for ubiquitination and degradation in a proteasome-dependent manner resulting in tumor cell apoptosis [100].

The most extensive research to date has involved the examination of molecules that directly inhibit the catalytic activity of CDK/cyclin complexes [99,101–103]. Examples include olomoucine, flavopiridol, roscovitine, and UCN-01 (7-hydroxystaurosporine) and fall into several classes including staurosporines, flavonoids, paullones, polusulfates, indigoids, and purine analogs (reviewed in [104]).

Examples from each of the classes act similarly by occupying the ATP-binding pocket of the enzyme and competing with ATP.

TYPE I INHIBITORS: DIRECT INHIBITORS OF CDK CATALYTIC ACTIVITY

Olomoucine and the Purine Analogs

The purine-based olomoucine was one of the first CDK inhibitors to be developed. Both olomoucine and its analogs have proven to be highly specific for CDK1 and its related kinases (CDKs 2, 5, and 7) with no detectable effects on CDK4 or 6 [105]. Crystallization analysis showed specificity in the orientation of the molecule when bound to the pocket region of the enzyme. The adenine side chain of olomoucine was found to lie in a completely different orientation when compared to the adenine group from ATP. In addition, a substituent of olomoucine was found to bind to CDK outside of the binding pocket making contacts with the protein that ATP was unable to achieve [106]. Roscovitine is a purine analog with binding affinity for CDKs similar to that of olomoucine. However, roscovitine was found to be 10 times more potent in inhibiting CDK1 activity than olomoucine and it has shown increased anti-tumor activity [101,107]. The crystal structure of roscovitine with CDK2 indicated that the drug was capable of similar binding outside of the pocket region [101]. Purvalanol B has been shown to be even more potent than either olomoucine or roscovitine with similar binding activity and selective inhibition towards CDKs 1 and 2 [108]. Purvalanol's ability to induce G2/M arrest and inhibit cell growth has been demonstrated and is linked to inhibition of CDK1 and p42/44 mitogen activated kinase (MAPK) [104].

In vitro studies have shown that both olomoucine and roscovitine are capable of inducing cell cycle arrest in a number of different cell lines [102,109]. Studies showed that treatment of cells with these drugs blocked phosphorylation of downstream targets of CDK1 such as protein phosphatase I and vimentin [102,110]. The potential of these agents in chemotherapeutic regimens has been suggested by recent *in vitro* data demonstrating induction of cytotoxicity in tumor cells. Kim *et al.* recently showed that treatment with tumor necrosis factor-related apoptosis-inducing ligand (TRAIL) in combination with roscovitine significantly induced the activation of caspase-dependent apoptosis in TRAIL-resistant glioma cells [111]. Both olomoucine and roscovitine

have also been shown to effectively stabilize and activate p53 by suppressing HDM2 expression which suggests these drugs could function as sensitizing agents for DNA damaging drugs [112,113]. In addition, the ability of roscovitine to enhance the efficacy of fractionated radiation in glioma cell spheroids suggests that disruption of CDK-mediated pathways may potentially enhance the effectiveness of radiation therapy in malignant gliomas [114].

Flavopiridol

Flavopiridol, a synthetic analog of a natural alkaloid, was shown to have potent CDK4-blocking activity [115]. Initial studies with this flavonoid demonstrated its ability to induce G1/S as well as G2/M cell cycle arrest due to loss of CDK1 and CDK2 activity [116–118]. Flavopiridol was shown to bind to the ATP-binding pocket of CDK2 and, thus, competes with ATP for binding [115] (Fig. 9.7). Unlike olomoucine and roscovitine which show some CDK specificity, flavopiridol inhibits all CDKs as well as other protein kinases [118,119] with the greatest activity towards CDKs (reviewed in [120]).

Administration of flavopiridol has been shown to have multiple effects on cells including transcriptional inhibition, induction of apoptosis, and antiangiogenesis [121–123]. Direct cell cycle arrest at both the G1/S and G2/M boundaries is induced by flavopiridol's direct inhibition of CDKs 1, 2, and 4 [124,125]. Its effects on RNA polymerase II transcription are *via* inhibition of CDK9/cyclin T [121]. Of interest, is flavopiridol's inhibition of CDK9 activity, which is non-competitive with ATP binding unlike its activity towards other CDKs [122]. Microarray data suggest that the broad-range effects of flavopiridol-induced transcriptional arrest are similar to those of actinomycin D and DRB (5,6-dichloro-beta-D-ribofuranosylbenzimidazole), global inhibitors of transcription [126]. It has been demonstrated that flavopiridol is capable of inhibiting transcription at concentrations much lower than those required of other CDK inhibitors even in the presence of ATP. One target of interest with respect to transcriptional downregulation is cyclin D1. The transcriptional inhibition of cyclin D1 has been associated with cell cycle arrest and, in some cases, cell death following treatment with flavopiridol. These results are consistent with observations that flavopiridol can delay disease progression in 84 per cent of patients with mantle cell lymphoma, a tumor associated with overexpressed cyclin D1 in 95 per cent of cases [127,128].

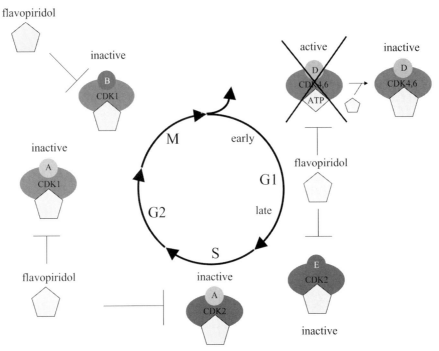

FIGURE 9.7   Inactivation of CDK/cyclin kinase activity by flavopiridol. By binding to the ATP-binding site of CDKs, flavopiridol is capable of inhibiting various CDKs throughout the cell cycle. Inhibition of ATP binding prevents kinase complex activation. See Plate 9.7 in Color Plate Section.

A variety of tumor cell lines have been found to be sensitive to flavopiridol-induced apoptosis *in vitro* (reviewed in [120]). Glioma cell lines were shown to undergo apoptosis in response to flavopiridol regardless of pRB or p53 status [129]. Despite the fact that flavopiridol is able to inhibit the expression of bcl-2 (an anti-apoptotic factor), downregulation of bcl-2 does not play a role in flavopiridol-mediated apoptosis [130,131]. It is still unclear whether or not inhibition of CDK activity is required for apoptosis.

Several clinical trials combining flavopiridol and various chemotherapeutic agents have been initiated [132,133]. It has been shown that coadministration of flavopiridol with cytostatic drugs, particularly taxanes, significantly enhanced the cytotoxic effects in a variety of tumor cells. The enhanced synergistic effects of flavopiridol and cytostatic drugs appeared to depend on the order of administration. In lung cancer cells, pre-treament with gemcitabine or cisplatin sensitizes cells to the cytotoxicity of flavopiridol whereas in leukemic cells, pre-treatment with flavopiridol enhances the cytotoxicity of nucleoside analogs such as ara-C [134–136]. Cells with defective pRB signaling undergo enhanced doxorubicin-mediated cell death in conjunction with flavopiridol [131]. Since a large percentage of glial tumors harbour defects in the pRB pathway, treatment with flavoripirdol or

flavopiridol/combination chemotherapy might be a therapeutic possibility.

Additional synergism between flavopiridol and signaling modulators has been detected. Co-treatment of leukemic cells with flavopiridol and the phorbol ester PMA, a potent inducer of differentiation, was found to induce apoptosis and overcome resistance mediated by bcl-2 expression [137]. Studies of flavopiridol synergism with additional chemotherapeutic agents have been extended to include a range of additional differentiation-inducing agents including Histone DeACetylase (HDAC) inhibitors. The downstream effects of flavopiridol and PMA or HDAC co-treatments include downregulation of cyclin D1, cleavage of pRB and bcl-2 and activation of E2F1. The identification of cytotoxic synergy between flavopiridol and other agents could produce more effective chemotherapeutic strategies in the future. The fact that flavopiridol does not induce resistance to most conventional cytotoxic agents including etoposide, doxorubicin, vinblastine, paclitaxel, cisplatin, topotecan, and 5-Fluorouracil (5FU) is promising. In addition, overexpression of the multidrug resistance proteins mdr-1 or mrp-1 did not cause appreciable resistance to flavopiridol [138–140].

Both phase I and phase II trials for flavopiridol have been completed. Clinical doses higher than those concentrations required to inhibit CDKs and cell growth were achieved without severe side effects. In addition, there was evidence of anti-tumor activity in patients with renal cell carcinoma, colon carcinoma, non-Hodgkin's lymphoma, gastric carcinoma, and mantle cell lymphoma [141–143] (reviewed in [132]). The effects of flavopiridol administered singly in clinical settings have not equaled those observed in *in vitro* settings. Therefore, combinatorial therapies are in development that would couple flavopiridol with FDA-approved anti-cancer agents such as paclitaxel, cytosine arabinoside, and irinotecan. A phase I trial of flavopiridol in combination with paclitaxel has demonstrated good tolerance and follow up studies are underway [135].

UCN-01

7-hydroxystaurosporine/UCN-01 was initially developed as a protein kinase C (PKC) inhibitor (reviewed in [144]). UCN-01 was subsequently shown to have anti-proliferative activity in several tumor cell lines [145–149], an effect that appears to be unrelated to the inhibition of PKC signaling [148]. Effects of UCN-01 included induction of apoptosis and cell cycle arrest at G1/S and G2/M which cannot be explained by UCN-01's effects on PKC signaling pathways [148, 150]. Yamasaki *et al.* recently demonstrated that human glioma cell lines treated with low doses of staurosporine underwent G1 arrest whereas higher doses of the drug induced G2/M arrest and apoptosis [151].

The mechanisms by which UCN-01 inhibits progression through the cell cycle remain poorly understood. At higher concentrations, UCN-01 was found to inhibit CDK1 and CDK2 *in vitro*, however, this activity required concentrations 1- to 2-fold higher than those necessary for cell cycle arrest [148,150]. It has been shown that UCN-01 treatment results in the upregulation of CDK inhibitors p21 and p27 with corresponding decrease in G1 CDK activity and pRB dephosphorylation [149,150]. Facchinetti *et al.* recently demonstrated that induction of p21 is critical to UCN-01's ability to arrest cells [152]. The effects of UCN-01 on p21 transcription were shown to be both p53- and PKC-independent, however, activation of the mitogen-activated protein/extracellular signal-regulated kinase kinase (MEK)/extracellular signal-regulated kinase (ERK) pathway was required.

UCN-01 was found to be particularly effective in inducing a cytotoxic response in cells expressing mutant p53 or lacking p53 [153]. These results suggested that UCN-01 might be of therapeutic use in tumors normally associated with poor prognosis and nonresponsive to standard chemotherapeutic regimens. The first phase I clinical trial of UCN-01 has been completed, which showed the drug to have a long half-life following administration [154,155]. Phase II trials indicated that UCN-01 was effective in stabilizing disease in patients with leiomyosarcoma, non-Hodgkin's lymphoma, and lung cancer.

As with flavopiridol, synergism between UCN-01 and other chemotherapeutic agents such as mitomycin C, 5-FU and camptothecin have been observed *in vitro* [154,156–163]. Interference with G1 and G2 cell cycle checkpoints appears to be responsible for the increased efficacy of UCN-01 in combination with conventional chemotherapeutic agents [164,165]. Clinical trials have commenced exploring the use of UCN-01 in combination with gemcitabine, cisplatin, 5-FU as well as others.

E7070, a Sulfonamide

E7070 has been shown to target the G1 phase of the cell cycle [166]. Inhibition of G1/S transition is achieved by blocking the activation of both CDK2 and cyclin E. This synthetic sulfonamide has been developed for use in clinical trials for the treatment of solid tumors including those of the colon and lung. Phase I trials with E7070 evaluated different treatment regimens including combination therapy with irinotecan (http://www.clinicaltrials.gov identifier NCT00060567)[167]. While administration of E7070 singly had limited anti-tumor effects, consequent studies have evaluated its effects in combination with other chemotherapeutic agents. Phase II clinical trials have been completed in patients with stage IV melanoma (NCI clinical trials, http://www.cancer.gov/clinical trials/EORTC-16005).

RGB-286199

In June 2004, a new cell cycle inhibitor was identified as a potent inhibitor of CDKs 1, 2, 4, and 6, capable of inhibiting cell division and promoting cell death in cancer cells (http://www.gpc-biotech.com/en/anticancer_programs/cdk/references/index.html). The antitumor activity of RGB-286199 has been demonstrated in animal models of ovarian, prostate, and colon cancer. While the compound is still undergoing pre-clinical workup, its ability to target a range of CDKs involved in the various phases of the cell cycle makes it a promising candidate for future cancer therapeutics.

TYPE II INHIBITORS: INHIBITORS OF CDK REGULATORY MECHANISMS

Proteasome Inhibitors

Regulation of the cyclins and CDKIs are critical to maintaining appropriate cell cycle progression or arrest. Cyclins and CDKIs are regulated at both the transcriptional and protein stability levels. As discussed above, ubiquitin-mediated proteolysis plays an important role in regulating cyclin expression and, to a lesser extent, expression of CDKIs (reviewed in [5]). In light of these data, the proteasome machinery is an attractive target in the treatment of cancer.

It has been shown that inhibition of the proteasome results in cell cycle arrest and apoptosis. Pharmacological inhibitors of the proteasome such as the calpain inhibitor acetyl-leucinyl-leucinyl-norleucinal (ALLN) and lactacystin were shown to sensitize cells to apoptosis [168,169]. Both ALLN and lactacystin were shown to induce p53-mediated apoptosis in glioma cell lines, an effect accompanied by upregulation of p53, p21^{CIP1}, p27^{KIP1}, and HDM2, a negative regulator of p53, as well as p53-independent apoptosis [170,171]. In addition, these proteasome inhibitors have also been shown to mediate accumulation of c-Myc stimulating FAS-activated apoptosis in glioma cell lines [172].

The first *in vivo* demonstration of the antitumor activity of proteasome inhibitors involved treatment of Burkitt's lymphoma xenografts with PS-341, since renamed Bortezomib [173]. A screen of tumor cell lines isolated from 9 different cancer types (brain, colon, melanoma, prostate, lung, breast, renal, and ovarian) showed that Bortezomib had broad antitumor activity both *in vitro* and *in vivo* [174,175]. Additional studies found that transformed cells were selectively sensitive to the cytotoxic effects of proteasome inhibition when compared to normal cells in culture [173,176–179]. *In vivo* effects were detected in prostate tumor xenograft and murine mammary carcinoma models [180].

Bortezomib (velcade) is the first proteasome inhibitor with demonstrated anti-cancer activity. Several important regulators of the cell cycle undergo proteasome-mediated degradation including p53, p21^{CIP1}, and p27^{KIP1}. Proteasome inhibition results in increased expression of p53 targets important to apoptosis and negative regulation of the cell cycle such as p21 [181,182]. In addition, proteasome inhibition stabilizes pools of p21 and p27 which mediate cell cycle arrest [183,184]. Currently, clinical trials are underway to evaluate the anti-cancer activity of bortezomib in patients (reviewed in [185]). Phase I

trials evaluated the use of bortezomib in the treatment of both chemoresistant and recurring solid tumors (hepatocellular carcinoma, melanoma, metastatic breast cancer, metastatic colorectal carcinoma, metastatic renal cell carcinoma, neuroendocrine tumors, ovarian carcinoma, non-small cell lung carcinoma, and sarcoma) and hematological malignancies [186,187]. Bortezomib was administered to multiple myeloma patients who were not responsive to current chemotherapies in phase II clinical trials with encouraging results [188]. Additional phase II studies include evaluation of effects in follicular lymphoma, mantle cell lymphoma, and chronic lymphocytic leukemia patients [189].

As with the CDK/cyclin enzymatic inhibitors, bortezomib is undergoing evaluation in combination therapies. A number of phase I studies are completed or ongoing including bortezomib in conjunction with irinotecan [190], doxorubicin [191,192], paclitaxel [193], 5-FU [190], and gemcitabine [194]. Phase II trials targeting specific solid tumors are underway at a number of institutions with use of bortezomib as either a single agent in patients with metastatic or recurrent sarcoma and metastatic neuroendocrine tumors. While clinical studies have not yet explored the use of proteasome inhibitors in the treatment of brain malignancies, early *in vitro* observations suggest that it might be useful in the treatment of these cancers.

SUMMARY

The events of the cell division cycle can be summarized as follows: the G1 cyclins prepare the chromosomes for replication *via* interactions with corresponding CDKs. A consequent increase in S-phase promoting factors prepares the cell to enter S-phase and duplicate its DNA. As replication continues, one of the cyclins shared by both the G1- and the S-phase CDKs, cyclin E, is destroyed and an increase in M-phase cyclins (cyclins A, B) ensues. M-phase-promoting factors such as cyclins A and B and CDK1 lead to mitotic spindle assembly, breakdown of the nuclear envelope, and chromatin condensation, each of which takes place prior to mitotic anaphase. At this point, the MPF (CDK1/cyclin B) promotes APC assembly which induces sister chromatid separation and movement to the poles. In order for cells to continue cyclin, phase-specific cyclins must be degraded by the ubiquitin proteasome pathway which targets proteins for destruction by covalent bonding of small ubiquitin molecules permitting their recognition by the proteasome. The INK4 and CIP/KIP families of CDKIs add another level of regulation to the cell cycle. Inhibition of CDK activation through the administration of small-molecule inhibitors is showing promise as a clinical therapeutic strategy through the ability to interrupt CDK enzymatic activity either directly or indirectly.

References

1. Fattaey, A., and Booher, R. N. (1997). Myt1: a Wee1-type kinase that phosphorylates Cdc2 on residue Thr14. *Prog Cell Cycle Res* **3**, 233–240.

2. Pines, J., and Hunter T. (1989). Isolation of a human cyclin cDNA: evidence for cyclin mRNA and protein regulation in the cell cycle and for interaction with p34cdc2. *Cell* **58**, 833–846.

3. Muratani, M., and Tansey, W. P. (2003). How the ubiquitin-proteasome system controls transcription. *Nat Rev Mol Cell Biol* **4**, 192–201.

4. Polyak, K., Lee M-H, Erdjument-Bromage, H. *et al.* (1994). Cloning of p21Kip1, a cyclin-dependent kinase inhibitor and a potential mediator of extracellular antimitogenic signals. *Cell* **8**, 59–66.

5. Murray, A. W. (2004). Recycling the cell cycle: cyclins revisited. *Cell* **116**, 221–234.

6. Lavoie, J. N., L'Allemain, G., Brunet, A., Muller, R., and Pouyssegur, J. (1996). Cyclin D1 expression is regulated positively by the p42/p44MAPK and negatively by the p38/HOGMAPK pathway. *J Biol Chem* **271**, 20608–20616.

7. Page, K., Li, J., Hodge, J. A. *et al.* (1999). Characterization of a Rac1 signaling pathway to cyclin D(1) expression in airway smooth muscle cells. *J Biol Chem* **274**, 22065–22071.

8. Lee, R. J., Albanese, C., Fu, M. *et al.* (2000). Cyclin D1 is required for transformation by activated Neu and is induced through an E2F-dependent signaling pathway. *Mol Cell Biol* **20**, 672–683.

9. Lenferink, A. E., Busse, D., Flanagan, W. M., Yakes, F. M., and Arteaga, C. L. (2001). ErbB2/neu kinase modulates cellular p27(Kip1) and cyclin D1 through multiple signaling pathways. *Cancer Res* **61**, 6583–6591.

10. Sclafani, R. A. (1996). Cyclin dependent kinase activating kinases. *Curr Opin Cell Biol* **8**, 788–794.

11. Coleman, T. R., and Dunphy, W. G. (1994). Cdc2 regulatory factors. *Curr Opin Cell Biol* **6**, 877–882.

12. Buschges, R., Weber, R. G., Actor, B. *et al.* (1999). Amplification and expression of cyclin D genes (CCND1, CCND2 and CCND3) in human malignant gliomas. *Brain Pathol* **9**, 435–442; discussion 432–433.

13. Lukas, J., Parry, D., Aagaard, L. *et al.* (1995). Retinoblastoma-protein-dependent cell-cycle inhibition by the tumor suppressor p16. *Nature* **375**, 503–506.

14. Draetta, G. F. (1994). Mammalian G1 cyclins. *Curr Opin Cell Biol* **6**, 842–846.

15. Rane, S. G., Dubus, P., Mettus, R. V. *et al.* (1999). Loss of Cdk4 expression causes insulin-deficient diabetes and Cdk4 activation results in beta-islet cell hyperplasia. *Nat Genet* **22**, 44–52.

16. Tsutsui, T., Hesabi, B., Moons, D. S. *et al.* (1999). Targeted disruption of CDK4 delays cell cycle entry with enhanced p27(Kip1) activity. *Mol Cell Biol* **19**, 7011–7019.

17. Osborne, R. H., Houben, M. P., Tijssen, C. C., Coebergh, J. W., and van Duijn, C. M. (2001). The genetic epidemiology of glioma. *Neurology* **57**, 1751–1755.

18. Behin, A., Hoang-Xuan, K., Carpentier, A. F., and Delattre, J. Y. (2003). Primary brain tumors in adults. *Lancet* **361**, 323–331.

19. Arato-Ohshima, T., and Sawa, H. (1999). Over-expression of cyclin D1 induces glioma invasion by increasing matrix metalloproteinase activity and cell motility. *Int J Cancer* **83**, 387–392.

20. Sallinen, S. L., Sallinen, P. K., Kononen, J. T. *et al.* (1999). Cyclin D1 expression in astrocytomas is associated with cell proliferation activity and patient prognosis. *J Pathol* **188**, 289–293.

21. Allan, K., Jordan, R. C., Ang, L. C., Taylor, M., and Young, B. (2000). Overexpression of cyclin A and cyclin B1 proteins in astrocytomas. *Arch Pathol Lab Med* **124**, 216–220.

22. Dyson N. (1994). pRB, p107 and the regulation of the E2F transcription factor. *J Cell Sci Suppl* **18**, 81–87.

23. Flemington, E. K., Speck, S. H., and Kaelin, W. G. Jr. (1993). E2F-1-mediated transactivation is inhibited by complex formation with the retinoblastoma susceptibility gene product. *Proc Natl Acad Sci U S A* **90**, 6914–6918.

24. Ortega, S., Malumbres, M., and Barbacid, M. (2002). Cyclin D-dependent kinases, INK4 inhibitors and cancer. *Biochim Biophys Acta* **1602**, 73–87.

25. Hannon, G. J., and Beach, D. (1994). p15 is a potential effector of cell cycle arrest mediated by TGF beta. *Nature* **371**, 257–261.

26. Watanabe, T., Nakamura, M., Yonekawa, Y., Kleihues, P., and Ohgaki, H. (2001). Promoter hypermethylation and homozygous deletion of the p14ARF and p16INK4a genes in oligodendrogliomas. *Acta Neuropathol (Berl)* **101**, 185–189.

27. Watanabe, T., Yokoo, H., Yokoo, M., Yonekawa, Y., Kleihues, P., and Ohgaki, H. (2001). Concurrent inactivation of RB1 and TP53 pathways in anaplastic oligodendrogliomas. *J Neuropathol Exp Neurol* **60**, 1181–1189.

28. Herman, J. G., Jen, J., Merlo, A., and Baylin, S. B. (1996). Hypermethylation-associated inactivation indicates a tumor suppressor role for p15INK4B. *Cancer Res* **56**, 722–727.

29. Fuxe, J., Akusjarvi, G., Goike, H. M., Roos, G., Collins, V. P., and Pettersson, R. F. (2000). Adenovirus-mediated overexpression of p15INK4B inhibits human glioma cell growth, induces replicative senescence, and inhibits telomerase activity similarly to p16INK4A. *Cell Growth Differ* **11**, 373–384.

30. Duro, D., Bernard, O., Della Valle, V., Berger, R., and Larsen, C. J. (1995). A new type of p16INK4/MTS1 gene transcript expressed in B-cell malignancies. *Oncogene* **11**, 21–29.

31. Quelle, D. E., Zindy, F., Ashmun, R. A., and Sherr, C. J. (1995). Alternative reading frames of the INK4a tumor suppressor gene encode two unrelated proteins capable of inducing cell cycle arrest. *Cell* **83**, 993–1000.

32. Stone, S., Jiang, P., Dayananth, P., Tavtigian, S. V., Katcher, H., Parry, D., Peters, G., and Kamb, A. (1995). Complex structure and regulation of the P16 (MTS1) locus. *Cancer Res* **55**, 2988–2994.

33. Serrano, M., Lee, H. W., Chin, L., and Cordon-Cardo, C. (1996). Role of the INK4 locus in tumor suppression and cell mortality. *Cell* **85**, 27–37.

34. Kamijo, T., Zindy, F., Roussel, M. F. *et al.* (1997). Tumor suppression at the mouse INK4a locus mediated by the alternative reading frame product p19ARF. *Cell* **91**, 649–659.

35. Krimpenfort, P., Quon, K. C., Mooi, W. J., Loonstra, A., and Berns, A. (2001). Loss of p16Ink4a confers susceptibility to metastatic melanoma in mice. *Nature* **413**, 83–86.

36. Sharpless, N. E., Bardeesy, N., Lee, K. H. *et al.* (2001). Loss of p16Ink4a with retention of p19Arf predisposes mice to tumorigenesis. *Nature* **413**, 86–91.

37. Kamijo, T., Bodner, S., van de Kamp, E., Randle, D. H., and Sherr, C. J. (1999). Tumor spectrum in ARF-deficient mice. *Cancer Res* **59**, 2217–2222.

38. Sharpless, N. E., Ramsey, M. R., Balasubramanian, P., Castrillon, D. H., and DePinho, R. A. (2004). The differential impact of p16(INK4a) or p19(ARF) deficiency on cell growth and tumorigenesis. *Oncogene* **23**, 379–385.

39. Holland, E. C., Hively, W. P., DePinho, R. A., and Varmus, H. E. (1998). A constitutively active epidermal growth factor receptor cooperates with disruption of G1 cell-cycle arrest pathways to induce glioma-like lesions in mice. *Genes Dev* **12**, 3675–3685.

40. Holland, E. C., Celestino, J., Dai, C., Schaefer, L., Sawaya, R. E., and Fuller, G. N. (2000). Combined activation of Ras and Akt in neural progenitors induces glioblastoma formation in mice. *Nat Genet* **25**, 55–57.

41. Zindy, F., van Deursen, J., Grosveld, G., Sherr, C. J., and Roussel, M. F. (2000). INK4d-deficient mice are fertile despite testicular atrophy. *Mol Cell Biol* **20**, 372–378.

42. Franklin, D. S., Godfrey, V. L., Lee, H. *et al.* (1998). CDK inhibitors p18(INK4c) and p27(Kip1) mediate two separate pathways to collaboratively suppress pituitary tumorigenesis. *Genes Dev* **12**, 2899–2911.

43. Pei, X. H., Bai, F., Tsutsui, T., Kiyokawa, H., and Xiong, Y. (2004). Genetic evidence for functional dependency of p18Ink4c on Cdk4. *Mol Cell Biol* **24**, 6653–6664.

44. Harper, J. W., Adami, G. R., Wei, N., Keyomarsi, K., and Elledge, S. J. (1993). The p21 Cdk-interacting protein Cip1 is a potent inhibitor of G1 cyclin-dependent kinases. *Cell* **75**, 805–816.

45. el-Deiry, W. S., Tokino, T., Velculescu, V. E. *et al.* (1993). WAF1, a potential mediator of p53 tumor suppression. *Cell* **75**, 817–825.

46. Cox, L. S. (1997). Multiple pathways control cell growth and transformation: overlapping and independent activities of p53 and p21Cip1/WAF1/Sdi1. *J Pathol* **183**, 134–140.

47. LaBaer, J., Garrett, M. D., Stevenson, L. F. *et al.* (1997). New functional activities for the p21 family of CDK inhibitors. *Genes Dev* **11**, 847–862.

48. Cheng, M., Olivier, P., Diehl, J. A. *et al.* (1999). The p21(Cip1) and p27(Kip1) CDK "inhibitors" are essential activators of cyclin D-dependent kinases in murine fibroblasts. *Embo J* **18**, 1571–1583.

49. Cheng, M., Sexl, V., Sherr, C. J., and Roussel, M. F. (1998). Assembly of cyclin D-dependent kinase and titration of p27Kip1 regulated by mitogen-activated protein kinase kinase (MEK1). *Proc Natl Acad Sci U S A* **95**, 1091–1096.

50. Deng, C., Zhang, P., Harper, J. W., Elledge, S. J., and Leder, P. (1995). Mice lacking p21CIP1/WAF1 undergo normal development, but are defective in G1 checkpoint control. *Cell* **82**, 675–684.

51. Kiyokawa, H., Kineman, R. D., Manova-Todorova, K. O. *et al.* (1996). Enhanced growth of mice lacking the cyclin-dependent kinase inhibitor function of p27(Kip1). *Cell* **85**, 721–732.

52. Fero, M. L., Rivkin, M., Tasch, M. *et al.* (1996). A syndrome of multiorgan hyperplasia with features of gigantism, tumorigenesis, and female sterility in p27(Kip1)-deficient mice. *Cell* **85**, 733–744.

53. Nakayama, K., Ishida, N., Shirane, M. *et al.* (1996). Mice lacking p27(Kip1) display increased body size, multiple organ hyperplasia, retinal dysplasia, and pituitary tumors. *Cell* **85**, 707–720.

54. Zindy, F., den Besten, W., Chen, B. *et al.* (2001). Control of spermatogenesis in mice by the cyclin D-dependent kinase inhibitors p18(Ink4c) and p19(Ink4d). *Mol Cell Biol* **21**, 3244–3255.

55. Alt, J. R., Gladden, A. B., and Diehl, J. A. (2002). p21(Cip1) Promotes cyclin D1 nuclear accumulation via direct inhibition of nuclear export. *J Biol Chem* **277**, 8517–8523.

56. Alt, J. R., Cleveland, J. L., Hannink, M., and Diehl, J. A. (2000). Phosphorylation-dependent regulation of cyclin D1 nuclear export and cyclin D1-dependent cellular transformation. *Genes Dev* **14**, 3102–3114.

57. Ohtsubo, M., Theodoras, A. M., Schumacher, J., Roberts, J. M., and Pagano, M. (1995). Human cyclin, E., a nuclear protein essential for the G1-to-S-phase transition. *Mol Cell Biol* **15**, 2612–2624.

58. Montagnoli, A., Fiore, F., Eytan, E. *et al.* (1999). Ubiquitination of p27 is regulated by Cdk-dependent phosphorylation and trimeric complex formation. *Genes Dev* **13**, 1181–1189.

59. Shirane, M., Harumiya, Y., Ishida, N. *et al.* (1999). Down-regulation of p27(Kip1) by two mechanisms, ubiquitin-mediated degradation and proteolytic processing. *J Biol Chem* **274**, 13886–13893.

60. Tsvetkov, L. M., Yeh, K. H., Lee, S. J., Sun, H., and Zhang, H. (1999). p27(Kip1) ubiquitination and degradation is regulated by the SCF(Skp2) complex through phosphorylated Thr187 in p27. *Curr Biol* **9**, 661–664.

61. Deng, X., Mercer, S. E., Shah, S., Ewton, D. Z., and Friedman, E. (2004). The cyclin-dependent kinase inhibitor p27Kip1 is stabilized in G(0) by Mirk/dyrk1B kinase. *J Biol Chem* **279**, 22498–22504.

62. Zeng, Y., Hirano, K., Hirano, M., Nishimura, J., and Kanaide, H. (2000). Minimal requirements for the nuclear localization of p27(Kip1), a cyclin-dependent kinase inhibitor. *Biochem Biophys Res Commun* **274**, 37–42.

63. Sheaff, R. J., Groudine, M., Gordon, M., Roberts, J. M., and Clurman, B. E. (1997). Cyclin E-CDK2 is a regulator of p27Kip1. *Genes Dev* **11**, 1464–1478.

64. Malek, N. P., Sundberg, H., McGrew, S., Nakayama, K., Kyriakides, T. R., Roberts, J. M., and Kyriakidis, T. R. (2001). A mouse knock-in model exposes sequential proteolytic pathways that regulate p27Kip1 in G1 and S-phase. *Nature* **413**, 323–327.

65. Brugarolas, J., Chandrasekaran, C., Gordon, J. I. *et al.* (1995). Radiation-induced cell cycle arrest compromised by p21 deficiency. *Nature* **377**, 552–557.

66. Kirla, R. M., Haapasalo, H. K., Kalimo, H., and Salminen, E. K. (2003). Low expression of p27 indicates a poor prognosis in patients with high-grade astrocytomas. *Cancer* **97**, 644–648.

67. Tamiya, T., Mizumatsu, S., Ono, Y. *et al.* (2001). High cyclin E/low p27Kip1 expression is associated with poor prognosis in astrocytomas. *Acta Neuropathol (Berl)* **101**, 334–340.

68. Sherr, C. J., and Weber, J. D. (2000). The ARF/p53 pathway. *Curr Opin Genet Dev* **10**, 94–99.

69. Taylor, W. R., and Stark, G. R. (2001). Regulation of the G2/M transition by p53. *Oncogene* **20**, 1803–1815.

70. Schwartz, D., and Rotter, V. (1998). p53-dependent cell cycle control: response to genotoxic stress. *Semin Cancer Biol* **8**, 325–336.

71. Kamijo, T., Weber, J. D., Zambetti, G. *et al.* (1998). Functional and physical interactions of the ARF tumor suppressor with p53 and Mdm2. *Proc Natl Acad Sci USA* **95**, 8292–8297.

72. Sherr, C. J., and Roberts, J. M. (1999). CDK inhibitors: positive and negative regulators of G1-phase progression. *Genes Dev* **13**, 1501–1512.

73. Hatakeyama, M., Brill, J. A., Fink, G. R., and Weinberg, R. A. (1994). Collaboration of G1 cyclins in the functional inactivation of the retinoblastoma protein. *Genes Dev* **8**, 1759–1771.

74. Mittnacht, S., Lees, J. A., Desai, D., Harlow, E., Morgan, D. O., and Weinberg, R. A. (1994). Distinct sub-populations of the retinoblastoma protein show a distinct pattern of phosphorylation. *Embo J* 13, 118–127.

75. Alexander, K., and Hinds, P. W. (2001). Requirement for p27(KIP1) in retinoblastoma protein-mediated senescence. *Mol Cell Biol* 21, 3616–3631.

76. Koff, A., Giordano, A., Desai, D. *et al.* (1992). Formation and activation of a cyclin E-cdk2 complex during the G1 phase of the human cell cycle. *Science* 257, 1689–1694.

77. van den Heuvel, S., and Harlow, E. (1993). Distinct roles for cyclin-dependent kinases in cell cycle control. *Science* 262, 2050–2054.

78. Jiang, H., Chou, H. S., and Zhu L. (1998). Requirement of cyclin E-Cdk2 inhibition in p16(INK4a)-mediated growth suppression. *Mol Cell Biol* 18, 5284–5290.

79. Nigg, E. A. (2001). Cell cycle regulation by protein kinases and phosphatases. *Ernst Schering Res Found Workshop*, 19–46.

80. Kramer, A., Mailand, N., Lukas, C. *et al.* (2004). Centrosome-associated Chk1 prevents premature activation of cyclin-B-Cdk1 kinase. *Nat Cell Biol* 6, 884–891.

81. Castedo, M., Perfettini, J. L., Roumier, T., and Kroemer, G. (2002). Cyclin-dependent kinase-1: linking apoptosis to cell cycle and mitotic catastrophe. *Cell Death Differ* 9, 1287–1293.

82. Waizenegger, I., Gimenez-Abian, J. F., Wernic, D., and Peters, J. M. (2002). Regulation of human separase by securin binding and autocleavage. *Curr Biol* 12, 1368–1378.

83. Kotani, S., Tanaka, H., Yasuda, H., and Todokoro, K. (1999). Regulation of APC activity by phosphorylation and regulatory factors. *J Cell Biol* 146, 791–800.

84. Kraft, C., Herzog, F., Gieffers, C. *et al.* (2003). Mitotic regulation of the human anaphase-promoting complex by phosphorylation. *Embo J* 22, 6598–6609.

85. Peters, J. M. (2002). The anaphase-promoting complex: proteolysis in mitosis and beyond. *Mol Cell* 9, 931–943.

86. Nasmyth K. (2002). Segregating sister genomes: the molecular biology of chromosome separation. *Science* 297, 559–565.

87. Reimann, J. D., Gardner, B. E., Margottin-Goguet, F., and Jackson, P. K. (2001). Emi1 regulates the anaphase-promoting complex by a different mechanism than Mad2 proteins. *Genes Dev* 15, 3278–3285.

88. Reimann, J. D., Freed, E., Hsu, J. Y., Kramer, E. R., Peters, J. M., and Jackson, P. K. (2001). Emi1 is a mitotic regulator that interacts with Cdc20 and inhibits the anaphase promoting complex. *Cell* 105, 645–655.

89. Hsu, J. Y., Reimann, J. D., Sorensen, C. S., Lukas, J., and Jackson, P. K. (2002). E2F-dependent accumulation of hEmi1 regulates S-phase entry by inhibiting APC(Cdh1). *Nat Cell Biol* 4, 358–366.

90. Margottin-Goguet, F., Hsu, J. Y., Loktev, A., Hsieh, H. M., Reimann, J. D., and Jackson, P. K. (2003). Prophase destruction of Emi1 by the SCF(betaTrCP/Slimb) ubiquitin ligase activates the anaphase promoting complex to allow progression beyond prometaphase. *Dev Cell* 4, 813–826.

91. Zachariae, W., Shevchenko, A., Andrews, P. D. *et al.* (1998). Mass spectrometric analysis of the anaphase-promoting complex from yeast: identification of a subunit related to cullins. *Science* 279, 1216–1219.

92. Jaspersen, S. L., Charles, J. F., and Morgan, D. O. (1999). Inhibitory phosphorylation of the APC regulator Hct1 is controlled by the kinase Cdc28 and the phosphatase Cdc14. *Curr Biol* 9, 227–236.

93. Acquaviva, C., Herzog, F., Kraft, C., and Pines, J. (2004). The anaphase promoting complex/cyclosome is recruited to centromeres by the spindle assembly checkpoint. *Nat Cell Biol* 6, 892–898.

94. Song, M. S., and Lim, D. S. (2004). Control of APC-Cdc20 by the tumor suppressor RASSF1A. *Cell Cycle* 3, 574–576.

95. Hoque, M. T., and Ishikawa, F. (2001). Human chromatid cohesin component hRad21 is phosphorylated in M-phase and associated with metaphase centromeres. *J Biol Chem* 276, 5059–5067.

96. Blanco, M. A., Sanchez-Diaz, A., de Prada, J. M., and Moreno, S. (2000). APC(ste9/srw1) promotes degradation of mitotic cyclins in G(1) and is inhibited by cdc2 phosphorylation. *Embo J* 19, 3945–3955.

97. Pfleger, C. M., Lee, E., and Kirschner, M. W. (2001). Substrate recognition by the Cdc20 and Cdh1 components of the anaphase-promoting complex. *Genes Dev* 15, 2396–2407.

98. Gieffers, C., Peters, B. H., Kramer, E. R., Dotti, C. G., and Peters, J. M. (1999). Expression of the CDH1-associated form of the anaphase-promoting complex in postmitotic neurons. *Proc Natl Acad Sci USA* 96, 11317–11322.

99. Senderowicz, A. M., and Sausville, E. A. (2000). RESPONSE: re: preclinical and clinical development of cyclin-dependent kinase modulators. *J Natl Cancer Inst* 92, 1185.

100. Chen, W., Lee, J., Cho, S. Y., and Fine, H. A. (2004). Proteasome-mediated destruction of the cyclin a/cyclin-dependent kinase 2 complex suppresses tumor cell growth in vitro and in vivo. *Cancer Res* 64, 3949–3957.

101. De Azevedo, W. F., Leclerc, S., Meijer, L., Havlicek, L., Strnad, M., and Kim, S. H. (1997). Inhibition of cyclin-dependent kinases by purine analogues: crystal structure of human cdk2 complexed with roscovitine. *Eur J Biochem* 243, 518–526.

102. Meijer, L., and Kim, S. H. (1997). Chemical inhibitors of cyclin-dependent kinases. *Methods Enzymol* 283, 113–128.

103. Zaharevitz, D. W., Gussio, R., Leost, M. *et al.* (1999). Discovery and initial characterization of the paullones, a novel class of small-molecule inhibitors of cyclin-dependent kinases. *Cancer Res* 59, 2566–2569.

104. Knockaert, M., Greengard, P., and Meijer, L. (2002). Pharmacological inhibitors of cyclin-dependent kinases. *Trends Pharmacol Sci* 23, 417–425.

105. Vesely, J., Havlicek, L., Strnad, M. *et al.* (1994). Inhibition of cyclin-dependent kinases by purine analogues. *Eur J Biochem* 224, 771–786.

106. Schulze-Gahmen, U., Brandsen, J., Jones, H. D. *et al.* (1995). Multiple modes of ligand recognition: crystal structures of cyclin-dependent protein kinase 2 in complex with ATP and two inhibitors, olomoucine and isopentenyladenine. *Proteins* 22, 378–391.

107. Havlicek, L., Hanus, J., Vesely, J. *et al.* (1997). Cytokinin-derived cyclin-dependent kinase inhibitors: synthesis and cdc2 inhibitory activity of olomoucine and related compounds. *J Med Chem* 40, 408–412.

108. Gray, N. S., Wodicka, L., Thunnissen, A. M. *et al.* (1998). Exploiting chemical libraries, structure, and genomics in the search for kinase inhibitors. *Science* 281, 533–538.

109. Abraham, R. T., Acquarone, M., Andersen, A. *et al.* (1995). Cellular effects of olomoucine, an inhibitor of cyclin-dependent kinases. *Biol Cell* 83, 105–120.

110. Kwon, Y. G., Lee, S. Y., Choi, Y., Greengard, P., and Nairn, A. C. (1997). Cell cycle-dependent phosphorylation of mammalian protein phosphatase 1 by cdc2 kinase. *Proc Natl Acad Sci U S A* 94, 2168–2173.

111. Kim, E. H., Kim, S. U., Shin, D. Y., and Choi, K. S. (2004). Roscovitine sensitizes glioma cells to TRAIL-mediated

apoptosis by downregulation of survivin and XIAP. *Oncogene* **23**, 446–456.

112. Ljungman, M., and Paulsen, M. T. (2001). The cyclin-dependent kinase inhibitor roscovitine inhibits RNA synthesis and triggers nuclear accumulation of p53 that is unmodified at Ser15 and Lys382. *Mol Pharmacol* **60**, 785–789.

113. Lu, W., Chen, L., Peng, Y., and Chen, J. (2001). Activation of p53 by roscovitine-mediated suppression of MDM2 expression. *Oncogene* **20**, 3206–3216.

114. Eshleman, J. S., Carlson, B. L., Mladek, A. C., Kastner, B. D., Shide, K. L, and Sarkaria, J. N. (2002). Inhibition of the mammalian target of rapamycin sensitizes U87 xenografts to fractionated radiation therapy. *Cancer Res* **62**, 7291–7297.

115. De Azevedo, W. F., Jr., Mueller-Dieckmann, H. J., Schulze-Gahmen, U., Worland, P. J., Sausville, E., and Kim, S. H. (1996). Structural basis for specificity and potency of a flavonoid inhibitor of human CDK2, a cell cycle kinase. *Proc Natl Acad Sci U S A* **93**, 2735–2740.

116. Kaur, G., Stetler-Stevenson, M., Sebers, S. *et al.* (1992). Growth inhibition with reversible cell cycle arrest of carcinoma cells by flavone L86–8275. *J Natl Cancer Inst* **84**, 1736–1740.

117. Worland, P. J., Kaur, G., Stetler-Stevenson, M., Sebers, S., Sartor, O., and Sausville, E. A. (1993). Alteration of the phosphorylation state of p34cdc2 kinase by the flavone L86–8275 in breast carcinoma cells. Correlation with decreased H1 kinase activity. *Biochem Pharmacol* **46**, 1831–1840.

118. Losiewicz, M. D., Carlson, B. A., Kaur, G., Sausville, E. A., and Worland, P. J. (1994). Potent inhibition of CDC2 kinase activity by the flavonoid L86-8275. *Biochem Biophys Res Commun* **201**, 589–595.

119. Carlson, B. A., Dubay, M. M., Sausville, E. A., Brizuela, L., and Worland, P. J. (1996). Flavopiridol induces G1 arrest with inhibition of cyclin-dependent kinase (CDK) 2 and CDK4 in human breast carcinoma cells. *Cancer Res* **56**, 2973–2978.

120. Dai, Y., and Grant S. (2003). Cyclin-dependent kinase inhibitors. *Curr Opin Pharmacol* **3**, 362–370.

121. Chao, S. H., Fujinaga, K., Marion, J. E. *et al.* (2000). Flavopiridol inhibits P-TEFb and blocks HIV-1 replication. *J Biol Chem* **275**, 28345–28348.

122. Chao, S. H., and Price, D. H. (2001). Flavopiridol inactivates P-TEFb and blocks most RNA polymerase II transcription in vivo. *J Biol Chem* **276**, 31793–31799.

123. de Azevedo, W. F., Jr., Canduri, F., and da Silveira, N. J. (2002). Structural basis for inhibition of cyclin-dependent kinase 9 by flavopiridol. *Biochem Biophys Res Commun* **293**, 566–571.

124. Senderowicz, A. M. (2001). Development of cyclin-dependent kinase modulators as novel therapeutic approaches for hematological malignancies. *Leukemia* **15**, 1–9.

125. Zhai, S., Senderowicz, A. M., Sausville, E. A., and Figg, W. D. (2002). Flavopiridol, a novel cyclin-dependent kinase inhibitor, in clinical development. *Ann Pharmacother* **36**, 905–911.

126. Lam, L. T., Pickeral, O. K., Peng, A. C. *et al.* (2001). Genomic-scale measurement of mRNA turnover and the mechanisms of action of the anti-cancer drug flavopiridol. *Genome Biol* **2**, RESEARCH0041.

127. Burdette-Radoux, S., Tozer, R. G., Lohmann, R. C. *et al.* (2004). Phase II trial of flavopiridol, a cyclin dependent kinase inhibitor, in untreated metastatic malignant melanoma. *Invest New Drugs* **22**, 315–322.

128. Liu, G., Gandara, D. R., Lara, P. N. *et al.* (2004). A Phase II trial of flavopiridol (NSC #649890) in patients with previously untreated metastatic androgen-independent prostate cancer. *Clin Cancer Res* **10**, 924–928.

129. Alonso, M., Tamasdan, C., Miller, D. C., and Newcomb, E. W. (2003). Flavopiridol induces apoptosis in glioma cell lines independent of retinoblastoma and p53 tumor suppressor pathway alterations by a caspase-independent pathway. *Mol Cancer Ther* **2**, 139–150.

130. Pepper, C., Thomas, A., Hoy, T., Fegan, C., and Bentley, P. (2001). Flavopiridol circumvents Bcl-2 family mediated inhibition of apoptosis and drug resistance in B-cell chronic lymphocytic leukaemia. *Br J Haematol* **114**, 70–77.

131. Li, W., Fan, J., and Bertino, J. R. (2001). Selective sensitization of retinoblastoma protein-deficient sarcoma cells to doxorubicin by flavopiridol-mediated inhibition of cyclin-dependent kinase 2 kinase activity. *Cancer Res* **61**, 2579–2582.

132. Colevas, D., Blaylock, B., and Gravell A. (2002). Clinical trials referral resource. Flavopiridol. *Oncology (Huntingt)* **16**, 1204–1205, 1210–1212, 1214.

133. Schwartz, G. K., Ilson, D., Saltz, L. *et al.* (2001). Phase II study of the cyclin-dependent kinase inhibitor flavopiridol administered to patients with advanced gastric carcinoma. *J Clin Oncol* **19**, 1985–1992.

134. Karp, J. E., Ross, D. D., Yang, W. *et al.* (2003). Timed sequential therapy of acute leukemia with flavopiridol: in vitro model for a phase I clinical trial. *Clin Cancer Res* **9**, 307–315.

135. Schwartz, G. K., O'Reilly, E., Ilson, D. *et al.* (2002). Phase I study of the cyclin-dependent kinase inhibitor flavopiridol in combination with paclitaxel in patients with advanced solid tumors. *J Clin Oncol* **20**, 2157–2170.

136. Matranga, C. B., and Shapiro, G. I. (2002). Selective sensitization of transformed cells to flavopiridol-induced apoptosis following recruitment to S-phase. *Cancer Res* **62**, 1707–1717.

137. Cartee, L., Wang, Z., Decker, R. H. *et al.* (2001). The cyclin-dependent kinase inhibitor (CDKI) flavopiridol disrupts phorbol 12-myristate 13-acetate-induced differentiation and CDKI expression while enhancing apoptosis in human myeloid leukemia cells. *Cancer Res* **61**, 2583–2591.

138. Robey, R. W., Medina-Perez, W. Y., Nishiyama, K. *et al.* (2001). Overexpression of the ATP-binding cassette half-transporter, ABCG2 (Mxr/BCrp/ABCP1), in flavopiridol-resistant human breast cancer cells. *Clin Cancer Res* **7**, 145–152.

139. Smith, V., Raynaud, F., Workman, P., and Kelland, L. R. (2001). Characterization of a human colorectal carcinoma cell line with acquired resistance to flavopiridol. *Mol Pharmacol* **60**, 885–893.

140. Boerner, S. A., Tourne, M. E., Kaufmann, S. H., and Bible, K. C. (2001). Effect of P-glycoprotein on flavopiridol sensitivity. *Br J Cancer* **84**, 1391–1396.

141. Senderowicz, A. M., Headlee, D., Stinson, S. F. *et al.* (1998). Phase I trial of continuous infusion flavopiridol, a novel cyclin-dependent kinase inhibitor, in patients with refractory neoplasms. *J Clin Oncol* **16**, 2986–2999.

142. Tan, A. R., Headlee, D., Messmann, R. *et al.* (2002). Phase I clinical and pharmacokinetic study of flavopiridol administered as a daily 1-hour infusion in patients with advanced neoplasms. *J Clin Oncol* **20**, 4074–4082.

143. Tan, A. R., Yang, X., Berman, A. *et al.* (2004). Phase I trial of the cyclin-dependent kinase inhibitor flavopiridol in combination with docetaxel in patients with metastatic breast cancer. *Clin Cancer Res* **10**, 5038–5047.

144. Komander, D., Kular, G. S., Bain, J., Elliott, M., Alessi, D. R., and Van Aalten, D. M. (2003). Structural basis for UCN-01 (7-hydroxystaurosporine) specificity and PDK1 (3-phosphoinositide-dependent protein kinase-1) inhibition. *Biochem J* **375**, 255–262.

145. Akinaga, S., Gomi, K., Morimoto, M., Tamaoki, T., and Okabe, M. (1991). Antitumor activity of UCN-01, a selective inhibitor of protein kinase, C., in murine and human tumor models. *Cancer Res* **51**, 4888–4892.

146. Akinaga, S., Nomura, K., Gomi, K., and Okabe, M. (1994). Effect of UCN-01, a selective inhibitor of protein kinase, C., on the cell-cycle distribution of human epidermoid carcinoma, A431 cells. *Cancer Chemother Pharmacol* **33**, 273–280.

147. Seynaeve, C. M., Stetler-Stevenson, M., Sebers, S., Kaur, G., Sausville, E. A., and Worland, P. J. (1993). Cell cycle arrest and growth inhibition by the protein kinase antagonist UCN-01 in human breast carcinoma cells. *Cancer Res* **53**, 2081–2086.

148. Wang, Q., Worland, P. J., Clark, J. L., Carlson, B. A., and Sausville, E. A. (1995). Apoptosis in 7-hydroxystaurosporine-treated T lymphoblasts correlates with activation of cyclin-dependent kinases 1 and 2. *Cell Growth Differ* **6**, 927–936.

149. Akiyama, T., Yoshida, T., Tsujita, T. *et al.* (1997). G1 phase accumulation induced by UCN-01 is associated with dephosphorylation of Rb and CDK2 proteins as well as induction of CDK inhibitor p21/Cip1/WAF1/Sdi1 in p53-mutated human epidermoid carcinoma A431 cells. *Cancer Res* **57**, 1495–1501.

150. Patel, V., Lahusen, T., Leethanakul, C. *et al.* (2002). Antitumor activity of UCN-01 in carcinomas of the head and neck is associated with altered expression of cyclin D3 and p27(KIP1). *Clin Cancer Res* **8**, 3549–3560.

151. Yamasaki, F., Hama, S., Yoshioka, H. *et al.* (2003). Staurosporine-induced apoptosis is independent of p16 and p21 and achieved via arrest at G2/M and at G1 in U251MG human glioma cell line. *Cancer Chemother Pharmacol* **51**, 271–283.

152. Facchinetti, M. M., De Siervi, A., Toskos, D., and Senderowicz, A. M. (2004). UCN-01-induced cell cycle arrest requires the transcriptional induction of p21(waf1/cip1) by activation of mitogen-activated protein/extracellular signal-regulated kinase kinase/extracellular signal-regulated kinase pathway. *Cancer Res* **64**, 3629–3637.

153. Wang, Q., Fan, S., Eastman, A., Worland, P. J., Sausville, E. A., and O'Connor, P. M. (1996). UCN-01: a potent abrogator of G2 checkpoint function in cancer cells with disrupted p53. *J Natl Cancer Inst* **88**, 956–965.

154. Sausville, E. A., Arbuck, S. G., Messmann, R. *et al.* (2001). Phase I trial of 72-hour continuous infusion UCN-01 in patients with refractory neoplasms. *J Clin Oncol* **19**, 2319–2333.

155. Sausville, E. A. (2003). Cyclin-dependent kinase modulators studied at the NCI: pre-clinical and clinical studies. *Curr Med Chem Anti-Canc Agents* **3**, 47–56.

156. Bunch, R. T., and Eastman, A. (1996). Enhancement of cisplatin-induced cytotoxicity by 7-hydroxystaurosporine (UCN-01), a new G2-checkpoint inhibitor. *Clin Cancer Res* **2**, 791–797.

157. Pollack, I. F., Kawecki, S., and Lazo, J. S. (1996). Blocking of glioma proliferation in vitro and in vivo and potentiating the effects of BCNU and cisplatin: UCN-01, a selective protein kinase C inhibitor. *J Neurosurg* **84**, 1024–1032.

158. Husain, A., Yan, X. J., Rosales, N., Aghajanian, C., Schwartz, G. K., and Spriggs, D. R. (1997). UCN-01 in ovary cancer cells: effective as a single agent and in combination with cis-diamminedichloroplatinum(II)independent of p53 status. *Clin Cancer Res* **3**, 2089–2097.

159. Shao, R. G., Shimizu, T., and Pommier, Y. (1997). 7-Hydroxystaurosporine (UCN-01) induces apoptosis in human colon carcinoma and leukemia cells independently of p53. *Exp Cell Res* **234**, 388–397.

160. Tsuchida, E., and Urano, M. (1997). The effect of UCN-01 (7-hydroxystaurosporine), a potent inhibitor of protein kinase, C., on fractionated radiotherapy or daily chemotherapy of a murine fibrosarcoma. *Int J Radiat Oncol Biol Phys* **39**, 1153–1161.

161. Hsueh, C. T., Kelsen, D., and Schwartz, G. K. (1998). UCN-01 suppresses thymidylate synthase gene expression and enhances 5-fluorouracil-induced apoptosis in a sequence-dependent manner. *Clin Cancer Res* **4**, 2201–2206.

162. Jones, C. B., Clements, M. K., Redkar, A., and Daoud, S. S. (2000). UCN-01 and camptothecin induce DNA double-strand breaks in p53 mutant tumor cells, but not in normal or p53 negative epithelial cells. *Int J Oncol* **17**, 1043–1051.

163. Sugiyama, K., Shimizu, M., Akiyama, T. *et al.* (2000). UCN-01 selectively enhances mitomycin C cytotoxicity in p53 defective cells which is mediated through S and/or G(2) checkpoint abrogation. *Int J Cancer* **85**, 703–709.

164. Monks, A., Harris, E. D., Vaigro-Wolff, A., Hose, C. D., Connelly, J. W., and Sausville, E. A. (2000). UCN-01 enhances the in vitro toxicity of clinical agents in human tumor cell lines. *Invest New Drugs* **18**, 95–107.

165. Harvey, S., Decker, R., Dai, Y. *et al.* (2001). Interactions between 2-fluoroadenine 9-beta-D-arabinofuranoside and the kinase inhibitor UCN-01 in human leukemia and lymphoma cells. *Clin Cancer Res* **7**, 320–330.

166. Scovassi, A. I., Stivala, L. A., Rossi, L., Bianchi, L., and Prosperi, E. (1997). Nuclear association of cyclin D1 in human fibroblasts: tight binding to nuclear structures and modulation by protein kinase inhibitors. *Exp Cell Res* **237**, 127–134.

167. Haddad, R. I., Weinstein, L. J., Wieczorek, T. J. *et al.* (2004). A phase II clinical and pharmacodynamic study of E7070 in patients with metastatic, recurrent, or refractory squamous cell carcinoma of the head and neck: modulation of retinoblastoma protein phosphorylation by a novel chloroindolyl sulfonamide cell cycle inhibitor. *Clin Cancer Res* **10**, 4680–4687.

168. Imajoh-Ohmi, S., Kawaguchi, T., Sugiyama, S. *et al.* (1995). Lactacystin, a specific inhibitor of the proteasome, induces apoptosis in human monoblast U937 cells. *Biochem Biophys Res Commun* **217**, 1070–1077.

169. Fujita, E., Mukasa, T., Tsukahara, T., Arahata, K., Omura, S., and Momoi, T. (1996). Enhancement of CPP32-like activity in the TNF-treated U937 cells by the proteasome inhibitors. *Biochem Biophys Res Commun* **224**, 74–79.

170. Kitagawa, H., Tani, E., Ikemoto, H., Ozaki, I., Nakano, A., and Omura, S. (1999). Proteasome inhibitors induce mitochondria-independent apoptosis in human glioma cells. *FEBS Lett* **443**, 181–186.

171. Wagenknecht, B., Hermisson, M., Eitel, K., and Weller, M. (1999). Proteasome inhibitors induce p53/p21-independent apoptosis in human glioma cells. *Cell Physiol Biochem* **9**, 117–125.

172. Tani, E., Kitagawa, H., Ikemoto, H., and Matsumoto, T. (2001). Proteasome inhibitors induce Fas-mediated apoptosis by c-Myc accumulation and subsequent induction of FasL message in human glioma cells. *FEBS Lett* **504**, 53–58.

173. Orlowski, R. Z., Eswara, J. R., Lafond-Walker, A. *et al.* (1998). Tumor growth inhibition induced in a murine model of human Burkitt's lymphoma by a proteasome inhibitor. *Cancer Res* **58**, 4342–4348.

174. Adams, J., and Elliott, P. J. (2000). New agents in cancer clinical trials. *Oncogene* **19**, 6687–6692.

175. Sunwoo, J. B., Chen, Z., Dong, G. *et al.* (2001). Novel proteasome inhibitor PS-341 inhibits activation of nuclear factor-kappa, B., cell survival, tumor growth, and angiogenesis in squamous cell carcinoma. *Clin Cancer Res* **7**, 1419–1428.

176. Masdehors, P., Omura, S., Merle-Beral, H. *et al.* (1999). Increased sensitivity of CLL-derived lymphocytes to apoptotic death activation by the proteasome-specific inhibitor lactacystin. *Br J Haematol* **105**, 752–757.

177. Soligo, D., Servida, F., Delia, D., *et al.* (2001). The apoptogenic response of human myeloid leukaemia cell lines and of normal and malignant haematopoietic progenitor cells to the proteasome inhibitor PSI. *Br J Haematol* **113**, 126–135.

178. Hideshima, T., Chauhan, D., Podar, K., Schlossman, R. L., Richardson, P., and Anderson, K. C. (2001). Novel therapies targeting the myeloma cell and its bone marrow microenvironment. *Semin Oncol* **28**, 607–612.

179. Guzman, M. L., Swiderski, C. F., Howard, D. S. *et al.* (2002). Preferential induction of apoptosis for primary human leukemic stem cells. *Proc Natl Acad Sci U S A* **99**, 16220–16225.

180. Teicher, B. A., Ara, G., Herbst, R., Palombella, V. J., and Adams, J. (1999). The proteasome inhibitor PS-341 in cancer therapy. *Clin Cancer Res* **5**, 2638–2645.

181. Maki, C. G., Huibregtse, J. M., and Howley, P. M. (1996). In vivo ubiquitination and proteasome-mediated degradation of p53(1). *Cancer Res* **56**, 2649–2654.

182. Hateboer, G., Kerkhoven, R. M., Shvarts, A., Bernards, R., Beijersbergen, R. L. (1996). Degradation of E2F by the ubiquitin-proteasome pathway: regulation by retinoblastoma family proteins and adenovirus transforming proteins. *Genes Dev* **10**, 2960–2970.

183. Wagenknecht, B., Hermisson, M., Groscurth, P., Liston, P., Krammer, P. H., and Weller, M. (2000). Proteasome inhibitor-induced apoptosis of glioma cells involves the processing of multiple caspases and cytochrome c release. *J Neurochem* **75**, 2288–2297.

184. Kudo, Y., Takata, T., Ogawa, I. *et al.* (2000). p27Kip1 accumulation by inhibition of proteasome function induces apoptosis in oral squamous cell carcinoma cells. *Clin Cancer Res* **6**, 916–923.

185. Park, D. J., and Lenz, H. J. (2004). The role of proteasome inhibitors in solid tumors. *Ann Med* **36**, 296–303.

186. Aghajanian, C., Soignet, S., Dizon, D. S. *et al.* (2002). A phase I trial of the novel proteasome inhibitor PS341 in advanced solid tumor malignancies. *Clin Cancer Res* **8**, 2505–2511.

187. Orlowski, R. Z., Stinchcombe, T. E., Mitchell, B. S. *et al.* (2002). Phase I trial of the proteasome inhibitor PS-341 in patients with refractory hematologic malignancies. *J Clin Oncol* **20**, 4420–4427.

188. Richardson, P. G., Barlogie, B., Berenson, J. *et al.* (2003). A phase 2 study of bortezomib in relapsed, refractory myeloma. *N Engl J Med* **348**, 2609–2617.

189. Goy, A., and Gilles, F. (2004). Update on the proteasome inhibitor bortezomib in hematologic malignancies. *Clin Lymphoma* **4**, 230–237.

190. Lenz, H. J. (2003). Clinical update: proteasome inhibitors in solid tumors. *Cancer Treat Rev.* **29 Suppl 1**, 41–48.

191. Chauhan, D., Li, G., Podar, K. *et al.* (2004). The bortezomib/proteasome inhibitor PS-341 and triterpenoid CDDO-Im induce synergistic anti-multiple myeloma (MM) activity and overcome bortezomib resistance. *Blood* **103**, 3158–3166.

192. Mitsiades, N., Mitsiades, C. S., Richardson, P. G. *et al.* (2003). The proteasome inhibitor PS-341 potentiates sensitivity of multiple myeloma cells to conventional chemotherapeutic agents: therapeutic applications. *Blood* **101**, 2377–2380.

193. Bayes, M., Rabasseda, X., and Prous, J. R. (2004). Gateways to clinical trials. *Methods Find Exp Clin Pharmacol* **26**, 211–244.

194. Kamat, A. M., Karashima, T., Davis, D. W. *et al.* (2004). The proteasome inhibitor bortezomib synergizes with gemcitabine to block the growth of human 253JB-V bladder tumors in vivo. *Mol Cancer Ther* **3**, 279–290.

Apoptosis Pathways and Chemotherapy

Joachim P. Steinbach and Michael Weller

ABSTRACT: Tumor cell death is the ultimate goal of chemotherapy. In particular, the induction of apoptosis has been suggested as a strategy for efficient and selective elimination of tumor cells. However, in malignant gliomas, the most common malignant intrinsic brain tumors, there is little evidence that the currently available chemotherapeutic drugs mediate their effects *via* induction of apoptosis, with the possible exception of anaplastic oligodendrogliomas which often show striking tumor regression on neuroimaging. Nevertheless, the induction of apoptosis plays a major conceptual role in the majority of novel experimental approaches to brain tumors.

In this chapter, we review the different components of the apoptotic pathways and explore how tumors can acquire antiapoptotic properties. We then analyze the mechanism of action of the currently available chemotherapeutic drugs and their impact on the cell death programs. Finally, we discuss novel approaches for the induction of apoptosis including death ligands, p53-based therapy and others, with a special focus on malignant gliomas.

ARCHITECTURE OF THE MAJOR APOPTOTIC PATHWAYS

Introduction to the Concept of Apoptotic Cell Death

It has long been known that the development of mammalian organisms involves the loss of numerous cells in various organs at predetermined phases of ontogenesis. This type of cell death, which plays a major role, for example in shaping the immune and central nervous systems, is a physiological process referred to as *developmental* or *programmed cell death*. This concept of programmed cell death has been developed in contradistinction to various other instances of cell death which are unscheduled and which are caused, for example by mechanical trauma, inflammation, or ischemia. Many types of programmed cell death are characterized by typical morphological changes which are collectively referred to as apoptosis [1]. These include condensation and fragmentation of chromatin and nuclear remnants and cytoplasmic compartmentalization, resulting in membrane blebbing and the formation of apoptotic bodies. The latter are efficiently cleared from the tissue by nonprofessional phagocytes which recognize signals displayed by apoptotic cells and cell remnants. The clearance of apoptotic cells accounts for the lack of inflammatory responses and of macrophage activation in response to cell death during ontogeny. Apoptotic cell death tends to eliminate solitary cells within a given target tissue. In contrast, tissue loss by necrosis, a process considered an *accidental cell death*, commonly affects continuous parts of a tissue.

The features of necrosis include disturbances of energy metabolism, cell swelling, osmotic lysis, degeneration of organelles, activation of catabolic enzymes, such as proteases and endonucleases, and the release of proinflammatory mediators.

There are also a number of other types of cell death distinct from apoptosis and necrosis. In particular, mitotic catastrophe deserves mentioning. This type of cell death occurs when cells with unbalanced genetic lesions enter mitosis due to deficient G_2 checkpoint systems. It is characterized by multiple micronuclei and nuclear fragmentation. Autophagy is another specific type of cell death which features an increased number of autophagic vesicles. Of note, when apoptosis is inhibited, the fraction of tumor cells that undergo nonapoptotic cell death is increased [2]. In addition, permanent cellular senescence may be induced by sublethal damage.

Most, but not all, types of apoptotic cell death share the biochemical features of caspase activation and DNA fragmentation. Caspases are proteases which degrade proteins in a sequence-specific manner during the apoptotic process and are thus executioners of apoptosis. The DNA lesions of apoptosis may be restricted to the formation of large DNA fragments of up to 300 kb. The classical nucleosomal size, ladder-like pattern of DNA fragmentation, that was initially characterized and is commonly seen in lymphoid cells undergoing apoptosis, may be lacking or at least difficult to detect in apoptosis of nonlymphoid cells.

Programmed cell death is an active type of cell death. There is evidence for the participation of the dying cell during most of the instances of programmed cell death, e.g., the morphological process of apoptosis is energy consuming. Further, inhibition of RNA and protein synthesis prevents cell death in many paradigms of apoptosis, notably in nontransformed cells. This suggests that cells induced to undergo apoptosis may synthesize mRNA encoded by *killer genes* which is translated into *killer proteins* which execute cellular suicide from within. However, it is important to note that inhibitors of RNA and protein synthesis are themselves potent inducers of death in many tumor-cell-types, presumably because such drugs interfere with the synthesis of proteins essential for survival.

The major molecular players of the apoptotic process are largely the same in glioma cells, other cancer cells and nontransformed glial and non-glial cells (Table 10.1):

The Extrinsic Pathway

The detection of endogenous death ligands mediating the death of distinct target cell populations *via* the activation of specific death receptors at the cell surface has led to the concept of what is now commonly called the extrinsic apoptotic pathway (Fig. 10.1). The most important death ligands are tumor necrosis factor alpha (TNF-α), CD95 ligand (CD95L), also known as Fas ligand or Apo-1 ligand, and Apo-2 ligand/TNF-related apoptosis-inducing ligand (Apo2L/TRAIL) [3]. CD95L and Apo2L/TRAIL act on target cells through binding their specific receptors, CD95 and death receptors DR4/DR5. Following ligation to their cognate receptors,

death ligands induce the formation of the death-inducing signaling complex (DISC) which then activates procaspases 8 (and 10). A mechanism of "induced proximity" has been proposed. This involves clustering of procaspase zymogens around aggregated receptor complexes at the DISC, inducing dimerization of caspase 8, resulting in its activation. Cleavage of caspase 8 then stabilizes its active form. Following release of caspases 8 and 10 from the DISC into the cytosol, they cleave and thereby activate effector caspases (e.g., caspases 3 and 7). However, in most of the cell lines and paradigms, an amplification of the death signal *via* the mitochondrial pathway is required for apoptosis (intrinsic pathway or mitochondrial pathway).

The Intrinsic Pathway

This pathway may either be cross-activated as a result of activation of the extrinsic pathway through caspase 8-induced cleavage of BID or by other stimuli resulting in mitochondrial injury. Mitochondrial membrane permeabilization is essentially regulated by the opposing actions of pro- and anti-apoptotic members of the BCL-2 family. Stimuli that result in a shift in the ratio of proapoptotic *versus* antiapoptotic BCL-2 family members towards proapoptotic proteins induce this pathway. Multidomain proapoptotic BCL-2 members (BAK, BAX) can be activated directly following interaction with the BH3-only BCL-2 member protein BID. Binding of other proapoptotic BH3-only members (NOXA, PUMA, BAD, BIM) to antiapoptotic BCL-2 members (BCL-2, BCL-X_L) also results in the activation of BAK and BAX [4]. Whether BAK and BAX directly form pores in the mitochondrial membrane is still debated. Ultimately, they induce mitochondrial membrane permeabilization, resulting in the release of proapoptotic molecules such as cytochrome c, second mitochondria-derived activator of caspase (Smac) and apoptosis-inducing factor (AIF), and loss of mitochondrial function. In the presence of dATP, cytochrome c promotes the assembly of apoptotic protease activating factor 1 (APAF-1) to form the apoptosome and cleavage of pro-caspase 9 which in turn cleaves caspases 3 and 7. Smac indirectly promotes apoptosis by blocking the inhibitory action of inhibitor of apoptosis proteins (IAPs) on caspases 3, 7, and 9. The final steps of the cell death program are executed by the effector caspases 3 and 7 and include DNA fragmentation and cleavage of a large number of essential proteins, resulting in the biochemical and morphological features of apoptosis.

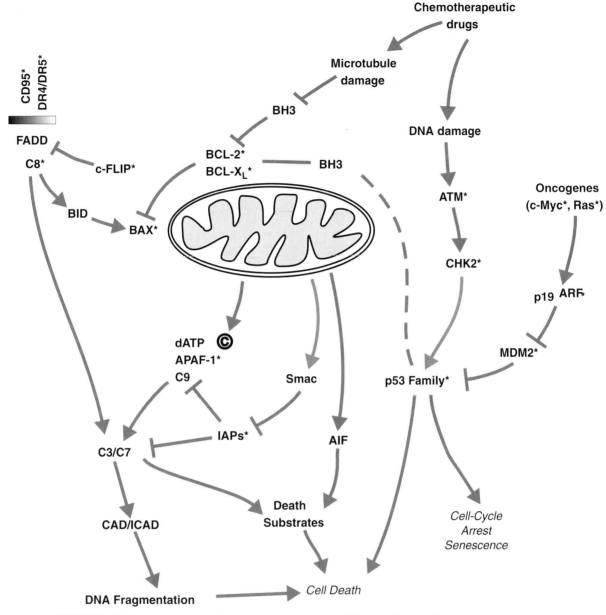

FIGURE 10.1 Current model of apoptotic pathways. This model visualizes the most important components of the cell death machinery. Chemotherapeutic drugs act on the intrinsic pathway through p53-dependent and -independent mechanisms converging on altered balance of proapoptotic/antiapoptotic BCL-2 family member proteins. BAX activation then results in the release of cytochrome c from the mitochondria and activation caspase 9 at the apoptosome. Caspase 9 in turn activates effector caspases, resulting in DNA and target protein cleavage and ultimately cell death. In contrast, the extrinsic pathway downstream of death receptors is of little importance for drug-induced apoptosis, but holds promise for therapeutic induction of cell death. * Apoptosis-related proteins with frequent mutation/deregulation in cancers. ©, cytochrome c; C3/8/9, caspases 3/8/9.

Intrinsic stresses in tumor cells can activate the mitochondrial pathway. These include hypoxia, survival factor deprivation and the action of deregulated oncogenes. Genotoxic stress also acts *via* the intrinsic pathway. One major mechanism is through

p53, a cell cycle control protein which senses genotoxic stress, leading to DNA damage and arrests cells in G_0/G_1 or G_2/M, or induces their apoptosis. P53 is activated through proteins that sense DNA damage such as the ataxia-telangiectasia-mutated

TABLE 10.1   Molecules with Key Roles in Tumorigenesis and Apoptosis (Adapted from [106]).

Tumor suppressors and death-inducing factors	Oncogenes and antiapoptotic factors
AIF Released from injured mitochondria. Induces caspase-independent cell death during development.	**BCL-2/BCL-X$_L$** Antagonize BAX/BAK and inhibit mitochondrial membrane disruption. Frequently overexpressed in tumors. Potently inhibit drug-induced apoptosis.
ATM Senses DNA double strand breaks and stabilizes p53. Mutated in ataxia-talangiectasia syndrome. Deficiencies increase risk of developing haematological malignancies and breast cancer.	**c-FLIP** Endogenous inhibitor of the extrinsic pathway. Overexpressed in some cancers. Prevents activation of caspase 8 and apoptosis induced by some chemotherapeutic drugs.
APAF-1 Necessary for activation of caspase-9 following cytochrome c release. Mutated and transcriptionally silenced in melanoma and leukemia cell lines. APAF-1-/- cells are chemoresistant.	**c-Myc** Induces proliferation in the presence of survival factors, such as BCL-2, and apoptosis in the absence of survival factors. Deregulated expression in many cancers. Can sensitize cells to drug-induced apoptosis.
BAK/BAX Mediate mitochondrial membrane damage. Mutated or decreased expression in some tumors. Key mediators of drug-induced apoptosis.	**EGFR** Growth and survival factor activating the MAPK and PI3K/Akt pathways. Amplified/overexpressed in the majority of primary glioblastomas. Activation inhibits death receptor- and drug-induced apoptosis.
Caspase 8 Initiator caspase activated by death ligands. Gene silenced in neuroblastomas, resulting in resistance to drug-induced apoptosis.	**IAPs** Inhibit effector caspase activation. Frequently overexpressed in cancer. Overexpression inhibits drug-induced apoptosis.
CD95L/ CD95 Initiate the extrinsic apoptotic pathway. CD95 is mutated and down-regulated in some lymphoid and solid tumors. In glioma cells, CD95L/CD95 function is not necessary for drug-induced apoptosis.	**MDM2** Negative regulator of p53. Overexpressed in some tumors. Inhibits drug-induced p53 activation.
CHK2 Senses DNA double strand breaks and phosphorylates and stabilizes p53. Mutated in cases of Li-Fraumeni syndrome without p53 mutations.	**mTOR** Target of PKB/Akt and key regulator of translational responses to nutrient and growth factor availability. Overexpressed in glioblastomas. May suppress apoptosis.
Cytochrome c Released from mitochondria following activation of BAK/BAX. Activates the assembly of the apoptosome from APAF-1 and caspase 9 in the presence of ATP.	**NF-κB** Survival pathway transcriptionally inducing expression of members of the BCL-2 and IAP families. Deregulated activity in many cancers. Can inhibit both the extrinsic and intrinsic death pathways and induce drug-resistance.
DR4/DR5 Agonistic receptors for Apo2L/TRAIL that Initiate the extrinsic apoptotic pathway. Preferential expression of DR4/5 over the antagonistic receptor DR2 may explain preferential sensitivity of glioma cells to Apo2L/TRAIL.	**PI3K** Mediates survival signaling downstream of RTKs, e.g., phosphorylation of PKB/Akt. Overexpressed or deregulated in some cancers. Inhibition of PI3K enhances chemotherapeutic drug-induced apoptosis.
p53 Initiates the intrinsic apoptotic pathway. Mutated or altered expression in many cancers. p53-deficient cells may be resistant to drug-induced apoptosis.	**PKB/Akt** Key antiapoptotic signaling molecule, phosphorylates many targets with antiapoptotic function. Frequently overexpressed/activated in solid tumors. Induces resistance to a wide range of apoptotic stimuli including drugs.
p19^{ARF} Blocks MDM2 inhibition of p53. Mutated or altered expression in many cancers. Enhances drug-induced apoptosis through p53-dependent mechanisms.	**Ras** Mutated or deregulated in many cancers. Activates PI3K and downstream pathways. Induces proliferation and inhibits c-Myc and drug-induced apoptosis.
PTEN Dephosphorylates PI3P and thus negatively regulates PKB/Akt. Mutated or altered expression in many cancers. Loss of PTEN results in resistance to many apoptotic stimuli.	
Rb Controls cell-cycle entry and inhibits E2F-medidated transcription. Mutated in some and functionally disrupted in many cancers. Loss of Rb function induces p53-dependent and -independent apoptosis.	
Smac Mitochondrial death factor inhibiting IAPs. Smac agonists sensitize tumor cells to drug-induced apoptosis.	

kinase ATM and CHK1 which phosphorylate and stabilize p53 and prevent its MDM2-induced ubiquitin-dependent degradation [5]. Mitogenic oncogenes such as p19^ARF can also activate p53 by binding and inactivating MDM2 [6]. The molecular consequences of p53 activation include changes in the transcription of various p53 response genes, including BAX, PUMA, NOXA, APAF-1, and p21, as well as transcription-independent effects. The latter have recently been better defined [7]. In brief, p53 can also localize to the cytoplasm and induce BAX oligomerization, resulting in mitochondrial membrane depolarization and cytochrome c release. Direct binding of p53 to BCL-X_L *via* the polyproline domain may be important for this, leading to the liberation of BH3-only proteins from BCL-X_L, which in turn induce BAX oligomerization. The relative importance of transcription-dependent *versus* transcription-independent functions in human cancer cells remains to be defined.

In addition to p53-dependent mechanisms, components of the stress response include activation of the c-Jun N-terminal kinase (JNK) and nuclear factor kappa B (NF-κB) pathways, endoplasmatic reticulum (ER) damage, elevation of ceramide levels, and others [8].

DEREGULATION OF APOPTOTIC PATHWAYS IN BRAIN TUMORS

Alterations of specific components of the cell death machinery are highly relevant for both tumorigenesis and resistance to therapy. Selective pressures for apoptosis-deficient clones are apparent already at early stages of tumorigenesis. One major proapoptotic stress is hypoxia. Recent progress in the understanding of tumor evolution has demonstrated an initial stage of vessel cooption followed by regression, leading to a secondarily avascular tumor and massive tumor cell loss [9]. Ultimately, the remaining tumor is rescued by robust angiogenesis at the tumor margin. Other factors implicated in the development of apoptosis include deprivation of survival factors, the acquisition of imbalanced DNA lesions, reactive oxygen intermediates generated by mitochondrial damage, and arachidonic acid metabolites released from necrotic cells. Coupling of apoptosis to unconstrained proliferation has also been proposed, mediated for instance by the activation of c-Myc which results in a substantial increase in the number of apoptotic tumor cells predominantly in the regions of tumor hypoxia [10]. Tumor cells must also evade the immune system, but there is no evidence for

a major role of inflammatory cells in inducing apoptosis of glioma cells *in vivo*. All these factors may be driving forces for the aquisition of genetic defects by tumor cells that result in the disruption of apoptotic pathways. The most important of these alterations are discussed below.

In established malignant gliomas, there is also extensive cell death. However, the type of cell death in these tumors is predominantly necrotic [11]. Central fields of necrosis which are prominent in the primary type of glioblastoma [12] can be distinguished from small foci of necrosis often surrounded by pseudopalisading tumor cells. Yet, apoptosis does also occur in malignant gliomas on the basis of light microscopic and ultrastructural criteria, including nuclear condensation and pyknosis and formation of apoptotic bodies, and by *in situ* detection of DNA fragmentation using *in situ* terminal transferase-mediated dUTP nick end labeling (TUNEL) [13]. TUNEL-positive tumor cells are preferentially located among these perinecrotic pseudopalisading cells [13,14]. Hence, specific properties of the microenvironment may account for the induction of the different types of cell death. In particular, gradients of *tumor hypoxia* combined with reduced supply of nutrients, such as glucose and amino acids are relevant for this process. The major cause of these changes appears to be deregulated tumor growth outscaling neoangiogenesis, resulting in chronic hypoxia caused by increased distance of tumor cells from capillaries and an acute, sometimes reversible, type of hypoxia caused by malfunction of the tumor vasculature, e.g., venous thrombosis and, collapse or compression of vessels, and even reversal of blood flow in tumor veins. Perinecrotic cells are still viable, yet, they express vascular endothelial growth factor (VEGF) [15], signifying sublethal exposure to hypoxia. Since apoptosis requires ATP [16], perinecrotic cells may die by apoptosis while ATP depletion caused by a steeper decline of O_2 partial pressure and nutrient supply in the central area may cause necrotic cell death. In a paradigm of hypoxia with partial glucose depletion, malignant glioma cells are resistant to hypoxia *per se*, and die only after their glucose supply is exhausted, resulting in ATP depletion. Although the release of cytochrome c is a prominent feature in this model, the mode of cell death is largely necrotic due to the decoupling of caspase activation from mitochondrial injury [17]. These results are compatible with clonal evolution of a tumor cell phenotype that tolerates hypoxia as long as energy metabolism can be maintained, and (indirectly) implies the development of resistance to hypoxia-induced apoptosis.

The p53 and pRb Pathways

p53 may be the most common target gene for mutational inactivation in human cancers. p53 mutations are rather common (65 per cent) in secondary glioblastomas thought to be derived through the malignant progression from grade II or III astrocytomas. In approximately 90 per cent of these patients, the same p53 mutations are already found in the less malignant precursor lesion [18,19]. In contrast, only 10 per cent of primary glioblastomas exhibit p53 mutations. Interestingly, p53 mutations and amplification of the epidermal growth factor receptor (EGFR) gene appear to be mutually exclusive. The molecular basis for this phenomenon remains to be identified.

In untransformed cells, the loss of p53 may enhance rather than decrease the vulnerability to apoptosis. However, within the process of neoplastic transformation, the loss of p53 probably allows the cell to accumulate random genetic and chromosomal aberrations without triggering the endogenous p53-controlled cell death pathway. Further, naturally occurring mutations of p53 often result in unpredictable biological effects which include gain-of-function properties of mutant p53 [20]. In human malignant glioma cell lines, there is no apparent correlation between the sensitivity to cytotoxic therapy and genetic or functional p53 status or expression of p53 response genes [21]. The Rb pathway is also deregulated in the majority of human cancers, including glioma [22,23]. Although pRb acts as a tumor suppressor by gating cell-cycle entry, its (isolated) loss may actually promote apoptosis [24,25]; thus tumor cells need to acquire other mutations to counteract this.

BCL-2 Family Proteins

Human glioma cell lines express a variety of antiapoptotic and proapoptotic BCL-2 family proteins [21], and it has been shown that overexpression of BCL-2 and BCL-X$_L$ can protect these cells from apoptosis induced by diverse stimuli [26]. In contrast to the expectation that the expression of antiapoptotic members of the BCL-2 family increases with malignancy, stronger staining in astrocytoma and anaplastic astrocytoma compared with glioblastoma has been observed in most [27], but not all the studies [26]. An up-regulation of BCL-2 and BCL-X$_L$, but a down-regulation of BAX has been described in recurrent glioblastoma independent from treatment [28], suggesting therapy-independent pressures for

the development of an apoptosis-resistant phenotype. Overexpression of BCL-2 or BCL-X$_L$ induces complex changes of the glioma cell phenotype in that it not only protects glioma cells from various pro-apoptotic stimuli [26,29,30], but also enhances their motility [31]. While the molecular pathway mediating enhanced motility has not been clarified, inhibition of apoptosis most likely involves the preservation of mitochondrial integrity in response to apoptotic stimuli.

Regarding cytotoxicity, the mRNA expression levels of BCL-X$_L$ had the highest correlation with drug resistance in a screen of the 60 tumor cell lines of the National Cancer Institute's Anticancer Drug Screen (NCI-ADS) [32].

IAP Proteins

IAPs are potent inhibitors of caspases 3, 7, and 9 [33]. However, they also inhibit cell death by modulating cell-cycle progression, cell division, and signal transduction pathways. Members of this family include X-linked IAP (XIAP), which inhibits caspases 3, 7, and 9, but not caspase 8. The cIAPs 1 and 2 inhibit caspases 3 and 7. Survivin is a member of another subclass of IAPs and is preferentially expressed fetally. Survivin is frequently overexpressed in many different malignancies. The precise mechanism of its action is debated; it may primarily consist in the inhibition of caspase 9, which depends upon an unidentified co-factor [34]. However, it has also been proposed that the primary function of survivin is the regulation of mitotic progression and that it is necessary for the maintenance of the spindle checkpoint [2]. Loss of survivin induces cell-cycle arrest and cell death by mitotic catastrophe in a p53 and BCL-2 independent manner. Thus, targeting survivin has been suggested as a novel approach for the therapy of p53-deficient tumors.

In human gliomas, XIAP is overexpressed and counteracts apoptosis induced by a variety of stimuli [35]. Another IAP member, IAP-2 is prominently upregulated by hypoxia [36]. A recent report suggests that upregulation of IAP expression is necessary to counteract high levels of constitutive caspase 3 activity detected in cancer cell lines which surprisingly do not lead to apoptosis [37].

A major breakthrough has been achieved by the identification of endogenous inhibitors of IAP function. Smac and Omi/HTRA2 are released from the mitochondria similar to cytochrome c and enhance caspase cleavage by disrupting IAP function.

Survival Pathways

In addition to deletions and mutations in the above-mentioned apoptosis-relevant proteins and others that can not be reviewed in more detail, e.g., heat shock proteins, phosphorylation by deregulated signaling can also induce important antiapoptotic effects. In particular, signaling through growth factor receptor tyrosine kinases, notably EGFR, *via* phosphatidylinositol-3-kinase (PI3K) and PKB/Akt mediates pleiotropic antiapoptotic effects [38,39]. Examples include inactivation of the BCL-2 family member BAD, caspase 9, and members of the Forkhead-class of transcription factors (FOX01, FOX03a) and nuclear translocation of MDM2 through Akt-dependent phosphorylation. Transcriptional and translational consequences of PKB/Akt signaling are also relevant, e.g., increased expression of c-FLIP and BCL-X_L [40,41].

Thus survival signals originating from the EGFR–PI3K–PKB/Akt pathway are linked mainly to the intrinsic pathway. Other important survival pathways include the NFκB and JNK pathways. However, depending on cell type and stimulus, these can also contribute to death signaling.

In summary, tumor cells evade apoptosis by pleiotropic mechanisms, most of which act on the intrinsic pathway, reflecting the profile of selective pressures. Specific alterations in the extrinsic pathway, e.g., overexpression of c-FLIP are rare in comparison. This may indicate that immune attack is overall of less importance. However, overexpression of the perforin/granzyme B inhibitor PI-9/SPI-6 in many tumor cell lines suggests a role for this pathway in immune evasion [42].

CHEMOTHERAPEUTIC DRUGS: MECHANISM OF ACTION AND IMPACT ON BRAIN TUMORS

Although clinically apparent tumors have already established antiapoptotic mechanisms in response to selective pressures when they are exposed to chemotherapy as outlined above, apoptotic pathways contribute to the cytotoxic actions of most chemotherapeutic drugs [6]. This is, in principle, also true for human malignant glioma cells [29,30,43].

However, the mechanisms through which apoptosis is induced by chemotherapy has been a subject of considerable debate. One issue was the contribution of death ligand–death receptor interactions to chemotherapy induced apoptosis. Data in support of a role for this phenomenon were reported for T-cell leukemia, hepatoma, neuroblastoma, and other solid tumor cell lines [44–46]. Further, cross-resistance to CD95-mediated apoptosis and chemotherapy has been reported, implicating common mechanisms of cell death [47].

In glioma cell lines, however, neither CD95/CD95L nor Apo2L/TRAIL receptor interactions are responsible for drug-induced cell death even though these cell lines co-express ligands and receptors. This has been demonstrated using crm-A as a tool, which preferentially inhibits caspase 8. Transfection with crm-A abolishes death ligand-induced cell death, but does not affect chemotherapy induced cell death [29]. In addition, treatment with inhibitors of RNA and protein synthesis, while sensitizing glioma cells towards death ligands, does not enhance chemotherapy induced apoptosis. However, chemotherapy induces an apoptotic type of cell death, as demonstrated by typical ultrastructural changes and DNA fragmentation. Chemotherapy also results in the activation of caspases other than caspase 8, notably caspases 7 and 9 [30]. Further, the pan-specific caspase inhibitor zVAD-fmk efficiently suppresses both death ligand- and chemotherapy-induced apoptosis. Therefore, downstream caspases represent a common effector pathway, but the upstream events are stimulus-specific. This conclusion is also supported by the fact that glioma cells selected for death ligand resistance did not acquire cross-resistance to chemotherapy [29]. Regarding the mechanism of chemotherapy induced cell death, the activation of caspase 9 by chemotherapy pointed to mitochondrial injury. Indeed, caspase activation was preceded by the release of cytochrome c into the cytosol [30]. The importance of mitochondrial damage for chemotherapy-induced cell death is underscored by the finding that overexpression of BCL-X_L protects against this insult and prevents cytochrome c release, caspase activation, and cell death. Largely similar results were observed utilizing a panel of different chemotherapeutic drugs including vincristine, teniposide, BCNU (carmustine), doxorubicine, and camptothecin, suggesting that the mitochondria are a uniform target of the proximate injury conferred by these agents, i.e., DNA and microtubular damage. Combinations of chemotherapeutic agents with different targets have in general not fulfilled their promise of synergistic efficacy in the treatment of malignant glioma cells. One likely explanation is that defects in common downstream death pathways rather than individual mutations in the drug targets mediate resistance [48].

A number of new chemotherapeutic agents with potentially better efficacy and clinical tolerability are in preclinical and clinical development; their impact on the cellular death pathways remains to be determined.

The limited impact of chemotherapy may be enhanced by the combination with inhibitors of survival pathways, such as EGFR [41], PI3K [49], protein kinase C [50], and NF-κB [51]. It will also be important to investigate the efficacy of chemotherapy under the specific conditions of the tumor microenvironment (e.g., hypoxia, limited supply of nutrients, and acidosis), since nongenetic short-term effects of hypoxia such as down-regulation of BID and BAX have been shown to mediate resistance to chemotherapy [52]. In this regard, the improved stability of CCNU (lomustine) under acidotic conditions may contribute to its activity against some types of malignant glioma [53]. Agents with selective cytotoxicity in hypoxic conditions such as tirapazamine may also be valuable for combination therapy [54].

The induction of nonapoptotic cell death by chemotherapy is also an important focus of interest [2]. One example is a regulated form of necrotic cell death induced by alkylating chemotherapeutic drugs which occurs when apoptotic pathways are disabled and results in the activation of the repair enzyme poly (ADP-ribose) polymerase (PARP). Cells utilizing glycolysis as their major source of energy production undergo cell death in response to PARP activation because of NAD depletion in the cytoplasm, but not in the mitochondria. This mechanism may explain at least in part the relative selectivity of this class of drugs for tumor cells [55]. The microtubule-damaging agent paclitaxel induces cell death with features of mitotic catastrophe by inducing an abnormal metaphase in which sister chromatides fail to segregate properly [56].

Deamidation of BCL-X$_L$, a posttranslational modification in which asparagines are converted to a mixture of aspartates and isoaspartates, is another recently identified mechanism of tumor-specific cytotoxicity [25]. Disruption of pRb function is the critical determinant of this process. In response to alkylating agents deamidation occurs in pRb- or p53-deficient cells, disrupting the interaction of BCL-X$_L$ with BH3-only proteins, thus activating the intrinsic pathway. In contrast, in normal tissues, DNA-damaging agents induce activation (dephosphorylation) of pRb, which suppresses BCL-X$_L$ deamidation through an as yet unknown mechanism.

NOVEL APPROACHES FOR THE INDUCTION OF APOPTOSIS

Death Ligands / Receptors

The presence of both agonistic and antagonistic (decoy) receptors for Apo2L/TRAIL may be of special importance since the differential distribution of these receptors in tumor cells and normal tissue has been proposed to underlie the selective activity of Apo2L/ TRAIL against tumor cells. A preferential expression of agonistic Apo2L/TRAIL receptors is also seen in malignant glioma cells [57,58]. A soluble decoy receptor (DcR3) for CD95L has been characterized as well. DcR3 is released by malignant glioma cells and protects them from CD95L-induced apoptosis [59]. In contrast, release of the soluble receptor for Apo2L/ TRAIL, osteoprotegerin, may be an exception in gliomas [60]. The majority of human glioma cell lines are partially sensitive to apoptosis induced by agonistic CD95 antibodies [61], or soluble CD95L [43], or viral vectors encoding CD95L [62]. CD95L has also been shown to induce cell death in primary human glioma-derived cell cultures which are resistant to the chemotherapeutic agent, lomustine [63]. CD95-mediated apoptosis is enhanced when RNA or protein synthesis are inhibited, suggesting that glioma cells express cytoprotective proteins which block the killing pathway [61]. One candidate gene product rapidly lost upon inhibition of protein synthesis in CD95L-treated glioma cells is the cell-cycle inhibitor p21 [64]. Further, suppression of p21$^{WAF/Cip1}$ by antisense oligonucleotides also enhanced apoptosis induced by irradiation in radioresistant glioma cells [65], and attenuation of p21$^{WAF1/Cip1}$ expression by an antisense adenovirus expression vector sensitized malignant glioma cells to apoptosis induced by the chemotherapeutic agents carmustine and cisplatin [66]. p21$^{WAF/Cip1}$ inhibition may, therefore, evolve to be a major target for strategies aiming at sensitizing malignant glioma cells to radiotherapy, chemotherapy, and death ligand therapy.

CD95L is probably too toxic for a systemic application in human patients because the systemic activation of CD95 results in liver failure within a few hours [67]. The local application of adenoviral vectors expressing CD95L may be a strategy to circumvent systemic side effects in gliomas [68]. Transferring the gene encoding an adaptor protein of the CD95-dependent killing cascade, Fas-associated protein with death domain (FADD), into glioma cells also inhibits glioma growth *in vitro* and *in vivo* [69]. Broad-scale gene expression analysis showed that the TNF receptor-1-associated death domain protein (TRADD) mediates p53-independent radiation-induced apoptosis of glioma cells and that overexpression of TRADD sensitized glioma cells to radiotherapy [70]. Finally, even more downstream effectors of death receptor-mediated apoptosis, the caspases, have been successfully employed to promote glioma cell death *in vitro* and *in vivo* [71]. Yet, targeting potent intracellular mediators of apoptosis is only feasible clinically if major problems of therapeutic gene delivery will be solved, involving both efficacy and specificity. Thus caspase 3 gene transfer will only kill cancer cells that are transduced, but there will be no bystander effect, and caspase 3 is also a potent inducer of apoptosis in neurons, suggesting that neurotoxicity might become a problem *in vivo*. Therefore, activating signaling pathways that are selectively active in tumor cells is still most promising.

In the field of death ligands and receptors, inducing cancer cell apoptosis *via* local or systemic application of Apo2L/TRAIL is one of the most promising strategies. The growth of intracranial human glioma xenografts in nude mice is inhibited by the locoregional administration of Apo2L/TRAIL [72]. Further, Apo2L/TRAIL acts synergistically with lomustine to induce apoptosis in glioma cell lines [58], and systemic co-treatment of nude mice bearing intracranial glioma xenografts with Apo2L/TRAIL and cisplatin significantly extended survival [73]. Convection-enhanced delivery of Apo2L/TRAIL can also enhance the effect of systemic treatment of temozolomide in an intracranial glioma model [74]. Glioma cell lines exhibiting resistance to Apo2L/TRAIL may be sensitized by pretreatment with cisplatin, camptothecin, or etoposide [75] by the induced down-regulation of the caspase 8 inhibitor c-FLIP(S) (cellular Fas-associated death domain-like interleukin 1-converting enzyme-inhibitory protein) and the up-regulation of the pro-apoptotic BCL-2 family member BAK, independent of the p53 status. These authors further demonstrated that Apo2L/TRAIL alone or in combination with chemotherapeutic agents induced apoptosis not only in established cell lines, but also in primary glioma tumor cultures. Up-regulation of DR5 by cisplatin, doxorubicin, or camptothecin has also been reported as a mechanism of sensitization of glioma cells for Apo2L/TRAIL-induced apoptosis [76]. Resistance to Apo2L/TRAIL may be caused by the deletion of BAX, a proapoptotic BCL-2 family member, in colon cancer cells [77]. In this model, challenge with Apo2L/TRAIL in a mutation-prone genetic environment created by the mutation of a mismatch-repair enzyme resulted in BAX mutations. Similar to the findings by Arizono *et al.* [76], exposure to the chemotherapeutic agents etoposide and camptothecin rescued Apo2L/TRAIL sensitivity by the up-regulation of DR5 and the BAX homolog BAK [77]. Specific targeting of DR5 with an agonistic antibody might be an even more selective strategy to efficiently kill tumor cells in the absence of toxicity [78]. Recent progress in the understanding of the varying susceptibility of glioma cell lines to Apo2L/TRAIL-induced apoptosis has revealed that resistant cell lines expressed 2-fold higher levels of the apoptosis inhibitor phosphoprotein enriched in diabetes/phosphoprotein enriched in astrocytes-15 kDa (PED/PEA-15) [79]. Susceptibility towards TRAIL-induced apoptosis may also be modulated by calcium/calmodulin-dependent protein kinase, since treatment with a specific inhibitor rescued TRAIL sensitivity [80]. In contrast, the levels of the surface receptors for Apo2L/TRAIL do not correlate well with the sensitivity of cell lines to apoptosis [58,81]. Differences in the intracellular actions of CD95L and Apo2L/TRAIL can account for selective sensitivity of glioma cell lines to either ligand [82,83], and combined treatment with adenovirally delivered Apo2L/TRAIL and CD95L may have synergistic effects in some cell lines [81].

A potentially clinically applicable strategy to promote glioma cell apoptosis consists of the combination of peptides derived from Smac protein with chemotherapy or Apo2L/TRAIL. Smac is a potent inhibitor of members of the IAP family of caspase inhibitors, and overexpression of Smac or the treatment with cell-permeable Smac peptides bypasses the BCL-2 block to apoptosis and greatly facilitates chemotherapy and Apo2L/TRAIL-mediated apoptosis *in vitro*. Even more impressive, intracranially grafted LN-229 human malignant glioma cells were eradicated by local combination treatment with Smac peptides and Apo2L/TRAIL in a mouse model [84]. While the application of Smac peptides is hampered by technical problems, nonpeptidyl small molecule drugs which mimic the inhibitory action of Smac on IAPs are already tested in preclinical studies [33].

Inactivation of XIAP by an antisense construct is another strategy, and there is a phase I clinical study with the XIAP antisense molecule AEG35165 [33]. For survivin, proapoptotic effects of either antisense oligonucleotide-mediated down-regulation or functional suppression through transduction with a dominant negative mutant have been reported

[85,86]. A clinical trial with survivin antisense oligonucleotides has also been announced.

p53-based Therapy

An adenovirus encoding an artificial p53-based gene designed to bypass various pathways of p53 inactivation (chimeric tumor suppressor 1, CTS1) induced cell death in all the glioma cell lines examined. In contrast, wild-type p53 did not consistently induce cell death in the same cell lines [87]. In a different study, treatment of intracerebral p53 wild-type U87MG xenografts with adenovirus encoding p53 resulted in apoptosis of tumor cells *in vivo* and prolonged survival [88].

A phase I study of adenoviral p53 gene therapy in patients with malignant glioma demonstrated that ectopic expression of p53 can induce apoptosis *in vivo*, but transduction efficiency was too low for clinically relevant impact of this strategy [89]. The stabilization of mutant p53 proteins by small molecules that act in a chaperone-like fashion and promote wild-type p53 activity is another promising approach to cancer therapy [90]. In human malignant glioma cells, the compound CP-31398 induces cell death in both p53 mutant and p53 wild-type cell lines, but not in the p53-null LN-308 cell line [91]. Cell death in this paradigm had some features of apoptosis, namely phosphatidylserine exposure on the outside of the cell membrane, but was independent of BCL-X_L and lacked caspase activation. The proximate cause of death thus remained obscure. In addition, p53-independent cytotoxic effects of CP-31398 may limit the usefulness of this specific compound. However, a number of related agents with supposedly more specificity are currently being developed [92].

Targeting Members of the BCL-2 Family

The neutralization of antiapoptotic BCL-2 family proteins by antisense technology or by overexpression of proapoptotic BCL-2 family proteins has remained an active area of apoptosis research for more than a decade. Natural born killer (NBK) is a prototype member of the proapoptotic BH-3 only BCL-2 family members which heterodimerizes with BCL-2 and BCL-X_L. Adenoviral transfer of NBK induces cell death independent of activation of caspases 3, 7, 8, 9, and 10. However, cell death can be suppressed by the overexpression of BCL-X_L or by XIAP. While NBK expression uniformly induces cell death independent of p53 status in all 12 glioma cell lines investigated

in vitro and abrogated the tumorigenicity of LN-229 cells *in vivo*, selective targeting to glioma cells appears necessary since NBK expression is also toxic to nontransformed (rat) glial cells and neurons [93].

Other Novel Approaches of Inducing Apoptosis in Tumor Cells or Sensitizing Tumor Cells Towards Cell Death

In addition to the direct targeting of classical players involved in apoptosis such as p53 or death receptors, there are some novel approaches to induce apoptosis in glioma cells which appear to be close to a clinical assessment. A report of antitumoral actions of cannabinoids, thought to act *via* the ceramide pathway, in a rat glioma model *in vivo* in the absence of neurotoxicity is provoking in this regard [94]. However, lower levels of cannabinoids transactivate the EGFR and enhance tumor cell proliferation [95]. Also, the suppression of Rac1, a small GTP-binding protein, inhibited survival and produced apoptosis in three human glioma cell lines and 19 of 21 short-term cultures of human gliomas that varied in p53, EGFR, the human analogue of the mouse double minute-2 (MDM2), and p16/p19 mutational or amplification status. In contrast, the inhibition of Rac1 activity did not induce apoptosis in normal primary human adult astrocytes. The mechanism may involve inhibition of mitogen-activated protein kinase kinase 1, an activator of JNK, suggesting that JNK functions downstream of Rac1 in glioma cells [96].

Inhibition of growth factor receptor tyrosine kinases is also an important strategy, since many of these, notably EGFR, mediate antiapoptotic effects and enhance survival of glioma cells *in vivo* [41,97]. Accordingly, pharmacological inhibition of EGFR inhibition sensitizes glioma cells to CD95L- and Apo2L/TRAIL-induced apoptosis [98]. In some cell lines, the inhibition of EGFR may be compensated by signaling through other receptor tyrosine kinases such as insulin-like growth factor receptor I [99]. The combination of EGFR inhibitors with inhibitors of other receptor tyrosine kinases may, therefore, improve the efficacy of this approach [100,101]. Recent work has offered an explanation for the efficacy of co-inhibition: the kinase c-Src downstream of the EGFR phosphorylates and inhibits the tumor suppressor phosphatase and tensin homolog deleted on chromosome ten (PTEN). Inhibition of EGFR may restore the function of PTEN, and thus put a break on PI3K signaling [101]. Glioma cells with deregulated activity of the EGFR pathway may also display enhanced sensitivity for Ras inhibition by farnesyltransferase inhibitors [102]. However, we have recently found that inhibition of EGFR signaling may also mimic a "starvation signal" and thus protect glioma cells from acute hypoxia by decreasing energy demand [103]. While the importance of this phenomenon for the therapy of human tumors is not yet known, it may be prudent to consider ambiguous effects of therapies targeting the EGFR when pronounced hypoxia is present, e.g., concurrent treatment with angiogenesis inhibitors.

A large body of research has delineated targets in the signaling pathways downstream of receptor tyrosine kinases in gliomas and other malignancies, including PI3K, PKB/Akt, mitogen-activated protein kinase (p42/44 MAPK), and mammalian target of rapamycin (mTOR). These—alone or in combination—are therefore promising new targets for the experimental therapy of gliomas. Of note, inactivation of PTEN—a common event in malignant gliomas that contributes to resistance to apoptosis [104]—renders these cells particularly vulnerable to inhibition of the mTOR pathway by rapamycin and its homologs [105].

NF-κB is activated by chemotherapeutic agents and may mediate antiapoptotic transcriptional responses in malignant glioma cells. Expression of dominant negative I-κB inhibits the nuclear translocation of NF-κB and augments the cytotoxicity of carmustine, carboplatin, and SN-38 [51]. Exposure to sulfasalazine, a potent inhibitor of NF-κB, has differential effects on CD95L and Apo2L/TRAIL-mediated apoptosis: the former is inhibited while the latter is enhanced [82]. Intriguingly, these effects appear to be independent from the inhibitory action on NF-κB, but require protein synthesis and possibly p21.

PERSPECTIVE

In summary, the intrinsic pathway is the key target for drugs to induce apoptosis in brain tumors. Both the p53-dependent and –independent mechanisms converge on an altered balance of proapoptotic/antiapoptotic BCL-2 protein family members. However, the efficacy of cell death induction is limited by mutation and deregulation of antiapoptotic proteins.

While our understanding of cellular death pathways in cancer has increased greatly in the past years, translation of this knowledge into clinically applicable strategies for selective interference with molecules involved in apoptosis has only just begun.

References

1. Kerr, J. F., Wyllie, A. H., and Currie, A. R. (1972). Apoptosis: a basic biological phenomenon with wide-ranging implications in tissue kinetics. *Br J Cancer* **26** 239–257.

2. Okada, H., and Mak, T. W. (2004). Pathways of apoptotic and non-apoptotic death in tumour cells. *Nat Rev Cancer* **4**, 592–603.

3. Weller, M., Kleihues, P., Dichgans, J., and Ohgaki, H. (1998). CD95 ligand: lethal weapon against malignant glioma? *Brain Pathol* **8**, 285–293.

4. Huang, D. C., and Strasser, A. (2000). BH3-Only proteins-essential initiators of apoptotic cell death. *Cell* **103**, 839–842.

5. Khanna, K. K., and Jackson, S. P. (2001). DNA double-strand breaks: signaling, repair and the cancer connection. *Nat Genet* **27**, 247–254.

6. Lowe, S. W., and Lin, A. W. (2000). Apoptosis in cancer. *Carcinogenesis* **21**, 485–495.

7. Baptiste, N., and Prives, C. (2004). p53 in the cytoplasm: a question of overkill? *Cell* **116**, 487–489.

8. Herr, I., and Debatin, K. M. (2001). Cellular stress response and apoptosis in cancer therapy. *Blood* **98**, 2603–2614.

9. Holash, J., Maisonpierre, P. C., Compton, D. *et al.* (1999). Vessel cooption, regression, and growth in tumors mediated by angiopoietins and VEGF. *Science* **284**, 1994–1998.

10. Alarcon, R. M., Rupnow, B. A., Graeber, T. G. Knox, S. J., and Giaccia, A.J. (1996). Modulation of c-Myc activity and apoptosis in vivo. *Cancer Res* **56**, 4315–4319.

11. Kleihues, P., Burger, P. C., Collins, V. P., Newcomb, E. W., Ohgaki, H., and Cavenee, W. K. (2000). Glioblastoma. In *Pathology and Genetics of Tumours of the Nervous System* (P. Kleihues, and W. K. Cavence, Ed.), 2nd ed., pp 29–39 IARC press, Lyon.

12. Tohma, Y., Gratas, C., Van Meir, E. G. *et al.* (1998). Necrogenesis and Fas/APO-1 (CD95) expression in primary (de novo) and secondary glioblastomas. *J Neuropathol Exp Neurol* **57**, 239–245.

13. Schiffer, D., Cavalla, P., Migheli, A. *et al.* (1995). Apoptosis and cell proliferation in human neuroepithelial tumors. *Neurosci Lett* **195**, 81–84.

14. Tachibana, O., Lampe, J., Kleihues, P., and Ohgaki, H. (1996). Preferential expression of Fas/APO1 (CD95) and apoptotic cell death in perinecrotic cells of glioblastoma multiforme. *Acta Neuropathol* **92**, 431–434.

15. Plate, K. H., Breier, G., Weich, H. A., and Risau, W. (1992). Vascular endothelial growth factor is a potential tumour angiogenesis factor in human gliomas in vivo. *Nature* **359**, 845–848.

16. Leist, M., Single, B., Castoldi, A. F., Kuhnle, S., and Nicotera, P. (1997). Intracellular adenosine triphosphate (ATP) concentration: a switch in the decision between apoptosis and necrosis. *J Exp Med* **185**, 1481–1486.

17. Steinbach, J. P., Wolburg, H., Klumpp, A., Probst, H., and Weller, M. (2003). Hypoxia-induced cell death in human malignant glioma cells: energy deprivation promotes decoupling of mitochondrial cytochrome c release from caspase processing and necrotic cell death. *Cell Death Differ* **10**, 823–832.

18. Watanabe, K., Tachibana, O., Sata, K., Yonekawa, Y., Kleihues, P., and Ohgaki, H. (1996). Overexpression of the EGF receptor and p53 mutations are mutually exclusive in the evolution of primary and secondary glioblastomas. *Brain Pathol* **6**, 217–223.

19. Watanabe, K., Sato, K., Biernat, W. *et al.* (1997). Incidence and timing of p53 mutations during astrocytoma progression in patients with multiple biopsies. *Clin Cancer Res* **3**, 523–530.

20. Weller, M. (1998). Predicting response to cancer chemotherapy: the role of p53. *Cell Tissue Res* **292**, 435–445.

21. Weller, M., Rieger, J., Grimmel, C. *et al.* (1998). Predicting chemoresistance in human malignant glioma cells: the role of molecular genetic analyses. *Int J Cancer* **79**, 640–644.

22. Ichimura, K., Schmidt, E. E., Goike, H. M., and Collins, V. P. (1996). Human glioblastomas with no alterations of the CDKN2A (p16INK4A, MTS1) and CDK4 genes have frequent mutations of the retinoblastoma gene. *Oncogene* **13**, 1065–1072.

23. Nevins, J. R. (2001). The Rb/E2F pathway and cancer. *Hum Mol Genet* **10**, 699–703.

24. Almasan, A., Yin, Y., Kelly, R. E. *et al.* (1995). Deficiency of retinoblastoma protein leads to inappropriate S-phase entry, activation of E2F-responsive genes, and apoptosis. *Proc Natl Acad Sci USA* **92**, 5436–5440.

25. Deverman, B. E., Cook, B. L., Manson, S. R. *et al.* (2002). Bcl-xL deamidation is a critical switch in the regulation of the response to DNA damage. *Cell* **111**, 51–62.

26. Weller, M., Malipiero, U., Aguzzi, A., Reed, J. C., and Fontana, A. (1995). Protooncogene bcl-2 gene transfer abrogates Fas/APO-1 antibody-mediated apoptosis of human malignant glioma cells and confers resistance to chemotherapeutic drugs and therapeutic irradiation. *J Clin Invest* **95**, 2633–2643.

27. Krajewski, S., Krajewska, M., Ehrmann, J. *et al.* (1997). Immunohistochemical analysis of Bcl-2, Bcl-X, Mcl-1, and Bax in tumors of central and peripheral nervous system origin. *Am J Pathol* **150**, 805–814.

28. Strik, H., Deininger, M., Streffer, J. *et al.* (1999). BCL-2 family protein expression in initial and recurrent glioblastomas: modulation by radiochemotherapy. *J Neurol Neurosurg Psychiatry* **67**, 763–768.

29. Glaser, T., Wagenknecht, B., Groscurth, P., Krammer, P. H., and Weller, M. (1999). Death ligand/receptor-independent caspase activation mediates drug- induced cytotoxic cell death in human malignant glioma cells. *Oncogene* **18**, 5044–5053.

30. Glaser, T., and Weller, M. (2001). Caspase-dependent chemotherapy-induced death of glioma cells requires mitochondrial cytochrome c release. *Biochem Biophys Res Commun* **281**, 322–327.

31. Wick, W., Wagner, S., Kerkau, S., Dichgans, J., John, J. C., and Weller, M. (1998). BCL-2 promotes migration and invasiveness of human glioma cells. *FEBS Lett* **440**, 419–424.

32. Amundson, S. A., Myers, T. G., Scudiero, D., Kitada, S., Reed, J. C., and Fornace, A. J., Jr. (2000). An informatics approach identifying markers of chemosensitivity in human cancer cell lines. *Cancer Res* **60**, 6101–6110.

33. Schimmer, A. D. (2004). Inhibitor of apoptosis proteins: translating basic knowledge into clinical practice. *Cancer Res* **64**, 7183–7190.

34. Marusawa, H., Matsuzawa, S., Welsh, K. *et al.* (2003). HBXIP functions as a cofactor of survivin in apoptosis suppression. *Embo J* **22**, 2729–2740.

35. Wagenknecht, B., Glaser, T., Naumann, U. *et al.* (1999). Expression and biological activity of X-linked inhibitor of apoptosis (XIAP) in human malignant glioma. *Cell Death Differ* **6**, 370–376.

36. Dong, Z., Venkatachalam, M. A., Wang, J. *et al.* (2001). Up-regulation of apoptosis inhibitory protein IAP-2 by hypoxia. Hif-1-independent mechanisms. *J Biol Chem* **276**, 18702–18709.

37. Yang, L., Cao, Z., Yan, H., and Wood, W. C. (2003). Coexistence of high levels of apoptotic signaling and inhibitor of apoptosis proteins in human tumor cells: implication for cancer specific therapy. *Cancer Res* **63**, 6815–6824.

38. Sibilia, M., Fleischmann, A., Behrens, A. *et al.* (2000). The EGF receptor provides an essential survival signal for SOS-dependent skin tumor development. *Cell* **102**, 211–220.

39. Holland, E. C., Celestino, J., Dai, C., Schaefer, L., Sawaya, R. E., and Fuller, G. N. (2000). Combined activation of Ras and Akt in neural progenitors induces glioblastoma formation in mice. *Nat Genet* **25**, 55–57.

40. Panka, D. J., Mano, T., Suhara, T., Walsh, K., and Mier, J. W. (2001). Phosphatidylinositol 3-kinase/Akt activity regulates c-FLIP expression in tumor cells. *J Biol Chem* **276**, 6893–6896.

41. Nagane, M., Levitzki, A., Gazit, A., Cavenee, W. K., and Huang, H. J. (1998). Drug resistance of human glioblastoma cells conferred by a tumor- specific mutant epidermal growth factor receptor through modulation of Bcl-XL and caspase-3-like proteases. *Proc Natl Acad Sci USA* **95**, 5724–5729.

42. Medema, J. P., de Jong, J., Peltenburg, L. T. *et al.* (2001). Blockade of the granzyme B/perforin pathway through overexpression of the serine protease inhibitor PI-9/SPI-6 constitutes a mechanism for immune escape by tumors. *Proc Natl Acad Sci USA* **98**, 11515–11520.

43. Roth, W., Fontana, A., Trepel, M., Reed, J. C., Dichgans, J., and Weller, M. (1997). Immunochemotherapy of malignant glioma: synergistic activity of CD95 ligand and chemotherapeutics. *Cancer Immunol Immunother* **44**, 55–63.

44. Friesen, C., Herr, I., Krammer, P. H., and Debatin, K. M. (1996). Involvement of the CD95 (APO-1/FAS) receptor/ligand system in drug-induced apoptosis in leukemia cells. *Nat Med* **2**, 574–577.

45. Müller, M., Strand, S., Hug, H. *et al.* (1997). Drug-induced apoptosis in hepatoma cells is mediated by the CD95 (APO-1/Fas) receptor/ligand system and involves activation of wild-type p53. *J Clin Invest* **99**, 403–413.

46. Fulda, S., Scaffidi, C., Pietsch, T., Krammer, P. H., Peter, M. E., and Debatin, K. M. (1998). Activation of the CD95 (APO-1/Fas) pathway in drug- and gamma-irradiation-induced apoptosis of brain tumor cells. *Cell Death Differ* **5**, 884–893.

47. Los, M., Herr, I., Friesen, C., Fulda, S., Schulze-Osthoff, K., and Debatin, K. M. (1997). Cross-resistance of CD95- and drug-induced apoptosis as a consequence of deficient activation of caspases (ICE/Ced-3 proteases). *Blood* **90**, 3118–3129.

48. Brown, J. M., and Wouters, B. G. (1999). Apoptosis, p53, and tumor cell sensitivity to anticancer agents. *Cancer Res* **59**, 1391–1399.

49. Shingu, T., Yamada, K., Hara, N. *et al.* (2003). Synergistic augmentation of antimicrotubule agent-induced cytotoxicity by a phosphoinositide 3-kinase inhibitor in human malignant glioma cells. *Cancer Res* **63**, 4044–4047.

50. Chen, T. C., Su, S., Fry, D., and Liebes, L. (2003). Combination therapy with irinotecan and protein kinase C inhibitors in malignant glioma. *Cancer* **97**, 2363–2373.

51. Weaver, K. D., Yeyeodu, S., Cusack, J. C. Jr., Baldwin, A. S. Jr., and Ewend, M. G. (2003). Potentiation of chemotherapeutic agents following antagonism of nuclear factor kappa B in human gliomas. *J Neurooncol* **61**, 187–196.

52. Erler, J. T., Cawthorne, C. J., Williams, K. J. *et al.* (2004). Hypoxia-mediated down-regulation of Bid and Bax in tumors occurs via hypoxia-inducible factor 1-dependent and -independent mechanisms and contributes to drug resistance. *Mol Cell Biol* **24**, 2875–2889.

53. Reichert, M., Steinbach, J. P., Supra, P., and Weller, M. (2002). Modulation of growth and radiochemosensitivity of human malignant glioma cells by acidosis. *Cancer* **95**, 1113–1119.

54. Brown, J. M. (1999). The hypoxic cell: a target for selective cancer therapy–eighteenth Bruce F. Cain Memorial Award lecture. *Cancer Res* **59**, 5863–5870.

55. Zong, W. X., Ditsworth, D., Bauer, D. E., Wang, Z. Q., and Thompson, C. B. (2004). Alkylating DNA damage stimulates a regulated form of necrotic cell death. *Genes Dev* **18**, 1272–1282.

56. Jordan, M. A., Wendell, K., Gardiner, S., Derry, W. B., Copp, H., and Wilson, L. (1996). Mitotic block induced in HeLa cells by low concentrations of paclitaxel (Taxol) results in abnormal mitotic exit and apoptotic cell death. *Cancer Res* **56**, 816–825.

57. Rieger, J., Naumann, U., Glaser, T., Ashkenazi, A., and Weller, M. (1998). APO2 ligand: a novel lethal weapon against malignant glioma? *FEBS Lett* **427**, 124–128.

58. Röhn, T. A., Wagenknecht, B., Roth, W. *et al.* (2001). CCNU-dependent potentiation of TRAIL/Apo2L-induced apoptosis in human glioma cells is p53-independent but may involve enhanced cytochrome c release. *Oncogene* **20**, 4128–4137.

59. Roth, W., Isenmann, S., Nakamura, M. *et al.* (2001). Soluble decoy receptor 3 is expressed by malignant gliomas and suppresses CD95 ligand-induced apoptosis and chemotaxis. *Cancer Res* **61**, 2759–2765.

60. Naumann, U., Wick, W., Beschorner, R., Meyermann, R., and Weller, M. (2004). Expression and functional activity of osteoprotegerin in human malignant gliomas. *Acta Neuropathol* **107**, 17–22.

61. Weller, M., Frei, K., Groscurth, P., Krammer, P. H., Yonekawa, Y., and Fontana, A. (1994). Anti-Fas/APO-1 antibody-mediated apoptosis of cultured human glioma cells. Induction and modulation of sensitivity by cytokines. *J Clin Invest* **94**, 954–964.

62. Shinoura, N., Yoshida, Y., Sadata, A. *et al.* (1998). Apoptosis by retrovirus- and adenovirus-mediated gene transfer of Fas ligand to glioma cells: implications for gene therapy. *Hum Gene Ther* **9**, 1983–1993.

63. Maleniak, T. C., Darling, J. L., Lowenstein, P. R., and Castro, M. G. (2001). Adenovirus-mediated expression of HSV1-TK or Fas ligand induces cell death in primary human glioma-derived cell cultures that are resistant to the chemotherapeutic agent CCNU. *Cancer Gene Ther* **8**, 589–598.

64. Glaser, T., Wagenknecht, B., and Weller, M. (2001). Identification of p21 as a target of cycloheximide-mediated facilitation of CD95-mediated apoptosis in human malignant glioma cells. *Oncogene* **20**, 4757–4767.

65. Kokunai, T., Urui, S., Tomita, H., and Tamaki, N. (2001). Overcoming of radioresistance in human gliomas by p21WAF1/CIP1 antisense oligonucleotide. *J Neurooncol* **51**, 111–119.

66. Ruan, S., Okcu, M. F., Pong, R. C. *et al.* (1999). Attenuation of WAF1/Cip1 expression by an antisense adenovirus expression vector sensitizes glioblastoma cells to apoptosis induced by chemotherapeutic agents 1,3-bis(2-chloroethyl)-1-nitrosourea and cisplatin. *Clin Cancer Res* **5**, 197–202.

67. Ogasawara, J., Watanabe-Fukunaga, R., Adachi, M. *et al.* (1993). Lethal effect of the anti-Fas antibody in mice. *Nature* **364**, 806–809.

68. Ambar, B. B., Frei, K., Malipiero, U. *et al.* (1999). Treatment of experimental glioma by administration of adenoviral vectors expressing Fas ligand. *Hum Gene Ther* **10**, 1641–1648.

69. Kondo, S., Ishizaka, Y., Okada, T. *et al.* (1998). FADD gene therapy for malignant gliomas in vitro and in vivo. *Hum Gene Ther* **9**, 1599–1608.

70. Yount, G. L., Afshar, G., Ries, S. *et al.* (2001). Transcriptional activation of TRADD mediates p53-independent radiation-induced apoptosis of glioma cells. *Oncogene* **20**, 2826–2835.

71. Yu, J. S., Sena-Esteves, M., Paulus, W., Breakefield, X. O., and Reeves, S. A. (1996). Retroviral delivery and tetracycline-dependent expression of IL-1beta- converting enzyme (ICE) in a rat glioma model provides controlled induction of apoptotic death in tumor cells. *Cancer Res* **56**, 5423–5427.

72. Roth, W., Isenmann, S., Naumann, U. *et al.* (1999). Locoregional Apo2L/TRAIL eradicates intracranial human malignant glioma xenografts in athymic mice in the absence of neuro-toxicity. *Biochem Biophys Res Commun* **265**, 479–483.

73. Nagane, M., Pan, G., Weddle, J. J., Dixit, V. M., Cavenee, W. K., and Huang, H. J. (2000). Increased death receptor 5 expression by chemotherapeutic agents in human gliomas causes synergistic cytotoxicity with tumor necrosis factor-related apoptosis-inducing ligand in vitro and in vivo. *Cancer Res* **60**, 847–853.

74. Saito, R., Bringas, J. R., Panner, A. *et al.* (2004). Convection-enhanced delivery of tumor necrosis factor-related apoptosis-inducing ligand with systemic administration of temozolomide prolongs survival in an intracranial glioblastoma xenograft model. *Cancer Res* **64**, 6858–6862.

75. Song, J. H., Song, D. K., Pyrzynska, B., Petruk, K. C., Van Meir, E. G., and Hao, C. (2003). TRAIL triggers apoptosis in human malignant glioma cells through extrinsic and intrinsic pathways. *Brain Pathol* **13**, 539–553.

76. Arizono, Y., Yoshikawa, H., Naganuma, H., Hamada, Y., Nakajima, Y., and Tasaka, K. (2003). A mechanism of resistance to TRAIL/Apo2L-induced apoptosis of newly established glioma cell line and sensitisation to TRAIL by genotoxic agents. *Br J Cancer* **88**, 298–306.

77. LeBlanc, H., Lawrence, D., Varfolomeev, E. *et al.* (2002). Tumor-cell resistance to death receptor–induced apoptosis through mutational inactivation of the proapoptotic Bcl-2 homolog Bax. *Nat Med* **8**, 274–281.

78. Ichikawa, K., Liu, W., Zhao, L. *et al.* (2001). Tumoricidal activity of a novel anti-human DR5 monoclonal antibody without hepatocyte cytotoxicity. *Nat Med* **7**, 954–960.

79. Hao, C., Beguinot, F., Condorelli, G. *et al.* (2001). Induction and intracellular regulation of tumor necrosis factor-related apoptosis-inducing ligand (TRAIL) mediated apotosis in human malignant glioma cells. *Cancer Res* **61**, 1162–1170.

80. Xiao, C., Yang, B. F., Asadi, N., Beguinot, F., and Hao, C. (2002). Tumor necrosis factor-related apoptosis-inducing ligand-induced death-inducing signaling complex and its modulation by c-FLIP and PED/PEA-15 in glioma cells. *J Biol Chem* **277**, 25020–25025.

81. Rubinchik, S., Yu, H., Woraratanadharm, J., Voelkel-Johnson, C., Norris, J. S., and Dong, J. Y. (2003). Enhanced apoptosis of glioma cell lines is achieved by co-delivering FasL-GFP and TRAIL with a complex Ad5 vector. *Cancer Gene Ther* **10**, 814–822.

82. Hermisson, M., and Weller, M. (2003). NF-kappaB-independent actions of sulfasalazine dissociate the CD95L- and Apo2L/TRAIL-dependent death signaling pathways in human malignant glioma cells. *Cell Death Differ* **10**, 1078–1089.

83. Knight, M. J., Riffkin, C. D., Muscat, A. M. Ashley, D. M., and Hawkins, C. J. (2001). Analysis of FasL and TRAIL induced apoptosis pathways in glioma cells. *Oncogene* **20**, 5789–5798.

84. Fulda, S., Wick, W., Weller, M., and Debatin, K. M. (2002). Smac agonists sensitize for Apo2L/TRAIL- or anticancer drug-induced apoptosis and induce regression of malignant glioma in vivo. *Nat Med* **8**, 808–815.

85. Olie, R. A., Simoes-Wust, A. P., Baumann, B. *et al.* (2000). A novel antisense oligonucleotide targeting survivin expression induces apoptosis and sensitizes lung cancer cells to chemotherapy. *Cancer Res* **60**, 2805–2809.

86. Mesri, M., Wall, N. R., Li, J., Kim, R. W., and Altieri, D. C. (2001). Cancer gene therapy using a survivin mutant adenovirus. *J Clin Invest* **108**, 981–990.

87. Naumann, U., Kügler, S., Wolburg, H. *et al.* (2001). Chimeric tumor suppressor 1, a p53-derived chimeric tumor suppressor gene, kills p53 mutant and p53 wild-type glioma cells in synergy with irradiation and CD95 ligand. *Cancer Res* **61**, 5833–5842.

88. Li, H., Alonso-Vanegas, M., Colicos, M. A. *et al.* (1999). Intracerebral adenovirus-mediated p53 tumor suppressor gene therapy for experimental human glioma. *Clin Cancer Res* **5**, 637–642.

89. Lang, F. F., Bruner, J. M., Fuller, G. N. *et al.* (2003). Phase I trial of adenovirus-mediated p53 gene therapy for recurrent glioma: biological and clinical results. *J Clin Oncol* **21**, 2508–2518.

90. Foster, B. A., Coffey, H. A., Morin, M. J., and Rastinejad, F. (1999). Pharmacological rescue of mutant p53 conformation and function. *Science* **286**, 2507–2510.

91. Wischhusen, J., Naumann, U., Ohgaki, H., Rastinejad, F., and Weller, M. (2003). CP-31398, a novel p53-stabilizing agent, induces p53-dependent and p53-independent glioma cell death. *Oncogene* **22**, 8233–8245.

92. Bykov, V. J., Issaeva, N., Shilov, A. *et al.* (2002). Restoration of the tumor suppressor function to mutant p53 by a low-molecular-weight compound. *Nat Med* **8**, 282–288.

93. Naumann, U., Schmidt, F., Wick, W. *et al.* (2003). Adenoviral natural born killer gene therapy for malignant glioma. *Hum Gene Ther* **14**, 1235–1246.

94. Galve-Roperh, I., Sanchez, C., Cortes, M. L., del Pulgar, T. G., Izquierdo, M., and Guzman, M. (2000). Anti-tumoral action of cannabinoids: involvement of sustained ceramide accumulation and extracellular signal-regulated kinase activation. *Nat Med* **6**, 313–319.

95. Hart, S., Fischer, O. M., and Ullrich, A. (2004). Cannabinoids induce cancer cell proliferation via tumor necrosis factor alpha-converting enzyme (TACE/ADAM17)-mediated

96. Senger, D. L., Tudan, C., Guiot, M. C. *et al.* (2002). Suppression of Rac activity induces apoptosis of human glioma cells but not normal human astrocytes. *Cancer Res* **62**, 2131–2140.

97. Nishikawa, R., Ji, X. D., Harmon, R. C. *et al.* (1994). A mutant epidermal growth factor receptor common in human glioma confers enhanced tumorigenicity. *Cancer Res* **91**, 7727–7731.

98. Steinbach, J. P., Supra, P., Huang, H. J., Cavenee, W. K., and Weller, M. (2002). CD95-mediated apoptosis of human malignant glioma cells: modulation by epidermal growth factor receptor activity. *Brain Pathol* **12**, 12–20.

99. Chakravarti, A., Loeffler, J. S., and Dyson, N. J. (2002). Insulin-like growth factor receptor I mediates resistance to anti-epidermal growth factor receptor therapy in primary human glioblastoma cells through continued activation of phosphoinositide 3-kinase signaling. *Cancer Res* **62**, 200–207.

100. Steinbach, J. P., Eisenmann, C., Klumpp, A., and Weller M. (2004). Co-inhibition of epidermal growth factor receptor and type 1 insulin-like growth factor receptor synergistically sensitizes human malignant glioma cells to CD95L-induced apoptosis. *Biochem Biophys Res Commun* **321**, 524–530.

101. Nagata, Y., Lan, K. H., Zhou, X. *et al.* (2004). PTEN activation contributes to tumor inhibition by trastuzumab, and loss of PTEN predicts trastuzumab resistance in patients. *Cancer Cell* **6**, 117–127.

102. Feldkamp, M. M., Lau, N., and Guha, A. (1999). Growth inhibition of astrocytoma cells by farnesyl transferase inhibitors is mediated by a combination of anti-proliferative, pro-apoptotic and anti-angiogenic effects. *Oncogene* **18**, 7514–7526.

103. Steinbach, J. P., Klumpp, A., Wolburg, H., and Weller, M. (2004). Inhibition of epidermal growth factor receptor signaling protects human malignant glioma cells from hypoxia-induced cell death. *Cancer Res* **64**, 1570–1574.

104. Wick, W., Furnari, F. B., Naumann, U., Cavenee, W. K., and Weller, M. (1999). PTEN gene transfer in human malignant glioma: sensitization to irradiation and CD95L-induced apoptosis. *Oncogene* **18**, 3936–3943.

105. Neshat, M. S., Mellinghoff, I. K., Tran, C. *et al.* (2001). Enhanced sensitivity of PTEN-deficient tumors to inhibition of FRAP/mTOR. *Proc Natl Acad Sci U S A* **98**, 10314–10319.

106. Johnstone, R. W., Ruefli, A. A., and Lowe, S. W. (2002). Apoptosis: a link between cancer genetics and chemotherapy. *Cell* **108**, 153–164.

transactivation of the epidermal growth factor receptor. *Cancer Res* **64**, 1943–1950.

11

Growth Factor Signaling Pathways and Receptor Tyrosine Kinase Inhibitors

Ian F. Pollack

INTRODUCTION

Astrocytomas constitute the largest group of central nervous system (CNS) tumors in both the pediatric [1] and adult [2] age groups. Despite recent refinements in surgery, radiotherapy, and conventional chemotherapy, the prognosis remains poor for patients with malignant gliomas, such as glioblastoma multiforme and anaplastic astrocytoma, which account for the majority of astrocytomas and generally lead to death within several years after diagnosis [1,2]. The poor response of these tumors to conventional chemotherapy and radiotherapy in part reflects a resistance of malignant glioma cells to undergoing apoptosis in response to DNA damage or depletion of exogenous mitogens. This resistance may result from mutations of tumor suppressor and cell cycle control genes and aberrant activation of growth and survival signaling pathways, as a consequence of autocrine and paracrine stimulation through tyrosine kinase growth factor receptors [3]. Although more than 20 members of the receptor tyrosine kinase family have been identified [4], and a variety of such receptors have been implicated in glial tumorigenesis, previous studies have demonstrated that receptors for platelet-derived growth factor (PDGFR) and epidermal growth factor (EGFR) may represent particularly important contributors to this dysregulated proliferation. This notion is supported by recent observations that antibody- and antisense-mediated neutralization of PDGFR and EGFR can substantially inhibit glioma growth *in vitro*. However, significant limitations to the use of these strategies for the *in vivo* treatment of human gliomas have precluded their clinical application. Recently, pharmacological inhibitors of PDGFR (e.g., STI571) and EGFR (e.g., ZD1839) have been developed, which demonstrate potent inhibition of receptor-dependent signaling. However, the appropriate subgroups of gliomas to be treated with these agents remains to be defined, calling attention to the need for identifying biologically relevant surrogates that predict response. Because these agents will likely be independently effective in only a subset of tumors, optimal strategies for combining them with other therapeutic approaches need to be determined and validated.

GLIOMA CELL RESISTANCE TO CONVENTIONAL THERAPEUTIC MODALITIES

The poor response of malignant glioma cells to radiotherapy and conventional chemotherapy results in part from their resistance to undergoing apoptosis in response to either DNA damage [5] or depletion of exogenous mitogens. A number of factors have been implicated in mediating this resistance. For example, the p53 protein, a critical facilitator of DNA damage-induced apoptosis [6], is mutated or deleted in 50 per cent of high-grade gliomas [7,8]. In a variety of tumor systems [6,9–11], such mutations have been associated with poor responsiveness to radiotherapy or chemotherapy, and have correlated with an adverse outcome. Mutations of other tumor suppressor genes, such as *PTEN*, which encodes a phosphoprotein and phospholipid phosphatase that inhibits Akt/PKB, have also been implicated in mediating resistance to apoptosis in these neoplasms [12–14].

In addition to their resistance to DNA damage-induced apoptosis, glioma cells are able to avoid undergoing apoptosis in response to mitogen depletion, as a result of aberrant expression of growth factors or their receptors. This leads to constitutive activation of downstream signaling elements, such as Ras, protein kinase C, and Akt, which not only drive cell proliferation, but also render these cells resistant to apoptosis induction. Receptors for platelet-derived growth factor and epidermal growth factor appear to be particularly important for promoting glioma cell proliferation and viability, and both constitute potential therapeutic targets that will be discussed in this chapter.

PLATELET-DERIVED GROWTH FACTOR AS A DOMINANT GLIOMA MITOGEN

Platelet-derived growth factor (PDGF) was originally identified as a potent mitogen for fibroblasts, glial cells, and smooth muscle. The observation that many malignant gliomas not only produced PDGF, but also contained its receptors, raised suspicion that this might be an important mediator of glioma growth [15]. Further studies demonstrated that PDGF was a disulfide-linked dimer composed of A, B, C, and D polypeptide chains, with the B chain being homologous to the *v-sis* oncogene isolated from simian sarcoma virus-transformed cells [16]. The various PDGF isoforms (AA, AB, BB, CC, DD) bind with differential affinity to two cell-surface PDGF receptors [17–20]. Concurrent expression of one or more of these ligands and their receptors has been observed in a significant percentage of high-grade gliomas [17,18,21–24] and the acquisition of PDGF receptor overexpression, in some cases by gene amplification, represents an important step in the transition from grade II to grade III gliomas [25,26]. Recent studies in a transgenic mouse model have shown that PDGF overexpression in either neural progenitor cells or differentiated astrocytes led to the formation of gliomas with oligodendroglial features [27], suggesting a role for PDGF and PDGFR in the early stages of tumor development. Combining PDGF overexpression with other genetic changes leads to a more malignant phenotype, highlighting the likely contribution of multiple genetic lesions to the development of glioblastoma. Recent studies have also indicated a role for the PDGFR in the development of a metastatic phenotype in medulloblastoma, the most common malignant brain tumor of childhood [28,29].

From a mechanistic perspective, PDGF and its receptors have been implicated in autocrine stimulation (by which a tumor cell promotes its own proliferation) as well as paracrine stimulation (by which a tumor cell promotes proliferation of neighboring cells, including PDGFR-expressing endothelial cells) in malignant gliomas [17,18,21,30,31]. Ligand binding to the receptor leads to a process of receptor dimerization, autophosphorylation of tyrosine residues, and phosphorylation of a series of signaling intermediaries, such as Ras, phospholipase C-gamma (PLCγ), phosphatidylinositol 3-kinase (PI3K), and Src family members, as described below.

Although previous studies in several laboratories [32–34] have indicated that a number of growth factors can stimulate the proliferation of malignant glioma cells, comparative analyses demonstrated that PDGF was a dominant mitogen [33], stimulating not only the most vigorous proliferation in individual established and low-passage cell lines, but also exhibiting the most consistent growth stimulatory effects in the panel of cell lines examined. Conversely, antibody- and antisense-mediated inhibition of PDGFR activation have been observed to inhibit glioma growth *in vitro* and *in vivo* [35–37]. In addition, because of the role of PDGF in supporting glioma-induced angiogenesis [38,39], inhibition of PDGFR may provide a means for simultaneously blocking tumor growth and angiogenic activation.

EGFR AMPLIFICATION/ OVEREXPRESSION IN GLIOMA PROGRESSION

EGFR, also known as ErbB1, the cellular homolog of the *v-erbB* oncogene, is one of a family of receptors that includes ErbB2 (HER2/neu), ErbB3 (HER3), and ErbB4 (HER4). The ligands for these receptors include EGF, TGFα, and amphiregulin, which bind exclusively to EGFR, heparin-binding EGF and beta-cellulin, which bind to EGFR and ErbB4, epiregulin, which binds to all receptors except ErbB2 homodimers, and the heregulins, which bind to ErbB3 and ErbB4, as well as heterodimers with ErbB2 [40–42]. *EGFR*, in particular, is amplified in 40–50 per cent of adult glioblastomas [43,44], resulting in overexpression of the epidermal growth factor receptor [45]. This transmembrane protein is normally activated by ligand binding to its extracellular receptor domain [46]. However, in some tumors with *EGFR* amplification, the gene is also rearranged, leading to constitutively active mutants [47–49]. The most common mutant, the so-called *EGFRvIII*, is caused by deletions of exons 2 to 7 [50,51], although a number of other common variants have been identified [49]. Because activated EGFR induces tyrosine phosphorylation of substrates that contribute to cell proliferation, as noted above for PDGFRs, excessive activation of this protein, either by ligand binding or mutation-induced constitutive signaling, may provide cells with a growth advantage under certain conditions. The observation that *EGFR* amplification occurs in almost 50 per cent of adult glioblastomas, but rarely in anaplastic astrocytomas [52–54], suggests that this change may contribute to the end stages of neoplastic progression by favoring cell proliferation under conditions in which exogenous substrate availability becomes limiting. However, amplification of this gene is not a requirement for neoplastic progression. Although most "primary" adult glioblastomas (which are histologically malignant at diagnosis and generally arise in older patients) show *EGFR* amplification, this change is less common in secondary glioblastomas (which evolve from lower grade gliomas, usually in younger adults) [52–54], and in pediatric glioblastomas [55]. However, the vast majority of childhood and secondary adult high-grade gliomas have high levels of EGFR expression, despite the absence of gene amplification, which suggests that excessive EGFR signaling may contribute to cell proliferation in these tumors as well [55].

The potential role of EGFR overexpression in glioma development is also supported by the observation that overexpression of mutant EGFR in neural progenitors or astrocytes induces the formation of tumors with histological features of glioblastoma in transgenic models [56]. Other EGFR family members, such as ErbB2, have been strongly implicated in medulloblastoma development, and appear to constitute a prognostic factor adversely associated with outcome and metastases [57–59]. Co-expression of ErbB4 was also commonly noted and heterodimerization between these receptors was postulated to be an important contributor to the adverse effects observed. Subsequent analysis of the prognostic significance of this marker confirmed that it had independent prognostic utility, even in otherwise favorable risk patients [59]. Recently, this marker has also been associated with outcome in a cohort of pediatric patients with ependymoma [60].

The important role of Erb family members in brain tumor development is supported by the fact that targeted inhibition of EGFR, using dominant negative truncated receptor constructs, interfered with glioma proliferation [61]. Similarly, antibody-based therapies targeted at wild-type or mutant EGFR

family members have been shown to have efficacy in preclinical glioma models [62,63] and ErbB2-targeted approaches have shown activity in medulloblastoma models [64], which provides a rationale for the small molecule approaches discussed below.

Cell-Surface Interactions

As noted above, the response of cells to both EGF and PDGF is mediated by cell-surface receptors that contain an extracellular domain that interacts with a ligand, a transmembrane domain that anchors the receptor to the cell membrane, and an intracytoplasmic domain that interacts with downstream signaling components. Binding of a ligand to its receptor induces conformational changes within the receptor and/or receptor cross-linking, and phosphorylation of tyrosine residues within their cytoplasmic domains [46,65]. Phosphorylation further activates the kinase and exposes binding sites for intracellular adapter molecules, such as Grb2 (growth factor receptor binding protein 2) and Shc [40,46,66] (Fig. 11.1). These binding sites are so-called Src homology 2 (SH2) domains, which consist of phosphorylated tyrosine residues presented in the context of specific adjoining amino acid sequences. The adapter proteins in turn associate with guanine nucleotide exchange factors, such as SOS (son of sevenless), *via* Src homology 3 (SH3) domains, which recognize proline-rich regions at the carboxyl

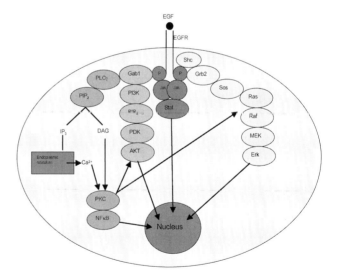

FIGURE 11.1 Schematic depiction of receptor-mediated signaling in response to EGF binding to EGFR, highlighting the complexity of parallel pathways transmitting signals for growth, gene transcription, and survival, which are then mediated by effectors within the nucleus. See Plate 11.1 in Color Plate Section.

terminus of Sos [67–70]. The exchange factors facilitate activation of the Ras family of guanine nucleotide triphosphatases by exchanging GDP for GTP. To promote this process, Sos binds to GDP-Ras, which leads to dissociation of GDP, allowing GTP to bind to the Ras-Sos complex. Sos then dissociates, leaving Ras in its active GTP-bound form. Excessive activation of these pathways has been implicated in the development of a variety of human tumors. In certain tumor types, mutation of one of the *ras* genes to a constitutively active protein has been associated with tumorigenesis [71]. Although such mutations are rare in gliomas [72,73], Ras activity is nonetheless markedly elevated [73–75] as a result of deregulation of upstream signaling elements, such as growth factor receptors [76].

In addition to promoting activation of Ras *via* the Grb2-Sos pathway, receptor tyrosine kinases activate a number of parallel signaling pathways by interactions between phosphotyrosine domains and SH2 domains of other cytoplasmic intermediates (Fig. 11.1). For example, PLCγ and PI3K, which play critical roles in phospholipid metabolism, are recruited to the activated receptor by adapter proteins, such as Gab1 [77]. PI3K stimulates inositol phosphorylation that in turn leads to activation of Akt, which has multiple functions, among which are cell survival. Activation of PLCγ produces other important lipid intermediates, such as diacylglycerol (DAG) and inositol 1,4,5-triphosphate (IP3). IP3 leads to release of intracellular calcium, thereby activating calmodulin-dependent protein kinases. The calcium, in combination with DAG, also activates protein kinase C (PKC), which then stimulates many of the same downstream elements that are activated by Ras [78–80]. PKC, in particular, appears to also activate Ras directly, and by employing a mechanism that is distinct from the aforementioned Grb2–Sos interaction [78,81], provides an additional pathway by which receptor-mediated mitogenic signaling can stimulate downstream proliferative and cell survival cascades.

Receptors that lack tyrosine kinase activity but contain sites for tyrosine phosphorylation may become transactivated by association with a soluble tyrosine kinase, such as a member of the Src family [82]. Following phosphorylation, these proteins can interact with adapter proteins, such as Shc, to activate G-protein-mediated signaling pathways and can also directly stimulate downstream effectors. Finally, both PDGFR and EGFR have also been found to activate members of the signal transducers and activators of transcription (STAT) family, which translocate to the nucleus to directly influence the transcription of genes involved in cell cycle progression and cell survival [83–88].

Signal Transduction

MAPK Cascade

Following activation of the membrane-associated components of the various signaling pathways, downstream signals reach the nucleus by a variety of mechanisms (Fig. 11.1). One of the most critical pathways involves the mitogen-activated protein kinase (MAPK) cascade. This cascade involves at least three separate protein kinases. The most proximal kinase in the cascade is the Raf (MAPKKK) family, which includes at least three members. Raf, a serine/threonine kinase, is the cellular homolog of a viral oncogene, capable of inducing transformation if constitutively overexpressed. Raf is recruited to the cell membrane, stabilized by interaction with other proteins, such as members of the 14-3-3 family, and possibly by dimerization [89,90], and phosphorylated to an active form. Activated Raf phosphorylates and activates MAP/ERK kinase (MEK, also known as MAPKK), a dual-specificity (tyrosine and serine/threonine) kinase, which subsequently activates MAPK (also known as ERK (extracellular signal-regulated kinase)) by phosphorylating tyrosine and threonine residues [91–93]. MAPKs are serine/threonine kinases that activate a number of additional cytoplasmic downstream mediators that regulate transcription, protein translation, and cytoskeletal rearrangement. These cytoplasmic mediators include other kinases, such as p-90RSK (ribosomal S6kinase) and p70-S6 kinase, which are broadly grouped as MAPK-activated protein kinases (MAPKAPKs), proteins that can impact directly on the nucleus by activating transcription regulatory factors, such as Fos, Jun, Myc, and Elk-1 [94–96]. In addition, these proteins can directly stimulate translation by activating ribosomal S6 protein, or down-regulate transcription by activating GSK3 (glycogen synthase kinase 3).

In parallel with the MAPK pathway, which transmits signals for cell growth and proliferation, are other similar pathways that convey signals that respond to injury or stress [97,98]. These include the JNK (Jun terminal kinase) pathway, which may be activated directly or indirectly *via* receptor tyrosine kinase signaling *via* Ras-related G-proteins, such as Rac, followed by activation of PAK (p21Ras-related protein activated kinase), which activates a MAPKKK molecule. This in turn activates the dual specificity kinase JNKK (also known as SEK (SAPK/ERK kinase)), which is structurally similar to MEK. JNKK activates JNK (also known as SAPK (stress-activated protein kinase)). These intermediates subsequently activate MAPKAPK-related kinases and various nuclear effectors, such as Jun, to modulate DNA synthesis and transcription. Because certain Jnk isoforms are overexpressed and excessively activated in gliomas [99,100] it is conceivable that they contribute along with MAPK signaling to glioma cell survival. In addition, the involvement of other G-proteins, such as Rac and Rho, may contribute to changes in cytoskeletal organization and motility that are associated with Ras activation.

Although early studies suggested that these individual cascades might function as discrete pathways, more recent studies clearly show substantial cross-talk. For example, JNKK, an activator of JNK/SAPK, can also activate p38, a constituent of a third MAPK-like cascade [97]. All three kinase cascades can activate similar downstream targets, such as cytoplasmic kinases, transcription factors, and cell-cycle checkpoint control elements [101,102]. For example, Mnk1 (MAPK-interacting kinase 1) appears to be activated by MAPK and p38 pathways to phosphorylate the initiation factor eIF-4E, which suggests that Mnk1 may regulate the rate of protein synthesis in cells that are stressed or stimulated to proliferate by growth factors [102]. The MAPKAPK-related protein 3pK appears to be activated by all three kinase cascades (MAPK, SAPK, and p38) [101]. In addition, various transcription factors, such as ATF-2 and Sap-1a can be phosphorylated and activated by members of the p38, JNK, and MAPK families, and Jun and Elk-1 are phosphorylated by MAPK as well as JNK/SAPK. Sap-1a forms a complex with serum response factor, binds to serum response element, and activates the expression of Fos, which serves as a transcription factor for a host of other genes. These pathways also interact with cAmp-dependent protein kinases, with cell-cycle regulatory kinases, and with apoptotic signaling cascades to influence other regulators of cell homeostasis, proliferation, and death [103,104]. Thus, although the various kinase cascades seem to serve discrete functions, with MAPK transducing proliferative signals and JNK/SAPK and p38 mediating stress-related processes, the different cascades converge on similar downstream targets to modulate cellular homeostasis. The response of a given cell to exogenous and endogenous stimuli may therefore depend on the profile of signaling elements that are activated, which in turn stimulate mechanisms for protein synthesis, proliferation, cytoskeletal rearrangement, differentiation, cell death, or a combination thereof in a cell-type-specific manner.

Akt-mediated Signaling

A second major downstream pathway involves activation of Akt *via* PI3K (Fig. 11.1). PI3K is a phospholipid kinase that contains both a regulatory subunit, p85, and a catalytic subunit, p110. Upon cell-surface activation, PI3K phosphorylates phosphatidylinositol 4,5-biphosphate ($PI4,5P_2, PIP_2$) to form $PI3,4P_2$, and PIP_3. PIP_3 leads to translocation of PDK1 and Akt. Akt is a serine/threonine kinase that is phosphorylated at threonine 308 by PDK1 and at serine 473 by a second kinase. Activated Akt phosphorylates several proteins involved in cell-survival signaling, such as Bad [105], mTOR, forkhead transcription factor [106], and glycogen synthase kinase [107].

Under normal conditions, Akt activation status is negatively regulated by the tumor suppressor PTEN, which is a lipid phosphatase that converts PIP_3 to PIP_2 [13,108]. However, mutations of *PTEN* are observed in at least 40 per cent of grade IV gliomas [12–14,109], and are particularly common among primary glioblastomas. The potential importance of this pathway in glioma development is highlighted by the fact that transfer of a wild-type *PTEN* gene to the *PTEN*-deleted U87 malignant glioma cell line suppressed tumor growth, leading to cell cycle arrest [108]. Introduction of the chromosomal region of the *PTEN* gene also suppressed soft agar colony formation and tumor growth in nude mouse models [109].

PLCγ/PKC-mediated Signaling

A third important signaling pathway involves activation of PKC *via* PLCγ-mediated production of calcium and DAG [78–80] (Fig. 11.1). PKC comprises a family of serine/threonine kinases that constitute an important component of the signaling cascade of several growth factors that stimulate glioma cell proliferation, such as EGF and PDGF [110–114]. These isoforms can be subdivided into three main groups. The first group consists of four "conventional" (c)PKC isoenzymes (α, β_I, β_{II}, and γ), which require Ca^{2+}, phosphatidylserine (PS) and DAG or tumor promoting phorbol esters for their activation [110,115]. These isoenzymes share a similar structural organization, with four highly conserved regions (C1–C4) and five variable regions (V1–V5). The second group includes five "novel" (n)PKC isoforms (δ, ε, η(L), θ, μ), which lack the C2 region, a putative

Ca^{2+}-binding domain, and thus do not require Ca^{2+} for activation. The third group is comprised of two "atypical" (a)PKC isoenzymes (ζ and i), which lack the Ca^{2+}-binding C2 region and contain an atypical C1 region. These isoenzymes require neither Ca^{2+} nor DAG or phorbol esters, but only PS for activation [110,115].

Activation of PKC leads to diverse biological effects, depending upon which isoforms are activated. These include phosphorylation of other effectors, such as Raf [116] and MAPK [117], and influencing the activation status of Ras [78], which together contribute to the transduction of a proliferative signal to the nucleus [94,114]. Levels of PKC expression and activity have been noted to correlate with proliferative status in normal and neoplastic astrocytes [118–121]. Astrocytoma cells express levels of PKCα and ε that are up to ten-fold higher than in normal astrocytes [119,120]. Malignant glioma cells also have levels of PKC activity several orders of magnitude greater than those of non-neoplastic astrocytes [121]. Glioma mitogens, such as PDGF and EGF, produce elevations in PKC activity that parallel increases in DNA synthesis [122]. In addition, selective pharmacological and antisense agents that target PKC isoforms overexpressed in astrocytomas, particularly PKCα, diminish glioma proliferation *in vitro* [123–126]. Recent results indicate that interference with PKC-mediated signaling not only blocks glioma proliferation, but also induces apoptosis [124,126].

STAT-mediated Signaling

A fourth signaling pathway that has recently been implicated in EGFR and PDGFR-mediated signaling involves the STATs (Fig. 11.1). These intermediates are stimulated by a variety of receptors, in some cases *via* Janus kinase or Src family members as intermediaries [83,87,88,127]. JAK proteins appear to be particularly important in mediating direct nuclear signaling. Their association with the receptor promotes receptor aggregation, which leads to homotypic and heterotypic association of multiple JAKs. The activated STATs then are translocated into the nucleus to influence transcription of genes involved in cell-cycle regulation and apoptotic signaling. Given the multiple isoforms present, these proteins have diverse effects on cell survival. STAT3, in particular, appears to confer proliferative and anti-apoptotic effects [128], and constitutive activation of this protein has been observed in both gliomas and medulloblastomas [129]. Recent studies in squamous cell carcinoma cells indicate selective EGFR signal transduction through STAT3 in tumor cells, but not in non-neoplastic immortalized keratinocytes engineered to overexpress EGFR, suggesting the complexity of this signaling pathway in tumor cells [86].

STRATEGIES FOR DIRECT INHIBITION OF GROWTH FACTOR PATHWAY ACTIVATION

Inhibition of intermediate and downstream components of growth factor signaling pathways, such as Ras, PI3K, PKC, and mTOR, is a promising strategy for interfering with the proliferation of malignant gliomas and other brain tumors [114,124,130], which will be discussed in detail in other chapters. One theoretical limitation to the activity of these agents is that they target pathways critical to the functioning of normal as well as tumor cells, and therefore can cause significant nonspecific toxicity, which has been a limiting factor for several of these approaches. The fact that constitutive activation of these pathways generally represents an indirect consequence of a primary abnormality in growth factor receptor signaling implies that greater selectivity and efficacy can potentially be achieved by directly targeting the growth factor receptor.

A variety of strategies have been effectively used for growth factor signaling inhibition, including neutralizing antibodies, antisense techniques, and dominant negative mutants, in pilot studies of individual glioma cell lines [35]. For example, Nitta and Sato [37] observed that c-sis antisense oligonucleotides inhibited growth *in vitro* of A172 glioma cells, which express large quantities of PDGF-BB and the β receptor. Similarly, growth of U87 and U343 glioma cells transfected with a dominant negative construct for PDGF-A, designed to form unstable dimers with normal A and B chains, was suppressed *in vitro*, and these cells exhibited diminished tumorigenicity in a nude mouse subcutaneous model [131]. Our previous studies have also demonstrated significant antiproliferative activity *in vitro* using a PDGF neutralizing antibody [33]. Other groups have noted significant activity of antibodies designed to neutralize EGFR or to selectively target the most common, constitutively active EGFR mutant (EGFR vIII) [62,63]. Similarly, in non-CNS tumors, monoclonal antibodies directed against EGFR (e.g., C225, Cetuximab, ImClone Systems, New York, NY) and EGFR family members (e.g., Herceptin, Trastuzumab, Genentech, South San Francisco, CA and ABX-EGF, Abgenix, Inc) have demonstrated efficacy in preclinical models as well as initial clinical studies [132–137].

Although the above results confirm that blocking critical growth factor pathways can interfere with cell proliferation in appropriately selected tumor cell lines and clinical situations, the applicability of these results for glioma therapeutics is limited by the challenges inherent in inhibiting growth factor receptor signaling within a CNS tumor. For example, antibody- and antisense-mediated approaches would have to overcome difficulties imposed by the blood–brain barrier. These limitations also apply to transfection-based dominant negative mutant approaches, with the added caveat that diffuse expression of the gene within the tumor would be required to achieve meaningful inhibition of tumor growth. One strategy for circumventing blood–brain barrier limitations has involved the use of convection-enhanced delivery, which employs slow intraparenchymal infusion of microliter volumes to achieve bulk flow of a large molecular weight protein directly into the brain tumor or peritumoral brain. Most applications to date have focused on the use of immunotoxin conjugates, in which a receptor ligand is conjugated to a mutated toxin gene, such as pseudomonas exotoxin lacking the cell membrane translocation domain [138,139]. Clinical trials are in progress using an EGFR-targeted approach, TGFα-PEQQR (TP-38, IVAX Research, Inc.), in which the ligand interactions between TGFα and EGFR provides the basis for internalization of the toxin. In contrast to the above approaches, the therapeutic effect is provided not by inhibiting the receptor, but rather by using the receptor as a way of selectively delivering a toxin gene into the tumor cells.

An alternative approach for growth-factor-targeted inhibition involves the use of pharmacological inhibitors of receptor activation. Until recently, most such applications of this approach used natural products with activity against a given receptor target, but minimal selectivity, such as genistein or staurosporine. This nonselectivity limited the clinical applicability of these agents. However, during the last several years, molecular modeling approaches have allowed the development of "designer inhibitors", most from the quinazoline and pyrrolopyrimidine classes of chemicals, that selectively target one or more tyrosine kinase receptors by competitive inhibition of the ATP-binding sites involved in receptor activation. Because these are "small molecule" inhibitors with a high degree of receptor selectivity, there is a reasonable expectation that systemic delivery could achieve penetrance into the brain tumor microenvironment and tumor-selective therapeutic effects. A brief summary of several promising agents is provided below.

STI571 AS A POTENTIAL PDGFR INHIBITOR

Although a number of pharmaceutical companies have developed PDGFR inhibitors that are in various phases of clinical development, the agent that has progressed by far the furthest is STI571 (CGP57148B, also known as Gleevec or Imatinib, Novartis Pharmaceutical Corp., East Hanover, NJ). This compound is a low-molecular-weight synthetic 2-phenylaminopyrimidine that initially became a focus of clinical interest, not because of its PDGFR-inhibitory effects, but because of its potent inhibition of Bcr-Abl, a fusion protein produced by Philadelphia chromosome-positive leukemias [140–142]. Significant activity of this agent was demonstrated in preclinical models of Bcr-Abl+ tumors, which formed an impetus for phase I, and subsequently, phase II studies in patients with Philadelphia chromosome-positive leukemias. Because of the dramatic hematologic and cytologic responses that were observed [141–143], phase III studies of this agent have been initiated, which demonstrated superiority of this agent in terms of complete cytogenetic responses and freedom from disease progression in comparison to conventional therapy with interferon and cytarabine [144], and a high frequency of long-term disease control in the subset of patients who had significant reduction in Bcr-Abl transcript levels [145]. The importance of this target in response is highlighted by the fact that acquired drug resistance in such patients is associated with mutation or amplification of the Bcr-Abl gene [146,147], thereby circumventing drug efficacy.

A second target for STI571, based on its receptor inhibitory profile, was c-kit, which is mutated to a constitutively activated form in a substantial percentage of gastrointestinal stromal tumors. Initial studies demonstrated striking efficacy in patients with advanced disease [148,149] and, as with BCR–Abl+ leukemias, this agent has constituted a molecular "magic bullet" for these tumors in which a single genetic target is a predominant contributor to tumor development.

Because subsequent studies also demonstrated that this agent was extremely potent in blocking PDGFR signaling by both the α and β receptors, and disrupting PDGF/PDGFR ligand-receptor autocrine and paracrine loops [150–153], attention was also focused on the potential application of this agent in tumors with PDGFR-driven proliferation. Preliminary studies in glioma cell lines demonstrated inhibition of proliferation *in vitro* and delay of tumor growth *in vivo* [152], although the results were less striking than those noted earlier for c-kit and Bcr-Abl-dependent tumors. However, this likely relates to the fact the multiple pathways are known to contribute to glioma proliferation, rather than a single overriding pathway, not to the limitations of PDGFR as a target in general. In fact, particularly striking effects were observed with this agent in the subset of myeloproliferative disorders involving rearrangements of the *PGFRβ* gene, leading to constitutive receptor activation [154].

Previous phase I studies with STI571 used oral daily dosing, based on the long half-life of this agent, and demonstrated minimal toxicity at doses up to 1000 mg/day in adults [141,142], which achieved serum concentrations in the low micromolar range. In contrast, doses of 300 mg/day led to complete hematological remissions in Philadelphia chromosome-positive leukemias, with loss of constitutively activated fusion protein, and inhibition of Bcr-Abl-induced tyrosine phosphorylation [141,142]. Thus, despite the fact that the tyrosine kinase receptors targeted by STI571 are expressed on a variety of normal cells throughout the body, preclinical and clinical studies with this agent seemed to indicate that antiproliferative effects were mediated selectively on neoplastic cells [141,142,149,154], with relatively little systemic toxicity.

Based on these promising results, initial dose escalation studies in children and adults with malignant gliomas were initiated by several cooperative groups, including the North American Brain Tumor Consortium and Pediatric Brain Tumor Consortium. An unexpected finding from several of these studies, which have yet to fully analyzed, is that a subset of patients experienced intratumoral hemorrhage during treatment. The etiology of this effect remains to be determined, but one potential mechanism relates to the known effects of this agent on perivascular cell permeability [153,155]. Correlative imaging studies, including MR spectroscopy and PET, have been performed on a significant percentage of the patients on the pediatric trial, and it is hoped that this will provide insights into this issue. Because objective responses and long-term survivors have been observed in the pediatric phase I trial, a phase II study has been considered for patients with brainstem malignant gliomas, although care has been taken to include stopping rules for an intolerable frequency of

hemorrhagic events. In contrast to these observations, an intergroup study from Europe observed intratumoral hemorrhage in only one of 51 patients. Partial radiological responses were observed in 3 of 19 patients treated with 600 mg/day and 1 of 33 patients treated with 800 mg/day, and an additional 1 and 4 patients in these respective subsets had stable disease for more than 6 months [156]. An intriguing observation from a second, smaller, study was that prolonged disease control was achieved in 3 of 6 high-grade glioma patients younger than 45 years of age, *versus* only one of 9 older than 45 years [157]. Combinations of STI571 and conventional chemotherapeutic agents have also been examined. Although the optimal agent to employ in this regard remains uncertain, promising results have been reported with hydroxyurea (1000 mg/day) combined with 400 mg/day of STI571, with objective disease regression in 5 of 26 patients [158].

ZD1839 AS A POTENTIAL EGFR INHIBITOR

ZD1839 (Iressa, Gefitinib, AstraZeneca, Wilmington, DE), a low molecular weight synthetic quinazoline, was designed as a potent and selective inhibitor of the epidermal growth factor receptor (EGFR) tyrosine kinase. It is active against EGFR at nanomolar concentrations in cell-free systems, with 100-fold less activity against other EGFR family members, such as erbB2, and has little or no enzyme inhibitory activity against other tyrosine and serine-threonine kinases [159–162]. It is effective in blocking EGFR autophosphorylation and inhibiting EGFR-dependent cell signaling *in vitro* in cell lines that rely heavily on EGFR activation for proliferative stimulation, but has virtually no effect on EGFR-independent proliferation. Tumor growth inhibition is associated with cell cycle arrest in the G1 phase of the cell cycle. ZD1839 has also demonstrated antitumor activity *in vivo* in a number of human tumor xenograft models after oral administration, particularly in model systems in which proliferation is dependent on EGFR-mediated signaling [162]. For example, this agent produced significant tumor growth inhibition and tumor regression in animals harboring xenografts of the A431 squamous carcinoma cell line, which is characterized by marked EGFR overexpression as a result of *EGFR* gene amplification. In this EGFR-dependent tumor model, low-dose therapy (10 mg/kg/day) achieved tumor growth inhibition, whereas higher doses (200 mg/kg/day) were observed to produce tumor regression [160–162], although tumors eventually recurred if therapy was stopped.

Initial clinical data indicated that ZD1839 has good oral bioavailability, a long half-life [163], and was well tolerated at effective doses [164,165], with the most common toxicity being an acneiform skin rash, and diarrhea being a common dose-limiting toxicity on clinical trials. Daily dosing achieved a Cmax of 5 µM at a dose of 500 mg/day, and even trough concentrations at this dose were in the 1 to 2 µM range that produced complete or nearly complete inhibition of EGFR signaling in many cell lines [166]. Results from several early clinical studies for non-central nervous system solid tumors have demonstrated activity of ZD1839 as a single agent [167–169], although large phase III studies in patients with advanced small cell lung cancer failed to demonstrate a convincing benefit of adding this agent to regimens including either gemcitabine/cisplatin or paclitaxel/carboplatin alone [170,171].

Although these results were initially a source of disappointment [172], recent landmark reports have demonstrated that response to ZD1839 is strongly influenced by tumor EGFR status, confirming that this agent may well be an exquisitively selective inhibitor in appropriate tumors [173]. In one institutional study, 25 of 275 patients treated with Iressa were noted to have objective therapeutic responses; eight of the nine who had tumor specimens available for mutational analysis were observed to have mutations within the EGFR gene, generally involving gain-of-function changes close to the ATP-binding pocket of the tyrosine kinase domain [173]. In contrast, none of the seven patients with no response to this agent who were analyzed in parallel were found to have mutations. In two cases in which the functional effect of the mutation on gefitinib sensitivity was examined, it was observed that this was increased ten-fold, with complete inhibition of protein activity at a concentration of 0.2 µM *versus* 2 µM with the wild-type receptor. These observations emphasize the importance of studies of tumor genotype and phenotype in considerations of both study design and response analysis.

It is likely that it is not only the presence of such mutations, but also the type, that may influence the effect of this agent. In contrast to the above results showing increased therapeutic efficacy with certain types of mutations, other groups have observed that the most common mutational phenotype in gliomas (EGFRvIII) may confer reduced sensitivity to ZD1839 in *in vivo* brain tumor models [174], because this receptor is active independent of ligand binding. An additional consideration that may be relevant to the application of these agents in glioma therapeutics is that involvement of other key signaling pathways in driving tumor proliferation may function to counteract the effects of this agent [175]. For example, *PTEN* mutations, which are common in these tumors, appear to counteract the effect of EGFR inhibitors, such as ZD1839, on downstream growth and survival signaling [176], necessitating the use of multiple signaling inhibitors in order to achieve a growth inhibitory effect.

With these issues in mind, phase I/II studies of ZD1839 have been initiated within the North American Brain Tumor Consortium (for adults with malignant glioma and meningioma) and the Pediatric Brain Tumor Consortium (for children with recurrent malignant glioma and newly diagnosed brainstem malignant glioma). These studies are including analyses of receptor expression and mutational status, in an effort to correlate treatment responses, which have been observed, with tumor genotype and phenotype. In the NABTC study, patients who were not receiving enzyme-inducing anticonvulsants or corticosteroids received a daily dose of 500 mg and those receiving corticosteroids were escalated incrementally to a dose of 1000 mg/day. A phase I study was conducted in those patients receiving enzyme-inducing anticonvulsants, which established a maximal tolerated dose of 1500 mg/day. Among 55 patients with malignant gliomas who were in the first two groups, a phase II evaluation of activity identified partial responses in seven, although the median times to progression in patients with glioblastoma and anaplastic glioma were not superior to historical controls (8 and 12 weeks, respectively) [177]. Given the fact that responses were observed, but only in a small subset of patients, efforts are in progress to determine whether a consistent molecular signaling profile will help to distinguish responders from non-responders. Because a study of ZD1839 in newly diagnosed patients with malignant gliomas failed to demonstrate a significant improvement in survival, or an association between EGFR amplification status as assessed by FISH and outcome [178], it is conceivable that a more detailed analysis of mutations and signaling pathway phenotype will be required to identify an appropriate subset of patients for treatment with this agent. Studies are also in progress to determine whether combinations of ZD1839 with conventional chemotherapeutic agents, such as temozolomide, may enhance efficacy [179].

OTHER SMALL MOLECULE EGFR INHIBITORS

Although studies with Iressa are perhaps furthest along, a number of other small molecule inhibitors of EGFR are currently or soon to begin clinical trials in patients with brain tumors. OSI-774 (Erlotinib, Tarceva, Genentech, South San Francisco, CA) exhibits reversible inhibition of EGFR, by competition with the ATP-binding site, similar to ZD1839, with a median inhibitory concentration in the low nanomolar range [132,180]. This agent inhibits EGFR autophosphorylation at concentrations in the range of 20 nM, and in cell lines dependent on EGFR for proliferation, inhibition of cell growth was observed in a comparable range [180,181]. Higher concentrations were capable of inducing apoptosis. As with ZD1839, this agent induces cell cycle arrest at G1, with accumulation of p27^{Kip1} [182]. In preclinical models, activity has been observed at concentrations in the range of 10 to 200 mg/kg/day, and early clinical studies confirmed good oral bioavailability. In phase I clinical studies, toxicities were comparable to ZD1839, with rash and diarrhea as common events [183,184]. At the maximal tolerated dose, steady-state plasma concentrations exceeded 1 µg/ml, above the dose needed to inhibit tumor growth in preclinical models. Phase II studies in patients with squamous cell carcinoma, non-small cell lung cancer, and ovarian cancer have suggested activity [185–188], as has a recently completed trial in adults with malignant glioma, conducted by the North American Brain Tumor Consortium [189]. In the phase II component of the latter study, patients were treated with 150 mg/day; in a recent report of that study, only one of 45 patients had an objective response, with median time to progression for the cohort no better than historical control data [190]. Although a higher rate of responses was observed in a single-institution study using the same dose [4 of 16 patients], these responses were not durable and median time to progression was not improved compared to historical controls [191]. An attempt to correlate response with *EGFR* amplification, EGFR expression, and EGFRvIII status is in progress [192]. In addition, a pediatric trial, combining OSI-774 with temozolomide in patients with recurrent tumors, has recently opened, although no results are available to date.

Two other selective, reversible EGFR inhibitors that are in early phase trials are GW2016 (Lapatinib, GlaxoSmithKline, Research Triangle Park, NC) and PKI166 (Novartis), both of which have demonstrated preclinical activity at submicromolar concentrations

[193,194]. GW2016 has also demonstrated activity against erbB2-expressing tumors [194,195], and is accordingly soon to begin clinical testing in childhood medulloblastoma and ependymoma, tumors for which ErbB2 overexpression has been associated with an adverse prognosis [57,59,60], as well as recurrent malignant glioma.

Several irreversible EGFR inhibitors, with activity against other ErbB family members are also beginning clinical trials following promising preclinical results. These include CI-1033 (Pfizer Inc, Groton, CT) and EKB-569 (Wyeth-Ayerst Laboratories, St. Davids, PA) [196–200]. CI-1033 has been observed to induce regression of A431 tumors in rodent models, maintaining tumor suppression for extended intervals with as little as once-weekly dosing [196]. In early clinical trials, the principal toxicities were identical to those with other EGFR inhibitors, specifically diarrhea and rash [197,198].

INHIBITION OF GROWTH FACTOR SIGNALING MAY INTERFERE WITH ANGIOGENESIS AND TUMOR INVASION

In addition to direct effects of growth factor receptor inhibition on tumor proliferation, there is compelling evidence that these agents may have indirect effects on the tumor microenvironment, which may be relevant for brain tumor therapeutics. Recent studies have confirmed that the secretion by tumor cells of vascular endothelial growth factor (VEGF), the most potent endothelial cell mitogen, depends heavily on EGFR-mediated signaling [201–203] and that a significant component of the therapeutic efficacy of EGFR-targeting agents, such as monoclonal antibodies or small molecule inhibitors, reflects this secondary effect on tumor angiogenesis [159,204,205]. Similarly, PDGF has been demonstrated to stimulate tumor angiogenesis [39,206] in addition to supporting the growth and survival of vascular pericytes [207] and promoting glioma VEGF secretion [38,201]. In view of these potent paracrine interactions, an important element in evaluating the therapeutic utility of EGFR- and PDGFR-targeted inhibitors must focus on understanding the potential effects of these agents on not only the tumor cells themselves, but also on the surrounding vasculature [153,159,208].

EGFR and PDGFR-mediated signaling also has effects on cell motility and invasion, which may relate to effects on G-proteins, such as Rac, that influence cytoskeletal organization, as well as matrix metalloproteinases [209]. The EGFRvIII variant, in particular has been associated in *in vivo* models with enhanced invasiveness, which is blocked by EGFR inhibition by OSI-774 [210], suggesting the importance of constitutive activation in this phenotypic effect.

INTERFERENCE WITH GROWTH FACTOR RECEPTOR SIGNALING MAY POTENTIATE OTHER THERAPIES

Although signal transduction inhibition may have independent activity in decreasing tumor cell proliferation and inducing apoptosis, the utility of these approaches may be amplified by combining them with other therapeutic strategies. Preclinical data suggest potentiation of the activity of cytotoxic drugs by growth factor receptor inhibitors. EGFR-targeted monoclonal antibodies have been noted to enhance the effects of cisplatin [135, 136], topotecan [211], gemcitabine [212], and taxol [213], as have small molecule inhibitors, such as ZD1839 [135,160,214] and CI-1033 [196]. In addition, marked radiosensitization has been achieved by EGFR-specific monoclonal antibodies [132,137,215–217]. *In vivo* studies have demonstrated enhancement of necrosis and vascular thrombosis in the context of decreased angiogenesis [216]. A similar potentiation has been observed in glioma cell lines using a dominant negative EGFR transfectant in the U87 model system [218]. Increased G1 cell cycle arrest and higher levels of apoptosis were also observed. Recent studies also suggest that small molecule inhibitors of EGFR kinase activity, such as ZD1839 [164,205,214,219–221], and PDGFR kinase activity, such as STI571 [222–224], may also achieve potentiation of conventional therapies. The ability of STI571 to potentiate the efficacy of conventional chemotherapeutic agents, such as taxol and 5-fluorouracil, may in part result from a reduction of interstitial pressure through effects on tumor vasculature [155]. The application of such combinatorial approaches in brain tumor therapeutics is already under way. Both adult and pediatric brain tumor trials are examining the combination of OSI774 and temozolomide, and ongoing pediatric studies in brainstem glioma are evaluating the combination of ZD1839 and concurrent irradiation, and

STI571 immediately following irradiation in newly diagnosed patients, with outcome comparisons *versus* historical control groups treated with irradiation alone.

In addition to the applicability of combining growth factor receptor-targeted inhibitory strategies with conventional chemotherapy, recent studies indicate that combinations of growth factor receptor inhibitors with other molecularly targeted approaches may dramatically potentiate efficacy by circumventing intrinsic resistance mechanisms [176]. For example, Bianco *et al.* [176] observed that treatment resistance in tumors with PTEN deletions could be counteracted by independent inhibition of Akt signaling, suggesting a role for agents, such as rapamycin, that block downstream Akt signaling, in conjunction with EGFR-targeted therapies. The optimal agent(s) to combine with individual growth factor receptor inhibitors is a topic of intense research interest.

SUMMARY

In view of the limited responsiveness of malignant gliomas to conventional therapies, there is a strong need to identify and exploit novel strategies to improve the outcome of patients with these tumors. Previous studies in many laboratories have demonstrated that the proliferation and survival of malignant glioma cells is strongly influenced by a number of growth factor receptor-mediated signaling pathways. These have therefore emerged as promising targets, which have been effectively exploited by small molecule inhibitors. The major ongoing challenges are the identification of tumor genotypic and phenotypic features that predict treatment response, the detection of physiological imaging correlates, including MR spectroscopy and PET, which may provide early insights into treatment efficacy, and the determination of optimal agents to combine with receptor-targeted therapeutics. In view of the close association between genotypic features and response to ZD1839 in non-small-cell lung cancer and to STI571 in chronic myelogenous leukemia and gastrointestinal stromal tumor, it is unlikely that any single agent will have broad applicability in all brain tumors, given their diversity. Rather, the appropriate use of these agents will likely require transitioning from the current "one-size-fits-all" approach to a more refined tumor-tailored therapy paradigm in future studies.

ACKNOWLEDGMENT

This work was supported in part by NIH grant NSP0140923 and a grant from the Wichmann Foundation.

References

1. Pollack, I. F. (1994). Current Concepts: Brain tumors in children. *N Engl J Med* **331**, 1500–1507.

2. DeAngelis, L. M. (2001). Brain tumors. *N Engl J Med* **344**, 114–123.

3. Hanahan, D., and Weinberg, R. A. (2000). The hallmarks of cancer. *Cell* **100**, 57–70.

4. Kapoor, G. S., and O'Rourke, D. M. (2003). Mitogenic signaling cascades in glial tumors. *Neurosurgery* **52**, 1425–1435.

5. Yount, G. L., Levine, K. S., Kuriyama, H., Haas-Kogan, D. A., and Israel, M. A. (1999). Fas (APO-1/CD95) signaling pathway is intact in radioresistant human glioma cells. *Cancer Res* **59**, 1362–1365.

6. Lowe, S., Bodis, S., McClatchy, A. *et al.* (1994). Jacks T. p53 status and the efficacy of cancer therapy in vivo. *Science* **266**, 807–810.

7. Pollack, I. F., Finkelstein, S. D., Burnham, J. *et al.* (2001). Age and TP53 mutation frequency in childhood gliomas. Results in a multi-institutional cohort. *Cancer Res* **61**, 7404–7407.

8. Sidransky, D., Mikkelsen, T., Schechheimer, K., Rosenblum, M. L., Cavenee, W., and Vogelstein, B. (1992). Clonal expansion of p53 mutant cells is associated with brain tumor progression. *Nature* **355**, 846–847.

9. Kinzler, K. W., and Vogelstein, B. (1994). Cancer therapy meets p53. *N Engl J Med* **331**, 49–50.

10. Goh, H. -S., Yao, J., and Smith, D. R. (1995). p53 point mutation and survival in colorectal cancer patients. *Cancer Res* **55**, 5217–5221.

11. Pollack, I. F., Finkelstein, S. D., Woods, J. *et al.* (2002). Expression of p53 and prognosis in malignant gliomas in children. *N Engl J Med* **346**, 420–427.

12. Li, J., Yen, C., Liaw, D. *et al.* (1997). PTEN, a putative protein tyrosine phosphatase gene mutated in human brain, breast, and prostate cancer. *Science* **275**, 1943–1947.

13. Davies, M. A., Lu, Y., Sano, T. *et al.* (1998). Adenoviral transgene expression of MMAC/PTEN in human glioma cells inhibits Akt activation and induces anoikis. *Cancer Res* **58**, 5285–5290.

14. Cantley, L. C., and Neel, B. G. (1999). New insights into tumor suppression: PTEN suppresses tumor formation by restraining the phosphoinositide 3-kinase/AKT pathway. *Proc Natl Acad Sci USA* **96**, 4240–4245.

15. Nister, M., Libermann, T. A., Betscholz, C. *et al.* (1988). Expression of messenger RNAs for platelet-derived growth factor and transforming growth factor α and their receptors in human malignant glioma cell lines. *Cancer Res* **48**, 3910–3918.

16. Waterfield, M. D., Scrace, G. T., Whittle, N. *et al.* (1989). Platelet-derived growth factor is structurally related to the putative transforming protein p28sis of simian sarcoma virus. *Nature* **304**, 35–39.

17. Nistér, M., Claesson-Welch, L., Erikssonm, A., Heldin, C.-H., and Westermark, B. (1991). Differential expression of platelet-derived growth factor receptors in human malignant glioma cell lines. *J Biol Chem* **266**, 16755–16763.

18. Lokker, N. A., Sullivan, C. M., Hollenbach, S. J., Israel, M. A., and Giese, N. A. (2002). Platelet-derived growth factor (PDGF) autocrine signaling regulates survival and mitogenic pathways in glioblastoma cells: Evidence that the novel PDGF-C and PDGF-D ligands may play a role in the development of brain tumors. *Cancer Res* **62**, 3729–3735.

19. Bergsten, E., Uutela, M., Li, X. *et al.* (2001). PDGF-D is a specific, protease-activated ligand for the PDGF beta receptor. *Nat Cell Biol* **3**, 512–516.

20. Gilbertson, D. G., Duff, M. E., West, J. W. *et al.* (2001). Platelet-derived growth factor C (PDGF-C), a novel growth factor that binds to PDGF alpha and beta receptor. *J Biol Chem* **276**, 27406–27414.

21. Harsh, G. R., Keating, M. T., and Escobedo, J. A. (1990). Williams LT. Platelet-derived growth factor (PDGF) autocrine components in human tumor cell lines. *J Neuro-Oncol* **8**, 1–12.

22. Maxwell, M., Naber, S. P., Wolfe, H. J., Galanopoulos, T., Hedley-Whyte, E. T., and Black, P. M. (1990). Antoniades HN. Coexpression of platelet-derived growth factor (PDGF) and PDGF-receptor genes by primary human astrocytomas may contribute to their development and maintenance. *J Clin Invest* **86**, 131–140.

23. Mauro, A., Bulfone, A., Turco, E., and Schiffer, D. (1991). Co-expression of platelet-derived growth factor (PDGF) B chain and PDGF B-type receptor in human gliomas. *Child's Nerv Syst* **7**, 432–436.

24. Hermanson, M., Funa, K., Hartman, M. *et al.* (1992). Platelet-derived growth factor and its receptors in human glioma tissue: expression of messenger RNA and protein suggests the presence of autocrine and paracrine loops. *Cancer Res* **52**, 3213–3219.

25. Hermanson, M., Funa, K., Koopman, J. *et al.* (1996). Association of loss of heterozygosity on chromosome 17p with high platelet-derived growth factor alpha receptor expression in human malignant gliomas. *Cancer Res* **56**, 164–171.

26. Fleming, T. P., Saxena, A., Clark, W. C. *et al.* (1992). Amplification and/or overexpression of platelet-derived growth factor receptors in human glial tumors. *Cancer Res* **52**, 4550–4553.

27. Dai, C., Celestino, J. C., Okada, Y., Louis, D. N., Fullerm, G. N., and Holland, E. C. (2001). PDGF autocrine stimulation dedifferentiates cultured astrocytes and induces oligodendro-gliomas and oligoastrocytomas from neural progenitors and astrocytes in vivo. *Genes Dev* **15**, 1913–1925.

28. Black, P., Carroll, R., and Glowacka, D. (1996). Expression of platelet-derived growth factor transcripts in medullo-blastomas and ependymomas. *Pediatr Neurosurg* **24**, 74–78.

29. MacDonald, T. J., Brown, K. M., LaFleur, B. *et al.* (2001). Expression profiling of medulloblastoma: PDGFRA and the RAS/MAPK pathway as therapeutic targets for metastatic disease. *Nature Genet* **29**, 143–52.

30. Vassbotn, F. S., Ostman, A., Langeland, N. *et al.* (1994). Activated platelet-derived growth factor autocrine pathway drives the transformed phenotype of a human glioblastoma cell line. *J Cell Physiol* **158**, 381–389.

31. Guha, A., Dashner, K., Black, P. M., Wagner, J. A., and Stiles, C. D. (1995). Expression of PDGF and PDGF receptors in human astrocytoma operation specimens supports the existence of an autocrine loop. *Int J Cancer* **60**, 168–173.

32. Pollack, I. F., Randall, M. S., Kristofik, M. P., Kelly, R. H., Selker, R. G., and Vertosick, F. T., Jr. (1990). Response of malignant glioma cell lines to epidermal growth factor and platelet-derived growth factor in a serum-free medium. *J Neurosurg* **73**, 106–112.

33. Pollack, I. F., Randall, M. S., Kristofik, M. P., Kelly, R. H., Selker, R. G., and Vertosick, F. T., Jr. (1991). Response of low-passage human malignant gliomas in vitro to stimulation and selective inhibition of growth factor-mediated pathways. *J Neurosurg* **75**, 284–293.

34. Westphal, M., Brunken, M., Rohde, E., and Herrmann, H.-D. (1988). Growth factors in cultured human glioma cells: differential effects of FGF, EGF, and PDGF. *Canc Lett* **38**, 283–296.

35. Campbell, J., and Pollack, I. F. (1997). Growth factors in gliomas: Antisense and dominant negative mutant strategies. *J Neuro-oncol* **35**, 275–285.

36. Kovalenko, M., Gazit, A., Böhmer, A. *et al.* Selective platelet-derived growth factor receptor kinase blockers reverse sis-transformation. *Cancer Res* **54**, 6106–6114.

37. Nitta, T., and Sato, K. (1994). Specific inhibition of c-sis protein synthesis and cell proliferation with antisense oligodeoxynucleotides in human glioma cells. *Neurosurgery* **34**, 309–314.

38. Wang, D., Huang, H. J., Kazlauskas, A., and Cavenee, W. K. (1999). Induction of vascular endothelial growth factor expression in endothelial cells by platelet-derived growth factor through the activation of phosphatidylinositol 3-kinase. *Cancer Res* **59**, 1464–1472.

39. Carmeliet, P., and Jain, R. K. (2000). Angiogenesis in cancer and other diseases. *Nature* **407**, 249–257.

40. Schlessinger, J. (2000). Cell signaling by receptor tyrosine kinases. *Cell* **103**, 211–225.

41. Shelly, M., Pinkas-Kramarski, R., Guarino, B. C. *et al.* (1998). Epiregulin is a potent pan-ErbB ligand that preferentially activates heterodimeric receptor complexes. *J Biol Chem* **273**, 10496–10505.

42. Tzahar, E., Waterman, H., Chen, X. *et al.* (1996). A hierarchical network of interreceptor interactions determines signal transduction by Neu differentiation factor/neuregulin and epidermal growth factor. *Mol Cell Biol* **16**, 5276–5287.

43. Libermann, T. A., Nusbaum, H. R., Razon, N. *et al.* (1985). Amplification, enhanced expression and possible rearrangement of EGF receptor gene in primary human brain tumors of glial origin. *Nature* **313**, 44–147.

44. von Deimling, A., Louis, D. N., von Ammon, K. *et al.* (1993b). Association of epidermal growth factor receptor gene amplification with loss of chromosome 10 in human glioblastoma multiforme. *J Neurosurg* **77**, 95–301.

45. Ekstrand, A. H., James, C. D., Cavenee, W. K., Seliger, B., Petterson, R. F., and Collins, V. P. (1991). Genes for epidermal growth factor receptor, transforming growth factor α, and epidermal growth factor and their expression in human gliomas in vivo. *Cancer Res* **51**, 2164–2172.

46. Ullrich, A., and Schlessinger, J. (1990). Signal transduction by receptor with tyrosine kinase activity. *Cell* **61**, 203–212.

47. Wong, A. J., Ruppert, J. M., Bigner, S. H. *et al.* (1992). Structural alterations of the epidermal growth factor receptor gene in human gliomas. *Proc Natl Acad Sci USA* **89**, 2965–2969.

48. Schwechheimer, K., Huang, S., and Cavenee, W. K. (1995). EGFR gene amplification-rearrangement in human glioblastomas. *Int J Cancer* **62**, 145–148.

49. Frederick, L., Wang, X.-Y., Eley, G., and James, C. D. (2000). Diversity and frequency of epidermal growth factor receptor mutations in human glioblastomas. *Cancer Res* **60**, 1383–1387.

50. Moscatello, D. K., Holgado-Madruga, M., Emlet, D. R. *et al.* (1998). Constitutive activation of phosphatidyl 3-kinase by a naturally occurring mutant epidermal growth factor receptor. *J Biol Chem* **273**, 200–206.

51. Chu, C. T., Everiss, K. D., Wikstrand, C. J. *et al.* (1997). Receptor dimerization is not a factor in the signaling activity of a transforming variant epidermal growth factor receptor (EGFRvIII). *Biochem J* **324**, 855–861.

52. Louis, D. N. (1997). A molecular genetic model of astrocytoma histopathology. *Brain Pathol* **7**, 755–764.

53. Watanabe, K., Tachibana, O., Sato, K., Yonekawa, Y., Kleihues, P., and Ohgaki, H. (1996). Overexpression of the EGF receptor and p53 mutations are mutually exclusive in the evolution of primary and secondary glioblastomas. *Brain Pathol* **6**, 217–224.

54. von Deimling, A., von Ammon, K., Schoenfeld, D., Wiestler, O. D., Seizinger, B. R., and Louis, D. N. (1993a). Subsets of glioblastoma multiforme defined by molecular genetic analysis. *Brain Pathol* **3**, 19–26.

55. Bredel, M., Pollack, I. F., Hamilton, R. L., and James, C. D. (1999). Epidermal growth factor receptor (EGFR) expression in high-grade non-brainstem gliomas of childhood. *Clin Cancer Res* **5**, 1786–1792.

56. Holland, E. C., Hively, W. P., DePinho, R. A., and Varmus, H. E. (1998). A constitutively active epidermal growth factor receptor cooperates with disruption of G1 cell-cycle arrest pathways to induce glioma-like lesions in mice. *Genes Dev* **12**, 3675–3685.

57. Gilbertson, R. J., Perry, R. H., Kelly, P. J., Pearson, A. D., and Lunec, J. (1997). Prognostic significance of HER2 and HER4 coexpression in childhood medulloblastoma. *Cancer Res* **57**, 3272–3280.

58. Gilbertson, R. J., Clifford, S. C., Meekin, W. *et al.* (1998). Expression of the ErbB-neuregulin signaling network during human cerebellar development: Implications for the biology of medulloblastoma. *Cancer Res* **58**, 3932–3941.

59. Gilbertson, R., Wickramasinghe, C., Hernan, R. *et al.* (2001). Clinical and molecular stratification of disease risk in medulloblastoma. *Br J Cancer* **85**, 705–712.

60. Gilbertson, R. J., Bentley, L., Hernan, R. *et al.* (2002). ErbB signaling promotes ependymoma cell proliferation and represents a potential novel therapeutic target for this disease. *Clin Cancer Res* **8**, 3054–3064.

61. O'Rourke, D. M., Qian, X., Zhang, H. T. *et al.* (1997). Trans receptor inhibition of human glioblastoma cells by erbB family ectodomains. *Proc Natl Acad Sci USA* **94**, 3250–3255.

62. Mishima, K., Johns, T. G., Luwor, R. B. *et al.* (2001). Growth suppression of intracranial xenografted glioblastomas overexpressing mutant epidermal growth factor receptors by systemic administration of monoclonal antibody (mAb) 806, a novel monoclonal antibody directed to the receptor. *Cancer Res* **61**, 5349–5354.

63. Luwor, R. B., Johns, T. G., Murone, C. *et al.* (2001). Monoclonal antibody 806 inhibits the growth of tumor xenografts expressing either the de2–7 or amplified epidermal growth factor receptor (EGFR) but not wild-type EGFR. *Cancer Res* **61**, 5355–5361.

64. Hernan, R., Fasheh, R., Calabrese, C. *et al.* (2003). ErbB2 up-regulates S100A4 and several other prometastatic genes in medulloblastoma. *Cancer Res* **63**, 140–148.

65. Weiner, H. L. (1995). The role of growth factor receptors in central nervous system development and neoplasia. *Neurosurgery* **37**, 179–194.

66. McCormick, F. (1993). How receptors turn Ras on. *Nature* **363**, 15–16.

67. Pawson, T. (1995). Protein modules and signalling networks. *Nature* **373**, 573–580.

68. Boguski , M. S., and McCormick, F. (1993). Proteins regulating Ras and its relatives. *Nature* **366**, 643–654.

69. Li, N., Batzer, A., Daly, R. et al. (1993). Guanine-nucleotide-releasing factor hSos1 binds to Grb2 and links receptor tyrosine kinases to Ras signaling. Nature 363, 85–88.

70. Rozakis-Adcock, M., Fernley, R., Wade, J., Pawson, T., and Bowtell, D. (1993). The SH2 and SH3 domains of mammalian Grb2 couple the EGF receptor to the Ras activator mSos1. Nature 363, 83–85.

71. Peddanna, N., Mendis, R., Holt, S., and Verma, R. S. (1996). Genetics of colorectal cancer. Int J Oncol 9, 327–335.

72. Bos, J. L. (1989). Ras oncogenes in human cancer: A review. Cancer Res 49, 4682–4689.

73. Guha, A., Feldkamp, M. M., Lau, N., Boss, G., and Pawson, A. (1997). Proliferation of human malignant astrocytomas is dependent on Ras activation. Oncogene 15, 2755–2765.

74. Orian, J. M., Vasilopoulos, K., Yoshida, S., Kaye, A. H., Chow, C. W., and Gonzales, M. F. (1992). Overexpression of multiple oncogenes related to histopathological grade of astrocytic glioma. Br J Cancer 66, 106–112.

75. Riccardi, A., Danova, M., Giordano, M. et al. (1991). Proto-oncogene expression and proliferative activity in human malignant gliomas. Dev Oncol 66, 81–84.

76. Prigent, S. A., Nagane, M., Lin, H. et al. (1996). Enhanced tumorigenic behavior of glioblastoma cells expressing a truncated epidermal growth factor receptor is mediated through the Ras-SHC-GRB2 pathway. J Biol Chem 271, 25639–25645.

77. Rodrigues, G. A., Falasca, M., Zhang, Z., Ong, S.H., and Schlessinger, J. (2000). A novel positive feedback loop mediated by the docking protein Gab1 and phosphotidylinositol 3-kinase in epidermal growth factor receptor signaling. Molec Cell Biol 20, 1448–1459.

78. Marais, R., Light, Y., Mason, C., Paterson, H., Olson, M. F., and Marshall, C. J. (1998). Requirement of Ras-GTP-Raf complexes for activation of Raf-1 by protein kinase C. Science 280, 109–112.

79. Meisenhelder, J., Suh, P.-G., Rhee, S.-G., and Hunter, T. (1989). Phospholipase C-γ is a substitute for the PDGR and EGF receptor protein-tyrosine kinases in vivo and in vitro. Cell 57, 1109–1122.

80. Kauffmann-Zeh, A., Thomas, G. M. H., Ball, A. et al. (1995). Requirement for phosphatidylinositol transfer protein in epidermal growth factor signaling. Science 368, 1188–1190.

81. Luttrell, L. M., Daaka, Y., and Lefkowitz, R. J. (1999). Regulation of tyrosine kinase cascades by G-protein-coupled receptors. Curr Opin Cell Biol 11, 177–183.

82. Dikic, I., Tokiwa, G., Lev, S., Courtneidge, S. A., and Schlessinger, J. (1996). A role of Pyk2 and Src in linking G-protein-coupled receptors with MAP kinase activation. Nature 383, 547–550.

83. Bromberg, J., and Chen, X. (2001). STAT proteins: Signal transducers and activators of transcription. Methods Enzymol 333, 138–151.

84. Heldin, C. H., Ostman, A., and Ronnstrand, L (1998). Signal transduction via platelet-derived growth factor receptors. Biochim Biophys Acta 1378, F79–F113.

85. Valgeirsdottir, S., Paukku, K., Silvennoinen, O., Heldin, C. H., and Claesson-Welsh, L. (1998). Activation of Stat5 by platelet-derived growth factor (PDGF) is dependent on phosphorylation sites in PDGF-beta receptor juxtamembrane and kinase insert domains. Oncogene 16, 505–515.

86. Quadros, M. R. D., Peruzzi, F., Kar, C., and Rodeck, U. (2004). Complex regulation of signal transducers and activators of transcription 3 activation in normal and malignant keratinocytes. Cancer Res 64, 3934–3939

87. Zhong, Z., Wen, Z., and Darnell, J. E., Jr. (1994). Stats: A STAT family member activated by tyrosine phosphorylation in response to epidermal growth factor and interleukin-6. Science 264, 95–98.

88. Shual, K., Ziemiecki, A., Wilks, A. F. et al. (1993). Polypeptide signalling to the nucleus through tyrosine phosphorylation of Jak and Stat proteins. Nature 366, 580–585.

89. Luo, Z., Tzivion, G., Belshaw, P. J., Vavvas, D., Marshall, M., and Avruch, J. (1996). Oligomerization activates c-Raf-1 through a Ras-dependent mechanism. Nature 383, 181–185.

90. Freed, E., Symons, M., MacDonald, S. G., McCormick, F., and Ruggieri, R. (1994). Binding of 14–3-3 proteins to the protein kinase Raf and effects on its activation. Science 265, 1713–1716.

91. Howe, L. R., Leevers, S. J., Gomez, N., Nakielny, S., Cohen, P., and Marshall, C. J. (1992). Activation of the MAP kinase pathway by the protein kinase raf. Cell 71, 335–342.

92. Kyriakis, J. M., Force, T. L., Rapp, U. R., Bonventre, J. V., and Avruch, J. (1993). Mitogen regulation of c-raf-1 protein kinase activity toward mitogen-activated protein kinase-kinase. J Biol Chem 268, 16009–16019.

93. Derijard, B., Raingeaud, J., Barrett, T., Wu, I.-H., Han, J., Ulevitch, R. J., and Davis, R. J. (1995). Independent human MAP kinase signal transduction pathways defined by MEK and MKK isoforms. Science 367, 682–684.

94. Blenis, J. (1993). Signal transduction via the MAP kinases: proceed at your own RSK. Proc Natl Acad Sci USA 90, 5889–5892.

95. Gille, H., Kortenjahn, M., Thomae, O. et al. (1995). ERK phosphorylation potentiates Elk-1-mediated ternary complex formation and transactivation. EMBO J 14, 951–962.

96. Marais, R., Wynne, J., and Treisman, R. (1993). The SRF accessory protein Elk-1 contains a growth factor-regulated transcriptional activation domain. Cell 73, 381–393.

97. Lin, A., Minden, A., Martinetto, H. et al. (1995). Identification of a dual specificity kinase that activates the Jun kinases and p38-Mpk2. Science 268, 286–290.

98. Behrens, A., Jochum, W., Sibilia, M., and Wagner, E. F. (2000). Oncogenic transformation by ras and fos is mediated by c-Jun N-terminal phosphorylation. Oncogene 19, 2657–2663.

99. Wu, C. J., Qian, X., and O'Rourke, D. M. (1999). Sustained mitogen-activated protein kinase activation is induced by transforming erbB receptor complexes. DNA Cell Biol 18, 731–741.

100. Tsuiki, H., Tnani, M., Okamoto, I. et al. (2003). Constitutively active forms of c-Jun NH2-terminal kinase are expressed in primary glial tumors. Cancer Research 63, 250–255.

101. Ludwig, S., Engel, K., Hoffmeyer, A. et al. (1996). 3pK, a novel mitogen-activated protein (MAP) kinase-activated protein kinase, is targeted by three MAP kinase pathways. Molec Cell Biol 16, 6687–6697.

102. Waskiewicz, A. J., Flynn, A., Proud, C. G., and Cooper, J. A. (1997). Mitogen-activated protein kinases activate the serine/threonine kinases Mnk1 and Mnk2. Embo J 16, 1909–1920.

103. Ashkenazi, A., and Dixit, V. M. (1998). Death receptors: signaling and modulation. Science 281, 1305–1308.

104. Wu, J., Dent, P., Jelinek, T., Wolfman, A., Weber, M. J., and Sturgill, T. W. (1993). Inhibition of the EGF-activated MAP kinase signaling pathway by adenosine 3',5'-monophosphate. Science 262, 1065–1069.

105. Cardone, M. H., Roy, N., Stennicke, H. R. et al. (1998). Regulation of cell death protease caspase-9 by phosphorylation. Science 282, 1318–1321.

106. Brunet, A., Bonni, A., Zigmond, M. J., Lin, M. Z., Juo, P., Hu, L. S., et al. (1999). Akt promotes cell survival by

phosphorylating and inhibiting a Forkhead transcription factor. *Cell* **96**, 857–868.

107. Cross, D. A., Alessi, D. R., Cohen, P., Andjelkovich, M., and Hemmings, B. A. (1995). Inhibition of glycogen synthase kinase-3 by insulin mediated by protein kinase B. *Nature* **378**, 785–789.

108. Li, D. M., and Sun, H. (1998). PTEN/MMAC1/TEP1 suppressed the tumorigenicity and induces G1 cell cycle arrest in human glioblastoma cells. *Proc Natl Acad Sci USA* **95**, 15406–15411.

109. Pershouse, M. A., Stubblefield, E., Hadi, A., Killary, A. M., Yung, W. K., and Steck, P. A. (1993). Analysis of the functional role of chromosome 10 loss in human glioblastomas. *Cancer Res* **53**, 5043–5050.

110. Nishizuka, Y. (1988). The molecular heterogeneity of protein kinase C and its implications for cellular regulation. *Nature* **334**, 661–665.

111. Nishizuka, Y. (1992). Intracellular signaling by hydrolysis of phospholipids and activation of protein kinase C. *Science* **258**, 607–613.

112. Susa, M., Olivier, A. R., Fabbro, D., and Thomas, G. (1989). EGF induces biphasic S6 kinase activation: late phase is protein kinase C-dependent and contributes to mitogenicity. *Cell* **57**, 817–824.

113. Fields, A. P., Tyler, G., Kraft, A. S., and May, W. S. (1990). Role of nuclear protein kinase C in the mitogenic response to platelet-derived growth factor. *J Cell Science* **96**, 107–114.

114. Pollack, I. F., Bredel, M., and Erff, M. (1998). The application of signal transduction inhibition as a therapeutic strategy for central nervous system tumors. *Pediatr Neurosurg* **29**, 228–244.

115. Basu, A., and Lazo, J. S. (1994). Protein kinase C In: New Molecular Targets for Cancer Chemotherapy. Kerr, D. J., and Workman, P. (eds), CRC Press, 121–141.

116. Kolch, W., Heldecker, G., Kochs, G. *et al.* (1993). Protein kinase Cα activates Raf-1 by direct phosphorylation. *Nature* **364**, 249–252.

117. Chao, T.-S. O., Foster, D. A., Rapp, U. R., and Rosner, M. R. (1994). Differential raf requirement for activation of mitogen-activated protein kinase by growth factors, phorbol esters, and calcium. *J Biol Chem* **269**, 7337–7341.

118. Neary, J. T., Norenberg, O. B., and Norenberg, M. D. (1988). Protein kinase C in primary astrocyte cultures: cytoplasmic localization and translocation by a phorbol ester. *J Neurochem* **50**, 1179–1184.

119. Reifenberger, G., Deckert, M., and Wechsler, W. (1989). Immunohistochemical determination of protein kinase C expression and proliferative activity in human brain tumors. *Acta Neuropathol* **78**, 166–175.

120. Benzil, D. L., Finkelstein, S. D., Epstein, M. H., and Finch, P. W. (1992). Expression pattern of α-protein kinase C in human astrocytomas indicates a role in malignant progression. *Cancer Res* **52**, 2951–2956.

121. Couldwell, W. T., Uhm, J. H., Antel, J. P., and Yong, V. W. (1991). Enhanced protein kinase C activity correlates with the growth rate of malignant gliomas in vitro. *Neurosurg* **29**, 880–887.

122. Couldwell, W. T., Antel, J. P., and Yong, V. W. (1992). Protein kinase C activity correlates with the growth rate of malignant gliomas: Part II. Effects of glioma mitogens and modulators of protein kinase C. *Neurosurgery* **31**, 717–724.

123. Ahmad, S., Mineta, T., Martuza, R. L., and Glazer, R. I. (1994). Antisense expression of protein kinase C-alpha inhibits the growth and tumorigenicity of human glioblastoma cells. *Neurosurgery* **35**, 904–909.

124. Pollack, I. F., Kawecki, S., and Lazo, J. S. (1996). 7-hydroxystaurosporine (UCN-01), a selective protein kinase C inhibitor, exhibits cytotoxicity against cultured glioma cells and potentiates the antiproliferative effects of BCNU and cisplatin. *J Neurosurg* **84**, 1024–1032.

125. Yazaki, T., Ahmad, S., Chahlavi, A. *et al.* (1996). Treatment of glioblastoma U-87 by systemic administration of an antisense protein kinase C-α phosphorothioate oligodeoxynucleotide. *Molec Pharmacol* **50**, 236–242.

126. Bredel, M., Pollack, I. F., Freund, J. M., Rusnak, J., and Lazo, J. S. (1999a). Protein kinase C inhibition by UCN-01 induces apoptosis in human glioma cells in a time-dependent fashion. *J Neuro-oncol* **41**, 9–20.

127. Olayioye, M. A., Beuvink, I., Horsch, K., Daly, J. M., and Hynes, N. E. (1999). ErbB receptor-induced activation of stat transcription factors in mediated by Src tyrosine kinases. *J Biol Chem* **274**, 17209–17218.

128. Bienvenu, F., Gascan, H., and Coqueret, O. (2001). Cyclin D1 represses STAT3 activation through a Cdk4-independent mechanism. *J Biol Chem* **276**, 16840–16847.

129. Schaefer, L. K., Ren, Z., Fuller, G. N., and Schaefer, T. S. (2002). Constitutive activation of Stat3α in brain tumors: Localization to tumor endothelial cells and activation by the endothelial tyrosine kinase receptor (VEGFR-2). *Oncogene* **21**, 2058–2065.

130. Pollack, I. F., Bredel, M., Erff, M., and Sebti, S. M. (1999a). Inhibition of Ras and related G-proteins as a novel therapeutic strategy for blocking malignant glioma growth. II. In vivo results in a nude mouse model. *Neurosurgery* **45**, 1208–1214.

131. Shamah, S. M., Stiles, C. D., and Guha, A. (1993). Dominant-negative mutants of platelet-derived growth factor revert the transformed phenotype of human astrocytoma cells. *Molec Cell Biol* **13**, 7203–7212.

132. Huang, S. M., and Harari, P. M. (1999). Epidermal growth factor receptor inhibition in cancer therapy: biology, rationale and preliminary clinical results. *Invest New Drugs* **17**, 259–269.

133. Overholser, J. P., Prewett, M. C., Hooper, A. T., Waksal, H. W., and Hicklin, D. J. (2000). Epidermal growth factor receptor blockade by antibody IMC-C225 inhibits growth of a human pancreatic carcinoma xenograft in nude mice. *Cancer* **89**, 74–82.

134. Pegram, M. D., Lipton, A., Hayes, D. F. *et al.* (1998). Phase II study of receptor-enhanced chemosensitivity using recombinant humanized anti-p185HER2/neu monoclonal antibody plus cisplatin in patients with HER2/neu-overexpressing metastatic breast cancer refractory to chemotherapy treatment. *J Clin Oncol* **16**, 2659–2671.

135. Baselga, J., Pfister, D., Cooper, M. R. *et al.* (2000). Phase I studies of anti-epidermal growth factor receptor chimeric antibody C225 alone and in combination with cisplatin. *J Clin Oncol* **18**, 904–914

136. Shin, D. M., Donato, N. J., Perez-Soler, R. *et al.* (2001). Epidermal growth factor receptor-targeted therapy with C225 and cisplatin in patients with head and neck cancer. *Clin Cancer Res* **7**, 1204–1213.

137. Huang, S. M., Bock, J. M., and Harari, P. M. (1999). Epidermal growth factor receptor blockade with C225 modulates proliferation, apoptosis, and radiosensitivity in squamous cell carcinomas of the head and neck. *Cancer Res* **59**, 1935–1940.

138. Laske, D. W., Ilercil, O., Akbasak, A., Youle, R. J., and Oldfield, E.H. (1994). Efficacy of direct intratumoral therapy

with targeted protein toxins for solid human tumors in nude mice. *J Neurosurg* **80**, 520–526.

139. Laske, D. W., Youle, R. J., and Oldfield, E. H. (1997). Tumor regression with regional distribution of targeted toxin TF-CRM107 in patients with malignant brain tumors. *Nature Med* **3**, 1362–1368.

140. Druker, B. J., Tamura, S., Buchdunger, E. *et al.* (1996). Effects of a selective inhibitor of the Abl tyrosine kinase on the growth of Bcr-Abl positive cells. *Nat Med* **2**, 561–566.

141. Druker, B. J., Sawyers, C. L., Kantarjian, H. *et al.* (2001). Activity of a specific inhibitor of the BCR-ABL tyrosine kinase in the blast crisis of chronic myeloid leukemia and acute lymphoblastic leukemia with the Philadelphia chromosome. *N Engl J Med* **344**, 1038–1042.

142. Druker, B. J., Talpaz, M., Resta, D. J. *et al.* (2001b). Efficacy and safety of a specific inhibitor of the BCR-ABL tyrosine kinase in chronic myeloid leukemia. *N Engl J Med* **344**, 1031–1037.

143. Sawyers, C. L., Hochhaus, A., Feldman, E. *et al.* (2002). Imatinib induces hematologic and cytogenetic responses in patients with chronic myelogenous leukemia in myeloid blast crisis: results in a phase II study. *Blood* **99**, 3530–3539.

144. O'Brien, S. G., Guilhot, F., Larson, R. A. *et al.* (2003). Imatinib compared with interferon and low-dose cytarabine for newly diagnosed chronic-phase chronic myeloid leukemia. *New Engl J Med* **348**, 994–1004.

145. Hughes, T. P., Kaeda, J., Branford, S. *et al.* (2003). Frequency of major molecular responses to imatinib or interferon alfa plus cytarabine in newly diagnosed chronic myeloid leukemia. *New Engl J Med* **349**, 1423–1432.

146. Gorre, M. E., Mohammed, M., Ellwood, K. *et al.* (2001). Clinical resistance to STI-571 cancer therapy caused by BCR-ABL gene mutation or amplification. *Science* **293**, 876–880.

147. Branford, S., Rudzki, Z., Walsh, S. *et al.* (2003). Detection of BCR-ABL mutations in patients with CML treated with imatinib is virtually always accompanied by clinical resistance, and mutations in the ATP phosphate-binding loop (P-loop) are associated with a poor prognosis. *Blood* **102**, 276–283.

148. Heinrich, M. C., Griffith, D. J., Druker, B. J., Wait, C. L., Ott, K. A., and Zigler, A. J. (2000). Inhibition of c-kit receptor tyrosine kinase activity by STI 571, a selective tyrosine kinase inhibitor. *Blood* **96**, 925–932.

149. Joensuu, H., Roberts, P. J., Sarlomo-Rikala, M. *et al.* (2001). Effect of the tyrosine kinase inhibitor STTI571 in a patient with a metastatic gastrointestinal stromal tumor. *N Engl J Med* **344**, 1052–1056.

150. Buchdunger, E., Cioffi, C. L., Law, N. *et al.* (2000). Abl protein-tyrosine kinase inhibitor STI571 inhibits in vitro signal transduction mediated by c-kit and platelet-derived growth factor receptors. *J Pharmacol Exper Ther* **295**, 139–145.

151. Uhrbom, L., Hesselager, G., Ostman, A., Nister, M., and Westermark, B. (2000). Dependence of autocrine growth factor stimulation in platelet-derived growth factor-B-induced mouse brain tumor cells. *Int J Cancer* **85**, 398–406.

152. Kilic, T., Alberta, J. A., Zdunek, P. R. *et al.* (2000). Intracranial inhibition of platelet-derived growth factor-mediated glioblastoma cell growth by an orally active kinase inhibitor of the 2-phenylaminopyrimidine class. *Cancer Res* **60**, 143–150.

153. Pietras, K., Ostman, A., Sjoquist, M. *et al.* (2001). Inhibition of platelet-derived growth factor receptors reduces interstitial hypertension and increases transcapillary transport in tumors. *Cancer Res* **61**, 2929–2934.

154. Apperly, J. F., Gardembas, M., Melo, J. V. *et al.* (2002). Response to imatinib mesylate in patients with chronic myeloproliferative diseases with rearrangements of the platelet-derived growth factor receptor beta. *New Engl J Med* **347**, 481–487.

155. Pietras, K., Rubin, K, Sjoblom, T. *et al.* (2002). Inhibition of PDGF receptor signaling in tumor stroma enhances antitumor effect of chemotherapy. *Canc Res* **62**, 5476–5484.

156. Raymond, E., Brandes, A., Van Oosterom, A. *et al.* (2004). Multicentre phase II study of imatinib mesylate in patients with recurrent glioblastoma: An EORTC:NDDG/BTG intergroup study. *Proc ASCO* 107.

157. Katz, A., Barrios, C. H., Abramoff, R., Simon, S. D., Tabacof, J., and Gansl, R. C. (2004). Imatinib (STI 571) is active in patients with high-grade gliomas progressing on standard therapy. *Proc ASCO* 117.

158. Dresemann, G. (2004). Imatinib (STI571) plus hydroxyurea: Safety and efficacy in pretreated progressive glioblastoma multiforme patients. *Proc ASCO* 119.

159. Ciardiello, F., Caputo, R., Bianco, R. *et al.* (2001). Inhibition of growth factor production and angiogenesis in human cancer cells by ZD1839 (Iressa), a selective epidermal growth factor receptor tyrosine kinase inhibitor. *Clin Cancer Res* **7**, 1459–1465.

160. Ciardiello, F., Caputo, R., Bianco, R. *et al.* (2000). Antitumor effect and potentiation of cytotoxic drugs activity in human cancer cells by ZD-1839 (Iressa), an epidermal growth factor receptor-selective tyrosine kinase inhibitor. *Clin Cancer Res* **6**, 2053–2063.

161. Baselga, J., and Averbuch, S. D. (2000). ZD1839 ('Iressa') as an anticancer agent. *Drugs* **60 (suppl 1)**, 33–40.

162. Wakeling, A. E., Guy, S. P., Woodburn, J. R. *et al.* (2002). ZD1839 (Iressa): An orally active inhibitor of epidermal growth factor signaling with potential for cancer therapy. *Cancer Res* **62**, 5749–5754.

163. Swaisland, H., Laight, A., Stafford, L. *et al.* (2001). Pharmacokinetics and tolerability of the orally active selective epidermal growth factor receptor tyrosine kinase inhibitor ZD1839 in healthy volunteers. *Clin Pharmacokinetics* **40**, 297–306.

164. Raymond, E., Faivre, S., and Armand, J. P. (2000). Epidermal growth factor receptor tyrosine kinase as a target for anticancer therapy. *Drugs* **60 Suppl 1**, 15–23.

165. Lorusso, P.M. (2003). Phase I studies of ZD1839 in patients with common solid tumors. *Sem Oncol* **30 (Suppl 1)**, 21–29.

166. Albanell, J., Royo, F., Auerbach, S. *et al.* (2002). Pharmacokinetic studies of the epidermal growth factor receptor inhibitor ZD1839 in skin from cancer patients: histological and molecular consequences of receptor inhibition. *J Clin Oncol* **20**, 110–124.

167. Baselga, J., Rischin, D., Ranson, M. *et al.* (2002). Phase I safety, pharmacokinetic, and pharmacodynamic trial of ZD1839, a selective oral epidermal growth factor receptor tyrosine kinase inhibitor, in patients with five selected solid tumor types. *J Clin Oncol* **20**, 4292–4302.

168. Fukuoka, M., Yano, S., Giaccone, G. *et al.* (2003). Multi-institutional randomized phase II trial of gefitinib for previously treated patients with advanced non-small-cell lung cancer. *J Clin Oncol* **21**, 2237–2246.

169. Kris, M. G., Natale, R. B., Herbst, R. S. *et al.* (2003). Efficacy of gefitinib, an inhibitor of the epidermal growth factor receptor tyrosine kinase, in symptomatic patients with non-small cell lung cancer: a randomized trial. *JAMA* **290**, 2149–2158.

170. Herbst, R. S., Giaccone, G., Schiller, J. H., Natale, R. B., Miller, V., Manegold, C., *et al.* (2004). Gefinib in combination with paclitaxel and carboplatin in advanced non-small-cell lung cancer: a phase III trial—Intact 2. *J Clin Oncol* **22**, 785–94.

171. Giaccone, G., Herbst, R. S., Manegold, C., Scagliotti, G., Rosell, R., Miller, V., *et al.* (2004). Gefinib in combination with gemcitabine and cisplatin in advanced non-small-cell lung cancer: a phase III trial—Intact 1. *J Clin Oncol* **22**, 777–84.

172. Dancey, J. E., and Freidlin, B. (2003). Targeting epidermal growth factor receptor – are we missing the mark. *Lancet* **362**, 62–64.

173. Lynch, T. J., Bell, D. W., Sordella, R., Gurubhagavatula, S., Okimoto, R. A., Brannigan, B. W., *et al.* (2004). Activating mutations in the epidermal growth factor receptor underlying responsiveness of non-small-cell lung cancer to gefitinib. *New Engl J Med* **350**, 2129–2139.

174. Heimberger, A. B., Learn, C. A., Archer, G. E., McLendon, R. E., and Chewning, T. A., Tuck F. L., *et al.* (2002). Brain tumors in mice are susceptible to blockade of epidermal growth factor receptor (EGFR) with the oral, specific, EGFR-tyrosine kinase inhibitor ZD1839 (Iressa). *Clin Cancer Res* **8**, 3496–3502.

175. Li, B., Chang, C.-M., Yuan, M., McKenna, G., and Shu, H.-K. (2003). Resistance to small molecule inhibitors of epidermal growth factor receptor in malignant gliomas. *Cancer Res* **63**, 7443–7450.

176. Bianco, R., Shin, I., Ritter, C. A. *et al.* (2003). Loss of PTEN/MMAC1/TEP in EGF receptor-expressing tumor cells counteracts the antitumor action of EGFR tyrosine kinase inhibitors. *Oncogene* **22**, 2812–2822.

177. Lieberman, F. S., Cloughesy, T., Fine, H., Kuhn, J., Lamborn, K., Malkin, M., *et al.* (2004). NABTC phase I/II trial of ZD-1839 for recurrent malignant gliomas and unresectable meningiomas. *Proc ASCO* 109.

178. Uhm, J. H., Ballman, K. V., Giannini, C. *et al.* (2004). Phase II study of ZD 1839 in patients with newly diagnosed grade 4 astrocytomas. *Proc ASCO* 108.

179. Prados, M., Yung, W., Wen, P. *et al.* (2004). Phase I study of ZD1839 plus temozolomide in patients with malignant glioma. *Proc ASCO* 108.

180. Moyer, J. D., Barbacci, EG., Iwata, K. K. *et al.* (1997). Induction of apoptosis and cell cycle arrest by CP-358,774, an inhibitor of epidermal growth factor receptor tyrosine kinase. *Cancer Res* **57**, 4838–4848.

181. Pollack, V. A., Savage, D., Baker, D. *et al.* (1999). Inhibition of epidermal growth factor receptor-associated tyrosine phosphorylation in human carcinomas with CP-358,774. Dynamics of receptor inhibition in situ and antitumor effects in athymic mice. *J Pharmacol Exp Ther* **291**, 739–748.

182. Malik, S. N., Siu, L. L., Rowinsky, E. K. *et al.* (2003). Pharmacodynamic evaluation of the epidermal growth factor receptor inhibitor OSI-774 in human epidermis of cancer patients. *Clin Cancer Res* **9**, 2478–2486.

183. Hidalgo, M., Siu, L. L., Nemunaitis, J. *et al.* (2001). Phase I and pharmacologic study of OSI-774, an epidermal growth factor receptor tyrosine kinase inhibitor, in patients with advanced solid malignancies. *J Clin Oncol* **19**, 3267–3279.

184. Rowinsky, E. K., Hammond, L., Siu, L. *et al.* (2001). Dose-schedule-finding, pharmacokinetic, biologic, and functional imaging studies of OSI-774, a selective epidermal growth factor receptor (EGFR) tyrosine kinase inhibitor. *Proc Am Soc Clin Oncol* **20**, 2a (abstract 5).

185. Perez-Soler, R., Chachoua, A., Huberman, M. *et al.* (2001). A phase II trial of the epidermal growth factor receptor tyrosine kinase inhibitor OSI-774, following platinum-based chemotherapy, in patients with advanced, EGFR-expressing non-small cell lung cancer. *Proc Am Soc Clin Oncol* **20**, 310a, abstract 1235.

186. Finkler, N., Gordon, A., Crozier, M. *et al.* (2001). Phase 2 evaluation of OSI-774, a potent oral antagonist of the EGFR-TK in patients with advanced ovarian cancer. *Proc Am Soc Clin Oncol* **20**, 208a, abstract 831.

187. Senzer, N. N., Soulieres, D., Siu, L. *et al.* (2001). Phase 2 evaluation of OSI-774, a potent oral antagonist of the EGFR-TK in patients with advanced squamous cell carcinoma of the head and neck. *Proc Am Soc Clin Oncol* **20**, 2a, abstract 6.

188. Perez-Soler, R. (2004). The role of erlotinib (Tarceva, OSI 774) in the treatment of non-small cell lung cancer. *Clinical Cancer Research.* **10**, 4238s-4240s.

189. Prados, M., Chang, S., Burton, E. *et al.* (2003). Phase I study of OSI-774 alone or with temozolomide in patients with malignant glioma. *Proc ASCO,* Abstract 394.

190. Raizer, J. J., Abrey L.E., Wen, P., Cloughesy, T., Robins, I. A., Fine, H. A., *et al.* (2004). A phase II trial of erlotinib (OSI-774) in patients with recurrent malignant gliomas not on EIACDs. *Proc ASCO* 107.

191. Vogelbaum, M. A., Peereboom, D., Stevens, G., Barnett, G., and Brewer, C. (2004). Phase II trial of the EGFR tyrosine kinase inhibitor erlotinib for single agent therapy of recurrent glioblastoma multiforme: Interim results. *Proc ASCO* 121.

192. Yung, A., Vredenburgh, J., Cloughesy, T., Klencke, B. J., Mischel, P. S., Bigner, D. D., *et al.* (2004). Erlotinib HCL for glioblastoma multiforme in first relapse, a phase II trial. *Proc ASCO* 120.

193. Lydon, N. B., Mett, H., Mueller, M. *et al.* (1998). A potent protein-tyrosine kinase inhibitor which selectively blocks proliferation of epidermal growth factor receptor-expressing tumor cells in vitro and in vivo. *Int J Cancer* **76**, 154–163.

194. Burris, H. A., III. (2004). Dual kinase inhibition in the treatment of breast cancer: Initial experience with the EGFR/ErbB-2 inhibitor lapitinib. *The Oncologist* **9** (suppl 3): 10–15.

195. Rusnak, D. W., Affleck, K., Cockerill, S. G. *et al.* (2001). The characterization of novel, dual ErbB-2/EGFR, tyrosine kinase inhibitors: potential therapy for cancer. *Cancer Res* **61**, 7196–7203.

196. Erlichman, C., Boerner, S. A., Hallgren, C. G. *et al.* (2001). The HER tyrosine kinase inhibitor CI1033 enhances cytotoxicity of 7-ethyl-10-hydroxy-camptothecin and topotecan by inhibiting breast cancer resistance protein-mediated drug efflux. *Cancer Res* **61**, 739–748.

197. Garrison, M., Tolcher, A., and McCreery, H. (2001). A phase I and pharmacokinetic study of CI-1033, a pan-ErbB tyrosine kinase inhibitor, given orally on days 1, 8, 15, every 28 days to patients with solid tumors. *Proc Am Soc Clin Oncol* **20**, 72a, abstract 283.

198. Shin, D., Nemunaitis, J., and Zinner, R. (2001). A phase I clinical and biomarker study of CI-1033, a novel pan-ErbB tyrosine kinase inhibitor in patients with solid tumors. *Proc Am Soc Clin Oncol* **20**, 82a, abstract 324.

199. Greenberger, L. M., Discafani, C., Wang, Y-F. *et al.* (2000). EKB-569: a new irreversible inhibitor of EGFR tyrosine kinase for the treatment of cancer. *Clin Cancer Res* **6**, 4544s.

200. Torrance, C. J., Jackson, P. E., Montgomery, E. *et al.* Combinatorial chemoprevention of intestinal neoplasia. *Nat Med* **6**, 1024–1028.

201. Tsai, J. C., Goldman, C. K., and Gillespie, G. Y. (1995). Vascular endothelial growth factor in human glioma cell lines: induced secretion by EGF, PDGF-BB, and bFGF. *J Neurosurg* **82**, 864–873.

202. Maity, A., Pore, N., Lee, J., Solomon, D., and O'Rourke, D. M. (2000). Epidermal growth factor transcriptionally up-regulated vascular endothelial growth factor expression in human glioblastoma cells via a pathway involving phosphatidylinositol 3'-kinase and distinct from that induced by hypoxia. *Cancer Res* **60**, 5879–5886.

203. Clark, K., Smith, K., Gullick, W. J., and Harris, A. J. (2001). Mutant epidermal growth factor receptor enhances induction of vascular endothelial growth factor by hypoxia and insulin-like growth factor-1 by a PI3 kinase dependent pathway. *Br J Cancer* **84**, 1322–1329.

204. Viloria-Petit, A., Crombet, T., Jothy, S. *et al.* (2001). Acquired resistance to the antitumor effect of epidermal growth factor receptor-blocking antibodies in vivo: a role for altered tumor angiogenesis. *Cancer Res* **61**, 5090–5101.

205. Huang, S.-M., Li, E., Armstrong, E. A., and Harari, P. M. (2002). Modulation of radiation response and tumor-induced angiogenesis after epidermal growth factor receptor inhibition by ZD1839 (Iressa). *Cancer Res* **62**, 4300–4306.

206. Yancopoulos, G. D., Davis, S., Gale, N. W., Rudge, J. S., Wiegand, S. J., and Holash, J. (2000). Vascular-specific growth factors and blood vessel formation. *Nature* **407**, 242–248.

207. Jensen, R.L. (1998). Growth factor-mediated angiogenesis in the malignant progression of glial tumors: a review. *Surg Neurol* **49**, 189–195.

208. Hirata, A., Ogawa, S.-I., Kometani, T. *et al.* (2002). ZD1839 (Iressa) induces antiangiogenic effects through inhibition of epidermal growth factor receptor tyrosine kinase. *Cancer Res* **62**, 2554–2560.

209. Kondapaka, S. B., Fridman, R., and Reddy, K. B. Epidermal growth factor and amphiregulin up-regulate matrix metallo-proteinase-9 (MMP-9) in human breast cancer cells. *Int J Cancer* **17**, 722–726.

210. Lal, A., Glazer, C. A., Martinson, H. M. *et al.* (2002). Mutant epidermal growth factor receptor up-regulates molecular effectors of tumor invasion. *Cancer Res* **62**, 3335–3339.

211. Ciardiello, F., Bianco, R., Damiano, V. *et al.* (1999). Antitumor activity of sequential treatment with topotecan and anti-epidermal growth factor receptor monoclonal antibody C225. *Clin Cancer Res* **5**, 909–916.

212. Bruns, C. J., Harbison, M. T., Davis, D. W. *et al.* (2000). Epidermal growth factor receptor blockade with C225 plus gemcitabine results in regression of human pancreatic carcinoma growing orthotopically in nude mice by antiangiogenic mechanisms. *Clin Cancer Res* **6**, 1936–1948.

213. Inoue, K., Slaton, J. W., Perrotte, P. *et al.* (2000). Paclitaxel enhances the effects of the anti-epidermal growth factor receptor monoclonal antibody ImClone C225 in mice with metastatic human bladder transitional cell carcinoma. *Clin Cancer Res* **6**, 4874–4884.

214. Sirotnak, F.M. (2003). Studies with ZD1839 in preclinical models. *Sem Oncol* **30**, Suppl 1: 12–20.

215. Bonner, J. A., Raisch, K. P., Trummell, H. Q. *et al.* (2000). Enhanced apoptosis with combination C225/radiation treatment serves as the impetus for clinical investigation in head and neck cancers. *J Clin Oncol* **18**, 47S-53S.

216. Milas, L., Mason, K., and Hunter, N., (2000). In vivo enhancement of tumor radioresponse by C225 antiepidermal growth factor receptor antibody. *Clin Canc Res* **6**, 323–325.

217. Robert, F., Ezekiel, M. P., Spencer, S. A. *et al.* (2001). Phase I study of anti-epidermal growth factor receptor antibody cetuximab in combination with radiation therapy in patients with advanced head and neck cancer. *J Clin Oncol* **19**, 3234–3243.

218. O'Rourke, D. M., Kao, G. D., Singh, N. *et al.* (1998). Conversion of a radioresistant phenotype to a more sensitive one by disabling erbB receptor signaling in human cancer cells. *Proc Natl Acad Sci USA* **95**, (18):10842–7.

219. Bianco, C., Tortora, G., Bianco, R. *et al.* (2002). Enhancement of antitumor activity of ionizing radiation by combined treatment with the selective epidermal growth factor receptor-tyrosine kinase inhibitor ZD1839 (Iressa). *Clin Cancer Res* **8**, 3250–3258.

220. Gee, J. M., and Nicholson, R. I. (2003). Expanding the therapeutic repertoire of epidermal growth factor receptor blockade: Radiosensitization. *Breast Cancer Res* **5**, 126–129.

221. Solomon, B., Hagekyriakou, J., Trivett, M. K., Stacker, S. A., McArthur, G. A., and Cullinane, C. (2003). EGFR blockade with ZD1839 (Iressa) potentiates the antitumor effects of single and multiple fractions of ionizing irradiation in human A431 squamous cell carcinoma. *Int J Radiat Oncol Biol Phys* **55**, 713–723.

222. Thiesing, J. T., Ohno-Jones, S., Kolibaba, K. S., and Druker, B. J. (2000). Efficacy of STI571, an abl tyrosine kinase inhibitor, in conjunction with other antileukemic agents against bcr-abl-positive cells. *Blood* **96**, 3195–3199.

223. Kano, Y., Akutsu, M., Tsunoda, S. *et al.* (2001). In vitro cytotoxic effects of a tyrosine kinase inhibitor STI571 in combination with commonly used antileukemia agents. *Blood* **97**, 1999–2007.

224. Topaly, J., Zeller, W. J., and Fruehauf, S. (2001). Synergistic activity of the new ABL-specific tyrosine kinase inhibitor STI571 and chemotherapeutic drugs on the BCR-ABL-positive chronic myelogenous leukemia cells. *Leukemia* **15**, 342–347.

CHAPTER

12

Ras Signaling Pathways and Farnesyltransferase Inhibitors

Joydeep Mukherjee and Abhijit Guha

ABSTRACT: Activating mutations of p21-Ras is the most common oncogene in human cancers. Secondary activation, without oncogenic mutations, of this major mitogenic signaling pathway is also prevalent in breast and malignant astrocytomas. Hence, development of p21-Ras inhibitors, such as farnesyltransferase inhibitors (FTI), as potential anti-cancer agents is a very promising but yet unproven strategy. FTIs inhibit a key post-translational modification required for normal or oncogenic p21-Ras to attach to the inner cell membrane, become activated and transmit mitogenic signals to the nucleus by activating several downstream effector pathways. However, this obligatory role of farnesylation inhibited by FTIs, is restricted to only the Ha-Ras isoform, while the other human p21-Ras isoforms can shift to other post-translational pathways when farnesylation is inhibited. Second, in addition to p21-Ras a large number of additional normal cellular proteins, also undergoes farnesylation. However, the concern of FTI-mediated toxicity was not demonstrated in pre-clinical studies, leading to the use of these agents in clinical trials. Majority of these clinical trials have been undertaken in human cancers where oncogenic p21-Ras mutations are common, such as in lung, colon, and pancreas. Minimal efficacy has been demonstrated with FTIs as a single agent, as with many other biological anti-cancer therapies. Pre-clinical *in vitro* and *in vivo* results in human GBMs, where p21-Ras activation is secondarily increased, are promising. Early clinical trials are being undertaken in GBMs, though as single agents, their chance of success remains doubtful. Combinations of FTIs with non-specific and other targeted biological therapies, are likely to have a better chance of success.

INTRODUCTION

Aberrant function of genes controlling cell proliferation, differentiation, and apoptosis underlie neoplastic transformation. Broadly, these include aberrant activation of gain-of-function oncogenes and inactivation of loss-of-function tumor suppressor genes. Understanding the nature of these genetic alterations, their interactions, and their functional relevance, is the rational basis for evolving "biologically targeted therapies" against biochemical processes that are essential to the malignant phenotype of cancer cells. Amongst these targets is p21-Ras, which has attracted attention due to its central role in many normal intracellular signaling pathways and due to the prevalence of primary activating p21-Ras mutations in approximately 30 per cent of all human cancers. In addition, secondary activation of p21-Ras-regulated pathways has also been implicated to contribute to the growth of human breast cancers and malignant gliomas.

p21-Ras is a member of the small guanine nucleotide-binding proteins (G-proteins), which play a pivotal role in signal transduction, since its activity can be modulated by a variety of cytokines and growth factors. Mutated or normally activated p21-Ras undergoes various post-translational modifications, critical towards its attachment and activation in the inner plasma membrane. Of these, the first and most important modification is the addition of a farnesyl isoprenoid moiety to the C-terminal region of p21-Ras, requiring the enzyme protein farnesyltransferase (FTase). Amongst the four prevalent human p21-Ras isoforms, Ha-Ras activation is fully dependent on FTase, while the other isoforms, mutations of which are more common in human cancers, can also undergo geranylation by geranyl-geranyl-transferase (GGTase) if FTase activity is inhibited.

FTase inhibitors (FTIs) are currently of clinical interest in human cancers, with pre-clinical and phase-1 clinical results demonstrating acceptable toxicities, despite farnesylation of many normal human proteins other than p21-Ras. Initial focus has been in common human cancers which harbor activating p21-Ras mutations, such as lung, colon, and pancreas. However, interest also exists in the potential use of FTIs in tumors where secondary activation of p21-Ras has been demonstrated such as human malignant gliomas, with encouraging pre-clinical data. However, FTIs have not been fully evaluated in clinical trials of patients with malignant gliomas, perhaps as a result of the disappointing initial results in the more common systemic cancers. These tempered results likely represent the ability of the commonly mutated p21-Ras isoforms to shuttle their processing to geranylation and lack of clear understanding of what is actually being targeted by FTIs. Nevertheless, targeting p21-Ras in gliomas still remains to be fully explored, with its likely role if any, to be in combinatorial therapy with other conventional and biological therapies.

UNDERSTANDING p21-RAS STRUCTURE AND FUNCTION

p21-Ras Proto-oncogene and p21-Ras Protein

After ligand–receptor interaction, receptor activation leads to a large variety of biochemical events in which small G-proteins such as p21-Ras are very important. Small G-proteins are a superfamily of regulatory GTP hydrolases that cycle between two conformational states, induced by the binding of either guanosine diphosphate (GDP: inactive) or guanosine triphosphate (GTP: active) [1,2]. Three p21-Ras proto-oncogenes have been identified: the Ha-Ras gene (homologous to the Harvey murine sarcoma virus oncogenes), the K-Ras gene (homologous to Kristen murine sarcoma virus oncogenes), and the N-Ras gene (which was first isolated from a neuroblastoma cell line and does not have a retroviral homologue) [3–7]. The p21-Ras oncogenes encode four 21-Kd proteins between 188 and 189 amino acids long, called Ha-Ras, N-Ras, K-Ras4A, and K-Ras4B, the latter two resulting from two alternatively spliced

K-Ras gene products. There is high sequence homology between these four p21-Ras proteins, with the first 86 amino acids being identical, the next 78 having 79 per cent homology, and the remaining 25 amino acids being highly variable.

Post-translational Modification of p21-Ras

p21-Ras proteins are produced as cytoplasmic precursor pro-Ras proteins and require several post-translational modifications to become fully biologically active (see Fig. 12.1). Pro-Ras is sequentially modified by:

1. prenylation of the cysteine residue,
2. proteolytic cleavage of the AAX peptide by proteases,
3. carboxymethylation of the new C-terminal carboxylate by carboxy-methyl transferase, and
4. palmitoylation [8–10].

Three different enzymes catalyze prenylation of protein intermediates: (A) farnesyltransferase (FTase), (B) geranyl-geranyltransferase type-I (GGTase-I), and (C) geranyl-geranyltransferase type-II (GGTase-II) [8–10]. FTase transfers a farnesyl diphosphate (FDP), while GGTase-I transfers a geranyl-geranyl diphosphate (GGDP) to the cysteine residue of the CAAX motif. GGTase-II transfers geranyl-geranyl groups from GGPPs to both cysteine residues of CC or CXC motifs.

Farnesyltransferase (FTase)

FTase is a heterodimeric enzyme, which catalyzes farnesylation, resulting in prenylation of the cysteine in the C-terminal CAAX motif of p21-Ras. [11–13]. Farnesylation involves attachment of a 15-carbon farnesyl moiety in a thioether covalent linkage to the cysteine. FTase is composed of 48Kd-α and 45Kd-β subunits [14–16], with Zn^{+2} and Mg^{+2} required for function. The terminal residue of the C-terminal CAAX motifs in proteins directs differential affinity to FTase [17,18]. For example, if X is methionine (e.g., K-Ras4B) then there is a 10–30X increased affinity to FTase, compared to proteins where X is either serine or glycine (e.g., Ha-Ras). From the crystalline structure, FTase contains two clefts, which may represent the FDP and CAAX binding sites [19]. A Zn^{+2} atom lies in the junction between the two clefts, with both FDP and CAAX probably binding with the β-subunit of FTase, whereas the α-subunit may stabilize the β-subunit and catalyze transfer of the farnesyl isoprenoid moiety [20].

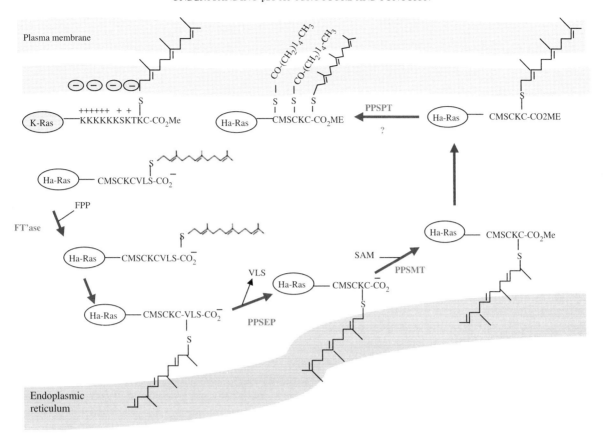

FIGURE 12.1 Post-translational modification of p21-Ras p21-Ras is synthesized as a pro-peptide, which undergoes a series of post-translational modifications, resulting in attachment to the inner surface of the plasma membrane, where it can be cycled from inactive GDP- to active GTP-bound state. First, FTase or GGT'ase transfers a farnesyl or a geranyl-geranyl group from FPP or GGPP to the thiol group of the cysteine residue in the CAAX motif of p21-Ras. The C-terminal tri-peptide is then removed by a CAAX-specific endoprotease (PPSEP) in the endoplasmic reticulum. A PPSMTase attaches the methyl group from S-adenosylmethionine (SAM) to the C-terminal cysteine. Finally, a prenyl-protein-specific palmitoyltranferase (PPSPTase) attaches palmitoyl groups to cysteines near the farnesylated C-terminus. See Plate 12.1 in Color Plate Section.

The α-subunit undergoes phosphorylation, which regulates the activity of FTase [21]. Of importance towards potential complications of inhibiting FTase, there are over 200 normal proteins, which also undergo farnesylation in addition to p21-Ras. These include nuclear laminins A and B, the γ-subunit of the retinal trimeric G-Protein Transducin, Rhodopsin Kinase, and a peroxisomal protein termed PxF [9,10].

After farnesylation, there is endoproteolytic removal of the three carboxy-terminal amino acids (AAX) by a cellular thiol-dependent zinc metallo-peptidase [22,23] and subsequent methylation of the carboxyl group of the prenylated cysteine residue by an uncharacterized methyltransferase. All the p21-Ras proteins, except K-Ras, undergo further palmitoylation at 1–2 cysteines near the farnesylated carboxy-terminus [8–10,24–26]. In contrast to farnesylation and proteolysis, palmitoylation and methylation of p21-Ras are thought to be reversible and may have a regulatory role. Although each of the post-translational modifications increases the hydrophobicity of p21-Ras and contributes to its association with the plasma membrane, the initial farnesylation step alone is sufficient to promote substantial membrane association and confer transforming potentials [26–28].

Geranyl-Geranyl Transferase (GGTase)

There are two other protein prenyltransferases, GGTase-I and GGTase-II, which can prenylate the C-terminal ends of proteins like FTase, by attaching either one or two 20-carbon geranyl-geranyl isoprenoid lipid moieties [9,26]. GGTase-I and FTase share an identical α-subunit, but have distinct β-subunits. Like FTase, GGTase-I is a Zn^{+2} metalloenzyme that requires Mg^{+2} for catalysis, however, in contrast to FTase which binds FDP with approximately 30X greater affinity than GGDP, GGTase-I binds GGDP approximately 300X greater than FDP. GGTase-I consists of three subunits, of which the catalytic unit is the β-heterodimer, however, GGTase-II requires an additional protein known as the Rab escort protein to facilitate interaction of its substrate proteins with the enzyme. GGTase-I preferentially prenylates proteins in which the X residue is leucine, but their substrate specificities are not absolute.

The cumulative results of studies indicate, that although membrane localization is critical for p21-Ras function modification, with which specific isoprenoid lipid moiety (FDP or GGDP) this is achieved, is not a critical issue. The potential for cross-prenylation of FTase and GGTase-I implies that GGTase-I might be able to restore the function of

p21-Ras and other proteins after FTase inhibition, which probably has implications for the development of resistance to FTIs.

Activation and Inactivation of p21-Ras

p21-Ras functions as a membrane-associated biologic switch that relays signals from ligand-stimulated receptors to cytoplasmic effector signaling cascades (see Fig. 12.2). Ligand binding to the extracellular domain of receptor tyrosine kinases (RTKs) causes receptor dimerization, stimulation of protein tyrosine kinase activity, and autophosphorylation [28–30]. Specific phosphorylated tyrosines on the cytoplasmic domains of receptors act as docking sites to recruit adapter proteins, such as Shc and Grb2 to the inner cell membrane, through interactions of specific

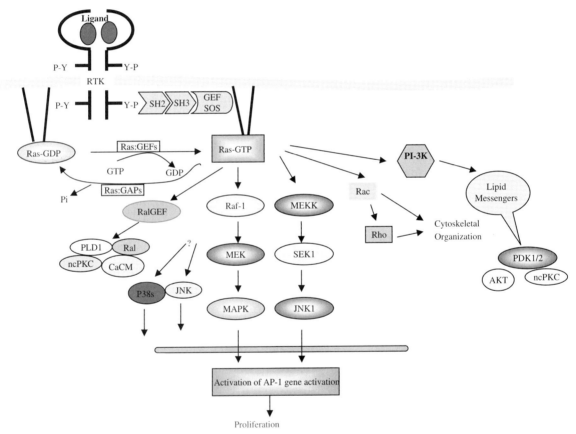

FIGURE 12.2   Activation and inactivation of p21-Ras. In response to growth factor receptor activation and tyrosine (Y) phosphorylation, Grb2 which is complexed to p21-Ras guanine nucleotide exchange factor Sos, is recruited in proximity to mature p21-Ras attached to the inner cell membrane. Sos catalyzes exchange of GTP for GDP to form activated p21-Ras-GTP. Activated p21-Ras-GTP interacts and activates several effector pathways, including Raf-MAPK, PI-3K, Rac and Rho, to transmit signals to other cytoplasmic regions and the nucleus. Although there is a small intrinsic GTPase activity to revert activated p21-Ras-GTP to inactive p21-Ras-GDP, this hydrolysis is catalyzed by a family of proteins termed Ras:GAPs (GTPase Activating Proteins), among which is neurofibromin, the gene product lost in the cancer pre-disposition syndrome, Neurofibromatosis-1. See Plate 12.2 in Color Plate Section.

protein modules, such as SH2 and PTB domains in these proteins. Grb2, is constitutively complexed with the enzyme Sos, a p21-Ras guanine-exchange-factor or Ras-GEF, by its SH3 domain interacting with the proline-rich region of Sos. Bringing this Grb2–Sos complex in proximity to the post-translationally modified p21-Ras in the inner cell membrane, allows activation of p21-Ras by exchanging GDP for GTP [28–30]. In addition to RTKs, a variety of non-receptor tyrosine kinases, such as Lyn and Fes and the Janus kinase JAK2, also can bind to docking adapter proteins such as Grb2 resulting in the activation of p21-Ras to its GTP-bound state.

Inactivation of p21-Ras requires hydrolysis of GTP and binding of GDP, a reaction known as GTPase. Small G-proteins such as p21-Ras have slow intrinsic GTPase function, requiring catalysis by another family of proteins known as Ras-GAPs. The two best known mammalian Ras-GAPs are p120GAP and neurofibromin, the latter being the protein product which is lost in the cancer pre-disposition syndrome Neurofibromatosis-1 (NF-1).

p21-Ras Signal Transduction Pathways

In its GTP-bound state, p21-Ras can activate several downstream effector pathways, of which activation of the serine-threonine kinase Raf-1 and subsequently, the MAPK (mitogen-activated protein kinase) pathway to the nucleus, has been most thoroughly elucidated (see Fig. 12.2) [31–50]. Other effector pathways include those mediated by phosphatidylinositol-3′-kinase (PI3K) and small G-proteins Rac and Rho. Two mechanisms have been proposed to explain how p21-Ras-GTP activates its downstream effectors:

(A) Recruitment model: Activated p21-Ras bound to GTP and anchored to the inner cell membrane binds to the cytoplasmic effectors, facilitating their activation by other membrane proteins;
(B) Allosteric model: Activated p21-Ras binding to the effectors induces their conformational change, resulting in their activation.

Both mechanisms are likely involved, depending on the particular effector protein being activated.

p21-Ras-Raf-MAPK Signaling Cascade

Several lines of evidence indicate that Raf serine-threonine kinases (A-Raf, B-Raf, and C-Raf) are critical effectors of p21-Ras function: (A) Dominant-negative mutants of Raf can impair p21-Ras-transforming

activity [31–50]; (B) Constitutively activated forms of Raf exhibit transforming activity comparable to that of p21-Ras, and are themselves sufficient to transform some of the murine cells. The activation of Raf occurs after it is recruited to the cell membrane, however, the precise mechanism(s) by which p21-Ras activates Raf is unknown. Once activated, Raf phosphorylates two MAP kinase kinases (MAPKK), MEK1 and MEK2, which are also serine-threonine kinases that phosphorylate the mitogen-activated protein kinases (MAPK), p44MAPK and p42MAPK (also known as extracellular signal-regulated kinases, ERK-1 and ERK-2). The MAPK pathways are well-conserved signaling pathways, with at least six mammalian MAPK cascades, of which those mediated by ERKs are best characterized. ERK-1 and ERK-2 are proline-directed protein kinases that phosphorylate Ser–Thr–Pro motifs of several cytoplasmic and nuclear proteins, such as phospholipase A2 (PLA2), ribosomal protein S6 kinases (RSKs), and most importantly the transcription factor Elk-1 to alter gene transcription. Activation of the ERK pathway is responsive for many mitogenic signals from protein tyrosine kinases, hematopoietic- and G-protein-coupled growth factor receptors.

Activation of PI-3K Signaling Cascade

p21-Ras-GTP binds to and activates the catalytic p110 subunit of PI-3K, a member of a family of lipid kinases that phosphorylate phosphoinositides [51–55], which activate the phosphoinositide-dependent kinases PDK-1 and PDK-2, leading to the activation of Akt and non-conventional isoforms of protein kinase C (ncPKC). PI-3K has been implicated in many distinct cellular functions, including mitogenic signaling, inhibition of apoptosis, regulation of cell size, intracellular vesicle trafficking and secretion, and regulation of actin and integrin functions. In addition to p21-Ras-dependent activation, PI3K can also directly be activated by several growth receptors, by binding the SH2 domains of the p85 regulatory subunit of PI3K directly to specific phosphorylated tyrosine residues on activated receptors.

Activation of Rho and Rac

Activated p21-Ras regulates several small G-proteins, including Rho, Rac, and CDC42, involved in regulating cell morphology [56–60]. Like p21-Ras, these proteins cycle between GDP- and GTP-bound states, activated by GEFs similar to Sos, and inactivated by GAPs similar to Ras-GAPs. Rho modulates the actin cytoskeleton, involved in regulating membrane ruffling, formation of stress fibers, focal adhesions, and filopodia, integral to maintaining cell shape, invasion, and transformation. CDC42 and Rac are two GTPases involved in regulation of the actin cytoskeleton, the SAPK/JNK pathway and the p38 pathway.

ROLE OF p21-RAS MUTATIONS IN DIFFERENT MALIGNANCIES

General

Mutated p21-Ras oncogenes were first identified by their ability to transform NIH 3T3 cells [61–64], with mutations subsequently identified in a variety of human tumors. The majority of p21-Ras mutations in human tumors occur in K-Ras, some in N-Ras and rarely Ha-Ras, with the type of p21-Ras mutations correlating with tumor type. Mutations occur in amino acids #12, 13, or 61, making p21-Ras-GTP unable to bind and be inactivated by Ras-GAPs. Mutations at amino acid #12, with the normal glycine mutated to serine, cysteine, arginine, asparagines, alanine, or valine [65–68] are most common. Although all the activating mutations result in aberrant increased levels of p21-Ras activity, there likely are varying biochemical consequences between them, which are not well understood, but are of likely therapeutic importance. In addition to transformation, aberrant activation of p21-Ras also leads to both radiation and chemo-resistance towards the therapy of human tumors, including gliomas. For example, inhibition of p21-Ras activity by dominant-negative mutants leads to enhanced sensitivity of GBM (glioblastoma multiforme) cells to cisplatin *in vitro* and *in vivo* [69]. Second, antagonistic effects were observed between sequential administration of radiation and chemotherapy in human GBM cells, with increased expression of EGFR and thereby activation of p21-Ras and the PI3-K/Akt signaling pathways [70].

Brain Tumor

Oncogenic mutations in p21-Ras are not prevalent in human primary glial or non-glial tumors. However, several direct and indirect pieces from pre-clinical and clinical experiments demonstrate that activation of p21-Ras and its downstream effector pathways are an important growth promoting signaling pathway in human malignant astrocytomas: First, research

in our lab initially described increased levels of activated p21-Ras in human malignant astrocytoma cell lines and operative specimens, compared to normal astrocytes or brain [71,72]. Sequencing of endogenous p21-Ras confirmed that primary mutations were not present in the malignant astrocytoma specimens, suggesting that p21-Ras was activated secondary to aberrantly activated receptors in these tumors; Second, our group and subsequently others demonstrated that the activated levels of p21-Ras are of functional importance for proliferation and angiogenesis-mediated growth of human astrocytomas both *in vitro* and *in vivo* [73–75]. These and other pre-clinical experimental findings are the basis for consideration of inhibiting the p21-Ras pathway with reagents such as FTIs in the management of human gliomas; Third, we, along with others have also demonstrated that increased p21-Ras activity is present in gliomas associated with NF-1, a result of decreased Ras:GAP activity [75–76]; Fourth, we have developed a germline transgenic mouse model with GFAP-regulated expression of oncogenic activated ^{12}V-Ha-Ras [77–78]. These mice develop high-grade astrocytomas that have pathological similarities to human tumors after 3–4 months of age. A non-germline glioma model also demonstrates that the combination of activated K-Ras and activated-Akt signaling in the context of neuroglial nestin positive precursors, or in Ink4a-Arf-deficient differentiated GFAP-positive astrocytes, developed gliomas [79].

FARNESYLTRANSFERASE INHIBITORS (FTIs)

Farnesylated Proteins other than p21-Ras

The importance of farnesylation towards the post-translational maturation of p21-Ras has been discussed above. However, there are ≈200 other farnesylated human proteins, including several nuclear laminins and ocular lens proteins [80]. The best characterized non-p21-Ras-farnesylated protein is RhoB [81,82], which can be processed by both farnesylation and geranyl-geranylation like the various p21-Ras sub-types discussed above. Similar to K-Ras and N-Ras, in the presence of FTIs, RhoB becomes exclusively geranyl-geranylated to exhibit growth inhibitory activity. Another farnesylated protein is the p21-Ras-related protein family designated Rig (Ras-related inhibitor of cell growth). Rig actually inhibits growth, survival, and transformation, with frequent loss noted in primary human astrocytomas [83]. Examination of these and other farnesylated proteins are of interest in the development and understanding of the biological effects and toxicities of FTIs in neoplastic and normal tissues. Although significant toxicities were initially predicted with FTIs, since protein farnesylation is so common, this was not found in pre-clinical studies, thereby leading to the human clinical trials discussed below. The explanation for this relative lack of toxicities is not entirely clear, but probably involves redundancy of farnesylation-related post-translational modification by geranylation and other pathways (e.g., p21-Ras). Furthermore, the anti-tumor effects of FTIs are often not predictable or even predicated on the presence of increased p21-Ras activity, suggesting that they actually target additional tumor promoting proteins, which also are farnesylated, such as enzymatic modification of RhoB. Understanding farnesylation in p21-Ras and other proteins, therefore, is an ongoing and important research area, as FTIs become commonly used in the clinic.

FTIs in Pre-clinical and Clinical Studies

Mechanisms

Pre-clinical studies utilizing several strategies to inhibit p21-Ras signaling, such as with antisense RNA, homologous recombination, dominant-negatives, and FTIs, have demonstrated anti-tumor effects *in vitro* and *in vivo* for several human tumors [84–91]. Numerous sub-types of FTIs have been synthesized, which can be grouped into three classes:

1. Analogs of the FPP substrate such as (-hydroxy-farnesyl) phosphonic acid, -ketophosphonic and -hydroxyphosphonic acid derivatives
2. CAAX peptide analogs and non-peptidic tricyclic FTIs
3. Bi-substrate inhibitors, such as phosphonic acid and hydroxamine acid analogs.

In addition to chemically synthesized compounds, several naturally occurring FTIs have been evaluated [92]. *In vivo* efficacy of FTIs has been demonstrated in a variety of xenograft, as well as transgenic mouse tumor models, with limited toxicity to the mice [93,94]. In the latter, mammary and salivary carcinomas which develop in the MMTV-^{12}VHa-Ras mice were quite sensitive to FTIs [95,96]. In contrast, tumors from N-Ras or K-Ras transgenics were more resistant, in keeping with these two p21-Ras isoforms being able to switch to geranylation when farnesylation is inhibited. In addition to these predicted effects

on tumors that develop in p21-Ras-driven transgenics, FTIs have also shown efficacy in mammary tumors from MMTV-TGF-α and MMTV-neu transgenics [97]. This point is of interest, as similar to malignant gliomas, activating p21-Ras mutations are not prevalent in breast carcinomas, though levels of activated p21-Ras-GTP is increased, likely from aberrant activation of receptors, similar to GBMs.

These pre-clinical studies demonstrate that the anti-tumorigenic effects of FTIs involve several mechanisms in addition to inhibiting proliferation, such as being pro-apoptotic, anti-angiogenic, and also sensitizing tumors to radiation and chemotherapy. Anti-proliferative effects of FTIs involve progressive dose-dependent cytoplasmic accumulation of unprocessed p21-Ras and inability to activate downstream effector mitogenic pathways, as discussed above. The main proliferative signals likely involve silencing the Raf/MAPKK/MAPK signaling pathway. Pro-apoptotic effects of FTIs may be mediated by up-regulating Bax and Bcl-Xs, and activation of caspases [98–100]. Ant-angiogenic effects of FTIs have been directly demonstrated [72–75] by inhibiting expression of pro-angiogenic molecules such as VEGF (Vascular Endothelial Growth Factor), providing an explanation why greater *in vivo* effects are sometimes observed than those predicted by the *in vitro* results. The mechanism(s) of how FTIs modulate this anti-angiogenic effect is not clear, and likely involves both p21-Ras-dependent and -independent signaling pathways. Synergy of FTIs with established anti-cancer treatments, such as radiation and chemotherapeutic treatment is of therapeutic interest. Anti-microtubules, such as taxols, synergize with FTIs to induce metaphase block [101,102], with FTIs increasing radiosensitivity of p21-Ras mutant harboring human tumor cells [103].

Although FTI mediated inhibition of p21-Ras farnesylation and p21-Ras induced transformation is an established mechanism, other targets for FTIs anti-tumor effects are likely present [104–107]. Evidence for this comes from observations that there is little correlation between sensitivity to FTIs and presence of activating p21-Ras mutations in a variety of human cancers [93,94]. As discussed earlier, another key farnesylated protein that may mediate the anti-tumorigenic effects of FTIs is RhoB [104–107]. In contrast to Ras proteins, RhoB exists normally *in vivo* in a farnesylated (RhoB-FF) and a geranyl-geranylated version (RhoB-GG). RhoB-GG is essential for the degradation of p27KIP1 and facilitates the progression of cells from G1- to S- phase. Treatment with FTIs results in a loss of RhoB-FF and a gain of RhoB-GG, which reverts transformation by activating the cell-cycle kinase inhibitor p21WAF1 and hence inducing G1 arrest.

Brain Tumor

Initial work on proof that inhibiting p21-Ras activity in human malignant astrocytoma cell lines, which do not harbor activating p21-Ras mutations, results in anti-proliferative effects, was undertaken by our lab with dominant-negative mutants. These results were subsequently supported by experiments with FTIs by our lab as well as by other labs, in collaboration with pharmaceutical companies [71–76]. Of interest, GBM lines with aberrant over expression of mutant EGFRs (epidermal growth factor receptor), prevalent in a large proportion of human malignant gliomas, were more sensitive to FTIs. These in vitro studies were extended to several subcutaneous xenograft human malignant glioma explant models, with the majority but not all demonstrating efficacy, by our group. The anti-tumor effects were not just a result of decreased proliferation, but also reduced vascularization with increased apoptosis of tumor-associated vasculature. Since not all the human glioma cell lines or explant xenografts were sensitive to FTIs, isoform specific p21-Ras-GTP was measured, to determine if this could be predictive. A close correlationship was found between both in vitro and in vivo sensitivity and those tumors with high levels of activated Ha-Ras versus those with high N-Ras or K-Ras levels. Recently, use of a FTI (R115777) in U87 xenograft model showed reduction in tumor hypoxia and MMP-2 (matrix-mettaloproteases) expression [108]. Inhibition of MMPs also is suggestive of a potential anti-angiogenic and anti-invasion role of FTIs in gliomas.

Based on the above pre-clinical evidence, early phase I/II clinical trials with FTIs have been/are being undertaken in brain tumors. A Pediatric Brain Tumor Consortium (PBTC) sponsored dose-escalation study with the Schering-Plough FTI (Ionafarnib-SCH66336- Sarasar), was undertaken in 39 children [109]. The FTI, R115777 (Tipifarnib/Zarnestra) was evaluated in phase I/II trials of recurrent malignant glioma. The phase I findings demonstrated a difference in the MTD depending upon the use of enzyme-inducing anticonvulsants. Phase II evaluations in the same patient populations utilizing the defined MTDs showed encouraging activity with 23 per cent of patients progression-free at 6 months [110]. A second phase I study with R115777 combined with Temodar, a combination which will likely be the standard of future trials in GBMs, showed minimal toxicity, leading to a planned phase II trial

[111]. Additional trials with other FTIs using such combinations with radiation and Temodar are being planned by RTOG (Radiation Therapy Oncology Group) and other such consortiums.

In addition to specific FTIs, other non-specific inhibitors of farnesylation, such as Lovastatin and Manumycin, have also demonstrated efficacy against human glioma cells in vitro, by the induction of apoptosis [112,113]. Diallyl disulfide (DADS), a non-specific FTI, demonstrated inhibition of Ha-Ras activity in a rat glioma model, though there was no increased survival of the animals [114].

Resistance Against FTIs in Tumor Cells

Development of tumor resistance to FTIs is an important issue, though the relative frequency and mechanism(s) remain somewhat unclear. First, it is well known that K-Ras and N-Ras transformants are more resistant to FTIs, than Ha-Ras-transformed cells, in part due to 10–50X higher affinity of FTase for these Ras isoforms [115–118]. Second, although normally all Ras isoforms are only farnesylated, K-Ras and N-Ras (but not Ha-Ras) can become geranyl-geranylated by GGTase-I in a dose-dependent manner, when farnesylation is inhibited by FTIs. Surprisingly, FTI treatment can still significantly inhibit the growth of tumors containing mutated K-Ras4B. This is in support of the theory that non-p21-Ras farnesylated proteins, such as Rho, which are involved in transformation, are targeted by FTIs. Additional mechanisms of FTI resistance include development of resistant mutations of the FTase subunits. For example, the Y361L FTase mutant is resistant to FTIs while maintaining FTase activity toward substrates [119,120].

CONCLUSION AND FUTURE DIRECTIONS

FTIs represent a growing arena of biologically targeted anti-tumor strategies, based on sound pre-clinical data, but for the most part awaiting clinical validation. For FTIs, promise at the clinical level in oncogenic p21-Ras mutant harboring tumors, such as lung, colon, and pancreatic cancers, has not yielded the predicted favorable outcomes from pre-clinical results. Evaluation of nervous system tumors, without oncogenic p21-Ras mutations but where pre-clinical data is promising, such as malignant gliomas and neurofibromas, is still awaiting formal clinical evaluation. Specific issues to the brain such as blood–brain barrier penetration and neural toxicity are of importance and requires study at the pre-clinical level, as for any therapeutic agent intended for gliomas. The variety of FTIs, with different pharmacokinetic profiles and limited neural toxicity even after long-term use in pre-clinical models, gives hope that these obstacles are surmountable. However, FTIs will likely not suffice as a single agent in gliomas and other human cancers, requiring evaluation in a combinatorial manner, which although, a desired stated goal is extremely hard to evaluate in the clinical setting. Furthermore, there is heterogeneity in which Ras isoform is differentially activated, as per our results on isotype sensitivity in gliomas [74]. Hence, "biological tailoring" of tumors sensitive to FTIs, is desirable to select the sub-population of tumors that may benefit from FTIs. The resistance of several Ras isoforms to FTIs is another area of ongoing drug and pre-clinical development. Although the emphasis has been on designing specific FTIs to avoid possible toxicities from inhibition of GGTase, the issue of resistance has led towards the search of GGTase inhibitors with acceptable toxicity profiles, with some showing promise at pre-clinical and early clinical levels. In summary, development and testing of biological therapeutics aimed at key signaling pathways such as p21-Ras is promising. FTIs likely will not be the magic bullet in our research to help patients afflicted with gliomas, however, they may form part of the answer. Certainly strategies aimed at modulating key signaling pathways, such as p21-Ras and others in gliomas, is worthy of pursuit and careful evaluation in the laboratory and clinic.

References

1. Sprang, S. R. (1997). G protein mechanisms: insights from structural analysis. *Annu Rev Biochem* **66**, 639–678.
2. Rebollo, A., and Martinez, C. A. (1999). Ras proteins: recent advances and new functions. *Blood* **94**, 2971–2980.
3. Boguski, M. S., and McCormick, F. (1993). Proteins regulating Ras and its relatives. *Nature* **366**, 643–654.
4. Lowy, D. R., and Willumsen, B. M. (1993). Function and regulation of Ras. *Ann Rev Biochem* **62**, 851–891.
5. Ellis, R. W., Defeo, D., Shih, T. Y. *et al.* (1981). The p21 src genes of Harvey and Kirsten sarcoma viruses originate from divergent members of a family of normal vertebrate genes. *Nature* **292**, 506–511.
6. Shimizu, K., Goldfarb, M., Suard, Y. *et al.* (1983). Three human transforming genes are related to the viral oncogenes. *Proc Natl Acad Sci U S A* **80**, 2112–2116.
7. Pells, S., Divjak, M., Romanowski, P. *et al.* (1997). Developmentally-regulated expression of murine K-ras isoforms. *Oncogene* **15**, 1781–1786.
8. Glomset, J. A., and Farnsworth, C. C. (1994). Role of protein modification reactions in programming interactions between Ras-related GTPases and cell membranes. *Annu Rev Cell Biol* **10**, 181–205.
9. Zhang, F. L., and Casey, P. J. (1996). Protein prenylation: molecular mechanisms and functional consequences. *Annu Rev Biochem* **65**, 241–269.
10. Gelb, M. H. (1997). Protein prenylation, et cetera: signal transduction in two dimensions. *Science* **275**, 1750–1751.
11. Pellicena, P., Scholten, J. D., Zimmerman, K., Creswell, M., Huang, C. C., and Miller, W. T. (1996). Involvement of the alpha subunit of farnesyl-protein transferase in substrate recognition. *Biochemistry* **35**, 13494–13500.
12. Trueblood, C. E., Boyartchuk, V. L, and Rine, J. (1997). Substrate specificity determinants in the farnesyltransferase -subunit. *Proc Natl Acad Sci U S A* **94**, 10774–10779.
13. Park, H.-W., Boduluri, S. R., Moomaw, J. F., Casey, P. J., and Beese, L. S. (1997). Crystal structure of protein farnesyltransferase at 2.25 Angstrom resolution. *Science* **275**, 1800–1804.
14. Gibbs, J. B., and Oliff, A. (1997). The potential of farnesyltransferase inhibitors as cancer chemotherapeutics. *Ann Rev Pharmacol Toxicol* **37**, 143–166.
15. Leonard, D. M. (1997). Ras farnesyltransferase: A new therapeutic target. *J Med Chem* **40**, 2971–2990.
16. Zhang, F. L., and Casey, P. J. (1996). Protein prenylation: Molecular mechanisms and functional consequences. *Ann Rev Biochem* **65**, 241–269.
17. Reiss, Y., Goldstein, J. L., Seabra, M. C., Casey, P. J., and Brown, M. S. (1990). Inhibitors of purified p21ras farnesyl: Protein transferase by cys-AAX tetrapeptides. *Cell* **62**, 81–88.
18. Reiss, Y., Stradley, S. J., Gierasch, L. M., Brown, M.S., and Goldstein, J. L. (1991). Sequence requirement for peptide recognition by rat brain p21ras protein farnesyltransferase. *Proc Natl Acad Sci U S A* **88**, 732–736.
19. Park, H. W., Boduluri, S. R., Moomaw, J. F., Casey, P. J., and Beese, L. S. (1997). Crystal structure of protein farnesyltransferase at 2.25 angstrom resolution. *Science* **275**, 1800–1804.
20. Andres, D. A., Goldstein, J. L., Ho, Y. K., and Brown, M. S. (1993). Mutational analysis of alpha-subunit of protein farnesyltransferase: Evidence for a catalytic role. *J Biol Chem* **268**, 1383–1390.
21. Kumar, A., Beresini, M. H., Dhawan, P., and Mehta, K. D. (1996). Alpha-subunit of farnesyltransferase is phosphorylated in vivo: Effect of protein phosphatase-1 on enzymatic activity. *Biochem Biophys Res Commun* **222**, 445–452, 1996.
22. Akopyan, T. N., Couedel, Y., Orlowski, M., Fournie-Zaluski, M. C., and Roques, B. P. (1994). Proteolytic processing of farnesylated peptides: assay and partial purification from pig brain membranes of an endopeptidase which has the characteristics of E.C. 3.4.24.15. *Biochem Biophys Res Commun* **198**, 787–794.
23. Boyartchuk, V. L., Ashby, M. N., and Rine, J. (1997). Modulation of Ras and a-factor function by carboxyl-terminal proteolysis. *Science* **275**, 1796–1800.
24. Hancock, J., Magee, A., Childs, J., and Marshall, C. (1989). All ras proteins are polyisoprenylated but only some are palmitoylated. *Cell* **57**, 1167 1177.
25. Ross, E. M. (1995). Palmitoylation in G-protein signaling pathways. *Curr Biol* **5**, 107–109.
26. Gibbs, J. B. (1993). Lipid modifications of proteins in the Ras superfamily. *In GTPases in Biology* (L. Birnbaumer and B. Dickey, Eds.), pp. 335–344. Springer-Verlag, New York, NY.
27. Casey, P. J. (1989). p21 Ras is modified by a farnesyl isoprenoid. *Proc Natl Acad Sci U S A* **86**, 8323–8327.
28. McCormick, F. (1993). How receptors turn Ras on. *Nature* **363**, 15–17.
29. Schlessinger, J. (1993). How receptor tyrosine kinases activate Ras. *Trends Biol Sci* **18**, 273–275.

30. Marshall, C. J. (1995). Specificity of receptor tyrosine kinase signaling: transient versus sustained extracellular signal-regulated kinase activation. *Cell* **80**, 179–185.

31. Marshall, C. J. (1996). Raf gets it together. *Nature* **383**, 127–128.

32. Pawson, T., and Saxton, T. M. (1999). Signaling networks do all roads lead to the same genes? *Cell* **97**, 675–678.

33. Bos, J. L. (1998). All in the family? New insights and questions regarding interconnectivity of Ras, Rap1 and Ral. *EMBO J* **17**, 6776–6782.

34. Wittinghofer, A. (1998). Signal transduction via Ras. *Biol Chem* **379**, 933–937.

35. Van Aelst, L., White, M., and Wigler, M. H. (1994). Ras partners. *Cold Spring Harbor Symp Quant Biol* **59**, 181–186.

36. Marshall, C. J. (1996) Ras effectors. *Curr Opin Cell Biol* **8**, 197–204.

37. Katz, M. E., and McCormick, F. (1997). Signal transduction from multiple Ras effectors. *Curr Opin Genet Dev* **7**, 75–79.

38. Kolch, W., Heidecker, G., Lloyd, P., and Rapp, U. R. (1991). Raf-1 protein kinase is required for growth of induced NIH/3T3 cells. *Nature* **349**, 426–428.

39. Cowley, S., Paterson, H., Kemp, P., and Marshall, C. J. (1994). Activation of MAP kinase kinase is necessary and sufficient for PC12 differentiation and for transformation of NIH 3T3 cells. *Cell* **77**, 841–852.

40. Mansour, S. J., Matten, W. T., Hermann, A. S. *et al.* (1994). Transformation of mammalian cells by constitutively active MAP kinase kinase. *Science* **265**, 966–970.

41. Stokoe, D., Macdonald, S. G., Cadwallader, K., Symons, M., and Hancock, J. F. (1994). Activation of Raf as a result of recruitment to the plasma membrane. *Science* **264**, 1463–1467.

42. Leevers, S. J., Paterson, H. F., and Marshall, C. J. (1994). Requirement for Ras in Raf activation is overcome by targeting Raf to the plasma membrane. *Nature* **369**, 411–414.

43. Pritchard, C., and McMahon, M. (1997). Raf revealed in life-or-death decisions. *Nature Genet* **16**, 214–215.

44. Tamada, M., Hu, C. D., Kariya, K. *et al.* (1997). Membrane recruitment of Raf-1 is not the only function of Ras in Raf-1 activation. *Oncogene* **15**, 2959–2964.

45. Morrison, D. K., and Cutler, R. E. Jr. (1997). The complexity of Raf-1 regulation. *Curr Opin Cell Biol* **9**, 174–179.

46. Marshall, C. J. (1996). Cell signalling: Raf gets it together. *Nature* **383**, 127–128.

47. Fanger, G. R., Gerwins, P., Widmann, C., Jarpe, M. B., and Johnson, G. L. (1997). MEKKs, GCKs, MLKs, PAKs, TAKs, and Tpls: upstream regulators of the c-Jun amino-terminal kinases? *Curr Opin Genet Dev* **7**, 67–74.

48. Robinson, M. J., and Cobb, M. H. (1997). Mitogen-activated protein kinase pathways. *Curr Opin Cell Biol* **9**, 180–186.

49. Elion, E. A. (1998). Routing MAP kinase cascades. *Science* **281**, 1625–1626.

50. Jaaro, H., Rubinfeld, H., Hanoch, T., and Seger, R. (1997). Nuclear translocation of mitogen-activated protein kinase kinase (MEK1) in response to mitogenic stimulation. *Proc Natl Acad Sci U S A* **94**, 3742–3747.

51. Carpenter, C. L., and Cantley, L. C. (1996). Phosphoinositide kinases. *Curr Opin Cell Biol* **8**, 153–158.

52. Yan, J., Roy, S., Apolloni, A., Lane, A., and Hancock, J. F. (1998). Ras isoforms vary in their ability to activate Raf-1 and phosphoinositide 3-kinase. *J Biol Chem* **273**, 24052–24056.

53. Kodaki, T., Woscholski, R., Hallberg, B., Rodriguez-Viciana, P., Downward, J., and Parker, P. J. (1994). The activation of phosphatidylinositol 3-kinase by Ras. *Curr Biol* **4**, 798–806.

54. Rodriguez-Viciana, P., Warne, P. H, Vanhaesebroeck, B., Waterfield, M. D., and Downward, J. (1996). Activation of phosphoinositide 3-kinase by interaction with Ras and by point mutation. *EMBO J* **15**, 2442–2451.

55. Rodriguez-Viciana, P., Warne, P. H., Khwaja, A. *et al.* (1997). Role of phosphoinositide 3-OH kinase in cell transformation and control of the actin cytoskeleton by Ras. *Cell* **89**, 457–467.

56. Keely, P. J., Westwick, J. K., Whitehead, I. P., Der, C. J., and Parise, L. V. (1997). Cdc42 and Rac1 induce integrin-mediated cell motility and invasiveness through PI(3)K. *Nature* **390**, 632–636.

57. Prendergast, G. C., and Gibbs, J. B. (1993). Pathways of Ras function: Connections to the actin cytoskeleton. *Adv Cancer Res* **62**, 19–64.

58. Joneson, T., White, M. A., Wigler, M. H. *et al.* (1996). Stimulation of membrane ruffling and MAP kinase activation by distinct effectors of Ras. *Science* **271**, 810–812.

59. Ridley, A. J., Paterson, H. F., Johnston, C. L., Diekmann, D., and Hall, A. (1992). The small GTP-binding protein Rac regulates growth-factor induced membrane ruffling. *Cell* **70**, 401–410.

60. Ridley, A. J., and Hall, A. (1992). The small GTP-binding protein Rho regulates the assembly of focal adhesions and actin stress fibers in response to growth factors. *Cell* **70**, 389–399.

61. Bos, J. L. (1989). ras oncogenes in human cancer: A review. *Cancer Res* **49**, 4682–4689.

62. Shih, C., and Weinberg, R. A. (1982). Isolation of transforming sequence from a human bladder carcinoma cell line. *Cell* **29**, 161–169.

63. Krontiris, T., and Cooper, G. M. (1981). Transforming activity in human tumor DNAs. *Proc Natl Acad Sci U S A* **78**, 1181–1184.

64. Perucho, M., Goldfarb, M., Shimizu, K., Lama, C., Fogh, J., and Wigler, M. (1981). Human tumor-derived cell lines contain common and different transforming genes. *Cell* **27**, 467–476.

65. Lowy, D. R., and Willumsen, B. M. (1993). Function and regulation of Ras. *Ann Rev Biochem* **62**, 851–891.

66. Bollag, G., and McCormick, F. (1991). Regulators and effectors of ras proteins. *Ann Rev Cell Biol* **7**, 601–632.

67. Barbacid, M. (1987). Ras genes. *Ann Rev Biochem* **56**, 779–827.

68. Boguski, M. S., and McCormick, F. (1993). Proteins regulating Ras and its relatives. *Nature* **366**, 643–654.

69. Messina, S., Leonetti, C., De Gregorio, G. *et al.* (2004). Ras inhibition amplifies cisplatin sensitivity of human glioblastoma. *Biochem Biophys Res Commun* **320**(2), 493–500.

70. Feldkamp, M. M., Lala, P., Lau, N., Roncari, L., and Guha, A. (1999). Expression of activated epidermal growth factor receptors, Ras-guanosine triphosphate, and mitogen-activated protein kinase in human glioblastoma multiforme specimens. *Neurosurgery* **45**(6), 1442–1453.

71. Guha, A., Feldkamp, M. M., Lau, N., Boss, G., and Pawson, A. (1997). Proliferation of human malignant astrocytomas is dependent on Ras activation. *Oncogene* **15**(23), 2755–2765.

72. Feldkamp, M. M, Lau, N., and Guha, A. (1999). Growth inhibition of astrocytoma cells by farnesyl transferase inhibitors is mediated by a combination of anti-proliferative, pro-apoptotic and anti-angiogenic effects. *Oncogene* **18**(52), 7514–7526.

73. Feldkamp, M. M. Lau, N., Rak, J., Kerbel, R. S., and Guha, A. (1999). Normoxic and hypoxic regulation of vascular endothelial growth factor (VEGF) by astrocytoma cells is mediated by Ras. *Int J Cancer* **81**(1), 118–124.

74. Feldkamp, M. M., Lau, N., Roncari, L., and Guha, A. (2001). Isotype-specific Ras.GTP-levels predict the efficacy of farnesyl transferase inhibitors against human astrocytomas regardless of Ras mutational status. *Cancer Res* **61**(11), 4425–4431.

75. Woods, S. A., Marmor, E., Feldkamp, M. *et al.* (2002). Aberrant G protein signaling in nervous system tumors. *J Neurosurg* **97**(3), 627–642.

76. Lau, N., Feldkamp, M., Roncari, L. *et al.* (2000). Loss of NF1 causes activation of Ras/MAPK and PI-3K/Akt signaling in an NF1-associated astrocytoma. *J Neuropathology Expt Neurology* **59**(9), 759–767.

77. Ding, H., Roncari, L., Shannon, P. *et al.* (2001). Astrocyte-specific expression of activated p21-ras results in malignant astrocytoma formation in a transgenic mouse model of human gliomas. *Cancer Res* **1, 61**(9), 3826–3836.

78. Ding, H., Shannon, P., Lau, N. *et al.* (2003). Astrocyte-specific expression of activated p21 ras and constitutively active form of EGFR results in malignant oligodendroglioma formation. *Cancer Res* **63**, 1106–1113.

79. Uhrbom, L., Dai, C., Celestino, J. C., Rosenblum, M. K., Fuller, G. N., and Holland, E. C. (2002). Ink4a-Arf loss cooperates with KRas activation in astrocytes and neural progenitors to generate glioblastomas of various morphologies depending on activated Akt. *Cancer Res* **62**(19), 5551–5558.

80. Tamanoi, F., Kato-Stankiewicz, J., Jiang, C., Machado, I., and Thapar, N. (2001). Farnesylated proteins and cell cycle progression. *J Cell Biochem Suppl* **37**, 64–70.

81. Du, W., and Prendergast, G. C. (1999). Geranylgeranylated RhoB mediates suppression of human tumor cell growth by farnesyltransferase inhibitors. *Cancer Res* **59**, 5492–5496.

82. Prendergast, G. C., and Rane, N. (2001). Farnesyltransferase inhibitors: mechanism and applications. *Expert Opin Investig Drugs* **10**, 2105–2116.

83. Ellis, C. A., Vos M. D., Howell, H., Vallecorsa, T., Fults, D. W., and Clark, G. J. (2002). Rig is a novel Ras-related protein and potential neural tumor suppressor. *Proc Natl Acad Sci U S A* **99**(15), 9876–9881

84. Gibbs, J. B. (1991). Ras C-terminal processing enzymes: new drug targets? *Cell* **65**, 1–4.

85. Tamanoi, F. (1993). Inhibitors of Ras farnesyltransferases. *Trends Biochem Sci* **18**, 349–353.

86. Gibbs, J. B., and Oliff, A. (1994). Pharmaceutical research in molecular oncology. *Cell* **79**, 193–198.

87. Gibbs, J. B., Oliff, A., and Kohl, N. E. (1994). Farnesyltransferase inhibitors: Ras research yields a potential cancer therapeutic. *Cell* **77**, 175–178.

88. Lowy, D. R., and Willumsen, B. M. (1995). Rational cancer therapy. *Nat Med* **1**, 747–748.

89. Gibbs, J. B., and Oliff, A. (1997). The potential of farnesyl-transferase inhibitors as cancer chemotherapeutics. *Annu Rev Pharmacol Toxicol* **37**, 143–166.

90. Cox, A. D., and Der, C. J. (1997). Farnesyltransferase inhibitors and cancer treatment: Targeting simply Ras? *Biochim Biophys Acta* **1333**, F51–F71.

91. Omer, C. A., Anthony, N. J., Buser-Doepner, C. A. *et al.* (1997). Farnesyl: Proteintransferase inhibitors as agents to inhibit tumor growth. *Biofactors* **6**, 359–366.

92. Lee, S., Park, S., Oh, J. W. *et al.* (1998). Natural inhibitors for protein prenyltransferase. *Planta Med.* **64**, 303–308.

93. Nagasu, T., Yoshimatsu, K., Rowell, C., Lewis, M. D., and Garcia, A. M. (1995). Inhibition of human tumor xenograft growth by treatment with the farnesyl transferase inhibitor B956. *Cancer Res.* **55**, 5310–5314.

94. Sepp-Lorenzino, L., Ma, Z., Rands, E. *et al.* (1995). A peptidomimetic inhibitor of farnesyl: protein transferase blocks the anchorage-dependent and -independent growth of human tumor cell lines. *Cancer Res* **55**, 5302–5309.

95. Kohl, N. E., Omer, C. A., Conner, M. W. *et al.* (1995). Inhibition of farnesyltransferase induces regression of mammary and salivary carcinomas in ras transgenic mice. *Nat Med* **1**, 792–797.

96. Mangues, R., Corral, T., Kohl, N. E. *et al.* (1998). Antitumor effect of a farnesyl protein transferase inhibitor in mammary and lymphoid tumors overexpressing N-Ras in transgenic mice. *Cancer Res* **58**, 1253–1259.

97. Norgaard, P., Law, B., Joseph, H. *et al.* (1999). Treatment with farnesyl-protein transferase inhibitor induces regression of mammary tumors in transforming growth factor (TGF) alpha and TGF alpha/neu transgenic mice by inhibition of mitogenic activity and induction of apoptosis. *Clin Cancer Res* **5**, 35–42.

98. Lebowitz, P. F., Sakamuro, D., and Prendergast, G. C. (1997). Farnesyl transferase inhibitors induce apoptosis of Ras-transformed cells denied substratum attachment. *Cancer Res* **57**, 708–713.

99. Hung, W. C., and Chuang, L. Y. (1998). Involvement of caspase family proteases in FPT inhibitor II-induced apoptosis in human ovarian cancer cells. *Int J Cancer* **12**, 1339–1342.

100. Hung, W. C., and Chuang, L. Y. (1998). The farnesyltransferase inhibitor, FPT inhibitor III upregulates Bax and Bcl-xs expression and induces apoptosis in human ovarian cancer cells. *Int J Oncol* **12**, 137–140.

101. Ashar, H. R., James, L., Gray, K. *et al.* (2000). Farnesyl transferase inhibitors block the farnesylation of CENP-E and CENP-F and alter the association of CENP-E with the microtubules. *J Biol Chem* **275**, 30451–30457.

102. Suzuki, N., Del Villar, K., and Tamanoi, F. (1998). Farnesyl-transferase inhibitors induce dramatic morphological changes of KNRK cells that are blocked by microtubule interfering agents. *Proc Natl Acad Sci U S A* **95**, 10499–10504

103. Bernhard, E. J., McKenna, W. G., Hamilton, A. D. *et al.* (1998). Inhibiting Ras prenylation increases the radiosensitivity of human tumor cell lines with activating mutations of ras oncogenes. *Cancer Res* **58**, 1754–1761.

104. Lebowitz, P. F., Davide, J. P., and Prendergast, G. C. (1995). Evidence that farnesyltransferase inhibitors suppress Ras transformation by interfering with Rho activity. *Mol Cell Biol* **15**, 6613–6622.

105. Lebowitz, P. F., and Prendergast, G. C. (1998). Non-Ras targets of farnesyltransferase inhibitors: focus on Rho. *Oncogene* **17**, 1439–1445.

106. Du, W., Lebowitz, P. F., and Prendergast, G. C. (1999). Cell growth inhibition by farnesyltransferase inhibitors is mediated by gain of geranylgeranylated RhoB. *Mol Cell Biol.* **19**, 1831–1840.

107. Adamson, P., Marshall, C. J., Hall, A. *et al.* (1992). Post-translational modifications of p21rho proteins. *J Biol Chem* **267**, 20033–20038.

108. Delmas, C., End, D., Rochaix, P., Favre, G., Toulas, C., and Cohen-Jonathan, E. (2003). The farnesyltransferase inhibitor R115777 reduces hypoxia and matrix metalloproteinase 2 expression in human glioma xenograft. *Clin Cancer Res* **1, 9**, 6062–6068.

109. Kieran, M. W., Packer, R., Boyett, J., Sugrue, M., and Kun, L. (2004). Phase I trial of the oral farnesyl protein transferase inhibitor Ionafarnib (SCH66336): A Pediatric Brain Tumor Consortium (PBTC) study. *J Clin Oncol* **22** (14S), 15–17.

110. Nabors, L. B. (2004). Targeted Molecular Therapy for Malignant Gliomas. *Curr Treat Options in Oncol* **5**, 519–526.

111. Gilbert, M. R., Hess, K., Gaupp, P. *et al.* (2004). A phase I study of Temozolomide (TMZ) and the farnesyltransferase inhibitor

(FTI), Tipifarnib (Zanestra, R115777) in recurrent glioblastoma: A dose and schedule intensive regimen. *Neuro-Oncol* **6**, 375.

112. Bouterfa, H. L., Sattelmeyer, V., Czub, S., Vordermark, D., Roosen, K., and Tonn, J. C. (2000). Inhibition of Ras farnesylation by lovastatin leads to downregulation of proliferation and migration in primary cultured human glioblastoma cells. *Anticancer Res* **20**(4), 2761–71.

113. Wang, W., and Macaulay, R. J. (1999). Apoptosis of medulloblastoma cells in vitro follows inhibition of farnesylation using manumycin A. *Int J Cancer* **82**(3), 430–443.

114. Perkins, E., Calvert, J., Lancon, J. A., Parent, A. D., and Zhang, J. (2003). Inhibition of H-ras as a treatment for experimental brain C6 glioma. *Brain Res Mol Brain Res* **111**(1–2), 42–51.

115. Whyte, D. B., Kirschmeier, P., Hockenberry, T. N. *et al.* (1997). K-Ras and N-Ras are geranylgeranylated in cells treated with farnesyl protein transferase inhibitors. *J Biol Chem* **272**, 14459–14464.

116. Zhang, F. L., Kirschmeier, P., Carr, D. *et al.* (1997). Characterization of Ha-ras, N-ras, Ki-Ras4A, and Ki-Ras4B as in vitro substrates for farnesyl protein transferase and geranylgeranyl protein transferase type I. *J Biol Chem* **272**, 10232–10239.

117. Rowell, C. A., Kowalczyk, J. J., Lewis, M. D., and Garcia, A. M. (1997). Direct demonstration of geranylgeranylation and farnesylation of Ki-Ras in vivo. *J Biol Chem* **272**, 14093–14097.

118. James, G. L., Goldstein, J. L., and Brown, M. S. (1995). Polylysine and CVIM sequences of K-RasB dictate specificity of prenylation and confer resistance to benzodiazepine peptidomimetic in vitro. *J Biol Chem* **270**, 6221–6226.

119. Prendergast, G. C., Davide, J. P., Lebowitz, P. F., Wechsler-Reya, R., and Kohl, N. E. (1996). Resistance of a variant ras-transformed cell line to phenotypic reversion by farnesyl transferase inhibitors. *Cancer Res* **56**, 2626–2632.

120. Del Villar, K., Urano, J., Guo, L., and Tamanoi, F. (1999). A mutant form of human protein farnesyltransferase exhibits increased resistance to farnesyltransferase inhibitors. *J Biol Chem* **274**, 27010–27017.

13

PI3-Kinase, PKB/Akt, mTOR, and Internal Signaling Pathways

Jean L. Nakamura and Daphne A. Haas-Kogan

ABSTRACT: Best known for its normal physiologic function in insulin-mediated signaling and glycolysis, PI3-kinase and its effectors PKB/Akt and mTOR are of particular interest to the oncologist due to their roles in tumorigenesis and response to therapy. Dysregulation of PI3-kinase signaling is implicated in mediating many features of malignancy, including aberrant growth, invasiveness, angiogenesis, and cell survival. The effective incidence of PI3-kinase dysregulation in malignant gliomas may be as high as 80 per cent, when one considers the number of aberrant signaling pathways converging through PI3-kinase and PKB/Akt. In addition to contributing to gliomagenesis, constitutively activated PKB/Akt is implicated in the development of resistance to both chemotherapy and radiotherapy. Although the mechanisms by which PI3-kinase, PKB/Akt, and mTOR modulate responses to therapy are unclear, the signaling pathway outlined by these effectors may prove relevant in current and developing molecular therapeutics. Signaling inhibitors currently in clinical trials target mTOR, farnesyl transferase, EGFR, and PDGFR. PI3-kinase is positioned either upstream or downstream of these targets, highlighting the importance of accounting for PI3-kinase activity in tumor biology. Emerging *in vitro* evidence indicates that multiple signaling aberrations in a single tumor may necessitate inhibition at multiple levels in signaling pathways. A major challenge remains to logically and intelligently coordinate which levels within signaling cascades to inhibit in the treatment of any given tumor.

Abbreviations: mTOR – mammalian target of rapamycin; PI3-kinase – phosphatidylinositol-3′ kinase; PDK-1 – 3-phosphoinositide-dependent protein kinase-1; PKB/Akt – protein kinase B; p70^{S6K} – p70 ribosomal S6 kinase; EGFR – epidermal growth factor receptor.

OVERVIEW OF THE BIOLOGY OF PI3-KINASE AND ITS EFFECTORS

PI3-kinase Function and Regulation

Phosphatidylinositol-3′ kinase (PI3-kinase) is best known for its integral function in a number of physiologic processes, including insulin and growth factor receptor signaling, cell proliferation, and motility. Many lines of evidence demonstrate that PI3-kinase roles in proliferation and motility contribute to oncogenesis and tumor resistance to therapy. PI3-kinase is a lipid kinase composed of a p85 regulatory subunit and a p110 catalytic subunit. Upon activation, typically from membrane-based inputs such as growth factor receptors, the catalytic subunit phosphorylates phosphoinositides at the 3′ position of the inositol ring to generate PI(3,4)P2 and PI(3,4,5)P3. These phosphorylated lipid products recruit PKB/Akt to the membrane and activate it, resulting in the phosphorylation of PKB at its Ser473 and Thr308 residues [1]. As a major intracellular signaling effector, PI3-kinase transduces membrane-based activation from several sources, including G proteins such as Ras and receptor tyrosine kinases such as epidermal growth factor receptor (EGFR) (Fig. 13.1).

PI3-kinase activity is opposed by PTEN (phosphatase and tensin homologue deleted on chromosome ten), a tumor suppressor located on chromosome

10q23. Also known as MMAC1 (mutated in multiple advanced cancers), PTEN is a dual function lipid and protein phosphatase composed of an N-terminal phosphatase domain, a C2 domain, and a C-terminal tail region containing a PSD-95/Dlg/ZO-1 (PDZ) homology domain. Its lipid phosphatase function negatively regulates PKB/Akt phosphorylation by dephosphorylating phosphatidylinositol (3,4,5) tris-phosphate (PtdIns(3,4,5)P3) and phosphatidyli-nositol (3,4) bisphosphate (PtdIns(3,4)P2), thereby restraining the PI3-kinase pathway. PTEN plays a critical role in the development of the central nervous system in mammals; early embryonic lethality results from homozygous *PTEN* deletions in transgenic mice [2].

An alternative lipid phosphatase regulating PKB/Akt is SH2-containing inositol-5-phosphatase (SHIP), an inositol 5′ phosphatase that hydrolyzes PI(3,4,5)P3 to PI(3,4)P2. SHIP-2 is widely expressed in non-hematopoietic cells, in contrast to SHIP-1, which is found primarily in hematopoeitic cells [3,4]. To date, SHIP has not been implicated in glioma-genesis and studies focusing on the relative roles of SHIP and PTEN as suppressors of the PI3-kinase pathway have identified PTEN as negatively regula-ting the PI-kinase/PKB/Akt cascade in leukemo-genesis [5].

Downstream Effectors

PKB/Akt

PKB/Akt is a serine/threonine kinase that exists as three isoforms (α, β, and γ). All three PKB/Akt isoforms are ubiquitously expressed in mammals, although levels of expression vary among tissues, with PKBγ displaying the most limited expression of all isoforms, expressed preferentially in lymphocytes. PtdIns(3,4)P2 and PtdIns(3,4,5)P3 produced by PI3-kinase bind to the pleckstrin homology domain of PKB/Akt, translocate PKB/Akt to the plasma membrane, and activate it [6]. PKB/Akt links PI3-kinase to effectors of cell-cycle regulation, meta-bolism, and invasion through the phosphorylation of many targets, some of which will be described more fully in this chapter. The cyclin-dependent kinase (cdk) inhibitor p27^{Kip1} [7], Bcl-2 family member Bad [8], caspase-9, FHKR, glycogen synthase kinase 3 (GSK3), TSC2, and RhoB are prominent targets directly phosphorylated by PKB/Akt.

With respect to metabolism, PKB/Akt transmits signals from IGF-1 to intracellular targets. In response to insulin stimulation, GSK3 is phosphorylated

and inactivated by PKB in a PI3-kinase-dependent process, resulting in GSK3 phosphorylating and inactivating glycogen synthase.

PKB/Akt effects on cell-cycle progression are best understood at the level of G1 arrest; PKB/Akt-mediated phosphorylation of p27^{Kip1} results in cyto-plasmic localization of p27^{Kip1} and loss of G1 arrest [7]. In addition to their documented effects on the G1 checkpoint, PI3-kinase and activated PKB/Akt can override the G2/M checkpoint [9,10] although the mechanisms underlying this response are to date unclear.

PKB/Akt shapes the anti-apoptotic response through its phosphorylation of Bad and caspase-9. Bad, a member of the Bcl-2 family, binds and anta-gonizes Bcl-2 and Bcl-X, inhibiting their anti-apoptotic potential [11]. Caspase-9, an apoptotic effector, is phosphorylated by PKB/Akt, resulting in the inhibi-tion of cytochrome C-induced cleavage [12].

Transcriptional regulation is a mechanism by which PKB/Akt promotes survival. In response to IGF-1 signaling, the Forkhead family of transcription factors (FH or FoxO) are phosphorylated by PKB/Akt, which results in nuclear exclusion and reduced transcriptional activity needed for promoting apoptosis [13–16].

The identification of new PKB/Akt-binding proteins adds to the apparent complexity of PKB/Akt roles, and identifies potential therapeutic targets. Among these is Ft1, a protein shown to interact with

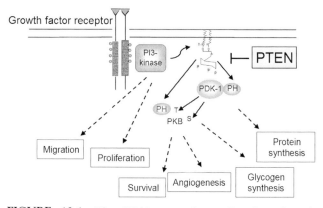

FIGURE 13.1 The PI3-kinase pathway. Signaling through growth factor receptors activates PI3-kinase, which produces PI(3,4,5)P$_3$, a signaling messenger that recruits PKB/Akt to the membrane and, through activation of PDK-1, activates PKB/Akt. PKB/Akt, in turn, phosphorylates a number of targets, resulting in the wide range of cellular processes affected. These phosphorylation cascades are restrained by the activity of the tumor suppressor PTEN, a lipid and protein phosphatase that opposes PI3-kinase activity by dephosphorylating PI(3,4,5)P3. See Plate 13.1 in Color Plate Section.

PKB/Akt, increasing its activation and increasing dexamethasone-induced apoptosis in T lymphocytes by increasing production of the pro-apoptotic Fas ligand [17]. Carboxyl-terminal modulator protein (CTMP) is another recently identified PKB/Akt-binding protein that interacts with PKB/Akt at the plasma membrane. In contrast to Ft1, CTMP reduces PKB/Akt phosphorylation at Thr308 and Ser473. CTMP expression reverts the phenotype of v-Akt-transformed cells, suggesting a possible role for CTMP in modulating PKB/Akt contributions to tumorigenesis [18].

mTOR

Mammalian target of rapamycin (mTOR) is a 290 kDa protein whose two effectors eIF4E and p70^{S6K} promote protein translation of capped messages and messages possessing 5′ terminal oligopyrimidine tracts (TOP) [19,20], although the role of p70^{S6K} in TOP mRNA translation has been disputed [21–23]. Activated mTOR directly phosphorylates p70^{S6K} and the translation factor inhibitor 4EBP1 that releases eIF4E, allowing its association with other translation initiation factors such as eIF4G to promote cap-dependent translation [24]. These effectors appear to mediate mTOR functions, as *in vitro* mTOR inhibition by rapamycin generally results in G1 arrest, with control of cell-cycle progression occurring through p70^{S6K} and the 4EBP1/eIF4E complex [25].

mTOR function is important in propagating growth signals and is capable of regulating intracellular ATP, independent of amino acid availability. Loss of mTOR function results in embryonic lethality [26–28]. mTOR is negatively regulated by TSC1/TSC2 through inactivation of Rheb (Ras homolog enriched in brain), a positive regulator of mTOR. TSC2 is a GTPase activating protein for Rheb, leading to phosphorylation of ribosomal S6 and 4EBP1 [29]. This function is rapamycin sensitive, and Rheb is proposed to regulate ribosomal S6- and 4EBP1-based responses to nutrients by stimulating the phosphorylation of mTOR. PKB/Akt regulates mTOR through phosphorylation of TSC2, resulting in the release of mTOR from inhibition by TSC2.

mTOR effectors eIF4E and S6 Kinase mediate mTOR's control of cell size and proliferation. The loss of the drosophila homolog of TOR, dTOR, results in smaller eyes and heads due to reduced cell size [30]. p70S6 Kinase loss in mice results in a smaller, but viable and fertile, phenotype [21].

MULTI-LEVEL DYSREGULATION OF PI3-KINASE: IMPLICATIONS FOR GLIOMAGENESIS

Constitutively activated PI3-kinase, PKB/Akt, and mTOR appear to contribute, within select backgrounds, to the development of high-grade gliomas. Dysregulation of PI3-kinase occurs through several of the most commonly observed genetic aberrations in malignant gliomas. These include loss of *PTEN* and amplification or overexpression of growth factor receptors, such as PDGFR and EGFR.

PTEN loss has been described in several solid tumors, notably glioblastoma multiforme and endometrial and prostate cancers, with 20–40 per cent of malignant gliomas harboring *PTEN* mutations [31]. However, when one considers the fact that PI3-kinase signaling is shared by several growth pathways, including EGFR, which alone is amplified or overexpressed in 40–90 per cent of malignant gliomas, the effective incidence of dysregulated PI3-kinase signaling is in fact substantially higher. Mutations in or deletions of *PTEN* have been shown to result in elevated, constitutive PKB/Akt activity in glioblastoma multiforme cell lines [32]. In glioma patients, PI3-kinase activity may prove to be of prognostic significance [33], as PTEN levels in human glioma specimens correlate with tumor grade and survival [34,35].

Germline mutations in *PTEN* result in three rare autosomal dominant inherited cancer syndromes: Cowden's disease, L'hermitte-Duclos, and Bannayan-Zonana. All three syndromes are characterized by the propensity to develop multiple solid tumors, most commonly hamartomas, although central nervous system dysplasias and tumors such as meningiomas also occur. Heterozygosity of *PTEN* in mice results in an increased incidence of tumor formation in a spectrum reminiscent of Cowden's disease [2], consistent with PTEN's role as a tumor suppressor.

Much attention has been paid to PTEN lipid phosphatase activity, as loss of this function abrogates PKB/Akt-mediated control of the cell cycle [36]. PTEN regulates G1 progression *via* inhibition of PKB/Akt [37] which increases expression of p27^{Kip1} [38]; Re-introduction of PTEN into human glioma cell lines results in PKB/Akt-mediated G1 cell-cycle arrest, astrocytic differentiation, and anoikis [39,40]. PTEN also plays a role in tumor angiogenesis, as Pore *et al.* documented PTEN and PI3-kinase-mediated transcriptional up-regulation of VEGF in human glioblastoma cell lines [41].

Few lipid-phosphatase-independent functions of PTEN have been described, but PTEN may regulate cell migration and invasion through its protein phosphatase activity and its interactions with focal adhesion kinase [42–44].

An alternative mechanism of lipid phosphatase regulation of PKB/Akt is SHIP, an inositol 5'-phosphatase that hydrolyzes PtdIns(3,4,5)P3 and PtdIns(3,4)P2. SHIP-2, which is widely expressed in non-hematopoietic cells [3,4], has been shown to result in PKB/Akt inactivation and cell-cycle arrest in glioblastoma multiforme through enhanced stability of the cell-cycle inhibitor p27^{Kip1} [45]. Although dysregulated SHIP activity has not been implicated in gliomagenesis, these data underscore the importance of PKB/Akt in mediating PI3-kinase-driven effects on cell-cycle progression.

PI3-kinase activity, although not directly transforming, appears in certain cell types to mediate transformation driven by prominent oncogenes such as Ras. Activated mutant Ras-driven transformation of NIH3T3 is abrogated by expression of dominant negative p85, the regulatory subunit of PI3-kinase [46]. To some extent, these effects are cell-type-specific, as isolated activation of PKB/Akt in immortalized human astrocytes fails to transform *in vitro* [47]. One link between Ras-driven transformation and PI3-kinase has been suggested to occur through RhoB, a GDP/GTP binding GTPase shown to inhibit proliferation and tumor growth [48].

Although PKB/Akt dysregulation alone is not sufficient to transform astrocytes, several models of gliomagenesis show that constitutively activated PKB/Akt transforms moderately malignant astrocytic histologies into tumors with features of glioblastoma multiforme *in vitro* [49,50]. Similarly, within *in vivo* mouse models of gliomagenesis, PKB/Akt in conjunction with activated Ras produces high-grade gliomas consistent with human glioblastoma multiforme [51].

In addition to growth factor receptor mediated activation, PKB/Akt may be activated as a consequence of interactions between gliomas and their microenvironment. Recently, secreted protein acidic, rich in cysteine (SPARC), an extracellular matrix protein expressed in malignant gliomas, has been shown to activate PKB/Akt *in vitro* [52], suggesting that PKB/Akt activation induced by glioma–microenvironmental interactions may support glioma invasiveness.

While the mechanisms by which PKB/Akt confers an *in vivo* growth advantage are not fully understood, activation of mTOR appears to be an important effector in defining tumorigenesis. Ras- and PKB/Akt-transformed glial progenitor cells demonstrate differential recruitment of mRNAs to polysomes, with preferential recruitment of proteins involved in growth and transcriptional regulation [53].

With respect to tumorigenesis, activated PKB/Akt may confer a proliferative advantage to tumors by permitting a means of averting cell-cycle arrest by cyclin-dependent kinases. A study of gliomas *in vitro* indicates that PTEN-induced G1 arrest is p27^{Kip1}-dependent [38]. In primary breast cancers, it has been proposed that one mechanism by which activated PKB/Akt opposes G1 arrest is by mediating cytoplasmic localization of p27^{Kip1} [54]. Immunohistochemical studies of primary breast tumor specimen have correlated reduced p27^{Kip1} levels with worse prognosis in breast cancer patients [55]. It was subsequently shown that p27^{Kip1} possesses a PKB/Akt consensus site within the nuclear localization signal, and constitutively activated PKB/Akt has been shown to result in cytoplasmic mislocalization of p27^{Kip1} [7].

Emerging data implicate mTOR as an important mediator of PI3-kinase-induced tumorigenesis. *In vivo* PKB/Akt-dependent expansion in prostatic epithelium is mTOR-dependent [25], and may occur through propagation of PKB/Akt signaling to downstream effectors such as HIF-1α. Expression of eIF4E in human mammary epithelial cells permits anchorage-independent growth [56]. Evidence from Riesterer *et al.* indicates that mTOR plays an important role in maintaining PKB/Akt protein stability in response to VEGF stimulation in endothelial cells [57].

mTOR may also promote tumorigenesis through the activation of its downstream effector eIF4E, as eIF4E can transform rat fibroblasts *in vitro* [58], and promote tumor formation *in vivo* [59]. eIF4E increases expression of growth promoting proteins such as cyclin D1 [60], by increasing cyclin D1 associated with polysomes [61], and promoting mRNA nucleo-cytoplasmic transport [62,63].

THERAPEUTIC IMPLICATIONS

The marked chemotherapy and radiation resistance of gliomas has focused attention on the means by which signaling aberrations underlying gliomagenesis contribute to resistance to cell death. In this regard, the PI3-kinase pathway is recognized for supporting cellular proliferation and survival, thereby promoting both malignant transformation and resistance to therapy.

Several lines of investigation identify PI3-kinase as a regulator of cellular responses to ionizing radiation. Biochemical inhibitors of PI3-kinase, LY294002, and wortmannin, enhance the anti-neoplastic effects of radiation [64–66], and recent data indicate that PKB/Akt mediates LY294002-mediated radiosensitization [82]. Ionizing radiation has been shown to activate PKB/Akt and p70^{S6K} in epithelial tumors *in vitro*, however this has not been shown to be the case in malignant glioma cell lines (Nakamura and Haas-Kogan, unpublished data).

In addition to direct effects on tumor cells, PKB/Akt mediates responses of vascular endothelium to ionizing radiation [67]. PI3-kinase inhibition may provide a means of targeting elements in the tumor microenvironment such as vascular endothelium [68]. Thus, PI3-kinase inhibition may have anti-neoplastic effects through direct tumor killing as well as through disruption of the supportive microenvironmental components such as vascularization [69].

The importance of PKB/Akt function in survival after cytotoxic therapy is highlighted by the evidence that mutant receptor tyrosine kinases such as EGFRvIII exert cytoprotective effects through PI3-kinase [70] (Fig. 13.2). However, the effectiveness of EGFR tyrosine kinase inhibitors may be compromised by compensatory signaling through PTEN loss and/or PKB/Akt activation [71,72]. Combinations of signaling inhibitors, based on growing insights into glioma genetic pathogenesis, may help circumvent these escape mechanisms. Although such combinations may appear redundant, the ability of dysregulated PI3-kinase signaling to rescue cells from cytotoxic therapies may be one molecular explanation for the modest clinical efficacy of signaling inhibitors [73].

Redundant activation of PI3-kinase from alternate growth factor receptors may also provide a means of rescuing malignant gliomas from EGFR inhibition. For example, insulin-like growth factor receptor-I (IGFR-I) has been shown to maintain PI3-kinase signaling in the face of AG1478, the EGFR inhibitor and in fact co-suppression of IGFR-I and EGFR produced greater cytotoxicity compared to individual receptor inhibition [74]. These data further confirm that therapeutic signaling inhibition will be delivered as polychemotherapy, and increasing recognition of the molecular basis of resistance will improve our ability to intelligently incorporate signaling inhibitors into multimodality therapy. The same complexity and cross talk of signaling cascades contributing to drug resistance may also protect tumors from alternate mechanisms of cell death. EGFR inhibitors AG1478 and PD153035 can protect malignant gliomas from hypoxia-induced cell death [75].

Signaling through mTOR represents another potential target for molecular-based therapeutics. Phase I studies of CCI-779, an ester of rapamycin that inhibits mTOR, are ongoing for malignant gliomas [76]. mTOR and its downstream effector eIF4E may be particularly important targets because recent data indicate that they mediate resistance to therapy and treatment with rapamycin sensitizes lymphomas to chemotherapy [77]. Furthermore, tumors with activated PKB/Akt display particular sensitivity to mTOR inhibition [78]. In glioma, mTOR inhibition may have multiple inhibitory effects. Rapamycin radiosensitizes U87 *in vivo* [79], a finding that suggests improved efficacy when combining mTOR inhibition with radiotherapy. Recent data also indicate that mTOR inhibition alone using the rapamycin derivative RAD001 reduces glioma invasiveness and VEGF secretion [80].

Recent data suggest the possibility of targeting eIF4E as a strategy. Kentsis, *et al.* described the suppression of eIF4E-mediated transformation by the guanosine ribonucleoside analog ribavirin [81].

Increasing recognition of the redundancy of inputs into the PI3-kinase pathway and the cross talk between signaling pathways emphasizes the need to rationally coordinate signaling inhibitors for maximal efficacy. *In vitro* and *in vivo* data defining compensatory mechanisms after single-agent signaling inhibition should be used to offer more appropriate combinatorial therapy.

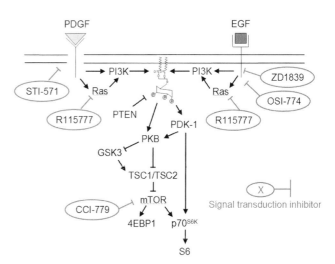

FIGURE 13.2 Schema of some signaling inhibitors currently in clinical trials. A variety of mechanisms are available to suppress aberrant signaling through growth factor receptors and the PI3-kinase pathway. STI-571, ZD1839, and OSI-774 inhibit tyrosine kinase activity, while R115777 is a farnesyl transferase inhibitor suppressing the necessary post-translational processing of Ras, and CCI-779, a rapamycin analog, inhibits mTOR. See Plate 13.2 in Color Plate Section.

References

1. Alessi, D. R., Andjelkovic, M., Caudwell, B. *et al.* (1996). Mechanism of activation of protein kinase B by insulin and IGF-1. *Embo J* **15**, 6541–6551.

2. Di Cristofano, A., Pesce, B., Cordon-Cardo, C., and Pandolfi, P. P. (1998). PTEN is essential for embryonic development and tumour suppression. *Nat Genet* **19**, 348–355.

3. Habib, T., Hejna, J. A., Moses, R. E., and Decker, S. J. (1998). Growth factors and insulin stimulate tyrosine phosphorylation of the 51C/SHIP2 protein. *J Biol Chem* **273**, 18605–18609.

4. Ishihara, H., Sasaoka, T., Hori, H. *et al.* (1999). Molecular cloning of rat SH2-containing inositol phosphatase 2 (SHIP2) and its role in the regulation of insulin signaling. *Biochem Biophys Res Commun* **260**, 265–272.

5. Choi, Y., Zhang, J., Murga, C. *et al.* (2002). PTEN, but not SHIP and SHIP2, suppresses the PI3K/Akt pathway and induces growth inhibition and apoptosis of myeloma cells. *Oncogene* **21**, 5289–5300.

6. Stokoe, D., Stephens, L. R., Copeland, T. *et al.* (1997). Dual role of phosphatidylinositol-3,4,5-trisphosphate in the activation of protein kinase B. *Science* **277**, 567–570.

7. Liang, J., Zubovitz, J., Petrocelli, T. *et al.* (2002). PKB/Akt phosphorylates p27, impairs nuclear import of p27 and opposes p27-mediated G1 arrest. *Nat Med* **8**, 1153–1160.

8. Datta, S. R., Dudek, H., Tao, X. *et al.* (1997). Akt phosphorylation of BAD couples survival signals to the cell-intrinsic death machinery. *Cell* **91**, 231–241.

9. Shtivelman, E., Sussman, J., and Stokoe, D. (2002). A role for PI3-kinase and PKB activity in the G2/M phase of the cell cycle. *Curr Biol* **12**, 919–924.

10. Kandel, E. S., Skeen, J., Majewski, N. *et al.* (2002). Activation of Akt/protein kinase B overcomes a G(2)/m cell cycle checkpoint induced by DNA damage. *Mol Cell Biol* **22**, 7831–7841.

11. Downward, J. (1999). How BAD phosphorylation is good for survival. *Nat Cell Biol* **1**, E33–35.

12. Cardone, M. H., Roy, N., Stennicke, H. R. *et al.* (1998). Regulation of cell death protease caspase-9 by phosphorylation. *Science* **282**, 1318–1321.

13. Rena, G., Guo, S., Cichy, S. C., Unterman, T. G., and Cohen, P. (1999). Phosphorylation of the transcription factor forkhead family member FKHR by protein kinase B. *J Biol Chem* **274**, 17179–17183.

14. Biggs, W. H., 3rd, Meisenhelder, J., Hunter, T., Cavenee, W. K., and Arden, K. (1999). Protein kinase B/Akt-mediated phosphorylation promotes nuclear exclusion of the winged helix transcription factor FKHR1. *Proc Natl Acad Sci USA* **96**, 7421–7426.

15. Wolfrum, C., Besser, D., Luca, E., and Stoffel, M. (2003). Insulin regulates the activity of forkhead transcription factor Hnf-3beta/Foxa-2 by Akt-mediated phosphorylation and nuclear/cytosolic localization. *Proc Natl Acad Sci USA* **100**, 11624–11629.

16. Brunet, A., Bonni, A., Zigmond, M. J. *et al.* (1999). Akt promotes cell survival by phosphorylating and inhibiting a Forkhead transcription factor. *Cell* **96**, 857–868.

17. Remy, I., and Michnick, S. W. (2004). Regulation of apoptosis by the Ft1 protein, a new modulator of protein kinase B/Akt. *Mol Cell Biol* **24**, 1493–1504.

18. Maira, S. M., Galetic, I., Brazil, D. P. *et al.* (2001). Carboxyl-terminal modulator protein (CTMP), a negative regulator of PKB/Akt and v-Akt at the plasma membrane. *Science* **294**, 374–380.

19. Jefferies, H. B., Reinhard, C., Kozma, S. C., and Thomas, G. (1994). Rapamycin selectively represses translation of the "polypyrimidine tract" mRNA family. *Proc Natl Acad Sci USA* **91**, 4441–4445.

20. Jefferies, H. B., Fumagalli, S., Dennis, P. B., Reinhard, C., Pearson, R. B., and Thomas, G. (1997). Rapamycin suppresses 5′TOP mRNA translation through inhibition of p70s6k. *Embo J* **16**, 3693–3704.

21. Shima, H., Pende, M., Chen, Y., Fumagalli, S., Thomas, G., and Kozma, S. C. (1998). Disruption of the p70(s6k)/p85(s6k) gene reveals a small mouse phenotype and a new functional S6 kinase. *Embo J* **17**, 6649–6659.

22. Stolovich, M., Tang, H., Hornstein, E. *et al.* (2002). Transduction of growth or mitogenic signals into translational activation of TOP mRNAs is fully reliant on the phosphatidylinositol 3-kinase-mediated pathway but requires neither S6K1 nor rpS6 phosphorylation. *Mol Cell Biol* **22**, 8101–8113.

23. Tang, H., Hornstein, E., Stolovich, M. *et al.* (2001). Amino acid-induced translation of TOP mRNAs is fully dependent on phosphatidylinositol 3-kinase-mediated signaling, is partially inhibited by rapamycin, and is independent of S6K1 and rpS6 phosphorylation. *Mol Cell Biol* **21**, 8671–8683.

24. Burnett, P. E., Barrow, R. K., Cohen, N. A. Snyder, S. H., and Sabatini, D. M. (1998). RAFT1 phosphorylation of the translational regulators p70 S6 kinase and 4E-BP1. *Proc Natl Acad Sci USA* **95**, 1432–1437.

25. Fingar, D. C., Richardson, C. J., Tee, A. R. Cheatham, L., Tsou, C., and Blenis, J. (2004). mTOR controls cell cycle progression through its cell growth effectors S6K1 and 4E-BP1/eukaryotic translation initiation factor 4E. *Mol Cell Biol* **24**, 200–216.

26. Dennis, P. B., Jaeschke, A., Saitoh, M., Fowler, B., Kozma, S. C., and Thomas, G. (2001). Mammalian TOR: a homeostatic ATP sensor. *Science* **294**, 1102–1105.

27. Murakami, M., Ichisaka, T., Maeda, M. *et al.* (2004). mTOR is essential for growth and proliferation in early mouse embryos and embryonic stem cells. *Mol Cell Biol* **24**, 6710–6718.

28. Hentges, K. E., Sirry, B., Gingeras, A. C. *et al.* (2001). FRAP/mTOR is required for proliferation and patterning during embryonic development in the mouse. *Proc Natl Acad Sci USA* **98**, 13796–13801.

29. Inoki, K., Li, Y., Xu, T., and Guan, K. L. (2003). Rheb GTPase is a direct target of TSC2 GAP activity and regulates mTOR signaling. *Genes Dev* **17**, 1829–1834.

30. Oldham, S., Montagne, J., Radimerski, T., Thomas, G., and Hafen, E. (2000). Genetic and biochemical characterization of dTOR, the Drosophila homolog of the target of rapamycin. *Genes Dev* **14**, 2689–2694.

31. Ishii, N., Maier, D., Merlo, A. *et al.* (1999). Frequent co-alterations of TP53, p16/CDKN2A, p14ARF, PTEN tumor suppressor genes in human glioma cell lines. *Brain Pathol* **9**, 469–479.

32. Haas-Kogan, D., Shalev, N., Wong, M., Mills, G., Yount, G., and Stokoe, D. (1998). Protein kinase B (PKB/Akt) activity is elevated in glioblastoma cells due to mutation of the tumor suppressor PTEN/MMAC. *Curr Biol* **8**, 1195–1198.

33. Chakravarti, A., Zhai, G., Suzuki, Y. *et al.* (2004). The prognostic significance of phosphatidylinositol 3-kinase pathway activation in human gliomas. *J Clin Oncol* **22**, 1926–1933.

34. Smith, J. S., Tachibana, I., Passe, S. M. *et al.* (2001). PTEN mutation, EGFR amplification, and outcome in patients with anaplastic astrocytoma and glioblastoma multiforme. *J Natl Cancer Inst* **93**, 1246–1256.

35. Ermoian, R. P., Furniss, C. S., Lamborn, K. R. *et al.* (2002). Dysregulation of PTEN and protein kinase B is associated with glioma histology and patient survival. *Clin Cancer Res* **8**, 1100–1106.

36. Myers, M. P., Pass, I., Batty, I. H. *et al.* (1998). The lipid phosphatase activity of PTEN is critical for its tumor supressor function. *Proc Natl Acad Sci USA* **95**, 13513–13518.

37. Ramaswamy, S., Nakamura, N., Vazquez, F. *et al.* (1999). Regulation of G1 progression by the PTEN tumor suppressor protein is linked to inhibition of the phosphatidylinositol 3-kinase/Akt pathway. *Proc Natl Acad Sci USA* **96**, 2110–2115.

38. Gottschalk, A. R., Basila, D., Wong, M. *et al.* (2001). p27Kip1 is required for PTEN-induced G1 growth arrest. *Cancer Res* **61**, 2105–2111.

39. Adachi, J., Ohbayashi, K., Suzuki, T., and Sasaki, T. (1999). Cell cycle arrest and astrocytic differentiation resulting from PTEN expression in glioma cells. *J Neurosurg* **91**, 822–830.

40. Davies, M. A., Koul, D., Dhesi, H. *et al.* (1999). Regulation of Akt/PKB activity, cellular growth, and apoptosis in prostate carcinoma cells by MMAC/PTEN. *Cancer Res* **59**, 2551–2556.

41. Pore, N., Liu, S., Haas-Kogan, D. A., O'Rourke, D. M., and Maity, A. (2003). PTEN mutation and epidermal growth factor receptor activation regulate vascular endothelial growth factor (VEGF) mRNA expression in human glioblastoma cells by transactivating the proximal VEGF promoter. *Cancer Res* **63**, 236–241.

42. Tamura, M., Gu, J., Matsumoto, K., Aota, S., Parsons, R., and Yamada, K. M. (1998). Inhibition of cell migration, spreading, and focal adhesions by tumor suppressor PTEN. *Science* **280**, 1614–1617.

43. Maier, D., Jones, G., Li, X. *et al.* (1999). The PTEN lipid phosphatase domain is not required to inhibit invasion of glioma cells. *Cancer Res* **59**, 5479–5482.

44. Raftopoulou, M., Etienne-Manneville, S., Self, A., Nicholls, S., and Hall, A. (2004). Regulation of cell migration by the C2 domain of the tumor suppressor PTEN. *Science* **303**, 1179–1181.

45. Taylor, V., Wong, M., Brandts, C. *et al.* (2000). 5' phospholipid phosphatase SHIP-2 causes protein kinase B inactivation and cell cycle arrest in glioblastoma cells. *Mol Cell Biol* **20**, 6860–6871.

46. Rodriguez-Viciana, P., Warne, P. H., Khwaja, A. *et al.* (1997). Role of phosphoinositide 3-OH kinase in cell transformation and control of the actin cytoskeleton by Ras. *Cell* **89**, 457–467.

47. Sonoda, Y., Ozawa, T., Hirose, Y. *et al.* (2001). Formation of intracranial tumors by genetically modified human astrocytes defines four pathways critical in the development of human anaplastic astrocytoma. *Cancer Res* **61**, 4956–4960.

48. Jiang, K., Sun, J., Cheng, J., Djeu, J. Y., Wei, S., and Sebti, S. (2004). Akt mediates Ras downregulation of RhoB, a suppressor of transformation, invasion, and metastasis. *Mol Cell Biol* **24**, 5565–5576.

49. Sonoda, Y., Ozawa, T., Aldape, K. D., Deen, D. F., Berger, M. S., and Pieper, R. O. (2001). Akt pathway activation converts anaplastic astrocytoma to glioblastoma multiforme in a human astrocyte model of glioma. *Cancer Res* **61**, 6674–6678.

50. Rich, J. N., Guo, C., McLendon, R. E., Bigner, D. D., Wang, X. F., and Counter, C. M. (2001). A genetically tractable model of human glioma formation. *Cancer Res* **61**, 3556–3560.

51. Holland, E. C., Celestino, J., Dai, C., Schaefer, L., Sawaya, R. E., and Fuller, G. N. (2000). Combined activation of Ras and Akt in neural progenitors induces glioblastoma formation in mice. *Nat Genet* **25**, 55–57.

52. Shi, Q., Bao, S., Maxwell, J. A. *et al.* (2004). Secreted protein acidic, rich in cysteine (SPARC) mediates cellular survival of gliomas through AKT activation. *J Biol Chem*.

53. Rajasekhar, V. K., Viale, A., Socci, N. D., Wiedmann, M., Hu, X., and Holland, E. C. (2003). Oncogenic Ras and Akt signaling contribute to glioblastoma formation by differential recruitment of existing mRNAs to polysomes. *Mol Cell* **12**, 889–901.

54. Alkarain, A., and Slingerland, J. (2004). Deregulation of p27 by oncogenic signaling and its prognostic significance in breast cancer. *Breast Cancer Res* **6**, 13–21.

55. Catzavelos, C., Bhattacharya, N., Ung, Y. C. *et al.* (1997). Decreased levels of the cell-cycle inhibitor p27Kip1 protein: prognostic implications in primary breast cancer. *Nat Med* **3**, 227–230.

56. Avdulov, S., Li, S., Michalek, V. *et al.* (2004). Activation of translation complex eIF4F is essential for the genesis and maintenance of the malignant phenotype in human mammary epithelial cells. *Cancer Cell* **5**, 553–563.

57. Riesterer, O., Zingg, D., Hummerjohann, J., Bodis, S., and Pruschy, M. (2004). Degradation of PKB/Akt protein by inhibition of the VEGF receptor/mTOR pathway in endothelial cells. *Oncogene* **23**, 4624–4635.

58. Lazaris-Karatzas, A., Montine, K. S., and Sonenberg, N. (1990). Malignant transformation by a eukaryotic initiation factor subunit that binds to mRNA 5' cap. *Nature* **345**, 544–547.

59. Ruggero, D., Montanaro, L., Ma, L. *et al.* (2004). The translation factor eIF-4E promotes tumor formation and cooperates with c-Myc in lymphomagenesis. *Nat Med* **10**, 484–486.

60. Rosenwald, I. B., Lazaris-Karatzas, A., Sonenberg, N., and Schmidt, E. V. (1993). Elevated levels of cyclin D1 protein in response to increased expression of eukaryotic initiation factor 4E. *Mol Cell Biol* **13**, 7358–7363.

61. Rosenwald, I. B., Kaspar, R., Rousseau, D. *et al.* (1995). Eukaryotic translation initiation factor 4E regulates expression of cyclin D1 at transcriptional and post-transcriptional levels. *J Biol Chem* **270**, 21176–21180.

62. Rousseau, D., Kaspar, R., Rosenwald, I., Gehrke, L., and Sonenberg, N. (1996). Translation initiation of ornithine decarboxylase and nucleocytoplasmic transport of cyclin D1 mRNA are increased in cells overexpressing eukaryotic initiation factor 4E. *Proc Natl Acad Sci USA* **93**, 1065–1070.

63. Topisirovic, I., Guzman, M. L., McConnell, M. J. *et al.* (2003). Aberrant eukaryotic translation initiation factor 4E-dependent mRNA transport impedes hematopoietic differentiation and contributes to leukemogenesis. *Mol Cell Biol* **23**, 8992–9002.

64. Gupta, A. K., Cerniglia, G. J., Mick, R. *et al.* (2003). Radiation sensitization of human cancer cells in vivo by inhibiting the activity of PI3K using LY294002. *Int J Radiat Oncol Biol Phys* **56**, 846–853.

65. Rosenzweig, K. E., Youmell, M. B., Palayoor, S. T., and Price, B. D. (1997). Radiosensitization of human tumor cells by the phosphatidylinositol3-kinase inhibitors wortmannin and LY294002 correlates with inhibition of DNA-dependent protein kinase and prolonged G2-M delay. *Clin Cancer Res* **3**, 1149–1156.

66. Shi, Y. Q., Blattmann, H., and Crompton, N. E. (2001). Wortmannin selectively enhances radiation-induced apoptosis in proliferative but not quiescent cells. *Int J Radiat Oncol Biol Phys* **49**, 421–425

67. Zingg, D., Riesterer, O., Fabbro, D., Glanzmann, C., Bodis, S., and Pruschy, M. (2004). Differential activation of the phosphatidylinositol 3'-kinase/Akt survival pathway by ionizing radiation in tumor and primary endothelial cells. *Cancer Res* **64**, 5398–5406.

68. Geng, L., Tan, J., Himmelfarb, E. *et al.* (2004). A specific antagonist of the p110delta catalytic component of

phosphatidylinositol 3′-kinase, IC486068, enhances radiation-induced tumor vascular destruction. *Cancer Res* **64**, 4893–4899.

69. Edwards, E., Geng, L., Tan, J., Onishko, H., Donnelly, E., and Hallahan, D. E. (2002). Phosphatidylinositol 3-kinase/Akt signaling in the response of vascular endothelium to ionizing radiation. *Cancer Res* **62**, 4671–4677.

70. Li, B., Yuan, M., Kim, I. A., Chang, C. M., Bernhard, E. J., and Shu, H. K. (2004). Mutant epidermal growth factor receptor displays increased signaling through the phosphatidylinositol-3 kinase/AKT pathway and promotes radioresistance in cells of astrocytic origin. *Oncogene* **23**, 4594–4602

71. Bianco, R., Shin, I., Ritter, C. A. *et al.* (2003). Loss of PTEN/MMAC1/TEP in EGF receptor-expressing tumor cells counteracts the antitumor action of EGFR tyrosine kinase inhibitors. *Oncogene* **22**, 2812–2822.

72. Li, B., Chang, C. M., Yuan, M., McKenna, W. G., and Shu, H. K. (2003). Resistance to small molecule inhibitors of epidermal growth factor receptor in malignant gliomas. *Cancer Res* **63**, 7443–7450.

73. Fan, Q. W., Specht, K. M., Zhang, C., Goldenberg, D. D., Shokat, K. M., and Weiss, W. A. (2003). Combinatorial efficacy achieved through two-point blockade within a signaling pathway-a chemical genetic approach. *Cancer Res* **63**, 8930–8938.

74. Chakravarti, A., Loeffler, J. S., and Dyson, N. J. (2002). Insulin-like growth factor receptor I mediates resistance to anti-epidermal growth factor receptor therapy in primary human glioblastoma cells through continued activation of phosphoinositide 3-kinase signaling. *Cancer Res* **62**, 200–207.

75. Steinbach, J. P., Klumpp, A., Wolburg, H., and Weller, M. (2004). Inhibition of epidermal growth factor receptor signaling protects human malignant glioma cells from hypoxia-induced cell death. *Cancer Res* **64**, 1575–1578.

76. Chang, S. M., Kuhn, J., Wen, P. *et al.* (2004). Phase I/pharmacokinetic study of CCI-779 in patients with recurrent malignant glioma on enzyme-inducing antiepileptic drugs. *Invest New Drugs* **22**, 427–435.

77. Wendel, H. G., De Stanchina, E., Fridman, J. S. *et al.* (2004). Survival signalling by Akt and eIF4E in oncogenesis and cancer therapy. *Nature* **428**, 332–337.

78. Podsypanina, K., Lee, R. T., Politis, C. *et al.* (2001). An inhibitor of mTOR reduces neoplasia and normalizes p70/S6 kinase activity in Pten+/- mice. *Proc Natl Acad Sci USA* **98**, 10320–10325.

79. Eshleman, J. S., Carlson, B. L., Mladek, A. C., Kastner, B. D., Shide, K. L., and Sarkaria, J. N. (2002). Inhibition of the mammalian target of rapamycin sensitizes U87 xenografts to fractionated radiation therapy. *Cancer Res* **62**, 7291–7297.

80. LaFortune, T., Liu, T.J., Yung, W.K.A. (1994). The anticancer effects of RAD001 on glioma cell lines. *Neuro-Oncol* **6**, 332.

81. Kentsis, A., Topisirovic, I., Culjkovic, B., Shao, L., and Borden, K. L. (2004). Ribavirin suppresses eIF4E-mediated oncogenic transformation by physical mimicry of the 7-methyl guanosine mRNA cap. *Proc Natl Acad Sci USA* **101**, 18105–18110.

82. Nakamura, J. L., Karlsson, A., Arvold, N. D. *et al.* (2005). PKB/Akt mediates radiosensitization by the signaling inhibitor LY294002 in human malignant gliomas. *J Neurooncol* **71**, 215–222.

14

Tumor Invasiveness and Anti-invasion Strategies

Sandra A. Rempel and Tom Mikkelsen

PATHOLOGIC FEATURES OF MALIGNANT BRAIN TUMOR INVASION

Gliomas in general, and anaplastic gliomas in particular, invade the surrounding brain to a great extent. The regional infiltration during tumor progression has been shown most strikingly in whole-mount studies of patients untreated for their malignant gliomas [1,2]. In these specimens, glioblastoma cells appear to arise within a bed of better-differentiated tumor that is associated with regional brain infiltration. In standard histological sections, most glioblastomas contain a central area of necrosis, a highly cellular rim of tumor, and a peripheral zone of infiltrating cells that can access several routes of dissemination. Regional infiltration can proceed as single-cell or small cell clusters infiltrating into the adjacent brain parenchyma, along white matter tracts, around nerve cells, along the perivascular spaces, or along subependymal and subpial basement membranes, depending on the location of the original tumor mass. Importantly, these studies have shown that, in the majority of cases, tumor cells have not only infiltrated regionally, but have also migrated to distant sites from the primary site of malignant gliomas by the time of diagnosis. Significantly, it is these infiltrating cells that are responsible for both the local and distant recurrences and tumor progression seen clinically.

CLINICAL AND PATHOLOGICAL EVALUATION OF TUMOR INFILTRATION

Clinical recognition of tumor infiltration depends on the accuracy and specificity of neuroimaging to identify the regional and distant infiltrating tumor cells. To date, standard MRI and CT scans are insufficient to detect these infiltrating cells. For example, biopsy studies have shown that infiltrating tumor cells appear to extend as far as the most distant area of abnormality seen by prolonged T2-weighted signal in MRI [3]. However, using post-mortem MRI of formalin-fixed brains, MRI was unable to detect tumor cells up to or beyond the MRI-abnormal areas [4]. Unfortunately, both neuroimaging and pathological assessment are still required, and together have provided important information regarding tumor progression and invasion. For instance, Burger et al. detailed the clinical appearance of malignant glioma in whole-mount brain sections and its relation to CT images [2,5–7]. In these studies, investigators were able to deduce the natural history of malignant glioma by examining untreated cases and the anatomic substrates upon which tumor cells migrate in the process of tumor progression [8]. The distant spread of tumor cells was shown not to be random, but to follow distinct pathways, such as the corpus callosum, fornix, anterior and posterior commisures, optic radiations, and association fibers.

Schiffer examined a series of 90 malignant gliomas and tumor cell infiltration could be detected histologically at a distance of >2 cm from the tumor edge, with finger-like extensions in the white matter [9]. Foci of infiltrating cells were seen in some cases, whereas, in others, single-cell mitoses were found in the normal white matter far from the tumor edge. In this study, CT could not distinguish edematous areas from underlying low-grade malignancy or small foci of glioblastoma cells, either in the parenchyma or in subarachnoid deposits. Furthermore, T2-weighted MRI also underestimated the margins of lower grade tumors and small foci of tumor cells. This inability to clinically detect invading tumor cells leads to a high therapeutic failure rate.

PATTERNS OF FAILURE

Studies indicate that local recurrence is the major contributing factor to therapeutic failure. In one clinical series, more than 90 per cent of tumors recurred within 2 cm of the original tumor [10]. Multicentric tumors were uncommon, occurring in only 1–5 per cent of cases, and dissemination throughout the leptomeninges or spinal cord rarely occurred. In a pathologic series describing the pattern of therapeutic failure, these values were confirmed, and this illustrated that recurrence is more likely to be local because of the density of residual cells remaining at resection margins [11–16].

Regional and remote recurrences appear to occur less frequently since there is a rapid fall-off in the number of residual tumor cells at a distance further than 2 cm. However, when occurrence of new lesions remote from the original tumor bed was observed, recurrence was almost always accompanied by failure at the original site [17].

Since tumor infiltration is a major reason for patient failures, therapeutic strategies aimed at limiting or eliminating the invasive phenotype of these tumors could have a profound effect on patient outcome.

THERAPEUTIC CONSIDERATIONS

Despite the fact that malignant gliomas are widely infiltrative at the time of diagnosis, there remains a solid rationale for employing therapies at the phenotype of invasion [18]. Pathologically, the patterns of tumor cell spread in the brain indicate those substrates that likely are critical in the dissemination of tumor cells. Adjacent parenchymal infiltration of tumor cells into normal brain substance, which may

occur as single-cells or small cell clusters, would appear to require the dissolution of the cell-to-cell contacts and scaffolding of the brain extracellular matrix (ECM). The extension of cells along (but not through) perivascular basement membrane, such as that in the Virchow-Robin spaces, and along subpial and subependymal basement membrane requires modulation of cell-to-cell and cell-to-substrate adhesion, in addition to motility factors. The widespread infiltration of tumor cells along compact white matter pathways, such as the corpus callosum, the optic radiations, fornix, and association fibers, suggests the involvement of cell-to-cell and cell-to-matrix recognition molecules and motility factors, perhaps associated with a modification of the local substrate permissive for cell motility. Thus, the signaling mechanisms (and thereby the underlying molecules) involved in regulating these patterns of invasion may prove to be valuable therapeutic targets.

In addition, there are a number of studies suggesting that the population of migrating glioma cells has a temporarily limited capability for proliferation. The ability to reverse this relationship could sensitize the once migratory invasive tumor cells to antiproliferative therapies (see Chapter 9 on Cell Cycle Signaling Pathways).

Furthermore, there is increasing evidence that those invasive glioma cells with a low propensity to proliferate may also be resistant to apoptosis, precluding their treatment by proapoptotic agents (see Chapter 10 on Apoptosis Signaling Pathways). For example, Joy et al. [19] showed that an increased expression of PI-3 kinase occurred in infiltrating cells compared with stationary tumor core cells, which suggests that these cells might also be relatively resistant to conventional cytotoxic treatments that target proliferating cells. Since the infiltrating cells are often the predominant cells remaining after successful debulking surgery, their relative resistance to the pro-apoptotic effects of cytotoxic chemotherapy or radiation could be favorably impacted by targeting their infiltrating behavior. Novel agents that inhibit the migratory phenotype may, therefore, re-initiate cell cycling and, in effect, restore sensitivity to pro-apoptotic therapy.

Molecules active in the process of glioma cell motility and invasion, therefore, may be viable targets for drug development. As discussed below, though, there are a number of interactions within and between biological pathways and multiple complex mechanisms acting simultaneously which are associated in the process of glioma infiltration. To further complicate matters, the tumor cells do not act in isolation, but are also influenced by the molecules in their surrounding brain matrix. Thus, while characterizing the underlying genetic changes to the tumor, the environmental influences must also be considered.

ENVIRONMENTAL INFLUENCES

The external environment surrounding cells is different in each tissue type. This tissue-specific external environment is comprised of proteins that play important roles in either physically supporting the cells (the ECM proteins) or those directly modulating the behavior of the cells (matricellular proteins, soluble growth factors, proteases) through the interaction with membrane receptors (growth factor receptor, proteases) and adhesion proteins (such as integrins). These cell surface interactions regulate intracellular signaling pathways and cytoskeletal changes, which in turn influence a number of biological functions including ECM degradation, proliferation, apoptosis, angiogenesis, and motility.

Invasion itself is a complex biological phenotype, and many of the biological signals that modulate invasion emanate from stromal–tumor cell interactions that (1) promote the loss of tumor cell–cell adhesion, (2) induce an appropriate state of cell–ECM adhesion with concomitant cytoskeletal changes to increase cell motility, (3) increase degradation of matrix, and (4) inhibit proliferation. It is important then to understand the tumor–stromal interactions in each of these complex biological interactions as potential molecular therapeutic targets may be tumor and/or stromal derived.

MOLECULAR TARGETS

Molecular targets for the treatment of glioma invasion may consist of the stromal- or tumor-derived ECM molecules, adhesion molecules, matricellular proteins, proteases, and growth factors and their receptors. As illustrated in Fig. 14.1, the ECM molecules, consisting of proteoglycans, glycosaminoglycans (GAG), and structural glycoproteins, can interact independently with the tumor cell-associated adhesion molecules, secreted matricellular proteins, and proteases. Figure 14.2 highlights the interconnectedness between these groups of molecules in modulating the invasive phenotype and its impact on the proliferative phenotype. Figure 14.3 focuses on the impact of the ECM glycoproteins and the matricellular proteins on integrin action, demonstrating their pivotal role in the regulation of tumor cell adhesion and motility.

While each major category of molecular targets is considered in turn, the complex interactions between them must also be considered and appreciated.

ECM Molecules

The tumor-associated ECM is comprised of adjacent brain-derived and tumor-secreted ECM molecules. The major structural ECM constituents confronted by the tumor cells are members of the proteoglycan, GAG, and glycoprotein superfamilies [20] (Fig. 14.1, Table 14.1).

Proteoglycans

A proteoglycan consists of a core protein having a variable number of GAG side chains [21] (Table 14.1). The ECM proteoglycan superfamily includes the small leucine-rich repeat (LRR) family, the lectican family, the collagen family, and others. Members of these families present in the brain ECM include phosphacan, agrin, versican (also known as PG-M), neurocan, and brain-enriched hyaluronan-binding (BEHAB [22]; also known as brevican), and the vascular basement membrane proteoglycan perlecan. In brain tumors, changes in the expression of several ECM proteoglycans are observed within the tumor-brain adjacent ECM and the tumor vasculature.

Three of the lectican family members are implicated in invasion. BEHAB expression is upregulated in astrocytic tumors *in vivo* [23], and either protein cleavage [24] or specific isoform expression and upregulation [25] appear to contribute to glioma invasion. Versican is alternatively spliced into 4 different isoforms, and the V2 isoform may be brain-specific. Its expression is decreased in glioma tumor ECM compared to levels seen in normal neuropil, whereas its expression appears to be upregulated in small blood vessels of gliomas *in vivo* [26]. Its role in glioma cells remains to be deciphered; however, transfection of a dominant-negative versican mutant that inhibits the secretion and binding of endogenous versican inhibited tumor formation, suggesting that versican can play a role in tumor formation [27]. Neurocan, another member of this family can bind to the adhesion molecules neural cell adhesion molecule (NCAM) and neuronglia cell adhesion molecule (NgCAM) (Table 14.2), inhibiting both homophilic cell–cell binding and adhesion [20]. It is present in normal brain and oligodendrogliomas; however, its expression is lost in astrocytic tumors [28].

Phosphacan is a chrondroitin sulfate or keratin sulfate (phosphacan-KS) proteoglycan of nervous

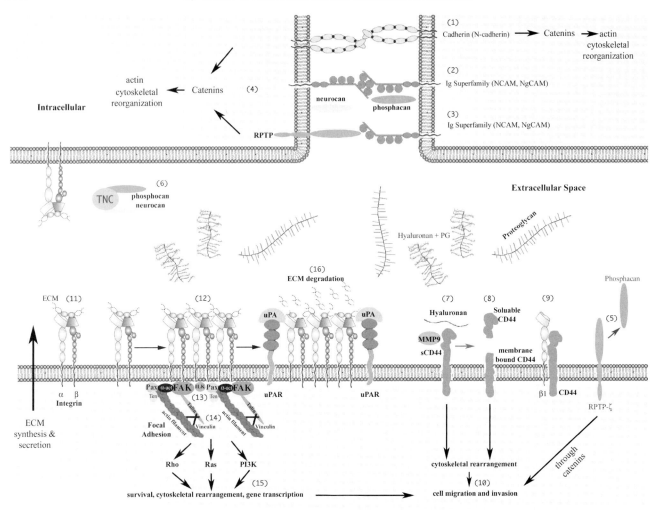

FIGURE 14.1 Tumor Adjacent External Environment. A schematic representation of the external environment, with an emphasis on the adhesion and ECM molecules involved in glioma invasion. Cell–cell interactions are modulated by the homophilic cell–cell adhesion of cadherins (1) and Ig-superfamily proteins (2), the latter of which can engage in heterophilic cell–cell interactions with members of the RPTP family (3). Both cadherins and RPTP interact with catenins to induce intracellular signaling and cytoskeletal changes (4). The cytoplasmic domain of RPTP can be cleaved to release phosphacan (5), which can interact with the matricellular protein TN-C (6), or the Ig-superfamily molecules (2). Cell–ECM interactions are modulated by CD44, which can bind to its ligand (7), become cleaved (8), or associate with β1 integrin (9), and thereby initiate intracellular signaling, induce cytoskeletal rearrangement, and promote migration and invasion (10). Cell–ECM interactions are also modulated by integrins (11). Ligation of ECM promotes clustering of integrins to focal adhesions (12), which contain actin-associated proteins (13), and which link the integrin to the cytoskeleton (14) thereby triggering activation of FAK and ILK (12) and downstream events involved in cytoskeleton rearrangement and cell migration and invasion (15,10). Association of integrins with uPA/uPAR induces ECM degradation (16). (Adapted from International Journal of Biochemistry and Cell Biology. Vol 36. Bellail, A. C., Hunter, S. B., Brat, D. J., Tan, C., Van Meir, E. G., Microregional extracellular matrix heterogeneity in brain modulates glioma cell invasion. pp 1046–1069, Figures 4 and 5, 2004 with permission from Elsevier) Pax — paxillin, α-act — α-actinin, Ten — tensin, FAK — focal adhesion kinase, ILK — integrin linked kinase, uPA — urokinase plasminogen activator, uPAR — uPA receptor. See Plate 14.1 in Color Plate Section.

tissue, grouped in the "Other" category (Table 14.1). It is the extracellular domain of the receptor-type tyrosine phosphatase adhesion molecule (Table 14.2). Like neurocan, it is present in normal brain and oligodendrogliomas; but expression is lost in astrocytic tumors [28]. It has a high affinity for

fibroblast growth factor-2 (FGF2) [21], and like phosphacan, can bind NCAM and NgCAM, suggesting influences on growth and adhesion, respectively. In addition, it can bind to TN-C, and inhibits adhesion of C6 glioma cells to this matricellular protein [29].

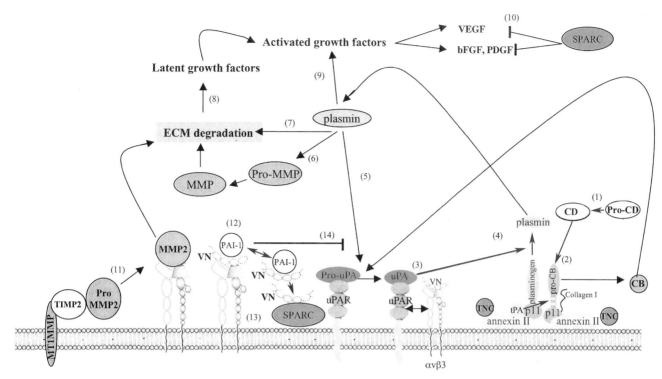

FIGURE 14.2 Interactions Between Major Signaling Pathways in Invasion and Subsequent Impact on Proliferation. This is a schematic representation of the interaction between the proteolytic cascade, integrins, ECM molecules, and the matricellular proteins. Pro-cathepsin D is autocatalytically cleaved to generate cathepsin D (1). Cathepsin D and tPA activate procathepsin B (2). Cathepsin B initiates a sequence of events, including the activation of uPA from pro-uPA (3), which then converts plasminogen to plasmin (4). Plasmin reinforces the generation of uPA (5), activates pro-MMPs (6), and degrades ECM (7). This releases latent growth factors (8), permitting their activation by plasmin (9). Growth factors can in turn initiate intracellular signaling by binding to their receptors. They can be inhibited though by binding to the matricellular protein SPARC (10). Membrane-bound MMPs and TIMP2 activate pro-MMP2 to active MMP-2, a step requiring integrins (11). This can be mediated by integrin binding to ECM alone or with other proteins such as PAI-1 (12). The PAI-1 and VN binding to integrin can be modulated by the matricellular protein SPARC (13). The balance of these interactions leads to the modulation of pro-uPA activation (14) and the subsequent ECM degradation of plasmin, as described above. See Plate 14.2 in Color Plate Section.

Glycosaminoglycans (GAGs)

The GAG side-chains of the proteoglycans are the most abundant heteropolysaccharides in the body (Table 14.1). These side chains include chrondroitin sulfate, heparin sulfate, keratin sulfate, and dermatan sulfates. GAGs provide structural integrity to cells and provide passageways between cells, allowing for cell migration. The content of GAGs increases in gliomas, mainly due to increases in heparin sulfate and dermatan sulfate [30]. Hyaluronan is a major GAG constituent of brain ECM often found non-covalently attached to other proteins including members of the lectican proteoglycan family (Table 14.1). In normal adult brain, hyaluronan constitutes 36 and 54 per cent of grey and white matter GAG, respectively. In brain tumors, the percentage of hyaluronan is highest in astrocytoma (56 per cent), decreasing to 32 per cent in anaplastic

astrocytomas, and 29 per cent in glioblastoma [20]. Hyaluronan can induce intracellular signaling by binding to its cell surface proteoglycan receptor, the adhesion molecule CD44 (Table 14.2, Fig. 14.1; see CD44 below). This GAG can therefore contribute structurally to the ECM, as well as serve as a regulatory signal. The binding of hyaluronan to its receptor is enhanced by the ECM glycoprotein collagen 1 [31], demonstrating the ability of other ECM components to influence hyaluronan-induced intracellular signaling. Its role in invasion has been substantiated by perturbation of hyaluronan–CD44 interactions using either small hyaluronan oligosaccharides that compete for endogenous hyaluronan polymer interactions, or by overexpression of soluble hyaluronan-binding proteins, which leads to inhibition of anchorage-dependent growth and decreased invasion [32]. These experiments suggest that

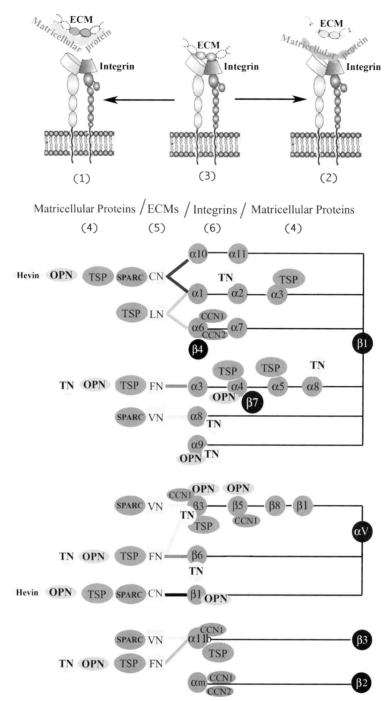

FIGURE 14.3 Interactions Between Integrins, Matricellular and Extracellular (ECM) Proteins. The matricellular proteins have a significant impact on integrin signaling. They can either bind ECM proteins (1) or integrins (2) to disrupt integrin-ECM interactions (3), and thereby influence adhesion, motility and signaling. Colored circles to the left indicate the matricellular proteins that can bind to the same ECM protein (4). The ECM proteins can often bind to more than one integrin (5). The grey and black circles connected by the black lines depict the major β, α, and αv integrins involved in glioma adhesion and invasion (6). Integrins can bind to more than one ECM. Integrins can also directly bind to matricellular proteins, the colored circles depicted on the right. Therefore, there are many potential interactions that govern the state of adhesion, and a change to just one component, the integrin, ECM, or matricellular protein could significantly impact the state of adhesion and consequently migration and invasion. OPN — osteopontin, TSP — thrombospondin, TN — Tenascin C, CN — collagen, LN — laminin, FN — fibronectin, and VN — vitronectin. See Plate 14.3 in Color Plate Section.

TABLE 14.1 ECM Molecules

Proteoglycan/GAG Superfamily			GAGs	Normal
ECM Proteoglycans	Leucine-rich repeat (LRR) family	Biglycan (PG-I)	**CS, DS**	Ca, Bn
		Decorin (PG-II)	CS, DS	CT
		Fibromodulin	**KSI**	Ca, T, S
		Lumican	KSI	Co, Mu, K, L
		PRELP	KSI	Ca
		Keratocan	KSI	Co
		Osteoadherin	KSI	Bn
		Epiphycan (PG-Lb)	CS, DS	Ca
		Osteoglycin (mimecan)	KSI	Co, Br, L
Large aggregating chondroitin sulfates	Lectican family	Aggrecan	CS, KSI, KSII	Ca, Br, A, T
		Versican (PG-M)	CS, DS	F
		Neurocan	CS, DS	Br, R
		Brevican (BEHAB)	CS	Br, NS
	Collagen Family	Type IX collagen	C	Ca
		Type XII collagen	CS, DS	
		Type XIV collagen	CS, DS	
	Others	Perlecan	**HS**	BM
		Bamacan	CS	BM
		Agrin	HS	NMJ
		Serglycin	Hep, CS	MC
		Phosphacan	CS, KS	Br
		Hyaluronan		Br
Cell-Surface Proteoglycans	Syndecan family (transmembrane HSPG)	Syndecan-1 (syndecan)	HS, CS, DS	Epi
		Syndecan-2 (fibroglycan)	HS	F
		Syndecan-3 (N-syndecan)	HS	NS
		Syndecan-4 (ryudocan)	HS	
	Glypican (GPI-anchored HSPG)	Glypican	HS	Bn, M, K
		Glypican-2 (cerebroglycan)	HS	NS
		Glypican-3 (OCI-5)	HS	O, L, K
		Glypican-4 (K-glypican)	HS	K, Br, L
		Glypican-5 (GCP5)	HS	Br, K, Lb, Lv
		Glypican-6	HS	Lv, K, O
	Others	Betaglycan	CS, DS, HS	
		250 kDa CSPG (NG2 rat homolog)	CS, DS	Ca, NS (pericytes)
		CD44	CS, DS, HS	Br
Glycoproteins		**Fibronectin**		BM
		Collagen I, III, **IV**, V		BM
		Laminin		BM
		Vitronectin		BM
		Entactin		BM

Glycosaminoglycans (GAGs): Chrondroitin sulfate (CS), dermatan sufate (DS), Heparan sulfate (HS), Keratin sulfate 1 (KS1), keratin sulfate II (KSII), chrondroitin (C), heparin (Hep). Normal Localization (Normal): Epithelial cells (Epi), fibroblast (F), nervous system (NS), bone (Bn), marrow (M), kidney (K), ovary (O), lung (L), brain (Br), limb (Lb), liver (Lv), cartilage (Ca), connective tissue (CT), tendon (T), skin (S), cornea (Co), muscle (Mu), aorta (A), retina (R), basement membrane (BM), mast cell (MC), neuromuscular junction (NMJ). **Bold**: Either overexpression or loss of expression in brain tumors (see text). (file://localhost/Users/apple/Desktop/review%20&%20papers/Proteoglycan%20Super family.html.)

TABLE 14.2   Cell Adhesion Molecules

Cell–Cell Adhesion			Ligands	Normal
Cadherins		**Cadherin N (2)**	Homophilic	N
		Cadherin BR (12)		Br
		Cadherin VE (5)		Br, EC
		Cadherin P (3)		Pla
		Cadherin R (4)		R
		Cadherin E (1)		Epi
		Cadherin M (15)		Mu
		Cadherin T & H (13)		H, astrocytes
		Cadherin OB (11)		Osteo
		Cadherin K (6)		Br, K
		Cadherin 7		
		Cadherin 8		Br
		Cadherin KSP (16)		K
		Cadherin LI (17)		GI, Pa
		Cadherin 18		CNS
		Cadherin fibroblast 1 (19)		F
		Cadherin fibroblast 2 (20)		F
		Cadherin fibroblast 3 (21)		F
		Cadherin 23		E
	Desmosomes	Desmocollin 1		S
		Desmocollin 2		Epi, Muc, Myo, LN
		Desmocollin 3		
		Desmoglein 1		EpiD, T
		Desmoglein 2		All
		Desmoglein 3		EpiD, T
		Protocadherin 1, 2, 3, 7, 8, 9		
Ig superfamily	Neural specific Homophillic	Adhesion molecule on glia (AMOG)		Glia, N
		L1CAM	Axonin	N
		Myelin-associated glycoprotein (MAG)	MAG	My
		Myelin-oligodendrocyte glycoprotein (MOG)		My, oligo
		NCAM-1 (CD56)	NCAM-1 via polysialic acid	N
		NrCAM	Ig superfamily	N
		OBCAM	Opiods, acidic lipids	Br
		P_O protein	P_O	My
		PMP-22 protein	PMP-22	My
		SynCAM	Ig superfamily	N, Sy
		neurofascin		N
		NgCAM	Neurocan, phosphacan	N
	Systemic Heterophillic	ALCAM (CD166)	CD6, CD166, NgCAM, 35 kD protein	N, Lk
		Basigin (CD147)		Lk, RBC, Pl, EC
		BL-CAM (CD22)	Sialated glycoproteins LCA (CD45)	B lymph
		CD44	Hyal, ankyrin, FN, MIP1β, OPN	Lymph, Epi, perivascular astro

(Continues)

TABLE 14.2 *(Continued)*

Cell–Cell Adhesion			Ligands	Normal
		ICAM-1 (CD54)	$\alpha L\beta2$, LFA-1	Lk, E DC, F, Epi, SC
		ICAM-2 (CD102)	$\alpha L\beta2$, LFA-1	EC, lymph, mono
		ICAM-3 (CD50)	$\alpha L\beta2$	Lk
		Lymphocyte function antigen-2 (LFA-2) (CD2)	LFA-3	Lymph, thymo
		LFA-3 (CD58)	LFA-2	Lk, stroma, EC
		Major histocompatibility complex (MHC) molecules		
		MAdCAM-1	$\alpha4\beta7$, Selectin	Mucosal EC
		PECAM (CD31)	$\alpha v\beta3$, CD31	Lk, SC, EC
		T-cell receptor (C-region)		
		VCAM-1	$\alpha4\beta1$, $\alpha4\beta7$	Mono, SC, act. EC
Selectins		Selectin L (CD62L)	GlyCAM-1, CD34, MAdCAM-1	Lk
		Selectin P (CD62P)	Sialyl-Lewis^x, **P-selectin GL-1**	
		Selectin E (CD62e)	Sialyl-Lewis^x, Sialyl-Lewis^a, (CLAA), **GSL1**	EC

Cell–Cell and Cell–ECM Adhesion

Receptor protein tyrosine phosphatases (PTPs)	Type I/VI	***CD44**	Hyal, ankyrin, FN, MIP1β, OPN
	Type IIa	*PTP LAR	Laminin-nidogen complex
		*PTP-σ	Agrin, CNXVIII
		*PTP-δ	
		*HmLAR *(Hirudo medicinalis)*	
		*HmLAR2 *(Hirudo medicinalis)*	
		*DPTP69D *(Drosophila)*	
		*Dlar *(Drosophila)*	
	Type IIb	PTP3 *(c.elegans)*	
		*PTP-μ	
		*PTP-κ	
		*PTP-ρ	
		*PTP-λ	
		*PTP-ψ	
	Type III	*PTPRO	
		*DPT99A *(Drosophila)*	
		*DPT99A *(Drosophila)*	
		*DPTP10D *(Drosophila)*	
		*DPTP52F *(Drosophila)*	
		DPTP4E *(Drosophila)*	
		PTP-β_1	
		DEP1	
		SAP1	
		PTPS31	
	Type IV	PTP-α	
		PTP-ε	

(Continues)

TABLE 14.2 *(Continued)*

Cell–Cell and Cell–ECM Adhesion			Ligands	Normal
	Type V	*PTP-ζ (extracellular domain- phosphacan)	TN, axonin-1, contactin, F3, NCAM, NrCAM, NgCAM, pleiotrophin, midkine	
		*PTP-γ		
	Type VII	PCPTP1		
		HePTP		
		STEP		
	Type VIII	IA2		
		IA2β		
Galectins		Galectin-1		Mu, H, Lu, Lv, LN, Ty, Pr, C, Br
		Galectin-2		SI,
		Galectin-3		C, Ma,
		Galectin-4		AT
		Galectin-5		Er
		Galectin-6		GI S
		Galectin-7		S
		Galectin-8		Lv, K, Lu
		Galectin-9		Ty, K
		Galectin-10		Eos, baso

CEll–ECM Adhesion				
Integrins		α1β1	LN, CN	NK, B lymph, (act) T lymph, F, glial peri, Schwann, EC
		α2β1	LN, **CN, TN**	NK, B lymph, (act) T lymph, Pl, EC, F, Epi, astro, Schwann, ependymal
		α3β1	LN, **CN, TN**, FN, TSP	(act) T lymph, thymo, EC, F, Epi, astro
		α4β1	FN, TSP-1, α4β1, α4β7, VCAM-1, MAdCAM-1	NK, B & T lymph, Eos, endo, Mu, NC-derived
		α5β1	**FN**, TSP, Murine L1 VN cell-line specific)	(act) B & T lymph; memory T lymph; Thymo, F, Epi, Pl, EC, Astro
		α6β1	**LN**, CCN1, CCN2	Lk, Thymo, Epi, T lymph (memory & activated), Glial, F, EC
		α7β1	LN	Sk & H Mu
		α8β1	FN, **VN**, TN	Epi, N, oligo
		α9β1	TN	Epi (airway), Mu
		α10β1	CN	Mu, H, Ca
		α11β1	CN	Bn, Br, Mu, Colon, Bn, Ma, Lv, Ov, Pla, S, Pa, MG, Te, E

(Continues)

TABLE 14.2 *(Continued)*

Cell–ECM Adhesion		Ligands	Normal
	$\alpha\beta1$	**VN**, FN, CN, von Willebrand's factor, fibrinogen	oligo
	$\alpha L\beta2$ (LFA-1α; CD11a)	ICAM-1, ICAM-2, ICAM-3	Lk, thymo, macro, T lymph, micro
	$\alpha M\beta2$ (CD11b)	ICAM-1, factor X, iC3b, fibrinogen	(act) myeloid B lymph, NK cells, macro, micro
	$\alpha X\beta2$ (CD11c)	IC3b, fibrinogen	Myeloid, dendridic cells, (act) B lymph, macro, micro
	$\alpha11b\beta3$	FN, VN, TSP-1, von Willebrand's factor	Pl
	$\alpha v\beta3$ (CD51/CD61)	**VN, FN, CN, TN**, OPN, TSP, von Willebrand's factor, PECAM-1, fibrinogen, human L1	(act) B & T lymph, EC, mono, glia, Schwann cells
	$\alpha6\beta4$	LN	Schwann cells, peri, EC, Epi, F
	$\alpha v\beta5$ (CD51/CD-)	**VN, FN**, OPN, Fibrinogen, CCN1	F, mono, macro, Epi, oligo
	$\alpha v\beta6$	FN, TN	
	$\alpha v\beta8$	FN, VN	Oligo, Schwann cells, brain synapses
	$\alpha4\beta7$	FN, VCAM-1, MAdCAM-1	NK, B & T lymph
	$\alpha IEL\beta7$	E-cadherin	Intraepithelial T lymph (IEL) intestinal
	$\alpha11$		U, H, Sk Mu, SM Mu-containing tissues
Cell surface proteoglycan	**250 kDa CSPG (NG2 rat homolog)**	CS, DS, FGF-2, PDGF-AA	Br
Hyaluronan receptor	**CD44**	Hyal, ankyrin, FN, MIP1β, OPN	Br

Ligands: laminin (LN), vitronectin (VN), fibronectin (FN), tenascin (TN), thrombospondin-1 (TSP-1), hyaluronin (Hyal), osteopontin (OPN) cutaneous lymphocyte-associated antigen (CLAA), P-selectin glycoprotein ligand-1 (P-selectin GL-1). Normal Localization (Normal): alimentary tract (AT), astrocytes (astro), basophils (baso), brain (Br), colon (C), dendritic cells (DC), ear (E), endothelial cells (EC), eosinophils (eos), epithelial cells (Epi), erythrocytes (Er), fibroblasts (F), GI tract (GI), heart (H), kidney (K), leukocytes (Lk), lung (Lu), lymph nodes (LN), lymphocytes (lymph), macrophages (Ma), mammary gland (MG), microglia (micro), monocytes (mono), mucosa (Muc), muscle (Mu), myelin (My), myocardium (myo), neural (N), neural-crest-derived (NC-derived), neural synapse (N Sy), oligodendrocytes (oligo), osteoblast (O), pancreas (Pa), perineurium (peri), placenta (Pla), platelets (Pl), prostate (Pr), red blood cells (RBC), retina (R), skeletal (Sk), skin (S), small intestine (SI), synovial cells (SC), testes (Te), thymocytes (thymo), thymus (Ty), tongue (T), http://www.neuro.wustl.edu/neuromuscular/lab/adhesion.htm. Bold cells: proteins implicated in gliomas; see text. * — expression in CNS.

targeting the ligand may be a viable means to affect glioma invasion.

Glycoproteins

The brain matrix is unique among the organs in that it does not express collagens I, III, IV, V, or laminin, fibronectin, and vitronectin. Instead, they are absent or are restricted to perivascular and vascular basement membranes [20,33]. Abnormal expression of these glycoproteins does occur in gliomas (Table 14.1), and these changes may also have profound effects on tumor cell adhesion, migration, and invasion.

Vitronectin synthesis is upregulated at the tumor–brain interface [34,35]. Vitronectin contains an RGD-peptide cell adhesion domain, and thereby modulates cell adhesion and migration through integrin binding [20] (Table 14.2). In addition, it facilitates invasion by binding to and concentrating the uPA/uPAR complex, which leads to the activation of the plasminogen–plasmin cascade of matrix degradation [36] (Fig. 14.2, and see uPA/uPAR later). Furthermore, PAI-1 promotes vitronectin multimer formation, which promotes vitronectin binding to the matricellular protein SPARC (Table 14.3, Fig. 14.2, and see Matricellular Proteins) [37]. Thus, through multiple

TABLE 14.3 Matricellular Proteins

		Matricellular Protein Integrin Receptors,	Other interactions
TSP1*	Thrombospondin-1	$\alpha3\beta1$, $\alpha5\beta1$, $\alpha v\beta3$, $\alpha2\beta1$, $\alpha4\beta1$, $\alpha11\beta\beta3$	syndecan, HSPG, TGFβ, CD36, IAP, calreticulin decorin
TSP2	Thrombospondin-2	$\alpha v\beta3$	heparin sulfate, LRP, TGFβ, CD36
SPARC*	Secreted protein acidic and rich in cysteine, osteonectin, BM-40	Unknown	thrombospondin, collagens I, II, III, IV, V, VIII, PDGF, VEGF
Hevin	SC1, SPARC-like 1, MAST9, RAGs1, ECM2	Unknown	
TN-C*	Tenascin-C, hexabrachion, cytotactin, neuronectin, myotendinous antigen, glial/mesenchyme extracellular matrix protein (GMEM), GP 150–225, (h, p)	$\alpha2\beta1$, $\alpha3\beta1$, $\alpha8\beta1$, $\alpha v\beta3$, $\alpha8\beta1$, $\alpha v\beta6$	Annexin II, FN
TN-X	Fibrinogen, hexabrachion-like (h), G13 protein (h), ATF-6 (h, m), creb-r (h, m)		
OPN	Bone sialoprotein I, early T-lymphocyte activation 1, secreted phosphoprotein 1	$\alpha v\beta3$, $\alpha v\beta1$, $\alpha v\beta5$, $\alpha9\beta1$, $\alpha4\beta1$	Cd44v3-v6, CN, FN, osteocalcin
CCN1*	CYR-61 (cysteine-rich protein; h,m,x), CEF10 (c), IGFBP-rP4 (h), β–IG-IG (m), CTGF-2 (h), IGFBP10 (h), angiopro	$\alpha6\beta3$, $\alpha v\beta3$, $\alpha v\beta5$, $\alpha11\beta3$, $\alpha_M\beta2$	HSPG (concomitant with integrins), CX43
CCN2*	CTGF (connective tissue growth factor; h,m,c,x), β–IG-M2 (m), FISP12 (m), IGFBP-rP2 (h), Hcs24 (h), IGFBP8 (H), HBGF-0.8 (h), ecogenin (h)	$\alpha6\beta1$, $\alpha_M\beta2$ (monocyte)?	LRP, heparin, HSPG (concomitant with integrins)
CCN3*	NOV (neuroblastoma overexpressed gene; h,r,c,m,q), IFGBP-Rp3 (h), IGFBP9 (h), NOVH (h), NOVm (m), ccmNOV (m), xNOV (x)		Fibulin C, notch I, S100A4
CCN4	WISP-1 (Wnt-1 inducible gene; h), ELM-1		Decorin, biglycan
CCN5	WISP-2 (h), CTGF-1 (r), CTGF-3 (r), HICP (r), rCOP-1 (r)		
CCN6	WISP-3 (h)		

* — Genes implicated in brain tumors, all are upregulated except for TSP1, which decreases with tumor grade. Note: IAP — integrin-associated protein, h — human, m — mouse, r — rat, c — chicken, q — quail, p — pig, x — xenopus. http://www.ebi.uniprot.org/index.shtml

binding partners consisting of integrins, matricellular proteins, and proteases, this one ECM is capable of modulating adhesion, migration, and matrix degradation, and ultimately tumor cell invasion.

Laminins are expressed in normal vasculature (Table 14.1), and their expression remains unchanged in tumor vasculature. However, laminin expression is upregulated in glioma tumor cells [38–40], and may similarly contribute to the regulation of adhesion and migration through integrin binding. Recently, laminin-5 was found to be a potent promoter of glioma cell adhesion, migration, and invasion, and its action is mediated through integrin $\alpha3\beta1$ [41]. Conversely, invasion was inhibited using anti-$\alpha3$ and anti-$\beta1$ integrin antibodies [42].

Fibronectin is also expressed in normal blood vessel basement membranes. Like laminin, it is upregulated in glioma ECM [40,43], and although it is still expressed in the vasculature, its expression is reduced to two-thirds of the vessels [44].

Interestingly, its expression pattern in the vessels is mutually exclusive with that of matricellular protein TN-C (see Matricellular Proteins; [45]), which is upregulated in the tumor vessel walls. Since fibronectin, like other ECM glycoproteins is considered to be adhesive, and TN-C anti-adhesive, their respective decrease and increase may facilitate tumor cell migration along the blood vessel basement membranes (20).

Therefore, the complex tumor adjacent ECM environment is comprised of proteoglycans, GAGs, and glycoproteins that were initially resident in the brain ECM or were tumor-secreted. As discussed, changes in the ECM environment alter tumor cell adhesion, migration, and invasion. As discussed below, this is brought about *via* the ECM environmental influences on the functions of adhesion molecules, matricellular proteins, proteases, and growth factor/growth factor receptor complexes (Figs. 14.1 and 14.2).

Adhesion Molecules

In normal adult tissues, there are several classes of adhesion molecules that govern cell–cell and cell–ECM interactions (Fig. 14.1, Table 14.2). The adhesion molecules governing cell–cell interactions are less well characterized. Those mediating cell–ECM interactions are better characterized, and these play a significant role.

Cell–Cell Adhesion Molecules

Included in this group are the cadherins, Ig-superfamily, and the selectins. Of these, the Ig-superfamily is better characterized with regard to gliomas. This superfamily consists of adhesion molecules that were originally grouped according to their abilities to mediate either homophilic, neural-specific cell–cell interactions, or systemic, hetero-phillic cell–cell interactions [46] (Table 14.2). Of the neural-specific proteins, neural cell adhesion molecule (NCAM) and neuroglia related cell adhesion molecule (Nr-CAM) have been implicated in brain tumor invasion (Table 14.2, Fig. 14.1). Four NCAM isoforms (180, 170, 140, and 120 kDa) are expressed in both adult brain and astrocytomas [47]. Increasing the amount of the 140-kDa isoform reduced glioma cell migration in vitro [48,49], and invasion in vivo [49]. Expression of this isoform was also inversely correlated with the expression of matrix metalloproteinases MMP-2 and MMP-9 [50] - enzymes that are involved in matrix degradation (see Proteases). Thus, the converse and complementary reduction of NCAM and the increase in MMP expression together promote both loss of cell–cell mediated and cell–matrix mediated constraints. Nr-CAM is also overexpressed in gliomas [51], and anti-sense Nr-CAM reduced both tumor growth and invasion in vitro [52], suggesting it plays a role in promoting invasion.

Of the heterophillic, systemic adhesion molecules (Table 14.2), several are expressed in astrocytic tumors and glioma cells. VCAM was found to be an astrocytoma cell surface marker by SAGE analysis [53], and to be expressed immunohistochemically in 15 of 25 gliomas, but not in normal brain [54]. In particular, LFA-3 (CD58) and ICAMs were highly expressed in glioblastomas [54,55], whereas LFA-3 was also increased in the tumor vessels [55]. Thus, the increased expression of these adhesion molecules may account for lymphocyte infiltration. This suggests that antibodies against these adhesion molecules, such as anti-VCAM1, may be of potential therapeutic benefit [54].

Less well characterized in gliomas are the 23 cadherin proteins (Table 14.2), which are transmembrane cell–cell linker molecules [46]. Their cytoplasmic domains are linked to the actin cytoskeleton via the cytoplasmic α, β, and γ catenins (Fig. 14.1). Both E and N cadherins are expressed in normal brain tissue. However, their role in glioma invasion remains controversial [56–58]. While it is reasonable to expect that the loss of cadherin expression, and therefore cell–cell contact, would promote glioma invasion, it was the overexpression of a truncated cadherin (T-cadherin) that decreased tumor cell motility by mechanisms presently unknown [59]. Clearly, the involvement of cadherins remains to be more clearly delineated.

The selectins L- (lymphocyte), E- (endothelial), and P- (platelet) mediate the adhesion of lymphocytes to endothelial cells, inflammatory responses, and adhesion between leukocytes and platelets, respectively, via their lectin-like carbohydrate binding extracellular domains [46] (Table 14.2). The selectins have not been implicated in tumor invasion per se, as they are not expressed on the tumor cells. However, they appear to play a role in angiogenesis, which can impact tumor invasion. It was reported that embryonic endothelial progenitor cells (eEPCs) that express E-and P-selectin preferentially homed to C6 tumor endothelium, and contributed to tumor angiogenesis. Since these incorporated eEPCs did not increase the overall vessel density of tumor size, this homing mechanism could be used to recruit therapeutically manipulated eEPCs to treat tumors without stimulating tumor growth [60].

Cell–Cell and Cell–ECM Adhesion Molecules

These adhesion molecules are particularly effective because of their diverse homophilic and heterophilic interactions.

Receptor protein tyrosine phosphatases (RPTPs) These molecules have been grouped into 6 subfamilies based on the diversity of their extracellular domains [61] (Table 14.2, Fig. 14.1). They regulate not only cell–cell, but also cell–matrix adhesion and motility [61]. Because they are transmembrane enzymes having intracellular domains possessing phosphatase activity, they do not require adaptor molecules. Rather they directly transduce extracellular interactions into intracellular events (Fig. 14.1). Several types of RPTPs are expressed in CNS tissues. RPTPzeta/beta and its extracellular domain phosphacan are increased in gliomas and implicated in glioma cell motility [62–64]. Although they partake in homophilic interactions, they also engage in heterophilic interactions with members of the Ig-Superfamily (Fig. 14.1). Furthermore, the receptor also binds to other secreted molecules and one report demonstrated that RPTPβ also mediated glioma cell adhesion to the matricellular protein TN-C [65].

Galectins   When these lectins are secreted into the external environment, they modulate both homophilic and heterophilic cell–cell adhesion interactions with NCAM and integrins [66]. In addition, these molecules also interact with the matricellular protein TN-C [67]. Galectin-1 is the most predominant galectin in normal brain [68]. Galectins 1, 3, and 8 are upregulated in gliomas [69], and galectins 1 and 3 are implicated in glioma migration [68,69]. Further, galectin-1 modulates glioma invasion *via* intracellular signaling with resultant modifications to the actin cytoskeleton and changes in the expression levels of small GTPases [68].

Cell–ECM Adhesion Molecules

The ability of a normal cell to move through the matrix during physiological development, or during wound healing in response to injury, depends upon the level of cell–ECM adherence. Adhesion is controlled through the interaction of cell-surface receptors (cell-surface proteoglycans, integrins) with ECM proteins. Many of these proteins are also upregulated in gliomas and contribute to their invasive phenotype.

Cell-surface proteoglycans.   The major cell surface proteoglycans include the syndecans and glypicans, and "others." To date, the syndecans and glypicans are not implicated in gliomas. However, both the 250 kDa chrondroitin sulfate proteoglycan (CSPG) and CD44 (Table 14.1), belonging to the "others" category, are implicated in brain tumors. The 250 kDa CSPG (also known as the rat homolog NG2) is not expressed in normal human brain parenchyma, but is expressed in pericytes in small arteries [70]. In astrocytic tumors, it is expressed in both the perivascular ECM as well as on the surface of perivascular tumor cells [70]. These alterations in protein level likely affect adhesive function. In addition, like phosphacan, the 250 kDa CSPG, exhibits high binding affinity for FGF2 and PDGF-AA [21], suggesting a further role in the regulation of tumor cell proliferation.

CD44 is the principal receptor for hyaluronan, one of the major brain ECM constituents. Discussed here as a cell surface proteoglycan, it is also characterized as belonging to the Ig superfamily and the RTPT family (Table 14.2, Fig. 14.1). Its many and diverse interactions make it a pivotal player in glioma migration. Engagement of the CD44 receptor with hyaluronan induces intracellular signaling resulting in the activation of Rac1, actin cytoskeletal reorganization, and redistribution of CD44 to regions of membrane ruffling [71,72] (Fig. 14.1), supporting a role for CD44 in glioma migration and invasion. Overexpression of one variant in particular, the CD44s variant, has been implicated in enhanced motility [72,73]. In addition, CD44 furthers this invasion by associating with MMP-9, a matrix metalloproteinase upregulated in gliomas, which is capable of degrading the ECM [74]. In turn, engagement of CD44 results in its cleavage by either MMP9 or ADAM10 (a disintegrin and metalloproteinase-10) [71]. Thus, the concomitant engagement by ligand, induced cytoskeletal changes and cleavage are all needed to promote motility. In addition, the growth factor EGF increases CD44 transcript and protein levels, lending insight into the mechanism as to how a factor involved in cell growth might also contribute to invasion (see Growth Factors/Growth Factor Receptors) [75]. Furthermore, evidence suggests that CD44 can bind directly to EGFR and modulate downstream gene expression of genes involved in glioma invasion, including urokinase-type plasminogen activator (uPA), plasminogen activator inhibitor-1 (PAI-1), and tissue inhibitor of mellatoproteinase-1 (TIMP-1) [76]. Since CD44 and its ligand hyaluronan are both upregulated in gliomas, the many interactions that CD44 has with its ligands, proteases, growth factor receptors, and subsequent intracellular signaling mechanisms most likely govern much of the observed glioma invasion into brain parenchyma.

Integrins   Integrins are a family of transmembrane heterodimer (α and β subunits) receptor proteins that link the ECM to the cortical cytoskeleton of the cell (Table 14.2, Fig. 14.1). Integrins are subclassified according to either the αv or the β integrin involved in the heterodimer. Of the three major beta chains (β1–β3), β1 and β2 subunits are primarily involved in cell–ECM interactions. While all cells express integrins, some integrin heterodimers are cell-type specific. In the normal brain, the neurons, glial cells, meningeal cells, and endothelial cells express integrin members of the β1 and αv heterodimers.

Most integrin heterodimers recognize more than one ligand (Table 14.2), and it is the strength of integrin binding to ECM glycoproteins that is pivotal to regulating cell motility. Weakly adherent cells do not have a sufficient level of adhesion to migrate, while strongly adherent cells are too tightly attached. Rather, an intermediate state of attachment is necessary for the cell to be able to migrate [77]. This level of attachment is, of course, dependent upon integrin expression levels and the expression levels of cognate ECM ligands in a given environment. When the integrins bind to their ECM glycoprotein ligands, integrin reconfiguration induces a clustering of the integrins into focal adhesions (Fig. 14.1). These focal adhesions are necessary to generate traction and forces required at the cell's leading edge during motility. However, as a cell moves forward, it is necessary to release adhesion at the rear of the cell. Thus, there is a dynamic turnover of the focal adhesions.

In brain tumors, the integrin heterodimers $\alpha2\beta1$, $\alpha3\beta1$, $\alpha5\beta1$, $\alpha6\beta1$, as well as $\alpha v\beta3$ and $\alpha v\beta5$ are highly expressed [20,33]. The integrins $\alpha2\beta1$, $\alpha3\beta1$, $\alpha5\beta1$, and $\alpha6\beta1$ are receptors for the ECM components collagen, fibronectin, and laminin, and by their localization, their overexpression appears to promote tumor cell migration along blood vessels [33]. Integrin $\alpha v\beta3$ is over-expressed in both glial tumor cells and endothelial cells, and its ability to bind to ECM proteins and matricellular proteins implicates it in the regulation of both angiogenesis and the migration of tumor cells [33]. Further, this integrin may play a pivotal role in regulating ECM degradation through the activation of proteases. Colocalization of receptor integrin $\alpha v\beta3$ and the urokinase plasminogen activator receptor (uPAR), through interaction with the upregulated ECM protein vitronectin appears to activate the conversion of plasminogen to plasmin, which in turn promotes the pro-MMP/MMP proteolytic cascade (Fig. 14.2).

Concomitantly, the integrin clustering induces a cytoplasmic association and/or phosphorylation of molecules such as α-actinin, vinculin, tensin, paxillin, and focal adhesion kinase (FAK), which in turn activates a cascade of kinases including ILK, MAPK, and PI3K, and the restructuring of the cytoskeleton, which promotes cell motility (Fig. 14.1) [33]. These downstream effectors may also serve as therapeutic targets. For example, inhibition of ILK activity by a small molecule inhibitor [78] or by the COX-2 specific inhibitor NS-398 [79] reduced *in vitro* invasiveness and cell proliferation, respectively.

The crucial role that integrins play in mediating adhesion and migration makes them attractive as potential therapeutic targets. This approach is substantiated by *in vitro* and *in vivo* studies. For example, αv integrin antagonist EMD 121974 induces apoptosis in brain tumor cells by preventing interaction with the ligands vitronectin and tenascin [80]. Studies such as these have led to the use of integrin antagonists in clinical trials (see Therapeutic Strategies below).

The integrins are also important in coordinating the recruitment of other molecules to focal adhesions. These other ligands include cadherins (Table 14.2) and cell surface counter-receptors of the Ig-superfamily previously discussed (Table 14.2), which also contribute to the overall level of cell adhesion. Thus, integrin–ECM interactions are pivotal to the regulation of cell adhesion and motility, but, as discussed below, they are also mediated or augmented by the interaction with secreted matricellular proteins, proteases, and growth factor receptors. The influences of these molecules on integrin–ECM interactions relative to tumor invasion are also important to consider.

Matricellular Proteins

The matricellular proteins include thrombospondins 1 and 2, Secreted Protein Acidic and Rich in Cysteine (SPARC/osteonectin/BM40), likely hevin/SC1 (a member of the SPARC family of proteins), Tenascins-C and -X, osteopontin and the CCN1-6 family of proteins [81,82] (Table 14.3). The cognate cell membrane receptors have been identified for some, but not all, of these proteins (Table 14.3). This group of proteins consists of developmentally regulated glycoproteins that are secreted into the ECM. These proteins are not grouped as a consequence of their structural similarity. Rather, they are grouped as a result of their shared function as regulators of cell–ECM interactions that, in turn, influence cell function. The features used to distinguish these proteins as a separate group from the structural ECMs have been summarized and include the following [81]:

(1) Matricellular proteins are expressed at high levels during development and in response to injury;
(2) They do not subserve structural roles but function contextually as modulators of cell-matrix interactions;
(3) They bind to many cell surface receptors, ECM proteins, growth factors, cytokines and proteases;
(4) They generally induce de-adhesion of normal cells, which contrasts the function of adhesivity of most ECM proteins; and
(5) In most cases, targeted gene disruption in mice produces either a grossly normal or a subtle phenotype that is exacerbated upon injury.

These proteins are believed to have profound effects on cell behavior such as proliferation, migration, and apoptosis by modulating ECM–integrin, integrin–MMP, and growth factor–growth factor receptor interactions [81]; either by directly binding to the integrin (Fig. 14.3), by binding to the ECM protein (Fig. 14.3), by binding to growth factors (Fig. 14.2), or modulating proteinase inhibitor activity (Fig. 14.2).

In the normal adult brain, the matricellular proteins thrombospondin and Tenascin-C are largely restricted to the vascular basement membrane or endothelial cells where they regulate angiogenesis (Table 14.3). In gliomas, thrombospondin-1 (TSP-1, [83]) is expressed in the walls of small blood vessels in high-grade astrocytic tumors [20]. Tenascin-C (also known as glial/mesenchyme extracellular matrix protein) is expressed in tumor cells and blood vessel walls in regions of microvascular proliferation [84,85], SPARC [86,87], and osteopontin [88,89] are highly expressed in tumor and endothelial cells [86,87,89], and CCNs1-3 are upregulated in tumor cells [90–93].

Several of these proteins are implicated in the regulation of glioma adhesion, migration and/or invasion. TSP-1's affects on adhesion and migration are mediated through multiple integrin interactions, including $\alpha v\beta 3$, $\alpha II\beta 3$, $\alpha 3\beta 1$, $\alpha 4\beta 1$, $\alpha 5\beta 1$, and with the proteoglycans syndecan and HSPG [83]. Incorporation of TN-C into the ECM requires interaction with integrins $\alpha 5$, αv, and $\beta 3$, fibronectin and perlecan [94]. Additional interactions occur with the integrins $\alpha 2\beta 1$, $\alpha 8\beta 1$, $\alpha 9\beta 1$, $\alpha v\beta 3$, and $\alpha v\beta 6$. TN-C modulates glioma migration; however, it has opposing effects on glioma migration, depending on the density of TN-C and that of its integrin ligands. At low density, TN-C promotes pro-adhesive and pro-migratory responses; at high density, it supports anti-adhesive and anti-migratory responses [95], which are modulated by binding to phosphacan [29]. Increased SPARC expression is accompanied by reduced expression of proliferation-associated genes and an upregulation in the expression of matrix proteases and matrix turnover-associated genes [96]. These gene expression changes were consistent with decreased growth and increased invasion observed *in vitro* [97] and *in vivo* [98,99]. Thus, an increase in SPARC expression appears to contribute to the invasive phenotype; however, it should be noted that the extent and route of invasion were dependent upon the amount of SPARC secreted. As noted above for TN-C, these biphasic effects are common to the matricellular proteins. CCNs 1–3 mediate their actions in part through the binding of integrins. CCN3 increases cell adhesion and migration, activates FAK, and binds to integrins at focal adhesion [91]. It also increases migration and MMP3 expression in a PDGF-a-dependent mechanism [92]. In gliomas, CCN3 is implicated in the regulation of proliferation and CCN1 expression is highly upregulated [93].

The complexity of signaling pathways attributed to the matricellular proteins is illustrated in Fig. 14.3. All matricellular proteins can bind independently to specific integrins or ECM proteins. Binding to the ECM proteins may attenuate the downstream signaling induced by integrin–ECM binding; whereas, specific binding to the integrins may initiate different signaling pathways. Either interaction could result in de-adhesion. Further complexity arises in that the different matricellular proteins share binding affinity for the same ECMs, and the downstream signaling may be regulated by the relative levels of overexpression of the matricellular proteins to one another. Such competition in specific cell type backgrounds may be responsible for different observations for individual matricellular proteins regarding their biphasic roles in regulating such diverse biological phenotypes as proliferation, adhesion, and migration.

A further complication is that the function of matricellular proteins may be regulated by their ECM environment. For example, TN-C incorporation into the matrix requires fibronectin and its integrin receptors. It has been suggested that since TN-C also binds to the same integrin receptors, its incorporation into the ECM stabilizes the fibronectin template. This is supported by the observation that fibronectin deposition is reduced in TN-C knockout mice [94]. Similarly, lack of collagen in MOV- null mice results in loss of SPARC retention in the ECM [100].

Considering the pivotal roles that the matricellular proteins play in glioma migration and invasion, they are considered to be attractive therapeutic targets. While the use of full-length TSP-1 is prohibitive as an anti-angiogenic therapy due to its large size, an angiostatic fragment derived from TSP-1, called TSP1ang, was evaluated as a potential agent for C6 gliomas in nude mice. Unfortunately, while the tumors were less vascularized, the tumors were more invasive [101], precluding its use as a therapeutic agent. This experiment underscores the difficulty in predicting the efficacy and outcome of using smaller fragments of multimodular proteins in glioma treatment, a concern that is relevant for many of the matricellular proteins and other molecular targets. The most promising therapy targeting a matricellular protein derives from an antibody against Tenascin-C has been developed and is currently being used in clinical trial (see Therapeutic Strategies). As discussed previously, ILK plays an important in role in focal adhesion signaling, and ILK is currently being evaluated as a therapeutic target in clinical trials (see Therapeutic Strategies). Of note, increased expression of CCN-1 induces increased the activity of ILK [93].

In summary, the astrocytic tumors have remodeled ECM, alterations in adhesion molecules, and aberrant expression of regulatory matricellular proteins including, thrombospondin, Tenascin-C, and SPARC, osteopontin, and CCNs. Their contributions to the migratory and invasive phenotypes make them attractive therapeutic targets. However, the presence of matricellular proteins also influences other integrin interactions that modulate protease extracellular matrix degradation, a process that also has a profound impact on glioma invasion. Thus, studies of the matricellular protein functions and their potential as therapeutic targets should also consider other interacting proteins.

Proteases

Proteases play a major role in promoting brain tumor invasion. Increased expression and activity results in increased proteolysis of the peritumoral ECM, which allows cells detached from their neighboring cells to invade adjacent brain tissue, either intraparenchymally or along nearby blood vessel basement membranes. There are three major classes of proteinases that influence brain tumor invasion (Table 14.4). These include the cathepsins, the serine proteases, and the matrix metalloproteinases (MMPs). The overall effect of activation of these enzymes is a redundancy in and amplification of the degradative process [102].

Cathepsins

The cathepsins are a family of lysosomal proteases that contain cysteine in their active site. They exhibit strict or partial endopeptidase activity and cleave proteins through a redox mechanism. There are over 20 family members denoted A–Z, of which 11 are known human enzymes [103] (Table 14.4). These proteases are synthesized as inactive proenzymes that are either autocatalytically activated (by low pH; Fig. 14.3), or catalytically activated by other enzymes (trypsin, uPA, cathepsin D). The cathepsins are regulated by their cystatin inhibitors.

In normal cells, the cathepsins are highly expressed in the lysosomes, but are also secreted into the matrix where they participate in the degradation of the ECM. In brain tumors, cathepsins B [104–106], D [67,107], H [108], L [109] and S [110] expression and secretion are correlated with either higher expression in tumors compared to normal brains, or demonstrate increased expression during glioma progression [111]. Of these, the lysosomal cathepsin B and non-lysosomal cathepsin D have been more extensively studied in gliomas

and both play a pivotal role in the cascade of proteolytic enzyme activation of plasminogen to plasmin by the uPA/uPAR system and pro-MMPs to active MMPs, which results in broad ECM degradation and the release and activation of ECM constituents such as growth factors VEGF, PDGF, and bFGF [102] (Fig. 14.2). At the cell surface, tissue plasminogen activator (tPA), collagen I, and pro-cathepsin B are sequestered together through the binding of pII and the annexin II tetramer complex [102]. In the ECM, pro-cathepsin D autoactivates and the active cathepsin D together with tPA activates cathepsin B. The matricellular protein TN-C is also a part of this complex, binding to annexin II (Fig. 14.2). However, its influence on cathepsin B activation is as yet unknown. Once active and present in the ECM, cathepsin B can promote the activation of pro-urokinase-like plasminogen activator (pro-uPA; also known as urokinase; urokinase-type) to active uPA. That its expression contributes to tumor invasion was supported by the ability of antisense to cathepsin B to impair invasion of SNB19 cells [112].

uPA and uPAR

Inactive uPA is secreted into the ECM where it is activated primarily by plasmin (but also by other proteinases such as cathepsin B) upon binding to its 55–60 kDa receptor uPAR, which is tethered by a glycosyl phosphatydylinositol (GPI) anchor to the cell membrane (Fig. 14.2). At the cell surface, uPA then converts plasminogen to plasmin, which is a potent and non-specific protein that will hydrolyze virtually any protein. In normal cells, this system is kept in check, in part by plasminogen activator inhibitor 1 (PAI-1), which can induce internalization and degradation of the uPA–uPAR complex, a process that requires the low-density lipoprotein receptor-related protein (LRP) [102]. The increase in plasmin promotes further activation of uPAR-bound uPA, which further reinforces ECM degradation [102] (Fig. 14.2). While this system certainly provides the space for the movement of tumor cells away from the mass into the adjacent brain, the uPA receptor is also implicated in adhesion and migration. It can bind to integrins, such as $\alpha v \beta 3$, which can then serve as transducers of uPAR signaling, and it can bind to ECM proteins such as vitronectin, and thereby alter integrin/ECM interactions and signaling. Both interactions can result in changes to the cytoskeleton and consequently to migration. The situation becomes more complex, with additional layers of control. The $\alpha v \beta 3$ integrin–vitronectin complex can bind PAI-1, which reduces the internalization of uPA–uPAR complex. In turn, this

TABLE 14.4   Human proteinases involved in ECM turnover and degradation

Proteinase			Substrate
Serine	**uPA**		Plasminogen, pro-HGF/SF, FN, PAI-1
	tPA		Plasminogen, pro-cathepsin B
	Plasmin		Broad spectrum
	Cathepsins A, G*		Broad-endopeptidases
Aspartic	Cathepsins **D**, E*		Broad-endopeptidases
Cysteine	Cathepsins **B**, C, F, **H**, K, **L**, O, **S**, V, W, X, Z		Broad-endopeptidases
Metallo-			
Collagenases	Interstitial	MMP-1	CNs 1, II, III, VII, X; gelatin, aggrecan
	Neurophil	MMP-8	CNs 1, II, III, aggrecan
	Collagenase-3	MMP-13	CNs 1, II, III, gelatin, FN, LN, TN
	Collagenase-4	MMP-18	Unknown
Gelatinases	**Gelatinase-A**	**MMP-2**	CNs I, IV, V, VII, X; FN, LN, TN, VN, aggrecan
	Gelatinse-B	**MMP-9**	CNs IV, V, XIV; TN, VN, aggrecan, elastin
Stromelysins	**Stromelysin-1**	**MMP-3**	CNs III, IV, IX, X; FN, LN, TN, VN, gelatin
	Stromelysin-2	MMP-10	CN IV, FN, aggrecan
	Stromelysin-3	MMP-11	CN IV, FN, LN, aggrecan, gelatin
	Matrilysin	**MMP-7**	CN IV, FN, LN, TN, VN, aggrecan, gelatin, elastin
	Endometase	MMP-26	Unknown
Membrane-bound MMPs	**MT1**	**MMP-14**	CNs 1, II, III, FN, LN, VN, proteoglycans, activates pro-MMP2, pro-MMP-9
	MT2	MMP-15	Activates pro-MMP2
	MT3	MMP-16	Activates pro-MMP2
	MT4	MMP-17	Unknown
	MT5	MMP-24	Activates pro-MMP2
	MT6	MMP-25	Unknown
Others	Metalloelastase	MMP-12	Elastin
	Unmanned	MMP-19	Unknown
	Enamelysin	MMP-20	Aggrecan

Note: * — not lysosomal, FN — fibronectin, LN — laminin, TN — tenascin, VN — vitronectin. pro-HGF/SF — pro-hepatocyte growth factor/scatter factor, PAI-1 — plasminogen activator inhibitor-1, **Bold** — upregulated in gliomas.

sequestering of PAI-1 by integrin–ECM can be influenced by the binding of the matricellular protein SPARC to vitronectin [37] (Fig. 14.2).

Thus, there are many levels of activation and deactivation, and a change in one of these components could have a huge impact on overall ECM degradation and tumor cell migration and invasion. In gliomas, uPA, uPAR, LRP, PAI-1, SPARC, VN, and cathepsin B are upregulated in tumors compared to normal brain. Each may be a therapeutic target, but targeting more than one of these may prove to be more beneficial for treatment. For example, co-treatment with adenovirus-mediated expression

of antisense uPAR and antisense cathepsin B inhibits glioma invasion to a greater extent that treatment with either antisense alone [113]. One of the major consequences of plasminogen to plasmin conversion is the amplification of the degradation of ECM *via* activation of pro-matrix metalloproteinases (pro-MMPs) to their active forms.

MMPs

The matrix metalloproteinases (MMPs) [33,102] constitute a family of secreted and membrane-type proteases involved in matrix remodeling. The class of secreted MMPs includes the collagenases that cleave collagens at specific sites, and the gelatinases and the stromelysins that cleave the collagen fragments into smaller peptides (Table 14.4). These smaller peptides can be internalized and degraded in the lysosomes by the cathepsins (described above). Their substrate specificity however is not rigid, and many have overlapping functions. These enzymes are present in normal tissues where their activity is also tightly controlled. They are secreted as proenzymes that must be activated by proteolytic cleavage by other MMPs or plasmin. They are regulated by the TIMPs. A major function of the membrane-type matrix metalloproteinases 1–6 (MT-MMPs 1–6) appears to be the activation of pro-MMP2 to MMP2 (Table 14.4).

In gliomas, there is growing and substantial evidence that MMP-2, MMP-9, MT1-MMP, and perhaps MMP-3 [114] and matrilysin, contribute to ECM degradation and tumor cell migration and invasion [102]. Of these, MMP-2 and MMP-9 are best characterized. MMP-2, MMP-9, and MT1-MMP have been detected in human tumor specimens, and all increase during glioma progression [115,116]. Their role in invasion has been demonstrated by their inhibition of expression. For example, studies of MMP-9 inhibition by antisense significantly decreased the invasion *in vitro* [117,118], and inhibited tumor formation in nude mice [117,118]. Angiopoietin-2 also appears to induce human glioma invasion through the activation of MMP-2 [119].

Therapeutic strategies targeting more than one of these components will likely be more successful. A recent study demonstrated that simultaneous RNAi-mediated targeting of both MMP-9 and cathepsin B significantly inhibited invasion of established SNB19 tumors *in vivo*, compared to RNAi treatment for either MMP-9 or Cathepsin B alone [120]. Also, the synergistic downregulation of uPAR and MMP-9 in SNB19 glioblastoma cells efficiently inhibits glioma invasion [121]. These studies support the concept that multiple targets are a better treatment approach.

In summary, these different classes of proteases not only produce their own specific effects, but they also influence one another. Since brain tumors utilize several protease pathways, there is a means of greatly amplifying the invasive capability of the tumor cells. This is further augmented by the interaction of proteases with adhesion proteins such as integrins, and the resultant modulation of integrin/ECM/matricellular protein interactions (Fig. 14.3). In addition to creating physical space for invading tumor cells, the attack on the matrix also releases inactive growth factors which can in turn be activated to stimulate many biological functions, including cell migration.

Growth Factor/Growth Factor Receptor

Growth factors bind to their cognate growth factor receptors to induce changes in cellular growth. Many growth factors and/or their receptors are upregulated in glioma growth (reviewed in detail in Chapter 11). Interestingly, many growth factors and their receptors have been shown to influence glioma cell migration and invasion [122], including vascular endothelial growth factor (VEGF) and its receptors [122–125], hepatocyte growth factor/scatter factor (HGF/SF) and its receptor c-met [129,126–132], epidermal growth factor receptor (EGFR) and ligands [75,122, 133,134], basic fibroblast growth factor (bFGF) and its receptors [122, 133, 135–138], transforming growth factor-beta (TGF-β) [139], nerve growth factor (NGF) [140–142], insulin-like growth factors-I and II (IGFs-I and II), their type I and type II receptors [141,143,144]. The signaling pathways involved in modulating invasion have been delineated for many of these growth factor–receptor interactions, and most involve an increase in the expression and activation of proteases including MT1-MMP [129,130], MMP-2 [129,130,133,141], MMP-9 [133,142], uPA/uPAR [131,137], PAI-1 [134] other receptors including CD44 [75], other growth factors [136], integrins [146,147], and changes in ECM production [133,148–150]. Of these, VEGF has been a major focus as a therapeutic target.

VEGF is not only a potent factor for angiogenesis in gliomas, but it also contributes to invasion through the upregulation of MMP-2 and MMP-9 [122]. Initially envisioned as a treatment option to inhibit angiogenesis and therefore tumor growth, VEGF has also become a target for anti-invasive strategies since inhibition of VEGF can also inhibit invasion. For example, administration of SU5416 both inhibited migration and reduced expression of the pro-invasive matricellular protein SPARC [123]. However, since the discovery that some tumor cells also express the VEGFR-1 receptor, the use of VEGF as a target should be considered carefully for the following reasons. When VEGF binds its receptor on glioma cells, it inhibits their invasion and proliferation. However, this negative effect on the tumor cells is accompanied by an angiogenic response of endothelial cells. Once sufficient angiogenesis provides adequate oxygen and nutrients, VEGF is downregulated, and the tumor cells are released from its negative regulation and proceed to grow and invade. Thus, targeting VEGF may stop angiogenesis, but in those tumors expressing the receptor, inhibition of VEGF may actually promote invasion [124]. VEGF is also inhibited from binding to its receptor by the matricellular protein SPARC, and therefore the amount of SPARC secreted into the external environment can also attenuate the action of VEGF [125; Fig. 14.3].

Since so many growth factors and/or their receptors are upregulated in gliomas, therapy may be more successful if more than one signaling pathway is targeted. For example, with respect to IGF/IGFR signaling and invasion, a recent study suggests that this signaling mechanism can compensate for the targeted loss of EGFR function by signaling through PI 3-Kinase [144]. This suggests that any therapeutic approach targeting EGFR should also consider targeting the IGF system. In support of this, a recent study demonstrated that co-inhibition of EGFR and type-I IGFR synergistically sensitized human glioma cells to CD95L-induced apoptosis [145].

In summary, many of the growth factor/growth factor receptor signaling pathways essential for normal development are overexpressed in gliomas. Most initiate tyrosine receptor kinase pathways that potentially cross talk or compensate, partially, for the loss of another signaling pathway. Therefore, the potential of each as therapeutic targets should also consider these compensatory mechanisms. In addition, interactions with other matricellular proteins and integrins provide a means of altering and/or enhancing the invasive phenotype.

THERAPEUTIC STRATEGIES

As discussed above, molecular and biological characterization of the gliomas has identified a number of molecules, or classes of molecules, that could be evaluated as therapeutic targets. Only a few have been examined to date.

Integrin Antagonists

One of the best examples of employing anti-invasion therapy may be targeting the $\alpha_v\beta_{3/5}$ integrins. As described previously, the integrins are mediators of cell–matrix interactions, which have important clinical consequences. For example, Uhm et al. showed that vitronectin, a glioma-derived ECM protein and $\alpha_v\beta_{3/5}$ integrin ligand, protects tumor cells from apoptotic death [151]. The $\alpha_v\beta_{3/5}$ integrin is highly expressed in glioma periphery [152] in invading cells which have, in turn, been shown to have upregulated genes including PI3kinase, resulting in the aforementioned apoptosis resistance [19]. The integrin antagonist EMD 121974 (Cilengitide)

induces apoptosis in brain tumor cells growing on vitronectin and tenascin [80] and show efficacy in vivo using orthotopic glioma models [153]. The agent is a cyclic pentapeptide, which binds to the RGD sequence of the integrin, interrupting ligand interaction and blocking downstream signaling including focal adhesion kinase-dependent cell survival pathways [154].

Clinical trials with the $\alpha_v\beta_{3/5}$ antagonist Cilengitide by the NABTT brain tumor consortium have shown excellent safety, occasional objective response, and instances of long-term disease stabilization when used as a single agent [155,156]. Additional studies from this trial have also demonstrated the utility of perfusion-sensitive MRI studies as potential measures of disease efficacy [157].

The lack of more striking activity as a single agent could be explained by the fact that the agent is not highly cytotoxic and that its effects in restoring sensitivity to pro-apoptotic stimuli require co-treatment for full effect. Studies in breast cancer xenografts have demonstrated that Cilengitide targeting of v3 integrin receptor synergizes effectively with radioimmunotherapy to increase efficacy and apoptosis [158]. Studies in our own laboratory with orthotopic glioma have shown striking synergy with radiation [159] and current generation brain tumor clinical trials will test concurrent Cilengitide with pro-apoptotic radiation and chemotherapy.

Integrin signaling through the PI3kinase/AKT pathway has also been targeted. Integrin-linked kinase (ILK) inhibitors effectively inhibit signaling via the ILK-AKT cascade, blocking both basal and EGF-induced phosphorylation of AKT downstream targets, including p70S6K and GSK3a/b. ILK inhibitors reduce growth and invasion in U87 glioma cells and further development as a target for therapy in animal models is underway [160].

Matricellular Proteins

As discussed previously, designing therapeutic peptides to circumvent these proteins may be difficult. To date, the only matricellular protein target taken to clinical trial is Tenascin-C. After demonstrating that anti-tenascin monoclonal antibody 81C6 exhibited therapeutic potential in s.c. and intracranial human xenografts in athymic mice, a study demonstrated that the ^{131}I-labeled antibody demonstrated selective intracranial localization [161]. Currently, a clinical study assessing the efficacy and maximum tolerated dose of 81C6 by bolus injection or micro-infusion is underway [162].

Protease Inhibitors

Marimastat is a low-molecular-weight peptide mimetic inhibitor matrix metalloproteinase (MMP). *In vitro* studies of Marimastat demonstrated significant inhibition of invasion of glioma cell lines, suggesting that MMP inhibitors (MMPIs) may have a role in the treatment of GBM. Marimastat administered adjuvantly to patients with GBM after radiotherapy did not show significant efficacy (unpublished data). In another trial, the combination of temozolomide and Marimastat did result in a benefit [163]. As suggested above, this supports the hypothesis that anti-invasion therapy has little utility alone, but significantly enhances the utility of pro-apoptotic therapies. Another MMP inhibitor, Col-3 has been used in a single agent phase I/II study of using a continuous oral schedule in patients with recurrent high-grade glioma with good safety, but limited efficacy [164].

Cathepsin B and urokinase plasminogen activator and receptor (uPA/uPAR) inhibition has shown some efficacy in pre-clinical models [120,121], but no inhibitors have demonstrated the same in clinical use.

Growth Factor/Growth Factor Receptor Inhibitors

As has been suggested above, growth factor signaling is implicated in the process of glioma proliferation and, possibly, invasion (see Chapter 11 on Growth Factor Signaling Pathways). While specific inhibitors of a number of steps in the signaling cascade have become available, none specifically targets the invasion phenotype *per se*.

OTHER CONSIDERATIONS

So far, these clinical trial examples demonstrate the limited efficacy of targeting single molecules. As previously discussed, *in vitro* data and *in vivo* animal data suggest that a more successful approach in eradicating invasion will likely result when key molecules from the various biological pathways that promote invasion are targeted.

Complicating the issue is the fact that treatments themselves may inadvertently promote invasion. One provocative observation, which could have clinical consequences for invasion-directed therapy, is the finding that radiation itself may promote migration and invasiveness [165]. One might imagine that one strategy tumor cells might employ in the face of attempted therapeutic eradication of cells would be to switch from a proliferative and, therefore, apoptosis-receptive state, to a migratory, apoptosis-resistant state; however, the molecular mechanism for this response is not clearly understood. In a study intended to examine this mechanism, Wick showed that temozolomide was involved with the cleavage of FAK and restored the induction of the apoptosis effector caspase-3 by radiation [166].

SUMMARY

Considerable progress has been made in identifying the molecules involved in promoting glioma migration and invasion. Although many anti-proliferative and proapoptotic therapeutic strategies have been employed, less headway has been made with targeting the molecules involved in invasion. Those that have been used have not been very successful. Importantly, the complex interactions between the different classes of proteins involved in invasion are better delineated and this will hopefully provide the basis for a more successful, multi-targeted approach to the inhibition of this phenotype.

References

1. Scherer, H. D. (1940). Cerebral astrocytomas and their derivatives. *Am J Cancer* **1**, 9–198.
2. Burger, P. C., Heinz, E. R., Shibata, T., and Kleihues, P. (1988). Topographic anatomy and CT correlations in the untreated glioblastoma multiforme. *J Neurosurg* **68**, 698–704.
3. Kelly, P. J., Daumas-Duport, C., Scheithauer, B. W., Kall, B. A., and Kispert, D. B. (1987). Stereotactic histologic correlations of computed tomography and magnetic resonance imaging-defined abnormalities in patients with glial neoplasms. *Mayo Clin Proc* **62**, 450–459.
4. Johnson, P. C., Hunt, S. J., and Drayer, B. P. (1989). Human cerebral gliomas. Correlation of postmortem MR imaging and neuropathologic findings. *Radiology* **170**, 211–217.
5. Burger, P. C. (1987). The anatomy of astrocytomas. *Mayo Clin Proc* **62**, 527–529.
6. Burger, P. C., and Kleihues, P. (1989). Cytologic composition of the untreated glioblastoma with implications for evaluation and needle biopsies. *Cancer* **63**, 2014–2023.
7. Giangaspero, F., and Burger, P. C. (1983). Correlations between cytologic composition and biologic behavior in the glioblastoma multiforme: A postmortem study of 50 cases. *Cancer* **52**, 2320–2333.
8. Burger, P. C. (1990). Classification, grading, and patterns of spread of malignant gliomas. *In Malignant cerebral glioma* (M. V. J. Apuzzo, Ed.), pp. 3–17. American Association of Neurosurgeons.
9. Schiffer, D. (1986). Neuropathology and imaging: the ways in which glioma spreads and varies in its histological aspect.

In Biology of brain tumour (M. D. Walker and D. G. T. Thomas, Eds.), pp. 163–172. Boston: Nijhoff.

10. Choucair, A. K., Levin, V. A., Gutin, P. H. *et al.* (1986). Development of multiple lesions during radiation therapy and chemotherapy in patients with gliomas. *J Neurosurg* **65**, 654–658.

11. Hochberg, F. H., and Pruitt, A. (1980). Assumptions in the radiotherapy of glioblastoma. *Neurology* **30**, 907–911.

12. Awad, I. A. (1987). Spread of malignant gliomas [letter]. *J Neurosurg* **66**, 946–947.

13. Bashir, R., Hochberg, F., and Oot, R. (1988). Regrowth patterns of glioblastoma multiforme related to planning of interstitial brachytherapy radiation fields. *Neurosurgery* **23**, 27–30.

14. Wallner, K. E., Galicich, J. H., Krol, G., Arbit, E., and Malkin, M. G. (1989). Patterns of failure following treatment for glioblastoma multiforme and anaplastic astrocytoma. *Int J Radiat Oncol Biol Phys* **16**, 1405–1409.

15. Liang, B. C., Thornton, A. F., Jr., Sandler, H. M., and Greenberg, H. S. (1991). Malignant astrocytomas: Focal tumor recurrence after focal external beam radiation therapy. *J Neurosurg* **25**, 559–563.

16. Garden, A. S., Maor, M. H., Yung, W. K. A. *et al.* (1991). Outcome and patterns of failure following limited-volume irradiation for malignant astrocytomas. *Radiother Oncol* **20**, 99–110.

17. Tsuboi, K., Yoshii, Y., Nakagawa, K., and Maki, Y. (1986). Regrowth patterns of supratentorial gliomas. Estimation from computed tomographic scans. *Neurosurgery* **19**, 946–951.

18. Giese, A., Bjerkvig, R., Berens, M. E., and Westphal, M. (2003). Cost of migration: Invasion of malignant gliomas and implications for treatment. *J Clin Oncol* **21, 1624–1636.**

19. Joy, A. M., Beaudry, C. E., Tran, N. L. *et al.* (2003). Migrating glioma cells activate the PI3-K pathway and display decreased susceptibility to apoptosis. *J Cell Sci* **116**, 4409–4417.

20. Gladson, C. L. (1999). The extracellular matrix of gliomas: modulation of cell function. *J Neuropathol Exp Neurol* **58**, 1029–1040.

21. Kresse, H., and Schönherr, E. (2001). Proteoglycans of the extracellular matrix and growth control. *J Cell Physiol* **189**, 266–274.

22. Jaworski, D. M., Kelly, G. M., and Hochfield, S. (1994). BEHAB, a new member of the proteoglycan tandem repeat family of hyaluronan-binding proteins that is restricted to the brain. *J Cell Biol* **125**, 495–509.

23. Jaworski, D. M., Kelly, G. M., Piepmeier, J. M., and Hochfield, S. (1996). BEHAB (brain-enriched hyaluronan binding) is expressed in surgical samples of glioma and in intracranial grafts of invasive glioma cell lines. *Cancer Res* **56**, 2293–2298.

24. Zhang, H., Kelly, G., Zerillo, C., Jaworski, D. M., and Hochfield, S. (1998). Expression of a cleaved brain-specific extracellular matrix protein mediates glioma cell invasion in vivo. *J Neurosci* **18**, 2370–2376.

25. Viapiano, M. S., Mattews, R. T., and Hockfield, S. (2003). A novel membrane-associated glycovariant of BEHAB/brevican is upregulated during rat brain development and in a rat model of invasive glioma. *J Biol Chem* **278**, 33239–33247.

26. Paulus, W., Baur, I., Dours-Zimmermann, M. T., and Zimmermann, D. R. (1996). Differential expression of versican isoforms in brain tumors. *J Neuropathol Exp Neurol* **55**, 528–533.

27. Wu, Y., Chen, L., Cao, L., Sheng, W., and Yang, B. B. (2004). Overexpression of the C-terminal PG-M/versican domain impairs growth of tumor cells by intervening in the interaction between epidermal growth factor receptor and β1-integrin. *J Cell Sci* **117**, 2227–2237.

28. Preobrazhensky, A. A., Oohira, A., Maier, G., Veronina, A. S., Vovk, T. S., and Barabanov, V. M. (1997). Identification of monoclonal antibody At5 as a new member of NHK-1 antibody family: the reactivity with myelin-associated glycoprotein and with two brain-specific proteoglycans, phosphacan and neurocan. *Neurochem Res* **22**, 133–140.

29. Grumet, M., Milev, P., Sakurai, T. *et al.* (1994). Interactions with tenascin and differential effects on cell adhesion of neurocan and phosphacan, two major chondroitin sulfate proteoglycans of the central nervous tissue. *J Cell Biol* **269**, 12142–12146.

30. Bertolotto, A., Magrassi, M. L., Orsi, L., and Schiffer, D. (1986). Glycosaminoglycan changes in human gliomas. *J Neurooncol* **4**, 43–48.

31. Annabi, B., Thibault, S., Moumdjian, R., and Béliveau, R. (2004). Hyaluronan cell surface binding is induced by type I collagen and regulated by caveolae in glioma cells. *J Biol Chem* **279**, 21888–21896.

32. Ward, J. A., Huang, L., Guo, H., Ghatak, S., and Toole, B. P. (2003). Perturbation of hyaluronan interactions inhibits malignant properties of gliomas cells. *Am J Pathol* **162**, 1403–1409.

33. Bellail, A. C., Hunter, S. B., Brat, D. J., Tan, C., and Van Meir, E. G. (2004). Microregional extracellular matrix heterogeneity in brain modulates glioma cell invasion. *Int J Biochem Cell Biol* **36**, 1046–1069.

34. Gladson, C. L., and Cherish, D. A. (1991). Glioblastoma expression of vitronectin and the αvβ3 integrin: adhesion mechanism for transformed glial cells. *J Clin Invest* **88**, 1924–1932.

35. Gladson, C. L., Wilcox, J. N., Sanders, L., Gillespie, G. Y., and Cherish, D. A. (1995). Cerebral microenvironment influences expression of the vitronectin gene in astrocytic tumors. *J Cell Sci* **108**, 947–956.

36. Chavakis, T., Kanse, S. M., Yutzy, B., Lijnen, R., and Preissner, K. T. (1998). Vitronectin concentrates proteolytic activity on the cell surface and extracellular matrix by trapping soluble urokinase receptor-urokinase complexes. *Blood* **91**, 2305–2312.

37. Rosenblatt, S., Bassuk, J. A., Alpers, C. E., Sage, E. H., Timpl, R., and Preissner, K. T. (1997). Differential modulation of cell adhesion by interaction between adhesive and counter-adhesive proteins: characterization of the binding of vitronectin to osteonectin (BM-40, SPARC). *Biochem J* **324**, 311–319.

38. Tysnes, B. B., Mahesparan, R., Thorsen, F. *et al.* (1999). Laminin expression by glial fibrillary acidic protein positive cells in human gliomas. *Int J Dev Neurosci* **17**, 531–539.

39. Ljubimova, J. Y., Fugita, M., Khazenzon, N. M. *et al.* (2004). Association between laminin-8 and glial tumor grade, recurrence, and patient survival. *Cancer* **101**, 604–612.

40. Mahesparan, R., Read, T.-A., Lund-Johansen, M., M., Skaftnesmo, K. O., Bjerkvig, R., and Engebraaten, O. (2003). Expression of extracellular matrix components in a highly infiltrative in vivo glioma model. *Acta Neuropathol (Berl)* **105**, 49–57.

41. Fukushima, Y., Ohnishi, T., Arita, N., Hayakawa, T., and Sekiguchi, K. (1998). Integrin α3β1-mediated interaction with laminin-5 stimulates adhesion, migration and invasion of malignant glioma cells. *Int J Cancer* **76**, 63–72.

42. Tysnes, B. B., Larsen, L. F., Ness, G. O. *et al.* (1996). Stimulation of glioma-cell migration by laminin and inhibition by anti-alpha3 and anti-beta1 integrin antibodies. *Int J Cancer* **67**, 777–784.

43. Chintala, S. K., Sawaya, R., Gokaslan, Z. L., Fuller, G., and Rao, J. S. (1996). Immunohistochemical localization of extracellular

matrix proteins in human glioma, both in vivo and in vitro. *Cancer Lett* **101**, 107–114.

44. McComb, R. D., Moul, J. M., and Bigner, D. D. (1987). Distribution of type IV collagen in human gliomas: comparison with fibronectin and glioma-mesenchymal matrix glycoprotein. *J Neuropathol Exp Neurol* **46**, 623–633.

45. Higuchi, M., Ohnishi, T., Arita, N., Hiraga, S., and Hasyakawa, T. (1993). Expression of tenascin in human gliomas: its relation to histological malignancy, tumor differentiation, and angiogenesis. *Acta Neuropathol (Berl)* **85**, 481–487.

46. Lukáš, Z., and Dvorák, K. (2004). Adhesion molecules in biology and oncology. *Acta Vet BRNO* **73**, 93–104.

47. Sasaki, H., Yoshida, K., Ikeda, E. *et al.* (1998). Expression of the neural cell adhesion molecule in astrocytic tumors: an inverse correlation with malignancy. *Cancer* **82**, 1921–1931.

48. Prag, S., Lepekhin, E. A., Kolkova, K. *et al.* (2002). NCAM regulates cell motility. *J Cell Sci* **115**, 283–292.

49. Edvardsen, K., Pedersen, P. H., Bjerkvig, R. *et al.* (1994). Transfection of glioma cells with the neural-cell adhesion molecule NCAM: effect on glioma-cell invasion and growth in vivo. *Int J Cancer* **58**, 116–122.

50. Maidment, S. L., Ruckledge, G. J., Roopai, H. K., and Pilkington, G. J. (1997). An inverse correlation between expression of NCAM-A and the matrix metalloproteinases gelatinase-A and gelatinase-B in human glioma cells in vitro. *Cancer Lett* **116**, 71–77.

51. Sehgal, A., Boynton, A. L., Young, R. F. *et al.* (1998). Cell adhesion molecule Nr-CAM is overexpressed in human brain tumors. *Int J Cancer* **76**, 451–458.

52. Sehgal, A., Ricks, S., Warrick, J., Boynton, A. L., and Murphy, G. P. (1999). Anti-sense human neuroglia related cell adhesion molecule hNr-CAM, reduces the tumorigenic properties of human glioblastoma cells. *Anticancer* **19**, 4947–4953.

53. Boon, K., Edwards, J. B., Eberhart, C. G., and Riggins, G. J. (2004). Identification of astrocytoma associated genes including cell surface markers. *BMC Cancer* **4**, 39.

54. Maepaa, A., Kovanen, P. E., Patau, A., Jaaskelainen, J., and Timonen, T. (1997). Lymphocyte adhesion molecule ligands and extracellular matrix proteins in gliomas and normal brain. *Acta Neuropathol (Berl)* **94**, 216–225.

55. Gringas, M. C., Roussel, E., Bruner, J. M., Branch, C. D., and Moser, R. P. (1995). Comparison of cell adhesion molecule expression between glioblastoma multiforme and autologous normal brain tissue. *J Neuroimmunol* **57**, 143–153.

56. Shinoura, N., Paradies, N. E., Warnick, R. E. *et al.* (1995). Expression of N-cadherin and alpha-catenin in astrocytomas and glioblastomas. *Br J Cancer* **72**, 627–633.

57. Asano, K., Kubo, O., Tajika, Y. *et al.* (1997). Expression and role of cadherins in astrocytic tumors. *Brain Tumor Pathol* **14**, 27–33.

58. Asano, K., Kubo, K., Tajika, Y., Takakur, K., and Suzuki, S. (2000). Expression of cadherin and CSF dissemination in malignant astrocytic tumors. *Neurosurg Rev* **23**, 39–44.

59. Huang, Z. Y., Wu, Y., Hedrick, N., and Gutmann, D. H. (2003). T-cadherin-mediated cell growth regulation involves G2 phase arrest and requires p21 (CIP/WAF1) expression. *Mol Cell Biol* **23**, 566–578.

60. Vajkoczy, P., Blum, S., Lamparter, M. *et al.* (2003). Multistep nature of microvascular recruitment of ex vivo-expanded embryonic endothelial progenitor cells during tumor angiogenesis. *J Exp Med* **197**, 1755–1765.

61. Johnson, K. G., and Van Vactor, D. (2003) Receptor protein tyrosine phosphatases in nervous system development. *Physiol Rev* **83**, 1–24.

62. Norman, S. A., Golfinos, J. G., and Scheck, A. C. (1998). Expression of a receptor protein tyrosine phosphatase in human glial tumors. *J Neurooncol* **36**, 209–217.

63. Ulbricht, U., Brockmann, M. A., Aiger, A. *et al.* (2003). Expression and function of the receptor protein tyrosine phosphatase zeta and its ligand pleiotrophin in human astrocytomas. *J Neuropathol Exp Neurol* **62**, 1265–1275.

64. Muller, S., Kunkel, P., Lamzsus, K. *et al.* (2003). A role for receptor tyrosine phosphatase zeta in glioma cell migration. *Oncogene* **22**, 6661–6668.

65. Adamsky, K., Schilling, J., Garwood, J., Faissner, A., and Peles, E. (2001). Glial cell adhesion is mediated by binding of the FNIII domain of receptor protein phosphatase β (RPTPβ) to Tenascin-C. *Oncogene* **20**, 609–618.

66. Tews, D. S. (2000). Adhesive and invasive features in gliomas. *Pathol* **196**, 701–711.

67. Tews, D. S., and Nissen, A. (1998–1999). Expression of adhesion factors and degrading proteins in primary and secondary glioblastomas and their precursor tumors. *Invasion Metastasis* **18**, 271–284.

68. Camby, I., Belot, N., Lefranc, F. *et al.* (2002). Galectin-1 modulates human glioblastoma cell migration into the brain through modifications to the actin cytoskeleton and levels of expression of small GTPases. *J Neuropathol Exp Neurol* **61**, 585–596.

69. Camby, I., Belot, N., Rorive, S. *et al.* (2001). Galectins are differentially expressed in supratentorial pilocytic astrocytomas, astrocytomas, anaplastic astrocytomas and glioblastomas, and significantly modulate tumor astrocyte migration. *Brain Pathol* **11**, 12–26.

70. Schrappe, M., Klier, F. G., Spiro, R. C., Waltz, T. A., Reisfeld, R. A., and Gladson, C. L. (1991). Correlation of chrondroitin sulfate proteoglycan expression on proliferating brain capillary endothelial cells with the malignant phenotype of astroglial cells. *Cancer Res* **51**, 4986–4993.

71. Murai, T., Miyazaki, Y., Nishinakamura, H. *et al.* (2004). Engagement of CD44 promotes Rac activation and CD44 cleavage during tumor cell migration. *J Biol Chem* **279**, 4541–4550.

72. Ranuncolo, S. M., Ladeda, V., Specterman, S. *et al.* (2002). CD44 expression in human gliomas. *J Surg Oncol* **79**, 30–36.

73. Akiyama, Y., Jung, S., Salhia, B. *et al.* (2001). Haluronate receptors mediating glioma cell migration and proliferation. *J Neurooncol* **53**, 115–127.

74. Yu, Q., and Stamenkovic, I. (1999). Localization of matrix metalloproteinase 9 to the cell surface provides a mechanism for CD44-mediated tumor invasion. *Genes Dev* **13**, 35–48.

75. Monaghan, M., Mulligan, K. A., Gillespie, H. *et al.* (2000). Epidermal growth factor up-regulates CD44-dependent astrocytoma invasion in vitro. *J Pathol* **192**, 519–525.

76. Tsatas, D., Kanagasundaram, V., Kaye, A., and Novak, U. (2002). EGF receptor modifies cellular responses to hyaluronan in glioblastoma cell lines. *J Clin Neurosci* **9**, 282–288.

77. Murphy-Ullrich, J. E. (2001). The de-adhesive activity of matricellular proteins: is intermediate cell adhesion an adaptive state? *J Clin Invest* **107**, 785–790.

78. Troussard, A. A., Costello, P., Yoganathan, T. N., Kumagai, S., Roskelley, C. D., and Dedhar, S. (2000). The integrin linked kinase (ILK) induces an invasive phenotype via AP-1 transcription factor-dependent upregulation of matrix metalloproteinase 9 (MMP-9). *Oncogene* **19**, 5444–5452.

79. Obara, S., Nakata, M., Takeshima, H. *et al.* (2004). Integrin-linked kinase (ILK) regulation of the cell viability in PTEN mutant glioblastoma and in vitro inhibition by the specific COX-2 inhibitor NS-398. *Cancer Letters* **208**, 115–122.

80. Taga, T., Suzuki, A., Gonzalez-Gomez, I. *et al.* (2002). αv-integrin antagonist EMD 121974 induces apoptosis in brain tumor cells growing on vitronectin and tenascin. *Int J Cancer* **98**, 690–697.

81. Bornstein, P., and Sage, E. H. (2002). Matricellular proteins: extracellular modulators of cell function. *Curr Opin Cell Biol* **14**, 608–616.

82. Sullivan, M. M., and Sage, E. H. (2004). Hevin/SC1, a matricellular glycoprotein and potential tumor-suppressor of the SPARC/BM-40/Osteonectin family. *Int J Biochem Cell Biol* **36**, 991–996.

83. Chen, H., Herndon, M. E., and Lawler, J. (2000). The cell biology of thrombospondin-1. *Matrix Biol* **19**, 597–614.

84. Zamecnik, J., Vargova, L., Homola, A., Kodet, R., and Sykova, E. (2004). Extracellular matrix glycoproteins and diffusion barriers in human astrocytic tumours. Neuropathol. *Appl Neurobiol* **30**, 338–350.

85. Zagzag, D., Friedlander, D. R., Miller, D. C. *et al.* (1995). Tenascin expression in astrocytomas correlates for angiogenesis. *Cancer Res* **55**, 907–914.

86. Rempel, S. A., Golembieski, W. A., Ge, S. *et al.* (1998). SPARC: a signal of astrocytic neoplastic transformation and reactive response in human primary and xenograft gliomas. *J Neuropathol Exp Neurol* **57**, 1112–1121.

87. Rempel, S. A., Ge, S., and Gutierrez, J. A. (1999). SPARC: a potential diagnostic marker of invasive meningiomas. *Clin Cancer Res* **5**, 237–241.

88. Saitoh, Y., Kuratsu, J., Takeshima, H., Yamamoto, S., and Ushio, Y. (1995). Expression of osteopontin in human glioma. Its correlation with the malignancy. *Lab Invest* **72**, 55–63.

89. Takano, S., Tsuboi, K., Tomono, Y., Mitsui, Y., and Nose, T. (2000). Tissue factor, osteopontin, alphavbeta3 integrin expression in microvasculature of gliomas associated with vascular endothelial growth factor expression. *Br J Cancer* **82**, 1967–1973.

90. Perbal, B. (2004). CCN proteins: multifunctional signaling regulators. *The Lancet* **363**, 62–64.

91. Perbal, B., Brigstock, D. R., and Lau, L. F. (2003). Report on the second international workshop on the CCN family of genes. *BMA* **56**, 80–85.

92. Laurent, M., Martinerie, C., Thibout, H. *et al.* (2003). NOVH increases MMP3 expression and cell migration in glioblastoma cells via a PDGFR-α-dependent mechanism. *FASEB J* **17**, 1919–1921.

93. Xie, D., Yin, D., Tong, X. *et al.* (2004a). Cyr61 is overexpressed in gliomas and involved in integrin-linked kinase-mediated Akt and β-catenin-TCF/Lef signaling pathways. *Cancer Res* **64**, 1987–1996.

94. Jones, P. L., and Jones, F. S. (2000). Tenascin-C in development and disease: gene regulation and cell function. *Matrix Biol* **19**, 581–596.

95. Giese, A., Loo, M. A., Norman, S. A., Treaurywala, S., and Berens, M. E. (1996). Contrasting migratory response of astrocytoma cells to tenascin mediated by different integrins. *J Cell Sci* **109**, 2161–2168.

96. Golembieski, W. A., and Rempel, S. A. (2002). cDNA array analysis of SPARC-modulated changes in glioma gene expression. *J Neurooncol* **60**, 213–226.

97. Golembieski, W. A., Ge, S., Nelson, K., Mikkelsen, T., and Rempel, S. A. (1999). Increased SPARC expression promotes U87 glioblastoma invasion in vitro. *Int J Dev Neurosci* **17**, 463–472.

98. Schultz, C., Lemke, N., Ge, S., Golembieski, W. A., and Rempel, S. A. (2002). Secreted protein acidic and rich in cysteine promotes glioma invasion and delays tumor growth in vivo. *Cancer Res* **62**, 6270–6277.

99. Rich, J. N., Shi, Q., Hjelmeland, M. *et al.* (2003). Bone-related genes expressed in advanced malignancies induce invasion and metastasis in a genetically defined human cancer model. *J Biol Chem* **278**, 15951–15957.

100. Iruela-Arispe, M. L., Vernon, R. B., Rudolf, J., and Sage, E. H. (1996). Type I collagen-deficient MOV-13 mice do not retain SPARC in the extracellular matrix: implications for fibroblast function. *Dev Dyn* **207**, 171–183.

101. De Fraipont, F., Keramidas, M., El Atifi, M., Chambaz, E. M., Berger, F., and Fiege, J.-J. (2004). Expression of the thrombospondin 1 fragment 167–569 in C6 glioma cells stimulates tumorigenicity despite reduced neovascularization. *Oncogene* **23**, 3642–3649.

102. Rao, J. S. (2003). Molecular mechanisms of glioma invasiveness: the role of proteases. *Nat Rev Cancer* **3**, 489–501.

103. Turk, V., Turk, B., and Turk, D. (2001). Lysosomal cysteine proteases: facts and opportunities. *EMBO J* **20**, 4629–4633.

104. Rempel, S. A., Rosenblum, M. L., Mikkelsen, T. *et al.* (1994). Cathepsin B expression and localization in glioma progression and invasion. *Cancer Res* **54**, 6027–6031.

105. Mikkelsen, T., Yan, P. S., Ho, K. L., Sameni, M., Sloane, B. F., and Rosenblum, M. L. Immunolocalization of cathepsin B in human glioma; implications for tumor invasion and angiogenesis. *J Neurosurg* **83**, 285–290.

106. Sivaparvathi, M., Sawaya, R., Wang, S. W. *et al.* (1995). Overexpression and localization of cathepsin B during the progression of human gliomas. *Clin Exp Metastasis* **13**, 49–56.

107. Sivaparvathi, M., Sawaya, R., Chintala, S. K., Go, Y., Gokaslan, Z. L., and Rao, J. S. (1996a). Expression of cathepsin D during the progression of human gliomas. *Neurosci Lett* **208**, 171–174.

108. Sivaparvathi, M., Sawaya, R., Gokaslan, Z. L., Chintala, S. K., Rao, J. S., and Chintala, K. S. (1996b). Expression and the role of cathepsin H in human glioma progression and invasion. *Cancer Letters* **104**, 121–126.

109. Sivaparvathi, M., Yamamoto, M., Nicolson, G. L., Gokaslan, Z. L., and Rao, J. S. (1996c). Expression and immunohistochemical localization of cathepsin L during the progression of human gliomas. *Clin Exp Metastasis* **14**, 27–34.

110. Flannery, T., Gibson, D., Mirakhur, M. *et al.* (2003). The clinical significance of cathepsin S expression in human astrocytomas. *Am J Pathol* **163**, 175–182.

111. Levicar, N., Strojnik, T., Kos, J., Dewey, T. A., Pilkington, G. J., and Lah, T. T. (2002). Lysosomal enzymes, cathepsins in brain tumour invasion. *J Neurooncol* **58**, 21–32.

112. Mohanam, S., Jasti, S. L., Kondrganti, S. R. *et al.* (2001). Down-regulation of cathepsin B expression impairs the invasive and tumorigenic potential of human glioblastoma cells. *Oncogene* **20**, 3663–3673.

113. Gondi, C. S., Lakka, S. S., Yanamandra, N. *et al.* Adenovirus-mediated expression of antisense urokinase plasminogen activator receptor and antisense cathepsin B inhibits tumor growth, invasion, and angiogenesis in gliomas. *Cancer Res* **64**, 4069–4077.

114. Mercapide, J., Lopez De Cicco, R., Castresana J. S., Castresana J. S., and Klein-Szanto, A. J. (2003). Stromelysin-1/ matrix metalloproteinase 3 (MMP-3) expression accounts from invasive properties of human astrocytoma cell lines. *Int J Cancer* **106**, 676–682.

115. Rao, J. S., Yamamoto, M., Mohanam, S. *et al.* (1996). Expression and localization of 92 kDa type IV collagenases/ gelatinase B (MMP-9) in human gliomas. *Clin Exp Metastasis* **14**, 21–28.

116. Forsyth, P. A., Wong, H., Laing, T. D. *et al.* (1999). Gelatinase-A (MMP-2), gelatinase B (MMP-9) and membrane type matrix metalloproteinase-1 (MT1-MMP) are involved in different aspects of the pathophysiology of malignant gliomas. *Br J Cancer* **79**, 1828–1835.

117. Kondraganti, S., Mohanam, S., Chintala, S. K. *et al.* (2000). Selective suppression of matrix metalloproteinase-9 in human glioma cells by antisense gene transfer impairs glioblastoma cell invasion. *Cancer Res* **60**, 6851–6855.

118. Lakka, S. S., Rajan, M., Gondi, C. *et al.* (2002). Adenovirus-mediated expression of antisense MMP-9 in glioma cells inhibits tumor growth and invasion. *Oncogene* **21**, 8011–8019.

119. Hu, B., Guo, P., Fang, Q. *et al.* (2003) Angiopoietin-2 induces human glioma invasion through the activation of matrix metalloprotease-2. *Proc Natl Acad Sci USA* **100**, 8904–8909.

120. Lakka, S. S., Gondi, C. S., Yanamandra, N. *et al.* (2004). Inhibition of cathepsin B and MMP-9 gene expression in glioblastoma cell line via RNA interference reduces tumor cell invasion, tumor growth and angiogenesis. *Oncogene* **23**, 4681–4689.

121. Lakka, S. S., Gondi, C. S., Yanamandra, N. *et al.* (2003). Synergistic down-regulation of urokinase plasminogen activator receptor and matrix metalloproteinase-9 in SNB19 glioblastoma cells efficiently inhibits glioma cell invasion, angiogenesis, and tumor growth. *Cancer Res* **63**, 2454–2461.

122. Mueller, M. M., Werbowetski, T., and del Maestro, R. F. (2003). Soluble factors involved in glioma invasion. *Acta Neurochir (Wein)* **145**, 999–1008.

123. Vajkoczy, P., Menger, M. D., Goldbrunner, R. *et al.* (2000). Targeting angiogenesis inhibits tumor infiltration and expression of the pro-invasive protein SPARC. *Int J Cancer* **87**, 261–268.

124. Harold-Mende, C., Steiner, H.-H., Andl, T. *et al.* (1999). Expression and functional role of VEGF receptors in human tumor cells. *Lab Invest* **79**, 1573–1582.

125. Kupprion, C., Motamed, K., and Sage, E. H. (1998). SPARC (BM-40, osteonectin) inhibits the mitogenic effect of vascular endothelial growth factor on microvascular endothelial cells. *J Biol Chem* **273**, 29635–29640.

126. Moriyama, T., Kataoka, H., Koono, M., and Wakisak, S. (1999). Expression of hepatocyte growth factor/scatter factor and its receptor c-met in brain tumors: evidence for its role in progression of astrocytic tumors (review). *Int J Med* **3**, 531–536.

127. Lamzus, K., Schmidt, N. O., Lin, J. *et al.* (1998). Scatter factor promotes motility of human glioma and neuromicrovascular endothelial cells. *Int J Cancer* **75**, 19–28.

128. Lamzus, K., Laterra, J., Westphal, M., and Rosen, E. M. (1999). Scatter factor/hepatocyte growth factor (SF/HGF) content and function in human gliomas. *Int J Cancer* **17**, 517–530.

129. Hamasuna, R., Kataoka, H., Moriyama, T., Itoh, H., Seiki, M., and Koono, M. (1999). Regulation of matrix metalloproteinase-2 (MMP-2) by hepatocyte growth factor/scatter factor (HGF/SF) in human glioma cells: HGF/SF enhances MMP-2 expression and activation accompanying up-regulation of membrane type-1 MMP. *Int J Cancer* **82**, 274–281.

130. Hirohito, Y., Hara, A., Murase, S., Nakano, S., Koono, M., and Wakisak, S. (2001). Expression of hepatocyte growth factor and matrix metalloproteinase-2 in human glioma. *Brain Tumor Pathol* **18**, 7–12.

131. Moriyama, T., Kataoka, H., Hamasuna, R., Nakano, S., Koono, M., and Wakisak, S. (1999b). Upregulation of urokinase type plasminogen activator and u-PA receptor by hepatocyte

132. Brockmann, M. A., Papadimitriou, A., Brandt, M., Fillbrandt, R., Westphal, M., and Lamzsus, K. (2003). Inhibition of intracerebral glioblastoma growth by local treatment with the scatter factor/hepatocyte growth factor-antagonist NK4. *Clin Cancer Res* **9**, 4578–4585.

133. Rooprai, H. K., Rucklidge, G. J., Panou, C., and Pilkington, G. J. (2000). The effect of exogenous growth factors on matrix metalloproteinase secretion by human brain tumor cells. *Br J Cancer* **82**, 52–55.

134. Kazsa, A., Kowanetz, M., Poslednik, K., Witek, B., Kordula, T., and Koj, A. (2001). Epidermal growth factor and pro-inflammatory cytokines regulate the expression of components of the plasminogen activation system in U737-MG astrocytoma cells. *Cytokine* **16**, 187–190.

135. Koocheckpour, S., Merzak, A., and Pilkington, G. J. (1995). Extracellular matrix proteins inhibit proliferation, upregulate migration and induce morphological changes in human glioma cell lines. *Eur J Cancer* **31**, 375–380.

136. Aguste, P., Gürsel, D. B., Lemiere, S. *et al.* (2001). Inhibition of fibroblast growth factor/fibroblast growth factor receptor activity in glioma cells impedes tumor growth both by angiogenesis-dependent and -independent mechanisms. *Cancer Res* **61**, 1717–1726.

137. Mori, T., Tatsuya, A., Wakabayashi, Y. *et al.* (2000). Up-regulation of urokinase-type plasminogen activator and its receptor correlates with enhanced invasion activity of human glioma cells mediated by transforming growth factor-α or basic fibroblast growth factor. *J Neurooncol* **46**, 115–123.

138. Ovstrovsky, O., Berman, B., Gallagher, J. *et al.* (2002). Differential effects of heparin saccharides on the formation of specific fibroblast growth factor (FGF) and FGF receptor complexes. *J Biol Chem* **277**, 2444–2453.

139. Yamada, N., Kato, M., Yamashita, H. *et al.* (1995). Enhanced expression of transforming growth factor-beta and its type-I and -II receptors in human glioblastoma. *Int J Cancer* **62**, 386–392.

140. Hamel, W., Westphal, M., Szonyi, E. *et al.* (1993). Neurotrophin gene expression by cell lines derived from human gliomas. *J Neurosci Res* **34**, 147–157.

141. Wang, H., Wang, H., Shen, W. *et al.* (2003). Insulin-like growth factor binding protein 2 enhances glioblastoma invasion by activating invasion-related genes. *Cancer Res* **63**, 4315–4321.

142. Kahn, K. M., Falcone, D. J., and Kraemer, R. (2002). Nerve growth factor activation of Erk-1 and Erk-2 induced matrix metalloproteinase-9 expression in vascular smooth muscle cells. *J Biol Chem* **277**, 2353–2359.

143. Zumkeller, W., and Westphal, M. (2001). The IGF/IGFBP system in CNS malignancy. *J Clin Pathol Mol Pathol* **54**, 227–229.

144. Chakravarti, A., Zhai, G., Suzuki, Y. *et al.* (2004). The prognostic significance of phosphatidylinositol 3-kinase pathway activation in human gliomas. *J Clin Oncol* **22**, 1926–1933.

145. Steinbach, J. P., Eisenmann, C., Klumpp, A., and Weller, M. (2004). Co-inhibition of epidermal growth factor receptor and type I insulin-like growth factor receptor synergistically sensitizes human malignant glioma cells to CD95L-induced apoptosis. *Biochem Biophys Res Commun* **321**, 524–530.

146. Miyake, K., Kimura, S., Nakanishi, M. *et al.* (2000). Transforming growth factor beta-1 stimulates contraction of

human glioblastoma cell-mediated collagen lattice through enhanced alpha2 integrin expression. *J Neuropathol Exp Neurol* **59**, 18–28.

147. Platten, M., Wick, W., Wild-Bode, C., Aulwurm, S., Dichgans, J., and Weller, M. (2000). Transforming growth factor beta-1 (TGF-beta1) and TGF-beta2 promote glioma cell migration of alphaV-beta3 integrin expression. *Biochem Biophy Res Commun* **268**, 607–611.

148. Neubauer, K., Krüger, M., Qondamatteo, F., Knittel, T., Siale, B., and Ramadori, G. (1999). Transforming growth factor-$\beta1$ stimulates he synthesis of basement membrane protein laminin, collagen type IV and entactin in rat liver sinusoidal endothelial cells. *J Hepathol* **31**, 692–702.

149. Paulus, W., Baur, I., Huettner, C. *et al.* (1995). Effects of transforming growth factor-beta 1 on collagen synthesis, integrin expression, adhesion and invasion of glioma cells. *J Neuropathol Exp. Neurol* **54**, 236–244.

150. Wick, W., Platten, M., and Weller, M. (2001). Glioma cell invasion: regulation of metalloproteinase activity by TGF-β. *J Neurooncol* **53**, 177–185.

151. Uhm, J. H., Dooley, N. P., Kyritsis, A. P., Rao, J. S., and Gladson, C. L. (1999). Vitronectin, a glioma-derived extracellular matrix protein, protects tumor cells from apoptotic death. *Clin Cancer Res* **5**, 1587–1594.

152. Bello, L., Francolini, M., Marthyn, P. *et al.* (2001) $\alpha v\beta3$ and $\alpha v\beta5$ integrin expression in glioma periphery. *Neurosurgery,* **49**, 380–390.

153. MacDonald, T. J., Taga, T., Shimada, H. *et al.* (2001). Preferential susceptibility of brain tumors to the antiangiogenic effects of an αv integrin antagonist. *Neurosurgery* **48**, 151–157.

154. Chatterjee, S., Hiemstra Brite, K., and Matsumura, A. (2001). Induction of apoptosis of Integrin-expressing human prostate cancer cells by cyclic arg-gly-asp peptides. *Clin Cancer Res* **7**, 3006–3011.

155. Nabors, L. B., Rosenfeld, S. S., Mikkelsen, T. *et al.* (2002). Phase I trial of EMD 121974 for treatment of patients with recurrent malignant gliomas A236. *Neuro-Oncology* **4**, 373.

156. Nabors, L. B., Rosenfeld, S. S., Mikkelsen, T. *et al.* (2004). NABTT 9911: A phase I trial of EMD 121974 for treatment of patients with recurrent malignant gliomas. TA-39. *Neuro-Oncology* **6**, 379.

157. Shastry, A. N., Twieg, D. B., Mikkelsen, T. *et al.* (2004). Assessment of brain tumor angiogenesis inhibitors using perfusion magnetic resonance imaging – quality and analysis results of a multi-institution phase I trial. *J Magn Reson Imag* **20**, 913–922.

158. Burke, P. A., DeNardo, S. J., Miers, L. A., Lamborn, K. R., Matzku, S., and DeNardo G. L. (2002) Cilengitide targeting of v 3 integrin receptor synergizes with radioimmunotherapy to increase efficacy and apoptosis in breast cancer xenografts. *Cancer Res* **62**, 4263–4272.

159. Mikkelsen, T., Nelson, K., and Brown, S. (2004). Cilengitide synergy with radiation therapy – molecular mechanisms. Proceedings of the AANS/CNS Section on Tumors Sixth Biennial Satellite Symposium. *J Neuro-Oncol* **69**.

160. Koul, D., Bergh, S., Shen, R., and Yung, W. K. A. (2004). Targeting integrin linked kinase (ILK) pathway in human glioblastoma. ET-10. *Neuro-oncology* **6**, 332.

161. Zalutsky, M. R., Moseley, R. P., Coakham, H. B., Coleman, R. E., and Bigner, D. D. (1989). Pharmacokinetics and tumor localization of 131I-labeled anti-tenascin monoclonal antibody 81C6 in patients with gliomas and other intracranial malignancies. *Cancer Res* **49**, 2807–2813.

162. Badruddoja, M. A., Reardon, D. A., Akabani, G. *et al.* (2004). Phase II trial of iodine 131-labeled murine anti-tenascin monoclonal anti-body 81C6 (M81C6) via surgically created resection cavity in the treatment of patients with recurrent malignant brain tumors. *J Clin Oncol* **22**, 4S, 1569.

163. Groves, M. D., Puduvalli, V. K., Hess, K. R. *et al.* (2002). Phase II trial of temozolomide plus the matrix metalloproteinase inhibitor, marimastat, in recurrent and progressive glioblastoma multiforme. *J Clin Oncol* **20**, 1383–1388.

164. New, P., Mikkelsen, T., Phuphanich, S. *et al.* (2002). A phase I/II study of col-3 administered on a continuous oral schedule in patients with recurrent high grade glioma: preliminary results of the NABTT 9809 clinical trial A237. *Neuro-Oncology* **4**, 373.

165. Wild-Bode, C., Weller, M., Rimner, A., Dichgans, J., and Wick, W. (2001). Sublethal irradiation promotes migration and invasiveness of glioma cells: Implications for radiotherapy of human glioblastoma. *Cancer Res* **61**, 2744–2750.

166. Wick, W., Wick, A., Schulz, J. B., Dichgans, J., Rodemann, H. P., and Weller, M. (2002). Prevention of radiation-induced glioma cell invasion by Temozolomide involves caspase 3 activity and cleavage of focal adhesion kinase. *Cancer Res* **62**, 1915–1919.

Mechanisms of Angiogenesis in Brain Tumors and their Translation into Therapeutic Anti-tumor Strategies

Till Acker and Karl H. Plate

SUMMARY

Despite an aggressive multimodal therapeutic approach, malignant gliomas have retained their dismal prognosis, which remained virtually unchanged over the last three decades, warranting the need for novel therapeutic modalities. The acquisition of a functional blood supply seems to be rate limiting for the tumor's ability to grow beyond a certain size and metastasize to other sites. Although vessel growth and maturation are complex and highly coordinated processes requiring the sequential activation of a multitude of factors, there is consensus that VEGF, angiopoietin, and possibly Eph/ephrin signaling represent crucial steps in tumor angiogenesis. In this review, we will present the accumulating evidence indicating that hypoxia and the key transcriptional system, HIF (hypoxia-inducible factor), are the major triggers for new blood vessel growth in malignant tumors. We will provide a model in which hypoxia, HIF, and several HIF-target genes participate in the coordinate collaboration between tumor, endothelial, inflammatory/hematopoietic, and circulating endothelial precursor cells to enhance and promote tumor vascularization. The current status and future perspectives of angiogenesis inhibition will be discussed. The current potential lies in the employment of an anti-angiogenic multimodal approach including radiotherapy or metronomically scheduled chemotherapy to counteract selection of less vascular-dependent tumor cell variants and prevent radio-resistance of endothelial cells conferred by tumor mediated production of cytoprotective factors. Integrated understanding of the influence of therapeutic approaches on the intricate tumor microenvironment may offer new opportunities for therapeutic intervention and tailoring of current treatments to enhance therapeutic efficacy.

Abbreviation: Ang – angiopoietin; ARNT – aryl hydrocarbon nuclear translocator; FIH – factor-inhibiting-HIF (asparagine hydroxylase); CA – carbonic anhydrase; CEP – circulating endothelial progenitor cell; EC – endothelial cell; HIF – hypoxia-inducible factor; HRE – hypoxia responsive element; MDM-2 – murine double minute -2; HPHD – HIF-prolyl hydroxylase; pVHL – von Hippel Lindau protein; PTEN – phosphatase and tensin homologue deleted on chromosome ten; TAM – Tumor-associated macrophage; VEGF – vascular endothelial growth factor

INTRODUCTION

In the United States, twenty thousand new primary central nervous system (CNS) neoplasms are diagnosed each year. CNS tumors are the second most frequent malignancy of childhood with their incidence in adults increasing with age. Importantly, cancers of the CNS are among the most debilitating of human malignancies as they affect the organ that defines the "self," often severely compromising quality of life. The CNS tumor classification by the

World Health Organization (WHO) recognizes a multitude of different neoplastic CNS entities of which malignant gliomas are the most common and most studied primary malignancies. Despite an aggressive multimodal therapeutic approach, malignant gliomas have retained their dismal prognosis, which remained virtually unchanged over the last three decades, warranting the need for novel therapeutic modalities.

Tumor growth depends on vascular supply to sustain the metabolic needs of the tumor tissue. Indeed, the acquisition of a functional blood supply seems to be rate limiting for the tumor's ability to grow beyond a certain size and metastasize to other sites. Glioblastomas are characterized by a prominent, proliferative vascular component [1] making them a suitable candidate for anti-angiogenic therapy approaches. The mechanisms by which the growing tumor tissue recruits new blood vessels has been the subject of intense investigations in the last few years, thereby providing novel targets for cancer therapy. Moreover, from a therapeutic point of view, the vascular architecture of the tumor highly determines the efficacy of delivery of anti-cancer drugs in effective quantities and to all regions within the tumor tissue. A mounting body of evidence suggests that hypoxia and the key transcriptional system, HIF (hypoxia-inducible factor), are the major triggers for new blood vessel growth in malignant tumors. Although vessel growth and maturation is a complex and coordinated process requiring the sequential activation of a multitude of factors, there is consensus that VEGF, angiopoietin, and possibly Eph/ephrin signaling represent crucial steps in tumor angiogenesis. The present review focuses on the important role of hypoxia in activating these key signaling molecules and in guiding the cross talk between different cellular components within the tumor microenvironment, both synergistically acting to govern tumor angiogenesis. It further discusses some therapeutic strategies to target the tumor vasculature and the hypoxic tumor cell population.

THE HYPOXIC TUMOR MICROENVIRONMENT AND HIF ACTIVATION

Tumor growth and progression occurs as a result of cumulative acquisition of genetic alterations affecting oncogenes or tumor suppressor genes selecting for tumor cell clones with enhanced proliferation and survival potential. However, high proliferating tumors frequently outstrip their vascular supply leading to a tumor microenvironment characterized by low oxygen tension, low glucose levels, and an acidic pH. Hypoxia is a common feature of solid tumor growth. Reduced pO_2 levels have been found in the majority of human tumors analyzed as compared to normal tissue of the corresponding organ [2,3]. A wide range of genes known to be involved in adaptive mechanisms to hypoxia such as angiogenic growth factors, enzymes of glucose metabolism, and pH regulation have classically been associated with tumors. Many of these genes have subsequently been shown to be regulated by HIF function (see below). To date, more than 60 genes have been identified as potential HIF-target genes providing further evidence for HIF as a key regulatory system of adaptive mechanisms [4,5]. HIF activation is a frequent finding in human tumors. Indirect experimental evidence for the induction of HIF activity by the hypoxic tumor environment came from studies showing perinecrotic expression patterns of HIF target genes or HRE-driven reporter genes [6–9]. Immunolabeling studies using monoclonal antibodies raised against the HIF-α subunit, which determines HIF activity, demonstrated increased HIF-α expression in a broad range of cancers as compared to the respective normal tissue [10,11]. When tumors develop, they often become more malignant with time, a process termed tumor progression. Overall, HIF expression is increased in more aggressive tumors and has recently been shown to correlate with tumor grade and tumor progression in a series of human brain tumors. While little or no HIF-α immunoreactivity could be detected in low-grade gliomas, glioblastomas revealed intense HIF-α immunostaining in perinecrotic tumor cells suggesting regulation by microenvironmental tumor hypoxia [12]. This data nicely correlates with previous studies showing increased vascularization and significantly higher levels of VEGF (vascular endothelial growth factor) in high grade gliomas [13]. Indeed, induction of blood vessel growth is one of the major adaptive mechanisms employed by the tumor to increase oxygen and nutrient supply. HIF serves as a key player in this process in response to hypoxia through transactivation of various angiogenic growth factors such as VEGF.

The Transcription Factor System HIF

Since its discovery in 1995, the HIF transcriptional complex has emerged as the key regulatory system of adaptive mechanisms in response to reductions in oxygen tension [14,15]. The HIF transcriptional complex is a heterodimer composed of HIF-α and HIF-β subunits belonging to the bHLH (basic helix loop helix)-PAS family of transcription factors These are conserved among mammalian cells and invertebrate model organisms such as *Drosophila melanogaster* and *Caenorhabditis elegans*. Both HIF-α and HIF-β proteins exist as homologs (HIF-1α [14,15], HIF-2α [16-19], HIF-3α [20] and ARNT, ARNT2, and ARTN3, respectively). Specificity to hypoxia-mediated responses is conferred by HIF-α subunits. At least two mammalian α subunits, HIF-α and HIF-2α, are regulated by oxygen in a similar fashion [21–23].

Regulation of HIF activity is complex involving multiple mechanisms of control at the level of mRNA expression, protein stability, nuclear translocation, and transactivation activity. These combine co-operatively to induce HIF activity to maximal levels under decreasing oxygen concentrations. On the molecular level, this is mediated by subjecting HIF-α subunits to multiple modes of modification including two different types of hydroxylation, acetylation and phosphorylation. Cellular O_2 concentrations regulate transcriptional activity of HIF-α subunits *via* influencing protein levels and transactivation domain functions. Under normoxic conditions HIF-α subunits are subject to rapid ubiquitination and proteosomal degradation [24–26]. Decreasing oxygen tensions dramatically reduce ubiquitination resulting in a rapid increase in protein levels [27]. In addition, deletion analysis revealed that HIF-α contains two transactivation regions, termed the amino- and carboxy-terminal transactivation domain (N-TAD, C-TAD), which upon reduced oxygen levels are relieved of a negative control [22,23,28,29]. HIF-α subunits subsequently translocate into the nucleus where they dimerize with HIF-β subunits, allowing binding to the conserved consensus DNA-binding motif RCGTG residing in the hypoxia-responsive elements (HRE) of many oxygen-regulated genes.

Oxygen-dependent HIF Activation

O_2-regulated degradation of HIF-α subunits is mediated by a functional domain of approximately 200 amino acids, termed oxygen-dependent degradation domain (ODD). This domain confers hypoxic stabilization to HIF-1α and HIF-2α, the feature of which is transferable to various fusion partner proteins [22,23,25,27,30–32]. pVHL (von Hippel Lindau protein), loss of function of which is implicated in VHL disease, acts as the recognition component of an E3 ubiquitin-protein ligase that binds to subsequences within the ODD, thus targeting

HIF-α subunits for proteosomal degradation [33-39]. HIF-α protein stability seems to be also regulated by ubiquitin ligases other than VHL, e.g., p53 has been implicated in promoting HIF degradation and decreasing transactivation of HRE-bearing genes possibly by promoting binding of the ubiquitin ligase MDM-2 and competing for the co-activator p300 [40,41].

Interaction of VHL with HIF-α requires an O_2- and iron-dependent hydroxylation of specific prolyl residues (Pro 402, Pro 564) within the HIF-α ODD. These posttranslational modifications are conferred by a distinct, conserved subclass of 2-oxoglutarate-dependent-oxygenases termed HIF-prolyl hydroxylase (HPHD). So far, four homologs of HPHD have been described (HPHD I-IV) [42–44]. A second oxygen-dependent switch involves an iron and 2-oxoglutarate-dependent hydroxylation of an asparagine residue within the C-TAD of HIF-α subunits by a recently identified HIF asparaginyl hydroxylase called factor-inhibiting HIF (FIH-1) [45]. Asparagine hydroxylation apparently interferes with recruitment of the coactivator p300 resulting in reduced transcriptional activity. Both, HPHD and FIH, belong to a superfamily of 2-oxoglutarate dependent hydroxylases which employ non-haem iron in the catalytic moiety [46]. They require oxygen in the form of dioxygen with one oxygen atom being incorporated in the prolyl or asparagyl residue, respectively, and the other into 2-oxoglutarate yielding succinate and CO_2 [47]. Thus, the hydroxylation reaction is inherently dependent on ambient oxygen pressure, providing a direct link between oxygen-dependent enzymatic activity and the regulation of hypoxia-inducible responses, a crucial criterion for an oxygen sensor (reviewed in [48]).

Oxygen-independent Mechanisms of HIF Activation

Apart from microenvironmental tumor hypoxia additional mechanisms have been identified which influence HIF function. HIF expression or activity is increased in response to genetic alterations, inactivating tumor suppressor genes or activating oncogenes, and in response to activation of various growth factor pathways [5,49,50]. This induction is generally less intense than that mediated by reductions in oxygen tension. [51] Recently, a novel mode of posttranslational modification of HIF-α subunits under normoxia has been identified which involves acetylation of a lysine residue (Lys 532) within the ODD domain by an acetyl-transferase termed ARD1. Lysyl acetylation has been shown to modulate HIF-α protein stability by promoting VHL-binding and subsequent proteosomal degradation. With decreasing oxygen tensions, acetylation is gradually reduced due to decreased ARD1 mRNA levels and decreased affinity of ARD1 to HIF-α subunits [52].

THE HYPOXIC TUMOR MICROENVIRONMENT, ANGIOGENESIS, AND EDEMA

Maintaining the body's pO_2 constant is absolutely essential for mammalian life. A highly developed, multilevel physiological system is devoted to the body's oxygen homeostasis. One of the first physiological systems to be established during development is the blood circulation. Failure in cardiac, erythroid, or vascular development leads to early embryonic lethality, suggesting that in multicellular organisms, oxygen as well as nutrients need to be delivered by a functional circulatory system. Mammalian cells are located within a distance of 100–200 μm (the diffusion limit of oxygen) from blood vessels. To grow beyond this size, multicellular organisms and tissues need to recruit new blood vessels by vasculogenesis and angiogenesis. The establishment of a highly branched vascular system is regulated by a balance between pro- and anti-angiogenic factors. This finely tuned balance is disturbed in various diseases including some of the leading causes of mortality in the West such as stroke, ischemic heart disease, and cancer.

In 1971, Folkman proposed that solid tumor growth is angiogenesis dependent. It is now widely accepted that tumors and metastases need to acquire a functional blood supply to grow beyond a volume of several mm^3. Thus, absence of angiogenesis can be considered as rate-limiting for tumor growth [53]. The angiogenic switch occurs when the balance between pro-angiogenic and anti-angiogenic molecules is shifted in favor of angiogenesis permitting rapid tumor growth and subsequent development of invasive and metastatic properties, characteristics that define the lethal cancer phenotype [54]. Indeed, a statistically significant correlation between vascular density as a parameter for tumor angiogenesis and patient survival has been established for a variety of tumors [55–57]. Various molecular players have been identified which are involved in orchestrating specific stages and mechanisms of vascular growth in response to developmental, physiological, and oncogenic stimuli. Among these, members of the VEGF, the angiopoietin (Ang) and ephrin family seem to

have a predominant role [58,59]. Other factors that can act as inducers or modulators of angiogenesis include acidic fibroblast growth factor (aFGF), basic fibroblast growth factor (bFGF), transforming growth factor alpha and beta (TGF-α and -β), tumor necrosis factor alpha (TNFα) and interleukin-8 (IL-8). In addition, factors have been reported which function as naturally occurring inhibitor of angiogenesis like angiostatin or endostatin being derived from proteolytic fragments of larger proteins such as plasminogen or collagen type XVIII, respectively [60].

The VEGF-family

VEGF and its tyrosine kinase receptors VEGFR-1 (flt-1) and VEGFR-2 (flk-1/KDR) are major regulators of vasculogenesis and angiogenesis. Gene-targeting studies suggest that signaling *via* VEGFR-2 mediates vascular permeabilty and endothelial cell (EC) growth, while VEGFR-1 (flt-1) plays a negative role by either acting as a decoy receptor or suppressing signaling through VEGFR-2. However, recent studies imply a positive regulatory role of VEGFR-1 in pathological angiogenesis *in vivo* (see below). VEGF is an endothelial cell-specific mitogen with vascular permeability inducing properties *in vivo* [61]. Interestingly, a fundamental link between microenvironmental tumor hypoxia and induction of angiogenesis could be established by several studies showing that VEGF expression is regulated by oxygen levels. VEGF is highly expressed in perinecrotic palisading cells but is downregulated in tumor cells adjacent to vessels, suggesting oxygen-dependent gene expression [6,7]. VEGF expression under hypoxia is subject to a multilevel regulation. The hypoxia-mediated response seems to depend on regulatory sequences in the 5′ and 3′ regions of the VEGF gene. It has been shown that (1) the 5′ HRE binding site for HIF is necessary for the hypoxic transactivation [8,62–64] and (2) that mRNA stabilization sites in the 3′ UTR of the VEGF gene restrict hypoxic gene expression to the perinecrotic palisading cells *in situ* [65–67]. mRNA stabilization seems to involve RNA–protein complexes in the 3′ UTR as formed by HuR, an RNA-binding protein [68,69], or hnRNP (heterogeneous nuclear ribonucleoprotein) L [70]. In addition, VEGF expression is regulated at the translational level by a functional IRES (internal ribosomal entry site) in the 5′ UTR which allows for efficient, cap-independent translation even under hypoxia [71]. Further, VEGF protein export and secretion seems to be controlled by oxygen tension [72].

VEGF, secreted by the hypoxic tumor cell compartment, is distributed throughout the tumor by diffusion generating a gradient. The diffusion capacity differs with the different splice-variants of VEGF depending on their heparin-binding affinity which determines their adherence to the extracellular matrix [73]. VEGF binding to its receptors (VEGFR) specifically expressed by EC (endothelial cell) enhances endothelial VEGFR expression in an auto-catalytic fashion [74]. VEGFR signaling leads to a cascade of events, including EC migration and proliferation as well as induction of fenestrae and vascular permeability in tumor vessels [75]. The importance of the VEGF family in regulating tumor angiogenesis, tumor growth, and progression was verified by several reports showing that inhibition of VEGF/VEGFR-2 signaling by VEGF neutralizing antibodies, low molecular VEGFR-2 inhibitors, or gene transfer of VEGFR-2 dominant-negative constructs lead to stunted tumor growth with reduced vascularization [76–78]. In an elegant approach using the well-established RIP1-Tag2 mouse model of multi-step tumorigenesis of pancreatic islet carcinoma [79], islet-specific VEGF-deletion by means of the Cre-lox system diminished angiogenic switching and tumor growth and progression [80]. The history of the identification of "VEGF and the quest for tumor angiogenesis factors" has been excellently depicted in a recent review by N. Ferrara [81].

PlGF (Placenta growth factor), another member of the VEGF family, which specifically binds to VEGFR-1, has been shown to exert important, possibly synergistic functions with VEGF in increasing tumor vascularization [82–85]. These findings have established a positive regulatory role of the VEGFR-1 signaling pathway in tumor angiogenesis. However, much of the biological function of the VEGFR-1 ligand system may rather lie in the regulation and recruitment of hematopoietic and inflammatory cells to the tumor site than by acting directly on EC [86]. Although a HIF binding site (HRE) has not yet been identified in the PlGF gene, several reports indicate oxygen-dependent PlGF expression in certain cell types [87–89]. Thus, tumor hypoxia may induce VEGFR-1 signaling synergistically by VEGF and PlGF upregulation.

The Angiopoietin/Tie Family

Angiopoietins, in particular Ang-1 and the naturally occurring antagonist Ang-2 are implicated in later stages of vascular development, i.e., during vascular remodeling and maturation [59]. In adult animals, Ang-2 induction is demonstrated in ECs undergoing active remodeling [90–92]. Hence, it was proposed that Ang-2 induced in the vascular endothelium blocked the constitutive stabilizing influence exerted by Ang-1. This would allow the EC to revert to a more plastic and unstable state. VEGF and hypoxia have been reported to increase Ang-2 expression in EC *in vitro* [93]. The observation that Ang-2 expression in tumor ECs can be seen in close vicinity to VEGF expressing tumor cells neighboring areas of necrosis suggests that similar mechanisms take place *in vivo* [94,95]. Moreover, tumor vessels are structurally and functionally abnormal with excessive branching, shunts, and leakiness resulting in regional heterogeneity in tumor perfusion [58]. As a consequence, tumor blood flow is chaotic leading to severely hypoxic regions within the tumor that even ECs of tumor vessel are subjected to hypoxia [96,97]. Acting in concert, EC-hypoxia and VEGF-mediated upregulation of Ang-2, would render EC more accessible to angiogenic inducers such as VEGF resulting in a strong angiogenic response [90,98,99]. In addition, angiopoietins have been implicated in vascular permeability. In particular, Ang-1 could counter the permeability inducing effects of VEGF [100]. In return, its antagonist Ang-2 even potentiated the VEGF-mediated increase in vascular permeability [101].

Together, these observations support the view that hypoxia and HIF are key regulators of blood vessel growth inducing both upregulation of pivotal angiogenic ligands and their cognate receptors. Tumor hypoxia and HIF mediated upregulation of VEGF and Ang-2 is most likely responsible for the two major culprits hindering effective cancer treatment: tumor angiogenesis and edema. In addition to its effects on the vascular system, HIF controls the activation of a range of other adaptive mechanisms important for tumor growth including the shift in energy metabolism from oxidative to glycolytic pathways (glucose transporters, glycolytic enzymes), pH regulation (CA IX), and cell survival and proliferation (IGF-2).

The Ephrin/Eph Family

Eph receptors (Ephs) and their membrane-tethered ligands, ephrins, have diverse biological roles including patterning and morphogenetic processes particularly in the nervous and vascular systems, involving the guidance of cell migration and positioning [102]. They are subdivided into an A-subclass (EphA1–EphA8; ephrinA1–ephrinA5) and B-subclass (EphB1–EphB4, EphB6, ephrinB1–ephrinB3) based on sequence similarities and binding characteristics. The above-mentioned receptor tyrosine kinases (RTK) bind soluble ligands, which are produced at some distance from the RTK-expressing cell, thus mediating long-range cell-to-cell communication. Eph receptors, however, bind membrane-bound ephrin ligands, which are expressed on neighboring cells, thus mediating short-range cell-to-cell communication. The influence of Eph–ephrin interaction on cell behavior depends on the cell type most commonly resulting in cell repulsion. But also opposite effects, i.e., increased adhesion or attraction, have been reported. The B-subclass of ephrins is unique among the RTK family as ephrinB ligands not only induce signaling downstream of Eph receptors (forward signaling) but function themselves as receptors as they possess intrinsic signaling properties (reverse signaling). *In vivo* evidence for the requirement of Eph/ephrins in vascular development came from studies in mouse embryos with targeted deletions of the EphB4 [103] or ephrinB2 gene [104–106]. These animals demonstrated defects in early angiogenic remodeling resulting in embryonic lethality around E10.0. Moreover, these studies implicated ephrinB2 (expression in primordial arterial vessels) and EphB4 (expression in primordial venous vessels) in establishing arterial *versus* venous identity in the developing embryo.

During tumor angiogenesis, Eph and ephrins are upregulated as blood vessels invade tumors. The arterial marker ephrinB2 is highly expressed in the endothelium of some angiogenic vessels and their sprouts, challenging the classically held view that tumor vessels arise from postcapillary venules [107,108]. In addition, EphA2 has been shown to be exclusively expressed at sites of neovascularization in the adult [109]. Functionally, the EphB/ephrinB system has been demonstrated to replicate various aspects of angiogenesis and sprouting *in vitro* [110, 111], while angiogenic requirement of EphA2 signaling was shown *in vitro* by blocking tube-like formation of human umbilical vein endothelial cells with a dominant-negative form of EphA2 [109]. Demonstrating decreased tumor vascularization and tumor growth using a soluble EphA2-FC chimeric receptor provided the first functional evidence for EphA receptor involvement in tumor angiogenesis [112]. Interestingly, *in vitro* EphA2-Fc has been shown to block various angiogenic activities of VEGF, suggesting a cooperative model of EphA/ephrinA function in tumor angiogenesis in the context of VEGF signaling [113]. This model is supported by studies suggesting that ephrinA1 [113] and ephrinB2 [114] may be downstream targets of VEGF.

It should be mentioned that Eph/ephrin action in tumor growth and tumor angiogenesis may not be restricted to signaling within the endothelial cell. Indeed, several Eph receptors and ephrins have been shown to be upregulated in tumors, particularly during invasive stages in tumor progression [102]. A recent study implicated EphB2 in promoting glioma migration and invasion [115]. In line with these observations, EphA2 overexpression was sufficient to confer malignant transformation and tumorigenic potential to non-transformed mammary epithelial cells [116]. The exact role of Eph/ephrins in tumorigenesis and tumor angiogenesis is, however, far from being understood. It is tempting to speculate, that the bi-directional signaling nature of Eph/ephrin system may indeed allow for a bi-directional cross talk between tumor and endothelial cells (see below). Taken together, these findings raise the possibility that Eph/ephrin molecules may serve as new prognostic tumor markers and potential targets for therapeutic intervention in cancer.

THE HYPOXIC TUMOR MICROENVIRONMENT AND ANGIOGENIC CELLULAR CROSS TALK

Apart from endothelial and perivascular cells, tumors attract a number of cell types which synergistically act to augment vascularization of the tumor including inflammatory/hematopoietic cells and circulating endothelial precursor cells (CEP). Current studies indicate that tumor hypoxia not only indirectly influences these cell types by tumor-cell-specific upregulation of various secreted, paracrine-acting factors but may in addition have direct cell-intrinsic effects (reviewed in [117]).

Endothelial Cells

Hypoxia-mediated induction of angiogenesis is thought to be mainly conferred by transactivation of VEGF in surrounding cells, thus acting in a paracrine fashion on EC (extrinsic pathway). However, exposure of EC to intermittent and chronic hypoxic conditions has been shown to occur *in vivo* as a result of the structural and functional abnormal tumor vasculature [96,97]. Recent studies suggest that hypoxia may operate as an intrinsic regulator of EC growth and function by stimulating receptor and ligand expression. HIF-α subunits have been reported to induce VEGFR-1, VEGFR-2, and tie2 [19,118–121]. Stabilization of HIF-1α by the peptide regulator (PR) 39 in different EC lines resulted in VEGF upregulation and accelerated formation of vascular structures. Moreover, *in vivo* expression of PR39 targeted to the myocardium increased myocardial vascularization, although it was not clear from that study how relevant HIF pathway activation in EC was as opposed to activation in the surrounding tissue [122]. Interestingly, hypoxic VEGF induction in EC was also shown to promote network formation *in vitro* [123]. Thus, hypoxia-driven autocrine stimulation of endothelial cells may enhance the angiogenic pathway and participate in the formation and reorganization of the vascular network in solid tumors. Tumor ECs are clearly morphologically and pathophysiologically distinct and exhibit a different expression profile than normal resting blood vessel ECs [124] as suggested by *in vivo* application of phage display libraries [125] or SAGE display [126]. It remains to be determined to what extent intrinsic EC hypoxia or hypoxia-induced angiogenic factors in the surrounding tissue contribute to the specialized phenotype of the tumor vasculature.

Apart from their metabolic function, ECs may provide inductive signals important for tumor development. Recent studies suggest that blood vessels, independent of their nourishing function, stimulate organ development and differentiation as shown for pancreas and liver formation [127,128], and enhance tumor proliferation as shown in co-culture studies with tumor spheroids and embryoid bodies [129]. Thus, signaling between tumor cells and ECs might indeed be bi-directional with tumor cells promoting blood vessel growth and ECs giving tumor cells specific differentiation and proliferation cues.

Inflammatory/hematopoietic Cells

When in the early 1970s, Folkman put forward the hypothesis that tumors induce blood vessel growth by secreting diffusible angiogenic factors, his concept was much criticized as common belief stated that tumor angiogenesis was directed by the inflammatory host response. His work inspired many researchers leading to the discovery of VEGF and its regulation by tumor hypoxia. Several decades later a mounting body of evidence suggests that tumors may indeed exploit the host-defense mechanism and attract inflammatory cells to further enhance vascularization. Interestingly, both concepts may finally be reconciled with the observation that tumor and host-cell-directed vascularization apparently makes use of similar mechanisms with analogous functions of the VEGF family and tumor hypoxia.

Induction of angiogenesis is known as a hallmark of various chronic inflammatory disorders such as rheumatoid arthritis and psoriasis. Tumors produce various cytokines and chemokines that attract inflammatory/hematopoietic cells [130]. Hematopoietic stem cells are pluripotent cells with the capacity for self-renewal and differentiation into specific lineages, e.g., when differentiating along the myeloid lineage, they give rise to neutrophils and macrophages. Immune surveillance has long been viewed as keeping tumor growth at bay. However, the inflammatory cell component may have dual roles in tumors. Apart from killing neoplastic cells, leukocytes such as tumor-associated macrophages (TAM) produce an array of potent angiogenic and lymphangiogenic growth factors, cytokines, and extracellular proteases which potentiate tumor growth and progression [53,131]. Failure to recruit TAMs has been demonstrated to significantly attenuate tumor progression and metastasis [132], partly due to impaired angiogenesis [133,134]. Hypoxia and the HIF transcriptional system may play an important role in these processes. It has been shown using transgenic mice expressing the green fluorescent protein under the control of the human VEGF promotor that the tumor environment is capable of inducing this HIF responsive promotor in stromal cells of host origin [135]. Moreover, hypoxia seems to reduce macrophage migration potentially leading to a preferential accumulation of macrophages in hypoxic tumor regions [136]. Interestingly, high HIF-2α expression levels were also found in tumor-associated macrophages [11]. In line with these observations, conditional gene targeting of HIF-1α in myeloid cells blunted the inflammatory cell response by attenuating macrophage/neutrophil invasion and migration [137]. Recent reports have further underlined the importance of several HIF-target genes such as VEGF, PlGF, and IL-8 either directly or by promoting the release of additional factors such as MMP-9, s-KitL in the indirect regulation of hematopoietic cell recruitment and function [82–85,138,139].

Inflammatory/hematopoietic cell recruitment from the blood stream occurs in sequential steps including rolling, adherence and transmigration. ECs represent a critical interface in controlling the emigration of inflammatory/hematopoietic cells in response to different stress signals [140]. Interestingly, hypoxia is known to activate ECs to initiate this cascade by upregulation of adhesion molecules and cytokines such as selectin, IL-8, and MCP (macrophage-chemotactic protein)-1, further underlining the function of oxygen tension in guiding cellular cross talk.

CEP

In addition to angiogenic sprouting and co-option of pre-existing vessels [141], new evidence indicates that tumor vascularization is enhanced by the mobilization and incorporation of circulating endothelial progenitor cells (CEP). CEPs are bone-marrow-derived cells with a high proliferation potential. Depending on the experimental system contribution to tumor vessels ranges from a few to up to 90 per cent of all vessels [130]. The VEGF family in line with its pivotal role in tumor angiogenesis has been also implicated in the mobilization of CEPs. While signaling in hematopoietic and inflammatory cells seems to be mediated mainly *via* the VEGFR-1/ligand system, VEGFR-2 signaling constitutes the prevailing pathway in CEPs to induce recruitment and proliferation. Interestingly in tumors, hemato-poetic cells and CEPs are found in close association suggesting cross talk between these two cell types. It remains to be determined whether hypoxia and the HIF pathway exert similar direct effects on CEPs.

ANTI-ANGIOGENESIS STRATEGIES AS NOVEL ANTI-TUMOR THERAPIES

The Rationale of Anti-angiogenic Therapy

The concept of anti-angiogenesis as an effective tumor treatment has been confirmed experimentally in various types of neoplasia including brain tumors. It has been demonstrated that gliomas overexpress VEGF and that this expression correlates with tumor vascularity and grade and inversely with prognosis making them a suitable candidate for anti-angiogenic therapy approaches [6,142]. Indeed, the value of anti-angiogenic approaches targeting VEGF/VEGFR and the angiopoietins/tie2 system has been validated experimentally [76–78,95,143] and is currently under clinical investigation [53,144]. Anti-angiogenic therapy may seem appealing for several reasons. It allows selective targeting based on distinct differences between quiescent and angiogenic vasculature. Since physiological angiogenesis is turned off in the adult, therapy should harbor minimal side effects. It may, moreover, have the potential to alleviate two major problems effectively as non-surgical tumor therapy is currently hampered by: (1) the development of drug resistance resulting from genetic and epigenetic changes which interfere with the effectiveness of available chemotherapeutics, (2) intratumoral drug delivery at effective concentration is impeded by the functional and structural heterogeneous blood supply and the increased interstitial pressure. The primary

target of anti-angiogenic approaches is the tumor EC that is considered to be genetically stable. Anti-angiogenic therapy may, in addition, help to prune the immature tumor blood vessels, thus reducing vascular permeability and normalizing tumor blood flow as suggested by studies interfering with the VEGF/VEGFR-2 system [145,146]. Improved blood flow may in turn change the tumor's sensitivity to current therapeutic approaches by potentiating the effects of chemo- and radiotherapy due to increased drug delivery and elevated oxygen levels providing a rationale for combination therapy (see below). Such combination therapy would obviously have to be optimized regarding schedule and dosage due to the potential drawback that increased blood flow sufficiency may enhance availability of oxygen and nutrients and thus further spur tumor cell growth [147]. Further concerns exist regarding potential side effects on tissue repair processes and the reproductive cycle as well as teratogenic effects. Moreover, anti-angiogenic compounds would have to be applied continuously to sustain tumor dormancy posing an increased risk of toxicity. These drawbacks have spurred a search for novel treatment modalities to enhance target specificity, delivery, and efficacy of angiogenic inhibitors (Fig. 15.1, Table 15.1).

From Bench to Bedside: Clinical Anti-angiogenesis Trials

Numerous clinical studies have been initiated to test anti-angiogenic strategies on various malignancies employing synthetic as well as naturally occurring anti-angiogenic compounds. Over 80 anti-angiogenic agents are currently evaluated in clinical trials having enrolled over 10 000 patients [148]. Several of these trials recruit patients with brain tumors (at the time of writing this article an internet search revealed 14 ongoing trials; sources: http://www.cancer.gov/clinical_trials; http://www.clinical-trials.gov; http://www.virtualtrials.org/index.cfm) targeting different pathways (see also Table 15.1) (see review [149]). Current reports from phase I and II clinical trials conclude that anti-angiogenic agents are well tolerated [150,151]. However, despite promising results obtained in experimental tumor models, none or only limited therapeutic effects have been reported from clinical trials. Differences in tumor vasculature in humans and animals as well as the use of tumor models established from cell lines which seldom resemble the molecular and morphological hetero-genous composition of primary tumors may be reasons for this discrepancy. The use of transgenic

Hypoxic Tumor Tissue

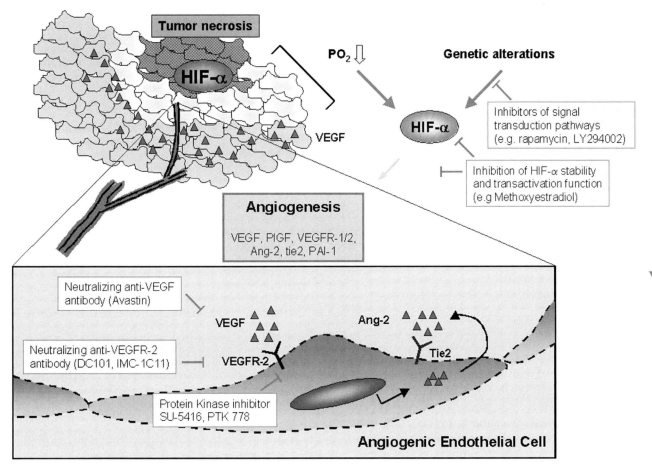

FIGURE 15.1 Schematic representation summarizing current strategies for anti-angiogenic tumor therapy. The most promising approaches some of them in clinical trial (see Table 15.1) target the hypoxic tumor cell population by inhibition of different steps in the HIF signaling pathway as well as by inhibition of different angiogenic pathways in endothelial cells. See Plate 15.1 in Color Plate Section.

tumor models may be one of the means to improve experimental testing of therapeutic approaches. Moreover, combination therapies with established conventional treatment regimens have been suggested to enhance therapeutic potential.

Combined Therapeutic Strategies in Anti-angiogenic Tumor Treatments

Several lines of evidence indicate that a multimodal approach is necessary to optimize clinical efficacy of anti-angiogenic treatments as resistance might develop over time in particular when used as mono-therapies [152]. As outlined above, the regulation of tumor-associated angiogenesis involves a complex and intricate interplay of different molecular and cellular players. To increase the efficacy of

anti-angiogenic strategies, these pathways need to be targeted simultaneously.

Moreover, although anti-angiogenic therapy targets genetically stable endothelial cells in the tumor vasculature, genetic alterations that decrease the vascular dependence of tumor cells can decisively influence the therapeutic response. Tumors represent heterogeneous cell populations as a consequence of genetic instability which is a hallmark of malignancy [153,154]. Accumulative acquisition of specific genetic alterations is implicated in tumor progression. Glioblastomas, for example, arise either *de novo* (primary glioblastoma) or by progression from a low-grade astrocytoma (secondary glioblastoma). Epidermal growth factor receptor overexpression and p53 mutations have been specifically associated with primary and secondary glioblastomas, respectively [155]. Interestingly, glioma progression *in vivo*

TABLE 15.1   Current Clinical Trials Employing Angiogenesis
Inhibitors for Treatment of Malignant Gliomas

Mechanism	Therapeutic compound	Phase	Comments
Chelators of copper	Penicillamine	II	
	Marimastat	III	
	Prinomastat	III	
Endothelial cell inhibitor	Endostatin	II	Not yet tested in glioma
	Angiostatin	I	Not yet tested in glioma
	Thalidomide	II	
	Farnesyl transferase inhibitor	I	
	TNP-470	II	
VEGF/VEGFR inhibitor	Suramin	II	
	Bevacizumab (Avastin)	III	Not yet tested in glioma
	SU-5416	I	
	PTK778/ZK22584	I	
HIF inhibitor	2-Methoxyestradiol	I	
Integrin inhibitor	EMD 121974	I/II	
MMP inhibitor	Col-3	I/II	

is associated with clonal expansion of p53 mutant cells [156]. These findings in mind, it is interesting to consider that genetic alterations found in tumor cells are not necessarily primary events, but may themselves be a consequence of tumor hypoxia. An elevated mutation frequency in hypoxic compared to normoxic tumor cells has been reported [157]. A diminished DNA repair capacity reported under hypoxic conditions may underlie this phenomenon [158].

Interestingly, hypoxia may not only induce mutations in tumor cells, but may select for malignant cell clones with increased resistance to hypoxia-mediated apoptosis [159]. Subjecting cells to hypoxia has been shown to induce p53 activity and apoptosis [160]. Hypoxia-induced acidosis has been implicated in this process [161]. When p53-deficient and p53 wt cells were mixed *in vitro*, several rounds of hypoxia led to an accumulation of p53-deficient cells. *In vivo*, highly apoptotic regions overlapped with hypoxic areas in wt p53 tumors, while only little apoptosis in hypoxic areas of p53-deficient tumors was reported [162]. These findings suggest that tumor cell apoptosis induced in hypoxic/hypoglycemic areas, possibly mediated by HIF activation, leads to selection of cell clones which have lost p53 activity and are less vulnerable to low oxygen tensions and, consequently, anti-angiogenic approaches. Indeed, mice with tumors derived from p53-deficient HCT116 human

colorectal cancer cells were less responsive to anti-angiogenic therapy than mice bearing p53 wt tumors [163]. Thus, the potential for selection of less vascular-dependent tumor cell variants throughout the course of disease progression may diminish the long-term efficacy of anti-angiogenic therapy. Anti-angiogenic therapies might even enhance this selective pressure by creating a more hypoxic micro-environment due to oxygen and nutrient supply deprivation.

Further, the susceptibility of endothelial cells to chemo- and radiotherapeutic treatments may critically influence the overall anti-tumor efficacy of such treatments. MCA/129 fibrosarcomas and B16F1 melanomas grown in apoptosis-resistant acid sphingomyelinase (asmase)-deficient or Bax-deficient mice displayed a 2–4 times increase in tumor growth compared to tumors grown on wild-type microvasculature markedly associated with reduced baseline microvascular endothelial apoptosis [164]. Further, tumor xenografts in apoptosis-resistant mice unlike tumors in wild-type mice were resistant to single-dose radiation up to 20 grays (Gy) concomitant with a reduced endothelial apoptosis response. Thus, endothelial apoptosis is an important mechanism regulating tumor growth. Interestingly, the combination of anti-angiogenic agents with radio- or chemotherapy greatly enhances therapy response. Combined treatment of radiotherapy and an anti-VEGF antibody in U87 glioma and colon adenocarcinoma (LS174T) xenografts [146,165], the VEGFR-2 specific inhibitor SU5416 or soluble VEGFR-2 in GL261 glioma [166] or the VEGF receptor tyrosine kinase inhibitor PTK787/ ZK22258 in p53-dysfunctional SW480 colon adeno-carcinomas [167] exerted a substantial tumor growth delay and reversal of radioresistance as compared to both treatments alone.

The mechanisms that appear to underlie these additive effects include chemotherapy- and radio-therapy-mediated upregulation of endothelial specific survival factors produced by tumor cells. These factors include VEGF and bFGF which have been shown to induce survival pathways in endothelial cells involving PI3-K/Akt signaling which diminish the sensitivity towards chemotherapy [168] and radiotherapy [169]. Tumor cells respond to radiation with upregulation of VEGF as part of a stress response [165,170]. A recent study identified HIF-1α as a critical determinant of this response regulating the radiosensitivity of the tumor vascula-ture by secretion of cytokines that control survival pathways in endothelial cells [171]. The authors showed that reoxygenation events after irradiation resulted in reactive oxygen species (ROS) formation inducing HIF-1α. HIF-1α, in turn, mediated the upregulation of known endothelium-protective factors such as VEGF and bFGF conferring increased radioresistance to EC. These studies indicate that the tumor microenvironment contains factors with endothelial-protective properties such as VEGF and bFGF that alter the susceptibility of endothelial cells to chemo- and radiotherapeutic treatments and may severely compromise the overall efficacy of current treatments. Thus, microvascular damage may be a decisive parameter in determining tumor response and sensitivity to current treatment strategies.

It is interesting to note that chemotherapeutic agents given in a continuous low-dose regimen, so called "metronomic dosing" may have anti-angiogenic properties and may thus be employed even to drug-resistant tumors [172]. When Lewis lung carcinoma and EMT-6 breast cancer were rendered drug resistant before therapy, metronomic dosing suppressed tumor growth 3-fold more effectively than the conventional schedule [173]. Anti-angiogenic effects seemed to be causative as endothelial cell apoptosis preceded the apoptosis of drug-resistant tumor cells. A recent study demonstrated that sensitivity to various chemotherapeutics clearly differed among tumor cells, non-transformed cells and endothelial cells with IC50 values range of 500 pM to >1 nM for fibroblasts and cancer cells and significantly lower IC50 in the range of 25–143 pM for endothelial cells [174]. Hence, it was proposed that an anti-angiogenic window for chemotherapeutic drugs may exist. This would allow administering chemotherapeutics at low doses that are selectively cytotoxic for proliferating endothelial cells, thereby reducing toxic side effects. In consequence, metro-nomic dosing may help to eradicate tumors that are considered refractory to conventional chemotherapies such as glioblastomas.

Taken together, these data clearly argue for a combination therapy. Anti-angiogenic compounds may act synergistically with conventional therapeutic regimens resulting, when administered concurrently, in an enhanced therapeutic responsiveness.

Novel Therapeutic Targets

Targeting the neovasculature by directing treat-ment to molecules that are exclusively expressed on tumor endothelial cells has the advantage of systemic instead of local delivery. However, certain approaches may require high therapeutic concentration of anti-angiogenic agents which may only be achieved by local delivery. Exploiting the interplay between tumor cells and the different stromal cells may con-stitute a different approach to improve drug delivery. Thus, macrophages, hematopoietic cells or CEPs which home to tumor tissues may be used as a vehicle to deliver or produce toxic substances within tumor [175].

The recent insight into the precise mechanisms of oxygen sensing and signaling may help to develop novel anti-tumor strategies which specifically target the HPHD–HIF–VHL pathway. Given the widespread HIF activation in tumors, the role of HIF in transactivating angiogenic factors and the role of angiogenic factors in tumor growth interfering with this pathway is particularly appealing. Its rationale lies in depriving the tumor cell of oxygen and nutrients by inhibiting angiogenesis while at the same time, disabling adaptive mechanisms that help the cell to survive in this microenvironment. The feasibility of this approach has been confirmed in different reports [176–179]. However, given their key regulatory role in various complex physiological pathways stretching from metabolism, proliferation, differentiation to apoptosis general manipulation of the HIF system is likely to show variable outcome depending on the cellular context and should for this reason be employed cautiously [180,181]. The recent finding that mice lacking the HRE motif in the VEGF promotor developed progressive motor neuron degeneration further warrants these safety considerations [182].

FUTURE PERSPECTIVES

Insights into the complex regulation of tumor-associated angiogenesis gained over the past years clearly indicate that multiple pathways in various cellular components have to be targeted to increase the effectiveness of anti-angiogenic strategies. In this context, the transcriptional complex HIF represents an intriguing candidate. HIF acts as a major regulator of the tumor's vascular network in particular by directing the cross talk between a multitude of molecular as well as cellular angiogenic players. In addition, recent attempts employing transcriptional and proteomic profiling techniques to identify differentially expressed genes between the host and the tumor vascular endothelium may not only provide tumor-specific vascular markers which can be used to specifically target the tumor but may also help to unravel novel functional gene and protein cluster networks operative in the tumor vasculature. Further understanding of these processes will facilitate the development, refinement, and cell-specific targeting of anti-tumor therapies.

ACKNOWLEDGMENTS

This research is supported by grants from the Deutsche Krebshilfe (Dr. Mildred-Scheel Stiftung), the Bundesministerium für Bildung und Forschung (01KV0102), and the Deutsche Forschungsgemeinschaft (AC110-1, PL158/3-1, PL158/4-1, 4-2, 4-3 and PL158/6-1).

References

1. Eberhard, A., Kahlert, S., Goede, V., Hemmerlein, B., Plate, K. H., Augustin, H. G. (2000). Heterogeneity of angiogenesis and blood vessel maturation in human tumors: implications for antiangiogenic tumor therapies. *Cancer Res* **60**, 1388–1393.
2. Brown, J. M., and Giaccia, A. J. (1998). The unique physiology of solid tumors: opportunities (and problems) for cancer therapy. *Cancer Res* **58**, 1408–1416.
3. Vaupel, P., Kallinowski, F., and Okunieff, P. (1989). Blood flow, oxygen and nutrient supply, and metabolic microenvironment of human tumors: a review. *Cancer Res* **49**, 6449–6465.
4. Wykoff, C. C., Pugh, C. W., Maxwell, P. H., Harris, A. L., Ratcliffe, P. J. (2000). Identification of novel hypoxia dependent and independent target genes of the von Hippel-Lindau (VHL) tumour suppressor by mRNA differential expression profiling. *Oncogene* **19**, 6297–6305.
5. Maxwell, P. H., Pugh, C. W., and Ratcliffe, P. J. (2001). Activation of the HIF pathway in cancer. *Curr Opin Genet Dev* **11**, 293–299.
6. Plate, K. H., Breier, G., Weich, H. A., Risau, W. (1992). Vascular endothelial growth factor is a potential tumour angiogenesis factor in human gliomas in vivo. *Nature* **359**, 845–848.
7. Shweiki, D., Itin, A., Soffer, D., Keshet, E. (1992). Vascular endothelial growth factor induced by hypoxia may mediate hypoxia-initiated angiogenesis. *Nature*. **359**, 843–845.
8. Damert, A., Machein, M., Breier, G. *et al.* (1997). Up-regulation of vascular endothelial growth factor expression in a rat glioma is conferred by two distinct hypoxia-driven mechanisms. *Cancer Res* **57**, 3860–3864.
9. Dachs, G. U., Patterson, A. V., Firth, J. D. *et al.* (1997). Targeting gene expression to hypoxic tumor cells. *Nat Med* **3**, 515–520.
10. Zhong, H., De Marzo, A. M., Laughner, E. *et al.* (1999). Overexpression of hypoxia-inducible factor 1alpha in common human cancers and their metastases. *Cancer Res* **59**, 5830–5835.
11. Talks, K. L., Turley, H., Gatter, K. C. *et al.* (2000). The expression and distribution of the hypoxia-inducible factors HIF-1alpha and HIF-2alpha in normal human tissues, cancers, and tumor-associated macrophages. *Am J Pathol* **157**, 411–421.
12. Zagzag, D., Zhong, H., Scalzitti, J. M., Laughner, E., Simon, J. W., Semenza, G. L. (2000). Expression of hypoxia-inducible factor 1alpha in brain tumors: association with angiogenesis, invasion, and progression. *Cancer* **88**, 2606–2618.
13. Plate, K. H. (1999). Mechanisms of angiogenesis in the brain. *J Neuropathol Exp Neurol* **58**, 313–320.
14. Wang, G. L., Jiang, B. H., Rue, E. A., Semenza, G. L. (1995). Hypoxia-inducible factor 1 is a basic-helix-loop-helix-PAS heterodimer regulated by cellular O2 tension. *Proc Natl Acad Sci U S A* **92**, 5510–5514.
15. Wang, G. L., and Semenza, G. L. (1995). Purification and characterization of hypoxia-inducible factor 1. *J Biol Chem* **270**, 1230–1237.
16. Ema, M., Taya, S., Yokotani, N., Sogawa, K., Matsuda, Y., Fujii-Kuriyama, Y. (1997). A novel bHLH-PAS factor with close sequence similarity to hypoxia-inducible factor 1alpha regulates the VEGF expression and is potentially involved in lung and vascular development. *Proc Natl Acad Sci U S A* **94**, 4273–4278.
17. Flamme, I., Frohlich, T., von Reutern, M., Kappel, A., Damert, A., Risau, W. (1997). HRF, a putative basic helix-loop-helix-PAS-domain transcription factor is closely related to hypoxia-inducible factor-1 alpha and developmentally expressed in blood vessels. *Mech Dev* **63**, 51–60.
18. Hogenesch, J. B., Chan, W. K., Jackiw, V. H. *et al.* (1997). Characterization of a subset of the basic-helix-loop-helix-PAS superfamily that interacts with components of the dioxin signaling pathway. *J Biol Chem* **272**, 8581–8593.
19. Tian, H., McKnight, S. L., and Russell, D. W. (1997). Endothelial PAS domain protein 1 (EPAS1), a transcription factor selectively expressed in endothelial cells. *Genes Dev* **11**, 72–82.
20. Gu, Y. Z., Moran, S. M., Hogenesch, J. B., Wartman, L., Bradfield, C. A. (1998). Molecular characterization and chromosomal localization of a third alpha-class hypoxia inducible factor subunit, HIF3alpha. *Gene Expr* **7**, 205–213.
21. Wiesener, M. S., Turley, H., Allen, W. E. *et al.* (1998). Induction of endothelial PAS domain protein-1 by hypoxia: characterization and comparison with hypoxia-inducible factor-1alpha. *Blood* **92**, 2260–2268.
22. Ema, M., Hirota, K., Mimura, J. *et al.* (1999). Molecular mechanisms of transcription activation by HLF and

HIF1alpha in response to hypoxia: their stabilization and redox signal-induced interaction with CBP/p300. *EMBO J* **18**, 1905–1914.

23. O'Rourke, J. F., Tian, Y. M., Ratcliffe, P. J., Pugh, C. W. (1999). Oxygen-regulated and transactivating domains in endothelial PAS protein 1: comparison with hypoxia-inducible factor-1alpha. *J Biol Chem* **274**, 2060–2071.

24. Salceda, S., and Caro, J. (1997). Hypoxia-inducible factor 1alpha (HIF-1alpha) protein is rapidly degraded by the ubiquitin-proteasome system under normoxic conditions. Its stabilization by hypoxia depends on redox-induced changes. *J Biol Chem* **272**, 22642–22647.

25. Huang, L. E., Gu, J., Schau, M., Bunn, H. F. (1998). Regulation of hypoxia-inducible factor 1alpha is mediated by an O2- dependent degradation domain via the ubiquitin-proteasome pathway. *Proc Natl Acad Sci U S A* **95**, 7987–7992.

26. Kallio, P. J., Wilson, W. J., O'Brien, S., Makino, Y., Poellinger, L. (1999). Regulation of the hypoxia-inducible transcription factor 1alpha by the ubiquitin-proteasome pathway. *J Biol Chem.* **274**, 6519–6525.

27. Sutter, C. H., Laughner, E., and Semenza, G. L. (2000). Hypoxia-inducible factor 1alpha protein expression is controlled by oxygen-regulated ubiquitination that is disrupted by deletions and missense mutations. *Proc Natl Acad Sci U S A* **97**, 4748–4753.

28. Jiang, B. H., Zheng, J. Z., Leung, S. W., Roe, R., Semenza, G. L. (1997). Transactivation and inhibitory domains of hypoxia-inducible factor 1alpha. Modulation of transcriptional activity by oxygen tension. *J Biol Chem* **272**, 19253–19260.

29. Gu, J., Milligan, J., and Huang, L. E. (2001). Molecular mechanism of hypoxia-inducible factor 1alpha -p300 interaction. A leucine-rich interface regulated by a single cysteine. *J Biol Chem* **276**, 3550–3554.

30. Pugh, C. W., O'Rourke, J. F., Nagao, M., Gleadle, J. M., Ratcliffe, P. J. (1997). Activation of hypoxia-inducible factor-1; definition of regulatory domains within the alpha subunit. *J Biol Chem* **272**, 11205–11214.

31. Srinivas, V., Zhang, L. P., Zhu, X. H., Caro, J. (1999). Characterization of an oxygen/redox-dependent degradation domain of hypoxia-inducible factor alpha (HIF-alpha) proteins. *Biochem Biophys Res Commun* **260**, 557–561.

32. Yu, F., White, S. B., Zhao, Q., Lee, F. S. (2001). Dynamic, site-specific interaction of hypoxia-inducible factor-1alpha with the von Hippel-Lindau tumor suppressor protein. *Cancer Res* **61**, 4136–4142.

33. Lisztwan, J., Imbert, G., Wirbelauer, C., Gstaiger, M., Krek, W. (1999). The von Hippel-Lindau tumor suppressor protein is a component of an E3 ubiquitin-protein ligase activity. *Genes Dev* **13**, 1822–1833.

34. Maxwell, P. H., Wiesener, M. S., Chang, G. W. *et al.* (1999). The tumour suppressor protein VHL targets hypoxia-inducible factors for oxygen-dependent proteolysis. *Nature* **399**, 271–275.

35. Stebbins, C. E., Kaelin W. G., Jr., and Pavletich, N. P. (1999). Structure of the VHL-ElonginC-ElonginB complex: implications for VHL tumor suppressor function. *Science* **284**, 455–461.

36. Cockman, M. E., Masson, N., Mole, D. R. *et al.* (2000). Hypoxia inducible factor-alpha binding and ubiquitylation by the von Hippel-Lindau tumor suppressor protein. *J Biol Chem* **275**, 25733–25741.

37. Ohh, M., Park, C. W., Ivan, M. *et al.* (2000). Ubiquitination of hypoxia-inducible factor requires direct binding to the beta-domain of the von Hippel-Lindau protein. *Nat Cell Biol* **2**, 423–427.

38. Kamura, T., Sato, S., Iwai, K., Czyzyk-Krzeska, M., Conaway, R. C, Conaway, J. W. (2000). Activation of HIF1alpha ubiquitination by a reconstituted von Hippel- Lindau (VHL) tumor suppressor complex. *Proc Natl Acad Sci U S A* **97**, 10430–10435.

39. Tanimoto, K., Makino, Y., Pereira, T., Poellinger, L. (2000). Mechanism of regulation of the hypoxia-inducible factor-1 alpha by the von Hippel-Lindau tumor suppressor protein. *EMBO J* **19**, 4298–4309.

40. Ravi, R., Mookerjee, B., Bhujwalla, Z. M. *et al.* (2000). Regulation of tumor angiogenesis by p53-induced degradation of hypoxia-inducible factor 1alpha. *Genes Dev* **14**, 34–44.

41. Blagosklonny, M. V., An, W. G., Romanova, L. Y., Trepel, J., Fojo, T., Neckers, L. (1998). p53 inhibits hypoxia-inducible factor-stimulated transcription. *J Biol Chem* **273**, 11995–11998.

42. Epstein, A. C., Gleadle, J. M., McNeill, L. A. *et al.* (2001). C. elegans EGL-9 and mammalian homologs define a family of dioxygenases that regulate HIF by prolyl hydroxylation. *Cell* **107**, 43–54.

43. Bruick, R. K., and McKnight, S. L. (2001). A conserved family of prolyl-4-hydroxylases that modify HIF. *Science* **294**, 1337–1340.

44. Oehme, F., Ellinghaus, P., Kolkhof, P. *et al.* (2002). Over-expression of PH-4, a novel putative proline 4-hydroxylase, modulates activity of hypoxia-inducible transcription factors. *Biochem Biophys Res Commun* **296**, 343–349.

45. Lando, D., Peet, D. J., Whelan, D. A., Gorman, J. J., Whitelaw, M. L. (2002). Asparagine hydroxylation of the HIF transactivation domain a hypoxic switch. *Science* **295**, 858–861.

46. Schofield, C. J., and Zhang, Z. (1999) Structural and mechanistic studies on 2-oxoglutarate-dependent oxygenases and related enzymes. *Curr Opin Struct Biol.* **9**, 722–731.

47. Lando, D., Gorman, J. J., Whitelaw, M. L., Peet, D. J. (2003). Oxygen-dependent regulation of hypoxia-inducible factors by prolyl and asparaginyl hydroxylation. *Eur J Biochem* **270**, 781–790.

48. Acker, T., and Acker, H. (2004). Cellular oxygen sensing need in CNS function: physiological and pathological implications. *J Exp Biol* **207**, 3171–3188.

49. Semenza, G. L. (2000). Hypoxia, clonal selection, and the role of HIF-1 in tumor progression. *Crit Rev Biochem Mol Biol* **35**, 71–103.

50. Acker, T., and Plate, K. H. (2002). A role for hypoxia and hypoxia-inducible transcription factors in tumor physiology. *J Mol Med* **80**, 562–575.

51. Bilton, R. L., and Booker, G. W. (2003). The subtle side to hypoxia inducible factor (HIFalpha) regulation. *Eur J Biochem* **270**, 791–798.

52. Jeong, J. W., Bae, M. K., Ahn, M. Y. *et al.* (2002). Regulation and destabilization of HIF-1alpha by ARD1-mediated acetylation. *Cell* **111**, 709–720.

53. Carmeliet, P., and Jain, R. K. (2000). Angiogenesis in cancer and other diseases. *Nature* **407**, 249–257.

54. Hanahan, D., and Folkman, J. (1996). Patterns and emerging mechanisms of the angiogenic switch during tumorigenesis. *Cell* **86**, 353–364.

55. Fox, S. B. (1997). Tumour angiogenesis and prognosis. *Histopathology* **30**, 294–301.

56. Ferrara, N., and Davis-Smyth, T. (1997). The biology of vascular endothelial growth factor. *Endocr Rev* **18**, 4–25.

57. Zetter, B. R. (1998). Angiogenesis and tumor metastasis. *Annu Rev Med* **49**, 407–424.

58. Carmeliet, P. (2000). Mechanisms of angiogenesis and arteriogenesis. *Nat Med* **6**, 389–395.

59. Yancopoulos, G. D., Davis, S., Gale, N. W., Rudge, J. S., Wiegand, S. J., Holash, J. (2000). Vascular-specific growth factors and blood vessel formation. *Nature* **407**, 242–248.

60. Jansen, M., Witt Hamer, P. C., Witmer, A. N., Troost, D., van Noorden, C. J. (2004). Current perspectives on antiangiogenesis strategies in the treatment of malignant gliomas. *Brain Res Brain Res Rev* **45**, 143–163.

61. Keck, P. J., Hauser, S. D., Krivi, G. *et al.* (1989). Vascular permeability factor, an endothelial cell mitogen related to PDGF. *Science* **246**, 1309–1312.

62. Levy, A. P., Levy, N. S., and Goldberg, M. A. (1996). Hypoxia-inducible protein binding to vascular endothelial growth factor mRNA and its modulation by the von Hippel-Lindau protein. *J Biol Chem* **271**, 25492–25497.

63. Liu, Y., Cox, S. R., Morita, T., Kourembanas, S. (1995). Hypoxia regulates vascular endothelial growth factor gene expression in endothelial cells. Identification of a 5′ enhancer. *Circ Res* **77**, 638–643.

64. Forsythe, J. A., Jiang, B. H., Iyer, N. V. *et al.* (1996). Activation of vascular endothelial growth factor gene transcription by hypoxia-inducible factor 1. *Mol Cell Biol* **16**, 4604–4613.

65. Ikeda, E., Achen, M. G., Breier, G., Risau, W. (1995). Hypoxia-induced transcriptional activation and increased mRNA stability of vascular endothelial growth factor in C6 glioma cells. *J Biol Chem* **270**, 19761–19766.

66. Stein, I., Neeman, M., Shweiki, D., Itin, A., Keshet, E. (1995). Stabilization of vascular endothelial growth factor mRNA by hypoxia and hypoglycemia and coregulation with other ischemia-induced genes. *Mol Cell Biol* **15**, 5363–5368.

67. Levy, A. P, Levy, N. S., and Goldberg, M. A. (1996). Post-transcriptional regulation of vascular endothelial growth factor by hypoxia. *J Biol Chem* **271**, 2746–2753.

68. Levy, N. S., Chung, S., Furneaux, H., Levy, A. P. (1998). Hypoxic stabilization of vascular endothelial growth factor mRNA by the RNA-binding protein HuR. *J Biol Chem* **273**, 6417–6423.

69. Goldberg, I., Furneaux, H., and Levy, A. P. (2002). A 40bp RNA element that mediates stabilization of VEGF mRNA by HuR. *J Biol Chem.*

70. Shih, S. C., and Claffey, K. P. (1999). Regulation of human vascular endothelial growth factor mRNA stability in hypoxia by heterogeneous nuclear ribonucleoprotein L. *J Biol Chem* **274**, 1359–1365.

71. Stein, I., Itin, A., Einat, P., Skaliter, R., Grossman, Z., Keshet, E. (1998). Translation of vascular endothelial growth factor mRNA by internal ribosome entry: implications for translation under hypoxia. *Mol Cell Biol* **18**, 3112–3119.

72. Ozawa, K., Kondo, T., Hori, O. *et al.* (2000). Expression of the oxygen-regulated protein ORP150 accelerates wound healing by modulating intracellular VEGF transport.

73. Grunstein, J., Masbad, J. J., Hickey, R., Giordano, F., Johnson. R. S. (1997). Isoforms of vascular endothelial growth factor act in a coordinate fashion to recruit and expand tumor vasculature. *Mol Cell Biol* **20**, 7282–7291.

74. Kremer, C., Breier, G., Risau, W., Plate, K. H. (1997). Up-regulation of flk-1/vascular endothelial growth factor receptor 2 by its ligand in a cerebral slice culture system. *Cancer Res* **57**, 3852–3859.

75. Roberts, W. G., and Palade, G. E. (1997). Neovasculature induced by vascular endothelial growth factor is fenestrated. *Cancer Res* **57**, 765–772.

76. Kim, K. J., Li, B., Winer, J. *et al.* (1993). Inhibition of vascular endothelial growth factor-induced angiogenesis suppresses tumour growth in vivo. *Nature* **362**, 841–844.

77. Millauer, B., Shawver, L. K., Plate, K. H., Risau, W., Ullrich, A. (1994). Glioblastoma growth inhibited in vivo by a dominant-negative Flk-1 mutant. *Nature* **367**, 576–579.

78. Strawn, L. M., McMahon, G., App, H. *et al.* (1996). Flk-1 as a target for tumor growth inhibition. *Cancer Res* **56**, 3540–3545.

79. Folkman, J., Watson, K., Ingber, D., Hanahan, D. (1989). Induction of angiogenesis during the transition from hyperplasia to neoplasia. *Nature* **339**, 58–61.

80. Inoue, M., Hager, J. H., Ferrara, N., Gerber, H. P., Hanahan, D. (2002). VEGF-A has a critical, nonredundant role in angiogenic switching and pancreatic beta cell carcinogenesis. *Cancer Cell* **1**, 193–202.

81. Ferrara, N. (2002). VEGF and the quest for tumour angiogenesis factors. *Nat Rev Cancer* **2**, 795–803.

82. Hiratsuka, S., Maru, Y., Okada, A., Seiki, M., Noda, T., Shibuya, M. (2001). Involvement of Flt-1 tyrosine kinase (vascular endothelial growth factor receptor-1) in pathological angiogenesis. *Cancer Res* **61**, 1207–1213.

83. Carmeliet, P., Moons, L., Luttun, A. *et al.* (2001). Synergism between vascular endothelial growth factor and placental growth factor contributes to angiogenesis and plasma extravasation in pathological conditions. *Nat Med* **7**, 575–583.

84. Luttun, A., Tjwa, M., Moons, L. *et al.* (2002). Revascularization of ischemic tissues by PlGF treatment, and inhibition of tumor angiogenesis, arthritis and atherosclerosis by anti-Flt1. *Nat Med* **8**, 831–840.

85. Hattori, K., Heissig, B., Wu, Y. *et al.* (2002). Placental growth factor reconstitutes hematopoiesis by recruiting VEGFR1(+) stem cells from bone-marrow microenvironment. *Nat Med* **8**, 841–849.

86. Eriksson, U., and Alitalo, K. (2002). VEGF receptor 1 stimulates stem-cell recruitment and new hope for angiogenesis therapies. *Nat Med* **8**, 775–777.

87. Nomura, M., Yamagishi, S., Harada, S., Yamashima, T., Yamashita, J., Yamamoto, H. (1998). Placenta growth factor (PlGF) mRNA expression in brain tumors. *J Neurooncol* **40**, 123–130.

88. Green, C. J., Lichtlen, P., Huynh, N. T. *et al.* (2001). Placenta growth factor gene expression is induced by hypoxia in fibroblasts: a central role for metal transcription factor-1. *Cancer Res* **61**, 2696–2703.

89. Beck, H., Acker, T., Puschel, A. W., Fujisawa, H., Carmeliet, P., Plate, K. H. (2002). Cell type-specific expression of neuropilins in an MCA-occlusion model in mice suggests a potential role in post-ischemic brain remodeling. *J Neuropathol Exp Neurol* **61**, 339–350.

90. Maisonpierre, P. C., Suri, C., Jones, P. F. *et al.* (1997). Angiopoietin-2, a natural antagonist for Tie2 that disrupts in vivo angiogenesis. *Science* **277**, 55–60.

91. Beck, H., Acker, T., Wiessner, C., Allegrini, P. R., Plate, K. H. (2000). Expression of angiopoietin-1, angiopoietin-2, and tie receptors after middle cerebral artery occlusion in the rat. *Am J Pathol* **157**, 1473–1483.

92. Acker, T, Beck, H., and Plate, K. H. (2001). Cell type specific expression of vascular endothelial growth factor and angiopoietin-1 and -2 suggests an important role of astrocytes in cerebellar vascularization. *Mech Dev* **108**, 45–57.

93. Oh, H., Takagi, H., Suzuma, K., Otani, A., Matsumura, M., Honda, Y. (1999). Hypoxia and vascular endothelial growth factor selectively up-regulate angiopoietin-2 in bovine microvascular endothelial cells. *J Biol Chem* **274**, 15732–15739.

94. Stratmann, A., Risau, W., and Plate, K. H. (1998). Cell type-specific expression of angiopoietin-1 and angiopoietin-2

suggests a role in glioblastoma angiogenesis. *Am J Pathol* **153**, 1459–1466.

95. Stratmann, A., Acker, T., Burger, A. M., Amann, K., Risau, W., Plate, K. H. (2001). Differential inhibition of tumor angiogenesis by tie2 and vascular endothelial growth factor receptor-2 dominant-negative receptor mutants. *Int J Cancer* **91**, 273–282.

96. Helmlinger, G., Yuan, F., Dellian, M., Jain, R. K. (1997). Interstitial pH and pO2 gradients in solid tumors in vivo: high- resolution measurements reveal a lack of correlation. *Nat Med* **3**, 177–182.

97. Kimura, H., Braun, R. D., Ong, E. T. *et al*. (1996). Fluctuations in red cell flux in tumor microvessels can lead to transient hypoxia and reoxygenation in tumor parenchyma. *Cancer Res* **56**, 5522–5528.

98. Hanahan, D. (1997). Signaling vascular morphogenesis and maintenance. *Science* **277**, 48–50.

99. Lauren, J, Gunji, Y., and Alitalo, K. (1998). Is angiopoietin-2 necessary for the initiation of tumor angiogenesis? *Am J Pathol* **153**, 1333–1339.

100. Thurston, G., Suri, C., Smith, K. *et al*. (1999). Leakage-resistant blood vessels in mice transgenically overexpressing angiopoietin-1. *Science* **286**, 2511–2514.

101. Jain, R. K., and Munn, L. L. (2000). Leaky vessels? Call Ang1! *Nat Med* **6**, 131–132.

102. Palmer, A., and Klein, R. (2003). Multiple roles of ephrins in morphogenesis, neuronal networking, and brain function. *Genes Dev* **17**, 1429–1450.

103. Gerety, S. S., Wang, H. U., Chen, Z. F., Anderson, D. J. (1999). Symmetrical mutant phenotypes of the receptor EphB4 and its specific transmembrane ligand ephrin-B2 in cardiovascular development. *Mol Cell* **4**, 403–414.

104. Adams, R. H., Wilkinson, G. A., Weiss, C. *et al*. (1999). Roles of ephrinB ligands and EphB receptors in cardiovascular development: demarcation of arterial/venous domains, vascular morphogenesis, and sprouting angiogenesis. *Genes Dev* **13**, 295–306.

105. Wang, H. U., Chen, Z. F., and Anderson, D. J. (1998). Molecular distinction and angiogenic interaction between embryonic arteries and veins revealed by ephrin-B2 and its receptor Eph-B4. *Cell* **93**, 741–753.

106. Gerety, S. S., and Anderson, D. J. (2002). Cardiovascular ephrinB2 function is essential for embryonic angiogenesis. *Development* **129**, 1397–1410.

107. Gale, N. W., Baluk, P., Pan, L. *et al*. (2001). Ephrin-B2 selectively marks arterial vessels and neovascularization sites in the adult, with expression in both endothelial and smooth-muscle cells. *Dev Biol* **230**, 151–160.

108. Shin, D., Garcia-Cardena, G., Hayashi, S. *et al*. (2001). Expression of ephrinB2 identifies a stable genetic difference between arterial and venous vascular smooth muscle as well as endothelial cells, and marks subsets of microvessels at sites of adult neovascularization. *Dev Biol* **230**, 139–150.

109. Ogawa, K., Pasqualini, R., Lindberg, R. A., Kain, R., Freeman, A. L., Pasquale, E. B. (2000). The ephrin-A1 ligand and its receptor, EphA2, are expressed during tumor neovascularization. *Oncogene* **19**, 6043–6052.

110. Adams, R. H., Diella, F., Hennig, S., Helmbacher, F., Deutsch, U., Klein, R. (2001). The cytoplasmic domain of the ligand ephrinB2 is required for vascular morphogenesis but not cranial neural crest migration. *Cell* **104**, 57–69.

111. Palmer, A., Zimmer, M., Erdmann, K. S. *et al*. (2002). EphrinB phosphorylation and reverse signaling: regulation by Src kinases and PTP-BL phosphatase. *Mol Cell* **9**, 725–737.

112. Brantley, D. M., Cheng, N., Thompson, E. J. *et al*. (2002). Soluble Eph A receptors inhibit tumor angiogenesis and progression in vivo. *Oncogene* **21**, 7011–7026.

113. Cheng, N., Brantley, D. M., Liu, H. *et al*. (2002). Blockade of EphA receptor tyrosine kinase activation inhibits vascular endothelial cell growth factor-induced angiogenesis. *Mol Cancer Res* **1**, 2–11.

114. Mukouyama, Y. S., Shin, D., Britsch, S., Taniguchi, M., Anderson, D. J. (2002). Sensory nerves determine the pattern of arterial differentiation and blood vessel branching in the skin. *Cell* **109**, 693–705.

115. Nakada, M., Niska, J. A., Miyamori, H. *et al*. (2004). The phosphorylation of EphB2 receptor regulates migration and invasion of human glioma cells. *Cancer Res* **64**, 3179–3185.

116. Zelinski, D. P., Zantek, N. D., Stewart, J. C., Irizarry, A. R., Kinch, M. S. (2001). EphA2 overexpression causes tumorigenesis of mammary epithelial cells. *Cancer Res* **61**, 2301–2306.

117. Acker, T., and Plate, K. H. (2003). Role of hypoxia in tumor angiogenesis – molecular and cellular angiogenic crosstalk. *Cell Tissue Res* **80**, 562–575.

118. Gerber, H. P., Condorelli, F., Park, J., Ferrara, N. (1997). Differential transcriptional regulation of the two vascular endothelial growth factor receptor genes. Flt-1, but not Flk-1/KDR, is up-regulated by hypoxia. *J Biol Chem.* **272**, 23659–23667.

119. Kappel, A., Ronicke, V., Damert, A., Flamme, I., Risau, W., Breier, G. (1999). Identification of vascular endothelial growth factor (VEGF) receptor-2 (Flk-1) promoter/enhancer sequences sufficient for angioblast and endothelial cell-specific transcription in transgenic mice. *Blood* **93**, 4284–4292.

120. Favier, J., Kempf, H., Corvol, P., Gasc, J. M. (2001). Coexpression of endothelial PAS protein 1 with essential angiogenic factors suggests its involvement in human vascular development. *Dev Dyn* **222**, 377–388.

121. Elvert, G., Kappel, A., Heidenreich, R. *et al*. (2003). Cooperative interaction of hypoxia-inducible factor-2alpha (HIF-2alpha) and Ets-1 in the transcriptional activation of vascular endothelial growth factor receptor-2 (Flk-1). *J Biol Chem* **278**, 7520–7530.

122. Li, J., Post, M., Volk, R. *et al*. (2000). PR39, a peptide regulator of angiogenesis. *Nat Med* **6**, 49–55.

123. Helmlinger, G., Endo, M., Ferrara, N., Hlatky, L., Jain, R. K. (2000). Formation of endothelial cell networks. *Nature* **405**, 139–141.

124. Ruoslahti, E. (2002). Specialization of tumour vasculature. *Nat Rev Cancer* **2**, 83–90.

125. Ruoslahti, E. (2000). Targeting tumor vasculature with homing peptides from phage display. *Semin Cancer Biol* **10**, 435–442.

126. St Croix, B., Rago, C., Velculescu, V. *et al*. (2000). Genes expressed in human tumor endothelium. *Science* **289**, 1197–1202.

127. Matsumoto, K., Yoshitomi, H., Rossant, J., Zaret, K. S. (2001). Liver organogenesis promoted by endothelial cells prior to vascular function. *Science* **294**, 559–563.

128. Lammert, E., Cleaver, O., and Melton, D. (2001). Induction of pancreatic differentiation by signals from blood vessels. *Science* **294**, 564–567.

129. Wartenberg, M., Donmez, F., Ling, F. C., Acker, H., Hescheler, J., Sauer, H. (2001). Tumor-induced angiogenesis studied in confrontation cultures of multicellular tumor spheroids and embryoid bodies grown from pluripotent embryonic stem cells. *FASEB J* **15**, 995–1005.

130. Rafii, S., Lyden, D., Benezra, R., Hattori, K., Heissig, B. (2002). Vascular and haematopoietic stem cells: novel targets for anti-angiogenesis therapy? *Nat Rev Cancer* **2**, 826–835.

131. Coussens, L. M., and Werb, Z. (2002). Inflammation and cancer. *Nature* **420**, 860–867.

132. Lin, E. Y., Nguyen, A. V., Russell, R. G. *et al.* (2001). Colony-stimulating factor 1 promotes progression of mammary tumors to malignancy. *J Exp Med* **193**, 727–740.

133. Takakura, N., Watanabe, T., Suenobu, S. *et al.* (2000). A role for hematopoietic stem cells in promoting angiogenesis. *Cell* **102**, 199–209.

134. Lyden, D., Hattori, K., Dias, S. *et al.* (2001). Impaired recruitment of bone-marrow-derived endothelial and hematopoietic precursor cells blocks tumor angiogenesis and growth. *Nat Med* **7**, 1194–1201.

135. Fukumura, D., Xavier, R., Sugiura, T. *et al.* (1998). Tumor induction of VEGF promoter activity in stromal cells. *Cell* **94**, 715–725.

136. Grimshaw, M. J., and Balkwill, F. R. (2001). Inhibition of monocyte and macrophage chemotaxis by hypoxia and inflammation–a potential mechanism. *Eur J Immunol* **31**, 480–489.

137. Cramer, T., Yamanishi, Y., Clausen, B. E. *et al.* (2003). HIF-1alpha is essential for myeloid cell-mediated inflammation. *Cell* **112**, 645–657.

138. Hiratsuka, S., Minowa, O., Kuno, J., Noda, T., Shibuya, M. (1998). Flt-1 lacking the tyrosine kinase domain is sufficient for normal development and angiogenesis in mice. *Proc Natl Acad Sci U S A* **95**, 9349–9354.

139. Desbaillets, I., Diserens, A. C., de Tribolet, N., Hamou, M. F., Van Meir, E. G. (1999). Regulation of interleukin-8 expression by reduced oxygen pressure in human glioblastoma. *Oncogene* **18**, 1447–1456.

140. Michiels, C., Arnould, T., and Remacle, J. (2000). Endothelial cell responses to hypoxia: initiation of a cascade of cellular interactions. *Biochim Biophys Acta* **1497**, 1–10.

141. Holash, J., Maisonpierre, P. C., Compton, D. *et al.* (1999). Vessel cooption, regression, and growth in tumors mediated by angiopoietins and VEGF. *Science* **284**, 1994–1998.

142. Samoto, K., Ikezaki, K., Ono, M. *et al.* (1995). Expression of vascular endothelial growth factor and its possible relation with neovascularization in human brain tumors. *Cancer Res* **55**, 1189–1193.

143. Lin, P., Polverini, P., Dewhirst, M., Shan, S., Rao, P. S., Peters, K. (1997). Inhibition of tumor angiogenesis using a soluble receptor establishes a role for Tie2 in pathologic vascular growth. *J Clin Invest* **100**, 2072–2078.

144. Folkman, J. (2002). Role of angiogenesis in tumor growth and metastasis. *Semin Oncol* **29**, 15–18.

145. Hansen-Algenstaedt, N., Stoll, B. R., Padera, T. P. *et al.* (2000). Tumor oxygenation in hormone-dependent tumors during vascular endothelial growth factor receptor-2 blockade, hormone ablation, and chemotherapy. *Cancer Res* **60**, 4556–4560.

146. Lee, C. G., Heijn, M., di Tomaso, E. *et al.* (2000). Anti-Vascular endothelial growth factor treatment augments tumor radiation response under normoxic or hypoxic conditions. *Cancer Res* **60**, 5565–5570.

147. Jain, R. K. (2001). Normalizing tumor vasculature with anti-angiogenic therapy: a new paradigm for combination therapy. *Nat Med* **7**, 987–989.

148. Novak, K. (2002). Angiogenesis inhibitors revised and revived at AACR. American Association for Cancer Research. *Nat Med* **8**, 427.

149. Newton, H. B. (2004). Molecular neuro-oncology and development of targeted therapeutic strategies for brain tumors. Part 2: PI3K/Akt/PTEN, mTOR, S. HH/PTCH and angiogenesis. *Expert Rev Anticancer Ther* **4**, 105–128.

150. Dahut, W. L., Gulley, J. L., Arlen, P. M. *et al.* (2004). Randomized phase II trial of docetaxel plus thalidomide in androgen-independent prostate cancer. *J Clin Oncol* **22**, 2532–2539.

151. Stadler, W. M., Kuzel, T., Shapiro, C., Sosman, J., Clark, J., Vogelzang, N. J. (1999). Multi-institutional study of the angiogenesis inhibitor TNP-470 in metastatic renal carcinoma. *J Clin Oncol* **17**, 2541–2545.

152. Kerbel, R. S., Yu, J., Tran, J. *et al.* (2001). Possible mechanisms of acquired resistance to anti-angiogenic drugs: implications for the use of combination therapy approaches. *Cancer Metastasis Rev* **20**, 79–86.

153. Kinzler, K. W., and Vogelstein, B. (1996). Life (and death) in a malignant tumour. *Nature* **379**, 19–20.

154. Hanahan, D., and Weinberg, R. A. (2000). The hallmarks of cancer. *Cell* **100**, 57–70.

155. Watanabe, K., Tachibana, O., Sata, K., Yonekawa, Y., Kleihues, P., Ohgaki, H. (1996). Overexpression of the EGF receptor and p53 mutations are mutually exclusive in the evolution of primary and secondary glioblastomas. *Brain Pathol* **6**, 217–223.

156. Sidransky, D., Mikkelsen, T., Schwechheimer, K. *et al.* (1992). Clonal expansion of p53 mutant cells is associated with brain tumour progression. *Nature* **355**, 846–847.

157. Reynolds, T. Y., Rockwell, S., and Glazer, P. M. (1996). Genetic instability induced by the tumor microenvironment. *Cancer Res* **56**, 5754–5757.

158. Yuan, J., Narayanan, L., Rockwell, S. *et al.* (2000). Diminished DNA repair and elevated mutagenesis in mammalian cells exposed to hypoxia and low pH. *Cancer Res* **60**, 4372–4376.

159. Giaccia, A. J. (1996). Hypoxic Stress Proteins: Survival of the Fittest. *Semin Radiat Oncol* **6**, 46–58.

160. Graeber, T. G., Peterson, J. F., Tsai, M., Monica, K., Fornace, A. J., Jr., Giaccia, A. J. (1994). Hypoxia induces accumulation of p53 protein, but activation of a G1-phase checkpoint by low-oxygen conditions is independent of p53 status. *Mol Cell Biol* **14**, 6264–6277.

161. Schmaltz, C., Hardenbergh, P. H., Wells, A., Fisher, D. E. (1998). Regulation of proliferation-survival decisions during tumor cell hypoxia. *Mol Cell Biol* **18**, 2845–2854.

162. Graeber, T. G., Osmanian, C., Jacks, T. *et al.* (1996). Hypoxia-mediated selection of cells with diminished apoptotic potential in solid tumours. *Nature* **379**, 88–91.

163. Yu, J. L., Rak, J. W., Coomber, B. L., Hicklin, D. J., Kerbel, R. S. (2002). Effect of p53 status on tumor response to antiangiogenic therapy. *Science* **295**, 1526–1528.

164. Garcia-Barros, M., Paris, F., Cordon-Cardo, C. *et al.* (2003). Tumor response to radiotherapy regulated by endothelial cell apoptosis. *Science* **300**, 1155–1159.

165. Gorski, D. H., Beckett, M. A., Jaskowiak, N. T. *et al.* (1999). Blockage of the vascular endothelial growth factor stress response increases the antitumor effects of ionizing radiation. *Cancer Res* **59**, 3374–3378.

166. Geng, L., Donnelly, E., McMahon, G. *et al.* (2001). Inhibition of vascular endothelial growth factor receptor signaling leads to reversal of tumor resistance to radiotherapy. *Cancer Res* **61**, 2413–2419.

167. Hess, C., Vuong, V., Hegyi, I. *et al.* (2001). Effect of VEGF receptor inhibitor PTK787/ZK222584 [correction of ZK222548]

combined with ionizing radiation on endothelial cells and tumour growth. *Br J Cancer* **85**, 2010–2016.

168. Tran, J., Master, Z., Yu, J. L., Rak, J., Dumont, D. J., Kerbel, R. S. (2002). A role for survivin in chemoresistance of endothelial cells mediated by VEGF. *Proc Natl Acad Sci U S A* **99**, 4349–4354.

169. Edwards, E., Geng, L., Tan, J., Onishko, H., Donnelly, E., Hallahan, D. E. (2002). Phosphatidylinositol 3-kinase/Akt signaling in the response of vascular endothelium to ionizing radiation. *Cancer Res* **62**, 4671–4677.

170. Brown, C. K., Khodarev, N. N., Yu, J. *et al.* (2004). Glioblastoma cells block radiation-induced programmed cell death of endothelial cells. *FEBS Lett* **565**, 167–170.

171. Moeller, B. J., Cao, Y., Li, C. Y., Dewhirst, M. W. (2004). Radiation activates HIF-1 to regulate vascular radiosensitivity in tumors: role of reoxygenation, free radicals, and stress granules. *Cancer Cell* **5**, 429–441.

172. Kerbel, R., and Folkman, J. (2002). Clinical translation of angiogenesis inhibitors. *Nat Rev Cancer* **2**, 727–739.

173. Browder, T., Butterfield, C. E., Kraling, B. M. *et al.* (2000). Antiangiogenic scheduling of chemotherapy improves efficacy against experimental drug-resistant cancer. *Cancer Res* **60**, 1878–1886.

174. Bocci, G., Nicolaou, K. C., and Kerbel, R. S. (2002). Protracted low-dose effects on human endothelial cell proliferation and survival in vitro reveal a selective antiangiogenic window for various chemotherapeutic drugs. *Cancer Res* **62**, 6938–6943.

175. Griffiths, L., Binley, K., Iqball, S. *et al.* (2000). The macrophage - a novel system to deliver gene therapy to pathological hypoxia. *Gene Ther* **7**, 255–262.

176. Maxwell, P. H., Dachs, G. U., Gleadle, J. M. *et al.* (1997). Hypoxia-inducible factor-1 modulates gene expression in solid tumors and influences both angiogenesis and tumor growth. *Proc Natl Acad Sci U S A* **94**, 8104–8109.

177. Ryan, H. E., Lo, J., and Johnson, R. S. (1998). HIF-1 alpha is required for solid tumor formation and embryonic vascularization. *EMBO J* **17**, 3005–3015.

178. Kung, A. L., Wang, S., Klco, J. M., Kaelin, W. G., Livingston, D. M. (2000). Suppression of tumor growth through disruption of hypoxia-inducible transcription. *Nat Med* **6**, 1335–1340.

179. Kim, M. S., Kwon, H. J., Lee, Y. M. *et al.* (2001). Histone deacetylases induce angiogenesis by negative regulation of tumor suppressor genes. *Nat Med* **7**, 437–443.

180. Carmeliet, P., Dor, Y., Herbert, J. M. *et al.* (1998). Role of HIF-1alpha in hypoxia-mediated apoptosis, cell proliferation and tumour angiogenesis. *Nature* **394**, 485–490.

181. Blancher, C., Moore, J. W., Talks, K. L., Houlbrook, S., Harris, A. L. (2000). Relationship of hypoxia-inducible factor (HIF)-1alpha and HIF-2alpha expression to vascular endothelial growth factor induction and hypoxia survival in human breast cancer cell lines. *Cancer Res* **60**, 7106–7113.

182. Oosthuyse, B., Moons, L., Storkebaum, E. *et al.* (2001). Deletion of the hypoxia-response element in the vascular endothelial growth factor promoter causes motor neuron degeneration. *Nat Genet* **28**, 131–138.

16

Biology of the Blood–Brain and "Blood–Brain Tumor" Barriers

Herbert H. Engelhard and Tibor Valyi-Nagy

THE NORMAL BLOOD–BRAIN BARRIER: ULTRASTRUCTURAL AND CELLULAR CONSIDERATIONS

The blood–brain barrier (BBB) was discovered by Paul Ehrlich in 1885 when he noted that dye injected into the bloodstream of a rabbit stained almost every bodily organ, but not the brain and spinal cord [1,2]. Intravascular dye introduction will also cause staining of the dura, but not affect the underlying leptomeninges or CSF [3,4]. At the cellular level, the BBB is composed of a continuous layer of endothelial cells, which constitutes the luminal surface of more than 99 per cent of the capillaries of the brain and spinal cord [1,5]. These capillary endothelial cells lack fenestrations and have high resistance tight junctions termed "zonulae occludentes," which limit the passage of even some small molecules into the CNS [1,6]. The BBB cells contain many mitochondria, indicating the large energy expenditures required for maintaining the integrity of the BBB [7]. The basal lamina surrounding the brain capillaries is more complex than elsewhere in the body, as is the extracellular matrix of the brain. Because of these features, movement of substances in and out of the CNS is quite limited in comparison to other organs.

The brain capillary endothelial cells which constitute the BBB are strongly influenced by astrocytes, pericytes, neurons and even the basement membrane (Fig. 16.1). The interactions between the various cells can be quite complex [1,8,9]. The foot processes of astrocytes ensheath more than 95 per cent of the capillary endothelium, and are thus interposed between the bloodstream and the majority of CNS neurons [1,6]. Astrocytes have been shown to induce BBB properties in developing brain during the angiogenic process, and orchestrate the tissue-specific gene expression of the brain capillary endothelium [1,5,10,11]. It is interesting that during the process of evolution, the "responsibility" for the barrier functions seems to have shifted more from the glial to the endothelial layer [11]. *In vitro* model systems of the BBB have been used to illustrate that astrocytes can also influence the "tightness" of the endothelial tight junctions [8]. Pericytes have the ability to contract, and have phagocytic activity and an antigen presenting function [1,5]. Lymphocytes may penetrate the BBB, and act to protect it [1,7].

The past several years have brought a significantly improved understanding of the molecular composition of tight junctions, including those formed between capillary endothelial cells in the brain [12,13]. It is now well established that endothelial BBB tight junctions differ from epithelial tight junctions in several morphological and functional aspects. Protein components of cerebral endothelial cell tight junctions include occludin, claudin-1, claudin-5, members of the junctional adhesion molecule (JAM) family, and endothelial cell-selective adhesion molecule (ESAM). Submembranous tight junction-associated proteins include zonula occludens (ZO) proteins ZO-1, -2, and -3—members of the membrane-associated guanylate kinase (MAGUK) family. These proteins are important for signal transduction and in anchoring

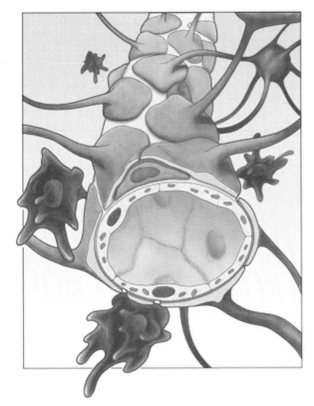

FIGURE 16.1 Three-dimensional artist's rendition of the BBB. The CNS capillary endothelial cells, with their abundant mitochondria and tight junctions (shown in yellow), constitute the structural basis for the blood–brain barrier. The capillary endothelium is 95 per cent covered by glial foot processes (red), and surrounded by the basal lamina, pericytes (orange), and perivascular phagocytes (blue). See Plate 16.1 in Color Plate Section.

the transmembrane tight junction proteins to the cytoskeleton. Adherens junctions are necessary for the primary contact between endothelial cells and formation of tight junctions. The cadherin–catenin system of adherens junction is also important for established tight junction morphology and function. Signaling pathways involved in tight junction regulation involve G-proteins, serine-, threonine-, and tyrosine kinases, calcium, cAMP, proteases and cytokines [12,13].

The BBB is absent at the pituitary and pineal glands, median eminence, area postrema, subfornical organ and lamina terminalis [2,14,15]. In these areas (the so-called "circumventricular organs"), the capillary endothelial cells are fenestrated, permitting molecules to pass more freely [1,14]. Substances which do not readily cross the BBB may reach the CNS by means of entry at these more vulnerable sites [8]. Another type of cell, the tanycyte, is found associated with the circumventricular organs, as well as in the region of the third ventricle, cerebral

aqueduct, floor of the fourth ventricle, and in the cervical spinal canal [3]. Tanycytes are somewhat like ependymal cells, but have few cilia. They may play a role in the transport of biogenic amines, which initiate the secretion of pituitary hormone release factors [16]. Recent evidence has also indicated that tanycytes may have a CNS regenerative function [17].

The blood–CSF barrier is the other major interface between the circulatory system and the CNS. By way of comparison however, the surface area available for exchange at the blood–CSF barrier is 1/5000th that of the BBB [18]. The blood–CSF barrier exists at the choroid plexus and in the subarachnoid space. The underlying vasculature of the choroid plexus consists of fenestrated, thin-walled, relatively large-diameter capillaries [3]. The epithelial cells of the choroid plexus, which constitute the blood–CSF barrier, have apical tight junctions only [1,8,13]. Fixed macrophages are associated with the choroid plexus. These cells are sometimes referred to as Kolmer or epiplexus cells. While most CSF is secreted by the choroid plexus, some is also produced by brain capillaries. The CSF flows through the ventricles of the brain, and a series of cisterns and subarachnoid spaces outside the brain and spinal cord, then is absorbed by bulk flow into the venous system at the arachnoid villi (or pacchionian granulations). The ependyma—which resembles renal tubular epithelium—constitutes the brain–CSF barrier [8].

While the ventricles are lined with the ependymal cells, the cisterns and spaces outside the brain are lined with arachnoid and pial cells [3]. Blood vessels run along the outer surface of the CSF-containing subarachnoid space [3]. It had been thought that the pia mater follows the arteries and arterioles for some short distance as they descend into the brain parenchyma. The perivascular space between the descending vessel and the pia, often referred to as the Virchow–Robin space, was thought to communicate with the subarachnoid space. Scanning electron microscopy, however, has revealed that the pia actually surrounds a vessel as it travels through the subarachnoid space, but does not accompany the vessel as it descends into the brain parenchyma. Instead, the pia surrounding the vessel spreads out over the pia which is covering the surface of the brain, effectively excluding the perivascular space from the subarachnoid space [3]. Thus, Virchow–Robin spaces actually communicate with the brain extracellular fluid space (ECF), rather than the subarachnoid space. Small solutes diffuse freely between the ECF and the CSF.

FUNCTIONS OF THE BLOOD–BRAIN BARRIER; GENERAL FACTORS AFFECTING OVERALL FUNCTION

The BBB acts to provide an optimal environment for the normal functioning of the brain, spinal cord and intradural nerve roots [5,7,15]. The usual serum concentrations of substances such as potassium, glycine and glutamate are toxic to the brain, necessitating the existence of such a barrier. On the other hand, in order to function normally, the brain requires several hydrophilic molecules, whose transfer is sharply curtailed by the BBB. Transportation of essential hydrophilic substances—including most metabolic substrates and inorganic ions—is therefore performed by carrier-mediated, energy-dependent transport systems [5,19,20]. Many CNS transport systems have been identified, including those for the glucose transporter (Glut-1 isoform) and hexoses; monocarboxy acids; amino acids and amines, nucleosides and purines; insulin; choline, transferrin and possibly cholesterol [7,20,21]. The glut-1 protein is present in virtually all of the brain microvascular endothelium having high resistance tight junctions, therefore it can be used histologically as a marker for the blood–brain and blood–CSF barriers [1,7]. Gene microarray studies have revealed that there are at least 50 genes that are specific for the BBB, most of which are related to various transport systems [15]. The function of BBB carrier-mediated transport systems may change with advancing age [22].

Other molecular systems transport molecules in the opposite direction, i.e. from the CNS into the bloodstream. Toxic metabolites, for instance, are actively pumped out of the brain [23]. Perhaps the best-known example of a "reverse BBB transport system" is the p-glycoprotein (multidrug resistance) efflux pump, which is highly expressed at the BBB [8,24]. Other examples are the multidrug-resistance-associated proteins (MRP 1 and 3) and organic anion transporters (OAT) [25]. Recently, a combination of *in vivo* and *in vitro* methods has been used to attempt to better characterize BBB efflux transport systems [26].

In short, the functions of the BBB include:

(1) protection of the brain by restricting the free exchange of water-soluble and/or large molecules between the blood and the CNS;

(2) selective (active) uptake of specific molecules across the endothelium by means of specialized transport systems; and

TABLE 16.1   Normal Composition of Cerebrospinal Fluid and Serum[1]

	CSF	Serum (arterial)
Osmolarity (mOsm/L)	295	295
Water content (%)	99	93
Sodium (mEq/L)	138	138
Potassium (mEq/L)	2.8	4.5
Chloride (mEq/L)	119	102
Bicarbonate (mEq/L)	22.0	24.0
Phosphorus (mg/dL)	1.6	4.0
Calcium (mEq/L)	2.1	4.8
Magnesium (mEq/L)	2.3	1.7
Iron (g/dL)	1.5	15.0
Urea (mmol/dL)	4.7	5.4
Creatinine (mg/dL)	1.2	1.8
Uric acid (mg/dL)	0.25	5.50
CO_2 tension (mmHg)	47.0	41.0
pH	7.33	7.41
Oxygen (mmHg)	43.0	104.0
Glucose (mg/dL)	60.0	90.0
Lactate (mEq/L)	1.6	1.0
Pyruvate (mEq/L)	0.08	0.11
Lactate:pyruvate ratio	26.0	17.6
Proteins (g/dL)	0.035	7.0

[1]Adapted from Barshes *et al.* [3] and Fishman [13].

(3) modification, metabolism and/or efflux transport of certain other substances from the blood and/or CNS [1,7,14].

The BBB therefore acts to provide the CNS with a privileged environment not only by blocking the entry of potentially toxic substances, but also by maintaining optimal concentrations of physiologic molecules, and actively eliminating unwanted substances such as metabolic byproducts and other molecules that may have somehow evaded its other defenses.

The composition of the "internal milieu" of the brain (i.e. the brain's ECF) is regulated not only by the BBB but also through modification by capillary-glial complexes, epithelia, and the neurons themselves [27]. The CSF (the composition of which is shown in Table 16.1) appears to have a function at least partially analogous to that of the lymphatics in other organs—namely, removing fat-soluble and toxic substances from the brain's ECF, across the ependyma [16]. Many fat-insoluble molecules are also removed from the brain ECF by the circulation of the CSF including urea, albumin, homovanillic acid, and norepinephrine [16]. The CSF certainly has a host of other functions as well, related to brain signaling and homeostasis. As a consequence of the functions of the BBB and CSF as summarized above, the CNS concentrations of most drugs given systemically, including most chemotherapeutic agents, will be relatively low.

Many conditions, most notably intrinsic CNS diseases (such as infection, neuroinflammatory disorders, tumors, or stroke) can dramatically affect the structure and function of the BBB. Disruption of the BBB causes vasogenic edema, which is defined as an increase in the volume of the ECF. Vasogenic edema is typically thought to involve white matter more than gray matter. BBB disruption may also occur due to hypertension, hypoxia-ischemia, hypercapnia or increased serum osmolarity [2,9]. Intravenous use of mannitol for the treatment of increased intracranial pressure may disrupt the BBB due to hypernatremia and high serum osmolarity. The BBB of older individuals may be relatively more susceptible to disruption [22]. Radiation therapy may increase the permeability of the BBB [28]. The effect of radiotherapy on the BBB provides a rationale for the co-administration of radiation and chemotherapy, for instance in the treatment of glioblastoma or impending brain metastases [28]. Corticosteroids, on the other hand, are thought to decrease the permeability of CNS capillaries, thereby protecting BBB function, and reversing the effect of influences acting to disrupt the BBB [19].

DRUG MOVEMENT ACROSS THE BBB AND DETERMINANTS OF CNS DRUG CONCENTRATION

The factors acting to determine the concentrations of drugs and other molecules at a particular location within the CNS are complex. As with other cell membranes, the cerebral endothelium is permeable to low molecular weight (< 500–600 Da), nonpolar, lipid-soluble molecules such as oxygen, carbon dioxide, ethanol, nicotine, diazepam and anesthetics [19,29]. For small molecules, a good correlation exists between the lipophilicity or lipid solubility of a drug (i.e. hydrophobicity), as approximated by the octanol/water partition coefficient, and its ability to cross the BBB [2,30,31]. Methotrexate, for instance, has a very low octanol/water partition coefficient and penetrates poorly into the brain when given systemically.

The more lipophilic compounds, however, can also be more vulnerable to cytochrome p450 metabolism [31]. Antiepileptic drugs such as phenytoin and phenobarbital may cause cytochrome p450 induction.

Knowledge of receptor/carrier systems has been exploited in order to increase the delivery of substances to the brain, including L-dopa, melphalan and baclofen [8]. In addition to molecular weight, charge, lipid solubility, and the presence of a carrier system, movement of compounds across the BBB also depends on physicochemical properties such as binding to plasma proteins, tertiary structure, extent of ionization and serum half-life [15,32]. When given intravascularly, all potential therapeutic substances face the drawback that the body acts as a "sink". Even in the best situations, only a small fraction of the administered drug actually reaches the tumor [33]. Larger molecules, such as blood-borne proteins, may enter CNS capillary endothelial cells through endocytosis, but are degraded in lysosomes before they can enter the ECF [19].

Since efflux transporters play an important role in the overall CNS exposure for many substrates [21,34], they too are an important factor that should be considered when planning therapeutic delivery of substances to the brain, and/or brain tumor cells [35]. Methotrexate, for instance, has been demonstrated to be (efflux) transported by MRP 1 and 3 [25]. P-glycoprotein inhibitors might be used clinically to increase the concentration of a drug within the brain [25,36]. *In vitro* cell systems—as well as purely mathematical models—aimed at predicting kinetic transfer rates across the BBB and drug concentrations within the brain have been described [7,37–39]. *In vivo*, the BBB penetration of specific drugs can be determined using radioactive tracers [14]. Animal studies are often still used to measure drug concentrations within the CNS [40], although improvements in radionuclide and MR imaging increasingly hold promise for obtaining CNS pharmacokinetic data directly from patients [25,31].

Since most chemotherapeutic agents achieve only low concentrations in the CNS when given intravenously, the BBB has been seen as an impediment to successful treatment of brain tumors by systemic chemotherapy. Systemic toxicity has usually been the dose-limiting factor in brain tumor chemotherapy [33]. While new strategies are constantly being developed, designing safe and effective drugs that cross the BBB continues to be a major challenge. Fortunately, temozolomide, like most nitrosoureas, *does* cross the blood–brain barrier [41]. Limitation of drug uptake into brain tumors could also be at least partially due to increased tissue pressure [2].

THE "BLOOD–BRAIN TUMOR" BARRIER; EFFECT OF TUMORS ON THE BBB

The capillary endothelium within brain tumors has sometimes been called the "blood–brain tumor barrier" [2,33]. Clinically, the permeability of this endothelium (and degree of vascularity) is indicated by the degree of tumor enhancement on CT and/or MRI scans. Meningiomas and brain metastases, as examples, would not be expected to have a true BBB, at least by the time they are visible by current neuroimaging studies. Most of the blood supply of meningiomas is derived from the dura; neither meningiomas nor metastases contain the astrocytes needed for the induction of the BBB during neovascularization. The structure of capillaries in malignant gliomas has been studied using light and electron microscopy [2,7,14,15,42]. Such studies have indicated that tumor capillaries are often surrounded by large collagen-filled extracellular spaces with gaps in the basal lamina and absent glial processes [2,42]; changes that are consistent with passage of large molecules across the endothelium. Molecular changes, such as alterations in claudins and occludin, have also been demonstrated [15].

Quantitative measurements of brain tumor capillary permeability in animal brain tumor models and human gliomas have indicated that capillary permeability in gliomas is regionally variable, and can be 10–30 times that of normal brain [7]. For gliomas and micrometastases, the real issue relates to the permeability of the endothelium in the region adjacent to, or even distant from, what is clearly demarcated as tumor by current imaging techniques. The enhancing (central) portion of a tumor can be addressed by techniques aimed at local control, such as surgery or radiosurgery. Unfortunately, malignant cells migrate into (or metastasize to) deeper brain regions where the BBB is at least initially intact, and thus may be shielded from the action of chemotherapeutic agents. Sometimes it has been assumed that the capillary permeability (and thus drug penetration) of the brain adjacent to tumor (BAT) is simply intermediate between that of the area of greatest BBB disruption, and that of normal brain. Such assumptions may not be accurate enough to plan therapy in a rational and successful manner. Better data pertaining to the delivery of specific drugs to brain tumors and the brain *in individual patients* is needed.

Brain tumor growth is often associated with BBB disruption and the development of tumor-associated (vasogenic) edema. Such edema has deleterious effects on cerebral blood flow, brain metabolism and intracranial pressure [2,7,42]. A variety of *in vitro* and animal models (C6, 9L, RG-2, and others) have been used to study the effects of brain tumor cells on the BBB, and the resultant edema formation [2,37]. Theoretically, tumor cells could affect BBB permeability in at least two ways:

(1) directly, through invasion or mechanical interference, and/or
(2) indirectly, by the production of diffusible molecules ("permeability factors") able to act at a distance from tumor cells [2,7].

In addition, glioma cells, unlike normal brain glia, probably lack the ability to induce the formation of a normal BBB. Mechanically, it has long been known that glioma cells often infiltrate brain parenchyma along "paths of least resistance" e.g. "Virchow–Robin" spaces and/or myelinated fiber tracts [43]. Proliferation of tumor cells adjacent to small vessels could directly affect capillary integrity and function [2,7]. It is likely that *"a critical threshold"* (in terms—as examples—of a percentage of brain tissue occupied by tumor cells, or the amount of "permeability factor" present) may need to be reached, before BBB disruption is detected, as illustrated in Fig. 16.2.

Observations regarding capillaries in brain tissue adjacent to tumors help explain the basis for BBB disruption. Studies have shown an increase in open junctions and pinocytoic vesicles in BAT capillaries, arguing for the importance of diffusible factors in the pathogenesis of BBB disruption [2,7,42]. Several BBB "permeability factors" have been identified for brain tumors. Vascular endothelial growth factor (VEGF) is a mitogen for endothelial cells and has extremely potent effects on microvascular permeability [14]. VEGF is expressed in malignant gliomas and its receptor is upregulated in tumor endothelial cells [7]. Degradation products of cell membrane phospholipids may also act as permeability factors, and be associated with changes in phospholipase A2 activity, arachidonic acid release, and increased production of leukotrienes and/or prostaglandins. Prostaglandin E and thromboxane B2 also act as vascular permeability factors [2,42]. Much research has focused on identifying proteolytic enzymes secreted by gliomas, which include:

(1) urokinase, a serine protease which is a plasminogen activator;
(2) matrix metalloproteinases, which include type IV and other collagenases, gelatinases and stromelysins, and

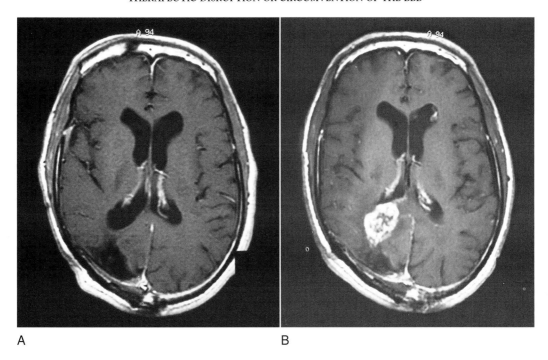

A   B

FIGURE 16.2 MRI scans illustrating the *critical threshold* concept for BBB disruption. Left: Gadolinium-enhanced MRI of a glioblastoma patient prior to tumor recurrence. Right: Similar slice, 12 weeks later. It is likely that tumor cells were radially distributed around the area of the occipital horn at the time of the scan at the left, but had not yet achieved the required density for BBB disruption, and thus detected by MRI.

(3) cathepsin B, a cysteine protease [2,7].

Such proteases, in addition to acting to degrade the extracellular matrix, and facilitate tumor cell migration, may be an important diffusible cause of BBB disruption.

THERAPEUTIC DISRUPTION OR CIRCUMVENTION OF THE BBB

If the BBB is indeed an impediment to successful brain tumor treatment, it would make sense to try to temporarily "open" it, or circumvent it altogether. Rapoport and colleagues first described the effect of intra-arterial infusions of hypertonic solutions on the integrity of the BBB, and hypothesized about the potential therapeutic applications [30,44]. Clinically, BBB disruption, with an increase in the delivery of drugs and/or macromolecules to the brain, has been most often produced by intra-carotid infusion of a hyperosmolar mannitol solution (see Chapter 18). Such treatment has been shown to transiently (10 minutes) increase drug delivery to normal brain and brain tumor, both through increased diffusion and bulk flow, with drug trapping within the BBB after it closes [44,45].

The mannitol is believed to cause shrinkage of the endothelial cells, with separation of the tight junctions [30]. The two parameters that are most important in mediating hyperosmolar disruption of the barrier are the osmolality of the solution and the infusion time [29]. Delivery is most effective when the drug is given 5–10 minutes after disruption [45]. Not unexpectedly, brain edema is a side effect of BBB disruption, and the procedure is performed under general anesthesia. Protocols for hyperosmolar BBB have been standardized, and the clinical studies well co-ordinated among multiple centers [29,46]. Other chemical agents used to modify the BBB have included dimethyl sulfoxide, arabinose, histamine, pentylenetetrazol, bradykinin, RMP-7 (a bradykinin B2 agonist), etoposide, and leukotrienes [2,30,36,47].

While use of therapeutic hyperosmotic BBB disruption (with infusion of various chemotherapeutic agents) has reported to be of benefit for treating primary CNS lymphoma [46,48], the respective roles of systemic chemotherapy, intrathecal chemotherapy and chemotherapy with BBB disruption for this disease may need to be further defined [49,50]. Antitumor activity of chemotherapy combined with BBB disruption for use against primitive neuroectodermal tumors and germinomas has also been shown [2,46].

Approaches designed to circumvent the BBB include the use of interstitial delivery of therapeutic molecules, or convection—enhanced delivery (see Chapter 20) [2,7,51–54]. Potential advantages include maximizing delivery of an effective drug/molecule or cell, while minimizing any systemic toxicity. However, local delivery can still produce variable and unpredictable concentration and spatial distribution of the therapeutic agent [33]. Depending on the type of delivery, many factors including infusion rate, initial drug concentration, depth of penetration and clearance across the BBB (efflux) and into the CSF will still be very important. The major dose-limiting factor of most local delivery methods is neurotoxicity [33].

Implantation of BCNU chemotherapy wafers into the tumor bed was approved by the U.S. Food and Drug Administration for use in the treatment of malignant glioma in early 1997, and has continued to be a treatment option for patients with these tumors [51] (see Chapter 19). The fact that results have been limited with this form of therapy may relate to the agent employed (BCNU), depth of penetration and drug concentration achieved, and/or the spatial distribution of the drug within the CNS [36,51]. A number of clinical trials evaluating the safety and/or efficacy of various local therapies have been undertaken, using either an injection/implantation strategy or convection-enhanced delivery (see www.virtualtrials.org). Many potential therapeutic agents have been studied in the laboratory and proposed for possible clinical use [33,52–55], as listed in Table 16.2.

CONCLUSIONS

The BBB provides protection and an optimal environment for the functioning of the CNS, strictly regulating the entry of molecules and cells into over 99 per cent of the brain and spinal cord. Astrocytes play an important role in the induction and maintenance of the BBB. When the BBB is compromised, vasogenic edema and neurologic deficits usually result. Potential BBB permeability factors have been identified for brain tumors. The capillary endothelium within brain tumors has been called the "blood–brain tumor barrier". For tumors such as meningiomas and clinically detectable brain metastases, a true BBB is absent. For patients with gliomas and (as yet undetected) micrometastases, the real issues relate to the presence of tumor cells, and permeability of the capillary endothelium, in what appears to be relatively normal brain tissue. The exact extent to which the BBB interferes with the delivery of

TABLE 16.2   Potential Therapeutic Agents for Direct CNS Implantation or Infusion

Monoclonal antibodies	Antisense oligodeoxynucleotides
Stem cells	Cytokines
Drug – protein conjugates	Radioligands
Drug polymers	Nanoparticles
Chemotherapeutic agents	Cytostatic drugs
Viral vectors	Vesicles / liposomes
Encapsulated cells	

chemotherapeutic agents will likely continue to be a topic of debate in the near future. Improved delivery of novel therapeutic agents to the CNS—whether by crossing or circumventing the BBB—should be of benefit for patients with a variety of CNS diseases, including malignant brain tumors.

References

1. Pardridge, W. M. (1991). Advances in cell biology of blood–brain barrier transport. *Semin Cell Biol* **2**, 419–426.
2. Engelhard, H. H. (2000). Brain tumors and the blood–brain barrier. *In Neuro-Oncology: The Essentials* (M. Bernstein, and M. Berger, Eds.), pp. 49–53, Thieme, New York.
3. Barshes, N., Demopoulos, A., and Engelhard, H. H. (2005). Anatomy and physiology of the leptomeninges and CSF space. *In Leptomeningeal Cancer* (L. Abrey, M. C. Chamberlain, and H. H. Engelhard, Eds.), Norwell, Kluwer Academic Publishers, Massachusetts.
4. Nabeshima, S., Reese, T. S., Landis, D. M. D., and Brightman, M. W. (1975). Junctions in the meninges and marginal glia. *J Comp Neur* **164**, 127–134.
5. Zinke, H., Mockel, B., Frey, A., Weiler-Guttler, H., Meckelein, B., and Gassen, H. G. (1992). Blood–brain Barrier: A molecular approach to its structural and functional characterization. *In Progress in Brain Research 91* (A. Ermisch, R. Landgraf, and H. J. Ruhle, Eds.), pp. 103–116, Elsevier Science Publishers.
6. Greig, N. H. (1989). Brain tumors and the blood-tumor barrier. *In Implications of the Blood–brain Barrier and Its Manipulation* (E. A. Neuwelt, Ed.), Vol. 2. pp 77–106. Plenum Medical Book Co. New York.
7. Engelhard, H. H., and Groothuis, D. G. (1998). The blood–brain barrier: Structure, function and response to neoplasia. *In The Gliomas* (M. Berger, and C. Wilson, Eds.), pp. 115–121. W. B. Saunders Co., Philadelphia.
8. DeBoer, A. G., and Breimer, D. D. (1994). The blood–brain barrier: Clinical implications for drug delivery to the brain. *J Royal College Physicians London* **28**, 502–506.
9. Neuwelt, E. A. (2004). Mechanisms of disease: the blood–brain barrier. *Neurosurgery* **54**, 141–2.
10. Abbott, N. J., Revest, P. A., and Romero, I. A. (1992). Astrocyte-endothelial interaction: physiology and pathology. *Neuropath Appl Neurobiol* **18**, 424–433.
11. Abbott, N. J. (2005). Dynamics of CNS barriers: Evolution, differentiation, and modulation. *Cell Mol Neurobiol* **25**, 5–23.
12. Wolburg, H., and Lippoldt, A. (2002). Tight junctions of the blood–brain barrier: development, composition and regulation. *Vascul Pharmacol* **38**, 323–337.

13. Vorbrodt, A. W., and Dobrogowska, D. H. (2003). Molecular anatomy of intercellular junctions in brain endothelial and epithelial barriers: electron microscopist's view. *Brain Res Brain Res Rev* **42**, 221–42.

14. Risau, W. (1994). Molecular biology of blood–brain barrier ontogenesis and function. *Acta Neurochir (Wien)* **60** Suppl 109–112.

15. DeVries, H. E., Montagne, L., Dijkstra, C. D., van der Valk, P. (2004). Molecular and cellular biology of the blood–brain barrier and its characteristics in brain tumors. *In Contemporary Cancer Research: Brain Tumors* (F. Ali-Osman. Ed.), pp. 157–174, Humana Press, New Jersey, Totowa.

16. Milhorat, T. H. (1975). The third circulation revisited. *J Neurosurg* **42**, 628–645.

17. Prieto, M., Chauvet, N., and Alonso, G. (2000). Tanycytes transplanted into the adult rat spinal cord support the regeneration of lesioned axons. *Exp Neurol* **161**, 27–37.

18. Pardridge, W. M. (1986). Blood–brain barrier transport of nutrients. *Fed Proc* **45**, 2047–9.

19. Salcman, M., and Broadwell, R. D. (1991). The blood–brain barrier. *In Neurobiology of Brain Tumors* (M. Salcman, Ed.), pp. 229–249. Williams & Wilkins. Baltimore.

20. Allen, D. D., and Lockman, P. R. (2003). The blood–brain barrier choline transporter as a brain drug delivery vector. *Life Sci* **73**, 1609–15.

21. Graff, C. L., and Pollack, G. M. (2004). Drug transport at the blood–brain barrier and the choroid plexus. *Cur Drug Metab* **5**, 95–108.

22. Shah, G. N., and Mooradian, A. D. (1997). Age-related changes in the blood–brain barrier. *Experimental Gerontology* **32**, 501–19.

23. Sun, H., Dai, H., Shaik, N., and Elmquist, W. F. (2003). Drug efflux transporters in the CNS. *Adv Drug Deliv Rev* **55**, 83–105.

24. Regina, A., Demeule, M., Laplante, A. *et al.* (2001). Multidrug resistance in brain tumors: roles of the blood–brain barrier. *Cancer Metastasis Rev* **20**, 13–25.

25. Bart, J., Groen, H. J., Hendrikse, N. H. *et al.* (2000). The blood–brain barrier and oncology: New insights into function and modulation. *Cancer Treat Rev* **26**, 449–62.

26. Hosoya, K., Ohtsuki, S., and Terasaki, T. (2002). Recent advances in the brain-to-blood efflux transport across the blood–brain barrier. *Int J Pharm* **248**, 15–29.

27. Fishman, R. A. (1992). *Cerebrospinal Fluid in Diseases of the Nervous System.* W.B.Saunders Co., Philadelphia.

28. Van Vulpen, M., Kal, H. B., Taphoorn, M. J., El-Sharouni, S. Y. (2002). Changes in blood–brain barrier permeability induced by radiotherapy: implications for timing of chemotherapy? *Oncol Rep* **9**, 683–8.

29. Fortin, D. (2004). The blood–brain barrier should not be underestimated in neuro-oncology. *Revue Neurologique* **160**, 523–32.

30. Kroll, R. A., and Neuwelt, E. A. (1998). Outwitting the blood–brain barrier for therapeutic purposes: osmotic opening and other means. *Neurosurgery* **42**, 1083–99.

31. Waterhouse, R. N. (2003). Determination of lipophilicity and its use as a predictor of blood–brain barrier penetration of molecular imaging agents. *Mol Imaging Biol* **5**, 376–89.

32. Engelhard, H. H., Duncan, H., Kim, S., Criswell, P. S., and Van Eldik, L. (2001) Therapeutic effects of sodium butyrate on glioma cells *in vitro* and in the rat C6 glioma model. *Neurosurgery* **48**, 616–625.

33. Groothuis, D. R. (2000). The blood–brain and blood-tumor barriers: A review of strategies for increasing drug delivery. *Neuro-oncol* **2**, 45–59.

34. Golden, P. L., and Pollack, G. M. (2003). Blood–brain barrier efflux transport. *J Pharm Sci* **92**, 1739–53.

35. Sun, H., Dai, H., Shaik, N., and Elmquist, W. F. (2003). Drug efflux transporters in the CNS. *Adv Drug Deliv Rev* **55**, 83–105.

36. Kemper, E. M., Boogerd, W., Thuis, *et al.* (2004). Modulation of the blood–brain barrier in oncology: Therapeutic opportunities for the treatment of brain tumors? *Cancer Treat Rev* **30**, 415–23.

37. Grabb, P. A., and Gilbert, M. R. (1995). Neoplastic and pharmacological influence on the permeability of an *in vitro* blood–brain barrier. *J Neurosurg* **82**, 1053–58.

38. Gratton, J. A., Abraham, M. H., Bradbury, M. W., and Chadha, H. S. (1997). Molecular factors influencing drug transfer across the blood–brain barrier. *J Pharm Pharmacol* **49**, 1211–6.

39. Chakrabarty, S. P., and Hanson, F. B. (2005). Optimal Control of Drug Delivery to Brain Tumors for a Distributed Parameters Model. Proceedings of 2005 American Control Conference, pp. 1–6.

40. Dash, A. K., and Elmquist, W. F. (2003). Separation methods that are capable of revealing blood–brain barrier permeability. *J Chromatogr B Analyt Technol in the Biomed Life Sci* **797**, 241–54.

41. DeAngelis, L. M. (2003). Benefits of adjuvant chemotherapy in high-grade gliomas. *Semin Oncol* **30** (6 Suppl 19), 15–18.

42. Schiffer, D. (1993). *Brain Tumors: Pathology and its Biological Correlates.* Springer-Verlag, New York.

43. Giese, A., and Westphal, M. (1996). Glioma invasion in the central nervous system. *Neurosurgery* **39**, 235–250.

44. Rapoport, S. I. (2001). Advances in osmotic opening of the blood–brain barrier to enhance CNS chemotherapy. *Expert Opin on Investing Drugs* **10**, 1809–18.

45. Zunkuler, B., Carson, R. E., Olson, J. *et al.* (1996). Quantification and pharmacokinetics of blood–brain barrier disruption in humans. *J Neurosurg* **85**, 1056–65.

46. Doolittle, N. D., Miner, M. E., Hall, W. A. *et al.* (2000). Safety and efficacy of a multicenter study using intraarterial chemotherapy in conjunction with osmotic opening of the blood-–brain barrier for the treatment of patients with malignant brain tumors. *Cancer* **88**, 637–47.

47. Cloughesy, T. F., and Black, K. L. (1995). Pharmacological blood–brain barrier modification for selective drug delivery. *J Neurooncol* **26**, 125–32.

48. Dahlborg, S. A., Henner, W. D., Crossen, J. R. *et al.* (1996). Non-AIDS primary CNS lymphoma: First example of a durable response in a primary brain tumor using enhanced chemotherapy delivery without cognitive loss and without radiotherapy. *Cancer J Sci Am* **2**, 166.

49. Park, D. M., and Abrey, L. E. (2002). Pharmacotherapy of primary CNS lymphoma. *Expert Opin Pharmacother* **3**, 39–19.

50. Plotkin, S. R., and Batchelor, T. T. (2002). Primary central nervous system lymphoma. *Curr Treatment Options Oncol* **3**, 525–35.

51. Engelhard, H. (2000). The role of interstitial BCNU chemotherapy in the treatment of malignant glioma. *Surg Neurol* **53**, 458–464.

52. Dunn, I. F., and Black, P. M. (2003). The neurosurgeon as local oncologist: cellular and molecular neurosurgery in malignant glioma therapy. *Neurosurgery* **52**, 1411–22.

53. Blomer, U., Ganser, A., and Scherr, M. (2002). Invasive drug delivery. *Adv Exp Med Biol* **513**, 431–51.

54. Bansal, K., and Engelhard, H. H. (2000) Gene therapy for brain tumors. *Cur Oncol Rep* **2**, 463–472.

55. Engelhard, H. H., Egli, M., and Rozental, J. (1998) Use of antisense vectors and oligodeoxynucleotides in neuro-oncology. Pediatr Neurosurg **29**, 1–10.

INNOVATIVE APPROACHES TO CHEMOTHERAPY DELIVERY

17

Intra-Arterial Chemotherapy

Herbert B. Newton

ABSTRACT: Intra-arterial (IA) chemotherapy is a form of regional delivery to brain tumors, designed to enhance the intra-tumoral concentrations of a given drug, in comparison to the intravenous route. Drugs that are likely to be benefit from IA delivery have a rapid systemic clearance and include carmustine and other nitrosoureas, cisplatin, carboplatin, etoposide, and methotrexate. Clinical studies have demonstrated activity of IA chemotherapy approaches for low- and high-grade gliomas, cerebral lymphoma, and brain metastases. However, a survival benefit for IA drug delivery, in comparison to intravenous administration, has not been proven in phase III trials. The technique is limited by the potential for significant vascular and neurological toxicity, including visual loss, stroke, and leukoencephalopathy. More recent studies suggest that toxicity can be reduced by the use of carboplatin- and methotrexate-based regimens. Further clinical studies will be needed to determine the appropriate role for IA chemotherapy in the treatment of brain tumors.

INTRODUCTION

Intra-arterial (IA) chemotherapy has been under investigation as a treatment modality for primary and metastatic brain tumors since the 1950s and 1960s, and continues to be evaluated for its proper place in the treatment armamentarium [1–3]. The goal of IA chemotherapy is to deliver a drug to tumor cells in a higher concentration for a given systemic exposure, thereby increasing the intracellular concentrations of drug and improving the likelihood of subsequent cell death. This regional approach is appropriate for most primary and metastatic brain tumors, since they are localized to the brain and

dependent on the cerebral arterial blood supply. Although approximately 2000 patients worldwide have received IA chemotherapy using a variety of different drugs, no definitive conclusions have been reached as to its efficacy against primary and metastatic brain tumors, procedural safety profile, or propensity for neurotoxicity [4–7]. This ambiguity results from several factors, including the large number of different drugs that have been used in IA studies, the wide variation in study methodology and definitions of patient and tumor responses, the fact that most of the reports have been small phase I and II studies, and the inconsistencies among studies in terms of patient demographics, especially patient age and tumor type. Despite these limitations in the data, there is a wealth of information that has been published regarding the use of IA chemotherapy for brain tumor treatment. An in-depth review of this literature will be the focus of the current chapter. In addition, other applications of IA chemotherapy can be found in Chapter 18 (Blood-Barrier Disruption Chemotherapy) and Chapter 35 (Chemotherapy of Brain Metastases).

PHARMACOLOGICAL RATIONALE

There are several important pharmacological considerations that must be taken into account when evaluating the IA route and comparing it to more conventional methods of administration (i.e., intravenous, IV) [4–11]. When a drug is administered IA, it results in an augmentation of the drug's local peak plasma concentration and the area under the concentration-time curve (AUC), when compared to an equivalent IV dose. Mathematical modeling of this comparison would predict up to a five-fold increase in drug delivery as defined by the concentration-time integral [8,9]. In addition, the peak effect can be quite significant and may result in a ten-fold increase in regional concentration of drug within the brain. As discussed in more detail below, the potential improvement in regional concentrations of drug achieved through IA administration can be affected by the arterial vessels chosen for infusion and the velocity of blood flow within those vessels. Local peak concentrations and the AUC will be maximized when the drug is delivered through a smaller artery with a slower flow rate. Administration *via* larger vessels with rapid blood flow results in dilution of the drug and less time for drug extraction by the tissues.

The pharmacological advantage of IA administration occurs in the "first pass" of drug through the tumor circulation [8–11]. After the "first pass", the drug continues to circulate through the bloodstream until it is cleared from the body, similar to an IV or orally administered drug. The "first pass" effect is maximized by drugs with certain characteristics, including a high extraction fraction within the target organ, high capillary permeability, which for a brain tumor would be equivalent to lipid solubility, and rapid systemic metabolism and/or excretion. As a class, nitrosourea drugs (e.g., carmustine) have many pharmacokinetic features that should make them ideal for IA administration. This has been supported by several different methods, including the IA infusion of ^{14}C-labeled carmustine into a monkey model, which provided a 2.5- to 5-fold delivery advantage of drug over IV infusion [12]. In another study using positron emission tomography (PET) and ^{11}C-labeled carmustine, superselective IA infusion into the middle cerebral artery of patients with recurrent gliomas resulted in up to a 50-fold increase in drug delivery compared to IV infusion [13]. In two of the patients, there was a correlation between high initial concentrations of intracellular ^{11}C-carmustine and a favorable clinical response to IA treatment. In a more recent study using a rabbit model, IA infusion of carmustine resulted in higher brain concentrations of drug than equivalent IV dosing [14]. However, the concentrations of measured carmustine were less than the levels predicted by pharmacokinetic modeling. A similar drug delivery advantage (approximately 2.5-fold) has been noted in both animal tumor models and clinical studies with the use of IA cisplatin [15,16]. A partial explanation of this effect may be the ability of IA cisplatin to disrupt the blood–brain barrier (BBB) [17]. In addition, an autopsy study of patients that had received cisplatin evaluated the levels of drug within tumor and noted higher levels

in the cohort that had received IA treatment [18]. IA delivery was found to be a significant independent factor related to intra-tumoral concentrations of cisplatin in a multiple linear regression analysis. Further evidence of the drug delivery advantage of IA administration has come from recent animal studies that have demonstrated an increase in intra-tumoral concentrations of methotrexate and boronated compounds by 3- to 5.5-fold in comparison to equivalent IV infusions [19,20].

An important consideration in the application of IA chemotherapy is choosing the proper drug or drugs to be infused. Not all drugs have the appropriate metabolism and pharmacokinetic profile for IA usage. As noted above, it is important that the drug have a rapid total body clearance, as defined in the *Regional Advantage* equation [8]:

$$R_a = 1 + \frac{CL(tb)}{F}$$

where F equals the blood flow (l/min) in the artery used for infusion, and CL(tb) is the total body clearance (l/min) of the infused drug. The R_a is the pharmacological advantage a drug has (or may not have) when administered IA *versus* the IV route; it is maximized by a rapid (i.e., large) total body clearance. With rapid systemic clearance, the amount of drug extracted by the tumor tissue after the "first pass" (i.e., during the venous recirculation phase) will be small relative to the amount of drug extracted during the "first pass" [8,9]. Conversely, drugs with slow systemic clearance will have a relatively large amount of drug extracted after the "first pass" in comparison to the amount extracted during the "first pass". The drugs with the most appropriate R_a for IA chemotherapy (ranked in descending order) include BCNU > cisplatin >> carboplatin > etoposide. The two drugs with the highest R_a (i.e., BCNU, cisplatin) are also known to have the most significant neuro-toxicity.

One theoretical limitation to the effectiveness of IA chemotherapy, and a potential cause of increased toxicity, is the concept of drug "streaming" [5,11]. It is hypothesized that streaming can occur after IA delivery of a drug, leading to inhomogeneous flow into the arterial branches and subsequent heterogeneous delivery into the tumor and surrounding brain tissue. Abnormalities of flow could result in inadequate delivery of drug to the tumor, low intra-tumoral drug concentrations, and a lack of response to treatment. In addition, the same anomalous flow dynamics could result in excessive drug concentrations being delivered to regions of normal brain, with a higher likelihood of neurotoxicity. Several investigators have studied the flow dynamics of IA chemotherapy, in an attempt to determine the incidence and severity of drug streaming. Early studies suggested that streaming might be a significant factor during IA infusion [21–23]. Blacklock and co-workers used quantitative autoradiography in a monkey model and noted significant non-uniform flow of ^{14}C-iodoantipyrine after IA infusion [21]. The streaming effect was more prominent at slower infusion rates and, in some cases, resulted in up to a ten-fold concentration difference between adjacent regions of cerebral cortex. Similar results have been noted by the same authors, during slow and medium rates of IA infusion using a rat model [22]. In these studies, the flow distribution became more uniform in distal vessels at fast infusion rates. Streaming has also been demonstrated during infra-ophthalmic infusion into an *in vitro* flow model of the human carotid arterial system [23]. The degree of streaming was greatly influenced by the positioning of the catheter tip during the IA infusion. In a similar study by the same authors using an *in vitro* flow model of the human vertebral arterial system, streaming was noted at slow intra-vertebral infusion rates (e.g., 2 ml/min) [24]. The streaming effect was minimized by placement of the catheter tip further away from the vertebral/basilar junction.

Data in humans would suggest that streaming is infrequent during IA infusion and that when it does occur, it is not very prominent, especially with infusions into the carotid artery. This is supported by several PET studies of patients with brain tumors [25,26]. In a study by Junck and associates, IA infusion of ^{15}O-H$_2$O at the carotid bifurcation resulted in homogeneous flow through the distal arterial system and uniform cerebral distribution over a 20-fold range of infusion rates (0.5–10 ml/min) [25]. These data suggested that mixing within the artery was complete or near complete and resulted from the pattern of blood flow within the artery, and not from jet effects at the catheter tip. A similar PET study by Saris and co-workers noted only a mild degree of flow heterogeneity within the ophthalmic artery and major cerebral vessels during continuous IA infusion in the carotid artery at the level of C2 [26]. In contrast, when the infusion occurred in the supraophthalmic portion of the carotid artery, there was more obvious heterogeneity of flow consistent with a streaming effect. The abnormalities of flow during carotid infusion were not affected by alterations of the infusion rate. A similar degree of streaming was noted during infusion into the supraophthalmic portion of the carotid artery in eight of ten patients

studied by Tc-99m hexamethyl-propyleneamine oxime (HMPAO) single-photon emission computed tomographic (SPECT) scanning [27]. The streaming effect was more prominent during slower infusion rates. A more recent study using the same Tc-99m HMPAO SPECT techniques studied six patients with high-grade glioma undergoing IA chemotherapy [28]. Flow dynamics and cerebral distribution were studied at slow and fast infusion rates into the infraophthalmic carotid artery. There was no evidence for significant streaming and no difference in flow dynamics or distribution between slow and fast infusion rates. The disparity of results between animal and human studies of IA infusion is probably explained by comparative differences in the anatomy of the carotid arteries, with a larger diameter, increased tortuosity, and increased distance between the infusion site and distal branch points in the human vessels. All of these aspects of the human carotid artery would lead to increased flow turbulence and mixing of drug, thereby reducing the potential for streaming.

Several authors have evaluated techniques to reduce drug streaming during IA infusion. The most well studied technique appears to be diastole-phased pulsatile infusion (DPPI) [26,29,30]. During DPPI the infusion is administered in a pulsatile manner, timed to the cardiac cycle so that the pulse occurs after the R wave, during "local diastole", when there is the lowest amount of flow at the injection site. Using this method in an *in vitro* internal carotid artery model, infusions resulted in a uniform distribution without streaming, even at slower infusion rates [29]. This approach requires the use of a specifically modified angiographic injector pump, but does provide excellent mixing of injectate with blood. In a monkey model, DPPI was shown to induce thorough mixing of ^{14}C-iodoantipyrine during IA infusion and was less likely to result in streaming when compared to more conventional continuous IA infusion methods [23]. Evaluation of DPPI has been extended to humans in a pilot study of ten patients with malignant gliomas [26]. In three patients with extensive streaming during IA infusion into the supraophthalmic carotid artery, the use of DPPI was able to substantially reduce or eliminate the streaming pattern.

SINGLE-AGENT IA CHEMOTHERAPY OF NEWLY DIAGNOSED GLIOMAS

The experience with IA chemotherapy has been extensive in patients with gliomas, both at initial diagnosis and at recurrence [4–7]. In patients with newly diagnosed tumors, IA chemotherapy has usually been administered just prior to, or in combination with, external beam irradiation. Single agents that have been used in this context include carmustine, nimustine, HeCNU, cisplatin, and 5-fluorouracil (see Table 17.1). The median time to progression (TTP) of newly diagnosed patients receiving single-agent IA chemotherapy has ranged from 12 to 32 weeks, with a median survival of approximately 1 year (range of 32–73 weeks). The nitrosourea class of drugs, especially carmustine, have been used most often in this group of patients. Initial experience began with single-agent carmustine in the 1980s, with doses ranging from 150 to 300 mg/m^2 IA every 6–8 weeks [31–34]. Greenberg and colleagues treated twelve patients using an intra-carotid approach and noted a 75 per cent response rate, with a median TTP of 25 weeks [31]. Toxicity was significant and included visual loss and seizure activity in 50 and 33 per cent of patients, respectively. Hochberg and co-workers treated 43 patients using both intra-carotid and supraophthalmic approaches, during and after irradiation [32]. The median survival for the two cohorts of patients ranged from 50 to 64 weeks. Neurological toxicity was less prominent, with deterioration noted in 17 per cent of patients and seizures in 8 per cent. Another report from the same group treated 28 patients with intra-carotid carmustine (400 mg every 4 weeks × 4 cycles) before the onset of irradiation [33]. Although there was a 44 per cent response rate, the responses were very brief and patients quickly developed progressive disease. The overall median survival of only 37 weeks was quite poor and not as good as their previous experience of using IA carmustine in combination with radiotherapy. In addition, up-front use of IA carmustine was more neurotoxic, with leukoencephalopathy noted in 7 per cent of patients, as well as visual loss (15 per cent) and neurological deterioration (7 per cent). In the report from Clayman and associates, IA carmustine (150 mg/m^2, q6 weeks) was administered to thirteen patients by selective intracerebral infusion [34]. The overall response rate was 38 per cent, with an impressive median survival of 73 weeks. Neurotoxicity appeared to be milder using this approach, with only 7 per cent of patients having neurological deterioration, and no evidence of visual loss or leukoencephalopathy.

Several investigators have treated newly diagnosed glioma patients with IA nimustine and HeCNU [35–40]. The initial experience was reported by Yamashita and co-workers in 1983, when they administered IA nimustine (2–3 mg/kg) to nine patients with glioblastoma multiforme (GBM), before and after irradiation [35]. Eight patients also

TABLE 17.1   Single-Agent IA Chemotherapy for Newly Diagnosed Gliomas

Author (ref.)	Drug	No. Pts.	% CR/PR/MR	Median TTP (weeks)	Median Survival (weeks)
Greenberg [31]	Carmustine, 200–300 mg/m^2 q6–8 wks	12	75	25	54
Hochberg [32]	Carmustine, 240 mg/m^2 q4–6 wks	43	NR	NR	50–64
Bashir [32]	Carmustine, 400 mg, q4 wk × 4 cycles	28	44	NR	37
Clayman [34]	Carmustine, 150 mg/m^2, q6 wks	13	38	NR	73
Yamashita [35]	Nimustine, 2–3 mg/kg, × 4 cycles	9	NR	NR	50
Roosen [36]	Nimustine, 100 mg, q6 wks	35	69 (includes SD)	NR	56
Vega [37]	Nimustine, 150 mg, q6–8 wks × 3	22	27	12	32
Chauveinc [38]	Nimustine, 150 mg, q6–8 wks × 3	27	52	NR	44
Kochi [39]	Nimustine, 80 mg/m^2, q6 wks × 3	42	NR	24	59
Fauchon [40]	HeCNU, 120 mg/m^2, q6–8 wks × 3	40	15	32	48
Calvo [41]	Cisplatin, 40–60 mg/m^2, q1 wk × 3–5	11	100	NR	42
Dropcho [42]	Cisplatin, 75 mg/m^2, q4 wk × 4	22	45	23	63
Mortimer [43]	Cisplatin, 150 mg, q3 wk × 2	27	22	NR	38.4–43.2
Larner [44]	5-FU, 200–600 mg, q1 wk × 4	25	NR	NR	60
Greenberg [45]	5-FU, 5 mg/m^2/day × 8 wks, with BudR	39	NR	NR	68

Derived from references [4–6,30–45]

Abbreviations: ref. — reference; No. — number; pts. — patients; CR — complete response; PR — partial response; MR — minor response; wks — weeks; 5-FU — 5-fluorouracil; NR — not reported; SD — stable disease.

received nimustine by the intravenous (IV) route. The median survival for patients receiving IA and IV nimustine showed a trend towards the IA cohort (50 *versus* 36 weeks). However, due to the small number of cases, the difference in survival did not reach statistical significance. No neurological toxicity was reported for either group of patients. In a larger study of 35 patients with GBM, Roosen and colleagues used intra-carotid nimustine (100 mg q6 weeks) before and after radiotherapy [36]. The overall response rate (including stable disease, SD) was 69 per cent, with a median survival of 56 weeks. The percentage of long-term survivors was impressive, with 18-month and 24-month survival rates of 41.0 per cent and 32.8 per cent, respectively. Neurological toxicity was noted in this study, including visual loss (6 per cent), neurological deterioration (9 per cent), and seizure activity (3 per cent). Vega and associates treated 22 patients (IA nimustine, 150 mg q6–8 weeks × 3 cycles) with newly diagnosed malignant gliomas before the onset of irradiation [37]. There was a 27 per cent objective response rate (6/22; GBM — 4, anaplastic astrocytoma, AA — 2), with a median TTP of only 12 weeks. The median survival was 32 weeks for the entire cohort, and was equivalent between GBM and AA patients. The only recorded neurotoxicity was visual loss in 9 per cent of patients. A more recent report by the same group described a cohort of 27 patients (10 GBM, 17 AA) treated by intra-carotid nimustine [38]. The objective response rate was 51.8 per cent, with a median survival of 44 weeks. The 24-month long-term survival rate for the entire cohort was 37 per cent. In the only randomized comparison of IA *versus* intravenous nimustine (80 mg/m^2 every 6 weeks), Kochi and associates treated a total of 84 patients with GBM [39]. The median TTP was longer in the intravenous cohort in comparison to the IA cohort (45 *versus* 24 weeks), although the difference was not significant. However, the median survival of the intravenous and IA groups were very similar (56 *versus* 59 weeks). Toxicity was equivalent between the two arms of the study. There were no episodes of leukoencephalopathy in the IA cohort. One study by Fauchon and co-workers administered intra-carotid HeCNU (120 mg/m^2, q6–8 weeks × 3 cycles) to a cohort of 40 patients before the onset of irradiation [40]. The objective response rate was only 15 per cent, with another 55 per cent experiencing SD. The median TTP was 32 weeks, with an overall median survival of 48 weeks. Neurological toxicity consisted of visual loss (15 per cent) and leukoencephalopathy (10 per cent).

Intra-arterial cisplatin has also been applied to newly diagnosed gliomas by several investigators [41–43]. Calvo and co-workers administered intra-carotid cisplatin (40–60 mg/m^2/week × 3–5 doses) to eleven evaluable patients with malignant glioma [41]. There was an excellent objective response rate of 100 per cent (5 complete responses, CR; 6 partial response, PR). However, the responses were not durable and the overall median survival was poor (42 weeks). Significant neurotoxicity was noted with IA cisplatin, including visual loss (9 per cent) and seizure activity (18 per cent). In a more extensive study by Dropcho and colleagues, 22 patients (13 GBM, 9 AA) received intra-carotid cisplatin (75 mg/m^2, q4 weeks x 4 cycles) before the onset of radiotherapy [42]. All patients underwent biopsy for histological verification, but did not undergo a resection, allowing for an assessment of the sensitivity of untreated gliomas to IA cisplatin. There was a 45 per cent objective response rate (i.e., PR 23 per cent, MR 22 per cent), as well as another 18 per cent of tumors that remained stable. The median TTP was 23 weeks, with 80 per cent of the tumors progressing locally within the distribution of the infused artery. The median survival for the cohort was 63 weeks, with a range of 18–217 weeks. Neurotoxicity consisted of visual loss (5 per cent), hearing loss (9 per cent), seizures (9 per cent), neurological deterioration (4 per cent), and cerebral herniation (4 per cent). Mortimer and associates evaluated the activity of intra-carotid cisplatin used before or in combination with radiotherapy [43]. In a series of 27 evaluable patients with malignant gliomas (23 GBM, 4 AA), half of the cohort was randomized to receive IA cisplatin (150 mg q3 week × 2) before irradiation, while the other half received it concurrently. Three patients had objective responses in each arm of the study (6/27; 22 per cent). The median survival was similar for the two arms of the study, 43.2 weeks for the concurrent cohort and 38.4 weeks for the pre-irradiation cohort. Neurological deterioration was noted in 5 per cent of the patients during IA cisplatin infusions.

Several investigators have administered single-agent 5-fluorouracil *via* the IA route [44,45]. In a combined phase I/II study by Larner and colleagues, superselective supraophthalmic 5-fluorouracil (200–600 mg, q1 week × 4) was administered to 25 patients with malignant gliomas (17 GBM, 8 AA) concomitantly with radiotherapy [44]. The median survival for the GBM cohort was 60 weeks. Neurotoxicity consisted of neurological deterioration in 16 per cent, including three patients with ischemic events and one with a cerebral bleed. Greenberg and co-workers also administered intra-carotid 5-fluorouracil

(5 mg/m^2/day by continuous infusion) over eight weeks during and after irradiation [45]. However, it was given in combination with bromodeoxyuridine (BUdR), which was used as a radiation sensitizer. A total of 39 patients with malignant gliomas were treated (30 GBM, 9 AA). The median survival was 68 weeks for the entire cohort. For the sub-groups of patients with GBM and AA, the median survivals were 54 and 124 weeks, respectively. The most common toxicities were focal dermatitis or iritis, noted in 77 per cent of patients. Neurological toxicity consisted of neurological deterioration (2 per cent) and visual loss (5 per cent).

COMBINATION IA CHEMOTHERAPY OF NEWLY DIAGNOSED GLIOMAS

Combination chemotherapy has also been attempted in patients with newly diagnosed gliomas, in which one or more drugs are administered *via* the IA route (see Table 17.2) [4–7]. The median TTP of combination IA regimens has ranged from 33 weeks to greater than 50 weeks in several studies. Similar to the reports of single-agent IA chemotherapy, the median survival for combination regimens is approximately one year, with a range of 40–228 weeks. The initial experience with combination IA treatment

involved carmustine and was reported in the early 1980s by West and colleagues [46]. They administered intra-carotid carmustine (100 mg/m^2) in combination with PCV (procarbazine, lomustine, vincristine) over four cycles to a cohort of 15 patients, before the onset of irradiation. There was a 75 per cent overall response rate; however, the median survival was only 50 weeks. Seizure activity was documented in 7 per cent of the cohort in association with IA carmustine. Similar results were noted in a study using selective supraophthalmic IA carmustine (300 mg) plus cisplatin (110–200 mg), administered to 19 patients with malignant gliomas [47]. In addition, selected patients also received lomustine (100 mg/m^2) once or twice per month. The overall response rate was 50 per cent, with a median survival of 50 weeks. Neurological deterioration following IA infusion was noted in 14 per cent of patients. Watne and co-workers reviewed a series of 173 patients to evaluate the efficacy of pre-irradiation intra-carotid carmustine (160 mg q4 weeks × 4 cycles) in combination with vincristine and procarbazine, as opposed to irradiation alone [48,49]. A total of 79 patients (60 GBM, 19 AA) received the IA chemotherapy regimen, while 94 patients received only radiotherapy. The response rate for the entire cohort was 54 per cent, with an overall median survival of 120 weeks. Patients in the GBM sub-group had a median survival of only

TABLE 17.2 Combination IA Chemotherapy for Newly Diagnosed Gliomas

Author (ref.)	Drug	No. Pts.	% CR/PR/MR	Median TTP (weeks)	Median Survival (weeks)
West [46]	Carmustine, 100 mg/m^2, plus PCV, × 4 cycles, pre-RT	15	75	NR	50
Bobo [47]	Carmustine, 300 mg plus IA cisplatin, 110–200 mg q8 wks	19	50	NR	50
Watne [48,49]	Carmustine, 160 mg plus VCR & PCB q4 wk × 4	79	54	NR	120 – overall 40 – GBM 228 – AA
Shapiro [50]	Carmustine, 200 mg/m^2, plus 5-FU, q8 wks	279	NR	NR	45
Lehane [51]	Cisplatin, 100 mg/m^2, q4 wks	22	NR	NR	91 projected
Recht [52]	Cisplatin, 90 mg/m^2 × 2, plus IV carmustine	25	16	53	60
Madajewicz [53]	Cisplatin, 40 mg/m^2, & etoposide, 20 mg/m^2 q3 wk × 2	20	60	NR	64 among responders
Madajewicz [54]	Cisplatin, 40 mg/m^2, & etoposide, 20 mg/m^2 q3 wk × 2; pre-RT or with RT	83	48	NR	pre-RT: GBM – 80 AA – 180; with RT: GBM – 28 AA – 48

Derived from references [4–6,46–54]

Abbreviations: ref. — reference; No. — number; pts. — patients; CR — complete response; PR — partial response; MR — minor response, wks — weeks; PCV — procarbazine, lomustine, vincristine; NR — not reported; SD — stable disease; VCR — vincristine; PCB — procarbazine; 5-FU — 5-fluorouracil; RT — radiotherapy.

40 weeks, while those in the AA sub-group were more responsive and had median survival of 228 weeks. No neurological toxicity data were reported for the regimen, except to mention a lack of leukoencephalopathy.

Shapiro and the Brain Tumor Cooperative Group reported the only phase III trial of IA *versus* intravenous chemotherapy for newly diagnosed patients with malignant gliomas [50]. A total of 279 patients were randomized to receive IA (N = 153) or intravenous (N = 126) carmustine 200 mg/m^2 every 8 weeks, with or without intravenous 5-fluorouracil (1 gm/m^2/day × 3 days, two weeks after each dose of carmustine). The median survival for the IA cohort was 11.2 months, with a 24-month survival rate of 13 per cent. In comparison, the median survival of the intravenous cohort was 14.0 months, significantly longer than the IA group (p = 0.008). 5-fluorouracil did not add any additional survival benefit to the IA or intravenous groups. The discrepancy in survival between the IA and intravenous cohorts was accounted for by the non-GBM cases, mostly among AA patients. It remains unclear why the AA cohort should have had a shorter survival. Significant neurological toxicity was noted in the IA cohort, including visual loss (17 per cent), leukoencephalopathy (10 per cent), stroke (4 per cent), and seizures, prompting early closure of the study.

Intra-arterial cisplatin has also been applied to newly diagnosed high-grade gliomas in combination with other drugs. The initial experience was reported by Lehane and associates, who used IA cisplatin (100 mg/m^2) in combination with lomustine and radiotherapy, for a series of 22 patients with malignant gliomas [51]. Survival for the cohort ranged from 2 to 104+ weeks, with a projected median survival of 91 weeks. Recht and colleagues administered intra-carotid cisplatin (90 mg/m^2 × 2) before irradiation, followed by intravenous carmustine, to a series of 25 patients [52]. There was a 16 per cent response rate, while another 52 per cent had stabilization of disease. The median TTP was 53 weeks, with a median survival of 60 weeks. Toxicity included hearing loss (8 per cent) and neurological deterioration (8 per cent). In a study of 20 patients, Madajewicz and co-workers administered intra-carotid cisplatin (40 mg/m^2) plus intra-carotid etoposide (20 mg/m^2) every three weeks × 2 doses before radiotherapy [53]. The objective response rate in patients with measurable tumors was 60 per cent (1 CR, 7 PR), with a median survival of 64 weeks among responders. The most significant toxicity was seizure activity, which was noted in 10 per cent of the patients. A follow-up study by the same group reported

a cohort of 83 patients with high-grade gliomas (GBM 63, AA 20), treated with the same regimen of IA cisplatin and etoposide, administered either before or during irradiation [54]. Among the 71 evaluable patients, the objective response rate was 48 per cent (CR 4, PR 30), with median survivals that were quite variable depending on the timing of chemotherapy. Patients with GBM and AA receiving IA chemotherapy before irradiation had median survival times of 20 and 45 months, respectively. In contrast, GBM and AA patients receiving IA chemotherapy during and after irradiation had median survivals of only 7 and 12 months, respectively. The main neurological toxicity included transient confusion, headache, focal seizures, and visual changes.

SINGLE-AGENT IA CHEMOTHERAPY OF RECURRENT GLIOMAS

There is an extensive literature regarding the use of single agents administered by the IA route for recurrent gliomas (see Table 17.3) [4–7]. The majority of studies have evaluated either a nitrosourea derivative or a platinum analog. Several investigators have evaluated the activity of single-agent IA carmustine [31,32,55,56]. Greenberg and co-workers evaluated the response of 24 patients with recurrent tumors to intra-carotid carmustine (200–300 mg/m^2, q6–8 weeks) [31]. There were objective responses noted in 54 per cent of the cohort (CR 3, PR 10), while another 16 per cent experienced stabilization of disease. The median TTP and survival were reported only for the responding cohort, and were noted to be 20 and 26 weeks, respectively. The neurotoxicity was similar to the experience in newly diagnosed patients and consisted of ipsilateral visual loss and leukoencephalopathy. The report by Hochberg and colleagues used a slightly more aggressive carmustine dosing regimen (240–600 mg/m^2 q6 weeks) for 30 patients with GBM, *via* the intra-carotid or supraophthalmic route [32]. The regimen appeared to be active, since the overall median survival was 50 weeks, which is quite good for patients with recurrent GBM. Toxicity was mainly ocular, with visual loss in 10 per cent of the cohort. Some investigators have used more conservative IA carmustine dosing regimens (120–180 mg/m^2 q6–8 weeks), with median survivals in recurrent patients ranging from 23 to 35 weeks [55,56]. Despite the use of lower doses, toxicity was still significant and included visual loss (10 per cent) and neurological deterioration (24 per cent). Other nitrosourea derivatives that have been administered IA to recurrent glioma

TABLE 17.3   Single-Agent IA Chemotherapy for Recurrent Gliomas

Author (ref.)	Drug	No. Pts.	% CR/PR/MR	Median TTP (weeks)	Median Survival (weeks)
Greenberg[31]	Carmustine, 200–300 mg/m^2, q6–8 wks	24	54	20 in responders	26 in responders
Hochberg [32]	Carmustine, 240–600 mg/m^2, q6 wks	30	NR	NR	50
Johnson [55]	Carmustine, 150 mg/m^2, q6 wks	20	40	NR	35
Bradac [56]	Carmustine, 120–180 mg/m^2, q6–8 wks	17	29	NR	23 in responders
Vega [37]	Nimustine, 150 mg, q6–8 wks	18	44	24	32
Stewart [57]	PCNU, 60–110 mg/m^2, q7 wks	16	44	20 in responders	NR
Poisson [58]	HeCNU, 120 mg/m^2, q6–8 wks	53	49	40 in responders	34
Lehane [51]	Cisplatin, 100 mg/m^2, q4 wks	10	80	NR	16
Feun [59]	Cisplatin, 60–120 mg/m^2, q4 wks	20	30	13 overall 33 in responders	NR
Newton [60]	Cisplatin, 60–100 mg/m^2, q4–6 wks	12	8	14	NR
Mahaley [61]	Cisplatin, 60 mg/m^2, q4 wks	34	34	20 in responders	35 in responders
Green [62]	Cisplatin, 60 mg/m^2, q4 wks	NR	NR	NR	38
Saris [26]	Cisplatin, 70–100 mg/m^2, q6 wks	10	10	NR	NR
Follezou [63]	Carboplatin, 400 mg/m^2, q4 wks	23	22	NR	NR
Stewart [64]	Carboplatin, 200–400 mg/m^2, q4 wks	10	10	NR	8
Qureshi [65]	Carboplatin, Mean dose 286 mg/m^2, q4–6 wks	16	2/10 evaluable patients	NR	Mean of 82.5
Greenberg [66]	Diaziquone, 10–30 mg/m^2, q4 wks	20	10	NR	NR
Feun [67]	Etoposide, 100–650 mg/m^2, q4 wks	15	7	NR	NR

Derived from references [4–6,26,31,32,37,55–67]

Abbreviations: ref. — reference; No. — number; pts. — patients; CR — complete response; PR — partial response; MR — minor response; wks — weeks; NR — not reported.

patients include nimustine, PCNU, and HeCNU [37,57,58]. In these studies, the objective response rates, median TTP, and median survival were all similar to the previously mentioned data for IA carmustine. However, neurological toxicity appeared to be more significant and included visual loss (6–19 per cent), neurological deterioration (7–71 per cent), and leukoencephalopathy (11 per cent).

Single-agent platinum analogs have also been widely applied to recurrent gliomas with IA approaches [4–7]. The most significant experience has been with cisplatin, with IA doses ranging from 60 to 120 mg/m^2, every 4–6 weeks [26,51,59–62]. The initial report was by Lehane and colleagues, who administered IA cisplatin (100 mg/m^2) to ten patients with recurrent malignant gliomas [51]. In this small cohort of patients, the objective response rate was very high (80 per cent). However, the response duration was brief, with an overall median survival of only 16 weeks. Feun and associates used a more aggressive dosing regimen (60–120 mg/m^2) in a cohort of 20 patients with recurrent tumors [59]. They noted a 30 per cent objective response rate, with another 25 per cent having stabilization of disease.

The overall median TTP was 13 weeks, with a median TTP of 33 weeks in the responding cohort. The neurological toxicity may have been worse than with IA carmustine, and included significant visual loss (20 per cent), hearing loss (15 per cent), seizure activity (20 per cent), herniation (5 per cent), and neurological deterioration (40 per cent). Mahaley and co-workers studied a larger cohort of 34 patients and used a more conservative dosing regimen (60 mg/m^2) [61]. They reported a 34 per cent objective response rate, with another 40 per cent of patients stabilizing while on IA treatment. Response criteria were only reported for the responding cohort, with a median TTP of 20 weeks and a median survival of 35 weeks. The regimen had less toxicity than the Feun study, with visual loss and herniation each noted in 3 per cent of patients. Other authors have reported less impressive results with IA cisplatin, noting poor objective response rates (8–10 per cent) and median TTP (14 weeks), along with significant neurotoxicity [26,60,62]. The median survival of 38 weeks reported by Green and colleagues was comparable to other studies of IA cisplatin and the nitrosoureas.

Carboplatin is a more recently developed platinum analog that has also been administered IA to patients with recurrent gliomas [4–7]. It has a lower R_a than cisplatin and clinical studies suggest it may be less neurotoxic. The initial investigation of IA carboplatin (400 mg/m^2) was by Follezou and associates, who treated 23 patients by the intracarotid route [63]. The objective response rate was 22 per cent, with another 22 per cent experiencing stabilization of disease. The PR and MR were maintained for 3–10 months and 2–8 months, respectively. Median TTP and overall survival data were not reported. Neurotoxicity consisted of visual loss (4 per cent) and neurological deterioration (4 per cent). A phase I study by Stewart and co-workers reported on the use of IA carboplatin (200–400 mg/m^2) in a series of 10 patients with recurrent gliomas [64]. The overall median survival was only 8 weeks; one patient with an anaplastic glioma had a brief MR. Toxicity was more pronounced than in the Follezou study, and included visual loss (50 per cent) and neurological deterioration (20 per cent). A more recent study applied superselective supraophthalmic delivery of carboplatin (mean dose 286 mg/m^2) to a cohort of 16 evaluable patients with high-grade gliomas (GBM 10, AA/AO 6) [65]. In a sub-group of ten patients available for evaluation of objective response, there was one CR and one MR. The mean overall survival of the cohort was 82.5 weeks. The procedure was relatively well tolerated, with neurological toxicity consisting of seizures (7 per cent), transient neurological deterioration (3 per cent), and an ischemic stroke (1 per cent).

Miscellaneous drugs that have been used for single-agent IA chemotherapy of recurrent gliomas include diaziquone and etoposide. Greenberg and associates evaluated IA diaziquone (10–30 mg/m^2) in a series of 20 patients with recurrent astrocytomas (GBM 6, AA 11, grade II 4) [66]. There was a 10 per cent objective response rate (PR — 2, 20, and 32+ weeks), with another 20 per cent of patients stabilizing for 12–32 weeks. Toxicity was similar to carmustine and cisplatin. The authors concluded that IA diaziquone was no more effective than using the drug intravenously. Feun and co-workers evaluated the efficacy of intra-carotid etoposide (100–650 mg/m^2) in a cohort of 15 evaluable patients with recurrent high-grade primary brain tumors [67]. The results were similar to diaziquone, with a low objective response rate (7 per cent), while another 33 per cent had stabilization of disease over 8 to 40 weeks. It was concluded that single-agent IA etoposide was no more effective than using the drug by the intravenous route.

COMBINATION IA CHEMOTHERAPY OF RECURRENT GLIOMAS

IA chemotherapy approaches utilizing multiple IA agents or the combination of an IA agent with an oral or intravenous drug have also been extensively reported in the literature (see Table 17.4). The majority of regimens have focused on the use of IA carmustine in combination with other drugs. The initial experience was reported by West and colleagues, when they utilized a regimen of IA carmustine (100 mg/m^2) plus PCV for nine patients with recurrent malignant gliomas [46]. The objective response rate was 22 per cent, with a median survival of 20 weeks. Neurotoxicity was similar to that reported above for IA carmustine. Another report combined IA carmustine (100–125 mg/m^2), IA cisplatin (60 mg/m^2), and IA teniposide (VM-26; 150–175 mg/m^2) in a series of 19 patients [68]. The response rate was 68 per cent, with another 6 per cent stabilizing while on treatment. However, the overall median TTP was only 15 weeks. The regimen was limited by retinal toxicity (19 per cent), hearing loss (5 per cent), and neurological deterioration (19 per cent); toxicity data included patients with brain metastases. A similar study by Kapp and co-workers combined IA carmustine (300 mg) and cisplatin (150–200 mg) with oral lomustine [69]. A series of 13 patients were treated, with an objective response rate of 62 per cent and another 23 per cent with SD. The median TTP was 36 weeks, with a median survival of 44 weeks. Although retinal toxicity was mild, there was a significant proportion of patients with neurological deterioration (38 per cent). In a follow-up study by Stewart and associates, a very aggressive regimen was attempted that included IA carmustin, cisplatin, and teniposide, as well as systemic bleomycin, vincristine, methotrexate, and procarbazine [70]. A cohort of 22 patients received the regimen, with an objective response rate of 52 per cent. There was a median TTP for responders of 19 weeks and an overall median survival of only 21 weeks. Neurological and systemic toxicity were considerable, and included neurological deterioration (31 per cent) and pulmonary fibrosis. The authors concluded that the addition of systemic chemotherapy did not improve efficacy in comparison to IA carmustine, cisplatin, and teniposide alone. In another aggressive approach, IA carmustine (300–400 mg) and IA cisplatin (150–200 mg) were administered in an alternating, sequential pattern (i.e., each drug × 2 doses, every 4–6 weeks) for a series of 43 patients with recurrent malignant gliomas [71]. Due to

TABLE 17.4   Combination IA Chemotherapy for Recurrent Gliomas

Author (ref.)	Drug	No. Pts.	% CR/PR/MR	Median TTP (weeks)	Median Survival (weeks)
West [46]	Carmustine, 100 mg/m^2, plus PCV, q6 wks	9	22	NR	20
Stewart [68]	Carmustine, 100–125 mg/m^2, cisplatin, 60 mg/m^2, teniposide, 150–175 mg/m^2	19	68	15	NR
Kapp [69]	Carmustine, 300 mg cisplatin, 150–200 mg, lomustine q8 wks	13	62	36	44
Stewart [70]	Carmustine, 100 mg/m^2, cisplatin, 60 mg/m^2, teniposide, 150 mg/m^2 IV bleomycin, VCR, MTX, PCB; q6 wks	22	52	19 in responders	21
Rogers [71]	Carmustine, 300–400 mg alternating with cisplatin, 150–200 mg, q4-6 wks	42 carmustine 27 cisplatin	10	NR	36
Watne [72]	Carmustine, 160 mg, vincristine, procarbazine 50 mg, q4 wks	79	60	NR	50 GBM – 26 AA – 80
Stewart [73]	Carmustine, 100 mg/m^2, cisplatin 60 mg/m^2, teniposide 150 mg/m^2 plus IV cisplatin, teniposide, & cytarabine	16	0	NR	14
Nakagawa [74]	Cisplatin, 50–100 mg/m^2, etoposide, 50–100 mg/m^2, q2 wks	13	23	14	24
Ashby and Shapiro [75]	Cisplatin, 60 mg/m^2, etoposide, 50 mg/m^2, q4 wks	20	10	18 in responders	56.5 in responders
Osztie [76]	Carboplatin, 400 mg/m^2, etoposide phos-phate, 400 mg/m^2, cytoxan, 660 mg/m^2	6	67	NR	NR Range 32–152
Newton [77]	Carboplatin, 200 mg/m^2, etoposide, 100 mg/m^2, q4 wks	25	20	24.2	34.2

Derived from rerferences [4–6,46,68–77]

Abbreviations: ref. — reference; No. — number; pts. — patients; CR — complete response; PR — partial response; MR — minor response; wks — weeks; PCV — procarbazine, lomustine, vincristine; PCB — procarbazine; VCR — vincristine; MTX — methotrexate; NR — not reported.

excessive toxicity, only 27 patients actually received the IA cisplatin component of the regimen. The objective response rate was only 10 per cent (PR 4, MR 4; associated with the carmustine component), although another 33–55 per cent had transient stabilization of disease. The overall median survival was 36 weeks for patients that received both carmustine and cisplatin. Neurotoxicity was common and included visual loss (38 per cent), hearing loss (12 per cent), seizure activity (2 per cent), and neurological deterioration (26 per cent). A similar but less aggressive study of combination IA carmustine and cisplatin evaluated 25 patients and noted a 52 per cent objective response rate [47]. No TTP data were reported, but the overall median survival from the

time of diagnosis was 108 weeks. In another attempt to combine IA carmustine (160 mg) and systemic chemotherapy, Watne and co-workers added intravenous vincristine and oral procarbazine (50 mg) to each cycle [72]. A total of 79 patients received the regimen, including 30 patients with GBM and AA. The response rate was 60 per cent, with an overall median survival of 50 weeks. In the sub-group analyses, the median survival for GBM and AA patients were 26 and 80 weeks, respectively. For the responding cohort, the median survival was 80 weeks, with a 30-month survival rate of 45 per cent. A more recent study by Stewart and colleagues combined IA carmustine, cisplatin, and teniposide with intravenous cisplatin, cytarabine, and teniposide in a series of 16 patients and then compared it to a separate cohort that received the same drugs, but only by the intravenous route [73]. The IA regimen demonstrated minimal activity, with no objective responses and only five patients with transient stabilization of disease. The overall median survival of 14 weeks was quite poor and was similar to the intravenous cohort (13 weeks). Neurological toxicity was extensive and included visual loss (12 per cent), hearing loss (8 per cent), seizures (19 per cent), and neurological deterioration (38 per cent). In addition, there was a 25 per cent rate of neutropenic sepsis.

Platinum-based combination IA regimens have also been attempted in patients with recurrent gliomas. Nakagawa and associates treated 13 patients with IA cisplatin (50–100 mg/m^2) and etoposide (50–100 mg/m^2), using a superselective supraophthalmic technique [74]. There were objective responses noted in 23 per cent of the cohort, with disease stabilization in the remaining patients. The median TTP was 14 weeks, with an overall median survival of 24 weeks. Toxicity consisted of seizure activity (23 per cent), neurological deterioration (23 per cent), and several episodes of stroke. In a variation on the Nakagawa approach, other authors have attempted to combine IA cisplatin (60 mg/m^2) and oral etoposide (50 mg/m^2/day, days 1–21) to a series of 20 evaluable patients [75]. The objective response rate was 10 per cent, while another 30 per cent had stabilization of disease. For the responding cohort, the median TTP and survival were 18 and 56.5 weeks, respectively. The most prominent neurological toxicity consisted of encephalopathy (45 per cent), headache, and seizure activity. Osztie and colleagues used IA carboplatin (400 mg/m^2) and etoposide phosphate (400 mg/m^2), in addition to intravenous cyclophosphamide (660 mg/m^2), for a series of six pediatric patients with progressive optic nerve and hypothalamic gliomas [76]. There were four patients with PR and another with stabilized disease, that lasted from 8 to 38 months. The regimen was well tolerated, with minimal neurological toxicity except for mild ototoxicity. In a more recent study, Newton and co-workers evaluated 25 patients with recurrent gliomas who received IA carboplatin (200 mg/m^2) and intravenous etoposide (100 mg/m^2) [77]. The objective response rate was 20 per cent (CR 1, PR 3, MR 1), with another 15 patients having stabilized disease. The overall median TTP was 24.2 weeks, with a TTP of 32 weeks in the responding cohort. Overall, the median survival was 34.2 weeks. The regimen was well tolerated, with minimal neurological toxicity.

IA CHEMOTHERAPY FOR PRIMARY CNS LYMPHOMA

Primary CNS lymphoma (PCNSL) has been shown to be a chemosensitive tumor, and can respond well to multi-agent chemotherapy, including methotrexate-based or carboplatin-based IA regimens [5,6,78,79]. In the majority of cases, IA chemotherapy for PCNSL is used in combination with IA mannitol and blood–brain barrier disruption, to further improve delivery of chemotherapy drugs to tumor cells [5,79,80]. Many patients are also treated "up-front", before any form of irradiation. The median survival with this approach ranges from 42 to 48 months, with an acceptable level of systemic and neurological toxicity. For an in-depth review of this topic, see Chapter 18.

IA CHEMOTHERAPY FOR BRAIN METASTASES

Intra-arterial chemotherapy techniques have been applied to the treatment of brain metastases for over 20 years, using various agents including carmustine, cisplatin, and carboplatin [4,81]. The more recent use of carboplatin-based combination regimens has demonstrated modest efficacy, with acceptable levels of neurological and systemic toxicity [82–84]. For an in-depth review of this topic, see the IA chemotherapy section of Chapter 35.

TOXICITY OF IA CHEMOTHERAPY

The most common toxicity of patients receiving IA chemotherapy is ipsilateral periorbital erythralgia and visual loss, which in many cases can be dose-limiting [4–7,85,86]. Ophthalmological examination of affected patients suggests a toxic retinal vasculitis, with retinal arteriolar narrowing, hemorrhages, nerve fiber layer infarctions, and arterial phase leaks [31]. An anterior ischemic optic neuropathy may also occur in some patients [58]. The incidence of visual loss among the cases cited in this review was approximately 10–12 per cent for patients receiving a nitrosurea-based regimen and 5–8 per cent for those receiving a cisplatin-based regimen. Intra-arterial PCNU and HeCNU appear to have a higher incidence of visual loss than carmustine. Overall, the potential for visual loss appears to be greater for patients receiving IA carmustine and other nitrosoureas than for patients receiving cisplatin, carboplatin, or methotrexate. Visual loss is less likely when patients receive IA carmustine doses less than 150 mg/m^2. Supraophthalmic delivery of IA drugs can reduce visual system injury, but cannot completely eradicate it. Visual loss can still occur in these patients with damage to the optic chiasm. For patients receiving IA cisplatin or carboplatin, no correlation has been noted for risk of visual loss and IA dose, infusion rate, or cumulative dose. In general, the potential for visual loss is less for IA carboplatin than for cisplatin or any of the nitrosourea drugs.

Hearing loss is a common form of toxicity in patients after IA cisplatin chemotherapy [4–7]. The deficit is often bilateral, but can be unilateral in some patients. In most studies, the incidence of symptomatic hearing loss is 5–15 per cent. However, the incidence is much higher (45–62 per cent) when serial audiometric testing is utilized [42]. Audiological testing usually demonstrates a bilateral, irreversible, dose-related loss in the 4000- to 8000-Hz frequency range. The affected frequencies are higher than what is required for conversational speech, so that most patients do not notice the deficit.

Neurological toxicity related to IA chemotherapy administration can be acute or chronic [4–7]. The most common acute neurological effects arise within 24–48 h of drug infusion and include seizure activity, confusional episodes, headaches, and focal deficits. For most studies, the risk of occurrence of an acute neurological event is 8–12 per cent. Supraophthalmic IA delivery of drug appears to be associated with a greater risk for acute neurological toxicity than infraophthalmic delivery [32,56]. In general, acute neurological toxicity is most likely to occur with IA carmustine (and other nitrosoureas) and cisplatin, and occurs less frequently with IA carboplatin and methotrexate.

Chronic neurological toxicity after IA chemotherapy is manifested by ipsilateral hemispheric white matter changes (i.e., leukoencephalopathy), and is noted in 8–10 per cent of patients [4–7,31,32,50,87–91]. This form of toxicity is most prevalent with carmustine, but has also been reported with other nitrosoureas and cisplatin. The leukoencephalopathy can develop rapidly in up to 20 per cent of patients (i.e., 2–4 weeks) or may evolve more indolently as the patient receives further cycles of IA treatment. In the majority of patients, the white matter damage remains subclinical or manifests as mild cognitive deficits. However, the course can be more malignant in some patients, with progressive hemiparesis, memory loss, confusion, and seizure activity. The syndrome does not respond well to corticosteroids and can negatively impact on survival [4,5,50]. Histopathological evaluation of the leukoencephalopathy reveals confluent regions of coagulative necrosis of edematous white matter, foci of petechial hemorrhages and axonal swelling, and fibrinoid necrosis of small vessels [87,89,91]. The probability of leukoencephalopathy appears to be higher in patients receiving IA chemotherapy before or coincident with irradiation. This suggests a possible synergistic toxic interaction between radiotherapy and the high concentrations of regionally infused drug.

CONCLUSIONS

The pharmacological goal of IA chemotherapy is to improve regional delivery of a given drug to the tumor, in an attempt to increase intra-tumoral concentrations of the agent. For several drugs with a favorable R_a (e.g., carmustine, cisplatin, carboplatin, methotrexate), IA chemotherapy has been proven in animal models to be capable of increasing intra-tumoral concentrations of drug in comparison to equivalent intravenous dosing. Intra-arterial chemotherapy approaches have demonstrated a therapeutic benefit for several tumor types, including low-grade and malignant astrocytomas, other gliomas (e.g., oligodendrogliomas), PCNSL, and brain metastases. However, even in the hands of an experienced treatment team, the technique is associated with a significant risk for vascular and neurological toxicity, and may negatively impact survival. This toxicity is most likely with the nitrosourea drugs, especially carmustine, as well as with cisplatin, and is much less prominent with carboplatin and methotrexate. In addition, for tumors with a significant degree of intrinsic chemotherapy resistance, the higher concentrations of intra-cellular drug may not translate to an improvement in tumor response or patient survival. It is also important to mention the paucity of phase III trials of IA chemotherapy approaches. The only phase III trials to date have evaluated nitrosoureas (carmustine and PCNU) and cisplatin, all of which have a prominent toxicity profile in the IA setting. Thus far, no phase III trials have been performed with carboplatin or methotrexate, which are better tolerated and have less significant toxicity. Further study will be required to determine the appropriate role of IA chemotherapy in the therapeutic armamentarium against brain tumors. These studies will need to focus on the current drugs, as well as newly developed agents, that have a more favorable toxicity profile and level of patient tolerance.

ACKNOWLEDGMENTS

The author would like to thank Ryan Smith for research assistance. Dr. Newton was supported in part by National Cancer Institute grant, CA 16058 and the Dardinger Neuro-Oncology Center Endowment Fund.

References

1. French, J. D., West, P. M., Ameraugh, F. K. *et al.* (1952). Effects of intracarotid administration of nitrogen mustard on normal brain and brain tumors. *J Neurosurg* **9**, 378–389.
2. Wilson, C. B. (1964). Chemotherapy of brain tumors by continuous arterial infusion. *Surgery* **55**, 640–653.
3. Owens, G. (1969). Intraarterial chemotherapy of primary brain tumors. Ann NY *Acad Sci* **159**, 603–607.
4. Stewart, D. J. (1991). Intraarterial chemotherapy of primary and metastatic brain tumors. *In Neurological Complications of Cancer Treatment* (D. A. Rottenberg, Ed.), 143–170. Butterworth-Heinimann, Boston.
5. Dropcho, E. J. (1999). Intra-arterial chemotherapy for malignant gliomas. *In The Gliomas* (M. S. Berger, and C. B. Wilson Eds.), 537–547. W. B. Saunders Co., Philadelphia.
6. Buckner, J. C. (2000). Intra-arterial chemotherapy. *In Neuroncology. The Essentials* (M. Bernstein and M. S. Berger Eds.), 234–239. Thieme Medical Publishers, Inc., New York.
7. Basso, U., Londardi, S., and Brandes, A. A. (2002). Is intra-arterial chemotherapy useful in high-grade gliomas? *Expert Rev Anticancer Ther* **2**, 507–519.
8. Eckman, W. W., Patlak, C. S., and Fenstermacher, J. D. (1974). A critical evaluation of the principles governing the advantages of intra-arterial infusion. *J Pharmacokinet Biopharmacokinet* **2**, 257–285.
9. Fenstermacher, J. D., and Cowles, A. L. (1977). Theoretic limitations of intracarotid infusions in brain tumor chemotherapy. *Cancer Treat Rep* **61**, 519–526.
10. Collins, J. M. (1984). Pharmacologic rationale for regional drug delivery. *J Clin Oncol* **2**, 498–504.
11. Stewart, D. J. (1989). Pros and cons of intra-arterial chemotherapy. *Oncol* **3**, 20–26.
12. Levin, V. A., Kabra, P. M., and Freeman-Dove, M. A. (1978). Pharmacokinetics of intracarotid artery ^{14}C-BCNU in the squirrel monkey. *J Neurosurg* **48**, 587–593.
13. Tyler, J. L., Yamamoto, L., Diksic, M. *et al.* (1986). Pharmacokinetics of superselective intra-arterial and intravenous ^{11}C-BCNU evaluated by PET *J Nucl Med* **27**, 775–780.
14. Hassenbusch, S. J., Anderson, J. H., and Colvin, O. M. (1996). Predicted and actual, BCNU concentrations in normal rabbit brain during intraarterial and intravenous infusions. *J Neurooncol* **30**, 7–18.
15. Stewart, D. J., Benjamin, R. S., Zimmerman, S. *et al.* (1983). Clinical pharmacology of intraarterial cisdiamminedichloroplatinum (II). *Cancer Res* **43**, 917–920.
16. Rottenberg, D. A., Dhawan, V., Cooper, A. J. *et al.* (1987). Assessment of the pharmacologic advantage of intra-arterial *versus* intravenous chemotherapy using ^{13}N-cisplatin and positron emission tomography (PET). *Neurol* **37(suppl 1)**, 335.
17. Ichimura, K., Ohno, K., Aoyagi, M., Tamaki, M., Suzuki, R., and Hirakawa, K. (1993). Capillary permeability in experimental rat glioma and effects of intracarotid CDDP administration on tumor drug delivery. *J Neurooncol* **16**, 211–215.

18. Stewart, D. J., Mikhael, N. Z., Nair, R. C. *et al.* (1988). Platinum concentrations in human autopsy tumor samples. *Am J Clin Oncol* **11**, 152–158.
19. Barth, R. F., Yang, W., Rotaru, J. H. *et al.* (1997). Boron neutron capture therapy of brain tumors: Enhanced survival following intracarotid injection of either sodium borocaptate or boronophenylalanine with or without blood-brain barrier disruption. *Cancer Res* **57**, 1129–1136.
20. Kroll, R. A., and Neuwelt, E. A. (1998). Outwitting the blood-brain barrier for therapeutic purposes: Osmotic opening and other means. *Neurosurg* **42**, 1083–1100.
21. Blacklock, J. B., Wright, D. C., Dedrick, R. L., *et al.* (1986). Drug streaming during intra-arterial chemotherapy. *J Neuorsurg* **64**, 284–291.
22. Saris, S. C., Wright, D. C., Oldfield, E. H. *et al.* (1988). Intravascular streaming and variable delivery to brain following intra-carotid artery infusions in the Sprague-Dawley rat. *J Cereb Blood Flow Metab* **8**, 116–120.
23. Lutz, R. J., Dedrick, R. L., Boretos, J. W., Oldfield, E. H., Blacklock, J. B., and Doppman, J. L. (1986). Mixing studies during intra-carotid artery infusions in an in vitro model. *J Neurosurg* **64**, 277–283.
24. Lutz, R. J., Warren, K., Balis, F., Patronas, N., and Dedrick, R. L. (2002). Mixing during intravertebral arterial infusions in an in vitro model. *J Neurooncol* **58**, 95–106.
25. Junck, L., Koeppe, R. A., and Greenberg, H. S. (1989). Mixing in the human carotid artery during carotid drug infusion studied with, PET. *J Cereb Blood Flow Metab* **9**, 681–689.
26. Saris, S. C., Blasberg, R. G., Carson, R. E. *et al.* (1991). Intravascular streaming during carotid artery infusions. Demonstration in humans and reduction using diastole-phased pulsatile administration. *J Neurosurg* **74**, 763–772.
27. Aoki, S., Terada, H., Kosuda, S. *et al.* (1993). Supraophthalmic chemotherapy with long tapered catheter: Distribution evaluated with intraarterial and intravenous Tc-99m, HMPAO. *Radiol* **188**, 347–350.
28. Agid, R., Rubinstein, R., Siegal, T. *et al.* (2002). Does streaming affect the cerebral distribution of infraophthalmic intra-carotid chemotherapy? *AJNR Am J Neuroradiol* **23**, 1732–1735.
29. Schook, D. R., Beaudet, L. M., and Doppman, J. L. (1987). Uniformity of intracarotid drug distribution with diastole-phased pulsed infusion. *J Neurosurg* **67**, 726–731.
30. Saris, S. C., Shook, D. R., Blasberg, R. G. *et al.* (1987). Carotid mixing with diastole-phased pulsed drug infusion. *J Neurosurg* **67**, 721–725.
31. Greenberg, H. S., Ensminger, W. D., Chandler, W. F. *et al.* (1984). Intra-arterial BCNU chemotherapy for treatment of malignant gliomas of the central nervous system. *J Neurosurg* **61**, 423–429.
32. Hochberg, F. H., Pruitt, A. A., Beck, D. O. *et al* (1985). The rationale and methodology for intra-arterial chemotherarpy with BCNU as treatment for glioblastoma. *J Neurosurg* **63**, 876–880.
33. Bashir, R., Hochberg, F. H., Linggood, R. M. *et al.* (1988). Pre-irradiation internal carotid artery BCNU in treatment of glioblastoma multiforme. *J Neurosurg* **68**, 917–919.
34. Clayman, D. A., Wolpert, S. M., and Heros, D. O. (1989). Superselective arterial BCNU infusion in the treatment of patients with malignant gliomas. *AJNR Am J Neuroradiol* **10**, 767–771.
35. Yamashita, J., Handa, H., Tokuriki, Y. *et al.* (1983). Intra-arterial ACNU therapy for malignant brain tumors. Experimental studies and preliminary clinical results. *J Neurosurg* **59**, 424–430.
36. Roosen, N., Kiwit, J. C. W., Lins, E., Schirmer, M., and Bock, W. J. (1989). Adjuvant intraarterial chemotherapy with nimustine

in the management of world health organization grade IV gliomas of the brain. Experience at the department of neurosurgery of Düsseldorf University. *Cancer* **64**, 1984–1994.

37. Vega, F., Davila, L., Chatellier, G. *et al.* (1992). Treatment of malignant gliomas with surgery, intraarterial chemotherapy with ACNU and radiation therapy. *J Neurooncol* **13**, 131–135.

38. Chauveinc, L., Sola-Martinez, M. T., Martin-Duverneuil, M. *et al.* (1996). Intra-arterial chemotherapy with ACNU and radiotherapy in inoperable malignant gliomas. *J Neurooncol* **27**, 141–147.

39. Kochi, M., Kitamura, I., Goto, T. *et al.* (2000). Randomized comparison of intra-arterial *versus* intravenous infusion of, ACNU for newly diagnosed patients with glioblastoma. *J Neurooncol* **49**, 63–70.

40. Fauchon, F., Davila, L., Chatellier, G. *et al.* (1990). Treatment of malignant gliomas with surgery, intraarterial infusions of 1-(2-hydroxyethyl) chloroethylnitrosourea and radiation therapy. A phase II study. *Neurosurg* **27**, 231–234.

41. Calvo, F. A., Dy, C., Henriquez, I., Hidalgo, V., Bilbao, I., and Santos, M. (1989). Postoperative radical radiotherapy with concurrent weekly intra-arterial cis-platinum for treatment of malignant glioma: A pilot study. *Radiother Oncol* **14**, 83–88.

42. Dropcho, E. J., Rosenfeld, S. S., Morawetz, R. B. *et al.* (1992). Pre-radiation intracarotid cisplatin treatment of newly diagnosed anaplastic gliomas. *J Clin Oncol* **10**, 452–458.

43. Mortimer, J. E., Crowley, J., Eyre, H., Weiden, P., Eltringham, J., and Stuckey, W. J. (1992). A phase II randomized study comparing sequential and combined intra-arterial cisplatin and radiation therapy in primary brain tumors. *Cancer* **69**, 1220–1223.

44. Larner, J. M., Kersh, C. R., Constable, W. C. *et al.* (1991). Phase I/II trial of superselective arterial 5-FU infusion with concomitant external beam radiation for patients with either anaplastic astrocytoma or glioblastoma multiforme. *Am J Clin Oncol* **14**, 514–518.

45. Greenberg, H. S., Chandler, W. F., Ensminger, W. D. *et al.* (1994). Radiosensitization with carotid intra-arterial bromodeoxyuridine ± 5-fluorouracil biomodulation for malignant gliomas. *Neurol* **44**, 1715–1720.

46. West, C. R., Avellanosa, A. M., Barua, N. R. *et al.* (1983). Intraarterial BCNU and systemic chemotherapy for malignant gliomas: a follow-up study. *Neurosurg* **13**, 420–426.

47. Bobo, H., Kapp, J. P., and Vance, R. (1992). Effect of intra-arterial cisplatin and 1,3-bis(2chloroethyl)-1-nitrosourea (BCNU) dosage on radiographic response and regional toxicity in malignant glioma patients: Proposal of a new method of intra-arterial dosage calculation. *J Neurooncol* **13**, 291–299.

48. Watne, K., Nome, O., Hager, B., and Hirschberg, H. (1991). Combined intra-arterial chemotherapy and irradiation of malignant gliomas. *Acta Oncol* **30**, 835–841.

49. Watne, K., Hannisdal, E., Nome, O. *et al.* (1993). Prognostic factors in malignant gliomas with special reference to intra-arterial chemotherapy. *Acta Oncol* **32**, 307–310.

50. Shapiro, W. R., Green, S. B., Burger, P. C. *et al.* (1992). A randomized comparison of intra-arterial *versus* intravenous, BCNU with or without intravenous 5-fluorouracil, for newly diagnosed patients with malignant glioma. *J Neurosurg* **76**, 772–781.

51. Lehane, D. E., Bryan, R. N., Horowitz, B. *et al.* (1983). Intraarterial cis-platinum chemotherapy for patients with primary and metastatic brain tumors. *Cancer Drug Del* **1**, 69–77.

52. Recht, L., Fram, R. J., Strauss, G. *et al.* (1990). Preirradiation chemotherapy of supratentorial malignant primary brain

tumors with intracarotid cisplatinum and IV BCNU. *Am J Clin Oncol* **13**, 125–131.

53. Madajewicz, S., Chowhan, N., Iliya, A. *et al.* (1991). Intracarotid chemotherapy with etoposide and cisplatin for malignant brain tumors. *Cancer* **67**, 2844–2849.

54. Madajewicz, S., Chowhan, N., Tfayli, A. *et al.* (2000). Therapy for patients with high grade astrocytoma using intraarterial chemotherapy and radiation therapy. *Cancer* **88**, 2350–2356.

55. Johnson, D. W., Parkinson, D., Wolpert, S. M. *et al.* (1987). Intracarotid chemotherapy with 1,3-bis(2-chloroethyl)-1-nitrosourea (BCNU) in 5 per cent dextrose in water in the treatment of malignant glioma. *Neurosurg* **20**, 577–583.

56. Bradac, G. B., Soffietti, R., Riva, A., Stura, G., Sales, S., and Schiffer, D. (1992). Selective intra-arterial chemotherapy with, BCNU in recurrent malignant gliomas. *Neuroradiol* **34**, 73–76.

57. Stewart, D. J., Grahovac, Z., Russel, N. A. *et al.* (1987). Phase I study of intracarotid, PCNU. *J Neurooncol* **5**, 245–250.

58. Poisson, M., Chiras, J., Fauchon, F., Debussche, C., and Delattre, J. Y. (1990). Treatment of malignant recurrent glioma by intra-arterial, infra-ophthalmic infusion of HECNU 1-(2-chloroethyl)-1-nitroso-3-(2-hydroxyethyl) urea. A phase II study. *J Neurooncol* **8**, 255–262.

59. Feun, L. G., Wallace, S., Stewart, D. J. *et al.* (1984). Intracarotid infusion of cis-diammine-dichloroplatinum in the treatment of recurrent malignant brain tumors. *Cancer* **54**, 794–799.

60. Newton, H. B., Page, M. A., Junck, L., and Greenberg, H. S. (1989). Intra-arterial cisplatin for the treatment of malignant gliomas. *J Neurooncol* **7**, 39–45.

61. Mahaley, M. S., Hipp, S. W., Dropcho, E. J. *et al.* (1989). Intracarotid cisplatin chemotherapy for recurrent gliomas. *J Neurosurg* **70**, 371–378.

62. Green, S. B., Shapiro, W. R., Burger, P. C. *et al.* (1989). Randomized comparison of intra-arterial cisplatin and intravenous, PCNU for the treatment of primary brain tumors (BTCG Study 8420A). *PROC Am Soc Clin Oncol* **8**, 86.

63. Follezou, J. Y., Fauchon, F., Chiras, J. (1989). Intraarterial infusion of carboplatin in the treatment of malignant gliomas: A phase II study. *Neoplasma* **36**, 349–352.

64. Stewart, D. J., Belanger, J. M., Grahovac, Z. *et al.* (1992). Phase I study of intracarotid administration of carboplatin. *Neurosurg* **30**, 512–516.

65. Qureshi, A. I., Suri, M. F. K., Khan, J. *et al.* (2001). Superselective intra-arterial carboplatin for treatment of intracranial neoplasms: experience in 100 procedures. *J Neurooncol* **51**, 151–158.

66. Greenberg, H. S., Ensminger, W. D., Layton, P. B. *et al.* (1986). Phase I-II evaluation of intra-arterial diaziquone for recurrent malignant astrocytomas. *Cancer Treat Rep* **70**, 353–357.

67. Feun, L. G., Lee, Y. Y., Yung, W. K. A., Savaraj, N., and Wallace, S. (1987). Intracarotid, VP-16 in malignant brain tumors. *J Neurooncol* **4**, 397–401.

68. Stewart, D. J., Grahovac, Z., Benoit, B. *et al.* (1984). Intracarotid chemotherapy with a combination of 1,3-bis(2-choroethyl)-1-nitrosourea (BCNU), cis-diaminedicholoroplatinum (cisplatin), and 4'-O-demethyl-1-O-(4,6-O-2-thenylidene-beta-D-glucopyranosyl) epipodophyllotoxin (VM-26) in the treatment of primary and metastatic brain tumors. *Neurosurg* **15**, 828–833.

69. Kapp, J. P., and Vance, R. B. (1985). Supraophthalmic carotid infusion for recurrent glioma: rationale, technique, and preliminary results for cisplatin and BCNU. *J Neurooncol* **3**, 5-11.

70. Stewart, D. J., Grahovac, Z., Hugenholtz, H., Russell, N., Richard, M., and Benoit, B. (1987). Combined intraarterial and

systemic chemotherapy for intracerebral tumors. *Neurosurg* **21**, 207–214.

71. Rogers, L. R., Purvis, J. B., Lederman, R. J. *et al.* (1991). Alternating sequential intracarotid BCNU and cisplatin in recurrent malignant glioma. *Cancer* **68**, 15–21.

72. Watne, K., Hannisdal, E., Nome, O., Hager, B., and Hirschberg, H. (1992). Combined intra-arterial and systemic chemotherapy for recurrent malignant brain tumors. *Neurosurg* **30**, 223–227.

73. Stewart, D. J., Grahovac, Z., Hugenholtz, H. *et al.* (1993). Feasibility study of intraarterial vs. intravenous cisplatin, BCNU, and teniposide combined with systemic cisplatin, teniposide, cytosine arabinoside, glycerol and mannitol in the treatment of primary and metastatic brain tumors. *J Neurooncol* **17**, 71–79.

74. Nakagawa, H., Fujita, T., Kubo, S. *et al.* (1994). Selective intra-arterial chemotherapy with a combination of etoposide and cisplatin for malignant gliomas: preliminary report. *Surg Neurol* **41**, 19–27.

75. Ashby, L. S., and Shapiro, W. R. (2001). Intra-arterial cisplatin plus oral etoposide for the treatment of recurrent malignant glioma: a phase II study. *J Neurooncol* **51**, 67–86.

76. Osztie, E., Várallyay, P., Doolittle, N. D. *et al.* (2001). Combined intraarterial carboplatin, intraarterial etoposide phosphate, and IV cytoxan chemotherapy for progressive optic-hypothalamic gliomas in young children. *AJNR Am J Neuroradiol* **22**, 818–823.

77. Newton, H. B., Slivka, M. A., Stevens, C. L. *et al.* (2002). Intra-arterial carboplatin and intravenous etoposide for the treatment of recurrent and progressive non-GBM gliomas. *J Neurooncol* **56**, 79–86.

78. El Kamar, F. G., and Abrey, L. E. (2004). Management of Primary Central Nervous System Lymphoma. *J Natl Cancer Comp Network* **2**, 341–349.

79. Doolittle, N. D., Miner, M. E., Hall, W. A. *et al.* (2000). Safety and efficacy of a multicenter study using intraarterial chemotherapy in conjunction with osmotic opening of the blood–brain barrier for the treatment of patients with malignant brain tumors. *Cancer* **88**, 637–647.

80. Kroll, R. A. and Neuwelt, E. A. (1998). Outwitting the blood–brain barrier for therapeutic purposes: osmotic opening and other means. *Neurosurgery* **42**, 1083–1100

81. Newton, H. B. (2002). Chemotherapy for the treatment of metastatic brain tumors. *Expert Rev Anticancer Ther* **2**, 495–506.

82. Newton, H. B., Stevens, C., and Santi, M. (2001). Brain metastases from fallopian tube carcinoma responsive to intra-arterial carboplatin and intravenous etoposide: A case report. *J Neurooncol* **55**, 179–184.

83. Newton, H. B., Snyder, M. A., Stevens, C. *et al.* (2003). Intra-arterial carboplatin and intravenous etoposide for the treatment of brain metastases. *J Neurooncol* **61**, 35–44.

84. Gelman, M., Chakares, D., and Newton, H. B. (1999). Brain tumors: Complications of cerebral angiography accompanied by intra-arterial chemotherapy. *Radiol* **213**, 135–140.

85. Gebarski, S. S., Greenberg, H. S., Gabrielson, T. O. *et al.* (1984). Orbital angiographic changes after intra-carotid, BCNU chemotherapy. *AJNR Am J Neuroradiol* **5**, 55–58.

86. Miller, D. F., Bay, J. W., Lederman, R. J. *et al.* (1985). Ocular and orbital toxicity following intracarotid injection of BCNU (carmustine) and cisplatinum for malignant gliomas. *Ophthalmol* **92**, 402–406.

87. Kleinschmidt-DeMasters, B. K. (1986). Intracarotid BCNU leukoencephalopathy. *Cancer* **57**, 1276–1280.

88. Lee, Y., Nauert, C., and Glass, P. (1986). Treatment-related white matter changes in cancer patients. *Cancer* **57**, 1473–1482.

89. Mahaley, M. S., Whaley, R. A., Blue, M. *et al.* (1986). Central neurotoxicity following intracarotid, BCNU chemotherapy for malignant gliomas. *J Neurooncol* **3**, 297–314.

90. Di Chiro, G., Oldfield, E., Wright, D. C. *et al.* (1988). Cerebral necrosis after radiotherapy and/or intra-arterial chemotherapy for brain tumors: PET and neuropathologic studies. *Am J Radiol* **150**, 189–197.

91. Rosenblum, M. K., Delattre, J. Y., Walker, R. W. *et al.* (1989). Fatal necrotizing encephalopathy complicating treatment of malignant gliomas with intra-arterial BCNU and irradiation: A pathological study. *J Neurooncol* **7**, 269–274.

18

Blood–Brain Barrier Disruption Chemotherapy

Nancy D. Doolittle, Tulio P. Murillo, and Edward A. Neuwelt

ABSTRACT: The goal of chemotherapy administered in conjunction with blood–brain barrier disruption (BBBD) is maximizing drug delivery to the brain, while preserving neurocognitive function and minimizing systemic toxicity. Translational blood–brain barrier (BBB) pre-clinical and clinical studies at Oregon Health & Science University (OHSU) evaluate the toxicity and antitumor efficacy of chemotherapeutics, chemoprotectants, and monoclonal antibodies (mAbs) as well as novel therapeutics and imaging agents. A current program focus is thiol chemoprotection against the hearing and bone marrow toxicity that is caused by platinum chemotherapy. Thiols may permit increased dose intensity of agents administered in conjunction with BBBD. In the clinic, BBBD has shown promising results in chemosensitive brain tumors such as primary central nervous system lymphoma (PCNSL) and offers a new strategy for global delivery of chemotherapy to tumors, such as anaplastic oligodendroglioma and central nervous system metastases. Multi-center clinical trials using BBBD are in progress at centers participating in the BBB Consortium. Current and future clinical studies include delivery of mAbs across the BBB.

INTRODUCTION

The efficacy of chemotherapy for malignant brain tumors has in general been disappointing, due in part to the limited passage of many systemically administered agents from the blood to the brain. The blood–brain barrier (BBB) excludes molecules from the brain based on electric charge, lipid solubility, and molecular weight. Molecules greater than $M_r 180$ are typically excluded. Investigators have previously reported the variability in brain tumor permeability within malignant brain tumors, and that the well-vascularized, actively proliferating edge of the tumor is particularly variable in terms of barrier integrity [1–3]. The current BBB transport paradigm is that movement across the BBB relates to the neurovascular unit which consists of endothelial cells, pericytes, glia, and neuronal elements [4].

The goal of blood–brain barrier disruption (BBBD) is maximizing delivery of agents to the brain while preserving neurocognitive function and quality of life, and minimizing systemic toxicity. BBBD is especially important in increasing the delivery of high molecular weight agents, such as proteins, antibodies, immunoconjugates, and viral vectors [4,5]. Pre-clinical and clinical BBB studies at Oregon Health & Science University (OHSU) evaluate

1. The toxicity of chemotherapeutics and chemoprotectants,
2. The potential for dose intensification of chemotherapy in combination with chemoprotectants, and
3. The antitumor efficacy of chemotherapeutics in combination with chemoprotectants and/or monoclonal antibodies (mAbs).

The technique of clinical BBBD used by Neuwelt *et al.* at OHSU is based on extensive pre-clinical toxicity and efficacy studies. BBBD involves administering the hyperosmolar solution mannitol intra-arterially in the carotid or the vertebral arteries. The infusion of mannitol is theorized to cause osmotic shrinkage of endothelial cells which line central nervous system (CNS) capillaries, with resultant separation of the tight junctions between the cells. Cellular messenger systems, such as calcium influx and nitrous oxide, as well as cytoskeletal changes, likely contribute to the transient opening of the barrier. To date, BBBD has shown promising clinical results in chemosensitive brain tumors such as primary central nervous system lymphoma (PCNSL).

The pre-clinical and clinical BBB program is based on a translational research model. Knowledge gained in the laboratory informs the clinical team, while clinical outcomes reciprocally inform the basic scientists. Translational studies conducted in the laboratory and in the clinic have led to innovative approaches such as a *two-compartment BBB model*, which is an exciting paradigm involving temporary and reversible osmotic opening of the BBB (see Figs. 18.1A,B). The two-compartment model is based on separating platinum chemotherapy and thiol chemoprotectants, by route and by timing of administration, and is further discussed at a later point in this chapter.

PRE-CLINICAL BBB DELIVERY STUDIES

Many important BBBD observations have been made in animal studies. For example,

1. A marked increase in brain and cerebrospinal (CSF) concentrations of methotrexate were documented after BBBD with intra-arterial chemotherapy administration [6–8],
2. Disruption of the BBB provides global delivery throughout the disrupted hemisphere, but is variable depending on the brain region and the type and size of the tumor,
3. Vascular permeability to small molecules such as methotrexate, as well as large molecules such as mAbs, is increased maximally by 15 min after mannitol, and
4. BBB permeability rapidly decreases, returning to pre-infusion levels within 2 h after BBBD.

Animal studies have also shown that antecedent cranial irradiation decreases agent delivery to the brain [9,10]. The studies evaluated long-term effects of

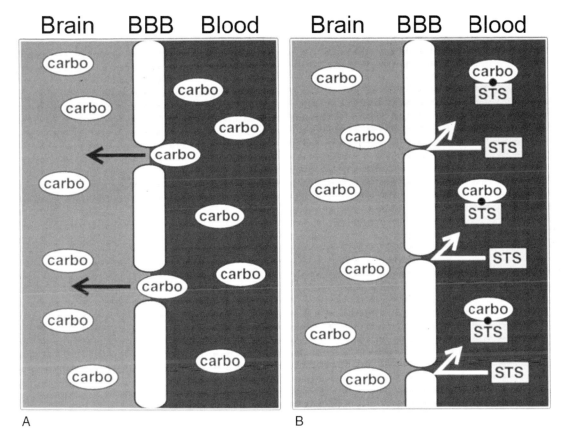

FIGURE 18.1 The two-compartment model. Carboplatin (carbo) is administered intra-arterially immediately after osmotic disruption of the BBB with hypertonic mannitol. **(A)** shows delivery of carboplatin across the BBB. Sodium thiosulfate (STS) is administered intravenously 4 (or 4 and 8) h after osmotic disruption, after BBB permeability has returned to baseline levels. **(B)** shows exclusion of STS from the brain.

various sequences of radiation therapy and BBBD chemotherapy in rodents. Drug delivery, acute toxicity and long-term (one year) neuropathological effects of methotrexate, or carboplatin plus etoposide, were evaluated. External beam radiation of 2000 cGy as a single fraction using parallel opposed portals, either 30 days before or concurrent with BBBD, resulted in a statistically significant decrease in drug delivery compared to animals not receiving cranial irradiation. Seizures were observed in 26 per cent of the animals that received irradiation before or concurrent with BBBD and methotrexate, but not carboplatin. The mortality rate for animals receiving radiotherapy 30 days prior to chemotherapy was significantly higher than the mortality rate for animals receiving only BBBD chemotherapy without irradiation [9,11].

Additional rodent studies evaluated whether prior irradiation influenced the efficacy of antibody targeted chemotherapy given with BBBD. Results showed that BR96-DOX, an antitumor mAb-doxorubicin immunoconjugate, administered prior to irradiation significantly increased survival compared to rodents receiving irradiation prior to chemotherapy, or compared to those receiving chemotherapy concurrently [10]. These findings were later supported in the clinic when subjects with PCNSL who received cranial irradiation before beginning BBBD chemotherapy, had significantly decreased median survival time compared to subjects who received initial BBBD chemotherapy [12].

The BBB pre-clinical and clinical teams carefully conduct toxicity studies, in animals and humans respectively, to determine which chemotherapy agents can be administered with BBBD with an acceptable safety and toxicity profile. However, extensive pre-clinical toxicity studies are always conducted prior to phase 1 clinical studies. For example, important knowledge was gained when laboratory studies showed severe neurotoxicity when adriamycin (intra-arterial) [13], cisplatinum (intra-arterial), or 5-FU (intra-arterial) [14] were administered as single agents after BBBD. Fortin [15] reported unexpected neurotoxicity when etoposide phosphate (intra-arterial) was administered in combination with melphalan (intra-arterial), methotrexate (intra-arterial), or carboplatin (intra-arterial) after BBBD, when propofol anesthesia was used.

Chemotherapy agents used most frequently in conjunction with BBBD in the clinical setting are methotrexate (intra-arterial, 2500 mg/day × two consecutive days), carboplatin (intra-arterial, 200 mg/m² per day × two consecutive days), melphalan (intra-arterial, a dose of 8 mg/m² per day × two consecutive days is currently under study), cyclophosphamide (intravenous, 500 mg/m² per day × two consecutive days when given with methotrexate; 330 mg/m² per day × two consecutive days when given with carboplatin), etoposide and etoposide phosphate (intravenous, 150 mg/m² per day × two consecutive days when given with methotrexate; 200 mg/m² per day × two consecutive days when given with carboplatin). Depending on brain tumor histology and according to the specific IRB-approved protocol, a combination of the above drugs are given with BBBD. These agents infused by the respective routes and doses have been routinely used in the clinical setting and have shown acceptable toxicity [12,16–19].

CLINICAL BBBD TECHNIQUE

The care of patients treated with BBBD requires a multi-disciplinary team approach. The team includes a neuro-oncologist, neurosurgeon, neuroradiologist, anesthesiologist, pharmacist, nurse coordinator, neuropsychologist, ophthalmologist, audiologist, physical therapist, and social worker. BBBD treatment is done on two consecutive days every four weeks for up to one year. Patients undergo baseline neuropsychologic evaluation, electrocardiogram, and port-a-cath placement. At baseline and prior to each monthly BBBD treatment patients undergo neurologic and KPS evaluation, brain MRI, chest X-ray, complete blood count, chemistry panel, and urinalysis. Patients are required to have adequate hematologic, renal, and hepatic function, and must have adequate pulmonary and cardiac function to tolerate general anesthesia. Ophthalmologic assessment is done if clinically indicated. Patients treated with carboplatin in conjunction with BBBD undergo monthly audiologic assessment.

In addition, all patients must meet neuroradiographic criteria prior to undergoing BBBD treatment. The radiographic criteria are:

1. An open quadrigeminal plate cistern,
2. Absence of dilatation of the contralateral frontal horn, and
3. Absence of uncal herniation.

In the setting of rapidly progressing brain disease with associated rapid neurologic deterioration, there is a risk of increasing mass effect following BBBD, thus BBBD is the safest before tumor burden becomes excessive [18].

BBBD is performed under general anesthesia to ensure patient comfort and safety during the rapid intra-arterial infusion of a large volume of hypertonic mannitol (25 per cent, warmed). Focal motor seizures occur during approximately 7 per cent of BBBD procedures (most often in conjunction with methotrexate), thus general anesthesia provides the capability of rapid control of seizures if necessary. A femoral artery is catheterized and a selected intracranial artery (either an internal carotid or a vertebral artery) is accessed. The transfemoral catheter is placed at cervical vertebrae 1–2 for a carotid artery infusion of chemotherapy or at cervical vertebrae 6–7 for a vertebral artery infusion. The transfemoral catheter placement level is confirmed by fluoroscopy prior to mannitol, and also prior to chemotherapy infusion. Mannitol is delivered *via* infusion device at a predetermined flow rate of 3–12 cc/s into the cannulated artery for 30 s. The precise flow rate is determined by fluoroscopy, to just exceed cerebral blood flow.

Following administration of mannitol, the intra-arterial chemotherapy agent(s) are infused, each over 10 min. If chemotherapy is administered *via* the intravenous route, it is begun as soon as the patient is under general anesthesia, to allow time for the chemotherapy to be delivered to tumor while the barrier is open. If intra-arterial chemotherapy without BBBD is used, patients undergo monitored anesthesia care instead of general anesthesia, and treatment is on one day instead of two consecutive days every four weeks.

Immediately following the mannitol, nonionic contrast dye is administered intravenously. Following completion of chemotherapy, the patient undergoes a computed tomography (CT) brain scan (see Figs. 18.2A,B) [20]. Contrast enhancement in the disrupted territory of the brain is compared to the nondisrupted territory. The degree of disruption is graded using the results reported by Roman-Goldstein *et al.* [21].

During each monthly treatment, one of the intra-cranial arteries (right or left internal carotid or a

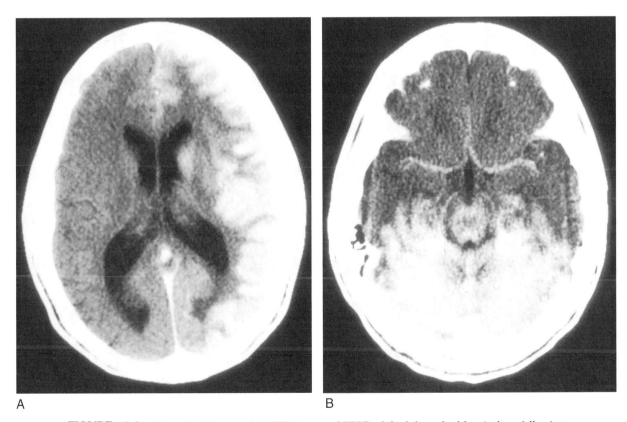

A   B

FIGURE 18.2 Computed tomographic (CT) images of BBBD of the left cerebral hemisphere following mannitol infusion in the left internal carotid artery **(A)**, and BBBD of the posterior fossa region following mannitol infusion in the vertebral artery **(B)** [20]. (Reprinted with permission from (1998): Doolittle, N. D., Petrillo, A., Bell, S., Cummings, P., and Eriksen, S. Blood–brain barrier disruption for the treatment of malignant brain tumors: The National Program. Journal of Neurosurgical Nursing **30(2)**, 80–91. Copyright 1998 by the American Association of Neuroscience Nurses.)

vertebral artery) is infused the first day of BBBD treatment, and a different artery is infused on the second day of BBBD, depending on the tumor type, extent, and location. In tumors that are not localized to one brain hemisphere or arterial territory, and/or tumors that have widespread microscopic infiltration of the brain such as PCNSL, infusion of the arteries is rotated so that during a year of BBBD treatment, each of the three intracranial arteries is infused eight times, thus providing global delivery to all cerebral circulations.

Following BBBD and during their hospital stay, patients undergo close observation including frequent monitoring of vital signs, neurologic status, and fluid balance. Fluid balance is meticulously maintained with diuretics or fluid boluses. In patients treated with methotrexate, sodium bicarbonate is added to intravenous fluids and titrated to achieve a urine pH greater than 6.5. Leucovorin rescue is used in methotrexate-based protocols, beginning 36 h after the first dose of methotrexate. Patients treated with methotrexate receive 80 mg of Leucovorin (intravenous) followed by 50 mg (intravenous or orally) every 6 h, for a total of 20 doses. Ganulocyte-colony stimulating factor (G-CSF) is given subcutaneously 48 h after the second day of BBBD chemotherapy. If filgrastim (Neupogen) is used, 5 μg/kg is given daily until the WBC is ≥ 5000/μl. If pegfilgrastim (Neulasta) is used, one dose of 6 mg is given. Following hospital discharge and between monthly BBBD treatments, complete blood counts are done twice a week for two weeks while the patient is receiving G-CSF. For the remaining two weeks, a complete blood count is done weekly.

Patients treated with BBBD may experience transient neurologic deficits after the BBBD procedure. In the setting of a good or excellent disruption, approximately 10 per cent of patients have decreased level of consciousness for up to 48 h, and this may be accompanied by temporary aphasia and/or weakness of the upper or lower extremities. In these instances patients are treated with dexamethasone, and usually return to baseline status within 48 h. The most common arterial injury that may occur during BBBD is a subintimal tear. These arterial injuries are usually asymptomatic and are noted during fluoroscopy of the carotid and vertebral arteries. If a subintimal tear occurs, mannitol and chemotherapy are not infused through the injured vessel, and the artery is re-assessed with angiography four weeks later, prior to resuming intra-arterial administration of mannitol or chemotherapy. There is a risk of deep venous thrombosis (DVT) in patients undergoing BBBD, thus patients routinely undergo Doppler monitoring of the extremities and those at high risk for DVT are placed on prophylactic anti-coagulation therapy.

Radiographic evidence of vascular injury, which may or may not be symptomatic, is seen in up to 5 per cent of patients after BBBD [12,16–18]. In the event of a stroke, in most cases the patient is asymptomatic with brain MRI changes consistent with a small infarction. In patients with MRI changes and associated neurologic deficits, the deficits usually occur within 24 h after BBBD, may include speech impairment and/or unilateral extremity weakness, and are usually caused by a small embolus resulting from the arterial catheter placement and infusion. The neurologic deficits may last greater than 48 h, however, most patients return to baseline neurologic status within 30 days.

Abnormal signal on cervical MRI has occurred in several patients treated with carboplatin-based chemotherapy with BBBD [22]. This toxicity requires immediate treatment with dexamethasone and very close observation of the patient. Carboplatin also causes high-frequency hearing loss when administered intra-arterially with BBBD [23]. This toxicity can be substantially decreased with delayed high-dose sodium thiosulfate (STS), a thiol chemoprotectant [24,25]. Additional side effects which are known to occur secondary to the chemotherapy drugs, such as nausea, fatigue, and myelosuppression, occur in patients in the BBBD program. Of note, the above side effects and toxicities can often be avoided if standard BBBD patient care guidelines developed by the BBB Consortium are closely followed.

The above technique of BBBD is performed at institutions participating in the BBB Consortium which include the Ohio State University in Columbus, University of Oklahoma Health Science Center in Oklahoma City, University of Minnesota in Minneapolis, Cleveland Clinic Foundation in Cleveland, University of Kentucky in Lexington, Hadassah-Hebrew University Medical Center in Jerusalem, Centre Hospital Universitaire de Sherbrooke in Quebec, and Oregon Health & Science University, in Portland (the coordinating center). Standard guidelines for anesthesia, transfemoral arterial catheterization, radiographic assessment of disruption and of tumor response, mannitol and chemotherapy infusion, and patient care guidelines, are used by participating BBBD centers.

Since 1995, the BBB Consortium has held an annual scientific meeting. The meetings are partially funded by an NIH R13 grant, and provide a forum for the BBB Consortium to share advances, results, and problems with current consortium clinical protocols, and to discuss future clinical trials. Following each meeting, a summary report is written by meeting participants and submitted for publication [26–28].

CLINICAL BBBD RESULTS

Since the goal of BBBD is enhanced delivery to the CNS, while preserving patient's neurocognitive functioning and quality of life, BBBD clinical protocols include an extensive neuropsychological test battery and quality of life questionnaire, which are completed at study entry and at follow-up. Excellent results have been obtained in patients with PCNSL, a highly chemosensitive tumor, using BBBD-enhanced delivery of methotrexate-based chemotherapy. It is well known that the prognosis for patients with PCNSL has improved dramatically with combined chemotherapy and radiation treatment. However, the risk of neurotoxicity associated particularly with whole brain radiation therapy (WBRT) is substantial, especially in older patients [29]. Enhanced delivery approaches such as BBBD offer the possibility of eliminating WBRT and thus the associated neurotoxicity.

A series of 74 PCNSL patients, with no prior radiotherapy, treated with methotrexate-based chemotherapy with osmotic BBBD, were reported by our clinical team [17]. The estimated 5-year survival of this series was 42 per cent with 86 per cent of patients in complete response at one year demonstrating no cognitive loss. Kraemer et al. assessed the association of total dose intensity and survival in the 74 PCNSL patients [30]. The number of BBB disruptions and the cumulative quality of disruption scores demonstrated longer survival with increased dose intensity. The relationship between magnetic resonance imaging changes and cognitive function was studied in a subset ($n = 16$) of the PCNSL patients treated with BBBD who were in complete response after a year of treatment. By the end of the treatment, all the patients' cognitive function improved, and T2 signal abnormalities associated with enhancing tumor were either stable, decreased, or resolved in 15 of 16 patients [31].

Our clinical team has reported sensitivity to carboplatin-based chemotherapy with BBBD in relapsed PCNSL [19]. It has been reported that rituximab (Rituxan), a mAb which reacts specifically with the B-cell antigen CD20, induces apoptosis and may synergize with platinum drugs [32]. Based on these findings we have begun studies using rituximab, prior to BBBD with carboplatin-based chemotherapy, thus delivering a mAb across the BBB. In patients with ocular involvement of PCNSL, intravitreal methotrexate has shown efficacy in inducing clinical remission with acceptable morbidity [33].

Anaplastic oligodendroglioma with allelic loss of heterozygosity on chromosome 1p are chemosensitive tumors, which respond well to alkylating agents. A study is in review, using carboplatin (intra-arterial), melphalan (intra-arterial), and etoposide or etoposide phosphate (intravenous) in conjunction with BBBD, in patients with anaplastic oligodendroglioma who have undergone prior chemotherapy or radiation. Molecular analysis for allelic loss of chromosomes 1p and 19q as well as p53 immunocytochemistry will be done as part of the protocol.

We reported clinical results in children and young adults with germ cell tumor ($n = 9$), PCNSL ($n = 9$), or primitive neuroectodermal tumor (PNET) ($n = 16$), who were treated with intra-arterial carboplatin or methotrexate-based chemotherapy with BBBD [16]. These patients underwent comprehensive neuropsychological evaluations at baseline prior to BBBD and at follow-up. Durable responses were seen in patients with germ cell tumor and with PCNSL. PNET required postchemotherapy radiotherapy for a durable response to be attained. The data obtained from the comprehensive neuropsychological evaluations at baseline prior to BBBD and at follow-up, showed a predominant pattern of stable or improved neuropsychological results after BBBD treatment in these patients [16].

Although CNS metastases occur much more frequently than primary brain tumors, treatment options are limited and chemotherapy in the setting of CNS metastases has not been widely studied (see Chapter 35 for more information). For example, platinum-based systemic chemotherapy is widely used in the treatment of ovarian cancer. Our clinical team treated five patients with CNS metastases from ovarian ($n = 4$) and endometrial ($n = 1$) cancer, who refused or failed to respond to conventional treatment, with carboplatin (intra-arterial) with or without BBBD, etoposide or etoposide phosphate (intravenous or intra-arterial) and cyclophosphamide (intravenous) (Murillo et al., unpublished data). The primary gynecologic cancer was in remission after systemic treatment, prior to the development of CNS metastases. If residual brain lesions were present after the intra-arterial chemotherapy, then focal radiation was given.

As a case example, one of the patients with ovarian cancer developed a single deep temporal-occipital

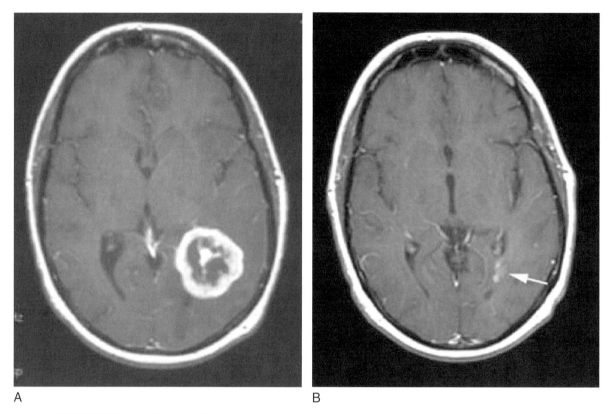

A   B

FIGURE 18.3   MRI with gadolinium shows a large deep temporal-occipital brain metastases in a patient with a history of ovarian carcinoma (A). The patient underwent five courses of carboplatin (intra-arterial), cyclophosphamide (intravenous), and etoposide phosphate (intravenous) with BBBD. MRI with gadolinium shows a dramatic response after BBBD chemotherapy (B). Radiosurgery was then given to the small residual enhancing lesion (arrow).

metastatic CNS lesion (see Figs. 18.3A,B). The patient refused brain surgery and was treated with carboplatin (intra-arterial), cyclophosphamide (intravenous), and etoposide phosphate (intravenous) with BBBD. MRI showed dramatic response to the chemotherapy. Radiosurgery was then given to the enhancing lesion. Following radiosurgery, the patient was in remission for 41 months and maintained full functional status (KPS = 90).

The overall survival from diagnosis of the gynecologic cancer ranged from 25 to 65 months, with an average of 47 months (see Table 18.1). The mean survival from the diagnosis of CNS metastases was 26 months. These results are among the best reported survivals for CNS metastases of ovarian cancer. Three of the five patients had KPS of 80 or above prior to starting the treatment; these three had the longest survivals and sustained a KPS of 80 or above during intra-arterial or BBBD treatment. These findings suggest that further investigation of the role of BBBD-enhanced chemotherapy in CNS metastases is warranted.

PRE-CLINICAL STUDIES OF THIOL CHEMOPROTECTION

Chemotherapy dose intensive strategies for the treatment of malignant brain tumors necessitate minimizing CNS and systemic toxicities. Carboplatin has shown efficacy in malignant brain tumors [23]. However, carboplatin causes myelosuppression including severe thrombocytopenia, often requiring platelet transfusions and dose reductions of subsequent carboplatin treatments. In addition, when administered in conjunction with BBBD, carboplatin causes irreversible hearing loss in a large proportion of subjects [24,25].

Thiols, such as sodium thiosulfate (STS) and N-acetylcysteine (NAC), which are small molecular weight agents with reactive sulfur groups, protect against in vitro chemotherapy cytotoxicity, and in vivo bone marrow toxicity and hearing toxicity in rat models. In vitro, STS binds directly to the electrophilic platinum, rendering it inactive [34–36],

TABLE 18.1 Location and Number of Brain Metastases, Additional Treatment for Brain Metastases

Patient	Location (and number) of brain metastases	Additional treatment for brain metastases	Survival (months) from diagnosis of gynecologic cancer	Survival (months) from diagnosis of brain metastases
1	Multiple deep and superficial supratentorial [4]	P: whole brain radiation	46	32
2	Left temporal-occipital [1]	A: stereotactic radiation	65	41
3	Right cerebellar hemisphere [1]	A: whole brain radiation	36	10
4	Left frontal [1], left parietal [1], right temporal [1]	None	25	9
5	Left cerebellar hemisphere [1]	A: stereotactic radiation	63	41

(P – prior to intra-arterial chemotherapy; A – after intra-arterial chemotherapy), and survival (months) for five patients with ovarian ($n = 4$) or endometrial ($n = 1$) cancer, from diagnosis of primary cancer and from diagnosis of brain metastases

while NAC may act more as an antioxidant and free radical scavenger. The potential effect of thiol agents on chemotherapy is a major concern. In a rat model of CNS malignancies this effect is minimized by the two-compartment model. Carboplatin (intra-arterial) is anatomically separated from the STS (intravenous), using osmotic opening of the BBB as well as the 4-h delay in administration of STS [36,37].

In a guinea pig model, high doses of STS when given up to 8 h after carboplatin, reduced carboplatin-mediated cochlear damage as confirmed by electrophysiological measurements and histology [38]. A further study in guinea pigs tested whether STS, given at time points that achieved hearing protection, could impact the chemotherapeutic efficacy of carboplatin. When STS administration was delayed until 8 h after carboplatin, there was no reduction in the anti-tumor cytotoxicity of the chemotherapy [36].

Our laboratory has also investigated whether NAC, a precursor of glutathione synthesis, was protective in animal studies against chemotherapy induced myelosuppression, when administered in the descending aorta to limit delivery to the brain [39]. L-buthionine-[S,R]-sulfoximine (BSO) was used to reduce cellular glutathione levels prior to intracarotid alkylator administration. BSO treatment significantly enhanced the toxicity of carboplatin, melphalan, and etoposide phosphate against granulocytes, total white cells, and platelets. When NAC, with or without STS was infused *via* aortic infusion prior to chemotherapy, the magnitude of the bone marrow toxicity nadir was minimized, even with BSO-enhanced myelosuppression.

Further studies have investigated whether an optimized bone marrow chemoprotection regimen impaired the efficacy of enhanced chemotherapy against rat brain tumors. Nude rats with human lung carcinoma xenografts were treated with carboplatin, melphalan, and etoposide phosphate (intra-arterially) with BBBD. NAC was given 60 min before chemotherapy and/or STS was given 4 and 8 h after chemotherapy. Pre-treatment with NAC combined with delayed STS protected against toxicity toward white cells, granulocytes, and platelets. Enhanced chemotherapy reduced intracerebral tumor volume compared to untreated animals ($p < 0.0001$). The data indicate that the efficacy of enhanced chemotherapy for rat brain tumors was not affected by thiol chemoprotection [40].

Our pre-clinical team reported the potential for hearing protection when NAC was administered prior to cisplatin, in rats. Cisplatin was administered 15 min after intravenous infusion of saline ($n = 8$) or NAC ($n = 8$). Subsequent groups were similarly treated with NAC 30 min before ($n = 7$) and 4 h after ($n = 7$) cisplatin. Auditory brainstem response thresholds were tested 7 days after the treatment and compared to baseline values. The NAC-treated rats exhibited no significant change from baseline hearing values at all time intervals, while the saline-treated rats showed marked hearing toxicity, especially at higher frequencies [41].

CLINICAL STUDIES OF THIOL CHEMOPROTECTION

Clinical studies have shown hearing protection when high-dose STS (16–20 g/m^2) was administered as part of a two-compartment model in patients with malignant brain tumors [24,25]. That is, carboplatin was administered intra-arterially immediately after BBBD. High-dose STS was administered intravenously in a delayed fashion, 4 (or 4 and 8) h after carboplatin, thus providing spatial and temporal

TABLE 18.2   Per cent Patients and Per cent Carboplatin Courses with Platelet Nadir Less than $20 \times 10^3/mm^3$, the Per cent of Patients and Carboplatin Courses Requiring Platelet Transfusion, and the Percent of Patients Requiring Carboplatin Dose Reduction

Treatment group	Platelet Nadir $<20 \times 10^3/mm^3$	Platelet transfusions	Dose reductions
Carboplatin/no STS			
$n = 24$ patients	25% patients	33% patients	33% patients
125 courses	7% courses	9% courses	
Carboplatin/STS			
$n = 29$ patients	4% patients	7% patients	0% patients
129 courses	1% courses	2% courses	

separation between chemotherapy and chemoprotection. The study showed a clear protective effect against carboplatin-induced hearing loss [25].

We reviewed hematologic data from patients with malignant brain tumors treated with carboplatin (intra-arterial), cyclophosphamide (intravenous), and etoposide or etoposide phosphate (intra-arterial or intravenous) BBBD with ($n = 29$) or without ($n = 24$) delayed high-dose STS for hearing protection (see Table 18.2) [42]. STS was administered 4 (or 4 and 8) h after carboplatin. The rate of grade 3 or 4 platelet toxicity (NCI Common Toxicity Criteria) without STS was 47.8 per cent and with STS was 17.2 per cent; there was a significant association of grade 3 or 4 platelet toxicity in patients without STS treatment ($p = 0.0018$). The rates of dose reduction of carboplatin, controlling for prior chemotherapy, were statistically significant between the two groups ($p = 0.0046$). These results suggest that STS may protect against severe thrombocytopenia, decreasing the number of platelet transfusions and dose reductions of carboplatin.

A randomized clinical trial is underway to more definitively determine the effect of delayed high-dose STS on platelet counts in patients with malignant brain tumors. The prospective clinical study includes carboplatin (intra-arterial), cyclophosphamide (intravenous), and etoposide phosphate (intravenous) without BBBD, in patients with high-grade glioma, with subjects randomized to receive delayed high-dose STS (intravenous) or no STS. Primary endpoints are per cent of subjects with platelet nadir $<20 \times 10^3/mm^3$, and per cent of subjects requiring platelet transfusion. Tumor response and duration of response are monitored to more definitively determine the effect of STS on clinical outcomes.

Cisplatin is a widely used and effective chemotherapeutic agent, however, it has a high incidence of associated toxicities, including hearing toxicity.

A prospective phase 3 Children's Oncology Group trial is under discussion, in which delayed high-dose STS will be evaluated in several pediatric histologies that are routinely treated with cisplatin (intravenous). This trial will evaluate whether hearing protection with delayed STS can be achieved in the pediatric population, without reducing anti-tumor efficacy.

A clinical phase 1 NAC dose-escalation study is underway to assess toxicity and to determine away maximum tolerated dose of NAC administered with carboplatin-based chemotherapy in patients with malignant brain tumors. The NAC is administered intravenously 60 min prior to BBBD and carboplatin (intra-arterial)-based chemotherapy. In a separate study under review, NAC is administered in the descending aorta 30 min prior to BBBD.

FUTURE BBBD DIRECTIONS

The occurrence of CNS metastases of systemic cancers far exceed the number of primary malignant brain tumors. Current therapies such as radiosurgery are effective for short-term palliation of CNS metastases, however, often do not provide long-term disease control. In many cases, WBRT used to treat CNS metastases has been associated with neurotoxicity. BBBD enables global delivery of chemotherapy to all cerebral circulations, and thus may offer a new treatment strategy for CNS metastases, administered before or after radiosurgery or WBRT. Based on the results of BBB pre-clinical studies, it is hypothesized that pre-irradiation BBBD chemotherapy may allow better drug penetration to CNS metastases in the clinical setting.

Tumor-specific mAbs can be used as anti-neoplastic agents or as delivery systems for anti-tumor therapies. A clinical protocol is under development for the use of mAb against the epidermal growth factor receptor (EGFR) in patients with breast cancer with CNS metastases. Chemotherapy will consist of monthly methotrexate and carboplatin, infused intra-arterially in conjunction with BBBD. HER-2 positive patients will receive trastuzamab (Herceptin) the evening prior to BBBD. Patient's neuropsychologic and quality of life status will be closely followed.

Translational studies have led to promising new targeted therapies such as the EGFR antagonists, used alone or as a part of combination therapy. Herstatin, a naturally occurring product of the HER-2 gene created by alternative mRNA splicing, blocks receptor dimerization and activation of EGFR [43]. Since glioblastoma is often driven by EGFR activation or mutation, we evaluated the potential for treatment with Herstatin in a rat glioblastoma model. Herstatin blocked the growth of intracerebral glioma expressing the normal EGFR, whereas *in vivo* and *in vitro* growth of cells expressing a mutant constitutively active EGFR were resistant to Herstatin. BBBD may provide a method to optimize CNS delivery of agents such as Herstatin and even viral vectors. As Herstatin can spread to and inhibit adjacent tumor cells, it may be effective as a transgene for gene therapy of CNS tumors [44].

In the pre-clinical setting, we have studied the BR96 mAb which binds to a Lewis^Y-related antigen expressed in lung, breast, ovarian, and colon cancer. The conjugate of BR96 mAb with adriamycin (BR96-DOX) is effective against a number of human carcinoma xenografts in rodents [45]. The efficacy of immunotherapy or immunoconjugates is limited by poor penetration in solid tumors. Additionally, antibodies do not cross the BBB. We are investigating the use of radiolabeled tumor-specific antibodies against brain tumors; an approach which could be further optimized by BBBD to achieve global delivery [46].

mAbs against the CD20 antigen, expressed by nearly all B-cell lymphomas, show promise. As previously mentioned, we are studying rituximab administered the evening prior to BBBD chemotherapy, in patients with relapsed PCNSL. On the horizon is the use of radiolabeled anti-B-cell antibodies in the treatment of relapsed PCNSL. We are interested in exploring the use of ibritumomab tiuxetan in conjunction with BBBD in relapsed PCNSL. In addition, a study is planned to assess the safety and efficacy of intravitreal rituximab in patients with intra-ocular lymphoma. As part of the study, apoptosis will be evaluated in vitreal cells, at multiple time points after injection of rituximab. There is evidence from the intravitreal injection of another immunoglobulin that

this class of therapeutic agents can be efficacious and administered intravitreally with acceptable toxicity.

Neuro-imaging techniques have become increasingly important in assessing biologic and physiologic aspects of brain tumors. The ability to image infiltrative disease and to better assess the extent of disease and actual tumor volume is critical. Our group has shown, using ultra small iron oxide particles (USPIOs), improvement in imaging tumor microvasculature and larger areas of tumor enhancement, in some cases [47]. Delayed imaging has shown that the USPIO is taken up into tumor macrophages and reactive astrocytes, visualized histochemically [48]. Due to the virus-like size of the USPIO, penetration of viral vectors may be monitored, when using gene therapy.

SUMMARY

BBB translational research evaluates toxicity and efficacy of global CNS delivery of chemotherapeutics in combination with chemoprotectants and/or mAbs. New clinical studies for chemotherapy responsive primary and metastatic CNS tumors have been developed to deliver mAbs across the BBB. The studies are conducted by participating BBB Consortium institutions. For additional information regarding our BBB pre-clinical and clinical translational program, as well as the BBB Consortium clinical studies, please visit www.ohsu.edu/bbb.

ACKNOWLEDGMENTS

We thank the OHSU pre-clinical and clinical BBB team for their commitment to BBB research and to the clinical care of brain tumor patients, as well as our colleagues in the BBB Consortium for critical dialogue and ongoing collaboration. Special thanks are also due to Leslie Muldoon Ph.D. for her thoughtful review of this chapter.

This work was supported by a Veterans Administration Merit Review Grant; National Institutes of Health R01 Grants NS 44687, NS 34608, and NS 33618 from the National Institute of Neurological Disorders and Stroke (NINDS); R13 CA 86959 from the National Cancer Institute, NINDS, and the National Institute of Deafness and Other Communication Disorders; and PHS Grant 5 M01 RR000334.

Dr. Neuwelt and the Oregon Health & Science University have significant financial interests in Adherex, a company that has a commercial interest in the results of this research and technology.

This potential conflict of interest has been reviewed and managed by the Oregon Health & Science University Conflict of Interest in Research Committee.

References

1. Groothuis, D. R., Vriesendorp, F. J., Kupfer, B. et al. (1991). Quantitative measurements of capillary transport in human brain tumors by computed tomography. Ann Neurol 30, 581–588.

2. Levin, V. A., Clancy, T. P., Ausman, J. I., and Rall, D. P. (1972). Uptake and distribution of 3H-methotrexate by the murine ependymoblastoma. J Natl Cancer Inst 48, 875–883.

3. Groothuis, D. R. (2000). The blood-brain and blood-tumor barriers: a review of strategies for increasing drug delivery. Neuro-oncol 2, 45–59.

4. Neuwelt, E. A. (2004). Mechanisms of disease: the blood-brain barrier. Neurosurgery 54, 131–142.

5. Neuwelt, E. A., Barnett, P. A., Hellstrom, K. E. et al. (1988). Delivery of melanoma-associated specific immunoglobulin monoclonal antibody and Fab fragments to normal brain utilizing osmotic blood-brain barrier disruption. Cancer Res 48, 4725–4729.

6. Neuwelt, E. A., Frenkel, E. P., Rapoport, S., and Barnett, P. (1980). Effect of osmotic blood-brain barrier disruption on methotrexate pharmacokinetics in the dog. Neurosurgery 7, 36–43.

7. Neuwelt, E. A. (Ed.) (1989). vol 2. Implications of the Blood-Brain Barrier and Its Manipulation. Plenum Publishing Corporation, New York.

8. Kroll, R. A., and Neuwelt, E. A. (1998). Outwitting the blood-brain barrier for therapeutic purposes: osmotic opening and other means. Neurosurgery 42, 1083–1100.

9. Remson, L. G., McCormick, C. I., Sexton, G., Pearse H. D., Garcia, R., and Neuwelt, E. A. (1995). Decreased delivery and acute toxicity of cranial irradiation and chemotherapy given with osmotic blood-brain barrier disruption in a rodent model: the issue of sequence. Clin Can Res 1, 731–739.

10. Remsen, L. G., Marquez, C., Garcia, R., Thrun, L. A., and Neuwelt, E. A. (2001). Efficacy after sequencing of brain radiotherapy and enhanced antibody targeted chemotherapy delivery in a rodent human lung cancer brain xenograft model. Int J Radiat Oncol Biol Phys 51, 1045–1049.

11. Remsen, L. G., McCormick, C. I., Sexton, G. et al. (1997). Long-term toxicity and neuropathology associated with the sequencing of cranial irradiation and enhanced chemotherapy delivery. Neurosurgery 40, 1034–1042.

12. Dahlborg, S. A., Henner, W. D., Crossen, J. R. et al. (1996). Non-AIDS primary cns lymphoma: first example of a durable response in a primary brain tumor using enhanced chemotherapy delivery without cognitive loss and without radiotherapy. The Cancer Journal from Scientific American 2, 166–174.

13. Neuwelt, E. A., Pagel, M., Barnett, P., Glasberg, M., and Frenkel, E. P. (1981). Pharmacology and toxicity of intracarotid adriamycin administration following osmotic blood-brain barrier modification. Cancer Res 41, 4466–4470.

14. Neuwelt, E. A., Barnett, P. A., Glasberg, M., and Frenkel, E. P. (1983). Pharmacology and neurotoxicity of cis-Diamminedichloroplatinum, bleomycin, 5-fluorouracil, and cyclophosphamide administration following osmotic blood-brain barrier modification. Cancer Res 43, 5278–5285.

15. Fortin, D., McCormick, C. I., Remsen, L. G., Nixon, R., and Neuwelt, E. A. (2000). Unexpected neurotoxicity of etoposide phosphate administered in combination with other chemotherapeutic agents after blood-brain barrier modification to enhance delivery, using propofol for general anesthesia, in a rat model. Neurosurgery 47, 199–207.

16. Dahlborg, S. A., Petrillo, A., Crossen, J. R. et al. (1998). The potential for complete and durable response in nonglial primary brain tumors in children and young adults with enhanced chemotherapy delivery. The Cancer Journal from Scientific American 4, 110–124.

17. McAllister, L. D., Doolittle, N. D., Guastadisegni, P. E. et al. (2000). Cognitive outcomes and long-term follow-up results after enhanced chemotherapy delivery for primary central nervous system lymphoma. Neurosurgery 46, 51–61.

18. Doolittle, N. D., Miner, M. E., Hall, W. A. et al. (2000). Safety and efficacy of a multicenter study using intraarterial chemotherapy in conjunction with osmotic opening of the blood-brain barrier for the treatment of patients with malignant brain tumors. Cancer 88, 637–647.

19. Tyson, R. M., Siegal, T., Doolittle, N. D., and Neuwelt, E. A. (2003). Current status and future of relapsed primary central nervous system lymphoma (PCNSL). Leuk Lymphoma 44, 627–633.

20. Doolittle, N. D., Petrillo, A., Bell, S., Cummings, P., and Eriksen, S. (1998). Blood-brain barrier disruption for the treatment of malignant brain tumors: the national program. J Neurosci Nurs 30, 81–90.

21. Roman-Goldstein, S., Clunie, D. A., Stevens, J. et al. (1994). Osmotic blood-brain barrier disruption: CT and radionuclide imaging. AJNR Am J Neuroradiol 15, 581–590.

22. Fortin, D., McAllister, L. D., Nesbit, G. et al. (1999). Unusual cervical spinal cord toxicity associated with intra-arterial carboplatin, intra-arterial or intravenous etoposide phosphate, and intravenous cyclophosphamide in conjunction with osmotic blood-brain barrier disruption in the vertebral artery. AJNR Am J Neuroradiol 20, 1794–1802.

23. Williams, P. C., Henner, D. W., Roman-Goldstein, S. et al. (1995). Toxicity and efficacy of carboplatin and etoposide in conjunction with disruption of the blood-brain barrier in the treatment of intracranial neoplasms. Neurosurgery 37, 17–28.

24. Neuwelt, E. A., Brummett, R. E., Doolittle, N. D. et al. (1998). First evidence of otoprotection against carboplatin-induced hearing loss with a two-compartment system in patients with central nervous system malignancy using sodium thiosulfate. J Pharmacol Exp Ther 286, 77–84.

25. Doolittle, N. D., Muldoon, L. L., Brummett, R. E. et al. (2001). Delayed sodium thiosulfate as an otoprotectant against carboplatin-induced hearing loss in patients with malignant brain tumors. Clin Cancer Res 7, 493–500.

26. Doolittle, N. D., Anderson, C. P., Bleyer, W. A. et al. (2001) for the Blood-Brain Barrier Disruption Consortium. Importance of dose intensity in neuro-oncology clinical trials: summary report of the sixth annual meeting of the Blood-Brain Barrier Disruption Consortium. Neuro-Oncology 3, 46–54. (Posted to Neuro-Oncology, Doc. 00-039, November 3, 2000 <neuro-oncology.mc.duke.edu>)

27. Doolittle, N. D., Abrey, L. E., Ferrari, N. et al. (2002). Targeted delivery in primary and metastatic brain tumors: summary report of the seventh annual meeting of the Blood-Brain Barrier Disruption Consortium. Clin Cancer Res 8, 1702–1709.

28. Doolittle, N. D., Abrey, L. E., Bleyer, W. A. et al. (2005). New frontiers in translational research in neuro-oncology

and the blood-brain barrier: report of the tenth annual Blood-Brain Barrier Disruption Consortium meeting. *Clin Cancer Res* **11**, 421–428.

29. Abrey, L. E., DeAngelis, L. M., and. Yahalom, J. (1998). Long-term survival in primary CNS lymphoma. *J Clin Oncol* **16**, 859–863.

30. Kraemer, D. F., Fortin, D., Doolittle, N. D., and Neuwelt, E. A. (2001). Association of total dose intensity of chemotherapy in primary central nervous system lymphoma (human non-acquired immunodeficiency syndrome) and survival. *Neurosurgery* **48**, 1033–1041.

31. Neuwelt, E. A., Guastadisegni, P. E., Varallyay, P., and Doolittle, N. D. (2005). In press. Imaging changes and cognitive outcome in primary CNS lymphoma after enhanced chemotherapy delivery. *AJNR Am J Neuroradiol* **26**, 258–265.

32. Alas, S., Ng, C. P., Bonavida, B. (2002). Rituximab modifies the cisplatin-mitochondrial signaling pathway, resulting in apoptosis in cisplatin-resistant non-Hodgkins lymphoma. *Clin Cancer Res* **8**, 836–845.

33. Smith, J. R., Rosenbaum, J. T., Wilson, D. J. *et al.* (2002). Role of intravitreal methotrexate in the management of primary central nervous system lymphoma with ocular involvement. *Ophthalmology* **109**, 1709–1716.

34. Dedon, P. C., and Borch, R. F. (1987). Characterization of the reactions of platinum antitumor agents with biologic and nonbiologic sulfur-containing nucleophiles. *Biochem Pharmacol* **36**, 1955–1964.

35. Elferink, F., van der Vijgh, W. J. F., Klein, I., and Pinedo, H. M. (1986). Interaction of cisplatin and carboplatin with sodium thiosulfate: reaction rates and protein binding. *Clin Chem* **32**, 641–645.

36. Muldoon, L. L., Pagel, M. A., Kroll, R. A. *et al.* (2000). Delayed administration of sodium thiosulfate in animal models reduces platinum ototoxicity without reduction of antitumor activity. *Clin Cancer Res* **6**, 309–315.

37. Muldoon, L. L., Walker-Rosenfeld, S. L., Hale, C., Purcell, S. E. Bennett, L. C., and Neuwelt, E. A. (2001). Rescue from enhanced alkylator-induced cell death with low molecular weight sulfur-containing chemoprotectants. *J Pharmacol Exp Ther* **296**, 797–805.

38. Neuwelt, E. A., Brummett, R. E., Remsen, L. G. *et al.* (1996). In vitro and animal studies of sodium thiosulfate as a potential chemoprotectant against carboplatin-induced ototoxicity. *Cancer Res* **56**, 706–709.

39. Neuwelt, E. A., Pagel, M. A., Hasler, B. P., Deloughery T. G., and Muldoon L. L. (2001). Therapeutic efficacy of aortic administration of N-acetylcysteine as a chemoprotectant against bone marrow toxicity after intracarotid administration of alkylators, with or without glutathione depletion in a rat model. *Cancer Res* **61**, 7868–7874.

40. Neuwelt, E. A., Pagel, M. A., Kraemer, D. F., Peterson, D. R., and Muldoon, L. L. (2004). Bone marrow chemoprotection without compromise of chemotherapy efficacy in a rat brain tumor model. *J Pharmacol Exp Ther* **309**, 594–599.

41. Dickey, D. T., Muldoon, L. L., Kraemer, D. F., and Neuwelt, E. A. (2004). Protection against cisplatin-induced ototoxicity by N-acetylcysteine in a rat model. *Hear Res* **193**, 25–30.

42. Doolittle, N. D., Tyson, R. M., Lacy, C. *et al.* (2001). Potential role of delayed sodium thiosulfate as protectant against severe carboplatin-induced thrombocytopenia in patients with malignant brain tumors. *Blood* **98(11)**, 37a–38a. Abstract 149.

43. Doherty, J. K., Bond, C., Jardim, A., Adelman, J. P., and Clinton, G. M. (1999). The HER-2/neu receptor tyrosine kinase gene encodes a secreted autoinhibitor. *Proc Natl Acad Sci U S A* **96**, 10869–74.

44. Staverosky, J. A., Muldoon, L. L., Guo, S., Evans A. J., Neuwelt, E. A., and Clinton G. M. (2005). Herstatin, an autoinhibitor of the EGF receptor family, blocks the intracranial growth of glioblastoma. *Clin Cancer Res* **11**, 335–340.

45. Trail, P. A., Willner, D., Lasch, S. J. *et al.* (1992). Antigen-specific activity of carcinoma-reactive BR64 doxorubicin conjugates evaluated in vitro and in human tumor xenograft models. *Cancer Res* **52**, 5693–5700.

46. Muldoon, L. L., and Neuwelt, E. A. (2003). BR96-DOX immuno-conjugate targeting of chemotherapy in brain tumor models. *J Neurooncol* **65**, 49-62.

47. Varallyay, P., Nesbit, G., Muldoon, L. L. *et al.* (2002). Comparison of two super paramagnetic viral-sized iron oxide particles ferumoxides and ferumoxtran-10 with a gadolinium chelate in imaging intracranial tumors. *AJNR Am J Neuroradiol* **23**, 510–519.

48. Neuwelt, E. A., Varallyay, P., Bago, A., Muldoon, L. L., Nesbit, G., and Nixon, R. (2004). Imaging of iron oxide nanoparticles by MR and light microscopy in patients with malignant brain tumours. *Neuropathol Appl Neurobiol* **30(5)**, 456–71.

Interstitial Chemotherapy and Polymer-Drug Delivery

Raqeeb M. Haque, Eric Amundson, Michael Dorsi, and Henry Brem

ABSTRACT: Malignancies of the central nervous system represent a formidable challenge and prognosis of patients remains dismal. Physiological barriers impede efficacious delivery of tumoricidal concentrations of anti-neoplastic agents to the CNS without incurring undesirable systemic toxicities. The development of local delivery devices for controlled administration of chemotherapeutics provides a new mechanism to effectively treat brain tumors while bypassing many of the harmful toxicities. In this chapter, we describe the rationale and history of the development and use of these polymers to deliver interstitial chemotherapy, and review new investigations of different drug-polymer combinations. Finally, we will briefly describe new local drug delivery systems.

INTRODUCTION

Treatment of gliomas remains one of the most challenging problems in clinical oncology today. Current brain tumor therapy consists of a triad of surgery, radiation, and chemotherapy. Despite vast improvements in the therapeutic armamentarium, the prognosis for most patients remains poor. Of the 16 800 people diagnosed with primary brain tumors every year, more than 13 000 die [1]. Thus, novel therapeutic approaches must be developed to improve treatment of these devastating and often terminal cancers.

Malignant brain tumors are the most commonly diagnosed adult primary tumors of the central nervous system. Over half of all primary brain tumors are glial cell malignancies, and more than 75 per cent of these arise from astrocytes. Malignant gliomas are classified as either primary or secondary tumors based on their progression from normal cells to malignancy. Primary glioblastomas arise *de novo*, without gradual progression from a low to a high-grade astrocytoma. Conversely, secondary gliomas arise from astrocytes that have undergone a stepwise progression of genetic changes (Fig. 19.1). Such genetic alterations correspond to the evolution of a low-grade astrocytoma into a glioblastoma. These genetic changes result in the up-regulation of certain soluble molecules and membrane receptors including platelet-derived growth factor, epidermal growth factor, and vascular endothelial growth factor [2].

There are numerous challenges to improve the treatment of gliomas. A surgical resection with "tumor free" margins would increase the risk of removing normal brain tissue, resulting in serious neurological sequelae. Similarly, doses of radiation that would be optimal for tumor eradication may cause damage to the surrounding brain. Systemic chemotherapy resulted in a small but significant improvement in the mortality rates of brain tumor patients [3,4]. The efficacy of systemic chemotherapy for brain tumors is limited by physiologic and pathological barriers within the central nervous system. Thus, systemically intolerable doses are required to deliver and maintain tumoricidal concentrations in the brain. Local delivery of chemotherapeutic agents directly to the tumor could potentially circumvent systemic side effects.

This chapter will focus on the development of interstitial chemotherapy and polymer-drug delivery.

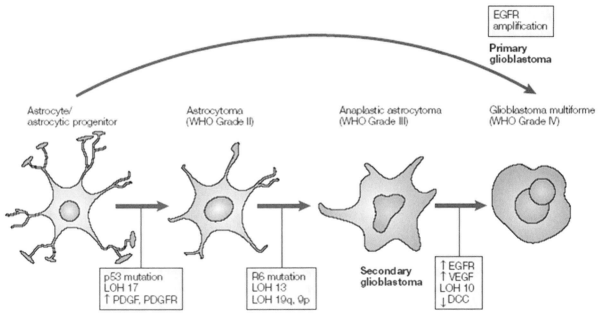

FIGURE 19.1 The stepwise genetic changes needed to progress from a low-grade astrocytoma to a glioblastoma multiforme. (Reprinted with permission from Nature Publishing Group) [5]. See Plate 19.1 in Color Plate Section.

We will describe the technologies being developed to circumvent the barriers encountered in delivering chemotherapeutics to the central nervous system. Special emphasis will be placed on the development of the FDA-approved BCNU-loaded polymer (Gliadel®) for the treatment of malignant tumors. This chapter will review the results and implications of pre-clinical and clinical studies of Gliadel® and other polymer-chemotherapeutic agent systems and describe the surgical technique used to implant Gliadel® wafers at the tumor site. Finally, we will address future directions of local drug delivery for treating brain tumors.

PHYSIOLOGIC AND PATHOLOGICAL BARRIERS TO ANTI-NEOPLASTICS IN THE CNS

The unique physiological and pathological barriers encountered in the central nervous system impede drug delivery for the treatment of brain lesions. Oral and intravenous injections of chemotherapeutics have difficulty reaching therapeutic concentrations at brain tumor sites because of these barriers, including the blood–brain barrier (BBB), blood–cerebrospinal fluid barrier (BCB), and blood–tumor barrier (BTB) [5].

The BBB is formed from tight junctions between cerebral capillary endothelial cells, serving as a

pharmacological barrier that restricts the influx of molecules from the vasculature into the brain. Of the three barriers mentioned above, the BBB is the most prominent in impeding drug transport. It is estimated to have a surface area 5000-fold greater than that of the BCB [6]. Only small, electrically neutral, lipid-soluble molecules are permeable across the BBB, leaving most chemotherapeutic drugs unable to penetrate the capillary endothelium (Fig. 19.2). In addition, the CNS has an active drug-efflux transport system that plays a significant role in preventing anti-neoplastics from entering the brain parenchyma, utilizing multiple organic anion transporters such as P-glycoprotein, MDR-1, and MDR-2 [6,7]. Lastly, the cerebral endothelial cells of the BBB lack pinocytotic vesicles seen in systemic endothelium, which normally helps with the transcytosis of molecules from the bloodstream into the brain [5].

The BCB is an additional obstacle that impedes drug delivery to the CNS [5]. Choroid epithelial cells, which line the ventricles and produce cerebrospinal fluid, form a tightly bound barrier that regulates the transfer of molecules into the interstitial fluid that surrounds the brain parenchyma. In this manner, the BCB prevents the penetration of chemotherapeutics. Also, analogous to the BBB, there exists an organic-acid transport system that actively removes molecules such as chemotherapeutics from the cerebrospinal fluid.

Finally, the BTB is a pathological barrier formed by the abnormally dilated and tortuous

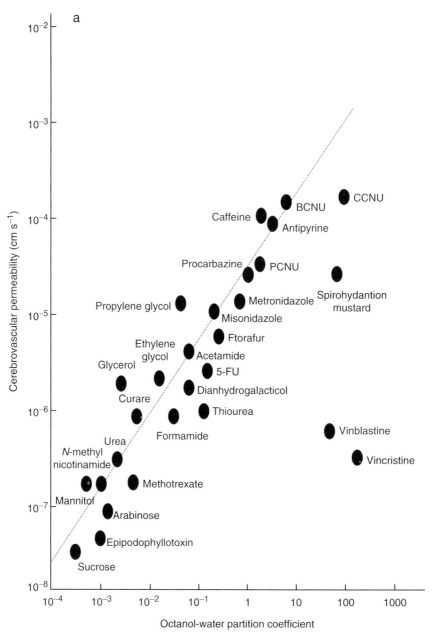

FIGURE 19.2 Cerebrovascular permeability versus octanol-water coefficient of selected chemicals and drugs. (Reprinted with permission from Elsevier Science) [118]

microvasculature of the brain tumor that limits the effectiveness of brain tumor therapy [8]. Leaky capillaries and decreased flow velocity also cause a concomitant increase in intratumoral interstitial pressure, further compromising drug delivery [9]. Also, the hypoxic environment caused by the erratic blood flow induces the expression of many bio-reductive enzymes, allowing the tumor to become more resistant to both radio- and chemotherapy [10]. Taken as a whole, these obstacles lend themselves to the development of local interstitial drug delivery methods.

DELIVERY SYSTEMS AND STRATEGIES FOR BRAIN TUMOR TARGETED DELIVERY

Several approaches have been implemented to address the limitations presented by both the natural and pathologic barriers in the CNS. The strategies used to bypass these boundaries can be divided into chemical and physical based techniques [7]. Chemical strategies focus on improving drug delivery by altering the structure of the drug or by disrupting the BBB. Physically based techniques include: (1) catheter, (2) convection-enhanced delivery, and (3) polymer or microchip systems which directly release therapeutics in the vicinity of the tumor. In this section, we will briefly describe some of these techniques, which are further illustrated in other chapters of this book.

As already described, the BBB is readily permeable only to small, electrically neutral, lipid-soluble molecules. Thus, altering the structure of chemotherapeutics may enhance their permeability. Such an approach was used to enhance delivery of carmustine (BCNU). Twenty lipophilic analogs of BCNU were studied, including lomustine (CCNU) and semustine (methyl-CCNU). However, clinical trials evaluating systemic administration of these agents showed no statistical improvement in efficacy over BCNU in treating glial tumors [11–13]. Another strategy involves linking a chemotherapeutic to a lipophilic carrier molecule that is capable of crossing the BBB. One such carrier molecule, dihydropyridine, has been implemented to enhance the penetrance of various neurotransmitters and chemotherapeutics [14]. Preliminary data indicate that this lipophilic carrier improves the drug delivery into the brain, and further trials are ongoing. In addition, techniques to target drugs to the brain parenchyma using transport vector receptor specific antibodies are currently being developed [15].

Disruption of the BBB is another option to enhance drug delivery. Hyperosmolar agents (e.g., mannitol) are used to create a concentration gradient across endothelial cells. Efflux of intracellular water to the intravascular compartment causes the endothelial cells to shrink, which loosens tight junctions holding adjacent cells together and creates fenestrations within the barrier. In a study by Williams *et al.*, intra-arterial co-administration of mannitol with carboplatin and etoposide resulted in clinical improvements in all of their patients with PNET (4/4) and half of the patients with lymphoma (2/4) [16]. Also, RMP-7, a bradykinin agonist, has been shown to enhance the concentration of anti-neoplastics within the CNS [17]. Yet, neither mannitol nor RMP-7 has been effective in clinical trials evaluating the treatment of low-grade astrocytomas, though both drugs increased the concentration of drug in the brain parenchyma [18,19].

A third strategy to bypass the various barriers of the CNS is to use a local delivery system that can be surgically implanted near the tumor site. With such an approach, a high local concentration of drug can be maintained without the toxicities associated with systemic therapy. Three such approaches exist, including direct infusion catheters, convection-enhanced drug delivery, and polymer-based drug delivery.

One of the first catheter delivery systems used was the Ommaya reservoir, which delivers intermittent bolus injections of chemotherapeutics such as BCNU, methotrexate, adriamycin, bleomycin, and cisplatin [20–24]. Interleukin-2 and interferon-gamma are two biological agents that have also been delivered through this intermittent catheter system [25,26]. There are anecdotal case reports indicating successful outcomes with using the Ommaya system to deliver various agents, but there have been no large-scale clinical trials proving their efficacy.

Recently, a new generation of implantable pumps has been developed that provide a constant infusion of drugs over an extended period of time. These pumps include the Infusaid pump (Infusaid), Mini-Med PIMS pump (Minimed), and the Medtronic SynchroMed system (Medtronic). Continuous drug infusion has proven to be beneficial in brain-tumor models [27]. Unfortunately, pump systems have many potential pitfalls, including the following: clot or tissue debris obstruction, mechanical failure, infection requiring further operations, and undesired alterations in the chemical properties of the drug during its transit from pump to the target site. Thus, other mechanisms of local delivery must also be explored.

Convection enhanced delivery is a promising technique for drug delivery to the CNS, especially in areas where surgical interventions are not possible [28,29]. This technique is based on a created pressure gradient, independent of the molecular weight of the delivered drug. Animal studies with gemcitabine and carboplatin have shown that convection enhanced delivery is safe and effective in the treatment of experimental gliomas [30]. Ongoing clinical studies using convection-enhanced delivery of taxol and certain immunotoxins, such as IL-13 and Pseudomonas exotoxin, are being conducted for the treatment of malignant gliomas [31–33].

A polymer-based drug delivery system has been developed to directly release chemotherapeutics to the tumor site. In this next section, we will describe the steps that were necessary to develop this novel approach, including the extensive pre-clinical and clinical studies using controlled release polymers and the development of Gliadel® (3.8 per cent BCNU), the first FDA-approved treatment for newly diagnosed and recurrent gliomas in 23 years.

DEVELOPMENT OF POLYMER-BASED LOCAL DELIVERY SYSTEMS

Controlled drug delivery of macromolecules utilizing implantable polymers was first described in 1976. Langer and Folkman introduced a novel method of sustained release of macromolecules from an ethylene vinyl acetate copolymer (EVAc) [34]. Using first order kinetics, this non-biodegradable polymer releases therapeutic agents by diffusion (Fig. 19.3A), based on the properties of the bound agent, including molecular weight, electrostatic charge, and liposolubility. Although EVAc polymers have proven useful in glaucoma treatment, contraception, and insulin therapy, they have not been successful in the brain. This is primarily due to their physical limitation of being inert. After the release of the drug, the intact matrix remains as a foreign body, requiring a subsequent invasive surgical removal [35].

This limitation led to the development of a new generation of biodegradable polymers, including a poly-lactide-co-glycolide (PLGA) matrix (Table 19.1). A variety of drugs were delivered using this system, including steroids, anti-inflammatory agents, and anti-neoplastics [36–43] through microspheres. A drawback of the PLGA polymers is that the bound drugs are released through bulk erosion, rather than surface erosion (Fig. 19.3B). As such, drugs are released inconsistently with sub-optimal tissue exposure profiles, resulting in possible toxicities [44]. An ideal polymer would combine the beneficial properties of EVAc and PLGA, specifically zero order release of drug *via* surface erosion.

Leong *et al.*, first described the development of the polyanhydride poly [1,3-bis(carboxyphenoxy)propane-co-sebacic-acid] (PCPP:SA) matrix, a biodegradable polymer that breaks down to dicarboxylic acids by spontaneous reaction with water [45]. Clinically, such a system has numerous advantages:

1. The hydrophobic properties shield embedded chemotherapeutic agents from hydrolysis and degradation,
2. The zero order kinetic degradation allows for prolonged drug release at a steady concentration,
3. The half-life of the drug when loaded in PCPP:SA is increased compared to systemic delivery, and

4. The breakdown of the polymer is limited to its surface. Furthermore, by altering the ratio of the monomers carboxyphenoxypropane (CPP) and sebacic acid (SA), the rate of breakdown and subsequent release of drugs can be manipulated appropriately.

Another advantage this polymer system has over EVAc and PLGA is that the high pressures and low temperatures needed for synthesis allows the PCPP-SA to be molded into various shapes, including rods, wafers, sheets, and microspheres, that can be loaded with different clinical compounds and implemented in various clinical settings [46–51]. PCPP:SA is a compatible polymer with the brain, and causes no signs of neurological change when implanted into cynomolgus monkeys (*Macaca fascicularis*) [52–54].

Recently, new types of polymers have been developed to improve drug delivery. In contrast to the PCPP:SA, a new fatty acid dimer-sebacic (FAD:SA) polymer can release chemotherapeutics with hydrophilic properties [55] and deliver hydrostatically unstable compounds such as carboplatinum. Gelatin-chondroitin sulfate coated microspheres have also been shown to release cytokines *in vivo* [56]. Finally, polyethylene glycol-coated liposomes have also been created to release drugs such as anthracyclines [57].

There have been many advances in polymer drug-delivery since the first polymers. In the next section, we will focus on Gliadel®, a PCPP:SA polymer loaded with carmustine. Specifically, we will chronicle the pre-clinical and clinical studies that were necessary for its FDA approval in the treatment of brain tumors. We will also describe many of the ongoing clinical trials with Gliadel®, as well as exciting new developments in polymer-based delivery of novel therapeutics.

BCNU (GLIADEL®)

Pre-Clinical Studies

Carmustine (BCNU), a low-molecular weight alkylating agent, has well known activity against malignant gliomas. As it is lipophilic, BCNU can partially cross the BBB, and has been the most widely used chemotherapeutic agent for the treatment of brain tumors to date. But its success as an antineoplastic has been limited primarily by significant systemic toxicity, such as bone marrow suppression and pulmonary fibrosis, and its short serum half-life of 15 min [58]. Local application of BCNU

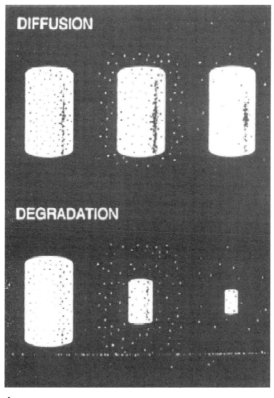

A

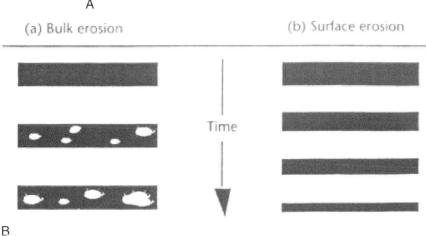

B

FIGURE 19.3 **(A)** Desired properties of polymer implants include drug release by degradation instead of diffusion. (Reprinted with permission) [44]. **(B)** Biodegradable controlled-release polymer implants are intended to release content at nearly constant rate (zero-order kinetics) as they dissolve in body water. Therefore, desired properties include **(a)** surface erosion instead of **(b)** bulk erosion. (Reprinted with permission) [44].

could potentially bypass both of these limitations. Consequently, BCNU was incorporated into a biodegradable polymer and evaluated for its use in the treatment of malignant gliomas.

The pre-clinical trials of BCNU-PCPP:SA began with experiments establishing the distribution and pharmacokinetics of active drug release both *in vitro* and in animal models. In both rabbit and monkey brain models, data showed that a controlled and sustained release of intact BCNU could be achieved with a PCPP:SA polymer system [59]. Furthermore, implementing quantitative autoradiography and thin-layer chromatography, tumoricidal levels of BCNU were detected at a distance of 4 cm from the site of polymer implantation one day after surgery and 1.3 cm on day 30 [60]. These results suggested that

TABLE 19.1 Clinical Use of Polymer Technology

Polymer	Chemical composition	Biodegradable	Clinical condition	Drug-polymer formulation
EVA	Ethylene/vinyl acetate	No	Glaucoma	Ocusert (EVA with Pilocarpine)
PCPP-SA	Poly[bis(p-carboxyphenoxy propane-sebacic acid)	Yes	Malignant glioma	Gliadel (PCPP-SA with Carmustine)
FAD-SA	Fatty acid dimmer-sebacic acid	Yes	—	—
PLG	Poly(lactide-co-glycolide)	Yes	Schizophrenia	Risperal Consta (PLG with Risperidone)
PEG	Polyethyleneglycol	Yes	SCID	Adagen (PEG with adenosine deaminase) Oncaspar (PEG with L-aspraginase)
Microspheres	Gelatine/chondroin-sulphate-coated microspheres	Yes	Periodontitis	Arestin (minocycline with microspheres)
Antibody impregnated shunt	Polydimethyl-siloxane	Yes	Hydrocephalus	Rifampin and Clindamycin impregnated shunts
Paclitaxel releasing cardiac stents	Translute^{TM} polymer	Yes	Coronary Artery Disease	Paclitaxel-Eluting Coronary Stent System

BCNU could persist in the adjacent parenchyma when delivered locally.

The next step was to test the efficacy of BCNU-polymers against a rat intracranial glioma model. Studies by Tamargo et al., consistently showed that local delivery of BCNU by polymer led to significant prolongation of survival in animals with malignant glioma when compared to systemic administration of BCNU or empty polymer [61]. Toxicity studies performed with BCNU-polymers in primates were well tolerated, and concurrent external beam radiotherapy did not result in any additional toxicities [62]. With the preclinical studies addressing the biocompatibility, biodistribution, toxicity, and efficacy of BCNU-loaded PCPP:SA polymer in place, translation into the clinical field was the next step.

Surgical Implantation of Gliadel®

Patients undergo a craniotomy for maximum resection of tumor. After removal of the tumor, up to eight discs are applied to the resection cavity surface. Sheets of oxidized regenerated cellulose (Surgicel, Johnson and Johnson, New Brunswick, NJ, USA) are used to secure the polymers against the brain (Fig. 19.4).

Clinical Trials Leading to Approval of Gliadel® by the FDA (Table 19.2)

The first clinical trial assessing the safety of 20:80 PCPP:SA BCNU-loaded polymer was a multicenter Phase I–II clinical trial in 1987 [63,64]. The study included 21 patients who were presented with recurrent malignant glioma. Enrollment criteria limited patients to those presenting with recurrent malignant gliomas that had previously undergone surgical debulking and in whom standard therapy had failed. Other inclusion criteria included: an indication for re-operation, a unilateral single tumor focus with greater than or equal to 1 cm^3 of enhancing volume on MRI or CT, completion of external beam radiotherapy, a Karnofsky Performance Scale (KPS) score greater than or equal to 60, and no exposure to nitrosureas during the 6 weeks prior to polymer implantation. In this study, three different doses of BCNU loaded in PCPP:SA polymer were tested: 1.93, 3.85 and 6.35% w/w. No evidence of systemic toxicity and no signs of neurological deterioration were seen in the treatment arms. The median survival of treated patients was 46 weeks after polymer implantation and 87 weeks after initial diagnosis. This Phase I–II trial demonstrated that implanted BCNU-PCPP:SA polymer was well tolerated by patients.

On the basis of earlier studies, a Phase III multi-centered, prospective, randomized, double-blinded, and placebo controlled clinical trial was initiated [64]. This study investigated the efficacy of 3.8 per cent BCNU-PCPP:SA polymer for the treatment of recurrent gliomas in 222 patients from 27 medical centers in the United States and Canada. The selection criteria were similar to those in the Phase I–II study with the additional provision that chemotherapy was not permitted 4 weeks pre-operatively and use of nitrosureas were not allowed for 6 weeks prior to polymer

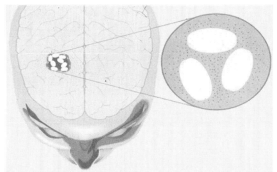

FIGURE 19.4 Up to eight polymer implants line the tumour resection cavity, where the loaded drug is gradually released as they dissolve. The inset shows conceptually how drug molecules diffuse away from these implants [44]. See Plate 19.4 in Color Plate Section.

TABLE 19.2 Clinical Trials Leading to Approval of Gliadel® by the FDA

Phase	Indications	Number of patients	References
I	Recurrent	21	Brem [120]
III	Recurrent	222	Brem [121]
I	Initial	22	Brem [122]
III	Initial	32	Valtonen [123]
IV	Recurrent	363	FDA Report, 1999
III	Initial	240	Westphal [124]

implantation. The patients were randomly divided into two groups, equally distributed for prognostic factors such as median age, histological grade, and neurological performance, and administered either BCNU ($n = 110$) or placebo ($n = 112$) polymers. The overall post-operative median survival was 31 weeks for the BCNU treatment arm and 23 weeks for the placebo group (hazard ratio = 0.67, $p = 0.0006$) (Fig. 19.5). Furthermore, the 6-month survival rate was 60 per cent in the treatment group *versus* 47 per cent in the placebo group. More significantly, in the GBM group, there was 50 per cent greater survival at six months in patients treated with BCNU polymer than placebo alone ($p - 0.02$). Consistent with the Phase I-II trials, there was no evidence of systemic toxicity. With the successful completion of the safety and efficacy clinical trials, the Food and Drug Administration (FDA) approved 3.85 per cent BCNU-loaded PCPP:SA wafer for the treatment of recurrent glioblastoma in 1996.

With the establishment of BCNU-PCPP:SA polymers as a safe and effective treatment for recurrent gliomas, attention naturally turned to its use in the treatment of primary gliomas. First, a Phase I trial with 22 patients newly diagnosed with malignant glioma was performed to test the safety of BCNU-PCPP:SA polymers and also the safety of concurrent standard external beam radiation therapy. Enrollment criteria consisted of: unilateral tumor focus greater than or equal to 1 cm^3, age > 18 years, KPS greater than or equal to 60, and intraoperative diagnosis of malignant gliomas. In addition, all of the patients received adjuvant standard radiation therapy after their surgical resection. There was no evidence of systemic or neurological toxicity. The median survival was 42 weeks, with four patients surviving greater than 18 months. Thus, BCNU-PCPP:SA polymers with conventional radiotherapy in the treatment of primary gliomas was established as safe.

Next, two Phase III trials were performed to establish the effectiveness of a 3.8 per cent BCNU-loaded polymer in the management of primary gliomas. In the first trial, [65] 32 patients with a histopathological diagnosis of grade III astrocytoma or GBM on intraoperative frozen sections were randomly assigned to receive BCNU-loaded polymer or placebo. The median survival was 58.1 weeks in the treatment group, compared to 39.3 weeks for the placebo group ($p = 0.012$) (Fig. 19.6A). The survival rates were significantly better for BCNU *versus* placebo at one (63 vs 19 per cent), two (31 vs 6 per cent), and three (25 vs 6 per cent) years post-implantation. Again, no signs of local or systemic toxicities were observed.

In the larger scale Phase III study by Westphal *et al.*, [66] 240 patients with newly diagnosed malignant glioma were randomized to receive either BCNU or placebo wafers during their surgical resection. Radiation therapy was instituted 14–21 days post-operatively. The primary endpoint for this study was survival. The median survival in the

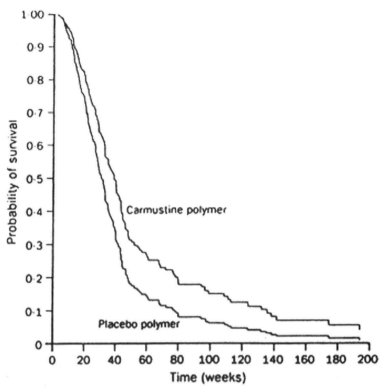

FIGURE 19.5 Overall survival curve for patients with initial therapy for Grade III and Grade IV gliomas treated with BCNU-polymer implants versus placebo. (Reprinted with permission from Elsevier Science) [64]

treatment group was 13.9 months with a 29 per cent risk reduction in death, while the placebo group had a median survival of 11.6 months ($p = 0.03$). Moreover, at three years post-implantaion, 11 of the patients who received Gliadel® were alive compared to 2 patients in the placebo group. This study demonstrated that BCNU-PCPP:SA polymer treatment reduced the overall risk of death during the 3–4 years post-treatment, as determined by a hazard ratio of 0.73 (95 per cent confidence interval: 0.56; $p < 0.05$). Time to decline in functional status was also significantly prolonged in the treatment group. Together with the Valtonen study, this large scale Phase III study of BCNU-PCPP:SA polymer as a treatment for primary malignant gliomas proved to be successful (Fig. 19.6B). As such, the FDA approved its use in initial resections of malignant gliomas in February of 2003.

New Clinical Trials Involving Gliadel® (Tables 19.3–19.6)

Since the approval of BCNU-PCPP:SA polymers for treatment of malignant gliomas, there have been multiple investigations to improve its therapeutic benefit. Twenty-eight additional clinical trials are underway evaluating other issues related to the BCNU-PCPP:SA polymer, including dosage, combination with systemic treatments, and combination with various forms of radiation and resistant modifiers, use in specialized populations, and use in those with cerebral metastases (Tables 19.3 and 19.4). Here we describe just a few of the many clinical trials involving BCNU-PCPP:SA.

Once the safety and efficacy of the 3.8 per cent BCNU-PCPP:SA was well established, a logical next step was to study the effect of increasing the BCNU-loading dose. Initial laboratory investigations indicated improved efficacy with higher doses of BCNU loaded polymers implanted into 9L gliosarcoma models [67]. In a recently completed Phase I–II dose escalation study, funded by the National Institutes of Health and involving 11 medical centers in the United States, Olivi *et al.*, found that doses ranging from 6.5 per cent to 20 per cent loaded into PCPP:SA polymers were well tolerated without significant adverse effects [68]. This Phase I–II escalation study established that the maximal non-toxic loading dose is 20 per cent [65]. A phase III efficacy trial is currently underway with PCPP:SA wafers loaded with this highest-tolerated dose.

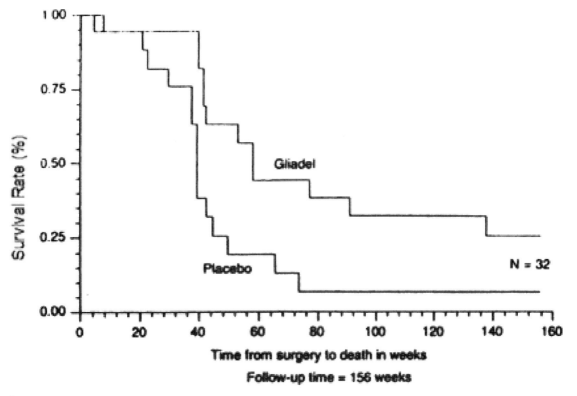

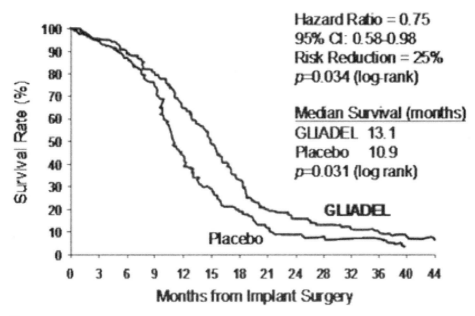

FIGURE 19.6 (A) Kaplan-Meier survival curve for patients with initial therapy for Grade II and Grade IV gliomas treated with BCNU-loaded polymer implants versus placebo. (Reprinted with permission from Lippincott Williams & Wilkins) [65]. (B) Kaplan-Meier survival curve depicting the results of the Valtonen and Westphal Phase III studies evaluating the local delivery of Gliadel® (3.8 per cent BCNU polymer) in patients with high grade malignant glioma.

Another strategy involves combining Gliadel® with other therapeutics to potentiate its anti-neoplastic effects. One such agent, O6-benzylguanine (O6-BG), inhibits the DNA repair enzyme O6-alkylguanine-DNA alkyltransferase (AGT). This enzyme has been reported to be responsible for the resistance of malignant glioma cell to BCNU by repairing carmustine-induced DNA damage [69]. O6-BG has been shown to be effective in animal models when co-administered with combination regimens such as BCNU/cyclophosphamide and temozolomide/irinotecan [70]. In another study by Schold et al., they found that AGT activity was decreased to less than detectable levels after a dose of 120 mg/m^2 when O6-BG was given 6 h prior to administration of an alkylating agent. With these results, a phase II trial was performed that combined systemic administration of O6-BG with systemic BCNU. But the

TABLE 19.3   Recently Completed Clinical Trials Using Controlled Release Polymers

Protocol	Phase	Indication	Number of patients	(Reference)
Gliadel® and i.v. O^6-BG	I	Recurrent	31	Rosenblum [125]
Gliadel®, Carboplatin, and Radiotherapy	I	Recurrent	13	Limentani [126,127]
Gliadel® and CPT-11	I	Recurrent	3	Colvin [128]
Gliadel® and CPT-11	I,II	Recurrent	10	Sungarian [129]
Gliadel®/Temodar	I/II	Recurrent	13	Rich [130]
Gliadel®/Temodar	II	Recurrent	10	Gururangan [131]
Gliadel® and I-125 Seeds	I	Recurrent	7	Albright [132]
Gliadel® and I-125 Seeds	II	Recurrent	34	Darakchiev [133]
Gliadel® and I-125 Seeds	II	New diagnosis	31	McPherson [134]
Gliadel® and CNS Metastasis	I/II	Metastatic tumors	10	Lowenthal [135]
Gliadel® and CNS Metastasis	II	Metastatic tumors	29	Brem [136]
Gliadel® with Aggressive Pituitary Adenomas and Craniopharyngiomas	I	Recurrent	10	Laws [137]
5-FU in PLGA polymer	I	Recurrent	10	Menei [138]
5-FU in PLGA polymer	II	Initial	95	Menei [88]
Gliadel® and Radiation Therapy	Retrospective Review	Initial	46	Kleinberg [139]

TABLE 19.4   Current Protocols for Brain Tumors Utilizing Gliadel®®

Protocol	Phase	PI, Site
Gliadel® and Irinotecan (CPT-11)	I	Friedman, Duke
Gliadel® for Low Grade Glioma	I	Friedman, Duke
Gliadel®, XRT, Penicillamine/Low Copper Diet	II	Brem, Moffit Cancer Center
Gliadel®, MR Spectroscopy	I	Pannullo, New York Presbyterian
Gliadel® and O^{6}BG in Pediatric Patients	I	Friedman, Duke
Gliadel® and i.v. Vincristine	I/II	Zeltzer, Cedars-Sinai
Gliadel® and CNS Metastasis	II	Asher, Carolina Neurosurgery
Gliadel® vs. IL-13/PE38QQR	III	PPD, Neopharm
Gliadel® and Temozolomide	II	Gururangan, Neuro-Oncol
Gliadel® with Gleevac	I/II	Sills, Memphis Regional

combination did not cause tumor regression, and was most likely due to the poor penetrance of systemic BCNU into the CNS. As such, systemic delivery of O6-BG could act in synergy with locally-delivered BCNU and potentiate the tumoricidal DNA damage produced by BCNU. This combination when administered in F98 rat glioma models resulted in a prolonged median survival compared to control (34 vs 25 days; $p = 0.0001$) [71]. These data indicate that simultaneous administration of O6-BG as a supplement to BCNU polymers may play an important role in the treatment of brain tumors resistant to standard chemotherapeutic modalities. Recently, a Phase I study demonstrated the safety of combining systemic delivery of O6-BG with local application of BCNU-PCPP:SA, [72] and further trials are currently in progress.

Gliadel® has also been used for the treatment of non-glial tumors. A pilot study has been initiated with BCNU-PCPP:SA for aggressive and recurrent pituitary adenomas and craniopharyngiomas [73]. A phase I study by Laws et al., concluded that Gliadel® implantation might have a role in patients with recurring aggressive pituitary adenomas and craniopharyngiomas. Similarly, for metastatic brain tumors, studies indicate that Gliadel® combined with post-operative systemic chemotherapy may prolong survival when compared to standard therapies alone. BCNU-PCPP:SA has been shown to prolong survival in animal models of metastatic malignancies, including lung carcinoma, renal cell carcinoma, colon carcinoma, and melanoma. Subsequent clinical studies have similarly been encouraging [74], and larger scale clinical trials are anxiously awaiting.

The effectiveness of Gliadel® used in combination with other chemotherapeutic agents, including temozolomide, carboplatin, irinotecan, vincristine, and Gleevac, is being evaluated in multiple clinical studies (Tables 19.3 and 19.4). Data from these clinical trials are pending. In addition, there have been new indications for Gliadel®, including: Gliadel® + radiotherapy for metastasis, low-grade gliomas, pituitary adenomas, and Gliadel® + O6-BG in pediatric patients (Tables 19.5). Even with such advances, BCNU is not the end point of local delivery, but rather the stepping-stone for other chemotherapies. The advantage of local delivery is the ability to load a variety of antineoplastics onto the polymer and deliver it to the tumor sites. In the next section, we will look at other therapeutics that are locally delivered to intracranial malignant gliomas, both in the clinical setting, and in the laboratory.

TABLE 19.5   New Indications for Gliadel®

Protocol	Phase	Indication	PI, Site
Metastasis: Gliadel® and Radiotherapy	II	Initial	Ewend, UNC
Low Grade Glioma	II	Recurrent	Friedman, Duke
Aggressive Pituitary Adenomas and Craniopharyngiomas: Gliadel®	I	Recurrent	Laws [137]
Gliadel® and O⁶BG in Pediatric Patients	I	Open	Friedman, Duke

OTHER CHEMOTHERAPEUTICS

Clinical Uses of Other Locally Delivered Chemotherapeutic Agents

Establishment of the clinical benefits of locally delivered BCNU has led to clinical trials investigating the local sustained delivery of other agents, including Taxol and 5-fluorouracil. Taxol is a naturally occurring microtubule stabilizer, that has already been FDA approved for the treatment of several human tumors. Although it has been shown to have activity against glioma cell lines *in vitro* [75], it has not been successful in patients when given systemically because it did not reach tumoricidal activity in the brain [76–78]. Implantation of a Taxol-loaded biodegradable polymer resulted in significant prolongation in survival in 9L glioma model [79], with therapeutic concentrations persisting up to 30 days after implantation [79]. Studies in the 9L gliosarcoma rat model showed that locally delivered taxol promoted survival two- to three-fold (38 days median survival with 40 per cent taxol and 61.5 days with 20 per cent taxol vs 19.5 days with placebo). However, taxol/PCPP:SA polymers showed a biphasic release profile that resulted in toxicity. Instead, taxol was loaded into a biodegradable polilactofate microspheres (Paclimer® delivery system, Guilford pharmaceutics, Baltimore, MD), and toxicity and efficacy studies were performed in a 9L gliosarcoma model [80]. The safety of implantable taxol-loaded microspheres has been established in a canine brain tumor model (unpublished data). Taxol has been delivered locally using convection enhanced delivery, and while a high anti-tumor response was obtained, significant complications were associated with the treatment [33].

Another established chemotherapeutic currently being assessed for its use in local delivery is 5-FU. This pyrimidine analog blocks the conversion of deoxyuridylic acid to thymidylic acid, a critical precursor for DNA synthesis. 5-FU has been effective in various cancers in the GI tract. However, its use in brain tumors has been limited by its systemic toxicity (e.g., myelosuppression) and its poor penetration of the BBB [81]. To bypass these limitations, 5-FU has been incorporated in a variety of polymer systems, including a matrix of glassified monomers and a polylactic acid-co-glycolic acid (PLGA) microsphere [82]. Implantation of 5-FU-loaded PLGA microspheres in both C6 rat glioma models [83] and F98 glioma models [84–86] resulted in decreased mortality without inducing any neurological toxicities. A small clinical trial by Menei *et al.*, showed that implantation of 5-FU loaded PLGA microspheres combined with surgical debulking and external beam therapy prolonged median survival to 98 weeks in 8 patients with newly diagnosed malignant gliomas [87]. Similar results were obtained following stereotactic placement of 5-FU PLGA microspheres in 10 patients with malignant gliomas (Table 19.6) [83]. A phase III study, however, failed to find a statistically significant prolongation of survival under the parameters tested [88].

Intracranial Local Delivery of Experimental Drugs in Rat Glioma Models (Table 19.7)

Advances in local drug delivery have stimulated interest in introducing a variety of chemotherapeutics into biodegradable polymers and observing their effectiveness in the treatment of malignant gliomas.

TABLE 19.6 New Polymer-Drug Combinations

Protocol	Phase	Indication/ Site	Number of patients	Site
BCNU Dose Escalation of 4%–28%	I	Recurrent Brain	44	Olivi [140]
5-FU in PLGA polymer	I	Recurrent Brain	10	Menei [138]
Taxol (10%) in Polyphosphoester	I	Ovarian, Lung	25	Armstrong, Hopkins, GOG, NCI

Many single agent therapeutics have been studied in the laboratory (see Table 19.7), and have demonstrated a proof of principle that local delivery has survival advantages compared to systemic therapy for experimental brain tumors.

Camptothecin, an inhibitor of the DNA-replicating enzyme topoisomerase I, initially appeared as a potent chemotherapeutic to combat cancer. Systemic delivery of the drug in clinical trial demonstrated significant toxicity, [89] preventing its use in the treatment of gliomas. To investigate the local polymeric delivery of camptothecin, sodium camptothecin was implanted into PCPP:SA polymer and tested against intracranial 9L gliosarcoma in rats [90]. In addition to causing no local or systemic toxicity, the median survival of rats implanted with 50 per cent loaded Na-camptothecin polymers were four times greater than control and three times greater than 3.8 per cent loaded BCNU polymers. New derivatives of camptothecin are also available, and data regarding their local delivery is under investigation. Clearly, the release, toxicity, biodistribution, and efficacy profile of camptothecin suggests that its local controlled delivery is a promising option in the treatment of malignant gliomas.

Mitoxantrone is both an intercalating agent and a potent inhibitor of topoisomerase II, the enzyme responsible for uncoiling and repairing damaged DNA. Its derivative, dihydroxyantracenedione, has already been used in the treatment of many malignancies, including advanced breast cancer, non-Hodgkins lymphoma, and ovarian cancer [91,92]. It has had similar cytotoxicity in glioma cell lines *in vitro* [93,94], but systemic delivery has failed to demonstrate efficacy against malignant gliomas because of its poor penetration to the CNS and dose-limiting myelosuppression. To bypass the BBB and side effects, mitoxantrone was incorporated into the PCPP:SA polymer. Subsequent efficacy studies with 10 per cent mitoxantrone-loaded polymer indicated that median survival was 2.5 times that of control rats [95]. Local delivery through Ommaya reservoirs have shown survival benefit with mitoxantrone in the treatment of malignant gliomas [96]. Thus, mitoxantrone delivered locally in a sustained release polymer may prove to be a useful option for the treatment of malignant gliomas.

Another group of drugs that could harness the technology of local delivery is anti-angiogenic agents. Such drugs include heparin, cortisone, minocycline, squalamine, and endostatin. Minocycline, a broad-spectrum antibiotic with anticollagenase activity, has been shown to inhibit induced neovascularization

TABLE 19.7 Summary of Intracranial Local Drug Delivery in Rat Glioma Models

Compound	Loading dose (%)	Polymer system*	Control	Treated	% increase survival	p-value	Ref.
Camptothecin	20	P	17	25	147	0.023	[141]
	50			69	406	<0.001	
Riluzole	10	P	13	18	138	0.028	[142]
SJG-136	1	P	17	58	341	0.0081	[143]
M4N	40	P	14	22	157	<0.02	[144]
IL-2	N/A	Cell	25	54	216	<0.05	[145]
IL-2	2.6	G	25	48	192	<0.05	[145]
IL-12	N/A	Cell	23	37	161	0.0002	[146]
IL-12	0.5 μg	G	22	43	195	<0.0016	[146]
	1.5 μg	G	22	41	186	<0.0002	
IL-12	5 μg	G	37	54	146	0.0025	[147]
Lactacystin	1	P	13	22	169	0.002	[148]
Lactacystin	1.5	P	11	80	727	0.0002	[149]
			11	28	255		
Minocycline	50	P	21	>200	>950	0.0001	[150]
Minocycline	50	E	15	30	200	0.0003	[151]
Chemically Modified Tetracycline-3	50	E	15	22	147	0.0004	[151]
Chemically Modified Tetracycline-8	50	E	15	>120	>800	0.0001	[151]
Chemically Modified Tetracycline-303	50	E	15	18	120	0.0473	[151]
Mitoxantrone	1	P	19	30	158	<0.0001	[152]
	5	P	19	34	179	<0.0001	
	10	P	19	50	263	<0.0001	
Paclitaxel	10	PPE	16	35	219	<0.0001	[153]
Temozolomide	50	P	11	23	209	<0.001	[154]
	40 mg/kg, D5–10	Oral	11	16	145	<0.0001	

in the rabbit cornea angiogenesis assay. Furthermore, laboratory data also indicate minocycline loaded polymers inhibit growth of glioma tumors in the flank and intracranially [97]. More recently, there have been reports of synergistic activity with systemic delivery of BCNU [97,98]. Squalamine, an inhibitor of proliferation and migration of endothelial cells (as demonstrated in a rat corneal assay [99]), is undergoing Phase II clinical trial evaluating its efficacy against advanced cancer when administered systemically [100]. If successful, local delivery of this drug could again bypass systemic toxicity and be used directly to treatment of gliomas. Finally, endostatin, an inhibitor of matrix metalloproteinase-2, also has anti-angiogenic properties [101–104]. It has been shown to improve survival in malignant glioma animal models, [105] and analogs have been synthesized and are currently being tested.

Cytokines and Immunotherapy

Local delivery of cytokines has the potential to offer another powerful therapeutic approach to treat malignant brain tumors. The reasoning behind using cytokines for the treatment of gliomas is based on their ability to initiate a strong inflammatory response, thereby enhancing the activation of tumor-specific lymphocytes. Two mechanisms of delivering specific cytokines were used: cells genetically engineered to produce cytokines and polymeric delivery.

The first strategy involves *ex vivo* gene transfer. B16/F10 melanoma cells were transfected with a cytokine gene and then injected intracranially into mice. IL-2 showed the most dramatic increases in survival when compared to IL-4 and GM-CSF, and emerged as the most promising candidate for use in paracrine immunotherapy [106]. IL-12, an antineoplastic cytokine with anti-angiogenic properties, has also been shown to increase median survival in rat glioma models when stereotactically injected with 9L gliosarcoma cells genetically engineered to express IL-12 [107]. Furthermore, Sampath et al., showed the efficacy of tumor cell engineered to produce IL-2 and local delivery of 10 per cent BCNU-PCPP:SA polymer in producing a synergistic increase in survival of mice intracranially challenged with B16-F10 tumor cells [108]. While experimentally efficacious in animal models, the technique of genetically engineered cells is much more labor-intensive and costly than polymeric delivery. As such, attention was turned to developing controlled release technology to accomplish paracrine cytokine production.

Consequently, an injectable polymeric microsphere system composed of gelatin and chondroin sulfate capable of releasing IL-2 in a sustained fashion has been developed [109]. These microspheres have been optimitized for size reproducibility, encapsulation efficiency, and cytokine release (over 2 weeks *in vitro*). Experiments in the 9L gliosarcoma model indicated that IL-2 microspheres were efficacious in the treatment of B16 melanoma metastases to the brain [109]. Also, recent data from the laboratory indicate a synergistic effect of local delivery of BCNU wafers in combination with IL-2 microspheres in treating 9L gliosarcoma in rats [110]. This study showed a median survival of animals receiving monotherapy was 24 days with IL-2 microspheres, 24 days with 3.8 per cent BCNU polymer, and 32.5 days with 10 per cent BCNU polymer. Research continues in the exciting field of local delivery of immunotherapy.

FUTURE DIRECTIONS

Microchip Drug Delivery

A new and potentially powerful method of drug delivery for more complex drug delivery approaches involves microchip technology [111, 112]. These solid-state silicon microchips provide controlled release of multiple microreservoirs, filled with solids, liquids, or gels and are then enveloped with an anode membrane (Figs. 19.7A,B). Dissolution of the membrane can be engineered to deliver drugs in a pulsatile fashion, and the independent release of each reservoir allows for an endless array of release profiles and therapy combinations. A microbattery, multiplexing circuitry, and memory can be installed into the device. Alternatively, a biodegradable polymeric microchip has been developed, in which the release mechanism is based on slow degradation of

a thin polymeric membrane covering each reservoir of anti-neoplastic agent. In preliminary laboratory studies with heparin, data indicated that heparin released over 142 days maintained approximately 96 per cent of its bioactivity [113]. These microchips were made with poly (L-lactic acid) and had poly (D,L-lactic-co-glycolic acid) membranes of different molecular masses on top of each reservoir. Potentially, 1000 different drug dosages or drugs can be delivered on demand, and this microchip technology makes it one of the most exciting developments in local drug delivery in the treatment of brain tumors.

Nanocarriers

As will be described in another chapter in this book, gene and viral therapies are innovative approaches to deliver chemotherapy (see Chapter 23). Unfortunately, these approaches have often been limited by the unsuccessful delivery of a gene to the nucleus, where it needs to be in order to ultimately express its gene product. Nanocarriers are synthetic gene carriers that allow the delivery of large DNA molecules efficiently into the nucleus. An example of a nanocarrier is polyethylenimine (PEI)DNA nano-complex, a synthetic vector capable of efficiently delivering DNA sequences to perinuclear regions *via* microtubule mediated mechanisms after being taken up through endocytosis by the target cell [114–117]. As shown by Sakhalkar et al., with adhesion molecules and inflamed respiratory endothelium, nano-carriers could also be targeted to cell surface receptors and allow for the delivery of genomic sequences to biologically modify neoplastic cells. Further studies are currently underway.

CONCLUSIONS

Treatment of malignant gliomas has proven to be very challenging. As discussed, difficulties with systemic chemotherapy and the high local recurrence rate of primary tumors have effectively driven the need for interstitial chemotherapy and polymer drug delivery. Biodegradable polymers can provide controlled, prolonged local delivery of anti-neoplastics, and these polymers allow for highly customizable release times and multiple formulation modalities. In addition to the ongoing clinical trials, exciting new data are emerging from the laboratory involving various therapeutics, combination therapeutics, and innovative delivery modalities to decrease mortality caused by malignant brain tumors.

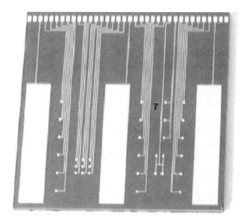
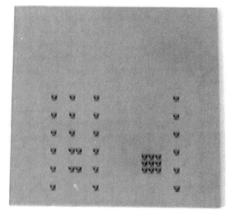

A

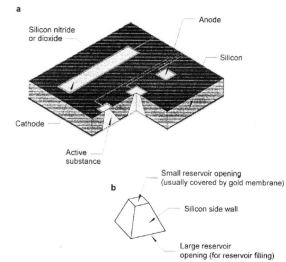

B

FIGURE 19.7 **(A)** Microchip on a quarter and compared to a pencil tip (photo by Carita Stubbe, courtesy of MicroCHIPS, inc.). The dots between the three large bars (cathodes) on the front are the caps (anodes) covering the reservoirs holding the chemicals. Electrical voltage applied between the cap and cathode causes a reaction that dissolves the cap, thus releasing the reservoir's contents. The back view shows the larger opening through which the content of the reservoirs are deposited. (Reprinted with permission) [119]. **(B)** Schematic of the passive microchip. Initial models will use PLGA and other existing polymer matrices for the substrate. The entire chip will be biodegradable. (Reprinted with permission) [119]. See Plate 19.7 in Color Plate Section.

DISCLOSURE

Under separate licensing agreements between the Johns Hopkins University and Guilford Pharmaceuticals, Inc. and the Johns Hopkins University and Angiotech Pharmaceuticals, Inc., Dr. Brem is entitled to a share of royalty received by the University on sales of products described in this work. Dr. Brem owns Guilford Pharmaceuticals, Inc. stock, which is subject to certain restrictions under University policy. Dr. Brem also is a paid consultant to Guilford Pharmaceuticals, Inc. The terms of this arrangement are being managed by the Johns Hopkins University in accordance with its conflict of interest policies.

References

1. Legler, J. M., Ries, L. A., Smith, M. A. et al. (1999). Cancer surveillance series [corrected]: Brain and other central nervous system cancers: Recent trends in incidence and mortality. *J Natl Cancer Inst* **91**, 1382–1390.
2. Guha, A., Dashner, K., Black, P. M., Wagner, J. A., Stiles C. D. (1995). Expression of pdgf and pdgf receptors in human astrocytoma operation specimens supports the existence of an autocrine loop. *Int J Cancer* **60**, 168–173.
3. Parker, S. L., Tong, T., Bolden S., Wingo, P. A. (1996). Cancer statistics, 1996. *CA Cancer J Clin* **46**, 5–27.
4. Black, P. M. (1991). Brain tumors. Part 1. *N Engl J Med* **324**, 1471–1476.
5. Lesniak, M. S., and Brem, H. (2004). Targeted therapy for brain tumours. *Nat Rev Drug Discov* **3**, 499–508.
6. Kusuhara, H., and Sugiyama, Y. (2001). Efflux transport systems for drugs at the blood-brain barrier and blood-cerebrospinal fluid barrier (part 2). *Drug Discov Today* **6**, 206–212.
7. Rautioa, J., and Chikhale, P. J. (2004). Drug delivery systems for brain tumor therapy. *Curr Pharm Des* **10**, 1341–1353.
8. Yuan, F., Salehi, H. A., Boucher, Y. et al. (1994). Vascular permeability and microcirculation of gliomas and mammary carcinomas transplanted in rat and mouse cranial windows. *Cancer Res* **54**, 4564–4568.
9. Jain, R. K. (1994). Barriers to drug delivery in solid tumors. *Sci Am* **271**, 58–65.
10. Rockwell, S. (1992). Use of hypoxia-directed drugs in the therapy of solid tumors. *Semin Oncol* **19**, 29–40.
11. Chin, H. W., Young, A. B., Maruyama, Y. (1981). Survival response of malignant gliomas to radiotherapy with or without bcnu or methyl-ccnu chemotherapy at the university of kentucky medical center. *Cancer Treat Rep* **65**, 45–51.
12. Kornblith, P. L., and Walker, M. (1988). Chemotherapy for malignant gliomas. *J Neurosurg* **68**, 1–17.
13. Medical Research Council Brain Tumor Working Party (2001). Randomized trial of procarbazine, lomustine, and vincristine in the adjuvant treatment of high-grade astrocytoma: A medical research council trial. *J Clin Oncol* **19**, 509–518.
14. Prokai, L., Prokai-Tatrai, K., and Bodor, N. (2000). Targeting drugs to the brain by redox chemical delivery systems. *Med Res Rev* **20**, 367–416.
15. Kurihara, A., and Pardridge, W. M. (1999). Imaging brain tumors by targeting peptide radiopharmaceuticals through the blood-brain barrier. *Cancer Res* **59**, 6159–6163.
16. Williams, P. C., Henner, W. D., Roman-Goldstein, S. et al. (1995). Toxicity and efficacy of carboplatin and etoposide in conjunction with disruption of the blood-brain barrier in the treatment of intracranial neoplasms. *Neurosurgery* **37**, 17–28.
17. Rapoport, S. I. (2001). Advances in osmotic opening of the blood-brain barrier to enhance cns chemotherapy. *Expert Opin Investig Drugs* **10**, 1809–1818.
18. Kobrinsky, N. L., Packer, R. J., Boyett J. M. et al. (1999). Etoposide with or without mannitol for the treatment of recurrent or primarily unresponsive brain tumors: A children's cancer group study, ccg-9881. *J Neurooncol* **45**, 47–54.
19. Prados, M. D., Schold, S. J. S., Fine, H. A. et al. (2003). A randomized, double-blind, placebo-controlled, phase 2 study of rmp-7 in combination with carboplatin administered intravenously for the treatment of recurrent malignant glioma. *Neuro-Oncol* **5**, 96–103.
20. Boiardi, A., Eoli, M., Pozzi, A. et al. (1999). Locally delivered chemotherapy and repeated surgery can improve survival in glioblastoma patients. *Ital J Neurol Sci* **20**, 43–48.
21. Morantz, R. A., Kimler, B. F., Vats, T. S., Henderson, S. D. (1983). Bleomycin and brain tumors. A review. *J Neurooncol* **1**, 249–255.
22. Patchell, R. A., Regine, W. F., Ashton, P. et al. (2002). A phase I trial of continuously infused intratumoral bleomycin for the treatment of recurrent glioblastoma multiforme. *J Neurooncol* **60**, 37–42.
23. Voulgaris, S., Partheni, M., Karamouzis, M. et al. (2002). Intratumoral doxorubicin in patients with malignant brain gliomas. *Am J Clin Oncol* **25**, 60–64.
24. Walter, K. A., Tamargo, R. J., Olivi, A., Buger, P. C., Brem, H. (1995). Intratumoral chemotherapy. *Neurosurgery* **37**, 1128–1145.
25. Huang, Y., Hayes, R. L., Wertheim, S., Arbit E., and Scheff, R. (2001). Treatment of refractory recurrent malignant glioma with adoptive cellular immunotherapy: A case report. *Crit Rev Oncol Hematol* **39**, 17–23.
26. Boiardi, A., Silvani, A., Milanesi, I. et al. (1991). Local immunotherapy (beta-ifn) and systemic chemotherapy in primary glial tumors. *Ital J Neurol Sci* **12**, 163–168.
27. Giussani, C., Carrabba, G., Pluderi, M. et al. (2003). Local intracerebral delivery of endogenous inhibitors by osmotic minipumps effectively suppresses glioma growth in vivo. *Cancer Res* **63**, 2499–2505.
28. Bobo, R. H., Laske, D. W., Akbasak, A. et al. (1994). Convection-enhanced delivery of macromolecules in the brain. *Proc Natl Acad Sci U S A* **91**, 2076–2080.
29. Chen, M. Y., Lonser, R. R., Morrison, P. F., Governale, L. S., Oldfield, E. H. (1999). Variables affecting convection-enhanced delivery to the striatum: A systematic examination of rate of infusion, cannula size, infusate concentration, and tissue-cannula sealing time. *J Neurosurg* **90**, 315–320.
30. Degen, J. W., Walbridge, S., Vortmeyer, A. O. Oldfield, E. H., Lonser, R. R. (2003). Safety and efficacy of convection-enhanced delivery of gemcitabine or carboplatin in a malignant glioma model in rats. *J Neurosurg* **99**, 893–898.
31. Husain, S. R., Joshi, B. H., Puri, R. K. et al. (2001). Interleukin-13 receptor as a unique target for anti-glioblastoma therapy. *Int J Cancer* **92**, 168–175.
32. Mardor, Y., Roth, Y., Lidar, Z. et al. (2001). Monitoring response to convection-enhanced taxol delivery in brain tumor patients using diffusion-weighted magnetic resonance imaging. *Cancer Res* **61**, 4971–4973.

33. Lidar, Z., Mardor, Y., Jonas, T. *et al.* (2004). Convection-enhanced delivery of paclitaxel for the treatment of recurrent malignant glioma: A phase I/II clinical study. *J Neurosurg* **100**, 472–479.

34. Langer, R., and Folkman, J. (1976). Polymers for the sustained release of proteins and other macromolecules. *Nature* **263**, 797–800.

35. Langer, R. S., and Wise, D. L. (1984). *Medical applications of controlled release.* CRC Press, Boca Raton, Fla.

36. Lewis, D. H. (1990). Controlled release of bioactive agents from lactide/glycolide polymers. In *Biodegradable polymers as drug delivery systems* (M. Chasin, R. and Langer, Eds.), pp. 1–41, Marcel Dekker, New York, NY.

37. Burt, H. M., Jackson, J. K., Bains, S. K. *et al.* (1995). Controlled delivery of taxol from microspheres composed of a blend of ethylene-vinyl acetate copolymer and poly (d,l-lactic acid). *Cancer Lett* **88**, 73–79.

38. Sato, H., Wang, Y. M., Adachi, I., Horikoshi, I. (1996). Pharmacokinetic study of taxol-loaded poly(lactic-co-glycolic acid) microspheres containing isopropyl myristate after targeted delivery to the lung in mice. *Biol Pharm Bull* **19**, 1596–1601.

39. Peyman, G. A., Conway, M., Khoobehi, B. Soike, K. (1992). Clearance of microsphere-entrapped 5-fluorouracil and cytosine arabinoside from the vitreous of primates. *Int Ophthalmol* **16**, 109–113.

40. Torres, A. I., Boisdron-Celle, M., and Benoit, J. P. (1996). Formulation of bcnu-loaded microspheres: Influence of drug stability and solubility on the design of the microencapsulation procedure. *J Microencapsul* **13**, 41–51.

41. Williams, R. C., Paquette, D. W., Offenbacher, S. *et al.* (2001). Treatment of periodontitis by local administration of minocycline microspheres: A controlled trial. *J Periodontol* **72**, 1535–1544.

42. Lamprecht, A., Ubrich, N., Yamamoto, H. *et al.* (2001). Biodegradable nanoparticles for targeted drug delivery in treatment of inflammatory bowel disease. *J Pharmacol Exp Ther* **299**, 775–781.

43. Lee, M., Browneller, R., Wu, Z. *et al.* (1997). Therapeutic effects of leuprorelin microspheres in prostate cancer. *Adv Drug Deliv Rev* **28**, 121–138.

44. Brem, H., and Langer, R. (1996). Polymer-based drug delivery to the brain. *Sci Med* July/August, 52–61.

45. Leong, K. W., Brott, B. C., and Langer, R. (1985). Bioerodible polyanhydrides as drug-carrier matrices. I: Characterization, degradation, and release characteristics. *J Biomed Mater Res* **19**, 941–955.

46. Chasin, M., Domb, A., and Ron, E. (1990). Polyanhydrides as drug delivery systems. In *Biodegradable polymers as drug delivery systems* (M. Chasin, and R. Langer Eds.), pp. 43–70. Marcel Dekker, New York, NY.

47. Mathiowitz, E., and Langer, R. (1987). Polyanhydride microspheres as drug carriers. I. Hot-melt microencapsulation. *J Control Rel* **5**, 13–22.

48. Mathiowitz, E., Saltzman, M., and Domb, A. (1988). Polyanhydride microspheres as drug carriers. Ii. Microencapsulation by solvent removal. *J Appl Polym Sci* **35**, 755–774.

49. Bindschaedler, C., Leong, K., Mathiowitz, E. Langer, R. (1988). Polyanhydride microsphere formulation by solvent extraction. *J Pharm Sci* **77**, 696–698.

50. Howard, M. A., III, Gross, A., Grady, M. S. *et al.* (1989). Intracerebral drug delivery in rats with lesion-induced memory deficits. *J Neurosurg* **71**, 105–112.

51. Mathiowitz, E., Kline, D., and Langer, R. (1990). Morphology of polyanhydride microsphere delivery systems. *Scanning Microsc* **4**, 329–340.

52. Langer, R., Brem, H., and Tapper, D. (1981). Biocompatibility of polymeric delivery systems for macromolecules. *J Biomed Mater Res* **15**, 267–277.

53. Brem, H., Kader, A., Epstein, J. I. *et al.* (1989). Biocompatibility of a biodegradable, controlled-release polymer in the rabbit brain. *Sel Cancer Ther* **5**, 55–65.

54. Tamargo, R. J., Epstein, J. I., Reinhard, C. S. Chasin, M., Brem, H. (1989). Brain biocompatibility of a biodegradable, controlled-release polymer in rats. *J Biomed Mater Res* **23**, 253–266.

55. Shieh, L., Tamada, J., Chen, I. *et al.* (1994). Erosion of a new family of biodegradable polyanhydrides. *J Biomed Mater Res* **28**, 1465–1475.

56. Golumbek, P. T., Azhari, R., Jaffee, E. M. *et al.* (1993). Controlled release, biodegradable cytokine depots: A new approach in cancer vaccine design. *Cancer Res* **53**, 5841–5844.

57. Gabizon, A. A. (1994). Liposomal anthracyclines. *Hematol Oncol Clin North Am* **8**, 431–450.

58. Loo, T. L., Dion, R. L., Dixon, R. L., Rall, D. P. (1966). The antitumor agent, 1,3-bis(2-chloroethyl)-1-nitrosourea. *J Pharm Sci* **55**, 492–497.

59. Grossman, S. A., Reinhard, C., Colvin, O. M. *et al.* (1992). The intracerebral distribution of bcnu delivered by surgically implanted biodegradable polymers. *J Neurosurg* **76**, 640–647.

60. Fung, L. K., Ewend, M. G., Sills, A. *et al.* (1998). Pharmacokinetics of interstitial delivery of carmustine, 4-hydroperoxycyclophosphamide, and paclitaxel from a biodegradable polymer implant in the monkey brain. *Cancer Res* **58**, 672–684.

61. Tamargo, R. J., Myseros, J. S., Epstein, J. I. *et al.* (1993). Interstitial chemotherapy of the 9l gliosarcoma: Controlled release polymers for drug delivery in the brain. *Cancer Res* **53**, 329–333.

62. Brem, H., Tamargo, R. J., Olivi, A. *et al.* (1994). Biodegradable polymers for controlled delivery of chemotherapy with and without radiation therapy in the monkey brain. *J Neurosurg* **80**, 283–290.

63. Brem, H., Mahaley, M. S., Jr, Vick, N. A. *et al.* (1991). Interstitial chemotherapy with drug polymer implants for the treatment of recurrent gliomas. *J Neurosurg* **74**, 441–446.

64. Brem, H., Piantadosi, S., Burger, P. C. *et al.* (1995). Placebo-controlled trial of safety and efficacy of intraoperative controlled delivery by biodegradable polymers of chemotherapy for recurrent gliomas. The polymer-brain tumour treatment group. *Lancet* **345**, 1008–1012.

65. Valtonen, S., Timonen, U., Toivanen, P. *et al.* (1997). Interstitial chemotherapy with carmustine-loaded polymers for high-grade gliomas: A randomized double-blind study. *Neurosurgery* **41**, 44–48.

66. Westphal, M., Hilt, D. C., Bortey, E. *et al.* (2003). A phase 3 trial of local chemotherapy with biodegradable carmustine (bcnu) wafers (gliadel wafers) in patients with primary malignant glioma. *Neuro-Oncol* **5**, 79–88

67. Sipos, E. P., Tyler, B., Piantadosi, S. Burger, P. C., Brem, H. (1997). Optimizing interstitial delivery of bcnu from controlled release polymers for the treatment of brain tumors. *Cancer Chemother Pharmacol* **39**, 383–389.

68. Olivi, A., Grossman, S. A., Tatter, S. *et al.* (2003). Dose escalation of carmustine in surgically implanted polymers in patients with recurrent malignant glioma: A new approaches to brain tumour therapy CNS consortium trial. *J Clin Oncol* **21**, 1845–1849.

69. Silber, J. R., Bobola, M. S., Ghatan, S. *et al.* (1998). O^6-methylguanine-DNA methyltransferase activity in adult gliomas: Relation to patient and tumour characteristics. *Cancer Res* **58**, 1068–1073.

70. Friedman, H. S., Keir, S., Pegg, A. E. *et al.* (2002). O^6-benzylguanine-mediated enhancement of chemotherapy. *Mol Cancer Ther* **1**, 943–948.

71. Rhines, L. D., Sampath, P., Dolan, M. E. *et al.* (2000). O^6-benzylguanine potentiates the antitumor effect of locally delivered carmustine against an intracranial rat glioma. *Cancer Res* **60**, 6307–6310.

72. Rosenblum, M., Weingart, J., Dolan, M. E. *et al.* (2002). Phase I study of gliadel combined with a continuous intravenous infusion of o6-benzylguanine in patients with recurrent malignant glioma. *Proc American Soc Clin Oncologists* **21**.

73. Lee, J. H., Chai, Y. G., and Hersh, L. B. (2000). Expression patterns of mouse repressor element-1 silencing transcription factor 4 (rest4) and its possible function in neuroblastoma. *J Mol Neurosci* **15**, 205–214.

74. Ewend, M. G., Brem, S., Gilber, M., Goodkin, R. and Penar, P. (2001) Treating single brian metastasis with resection, placement of bcnu-polymer wafers, and radiation therapy. *Am Assoc Neurol Surgeons*. Abstract in CD.

75. Cahan, M. A., Walter, K. A., Colvin, O. M. *et al.* (1994). Cytotoxicity of taxol in vitro against human and rat malignant brain tumors. *Cancer Chemother Pharmacol* **33**, 441–444.

76. Forsyth, P., Cairncross, G., Stewart, D. *et al.* (1996). Phase II trial of docetaxel in patients with recurrent malignant glioma: A study of the national cancer institute of canada clinical trials group. *Invest New Drugs* **14**, 203–206.

77. Freilich, R. J., Seidman, A. D., DeAngelis, L. M. (1995). Central nervous system progression of metastatic breast cancer in patients treated with paclitaxel. *Cancer* **76**, 232–236.

78. Glantz, M. J., Chamberlain, M. C., Chang, S. M., Prados, M. D., Cole, B. F. (1999). The role of paclitaxel in the treatment of primary and metastatic brain tumors. *Semin Radiat Oncol* **9**, 27–33.

79. Menei, P., Benoit, J. P., Boisdron-Celle, M., *et al.* (1994). Drug targeting into the central nervous system by stereotactic implantation of biodegradable microspheres. *Neurosurgery* **34**, 1058–1064.

80. Li, K. W., Dang, W., Tyler, B. M. *et al.* (2003). Polilactofate microspheres for paclitaxel delivery to central nervous system malignancies. *Clin Cancer Res* **9**, 3441–3447.

81. Goodman, L. S., Hardman, J. G., Limbird, L. E., Gilman, A. G., Goodman and Gilman's: (2001). *The pharmacological basis of therapeutics*. McGraw-Hill, New York.

82. Boisdron-Celle, M., Menei, P., and Benoit, J. P. (1995). Preparation and characterization of 5-fluorouracil-loaded microparticles as biodegradable anticancer drug carriers. *J Pharm Pharmacol* **47**, 108–114.

83. Menei, P., Boisdron-Celle, M., Croue, A. Guy, G., Benoit, J. P. (1996). Effect of stereotactic implantation of biodegradable 5-fluorouracil-loaded microspheres in healthy and c6 glioma-bearing rats. *Neurosurgery* **39**, 117–124.

84. Lemaire, L., Roullin, V. G., Franconi, F. *et al.* (2001). Therapeutic efficacy of 5-fluorouracil-loaded microspheres on rat glioma: A magnetic resonance imaging study. *NMR Biomed* **14**, 360–366.

85. Fournier, E., Passirani, C., Montero-Menei, C, *et al.* (2003). Therapeutic effectiveness of novel 5-fluorouracil-loaded poly(methylidene malonate 2.1.2)-based microspheres on f98 glioma-bearing rats. *Cancer* **97**, 2822–2829.

86. Fournier, E., Passirani, C., Vonarbourg, A. *et al.* (2003). Therapeutic efficacy study of novel 5-fu-loaded pmm 2.1.2-based microspheres on c6 glioma. *Int J Pharm* **268**, 31–35.

87. Menei, P., Venier, M. C., Gamelin, E. *et al.* (1999). Local and sustained delivery of 5-fluorouracil from biodegradable microspheres for the radiosensitization of glioblastoma: A pilot study. *Cancer* **86**, 325–330.

88. Menei, P., Capelle, L., Guyotat, J. *et al.* (2005). Local and sustained delivery of 5-fluorouracil from biodegradable microspheres for the radiosensitization of malignant glioma: A randomized phase II trial. *Neurosurgery* **56**, 242–248; discussion 242–248.

89. Slichenmyer, W. J., Rowinsky, E. K., Donehower, R. C., Kaufmann, S. H. (1993). The current status of camptothecin analogues as antitumor agents. *J Natl Cancer Inst* **85**, 271–291.

90. Storm, P. B., Moriarity, J. L., Tyler, B. *et al.* (2002). Polymer delivery of camptothecin against 9l gliosarcoma: Release, distribution, and efficacy. *J Neurooncol* **56**, 209–217.

91. Facoetti, A., Capelli, E., and Nano, R. (2001). HLA class I molecules expression: Evaluation of different immunocytochemical methods in malignant lesions. *Anticancer Res* **21**, 2435–2440.

92. Ehninger, G., Schuler, U., Proksch, B., Zeller, K. P., Blanz, J. (1990). Pharmacokinetics and metabolism of mitoxantrone. A review. *Clin Pharmacokinet* **18**, 365–380.

93. Wolff, J. E., Trilling, T., Molenkamp, G., Egeler, R. M., Jurgens, H. (1999). Chemosensitivity of glioma cells in vitro: A meta analysis. *J Cancer Res Clin Oncol* **125**, 481–486.

94. Senkal, M., Tonn, J. C., Schonmayr, R. *et al.* (1997). Mitoxantrone-induced DNA strand breaks in cell-cultures of malignant human astrocytoma and glioblastoma tumors. *J Neurooncol* **32**, 203–208.

95. DiMeco, F., Li, K. W., Tyler, B. M. *et al.* (2002). Local delivery of mitoxantrone for the treatment of malignant brain tumors in rats. *J Neurosurg* **97**, 1173–1178.

96. Boiardi, A., Eoli, M., Salmaggi, A. *et al.* (2001). Efficacy of intratumoral delivery of mitoxantrone in recurrent malignant glial tumours. *J Neurooncol* **54**, 39–47.

97. Weingart, J. D., Sipos, E. P., and Brem, H. (1995). The role of minocycline in the treatment of intracranial 9l glioma. *J Neurosurg* **82**, 635–640.

98. Frazier, J. L., Wang, P. P., Case, D. *et al.* (2003). Local delivery of minocycline and systemic bcnu have synergistic activity in the treatment of intracranial glioma. *J Neurooncol* **64**, 203–209.

99. Sills, A. K., Jr, Williams, J. I., Tyler, B. M. *et al.* (1998). Squalamine inhibits angiogenesis and solid tumor growth in vivo and perturbs embryonic vasculature. *Cancer Res* **58**, 2784–2792.

100. Bhargava, P., Marshall, J. L., Dahut, W. *et al.* (2001). A phase I and pharmacokinetic study of squalamine, a novel antiangiogenic agent, in patients with advanced cancers. *Clin Cancer Res* **7**, 3912–3919.

101. Kim, Y. M., Jang, J. W., Lee, O. H. *et al.* (2000). Endostatin inhibits endothelial and tumor cellular invasion by blocking the activation and catalytic activity of matrix metalloproteinase. *Cancer Res* **60**, 5410–5413.

102. Morbidelli, L., Donnini, S., Chillemi, F. Giachetti, A., Ziche, M. (2003). Angiosuppressive and angiostimulatory effects exerted by synthetic partial sequences of endostatin. *Clin Cancer Res* **9**, 5358–5369.

103. Chillemi, F., Francescato, P., Ragg, E. *et al.* (2003). Studies on the structure-activity relationship of endostatin: Synthesis of

human endostatin peptides exhibiting potent antiangiogenic activities. *J Med Chem* **46**, 4165–4172.

104. Cattaneo, M. G., Pola, S., Francescato, P. Chillemi, F., Vicentini, L. M. (2003). Human endostatin-derived synthetic peptides possess potent antiangiogenic properties in vitro and in vivo. *Exp Cell Res* **283**, 230–236.

105. Peroulis, I., Jonas, N., and Saleh, M. (2002). Antiangiogenic activity of endostatin inhibits c6 glioma growth. *Int J Cancer* **97**, 839–845.

106. Thompson, R. C., Pardoll, D. M., Jaffee, E. M. *et al.* (1996). Systemic and local paracrine cytokine therapies using transduced tumour cells are synergistic in treating intracranial tumors. *J Immunother Emphasis Tumor Immunol* **19**, 405–413.

107. DiMeco, F., Rhines, L. D., Hanes, J. *et al.* (2000). Paracrine delivery of il-12 against intracranial 9l gliosarcoma in rats. *J Neurosurg* **92**, 419–427.

108. Sampath, P., Hanes, J., DiMeco, F. *et al.* (1999). Paracrine immunotherapy with interleukin-2 and local chemotherapy is synergistic in the treatment of experimental brain tumors. *Cancer Res* **59**, 2107–2114.

109. Hanes, J., Sills, A., Zhao, Z. *et al.* (2001). Controlled local delivery of interleukin-2 by biodegradable polymers protects animals from experimental brain tumors and liver tumors. *Pharm Res* **18**, 899–906.

110. Rhines, L. D., Sampath, P., DiMeco, F. *et al.* (2003). Local immunotherapy with interleukin-2 delivered from biodegradable polymer microspheres combined with interstitial chemotherapy: A novel treatment for experimental malignant glioma. *Neurosurgery* **52**, 872–879; discussion 879–880.

111. Santini, J. T., Jr., Cima, M. J., and Langer, R. (1999). A controlled-release microchip. *Nature* **397**, 335–338.

112. Santini, J. T., Jr., Richards, A. C., Scheidt, R. A. *et al.* (2000). Microchip technology in drug delivery. *Ann Med* **32**, 377–379.

113. Richards Grayson, A. C., Choi, I. S., Tyler, B. M. *et al.* (2003). Multi-pulse drug delivery from a resorbable polymeric microchip device. *Nat Mater* **2**, 767–772.

114. Zanta, M. A., Boussif, O., Adib, A. Behr, J. P. (1997). In vitro gene delivery to hepatocytes with galactosylated polyethylenimine. *Bioconjug Chem* **8**, 839–844.

115. Boussif, O., Lezoualc'h, F., Zanta, M. A. *et al.* (1995). A versatile vector for gene and oligonucleotide transfer into cells in culture and in vivo: Polyethylenimine. *Proc Natl Acad Sci U S A* **92**, 7297–7301.

116. Demeneix, B., Behr, J., Boussif, O. *et al.* (1998). Gene transfer with lipospermines and polyethylenimines. *Adv Drug Deliv Rev* **30**, 85–95.

117. Suh, J., Wirtz, D., and Hanes, J. (2003). Efficient active transport of gene nanocarriers to the cell nucleus. *Proc Natl Acad Sci U S A* **100**, 3878–3882.

118. Abbott, N. J., and Romero, I. A. (1996). Transporting therapeutics across the blood-brain barrier. *Mol Med Today* **2**, 106–113.

119. Wang, P. P., Frazier, J., and Brem, H. (2002). Local drug delivery to the brain. *Adv Drug Deliv Rev* **54**, 987–1013.

120. Brem, H., Mahaley M. S., Jr., Vick, N. A. *et al.* (1991). Interstitial chemotherapy with drug polymer implants for the treatment of recurrent gliomas. *J Neurosurg* **74**, 441–446.

121. Brem, H., Piantadosi, S., Burger, P. C. *et al.* (1995). Placebo-controlled trial of safety and efficacy of intraoperative controlled delivery by biodegradable polymers of chemotherapy for recurrent gliomas. The polymer-brain tumor treatment group. *Lance* **345**, 1008–1012.

122. Brem, H., Ewend, M. G., Piantadosi, S. *et al.* (1995). The safety of interstitial chemotherapy with bcnu-loaded polymer followed by radiation therapy in the treatment of newly diagnosed malignant gliomas: Phase I Trial. *J Neurooncol* **26**, 111–123.

123. Valtonen, S., Timonen, U., Toivanen, P. *et al.* (1997). Interstitial chemotherapy with carmustine-loaded polymers for high-grade gliomas: A randomized double-blind study. *Neurosurgery* **41**, 44–48.

124. Westphal, M., Hilt, D. C., Bortey, E. *et al.* (2003). A phase 3 trial of local chemotherapy with biodegradable carmustine (BCNU) wafers (gliadel wafers) in patients with primary malignant glioma. *Journal of Neuro-Oncology* **5**, 79–88.

125. Rosenblum, M., Weingart, J., Dolan, M. E. *et al.* (2002). Phase I study of gliadel combined with a continuous intravenous infusion of o6-benzylguanine in patients with recurrent malignant glioma. *Proc American Soc Clin Oncologists* **21**.

126. Limentani, S., Asher, A., Fraser, R. *et al.* (1999). A phase I trial of surgery, gliadel, and carboplatin in combination with radiation therapy for anaplastic astrocytoma or glioblastoma multiforme. *ASCO Annual Meeting* Abstract in CD.

127. Limentani, S., Asher, A., and Fraser, R. (2000). A phase I trial of surgery, gliadel, and carboplatin in combination with radiation therapy for anaplastic astrocytoma or glioblastoma multiforme. *American Association of Neurological Surgeons.* Abstract in CD.

128. Colvin, O. M., Cogkor, I., Edwards, S. *et al.* (2000). Phase I trials of gliadel plus CPT-11 or temodar. *ASCO Annual Meeting* Abstract in CD.

129. Sungarian, A., Sampath, P., Maestri, X. *et al* (2001). Cpt-11 used in combination with bcnu-impregnated polymer (gliadel) in the treatment of recurrent glioblastoma multiforme: A preliminary clinical study. *Congress of Neurological Surgeons.* Abstract in CD.

130. Rich, J., Reardon, D. A., Allen, D. *et al.* (2001). Phase I/II trial of gliadel plus temodar for adult patients with recurrent high-grade glioma. *ASCO Annual Meeting.*

131. Gururangan, S., Cokgor, L., Rich, J. *et al.* (2001). Phase I study of gliadel wafers plus temozolomide in adults with recurrent supratentorial high-grade glioma. *J Neurooncol* **3**, 246–250.

132. Albright, R., Breneman, J., Warnick, R. *et al.* (2000). A phase I study of treatment for patients with relapsed malignant glioma using concurrent permanent I-125 seeds and dose escalation of gliadel bcu polymers. *American Association of Neurological Surgeons.* Abstract in CD.

133. Darakchiev, B., Albright, R., Breneman, J. *et al.* (2004). Prolonged survival in patients with recurrent glioblastoma multiforme treaed by resection with implantation of permannent I-125 seeds and bcnu -impregnated wafers. *American Association of Neurological Surgeons.* Abstract in CD.

134. McPherson, C., Breneman, J., Tobler, W. *et al.* (2004). Permanent low activity i-125 interstitial implants as an adjuvant therapy for the treatment of newly diagnosed glioblastoma multiforme. *American Association of Neurological Surgeons.* Abstract in CD.

135. Lowenthal, B., Sanan, A., Shannon, M. *et al.* (2003). A preliminary report on patients with cns metastatic disease enrolled in prolong outcomes study. *American Association of Neurological Surgeons.* Abstract in CD.

136. Brem, S., Snodgrass, S., Staller, A. *et al.* (2003). Implantation of bcnu chemotherapreutic wafers to prevent recurrence of brain metastases. *American Association of Neurological Surgeons.* Abstract in CD.

137. Laws, Jr. E. R., Morris, A. M., and Maartens, N. (2003). Gliadel for pituitary adenomas and craniopharyngiomas. *Neurosurgery* **53**, 255–269.

138. Menei, P., Jadaud, E., Faisant, N. *et al.* (2004). Stereotaxic implantation of 5-fluorouracil-releasing microspheres in malignant glioma. *Cancer* Jan 15, **100**, 405–410.

139. Kleinberg, L. R., Weingart, J., Burger, P. *et al.* (2004). Clinical course and pathologic findings after gliadel and radiotherapy for newly diagnosed malignant glioma: Implications for patient management. *Cancer Invest* **22**, 1–9.

140. Olivi, A., Grossman S. A., Tatter, S. *et al.* (2003). Dose escalation of carmustine in surgically implanted polymers in patients with recurrent malignant glioma: A new approaches to brain tumour therapy cns consortium trial. *J Clin Oncol* **21**, 1845–1849.

141. Storm, P. B., Moriarity, J. L., Tyler, B. *et al.* (2002). Polymer delivery of camptothecin against 9l gliosarcoma: Release, distribution, and efficacy. *J Neurooncol* **56**, 209–217.

142. Haque, R. (2004). Antagonism of brain-mediated excitotoxic destruction and invasion using a glutamate release inhibitor. *AANS.*

143. Amundson, E. (2003). SJG-136: A novel agent for the treatment of experimental gliomas: An in-vivo toxicity and efficacy study. *Congress of Neurological surgeons.* Abstract in CD.

144. Tyler, B. (2002). Tetra-o-methyl-nordihydroguaiaretic acid (m4m): A novel agent for the interstitial treatment of experimental brain tumors - an in vitro and in vivo assessment. *Congress of Neurological surgeons.* Abstract in CD.

145. Hanes, J., Sills, A., Zhao, Z. *et al.* (2001). Controlled local delivery of interleukin-2 by biodegradable polymers protects animals from experimental brain tumors and liver tumors. *Pharm Res* **18**, 899–906.

146. DiMeco, F. (2001). Interleukin-12 delivery by biodegradable microspheres (IL-12 ms) for the treatment of metastatic brain tumors. *AANS.*

147. DiMeco, F. (2000). Local delivery of interleukin-12 by biodegradable microspheres (IL-12ms) for the treatment of experimental brain tumors. *4th Congress of the European Association of Neuro-Oncology.*

148. Pradilla, G., Legnani, F., Hsu, W. Tyler, B., Brem, H. (2004). The protease inhibitor lactacystin prolongs survival in animals intracranially challenged with 9l glioksarcoma when delivered locally via controlled-release polymers. *Congress of Neurological surgeons.* Abstract in CD.

149. Hsu, W., Pradilla, G., Legnani, F., Tyler, B., Brem, H. (2004). Local edlivery of lactacystin via controlled-release polymers prolongs survival in a 9l gliosarcoma model. *AANS.*

150. Frazier, J. L., Wang, P. P., Case, D. *et al.* (2003). Local delivery of minocycline and systemic bcnu have synergistic activity in the treatment of intracranial glioma. *J Neurooncol* **64**, 203–209.

151. Haroun, R., Tyler, B., and Brem, H. (2000). Chemically modified tetracyclines inhibit solid tumour growth in brain tumor models in vivo. *Congress of Neurological surgeons.* Abstract in CD.

152. DiMeco, F., Li, K. W., Tyler, B. M. *et al.* (2002). Local delivery of mitoxantrone for the treatment of malignant brain tumors in rats. *J Neurosurg* **97**, 1173–1178.

153. Li, K. W., Dang, W., Tyler, B. M. *et al.* (2003). Polilactofate microspheres for paclitaxel delivery to central nervous system malignancies. *Clin Cancer Res* **9**, 3441–3447.

154. Brem, S., Tyler, B., Legnani, F. *et al.* (2004). Local delivery of temozolomide against an experimental malignant glioma model. *Congress of Neurological surgeons.* Abstract in CD.

20

Intratumoral Administration and Convection-Enhanced Delivery

Mike Yue Chen, Zhi-jian Chen, George T. Gillies, Peter J. Haar, and William C. Broaddus

ABSTRACT: Convection-enhanced delivery is a means of localized drug delivery to the central nervous system. The therapeutic agent, infused *via* a catheter, is carried (*convectus*, latin) by bulk flow though the interstitial space. This method has proved to be a useful laboratory technique for targeted, wide-spread distribution of a broad range of agents including small molecules and gene therapy vectors, such as viruses and liposomes. The success in the lab has translated into a significant number of clinical studies which combine the use of convection-enhanced delivery and elegantly devised molecular strategies. Additionally, several studies suggest that convection-enhanced delivery can be accurately predicted with mathematical models, allowing precise planning of dosage and distribution in the clinical setting. Taken together the results of these investigations hint at the promise of an expanded armantarium against brain tumors.

GENERAL OVERVIEW

Surgery, radiation therapy, and local or systemic chemotherapy have only a modest success rate in the treatment of malignant brain tumors. One reason for the failure of chemotherapeutic approaches is the difficulty of drug delivery to central nervous system tumors which have generally proved to be chemoresistant, the basis of which may be due to a variably intact blood–brain barrier and the relatively small window separating therapeutic efficacy from toxicity.

Ideally, an anti-tumor agent could be administered systemically, cross the blood–brain barrier as needed, and specifically target the tumor cells. Unfortunately, current techniques do not achieve this level of sophistication. One method, pioneered by Oldfield and colleagues at the National Institutes of Health, known as convection-enhanced delivery (CED), or also referred to as "intracerebral clysis" and "high-flow microinfusion", has recently gained significant momentum as a central nervous system drug delivery technique that bypasses the blood–brain barrier and allows spatially accurate targeting. The method is conceptually straightforward. A catheter is placed directly into the brain or tumor parenchyma and the drug solution is infused at a rate that allows bulk flow through the interstitial space (see Fig. 20.1). This technique then allows highly concentrated drug delivery to the tumor, and minimizes systemic and normal brain tissue exposure.

Currently, enthusiasm for CED has generated a large number of studies detailing the use of "molecular neurosurgery" in the laboratory and clinical neuro oncology environments. The goal of this chapter is to first discuss the underlying science, then to review important animal and human studies that are completed or ongoing, and finally to understand the obstacles between the current applications of CED and widespread clinical use.

PHYSICAL PRINCIPLES

In its simplest form, CED consists of using a pump to drive the flow of the desired agent through a delivery device, such as a cannula or catheter, the tip of which is stereotactically placed at the target location within the brain. As the agent is delivered

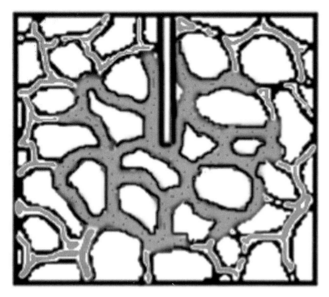

FIGURE 20.1 Drawing illustrating the principle of convection-enhanced delivery. After exiting the cannula tip, the infusate flows by bulk flow through channels in the extracellular space which are widened in the process.

through the tip and directly into the parenchymal tissues, the resulting pressure gradient then generates bulk flow of the fluid through the interstitial space. Drug solution permeates the targeted region at a final tissue concentration and volume of distribution governed by the infusion parameters, the flow resistance or hydraulic conductivity of the tissue, and the duration of the treatment. In this way, CED can avoid the screening effects of the blood–brain barrier and deliver anti-tumoral compounds and agents to specific locations, at desired concentrations, and over time-periods consistent with pharmacologic tolerance, pharmacokinetic stability of the agent, and central nervous system metabolic activity. Subsequent spread of the infusate within wider volumes can occur due to diffusion of the agent along the concentration gradient and to bulk flow along the pathways of tissue anisotropy, but metabolic uptake, vascular clearance, and sumping within the ventricular system ultimately provide pharmacodynamic and physiological sinks.

From a physical perspective, the CED process is governed in part by Darcy's Law for flow through a porous medium during the early (advective) component of the flow, and by Fick's Laws of diffusion during the latter part of the flow. The mathematical expressions of these laws are typically the starting points for analyses of transport within the interstitium. The biophysical basis of the diffusion process is discussed by Nicholson and Phillips and Nicholson, while models of the infusion process have been

developed by Basser and Morrison *et al.* [1–5]. All these are analytical approaches to the transport problem (from which numerical estimates have been made), but computational simulations using finite-element methods have been developed as well, with the latter taking into account both the diffusion and infusion components of the flow [6,7]. Still another model, based on the poroelastic characteristics of tissue, has been developed by Chen *et al.* [8]. Its predictions regarding the volumes of distribution and pressure profiles of an infusion compare well against experimental data obtained during the delivery of dye into an agarose brain phantom gel that has been shown to be a useful surrogate of living mammalian brain tissue [9].

There is one newly emerging model of the flow patterns that occur during CED in which the predictive capability of the technique has advanced to the point where accurate treatment planning has become possible (see Fig. 20.2). This is the work of Raghavan *et al.*, wherein the model is designed to be broadly applicable to most clinical situations [10]. It incorporates various provisions for sources and sinks of the infusate, e.g., backflow, leakage into the subarachnoid space, drainage into cavities, fluctuations in clearance rates resulting from variations in capillary permeability, etc. Their model uses anatomical and diffusion tensor magnetic resonance images as input, and it allows for anisotropic variations in hydraulic conductivity and diffusivity. The movement of the infusate is described by a stochastic differential equation with advective and diffusive terms. In their approach, the stochastic process is implemented as a simulation of the motion of individual particles, in such a way that local particle density is proportional to local concentration. Drug concentrations are predicted at a particular location and time, and the simulation steps are repeated for many particles. The approach allows re-sampling of particles for any intermediate time point and can be optimized for increased resolution or accuracy. They have shown in human trials that a drug injected at a plausible point near a tumor might flow in such a way as to be clinically useless, but such flow patterns can be predicted within enough time intraoperatively for the injection site to be corrected.

In general, CED bears resemblance to standard direct-injection delivery, but with important differences. For instance, the flow rates through the cannula or catheter are very low, typically between 0.5 and 5.0 μl min^{-1} (a presently ongoing phase III human clinical trial for recurrent glioblastoma multiforme infuses TransMID^{TM} at 0.2 ml h^{-1} = 3.33 μl min^{-1}). The goal for CED is one of introducing the infusate

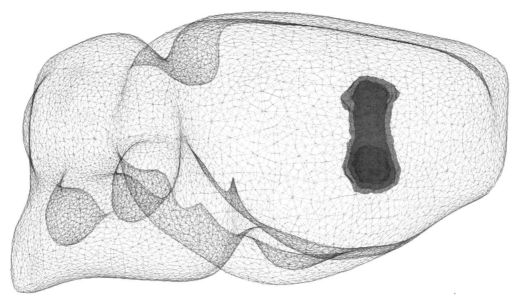

FIGURE 20.2 Illustration of the use of a finite-element approach to model convection-enhanced delivery in an experimental setting. Equations of convection and diffusion in a poroelastic media have been solved over a rat brain hexahedral grid. Several concentration isosurfaces are shown following an infusion of 50 min and diffusion for 20 min.

in a slow, controlled manner so that the agent gently swells the extracellular matrix and flows interstitially as opposed to creating rupture paths in the tissues along which essentially uncontrolled bulk flow might instead occur. Concomitant with this is maintaining an interstitial fluid pressure at the outlet port(s) of the delivery device that is no more than two to three times the background intracranial pressure. This is enough of a differential to cause flow along the resulting pressure gradient, but not so high that hydrodynamically driven damage to the tissue would occur.

The presence of ultra-structural fenestrations in the pia mater might logically suggest that an intrathecal means of drug delivery across the pia might be another way of delivering drugs into the parenchymal tissues. However, in their review of modalities for the delivery of neurotrophic factors into the brain, Thorne and Frey reiterate the well-known fact that delivery *via* the cerebrospinal fluid is indeed able to achieve a wide area of coverage, but is limited in efficacy by cellular barriers to tissue penetration [11]. For instance, in the cerebellum, the Purkinje cell layer in particular appears to block diffusive traversal of large molecules through the pia, and this general class of findings argues strongly that any natural fenestration that might exist in the pial covering of the cortical surfaces, as opposed to that surrounding the spinal cord, is insufficient to mediate a direct transpial transport process of anything but very small molecules [12].

As the ventricles are also one of the principal fluid-carrying compartments within the central nervous system, intraventricular delivery with the goal of obtaining eventual intraparenchymal distributions would also seem to be a potential avenue of approach. In fact, the volumes and flows of the cerebrospinal fluid and interstitial fluid pressure are interrelated, with the coupling between them arising at least in part due to the perivascular spaces of the brain's surface vessels. However, the blood–cerebrospinal fluid barrier at the ventricular walls plays a substantial role in regulating all aspects of exchange between the various compartments, and this in turn places limits on transport into the parenchymal tissues that are just as stringent as those dictated by the blood–brain barrier of the endothelial layer.

For all these reasons, direct intraparenchymal delivery *via* CED plays a unique role in achieving therapeutic concentrations of anti-tumoral agents on a regional basis within the brain. The technique is also being studied for cell delivery for neuro-degenerative diseases and will likely find several other clinical applications as well [13].

ANIMAL STUDIES

An extensive number of animal studies have established the safety and efficacy of convective delivery to normal brain such that calling it a novel technique, presently, is difficult. Trials have been performed in cats, rats, non-human primates, and pigs (unpublished data) [14–31]. Once anesthetized, the animal's head is secured and the cannula is placed with stereotactic guidance. Popular sites of infusion include the corona radiata, the striatum, and the brainstem [14,17,18,24,25,27,28,32]. Given the propensity of malignant brain tumors to diffusely infiltrate overtly normal tissue, characterization of bulk flow in non-tumoral tissues is pertinent.

The pressure gradient for the infusion is typically supplied by an external pump; however, use of an internally implanted pump for chronic infusion has been reported [29]. The infusates used in the studies are often potentially therapeutic agents or tracer solutions, the latter of which are commonly labeled or radioactive, allowing researchers to study flow patterns and concentration gradients. The size of the solute can also vary by orders of magnitude, ranging from inert sucrose which is less than 1 nm in diameter to polystyrene or viral particles 200 nm in diameter, though the properties of convective delivery at this upper boundary remain ill-defined [19,26,33]. Nevertheless, it is clear from laboratory studies and human trials that protein and oligonucleotide solutes, with molecular diameters in the 5–20 nm range, are easily delivered and distributed in brain and tumor parenchyma using this approach.

Animal studies have further clarified the important relationship between CED and the additional intra-cranial pressure that is generated. Measurements of interest include interstitial pressure, intracranial pressure, cisterna magna pressure, and line pressure, all common parameters that have been studied to optimize the safety and efficacy of the technique [14,16,26]. Bruce et al. determined that intracranial pressure during infusion in rats can vary between 5 and 21 mm Hg in direct correlation with the infusion volume and the rate of infusion which ranged from 0.5 to 4.0 μl/min in their study [16]. Also noted in those experiments was the absence of adverse effects with intracranial pressure changes less than 20 mm Hg, an observation that is con-cordant with data from the traumatic brain injury experience. Intuition is confirmed with these results that link CED to changes in cerebral pressure, but other, earlier findings are mere speculations. Bobo et al. and subsequent investigators have noted that infusion pressure generally rises at the beginning of the infusion, reaches a peak, and then falls to a plateau for the duration of the infusion [14,26]. This phenomenon is thought to possibly result from a tissue plug in the cannula, or possibly from the work required to open fluid pathways in the surrounding parenchyma [14,26].

CONVECTION-ENHANCED DELIVERY IN ANIMAL BRAIN TUMOR MODELS

Convective delivery of different agents into a specific area of normal brain has been extensively studied with a resultant understanding that allows a fairly accurate prediction of how the infusate will distribute. However, high-flow microinfusion into tumor tissues involves additional levels of complexity that are only beginning to be addressed. Additional variables include the lack of a blood–brain barrier, the inhomogeneity of tumor tissue, the increased interstitial pressure within a tumor, and the different biochemical and physical characteristics of the extra-cellular space. Despite these challenges, the feasibility of precisely targeted convective delivery to tumors appears promising.

In a prelude to human studies, Kaiser et al. used CED to locally distribute a chemotherapeutic agent, topotecan, a topoisomerase I inhibitor that has limited access to the central nervous system due to the blood–brain barrier [22]. The C6 rat intracerebral glioma model was used comparing survival among control, systemically treated, and convectively treated groups. Neither the control nor the conventionally treated groups produced a subject that survived longer than 26 days. However, in animals treated using the infusion technique, no residual tumors were detected, and all but one rat survived until 120 days, at which point all subjects were sacrificed.

In another study demonstrating the efficacy of CED in animal tumor models, Yang et al. performed a comparative study of the distribution of iodine[125]-boronated epithelial growth factor in normal rats and a rat glioma model using either high-flow microinfusion or rapid intratumoral bolus injections [34]. The uptake and biodistribution of iodine[125]-boronated epithelial growth factor in tumor or brain was studied by quantitative autoradiography and gamma-scintillation counting. They found that the volume of distribution with CED was 64.8 ± 13.4 μl which surpassed the spread of the rapidly injected agent by 620 per cent. Moreover, 24 h after treatment, 47.4 per cent of the convected dose remained in the gliomas whereas only 33.2 per cent of the rapidly injected dose was present. These results led the authors to conclude that high flow microinfusion could be especially useful for the administration of high molecular weight, receptor-targeted macromolecules, such as monoclonal anti-bodies and growth factors.

Utilizing a similar-sized macromolecule that has well-described surface receptors, Saito et al.

convectively delivered the highly potent tumor-necrosis-factor-related apoptosis inducing ligand (TRAIL) in combination with systemic administration of temozolomide [35]. In treatment groups, prolonged survival of animals with intracranial glioblastoma xenografts was attributed to effective distribution of the TRAIL ligand throughout the brain tumor mass that was synergistic with chemotherapy. Furthermore, the TRAIL ligand was safely and reliably spread using CED through both normal brain and the U87MG xenografts.

In a study that highlights the potential and inherent difficulties of using high-flow microinfusion for the distribution of viral vectors, Brust *et al.* studied radiosensitization of rat RT2 gliomas with bromo-deoxycytidine and adenovirus constructs expressing the herpes simplex virus-thymidine kinase gene [36]. They found that large volumes (100–150 μl) of virus delivered at rates of less than or equal to 1 μl/min were optimal for uniform, reproducible results. Using these infusion conditions, the authors achieved a 40 per cent rate of adenovirus infection in the tumors. Infection of the RT-2 tumors with adenovirus construct in combination with continuous adminis-tration of bromodeoxycytidine from an osmotic pump resulted in significant tumor regression six days after radiation (30 Gy delivered as 2×5 Gy over three days). The tremendous amount of virus needed to achieve that effect demonstrates the need to further optimize viral vectors required for gene therapy of brain tumors.

A study carried out by Mamot *et al.* is considered to be the first report of convective delivery into the central nervous system and a brain tumor model of liposomes [37]. In their experiment, liposomes labeled with a varying mixture of fluorochromes and gold particles were infused *via* CED into glioma xenograft models in rodent hosts. In the intracranial U-87 glioma xenografts, liposomes were robustly distributed throughout the tumor tissue and to some extent into surrounding normal tissue. Greater penetration was observed using liposomes with a diameter of 40 *versus* 90 nm. Co-infusion with mannitol was found to be synergistic as well. This study also demonstrated that gadodiamide, loaded into liposomes, provided a non-invasive method to track the infused liposomes.

Not all the studies have been positive regarding the feasibility of using CED in tumors. Vavra *et al.* compared tissue and plasma pharmacokinetics of C^{14}-sucrose in subcutaneous RG-2 rat gliomas after administration by three routes, intravenous bolus (50 microCi over 30 s), continuous intravenous infusion (50 microCi at a constant rate), and CED

(CED, 5 microCi infused at a rate of 0.5 μl/min) [38]. Plasma, tumor, and other tissue samples were obtained at three time points (0.5, 2, and 4 h) to measure tissue radioactivity. Plasma radioactivity in the CED group increased exponentially and lagged only slightly behind the continuous intravenous infusion group, but after 90 min, plasma values were similar in all. Compared to the other methods, mean tumor radioactivity was 100–500 times higher in the CED group at each time point. However, radioactivity distribution was inhomogeneous at all three time points; highest concentrations occurred in tissue around tumor and in areas of necrosis, while viable tumor contained the lowest and sometimes negligible amounts of isotope. The intravenously treated groups did not fare much better, tumor radioactivity was homogeneous at 0.5 h but was subsequently inhomogeneous at 1 and 2 h.

In the same study, Vavra *et al.* also examined efflux of C^{14}-sucrose from intracerebral tumors at 0.5, 1, 2, and 4 h. Less than five per cent of the radioactivity remained in intracerebral tumors at each time point. Overall, these results may indicate rapid efflux of low molecular weight drugs as well as marked hetero-geneity of drug distribution from brain tumors after convective administration. The authors speculate that high interstitial fluid pressure gradients possibly account for the witnessed behavior. More effective CED to brain tumors may necessitate acute or chronic lowering of the tumor interstitial pressure, but further studies are required to confirm or deny this interesting hypothesis.

HUMAN STUDIES

The extensive number of animal studies partially detailed in the previous section has created a fairly comprehensive understanding of the principles and potential of CED. This data has in turn generated exciting clinical studies which may soon validate the utility of CED as a neurosurgical tool. In this section, completed and ongoing human neuro-oncology studies which utilize high-flow microinfusion as a strategy for drug delivery are reviewed.

In our opinion, the factor that will most limit the use of CED as a tool for the treatment of brain tumors is the effectiveness of the infused agent. So far, the substances that have been tried vary from the mundane to the exotic and include: chemotherapy drugs, designer toxins, radiotoxic conjugates, and liposomes bearing modified viral genomes. The most widely studied agents are the receptor-directed toxins, of which a modified diphtheria toxin

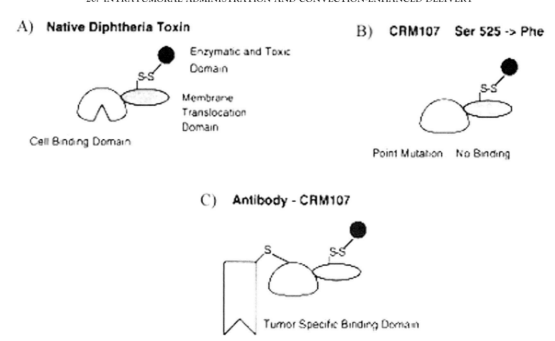

FIGURE 20.3   Tf-CRM107 protein-toxin conjugate components. **(A)** Native diphtheria toxin. **(B)** Point mutation disables cell binding domain. **(C)** Conjugation to transferrin or antibody confers specificity. Illustration reprinted with permission from Science, October 1987, volume 238.

(CRM107) conjugated to transferrin (Tf-CRM107, "TransMID") is the prototype and the subject of a phase III clinical trial (Xenova).

Details of the development of CRM107 for CED illustrate the concepts important to understanding the designer toxin therapies. Tf-CRM107 (140 000 molecular weight) is a genetically modified diphtheria toxin conjugated to the transferrin (Tf) ligand (see Fig. 20.3). Theoretically, transferrin binds to transferrin receptors which are upregulated in tumor cells but not normal brain, a phenomenon conferring relative specificity. After receptor-mediated endocytosis, the toxin, subjected to a lower pH, unfolds and is inserted into the endocytotic vesicle membrane. The alpha subunit is then released into the cytoplasm and catalyzes the ADP ribosylation of ribosomal elongation factor-2 (EF-2), a protein necessary for protein synthesis. Due to its capacity for catalytic inactivation of numerous ribosomal EF-2 molecules, diphtheria toxin is very potent. Only a single molecule is necessary to kill a cell.

The phase I trial was performed in 15 evaluable patients with malignant gliomas of which nine were glioblastomas, five were anaplastic astrocytomas, and one was an anaplastic oligodendroglioma [39]. Overall, median survival was greater than 75 weeks, median time to progression was 38 weeks, and in nine of the patients, tumor volumes were reduced by more than 50 per cent. Based on magnetic resonance imaging, there were two complete responders, though progression eventually occurred (see Fig. 20.4). Systemic toxicity was manifested by transient elevations of serum alanine and aspartate aminotransferases in fourteen patients. Mild hypoalbuminemia also occurred in twelve patients. Overall, the infusions were well tolerated, though peritumoral toxicity, associated with higher toxin concentrations, occurred in three patients one to four weeks after Tf-CRM107 introduction. Biopsied regions of magnetic resonance abnormalities in three patients who developed significant neurological deficits revealed thrombosed cortical venules and capillaries.

The encouraging results were further confirmed in a phase II trial (unpublished data presented in poster format at the Annual Meeting of American Association of Neurological Surgeons, 2000). Thirty-four out of forty-four patients were evaluated after two stereotactic intratumoral infusions, spaced four to ten weeks apart, of Tf-CRM107 using dual silastic catheters. The phase II toxin concentration was less than the concentration that was associated with peritumoral toxicity in the phase I trial. Five patients were complete responders (15 per cent), seven were partial responders (20 per cent), nine had stable disease (26 per cent), and thirteen had progressive disease (39 per cent). The median survival time was 37 weeks for all the patients and 13 were still alive one year after the initial infusion [40]. Overall, the

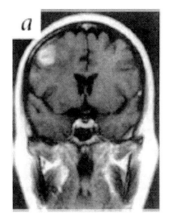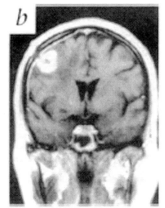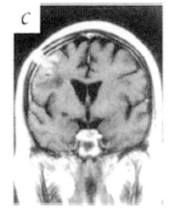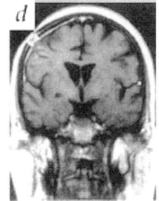

FIGURE 20.4 T1-weighted gadolinium-enhanced coronal MRI images depicting a complete response of a glioblastoma multiforme in a 48-year old female to convective delivery of Tf-CRM107. **(A)** Tumor prior to treatment. **(B)** Four days after treatment there is increased enhancement. **(C)** At seven months, the enhancement is resolving. **(D)** Complete resolution at 14 months with recurrence at 23 months (recurrence not shown). Figure reprinted with permission from Nature Medicine, volume 3, 1997: page 1365.

results of the phase II trial were comparable to those obtained in the phase I trial. Currently, a multicenter phase III trial is in progress to compare the efficacy of the Tf-CRM107 receptor directed toxin to the best standard of care in patients with non-operable, progressive or recurrent glioblastoma multiforme.

Following the encouraging results of the earlier Tf-CRM107 studies, other protein-toxin conjugates designed for local delivery are presently undergoing human trials. It should also be noted that a commonly used term for these agents, "immunotoxin", is a misnomer because this term would refer to a toxin conjugated to an antibody. Immunotoxins thus differ from Tf-CRM107 in terms of the targeted receptor and the nature of the toxin itself.

One example of another cytotoxin that is being delivered locally to glioblastomas in clinical trials is the recombinant fusion protein IL13-PE38QQR which is composed of the immune regulatory cytokine, interleukin-13 (IL-13), conjugated to a mutated form of Pseudomonas exotoxin [41]. At concentrations of 0.5–2.0 µg/ml, prolonged survival was observed among the 46 patients enrolled, as was histopathological tumor effect. This approach has been further extended by combining IL13-PE38QQR with the novel fusion protein DAB389EGF, with the combination showing enhanced cytotoxicity against glioblastoma cells *in vitro*, thus suggesting a synergistic effect of the combined agents. Ongoing phase I/II and III clinical trials are in progress to determine the effect of infusing the IL-13 cytotoxin into recurrent gliomas as a stand-alone therapy, prior to surgery, or after surgery [42].

A trial using a similar fusion protein, NBI-3001, an interleukin-4 Pseudomonas exotoxin combination, was performed in 31 patients with histologically verified supratentorial grade 3 and 4 astrocytomas [43]. Overall, the median survival was 8.2 months with a median survival of 5.8 months for the glioblastoma patients. Systemic toxicity was not evident and treatment-related adverse effects, occurring in 39 per cent of the patients, were limited to the central nervous system.

Some investigators have focused the use of high-flow microinfusion on conventional chemotherapeutic agents. For instance, an organic solvent-based BCNU solution is being convectively administered into glioblastoma multiforme tumor beds. The agent being used for this purpose is DTI-015 (Direct Therapeutics, Inc., San Bruno, California), which consists of BCNU in 100 per cent ethanol at a typical concentration of 45 mg/ml [44]. Diffusion-weighted magnetic resonance tracking performed in a phase I/II dose-escalation study revealed that the volume of distribution is up to 20 times the volume infused [45]. In patients who received equal or less than the maximum tolerated dose, median survival was 55 weeks. A phase III clinical trial of the technique is reported to be under planning, as are European trials.

Lidar *et al.* convectively administered another chemotherapeutic agent that is typically given systemically, paclitaxel, to fifteen patients with glioblastomas and anaplastic astrocytomas [46]. The authors achieved a relatively good effect, five complete and six partial responders, but this result came at the expense of a high incidence of adverse effects. Complications included transient chemical meningitis, infectious complications, and transient neurological deterioration. Of interest particularly to future studies, successful distribution correlated with hyperintense signal changes on diffusion-weighted magnetic resonance imaging, whereas non-effective CED was associated with "leakage of the convected drug into the subarachnoid space, ventricles, and cavities formed by previous resections, and was seen in tumors containing widespread necrosis."

Yet one more application that is being investigated is the use of CED to distribute radiotoxic species designed to preferentially target glioblastoma multiforme cells. Merlo et al. infused the peptide vector ^{90}Y-labeled DOTA^0-D-Phe^1-Tyr^3-octreotide (DOTA-TOC) into two cohorts of patients, one with low-grade gliomas and one with anaplastic gliomas [47]. Six of the eleven patients showed disease stabilization and tumor shrinkage. Papanastassiou et al. treated seven patients by convective infusion of I^{131} labeled monoclonal antibodies, and Riva et al. similarly treated fifty patients [48,49]. Remarkably, in the latter study, median survival was 17 months in patients with bulky tumors, and 26 months in patients with minimal disease. Moreover, median time to progression was three months in recurrent and seven months in newly diagnosed gliomas. Based on these results, Riva et al. suggested that direct intraparenchymal infusion might be particularly useful in newly diagnosed brain tumor cases in which debulking surgery had already been performed.

Using CED to transport gene therapy vectors, such as viruses and liposomes to brain tumors is perhaps the most intriguing application. Feasibility of using bulk flow to distribute nano-sized particles through the extracellular space appears promising, and, in fact, human clinical trials are in progress [28,33,50]. In a phase I/II study Voges et al. convectively administered liposomes carrying the herpes simplex I genome incorporating the thymidine kinase gene to eight patients with recurrent glioblastoma multiforme [51]. Two of the patients had a 50 per cent reduction in tumor volume and six patients had a focal response.

CHALLENGES

For CED to further mature as a method to treat tumors, several challenges still need to be overcome, one of which is catheter design, a seemingly trivial but important consideration. Many of the existing neurocatheter designs are typically not optimized for use in CED therapies. Some of the difficulties associated with attempting to use them for that purpose include:

i. inaccurate dosing resulting from a combination of non-uniform distribution of the infusate and

its backflow or reflux along the catheter insertion track,

ii. collateral tissue damage resulting from multiple traumatic catheter insertions into the target tissues in attempts to obtain volume-contoured deliveries,

iii. tissue damage resulting from high-pressure infusions associated with inappropriate flow characteristics at the distal tip of the catheter, and

iv. inaccurate dosing resulting from the introduction of trapped air bubbles into brain tissue.

There are some designs being developed now to overcome this class of limitations, and testing them to evaluate their efficacy in CED applications is presently underway [52].

Current limitations of pump systems present a different area of concern. Infusion systems typically consist of an external pump to which a percutaneous catheter is attached. This situation results in some discomfort and considerable inconvenience for the patient, whose head is "anchored" by a catheter, and furthermore restricts the duration of infusion because of the potential for infection. The longest duration of the infusion in a current human protocol is five days, a length of time that, based on extensive experience with external ventricular devices, could be increased only with a progressive increase in the risk of infection. An implantable pump system, much like those that are available for intrathecal infusions of morphine and baclofen, would be of considerable use. The longer, slower infusion could allow for larger volumes of distributions, more uniform delivery, decreased pressure changes, and possibly increased effect of certain infused agents that require prolonged exposure for effectiveness.

Another problem that needs to be addressed is determining how to precisely match the distribution of the infused agent with the contours of the tumor. Currently, target coverage may be inadequate even with multiple catheters. Validated predictive models and computer programs are needed that will accurately predict the spread of the wave front of infused therapeutic agent based on an individual patient's tumor configuration. By improving conformity, the toxicity of certain anti-tumor agents to the normal brain may become less of an issue, much as it has in the radiation oncology field.

A problem that compounds the need for better predictive techniques for tailoring the delivery of therapeutic agents to individual patients' tumors is the current lack of useful strategies for imaging the delivery of the agents in real-time. Although Lidar et al. demonstrated promising results by monitoring

T2 images for increases in signal attributed to the spread of the infused paclitaxel solution, this imaging strategy is indirect, and remains to be validated for demonstrating the extent and tissue content of paclitaxel distribution [46]. Thus, techniques for incorporating imaging labels or tracers in the therapeutic agents would seem to have great potential future value for monitoring the delivery of the agent to the target volume. It would, therefore, behoove pharmaceutical developers to consider incorporating such labels into the therapeutic agents prior to entering clinical trials, if possible, in order to avoid the potential need for repeating toxicology and efficacy studies after modification of the agent by the addition of a label or tracer.

Finally, as previously mentioned, the factor that most limits the success of convective delivery to tumors is the nature of the therapeutic agent that is infused. While this method removes a major obstacle by circumventing the problems associated with systemic delivery, the biological complexity of malignant brain tumors is still daunting, and more sophisticated agents will likely be required.

Since Dr. Rickman Godlee first excised—in the modern medical era—a brain tumor in November 1884, the treatment of brain tumors has advanced significantly. Though the medical profession still struggles with malignant gliomas, CED provides another tool to take advantage of the rapid gains in molecular biology and portends a brighter future.

References

1. Nicholson, C., and Phillips, J. M. (1981). Ion diffusion modified by tortuosity and volume fraction in the extracellular microenvironment of the rat cerebellum. *J Physiol* **321**, 225–257.

2. Nicholson, C. (1985). Diffusion from an injected volume of a substance in brain tissue with arbitrary volume fraction and tortuosity. *Brain Res* **333(2)**, 325–329.

3. Nicholson, C. (1993). Ion-selective microelectrodes and diffusion measurements as tools to explore the brain cell microenvironment. *J Neurosci Methods* **48(3)**, 199–213.

4. Basser, P. J. (1992). Interstitial pressure, volume, and flow during infusion into brain tissue. *Microvasc Res* **44(2)**, 143–165.

5. Morrison, P. F., Laske, D. W., Bobo, H., and Dedrick, R. L. (1994). High-flow microinfusion: tissue penetration and pharmacodynamics. *Am J Physiol* **266(1)**, R292–R305.

6. Kalyanasundaram, S., Calhoun, V. D., and Leong, K. W. (1997). A finite element model for predicting the distribution of drugs delivered intracranially to the brain. *Am J Physiol* **273(5)**, R1810–R1821.

7. Nagashima, T., Tamaki, N., Matsumoto, S., Horwitz, B., and Seguchi, Y. (1987). Biomechanics of hydrocephalus: a new theoretical model. *Neurosurgery* **21(6)**, 898–904.

8. Chen, Z. J., Broaddus, W. C., Viswanathan, R. R., Raghavan, R., and Gillies, G. (2002). Intraparenchymal drug delivery via positive-pressure infusion: experimental and modeling studies of poroelasticity in brain phantom gels. *IEEE Trans Biomed Eng* **49(2)**, 85–96.

9. Chen, Z. J., Gillies, G. T., Broaddus, W. C. *et al.* (2004). A realistic brain tissue phantom for intraparenchymal infusion studies. *J Neurosurg* **101(2)**, 314–322.

10. Raghavan, R., Poston, T., and Viswanathan, R. R. (2005). A method and apparatus for targeting material delivery to tissue. U.S.Patent No.6,549,803, April 15, 2003.

11. Thorne, R. G. and Frey, W. H. (2001). Delivery of neurotrophic factors to the central nervous system: pharmacokinetic considerations. *Clin Pharmacokinet* **40(12)**, 907–946.

12. Kumari, R. R., Dhaliwal, J. S., Stoodley, M. A., and Jones, N. R. (1999). Perivascular CSF flow in the rat cerebellum. *J Clin Neurosci* **6(2)**, 143–146.

13. Broaddus, W. C., Gillies, G. T., and Kucharczyk, J. (2005). Image-guided intraparenchymal drug and cell therapy. *In Imaging of the Nervous System: Diagnostic and Therapeutic Applications* (R. E. Latchaw, J. Kucharczyk, and M. E. Moseley, Eds.), Vol. 2, pp. 1467–1476, W. B. Saunders, Philadelphia.

14. Bobo, R. H., Laske, D. W., Akbasak, A., Morrison, P. F., Dedrick, R. L., and Oldfield, E. H. (1994). Convection-enhanced delivery of macromolecules in the brain. *Proc Natl Acad Sci USA* **91(6)**, 2076–2080.

15. Broaddus, W. C., Prabhu, S. S., Gillies, G. T. *et al.* (1998). Distribution and stability of antisense phosphorothioate oligonucleotides in rodent brain following direct intraparenchymal controlled-rate infusion. *J Neurosurg* **88(4)**, 734–742.

16. Bruce, J. N., Falavigna, A., Johnson, J. P. *et al.* (2000). Intracerebral clysis in a rat glioma model. *Neurosurgery* **46(3)**, 683–691.

17. Chen, M. Y., Lonser, R. R., Morrison, P. F., Governale, L. S., and Oldfield, E. H. (1999). Variables affecting Convection-enhanced delivery to the striatum: a systematic examination of rate of infusion, cannula size, infusate concentration, and tissue-cannula sealing time. *J Neurosurg* **90(2)**, 315–320.

18. Cunningham, J., Oiwa, Y., Nagy, D., Podsakoff, G., Colosi, P., and Bankiewicz, K. S. (2000). Distribution of AAV-TK following intracranial Convection-enhanced delivery into rats. *Cell Transplant* **9(5)**, 585–594.

19. Groothuis, D. R., Ward, S., Itskovich, A. C. *et al.* (1999). Comparison of 14C-sucrose delivery to the brain by intravenous, intraventricular, and convection-enhanced intracerebral infusion. *J Neurosurg* **90(2)**, 321–331.

20. Groothuis, D. R., Benalcazar, H., Allen, C. V. *et al.* (2000). Comparison of cytosine arabinoside delivery to rat brain by intravenous, intrathecal, intraventricular and intraparenchymal routes of administration. *Brain Res* **856(1–2)**, 281–290.

21. Haar, P. J., Stewart, J. E., Gillies, G. T., Prabhu, S. S., and Broaddus, W. C. (2001). Quantitative three-dimensional analysis and diffusion modeling of oligonucleotide concentrations after direct intraparenchymal brain infusion. *IEEE Transactions on Biomedical Engineering* **48(5)**, 560–569.

22. Kaiser, M. G., Parsa, A. T., Fine, R. L., Hall, J. S., Chakrabarti, I., and Bruce, J. (2000). Tissue distribution and antitumor activity of topotecan delivered by intracerebral clysis in a rat glioma model. *Neurosurgery* **47(6)**, 1391–1398.

23. Kroll, R. A., Pagel, M. A., Muldoon, L. L., Roman-Goldstein, S., and Neuwelt, E. A. (1996). Increasing volume of distribution to the brain with interstitial infusion: dose, rather than convection, might be the most important factor. *Neurosurgery* **38(4)**, 746–752.

24. Lieberman, D. M., Laske, D. W., Morrison, P. F., Bankiewicz, K. S., and Oldfield, E. H. (1995). Convection-enhanced

distribution of large molecules in gray matter during interstitial drug infusion. *J Neurosurg* **82(6)**, 1021–1029.

25. Nguyen, J. B., Sanchez-Pernaute, R., Cunningham, J., and Bankiewicz, K. S. (2001). Convection-enhanced delivery of AAV-2 combined with heparin increases TK gene transfer in the rat brain. *Neuroreport* **12(9)**, 1961–1964.

26. Prabhu, S. S., Broaddus, W. C., Gillies, G. T. Loudon, W. G., Chen, Z. J., and Smith, B. (1998). Distribution of macromolecular dyes in brain using positive pressure infusion: a model for direct controlled delivery of therapeutic agents. *Surg Neurol* **50(4)**, 367–375.

27. Zirzow, G. C., Sanchez, O. A., Murray, G. J., Brady, R. O., and Oldfield, E. H. (1999). Delivery, distribution, and neuronal uptake of exogenous mannose-terminal glucocerebrosidase in the intact rat brain. *Neurochem Res* **24(2)**, 301–305.

28. Bankiewicz, K. S., Eberling, J. L., Kohutnicka, M. *et al.* (2000). Convection-enhanced delivery of AAV vector in parkinsonian monkeys; in vivo detection of gene expression and restoration of dopaminergic function using pro-drug approach. *Exp Neurol* **164(1)**, 2–14.

29. Laske, D. W., Morrison, P. F., Lieberman, D. M. *et al.* (1997). Chronic interstitial infusion of protein to primate brain: determination of drug distribution and clearance with single-photon emission computerized tomography imaging. *J Neurosurg* **87(4)**, 586–594.

30. Lieberman, D. M., Corthesy, M. E., Cummins, A., and Oldfield, E. H. (1999). Reversal of experimental parkinsonism by using selective chemical ablation of the medial globus pallidus. *J Neurosurg* **90(5)**, 928–934.

31. Lonser, R. R., Corthesy, M. E., Morrison, P. F., Gogate, N., and Oldfield, E. H. (1999). Convection-enhanced selective excitotoxic ablation of the neurons of the globus pallidus internus for treatment of parkinsonism in nonhuman primates. *J Neurosurg* **91(2)**, 294–302.

32. Lonser, R. R., Walbridge, S., Garmestani, K. *et al.* (2002). Successful and safe perfusion of the primate brainstem: in vivo magnetic resonance imaging of macromolecular distribution during infusion. *J Neurosurg* **97(4)**, 905–913.

33. Chen, M. Y., Hoffer, A., Morrison, P. F. *et al.* Surface Properties, more than size, limit convective distribution of viral-sized particles and viruses in the CNS. *J Neurosurg* in press.

34. Yang, W., Barth, R. F., Adams, D. M. *et al.* (2002). Convection-enhanced delivery of boronated epidermal growth factor for molecular targeting of EGF receptor-positive gliomas. *Cancer Res* **62(22)**, 6552–6558.

35. Saito, R., Bringas, J. R., Panner, A. *et al.* (2004). Convection-enhanced delivery of tumor necrosis factor-related apoptosis-inducing ligand with systemic administration of temozolomide prolongs survival in an intracranial glioblastoma xenograft model. *Cancer Res* **64(19)**, 6858–6862.

36. Brust, D., Feden, J., Farnsworth, J., Amir, C., Broaddus, W. C., and Valerie, K. (2000). Radiosensitization of rat glioma with bromodeoxycytidine and adenovirus expressing herpes simplex virus-thymidine kinase delivered by slow, rate-controlled positive pressure infusion. *Cancer Gene Ther* **7(5)**, 778–788.

37. Mamot, C., Nguyen, J. B., Pourdehnad, M. *et al.* (2004). Extensive distribution of liposomes in rodent brains and

brain tumors following Convection-enhanced delivery. *J Neurooncol* **68(1)**, 1–9.

38. Vavra, M., Ali, M. J., Kang, E. W. *et al.* (2004). Comparative pharmacokinetics of 14C-sucrose in RG-2 rat gliomas after intravenous and Convection-enhanced delivery. *Neuro-oncol* **6(2)**, 104–112.

39. Laske, D. W., Youle, R. J., and Oldfield, E. H. (1997). Tumor regression with regional distribution of the targeted toxin TF-CRM107 in patients with malignant brain tumors. *Nat Med* **3(12)**, 1362–1368.

40. Hall, W. A., Rustamzadeh, E., and Asher, A. L. (2003). CED in clinical trials. *Neurosurg Focus* **14(2)** 1–4.

41. Kunwar, S. (2003). Convection enhanced delivery of IL13-PE38QQR for treatment of recurrent malignant glioma: presentation of interim findings from ongoing phase 1 studies. *Acta Neurochir Suppl* **88**, 105–111.

42. Husain, S. R. and Puri, R. K. (2003). Interleukin-13 receptor-directed cytotoxin for malignant glioma therapy: from bench to bedside. *J Neurooncol* **65(1)**, 37–48.

43. Weber, F. W., Floeth, F., Asher, A. *et al.* (2003). Local convection enhanced delivery of IL4-Pseudomonas exotoxin (NBI-3001) for treatment of patients with recurrent malignant glioma. *Acta Neurochir Suppl* **88**, 93–103.

44. Pietronigro, D., Drnovsky, F., Cravioto, H., and Ransohoff, J. (2003). DTI-015 produces cures in T9 gliosarcoma. *Neoplasia* **5(1)**, 17–22.

45. Hassenbusch, S. J., Nardone, E. M., Levin, V. A., Leeds, N., and Pietronigro, D. (2003). Stereotactic injection of DTI-015 into recurrent malignant gliomas: phase I/II trial. *Neoplasia* **5(1)**, 9–16.

46. Lidar, Z., Mardor, Y., Jonas, T. *et al.* (2004). Convection-enhanced delivery of paclitaxel for the treatment of recurrent malignant glioma: a phase I/II clinical study. *J Neurosurg* **100(3)**, 472–479.

47. Merlo, A., Hausmann, O., Wasner, M. *et al.* (1999). Locoregional regulatory peptide receptor targeting with the diffusible somatostatin analogue 90Y-labeled DOTA0-D-Phe1-Tyr3-octreotide (DOTATOC): a pilot study in human gliomas. *Clin Cancer Res* **5(5)**, 1025–1033.

48. Papanastassiou, V., Pizer, B. L., Coakham, H. B., Bullimore, J., Zananiri, T., and Kemshead, J. T. (1993). Treatment of recurrent and cystic malignant gliomas by a single intracavity injection of 131I monoclonal antibody: feasibility, pharmacokinetics and dosimetry. *Br J Cancer* **67(1)**, 144–151.

49. Riva, P., Arista, A., Franceschi, G. *et al.* (1995). Local treatment of malignant gliomas by direct infusion of specific monoclonal antibodies labeled with 131I: comparison of the results obtained in recurrent and newly diagnosed tumors. *Cancer Res* **55(23)**, Suppl 5952s–5956s.

50. Hadaczek, P., Mirek, H., Berger, M. S., and Bankiewicz, K. (2005). Limited efficacy of gene transfer in herpes simplex virus-thymidine kinase/ganciclovir gene therapy for brain tumors. *J Neurosurg* **102(2)**, 328–335.

51. Voges, J., Reszka, R., Gossmann, A. *et al.* (2003). Imaging-guided Convection-enhanced delivery and gene therapy of glioblastoma. *Ann Neurol* **54(4)**, 479–487.

52. Kucharczyk, J., Broaddus, W. C., Fillmore, H. L., and Gillies, G. T. (2005). Cell Delivery Catheter and Method. U.S.Patent No.6,599,274, July 29, 2003.

Marrow Ablative Chemotherapy with Hematopoietic Stem Cell Rescue

Anna Butturini, Matthew Carabasi, and Jonathan L. Finlay

ABSTRACT: Marrow ablative chemotherapy with hematopoietic stem cell rescue (also referred to as hematopoietic stem cell transplantation) is a therapeutic strategy used to treat children and young adults with brain tumors whose conventional therapy is either too toxic (e.g., radiotherapy in infants) or ineffective (e.g., rhabdoid tumors or progressive malignant tumors). In this strategy, high-dose chemotherapy (typically including alkylating agents and/or other radiomimetic drugs) is given to patients after initial conventional therapy, including surgery and irradiation, if possible, and conventional-dose chemotherapy. The success of marrow ablative chemotherapy with hematopoietic stem cell rescue depends on several variables, including the histological type of tumor, extent of disease and of surgical resection, and response to prior chemotherapy. In this chapter, we shall review results of marrow ablative chemotherapy with hematopoietic stem cell rescue in brain tumors of differing histologies.

INTRODUCTION

High-dose myeloablative chemotherapy with hematopoietic stem cell rescue is a novel approach in the treatment of brain tumors. Until the mid-1970s, the standard treatment for malignant brain tumors was based on surgery and irradiation. Then the addition of chemotherapy, given as "maintenance" or "adjuvant" therapy, was shown to increase the cure rate of some brain tumors, particularly medulloblastoma and other primitive neuroectodermal tumors (PNET) [1]. At the same time, it became clear that the

long-term consequences of high-dose cranial irradiation in very young children were significantly and irreversibly deleterious [2]. This notion prompted the development of new strategies using high-dose chemotherapy to delay, decrease, or avoid the use of radiotherapy in infants and toddlers. In parallel, the same strategies were attempted to treat tumors considered "incurable" by conventional therapy, either for histology (e.g., rhabdoid tumors, glioblastoma multiforme) or because they had recurred after conventional treatment.

The rationale for increasing chemotherapy doses derived from *in vitro* studies showing a dose-response curve for several drugs (see Chapter 4). Higher doses of chemotherapy could overcome tumor drug-resistance and increase the passage through the blood–brain barrier, which contribute to chemotherapy failures in patients with brain tumors (see Chapters 2 and 6). Because the expected consequences of high-dose chemotherapy were severe and possibly irreversible myelosuppression, these therapies were followed by the infusion of autologous hematopoietic stem cells, in order to rescue hematopoiesis. Hematopoietic stem cells were harvested and cryopreserved during the early phases of therapy either from bone marrow or from the blood. This treatment strategy, also referred to as "autologous hematopoietic stem cell transplantation", "auto-transplantation", or "auto-transplant" was similar to that used in leukemia and in other solid tumors [3].

In brain tumors, auto-transplantations were initially evaluated in patients with recurrent tumors having failed initial irradiation and chemotherapy. At first auto-transplantation was attempted irrespectively to the disease status (bulky *versus* minimal disease), response to prior therapy or even to the adequacy of organ function [4,8,9]. In the decade between 1985 and 1995, several trials of high-dose chemotherapy and autologous hematopoietic stem cell rescue in brain tumor patients were conducted in North America and Europe (Table 21.1) [5–18]. In that period, the source of stem cells was more commonly bone marrow. When autotransplants were performed in patients after recurrence [5,6,8,18] the presence of either bulky tumor burden and/or organ dysfunctions increased the risk of failures and acute and often fatal toxicities. Toxicity was higher than after conventional therapy; treatment-related mortality varied from less than 10 per cent to greater than 30 per cent [5,18]. The likelihood of cure varied from less than 10 per cent to about 50 per cent and correlated with histological type, extent of surgical resection, presence of metastases, type of pre-transplant therapy and extent of disease at transplant. Overall results were considered promising, therefore high-dose chemotherapy was attempted to limit the use of radiotherapy in young children (either less than three or six years of age) [11,19]. Again, results were encouraging, so that in the late 1990s, high-dose chemotherapy and autologous hematopoietic stem cell rescue became a standard therapy in children with malignant brain tumors diagnosed at less than three years of age. It was

TABLE 21.1   Reports >15 Patients Receiving Myeloablative Therapy With Hematopoietic Stem Cell Rescue 1985–1995

Reference	N	At diagnosis	After relapse	PNET Medullobl.	HG glioma	Epend.	Others	Conditioning Regimen	Post-transplant Outcome
		Timing of Transplant		**Diagnosis**					
Dupuis Girod, 1996 [7]	20	0	20	20	0	0	0	Bu, TT	2 years EFS: 50%
Finlay, 1996 [5]	45	0	45	12	18	5	10	TT, VP16	2 years Surv: 16%
Mahoney, 1996 [6]	18	0	18	9	3	3	3	CPM, Melphalan	1 year PFS: 17% 1 year Surv: 39%
Grill, 1996 [8]	16	0	16	0	0	16	0	Bu, TT	No response >50%
Bouffet, 1997 [9]	22	13	9	0	22	0	0	TT, VP16	3/22 >4 years EF
Graham, 1997 [10]	49	16	33	29	12	5	3	CPM, Melphalan Bu, Melphalan Carbo, VP16	2 years EFS ≈ 40%
Mason, 1998 [11]	37	37	0	21	2	9	5	Carbo, TT, VP16	2 years EFS: 41% 2 years Surv: 73%
Mason, 1998b [12]	15	0	15	0	0	15	0	Carbo, TT, VP16	2 years EFS: <5%
Dunkel, 1998 [13]	23	0	23	23	0	0	0	Carbo, TT, VP16	3 years EFS: 34% 3 years Surv: 46%
Gururangan, 1998 [15]	20	0	20	13	5	1	1	Carbo, TT, VP16	3 years EFS: 47% 3 years Surv: 43%
Dunkel, 1998b [14]	16	6	10	0	0	0	16*	TT, VP16 BCNU, TT, VP16 Carbo, TT, VP16	No long term cure
Papadopoulos, 1998 [16]	31**	13	18	9	12	1	9	Carbo, TT, VP16 BCNU, TT, VP16	Median survival 12 months
Abrey, 1999 [18]	26**	9	17	0	26	0	0	BCNU, TT, VP16 Carbo, TT, VP16	2 years EFS: 33% in relapsed

Abbreviations: PNET — primitive neuroendodermal tumor; HG — high grade; Epend — ependymoma; Bu — Busulfan; TT — thiotepa; CPM — cyclophosphamide; Carbo — Carboplatin; VP16 — Etoposide; EFS — event-free survival; PFS — progression-free survival; EF — event free; Surv. — survival.

*diffuse pontine and brain stem tumors
**mostly or exclusively adult patients.

TABLE 21.2 Examples of Intermediate-Dose Regimens Used With Stem Cell Support

Drug#1	Dose/m² (mg)	Drug#2	Dose/m² (mg)	Drug#3	Dose/m² (mg)	Total cycles	Reference
Cyclophosphamide	7000	–	–	–		2	[21]
Vincristine	1.5	CCNU	130	Procarbazine	1050	4	[20]
Vincristine	3	Cisplatin	75	Cyclophosphamide	4000	4	[22]

also used in children and young adults with high risk or recurrent malignant brain tumors.

Overall more then 300 patients with brain tumors received autologous stem cell transplants every year. Presently, auto-transplantations are typically given to patients with minimal tumor burden (achieved by surgical re-operation, irradiation, chemotherapy, or some combination of these) and with adequate organ (e.g., liver, kidney, and lung) function.

Mirroring what has happened in other cancers, peripheral blood has become the most common source of stem cells in auto-transplantations for brain tumor patients [19]. The ability to mobilize hematopoietic stem cells from bone marrow to blood made their collection more feasible. The increased availability of hematopoietic stem cells allowed their use to explore different ways to intensify chemotherapy in brain tumors. Hematopoietic stem cells were given to both increase chemotherapy doses and/or to decrease the length of pancytopenia following non-myeloablative chemotherapy. This strategy-referred to as "mini-autotransplants" or "stem-cell-supported intermediate-dose chemotherapy", typically uses multiple cycles of chemotherapy at doses higher than those used conventionally (Table 21.2) [20–22]. Also, the sequential effect of two or more myeloablative regimens each followed by hematopoietic stem cell infusions (tandem transplants) has been studied (Table 21.3) [23–26].

In this chapter, we will review the most common intermediate-dose and high-dose drug combinations used in the context of auto-transplantation in patients with brain tumors (the last referred to as conditioning regimens). Because most studies suggest that tumor histology is a main determinant of efficacy of high-dose chemotherapy, we will review the results of increasing chemotherapy doses according to the different brain tumor types. Also, we will summarize results of transplants using hematopoietic cells from allogeneic donors.

MARROW ABLATIVE THERAPIES

Drugs

To be escalated in the context of an auto-transplant, a drug must have been shown to have a steep dose-response anti-cancer effect *in vitro* and a modest extra-medullary toxicity in clinical studies. Alkylating agents have both these characteristics and therefore they are commonly used in conditioning for several cancers. In particular one alkylator, thioTEPA, is highly lipophilic and has a high capability of penetrating into the central nervous system (see Chapter 2).

Conditioning regimens in brain tumors use alkylating agents either alone or in combination with other drugs (Table 21.4) [27–32]. The rationale for using multiple drugs is based on *in vitro* synergy (e.g., alkylating agents which damage DNA and drugs affecting DNA repair, thioTEPA and etoposide) or on the combination of drugs with similar anti-cancer mechanisms but different toxicity (e.g., two alkylating agents, melphalan and cyclophosphamide). On the contrary, the rationale to use single-drug conditioning regimens is to use the highest possible dose of one effective drug avoiding the combined toxicity of association. So far no conditioning regimen has proven superior in brain cancer or in any other cancer in the context of autotransplantation.

Many drugs used in conditioning regimens for brain tumors are predominantly excreted through the kidneys. In most trials, assessment of renal function has been mandatory to screen patients suitable for high-dose therapies and/or to adjust chemotherapy dose. In particular, doses of carboplatin are adjusted

TABLE 21.3 Examples of Tandem Transplant Regimens

First Transplant	Second Transplant	Reference
Cyclophosphamide	Carboplatin	[23]
Cyclophosphamide + Carboplatin + Etoposide	Cyclophoshamide + Thiotepa	[25]
Thiotepa1	Carboplatin	[26]
Melphalan	Melphalan	[24]

TABLE 21.4   Conditioning Regimens

Alkylator	Total Dose/m² (mg)	Drug#2	Total dose/m² (mg)	Drug#3	Total dose/m² (mg)	Reference
Thiotepa	600–900	–	–	–	–	[27]
Thiotepa	900	Carboplatin	1500 (or by Calvert formula AUC = 7 × 3 days)	Etoposide	750	[11–14]
Thiotepa	600	BCNU	600	Etoposide	750	[5,17]
Thiotepa	900	Carboplatin	by Calvert formula AUC = 7 × 3 days	Topotecan	10	[28]
–	–	Carboplatin	2100	Etoposide	1500	[10]
Thiotepa	900	–	–	Etoposide	1500	[9]
Thiotepa	900	–	–	Etoposide	750	[5]
Thiotepa	900	Busulfan	600	–	–	[8,29]
Thiotepa	900	Cyclophosphamide	6000	–	–	[30]
Melphalan	75–180	Cyclophosphamide	6000	–	–	[10]
BCNU	600–1400	–	–	–	–	[31]

to the glomerular filtration rate using different formulae (e.g., the Calvert formula) [33].

Short-Term Toxicity

The almost universal complications after high-dose chemotherapy are pancytopenia and mucositis. The duration and severity of pancytopenia is a consequence of several variables related to the hematopoietic stem cells reinfused (or persisting) after high-dose chemotherapy and the ability of the stroma to sustain hematopoiesis. They include the amount of chemotherapy, and dose as well as extent of irradiation (i.e., craniospinal radiotherapy) the patients had received before stem cell collection and before auto-transplantation, the conditioning regimen itself, and characteristics of the hematopoietic stem cells infused. Since the introduction of hematopoietic growth factors, myeloid recovery is accelerated and neutropenia usually lasts two weeks or less. In contrast, thrombocytopenia may last several months, especially in heavily pre-treated patients [34].

Mucositis is often severe, requiring narcotic analgesics and intravenous alimentation. This is especially the case after thioTEPA-containing regimens and in patients with prior spinal irradiation. Patients receiving thioTEPA may also develop generalized skin erythema and desquamation, secondary to the excretion of thioTEPA in sweat. Acute neurological dysfunction, including hallucinations, coma, seizures, headaches, ataxia-tremor-dysarthria syndrome, anorexia, and nausea syndrome are reported in

about 50 per cent of patients in the first three months after transplant. These symptoms are possibly associated with iatrogenic metabolic dysfunction or with a direct neurotoxicity of chemotherapy. They are usually reversible [35].

Other common and possibly severe toxicities relate to microvasculature damage: capillary leaking and veno-occlusive disease may develop. Nephrotoxicity is possible especially after platinum-containing regimens. Acute pulmonary toxicity is described in persons treated with busulfan, BCNU, and, less commonly, thioTEPA. These complications are usually transient, but especially if associated with infection secondary to neutropenia, they may lead to multi-organ system failure and death. As discussed above, toxic death rates were as high as 35 per cent in early studies. In more recent studies, the toxic mortality rate has declined to below 10 per cent. Again, life-threatening toxicities may be more frequent in heavily pre-treated patients.

Long-Term Toxicity

There are relatively few data on the long-term toxicity of high-dose chemotherapy for brain tumors. Furthermore it is difficult to differentiate the consequences of high-dose chemotherapy from those of the other components of therapy or from the brain tumors themselves. However, hearing loss has been reported in about half of children who received carboplatin-containing conditioning regimens [36]. Neurological and intellectual functions in children who received high-dose chemotherapy but not radiotherapy during their first years of life were either in the low average or only moderately impaired [37].

Second cancers are reported in persons treated with autologous transplant for cancers other than brain tumors. For example, the risk of a second cancer developing in persons who received high-dose chemotherapy for Hodgkin disease or breast cancer is about 10–20 per cent after 15 years [38]. Whether this will be true also for survivors of autotransplant for brain tumors is unknown. Similarly, reproductive and endocrine impairments have been reported in patients treated with high-dose chemotherapy for other cancers [39], as well as in children with brain tumors treated with conventional-dose chemotherapy and irradiation. The extent of these complications after high-dose chemotherapy in brain tumor patients remains unknown.

RESULTS OF HIGH-DOSE CHEMOTHERAPY AND AUTO-TRANSPLANTATION

Medulloblastoma and PNET

Conventional-dose chemotherapy is reportedly effective in medulloblastoma and other intracranial PNETs. In newly diagnosed patients with "standard risk" (i.e., non-metastatic) medulloblastoma, the addition of chemotherapy to surgery and craniospinal radiotherapy has increased the cure rate from about 50 to over 80 per cent, and chemotherapy is now part of the standard therapy for medulloblastoma and PNET diagnosed in patients older than three years of age (see Chapter 29). In contrast, conventional-dose chemotherapy is ineffective in producing prolonged disease-free survival following a recurrence.

The efficacy of conventional-dose chemotherapy prompted studies of high-dose chemotherapy in three patient settings: young children (usually less than six years of age), older patients with the highest risk tumors (e.g., disseminated medulloblastoma, pineoblastoma), and patients with recurrent tumors. Chronologically, the first patient population in which high-dose chemotherapy was attempted were patients failing radiotherapy, with or without maintenance chemotherapy [3,6,10,13]. Because this strategy was able to rescue about a third of the patients, autologous transplant was then used in infants failing conventional-dose-induction chemotherapy [7, 15]. About half of these infants were cured by high-dose chemotherapy and either focal [7] or reduced dose

craniospinal irradiation [15] post-transplant. These data suggested that high-dose chemotherapy is effective in medulloblastoma and PNET. Therefore it was used upfront in both young children and in older high-risk patients [40–42]. In these studies typically young children received no or only focal radiotherapy, while older patients typically received craniospinal irradiation.

Overall, 60–80 per cent of very young children are cured by using surgery, conventional chemotherapy with high-dose chemotherapy as consolidation and limited field or no radiotherapy [7,11,15, 40–45]. Results were better in medulloblastoma and in those who did not have residual tumor at time of transplant, either because their tumor was completely resected or because it showed complete response to conventional-dose chemotherapy. In older children and adults with high risk features, high-dose chemotherapy up front resulted in about 60 per cent probability of long-term response; in patients treated after relapse the probability of long-term response varied from 25 to 50 per cent [10,13,18,42]. Even in these settings results correlated with tumor site and extension of resection (poorer in pineal and/or incompletely resected tumors).

Atypical Teratoid/Rhabdoid Tumors of the Brain

Atypical teratoid/rhabdoid tumors (AT/RT) of the brain have poor prognosis after surgery and irradiation. Conventional-dose chemotherapy has resulted in minimal improvement in survival and irradiation—usually given irrespective of age—has been considered a major treatment contributing to cure. Some data now indicate that children with AT/RT treated with current "Baby" protocols incorporating marrow ablative chemotherapy with stem cell rescue, can be cured without recourse to irradiation [46,47]

Gliomas

Malignant Gliomas

The benefit of adding conventional-dose chemotherapy to surgical resection and irradiation in supratentorial malignant (or high-grade) gliomas (e.g., anaplastic astrocytoma and glioblastoma multiforme) is modest (see Chapter 24). A recent meta-analysis reported a better outcome in patients who received both chemotherapy and irradiation [48]. Most recently, the administration of temozolomide concurrent with and following irradiation for adults with high grade gliomas appears to be producing promising prolongation of time to recurrence [49]. However conventional-dose chemotherapy has not been shown to prolong survival significantly in patients with recurrent malignant gliomas.

Several studies have attempted to improve results by increasing doses, with or without stem rescue. Increasing doses of thiotepa (40–70 mg/m^2) [50] or cyclophosphamide (up to 7 g) were ineffective [51,52]. Similarly, the use of stem cell support to decrease intervals between cycles of procarbazine, CCNU, and vincristine (PCV) or to administer radiation therapy concurrently [21,53,54] had no observable impact on cure.

In adults, myeloablative chemotherapy using a single drug (BCNU, etoposide, or thiotepa) [32,55,56] or multiple drugs (Thiotepa and etoposide in combination, with the addition of either BCNU or carboplatin) [57] were attempted in patients treated either at diagnosis or following relapse. Clinical responses were reported but the effect on cure was questionable: some studies reported more than 20 per cent long-term response [32], others failed to find any significant impact on cure [58]. Factors possibly affecting results of high-dose therapy in adults were histology, extent of surgical resection, and patient age [56]. Results of pediatric studies have been more promising. In children, high-dose BCNU does not seem effective [9,59], but multi-drug conditioning regimen despite significant morbidity and mortality, in about 20–35 per cent, resulted in long-term response [17,60,61]. Analysis of factors affecting outcome after high-dose chemotherapy in children is complex: the extent of surgical resection and possibly the use of radiotherapy may affect cure rate [15,60,61].

Diffuse Brainstem Gliomas

The outcome of diffuse brainstem (pontine) tumors is extremely poor. Conventional-dose chemotherapy is considered ineffective: high-dose radiotherapy is the usual treatment but progression-free survival is less than 10 per cent after 2 years. Although in most cases the histological diagnosis is unknown, the overwhelming majority of diffuse pontine tumors are high grade gliomas. High-dose chemotherapy had been added to radiotherapy in several studies of children with either recurrent or newly diagnosed pontine tumors, without benefit [14,29].

Oligodendroglioma

Oligodendrogliomas usually respond to conventional-dose chemotherapy but tend to have late relapses. Relapses usually respond poorly to chemotherapy (see Chapter 26). High-dose myeloablative mono-chemotherapy (thiotepa, BCNU) has been used in adult patients with high grade or recurrent oligodendroglioma [56,62]. Results in patients transplanted after recurrence have been poor [63]. In contrast, about half of the patients treated before recurrence with marrow ablative chemotherapy and without irradiation survived for a median of at least three years without recurrence of tumor [62]. Whether these results were better then those achievable with irradiation and conventional-dose chemotherapy is unclear.

Ependymoma

Conventional-dose therapy has a modest but now proven role in ependymoma (see Chapter 30). About 40–50 per cent of these tumors are cured by surgery and local field irradiation; the addition of chemotherapy prior to irradiation now appears to have improved the survival for children with incompletely resected tumors [64]. However, chemotherapy is not effective in improving cure rates after recurrence. High-dose chemotherapy has been attempted in young children as consolidation therapy: about 60 per cent were reported to be long-term survivors, about half of whom avoided irradiation [65]. Results with marrow ablative chemotherapy after relapse have been disappointing [8,12].

Central Nervous System Germ Cell Tumors

Conventional-dose chemotherapy is effective in germ cell tumors. Germinoma are usually also very radiosensitive: chemotherapy might not be used up front. Non-germinomatous tumors are usually treated with both irradiation and chemotherapy. These conventional approaches cure about 85–95 per cent of geminoma and 30–65 per cent of non-germinomatous germ cell tumors of the brain. Conventional-dose chemotherapy is only rarely effective in tumors after recurrence following irradiation (see Chapter 33). High-dose chemotherapy seems very effective in germinoma after relapse, with more than 80 per cent cure, but less effective in non-germinomatous tumors following relapse, in whom fewer than 50 per cent of patients were long-term survivors [66]. The current North American national (Childrens Oncology Group) study for newly diagnosed non-germinomatous germ cell tumors now incorporates marrow ablative chemotherapy with autologous stem cell rescue for patients whose tumors respond only slowly to initial chemotherapy, in an approach drawn from the experience with adults with testicular cancer.

Results of High-Dose Chemotherapy and Allotransplantation

As discussed above, high-dose chemotherapy with autologous stem cell rescue can be an effective treatment for patients with otherwise incurable CNS tumors. However, autologous hematopoietic stem cells may be difficult or impossible to collect in patients intensely pre-treated with chemotherapy and spinal irradiation and in those (typically with medulloblastoma or PNET) whose iliac bone and bone marrow are heavily infiltrated by tumor. In these patients, transplantation can be attempted using hematopoietic cells collected from an HLA-identical donor (allogeneic transplant).

In patients with leukemia and other hematological malignancies, allogeneic transplantation using stem cells from a family or unrelated donor is often considered as a treatment alternative. The use of allogeneic stem cells has several advantages. First, it eliminates the risk of reinfusion of malignant cells with the stem cell product. Second, it allows for the generation of immune-mediated anti-tumor effect (graft-versus-tumor or GVT) capable of eliminating residual disease that has survived the conditioning regimen. More recently, the recognition that the amount of conditioning required for this procedure could be significantly reduced has also made this a potential option for patients who cannot tolerate the high-dose therapy usually administered with autologous transplants. The impact of these advantages on survival can sometimes be diminished by an undesirable immune response against normal tissues called graft-versus-host disease (GVHD) seen frequently in allogeneic stem cell transplant recipients. In addition, these patients have an increased risk of opportunistic infection such as CMV pneumonia compared to autologous transplant recipients.

Experience with allogeneic transplant in patients with CNS malignancies and occurrence of "graft-versus-brain tumor" is limited to a few case reports in ependymoma and medulloblastoma [67,68]. Thus, the potential utility of allogeneic transplant for CNS tumors can only be gauged by examining whether GVT is seen in other relevant scenarios. The first question is whether GVT can occur in the brain. Allogeneic transplant treated successfully patients with acute myeloid leukemia and brain chloroma [69] and prolonged response in a patient with melanoma brain metastases [70], but these patients received myeloablative therapy, therefore it is difficult to establish the role of allogeneic anti-tumor immune effects. There is anecdotal experience with reduced-conditioning allogeneic transplantation

as a treatment for patients with CNS lymphoma. Aoyama and colleagues recently reported the result of allogeneic transplant in a 33-year-old male with diffuse large B cell lymphoma involving the CNS [71]. He was treated with standard dose chemotherapy, intra-thecal therapy and an autologous transplant. When his CNS disease recurred as documented by CSF exam and MRI, an allogeneic transplant was performed. Because of the prior history of autologous transplant, the patient received a reduced intensity conditioning regimen followed by an infusion of peripheral blood stem cells from his HLA-identical sibling. A second infusion of lymphocytes from the donor was administered 40 days later to enhance GVT. The patient rapidly developed both GVT and GVHD with complete resolution of all measurable disease and remains in complete remission 15 months post-transplant. A second reported case involves a female patient who received peripheral blood stem cells from an HLA-matched sibling after reduced intensity conditioning [72]. She demonstrated both GVT and GVHD early post-transplant with resolution of all CNS disease by the third month and remains in complete remission after 30 months following transplantation. These results demonstrate that GVT can occur in the brain at least against primary CNS lymphoma.

The second question is whether GVT occurs in patients with non-hematological malignancies. Results in two published series suggests that it does. In the first series, 18 patients with advanced solid tumors were transplanted with stem cells collected from family members or unrelated donors after treatment with reduced intensity conditioning [73]. One patient with colon cancer responded but died of pneumonia early post-transplant. Several patients with renal cell or colon cancer showed mixed responses associated with GVHD. The second series includes 23 patients with less advanced renal cell or breast cancer [74]. Patients were treated with fludarabine and melphalan followed by infusion of HLA identical stem cells collected from family members or unrelated donors. An overall response rate of 45 per cent was seen, with three complete responses, two partial responses, and five minor responses. Non-relapse mortality was 22 per cent in both series. Together, these reports demonstrate that GVT can be seen against some solid tumor types, but, like in leukemia, is not necessarily associated to GVHD.

As an alternative to full transplantation, investigators are examining the utility of using specific allogeneic cell populations to treat or protect against CNS malignancies in animal models [75]. Specifically, mice with either Gl261 glioma or SB-5b breast car-cinoma brain masses treated with allogeneic fibro-blasts modified to secrete interleukin-2 (IL-2) injected intracerebrally showed prolonged survival compared to untreated animals. In addition, pre-treatment with these cells blocked subsequent tumor develop-ment in 50 per cent of animals injected with Gl261 glioma cells and 75 per cent of animals injected with SB-5b breast cancer cells. Finally, animals injected with fibroblasts in the absence of tumor demon-strated no CNS toxicity, suggesting that these res-ponses are tumor-specific. These observations raise the intriguing possibility that allogeneic cells alone may be an effective treatment for CNS malignancies even without pre-treatment with high-dose chemotherapy.

CONCLUSIONS

The data we have reviewed demonstrate that high-dose chemotherapy has a role in the treatment of certain types of brain tumors, although it raises a number of questions.

First, what is the advantage of high-dose chemo-therapy over conventional-dose chemotherapy? Many recurring tumors appear to respond to high dose but not to conventional-dose chemotherapy, suggesting that high-dose therapy has greater anti-cancer effect. However the patient selection applied to transplants bias the comparison between the two strategies. High-dose chemotherapy is associated with significantly higher morbidity and mortality. In contrast, so far there is no indication of any long-term toxicities that are qualitatively or quantitatively more severe than that due to conventional therapy. In particular, the few neuro-cognitive studies conducted in children who received high-dose therapy in infancy, with avoidance of irradiation, suggest that this approach might be less toxic than those incorporating brain irradiation. Because results of transplant are poorer in older and heavily pre-treated patients and in those with bulky disease unresponsive to chemotherapy, a possible strategy could be to reserve autotransplant to those who could more likely benefit from it, like younger patients with chemo-responsive tumors in the earlier phases of therapy, either up front or after the first recurrence.

Second, why does high-dose chemotherapy fail in a significant proportion of patients? High-dose conditioning has at least some degree of anti-cancer effects in different brain tumors. For example, partial or complete regression of tumor which recurred after conventional therapy is described in malignant

glioma, medulloblastoma/PNET, germ cell tumors, occasionally rhabdoid tumors. However in few instances (mostly medulloblastoma and germinoma), responses consistently translate into long-term cure. This might be because the drugs that have been escalated, at the doses to which one can safely escalate, can cytoreduce but not eradicate cancer. For example, hematopoietic stem cells are consi-dered partially resistant to busulfan because of their proliferative characteristics [76]. Whether the same principle applies to subsets of brain tumor cells is unknown. Also it is possible that the timing characteristics of high-dose therapy are inadequate. In many tumors, both doses and duration of therapy (dose intensity) are more important than the absolute dose. Again that is not known in brain tumors [77]. Another limitation of auto-transplantation in other cancers is the possibility of re-infusion of cancer cells with the graft. Despite some data suggesting that brain tumor cells can exist outside the brain [78], this is usually not considered an issue in brain tumors. Finally, in brain tumors, a possibility of failure is due to inadequate drug pharmacokinetics. Cancer-induced changes may alter the brain–blood barrier, reducing drug influx into the tumor (see Chapters 6 and 16). However, it is unknown whether this applies also to single-cell micro-metastases.

Several studies which are now in progress may give some answers. For example, conditioning regi-mens using novel drugs, like temozolamide, may indicate whether using drugs different from classic alkylating agents will improve results [79]. Also, the use of tandem transplants and multiple cycles of intermediate-dose chemotherapy will provide data on the role of longer and intermittent exposure to high-dose therapy. Presently, most of these new approaches are investigated in the context of phase I or II studies. Large randomized clinical studies will be necessary to prove whether any of those approaches is superior, so that these achievements could be moved to other settings. One such study in recurrent malignant gliomas of childhood is about to be opened throughout the North American Children's Oncology Group—this study will rando-mize patients in minimal tumor burden, between a single cycle auto-transplant (thioTEPA, etoposide and carboplatin) against triple mini-transplant (thioTEPA and carboplatin).

In conclusion, the use of high-dose chemotherapy has revamped the role of cytotoxic therapy in brain tumors. It has dramatically changed our way of treating some patients, such as young children with medulloblastoma, PNET ependymoma, and possibly

AT/R tumors. In other situations it has not yet proven effective. However in many of the last instances, high-dose chemotherapy was at least able to achieve transient responses. The notion that high-dose chemotherapy may effectively reduce the number of brain tumor cells, even if fail to cure, could be exploited in strategies using high-dose chemotherapy to obtain the maximum tumor debulking. This could be followed by other types of therapies potentially effective in eradicating minimal residual disease. This approach has been attempted in different cancers and it has proven successful in neuroblastoma [80]. Finally, even if limited, evidence suggests that allogeneic cells may have long-term anti-cancer effect in patients with CNS malignancies. Although the role of allogeneic therapy in the treatment of CNS malignancies still needs to be tested in clinical trials, the experience with allogeneic transplants in brain tumors may have paved the way to novel and exciting ways to treat brain tumors.

References

1. Packer, R. J., Goldwein, J., Nicholson, H. S. et al. (1999). Treatment of children with medulloblastoma with reduced dose craniospinal radian therapy and adjuvant chemotherapy. A Children's Cancer Study. J Clin Oncol 17, 2127–2138.
2. Finlay, J. L. (1996). The role of high dose chemotherapy and stem cell rescue in the treatment of malignant brain tumors. Bone Marrow Transplant 18 (suppl 3), S1–S5.
3. Gale, R. P., Butturini, A., and Horowitz, M. M. (1992). What is the best strategy fot autotransplants in cancer? Bone Marrow Transplant 9, 303–309.
4. Kalifa, C., Hartmann, O., Demeocq, F. et al. (1992). High dose busulfan and thiotepa with autologous bone marrow transplantation in childhood malignant brain tumors: a phase II study. Bone Marrow Transplant 9, 227–233.
5. Finlay, J. L., Goldman, S., Wong, M. C. et al. (1996). Pilot study of high-dose thiotepa and etoposide with autologous bone marrow rescue in children and young adults with recurrent CNS tumors. The Children's Cancer Group. J Clin Oncol 14, 2495–2503.
6. Mahoney, D. H., Jr, Strother, D., Camitta, B. et al. (1996). High-dose melphalan and cyclophosphamide with autologous bone marrow rescue for recurrent/progressive malignant brain tumors in children: a pilot pediatric oncology group study. J Clin Oncol 14, 382–388.
7. Dupuis Girod, S., Hartmann, O., Benhamou, E. et al. (1996). Will high-dose chemotherapy followed by autologous bone marrow transplantation supplant craniospinal irradiation in young children treated for medulloblastoma? Journal of Neurooncol 27, 87–98.
8. Grill, J., Kalifa, C., Doz, F. et al. (1996). A high-dose busulfan-thiotepa combination followed by autologous bone marrow transplantation in childhood recurrent ependymoma. A phase-II study. Pediatr Neurosurg 25, 7–12.
9. Bouffet, E., Mottolese, C., Jouvet, A. et al. (1997). Etoposide and thiotepa followed by ABMT (autologous bone marrow transplantation) in children and young adults with high-grade gliomas. Eur J Cancer 33, 91–95.
10. Graham, M. L., Herndon, J. E., 2nd, Casey, J. R. et al. (1997). High-dose chemotherapy with autologous stem-cell rescue in patients with recurrent and high-risk pediatric brain tumors. J Clin Oncol 15, 1814–1823.
11. Mason, W. P., Grovas, A., Halpern, S. et al. (1998). Intensive chemotherapy and bone marrow rescue for young children with newly diagnosed malignant brain tumors. J Clin Oncol 16, 210–221.
12. Mason, W. P., Goldman, S., Yates, A. J., Boyett J., Li H. and Finlay J. L. (1998b). Survival following intensive chemotherapy with bone marrow reconstitution for children with recurrent intracranial ependymoma- a report of the Children's Cancer Group. J Neurooncol 37, 135–143.
13. Dunkel, I. J., Boyett, J. M., Yates, A. et al. (1998). High dose carboplatin, thiotepa and etoposide with autologous stem cell rescue for children with recurrent medulloblastoma. Children's Cancer Group. J Clin Oncol 16, 222–228.
14. Dunkel, I. J., Garvin, J. H., Jr, Goldman, S. et al. (1998b). High dose chemotherapy with autologous bone marrow rescue for children with diffuse pontine brain stem tumors. J Neurooncol 37, 67–73.
15. Guruangan, S., Dunkel, I. J., Goldman, S. et al. (1998) Myeloablative chemotherapy with autologous bone marrow rescue in young children with recurrent malignant brain tumors. J Clin Oncol. 16, 2486–2493.
16. Papadopoulos, K. P., Garvin, J. H., Fetell, M. et al. (1998). High-dose thiotepa and etoposide-based regimens with autologous hematopoietic support for high-risk or recurrent CNS tumors in children and adults. Bone Marrow Transplant 22, 661–667.
17. Grovas, A. C., Boyett, J. M., Lindsey, K. et al. (1999). Regimen related toxicity of myeloablative chemotherapy with BCNU, thiotepa and etoposide followed by autologous stem cell rescue for children with newly diagnosed glioblastoma multiforme: report from the Children's Cancer Group. Med Ped Oncol 33, 83–87.
18. Abrey, L. E., Rosenblum, M. K., Papadopoulos, E. et al. (1999). High dose chemotherapy with autologous stem cell rescue in adults with malignant primary brain tumors. J Neurooncol 147–153.
19. Henon, P. R., Butturini, A., and Gale, R. P. (1991). Blood derived hematopoietic stem cell transplants: blood to blood? Lancet 337, 961–963.
20. Jakacki, R. I., Jamison, C., Mathews, V. P. et al. (1998). Dose-intensification of procarbazine, CCNU (lomustine), vincristine (PCV) with peripheral blood stem cell support in young patients with gliomas. Med Ped Oncol 31, 483–490.
21. Yule, S. M., Foreman, N. K., Mitchell, C., Gouldon N., May P., and McDowell H. P. (1997). High-dose cyclophosphamide for poor-prognosis and recurrent pediatric brain tumors: a dose-escalation study. J Clin Oncol 15, 3258–3265.
22. Gajjar, A., Chintagumpala, M., Kellie, S. et al. (2004). Excellent event free survival in newly diagnosed high risk medulloblastoma treated with craniospinal radiation therapy followed by 4 cycles of high dose chemotherapy and stem cell rescue: results of a prospective multicenter trial. Abstract, International Symposium of Pediatric Neurooncology (ISPNO) Meeting. Boston, June 2004 p 147.
23. Vassal, G., Tranchand, B., Valteau-Couanet, D. et al. (2001). Pharmacodynamics of tandem high-dose melphalan with peripheral blood stem cell transplantation in children with neuroblastoma and medulloblastoma. Bone Marrow Transplant 27, 471–477.

24. Foreman, N., Handler, Schissel, D. *et al.* (2004). A study of sequential high dose cyclophosphamide and high dose carboplatin with peripheral stem cell rescue in resistant or recurrent pediatric brain tumors. Astract, *International Symposium of Pediatric Neurooncology (ISPNO) Meeting.* Boston, June 2004 p 145.

25. Zia, M. I., Forsyth, P., Chaudhry, A., Russell, J., and Stewart, D. A. (2002). Possible benefits of high-dose chemotherapy and autologous stem cell transplantation for adults with recurrent medulloblastoma. *Bone Marrow Transplant* **30**, 565–569.

26. Picton, S., Robinson, K., Weston, C. *et al.* (2004). A UKCCSG study of the treatment of relapsed CNS primitive neuroectodermal tumors using high dose chemotherapy, Abstract, *International Symposium of Pediatric Neurooncology (ISPNO) Meeting.* Boston, June 2004 p 169.

27. Ascensao, J., Ahmed, T., Feldman, E. *et al.* (1989). High dose thiotepa with autologous bone marrow transplantation and localized radiotherapy for patients with astrocytoma grade II-IV: a promising approach. *Proc Am Soc Clin Oncol* **21**, 1023 abstract.

28. Kushner, B. H., Cheung, N. K., Kramer, K., Dunkel, I. J., Calleja, E., and Boulad, F. (2001). Topotecan combined with myeloablative doses of thiotepa and carboplatin for neuroblastoma, brain tumors, and other poor-risk solid tumors in children and young adults. *Bone Marrow Transplant* **28**, 551–556.

29. Bouffet, E., Raquin, M., Doz, F. *et al.* (2000). Radiotherapy followed by high dose busulfan and thiotepa: a prospective assessment of high dose chemotherapy in children with diffuse pontine gliomas. *Cancer* **88**, 685–692.

30. Kedar, A., Maria, B. L., Graham Pole, J. *et al.* (1994). High dose chemotherapy with marrow reinfusion and hyperfractionated irradiation for children with high risk brain tumors. *Bone Marrow Transplant* **23**, 428–436.

31. Phillips, G. l., Wolff, S. N., and Fay, J. W. (1986). Intensive 1,2 bis (2chloroethyl) 1nitrosorea (BCNU) monotherapy and autologous bone marrow transplantation for malignant glioma. *J Clin Oncol* **4**, 639–645.

32. Fernandez-Hidalgo, O. A., Vanaclocha, V., Vieitez, J. M. *et al.* (1996). High-dose BCNU and autologous progenitor cell transplantation given with intra-arterial cisplatinum and simultaneous radiotherapy in the treatment of high-grade gliomas: benefit for selected patients. *Bone Marrow Transplant* **18**, 143–149.

33. Calvert, A. H., Newell, D. R. Gumbrell, L. A. *et al.* (1989). Carboplatin dosage: prospective evaluation of a simple formula based on renal function. *J Clin Oncol* **71**, 1748–1756.

34. Faulkner, L. B., Lindsley, K. L., Kher, U., Heller, G., Black, P., and Finlay, J. L. (1996). High-dose chemotherapy with autologous marrow rescue for malignant brain tumors: analysis of the impact of prior chemotherapy and cranio-spinal irradiation on hematopoietic recovery. *Bone Marrow Transplant* **17**, 389–394.

35. Kramer, E. D., Packer, R. J., Ginsberg, J. *et al.* (1997). Acute neurologic dysfunction associated with high-dose chemotherapy and autologous bone marrow rescue for primary malignant brain tumors. *Pediatr Neurosurg* **27**, 230–237.

36. Freilich, R. J., Kraus, D. H., Budnick, A. S., Bayer, L. A., and Finlay, J. L. (1996). Hearing loss in children with brain tumors treated with cisplatin and carboplatin-based high-dose chemotherapy with autologous bone marrow rescue. *Med Ped Oncol* **26**, 95–100.

37. Sands, S. A., van Gorp, W. G. and Finlay, J. L. (1998). Pilot neuropsychological findings from a treatment regimen consisting of intensive chemotherapy and bone marrow rescue for young children with newly diagnosed malignant brain tumors. *Childs Nerv System* **14**, 87–89.

38. Baker, K. S., DeFor, T. E., Burns, L. J. Ramsay, N. K., Neglia, J. P., and Robison, L. L. (2003). New malignancies after blood or marrow stem-cell transplantation in children and adults: incidence and risk factors. *J Clin Oncol* **21**, 1352–1358.

39. Brennan, B. M., and Shalet, S. M. (2002). Endocrine late effects after bone marrow transplant. *Br J Haematol* **118**, 58–66.

40. Ashley, D. M., Longee, D., Tien, R. *et al.* (1996). Treatment of patients with pineoblastoma with high dose cyclophosphamide. *Med Pediatric Oncol* **26**, 387–392.

41. Fagioli, F., Biasin, E., Mastrodicasa, L. *et al.* (2004). High dose thiotepa and etoposide in children with poor prognosis brain tumors. *Cancer* **100**, 2215–2221.

42. Broniscer, A., Nicolaides, T. P., Dunkel, I. J. *et al.* (2004). High-dose chemotherapy with autologous stem cell rescue in the treatment of patients with recurrent non-cerebellar primitive neuroectodermal tumors. *Ped Blood Cancer* **4**, 261–267.

43. Ridola, V., Doz, F., and Frappaz, D. (2004). High dose chemotherapy and posterior fossa radiotherapy for children less then 3–5 years with localized high risk medulloblastoma at diagnosis or at relapse. The SFOP experience. Astract, *International Symposium of Pediatric Neurooncology (ISPNO) Meeting.* Boston, June 2004 p 171.

44. Finlay, J. L., Allen, J., Kellie, S. *et al.* (2003). Survival after intensive chemotherapy followed by consolidative myeloablative chemotherapy with autologous stem cell rescue in young children newly diagnosed with non-cerebellar PNET. The Head Start Regimen. *Proc Am Soc Clin Oncol* **22**, 804 abstract.

45. Grodman, H. M., Gardner, S., Dunkel, I. *et al.* (2003). Outcome of children less then three years old diagnosed with non metastatic medulloblastoma treated with chemotherapy on Head Start I and II protocols. *Proc Am Soc Clin Oncol* **22**, 804 abstract.

46. Hilden, J. M., Meerbaum, S., Burger, P. *et al.* (2004). Central nervous system atypical teratoid/rhabdoid tumor: results of therapy in children enrolled in a registry. *J Clin Oncol* **22**, 2877–2884.

47. Gardner, S., Diez, B., Green, A. *et al.* (2004) Intensive induction chemotherapy followed by high dose chemotherapy with autologous stem cell rescue (ASCR) in young children newly diagnosed with central nervous system (CNS) atypical teratoid rhabdoid tumors (ATT/RT). The head start regimen. Abstract, *International Symposium of Pediatric Neurooncology(ISPNO) Meeting.* Boston, June 2004 p 1148.

48. Stewart, L. A. (2002). Chemotherapy in adult high grade glioma: a systematic review and meta-analysis of individual patient data from 12 randomized trials. *Lancet* **359**, 1011–1018.

49. Jaeckle, K. A., Hess, K. R., Yung, W. K. *et al.* (2003). Phase II evaluation of temozolomide and 13-cis-retinoic acid for the treatment of recurrent and progressive malignant glioma: a North American Brain Tumor Consortium study. *J Clin Oncol* **21**, 2305–2311.

50. Balmaceda, C., Fetell, M. R. and Hesdorffer, C. (1997). Thiotepa and etoposide treatment of recurrent malignant gliomas: phase I study. *Cancer Chemother Pharmacol* **40**, 72–74.

51. McCowage, G. B., Friedman, H. S., Moghrabi, A. *et al.* (1998). Activity of high-dose cyclophosphamide in the treatment of childhood malignant gliomas. *Med Ped Oncol* **30**, 75–80.

52. Bottom, K. S., Ashley, D. M., Friedman, H. S., and Longee D. C. (2000) Evaluation of pre-radiotherapy cyclophosphamide in patients with newly diagnosed glioblastoma multiforme.

Writing Committee for the Brain Tumor Center at Duke. *J Neurooncol* **46**, 151–156.

53. Jakacki, R. I., Jamison, C., Heifetz, S. A., Caldemeyer, K., Hanna, M., and Sender, L. (1997). Feasibility of sequential high-dose chemotherapy and peripheral blood stem cell support for pediatric central nervous system malignancies. *Med Ped Oncol* **29**, 553–559.

54. Jakacki, R. I., Siffert, J., Jamison, C., Velasquez, L., and Allen, J. C. (1999). Dose-intensive, time-compressed procarbazine, CCNU, vincristine (PCV) with peripheral blood stem cell support and concurrent radiation in patients with newly diagnosed high-grade gliomas. *J Neurooncol* **44**, 77–83.

55. Biron, P., Vial, C., Chauvin, F. *et al.* (1985). Strategies including surgery, high dose BCNU followed by ABMT and radiotherapy in supratentorial high grade astrocytoma. A report of 98 patients. *In Autologous bone marrow transplantation: Proceeding of the First International Symposium* (K. A. Dicke, J.O. Spitzer, M. J. Dicke Evinger, Eds.), pp. 227–230. University of Texas 1985, Houston.

56. Durando, X., Lemaire, J. J., Tortochaux, J. *et al.* (2003). High-dose BCNU followed by autologous hematopoietic stem cell transplantation in supratentorial high-grade malignant gliomas: a retrospective analysis of 114 patients. *Bone Marrow Transplant* **31**, 559–564.

57. Linassier, C., Destrieux, C., Benboubker, L. (2001). Role of high-dose chemotherapy with hemopoietic stem-cell support in the treatment of adult patients with high-grade glioma. *Bull Cancer* **88**, 871–876.

58. Brandes, A. A., Palmisano, V., Pasetto, L. M., Basso, U., and Monfardini, S. (2001). High-dose chemotherapy with bone marrow rescue for high-grade gliomas in adults. *Cancer Invest* **19**, 41–48.

59. Bouffet, E., Khelfaoui, F., Philip, I., Biron, P., Brunat-Mentigny, M., and Philip, T. (1997). High-dose carmustine for high-grade gliomas in childhood. *Cancer Chemother Pharmacol* **39**, 376–379.

60. Finlay, J. L., August, C., Packer, R. *et al.* (1990). High dose multiagent chemotherapy followed by bone marrow rescue for malignant astrocytoma of childhood and adolescence. *J Neurooncol* **9**, 239–248.

61. Heideman, R. L., Douglass, E. C., Krance, R. A. *et al.* (1993). High dose chemotherapy and autologous bone marrow rescue followed by interstitial and external beam radiotherapy in new diagnosed pediatric malignant glioma. *J Clin Oncol* **11**, 1456–1465.

62. Abrey, L. E., Childs, B. H., Paleologos, N. *et al.* (2003). High-dose chemotherapy with stem cell rescue as initial therapy for anaplastic oligodendroglioma. *J Neurooncol* **65**, 127–134.

63. Cairncross, G., Macdonald, D., Ludwin, S. *et al.* (1994). Chemotherapy for anaplastic oligodedroglioma. *J Clin Oncol* **12**, 2013–2021.

64. Gumley, D., Phipps, K., Reynolds, J., and Michalski, A. (2004). Infants with ependymoma: outcome following surgery and baby brain chemotherapy. Astract, *International Symposium of Pediatric Neurooncology (ISPNO) Meeting*. Boston, June 2004 p 152.

65. Levy, A. S., Gardner, S., Brady, K. *et al.* (2003) Outcome of young children with newly diagnosed ependymoma treated with intensive induction chemotherapy followed by myeloablative consolidative chemotherapy with autologous stem cell rescue. *Proc Am Soc Clin Oncol* **22**, 804 abstract.

66. Modak, S., Gardner, S., Dunkel, I. J. *et al.* (2004). Thiotepa-based high-dose chemotherapy with autologous stem-cell rescue in patients with recurrent or progressive CNS germ cell tumors. *J Clin Oncol* **22**, 1934–1943.

67. Lundberg, J. H., Weissman, D. E., Beatty, P. A., and Ash, R. C. (1992). Treatment of recurrent metastatic medulloblastoma with intensive chemotherapy and allogeneic bone marrow transplantation. *J Neurooncol* **13**, 151–155.

68. Tanaka, M., Shibui, S., Kobayashi, Y., Nomura, K., and Nakanishi, Y. (2002). A graft-versus-tumor effect in a patient with ependymoma who received an allogeneic bone marrow transplant for therapy-related leukemia. Case report. *J Neurosurg* **97**, 474–476.

69. Takada, S., Ito, K., Sakura, T. *et al.* (1999). Three AML patients with existing or pre-existing intracerebral granulocytic sarcomas who were successfully treated with allogeneic bone marrow transplantations. *Bone Marrow Transplant* **23**, 731–734.

70. Kasow, K. A., Handgretinger, R., Krasin, M. *et al.* (2003). Possible allogeneic graft-versus-tumor effect in childhood melanoma. *J Pediatr Hematol Oncol* **25**, 982–986.

71. Aoyama, Y., Yamamura, R., Shima, E. *et al.* (2003). Successful treatment with reduced-intensity stem cell transplantation in a case of relapsed refractory central nervous system lymphoma. *Ann Hematol* **82**, 371–373.

72. Varadi, G., Or, R., Kapelushnik, J. *et al.* (1999). Graft-versus-lymphoma effect after allogeneic peripheral blood stem cell transplantation for primary central nervous system lymphoma. *Leuk Lymphoma* **34**, 185–190.

73. Hentschke, P., Barkholt, L., Uzunel, M. *et al.* (2003). Low-intensity conditioning and hematopoietic stem cell transplantation in patients with renal and colon carcinoma. *Bone Marrow Transplant* **31**, 253–261.

74. Ueno, N. T., Cheng, Y. C., Rondon, G. *et al.* (2003). Rapid induction of complete donor chimerism by the use of a reduced-intensity conditioning regimen composed of fludarabine and melphalan in allogeneic stem cell transplantation for metastatic solid tumors. *Blood* **102**, 3829–3836.

75. Glick, R. P., Lichtor, T., Panchal, R. *et al.* (2003). Treatment with allogeneic interleukin-2 secreting fibroblasts protects against the development of malignant brain tumors. *J Neurooncol* **64**, 139–146.

76. Blackett, N. M. and Millard, R. E. (1973). Differential effect of Myleran on two normal haemopoietic progenitor cell populations. *Nature* **244**, 300–301.

77. Ellis, G. K., Livingston, R. B., Gralow, J. R., Green, S. J., and Thompson, T. (2002). Dose-dense anthracycline-based chemotherapy for node-positive breast cancer. *J Clin Oncol* **20**, 3637–3643.

78. Frank, S., Muller, J., Bonk, C., Haroske, G., Schackert, H. K., and Schackert, G. (1998). Transmission of glioblastoma multiforme through liver transplantation. *Lancet* **352**, 31.

79. Gardner, S., Baker, D., Belasco, J. *et al.* (2004). Phase I dose escalation of temozolomide with thiotepa and carboplatin with autologous stem cell rescue in patients with recurrent/ refractory central nervous system tumors. Astract, *International Symposium of Pediatric Neurooncology (ISPNO) Meeting* Boston, June 2004 p 47.

80. Matthay, K. K., Villablanca, J. G., Seeger, R. C. *et al.* (1999). Treatment of high-risk neuroblastoma with intensive chemotherapy, radiotherapy, autologous bone marrow transplantation and 13-cis-retinoic acid. Children's Cancer Group. *N Engl J Med* **341**, 1165–1173.

22

CSF Dissemination of Primary Brain Tumors

Marc C. Chamberlain

INTRODUCTION

Neoplastic meningitis (NM) is a common problem in neuro-oncology occurring in approximately 1–2 per cent of all the patients with primary brain tumors. NM is a disease affecting the entire neuraxis and, therefore, clinical manifestations are pleomorphic affecting the spine, cranial nerves, and cerebral hemispheres. Due to the craniospinal disease involvement, staging and treatment need encompass all cerebrospinal fluid (CSF) compartments. Treatment of NM utilizes involved-field radiotherapy of bulky or symptomatic disease sites and intra-CSF drug therapy. The inclusion of concomitant systemic therapy is often necessary in that the majority of patients with primary brain tumors have both recurrence of disease at the primary site and NM. As a consequence, there is a paucity of studies of NM in patients with primary brain tumors addressing NM specifically and its treatment. At present, intra-CSF drug therapy is confined to three chemotherapeutic agents (i.e., methotrexate, cytosine arabinoside, and thio-TEPA) administered by a variety of schedules either by intralumbar or intraventricular drug delivery. Although treatment of NM is palliative with an expected median patient survival of 2–6 months, it may afford stabilization and protection from further neurologic deterioration in patients with NM.

GENERAL OVERVIEW

Clinical Presentation

NM classically presents with pleomorphic clinical manifestations encompassing symptoms and signs in three domains of neurological functions:

1. the cerebral hemispheres;
2. the cranial nerves; and
3. the spinal cord and associated roots [1–6].

Signs on examination generally exceed patient-reported symptoms.

The most common manifestations of cerebral hemisphere dysfunction are headache and mental status changes. Other signs include confusion, dementia, seizures, and hemiparesis. Diplopia is the most common symptom of cranial nerve dysfunction with the cranial nerve VI being the most frequently affected, followed by cranial nerve III and IV. Trigeminal sensory or motor loss, cochlear dysfunction, and optic neuropathy are also common findings. Spinal signs and symptoms include weakness (lower extremities more often than upper), dermatomal or segmental sensory loss and pain in the neck, back, or following radicular patterns. Nuchal rigidity is only present in 15 per cent of the cases [1–6].

A high index of suspicion needs to be entertained in order to make the diagnosis of NM. The finding of multifocal neuraxis disease in a patient with known malignancy is strongly suggestive of NM, but it is also common for patients with NM to present with isolated syndromes such as symptoms of raised intracranial pressure, cauda equina syndrome, or cranial neuropathy.

New neurological signs and symptoms may represent progression of NM but it needs to be distinguished from the manifestations of parenchymal disease, from the side effects of chemotherapy or radiation used for the treatment [1–6].

Diagnosis

CSF Examination

The most useful laboratory test in the diagnosis of NM is the CSF examination [7–19]. Abnormalities include increased opening pressure (>200 mm of H_2O), increased leukocytes (>4/mm^3), elevated protein (>50 mg/dl), or decreased glucose (<60 mg/dl), which though suggestive of NM are not diagnostic. The presence of malignant cells in the CSF is diagnostic of NM but in general, as is true for most of the cytological analysis, assignment to a particular tumor is not possible [7–19].

In patients with positive CSF cytology (see below), up to 45 per cent will be cytologically negative on initial examination [7–11]. The yield is increased to 80 per cent with a second CSF examination but little benefit is obtained from repeated lumbar punctures after two lumbar punctures [2].

Of the 90 patients reported by Wasserstrom et al. with carcinomatous meningitis, 5 per cent had positive CSF cytology only from either the ventricles or cisterna magna [2]. In a series of 60 patients with NM, positive lumbar CSF cytology at diagnosis and no evidence of CSF flow obstruction, ventricular and lumbar cytologies obtained simultaneously were discordant in 30 per cent of the cases [9]. The authors observed that in the presence of spinal signs or symptoms, the lumbar CSF was more likely to be positive and, conversely, in the presence of cranial signs or symptoms, the ventricular CSF was more likely to be positive. Not obtaining CSF from a site of symptomatic or radiographically demonstrated disease was found to correlate with false negative cytology results in a prospective evaluation of 39 patients, as did withdrawing small CSF volumes (<10.5 ml), delayed processing of specimens, and obtaining less than two samples [10]. Even after correcting for these factors, there remains a substantial group of patients with NM and persistently negative CSF cytology. Glass reported on a postmortem evaluation estimating the value of premortem CSF cytology [8]. He demonstrated that up to 40 per cent of the patients with clinically suspected NM proven at the time of autopsy are cytologically negative. This figure increased to >50 per cent in patients with focal NM.

The low sensitivity of CSF cytology makes it difficult not only to diagnose NM, but also to assess the response to treatment. Biochemical markers, immunohistochemistry, and molecular biology techniques applied to CSF have been explored in an attempt to find a reliable biological marker of disease. Numerous biochemical markers have been evaluated but in general, their use has been limited by poor sensitivity and specificity [11].

Use of monoclonal antibodies for immunohistochemical analysis in NM does not significantly increase the sensitivity of cytology alone [12–19]. However, in the case of leukemia and lymphoma, antibodies against surface markers can be used to distinguish between reactive and neoplastic lymphocytes in the CSF [15,19].

Cytogenetic studies have also been evaluated in an attempt to improve the diagnostic accuracy of NM. Flow cytometry and DNA single-cell cytometry, techniques that measure the chromosomal content of cells, and fluorescent in situ hybridization (FISH), that detects numerical and structural genetic aberrations as a sign of malignancy, can give additional diagnostic information, but still have a low sensitivity [16–18]. Polymerase chain reaction (PCR) can establish a correct diagnosis when cytology is inconclusive, but the genetic alteration of the neoplasia must be known for it to be amplified with this technique, and this is generally not the case, particularly in solid tumors [19].

In the cases where there is no evidence of systemic cancer and CSF examinations remain inconclusive, a meningeal biopsy may be diagnostic. The yield of this test increases if the biopsy is taken from an enhancing region on MRI (see below) and if posterior fossa or pterional approaches are used [20].

Neuroradiographic Studies

Magnetic resonance imaging with gadolinium enhancement (MR-Gd) is the technique of choice to evaluate patients with suspected leptomeningeal metastasis [21–27]. As NM involves the entire neuraxis, imaging of the entire CNS is required in patients considered for further treatment. T1-weighted sequences, with and without contrast, combined with fat suppression T2-weighted sequences constitute the standard examination [24]. MRI has been shown to have a higher sensitivity than cranial contrast enhanced computed tomography (CE-CT) in several series [22,23], and is similar to computerized tomographic myelography (CT-M) for the evaluation of the spine, but significantly better tolerated [25,26].

Any irritation of the leptomeninges (i.e., blood, infection, and cancer) will result in their enhancement on MRI, which is seen as a fine signal-intense layer that follows the gyri and superficial sulci. Subependymal involvement of the ventricles often results in ventricular enhancement. Some changes such as cranial nerve enhancement on cranial imaging and intradural extramedullary enhancing nodules on spinal MR (most frequently seen in the cauda equina) can be considered diagnostic of NM in patients with cancer [21,26]. Lumbar puncture itself can rarely cause a meningeal reaction leading to dural-arachnoidal enhancement so imaging should be obtained preferably prior to the procedure [27]. MR-Gd still has a 30 per cent incidence of false negative results so that a normal study does not exclude the diagnosis of NM. On the other hand, in cases with a typical clinical presentation, abnormal MR-Gd alone is adequate to establish the diagnosis of NM [21,22,25,26].

Radionuclide studies using either [111]Indium-diethylenetriamine-pentaacetic acid or ^{99}Tc macro-aggregated albumin, constitute the technique of choice to evaluate CSF flow dynamics [28]. Abnormal CSF circulation has been demonstrated in 30–70 per cent of the patients with NM, with blocks commonly occurring at the skull base, the spinal canal, and over the cerebral convexities [28–32]. Patients with interruption of CSF flow demonstrated by radionuclide ventriculography have been shown in three clinical series to have decreased survival when compared to those with normal CSF flow [28–30]. Involved-field radiotherapy to the site of CSF flow obstruction, restores flow in 30 per cent of the patients with spinal disease and 50 per cent of the patients with intracranial disease [32]. Re-establishment of CSF flow with involved-field radiotherapy followed by intrathecal chemotherapy led to longer survival, lower rates of treatment-related morbidity, and lower rate of death from progressive NM, compared to the group that had persistent CSF blocks [28,30]. These findings may reflect that CSF flow abnormalities prevent homogenous distribution of intrathecal chemotherapy, resulting in

1. protected sites where tumor can progress and
2. in accumulation of drug at other sites leading to neurotoxicity and systemic toxicity.

Based on this, many authors have recommended that intrathecal chemotherapy be preceded by a radionuclide flow study and if a block is found, that radiotherapy be administered in an attempt to re-establish normal flow [28,30–32].

PRIMARY BRAIN TUMORS

Glioma

The management of patients with gliomas and leptomeningeal gliomatosis (LG) is particularly challenging in that the majority of patients have compromised neurological function resulting both from the treatment and topography of their primary tumor [33–40].

The majority of studies regarding LG are primarily autopsy based [33–40]. Erlich and Davis performed an autopsy study of 25 patients with GBM, amongst whom 20 spinal cord examinations were performed [33]. Five of these patients (25 per cent) had evidence of LG. Yung subsequently reported on 52 patients with HGG who underwent autopsy in whom 11 (21 per cent) had evidence of LG [34]. One additional patient was diagnosed without autopsy from a total study group of 12 patients with HGG and LG. Eight of these patients were diagnosed antemortem by positive CSF cytology.

In an antemortem series, where the diagnosis of LG was made based on neuroradiographic imaging but not CSF cytology, Vertosick and Selker reported an incidence of LG of 2 per cent [35]. Awad et al. reported on 13 patients seen at the Cleveland Clinic with HGG and LG in whom 8 had premortem symptoms consistent with LG [36]. In all the patients, the appearance of LG was a preterminal event. Onda et al. performed autopsies on 51 patients who died of GBM and demonstrated evidence of LG in 14 (27 per cent) [37]. Subsequently, Grant et al. reported on a series of 11 patients with HGG and LG all diagnosed antemortem [38]. In three patients, LG was found at the time of initial tumor presentation. Amongst the patients with previously diagnosed and treated AA or GBM, treatment of LG had little effect. In summary, these various reports suggest that autopsy evidence of LG is ten-fold greater than clinically evident LG in patients with HGG. The reason for this antemortem/post-mortem discrepancy is enigmatic.

Most recently, Witham et al. reported on 14 patients [GBM: $n = 9$; AA: $n = 5$] with LG and treated with intra-CSF thio-TEPA [39]. In three patients, LG was the presenting symptom of the HGG. In only five patients were clinical symptoms and signs compatible with LG. CSF cytology was not performed in six patients and amongst the eight patients in whom CSF was examined, it was positive in only two. The neuroradiographic criteria permitted subependymal disease (seen in eight patients) to be sufficient to diagnose LG. Of note, overall median survival was 10 months (10 months for GBM; 19 months for AA).

Another recent report suggests that patients with HGG and LG defined by clinical symptoms and signs, positive CSF cytology and neuroradiographic findings compatible with LG respond poorly to an aggressive multimodal therapy [40]. The median survival following the diagnosis of LG in this report is similar to that of NM secondary to systemic solid cancers [41,42]. Like the majority of patients with NM, LG in patients with HGG is a late event in the natural history of their cancer and predicts for limited survival. At present there is very little evidence suggesting an advantage for any particular intra-CSF chemotherapy relative to another or when used in combination for the treatment of NM [41–46]. Furthermore, the majority of patients in the above-mentioned study died of complications secondary to LG and not as a result of progression of the primary brain tumor [40].

The therapy for leptomeningeal metastasis is necessarily multimodal as approximately 60 per cent of the patients required radiotherapy to treat symptomatic or neuroradiographically bulky subarachnoid disease (60 per cent in the Chamberlain study), and approximately one-third received systemic chemotherapy for primary tumor progression (70 per cent in the Chamberlain study) [40]. Therefore, standard treatment of carcinomatous meningitis utilizes clinically appropriate treatments (involved-field radiotherapy, systemic chemotherapy, or intra-CSF chemotherapy) as the disease affects the entire neuraxis in a variable and pleomorphic manner. Treatment results, therefore, reflect a composite of therapies not unlike the treatment of high-grade gliomas, which creates difficulties when attempting to assign benefit to any specific therapy, as for example, intra-CSF chemotherapy.

In conclusion, notwithstanding the modest treatment-related toxicity observed in various studies, the poor response to treatment and limited survival in patients with HGG and LG suggests that a less aggressive approach may be justified. Palliative treatment with radiotherapy directed to symptomatic site(s) of the disease, perhaps in conjunction with simple systemic chemotherapy regimens, may prove to be as effective as an aggressive multimodal approach and much less intrusive to the patient. This statement, however, is not meant to exclude the enrollment of such patients from clinical trials designed to explore new approaches to the treatment of LG, a disease for which new and effective therapies would be welcomed [47–52].

Meningioma

Meningiomas are extra-axial brain tumors of middle to late adult life and have a female predominance. Overall, 90 per cent of meningiomas are benign, 6 per cent atypical, and 2 per cent are malignant [53–59]. Most of the patients diagnosed with a meningioma decided to have it removed surgically and are advised to do so based on their neurological symptoms [53–59]. Complete surgical resection is usually curative. For incompletely resected or recurrent tumors not previously irradiated, radiotherapy is administered [53–59]. Radiotherapy may be administered as either conventional external beam irradiation or stereotactically. Stereotactic radiotherapy (SRT) either as LINAC or gamma knife radiosurgery is increasingly utilized. When the meningioma is unresectable or all other treatments (surgery, radiotherapy) have failed, immunochemotherapy may be considered [59]. Rarely meningiomas metastasize and when seen, most often to extraneural sites (cervical lymph nodes, pulmonary) [60–65].

Metastatic meningioma is most often associated with either aggressive (WHO Grade 2) or malignant meningioma (WHO Grade 3) with a range of occurrence of 10–25 per cent. Not clear from the literature, however, is the risk of CSF dissemination [60–65]. In a small series, arguably biased by referral to a neuro-oncology clinic, 4 per cent of the 200 consecutively seen patients with meningioma manifested CSF spread of disease defined both radiographically and cytologically [66]. In addition to CSF spread of disease, the majority of these patients (5 of 8; 63 per cent) also had evidence of extraneural metastasis. Furthermore, all the patients had extensive prior therapy comprised of surgery (often multiple), radiotherapy (often both external beam and stereotactic), and chemotherapy (all had received prior hydroxyurea and progressed). Aside from hydroxyurea, there is a paucity of chemotherapy agents with demonstrated activity against recurrent meningiomas complicating the management of these already complicated patients. Similar to patients with meningeal gliomatosis, these patients were treated in a multimodal manner including both systemic and regional (intraventricular) chemotherapy and involved-field radiotherapy to sites of symptomatic spine disease (three patients). Despite this aggressive approach, median survival was only 5.5 months with all but three patients dying of progressive disease. In the three patients alive, all have residual disease, and two continue to receive therapy. Furthermore, no patient demonstrated a treatment response (best response was stable disease) as assessed by neuroradiography. Based on the results of this study, two conclusions seem justified. First, meningiomas do metastasize both hematologically and by CSF dissemination and second, that there is a need for more effective chemotherapy for recurrent surgical and radiotherapy refractory meningiomas.

Primary CNS Lymphoma (PCNSL)

PCNSL is a disease that primarily involves the brain, both as the source of origination and in patients who relapse [67–90]. The leptomeninges may be involved both at disease presentation and at recurrence. Notwithstanding that isolated disease of the leptomeninges is rare in patients with PCNSL regardless if one considers patients at presentation or recurrence, simultaneous brain and leptomeningeal disease is common.

In a study of 86 patients with PCNSL, all the patients underwent a complete CSF/leptomeningeal evaluation [73]. Twenty-six per cent had a positive cytology and neuropathology and neuroradiography demonstrated disease consistent with leptomeningeal spread in another 15 per cent. In studies that define lymphomatous meningitis as a positive CSF cytopathology the results are similar. Balmaceda *et al.* and Lawler *et al.* each reported a prevalence of 26 per cent while other studies, including a larger series by Ferreri *et al.* reported prevalence of 12–16 per cent [67,73,83]. Other studies using different modalities, including PCR for a component of IgG heavy chain or contrast-enhanced brain and spinal imaging alone reported lower prevalence figures for lymphomatous meningitis, 13 per cent and 12.5 per cent respectively. In an autopsy study of patients with PCNSL, 77 per cent of the patients (21 of 26) had evidence of lymphomatous meningitis [83]. A similar figure (40 per cent incidence of PCNS-LM) was found in another autopsy study [73]. Both these studies are perhaps most relevant to patients with PCNSL at relapse and suggest combined brain and leptomeningeal disease is seen in the majority of such patients. Taken together, the above-mentioned antemortem and postmortem studies suggest that PCNSL is a disease of the neuraxis with frequent evidence of lymphomatous meningitis.

A large multicenter study conducted by Ferreri *et al.* of 370 patients with PCNSL evaluated the benefit of adjuvant intrathecal methotrexate [67]. They compared patients who received high-dose methotrexate-based regimens with or without intrathecal chemotherapy and were unable to demonstrate a survival benefit [67]. Furthermore, notwithstanding that some patients were treated with less than 3.0 g/m^2 of methotrexate (MTX), intrathecal chemotherapy had no impact regarding meningeal relapse or on survival, even in patients who had positive CSF cytology at diagnosis. It would appear, therefore, that even in patients with leptomeningeal disease at diagnosis, high-dose MTX systemic regimens provide sufficient CSF MTX levels to obviate the need for intra-CSF chemotherapy.

As lymphomatous meningitis is sufficiently common in PCNSL, leptomeningeal/CSF compartment therapy is indicated as part of the treatment regimen in patients with newly diagnosed PCNSL. Leptomeningeal/CSF compartment therapy can take several forms including intrathecal chemotherapy, radiation (i.e., craniospinal irradiation) or high-dose systemic chemotherapy (MTX, cytarabine, or thio-TEPA). The most effective drug utilized in patients with newly diagnosed PCNSL is high-dose MTX. When this drug is administered in gram quantities (high-dose), cytotoxic CSF levels are achieved. In a study by Glantz *et al.* of 16 solid tumor patients with leptomeningeal metastases, MTX was administered intravenously at 8 g/m^2 over four hours followed by serial sampling of CSF and blood [86]. In this study, there was in addition, a parallel group of patients who received intrathecal MTX at a standard dose and schedule. After a single intravenous dose, MTX levels of 1.0 μM are maintained in the CSF for, on average, 48 h, and 0.1 μM for up to 93 h. In the patients who received a single intrathecal MTX dose, levels of 1.0 μM are maintained for 35–48 h, and 0.1 μM levels are maintained for about 57 h. Therefore, the duration of cytotoxic drug exposure in the CSF is similar in patients who received intravenous or intrathecal MTX on these schedules and at these doses. In another study by Chamberlain *et al.* two groups of patients with recurrent PCNSL and isolated lymphomatous meningitis were compared [90]. One group was treated with intra-CSF chemotherapy whereas the other received high-dose (8 g/m^2) systemic methotrexate. Outcome was similar in both groups (overall survival, progression free survival, and lymphomatous meningitis response rates) and differences were seen only in toxicity and pharmacoeconomics (both favoring intra-CSF chemotherapy). These studies suggest that intra-CSF chemotherapy is not required in patients receiving systemic high-dose MTX, but should, however, be considered in patients receiving alternative therapies (i.e., PCV, rituximab, and Topotecan) and with evidence of CSF dissemination.

Ependymomas

CSF dissemination occurs in 3–12 per cent of all intracranial ependymomas and is most frequent in patients with infratentorial anaplastic ependymomas [91–125]. As a small but measurable risk for CSF dissemination exists for all patients with newly diagnosed ependymoma, an extent of disease evaluation including both CSF cytology and craniospinal

MRI is mandated following surgery. This staging permits stratification of patients into those with (M_{1+}) or without CSF metastasis (M_0) and patients with or without residual disease following surgery, the two most important clinical parameters affecting the outcome.

Treatment of ependymomas is primarily surgical, as essentially all analyses have determined that completeness of surgical resection is the most important covariant affecting progression-free and overall survival [98–101,104,106–111,124]. Radiotherapy represents the second most frequently utilized adjuvant treatment modality for ependymomas despite the lack of a randomized clinical trail showing benefit and the general consensus that ependymomas are radioresistant [100,101,104,106,109,117,124,125]. Furthermore, there are no data regarding a dose–response relationship in ependymomas and as such, total tumor dose has varied. Due to the possibility of CSF spread, one of the controversies regarding the radiotherapeutic management of ependymomas is the volume of brain that needs to be treated. Notwithstanding early enthusiasm for craniospinal irradiation (CSI), several recent studies support the application of limited-field radiotherapy for M_0 tumors and reserve CSI for M_{1+} tumors regardless of tumor histology and according to some authors for anaplastic ependymomas of the infratentorial compartment [106,107,109,110,114,125].

Robertson *et al.* reported a prospective trial involving 32 children with newly diagnosed intracranial ependymoma that examined the role of adjuvant chemotherapy (CCNU, vincristine, and prednisone *versus* so called "8 in 1" regimen) after initial surgery and CSI [97]. This multi-institutional Children's Cancer Group study demonstrated that both the extent of surgical resection and volume of residual disease on postoperative imaging predict for progression-free survival. Chemotherapy had no impact on progression-free survival with either chemotherapy drug regimen. Five-year progression-free survival and overall survival rates were 50 per cent and 64 per cent, respectively. The majority of relapses were local treatment failures (71 per cent) or concurrent local and distant CNS metastasis (21 per cent) and isolated metastatic relapse was uncommon (7 per cent) and occurred only in the setting of M_{1+} disease at diagnosis. The study concluded that involved-field radiotherapy results in similar outcomes as compared to CSI and, therefore, CSI is appropriately reserved for disseminated neuraxis disease (M_{1+}). Whether intra-CSF chemotherapy could replace CSI in patients with M_{1+} disease at presentation seems unlikely as adjuvant chemotherapy has neither improved survival nor are there studies comparing CSI to intra-CSF chemotherapy.

The management of recurrent ependymoma has not received much attention in the literature despite the fact that nearly 50 per cent of the patients will recur. There is consensus regarding the extent of disease evaluation (contrast-enhanced neuraxis MRI and CSF cytology) to be performed in all the patients with recurrent ependymoma as the discovery of disseminated disease changes management. Patients demonstrated to have disseminated disease (M_{1+}) are candidates for CSI (particularly if prior radiotherapy has not been administered), involved-field radiotherapy, or intra-CSF chemotherapy. Goldwein *et al.* reported on 36 patients with recurrent intracranial ependymoma in which 33 were treated with re-operation, 12 received conventional radiotherapy and all received chemotherapy [101]. Median time to recurrence was 2.8 years and in the majority, relapse was either local (78 per cent) or local with concomitant distant metastasis (14 per cent). Twenty-nine (79 per cent) of the initial cohorts had a second relapse in which a local component to the relapse was seen in 80 per cent. Two-year overall survival and progression-free survival were 29 per cent and 23 per cent, respectively. Considering only first relapse, 2-year actuarial survival was 39 per cent and median survival was 17 months. Median progression free survival was 12 months. Among 36 evaluable patients and 37 chemotherapy regimens, there was one partial response (3 per cent), seven stable disease patterns (20 per cent), and 29 disease progressions (77 per cent). In responding or stable disease patients, median duration of response was 9 months (range 3–23 months). Cisplatin was felt to be the most active agent amongst the four commonly used chemotherapeutics (cisplatin, procarbazine, CCNU, and vincristine). The role of intra-CSF chemotherapy for patients with recurrent M_{1+} ependymoma is uncertain, as there exists only case reports of patients with isolated CSF dissemination treated primarily with intra-CSF chemotherapy. In the author's opinion, intra-CSF chemotherapy should be offered to patients not otherwise treated with CSI for M_{1+} ependymoma recognizing that the dominant variable affecting survival is disease recurrence at the primary site.

Primitive Neuroectodermal Tumor (PNET)

PNET comprise both differentiated tumors (olfactory neuroblastoma, cerebral neuroblastoma, retinoblastoma, and pineoblastoma) and undifferentiated tumors (medulloblastoma) [126–141]. Due to the propensity for PNET and in particular, medulloblastoma to manifest CSF dissemination, standard initial treatment includes CSI regardless of the extent of disease (M_0 or M+ disease). As a consequence, there would appear to be no role for intra-CSF in the initial treatment of patients with CSF-disseminated PNET as CSF-based chemotherapy would be redundant to CSI. Furthermore, there are no adjuvant studies that in addition to CSI include intra-CSF chemotherapy for patients with M+ disease, however, this has been proposed to permit further dose reduction of CSI (see below). An argument, however, could be made for intra-CSF chemotherapy in patients with positive CSF cytology following the administration of CSI though the incidence of this clinical occurrence is not easily determined from the literature. Against this position are several studies that have demonstrated that M_1 disease (positive CSF cytology only i.e., metastatic microscopic disease) does not predict for reduced survival raising the question whether there exists a compelling role for intra-CSF chemotherapy in the initial treatment of medulloblastoma. Finally, a pilot study (CCG 9892) of average-risk medulloblastoma children ($n = 65$) suggests that reduced CSI dose (24 Gy) in conjunction with systemic chemotherapy (CCNU, cisplatin, and vincristine) results in both excellent outcomes (5-year event-free survival, 79 per cent) and no increased risk of neuraxis failures (i.e., CSF dissemination) [131]. Therefore, the best indication for intra-CSF chemotherapy is seen in patients with recurrent disease and evidence of CSF dissemination.

Data are most robust for medulloblastoma regarding patterns of relapse in which approximately one-third are local, one-third are local plus disseminated, and one-third metastatic only [132,134–141]. Amongst the metastatic group approximately 20 per cent will in addition have extraneural metastases and predominantly to bone. Overall, 50 per cent of the patients with recurrent medulloblastoma have evidence of CSF dissemination defined as either subarachnoid nodules (spinal or intracranial) seen neuroradiographically or by positive CSF cytology. In general, however, treatment of recurrent and disseminated medulloblastoma is with systemic chemotherapy regardless of CSF spread of disease [132, 134–141]. The only exception (similar to ependymoma discussed above) is seen in case reports of patients with isolated CSF dissemination and medulloblastoma treated with intra-CSF chemotherapy. How to extrapolate from the limited data regarding intra-CSF chemotherapy for PNET is problematic and any statements regarding treatment must be considered in this context.

Germ Cell Tumor (GCT)

GCT constitutes less than one per cent of all primary brain tumors; occur primarily in the first two decades of life (60–70 per cent) and principally in males (approximately 75 per cent) [142–164]. GCT like PCNSL and ependymoma has a proclivity to CSF dissemination and as a consequence, staging of disease (CSF cytology and oncofetal protein analysis, contrast-enhanced neuraxis MR) is performed at diagnosis [142–164]. CSF-disseminated disease is seen in approximately 20 per cent of germinomas and 35 per cent of nongerminomatous GCT [142–164].

Among patients with GCT, those with germinoma (seminoma) can be cured by radiotherapy alone or by combined platinum-based chemotherapy and radiotherapy [142,143,148,150–152,154,156,162]. Patients with mature teratoma are potentially curable by surgery alone. Patients with non-seminomatous GCT, though frequently recur, may be cured (approximately 65 per cent) by radiotherapy alone, high-dose thio-TEPA-based chemotherapy with autologous stem cell transplant or by combined platinum-based chemotherapy and radiotherapy [145,147,148, 160,161].

Similar to ependymoma, a point of controversy in the treatment of GCT remains with respect to the appropriate radiation field (localized *versus* whole ventricle *versus* whole brain *versus* CSI) [143,148–152,156,158,159] Increasingly, the volume of brain irradiated is determined by the extent of disease and secondarily by tumor histology. In general, disseminated disease is treated with CSI whereas localized disease is treated by either involved-field or whole ventricle radiotherapy. Additionally, there is increasing interest in utilizing combined modality therapy for GCT (e.g., platinum-based chemotherapy and radiotherapy) permitting a reduction in both radiation dose and volume. The role of intra-CSF chemotherapy for the treatment of disseminated GCT is unclear, as essentially all the studies have utilized systemic chemotherapy with radiotherapy. Patients with isolated CSF disease are rare as are reports of primary treatment with intra-CSF chemotherapy. Not unlike disseminated PNET, how to utilize intra-CSF chemotherapy in the treatment of disseminated GCT is uncertain and would reasonably be based on extrapolation of studies of carcinomatous meningitis.

TREATMENT

The evaluation of treatment of NM is complicated by the lack of standard treatments, the difficulty of

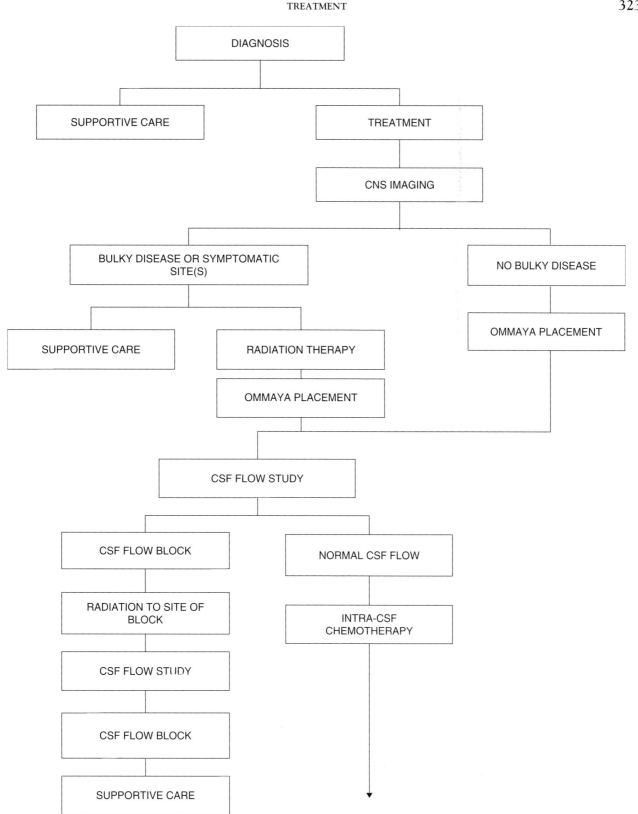

FIGURE 22.1   Treatment Algorithm of Neoplastic Meningitis.

determining the response to treatment given the suboptimal sensitivity of the diagnostic procedures and that most patients will die of systemic disease, and the fact that most of the studies are small, non-randomized, and retrospective [164]. However, it is clear that the treatment of NM can provide effective palliation and in some cases result in prolonged survival. The treatment requires the combination of surgery, radiation, and chemotherapy in most cases. Figure 22.1 outlines a treatment algorithm for NM.

Surgery

Surgery is used in the treatment of NM for the placement of

1. intraventricular catheter and subgaleal reservoir for the administration of cytotoxic drugs and
2. ventriculoperitoneal shunt in patients with symptomatic hydrocephalus.

Drugs can be instilled into the subarachnoid space by lumbar puncture or *via* an intraventricular reservoir system. The latter is the preferred approach because it is simpler, more comfortable for the patient and safer than repeated lumbar punctures. It also results in a more uniform distribution of the drug in the CSF space and produces the most consistent CSF levels. In up to 10 per cent of lumbar punctures drug is delivered to the epidural space, even if there is CSF return after placement of the needle, and drug distribution has been shown to be better after drug delivery through a reservoir [165].

NM often causes communicating hydrocephalus leading to symptoms of raised intracranial pressure. Relief of sites of CSF flow obstruction with involved-field radiation should be attempted to avoid the need for CSF shunting. If hydrocephalus persists, a ventriculoperitoneal shunt should be placed to relieve the pressure because relief of pressure often results in clinical improvement. If possible, an in-line on/off valve and reservoir should be used to permit the administration of intra-CSF chemotherapy, although some patients cannot tolerate having the shunt turned off to allow the circulation of the drug.

Additionally, in patients with a persistent blockage of ventricular CSF, a lumbar catheter and reservoir can be used in addition to a ventricular catheter, to allow treatment of the spine with intrathecal chemotherapy, although as discussed earlier, patients with persistent CSF flow blocks after radiation are probably best managed by supportive care alone.

Finally, occasional patients may undergo a meningeal biopsy so as to pathologically confirm neoplastic meningitis. However, in that most patients demonstrate MR leptomeningeal abnormalities, an abnormal CSF profile or a clinical examination consistent with NM, meningeal biopsies are rarely performed.

Radiotherapy

Radiotherapy is used in the treatment of NM for

1. palliation of symptoms, such as a cauda equina syndrome,
2. to decrease bulky disease such as co-existent parenchymal brain metastases, and
3. to correct CSF flow abnormalities demonstrated by radionuclide ventriculography.

Patients may have significant symptoms without radiographic evidence of bulky disease and still benefit from radiation. For example, patients with low back pain and leg weakness should be considered for radiation to the cauda equina, and those with cranial neuropathies should be offered whole-brain or base of skull radiotherapy [166].

Radiotherapy of bulky disease is indicated as intra-CSF chemotherapy is limited by diffusion to 2–3 mm penetration into tumor nodules. In addition, involved-field radiation can correct CSF flow abnormalities and this has been shown to improve patient outcome as discussed above. Whole neuraxis radiation is rarely indicated in the treatment of NM from solid tumors because it is associated with significant systemic toxicity (severe myelosuppression and mucositis among other complications) and is not curative.

Chemotherapy

Chemotherapy is the only treatment modality that can treat the entire neuraxis. Chemotherapy may be administered systemically or intrathecally [41–45, 165–170].

Intrathecal chemotherapy is the mainstay of treatment for NM in patients with solid tumors [41–45]. Retrospective analysis or comparison to historical series suggest that the administration of chemotherapy to the CSF improves the outcome of patients with NM and solid tumors. However, it is noted that most series will exclude patients that are too sick to receive any treatment, which may be up to one-third of the patients with NM. Three agents are routinely

TABLE 22.1 Regional Chemotherapy for Neoplastic Meningitis

Drugs	Induction regimens		Consolidation regimens		Maintenance regimens	
	Bolus regimen	CxT regimen	Bolus regimen	CxT regimen	Bolus regimen	CxT regimen
Methotrexate	10–15 mg twice weekly (Total 4 weeks)	2 mg/day every other week (Total 8 weeks)	10–15 mg once weekly (total 4 weeks)	2 mg/day for 5 days every other week (total 4 weeks)	10–15 mg once a month	2 mg/day for 5 days once a month
Cytarabine	25–100 mg 2 or 3 times weekly (Total 4 weeks)	25 mg/day for 3 days weekly (Total 4 weeks)	25–100 mg once weekly (Total 4 weeks)	25 mg/day for 3 days every other week (Total 4 weeks)	25–100 mg once a month	25 mg/day for 3 days once a month
DepoCyt®	50 mg every 2 weeks (Total 8 weeks)		50 mg every 4 weeks (Total 24 weeks)			
Thiotepa	10 mg 2 cr 3 times weekly (Total 4 weeks)	10 mg/day for 3 days weekly (total 4 weeks)	10 mg once weekly (Total 4 weeks)	10 mg/day for 3 days every other week (Total 4 weeks)	10 mg once a month	10 mg/day for 3 days once a month
α-Interferon	1×10^6 u 3 times weekly (Total 4 weeks)		1×10^6 u 3 times weekly every other week (Total 4 weeks)		1×10^6 u 3 times weekly one week per month	
Topotecan	400 μg twice weekly (total 6 weeks)		400 μg once weekly (total 6 weeks)		400 μg once a month	
Etoposide	0.5 mg twice weekly (total 6 weeks)		0.5 mg once weekly (total 6 weeks)		0.5 mg once a month	

used: methotrexate, cytarabine (including liposomal cytarabine or DepoCyt®), and thio-TEPA. No difference in response has been seen when comparing single-agent methotrexate with thio-TEPA or when using multiple agents (methotrexate, thio-TEPA, and cytarabine or methotrexate and cytarabine) *versus* single-agent methotrexate in adult randomized studies of NM and solid tumors [41–45]. Table 22.1 outlines the common treatment regimens for these drugs. A sustained-release liposomal form of cytarabine (DepoCyt®) results in cytotoxic cytarabine levels in the CSF for ≥ 10 days and when given bimonthly and compared to biweekly methotrexate, resulted in longer time to neurological progression in patients with NM due to solid tumors [41, 42]. Furthermore, quality of life and cause of death favored DepoCyt® over methotrexate. These findings were confirmed in a study of lymphomatous meningitis and in an open label study suggesting that DepoCyt® should be considered the drug of first choice in the treatment of NM when experimental therapies are unavailable [170].

The rationale to give intrathecal chemotherapy is based on the presumption that most chemotherapeutic agents when given systemically have poor CSF penetration and do not reach therapeutic levels. Exceptions to this would be systemic high-dose methotrexate, cytarabine, and thio-TEPA, all of which result in cytotoxic CSF levels [86–90]. Their systemic administration, however, is limited by systemic toxicity and the difficulty to integrate these regimens into other chemotherapeutic programs being used to manage the primary disease. Some authors argue that intrathecal chemotherapy does not add to improved outcome in the treatment of NM, since systemic therapy can obtain access to the subarachnoid deposits through their own vascular supply [166,169]. In a retrospective comparison of patients treated with systemic chemotherapy and radiation to involved areas, plus or minus intrathecal chemotherapy, Bokstein *et al.* did not find significant differences in response rates, median survival, or proportion of long-term survivors amongst the two groups but, of course, the group that did not receive the intrathecal treatment was spared the complications of this modality [167]. Glantz *et al.* treated 16 patients with high-dose intravenous methotrexate and compared their outcome with a reference group of 15 patients treated with intrathecal methotrexate [168]. They found that the response rates and survival were significantly better in the group treated with intravenous therapy. Finally, a recent report describes two patients with breast cancer in whom LM was controlled with systemic hormonal treatment [169].

In addition, there are anecdotal reports indicating that systemic chemotherapy is effective for meningeal gliomatosis.

Nonetheless, intrathecal chemotherapy remains the preferred treatment route for NM at this time [47]. New drugs are being evaluated for intrathecal administration, to determine tolerability and efficacy of treatment, including mafosfamide, diaziquone, topotecan, gemcitabine, interferon-α, and temozolomide are some of the new drugs being evaluated for intrathecal administration [48–52]. Immunotherapy, using IL-2 and IFN-α, ^{131}I-radiolabelled monoclonal antibodies, and gene therapy are other modalities that are being explored in clinical trials [50–52].

Supportive Care

Not all patients with NM are candidates for the aggressive treatment outlined above. Most authors agree that combined-modality therapy should be offered to patients with life expectancy greater than three months and a Karnofsky performance status of more than 60 per cent. Supportive care should be offered to every patient, regardless of whether they receive NM-directed therapy, including anticonvulsants for seizure control (seen in 10–15 per cent of the patients with NM), adequate analgesia with opioid drugs as needed as well as antidepressants and anxiolytics, if necessary. Corticosteroids have a limited use in NM-related neurological symptoms, but can be useful to treat vasogenic edema associated with intraparenchymal or epidural metastases, or for the symptomatic treatment of nausea and vomiting together with routine antiemetics. Decreased attention and somnolence secondary to whole-brain radiation can be treated with psychostimulants.

CONCLUSIONS

NM is a complicated disease for a variety of reasons. First, most reports concerning NM treat all the subtypes as equivalent with respect to CNS staging, treatment, and outcome. However, clinical trials in oncology are based on specific tumor histology. Comparing responses in patients with carcinomatous meningitis due to gliomas to patients with primary CNS lymphoma outside investigational new drug trials may be misleading. A general consensus is that lymphomatous meningitis is inherently more chemosensitive than meningeal gliomatosis or CSF-disseminated meningioma and, therefore, survival following chemotherapy is likely to be different. This observation has been substantiated in patients with primary disease recurrence though comparable data regarding CSF-disseminated disease, and in particular NM is meager.

A second feature of NM, which complicates therapy, is deciding whom to treat. Not all patients necessarily warrant aggressive CNS-directed therapy, however, few guidelines exist permitting appropriate choice of therapy. Based on the prognostic variables determined clinically and by the evaluation of the extent of disease, a sizable minority of patients will not be candidates for aggressive NM-directed therapy. Therefore, supportive comfort care (radiotherapy to symptomatic disease, antiemetics, and narcotics) is reasonably offered to patients with NM considered as poor candidates for aggressive therapy as seen in Fig. 22.1.

Third, optimal treatment of NM remains poorly defined. Given these constraints, the treatment of NM today is palliative and rarely curative with a median patient survival of several months based on data of the four prospective randomized trials (extrapolated to primary brain tumors). However, palliative therapy of NM often affords the patient protection from further neurological deterioration and consequently, an improved neurologic quality of life. No studies to date have attempted an economic assessment of the treatment of NM and, therefore, no information is available regarding a cost-benefit analysis as has been performed for other cancer-directed therapies.

Finally, in patients with NM, the response to treatment is primarily a function of CSF cytology and secondarily of clinical improvement of neurologic signs and symptoms. Aside from CSF cytology and perhaps biochemical markers, no other CSF parameters predict response. Furthermore, because CSF cytology may manifest a rostrocaudal disassociation, consecutive negative cytology (defined as a complete response to treatment) requires confirmation by both ventricular and lumbar CSF cytologies. In general, only pain-related neurologic symptoms improve with treatment. Neurologic signs such as confusion, cranial nerve deficit(s), ataxia, and segmental weakness minimally improve or stabilize with successful treatment.

References

1. Kaplan, J. G., DeSouza, T. G., Farkash, A. et al. (1990). Leptomeningeal metastases: comparison of clinical features and laboratory data of solid tumors, lymphomas and leukemia's. J Neurooncol 9(3), 225–229.
2. Wasserstrom, W. R., Glass, J. P., and Posner, J. B. (1982). Diagnosis and treatment of leptomeningeal metastases from solid tumors: experience with 90 patients. Cancer 49(4), 759–772.
3. Little, J. R., Dale, A. J., and Okazaki, H. (1974). Meningeal carcinomatosis. Clinical manifestations. Arch Neurol 30(2), 138–143.
4. van Oostenbrugge, R. J., and Twijnstra A. (1999). Presenting features and value of diagnostic procedures in leptomeningeal metastases. Neurology 53(2), 382–385.
5. Balm, M., and Hammack, J. (1996). Leptomeningeal carcinomatosis. Presenting features and prognostic factors. Arch Neurol 53(7), 626–632.
6. DeAngelis, L. M. (1998). Current diagnosis and treatment of leptomeningeal metastasis. J Neurooncol 38(2–3), 245–252.
7. Kolmel, H. W. (1998). Cytology of neoplastic meningosis. J Neurooncol 38(2–3), 121–125.
8. Glass, J. P., Melamed, M., Chernik, N. L. et al. (1979). Malignant cells in cerebrospinal fluid (CSF): the meaning of a positive CSF cytology. Neurology 29(10), 1369–1375.
9. Chamberlain, M. C., Kormanik, P. A., and Glantz, M. J. (2001). A comparison between ventricular and lumbar cerebrospinal fluid cytology in adult patients with leptomeningeal metastases. Neurooncol 3(1), 42–45.
10. Glantz, M. J., Cole, B. F., Glantz, L. K. et al. (1998). Cerebrospinal fluid cytology in patients with cancer: minimizing false-negative results. Cancer 82(4), 733–739.
11. Chamberlain, M. C. (1998). Cytologically negative carcinomatous meningitis: usefulness of CSF biochemical markers. Neurology 50(4), 1173–1175.
12. Garson, J. A., Coakham, H. B., Kemshead, J. T. et al. (1985). The role of monoclonal antibodies in brain tumour diagnosis and cerebrospinal fluid (CSF) cytology. J Neurooncol 3(2), 165–171.
13. Hovestadt, A., Henzen-Logmans, S. C., and Vecht, C. J. (1990). Immunohistochemical analysis of the cerebrospinal fluid for carcinomatous and lymphomatous leptomeningitis. Br J Cancer, 62(4), 653–654.
14. Boogerd, W., Vroom, T. M., van Heerde, P. et al. (1988). CSF cytology versus immunocytochemistry in meningeal carcinomatosis. J Neurol Neurosurg Psychiatry 51(1), 142–145.
15. van Oostenbrugge, R. J., Hopman, A. H., Ramaekers, F. C. et al. (1998). In situ hybridization: a possible diagnostic aid in leptomeningeal metastasis. J Neurooncol 38(2–3), 127–133.
16. Cibas, E. S., Malkin, M. G., Posner, J. B. et al. (1987). Detection of DNA abnormalities by flow cytometry in cells from cerebrospinal fluid. Am J Clin Pathol 88(5), 570–577.
17. Biesterfeld, S., Bernhard, B., Bamborschke, S. et al. (1993). DNA single cell cytometry in lymphocytic pleocytosis of the cerebrospinal fluid. Acta Neuropathol (Berl) 86(5), 428–432.
18. van Oostenbrugge, R. J., Hopman, A. H., Arends, J. W. et al. (1998). The value of interphase cytogenetics in cytology for the diagnosis of leptomeningeal metastases. Neurology 51(3), 906–908.
19. Rhodes, C. H., Glantz, M. J., Glantz, L. et al. (1996). A comparison of polymerase chain reaction examination of cerebrospinal fluid and conventional cytology in the diagnosis of lymphomatous meningitis. Cancer 77(3), 543–548.
20. Cheng, T. M., O'Neill, B. P., Scheithauer, B. W. et al. (1994). Chronic meningitis: the role of meningeal or cortical biopsy. Neurosurgery 34(4), 590–595.
21. Chamberlain, M. C., Sandy, A. D., and Press, G. A. (1990). Leptomeningeal metastasis: a comparison of gadolinium-enhanced MR and contrast-enhanced CT of the brain. Neurology 40(3 Pt 1), 435–438.
22. Schumacher, M., and Orszagh, M. (1998). Imaging techniques in neoplastic meningiosis. J Neurooncol 38(2–3), 111–120.
23. Sze, G., Soletsky, S., Bronen, R. et al. (1989). MR Imaging of the cranial meninges with emphasis on contrast enhancement and

meningeal carcinomatosis. *AJR Am J Roentgenol* **153**(5), 1039–1049.

24. Schuknecht, B., Huber, P., Buller, B. *et al.* (1992). Spinal leptomeningeal neoplastic disease. Evaluation by MR, myelography and CT myelography. *Eur Neurol* **32**(1), 11–16.

25. Chamberlain, M. C. (1995). Comparative spine imaging in leptomeningeal metastases. *J Neurooncol* **23**(3), 233–238.

26. Freilich, R. J., Krol, G., and DeAngelis, L. M. (1995). Neuroimaging and cerebrospinal fluid cytology in the diagnosis of leptomeningeal metastasis. *Ann Neurol* **38**(1), 51–57.

27. Mittl, R. L., Jr., and Yousem, D. M. (1994). Frequency of unexplained meningeal enhancement in the brain after lumbar puncture. *AJNR Am J Neuroradiol* **15**(4), 633–638.

28. Glantz, M. J., Hall, W. A., Cole, B. F. *et al.* (1995). Diagnosis, management, and survival of patients with leptomeningeal cancer based on cerebrospinal fluid-flow status. *Cancer* **75**(12), 2919–2931.

29. Trump, D. L., Grossman, S. A., Thompson, G. *et al.* (1982). CSF infections complicating the management of neoplastic meningitis. Clinical features and results of therapy. *Arch Intern Med* **142**(3), 583–586.

30. Chamberlain, M. C., and Kormanik, P. A. (1996). Prognostic significance of 111indium-DTPA CSF flow studies in leptomeningeal metastases. *Neurology* **46**(6), 1674–1677.

31. Mason, W. P., Yeh, S. D., and DeAngelis, L. M. (1998) 111Indium-diethylenetriamine pentaacetic acid cerebrospinal fluid flow studies predict distribution of intrathecally administered chemotherapy and outcome in patients with leptomeningeal metastases. *Neurology* **50**(2), 438–444.

32. Chamberlain, M. C., and Corey-Bloom, J. (1991). Leptomeningeal metastases: 111indium-DTPA CSF flow studies. *Neurology* **41**(11), 1765–1769.

33. Erlich, S., and Davis, R. L. (1978). Spinal subarachnoid metastasis from primary intracranial glioblastoma multiforme. *Cancer* **42**, 2854–2864.

34. Yung, W. A., Horten, B. C., and Shapiro, W. R. (1980). Meningeal Gliomatosis: A review of 12 cases. *Ann Neurol* **8**, 605–608.

35. Vertosick, F. T., and Selker, R. G. (1990). Brain stem and spinal metastases of supratentorial glioblastoma multiforme: a clinical series. *Neurosurgery* **27**, 516–522.

36. Awad, I., Bay, J. W., and Roger, L. (1986). Leptomeningeal metastasis from supratentorial malignant gliomas. *Neurosurgery* **19**, 247–251.

37. Onda, K., Tanaka, R., Takahashi, H., Takeda, N., and Ikuta, F. (1998). Cerebral glioblastoma with cerebrospinal fluid dissemination: A clinicopathological study of 14 cases examined by complete autopsy. *Neurosurgery* **25**, 533–540.

38. Grant, R., Naylor, B., Junck, L., and Greenberg, H. S. (1992). Clinical outcome in aggressively treated meningeal gliomatosis. *Neurology* **42**, 252–254.

39. Witham, T. F., Fukui, M. B., Meltzer, C. C., Burns, R., Kondziolka, D., and Bozik, M. E. (1999). Survival of patients with high-grade glioma treated with intrathecal thiotriethylenephosphoramide for ependymal or leptomeningeal gliomatosis. *Cancer* **86**, 1347–1353.

40. Chamberlain, M. C. (2003). Combined modality treatment of leptomeningeal gliomatosis. *Neurosurgery* **52**, 324–330.

41. Glantz, M., Jaeckle, K., Chamberlain, M. C. *et al.* (1999). A randomized trial comparing intrathecal sustained-release ara-C (DepoCyt) to intrathecal methotrexate in patients with neoplastic meningitis from solid tumors. *Clin Cancer Res* **11**, 3394–3402.

42. Jaeckle, K. A., Phuphanich, S., van den Bent, M. J. *et al.* (2001). Intrathecal treatment of neoplastic meningitis due to breast cancer with a slow release formulation of cytarabine. *Br J Cancer* **84**(2), 157–163.

43. Grossman, S. A., Finkelstein, D. M., Ruckdeschel, J. C., *et al.* (1993). Randomized prospective comparison of intraventricular methotrexate and thiotepa in patients with previously untreated neoplastic meningitis. *J Clin Oncol* **11**, 561–569.

44. Hitchens, R., Bell, D., Woods, R. *et al.* (1987). A prospective randomized trial of single- agent versus combination chemotherapy in meningeal carcinomatosis. *J Clin Oncol* **5**, 1655–1662.

45. Giannone, L., Greco, F. A., and Hainsworth, J. D. (1986). Combination intraventricular chemotherapy for meningeal neoplasia. *J Clin Oncol* **4**(1), 68–73.

46. Hildebrand, J. (1998). Prophylaxis and treatment of leptomeningeal carcinomatosis in solid tumors of adulthood. *J Neurooncol* **38**(2–3), 193–198.

47. Blaney, S. M., and Poplack, D. G. (1998). New cytotoxic drugs for intrathecal administration. *J Neurooncol* **38**, 219–223.

48. Chamberlain, M. C. (2002). Alpha–Interferon in the Treatment of Neoplastic Meningitis. *Cancer* **94**, 2675–2680.

49. Sampson, J. H., Archer, G. E., Villavicencio, A. T. *et al.* Treatment of Neoplastic Meningitis with Intrathecal Temozolomide. *Clin Cancer Res* **5**, 1183–1188, 1999.

50. Herrlinger, U., Weller, M., and Schabet, M. (1998). New aspects of immunotherapy of leptomeningeal metastasis. *J Neurooncol* **38**, 233–239.

51. Coakham, H. B., and Kemshead, J. T. (1998). Treatment of neoplastic meningitis by targeted radiation using ^{131}I-radiolabelled monoclonal antibodies. *J Neurooncol* **38**, 225–232.

52. Vrionis, F. D. (1998). Gene therapy of neoplastic meningiosis. *J Neurooncol* **38**, 241–244.

53. LeMay, D. R., Bucci, M. N., and Farhat, S. M. (1989). Malignant transformation of recurrent meningioma with pulmonary metastases. *Surg Neurol* **31**, 365–368.

54. Russell, T., and Moss, T. (1996). Metastasizing meningiomas. *Neurosurgery* **19**, 1028–1030.

55. Younis, G. A., Sawaya, R., DeMonte, F., Hess, K. R., Albrecht, S., and Bruner, J. M. (1995). Aggressive meningeal tumors: review of a series. *J Neurosurg* **82**, 17–27.

56. Stangl, A. P., Wellenreuther, R., Lenartz, D. *et al.* (1997). Clonality of multiple meningiomas. *J Neurosurg* **86**, 853–858.

57. Perry, A., Scheithauer, B. W., Strafford, S. L., Lohse, C. M., and Wollan, P. C. (1999). Malignancy in meningiomas. *Cancer* **85**, 2046–2056.

58. Enam, S. A., Abdulrauf, S., Mehta, B., Malik, G. M., and Mahood, A. (1996). Metastasis in meningioma. *Acta Neurochir (Wein)* **138**, 127–130.

59. Chamberlain, M. C., Tsao-Wei, D., and Groshen, S. (2004). Temozolomide for treatment resistant recurrent meningioma. *Neurology* **62**(7), 1210–1212.

60. Akimura, T., Orita, T., and Hayashida, O. (1992). Malignant meningioma metastasizing through the cerebrospinal pathway. *Acta Neurol Scand* **85**(5), 368–371.

61. Bigner, S. H., and Johnston W. W. (1981). The cytopathology of cerebrospinal fluid. II. Metastatic cancer, meningeal carcinomatosis and primary central nervous system neoplasms. *Acta Cytol* **25**(5), 461–479.

62. Kleinschmidt-DeMasters, B. K., and Avakian J. J. (1985). Wallenberg syndrome caused by CSF metastasis from malignant intraventricular meningioma. *Clin Neuropathol* **4**(5), 214–219.

63. Noterman, J., Depierreux, M., and Raftopoulos, C. *et al.* (1987). Metastases of meningioma; apropos of 2 cases. *Neurochirurgia* **33**(3), 184–189.

64. Ramakrishnamurthy, T. V., Murty, A. V., Purohit, A. K., and Sundaram, C. (2002). Benign meningioma metastasizing through CSF pathways: a case report and review of literature. *Neurol India* **50**(3), 326–329.

65. Satoh, T., Kageyama, T., Yoshimoto, Y., Kamata, I., Date, I., and Motoi, M. (1992). Intrathecal dissemination of meningiomas; a case report. *No Shinkei Geka* **20**(7), 805–808.

66. Chamberlain, M. C., and Krantz, M. J. (2004). Disseminated meningioma. *Cancer* (in press).

67. Ferreri, A. J. M., Reni, M., Pasini, F. *et al.* (2002). A multicenter study of treatment of primary CNS Lymphoma. *Neurology* **58**, 1513–1520.

68. Bataille, B., Delwail, V., Menet, E. *et al.* (2000). Primary intracerebral malignant lymphoma: report of 248 cases. *J Neurosurg* **92**, 261–266.

69. Batchelor, T., Carson, K., O'Neill, A. *et al.* (2003). Treatment of primary CNS lymphoma with methotrexate and deferred radiotherapy: a report of NABTT 96–07. *J Clin Oncol* **21**, 1044–1049.

70. Hoang-Xuan, K., Taillandier, L., Chinot, O. *et al.* (2003). Chemotherapy alone as initial treatment for primary CNS lymphoma in patients older than 60 years: A Multicenter Phase II Study (26952) of the European Organization for Research and Treatment of Cancer Brain Tumor Group. *J Clin Oncol* **21**, 2726–2731.

71. Poortmans, P. M. P., Kluin-Nelemans, H. C., Haaxma-Reiche, H. *et al.* (2003). High-dose methotrexate-based chemotherapy followed by consolidating radiotherapy in non-AIDS-related primary central nervous system lymphoma: European Organization for Research and Treatment of Cancer Lymphoma Group Phase II Trial 20962. *J Clin Oncol* **21**, 4483–4488.

72. Pels, H., Schmidt-Wolf, I. G. H., Glasmacher, A. *et al.* (2003). Primary central nervous system lymphoma: results of a pilot and phase II study of systemic and intraventricular chemotherapy with deferred radiotherapy. *J Clin Oncol* **21**, 4489–4495.

73. Balmaceda, C., Gaynor, J. J., Sun, M. *et al.* (1995). Leptomeningeal tumor in primary central nervous system lymphoma: recognition, significance, and implications. *Ann Neurol* **38**, 202–209.

74. Herrlinger, U., Schabet, M., Brugger, W. *et al.* (2002). German cancer society neuro-oncology working group NOA-03 multicenter trial of single-agent high-dose methotrexate for primary central nervous system lymphoma. *Ann Neurol* **51**, 247–252.

75. Gleissner, B., Siehl, J., Korfel, A. *et al.* (2002). CSF evaluation in primary CNS lymphoma patients by PCR of the CDR III IgG genes. *Neurology* **58**, 390–396.

76. Abrey, L. E., Yahalom, J., and DeAngelis, L. M. (2000). Treatment of primary CNS lymphoma: the next step. *J Clin Oncol* **18**, 3144–3150.

77. Fisher, B. J., Sieferheld, W., Chultz, C. *et al.* (2001). Secondary analysis of RTOG 9310, an intergroup phase II combined modality treatment of primary central nervous system lymphoma with chemotherapy and hyperfractionated radiotherapy. *Int J Radiat Oncol, Biol, Phys* (51) 3 (Suppl 1) 1850–59.

78. DeAngelis, L. M., Seiferheld, W., Schold, S. C. *et al.* (2002). Combination chemotherapy and radiotherapy for primary central nervous system lymphoma: Radiation Therapy Oncology Group Study 93–10. *J Clin Oncol* **20**, 4643–4648.

79. Freilich, R. J., Delattre, J-Y., Monjour, A., and De Angelis, L. M. (1996). Chemotherapy without radiation therapy as initial treatment for primary CNS lymphoma in older patients. *Neurology* **46**, 435–439.

80. DeAngelis, L. M., Yahalom, J., Heinemann, M-H. *et al.* (1990). Primary CNS lymphoma: Combined treatment with chemotherapy and radiotherapy. *Neurology* **40**, 80–86,.

81. Lai, R., Rosenberg, M., and DeAngelis, L. M. (2002). Primary CNS lymphoma: A whole brain disease? *Neurology* **59**, 1557–1562.

82. Onda, K., Wakabayashi, K., Tanaka, R., and Takahashi, H. (1999). Intracranial malignant lymphomas: clinicopathological study of 26 autopsy cases. *Brain Tumor Pathol* **16**, 29–35.

83. Lawler, B. E., Betensky, R., Hochberg, F., and Batchelor, T. (2002). Primary CNS lymphoma: the Massachusetts General Hospital experience, 1987–2001, Orlando, FL. Proceedings of ASCO 2002. Abstract No. 298.

84. Nakhleh, R. E., Manivel, J. C., Hurd, D. *et al.* (1989). Central nervous system lymphomas. Immunohistochemical and clinicopathologic study of 26 autopsy cases. *Arch Pathol Lab Med* **113**, 1050–1056

85. Kros, J. M., Bagdi, E. K., Zheng, P. *et al.* (2002). Analysis of *immunoglobulin H* gene rearrangement by polymerase chain reaction in primary central nervous system lymphoma. *J Neurosurg* **97**, 1390–1396.

86. Glantz, M. J., Cole, B. F., Recht, L. *et al.* (1998). High dose intravenous methotrexate for patients with leptomeningeal cancer: is intrathecal chemotherapy necessary? *J Clin Oncol* **16**, 1561–1567.

87. Buhring, U., Herrlinger, U., Krings, T., Thiex, R., Weller, M., and Kuker, W. (2001). MRI features of primary central nervous system lymphoma at presentation. *Neurology* **57**, 393–396.

88. Khan, R. B., Shi, W., Thaler, H. T., DeAngelis, L. M., and Abrey, L. E. (2002). Is intrathecal methotrexate necessary in the treatment of primary central nervous system lymphoma? *J Neurooncol* **58**, 175–178.

89. Gleissner, B., Siehl, J., Korfel, A., Reinhardt, R., and Thiel, E. (2002). CSF evaluation in primary central nervous system lymphoma patients by PCR of the CDR III IgG genes. *Neurology* **58**, 390–396.

90. Chamberlain, M. C., Kormanik, P., and Glantz, M. (1998). Recurrent primary central nervous system lymphoma complicated by lymphomatous meningitis. *Oncol Rep* **5**, 521–523.

91. Marks, J. E., and Adler, S. J. (1982). A comparative study of ependymomas by site of origin. *Int J Radiat Oncol Biol Phys* **8**, 37–43.

92. Dohrmann, G. J., Farwell, J. R., and Flannery, J. T. (1976). Ependymomas and ependymoblastomas in children. *J Neurosurg* **45**, 273–283.

93. Kovalic, J. J., Flaris, N., Grigsby, P. W. *et al.* (1993). Intracranial ependymoma long-term outcome, patterns of failure. *J Neurooncol* **15**, 125–131.

94. Goldwein, J. W., Glauser, T. A., Packer, R. J. *et al.* (1990). Recurrent intracranial ependymomas in children. Survival, patterns of failure, and prognostic factors. *Cancer* **66**, 557–563.

95. Chiu, J. K., Woo, S. Y., Ater, J. *et al.* (1992). Intracranial ependymoma in children: Analysis of prognostic factors. *J Neurooncol* **13**, 283–290.

96. Chamberlain, M. C. (2001). Recurrent intracranial ependymoma in children: salvage therapy with oral etoposide. *Pediatr Neurol* **24**, 117–121.

97. Robertson, P. L., Zeltzer, P. M., Boyeyy, J. M. *et al.* (1998). Survival and prognostic factors following radiation therapy and chemotherapy for ependymomas in children: A report on the Children's Cancer Group. *J Neurosurg* **88**, 695–703.

98. Duffner, P. K., Krischer, J. P., Sanford, R. A. *et al.* (1998). Prognostic factors in infants and very young children with intracranial ependymomas. *Pediatr Neurosurg* **28**, 215–222.

99. Grill, J., Kalifa, C., Doz, F. *et al.* (1996). A high dose busulfan-thiotepa combination followed by autologous bone marrow transplantation in childhood recurrent ependymoma. A phase 2 study. *Pediatr Neurosurg* **25**, 7–12.

100. Mason, W. P., Goldman, S., Yates, A. J., Boyett, J., Li, H., and Finaly, J. L. (1998). Survival following intensive chemotherapy with bone marrow reconstitution for children with recurrent intracranial ependymoma-A report of Children's Cancer Group. *J Neurooncol* **37**, 135–143.

101. Goldwein, J. W., Leahy, J. M., Packer, R. J. *et al.* (1990). Intracranial ependymomas in children. *Int J Radiat Oncol Phys* **19**, 1497–1502.

102. Good, C. D., Wade, A. M., Hayward, R. D. *et al.* (2001). Surveillance neuroimaging in childhood intracranial ependymoma: how effective, how often, and for how long? *J Neurosurg* **94**(1), 27–32.

103. Figarella-Branger, D., Civatte, M., Bouvier-Labit, C. *et al.* (2000). Prognostic factors in intracranial ependymomas in children. *J Neurosurg* **93**, 605–613.

104. Packer, R. J. (2000). Ependymomas in children. *J Neurosurg* **93**, 721–722.

105. Grill, J., Le Deley, M. C., Gambarelli, D. *et al.* (2001). Postoperative chemotherapy without irradiation for ependymoma in children under 5 years of age: A multicenter trial of the French Society of Pediatric Oncology. *J Clin Oncol* **19**, 1288–1296.

106. van Veelen-Vincent, M. R. C., Pierre-Kahn, A., Kalifa, C. *et al.* (2002). Ependymoma in childhood: prognostic factors, extent of surgery, and adjuvant therapy. *J Neurosurg* **97**, 827–835.

107. Comi, A. M., Backstrom, J. W., Burger, P. C., Duffner, P. K., and the Pediatric Oncology Group. (1998). Clinical and neuroradiological findings in infants with intracranial ependymomas. *Pediatr Neurol* **18**, 23–29.

108. Healey, E. A., Barnes, P. D., Kupsky, W. J. *et al.* (1991). The prognostic significance of postoperative residual tumor ependymoma. *Neurosurgery* **28**, 666–671.

109. Rezai, A. R., Woo, H. H., Lee, M. *et al.* (1996). Disseminated ependymomas of the central nervous system. *J Neurosurg* **85**, 618–624.

110. Needle, M. N., Goldwein, J. W., Grass, J. *et al.* (1997). Adjuvant chemotherapy for the treatment of intracranial ependymoma of childhood. *Cancer* **80**(2), 341–347.

111. Paulino, A. C., Wen, B. C., Buatti, J. M. *et al.* (2002). Intracranial ependymomas: an analysis of prognostic factors and patterns of failure. *Am J Clin Oncol* **25**(2), 117–122.

112. Vanuytsel, L., and Brada, M. (1991). The role of prophylactic spinal irradiation in localized intracranial ependymoma. *Int J Radiat Oncol Biol* **21**, 825–830.

113. Duffner, P. K., Horowitz, M. E., Krischer, J. P. *et al.* (1999). The treatment of malignant brain tumors in infants and very young children: An update of the Pediatric Oncology Group experience. *Neurooncol* **4**, 152–156.

114. Begemann, M., and DeAngelis, L. M. (2001). Chemotherapeutic treatment of ependymomas at Memorial Sloan-Kettering Cancer Center (MSKCC) from 1994 to 2000, Memorial Sloan-Kettering Cancer Center, New York, NY. Proceedings of ASCO 2001. 20:65a: 258 (abstract).

115. Aggarwal, R., Yeung, D., Kumar, P., Muhlbauer, M., and Kun, L. E. (1997). Efficacy and feasibility of stereotactic

116. Stafford, S. L., Pollock, B. E., Foote, R. L. *et al.* (2000). Stereotactic radiosurgery for recurrent ependymoma. *Cancer* 2,15 **88**(4), 870–875.

117. Schwartz, T. H., Kim, S., Glick, R. S. *et al.* (999). Supratentorial ependymomas in adult patients. *Neurosurgery* **44**(4), 721–731.

118. Merchant, T. E., Haida, T., Wang, M. H. *et al.* (1997). Anaplastic ependymoma: treatment of pediatric patients with or without craniospinal radiation therapy. *J Neurosurg* **86**, 943–949.

119. Ross, G. W., and Rubinstein, L. J. (1989). Lack of histopathological correlation of malignant ependymomas with postoperative survival. *J Neurosurg* **70**, 31–36.

120. Garret, P. G., and Simpson, W. J. K. (1983). Ependymomas: results of radiation treatment. *Int J Radiat Oncol Biol Phys* **9**, 1121–1124.

121. Awaad, Y. M., Allen, J. C., Miller, D. C. *et al.* (1996). Deferring adjuvant therapy for totally resected intracranial ependymoma. *Pediatr Neurol* **14**, 216–219.

122. Kovnar, E., Kun, L., Burger, P. *et al.* (1991). Patterns of dissemination and recurrence in childhood ependymoma: preliminary results of Pediatric Oncology Protocol #8532. *Ann Neurol* **30**, 457 (Abstract).

123. Duffner, P. K., Horowitz, M. E., Krischer, J. P. *et al.* (1993). Postoperative chemotherapy and delayed radiation in children less than three years of age with malignant brain tumors. *N Engl J Med* **328**, 1725–1731.

124. Needle, M. N., Molloy, P. T., Geyer, J. R. *et al.* (1997). Phase 2 study of daily oral etoposide in children with recurrent brain tumors and other solid tumors. *Med Pediatr Oncol* **29**, 28–32.

125. Newton, H. B., Henson, J., Walker, R. W. (1992). Extraneural metastases in ependymoma. *J Neurooncol* **14**, 135–142.

126. Fisher, P. G., Burger, P., and Eberhart, C. (2004). Biologic risk stratification of medulloblastoma. *J Clin Oncol* **22**(6), 971–974.

127. Gajjar, A., Hernan, R., Kocak, M. *et al.* (2004). Clinical, histopathological, and molecular markers of prognosis: toward a new disease risk stratification system for medulloblastoma. *J Clin Oncol* **22**(6), 984–993.

128. Le, Q., Huhn, S. L., Wara, W. M. *et al.* (2000). Prognostic factors in patients with relapsed medulloblastoma. *Int J Radiat Oncol Biol Phys* **48**(3), 258–259.

129. Chan, A. W., Tarbell, N., Black, P. *et al.* (2000). Adult Medulloblastoma: prognostic factors and patterns of relapse. *Neurosurgery* **47**(9), 623–632.

130. Thomas, P., Deutsch, M., Kepner, J. *et al.* (2000). Low-stage medulloblastoma: final analysis of trial comparing standard-dose with reduced-dose neuraxis irradiation. *J Clin Oncol* **18**(16), 3004–3011.

131. Dunkel, I., Bayett, J., Yates, A. *et al.* (1998). High dose carboplatin, thiotepa, and etoposide with autologous stem-cell rescue for patients with recurrent medulloblastoma. *J Clin Oncol* **16**(1), 222–228.

132. Ashley, D., Meier, L., Kerby, T. *et al.* (1996). Response of recurrent medulloblastoma to low-dose oral etoposide. *J Clin Oncol* **14**(6), 1922–1927.

133. Taylos, R., Bailey, C., Kath, R. *et al.* (2003). Results of a randomized study of preradiation chemotherapy versus radiotherapy alone for non metastatic medulloblastoma. *J Clin Oncol* **21**(8), 1581–1591.

134. Levin, V., Vestnys, P., Edwards, M. *et al.* (1983). Improvement in survival produced by sequential therapies in the treatment of recurrent medulloblastoma. *Cancer* **51**(8), 1364–1370.

135. Friedman, H., Mahaley, S., Schold, C. *et al.* (1986). Efficacy of vincristine and cyclophosphamide in the therapy of recurrent medulloblastoma. *Neurosurgery* **18**(3), 335–340.

136. Lefkowitz, I., Packer, R., Siegel, K. *et al.* (1990). Results of treatment of children with recurrent medulloblastoma/primitive neuroectodermal tumors with lomustine, cisplatin, and vincristine. *Cancer* **65**(3), 412–417.

137. Finlay, J., Goldman, S., Wang, M. *et al.* (1996). Pilot study of high-dose thiotepa and etoposide with autologous bone marrow rescue in children and young adults with recurrent CNS tumors. *J Clin Oncol* **14**(9), 2495–2503.

138. Mahoney, D., Strather, D., Camitta, B. *et al.* (1996). High-dose melphalan and cyclophosphamide with autologous bone marrow rescue for recurrent/progressive malignant brain tumors in children. *J Clin Oncol* **14**(2), 382–388.

139. Friedman, H., Schold, C., Mahaley, S. *et al.* (1989). Phase II treatment of medulloblastoma and pineoblastoma with melphalan. *J Clin Oncol* **7**(7), 904–911.

140. Mooney, C., Souhami, R., and Pritchard, J. (1983). Recurrent medulloblastoma lack of response to high-dose methotrexate. *Cancer Chemother Pharmacol* **10**, 135–136.

141. Chamberlain, M. C., and Kormanik, P. A. (1997). Chronic oral VP-16 for recurrent medulloblastoma. *Pediatr Neurol* **17**, 230–234.

142. Bloom, H. J. G. (1983). Primary intracranial germ cell tumors. *Clin Oncol* **2**(1), 233–257.

143. Haas-Kogan, D. A., Missett, B. T., Wara, W. M. *et al.* (2003). Radiation therapy for intracranial germ cell tumors. *Int J Radiat Oncol Biol Phys* **56**(2), 511–518.

144. Donadio, A. C., Motzer, R. J., Bajorian, D. F. *et al.* (2003). Chemotherapy for teratoma with malignant transformation. *J Clin Oncol* **21**(23), 4285–4291.

145. Kellie, S. J., Boyce, H., Dunkel, I. J. *et al.* (2004) Primary chemotherapy for intracranial nongerminomatous germ cell tumors: Results of the Second International CNS Germ Cell Study Group Protocol. *J Clin Oncol* **22**(5), 846–853.

146. Reddy, A. T., Wellons, J. C., Allen, J. C. *et al.* (2004). Refining the staging evaluation of pineal region germinoma using neuroendoscopy and the presence of preoperative diabetes insipidus. *Neuro-Oncol* **6**, 127–133.

147. Modak, S., Gardner, S., Dunkel, I. J. *et al.* (2004). Thiotepa-based high-dose chemotherapy with autologous stem-cell rescue in patients with recurrent or progressive CNS germ cell tumors. *J Clin Oncol* **22**, 1934–1943.

148. Matsutani, M. (2001). Combined chemotherapy and radiation therapy for CNS germ cell tumors-the japanese experience. *J Neurooncol* **54**(3), 311–316.

149. Kochi, M., Itoyama, Y., Shiraishi, S. *et al.* (2003). Successful treament of intracranial nongerminomatous malignant germ cell tumors by administering neoadjuvant chemotherapy and radiotherapy before excision of residual tumors. *J Neurosurg* **99**, 106–114.

150. Maity, A., Shu, H., Janss, A. *et al.* (2004). Craniospinal radiation in the treatment of biopsy-proven intracranial germinomas: twenty-five years' experience in a single center. *Int J Radiat Oncol Biol Phys* **58**(4), 1165–1170.

151. Shibamoto, Y., Mitsuyuki, A., Yamashita, J. *et al.* (1988). Treatment results of intracranial germinoma as a function of the irradiated volume. *Int J Radiat Oncol Biol Phys* **15**, 285–290.

152. Lindstat, D., Wara, W. M., Edwards, M. S. B. *et al.* (1988). Radiotherapy of primary intracranial germinomas: the case against routine craniospinal irradiation. *Int J Radiat Oncol Biol Phys* **15**, 291–297.

153. Shibamoto, Y., Takahashi, M., and Sasai, K. (1997). Prognosis of intracranial germinoma with syncytiotrophoblastic giant cells treated by radiation therapy. *Int J Radiat Oncol Ciol Phys* **37**(3), 505–510.

154. Missett, B. T., Le, E. Q. T., Wara, W. M. *et al.* (2001). The role of craniospinal radiation in the treatment of intracranial germ cell tumors. *Int J Radiat Oncol Biol Phys* **51**(3), 165–166.

155. Aoyama, H., Shirato, H., Ikeda, J. *et al.* (2002). Induction chemotherapy followed by low-dose involved-field radiotherapy for intracranial germ cell tumors. *J Clin Oncol* **20**(3), 857–865.

156. Bamberg, M., Kortmann, R., Calaminus, G. *et al.* (1999). radiation therapy for intracranial germinoma: results of the german cooperative prospective trials MAKEI 83/86/89. *J Clin Oncol* **17**(8), 2585–2592.

157. Miyanohara, O., Takeshima, H., Kaji, M. *et al.* (2002). Diagnostic significance of soluble C-Kit in the cerebrospinal fluid of patients with germ cell tumors. *J Neurosurg* **97**, 177–183.

158. Baranzelli, M. C., Patte, C., Bouffet, E. *et al.* (1998). An attempt to treat pediatric intracranial αFP and βHCG secreting germ cell tumors with chemotherapy alone. SFOP experience with 18 cases. *J Neurooncol* **37**, 229–239.

159. Sawamura, Y., Shirato, H., Ikeda, J. *et al.* (1998). Induction chemotherapy followed by reduced-volume radiation therapy for newly diagnosed central nervous system germinoma. *J Neurosurg* **88**, 66–72.

160. Kobayashi, T., Yoshida, J., Ishiyama, J. *et al.* (1989). Combination chemotherapy with cisplatin and etoposide for malignant intracranial germ-cell tumors. *J Neurosurg* **70**, 676–681.

161. Kida, Y., Kobayashi, T., Yoshida, J. *et al.* (1986). Chemotherapy with cisplatin for AFP-secreting germ-cell tumors for the central nervous system. *J Neurosurg* **65**, 470–475.

162. Matsutani, M., Sano, K., Takahura, K. *et al.* (1997). Primary intracranial germ cell tumors: a clinical analysis of 153 histologically verified cases. *J Neurosurg* **86**, 446–455.

163. Jennings, M., Gelman, R., Hochberg, F. Intracranial germ-cell tumors: natural history and pathogenesis. *J Neurosurg* **63**, 155–167.

164. Hoffman, H., Otsubo, H., Hendrick, E. B. *et al.* (1991). Intracranial germ-cell tumors in children. *J Neurosurg* **74**, 545–551.

165. Shapiro, W. R., Young, D. F., and Mehta, B. M. (1975). Methotrexate: distribution in cerebrospinal fluid after intravenous, ventricular and lumbar injections. *N Engl J Med* **293**(4), 161–166.

166. Siegal, T. (1998). Leptomeningeal metastases: rationale for systemic chemotherapy or what is the role of intra-CSF-chemotherapy? *J Neurooncol* **38**(2–3), 151–157.

167. Bokstein, F., Lossos, A., and Siegal, T. (1998). Leptomeningeal metastases from solid tumors: a comparison of two prospective series treated with and without intra-cerebrospinal fluid chemotherapy. *Cancer* **82**(9), 1756–1763.

168. Glantz, M. J., Cole, B. F., Recht, L. *et al.* (1998). High-dose intravenous methotrexate for patients with nonleukemic leptomeningeal cancer: is intrathecal chemotherapy necessary? *J Clin Oncol* **16**(4), 1561–1567.

169. Boogerd, W., Dorresteijn, L. D. A., van der Sande, J. J. *et al.* (2000). Response of leptomeningeal metastases from breast cancer to hormonal therapy. *Neurology* **55**, 117–119.

170. Glantz, M., Jaeckle, K. A., Chamberlain, M. C. *et al.* (1999). A randomized trial of a slow-release formulation of cytarabine for the treatment of lymphomatous meningitis. *J Clin Oncol* **17**, 3110–3116.

23

Chemotherapy-Activating Gene Therapy

Kaveh Asadi-Moghaddam and Antonio E. Chiocca

Malignant brain tumors, especially glioblastomas (GBM) remain among the most difficult cancers to treat, and their median survival time has not changed significantly over the past 50 years. Therefore, new treatments are needed. Brain tumors were among the first human malignancies to be targeted by a quite recent technology called gene therapy. In its simplest form, gene therapy is the process by which nucleic acids are transferred into tumor cells to bring about a therapeutic effect. The most effective way to transfer DNA into somatic cells is to use a viral host. To enhance their safety and antitumoral potency, these viruses are genetically modified and engineered with recombinant DNA technology. We review the basic concepts of gene therapy for brain tumors, thereby focusing on gene-directed enzyme-prodrug strategies, suicide gene therapy, and the common approaches that are being translated into clinical trials.

BASIC CONCEPTS OF GENE THERAPY

Any strategy that uses the transfer of genetic material into somatic cells to modulate intracellular activity can be used for gene therapy. This transfer of genetic material into cells can take place *in vitro*, where genetically modified cells are reimplanted into the target tissue [1], or more commonly gene transfer takes place *in vivo* by using a viral vector. In order to enable expression of the intended gene product inside the cell, the genetic material is linked to additional regulatory sequences, and viral genes are removed to minimize virus-mediated toxicity. Viral-based gene

therapy can be subdivided into two types:

1. those that use replication-defective viral vectors and, therefore, cannot grow in cells, are used to deliver an anticancer gene [2], and
2. those that use replication-competent (oncolytic) viruses [3].

The fact that malignant brain tumor cells represent islands of high mitotic activity on the background of a mostly postmitotic environment can be used to design viral vectors which have efficient gene expression only in dividing tumor cells, to achieve selective tumor toxic treatments [4].

ACTIVATION OF CHEMOTHERAPY WITH GENE THERAPY

The technique most commonly used in brain tumor clinical trials is gene-directed enzyme-prodrug therapy also known as suicide gene therapy. This approach is comprised of three components; the prodrug to be activated, the enzyme used for activation, and the delivery system for the corresponding gene [5]. With this strategy, the systemically administered prodrug is converted to the active chemotherapeutic agent only in cancer cells, thereby allowing a maximal therapeutic effect while limiting systemic toxicity. Several suicide gene therapy approaches for brain tumors are being explored:

A. Herpes simplex virus type 1 thymidine kinase/ ganciclovir;
B. cytosine deaminase/5-fluorocytosine;

C. cytochrome P450/cyclophosphamide or ifosfamide;
D. guanine phosphoribosyl-transferase/ 6-thioxantine;
E. nitroreductase/CB1954;
F. carboxylesterase/CPT-11;
G. *Escherichia coli* purine nucleoside phosphorylase/ purine analogs.

Herpes Simplex Virus Type 1 Thymidine Kinase (HSV*tk*)/Ganciclovir

The herpes simplex virus type 1 thymidine kinase enzyme (HSV*tk*) approach is the most widely studied suicide gene strategy. The HSV*tk* system was developed in 1986 [6], and was the first approach used in patients with malignant brain tumors in 1992 [7]. This approach has been conducted in combination with guanosine-based prodrugs, such as ganciclovir (GCV) and acyclovir, which were originally developed as antiviral agents [8]. These prodrugs are nontoxic nucleoside analogs, which are converted by HSV*tk* into phosphorylated compounds. Consequently these compounds directly inhibit DNA polymerase and render the formed DNA molecule unstable, leading to DNA synthesis arrest and cell death. The most commonly used prodrug, GCV is an acyclic analog of the natural nucleoside 2′-deoxyguanosine [9]. GCV is a specific substrate of the HSV*tk*, which is many orders of magnitudes more efficient than human nucleoside kinase at monophosphorylating GCV (see Fig. 23.1) [10].

The resulting GCV-monophosphate is then converted by cellular kinases into toxic GCV-triphosphate. GCV-triphosphate's structural resemblance to 2′-deoxyguanosine triphosphate (dGTP) makes it a substrate for DNA polymerase. Once bound to DNA polymerase, GCV-triphosphate inhibits the

polymerase or is incorporated into DNA, causing DNA chain elongation to terminate. This causes cell death by the inhibition of incorporation of dGTP into DNA, and also by the prevention of chain elongation [11].

Glioma cells transduced and selected to express HSV*tk* are 5000 times more sensitive to GCV than nontransduced cells [12], and a study using positron-emission tomography showed that the extent of gene expression correlates with the therapeutic response [13]. GCV's effects are limited to DNA and it targets replicating cells much like the S-phase-specific chemotherapeutics. *In vivo* efficacy of HSV*tk*/GCV was demonstrated in rats with intracranial gliosarcoma. The animals were treated with intratumoral implantation of a fibroblast packaging cell line secreting HSV*tk* retroviral vectors, followed by intraperitoneal GCV application. The treated animals survived more than twice as long as the controls [14]. Furthermore, preclinical experiments demonstrated marked tumor elimination, despite gene transfer into only a small fraction of the tumor cells [15]. This cytotoxic effect of transduced cells on adjacent nontransduced cells is termed the *bystander effect* [16]. The bystander effect is mediated mainly by the transfer of toxic phosphorylated forms of GCV to nontransduced cells, presumably *via* gap junctions [17]. Another presumed mechanism contributing to the bystander effect is the targeting of mitotically active endothelial cells in tumor vessels by the retroviral vector and the formation of zones of tumor infarction after their destruction by GCV administration [18]. An immune-associated response against a nonhuman protein, like HSV*tk*, leading to diffuse cell death that affects neighboring nontransduced cells has also been suggested [19]. In addition to the bystander effect, tumor cells transduced to express HSV*tk* and treated with the antiviral agent acyclovir display enhanced sensitivity to radiation

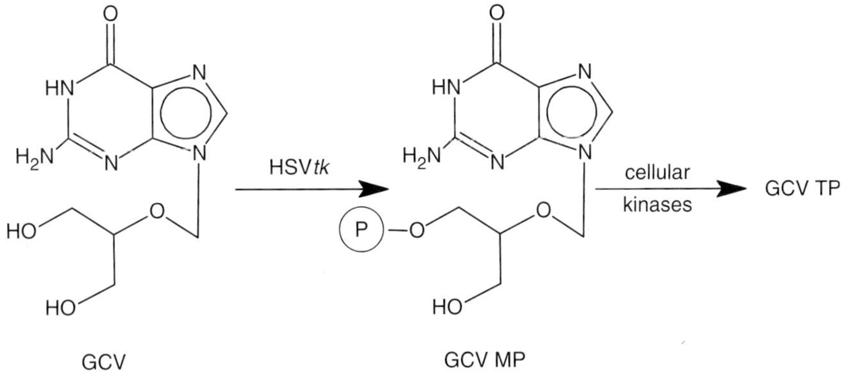

FIGURE 23.1 Herpes simplex virus thymidine kinase (HSV*tk*); ganciclovir (GCV), monophosphate (MP); triphosphate (TP).

in culture and *in vivo* [20]. Possible explanations for radiation enhancement is that DNA which has incorporated acyclovir may be susceptible to radiation-induced strand breakage, and/or acyclovir might sensitize cells by inhibiting polymerase activity required for the repair of radiation-induced DNA damage.

Several strategies have been developed in attempting to improve tumor cell killing. One method involves generating novel and enzymatically enhanced HSV*tk* mutants to induce an increased sensitivity to GCV in transfected cells [21]. The delivery of GCV was also improved by using biocompatible silicones that were directly implanted into the gliomas of experimental rodent models. The results revealed a hundredfold drug concentration over GCV that had been administered intraperitoneally [22]. One approach using a replication-defective herpes simplex virus type 1 vector (NUREL-C2) combines three strategies in order to improve tumor cell killing. The base vector expresses HSV*tk*, and human tumor necrosis factor alpha with coexpression of the gap-junction-forming protein connexin43. This vector has been shown to be effective in treating animal models of glioma [23,24]. The HSV*tk*/GCV "CLINICAL TRIALS" for brain tumors are discussed in section clinical trials of this chapter.

used to treat cancers like colon, pancreatic, and breast cancer. The cytotoxic effects of 5-FU occur following its conversion to 5-fluoro-2′-deoxyuridine-5′-monophosphate (5-FdUMP). 5-FdUMP is an irreversible inhibitor of thymidylate synthase and thus inhibits DNA synthesis by deoxythymidine triphosphate (dTTP) deprivation and causes DNA strand breakage, leading to cell death [25].

Rodent gliosarcoma cells expressing the *E. coli* CD gene become 77 times more sensitive to 5-FC in culture [26]. In addition, tumor cells expressing CD may present CD peptides on MHC class I, where they could lead to an immune response [27]. In order to improve tumor cell killing, a strategy has been developed, where the therapeutic vector encompassed two suicide genes to sensitize cells doubly to GCV and 5-FC [26,28]. In contrast to the HSV*tk*/GCV approach, 5-FU metabolites do not require cell–cell contact for a bystander effect. On cell lysis, 5-FU is released into the medium and is thus likely to be responsible for the bystander effect, and indeed the 5-FU levels in the medium correlated well with the degree of cytotoxicity [29]. Animal studies of CD/5-FC using adenoviral vectors for rodent and human glioma cell lines showed an increase in survival time compared to controls [30]. From the CD/5-FC clinical trials underway none is for brain tumors.

Cytosine Deaminase/5-Fluorocytosine

After HSV*tk*, the cytosine deaminase (CD)/5-Fluorocytosine (5-FC) approach is the next most widely studied suicide gene therapy approach. 5-FC (used to treat infections by fungi such as *Candida albicans* and *Cryptococcus neoformans*) is a prodrug converted into the active agent 5-Fluorouracil (5-FU) by CD (see Fig. 23.2), which is uniquely expressed in certain fungi and bacteria. While 5-FC is nontoxic to human cells because of the lack of CD, 5-FU is

Cytochrome P450/Cyclophosphamide or Ifosfamide

Gene therapy based on cytochrome P450 (CYP) enzymes uses prodrugs that are activated by one or more of the many CYP isozymes [31]. The large number of different isozymes, and the fact that many drugs are metabolized by them, makes the choice of prodrugs quite wide. Most of these enzymes are expressed in the liver rather than in tumor cells, so the goal of this strategy is to selectively increase tumor

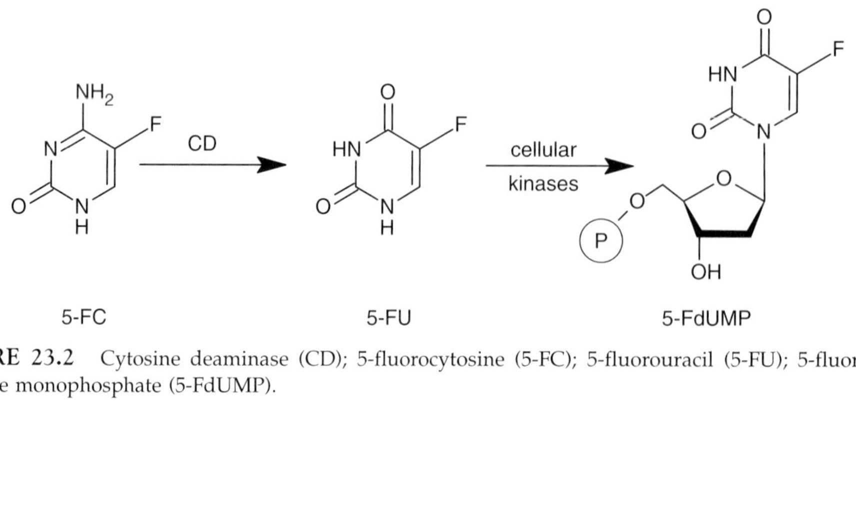

FIGURE 23.2 Cytosine deaminase (CD); 5-fluorocytosine (5-FC); 5-fluorouracil (5-FU); 5-fluorodeoxyuridine monophosphate (5-FdUMP).

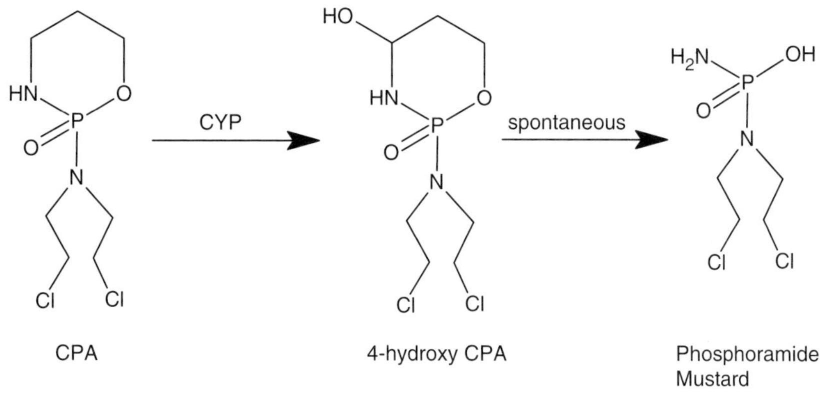

FIGURE 23.3 Cytochrome P450 (CYP); cyclophosphamide (CPA).

cell exposure to cytotoxic drug metabolites by targeting expression of enzymes to tumor cells. To date, this area has been dominated by two prodrugs, cyclophosphamide (CPA) and ifosfamide (IFA). CPA is a prodrug that is activated by liver-specific enzymes of the CYP family (see Fig. 23.3). The rat cytochrome P450 2B1 (CYP2B1), activates CPA with high efficiency [32]. The active form of CPA, phosphoramide mustard, is an alkylating agent that generates DNA cross-links and consecutively DNA strand breaks and results in cell death.

The efficacy of CPA in treating brain tumors has been limited by the fact that although CPA crosses the blood–brain barrier, its active metabolites can be generated only by liver P450, and these metabolites are poorly transported across the blood–brain barrier [33]. Gene therapy using CYP2B1 to activate CPA was designed primarily for use in brain tumors since other malignancies already have access to CPA's active metabolites. The implantation of CYP2B1 expressing retroviral vectors was shown to induce regression of intracerebral rat glioma cells after intratumoral or intrathecal CPA administration [34]. Like the CD/5-FC approach, CPA metabolites do not require cell–cell contact for a bystander effect, distributing by passive diffusion [35]. The transient immunosuppression provided by activated CPA metabolites has also been shown to favor viral replication and anticancer effects *in vivo* [36].

It has been reported that nucleoside analogs are synergistic in their anticancer action with alkylating agents [37]. While CPA metabolites are alkylating agents, GCV metabolites are nucleoside analogs. One approach uses the replacement of the large subunit of the HSV-1 genome with the CYP2B1 gene to generate a HSV-1 vector (rRp450) that is able to kill tumor cells through three modes: (1) using viral oncolysis and rendering infected cell sensitive to (2) CPA and (3) GCV [38]. Subcutaneous tumors established from

glioma cell lines in immunodeficient mice regress only when they are treated with rRp450, CPA, and GCV [26]. This has lead to the hypothesis that after DNA chain alkylation by CPA metabolites, DNA repair mediated by DNA polymerases, is affected by GCV metabolites.

The cytochrome P450 system actually comprises two polypeptide components, the P450 and the P450 reductase (RED). RED expression is required to provide full catalytic activity of the rat CYP2B1 for gene therapy. Studies performed with rat gliosarcoma cells stably transfected with CYP2B1, RED, or both cDNAs, showed that further supplementation of RED by gene transfer enhanced the CYP/CPA effects, although tumor cells express enough RED to fulfill the transferred CYP gene's capability of CPA conversion [39]. The addition of RED not only improves CYP2B1-mediated conversion of CPA but also provides the ability to convert other prodrugs such as tirapazamine into active anticancer agents [40]. Another method to improve the therapeutic index of CYP/CPA is the inhibition of hepatic metabolism of the prodrug [41]. This might decrease systemic toxicity of CPA metabolites as well as increasing prodrug availability to tumor cells expressing CYP. One of the essential steps in tumorigenesis is the active recruitment of a neovascular supply by the neoplasm. A modified CPA regime showed an anti-angiogenic effect which also increased the therapeutic index of the CYP/CPA approach [42, 43]. Three CYP/CPA clinical trials are underway, none of them in brain tumors.

Guanine phosphoribosyl-transferase/6-thioxantine

The *E. coli* gpt gene codes for the enzyme xanthine/guanine phosphoribosyl-transferase (XGPRT). Early generation retroviral vectors used this gene in

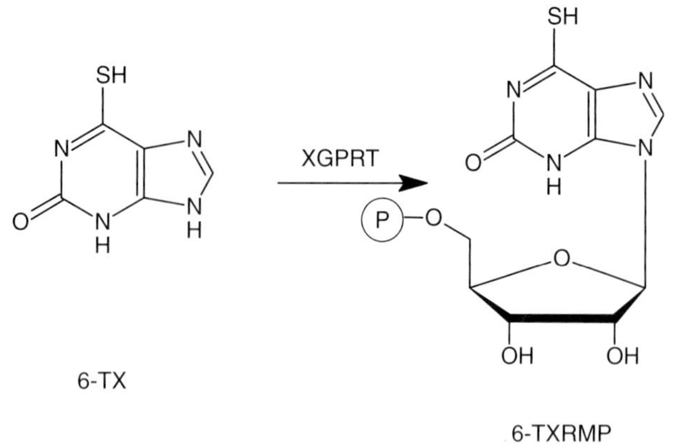

FIGURE 23.4 Xanthine/guanine phosphoribosyl-transferase (XGPRT); 6-thioxanthine (6-TX); 6-thioxanthine riboso-monophosphate (6-TXRMP).

mammalian cells, which do not efficiently use xanthine for purine nucleotide synthesis. After transfection cells producing XGPRT can be selectively grown with xanthine as the sole precursor for guanine nucleotide formation in a medium containing inhibitors (aminopterin and mycophenolic acid) that block *de novo* purine nucleotide synthesis [44,45]. XGPRT was also found to transform a xanthine analog, 6-thioxanthine (6-TX) to a toxic form for mammalian cells [46]. The weakly toxic purine analog 6-TX is phosphorylated to 6-thioxanthine monophosphate (6-XMP) by XGPRT (see Fig. 23.4). 6-XMP is subsequently converted to the highly toxic 6-thioguanine monophosphate (6-GMP).

A clonal line exhibited significant 6-TX susceptibility *in vitro*. In a "bystander" assay, tumor cells from the clonal line efficiently transferred 6-TX sensitivity to uninfected tumor cells. This *in vitro* bystander effect was abrogated when transduced and untransduced cells were separated by a microporous membrane, suggesting that it was not mediated by highly diffusible metabolites. *In vivo* both 6-TX and 6-thioguanine (6-TG) significantly inhibited the growth of subcutaneously transplanted XGPRT expressing clonal tumor cells. In an intracerebral

model, both 6-TX and 6-TG exhibited significant antiproliferative effects against transduced clonal tumors cells [47]. In a nude mouse model retrovirus-mediated transfer of the *gpt* gene into rat glioma cells without subsequent selection still inhibited the proliferation of this mixed polyclonal population upon treatment with 6-TX [48]. There are no GPT/6-TX clinical trial on brain tumors underway.

Nitroreductase/CB1954

Another suicide gene system is the *E. coli* enzyme nitroreductase (NTR) in combination with the prodrug CB1954 [5-(aziridin-1-yl)-2,4-dinitrobenzamide] (see Fig. 23.5). CB1954 is a synthesized, weak, alkylating agent [49,50]. The activating enzyme for CB1954 in mammals is DT-diaphorase (NAD(P)H dehydrogenase). This enzyme converts CB1954 to its 4-hydroxylamino derivative [51]. A further activation step then produces an alkylating agent, after acetylation *via* thioesters such as acetyl coenzyme A (CoA). The activated prodrug is then capable of forming poorly repaired DNA cross-links. The *E. coli* NTR is sensitized to CB1954 whereas human DT-diaphorase is poorly capable of performing this conversion, thereby limiting toxicity to transformed cells [52]. The advantage of the NTR/CB1954 suicide gene system is that killing mediated by activated CB1954 is not dependent on the cell-cycle phase, potentially allowing quiescent tumor cells to be killed.

In vitro studies using a retroviral vector expressing the *E. coli* NTR gene in human colorectal and pancreatic cencer cell lines showed selective killing of NTR-expressing cells following CB1954 administration [53]. To improve the delivery of the NTR gene a replication-defective adenoviral vector expressing NTR was constructed [54]. *In vivo* studies with the adenoviral vector in nude mice bearing human ovarian carcinoma showed promising results [55]. One study also demonstrated a synergistic effect when cells expressing both NTR and HSV*tk* were

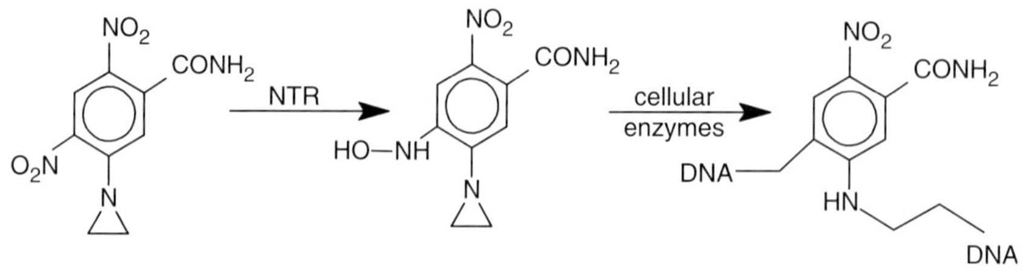

FIGURE 23.5 Nitroreductase (NTR).

treated with a combination of CB1954 and GCV [56]. In addition, a significant bystander effect is seen with the NTR/CB1954 approach analogous to the HSV*tk*/GCV system [54]. A very effective NTR vector is an oncolytic adenovirus. The combination of viral oncolysis and NTR expression results in significantly greater sensitization of colorectal cancer cells to the prodrug CB1954 *in vitro*. *In vivo*, the oncolytic adenoviral vector was shown to replicate in subcutaneous colorectal cancer tumor xenografts in immunodeficient mice, resulting in more NTR expression and greater sensitization to CB1954 than with replication-defective virus [57]. From the four NTR/CB1954 clinical trials underway, none is for brain tumors.

Carboxylesterase/CPT-11

The enzyme–prodrug combinations described so far illustrate the concept of introducing a viral or bacterial enzyme to provide an activity that is almost absent in mammalian cells, whereas in the carboxylesterase (CE) system a mammalian enzyme is used. CE converts the anticancer prodrug CPT-11 (irinotecan, 7-ethyl-10-[4-(1-piperidino)-1-piperidino]carbonyloxycamptothecin) to its active moiety, SN-38 (see Fig. 23.6). In this process, CPT-11 undergoes hydrolysis or deesterification to form the active metabolite SN-38, which is 100–1000 times as potent as CPT-11 as an inhibitor of topoisomerase I [58]. In addition, SN-38 freely passes though cell membranes, increasing the likelihood of a bystander effect. The activation of CPT-11 in humans has been thought to be mediated by the hepatic (hCE2) and human intestinal (hiCE) CEs. Evidence of *in vivo* activation of CPT-11 by endogenous human CEs of less than 5 per cent has led to the search of more efficient CEs from mammalian sources. *In vitro* assays with human tumor cell lines expressing different mammalian CEs found the rabbit CE (rCE) to be the most efficient CE so far.

CPT-11 has demonstrated remarkable antitumor activity both in animal models and in phase II/III trials. Investigations with more than 40 drugs in a CNS xenograft model using multiple adult and pediatric glioma cell lines found CPT-11 to be the most active agent tested [59]. An initial clinical trial of CPT-11 without viral-directed enzyme delivery in 60 patients with recurrent high-grade gliomas resulted in partial response in 10/49 (20 per cent) in GBM patients and 1/8 (12.5 per cent) in AA patients [60]. Currently experiments are ongoing to compare the efficacy and toxicity of hiCE and rCE with various viral constructs, including adenoviruses and retroviruses in which expression of intracellular or secreted forms of each enzyme are regulated by different tumor-cell-specific promoters.

E. coli PNP/Purine Analogs

The *E. coli* purine nucleoside phosphorylase (PNP) is involved in purine metabolism and causes cell death in the presence of nucleoside prodrugs. The prodrugs used with this strategy are 6-methyl purine-deoxyribose (6-MeP-dR), fluoro-deoxyadenosine (F-dAdo), and fluoro-arabinosyl adenosine monophosphate (F-araAMP). *E. coli* PNP cleaves these prodrugs to methyl purine (6-MeP) and 2-fluoroadenine (2-F-Ade), respectively (see Fig. 23.7). These two agents are converted to ATP analogs, which inhibit RNA and/or protein synthesis. The specific mechanism of action of these two agents is not known. It is possible that these agents inhibit dsRNA deaminases. However, RNA and/or protein synthesis inhibition makes these agents effective against proliferating and nonproliferating tumor cells [61]. Furthermore, the active metabolites have a high potency and a high bystander activity which does not require cell-to-cell contact, thereby distributing by passive diffusion [62]. Only a few (0.1–1 per cent) expressing cells are needed to kill

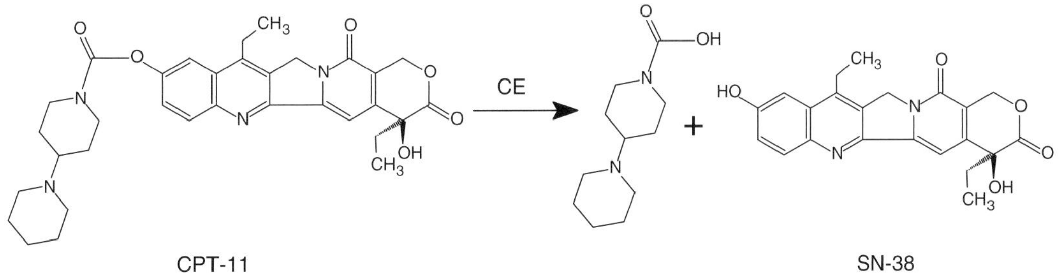

FIGURE 23.6 Carboxylesterase (CE).

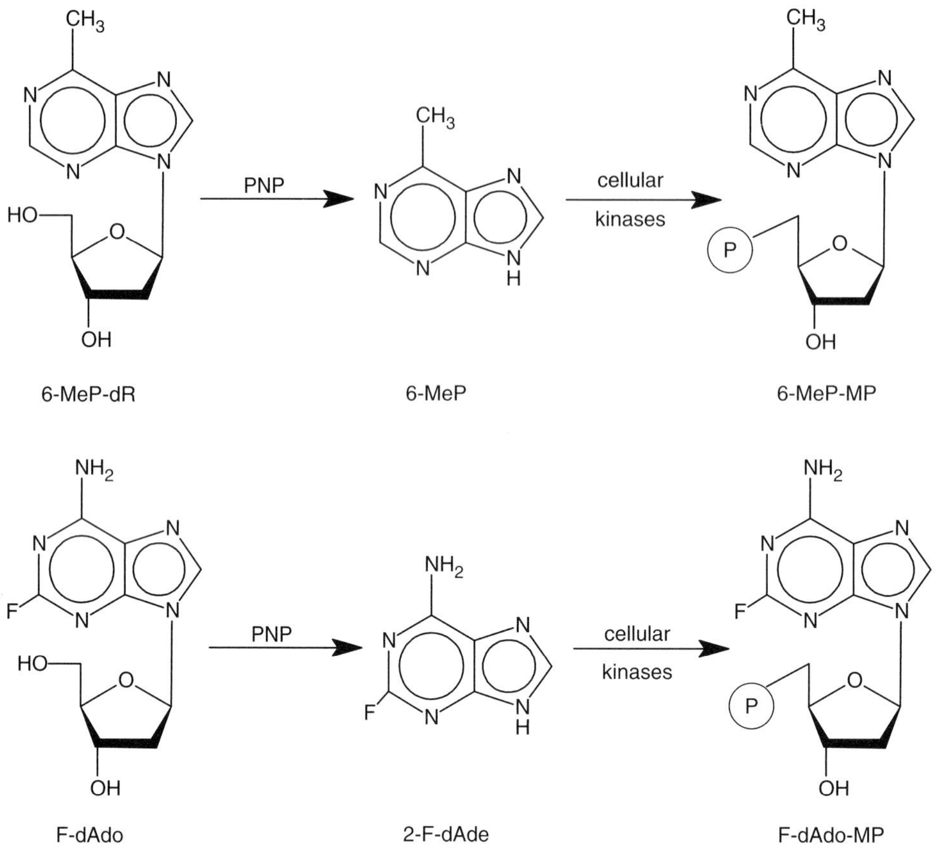

FIGURE 23.7 *E. coli* purine nucleoside phosphorylase (PNP); 6-methyl purine-deoxyribose (6-MeP-dR); 6-methyl purine (6-MeP); fluoro-deoxyadenosine (F-dAdo); fluoro adenine (2-F-dAde); monophosphate (MP).

the entire cell population [61]. Once metabolites such as 6-MeP are generated within a solid tumor, they appear to have a long half-life within the tumor (more than 24 h) and are very slowly released from tumor masses [63]. This long tumor half-life may be one explanation for the ability of these drugs to ablate tumors without profound systemic toxicity. Although, 6-MeP and 2-F-Ade are known to be too toxic for systemic administration (i.e., animals die at drug doses below those that could safely cause tumor regression) [64].

Several *in vitro* studies showed profound tumor cell killing across several cancer cell types by the PNP strategy [65–67]. A direct comparison with HSV*tk*/GCV or CD/5-FC demonstrated faster and more effective tumor cell killing by PNP/purine analogs [66,67]. Strong *in vivo* antitumor effects in mouse models have also been shown [68,69]. *In vivo* studies with human glioma cell lines developed for the *E. coli* PNP approach demonstrated strong antitumor effects after prodrug therapy with 6-MeP-dR and 2-F-araA [63]. There are no clinical trials for brain tumors underway using the *E. coli* PNP approach.

CLINICAL TRIALS

The only suicide gene therapy approach used so far in clinical trials for human brain tumors is the HSV*tk*/GCV strategy (see Table 23.1). Two viruses (retrovirus and adenovirus) have been genetically modified to express HSV*tk*. The first clinical study of brain tumor gene therapy in humans used stereotactic intratumoral inoculation of a retroviral vector carrying HSV*tk* in 15 patients with recurrent malignant brain tumors [70]. This study achieved some promising results in terms of antitumor efficacy. However, it was not a controlled randomized protocol. Subsequent phase I/II studies in patients with recurrent GBM, HSV*tk*/GCV gene therapy was performed by locally administered retroviral-vector-producing cells inoculated manually during open surgical resection of the tumor. One study involving 12 patients showed no adverse treatment-related effects, with an overall median survival of 7.4 months, and with one recurrence-free patient at 2.8 years after treatment [73]. A similar international, multicenter uncontrolled

TABLE 23.1   Selection of Closed HSV*tk*/GCV Gene Therapy Trials in Human Brain Tumors

Phase	Tumor type	Application	Vector	Patients(N)	Median survival
I	Recurrent GBM or metastasis	Stereotactic injection into tumor of RV-producer cells	RV	15	8 mo [70]
I	Recurrent GBM	Stereotactic injection of adenoviral vector	AV	13	4 mo [71]
I/II	Recurrent GBM	Ommaya reservoir injection of RV-producer cells	RV	30	8 mo [72]
I/II	Recurrent GBM	Freehand injection into resected tumor cavity of RV-producer cells	RV	12	7 mo [73]
I/II	Recurrent GBM	Freehand injection into resected tumor cavity of RV-producer cells	RV	48	9 mo [74]
I/II	Primary/recurrent GBM	Freehand injection into resected tumor cavity of RV-producer cells or adenoviral vector	RV or AV	21 (7 RV, 7 AV, 7 LacZ)	7 mo (RV) 15 mo (AV) 8 mo (LacZ) [75]
II/III	Primary/recurrent GBM	Freehand injection into resected tumor cavity of adenoviral vector	AV	36 (17 AV, 19 control)	16 mo (AV) 9 mo (control) [76]
III	Primary GBM	Freehand injection into resected tumor cavity, followed by radiotherapy	RV	248 (124 RV, 124 control)	365 d (RV) 354 d (control) [77]

Abbreviations: AV – adenovirus; d – days; GBM – glioblastoma multiforme; GCV – ganciclovir; HSV*tk* – herpes simplex virus thymidine kinase; LacZ – *E. coli* β-galactosidase gene; mo - months; RV – retrovirus.

study included 48 patients with recurrent GBM. The median survival time was 8.6 months. Tumor recurrence was absent on MRI in seven patients for at least 6 months, in two patients for at least 12 months, and one patient remained recurrence-free at 24 months [74]. A similar phase I study was performed in 12 children between ages 2 and 15 years with recurrent malignant supratentorial brain tumors. Disease progression occurred at a median time of 3 months after treatment [78]. A large controlled phase III study was conducted for an ultimate confirmation of the efficacy of the retroviral HSV*tk*/GCV approach. This study used an adjuvant gene therapy protocol to the standard therapy of maximum surgical resection and irradiation for newly diagnosed GBM. After four years of follow up of 248 patients, who were divided into gene therapy and control arms, survival analysis showed no advantage of gene therapy in terms of tumor progression and overall survival [77].

In order to identify the effectiveness and safety of adenoviral vectors bearing the HSV*tk* gene several phase I trials were conducted. In one study 13 malignant brain tumor patients were treated with a single intratumoral injection of a replication-defective adenoviral vector, followed by GCV treatment. Patients who received the highest vector dose showed central nervous system toxicity (confusion, seizures). Two patients survived for 2 years before lethal tumor progression. One patient survived 2.5 years after the treatment and remained in stable condition [71]. Retroviruses were compared with adenoviruses in another phase I/II trial, in which

21 patients with primary or recurrent GBM were randomly divided into three groups. Retroviruses were used for eight tumors in seven patients and adenoviruses were used for seven tumors in seven patients. At the time of surgical resection, the two experimental groups were treated with either retrovirus-vector-producing cells or adenoviruses expressing HSV*tk* (AdvHSV*tk*), whereas the control group received either adenovirus or retrovirus-vector-producing cells expressing *E. coli* β-galactosidase (lacZ), a marker gene. Four patients with adenovirus injections had a significant increase in anti-adenovirus antibodies and two of them had a short-term fever reaction. Frequency of epileptic seizures increased in two patients. After subsequent GCV treatment, the adenovirus group had significant improvement in mean survival time, with 15 month compared to 7.4 (retrovirus), and 8.3 months (control). The explanations offered for the greater efficacy of the adenovirus group were greater titer, a benefit from the inflammatory reaction to adenoviruses, and the ability of adenovirus to infect nonreplicating cells. In the retrovirus group, all the treated gliomas showed progression by MRI at the 3-month time point, whereas three of the seven patients treated with AdvHSV*tk* remained stable [75]. On the basis of these results a randomised controlled trial involving 36 patients with operable primary or recurrent malignant glioma was conducted. Seventeen patients were randomized to receive AdvHSV*tk* gene therapy by local injection into the wound bed after tumor resection, followed by intravenous GCV administration.

The control group of 19 patients received standard care consisting of radical excision followed by radiotherapy in those patients with primary tumors. AdvHSV*tk* treatment produced a significant increase in mean survival from 39 to 70.6 weeks. The median survival time increased from 37.7 to 62.4 weeks. Six patients had increased anti-adenovirus antibody titers, without adverse effects [76].

LIMITATIONS AND PROBLEMS

Some major problems remain to be solved before any larger scale clinical trials are conducted and gene therapy becomes routinely adopted in the clinic. One of the main challenges being the improvement of delivering genetic vectors into the tissue of interest (solid brain tumors) with efficient *in situ* gene transfer [79]. The efficiency of transduction could be improved by modifying vector producing cells (VPC), which have the ability to track even single tumor cells invading the surrounding brain tissue [80]. The currently used manual injection of VPCs, might be improved by the use of three-dimensional neuronavigation techniques and automated slow-speed injection devices [81]. Moreover, acute and chronic toxicity to normal tissue caused by the viral vector itself needs to be improved. In order to improve this hurdle several strategies have been used. One approach uses an oncolytic virus (OV) to deliver the cDNA into the tumor. OVs are genetically altered viruses with deletions that restrict viral replication in normal cells but permit it in tumor cells [82,83].

CONCLUSIONS

The rapid evolution of recombinant DNA technology enables us to envision new therapeutic modalities, including gene therapy. The completed clinical trials in brain tumor gene therapy have offered some promising results. However, the field of gene therapy is in its infancy. Screening of new approaches is mostly based on animal models that are far from being representative of the analogous clinical scenarios, as shown in the discrepancy between the experimental rodent studies and the clinical trials. Neither can the human response to various viral vectors be predicted in a reliable manner from animal experimentation nor do size, consistency, and extent of experimental brain tumors reflect the large, necrotic, and infiltrative nature of GBMs.

The next step in the development of brain tumor gene therapy will also require a multimodal approach that combines several of the ideas suggested in this book in order to improve the poor outcome of malignant brain tumors.

FURTHER INFORMATION

Official site about clinical trials, gene therapy trials, and human research studies: www.clinicaltrials.gov

The most comprehensive source on worldwide gene therapy clinical trials: www.wiley.co.uk/genmed/clinical.

References

1. Gage, F. H., Fisher, L. J., Jinnah, H. A. *et al.* (1990). Grafting genetically modified cells to the brain: Conceptual and technical issues. *Prog Brain Res* **82**, 1–10.
2. Kramm, C. M., Sena-Esteves, M., Barnett, F. H. *et al.* (1995). Gene therapy for brain tumors. *Brain Pathol* **5**, 345–381.
3. Chiocca, E. A. (2002). Oncolytic viruses. *Nat Rev Cancer* **2**, 938–950.
4. Lam, P. Y., and Breakefield, X. O. (2001). Potential of gene therapy for brain tumors. *Hum Mol Genet* **10**, 777–787.
5. Anderson, W. F. (2000). Gene therapy scores against cancer. *Nat Med* **6**, 862–863.
6. Moolten, F. L. (1986). Tumor chemosensitivity conferred by inserted herpes thymidine kinase genes: Paradigm for a prospective cancer control strategy. *Cancer Res* **46**, 5276–5281.
7. Oldfield, E. H., Ram, Z., Culver, K. W. *et al.* (1993). Gene therapy for the treatment of brain tumors using intra-tumoral transduction with the thymidine kinase gene and intravenous ganciclovir. *Hum Gene Ther* **4**, 39–69.
8. De Clercq, E. (2000). Guanosine analogues as anti-herpesvirus agents. *Nucleosides Nucleotides Nucleic Acids* **19**, 1531–1541.
9. Faulds, D., and Heel, R. C. (1990). Ganciclovir. A review of its antiviral activity, pharmacokinetic properties and therapeutic efficacy in cytomegalovirus infections. *Drugs* **39**, 597–638.
10. Elion, G. B., Furman, P. A., Fyfe, J. A. *et al.* (1977). Selectivity of action of an antiherpetic agent, 9-(2-hydroxyethoxymethyl) guanine. *Proc Natl Acad Sci USA* **74**, 5716–5720.
11. Mesnil, M., and Yamasaki, H. (2000). Bystander effect in herpes simplex virus-thymidine kinase/ganciclovir cancer gene therapy: Role of gap-junctional intercellular communication. *Cancer Res* **60**, 3989–3999.
12. Shewach, D. S., Zerbe, L. K., Hughes, T. L. *et al.* (1994). Enhanced cytotoxicity of antiviral drugs mediated by adenovirus directed transfer of the herpes simplex virus thymidine kinase gene in rat glioma cells. *Cancer Gene Ther* **1**, 107–112.
13. Jacobs, A., Voges, J., Reszka, R. *et al.* (2001). Positron-emission tomography of vector-mediated gene expression in gene therapy for gliomas. *Lancet* **358**, 727–729.
14. Culver, K. W., Ram, Z., Wallbridge, S. *et al.* (1992). In vivo gene transfer with retroviral vector-producer cells for treatment of experimental brain tumors. *Science* **256**, 1550–1552.
15. Ram, Z., Culver, K. W., Walbridge, S. *et al.* (1993). In situ retroviral-mediated gene transfer for the treatment of brain tumors in rats. *Cancer Res* **53**, 83–88.
16. Ishii-Morita, H., Agbaria, R., Mullen, C. A. *et al.* (1997). Mechanism of 'bystander effect' killing in the herpes simplex

thymidine kinase gene therapy model of cancer treatment. *Gene Ther* **4**, 244–251.

17. Fick, J., Barker, F. G., 2nd, Dazin, P. *et al.* (1995). The extent of heterocellular communication mediated by gap junctions is predictive of bystander tumor cytotoxicity in vitro. *Proc Natl Acad Sci U S A* **92**, 11071–11075.

18. Ram, Z., Walbridge, S., Shawker, T. *et al.* (1994). The effect of thymidine kinase transduction and ganciclovir therapy on tumor vasculature and growth of 9l gliomas in rats. *J Neurosurg* **81**, 256–260.

19. Barba, D., Hardin, J., Sadelain, M. *et al.* (1994). Development of anti-tumor immunity following thymidine kinase-mediated killing of experimental brain tumors. *Proc Natl Acad Sci USA* **91**, 4348–4352.

20. Kim, J. H., Kim, S. H., Kolozsvary, A. *et al.* (1995). Selective enhancement of radiation response of herpes simplex virus thymidine kinase transduced 9l gliosarcoma cells in vitro and in vivo by antiviral agents. *Int J Radiat Oncol Biol Phys* **33**, 861–868.

21. Wiewrodt, R., Amin, K., Kiefer, M. *et al.* (2003). Adenovirus-mediated gene transfer of enhanced herpes simplex virus thymidine kinase mutants improves prodrug-mediated tumor cell killing. *Cancer Gene Ther* **10**, 353–364.

22. Miura, F., Moriuchi, S., Maeda, M. *et al.* (2002). Sustained release of low-dose ganciclovir from a silicone formulation prolonged the survival of rats with gliosarcomas under herpes simplex virus thymidine kinase suicide gene therapy. *Gene Ther* **9**, 1653–1658.

23. Niranjan, A., Wolfe, D., Tamura, M. *et al.* (2003). Treatment of rat gliosarcoma brain tumors by hsv-based multigene therapy combined with radiosurgery. *Mol Ther* **8**, 530–542.

24. Wolfe, D., Niranjan, A., Trichel, A. *et al.* (2004). Safety and biodistribution studies of an hsv multigene vector following intracranial delivery to non-human primates. *Gene Ther* **11**, 1675–1684.

25. Grem, J. L. (1996). 5-fluoropyrimidines. In *Cancer chemotherapy and biotherapy: Principles and practice.* (B. A. Chabner, and D. L. Longo Eds.), pp. 149–212. Philadelphia, Lippincott.

26. Aghi, M., Kramm, C. M., Chou, T. C. *et al.* (1998). Synergistic anticancer effects of ganciclovir/thymidine kinase and 5-fluorocytosine/cytosine deaminase gene therapies. *J Natl Cancer Inst* **90**, 370–380.

27. Mullen, C. A., Petropoulos, D., and Lowe, R. M. (1996). Treatment of microscopic pulmonary metastases with recombinant autologous tumor vaccine expressing interleukin 6 and escherichia coli cytosine deaminase suicide genes. *Cancer Res* **56**, 1361–1366.

28. Desaknai, S., Lumniczky, K., Esik, O. *et al.* (2003). Local tumour irradiation enhances the anti-tumour effect of a double-suicide gene therapy system in a murine glioma model. *J Gene Med* **5**, 377–385.

29. Kuriyama, S., Masui, K., Sakamoto, T. *et al.* (1998). Bystander effect caused by cytosine deaminase gene and 5-fluorocytosine in vitro is substantially mediated by generated 5-fluorouracil. *Anticancer Res* **18**, 3399–3406.

30. Miller, C. R., Williams, C. R., Buchsbaum, D. J. *et al.* (2002). Intratumoral 5-fluorouracil produced by cytosine deaminase/5-fluorocytosine gene therapy is effective for experimental human glioblastomas. *Cancer Res* **62**, 773–780.

31. Waxman, D. J., Chen, L., Hecht, J. E. *et al.* (1999). Cytochrome p450-based cancer gene therapy: Recent advances and future prospects. *Drug Metab Rev* **31**, 503–522.

32. Clarke, L., and Waxman, D. J. (1989). Oxidative metabolism of cyclophosphamide: Identification of the hepatic monooxygenase catalysts of drug activation. *Cancer Res* **49**, 2344–2350.

33. Wei, M. X., Tamiya, T., Chase, M. *et al.* (1994). Experimental tumor therapy in mice using the cyclophosphamide-activating cytochrome p450 2b1 gene. *Hum Gene Ther* **5**, 969–978.

34. Manome, Y., Wen, P. Y., Chen, L. *et al.* (1996). Gene therapy for malignant gliomas using replication incompetent retroviral and adenoviral vectors encoding the cytochrome p450 2b1 gene together with cyclophosphamide. *Gene Ther* **3**, 513–520.

35. Wei, M. X., Tamiya, T., Rhee, R. J. *et al.* (1995). Diffusible cytotoxic metabolites contribute to the in vitro bystander effect associated with the cyclophosphamide/cytochrome p450 2b1 cancer gene therapy paradigm. *Clin Cancer Res* **1**, 1171–1177.

36. Ikeda, K., Ichikawa, T., Wakimoto, H. *et al.* (1999). Oncolytic virus therapy of multiple tumors in the brain requires suppression of innate and elicited antiviral responses. *Nat Med* **5**, 881–887.

37. Andersson, B. S., Sadeghi, T., Siciliano, M. J. *et al.* (1996). Nucleotide excision repair genes as determinants of cellular sensitivity to cyclophosphamide analogs. *Cancer Chemother Pharmacol* **38**, 406–416.

38. Chase, M., Chung, R. Y., and Chiocca, E. A. (1998). An oncolytic viral mutant that delivers the cyp2b1 transgene and augments cyclophosphamide chemotherapy. *Nat Biotechnol* **16**, 444–448.

39. Chen, L., Yu, L. J., and Waxman, D. J. (1997). Potentiation of cytochrome p450/cyclophosphamide-based cancer gene therapy by coexpression of the p450 reductase gene. *Cancer Res* **57**, 4830–4837.

40. Jounaidi, Y., and Waxman, D. J. (2000). Combination of the bioreductive drug tirapazamine with the chemotherapeutic prodrug cyclophosphamide for p450/p450-reductase-based cancer gene therapy. *Cancer Res* **60**, 3761–3769.

41. Huang, Z., Raychowdhury, M. K., and Waxman, D. J. (2000). Impact of liver p450 reductase suppression on cyclophosphamide activation, pharmacokinetics and antitumoral activity in a cytochrome p450-based cancer gene therapy model. *Cancer Gene Ther* **7**, 1034–1042.

42. Browder, T., Butterfield, C. E., Kraling, B. M. *et al.* (2000). Antiangiogenic scheduling of chemotherapy improves efficacy against experimental drug-resistant cancer. *Cancer Res* **60**, 1878–1886.

43. Jounaidi, Y., and Waxman, D. J. (2001). Frequent, moderate-dose cyclophosphamide administration improves the efficacy of cytochrome p-450/cytochrome p-450 reductase-based cancer gene therapy. *Cancer Res* **61**, 4437–4444.

44. Mulligan, R. C., and Berg, P. (1980). Expression of a bacterial gene in mammalian cells. *Science* **209**, 1422–1427.

45. Mulligan, R. C., and Berg, P. (1981). Selection for animal cells that express the escherichia coli gene coding for xanthine-guanine phosphoribosyltransferase. *Proc Natl Acad Sci USA* **78**, 2072–2076.

46. Besnard, C., Monthioux, E., and Jami, J. (1987). Selection against expression of the escherichia coli gene gpt in hprt+ mouse teratocarcinoma and hybrid cells. *Mol Cell Biol* **7**, 4139–4141.

47. Tamiya, T., Ono, Y., Wei, M. X. *et al.* (1996). Escherichia coli gpt gene sensitizes rat glioma cells to killing by 6-thioxanthine or 6-thioguanine. *Cancer Gene Ther* **3**, 155–162.

48. Ono, Y., Ikeda, K., Wei, M. X. *et al.* (1997). Regression of experimental brain tumors with 6-thioxanthine and escherichia coli gpt gene therapy. *Hum Gene Ther* **8**, 2043–2055.

49. Khan, A. H., and Ross, W. C. (1969). Tumour-growth inhibitory nitrophenylaziridines and related compounds: Structure-activity relationships. *Chem Biol Interact* **1**, 27–47.

50. Khan, A.H., and Ross, W. C. (1971). Tumour-growth inhibitory nitrophenylaziridines and related compounds: Structure-activity relationships. *Ii. Chem Biol Interact* **4**, 11–22.

51. Knox, R. J., Friedlos, F., Jarman, M. *et al.* (1988). A new cytotoxic, DNA interstrand crosslinking agent, 5-(aziridin-1-yl)-4-hydroxylamino-2-nitrobenzamide, is formed from 5-(aziridin-1-yl)-2,4-dinitrobenzamide (cb 1954) by a nitro-reductase enzyme in walker carcinoma cells. *Biochem Pharmacol* **37**, 4661–4669.

52. Boland, M. P., Knox, R. J., and Roberts, J. J. (1991). The differences in kinetics of rat and human dt diaphorase result in a differential sensitivity of derived cell lines to cb 1954 (5-(aziridin-1-yl)-2,4-dinitrobenzamide). *Biochem Pharmacol* **41**, 867–875.

53. Green, N. K., Youngs, D. J., Neoptolemos, J. P. *et al.* (1997). Sensitization of colorectal and pancreatic cancer cell lines to the prodrug 5-(aziridin-1-yl)-2,4-dinitrobenzamide (cb1954) by retroviral transduction and expression of the e. Coli nitror-eductase gene. *Cancer Gene Ther* **4**, 229–238.

54. Grove, J. I., Searle, P. F., Weedon, S. J. *et al.* (1999). Virus-directed enzyme prodrug therapy using cb1954. *Anticancer Drug Des* **14**, 461–472.

55. Weedon, S. J., Green, N. K., McNeish, I. A. *et al.* (2000). Sensitisation of human carcinoma cells to the prodrug cb1954 by adenovirus vector-mediated expression of e. Coli nitro-reductase. *Int J Cancer* **86**, 848–854.

56. Bridgewater, J. A., Springer, C. J., Knox, R. J. *et al.* (1995). Expression of the bacterial nitroreductase enzyme in mamma-lian cells renders them selectively sensitive to killing by the prodrug cb1954. *Eur J Cancer* **31A**, 2362–2370.

57. Chen, M. J., Green, N. K., Reynolds, G. M. *et al.* (2004). Enhanced efficacy of escherichia coli nitroreductase/cb1954 prodrug activation gene therapy using an e1b-55k-deleted oncolytic adenovirus vector. *Gene Ther* **11**, 1126–1136.

58. Rothenberg, M. L. (1997). Topoisomerase i inhibitors: Review and update. *Ann Oncol* **8**, 837–855.

59. Hare, C. B., Elion, G. B., Houghton, P. J. *et al.* (1997). Therapeutic efficacy of the topoisomerase i inhibitor 7-ethyl-10-(4-[1-piperidino]-1-piperidino)-carbonyloxy-camptothecin against pediatric and adult central nervous system tumor xenografts. *Cancer Chemother Pharmacol* **39**, 187–191.

60. Colvin, O. M., Cokgor, I., Ashley, D. M. *et al.* (1998). Irinotecan treatment of adults with recurrent or progressive malignant glioma. *Proc Am Soc Clin Oncol.* [Abstract 1493], 387a.

61. Parker, W. B., Allan, P. W., Shaddix, S. C. *et al.* (1998). Metabolism and metabolic actions of 6-methylpurine and 2-fluoroadenine in human cells. *Biochem Pharmacol* **55**, 1673–1681.

62. Hughes, B. W., King, S. A., Allan, P. W. *et al.* (1998). Cell to cell contact is not required for bystander cell killing by escherichia coli purine nucleoside phosphorylase. *J Biol Chem* **273**, 2322–2328.

63. Gadi, V. K., Alexander, S. D., Waud, W. R. *et al.* (2003). A long-acting suicide gene toxin, 6-methylpurine, inhibits slow growing tumors after a single administration. *J Pharmacol Exp Ther* **304**, 1280–1284.

64. Philips, F. S., Sternberg, S. S., Hamilton, S. *et al.* (1954). The toxic effects of 6-mercaptopurine and related compounds. *Ann N Y Acad Sci* **60**, 283–296.

65. Da Costa, L. T., Jen, J., He, T. C. *et al.* (1996). Converting cancer genes into killer genes. *Proc Natl Acad Sci USA* **93**, 4192–4196.

66. Nestler, U., Heinkelein, M., Lucke, M. *et al.* (1997). Foamy virus vectors for suicide gene therapy. *Gene Ther* **4**, 1270–1277.

67. Lockett, L. J., Molloy, P. L., Russell, P. J. *et al.* (1997). Relative efficiency of tumor cell killing in vitro by two enzyme-prodrug systems delivered by identical adenovirus vectors. *Clin Cancer Res* **3**, 2075–2080.

68. Martiniello-Wilks, R., Garcia-Aragon, J., Daja, M. M. *et al.* (1998). In vivo gene therapy for prostate cancer: Preclinical evaluation of two different enzyme-directed prodrug therapy systems delivered by identical adenovirus vectors. *Hum Gene Ther* **9**, 1617–1626.

69. Mohr, L., Shankara, S., Yoon, S. K. *et al.* (2000). Gene therapy of hepatocellular carcinoma in vitro and in vivo in nude mice by adenoviral transfer of the escherichia coli purine nucleoside phosphorylase gene. *Hepatology* **31**, 606–614.

70. Ram, Z., Culver, K. W., Oshiro, E. M. *et al.* (1997). Therapy of malignant brain tumors by intratumoral implantation of retroviral vector-producing cells. *Nat Med* **3**, 1354–1361.

71. Trask, T. W., Trask, R. P., Aguilar-Cordova, E. *et al.* (2000). Phase I study of adenoviral delivery of the hsv-tk gene and ganciclovir administration in patients with current malignant brain tumors. *Mol Ther* **1**, 195–203.

72. Prados, M. D., McDermott, M., Chang, S. M. *et al.* (2003). Treatment of progressive or recurrent glioblastoma multiforme in adults with herpes simplex virus thymidine kinase gene vector-producer cells followed by intravenous ganciclovir administration: A phase I/II multi-institutional trial. *J Neurooncol* **65**, 269–278.

73. Klatzmann, D., Valery, C. A., Bensimon, G. *et al.* (1998). A phase I/II study of herpes simplex virus type 1 thymidine kinase "suicide" gene therapy for recurrent glioblastoma. Study group on gene therapy for glioblastoma. *Hum Gene Ther* **9**, 2595–2604.

74. Shand, N., Weber, F., Mariani, L. *et al.* (1999). A phase 1–2 clinical trial of gene therapy for recurrent glioblastoma multiforme by tumor transduction with the herpes simplex thymidine kinase gene followed by ganciclovir. Gli328 european-canadian study group. *Hum Gene Ther* **10**, 2325–2335.

75. Sandmair, A. M., Loimas, S., Puranen, P. *et al.* (2000). Thymidine kinase gene therapy for human malignant glioma, using replication-deficient retroviruses or adenoviruses. *Hum Gene Ther* **11**, 2197–2205.

76. Immonen, A., Vapalahti, M., Tyynela, K. *et al.* (2004). Advhsv-tk gene therapy with intravenous ganciclovir improves survival in human malignant glioma: A randomised, controlled study. *Mol Ther* **10**, 967–972.

77. Rainov, N. G. (2000). A phase iii clinical evaluation of herpes simplex virus type 1 thymidine kinase and ganciclovir gene therapy as an adjuvant to surgical resection and radiation in adults with previously untreated glioblastoma multiforme. *Hum Gene Ther* **11**, 2389–2401.

78. Packer, R. J., Raffel, C., Villablanca, J. G. *et al.* (2000). Treatment of progressive or recurrent pediatric malignant supratentorial brain tumors with herpes simplex virus thymidine kinase gene

vector-producer cells followed by intravenous ganciclovir administration. *J Neurosurg* **92**, 249–254.

79. Puumalainen, A. M., Vapalahti, M., Agrawal, R. S. *et al.* (1998). Beta-galactosidase gene transfer to human malignant glioma in vivo using replication-deficient retroviruses and adenoviruses. *Hum Gene Ther* **9**, 1769–1774.

80. Herrlinger, U., Woiciechowski, C., Sena-Esteves, M. *et al.* (2000). Neural precursor cells for delivery of replication-conditional hsv-1 vectors to intracerebral gliomas. *Mol Ther* **1**, 347–357.

81. Rutka, J. T., Taylor, M., Mainprize, T. *et al.* (2000). Molecular biology and neurosurgery in the third millennium. *Neurosurgery* **46**, 1034–1051.

82. Smith, E. R., and Chiocca, E. A. (2000). Oncolytic viruses as novel anticancer agents: Turning one scourge against another. *Expert Opin Investig Drugs* **9**, 311–327.

83. Chung, R. Y., Saeki, Y., and Chiocca, E. A. (1999). B-myb promoter retargeting of herpes simplex virus gamma34.5 gene-mediated virulence toward tumor and cycling cells. *J Virol* **73**, 7556–7564.

CHEMOTHERAPY OF SPECIFIC TUMOR TYPES

Chemotherapy of High-Grade Astrocytomas

Herbert B. Newton

ABSTRACT: High-grade astrocytomas, including anaplastic astrocytoma (AA) and glioblastoma multiforme (GBM), are frequently diagnosed tumors that continue to have a poor prognosis and protracted survival. Chemotherapy has recently been demonstrated to provide a survival benefit in patients with AA and GBM, in addition to surgical resection and irradiation, by several meta-analyses. The improvement in survival has been modest and mainly associated with the use of nitrosourea drugs such as BCNU and CCNU, as well as platinum compounds. Recent clinical trial data also suggests that the alkylating agent, temozolomide (TZM), has activity against AA and GBM in the neoadjuvant, adjuvant, and recurrent settings. In addition, TZM has been shown in phase II and III clinical trials to improve survival of patients with GBM in combination with radiation therapy. Further drug discovery and clinical trial testing will be needed to make more substantial improvements in the survival of patients with high-grade astrocytomas.

INTRODUCTION

High-grade or malignant astrocytomas consist of anaplastic astrocytoma (AA) and glioblastoma multiforme (GBM), and represent the most commonly diagnosed brain tumors in adults [1–3]. These tumors account for 30 to 35 per cent of all newly diagnosed brain tumors in patients 18 years of age or older and affect an estimated 10 000 to 12 000 new patients each year in the United States. The overall incidence rates for AA and GBM are 0.13 and 3.24 per 100 000

person-years, respectively. The age of onset is different between the two forms of high-grade astrocytoma, with AA arising in patients between the ages of 45 and 55, and GBM typically affecting an older cohort between 55 and 65 years of age. There is usually a slight preponderance of male patients, with a male to female ratio of approximately 1.8:1 in most series [2]. The prognosis for patients with AA and GBM remains poor, with overall median survivals of 30 to 36 months and 12 to 15 months, respectively, for patients that have received standard treatment.

Anaplastic astrocytomas and GBM are classified by the World Health Organization (WHO) as grade III and grade IV astrocytomas, respectively (see Chapter 1) [4]. Pathologically, AA are characterized by the presence of high cellularity, prominent cellular and nuclear pleomorphism, nuclear atypia, and mitotic activity. The histological features of GBM are similar to AA, but with more pronounced cellular and nuclear anaplasia and the presence of microvascular proliferation and/or necrosis. Both AA and GBM are highly infiltrative tumors and commonly arise in the cerebral hemispheres or deep white matter.

OVERVIEW OF INITIAL TREATMENT

The most common form of initial treatment for a high-grade astrocytoma is surgical intervention. Indications for surgery include reducing tumor burden, alleviating mass effect, confirmation of the histological diagnosis, diversionary shunting procedures in selected cases, and the introduction of local antineoplastic agents [5,6]. Recent advances in neurosurgical technology offer new approaches to tumor removal, such as frame-based and frameless stereotactic biopsy, intra-operative cortical mapping, neuronavigation, and the use of intraoperative magnetic resonance imaging (MRI) [7–9]. These techniques allow the surgeon to more carefully delineate tumor margins and preserve surrounding regions of eloquent brain (e.g., Broca's area, primary motor cortex) and delicate vascular structures, while performing a more aggressive and thorough tumor resection. Although it remains debated in the literature, most neurosurgeons recommend a near-total or gross-total resection, whenever possible, of all enhancing tumor volume and regionally infiltrated brain as defined on T2 MRI images. Gross-total tumor resection has been associated with longer overall and progression-free survival in several studies [10,11]. For tumors that are diffusely infiltrative or multifocal, a stereotactic biopsy is more likely to preserve neurological function than an attempt at resection and, in most cases, will be able to provide a histological diagnosis to guide further treatment.

External beam fractionated radiation therapy is an appropriate form of treatment for virtually all patients with AA and GBM, and is similar for both tumor types [1,12–14]. Numerous randomized controlled trials have demonstrated a survival benefit for patients receiving surgical resection and irradiation in comparison to resection alone (approximately 34 to 38 weeks *versus* 14 to 18 weeks). The standard approach is administered in the early postoperative phase and uses initial radiation ports that encompass the T2-weighted target with a margin of 1 to 3 cm, using a dose of approximately 4500 to 4700 cGy in 180 to 200 cGy daily fractions. After this portion has been completed, a "cone down" is performed, targeting the T1-weighted contrast-enhancing volume of the tumor with a 1 to 3 cm margin, bringing the total dose to approximately 6000 cGy. Irradiation is performed over the course of 6 to 7 weeks, with the patient receiving treatment five days per week. Radiation therapy schedules can sometimes be modified with hypofractionation and/or an abbreviated treatment course for elderly patients or those with a low performance status, while maintaining a similar level of toxicity and overall survival [15,16]. More aggressive approaches to irradiation using hyperfractionation schemes have not been shown to improve tumor control and, in some reports, have been associated with worse outcomes [14]. Other techniques to increase localized radiation doses to the tumor resection cavity, such as brachytherapy with permanent or temporary radioactive seeds, have also had disappointing results in controlled trials [17].

Stereotactic radiosurgery (SRS), using a linear accelerator-based system or Gamma Knife® to deliver a single high-dose radiation fraction to a defined volume using stereotactic localization, is another method to boost radiation doses in the tumor bed of a newly diagnosed or recurrent high-grade astrocytoma [18,19]. Retrospective and single-armed uncontrolled trials suggest an improvement in local tumor control rates and survival when using either radiosurgical system. However, these results have not been confirmed in randomized, controlled trials, which are ongoing through the Radiation Therapy Oncology Group (RTOG).

CHEMOTHERAPY OF AA AND GBM—HISTORICAL OVERVIEW

In the 1960s and 1970s, surgical resection and postoperative external beam radiation therapy were established as the standard treatment approach for patients with AA and GBM [5,14]. However, in the late 1970s, investigators began to evaluate chemotherapy as a potential treatment modality when it became obvious that surgical resection and radiation therapy were not curative in the vast majority of patients, and that other forms of treatment were desperately needed to improve survival [20–24]. Over the next fifteen years numerous chemotherapy agents were tested, alone and in combination, for efficacy against newly diagnosed and recurrent high-grade gliomas. Single-agent nitrosoureas (i.e., BCNU) and nitrosourea-based combination regimens (i.e., PCV; procarbazine, CCNU, vincristine) were most effective (outlined in more detail below), although only to a modest degree. However, much of the literature regarding chemotherapy for patients with AA and GBM have methodological deficiencies that impair the ability to draw firm conclusions [25]. Overall, there are very few well-controlled, prospective, randomized clinical trials of chemotherapy in patients with AA and GBM. The reasons for this remain somewhat obscure, but include the relative rarity of these patients, the initial lack of clinical infrastructure to perform large multi-institutional clinical trials, the frequent presence of therapeutic nihilism in physicians caring for AA and GBM patients, and the fact that many of these patients were treated by different groups of sub-specialty physicians (e.g., neurosurgeons, neuro-oncologists, radiation oncologists, medical oncologists) that have varying levels of interest in referral to clinical trials for chemotherapy. Because of poor accrual and the small numbers of patients in many clinical trials, statistical power has been limited, further hampering efforts to draw meaningful conclusions from the available data. Another set of problems that have diluted the quality of published results are the inclusion of heterogeneous patient populations and the lack of uniformity of prognostic variables among those patients. Many of the studies have included patients with both AA and GBM, despite the fact that the prognosis and median survival differ significantly between these two histologic cohorts. As mentioned above, the median survival range for AA and GBM are 24 to 30 months and 10 to 14 months, respectively. The two-year survival is also very different between the two groups, with rates of approximately 47 per cent for AA and 10 per cent for GBM. Other prognostic factors of similar importance, that have not been well controlled in many studies, are age and performance status. In general, younger patients with either histology live longer, even when the rates are adjusted for other prognostic variables. In contrast, patients over the age of 60 tend to do poorly, with similar median survivals (27 *versus* 36 weeks) for both AA and GBM. Performance status has a similar effect on survival, with high performance patients consistently living longer than poor performance patients. The final factor that has negatively impacted on the quality of data from neuro-oncology chemotherapy clinical trials is the lack of uniformity in response criteria. The definitions for partial response, minor response, and stable disease have been quite variable between different studies, making it very difficult to compare results. This problem has been ameliorated with the almost universal adoption of the Macdonald criteria, a more rigorous and objective set of criteria for definition of response in clinical trials of brain tumor patients [26].

The use of chemotherapy for brain tumors began with several small studies in the late 1970s [27–29]. Several of these reports described the use of nitrosourea drugs (CCNU and BCNU) and were suggestive of a positive effect. These initial results led to the formation of the Brain Tumor Cooperative Group (BTCG) and the first large scale multi-center trial for brain tumor chemotherapy (BTCG 7201), which evaluated the possible benefit of adding nitrosourea drugs to radiotherapy [30]. A valid study cohort of 358 patients with malignant glioma was randomized to receive methyl-CCNU alone, radiotherapy alone, radiotherapy plus intravenous BCNU (80 mg/m^2/day × 3 days, every 8 weeks), or radiotherapy plus methyl-CCNU. The BCNU plus radiotherapy arm had the best median survival (51 weeks) and 18-month survival rate (27.2 per 7cent), although it did not reach full statistical

significance over the radiotherapy alone arm (36 weeks and 15.1 per cent, respectively). These results were the impetus for the BTCG 7501 trial, which randomized 527 patients with malignant glioma to receive radiation therapy in addition to either BCNU (80 mg/m^2/day × 3 days, every 8 weeks), oral procarbazine (150 mg/m^2/day × 28 days every 8 weeks), methylprednisolone (400 mg/m^2/day × 7 days, every 3 weeks), or BCNU plus methylprednisolone [31]. Patients on the BCNU and procarbazine arms (50 weeks, 47 weeks) had a statistically significant increase in median survival in comparison to the methylprednisolone or BCNU plus methylprednisolone arms (40 weeks, 41 weeks). Based on these results, the BTCG suggested that intravenous BCNU was the treatment of choice for patients with malignant glioma after the completion of radiation therapy. However, the results of the BTCG 7501 trial may be somewhat confounded because patients in the "control" arm receiving methylprednisolone may have had their survival negatively impacted by steroid-induced complications [25]. This supposition is consistent with the median survival in the BCNU plus methylprednisolone arm, which was significantly lower than the results for patients who received only BCNU. The results from subsequent phase III trials that have included intravenous BCNU (BTCG 77-02, BTCG 8001, BTCG 8301, RTOG 93-05) have been very consistent, with median survivals usually in the range of 10 to 13 months for patients with GBM (see Tables 24.1 and 24.2) [32–35]. The efficacy of BCNU was also tested using an intra-arterial route of administration in a phase III trial [34]. The median survival was significantly longer for the patients in the intravenous BCNU cohort (14.0 months) than for those in the intra-arterial cohort (11.2 months; $p = 0.03$) (see Chapter 17). The survival difference was most likely due to prominent neurotoxicity in the intra-arterial group of patients.

Another randomized trial performed in the 1980s by Levin and colleagues of the Northern California Oncology Group (NCOG) shaped the treatment approach to patients with AA for many years [36]. The NCOG 6G61 trial randomized 127 patients with GBM and anaplastic glioma to receive radiotherapy followed by chemotherapy with either BCNU or PCV. Patients in the PCV arm had improved survival in comparison to the BCNU arm for patients with anaplastic tumors (157.1 *versus* 82.1 weeks; $p = 0.021$), but not for those with GBM. Because of this apparent survival advantage for PCV over BCNU, which was specific only for AA patients, PCV became the standard chemotherapy approach for this group

TABLE 24.1   Selected Chemotherapy Regimens and Their Results for Patients with GBM

Regimen	Trial type	No. cases	Results	Reference #
RT + IV BCNU	Phase III	526	MST 40–56 weeks	[30–35]
RT + IA BCNU	Phase III	56	MST 44.8	[34]
RT + oral PCB	Phase III	114	MST 47	[31]
RT + PCV	Phase III	31	MST 50.4	[36]
RT + TZM	Phase II	64	MST 64	[65]
RT + TZM	Phase III	573	MST 58.4	[66]
RT + TZM	Phase II	138	MST 53.6	[70]
RT + TZM/Thal	Phase II	67	MST 73	[71]
IA DDP	Phase II	13	MNTP 39.8	[72]
IV GEM	Phase II	21	MNTP 11	[74]
IV GEM/Treo	Phase II	17	MNTP 12	[75]
Tam/Carbo	Phase II	50	MST 55	[76]
TZM + RT	Phase II	36	MST 53	[78]
TZM only	Phase II	84	MST 24	[79]
TZM only	Phase II	32	MST 25.6	[80]
TZM + DDP	Phase II	40	MST 50	[83]
PCV + DFMO	Phase III	272	MST 53.2	[87]
TZM	Phase II	13	MST 56	[89]
TZM + BCNU	Phase I	14	MST 69	[90]
TZM + Thal	Phase II	25	MST 103	[91]
oral PCB	Phase II	10	MTP 24	[92,93]
IV DDP	Phase II	10	MNTP 9.5	[94]
IV Carbo	Phase II	15	MST 32	[95]
IV Carbo + VP-16	Phase II	30	MST 43.5	[96]
oral VP-16	Phase II	21	MNTP 7.5	[98]
Irino	Phase II	40	MST 16	[99]
Irino	Phase II	12	MST 24	[100]
CTX	Phase II	40	MST 16	[101]
Tam	Phase II	20	MST 31	[102]
TZM	Phase II	225	MPFS 12.4	[107]
TZM	Phase II	138	MST 32	[110]
TZM	Phase II	72	MST 54	[111]
TZM	Phase II	28	MST 31	[112]
TZM + BCNU	Phase II	38	MST 34	[114]
TZM + Irino	Phase I	18	MNTP 24	[117]
TZM + DDP	Phase II	50	MST 48	[118]
TZM + Marim	Phase II	44	MST 45	[119]
TZM + CRA	Phase II	40	MST 35	[120]
TZM + LDox	Phase II	22	MST 32.8	[121]
TZM + Tam	Phase II	10	MNTP 10	[122]

Abbreviations: RT — radiation therapy; IV — intravenous; MST — median survival time; wks — weeks; IA — intra-artierial; PCB — procarbazine; PCV — procarbazine, CCNU, vincristine; Thal — thalidomide; MTP — median time to progression;

TABLE 24.2   Selected Chemotherapy Regimens and Their Results for Patients with AA

Regimen	Trial type	No. cases	Results	Reference #
RT + IV BCNU	Phase III	107	MST 100–125 weeks	[30–34]
RT + PCV	Phase III	36	MTP 125.6	[36]
IA DDP	Phase II	9	MNTP 39.8	[72]
TZM	Phase II	21	MST 94	[78]
PCV + DFMO	Phase III	181	MST 285	[88]
TZM + BCNU	Phase I	10	MST 132	[90]
oral PCB	Phase II	15	MTP 42	[91]
IV Carbo	Phase II	14	MST 32	[94]
oral VP-16	Phase II	15	MNTP 9.1	[97]
Tam	Phase II	12	MST 69	[101]
Tam	Phase II	24	MST 52	[102]
TZM	Phase II	97	MST 54.4	[105]
TZM	Phase II	53	MST 49	[109]
TZM	Phase II	13	MST 54	[110]
TZM + CRA	Phase II	28	MST 47	[120]

Abbreviations: RT — radiation therapy; IV — intravenous; MST — median survival time; wks — weeks; IA — intra-arterial; PCB — procarbazine; PCV — procarbazine, CCNU, vincristine; MTP — median time to progression; CTX — cyclophosphamide; MNTP — mean time to progression; TZM — temozolomide; DFMO — difluoromethylornithine; DDP — cisplatin; carbo — carboplatin; VP-16 — etoposide; Tam — tamoxifen; CRA — cis-retinoic acid

after irradiation. A more recent analysis of the RTOG database regarding chemotherapy for AA patients with PCV or BCNU did not corroborate an advantage for PCV [37]. All patients were treated with standard radiation therapy. After the completion of irradiation, a total of 257 patients had received BCNU and 175 had been treated with PCV. The BCNU and PCV cohorts were similar in terms of demographic and prognostic factors before treatment. Analysis of median and overall survival were equivalent between the two treatment arms, without an advantage for PCV ($p = 0.40$). The only variables with a significant effect on survival were age, Karnofsky performance status, and extent of surgical resection.

Although these initial studies suggested a modest benefit for the use of chemotherapy after the

GEM — gemcitabine; Treo — treosulfan; Tam — tamoxifen; DFMO — difluoromethylornithine; VP-16 — etoposide; Irino — irinotecan; Marim — marimostat; CRA — cis-retinoic acid; LDox — liposomal doxorubicin; CTX — cyclophosphamide; VCR — vincristine; MNTP — mean time to progression; MPFS — median progression-free survival; TZM — temozolomide; DDP — cisplatin; carbo — carboplatin completion of radiotherapy, the issue remained controversial within the neuro-oncological literature [20–25]. This prompted several meta-analyses of the use of chemotherapy in adults with high-grade gliomas. The first was performed by Fine and colleagues in 1993, with a review of 16 randomized trials that included more than 3000 patients, to compare the survival rates of patients that had received radiation therapy and chemotherapy *versus* irradiation alone [38]. The application of chemotherapy was associated with an absolute increase in survival of 10.1 per cent at 1 year and 8.6 per cent at 2 years. This improvement translated to a relative increase in survival of 23.4 per cent at 1 year and 52.4 per cent at 2 years. BCNU and CCNU accounted for most of the survival benefit, but other chemotherapy drugs appeared to contribute as well. When prognostic variables and histology were factored into the analysis, the survival benefit was present for both types of tumors and occurred earlier for patients with AA than for those with GBM. In a similar and more recent meta-analysis by Stewart and associates, the individual survival data of 3004 patients from 12 randomized trials were reviewed to compare the survival after radiation therapy plus chemotherapy to that of irradiation alone [39]. This analysis demonstrated a significant prolongation of survival from the use of chemotherapy, with a hazard ratio of 0.85 ($p < 0.0001$) and 15 per cent relative decrease in the risk of death. This effect translated to an absolute increase in 1-year survival of 6 per cent, with a 2-month increase in median survival time. In addition, the survival advantage related to chemotherapy was not affected by differences in histology (AA *versus* GBM), age, sex, performance status, or extent of resection.

CONCOMITANT CHEMOTHERAPY AND IRRADIATION APPROACHES

Overall, radiation therapy remains the most effective form of treatment for newly diagnosed high-grade astrocytomas and many other solid tumors [12–14]. Intensive research activity has led to an improved understanding of the cellular and molecular basis for radiation sensitivity and radioresistance [40]. In parallel to this research effort, many investigators have begun to evaluate agents with the ability to enhance the therapeutic effect of irradiation, while minimizing additional toxicity [41]. An effective radiosensitizing agent should have the following characteristics: non-toxic or tolerable side effects,

a potent radiosensitizing effect, independent activity against tumor cells, non-cell-cycle specific, and amenable to dose-intensive or prolonged administration schedules. The three most common mechanisms for radiosensitization are inhibition of radiation damage repair (e.g., purine and pyrimidine analogs), perturbation of the cell cycle in order to increase the fraction of tumor cells in the G_2/M phase (e.g., vinca alkaloids, paclitaxel), and alteration of hypoxic cell populations (e.g., RSR 13). Although numerous compounds have been tested for a radiation sensitizing effect, chemotherapy drugs have been most effective and form the basis for the evolving discipline of chemoradiation [42–45]. Chemoradiation has been applied to high-grade astrocytomas, initially with alkylating agents such as BCNU and methyl-CCNU, and halogenated pyrimidines (e.g., iododeoxyuridine) [45–49]. Other chemotherapy agents that have been used in chemoradiation regimens include cisplatin, fluorouracil, hydroxyurea, camptothecin derivatives (e.g., irinotecan, topotecan), etoposide, and paclitaxel [46,50,51]. An alternative strategy consists of using compounds with the ability to improve the oxygenation status of tumors, thereby augmenting the effectiveness of radiotherapy [46,52]. Agents in this group include RSR 13, tirapazamine, and gadolinium texaphyrin [46,52–54]. Although several of the drugs mentioned above have demonstrated modest improvements in overall and progression-free survival when used during chemoradiation in phase I and II trials, none of them has been effective enough to warrant further study in a phase III trial.

Temozolomide (TZM) is an imidazotetrazine derivative of the alkylating agent dacarbazine, which has demonstrated significant activity against high-grade astrocytomas (see Chapter 2) [55–61]. The drug undergoes chemical conversion at physiological pH to the active species 5-(3-methyl-1-triazeno)imidazole-4-carboxamide (MTIC), with subsequent methylation of DNA at N^7-guanine (70 per cent), N^3-adenine (9.2 per cent), and O^6-guanine (5 per cent). Temozolomide has also been evaluated for its potential as a sensitizing agent in combination with radiation therapy, based on the results of pre-clinical studies [60,62–64]. For example, when TZM was administered with concurrent radiotherapy, it demonstrated an additive cytotoxicity against U373MG glioma cells, especially when the expression levels of methylguanine-DNA methyltransferase (MGMT) were low [62]. Another study demonstrated that prolonged exposure to TZM during fractionated irradiation was able to enhance the cytotoxicity of treatment against D384 cells [63]. In addition, TZM has been shown to induce a G_2/M arrest in glioma cells, thereby synchronizing the cell cycle in a radiosensitive phase [64]. Based on these intriguing pre-clinical data, Stupp and colleagues performed a phase II trial of chemoradiation with TZM in a series of patients with newly diagnosed GBM [65]. Sixty-four patients were treated with oral TZM (75 mg/m^2/day × 7 days/week for 6 weeks) during the course of standard external beam radiotherapy (6000 cGy; 200 cGy/day × 5 days/week for 6 weeks). After the completion of irradiation, each patient received six cycles of single-agent TZM (200 mg/m^2/day × 5 days, every 28 days). The median overall survival of the cohort was 16 months, with 1- and 2-year survival rates of 58 per cent and 31 per cent, respectively. Prolonged survival was most likely in patients less than 50 years of age and in those who underwent a debulking surgical resection. The combined chemoradiation phase was well tolerated, with grade III/IV hematological toxicity occurring in only 6 per cent of patients, including two cases of pneumocystis carinii pneumonia. During the adjuvant TZM chemotherapy phase, grade III/IV neutropenia and thrombocytopenia were observed in 2 per cent and 6 per cent of treatment cycles, respectively.

Based on these encouraging phase II results, Stupp and associates organized an international, randomized phase III trial of chemoradiation with TZM in patients with newly diagnosed GBM, accruing to sites in Europe and Canada [66]. A total of 573 patients were randomly assigned to receive radiation alone (6000 cGy; 200 cGy/day × 5 days/week for 6 weeks) or radiotherapy in combination with daily TZM (75 mg/m^2/day × 7 days/week for 6 weeks). Similar to the phase II study, each patient received six cycles of adjuvant single-agent TZM (200 mg/m^2/day × 5 days, every 28 days) after the completion of irradiation. The overall median survival was 14.6 months for the radiotherapy plus TZM cohort and 12.1 months for the cohort that received irradiation alone, for an overall median survival benefit of 2.5 months. The unadjusted hazard ratio for death due to the GBM for the radiotherapy plus TZM cohort was 0.63 ($p < 0.001$, log-rank test). This indicates a 37 per cent relative reduction in risk of death in comparison to patients that received irradiation alone. The 2-year survival rate was 26.5 per cent for the chemoradiation cohort and 10.4 per cent for the cohort receiving radiotherapy alone. In addition, the median progression-free survival was also significantly different between the groups, measuring 6.9 months for the radiotherapy plus TZM cohort and 5.0 months for the irradiation alone group (hazard ratio of 0.54; $p < 0.001$, log-rank test). When a Cox proportional-hazard model analysis was performed using all prognostic

and treatment variables, the adjusted hazard ratio for death (0.62) was still significant for the radiation therapy plus TZM cohort, and was similar to the unadjusted hazard ratio. Concomitant radiotherapy and TZM resulted in grade III/IV hematologic toxicity in 7 per cent of patients. It remains unclear whether the improvement in overall and progression-free survival in newly diagnosed GBM patients was due to the effects of TZM in combination with radiotherapy or from the use of adjuvant TZM after the completion of irradiation. Nevertheless, based on the results of this study, the Food and Drug Administration (FDA) has approved the use of TZM for chemoradiation of newly diagnosed patients with GBM.

In parallel with the phase III trial mentioned above, another group of investigators was evaluating the effect of MGMT promoter methylation and gene silencing on the survival of patients enrolled in the two arms of the study [67]. The MGMT gene (located on chromosome 10q26) encodes a DNA-repair protein that removes alkyl groups from the O^6 position of guanine, a common site for DNA alkylation by chemotherapy drugs. Methylation of the promoter reduces expression of the MGMT protein and impairs DNA-repair capacity. Previously published reports had suggested that promoter methylation of MGMT and reduced expression of the protein positively correlated with clinical responses of high-grade glioma patients to TZM and other alkylating agents [68,69]. In order to verify this preliminary data, the promoter methylation status of the tumor was determined in a cohort of 206 assessable cases by methylation-specific polymerase-chain-reaction analysis, and then correlated with survival and prognostic value. The MGMT promoter was methylated in 45 per cent of the assessable cases and was an independently favorable prognostic factor ($p < 0.001$, log-rank test), regardless of the assigned treatment arm. Overall median survival was 18.2 months for the patients with methylated tumors *versus* 12.2 months for those without methylation. The associated hazard ratio for death in the sub-group with methylation was 0.45, which corresponds to a 55 per cent reduction of risk. This effect was most prominent for patients on the radiotherapy plus TZM arm of the study, with an overall median survival and 2-year survival rate of 21.7 months and 46 per cent, respectively. In contrast, patients on the irradiation alone arm had a significantly different overall median survival and 2-year survival rate of 15.3 months ($p = 0.007$, log-rank test) and 22.7 per cent, respectively. In the cohort without methylation of MGMT, there was a less significant overall survival advantage for patients in the radiotherapy plus TZM arm in comparison to those in the irradiation alone arm (12.7 months *versus* 11.8 months; $p = 0.06$, log-rank test). When a Cox proportional-hazard model analysis was performed using all prognostic and treatment variables, the adjusted hazard ratio for death (0.41) was still highly significant ($p < 0.001$) for MGMT promoter methylation status, and remained similar to the unadjusted hazard ratio.

A randomized phase II study of radiotherapy plus TZM *versus* irradiation alone was recently reported by Athanassiou and co-workers, which was very similar in design to the phase II and III studies by Stupp described above [70]. A total of 138 patients with newly diagnosed GBM were randomly assigned to receive radiation alone (6000 cGy in 30 fractions) or radiotherapy in combination with daily TZM (75 mg/m^2/day $\times$ 7 days/week for 6 weeks), followed by six cycles of a more dose-intensive adjuvant TZM regimen (200 mg/m^2/day on days 1 through 5 and 15 to 19, every 28 days). The median overall survival was significantly different between the groups ($p < 0.0001$), measuring 13.4 months for the radiotherapy plus TZM arm and 7.7 months for the irradiation alone arm. Similar results were noted for median time to progression (10.8 months *versus* 5.2 months; $p = 0.0001$), 1-year progression-free survival (36.6 per cent *versus* 7.7 per cent), and 1-year overall survival (56.2 per cent *versus* 15.7 per cent; $p < 0.0001$). As used in this study, dose intensification of TZM during the adjuvant treatment phase did not translate into a prolongation of median or progression-free survival.

In a variation on the above studies, Chang and co-workers have reported a chemoradiation trial using the combination of TZM and thalidomide during the course of radiotherapy [71]. A total of 67 patients with newly diagnosed GBM were enrolled to receive TZM (150–200 mg/m^2/day $\times$ 5 days every 4 weeks) plus daily thalidomide (200–1200 mg) concomitantly with irradiation (6000 cGy over six weeks). The TZM and thalidomide could be continued for one year on an adjuvant basis for stable patients, after the completion of radiation therapy. The median TTP was 22 weeks, with 6-month and 1-year progression-free survival rates of 45 and 29 per cent, respectively. The overall median survival was 73 weeks, with 1-year and 2-year survival rates of 64 and 27 per cent, respectively. In comparison to historical controls, this approach demonstrated significantly prolonged survival over groups of patients receiving irradiation alone (hazard ratio 0.58, $p < 0.001$) and was longer than that of patients receiving nitrosourea-based chemotherapy.

NEOADJUVANT CHEMOTHERAPY APPROACHES

In addition to BCNU and PCV (as outlined above), numerous other chemotherapy drugs, alone and in combination, have been used for neoadjuvant treatment of high-grade astrocytomas (see Tables 24.1 and 24.2) [20–24]. Dropcho and colleagues attempted neoadjuvant intra-arterial (IA; see Chapter 17) cis-platin (75 mg/m^2 every 4 weeks) in a series of 22 assessable patients with newly diagnosed high-grade astrocytomas (GBM 13, AA 9) [72]. There were five patients (23 per cent) with partial responses (PR) and five patients (23 per cent) with minor responses (MR) after IA therapy. The time to progression (TTP) ranged from 22 to 115 weeks, with a median of 39.8 weeks overall and 56.3 weeks in responders. Although hematologic, renal, and otic toxicity were mild, there were several patients with significant neurological toxicity. Other investigators have used high-dose intravenous cyclophosphamide (2 g/m^2/day for 2 doses every 28 days) as neoadjuvant treatment for 14 patients with newly diagnosed GBM [73]. There were 3 patients with CR and 3 with stable disease; TTP and survival data were limited. Toxicity was mainly hematological and included 13 admissions for neutropenia and fever, as well as transfusions for anemia and thrombocytopenia. Several reports have evaluated the use of intravenous gemcitabine, a cytosine arabinoside analog that functions as a pyrimidine antimetabolite, as a single agent or in combination for neoadjuvant chemotherapy of AA and GBM. Weller and co-workers administered gemcitabine (1000 mg/m^2 on days 1, 8, and 15) for one to four monthly cycles before irradiation in a series of 21 patients with newly diagnosed GBM [74]. Of the 17 patients with evaluable residual disease, there were no CR or PR, 14 patients had stabilization of disease. The median progression-free survival and 4-month progression-free survival rate were 11 weeks and 24 per cent, respectively. For the entire cohort, the median overall survival was 11 months. Although the regimen was safe and well tolerated, neoadjuvant single agent gemcitabine did not appear to confer any survival advantage in comparison to irradiation alone. The same group has also tested neoadjuvant gemcitabine in combination with treosulfan, a bifunctional alkylating agent [75]. In this study, 17 patients with newly diagnosed GBM received gemcitabine (1000 mg/m^2) and treosulfan (3500 mg/m^2) on days 1 and 8 every 28 days, for up to four cycles before irradiation. The median progression-free survival and 4-month progression-free survival rate were 12 weeks and 29 per cent, respectively. For the entire cohort, the median overall survival was 12 months. The regimen was associated with prohibitive hematological toxicity, including 18 per cent with grade IV events. The addition of treosulfan did not significantly improve the survival benefit of neoadjuvant gemcitabine and was not superior to irradiation alone. Puchner and co-workers used a combination regimen consisting of tamoxifen (200 mg/day) and intravenous carboplatin (300 mg/m^2 every 3 weeks × 3 cycles), followed by standard radiotherapy (5940 cGy over 6 weeks) for 50 patients with newly diagnosed GBM [76]. The median TTP was 30 weeks, with an overall median survival of 55 weeks. The 1- and 2-year survival rates were 58 and 18 per cent, respectively. It was concluded that this combination regimen did not impart any survival advantage over nitrosourea-based regimens.

Temozolomide has also been used in the neoadjuvant setting as a single agent and in combination with other cytotoxic drugs [55–61]. Friedman and associates treated 38 patients with newly diagnosed high-grade astrocytoma (GBM 33, AA 5) on a regimen of single agent TZM (200 mg/m^2 × 5 days every 28 days), for up to six cycles before the onset of radiotherapy or other treatment [77]. In the GBM cohort, there were 3 patients with CR, 14 with PR, and 4 with stable disease. For AA patients, there was one brief PR and 2 with stable disease. Overall, 18 of 38 patients responded (47 per cent) to TZM in the neoadjuvant setting. Tumor responses were inversely correlated with immunohistochemical staining for MGMT. A similar study by Gilbert and colleagues treated 57 patients with newly diagnosed malignant astrocytoma (GBM 36, AA 21) with neoadjuvant TZM, for up to four cycles before proceeding to radiotherapy [78]. Almost 50 per cent of both AA and GBM patients were able to complete all four cycles of TZM before the initiation of radiation therapy. Overall, there were 22 patients with objective responses (39 per cent), including 6 patients with CR and 16 with PR. Another 18 patients (32 per cent) had stabilization of disease. The median overall and progression-free survivals for the GBM patients were 13.2 months and 3.9 months, respectively. For the AA cohort, the median overall and progression-free survivals were 23.5 months and 7.6 months, respectively. Neoadjuvant TZM was well tolerated, with infrequent episodes of grade III (28 per cent) and IV (12 per cent) toxicity. Other investigators have attempted neoadjuvant TZM for elderly patients with AA and GBM, as an alternative to external beam radiotherapy [79,80]. In one report, 86 elderly patients (70 years of age or older) with malignant astrocytomas (GBM 84,

AA 2) were treated with either radiotherapy alone (N = 54; 6000 cGy over 6 weeks) or neoadjuvant TZM (N = 32; 150–200 mg/m^2/day × 5 days every 4 weeks) [79]. The median overall survival and 1-year survival rates were equivalent between the irradiation and TZM groups (4.1 *versus* 6.0 months and 9.26 per cent *versus* 11.88 per cent, respectively), with a slight advantage for the TZM treatment group which did not reach statistical significance. Temzolomide was well tolerated in this setting and appeared to be an excellent alternative to irradiation for elderly patients. A similar study by a French group evaluated the use of TZM in a cohort of 32 elderly GBM patients [80]. Objective responses were noted in nine patients (CR 0, PR 9; 31 per cent), while 12 patients had stabilization of disease. The median overall and progression-free survivals were 6.4 and 5.0 months, respectively, with a 1-year survival rate of 25 per cent.

A few reports have utilized neoadjuvant TZM in combination with other chemotherapy drugs, to determine if an additive survival benefit could be documented [55–61]. In a phase I study, TZM (200 mg/m^2/day × 5 days every 4 weeks) and procarbazine (50–125 mg/m^2/day × 5 days every 4 weeks, 1 hour before TZM) were administered to 28 chemotherapy-naïve patients with glioma (GBM 16, AA 7) [81]. All patients received TZM alone during cycle 1 and then had escalating doses of procarbazine added for subsequent cycles. The main toxicity from combination therapy was hematological, with lymphocytopenia and thrombocytopenia noted at dosing levels 3 and 4. At dose level 4, the dose-limiting toxicity was thrombocytopenia. The recommended dose for subsequent trials was level 3, with procarbazine at 100 mg/m^2/day. Objective responses were noted in 10 patients (36 per cent), nine of which had an AA (3) or GBM (6), with response duration ranging from 2 to 17+ months. Another variation of neoadjuvant combination chemotherapy used intravenous BCNU (150 mg/m^2), followed after two hours by TZM (550 mg/m^2) on day 1 of a 42-day cycle [82]. A total of 41 patients with anaplastic glioma were entered onto study, 33 of whom had an AA. Twenty-four patients (59 per cent) were able to complete four cycles of combined chemotherapy. Objective responses were noted in 9 patients (CR 1, PR 8; 29 per cent), all of which had AA histology. The regimen was associated with significant hematological toxicity, with grade III/IV granulocytopenia and thrombocytopenia noted in 21 and 46 per cent of patients, respectively. Overall, neoadjuvant TZM and BCNU was quite toxic and did not appear to provide a survival benefit in comparison to single-agent chemotherapy approaches. Neoadjuvant TZM

(200 mg/m^2/day × 5 days every 4 weeks) has also been used in combination with cisplatin (100 mg/m^2 on day 1) for newly diagnosed patients with GBM, in a report by a Spanish group [83]. Of 40 total patients, objective responses were noted in 45 per cent, including 3 CR (7.5 per cent) and 15 PR (37.5 per cent), as well as another 4 patients (10 per cent) with stabilized disease. The median TTP and overall survival were 6.3 and 12.5 months, respectively, with a 1-year survival rate of 55 per cent. The regimen was associated with significant hematological toxicity, including grade III/IV neutropenia and thrombocytopenia, which were noted in 19.2 and 8.7 per cent of patients, respectively.

ADJUVANT CHEMOTHERAPY APPROACHES

Many different chemotherapy drugs have been used in the adjuvant setting for high-grade astrocytomas, including the nitrosoureas, as outlined above. The vast majority of these drugs have either been ineffective or unable to demonstrate a clinical advantage (i.e., prolonged survival or TTP) over standard intravenous BCNU [20–24]. This conclusion has been further corroborated by a recent randomized phase III trial of irradiation plus adjuvant PCV *versus* radiation therapy alone for patients with high-grade astrocytoma (see Tables 24.1 and 24.2) [84]. A total of 674 patients were treated, including 449 with GBM and 113 with AA. The median overall survival for the radiation-PCV and radiation alone groups were 10 and 9.5 months, respectively (hazard ratio = 0.95; $p = 0.50$, log-rank test). The addition of PCV did not add a survival benefit and no difference could be detected in the efficacy of chemotherapy between the AA and GBM patients. However, a retrospective analysis of adjuvant chemotherapy in 133 patients with GBM does suggest that treatment confers a survival advantage over radiation therapy alone [85]. The addition of chemotherapy (most often BCNU or PCV) was able to increase the median survival from 11.5 to 15 months and extend 2-year survival rates from 12 to 28 per cent. In addition, during univariate and multivariate analyses, the use of chemotherapy was an independent prognostic factor for extended survival. No statistically significant difference in survival could be discerned among patients treated with BCNU, CCNU, or PCV.

In a phase II comparative trial of adjuvant chemotherapy, Jeremic and associates treated 67 patients (31 GBM, 36 AA) with a modified version of PCV (on a six week schedule) and 66 patients (GBM 29, AA 37) with CCNU (60 mg/m^2 days 3 and 4) and VM-26 (75 mg/m^2 days 1 and 2) every six weeks [86]. The TTP and survival data were superior for the PCV regimen in comparison to CCNU and VM-26, with a two-fold increase in survival at the 25th and 50th percentiles. However, this survival advantage was only noted for the AA patients. Although there was a higher percentage of long-term GBM survivors in the PCV cohort, the difference was not significant.

Levin and co-workers have reported several randomized studies evaluating the activity of adjuvant PCV *versus* PCV in combination with α-difluoromethylornithine (DFMO) for patients with malignant gliomas [87,88]. DFMO is an irreversible inhibitor of ornithine decarboxylase, which is capable of blocking polyamine accumulation and reducing cell proliferation rates in cell culture systems. *In vitro*, the addition of DFMO to a nitrosourea has been shown to augment antitumor activity against 9L gliosarcoma cell lines and rats with intracerebral 9L tumors. In the initial report, 272 patients with GBM were randomized to receive standard PCV or PCV plus DFMO (3.0 g/m^2 q8h × 14 days before and after CCNU) after the completion of radiotherapy [87]. The median overall survival (14.2 *versus* 13.3 months, respectively) and 5-year survival rates (8.7 *versus* 6.2 per cent, respectively) were equivalent between the two groups and did not support a therapeutic benefit for the addition of DFMO in GBM patients. In a similar phase III trial, 249 anaplastic glioma patients (181 AA) were randomized to receive PCV or PCV plus DFMO (3.0 gm/m^2 q8h on days 1–14 before CCNU and days 29–42 after CCNU) [88]. The PCV-DFMO cohort had longer median overall survival (71.2 *versus* 46.0 months) and progression-free survival (56.2 *versus* 22.2 months) in comparison to patients that received PCV alone. However, the differences were not statistically significant. A hazard ratio analysis did suggest a significant difference between the two treatment groups (hazard ratio 0.53, $p = 0.02$), but only during the first two years of the study.

Temozolomide has also been applied to high-grade astrocytoma patients in the adjuvant setting [55–61]. In a report by Chibbaro and colleagues, 42 patients with high-grade gliomas (GBM 28, AA 11) were treated with TZM (200 mg/m^2/day × 5 days) on a monthly schedule [89]. Of those 42 patients, 13 (all with GBM) received treatment in an adjuvant setting after the completion of irradiation. Objective responses were noted in four patients (36.3 per cent; 2 CR, 2 PR), while another two patients had disease stabilization. The median progression-free and overall survival (for all 42 patients) were 33.5 and 56 weeks, respectively. Using a more dose-intensive approach,

Raizer and co-workers performed a phase I study of TZM (50–90 mg/m^2/day × 28 days, every 8 weeks) in combination with intravenous BCNU (150 mg/m^2) for patients with malignant gliomas after radiotherapy [90]. A total of 24 patients were enrolled (GBM 14, AA 10), with the dose-limiting toxicity of neutropenia and thrombocytopenia at 90 mg/m^2/day of TZM. For the GBM cohort, the median overall survival and TTP were 69 and 14 weeks, respectively, with progression-free survival rates at 6- and 12-months of 21 and 14 per cent, respectively. The AA cohort had median overall survival and TTP of 132 and 82 weeks, respectively, with a progression-free survival rate of 70 per cent at 6- and 12-months. The maximum tolerated dose (MTD) of TZM, in combination with this dose of BCNU, was 80 mg/m^2/day. Other investigators have combined conventional TZM (200 mg/m^2) and thalidomide (200–600 mg/day) for adjuvant treatment of patients with GBM [91]. Responses included 2 patients with MR and 14 with stable disease. The median TTP and overall survival were 36 and 103 weeks, respectively. Overall, TZM and thalidomide was well tolerated and appeared to be an effective combination in the adjuvant setting.

CHEMOTHERAPY FOR RECURRENT HIGH-GRADE ASTROCYTOMAS

Similar to the experience with adjuvant treatment, chemotherapy for the majority of patients with recurrent high-grade astrocytomas has been disappointing. Over the past 25 years, numerous drugs have been tested, typically with minimal efficacy and no survival advantage over a nitrosourea-based regimen [20–24]. The use of BCNU, CCNU, or PCV at the time of recurrence has been associated with a 25 to 30 per cent response rate, with a survival benefit that is similar to that obtained when the treatment is used immediately after the completion of irradiation [22,24]. Among the nitrosourea-based approaches, single-agent BCNU appears to be the most active agent. Other drugs with activity against recurrent high-grade astrocytomas include single-agent procarbazine, an oral alkylating agent commonly used as part of the PCV regimen (see Tables 24.1 and 24.2). When used at the time of recurrence after BCNU failure, procarbazine (150 mg/m^2/day × 28 days every 8 weeks) is capable of inducing objective responses (CR 2, PR 7) and stabilization of disease, with a median TTP of approximately 24 weeks [92,93].

Platinum-based regimens have also shown activity against recurrent AA and GBM [22,24]. Single-agent intravenous cisplatin (100–200 mg/m^2 days 1 and 8,

every 4 weeks) was able to induce objective responses in 4 of 14 patients with recurrent malignant glioma (CR 0, PR 4) [94]. However, the responses were very brief, with a mean TTP for the GBM and AA cohorts of 9.5 and 32 weeks, respectively. More sustained responses have been noted for intravenous single-agent carboplatin (400–450 mg/m^2 every 4 weeks), with a response rate of 48 per cent (PR 2, MR 2, SD 10), median TTP of 26 weeks in responders, and an overall median survival of 32 weeks [95]. Carboplatin (300 mg/m^2 days 1 and 3 every 4 weeks) has also been used in combination with etoposide (100 mg/m^2 days 1 to 5 every 4 weeks) for patients with recurrent malignant glioma [96]. There were eight objective responses (CR 0, PR 8) and another 12 patients with stable disease. The median TTP and overall survival were 14 weeks and 43.5 weeks, respectively. Etoposide has also been administered as a single agent for patients with recurrent malignant astrocytomas. In one report, etoposide (50–100 mg/m^2/day IV × 5 days every 3 weeks) was administered as a continuous infusion to a cohort of 18 patients [97]. Although the overall response rate was 50 per cent, the duration of the responses were brief, with an overall median survival of only 18 weeks. Some investigators have attempted treatment with oral etoposide (50 mg/day), administered on a daily basis at a low dosage [98]. In the report by Fulton and associates, a total of 46 patients with recurrent malignant gliomas (GBM 21, AA 15) were treated. There were 8 patients with objective responses and another 11 with stabilization of disease. The overall median TTP was 8.8 weeks, with TTP in the AA and GBM sub-groups of 9.1 and 7.5 weeks, respectively.

Several investigators have reported on the use of irinotecan for patients with recurrent high-grade gliomas [99,100]. In a phase II study, 40 patients with recurrent GBM were treated with irinotecan (400–500 mg/m^2) every three weeks [99]. There were no objective responses and all patients had progressive disease within six weeks. The overall median survival was only 16 weeks. Slightly better results were reported by Cloughesy and colleagues, with a less intense regimen of irinotecan (300 mg/m^2 every three weeks) administered to 14 patients with recurrent high-grade glioma [100]. Two patients had PR and two had stable disease, with a median TTP and overall survival of 6 weeks and 24 weeks, respectively.

Other drugs used as single agents for patients with recurrent high-grade astrocytomas include cyclophosphamide (CTX) and tamoxifen. In a series of 40 patients with recurrent GBM that had all failed prior chemotherapy, including TZM, intravenous CTX

(750 mg/m^2 × 2 consecutive days) was administered every four weeks [101]. There were 7 patients with a PR and another 11 with stable disease. The median TTP and overall survival were 8 weeks and 16 weeks, respectively. Tamoxifen is an antiestrogenic agent and inhibitor of protein kinase C signal transduction that has been applied to patients with recurrent high-grade astrocytomas. In a series of 32 patients (20 GBM, 12 AA), tamoxifen (160–200 mg) was administered daily in a divided dosage [102]. Objective responses were noted in 8 patients (CR 0, PR 8), with an additional 6 patients with disease stabilization. Median overall survival from the initiation of tamoxifen was 44 weeks, with survival of 69 weeks in the AA cohort and 31 weeks for the GBM sub-group. In a similar study, tamoxifen (80 mg/m^2/day) was administered to 24 patients with recurrent AA [103]. Four patients had PR, while another 11 patients had stabilization of disease. The median TTP and overall survival were 48 and 52 weeks, respectively. Numerous other drugs have been tested, alone and in combination, and demonstrated low to intermediate activity in patients with recurrent high-grade astrocytoma [22,24]. These drugs include dacarbazine, thiotepa, melphalan, vindesine, mitoguazone, fludarabine, mitoxantrone, AZQ, PCNU, and many others.

Despite the modest results noted above for chemotherapeutic treatment of patients with recurrent high-grade astrocytomas, several meta-analyses of the earlier data do suggest a survival benefit [104,105]. The initial report evaluated the results of 40 clinical trials (32 chemotherapy trials, including four randomized trials) and 1415 high-grade astrocytoma patients [104]. Nitrosoureas were the most effective class of chemotherapy drugs, with a significantly longer TTP (26.9 weeks) in comparison to other compounds. Platinum drugs were also active and, along with the nitrosoureas, had the longest overall survival results (over 32 weeks). The combination of a nitrosourea and a platinum drug appeared to be an effective approach and was associated with the longest overall survival (40 weeks). A similar meta-analysis of 32 clinical trials focused on the activity of cisplatin and carboplatin in patients with high-grade astrocytomas [105]. Although the quality of the data in the analyzed trials was somewhat poor, the authors concluded that both compounds had activity against these tumors and warranted further study in more carefully designed clinical trials.

Over the past decade there has been extensive investigation of the use of TZM, alone and in combination with other cytotoxic and molecular agents, for the treatment of recurrent high-grade astrocytomas [55–61]. Overall, the drug has shown significant activity against recurrent AA and GBM, as demonstrated in several clinical trials by Yung and colleagues. The first study evaluated the use of TZM (150–200 mg/m^2/day × 5 days every 28 days) in a series of 162 patients with recurrent malignant gliomas, including 97 patients with AA [106]. In the AA cohort, there were 6 patients with CR, 27 with PR, and another 31 with stable disease (CR+PR+SD = 66 per cent). The response rate was similar in patients that had failed prior chemotherapy or were chemotherapy-naïve. Median overall PFS was 5.4 months, with 6- and 12-month PFS rates of 46 and 24 per cent, respectively. The median overall survival was 13.6 months, with 6- and 12-months survival rates of 75 cent and 56 per cent, respectively. A similar comparative phase II trial evaluated the activity of TZM *versus* procarbazine (125–150 mg/m^2/day × 28 days every 8 weeks) in a cohort of 225 patients with GBM at first relapse [107]. Overall response rates (PR+SD) were significantly higher for patients in the TZM cohort (45.6 *versus* 32.7 per cent; $p = 0.049$). Treatment with TZM resulted in a significant improvement in median PFS (12.4 weeks *versus* 8.32 weeks; $p = 0.0063$) and 6-month PFS (21 *versus* 8 per cent; $p = 0.008$) in comparison to procarbazine. In addition, the 6-month overall survival rate was significantly higher for patients in the TZM arm of the study (60 *versus* 44 per cent; $p = 0.019$). More in-depth analyses of the data from this trial, with a focus on neurological and health-related quality of life (HRQOL), have been reported by several groups [108,109]. The initial report by Osoba and associates demonstrated that the use of TZM resulted in an improvement in most measures of HRQOL, while the use of procarbazine typically lead to a deterioration of these measures [108]. This data was corroborated by a more recent report by Macdonald and co-workers, in which the use of TZM resulted in an improvement in the median time to neurological failure (4.18 *versus* 3.49 months; $p = 0.035$), time to deterioration of Karnofsky performance status (KPS) to less than or equal to 60 (5.56 *versus* 3.45 months; $p = 0.007$), and time to decline of the KPS by 30 points (6.71 *versus* 5.07 months; $p = 0.003$) in comparison to patients treated with procarbazine.

Other studies of single-agent TZM consistently demonstrate activity against recurrent high-grade astrocytomas. In one report, a series of 213 patients with recurrent malignant gliomas (GBM 138, AA 53) were treated with standard dose TZM [110]. For the entire cohort, there were 2 CR and 32 PR (objective response rate of 16 per cent), while another 79 patients had stabilization of disease. The median TTP and overall survival were 21 and 49 weeks, respectively, for the AA patient cohort. For the GBM patients, the median TTP and overall survival were 10 and 32 weeks, respectively. In a similar study, 117 patients with recurrent malignant glioma (GBM 72, AA 13) received TZM on the standard monthly schedule [111]. The objective response rate was 29 per cent, with another 34 per cent of patients having disease stabilization. For the entire cohort, the median PFS and overall survival were 26 and 54 weeks, respectively. In a phase II study of extended low-dose TZM (75 mg/m^2/day × 42 days every 70 days), Khan and colleagues treated 35 patients (GBM 28, AA 3) with recurrent malignant glioma [112]. Objective responses included 2 PR (AA) and 3 MR (GBM). For the GBM cohort, the median progression-free and overall survival were 2.3 and 7.7 months, respectively. Although the regimen was well tolerated, it did not provide a survival benefit in comparison to standard TZM dosing regimens.

In addition to all of the aforementioned single-agent studies, TZM has also been aggressively studied in combination with numerous cytotoxic and molecular chemotherapy drugs, for the treatment of recurrent high-grade astrocytomas [55–61]. Several studies have evaluated the combination of TZM and intravenous BCNU, including an initial phase I study of 45 patients (25 with GBM) [113]. In this study, both drugs were administered as a single dose on day 1 of a 42-day cycle. The sequence of drug administration was varied on two treatment arms during dose escalation (i.e., BCNU-TZM or TZM-BCNU). There were a total of 9 PR noted, including 5 in the cohort of patients with GBM. The final recommended dosing for subsequent phase II trials was BCNU 150 mg/m^2 followed in two hours by TZM 550 mg/m^2, repeated every 6 weeks. This dosing regimen was used in a phase II trial, which evaluated 38 patients with recurrent GBM [114]. The objective responses included 2 PR and 2 MR, with another 19 patients having stabilization of disease. The median PFS and overall survival were 11 and 34 weeks, respectively. At 6 months, the overall and progression-free survival were 68 and 21 per cent, respectively. This regimen was associated with significant hematological toxicity and did not impart a survival advantage over standard single agent TZM.

Other combination phase I trials with TZM have included the addition of oral etoposide, BCNU impregnated wafers, and irinotecan [115–117]. In a series of 29 patients with recurrent malignant gliomas (GBM 19, AA 5), the dose of TZM remained constant (150 mg/m^2/day × 5 days every 4 weeks), while the oral etoposide dose was escalated from 50 mg/m^2/day × 5 days to 50 mg/m^2/day × 20 days [115].

Responses included 1 PR, 1 MR, and 11 patients with stable disease. For the entire cohort, the median TTP was 16 weeks. Toxicity included hematological compromise and infections. The recommended dose of etoposide for use in combination with TZM was 50 mg/m^2/day × 12 days. BCNU impregnated wafers have also been used at the time of tumor resection, followed by escalating doses of TZM (100–200 mg/m^2/day × 5 days every 4 weeks), in 10 patients with recurrent high-grade gliomas (GBM 7, AA 3) [116]. The combination was well tolerated, with minimal toxicity. The recommended dose of TZM when used with the wafers is 200 mg/m^2 per day. Another phase I study evaluated the use of conventional TZM (200 mg/m^2) and irinotecan (125 mg/m^2 on days 6, 13, and 20 or 350 mg/m^2 on day 6) in a series of 32 patients with recurrent malignant glioma (GBM 18, AA 4) [117]. Objective responses included 3 CR (GBM 2, AA 1) and 3 PR (GBM 3), with another 13 patients demonstrating stable disease. The median TTP for GBM and AA patients was 24 and 31 weeks, respectively. The regimen was well tolerated, with mild hematological and gastrointestinal toxicity in most patients.

Additional drugs that have been used in combination with TZM in phase II studies include cisplatin, marimastat, cis-retinoic acid, liposomal doxorubicin, and tamoxifen [118–122]. A study by Brandes and co-workers administered cisplatin (75 mg/m^2 on day 1) followed by TZM (130 mg/m^2 bolus on day 2, then 70 mg/m^2 every 12 h × 9 doses) to a series of 50 chemotherapy-naïve patients with recurrent GBM [118]. Objective responses included 1 CR and 9 PR, with another 23 patients noting disease stabilization. The median TTP and overall survival were 18.4 and 48 weeks, respectively. The 6-month overall survival rate was 81 per cent, with a 6-month progression-free survival rate of 34 per cent. A similar study used conventional TZM (150–200 mg/m^2) in combination with marimastat (50 mg/day on days 8 to 28), a metalloproteinase inhibitor, to treat 44 patients with recurrent GBM [119]. The study was associated with significant toxicity in the form of joint and tendon pain (reported in 47 per cent of patients), including 5 patients in whom treatment was discontinued due to intolerable discomfort. Despite these limitations, the regimen appeared active, with objective responses in 6 patients and a median PFS and overall survival of 17 and 45 weeks, respectively. In addition, the 6- and 12-month PFS of 39 and 16 per cent, respectively, were favorable in comparison to historical controls. Another approach combined conventional TZM (150–200 mg/m^2) with 13-cis-retinoic acid (100 mg/m^2/day on days 1 to 21, every 28 days), a synthetic retinoid with the ability to induce apoptosis and differentiation [120]. A total of 88 patients were treated, including 40 with GBM and 28 with AA. Objective responses in the AA/GBM cohort included 1 patient with a CR and 4 with PR. The median PFS and overall survival were 19 weeks and 47 weeks, respectively (16 and 35 weeks in GBM patients). Six- and 12-month PFS rates were 43 and 16 per cent, respectively, with 6-month overall survival rate of 75 per cent. Similar to the study with marimastat, the combination of TZM and 13-cis-retinoic acid showed favorable PFS and overall survival data in comparison to historical controls. Chua and colleagues have reported the use of conventional TZM (200 mg/m^2) in combination with liposomal doxorubicin (40 mg/m^2 on day 1 every 4 weeks) for a series of 22 patients with recurrent GBM [121]. Responses included 1 patient with a CR, 3 with PR, and another 11 with stable disease. Seven patients remained progression free at 6 months. The median TTP and overall survival were 12.8 weeks and 32.8 weeks, respectively. Overall, the regimen was relatively well tolerated, with mainly hematological toxicity. Dose intensive continuous TZM (60–75 mg/m^2/day × 6 weeks, every 10 weeks) was combined with tamoxifen (100 mg bid) by Spence and associates for treatment of 14 patients with recurrent malignant astrocytoma (GBM 10, AA 4) [122]. The regimen was not very effective and only resulted in 1 patient with a PR and 1 with stable disease. Median TTP and overall survival from the time of chemotherapy initiation were 10 weeks and 26 weeks, respectively. Several patients had significant toxicity, including pancytopenia, transaminitis, and trigeminal herpes zoster.

CONCLUSIONS

Patients with AA and GBM continue to have a poor prognosis and protracted survival despite recent advances in neurosurgical techniques, the application of therapeutic radiation, and chemotherapy. Although several meta-analyses have demonstrated a survival benefit for chemotherapy with nitrosourea drugs and platinum compounds, the improvements have been modest. However, with the advent of TZM, the future of chemotherapy as a necessary component of multi-modality treatment is more secure. At the present time, TZM is the only chemotherapy drug with the ability to significantly impact on TTP and survival when used in the neoadjuvant, adjuvant, and recurrent settings, as well as in combination with radiation therapy. Further progress and clinical trial data are needed to define the most effective schedules and drug combinations to use with TZM. For example, after the completion of TZM chemoradiation, would 12 cycles of adjuvant TZM be more effective than 6 cycles, and extend survival at 12 and 24 months? In addition, which drugs are best to combine with TZM and maximize survival benefit, while maintaining quality of life? Preliminary results suggest that marimastat, cis-retinoic acid, cisplatin, and thalidomide are effective in combination with TZM. Other drugs will need to be tested as well, including the new molecular agents designed to "target" specific receptors and pathways within brain tumor cells that maintain the malignant phenotype, such as the epidermal growth factor receptor, platelet-derived growth factor receptor, ras pathways, Akt, mTOR, cell cycle-related proteins, and apoptotic pathways [123–125].

TZM has become an excellent starting point for the new era in chemotherapy of high-grade astrocytomas. Further progress will require aggressive drug dis-covery programs that can discern second and third generation compounds with more potent activity and better tolerability profiles. As these new agents are identified, they will require testing in the setting of well-designed, multi-institutional clinical trials.

ACKNOWLEDGMENTS

The author would like to thank Ryan Smith for research assistance. Dr. Newton was supported in part by National Cancer Institute grant, CA 16058 and the Dardinger Neuro-Oncology Center Endowment Fund.

References

1. Newton, H. B. (1994). Primary brain tumors: Review of etiology, diagnosis, and treatment. *Am Fam Phys* **49**, 787–797.
2. Wrensch, M., Minn, Y., Chew, T., Bondy M., Berger, M. S. (2002). Epidemiology of primary brain tumors: Current concepts and review of the literature. *Neurooncol* **4**, 278–299.
3. ACS (American Cancer Society) (2002). Cancer Facts and Figures 2002. Atlanta, American Cancer Society.
4. Kleihues, P., Louis, D. N., Scheithauer, B. W. *et al.* (2002). The WHO classification of tumors of the nervous system. *J Neuropathol Exp Neurol* **61**, 215–225.
5. Salcman, M. (1990). Malignant glioma management. *Neurosurg Clin N Am* **1**, 49–64.
6. Dunn, I. F., and Black, P. M. (2003). The neurosurgeon as local oncologist: cellular and molecular neurosurgery in malignant glioma therapy. *Neurosurg* **52**, 1411–1424.
7. Matz, P. G., Cobbs, C., and Berger, M. S. (1999). Intraoperative cortical mapping as a guide to the surgical resection of gliomas. *J Neurooncol* **42**, 233–245.

8. Lemole, G. M., Henn, J. S., Riina, H. A., and Spetzler R. F. (2001). Cranial application of frameless stereotaxy. *BNI Quart* **17**, 16–24.

9. Nimsky, C., Ganslandt, O., Kober, H., Buchfelder, M., and Fahlbusch, R. (2001). Intraoperative magnetic resonance imaging combined with neuronavigation: A new concept. *Neurosurg* **48**, 1082–1091.

10. DeVaux, B. C., O'Fallon, J. R., and Kelly, P. J. (2001). Resection, biopsy, and survival in malignant glial neoplasms. A retrospective study of clinical parameters, therapy, and outcome. *J Neurosurg* **78**, 767–775.

11. Lacroix, M., Abi-Said, D., Fourney, D. R. *et al.* (2001). A multivariate analysis of 416 patients with glioblastoma multiforme: prognosis, extent of resection, and survival. *J Neurosurg* **95**, 190–198.

12. Shrieve, D. C., and Loeffler, J. S. (1995). Advances in radiation therapy for brain tumors. *Neurol Clin* **13**, 773–794.

13. Miyamoto, C. (2001). Radiation therapy principles for high-grade gliomas. *In Combined Modality Therapy of Central Nervous System Tumors* (Z. Petrovich, L. W. Brady, M. L. Apuzzo, and M. Bamberg, Eds.), Vol. 18, pp. 345–364. Springer, Berlin.

14. Laperriere, N., Zuraw, L., and Cairncross, J. G. (2002). Radiotherapy for newly diagnosed malignant glioma in adults: a systematic review. *Radiother Oncol* **64**, 259–273.

15. Chang, E. L., Yi, W., Allen, P. K., Levin, V. A., Sawaya, R. E. and Maor, M. H. (2003). Hypofractionated radiotherapy for elderly or younger low-performance status glioblastoma patients: Outcome and prognostic factors. *Int J Rad Oncol Biol Phys* **56**, 519–528.

16. Roa, W., Brasher, P. M. A., Bauman, G. *et al.* (2004). Abbreviated course of radiation therapy in older patients with glioblastoma multiforme: A prospective randomized clinical trial. *J Clin Oncol* **22**, 1583–1588.

17. Selker, R. G., Shapiro, W. R., Burger, P. *et al.* (2002). The Brain Tumor Cooperative Group, N. I., H trial 87-01: A randomized comparison of surgery, external radiotherapy, and carmustine versus surgery, interstitial radiotherapy boost, external radiation therapy, and carmustine. *Neurosurg* **51**, 343–357.

18. Nwokedi, E., DiBiase, S. J., Jabbour, S., Herman, J., Amin, P., and Chin, L. S. (2002). Gamma knife stereotactic radiosurgery for patients with glioblastoma multiforme. *Neurosurg* **50**, 41–47.

19. Prisco, F. E., Weltman, E., Hanriot, R. M., and Brandt, R. A. (2002). Radiosurgical boost for primary high-grade gliomas. *J Neurooncol* **57**, 151–160.

20. Kornblith, P. L., and Walker, M. (1988). Chemotherapy for malignant gliomas. *J Neurosurg* **68**, 1–17.

21. Kyritsis, A. P. (1993). Chemotherapy for malignant gliomas. *Oncol* **7**, 93–100.

22. Brandes, A. A., and Fiorentino, M. V. (1996). The role of chemotherapy in recurrent malignant gliomas: An overview. *Cancer Investig* **14**, 551–559.

23. Pech, I. V., Peterson, K., and Cairncross, J. G. (1998). Chemotherapy for brain tumors. *Oncol* **12**, 537–547.

24. Cokgor, I., Friedman, H. S., and Friedman, A. H. (1999). Chemotherapy for adults with malignant glioma. *Cancer Investig* **17**, 264–272.

25. Fine, H. A. (1994). The basis for current treatment recommendations for malignant gliomas. *J Neurooncol* **20**, 111–120.

26. Macdonald, D. R., Cascino, T. L., Schold, S. C., and Cairncross, J. G. (1990). Response criteria for phase II studies of supratentorial malignant glioma. *J Clin Oncol* **8**, 1277–1280.

27. Garrett, M. J., Hughes, H. J., and Freeman, L. S. (1978). A comparison of radiotherapy alone with radiotherapy and CCNU in cerebral glioma. *Clin Oncol* **4**, 71–76.

28. Eagan, R. T., Childs, D. S., Layton, D. D. *et al.* (1979). Dianhydrogalactitol and radiation therapy: treatment of supratentorial glioma. *JAMA* **24**, 2046–2050.

29. Solero, C. L., Monfardini, S., Brambilla, C. *et al.* (1979). Controlled study with BCNU vs. CCNU as adjuvant chemotherapy following surgery plus radiotherapy for glioblastoma multiforme. *Cancer Clin Trials* **1979**, 43–48.

30. Walker, M. D., Green, S. B., Byar, D. P. *et al.* (1980). Randomized comparisons of radiotherapy and nitrosoureas for the treatment of malignant glioma after surgery. *N Engl J Med* **303**, 1323–1329.

31. Green, S. B., Byar, D. P., Walker, M. D. *et al.* (1983). Comparisons of carmustine, procarbazine, and high-dose methylprednisolone as additions to surgery and radiotherapy for the treatment of malignant glioma. *Cancer Treat Rep* **67**, 121–132.

32. Deutsch, M., Green, S. B., Strike, T. A. *et al.* (1989). Results of a randomized trial comparing BCNU plus radiotherapy, streptozotocin plus radiotherapy, BCNU plus hyperfractionated radiotherapy, and BCNU following misonidazole plus radiotherapy in the postoperative treatment of malignant glioma. *Int J Rad Oncol Biol Phys* **16**, 1389–1396.

33. Shapiro, W. R., Green, S. B., Burger, P. C. *et al.* (1989). Randomized trial of three chemotherapy regimens and two radiotherapy regimens in postoperative treatment of malignant glioma. Brain Tumor Cooperative Group trial 8001. *J Neurosurg* **71**, 1–9.

34. Shapiro, W. R., Green, S. B., Burger, P. C. *et al.* (1992). A randomized comparison of intra-arterial versus intravenous BCNU, with or without intravenous 5-fluorouracil, for newly diagnosed patients with malignant glioma. *J Neurosurg* **76**, 772–781.

35. Souhami, L., Seiferheld, W., Brachman, D. *et al.* (2004). Randomized comparison of stereotactic radiosurgery followed by conventional radiotherapy with carmustine to conventional radiotherapy with carmustine for patients with glioblastoma multiforme: Report of Radiation Therapy Oncology Group 93-05 protocol. *Int J Rad Oncol Biol Phys* **60**, 853–860.

36. Levin, V. A., Silver, P., Hannigan, J. *et al.* (1990). Superiority of post-radiotherapy adjuvant chemotherapy with CCNU, procarbazine, and vincristine (PCV) over BCNU for anaplastic gliomas: NCOG 6G61 final report. *Int J Rad Oncol Biol Phys* **18**, 321–324.

37. Prados, M. D., Scott, C., Curran, W. J., Nelson, D. F., Leibel S., and Karmer, S. (1999). Procarbazine, lomustine, and vincristine (PCV) chemotherapy for anaplastic astrocytoma: A retrospective review of radiation therapy oncology group protocols comparing survival with carmustine or PCV adjuvant chemotherapy. *J Clin Oncol* **17**, 3389–3395.

38. Fine, H. A., Dear, K. B. G., Loeffler, J. S., Black, P. M., and Canellos G. P. (1993). Meta-analysis of radiation therapy with and without adjuvant chemotherapy for malignant gliomas in adults. *Cancer* **71**, 2585–2597.

39. Steward, L. A., Burdett, S., Souhami, R. L., and Stenning, S. (2002). Chemotherapy in adult high-grade glioma: a systematic review and meta-analysis of individual patient data from 12 randomised trials. *Lancet* **359**, 1011–1018.

40. Rosen, E. M., Fan, S., Rockwell, S., and Goldberg, I. D. (1999). The molecular and cellular basis of radiosensitivity: Implications for understanding how normal tissues and tumors respond to therapeutic radiation. *Cancer Investig* **17**, 56–72.

41. Colevas, A. D., Brown, J. M., Hahn, S., Mitchell, J., Camphausen, K., and Coleman, C. N. (2003). Development of investigational radiation modifiers. *J Natl Cancer Inst* **95**, 646–651.

42. Vokes, E. E., and Weichselbaum, R. R. (1990). Concomitant chemoradiotherapy: rationale and clinical experience in patients with solid tumors. *J Clin Oncol* **8**, 911–934.

43. Devine, S. M., Vokes, E. E., and Weichselbaum, R. R. (1991). Chemotherapeutic and biologic radiation enhancement. *Curr Opin Oncol* **3**, 1087–1095.

44. Haraf, D. J., Weichselbaum, and R. R., Vokes, E. E. (1995). Timing and sequencing of chemoradiotherapy. *Cancer Treat Res* **74**, 173–198.

45. Masters, G. A., and Vokes, E. E. (1997). Radiotherapy and concomitant chemotherapy. *In Encyclopedia of Cancer.* (J. R. Bertino. Eds.), Volume III. pp. 1471–1480. Academic Press, San Diego.

46. Glantz, M. J., Kim, L., Choy, H., and Akerley, W. (1999). Concurrent chemotherapy and radiotherapy in patients with brain tumors. *Oncol* **13**, 78–82.

47. Goffman, T. E., Dachowski, L. J., Bobo, H. *et al.* (1992). Long-term follow-up on National Cancer Institute phase I/II study of glioblastoma multiforme treated with iododeoxyuridine and hyperfractionated irradiation. *J Clin Oncol* **10**, 264–268.

48. Werner-Wasik, M., Scott, C. B., Nelson, D. F. *et al.* (1996). Final report of a phase I/II trial of hyperfractionated and accelerated hyperfractionated radiation therapy with carmustine for adults with supratentorial malignant gliomas. Radiation Therapy Oncology Group study 83-02. *Cancer* **77**, 1535–1543.

49. Urtasun, R. C., Kinsella, T. J., Farnan, N., Del Rowe, J. D., Lester, S. G., and Fulton, D. S. (1996). Survival improvement in anaplastic astrocytoma, combining external radiation with halogenated pyrimidines: Final report of RTOG 86-12, phase I-II study. *Int J Rad Oncol Biol Phys* **36**, 1163–1167.

50. Chen, A. Y., Choy, H., and Rothenburg, M. L. (1999). DNA topoisomerase I-targeting drugs as radiation sensitizers. *Oncol* **13**, 39–46.

51. Grabenbauer, G. G., Anders, K., Fietkau, R. J. *et al.* (2002). Prolonged infusional topotecan and accelerated hyperfractinated 3d-conformal radiation in patients with newly diagnosed glioblastoma – a phase I study. *J Neurooncol* **60**, 269–275.

52. Rowinsky, E. K. (1999). Novel radiation sensitizers targeting tissue hypoxia. *Oncol* **13**, 61–70.

53. Del Rowe, J., Scott, C., Werner-Wasik, M. *et al.* (2000). Single-arm, open-label phase II study of intravenously administered tirapazamine and radiation therapy for glioblastoma multiforme. *J Clin Oncol* **18**, 1254–1259.

54. Kleinberg, L., Grossman, S. A., Carson, K. *et al.* (2002). Survival of patients with newly diagnosed glioblastoma multiforme treated with RSR13 and radiotherapy: Results of a phase II New Approaches to Brain Tumor Therapy CNS Consortium safety and efficacy study. *J Clin Oncol* **20**, 3149–3155.

55. Stevens, M. F. G., and Newlands, E. S. (1993). From triazines and triazenes to temozolomide. *Eur J Cancer* **20A**, 1045–1047.

56. Newlands, E. S., O'Reilly, S. M., Glaser, M. G. *et al.* (1996). The Charing Cross Hospital experience with temozolomide in patients with gliomas. *Eur J Cancer* **32A**, 2236–2241.

57. Newlands, E. S., Stevens, M. F. G., Wedge, S. R., Wheelhouse, R. T., and Brock, C. (1997). Temozolomide: a review of its discovery, chemical properties, pre-clinical development and clinical trials. *Cancer Treat Rev* **23**, 35–61.

58. Hvizdos, K. M., and Goa, K. L. (1999). Temozolomide. *CNS Drugs* **12**, 237–243.

59. Friedman, H. S., Kerby, T., and Calvert, H. (2000). Temozolomide and treatment of malignant glioma. *Clin Cancer Res* **6**, 2585–2597.

60. Stupp, R., Gander, M., Leyvraz, S., and Newlands, E. (2001). Current and future developments in the use of temozolomide for the treatment of brain tumours. *Lancet Oncol* **2**, 552–560.

61. Danson, S. J., and Middleton, M. R. (2001). Temozolomide: a novel oral alkylating agent. *Expert Rev Anticancer Ther* **1**, 13–19.

62. Wedge, S. R., Porteous, J. K., Glaser, M. G. *et al.* (1997). In vitro evaluation of temozolomide combined with x-irradiation. *Anticancer Drugs* **8**, 92–97.

63. van Rijn, J., Heimans, J. J., van den Berg, J. *et al.* (2000). Survival of human glioma cells treated with various combination of temozolomide and x-rays. *Int J Rad Oncol Biol Phys* **47**, 779–784.

64. Hirose, Y., Berger, M. S., and Pieper, R. O. (2001). P53 effects both the duration of G2/M arrest and the fate of temozolomide-treated human glioblastoma cells. *Cancer Res* **61**, 1957–1963.

65. Stupp, R., Dietrich, P. Y., Kraljevic, S. O. *et al.* (2002). Promising survival for patients with newly diagnosed glioblastoma treated with concomitant radiation plus temozolomide followed by adjuvant temozolomide. *J Clin Oncol* **20**, 1375–1382.

66. Stupp, R., Mason, W. P., van den Bent, M. J. *et al.* (2005). Radiotherapy plus concomitant and adjuvant temozolomide for glioblastoma. *New Engl J Med* **352**, 987–996.

67. Hegi, M. E., Diserens, A. C., Gorlia, T. *et al.* (2005). MGMT gene silencing and benefit from temozolomide in glioblastoma. *New Engl J Med* **352**, 997–1003.

68. Paz, M. F., Yaya-Tur, R., Rojas-Marcos, I. *et al.* (2004). CpG island hypermethylation of the DNA repair enzyme methyltransferase predicts response to temozolomide in primary gliomas. *Clin Cancer Res* **10**, 4933–4938.

69. Hegi, M. E., Diserens, A. C., Godard, S. *et al.* (2004). Clinical trial substantiates the predictive value of O-6-methylguanine-DNA methyltransferase promoter methylation in glioblastoma patients treated with temozolomide. *Clin Cancer Res* **10**, 1871–1874.

70. Athanassiou, H., Synodinou, M., Maragoudakis, E. *et al.* (2005). Randomized phase II study of temozolomide and radiotherapy compared with radiotherapy alone in newly diagnosed glioblastoma multiforme. *J Clin Oncol* **23**, 2372–2377.

71. Chang, S. M., Lamborn, K. R., Malec, M. *et al.* (2004). Phase II study of temozolomide and thalidomide with radiation therapy for newly diagnosed glioblastoma multiforme. *Int J Rad Oncol Biol Phys* **60**, 353–357.

72. Dropcho, E. J., Rosenfeld, S. S., Morawetz, R. B. *et al.* (1992). Preradiation intracarotid cisplatin treatment of newly diagnosed anaplastic gliomas. *J Clin Oncol* **10**, 452–458.

73. Bottom, K. S., Ashley, D. M., Friedman, H. S. *et al.* (2000). Evaluation of pre-radiotherapy cyclophosphamide in patients with newly diagnosed glioblastoma multiforme. *J Neurooncol* **46**, 151–156.

74. Weller, M., Streffer, J., Wick, W. *et al.* (2001). Preirradiation gemcitabine chemotherapy for newly diagnosed glioblastoma. A phase II study. *Cancer* **91**, 423–427.

75. Wick, W., Hermisson, M., Kortmann, R. D. *et al.* (2002). Neoadjuvant gemcitabine/treosulfan chemotherapy for newly diagnosed glioblastoma. A phase II study. *J Neurooncol* **59**, 151–155.

76. Puchner, M. J. A., Herrmann, H. D., Berger, J., and Cristante, L. (2000). Surgery, tamoxifen, carboplatin, and radiotherapy in the treatment of newly diagnosed glioblastoma patients. *J Neurooncol* **49**, 147–155.

77. Friedman, H. S., McLendon, R. E., Kerby, T. *et al.* (1998). DNA mismatch repair and O6-alkylguanine-DNA alkyltransferase

analysis and response to Temodal in newly diagnosed malignant glioma. *J Clin Oncol* **16**, 3851–3857.

78. Gilbert, M. R., Friedman, H. S., Kuttesch, J. F. *et al.* (2002). A phase II study of temozolomide in patients with newly diagnosed supratentorial malignant glioma before radiation therapy. *Neurooncol* **4**, 261–267.

79. Glantz, M., Chamberlain, M., Liu, Q., Litofsky, N. S., and Recht, L. D. (2003). Temozolomide as an alternative to irradiation for elderly patients with newly diagnosed malignant glioma. *Cancer* **97**, 2262–2266.

80. Chinot, O. L., Barrie, M., Frauger, E. *et al.* (2004). Phase II study of temozolomide without radiotherapy in newly diagnosed glioblastoma multiforme in an elderly populations. *Cancer* **100**, 2208–2214.

81. Newlands, E. S., Foster, T., and Zaknoen, S. (2003). Phase I study of temozolomide (TMZ) combined with procarbazine (PCB) in patients with gliomas. *Br J Cancer* **89**, 248–251.

82. Chang, S. M., Prados, M. D., Yung, W. K. A. *et al.* (2004). Phase II study of neoadjuvant 1,3-bis(2-chloroethyl)-1-nitrosourea and temozolomide for newly diagnosed anaplastic glioma. A North American Brain Tumor Consortium trial. *Cancer* **100**, 1712–1716.

83. Balaña, C., López-Pousa, A., Berrocal, A. *et al.* (2004). Phase II study of temozolomide and cisplatin as primary treatment prior to radiotherapy in newly diagnosed glioblastoma multiforme patients with measurable disease. A study of the Spanish Medical Neuro-Oncology Group (GENOM). *J Neurooncol* **70**, 359–369.

84. Medical Research Council Brain Tumour Working Party. Randomized trial of procarbazine, lomustine, and vincristine in the adjuvant treatment of high-grade astrocytoma: A Medical Research Council Trial. *J Clin Oncol* **19**, 509–518.

85. Reni, M., Cozzarini, C., Ferreri, A. J. M. *et al.* (2000). A retrospective analysis of postradiation chemotherapy in 133 patients with glioblastoma multiforme. *Cancer Investig* **18**, 510–515.

86. Jeremic, B., Jovanivic, D., Djuric, L. J., Jevremovic, S., and Mijatovic, L. J. (1992). Advantage of post-radiotherapy chemotherapy with CCNU, procarbazine, and vincristine (mPCV) over chemotherapy with VM-26 and CCNU for malignant gliomas. *J Chemother* **4**, 123–126.

87. Levin, V. A., Uhm, J. H., Jaeckle, K. A. *et al.* (2000). Phase III randomized study of postradiotherapy chemotherapy with alpha-difluoromethylornithine-procarbazine, N-(2-choroethyl)-N'-cyclohexyl-N-nitrosourea, vincristine (DFMO-PCV) versus PCV for glioblastoma multiforme. *Clin Cancer Res* **6**, 3878–3884.

88. Levin, V. A., Hess, K. R., Choucair, A. *et al.* (2003). Phase III randomized study of postradiotherapy chemotherapy with combination α–difluoromethylornithine-PCV versus PCV for anaplastic gliomas. *Clin Cancer Res* **9**, 981–990.

89. Chibbaro, S., Benvenuti, L., Caprio, A. *et al.* (2004). Temozolomide as first-line agent in treating high-grade gliomas: phase II study. *J Neurooncol* **67**, 77–81.

90. Raizer, J. J., Malkin, M. G., Kleber, M., and Abrey, L. E. (2004). Phase I study of 28-day, low-dose temozolomide and BCNU in the treatment of malignant gliomas after radiation therapy. *Neurooncol* **6**, 247–252.

91. Baumann, F., Bjeljac, M., Kollias, S. S. *et al.* (2004). Combined thalidomide and temozolomide treatment in patients with glioblastoma multiforme. *J Neurooncol* **67**, 191–200.

92. Newton, H. B., Junck, L., Bromberg, J., Page, M. A., and Greenberg, H. S. (1990). Procarbazine chemotherapy in the treatment of recurrent malignant astrocytomas after radiation and nitrosourea failure. *Neurol* **40**, 1743–1746.

93. Newton, H. B., Bromberg, J., Junck, L., Page, M. A., and Greenberg, H. S. (1993). Comparison between BCNU and procarbazine chemotherapy for treatment of gliomas. *J Neurooncol* **15**, 257–263.

94. Spence, A. M., Berger, M. S., Livinston, R. B., Ali-Osman, F., and Griffin, B. (1992). Phase II evaluation of high-dose cisplatin for treatment of adult malignant gliomas recurrent after chloroethylnitrosourea failure. *J Neurooncol* **12**, 187–191.

95. Yung, W. K. A., Mechtler, L., and Gleason, M. J. (1991). Intravenous carboplatin for recurrent malignant glioma: A phase II study. *J Clin Oncol* **9**, 860–864.

96. Jeremic, B., Grujicic, D., Jevremovic, S. *et al.* (1992). Carboplatin and etoposide chemotherapy regimen for recurrent malignant glioma: A phase II study. *J Clin Oncol* **10**, 1074–1077.

97. Tirelli, U., D'Incalci, M., Canetta, R. *et al.* (1984). Etoposide (VP-16-213) in malignant brain tumors: A phase II study. *J Clin Oncol* **2**, 432–437.

98. Fulton, D., Urtasun, R., and Forsyth, P. (1996). Phase II study of prolonged oral therapy with etoposide (VP16) for patients with recurrent malignant gliomas. *J Neurooncol* **27**, 149–155.

99. Chamberlain, M. C. (2002). Salvage chemotherapy with CPT-11 for recurrent glioblastoma multiforme. *J Neurooncol* **56**, 183–188.

100. Cloughesy, T. F., Filka, E., Nelson, G. *et al.* (2002). Irinotecan treatment for recurrent malignant glioma using an every-3-week regimen. *Am J Clin Oncol* **25**, 204–208.

101. Chamberlain, M. C., and Tsao-Wei, D. D. (2004). Salvage chemotherapy with cyclophosphamide for recurrent, temozolomide-refractory glioblastoma multiforme. *Cancer* **100**, 1213–1220.

102. Couldwell, W. T., Hinton, D. R., Surnock, A. A. *et al.* (1996). Treatment of recurrent malignant gliomas with chronic oral high-dose tamoxifen. *Clin Cancer Res* **2**, 619–622.

103. Chamberlain, M. C., and Kormanik, P. A. (1999). Salvage chemotherapy with tamoxifen for recurrent anaplastic astrocytomas. *Arch Neurol* **56**, 703–708.

104. Huncharek, M., and Muscat, J. (1998). Treatment of recurrent high grade astrocytoma; Results of a systematic review of 1,415 patients. *Anticancer Res* **18**, 1303–1312.

105. Huncharek, M., Kupelnick, B., and Bishop, D. (1998). Platinum analogues in the treatment of recurrent high grade astrocytoma. *Cancer Treat Rev* **24**, 307–316.

106. Yung, W. K. A., Prados, M. D., Yaya-Tur, R. *et al.* (1999). Multicenter phase II trial of temozolomide in patients with anaplastic astrocytoma or anaplastic oligoastrocytoma at first relapse. *J Clin Oncol* **17**, 2762–2771.

107. Yung, W. K. A., Albright, R. E., Olson, J. *et al.* (2000). A phase II study of temozolomide vs. procarbazine in patients with glioblastoma multiforme at first relapse. *Br J Cancer* **83**, 588–593.

108. Osoba, D., Brada, M., Yung, W. K. A., and Prados, M. (2000). Health-related quality of life in patients treated with temozolomide versus procarbazine for recurrent glioblastoma multiforme. *J Clin Oncol* **18**, 1481–1491.

109. Macdonald, D. R., Kiebert, G., Prados, M., Yung, W. K. A., and Olson, J., (2005). Benefit of temozolomide compared to procarbazine in treatment of glioblastoma multiforme at first relapse: Effect on neurological functioning, performance status, and health related quality of life. *Cancer Investig* **23**, 138–144.

110. Chang, S. M., Theodosopoulos, P., and Lamborn, K. *et al.* (2004). Temozolomide in the treatment of recurrent malignant glioma. *Cancer* **100**, 605–611.

111. Everaert, E., Neyns, B., Joosens, E., Strauven, T., Branle, F., and Menten, J. (2004). Temozolomide for the treatment of recurrent supratentorial glioma: results of a compassionate use program in Belgium. *J Neurooncol* **70**, 37–48.

112. Khan, R. B., Raizer, J. J., Malkin, M. G., Bazylewicz, K. A., Abrey, L. E. (2002). A phase II study of extended low-dose temozolomide in recurrent malignant gliomas. *Neurooncol* **4**, 39–43.

113. Schold, S. C., Kuhn, J. G., Chang, S. M. *et al.* (2000). A phase I trial of 1,3-bis(2-chloroethyl)-1-nitrosourea plus temozolomide: A North American Brain Tumor Consortium study. *Neurooncol* **2**, 34–39.

114. Prados, M. D., Yung, W. K. A., Fine, H. A. *et al.* (2004). Phase 2 study of BCNU and temozolomide for recurrent glioblastoma multiforme: North American Brain Tumor Consortium study. *Neurooncol* **6**, 33–37.

115. Korones, D. N., Benita-Weiss, M., Coyle, T. E. *et al.* (2003). Phase I study of temozolomide and escalating doses of oral etoposide for adults with recurrent malignant glioma. *Cancer* **97**, 1963–1968.

116. Gururangan, S., Cokgor, K., Rich, J. N. *et al.* (2001). Phase I study of Giadel wafers plus temozolomide in adults with recurrent supratentorial high-grade gliomas. *Neurooncol* **3**, 246–250.

117. Gruber, M. L., and Buster, W. P. (2004). Temozolomide in combination with irinotecan for treatment of recurrent malignant glioma. *Am J Clin Oncol* **27**, 33–38.

118. Brandes, A. A., Basso, U., Reni, M. *et al.* (2004). First-line chemotherapy with cisplatin plus fractionated temozolomide in recurrent glioblastoma multiforme: A phase II study of the Gruppo Italiano Cooperative di Neuro-Oncologia. *J Clin Oncol* **22**, 1598–1604.

119. Groves, M. D., Puduvalli, V. K., Hess, K. R. *et al.* (2002). Phase II trial of temozolomide plus the matrix metalloproteinase inhibitor, marimastat, in recurrent and progressive glioblastoma multiforme. *J Clin Oncol* **20**, 1383–1388.

120. Jaeckle, K. A., Hess, K. R., Yung, W. K. A. *et al.* (2003). Phase II evaluation of temozolomide and 13-cis-retinoic acid for the treatment of recurrent and progressive malignant glioma: A North American Brain Tumor Consortium study. *J Clin Oncol* **21**, 2305–2311.

121. Chua, S. L., Rosenthal, M. A., Wong, S. S. *et al.* (2004). Phase 2 study of temozolomide and Caelyx in patients with recurrent glioblastoma multiforme. *Neurooncol* **6**, 38–43.

122. Spence, A. M., Peterson, R. A., Scharnhorst, J. D., Selbergeld, D. L., and Rostomily, R. C. (2004). Phase II study of concurrent continuous temozolomide (TMZ) and tamoxifen (TMX) for recurrent malignant astrocytic gliomas. *J Neurooncol* **70**, 91–95.

123. Newton, H. B. (2003). Molecular neuro-oncology and the development of "targeted" therapeutic strategies for brain tumors. Part 1 – growth factor and ras signaling pathways. *Expert Rev Anticancer Ther* **3**, 595–614.

124. Newton, H. B. (2004). Molecular neuro-oncology and the development of "targeted" therapeutic strategies for brain tumors. Part 2 – PI3K/Akt/PTEN, mTOR, SHH/PTCH, and angiogenesis. *Expert Rev Anticancer Ther* **4**, 105–128.

125. Newton, H. B. (2005). Molecular neuro-oncology and the development of "targeted" therapeutic strategies for brain tumors. Part 5 – apoptosis and cell cycle. *Expert Rev Anticancer Ther* **5**, 355–378.

25

Chemotherapy of Low-Grade Astrocytomas

Mark G. Malkin

ABSTRACT: Rational decisions regarding the appropriate use of chemotherapy in patients with low-grade astrocytoma (LGA) will require further data from prospective randomized trials. Until such data becomes available, chemotherapy should probably still be used only as a salvage option in patients with recurrent or progressive LGA who have previously received radiation therapy (RT). Increasingly, due primarily to its low toxicity and possibly due to its modest efficacy, temozolomide has become the most popular chemotherapeutic agent in this setting. However, chemotherapy cannot yet be considered standard of care in patients with newly diagnosed tumors. When faced with an adult with a LGA, either at initial diagnosis or at recurrence, referral to an academic medical center with open clinical trials should be seriously considered.

BACKGROUND

LGA is a common brain tumor that occurs primarily in young adults. LGAs usually grow slowly, exhibit a high degree of cellular differentiation, and diffusely infiltrate surrounding brain. Although patients with LGA typically survive for several years longer than their counterparts with anaplastic astrocytoma (AA) or glioblastoma multiforme (GBM), most will ultimately die of tumor progression to those more malignant forms. Controversy continues to surround the optimal plan of management for these neoplasms. Treatment options include surgical resection, RT and chemotherapy [1]. The literature devoted to the topic of chemotherapy of LGA is sparse, indeed. Most series are retrospective

compilations of the experience at a single institution, and include a small number of patients with a mixture of low-grade histologies (i.e., LGA, low-grade oligodendroglioma, and low-grade mixed oligoastrocytoma), each with its unique range of prognosis. Frequently these studies include both patients who have and have not received prior RT, and patients who have been treated both at the time of initial diagnosis and for disease progression or recurrence. Most importantly, it is uncommon for patients treated with chemotherapy at the time of clinical or radiographic progression to have undergone biopsy or resection, especially if the tumor is located in an eloquent or high-risk anatomic area. Thus, it is impossible to confirm persistent low-grade pathology *versus* transformation to a higher grade astrocytoma in many of these patients. This is a critical issue since, at the time of recurrence, anywhere from 50–85 per cent of low-grade gliomas have progressed to a higher grade tumor [2,3]. The majority of published reports describing the efficacy of chemotherapy concern children with LGA whose tumors have recurred or progressed after surgery, with or without RT. That pediatric experience is summarized elsewhere in this book (see Chapter 36). Recent comprehensive review articles on the topic of chemotherapy for LGA have been published by Stieber [4], Lesser [5], and Stupp and Baumert [6].

PATHOLOGIC NOMENCLATURE

The term low-grade glioma encompasses not only LGA, but other neuroepithielial tumors including olgodendroglioma, mixed oligoastrocytoma, ependymoma, pilocytic astrocytoma, pleomorphic xanthoastrocytoma, and subependymal giant cell astrocytoma. Amongst these, LGA is the commonest low-grade glioma. The term astrocytoma, unless otherwise specified, usually refers to the low-grade diffuse astrocytoma (LGA) of adulthood. LGA corresponds to a WHO grade II astrocytoma. LGA must be differentiated from pilocytic astrocytoma, which has a different age distribution, location, and biology. Pilocytic astrocytomas most commonly occur in children, are situated in the cerebellum, and in the vast majority of cases, are treated adequately with surgical resection alone.

EPIDEMIOLOGY

LGA represents approximately 15 per cent of adult gliomas [7]. The average incidence of these tumors has been calculated at slightly less than 1 per 100 000 population per year [8,9]. The peak incidence is in young adults between the ages of 30 and 40 (25 per cent of all cases) [10]. Approximately, 10 per cent occur below the age of 20, 60 per cent between 20 and 45 years of age, and about 30 per cent over 45 years. For unclear reasons, there is a slight predominance in males, constituting approximately 60 per cent of cases [7,11].

LOCATION

LGAs may develop in any region of the central nervous system but most commonly arise in the cerebral hemispheres. The brainstem is the next most common site, while these tumors are distinctly uncommon in the cerebellum. Within the cerebrum, they arise roughly in proportion to the relative mass of the different lobes, hence the frontal lobe is the commonest location, followed by the temporal lobe [11].

CLINICAL PRESENTATION AND IMAGING FINDINGS

Common presenting symptoms of LGA include seizures, headache, altered mental status, and focal neurological deficit. Seizures are the presenting symptom in more than 50 per cent of all cases [12,13]. Headache and focal neurological deficit occur less frequently, and signs of raised intracranial pressure are uncommon [12]. Magnetic resonance imaging (MRI) is the most sensitive test available to diagnose LGA. The tumor typically appears as a low-intensity area on T1-weighted images, whereas there is almost always an increase in signal intensity corresponding to an increased relaxation time on T2-weighted images. The area of abnormal signal is usually homogeneous and well defined without evidence of hemorrhage or necrosis [14]. Enhancement after the administration of intravenous gadolinium occurs in between 8 and 15 per cent of cases [15–17]. Usually, the lack of enhancement coupled with the well-differentiated nature of the LGA will result in a "hypometabolic" or "cold" fluorodeoxyglucose positron emission tomography (FDG PET) scan. Often, but certainly not invariably, when dedifferentiation occurs to a more malignant state, the tumor may appear "hypermetabolic" or "hot" on FDG PET scan. This information may be of value in determining a relevant site for biopsy in a patient with a recurrent or progressive LGA, since astrocytomas tend to behave according to the highest grade of malignancy in a heterogeneous tumor, and the highest grade noted by the neuropathologist will usually determine the aggressiveness of subsequent therapy [18,19]. Magnetic resonance spectroscopy may also provide additional information with regard to the grade of astrocytic tumors [20].

PROGNOSTIC FACTORS

Almost all would agree that younger age at diagnosis is by far the most important factor that correlates with longer survival [10,21–26]. Seizures are correlated with a better prognosis while focal neurologic deficit and personality change indicate a worse prognosis [10,26,27]. The beneficial effect of good performance status at diagnosis is well documented [10,16,22]. More extensive surgical resections may predict longer survival, but there remains some disagreement about this issue. An important observation of a recent randomized trial on the appropriate dose of RT in the treatment of LGA has been that tumor volume is an important predictor both of overall survival and progression-free survival [28]. Several studies have determined that tumors with a higher mitotic activity have a poorer prognosis. LGAs with a bromodeoxyuridine (BUdR) labeling index <1 per cent, or a MIB-1 labeling index of >8 per cent, represent a group of patients with a poorer prognosis [29–31]. Lack of contrast enhancement on brain imaging predicts better progression-free and overall survival [32].

PROGRESSION AND OUTCOME OF LGA

These tumors slowly grow over months to years. Ultimately, conversion to a more malignant phenotype, and progressive neurologic dysfunction cause death in the majority of patients [33,15]. Progression to higher grades occurs more rapidly in older patients — in a recent study, age at diagnosis showed a strong negative correlation with interval to anaplastic progression [34]. Patients surviving 10 years or longer from diagnosis have been reported [10,35]. It is well recognized that the behavior of these tumors can be quite variable. Many patients, particularly younger individuals who present with seizures and no other neurological deficit, can survive for extended periods of time [10,36]. Median survival times of 5 to 7 years from diagnosis of a LGA are the norm.

A SINGLE RANDOMIZED TRIAL

The only published randomized trial to evaluate the role of chemotherapy in the treatment of LGA was published by Eyre, et al., in 1993 [37]. From February 1980 through March 1985, twenty-three institutions entered sixty adult patients with partially resected low-grade gliomas on study. According to the classification of Kernohan and Sayre, eligibility requirements included the histologic diagnosis of a grade I or II primary brain tumor, [38]. Grade II astrocytomas included pilocytic astrocytomas of the cerebral hemispheres, gemistocytic astrocytomas, mildly anaplastic astrocytomas, mixed gliomas, oligo-dendrogliomas, and gangliogliomas. Tumors were also classified using the three-tiered system of LGA, AA, and GBM. Patients with pilocytic cerebellar astrocytoma or GBM were not included. Patients were randomly assigned to receive RT alone versus RT with 1-(2-chloroethyl)-3-cyclohexyl-1-nitrosourea (CCNU). Patients were required to begin RT within six weeks of tumor resection. The target volume was defined as the primary tumor, as identified on computerized tomography (CT) scans, with a 2-cm margin. A total of 55 Gy was delivered to the target volume in 32 fractions, given 5 days per week over a total of 6½ to 7 weeks. In the chemotherapy arm, CCNU was begun two days prior to the onset of RT, at a dose of 100 mg/m^2 orally every six weeks. If the patient had a partial (PR) or complete response (CR), CCNU was continued for a total period not to exceed two years. Dexamethasone was prescribed at the discretion of the clinical investigator. Central neuro-pathologic review showed that six patients were ineligible for the study. Evaluation of patient age, extent of surgery, tumor grade, and performance status showed no significant differences between the treatment arms. Median patient age was 36 years (range 22 to 73 years) for the RT-only arm and 39 years (range 17 to 72 years) for the RT plus CCNU arm. The data safety monitoring committee closed the trial early. Slow accrual, and a rejection of the hypothesis of a 50 per cent improvement in the survival time of patients receiving RT plus CCNU, were the principal reasons for closure. The response rate, as judged by the disappearance or reduction in size of tumor on CT scans, was 79 per cent for RT alone versus 54 per cent for RT plus CCNU. This difference was not statistically significant. Median survival for patients who received RT alone was 4.5 years whereas that for patients who received RT plus CCNU was 7.4 years ($p = 0.7$). For the group as a whole, patient age and performance status were the most important prognostic parameters. Median survival in patients younger than 30 years of age was greater than 8 years, in patients between 30 and 50 years of age was 5.5 years, and in patients older than 50 years was 1.6 years ($p = 0.001$). Patients with an ECOG performance status of 0 to 1 experienced a median survival of 7.4 years, while those with a performance status of 2 to 4 had a median survival of 1.6 years ($p = 0.002$). Patients who underwent biopsy only had a median survival of 2.6 years, compared to 5.5 years for those who underwent partial resection ($p = 0.38$). Of the 32 evaluable patients receiving CCNU, 59 per cent experienced mild to moderate leukopenia and/or thrombocytopenia and 41 per cent experienced mild to moderate gastrointestinal upset. Several important conclusions can be drawn from this study:

1. The addition of CCNU did not improve upon the results of RT in the treatment of partially resected low-grade gliomas.
2. The important prognostic factors that need to be taken into account and controlled for in any future comparative trial of LGA include age, performance status, and extent of surgery.
3. Low-grade gliomas comprise a heterogeneous group of histopathologic entities, and it is inappropriate to extrapolate response to treatment for the group as a whole to the specific subset of LGA.

PHASE II TRIALS

Temozolomide is an imidazole tetrazinone that undergoes chemical conversion to the active methylating agent 5-(3-methyltriazen-1yl)imidazole-4-carboximide under physiologic conditions. In a phase II trial of temozolomide in patients with progressive low-grade glioma, temozolomide was administered orally once a day for five consecutive days (in a fasting state) at a starting dose of 200 mg/m^2/day [33]. Treatment cycles were repeated every 28 days following the first daily dose of temozolomide. The study objectives were two-fold: to determine the antitumor activity of temozolomide, including response and progression-free survival (PFS) in the treatment of adults and children with progressive low-grade glioma, including astrocytoma, oligodendroglioma, mixed glioma, and pilocytic astrocytoma; and to evaluate the toxicity of temozolomide in this patient population. Eligible patients were required to have a histologically confirmed diagnosis of primary intracranial, infratentorial, or supratentorial low-grade glioma (e.g., astrocytoma, oligodendroglioma, or mixed glioma). However, biopsy was not required for patients with an intrinsic chiasmatic mass or tumor infiltration along the posterior optic tracts. All patients were required to have measurable disease on MRI or CT scan. All patients, whether newly diagnosed or previously treated, were required to exhibit evidence of tumor progression while on a stable or increasing dose of corticosteroids. Progressive disease had to have been evident after all prior interventions with chemotherapy, RT, and/or surgical resection. However, if the sole intervention included biopsy or partial resection, where less than 50 per cent of the tumor was debulked, then progressive disease had to have been evident before or after the intervention. The patients were required to be at least 4 years of age, have a Karnofsky performance score (KPS) ≥ 70, and have an estimated life expectancy of greater than 12 weeks. In the absence of disease progression or unacceptable toxicity, patients continued to receive temozolomide for up to a maximum of 12 cycles. Forty-six patients with low-grade glioma were treated, of whom sixteen (35 per cent) had LGA. Thirty-nine of the 46 (85 per cent) had not received prior RT. Thirty-six of the 46 (78 per cent) had not received prior chemotherapy. Thirty-four of the 46 (70 per cent) had enhancing lesions on CT or MRI scan. For the 16 astrocytoma patients, five (31 per cent) had a CR, six (38 per cent) had a PR, and four (25 per cent) had stable disease (SD) for an overall response rate of 15/16 (94 per cent). Of the 16 LGA patients, four ultimately progressed. For the LGA patients, median PFS was not reached, 6-month PFS was 100 per cent and 12-month PFS was 73 per cent. Six low-grade glioma patients experienced notable toxicity during the study. Three patients experienced

grade 3 neutropenia, with a duration greater than three weeks in one patient, and two patients experienced grade 3 thrombocytopenia. One patient experienced $\geq$ grade 4 toxicity, with intracerebral hemorrhage, neutropenia, thrombocytopenia, sepsis, and death. Several conclusions regarding this study are worthy of consideration:

1. Children were treated, as well as adults, and younger age confers a better prognosis.
2. All patients were independent in activities of daily living, and higher performance status confers a better prognosis.
3. Newly diagnosed patients, and patients with recurrence or progression, were lumped together in the analysis.
4. The high response rates and PFS reported in this series have not yet been replicated by other investigators.

Brada *et al.* published a phase II study whose aim was to assess the response rate to temozolomide in patients with low-grade gliomas using imaging as the primary end point [39]. Eligible patients included adults (>18 years of age) with histologically confirmed grade II astrocytoma, grade II oligodendroglioma, or grade II mixed oligoastrocytoma, with evaluable disease on imaging, with no previous treatment other than surgery and with either stable or progressive disease. Patients with imaging or histologic evidence of transformation to high grade tumors were excluded. Patients felt to be in need of urgent decompressive surgery or RT due to rapidly evolving neurological deficit, or raised intracranial pressure with papilledema and corticosteroid dependence, were also excluded. Patients received temozolomide at 200 mg/m^2 orally daily for 5 days given every 28 days, and the dose and frequency were adjusted according to standard toxicity criteria. The study was closed prematurely when temozolomide became available commercially; presumably the investigators found it increasingly difficult to recruit patients to the study at that point. Thirty patients were entered into the study, 17 with LGA. All had previous surgery (biopsy alone, 18; surgical resection, 12); three patients had two or more attempts at excision. At the time of study entry, two patients had a KPS of 60, one a KPS of 80, 11 a KPS of 90, and 16 had a KPS of 100. Twenty-seven patients had a seizure history at the time of study entry. During chemotherapy, 14 (52 per cent) experienced improvement in seizure frequency. Twenty-seven of 28 patients (96 per cent) had an improvement in at least one health-related quality of life (QoL) domain. Of the LGA and mixed glioma patients, one (5 per cent) had a PR,

11 (58 per cent) had a mixed response (MR), six (32 per cent) had SD, and one (5 per cent) had progressive disease (PD). There were 11 episodes (3.5 per cent) of grade 3 or 4 hematologic toxicity seen in six patients; this presented as thrombocytopenia in six, neutropenia in three, and combined thrombocytopenia and neutropenia in two. Two patients had grade 3 constipation and one patient had grade 3 nausea and vomiting. While the study aimed to provide an objective assessment of the effectiveness of temozolomide, there is no uniformly agreed imaging end point in patients with WHO grade II tumors. While it is reasonable to interpret changes in the size of high signal abnormality as a reduction in tumor size, and therefore tumor cell mass, there is no definite imaging-pathological correlation to ensure that the changes represent purely a reduction in the number of tumor cells. FLAIR sequences detect alterations in free water in the soft tissues. In low-grade glioma, the areas of altered signal are postulated to reflect the combination of tumor cells and increased water (edema) in the adjacent tissues. Alteration in the size of the high signal region on FLAIR sequences may indicate both a reduction in the number of viable tumor cells and a reduction in edema. The area of signal change gives an approximation of the extent of tumor involvement, as tumor cells migrate along the widened edematous white matter tracts. Abnormalities seen on FLAIR sequences considered to represent the tumor can be measured on sequential imaging with relative ease, but remain subject to interobserver variation, particularly due to a gradation of signal change between the normal and abnormal areas. There are other pitfalls in the assessment of tumor size. They include lack of strict co-registration of images, variation due to changing slice angles, and different windowing of images. The correlation between imaging response, improvement in health-related QoL (74/112 domains [66 per cent] in responders; 41/95 domains [41 per cent] in non-responders; $p < 0.001$), and reduction in seizure frequency suggest a beneficial and clinically meaningful effect of the observed MRI response and probably reflects a worthwhile reduction in tumor size. The initial study design aim, which was to assess the effectiveness of temozolomide separately in astrocytoma and oligodendroglioma, was not achieved when the study was terminated prematurely, and the available information is on the whole group of patients with grade II glial tumors. The change in tumor size on imaging was slow, with a continued reduction over a number of months, which suggests that studies in patients with previously untreated LGA need to be performed over a considerable time. The high interobserver variation and difficulty in precise assessment of tumor size also argues for a study design with multiple observers. Patients received treatment up to 9 years after the initial diagnosis and were therefore at different stages of tumor evolution, ranging from indolent to more malignant variants. The median survival of this cohort from initial diagnosis was not reached. Five-year survival from initial diagnosis was 84 per cent at a median follow-up from initial diagnosis of 5 years, which suggests a selection of good prognosis patients from first diagnosis.

Pace *et al.* assessed the activity of temozolomide in patients with progressive low- grade glioma [40]. The primary endpoint of the study was response to treatment. Secondary endpoints included clinical benefits in terms of modifications of seizure frequency, QoL, and toxicity. Patients with WHO grade II histologically confirmed glioma (LGA, oligodendroglioma, and mixed oligoastrocytoma) were eligible for the study when clinical and radiographic progression were documented. All patients demonstrated measurable disease on pretreatment contrast-enhanced MRI scan and had a KPS ≥ 70. Temozolomide was administered orally, once daily, for five consecutive days every 28 days, at a starting dose of 200 mg/m^2/day if not pretreated, or at 150 mg/m^2/day in procarbazine-lomustine-vincristine (PCV) pretreated patients, with dose escalation to 200 mg/m^2/day in the absence of toxicity. Forty-three patients affected with low-grade glioma (29 LGA, 4 oligodendroglioma, and 10 mixed oligo-astrocytoma) were treated. Sixteen patients underwent re-operation after recurrence and histology showed anaplastic foci in eight cases and LGA in the other eight cases. Seventeen of the 43 patients (40 per cent) presented a non-enhancing lesion on MRI, while 26 (60 per cent) presented focal enhancing areas in the context of a non-enhancing tumor. Thirty patients (70 per cent) had previously received RT; 16 (37 per cent) were pretreated with a PCV regimen (usually oligodendroglioma or mixed oligoastrocytoma patients). Patients ranged in age from 21 to 62 years, with a median of 39 years. Median KPS was 90 (range 70–100). For the 29 LGA patients the median number of cycles of temozolomide was 11 (range 3–22). Of the 29 LGA patients there were 2 CRs, 12 PRs, 11 SDs, and 3 PDs. Three patients suffered grade 3 myelosuppression, and one patient experienced grade 3 gastrointestinal toxicity. For the group as a whole, median duration of response was 10 months, with a PFS rate of 76 per cent at 6 months and 39 per cent at 12 months. The response rate in patients presenting non-enhancing lesions on MRI (17 patients) was lower than that observed in patients presenting enhancing lesions on MRI (26 patients): 29 per cent (95 CI% 8 to 51 per cent) *versus* 58 per cent (95 CI% 38 to 76 per cent), respectively. The influence of temozolomide treatment on seizure frequency in the 31 patients presenting with uncontrolled epilepsy was remarkable, with complete seizure control in six patients and partial seizure control in nine, with steroid dose stable or reduced and no modification in anticonvulsant medication. Out of 16 patients receiving steroids at the beginning of chemotherapy, a reduction in steroid dose was achieved in nine (56 per cent). Eight patients (18 per cent) showed an increase in KPS. Responses to temozolomide were observed after a median of four cycles of therapy.

PENDING STUDY

The Radiation Therapy Oncology Group (RTOG) has recently closed to accrual a large study of low grade glioma in adults ≥ 18 years old with KPS ≥ 60 (RTOG 98-02). "Low risk" patients (age < 40 and gross total resection) are being observed. "High risk" patients (age ≥ 40 or subtotal resection or biopsy) were stratified by histology (LGA predominant or evenly mixed oligoastrocytoma *versus* oligodendroglioma predominant), by age (<40 *versus* $\geq$40), by KPS (60–80 *versus* 90–100) and by contrast enhancement (present *versus* absent). The high risk patients were then randomly assigned to RT alone *versus* RT followed by six cycles of PCV chemotherapy. The RT dose was 54 Gy in 30 fractions over six weeks, five days per week to a gross tumor volume defined by a T2 weighted post-operative MRI scan plus a 2 cm margin. The primary objective was to identify overall survival in the low risk and high risk patients. Secondary objectives included documentation of toxicity, comparison of the neurosurgeon's assessment of completeness of resection *versus* the neuroradiologist's opinion based upon a post-operative MRI scan, and the collection of brain tumor tissue and peripheral blood samples for future correlative studies. The trial, which was activated on October 31 1998, closed to accrual on June 27 2002. Given the low-grade nature of the tumors in patients on this study, it is anticipated that data concerning survival will not be available for at least another 2–3 years.

References

1. Kaye, A. H., and Walker, D. G. (2000). Low grade astrocytomas: controversies in management. *J Clin Neurosci* 7(6), 475–483.

2. Afra, D., Müller, W., Benoist, G., and Schröder, R. (1978). Supratentorial recurrences of gliomas, results of reoperations on astrocytomas and oligodendrogliomas. *Acta Neuroschir (Wien)* 43(3–4), 217–227.

3. Müller, W., Afra, D., and Schröder, R. (1977). Supratentorial recurrences of gliomas. Morphological studies in relation to time intervals with astrocytomas. *Acta Neurochir (Wien)* 37(1–2), 75–91.

4. Stieber, V. W. (2001). Low-grade gliomas. *Curr Treat Options Oncol* 2(6), 495–506.

5. Lesser, G. J. (2001). Chemotherapy of low-grade gliomas. *Semin Radiat Oncol* 11(2), 138–144.

6. Stupp, R., and Baumert, B. G. (2003). Promises and controversies in the management of low-grade glioma. *Ann Oncol* 14(12), 1695–1696.

7. Guthrie, B. L., and Laws, E. R., Jr. (1990). Supratentorial low-grade gliomas. *Neurosurg Clin North Am* 1(1), 37–48.

8. Radhakrishnan, K., Bohnen, N., and Kurland, L. T. (1993). Epidemiology of brain tumors. *In Brain tumors: a comprehensive text* (R. A. Morantz, and J. Walsh, Eds.), Marcel Dekker, New York.

9. Morantz, R. A. (1995). Low grade astrocytomas. *In Brain tumors* (A. H. Kaye, and E. R. Laws, Jr, Eds.), 433–446. Churchill Livingstone, Edinburgh.

10. Laws, E. R., Jr., Taylor, W. F., Clifton, M. B., and Okazaki, H. (1984). Neurosurgical management of low-grade astrocytoma of the cerebral hemispheres. *J Neurosurg* 61(4), 665–673.

11. Kleihues, P., Davis, R. L., Ohgaki, H., and Cavenee, W. K. (1997). Low-grade diffuse astrocytomas. *In Tumors of the nervous system: pathology and genetics* (P. Kleihues, and W. K. Cavenee, Eds.), 10–14. International Agency for Research on Cancer, Lyon.

12. Janny, P., Cure, H., Mohr, M. *et al.* (1994). Low grade supratentorial astrocytomas. Management and prognostic factors. *Cancer* 73(7), 1937–1945.

13. Afra, D., Osztie, E., Sipos, L., and Vitanovics, D. (1999). Preoperative history and postoperative survival of supratentorial low-grade astrocytomas. *Br J Neurosurg* 13(3), 299–305.

14. Morantz, R. A. (1996). Low-grade astrocytomas. *In Neurosurgery* (R. H. Wilkins, and S. S. Rengachary, Eds.), 789–798. McGraw-Hill, New York.

15. Piepmeier, J., Christopher, S., Spencer, D. *et al.* (1996). Variations in the natural history and survival of patients with supratentorial low-grade astrocytomas. *Neurosurgery* 38(5), 872–878.

16. Vertosick, F. T., Jr, Selker, R. G., and Arena, V. C. (1991). Survival of patients with well-differentiated astrocytomas diagnosed in the era of computed tomography. *Neurosurgery* 28(4), 496–501.

17. Lunsford, L. D., Somaza, S., Kondziolka, D., and Flickinger, J. C. (1995). Survival after stereotactic biopsy and irradiation of cerebral nonanaplastic, nonpilocytic astrocytoma. *J Neurosurg* 82(4), 523–529.

18. Francavilla, T. L., Miletich, R. S., Di Chiro, G., Patronas, N. J., Rizzoli, H. V., and Wright, D. C. (1989). Positron emission tomography in the detection of malignant degeneration of low-grade gliomas. *Neurosurgery* 24(1), 1–5.

19. Worthington, C., Tyler, J. L., and Villemure, J. G. (1987). Stereotaxic biopsy and positron emission tomography correlation of cerebral gliomas. *Surg Neurol* 27(1), 87–92.

20. Meyerand, M. E., Pipas, J. M., Mamourian, A., Tosteson, T. D., and Dunn, J. F. (1999). Classification of biopsy-confirmed brain tumors using single-voxel MR spectroscopy. *Am J Neuroradiol* 20(1), 117–123.

21. Franzini, A., Leocata, F., Cajola, L., Servello, D., Allegranza, A., and Broggi, G. (1994). Low-grade glial tumors in basal ganglia and thalamus: natural history and biological reappraisal. *Neurosurgery* **35**(5), 817–820.

22. Iwabuchi, S., Bishara, S., Herbison, P., Erasmus, A., and Samejima, H. (1999). Prognostic factors for supratentorial low grade astrocytomas in adults. *Neurol Med Chir (Tokyo)* **39**(4), 273–279.

23. McCormack, B. M., Miller, D. C., Budzilovich, G. N., Voorhees, G. J., and Ransohoff, J. (1992). Treatment and survival of low-grade astrocytomas in adults – 1977-1988. *Neurosurgery* **31**(4), 636–642.

24. North, C. A., North, R. B., Epstein, J. A., Piantadosi, S., and Wharam, M. D. (1990). Low-grade cerebral astrocytomas. Survival and quality of life after radiation therapy. *Cancer* **66**(1), 6–14.

25. Scanlon, P. W., and Taylor, W. F. (1979). Radiotherapy of intracranial astrocytomas: analysis of 417 cases treated from 1960 through 1969. *Neurosurgery* **5**(3), 301–308.

26. Soffietti, R., Chio, A., Giordana, M. T., Vasario, E., and Schiffer, D. (1989). Prognostic factors in well-differentiated cerebral astrocytomas in the adult. *Neurosurgery* **24**(5), 686–692.

27. Philippon, J. H., Clemenceau, S. H., Fauchon, F. H., and Foncin, J. F. (1993). Supratentorial low-grade astrocytomas in adults. *Neurosurgery* **32**(4), 554–559.

28. Karim, A. B., Maat, B., Hatlevoll, R. *et al.* (1996). A randomized trial on dose-response in radiation therapy of low-grade cerebral glioma: European Organization for Research and Treatment of Cancer (EORTC) Study 22844. *Int J Radiat Oncol Biol Phys* **36**(3), 549–556.

29. Ito, S., Chandler, K. L., Prados, M. D. *et al.* Proliferative potential and prognostic evaluation of two-grade astrocytomas. (1994). *J Neurooncol* **19**(1), 1–9.

30. Struikmans, H., Rutgers, D. H., Jansent, G. H., Tulleken, C. A. F., van der Tweel, I., and Batterman, J. J. (1998). Prognostic relevance of cell proliferation markers and DNA-ploidy in gliomas. *Acta Neurochir (Wien)* **140**(2), 140–147.

31. Schiffer, D., Cavalla, P., Chio, A., Richiardi, P., and Giordana, M. T. (1997). Proliferative activity and prognosis of low-grade astrocytomas. *J Neurooncol* **34**(1), 31–35.

32. Lote, K., Egeland, T., Hagar, B. *et al.* (1997). Survival, prognostic factors, and therapeutic efficacy in low-grade gliomas: a retrospective study in 379 patients. *J Clin Oncol* **15**(9), 3129–3140.

33. Quinn, J. A., Reardon, D. A., Friedman, A. H. *et al.* (2003). Phase II trial of temozolomide in patients with progressive low-grade glioma. *J Clin Oncol* **21**(4), 646–651.

34. Shafqat, S., Hedley-Whyte, E. T., and Henson, J. W. (1999). Age-dependent rate of anaplastic transformation in low-grade astrocytoma. *Neurology* **52**(4), 867–869.

35. Sheline, G. E. (1986). The role of radiation therapy in the treatment of low-grade gliomas. *Clin Neurosurg* **33**, 563–574.

36. Piepmeier, J. M. (1987). Observations of the current treatment of low-grade astrocytic tumors of the cerebral hemispheres. *J Neurosurg* **67**(2), 177–181.

37. Eyre, H. J., Crowley, J. J., Townsend, J. J. *et al.* (1993). A randomized trial of radiotherapy *versus* radiotherapy plus CCNU for incompletely resected low-grade gliomas: a Southwest Oncology Group study. *J Neurosurg* **78**(6), 909–914.

38. Kernohan, J. W., and Sayre, G. P. (1952). *Tumors of the central nervous system. In "Atlas of Tumor Pathology, Section X, Fascicle 35."* Armed Forces Institute of Pathology, Washington, DC.

39. Brada, M., Viviers, L., Abson, C. *et al.* (2003). Phase II study of primary temozolomide chemotherapy in patients with WHO grade II gliomas. *Ann Oncol* **14**(12), 1715–1721.

40. Pace, A., Vidiri, A., Galiè, *et al.* (2003). Temozolomide chemotherapy for progressive low-grade glioma: clinical benefits and radiological response. *Ann Oncol* **14**(12), 1723–1726.

26

Chemotherapy of Oligodendrogliomas

Nina A. Paleologos and Christopher Fahey

INTRODUCTION

Oligodendrogliomas are much less prevalent than astrocytomas, thought to comprise only 4 per cent of primary brain tumors [1]. Recently, others have suggested larger estimates [2]. Oligodendrogliomas were first appreciated as a distinct clinical and histological neoplasm in the 1920s [3–5]. The median age at diagnosis of low-grade oligodendroglioma is less than 40 years of age, while anaplastic tumors typically occur in patients greater than 40 years of age [6]. There is a wide survival range, which, in part, is related to tumor grade, with other factors contributing. Most low-grade tumors will eventually undergo transformation to more anaplastic forms, and the majority of patients with oligodendroglioma will ultimately die from the disease. Significant attention has been focused on these unique tumors following the seminal papers of Cairncross and Macdonald, which suggested that oligodendroglioma were uniquely sensitive to chemotherapy [7,8]. Advances in molecular genetic analyses have provided insight into which patients will be likely to respond to chemotherapy and have helped to identify patients with tumors that are likely to behave more aggressively. While many treatment modalities are useful in the successful treatment of oligodendrogliomas, this chapter will focus on chemotherapy.

GENETICS

The clinical observation that these tumors were chemosensitive, with the implication that they may be treated differently than astrocytomas, resulted in a heightened awareness of the importance of correct histopathologic diagnosis. Unfortunately, there are no reliable histochemical markers for oligodendrogliomas, and their histological features can, at times, be ambiguous with significant inter-observer variability, even among experienced neuropathologists. In an effort to "not miss" this diagnosis, many pathologists have loosened their criteria for the oligodendroglioma resulting in an increase in frequency of the diagnosis and the likelihood that some astrocytic tumors are being misclassified [9,10]. The need for objective methods to distinguish oligodendrogliomas is thus becoming increasingly important.

Oligodendrogliomas are frequently characterized by allelic deletions of chromosomes 1p and 19q. Low-grade (WHO grade II) oligodendrogliomas show losses of 1p and 19q in 80–90 per cent of cases, while anaplastic oligodendrogliomas show these losses in approximately 50–70 per cent of cases [9]. This compares with a loss of heterozygosity for 1p and 19q occurring at a frequency of only 11 per cent among astrocytomas [11] and 38 per cent among oligoastrocytomas [12]. Unlike losses on 1p, allelic losses on 19q are also commonly seen in high-grade astrocytic tumors [9]. Tumors with a classical histologic oligodendroglial appearance are more likely to have deletions of 1p and 19q as compared to those tumors which are atypical in appearance [10,13]. Further, anaplastic oligodendrogliomas demonstrate loss of heterozygosity at 1p or 19q more frequently when located in the frontal, parietal, and occipital lobes than when found in the temporal lobe, insula, or diencephalon [14]. Loss of 1p and 19q is also associated with bilateral tumor growth [14] and patients with these deletions more frequently present with seizures [15]. The specific tumor suppressor genes located on chromosome arms 1p and 19q remain unknown, but work to identify them is ongoing and several candidate regions have been identified [9,16].

Cairncross *et al.* were first to report that in patients with anaplastic oligodendroglioma, response to chemotherapy was linked to allelic loss of 1p or combined losses of 1p and 19q [17]. A second study by the same group reported a further analysis of fifty patients which confirmed these results and also reported that prolonged survival was predicted by combined allelic losses on 1p and 19q [18]. The initial study examined thirty-nine patients with anaplastic oligodendroglioma who were treated with chemotherapy and evaluated for 1p/19q status. Thirty-seven of these patients were treated with PCV (procarbazine, CCNU (1-(2-chloroethyl)-3-cyclo-hexyl-1-nitrosurea), and vincristine) therapy, one with carmustine, and one with cisplatin and etoposide. Of the twenty-four patients with loss of heterozygosity at 1p, all achieved a response to chemotherapy. In this study, loss of heterozygosity at 19q did not result in heightened chemosensitivity. However, combined loss of 1p and 19q did correlate with chemosensitivity in that each of the twenty-two patients with the combined loss responded to chemotherapy. Similar results have been reported by others. Smith *et al.* assessed tissue from 162 gliomas for 1p/19q status [19]. Tumor classification and grade were evaluated by three independent neuropathologists. The loss of both 1p and 19q was a positive predictor for longer survival in patients with oligodendroglioma. This relationship remained significant even after adjustment for tumor grade and patient age. Patients with individual losses of 1p or 19q trended towards improved survival, although this trend did not reach significance. Combined loss of 1p/19q was not a statistically significant predictor of survival in patients with mixed oligoastrocytoma in this study. Van den bent *et al.* [15], reported that of fifteen patients with oligodendroglial tumors with both 1p and 19q deletions, fourteen responded to chemotherapy, whereas only three of twelve without 1p or 19q deletions responded. All patients were treated with a standard PCV regimen. The chemosensitivity of oligodendroglial tumors with 1p/19q loss appears not to be exclusive to the PCV regimen. Nine of ten patients with loss of 1p followed by Chahlavi *et al.* [20] responded to temozolomide, while only two of six patients with an intact 1p responded to the same treatment. Overall, chemosensitivity of low-grade oligodendrogliomas may be better predicted by the loss of heterozygosity at these alleles than the presence of classical histological features [10].

Other individual genetic alterations have been reported as associated with progression in oligodendrogliomas, including EGFR (epidermal growth factor receptor) amplification, PDGFRA amplification, PTEN gene mutations, and homozygous deletion of the CDKN2a gene on chromosome 9p. To date, it appears that the most important progression-associated genetic changes which have been identified are deletions of chromosome arms 9p and 10q [9,18,17]. A recent report suggested that alterations of GLTSCR1 (or a closely linked gene) are associated with the development and progression of oligodendroglioma [21].

PROGNOSIS

As a whole, patients with anaplastic oligodendroglioma fare poorer than those with low-grade oligodendroglioma, with median survival times less than five years in the former [22–24] and potentially more than ten years in the latter [25,26]. The contribution of cellular heterogeneity and whether the percentage of cells in a tumor with 1p/19q deletions is relevant is not known. This could be particularly important in patients undergoing biopsy alone when there is only a small amount of tumor available for analysis. The utility of molecular analyses in patients with low-grade oligodendroglioma seems less certain currently than it is in anaplastic tumors. In addition to the genetic anomalies mentioned, there exist several other important prognostic factors. In a series of 81 patients with oligodendroglial tumors, Shaw *et al.* [24] reported factors which were associated with both improved and inferior survival rates. Better outcomes were associated with an age of less than twenty years, a tumor located in the parietal or frontal lobes, the existence of calcifications, the lack of enhancement on CT scan, lower grade, gross total resection, and a radiation dose of at least 5000 cGy. Poorer outcomes were associated with an age > 60 years and the occurrence of lobectomy at surgery. Puduvalli *et al.* [27], in a retrospective study of 106 patients with anaplastic oligodendroglioma, found that only younger age and better Karnofsky performance score correlated significantly with improved survival. The extent of tumor resection non-significantly trended towards improved outcome. Celli *et al.* [21] reported that of those with a low-grade tumor, total or large subtotal resection was associated with longer median survival than partial resection or biopsy alone. Additionally, patients who presented without focal neurological deficits fared better than those who presented with increased intracranial hypertension and/or focal neurological deficits. Olson *et al.* [26] followed 106 patients with either low-grade oligodendroglioma or oligoastrocytoma and found that none of the examined prognostic factors, including age, degree of resection, enhancement at time of diagnosis, and timing and choice of treatment, significantly affected ultimate survival.

ANAPLASTIC OLIGODENDROGLIOMA

In 1988, Cairncross and Macdonald [7] described eight consecutive patients with recurrent aggressive oligodendroglioma who durably responded to nitrosurea-based chemotherapy. Six of these patients were treated with PCV, a treatment regimen first described by Levin *et al.* [28]. PCV chemotherapy has subsequently been studied as first-line, second-line, adjuvant, and neoadjuvant (pre-radiation) therapy. Standard dose PCV is given at a frequency of once every eight weeks. CCNU is given by mouth at a dose of 110 mg/m^2 on day one, procarbazine is given by mouth daily on days eight through twenty-one at a dose of 60 mg/m^2/day and vincristine is given intravenously on days eight and twenty-nine at a dose of 1.4 mg/m^2 (maximum dose of 2.0 mg). Intensive PCV (IPCV) is administered once every six weeks. CCNU is given by mouth at a dose of 130 mg/m^2 on day one, procarbazine is given by mouth daily on days eight through twenty-one at a dose of 75 mg/m^2/day, and vincristine is given intravenously on days eight and twenty-nine at a dose of 1.4 mg/m^2 (without a maximum dose). Other agents, most notably temozolomide, have also been shown to be effective in treating oligodendroglioma but have not been studied as extensively.

First-line chemotherapy: Several studies have shown high response rates to PCV or IPCV when given at first recurrence, adjuvantly or neoadjuvantly. A prospective phase II multicenter trial of IPCV in 24 first relapsed patients demonstrated an overall response rate of 75 per cent, as well as a complete response rate of 38 per cent. Previous radiotherapy did not alter response to therapy. The median time to tumor progression was 6.8 months for stable patients, 14.2 months for partial responders, and at least 36.1 months for complete responders [28,29]. Myelosuppression was considerable in this trial with grade 3/4 neutropenia seen in 40 per cent of cycles and grade 3/4 thrombocytopenia seen in 15 per cent of cycles. Up to 85 per cent of all treatment cycles were complicated by at least some degree of myelosuppression. Distal numbness or paresthesias thought to be due to vincristine were present in 70 per cent of patients. Dose reductions and delays due to toxicity were frequent. Five patients had serious or adverse reactions possibly related to therapy including encephalopathy and intracranial hemorrhage. It remains uncertain whether the added toxicity of intensive PCV translates into improved efficacy.

Van den Bent *et al.* studied 43 patients with anaplastic oligodendroglioma and nine patients with anaplastic oligoastrocytoma (greater than 25 per cent oligodendroglial) [30]. All patients had received previous radiotherapy. PCV or IPCV was given at the time of first relapse. The overall response rate was 63 per cent. The median survival time for complete responders was 25 months and for partial responders was 12 months. Grade 3/4 hematologic toxicity occurred in 31 per cent of treated patients. Toxicity of IPCV tended to be more severe that that of standard PCV.

When used as adjuvant therapy in combination with surgery and radiotherapy, PCV therapy was thought to be effective against anaplastic oligodendrogliomas in a phase II study [31]. This response was durable as demonstrated by a ten-year follow-up report in which the median survival time in patients treated with this trimodality approach was 118 months. The five-year survival rate was 52 per cent, and the ten-year survival rate was 47 per cent [32].

Because of the possibility of deleterious delayed side effects of radiotherapy, the large radiation ports required for treatment in many patients, and the relatively long survival of those with oligodendroglial tumors, interest developed in using chemotherapy in a neoadjuvant role [8]. Paleologos *et al.* described thirty patients who received neoadjuvant PCV or IPCV after biopsy or tumor resection [33]. Twenty-one patients (70 per cent) demonstrated a response with twelve of the responses being complete. Twenty-five per cent of patients required early radiotherapy because of toxicity or progression. Cairncross *et al.* recently reported a large randomized trial of pre-radiation PCV plus radiotherapy *versus* radiotherapy alone in patients with anaplastic oligodendroglioma and anaplastic oligoastrocytoma [34]. Pre-radiation PCV did not impart an overall survival advantage, but progression free survival was improved in that group (2.6 years) compared to radiotherapy alone (1.9 years), but the difference seemed accounted for by patients with 1p/19q loss. Ninety-five of the 148 patients treated in the PCV arm experienced grade 3 or 4 toxicity with the dose intense regimen used, whereas radiation-induced toxicity was described as infrequent in both arms. Patients whose tumors lacked 1p and 19q lived longer regardless of treatment. While one might expect more benefit from adjuvant chemotherapy when treating patients with a chemosensitive tumor, this may not always be the case. If effective salvage treatment is available, overall survival may not be prolonged. It may be that the timing of chemotherapy treatment related to radiotherapy in this group of patients is what is of particular importance. One advantage of reserving chemotherapy until progression in patients with anaplastic oligodendroglioma who have received radiation is that response is quite clear quickly after one or two cycles (at least with PCV) allowing for discontinuation of that chemotherapy in patients who do not respond. Patients treated adjuvantly may end up receiving ineffective chemotherapy for many months thus exposing them to toxicity unnecessarily. Molecular analyses may lower the likelihood of that occurring, however whether it would be better to hold the chemotherapy in reserve in these patients is not known. Some question whether it is the radiotherapy that might be best held in reserve until time of tumor progression after chemotherapy, and whether 1p/19q status may be used to decide which patients may be effectively managed this way. Prospective trials are needed to answer those questions.

In a study of IPCV followed by high-dose thiotepa and stem cell rescue as neoadjuvant therapy for anaplastic oligodendrogliomas, nearly 75 per cent of patients who completed the intended therapy had their tumors controlled with chemotherapy alone for at least two years [35]. Twelve transplanted patients (31 per cent) ultimately relapsed. In the 39 patients who completed transplant, the estimated median progression-free survival was 69 months, and median survival time had not yet been reached. However, nearly half the patients enrolled did not complete the planned thiotepa treatment and stem cell rescue most commonly because of inadequate response to the IPCV induction regimen. Early relapse occurred in one-third of patients who did complete the planned treatment. Treatment was well tolerated as only two patients had toxicity from IPCV significant enough to preclude further therapy. This study did not incorporate molecular analyses of 1p and 19q which might better identify patients who are especially likely to benefit from such intensive chemotherapeutic approaches. A limitation of this study is the absence of a control arm, however long-term follow-up of the patients in this study may help to ascertain its significance.

There is no doubt that PCV regimens are effective and can be durable as first-line therapy. Response rates are high, ranging from 60 to 90 per cent, with complete response rates of 35–45 per cent. However, the adverse effects, particularly cumulative hematologic toxicity and gastrointestinal side effects, are significant, limiting treatment and resulting in patients frequently experiencing reduction in performance status during treatment. This is more problematic with the IPCV regimen [27,29,33].

Second- and third-line chemotherapy: Data regarding second- and third-line chemotherapy in anaplastic oligodendrogliomas is less encouraging regardless of the agent studied. The bulk of published reports describe patients treated initially with PCV or IPCV. This will change as more patients are treated with temozolomide as first-line, adjuvant, and neoadjuvant chemotherapy. While most patients initially respond to PCV, they ultimately experience progression of disease. Repeat use of PCV is limited because of cumulative myelosuppression and perhaps by acquired tumor resistance [33,36,37]. Not surprisingly, as second-line therapy, PCV may be most effective in those who have not undergone treatment with previous PCV or other nitrosurea-based chemotherapy. Peterson *et al.* reported a response rate of only 19 per cent to PCV in patients previously treated with PCV [36]. In patients treated with non-PCV therapy initially, the response rate to second-line PCV was much higher, with all seven patients responding. In this study, four of ten patients (40 per cent) treated initially with PCV responded to second-line treatment with etoposide plus cisplatin therapy. Of other agents used as second-line therapies in patients previously treated with PCV, temozolomide has been the most extensively studied, is the most useful, and will be addressed in detail later. Others agents have been disappointing.

Paclitaxel, at a dose of 175 mg/m^2 every three to four weeks, was studied in twenty patients previously treated with surgery, radiotherapy, and PCV. Only three (15 per cent) achieved a partial response, while seven had stable disease [38]. Median duration of responses and stable disease was nine months (range: 4–10 months), and all patients had progression of disease by one year. Median survival was 10 months (range: 5–14 months). Hematologic toxicity was common with 30 per cent experiencing grade 3/4 thrombocytopenia, and 15 per cent experiencing grade 3/4 neutropenia. The induction of cytochrome P450 by anticonvulsants further complicated treatment in some patients. Carboplatin has had similarly disappointing results with one phase II study demonstrating three of twenty-three patients achieving a response after second relapse [39]. Chamberlain reported a 13 per cent response rate to CPT-11 in 15 patients treated [40]. CPT-11 treatment is also complicated by increased drug metabolism due to induction of cytochrome P450 by many anticonvulsant drugs. Carboplatin and teniposide were given as third-line therapy to 23 patients after surgery, radiotherapy, PCV, and temozolomide, with two achieving a partial response [41]. Finally, in 16 patients with anaplastic oligodendroglioma or oligoastrocytoma treated with topotecan after relapse, none of 14 evaluable pati-ents achieved a response and this study was closed early [42].

Temozolomide

Temozolomide is an alkylating agent which is quickly well absorbed after oral administration, reaching peak plasma concentrations (C_{max}) one hour after administration. It is spontaneously converted to the cytotoxic mono-methyl 5 triazino imidazole carboxamide (MTIC), the active form of the drug. It does not require metabolic activation by the liver and crosses the blood–brain barrier, achieving excellent concentration within the central nervous system [43,44]. Its cytotoxicity is thought to be due to methylation of primarily DNA at the O6, N7, and N3 positions, inhibiting DNA replication. Dosage is 150–200 mg/m^2/day for 5 days per 28 day cycle. There is an absence of drug interaction between temozolomide and cytochrome-P450-inducing agents, an advantage in this population of patients, many of whom are on anticonvulsant drugs. Chemotherapy naïve patients are started at 200 mg/m^2/day. Patients previously treated with chemotherapy are started at 150 mg/m^2/day, but are usually allowed to escalate to 200 mg/m^2/day if they experience no or little toxicity. The drug is well tolerated in most patients. Yung et al. in a large pivotal study of temozolomide in recurrent anaplastic astrocytoma or anaplastic oligoastrocytoma, reported grade 3/4 thrombocytopenia in 6 per cent and grade 3/4 neutropenia in 2 per cent of patients [45]. Importantly, hematologic toxicity did not appear to be cumulative. Overall response in this trial, which included relatively few oligodendroglial tumors, was 35 per cent (14 of 111 were anaplastic oligoastrocytomas). Of the fourteen anaplastic oligoastrocytoma patients, six had partial or complete responses.

The efficacy of temozolomide in recurrent and newly diagnosed patients with malignant astrocytic tumors and in patients with mixed oligodendroglioma, as well as its high safety profile compared to PCV, has led to more focused investigations in patients with oligodendroglioma. The response rate to temozolomide when it is used as second-line chemotherapy in patients previously treated with PCV is about 20–30 per cent.

In a phase II trial, Chinot et al. treated 48 patients with recurrent anaplastic oligodendroglioma (39 patients) or anaplastic oligoastrocytoma (nine patients) with temozolomide [46]. All had been previously treated with radiation therapy and PCV chemotherapy. The overall response rate was 44 per cent (17 per cent achieving complete response and 27 per cent achieving partial response), and overall 12 month progression-free survival was 25 per cent. Responders experienced progression-free survival of 13 months and overall survival of 16 months, which was statistically improved compared to patients with stable disease or progressive disease. Of note, 83 per cent of patients in this study had previously responded to PCV, a number higher than that usually seen. In patients who had a previous complete response to PCV therapy, 48 per cent responded to temozolomide. In patients who had a partial response to previous PCV therapy, 46 per cent responded to temozolomide. A significant improvement in progression-free survival was seen with temozolomide in those patients with a previous response to PCV. Patients who failed to respond to PCV therapy also benefited from temozolomide. Patients with anaplastic oligodendroglioma had an objective response rate of 49 compared to 22 per cent for patients with anaplastic oligoastrocytoma. Their progression-free survival was 7.3 months (versus 5.6 for mixed tumors) and median overall survival was 9.9 months (versus 8.7 for mixed tumors). Hematologic toxicity was low with no patients experiencing grade 3/4 neutropenia and 6 per cent experiencing grade 3/4 thrombocytopenia. Grade 3/4 fatigue was experienced by 17 per cent.

van den Bent et al. reported a retrospective analysis of 30 patients with contrast enhancing recurrent or progressing oligodendroglioma (22 patients) or oligoastrocytoma (8 patients) treated with temozolomide [47]. All had been previously treated with radiotherapy, 27 had been previously treated with PCV, three patients were chemotherapy naïve, and three had received more than one line of chemotherapy prior to their treatment with temozolomide. The response rate in patients previously treated with chemotherapy was 26 per cent. Several of those patients had not responded to PCV. Median time to tumor progression in responders was 13 months. Grade 3/4 thrombocytopenia was seen in 13 per cent of patients and 2 patients discontinued treatment due to toxicity (hepatotoxicty and thrombocytopenia; both reversible). Costanza et al. reported a similar response rate of 22 per cent among thirty-two patients (twenty-five with pure oligodendroglioma and seven with oligoastrocytoma) treated with temozolomide as second-line therapy after initial PCV [48]. All of the responding patients had a pure oligodendroglioma. Grade 3 myelotoxicity was reported at 25 per cent.

More recently, van den Bent *et al.* reported results of a phase II study of 28 patients with confirmed recurrent oligodendroglioma (17 patients) or oligoastrocytoma (11 patients) treated with temozolomide at recurrence after first-line treatment with PCV [49]. Objective responses were achieved in 25 per cent of these patients with a median time to tumor progression for responders of 8 months. Toxicity was similar to that seen in previous studies. There was no significant difference in response rates to temozolomide between previous responders to PCV and those who had not responded.

In looking at these studies, it is clear that temozolomide is clearly effective and well tolerated as second-line chemotherapy after PCV. Both patients who have had previous responses to PCV and who were non-responders to PCV can respond to temozolomide. The 1p/19q status of patients not responding to PCV but responsive to temozolomide would be of great interest. Most studies show response rates for second-line temozolomide (after PCV) to be 22–26 per cent [47–49], with Chinot's study [46] showing a higher response rate of 44 per cent. More patients in the Chinot study had previously responded to first-line PCV (83 per cent) than in the other studies (38–50 per cent), which may account for the more favorable response rate in the Chinot study. Varying numbers of anaplastic oligoastrocytomas were included in these studies and response rates for those patients were not always separately reported, and the genetic characteristics of these tumors have not consistently been reported. In addition to higher response rates to second-line temozolomide, more patients appear to be free from tumor progression at one year after second-line temozolomide as opposed to those who had been treated with other second-line agents such as etoposide/cisplatin, paclitaxel, or carboplatin.

Less is known regarding first-line temozolomide treatment. Although its usage in that setting has become commonplace, due to ease of administration and a favorable toxicity profile, the first-line, neoadjuvant, and adjuvant response rates are still being evaluated and are not clearly documented. In one prospective study of 38 chemotherapy naïve, previously irradiated patients with recurrent enhancing oligodendroglial tumor (27 with oligodendroglioma, 6 with oligoastrocytoma, and 5 with other pathology after central review or no review), 54 per cent of the patients with confirmed oligodendroglial tumor achieved objective response (nine complete and ten partial responses). An additional eleven patients had stable disease. The median time to tumor progression for responders was 13.2 months compared to 10.4 months for all patients. Tumor-free progression

was 71 per cent at 6 months and 40 per cent at 12 months. There were similar response rates among patients with oligodendroglioma (14/27) and those with oligoastrocytoma (3/6). Three of the 5 patients with other tumors or unconfirmed pathology also responded [50]. Treatment was well tolerated. The response rate, progression-free survival, and median survival in this study were lower than those seen in most of the series evaluating PCV.

In a recent preliminary report of temozolomide as initial therapy in seventeen newly diagnosed patients with anaplastic oligodendroglioma, the response rate was 70 per cent [51]. Six patients had progressed from previously diagnosed grade II tumors but had not had prior treatment. Median time to tumor progression had not been reached at the time of the report with a median follow-up of 13.7 months (range 3–37 months).

Final data is awaited from an ongoing trial of neoadjuvant temozolomide chemotherapy for newly diagnosed anaplastic oligodendroglioma and anaplastic oligoastrocytoma. Treatment assignment is based on 1p/19q status. Patients who do not show loss of heterozygosity are treated with concurrent temozolomide and radiotherapy after initial temozolomide (while awaiting 1p/19q analysis results), followed by adjuvant temozolomide until progression. Patients with loss of heterozygosity are treated with temozolomide alone, deferring radiotherapy until time of progression. Interim analysis showed that 13 of 24 patients (54 per cent) accrued thus far have had a 50 per cent or greater decrease in gadolinium enhancement as seen on follow up MRI scans [52, personal communication].

In van den Bent *et al.*'s previously cited study of predominantly second-line treatment, three chemotherapy naïve patients were treated with temozolomide as first-line therapy and two achieved complete response [47]. Temozolomide may also be effective in the treatment of systemic metastases from oligodendroglioma [53] and in oligodendroglial gliomatosis cerebri [54].

Further studies are needed to evaluate whether the correlation between loss of heterozygosity of 1p/19q and PCV sensitivity also applies to temozolomide as well as studies to better delineate the role of temozolomide in the adjuvant, neoadjuvant, and first-line settings. There is a significant advantage to the use of temozolomide in regards to its superior toxicity profile, although it is unknown whether patients able to tolerate temozolomide for two or more years will face a higher risk for consequences secondary to prolonged myelosuppression. Furthermore, it remains uncertain whether the response rates will ultimately compare favorably to those of PCV. The response to temozolomide certainly appears to be promising thus far, and further evaluation is ongoing. Phase III studies incorporating 1p/19q status and comparing PCV with temozolomide would help to answer many of these questions. Also of interest is whether some subsets of patients may benefit from prolonged maintenance therapy with temozolomide.

LOW-GRADE OLIGODENDROGLIOMA

A number of issues complicate the management and study of low-grade oligodendrogliomas, which complicates the evaluation of the literature on these tumors. Appropriate management, including timing of treatment, especially at initial diagnosis, remains controversial. The available clinical data shows that low-grade oligodendrogliomas respond radiographically and clinically to both PCV and temozolomide regimens and that these responses can be durable. There have been no randomized studies comparing radiation to chemotherapy in these patients. The challenge of and inherent clinical subjectivity of the histopathologic diagnosis is a key issue that complicates both the literature and management of these patients. Sasaki et al. reported that only half of 44 tumors diagnosed as low-grade (WHO grade II) oligodendrogliomas by referral pathologists from sixteen different institutions were considered classic oligodendrogliomas at central review by an experienced neuropathologist [10]. The central reviewer's diagnosis of low-grade oligodendroglioma correlated closely with 1p/19q status, with 86 per cent showing 1p loss and 82 per cent showing 19q loss. Only 57 per cent of referral pathologist-diagnosed "oligodendroglioma" had 1p loss. There was a high correlation between classic oligodendroglioma appearance and loss of heterozygosity. The authors reported that in their hands the less classical-appearing tumors would have been diagnosed as astrocytomas. Whether these less classical-looking tumors represent astrocytomas, mixed tumors, or a subset of oligodendroglioma, and how one defines a low-grade oligodendroglioma, is still being debated. Some believe that genetic characteristics can define the histopathologic diagnosis, but whether this is true or not remains to be seen. Molecular analyses clearly play an important role in helping to define clinically relevant groups and may assist in individual patient management. In some cases, this might mean directing patients toward chemotherapy [17,18,19]. Possible under-grading of tumors due to sampling issues, especially in patients undergoing biopsy alone, further adds to the

complexity of the treatment decision-making process and to the critical review of the existing literature. A close review of the literature shows that some patients referred to as having a diagnosis of "low-grade oligodendroglioma" exhibit contrast enhancement on magnetic resonance imaging (MRI) studies, suggesting transformation to a higher grade in at least some portions of the tumor. Not surprisingly, these tumors behave more aggressively than do non-enhancing tumors. Other patients in published series are described as progressing low-grade tumors, but confirmation of pathology or comment on whether enhancement has or has not developed is not always made.

Finally, it is more difficult to evaluate radiographic response to chemotherapy in these non-enhancing tumors, and the usual method using extent of change in enhancement to demonstrate response does not apply. Response occurs more gradually than in anaplastic tumors and may not be seen until the completion of many cycles. Thus, actual response or progression may not be appreciated from scan to scan, given that only minor changes in fluid-attenuated inversion recovery (FLAIR) signal abnormality may exist. Only after a retrospective of serial scans is performed, can response or progression of low-grade tumor be recognized. The more indolent nature of these tumors adds to the difficulty in assessing response with years of follow-up needed to accurately measure time to tumor progression and survival. It appears clear that clinical improvement and improved seizure control occur in many patients with low-grade oligodendroglioma treated with chemotherapy, and that these clinical improvements can occur with minimal or no radiographic response.

The use of chemotherapy in low-grade oligodendroglioma has been explored for a variety of reasons, which include the lack of a clear survival benefit in patients undergoing radiotherapy and the desire to delay or avoid late radiation-induced neurotoxicity. Concern for eventual radiation-related side effects is especially relevant in patients with low-grade oligodendroglioma as they frequently have prolonged survival and may require large radiotherapy treatment ports. That being said, whether there is a substantial risk of delayed consequence from myelosuppression in patients with low-grade tumors treated with chemotherapy is not yet known. In addition to other significant toxicities, acute or cumulative myelosuppression frequently complicates the management of patients treated with PCV and can define the length of treatment. There is less known about temozolomide treatment of patients with low-grade oligodendroglioma, but its usage is becoming more common in this patient population due to its ease of administration and good tolerability. Length of treatment, response rate, and long-term safety will need to be defined for this drug, especially in this group of patients who may tolerate the drug for extended periods of time. Whether it is useful to treat small tumors with chemotherapy in order to avoid radiation is unclear. In addition, whether the toxicity and efficacy of modern radiotherapy using the limited fields needed in smaller tumors is greater or lesser than that of the toxicity and efficacy seen with chemotherapy is not known.

PCV

Paleologos *et al.* reported on five patients with low-grade oligodendroglioma, none of whom had received radiation or had required steroids [55]. The diagnosis was histologically confirmed by two independent pathologists. Three of the five had partial responses to PCV, one of whom experienced a significant decrease in seizure frequency which could only be ascribed to the chemotherapy. The remaining two patients had stable disease. Mason *et al.* reported on nine patients described as low-grade oligodendroglioma [56]. One of those patients was treated at time of recurrence after radiotherapy and had what the authors describe as "prominent enhancement" on pretreatment MRI. Two of eight patients treated at or near time of presentation also had enhancement on pretreatment MRI. Three of six patients in that series whose tumors did not enhance had objective responses to PCV as seen on MRI. Soffietti *et al.* evaluated PCV in thirteen newly diagnosed patients with low-grade, non-enhancing oligodendroglioma or oligoastrocytoma and seven patients with recurrent oligodendroglioma or oligoastrocytoma [57]. In the newly diagnosed group, three (23 per cent) achieved a partial response and ten (77 per cent) had stable disease. In the recurrent group, two (29 per cent) achieved a partial response and five (71 per cent) had stable disease. All three symptomatic patients improved (2 had partial responses and one had stable disease), and chemotherapy was well tolerated. Buckner *et al.* reported a 52 per cent radiographic response rate in evaluable "low-grade" oligodendroglioma or oligoastrocytoma, but this report is clouded by the presence of enhancement on 46 per cent of preoperative MRI scans [58]. See *et al.* reported three patients with non-enhancing symptomatic or progressing low-grade oligodendroglioma who responded to PCV [59].

In a phase II study, Soffietti *et al.* studied 26 patients with recurrent low-grade oligodendroglioma or oligoastrocytoma treated with up to six doses of PCV [37]. A total response rate of 62 per cent was reported. Three patients (12 per cent) had complete responses, 8 (31 per cent) had partial responses and 2 (8 per cent) had stable disease. It is important to note that in this report, 19 of 26 patients had enhancement on neuroimaging studies with 10 of those having had enhancement from the time of diagnosis and 9 who developed enhancement at time of progression. One question that arises in review of this series is whether these enhancing lesions were actually higher grade tumors. Of note, the response rate of enhancing tumors was markedly higher than that of non-enhancing tumors (74 per cent *versus* 29 per cent). It is unclear whether the patients without enhancement from this published report were the same ones or different ones as the ones described in the same author's 1999 abstract cited above.

Temozolomide

Several recent studies have investigated the use of temozolomide in low-grade oligodendroglial tumors. Hoang-Xuon *et al.* treated 60 patients with oligodendroglioma (49 patients) and oligoastrocytoma (11 patients) with temozolomide [60]. Fifty-nine were evaluable for response. At the time of analysis, the median number of temozolomide cycles delivered was eleven. Median follow-up was fourteen months (range 6–46 months). Of the 59 evaluable patients, 51 per cent clinically improved, particularly those with uncontrolled epilepsy. The overall objective radiologic response was 31 per cent, with ten patients (17 per cent) with a partial response and eight patients (14 per cent) with a minor or minimal response (between 25 and 50 per cent reduction in T2 signal). Sixty-one per cent had stable disease (36 patients) and 8 per cent (eight patients) had progressive disease. Seven patients (11 per cent) had contrast enhancement on MRI. The medium time to maximum tumor response was twelve months (range 5–20 months) with some responses only being apparent after up to ten cycles. Grade 3/4 hematologic toxicity was seen in 8 per cent of patients. One patient developed grade 4 thrombocytopenia after one cycle, stopped treatment, and was unevaluable for response. In all of the other patients who developed hematologic complications, treatment continued with a 25 per cent reduction in dose. Twelve of the 36 patients with stable disease improved neurologically. There was a significant association between loss of 1p and chemotherapy response in the 26 patients in whom molecular analyses were available.

Other studies of temozolomide have included a considerable number of patients with contrast-enhancing tumors among their study population of ostensibly low-grade tumors. Quinn *et al.* administered temozolomide daily for five days at 200 mg/m^2/day every twenty-eight days to patients with progressive low-grade gliomas [61]. Of the twenty patients with oligodendroglioma, 25 per cent demonstrated a complete response, and 35 per cent demonstrated a partial response. Of the entire group of 46 patients, 70 per cent demonstrated enhancement on neuroimaging studies suggesting these tumors were no longer low grade. Pace *et al.* assessed the use of temozolomide in patients with low-grade glioma at the time of radiographic and clinical progression [62]. Of the fourteen patients with oligodendroglial tumor (four with oligodendroglioma, ten with oligoastrocytoma), two had complete responses, while four had partial responses. Again, a high number of patients (60 per cent) entered into the study had contrast enhancement on MRI. Finally, Brada *et al.* treated patients with WHO grade II gliomas, eleven of which were oligodendroglioma, with temozolomide [63]. Two patients with oligodendroglioma had a partial response and three had a minimal response. The number of enhancing tumors is not directly commented upon within the study, although it is implied that the number is low.

With all histologies considered, the response rate of 31 per cent seen in the study by Hoang *et al.* is less than those seen by Quinn *et al.* and Pace *et al.* (61 and 47 per cent, respectively) [60–62]. However, the contrast enhancement was higher in the latter studies, and those studies included patients who had progressed after previous therapy (radiotherapy and/or chemotherapy), raising the possibility that they included higher-grade tumors.

In summary, it appears that both PCV and temozolomide have activity in low-grade oligodendrogliomas with temozolomide having less toxicity and greater ease of administration. Temozolomide may, therefore, be a reasonable treatment choice in these patients, especially those with 1p loss. Time to tumor response is long, averaging over many months. Consequently, length of treatment may need to be longer in this population of patients, with some patients responding at twenty months [60]. Clinical improvement, particularly in seizures, can be seen in patients treated with chemotherapy, even in the absence of objective response on MRI. Possible explanations for this include:

1. Improvement in tumor burden that does not reach the threshold of detectability on MRI;

2. Longer follow-up and longer duration of treatment may be required before objective changes are seen;

3. Careful serial review of multiple studies may be needed to appreciate small changes in FLAIR/T2 images; and

4. Changes that are less than the usual 25 per cent improvement required to be classified as a minimal response may be clinically important.

Considering the slow and delayed responses seen and the excellent tolerability of temozolomide, it also seems reasonable to recommend protracted treatment (up to 24 months) for some patients. Currently, the decisions on whether to treat a particular patient with a low-grade oligodendroglioma with chemotherapy and when to initiate that therapy is most commonly made on a case by case basis. Enrollment of these patients in clinical trials would hasten answers to many of the questions posed. It should be emphasized that the long-term effects of chemotherapy in this group of patients are still unknown with the most worrisome possibilities including myelodysplastic syndromes and myelogenous malignancies. It does appear that the data available thus far supports the need for a phase III trial comparing radiation therapy and chemotherapy with incorporation of molecular analyses and stratification for 1p loss.

References

1. Mork, S. J., Lindegaard, K. F., Halvorsen, T. B. *et al.* (1985). Oligodendroglioma: Incidence and biological behavior in a defined population. *J Neurosurg* **63**, 881–889.

2. Coons, S., Johnson, P., Scheithauer, B., Yates, A., and Pearl, D. (1997). Improving diagnostic accuracy and interobserver concordance in the classification and grading of primary gliomas. *Cancer* **79**, 1381–1393.

3. Bailey, P., and Bucy, P. C. (1929). Oligodendrogliomas of the brain. *J Pathol* **32**, 735–751.

4. Bailey, P., and Cushing, H. (1926). *Tumors of the Glioma Group.* Philadelphia, Lippincott.

5. Bailey, P., and Hiller, G. (1924). The interstitial tissues of central nervous system: A review. *J Nerv Ment Dis* 337–361.

6. van den Bent, M., Chinot, O. L., and Cairncross, J. G. (2003a). Recent developments in the molecular characterization and treatment of oligodendroglial tumors. *Neuro-oncol* **5**, 128–138.

7. Cairncross, J. G., and Macdonald, D. R. (1988). Successful chemotherapy for recurrent malignant oligodendroglioma. *Ann Neurol* **23**, 360–364.

8. Macdonald, D. R., Gaspar, L. E., and Cairncross, J. G. (1990). Successful chemotherapy for newly diagnosed aggressive oligodendroglioma. *Ann Neurol* **27**, 573–574.

9. Reifenberger, G., and Louis, D. N. (2003). Oligodendroglioma: Toward molecular definitions in diagnostic neuro-oncology. *J Neuropathol Exp Neurol* **62(2)**, 111–126.

10. Sasaki, H., Zlatescu, M., Betensky, R. *et al.* (2002). Histopathological-molecular genetic correlations in referral pathologist-diagnosed low-grade "oligodendroglioma." *J Neuropathol Exp Neurol* **61(1)**, 58–63.

11. Smith, J., Alderete, B., Minn, Y. *et al.* (1999). Localization of common deletion regions on 1p and 19q in human gliomas and their association with histological subtype. *Oncogene* **18**, 4144–4152.

12. Bigner, S., Matthews, M., Rasheed, B. K. A. *et al.* (1999). Molecular genetic aspects of oligodendrogliomas including analysis by comparative genomic hybridization. *Am J Pathol* **155(2)**, 275–386.

13. Watanabe, T., Nakamura, M., Kros, J. *et al.* (2002). Phenotype versus genotype correlation in oligodendrogliomas and low-grade diffuse astrocytomas. *Acta Neuropathol* **103**, 267–275.

14. Zlatescu, M., TehraniYazdi, A., Sasaki, H. *et al.* (2001). Tumor location and growth pattern correlate with genetic signature in oligodendroglial neoplasms. *Cancer Res* **61**, 6713–6715.

15. van den Bent, M. J., Looijenga, L. H., Langenberg, K. *et al.* (2003b). Chromosomal anomalies in oligodendroglial tumors are correlated with clinical features. *Cancer* **97(5)**, 1276–1284.

16. Felsberg, J., Erkwoh, A., Sabel, M. C. *et al.* (2004). Oligodendroglial tumors: refinement of candidate regions on chromosome arm 1p and correlation of 1p/19q status with survival. *Brain Pathol* **14(2)**, 121–130.

17. Cairncross, J. G., Ueki, K., Zlatescu, M. *et al.* (1998). Specific genetic predictors of chemotherapeutic response and survival in patients with anaplastic oligodendrogliomas. *J Natl Cancer Inst* **90**, 1473–1479.

18. Ino, Y., Betensky, R. A., Zlatescu, M. *et al.* (2001). Molecular subtypes of anaplastic oligodendroglioma: Implications for patient management at diagnosis. *Clin Cancer Res* **7**, 839–845.

19. Smith, J. S., Perry, A., Borell, T. J. *et al.* (2000). Alterations of chromosome arms 1p and 19q as predictors of survival in oligodendrogliomas, astrocytomas, and mixed oligoastrocytomas. *J Clin Oncol* **18**, 636–645.

20. Chahlavi, A., Kanner, A., Peereboom, D., Staugaitis, S. M., Elson, P., and Barnett, G. (2003). Impact of chromosome 1p status in response of oligodendroglioma to temozolomide: preliminary results. *J Neuro-oncol* **61(3)**, 267–273.

21. Kollmeyer, T. M., Yang, P., Buckner, K., Bamlet, W., Ballman, K. V., and Jenkins, R. B. (2004). A polymorphism in GLTSCR1 is associated with the development of oligodendrogliomas. *Neuro-oncol* **6(4)**, 327. (Abstract)

22. Celli, P., Nofrone, I., Palma, L., Cantore, G., and Fortuna, A. (1994). Cerebral oligodendroglioma: Prognostic factors and life history. *Neurosurgery* **35**, 1018–1034.

23. Smith, M. T., Ludwig, C. L., Godfrey, A. D., and Armbrustmacher, V. W. (1983). Grading of oligodendrogliomas. *Cancer* **52**, 2107–2114.

24. Shaw, E. G., Scheithauer, B. W., O'Fallon, J. R., Tazelaar, H. D., and David, D. H. (1992). Oligodendrogliomas: The Mayo Clinic experience. *J Neurosurg* **76**, 428–434.

25. Leighton, C., Fisher, B., Bauman, G. *et al.* (1997). Supratentorial low-grade glioma in adults: An analysis of prognostic factors and timing of radiation. *J Clin Oncol* **15**, 1294–1301.

26. Olson, J. D., Riedel, E., DeAngelis, L. M. (2000). Long-term outcome of low-grade oligodendroglioma and mixed gliomas. *Neurology* **54(7)**, 1442–1448.

27. Puduvalli, V. K., Hashmi, M., McAllister, L. D. *et al.* (2003). Anaplastic oligodendroliomas: Prognostic factors for tumor recurrence and survival. *Oncology* **65**, 259–266.

28. Cairncross, G., MacDonald, D., Ludwin, S. *et al.* (1994). Chemotherapy for anaplastic oligodendroglioma. National Cancer Institute of Canada Clinical Trials Group. *J Clin Oncol* **12**, 2013–2021.

29. Cairncross, J. G., and Eisenhauer, E. A. (1995). Response and control: Lessons from oligodendroglioma. *J Clin Oncol* **13**, 2475–2476. (Letter)

30. Van den Bent, M. J., Kros, J. M., Heimans, J. J. *et al.* (1998). Response rate and prognostic factors of recurrent oligodendroglioma treated with procarbazine, CCNU, and vincristine chemotherapy. Dutch Neuro-oncology Group. *Neurology* **51**, 1140–1145.

31. Jeremic, B., Shibamoto, Y., Grujicic, D. *et al.* (1999). Combined treatment modality for anaplastic oligodendroglioma: A phase II study. *J Neurooncol* **43**, 179–185.

32. Jeremic, B., Milicic, B., Grujicic, D. *et al.* (2004). Combined treatment modality for anaplastic oligodendroglioma and oligoastrocytoma: a 10-year update of a phase II study. *J Radiat Oncol Biol Phys* **59(2)**, 509–514.

33. Paleologos, N. A., Macdonald, D. R., Vick, N. A., and Cairncross, J. G. (1999a). Neoadjuvant procarbazapine, CCNU, and vincristine for anaplastic and aggressive oligodendroglioma. *Neurology* **53**, 1141–1143.

34. Cairncross, G., Seiferheld, W., Shaw, E. *et al.* (2004). An intergroup randomized controlled clinical trial (RCT) of chemotherapy plus radiotherapy (RT) versus RT alone for pure and mixed anaplastic oligodendroglioma: Initial report of RTOG 94–02. *Neuro-oncol* **6(4)**, 371. (Abstract)

35. Abrey, L. E., Childs, B. H., Paleologos, N. *et al.* (2003). High-dose chemotherapy with stem cell rescue as initial therapy for anaplastic oligodendroglioma. *J Neurooncol* **65(2)**, 127–134.

36. Peterson, K., Paleologos, N., Forsyth, P., Macdonald, D. R., and Cairncross, J. G. (1996). Salvage chemotherapy for oligodendroglioma. *J Neurosurg* **85**, 597–601.

37. Soffietti, R., Ruda, R., Bradac, G. B., and Schiffer, D. (1998). PCV chemotherapy for recurrent oligodendrogliomas and oligoastrocytomas. *Neurosurgery* **43**, 1066–1073.

38. Chamberlain, M. C., and Kormanik, P. A. (1997). Salvage chemotherapy with paclitaxel for recurrent oligodendrogliomas. *J Clin Oncol* **15**, 3427–3432.

39. Soffietti, R., Nobile, M., Ruda, R. *et al.* (2004). Second-line treatment with carboplatin for recurrent or progressive oligodendroglial tumors after PCV (procarbazine, lomustine, and vincristine) chemotherapy: a phase II study. *Cancer* **100(4)**, 807–813.

40. Chamberlain, M. C. (2001). A phase 1–2 trial of CPT-11 (Camptosar) in the treatment of recurrent intracranial oligodendrogliomas [oligos]. *Neuro-oncol* **3**, 356. (Abstract)

41. Brandes, A. A., Basso, U., Vastola, F. *et al.* (2003). Carboplatin and teniposide as third-line chemotherapy in patients with recurrent oligodendroglioma or oligoastrocytoma: a phase II study. *Ann Oncol* **14(12)**, 1727–1731.

42. Belanger, K., MacDonald, D., and Cairncross, G. (2003). A phase II study of topotecan in patients with anaplastic oligodendroglioma or anaplastic mixed oligoastrocytoma. *Invest New Drugs* **21(4)**, 473–480.

43. Marzolini, C., Decosterd, L. A., Shen, F. *et al.* (1998). Pharmocokinetics of temozolomide in association with fotemustine in malignant melanoma and malignant glioma patients: comparison of oral, intravenous, and hepatic intra-arterial administration. *Cancer Chemother Pharmacol* **42(6)**, 433–440.

44. Stupp, R., Ostermann, S., Leyvraz, S., Csajka, C., Buclin, T., and Decosterd, L. (2001). Cerebrospinal fluid levels of temozolomide as a surrogate marker for brain penetration. *Proc Am Soc Clin Oncol* **20**, 59. (Abstract)

45. Yung, W. K. A., Prados, M. D., Yaya-Tur, R. *et al.* (1999). Multicenter phase II trial of temozolomide in patients with anaplastic astrocytoma or oligoastrocytoma at first relapse. *J Clin Oncol* **17(9)**, 2762–2771.

46. Chinot, O. L., Honore, S., Dufour, H. *et al.* (2001). Safety and efficacy of temozolomide in patients with recurrent anaplastic oligodendrogliomas after standard radiotherapy and chemotherapy. *J Clin Oncol* **19**, 2449–2455.

47. van den Bent, M. J., Keime-Guibert, F., Brandes, A. A. *et al.* (2001). Temozolomide chemotherapy in recurrent oligodendroglioma. *Neurology* **57**, 340–342.

48. Costanza, A., Borgognone, M., Nobile, M., Ruda, R., Mutani, R., and Soffietti, R. (2001). Temozolomide in recurrent oligodendroglial tumors: A phase II study. *Neuro-oncol* **3 (Suppl. 1)**, 66. (Abstract)

49. van den Bent, M. J., Chinot, O., Boogerd, W. *et al.* (2003b). Second-line chemotherapy with temozolomide in recurrent oligodendroglioma after PCV (procarbazine, lomustine and vincristine) chemotherapy: EORTC Brain Tumor Group phase II study 26972. *Ann Oncol* **14(4)**, 599–602.

50. van den Bent, M. J., Taphoorn, M. J., Brandes, A. A. *et al.* (2003c). European Organization for Research and Treatment of Cancer Brain Tumor Group. Phase II study of first-line chemotherapy with temozolomide in recurrent oligodendroglial tumors: the European Organization for Research and Treatment of Cancer Brain Tumor Group Study 26971. *J Clin Oncol* **21(13)**, 2525–2528.

51. Taliansky, A., Bokstein, F., Lavon, I. *et al.* (2004). Temozolomide (TMZ) as initial treatment for newly diagnosed anaplastic oligodendroglioma: Preliminary results. *Neuro-oncol* **6(4)**, 383. (Abstract)

52. Mikkelsen, T., Doyle, T., Margolis, J. *et al.* (2004). Neoadjuvant temozolomide single-agent chemotherapy for newly diagnosed anaplastic oligodendroglioma and anaplastic oligoastrocytoma with or without radiation therapy. *Neuro-oncol* **6(4)**, 378. (Abstract)

53. Morrison, T., Bilbao, J. M., Yang, G., and Perry, J. R. (2004). Bony metasteses of anaplastic oligodendroglioma respond to temozolomide. *Can J Neurol Sci* **31**, 102–108.

54. Levin, N., Gomori, J. M., Siegal, T. (2003). Chemotherapy as initial treatment in gliomatosis cerebri: results with temozolomide. *Neurology* **63(2)**, 354–6.

55. Paleologos, N. A., Vick, N. A., Kachoris, J. P. (1994). Chemotherapy for low-grade oligodendrogliomas? *Ann Neurol* **36**, 294–295. (Abstract)

56. Mason, W. P., Krol, G. S., DeAngelis, L. M. (1996). Low-grade oligodendroglioma responds to chemotherapy. *Neurology* **46**, 203–207.

57. Soffietti, R., Ruba, R., Borgognone, M., and Schiffer, D. (1999). Chemotherapy with PCV for low grade nonenhancing oligodendrogliomas and oligoastrocytomas. *Neurology* **52(6): Supplement 2**, A423–A424.

58. Buckner, J. C., Gesme, D., Jr, O'Fallon, J. R. *et al.* (2003). Phase II trial of procarbazine, lomustine, and vincristine as initial therapy for patients with low-grade oligodendroglioma or oligoastrocytoma: efficacy and associations with chromosomal abnormalities. *J Clin Oncol* **21(2)**, 251–255.

59. See, S. J., Ty, A., and Wong, M. C. (2004). Low-grade oligodendroglioma and PCV chemotherapy: Pictures tell a tale. *Neuro-oncol* **6(4)**, 382. (Abstract)

60. Hoang-Xuan, K., Capelle, L., Kujas, M. *et al.* (2004). Temozolomide as initial treatment for adults with low-grade oligodendrogliomas or oligoastrocytomas and correlation with chromosome 1p deletions. *J Clin Oncol* **22**, 3133–3138.

61. Quinn, J. A., Reardon, D. A., Friedman, A. H. *et al.* (2003). Phase II trial of temozolomide in patients with progressive low-grade glioma. *J Clin Oncol* **21(4)**, 646–651.

62. Pace, A., Vidiri, A., Galie, E. *et al.* (2003). Temozolomide chemotherapy for progressive low-grade glioma: clinical benefits and radiological response. *Ann Oncol* **14**, 1722–1726.

63. Brada, M., Viviers, L., Abson, C. *et al.* (2003). Phase II study of primary temozolomide chemotherapy in patients with WHO grade II gliomas. *Ann Oncol* **14**, 1715–1721.

27

Chemotherapy of Oligoastrocytomas

Allison L. Weathers and Nina A. Paleologos

INTRODUCTION

Although Bailey and Cushing stated in 1926 "that gliomas are rarely found to be composed of a single type of cell should be no cause for surprise", the concept of oligoastrocytoma (OA) as a distinct tumor type remains controversial despite Hart's series of one hundred and two cases published in 1974 [1–3]. Oligoastrocytomas are tumors consisting of areas of cells with the morphologic characteristics of oligodendroglioma intermingled with areas where the morphology of the cells is astrocytic. In most, the cells are diffusely mixed. Tumors with separate areas of oligodendroglial and astrocytic cells are not common. Oligoastrocytomas have an incidence between 2 and 10 per cent of all gliomas [2,4,5]. In a population-based study of low-grade diffuse gliomas, Okamoto *et al.* determined that the incidence rate of low-grade oligoastrocytomas, adjusted to the World Standard Population, was 0.89 per million population per year [6].

There is considerable subjectivity in diagnosing these tumors by histologic criteria with significant inter-observer variability. The absence of standardized histologic diagnostic criteria for these tumors is in part responsible for the wide incidence range quoted in the literature [4]. In addition to inter-observer variability, there are many other factors which interfere with the reliability of histopathologic diagnoses. Two of the most significant may be sampling, especially in cases with significant intratumoral variation in tumor histology, and the degree of morphologic oligodendroglioma cells required for the designation. Though attempts have been made to quantify the percentage of cells with oligodendroglial morphology required for the determination of mixed histology, no specific guidelines exist as to what percentage of the second glial component needs to be present to make this diagnosis with the amount varying greatly across definitions [7–9]. This adds to the significant difficulty in critical review of the literature, which also is due in part to differences in inclusion criteria for the tumors studied, frequent inclusion of heterogeneous tumor populations, small numbers of subjects, and retrospective analyses of the data.

Oligoastrocytomas share many clinical characteristics, such as age and sex distribution, location, clinical presentation, and imaging features with oligodendrogliomas, and demonstrate behavior intermediate to that of astrocytomas and oligodendrogliomas in regards to survival and response rate to chemotherapy [2,4,8,10]. The median age at diagnosis of low-grade oligoastrocytoma is less than 40 years of age, there is a slight male predominance, the majority of patients present with seizures, and these tumors most frequently occur in the frontal lobe [2,3,6,9]. The median age at diagnosis is slightly higher for high-grade oligoastrocytoma, however other clinical characteristics are similar [7–9]. Hemiparesis and hemiplegia are the most common presenting signs on neurological examination [8]. A median Karnofsky Performance Score (KPS) at diagnosis of 90 per cent was reported in a study of 52 cases of oligoastrocytomas [8]. The presence of calcifications on computed tomography (CT) or in histopatholgoical specimens has been reported to be 14 to 60 per cent [2,3,8]. Contrast enhancement on pre-treatment CT has been reported to be as low as 52 per cent in one study and as high as 70 per cent in another, and when present is associated with a poorer prognosis [2,8].

At present, there seems to be two main categories of oligoastrocytoma. Those that have 1p and 19q deletions and those with TP53 mutations. The former, typical of pure oligodendroglioma, suggests oligodendroglial lineage. TP53 mutations are typical of astrocytomas, suggesting that those oligoastrocytomas may be derived from astrocytic precursor cells [11].

The controversy therefore continues; whether these tumors are a distinct entity or not remains a topic of discussion. Some wonder if the tumors with oligodendroglial-like molecular characteristics are in fact oligodendrogliomas and if the ones with TP53 mutations are astrocytomas [12]. Can there be two molecular subtypes of a single histologically described entity [12]? With time and further study, molecular analyses may lead to a better understanding of these tumors.

Another chapter (Chapter 26) in this book provides a review of the chemotherapeutic treatment of oligodendrogliomas. This chapter focuses on studies in which there has been an attempt to analyze oligoastrocytomas separately. Many studies which may have included oligoastrocytomas, but which did not evaluate this subgroup separately from pure oligodendrogliomas, are not included here and are discussed in the chapter on oligodendrogliomas. As with oligodendrogliomas, many treatment modalities are useful and may play an even more significant role in the treatment of oligoastrocytomas, however, this chapter will focus on chemotherapy.

GENETICS

The appreciation that oligodendrogliomas and oligoastrocytomas are chemosensitive tumors has led to a low threshold for diagnosing a tumor as oligodendroglial because of the difference in prognosis and implications for treatment [7,12–16]. In addition to the previously discussed lack of consensus on their defining histologic criteria, the histopathological diagnosis of oligoastrocytomas is further complicated by the absence of a reliable immuno histochemical marker [13,17]. As with oligodendrogliomas much effort has been put forth to determine objective methods for the diagnosis of these tumors.

Oligoastrocytomas are a genetically heterogeneous population, for the most part having the genetic alterations characteristic of either pure oligodendrogliomas or pure astrocytomas [11,17]. In the majority of cases the two glial components within a tumor share the same alterations, suggesting that these are monoclonal tumors arising from either an oligodendroglial lineage or an astrocytic lineage [10,17]. While oligodendrogliomas are frequently characterized by the deletion of chromosomes 1p and 19q, with losses of both the alleles in 80 to 90 per cent of cases, these deletions occur at a frequency of 26 to 50 per cent in oligoastrocytomas [6,17–20]. This frequency is intermediate between that of the pure glial types [20]. One-third of oligoastrocytomas will have intact 1p/19q but loss of heterozygosity on 17p and/or TP53 mutations [17]. Those alterations are usually associated with astrocytomas and only rarely are found in oligodendrogliomas [6,17]. An association exists between allelic loss of 19q and all glial tumors, but this is not the case for loss of 1p [18]. Oligoastrocytomas with a higher component of oligodendrocytes and those located outside the temporal lobe have been found to have a higher frequency of combined losses of 1p and 19q [17,18,21]. The presence of small areas of cells with oligodendroglial features in low-grade diffuse astrocytomas is not predictive of these alterations [22]. No genetic abnormalities clearly unique to oligoastrocytomas have yet been found.

Anaplastic oligoastrocytomas show progression associated genetic alterations similar to those seen in other malignant gliomas [17,18]. These changes include the loss of 9p, especially associated with the homozygous deletion of CDKN2A, losses of 10q (PTEN mutations), 11p, 13q, 17p, and the amplification of the CDK4 gene and of proto-oncogenes, such as the epidermal growth factor receptor precursor or the platelet-derived growth factor-alpha receptor precursor [17,18]. The progression-related alterations that occur have been shown to be related to the genetic aberrations of the low-grade oligoastrocytoma. Tumors with loss of 1p and 19q and those with TP53 mutations demonstrate changes seen in the progression of oligodendrogliomas and astrocytomas, respectively [18].

While valuable knowledge of the molecular characteristics of oligoastrocytomas has been gained, it must be realized that the lack of a universal definition of what constitutes an oligoastroctyoma extends to the methodology of the studies referenced here. These studies are additionally limited by sampling error and, therefore, the possible underestimation of the diagnosis of oligoastrocytoma that can occur if stereotactic biopsy is the method utilized to obtain tissue.

PROGNOSIS

The prognosis of mixed oligoastrocytomas, when not considering molecular characteristics, falls in between those of pure oligodendrogliomas and pure astrocytomas [2]. Many times these tumors are grouped together with the pure oligodendrogliomas, adding to the difficulty in interpreting the literature and in making specific comments about their epidemiology, prognosis, and response to treatment. The prognosis of these tumors being intermediate to that of pure tumors may be explained by their division into the two main molecular groups described above. The median survival time (MST) of low-grade oligoastrocytomas has been reported to be 6–7 years [2,6,23]. Although this value seems to be between that of astrocytomas and oligodendrogliomas, multiple studies have reported that the median difference in survival between oligoastrocytomas and oligodendrogliomas is not statistically significant [6,23–26].

In a survival analysis of 71 patients felt to have oligoastrocytomas based on a review of their histological material, Shaw et al. reported a median survival time of 2.8 years for those with anaplastic tumors [2]. The patients with low-grade oligoastrocytoma had a median survival time of 7.1 years. Lower grade, gross total resection, partial brain radiation, radiation dose of at least 5000 cGy, and age less than 37 years were all factors significantly associated with improved survival. In that study tumor grade was the factor with the strongest association with prolonged survival [2]. Variables of oligodendroglial tumors that have been shown to contribute toward a poorer prognosis include increased tumor grade, anaplastic histologic features (number of mitoses, Ki-67 labeling index, endothelial proliferation, nuclear pleomorphism, and necrosis), presence of enhancement on neuroimaging studies, older patient age, and poor performance status [27,28].

In their seminal review of 102 cases of gliomas (including 86 oligoastrocytomas) published in 1974, Hart et al. reported that survival was not significantly altered by the proportion of the individual glial components [3]. This finding has since been confirmed by studies in each of the subsequent decades [2,25,29]. In one study, a minimally better outcome was reported for patients whose tumors contained 30 per cent or more neoplastic oligodendrocytes, but a higher percentage did not further improve prognosis and furthermore, the percentage of neoplastic astrocytes was not correlated to survival [29]. This lack of correlation between histological predominance and prognosis may now be more understandable in light of the molecular data thus far which suggests that oligoastrocytomas are monoclonal tumors, whose genetic alterations have been found to be significant predictors of prognosis and are not always predicted by which glial component is in the majority [17].

For oligodendroglial tumors it appears that the combined loss of 1p and 19q may be even more important in predicting prognosis than either clinical or histological characteristics [17,30]. The presence of 1p loss in oligodendroglial tumors has been associated with long progression-free survival (PFS) in many but not all studies [6,19,20,30–33]. Whereas the increased survival time and responsiveness to chemotherapy of oligodendrogliomas with combined losses of 1p and 19q was initially reported by Cairncross et al. in 1998 and confirmed by the same group in 2001, oligoastrocytomas were not included in these landmark studies [30,33]. Smith et al. assessed 162 surgically resected diffuse gliomas for 1p/19q status [20]. Tumors were evaluated by three independent neuropathologists, and 19 tumors were classified as primary oligoastrocytomas, with 12 tumors classified as recurrent. While the combined loss of 1p and 19q was a statistically significant positive predictor for longer survival in patients with oligodendrogliomas, this was not found to be true for oligoastrocytomas. In a study of 34 oligoastrocytomas (grade II and grade III), Bissola et al. found that the loss of 1p was a statistically significant predictor of longer progression-free and overall survival [19]. In this study, loss of 10q was significantly associated with shorter survival. The authors of this study postulated that the discrepancy in the results of their study and the Smith et al. study may be due to a difference in the number of subjects and to the heterogeneity of the histological diagnosis [19]. Okamoto et al. found that losses on 1p/19q had no statistically significant predictive value for survival in any of the histologic subtypes in a study of 122 low-grade diffuse gliomas [6]. In this study, TP53 mutations were predictive of shorter survival only in patients with oligoastrocytoma. TP53 mutation status was also reported as an unfavorable prognostic factor by Ständer et al. in a study of 159 patients with low-grade astrocytoma or oligoastrocytoma [34]. The applicability of these results to oligoastrocytomas may be limited by the small number of patients (24) with these tumors in this study. Cairncross et al. recently reported a study of pre-radiation chemotherapy versus radiation alone in the treatment of anaplastic oligodendrogliomas (70 per cent) and anaplastic oligoastrocytomas (30 per cent) which showed that patients whose tumors had combined loss of 1p and 19q had a longer survival time than those whose tumors did not have 1p/19q loss [35]. It appears that mixed tumors with genetic lesions more characteristic of tumors of astrocytic lineage are less sensitive to treatment and have a poorer prognosis than those with the genetic characteristics more typical of oligodendroglial tumors. Some have shown that molecular analysis of nonclassical tumors is superior to histopathologic evaluation in predicting prognosis [36].

Molecular analyses may ultimately prove to be more helpful than histology in guiding clinical decision making, and assessment of 1p/19q status is becoming a standard diagnostic tool [17]. However much still needs to be done, including prospective clinical trials with stratification for genotype, before we can confidently know what the treatment implications of molecular analysis are. Identification of the critical genes and their products will shed even more light on these issues and hopefully lead to novel and more effective therapies.

ANAPLASTIC OLIGOASTROCYTOMAS

PCV

The use of procarbazine, lomustine [1,(2-chloro-ethyl)-3-cyclohexyl-1-nitrosurea, CCNU], and vincristine (PCV) was first described by Levin *et al.* and has been since studied as first-line, second-line, adjuvant, and neoadjuvant (pre-radiation) therapy in the treatment of malignant gliomas [37]. The landmark papers of Cairncross and Macdonald describing the unique chemosensitivity of oligodendrogliomas, most treated with PCV, did not include oligoastrocytomas [14,15]. These and other reports prompted the study of this regimen in oligoastrocytomas. Standard-dose PCV is administered once every eight weeks. CCNU is given orally at a dose of 110 mg/m^2 on day one, procarbazine is given orally daily on days eight through twenty-one at a dose of 60 mg/m^2 per day and vincristine is given intravenously on days eight and twenty-nine at a dose of 1.4 mg/m^2 with a maximum dose of 2.0 mg. Intensive PCV (IPCV) is administered once every six weeks. CCNU is given orally at a dose of 130 mg/m^2 on day one, procarbazine is given orally daily on days eight through twenty-one at a dose of 75 mg/m^2 per day, and vincristine is given intraveneously on days eight and twenty-nine at a dose of 1.4 mg/m^2 (with no maximum dose).

The first extensive study of the treatment of oligoastrocytomas with PCV was conducted by Glass *et al.* [16]. In this prospective study of 21 patients with oligodendroglial tumors 14 had oligoastrocytomas (1 low-grade oligoastrocytoma, 13 with at least one anaplastic feature). Ten of these patients received PCV as neoadjuvant (pre-radiation) therapy; at least three cycles were given. The objective response rate was 90 per cent. One patient had a complete response (CR) and eight patients had a partial response (PR). Partial responses were seen in two of the four patients with OA who received PCV

after radiation therapy (RT). Leukopenia occurred in 36 per cent of the cycles (including 2 cases of grade 3) and thrombocytopenia occurred in 24 per cent of cycles (including 2 cases of grade 3 and 1 case of grade 4). Five patients experienced distal paresthesias. The individual composition of the tumors (mix of glial subtypes) and the degree of malignancy were not found to be significant factors related to response.

In a retrospective review, Kyritsis *et al.* evaluated the response of 34 patients with anaplastic oligodendroglial tumors to chemotherapy [38]. Of these 34, 17 patients had oligoastrocytomas, 12 of whom were treated at the time of diagnosis. All 12 patients were treated with PCV and radiation therapy and received at least 2 cycles of PCV. Two of the twelve patients received the PCV prior to radiation therapy (neoadjuvantly). The overall response rate of this group of 12 patients was 41.6 per cent. Two patients had complete response, three patients had partial response and the remaining seven patients had stable disease (SD). The survival of the anaplastic oligoastrocytoma group treated at the time of diagnosis did not differ significantly from that of the pure anaplastic oligodendroglioma group. Of the five anaplastic oligoastrocytoma patients who received chemotherapy at the time of recurrence, only one was treated with PCV; this patient had a partial response. The other four patients were treated with a variety of agents, and none were reported as having a response, but all were reported as having stable disease. Survival of the recurrent oligoastrocytoma group was inferior to that of the recurrent oligodendroglioma group. Of note, the authors of this study did not use the Macdonald criteria for response, with partial response defined as "definite decrease in the size of the enhancing or nonenhancing tumor" [38]. The details of toxicity were not provided, however it was documented that chemotherapy toxicity occurred in three patients.

Kim *et al.* treated 32 patients with adjuvant or neoadjuvant PCV, 25 of these patients had anaplastic oligoastrocytoma (AOA) (19 grade III, 6 grade IV) [7]. Fourteen patients in the OA group received pre-radiotherapy PCV. The overall objective response rate of the entire study group was 91 per cent. The objective response rate of the oligoastrocytoma group was 89 per cent. The grade IV OA group had a median time to tumor progression (TTP) of 12.4 months and the grade III OA group had a median TTP of 13.8 months, compared to 63.4 months for the grade III oligodendroglioma group. Median survival of the grade III oligoastrocytoma group (49.8 months) was significantly less than the grade III pure oligodendroglioma group (76 months). There was a trend to improved TTP and survival with increasing percentage of the oligodendroglioma component. Within the oligoastrocytoma group the percentage of the oligodendroglioma component did not significantly affect response rate. Whether chemotherapy was administered neoadjuvantly or adjuvantly was not a statistically significant factor in TTP. Grade 3/4 hematologic toxicity occurred in 7 per cent of the 124 cycles of chemotherapy administered. No patient experienced hepatic or renal toxicity. Four patients discontinued PCV treatment due to toxicity.

Nine of the twenty-seven patients with oligodendroglial tumors treated by Streffer et al. with PCV either prior to radiation therapy, as adjuvant therapy with irradiation, or at the time of recurrence after radiotherapy had oligoastrocytoma (2 OA, 7 AOA) [39]. In the adjuvant group there were three complete responses and two patients with stable disease. The neoadjuvant group had a response rate of 45 per cent (2 CR, 3 PR) with the rest of the patients with stable disease. Response rates were no different between oligoastrocytomas and oligodendrogliomas.

As with patients with oligodendroglioma, the relatively long survival of patients with oligoastrocytoma and concern about delayed radiation-induced neurotoxicity has led to increased utilization of chemotherapy as neoadjuvant therapy, delaying or deferring radiation therapy. In the preceding studies, the neoadjuvant response rate to PCV ranged from 45 to 94.7 per cent [7,16,38,39]. The number of patients treated are small and there are no randomized studies comparing this approach (PCV alone) to PCV and radiation therapy. Cairncross et al. conducted a randomized controlled clinical trial comparing radiation therapy alone versus pre-radiation PCV plus radiotherapy [35]. In this study, 291 eligible patients with anaplastic oligodendroglial tumors were randomized; 30 per cent had mixed anaplastic oligoastrocytomas. A survival advantage was not obtained with pre-radiation PCV. The improvement in progression-free survival (PFS) of the patients who received both PCV and radiation therapy over those who received radiation therapy alone (2.6 versus 1.9 years) was of borderline significance. The patients whose tumors had deletions of 1p and 19q had longer survival times regardless of the treatment received, and these patients may account for the difference in PFS between the treatment groups. Grade 3 or 4 toxicity occurred in 64 per cent of the pre-radiation PCV-treated group. Radiation induced toxicities were reported as infrequent in both groups. As of yet the data of the individual histological subtypes (oligoastrocytomas versus oligodendrogliomas) has not been reported.

Abrey et al. enrolled 69 patients with AO and AOA in a study of neoadjuvant IPCV with stem cell rescue [40]. Sixteen of the initial 69 patients and 8 of the 39 transplanted patients had AOA. Histology (AOA versus AO) did not correlate with relapse. The two-year disease-free survival of the 39 transplanted patients was 78 per cent. The estimated median PFS was 69 months and median survival time had not yet been reached at the time of publication. Twelve of the transplanted patients (31 per cent) eventually relapsed. IPCV and transplantation was well tolerated (only two patients were unable to be transplanted due to toxicity from IPCV). Twenty-seven per cent of the patients whose response to IPCV induction was not adequate enough to receive transplant had oligoastrocytomas.

In a phase II study of PCV as adjuvant therapy in combination with surgery and radiotherapy for the treatment of anaplastic oligodendroglial tumors, Jeremic et al. reported an objective response rate of 83 per cent [41]. The 5 patients with oligoastrocytoma had a decreased 5-year survival compared to the 18 patients with pure oligodendroglioma (40 versus 56 per cent); however this difference was not statistically significant. The median survival time and median TTP had not yet been reached at the time the study was reported. In a ten-year follow-up report, response was shown to be durable and differences between the histologic groups were more evident with a median survival time of 35 months for the oligoastrocytoma group and 124 months for the oligodendroglioma group [42]. The 10-year survival rates were 0 and 56 per cent respectively. These differences almost reached statistical significance.

van den Bent et al. and Brandes et al. evaluated the use of PCV at the time of tumor recurrence after treatment with surgery and radiotherapy [43,44]. In the retrospective van den Bent study, 52 patients treated with PCV or IPCV for recurrent oligodendroglioma or oligoastrocytoma were evaluated [43]. Nine patients had oligoastrocytoma. A response rate of 63 per cent was observed for the entire group, median TTP was 10 months, and median overall survival was 20 months. Relapse within one year of initial treatment was negatively correlated with response rate, five of the twelve patients who relapsed early had oligoastrocytoma. The objective response rate of the oligoastrocytoma group was 33 per cent (three of the nine patients). Two of the four patients with oligoastrocytoma who had an initial disease-free interval of more than one year responded. Toxicity was common with 16 of the 52 patients studied (31 per cent) experiencing a grade 3 or 4 hematologic toxicity. Treatment was discontinued in seven due to toxicity (four hematologic, one intractable nausea and two hepatotoxicity). Toxicity was greater with the IPCV regimen and there seemed to be no benefit of this regimen over the standard PCV regimen

Brandes et al. conducted a phase II study in which 37 patients with anaplastic oligodendroglial tumors (14 AOA, 23 AO) were treated with PCV at recurrence [44]. None had received previous chemotherapy. There was a 59 per cent overall response rate. The only factor that significantly correlated with response on both univariate and multivariate analysis was histological subtype. The AOA group did worse with a response of 22.7 per cent compared to 77.2 per cent for the AO group. Median TTP for the AOA group was 6 versus 18.6 months for the AO group. This difference was of borderline significance. Median survival time was not statistically different between the two groups. Forty-three per cent of the patients without disease progression discontinued chemotherapy because of incomplete recovery from toxicity.

In many of these studies, when compared to the pure oligodendroglioma group, the oligoastrocytoma histological subtype had a negative association with the various measures of response (radiographic response, MST, and TTP) to PCV chemotherapy. This association was either significant or strongly trended towards significance. However, oligoastrocytomas were still shown to respond durably to treatment with PCV, with rates ranging from 42 to 90 per cent for neoadjuvant and adjuvant therapy and 22 to 33 per cent for use at recurrence. The analysis of the response of oligoastrocytomas to PCV in these studies is limited by their relatively small numbers. Additionally, due to the very small number of patients, the effectiveness of PCV as second-line chemotherapy in oligoastrocytomas can not be obtained from these studies. Adverse effects, especially cumulative hematologic toxicity, were significant and often limited treatment.

Most of these studies were done prior to recognition of the importance of molecular genetic analyses. Cairncross et al. was the only one of the preceding studies cited to include molecular analysis and assess its prognostic significance [35].

Temozolomide

Temozolomide (TMZ) is an imidazotetrazine derived alkylating agent that has excellent oral bioavailability, reaching peak plasma concentrations (C_{max}) one hour after administration. Temozolomide undergoes spontaneous hydrolysis to its active metabolite 3-methyl-(triasen-1-yl)imidazole-4-carboxamide (MTIC) and as such does not require metabolic activation by the liver [45]. The action and, therefore, cytotoxicity of this metabolite is thought to be methylation of DNA at the O6, N7, and N3 positions, which inhibits DNA replication [45,46]. Temozolomide achieves high measured concentrations in cerebrospinal fluid, approximately 35 to 39 per cent of that measured in plasma [45,47,48]. As temozolomide is not metabolized in the liver, there are no drug interactions between it and cytochrome P450 inducing agents. This imparts a significant advantage to this drug in the brain tumor population, as many patients are on antiepileptic drugs. Temozolomide is administered at a dose of 150–200 mg/m^2 per day for 5 days per 28 day cycle. Chemotherapy naïve patients are started at 200 mg/m^2 per day. Patients who have been previously exposed to chemotherapy are started at 150 mg/m^2 per day, and usually are advanced to the higher dose if the medication is tolerated with mild or no toxicity.

Temozolomide was shown to have clinical anti-glioma activity and was well tolerated in smaller studies [46,49]. This led to the major multicenter phase II study conducted by Yung et al. [45]. That study evaluated the efficacy and safety profile of TMZ in 111 patients with anaplastic astrocytoma (AA) or AOA [45]. Tumors with 80 to 90 per cent oligodendroglial component were classified as pure AO and tumors with less than 20 per cent oligodendroglial component were classified as AA. Anything else was considered mixed AOA. The objective response rate for all patients studied was 35 per cent. The response rate for the fourteen patients (9 per cent) with AOA was 42.9 per cent (2 CR, 4 PR) and PFS at 6 months was 46 per cent. Median overall survival in this group was 14.8 months. In the chemotherapy-naïve group, 43 per cent of the patients responded and 26 per cent had stable disease compared to a 30 per cent response rate and 28 per cent rate of stable disease in patients who received TMZ as second-line therapy. These differences were not statistically significant. There was no significant difference in efficacy between patients with AA and those with AOA. Toxicity was low with only 6 per cent developing grade 3/4 thrombocytopenia, 2 per cent developing grade 3/4 neutropenia, and 2 per cent developing grade 3/4 leukopenia. Thirty-three per cent of patients reported constipation and 34 per cent reported fatigue. Importantly, myelosuppresion did not appear cumulative and typically resolved with a one-dose level reduction.

Others have also evaluated the efficacy and safety of this agent as a second-line chemotherapy in patients with oligodendroglial tumors. Chinot et al. evaluated the response to TMZ of 48 patients with recurrent anaplastic oligodendroglial tumors after standard RT and chemotherapy [50]. The majority of the patients (93.6 per cent) had received PCV as first-line chemotherapy. A response rate of 43.8 per cent was achieved for the entire study population. Survival was 10 months for the entire group and 16 months for the responders. This overall survival and a PFS of 13 months in the responder group were significantly improved compared to those patients with stable or progressive disease. Nine of the forty-eight patients had AOA and this group had a response rate of 22.2 per cent, a median PFS of 5.6 months, and a median overall survival of 8.7 months. Although this response was less than that seen in the oligodendroglioma group (response rate 48.7 per cent, median PFS 7.3 months, median overall survival 9.9 months) it was still deemed by the authors to be clinically relevant. Whereas patients who had previously responded to PCV had a significant improvement in progression-free survival with TMZ, patients who failed to respond to PCV also benefited from this therapy. The 35 patients whose response to first-line chemotherapy was assessable had an objective response rate of 83 per cent, a much better response than that usually seen. Low-grade hematologic toxicity was the most common adverse event. Grade 3/4 thrombocytopenia occurred in 6 per cent of the patients and no patients experienced grade 3/4 neutropenia. Fatigue was reported as grade 3 or 4 in 17 per cent of the patients. Of note, some patients did not achieve a complete response until 3–5 cycles of TMZ had been administered.

There have been two European Organization for Research and Treatment of Cancer (EORTC) studies investigating TMZ in recurrent oligodendroglial tumors [51–53]. Prior to the start of official enrollment in these studies, a number of patients with oligodendroglial tumors had been treated with TMZ as salvage therapy in accordance with the EORTC protocol. A retrospective analysis of these patients was reported by van den Bent et al. in 2001 [51]. Thirty patients previously treated with radiation therapy met the study criteria of having a contrast enhancing recurrent or progressing oligodendroglial tumor. Twenty-seven patients had previously received PCV, three patients were chemotherapy naïve, and three had received more than one line of chemotherapy prior to treatment with temozolomide. Prior response to PCV was assessable in 24 patients. The objective response rate to PCV was 38 per cent. Patients who had been previously treated with PCV had a 26 per cent response rate to temozolomide. This included several patients who had not responded to first-line PCV. The median TTP in responders was 13 months.

Grade 3/4 thrombocytopenia occurred in 13 per cent of the patients. Two patients discontinued treatment for grade 4 toxicity (thrombocytopenia and hepatotoxicity) and both patients subsequently recovered. The authors noted that the low response rate to PCV as first-line chemotherapy and the relatively large number of patients with first relapse within two years of radiotherapy suggests their study population had clinically aggressive tumors. Eight patients had oligoastrocytoma, however their response rate was not reported separately. In another study of TMZ as second-line chemothereapy (after PCV), Costanza et al. reported a 22 per cent response rate in 32 patients with oligodendroglial tumors treated [54]. None of the seven patients with oligoastrocytoma responded.

In 2003 the results of the two EORTC trials were reported. One evaluated temozolomide as second-line chemotherapy and the other as first-line chemotherapy [52,53]. EORTC study 26972 was a phase II study of temozolomide as second-line chemotherapy in the treatment of 28 patients with recurrent oligodendroglial tumors who had received PCV as initial chemotherapy [52]. Eleven patients had oligoastrocytomas, their response rate and median TTP were not provided separately. Seven of the twenty-eight patients responded for a rate of 25 per cent. The median TTP of the responders was eight months, with mild toxicity. The patients who had responded to PCV did not have a significantly improved response rate.

EORTC study 26971 evaluated 38 previously irradiated patients with temozolomide as first-line chemotherapy, including 35 with confirmed oligodendroglial tumors [53]. All of the patients had recurrent, enhancing tumors. The response rate of the patients with confirmed oligodendroglial tumors was 54 per cent (10 CR, 9 PR). Eleven patients had stable disease. The median TTP was 10.4 months for all patients and 13.2 months for responders. Tumor-free progression was 71 per cent at 6 months and 40 per cent at 12 months. Median survival had not yet been reached at the time of publication. Three of the six confirmed patients with oligoastrocytoma responded. There was no difference between the response of the two histological subgroups. Treatment was well tolerated with grade 3/4 hematologic toxicity occurring in only 4 per cent of the patients.

Chahlavi et al. investigated the impact of chromosome 1p status in the response of 16 oligodendroglial tumors to temozolomide [55]. Although chromosome 1p loss was significantly associated with response to temozolomide, none of the three oligoastrocytomas had loss of 1p.

Currently ongoing is a trial of neoadjuvant (pre-radiation) temozolomide for newly diagnosed anaplastic oligodendroglial tumors, which is stratified for 1p/19q status. Patients with loss of 1p/19q receive temozolomide alone with radiotherapy being deferred until time of progression [56 and personal communication].

The efficacy and safety of temozolomide as second-line chemotherapy in the treatment of oligodendroglial tumors has been well demonstrated. Response to PCV does not appear to be a significant predictor of response to temozolomide, and patients who have not responded to PCV can benefit from this treatment [50–52]. In comparison there is a relative paucity of literature on the efficacy of temozolomide as adjuvant, neoadjuvant, and first-line chemotherapy. As is common to many studies on chemotherapy and oligodendroglial tumors, inclusion of oligoastrocytomas and the extent of histological subtype analysis varied greatly between studies. This considerably limits the conclusions about temozolomide in the treatment of oligoastrocytomas that can be drawn from their results. The excellent safety profile of temozolomide and its ease of administration make it very attractive. Treatment with PCV is limited by toxicity. Treatment with temozolomide is well tolerated in most and in some patients seems to be well tolerated over an extended period of time [57,58]. The delayed risks of the extended use of this agent are unknown, as is whether prolonged treatment adds any benefit. Still to be determined is whether the response to temozolomide is as great and durable as the response to PCV and if the considerably greater tolerance to this drug will outweigh any possible advantage of PCV. Further studies with specific attention to oligoastrocytomas are needed to correlate genetic alterations and responsiveness to temozolomide, to compare PCV to temozolomide, and to determine the safety and efficacy of maintenance therapy with temozolomide.

Other Chemotherapy Regimens

Limited data exists on the use of other regimens in the treatment of oligoastrocytomas. In a study of 32 newly diagnosed anaplastic oligoastrocytoma patients, Boiardi et al. evaluated adjuvant treatment with combined Cisplatin and BCNU [59]. The median TTP and median survival time were 54.6 and 70.1 months, respectively. Disease recurred in 14 patients. The others were described as being disease free during a median follow-up period of 62.3 months. There was a high incidence of hematologic toxicity with grade 3 leukopenia occurring in 25 patients, grade 4 in 8 patients, grade 3 thrombocytopenia in 18 patients, and grade 4 in 4 patients. Soffieti et al. studied the effectiveness of Carboplatin as second-line chemotherapy for recurrent or progressive oligodendroglial tumors after PCV [60]. Eight of the twenty-three patients had oligoastrocytoma (five grade II and three grade III). There were no responders in the oligoastrocytoma group in comparison to three partial responders in the oligodendroglioma group. In a study of Carboplatin and teniposide as third-line chemotherapy in 23 patients with recurrent oligodendroglial tumors (two low-grade and four anaplastic oligoastrocytomas), 33 per cent of the patients with oligoastrocytomas remained progression-free during treatment [61]. This was in comparison to 70 per cent of the patients with oligodendrogliomas. Median TTP of the entire study group was 19 weeks. The median survival of the entire 23 patients was 60.7 weeks. Oligoastrocytoma histology was a significant negative prognostic factor with a median TTP of 6 weeks and a MST of 15.4 weeks for those patients. A small phase II study of topotecan in 16 patients with progressive or recurrent anaplastic oligodendroglial tumors previously treated with PCV did not demonstrate efficacy [62].

LOW-GRADE OLIGOASTROCYTOMAS

One of the great dilemmas in neuro-oncology is the optimal management of low-grade glioma. Similar to other low-grade gliomas, the timing and type of treatment for low-grade oligoastrocytomas is controversial. Surgery and radiation therapy play important roles. This discussion will not address these modalities or whether they are more or less appropriate than chemotherapy, but will focus on what is known about the treatment of these tumors with chemotherapy. Appreciation of the chemosensitivity of anaplastic oligodendroglial tumors (including oligoastrocytomas) led to the use of chemotherapy in lower grade tumors. That low-grade oligodendrogliomas respond to chemotherapy is clear. Whether that is true of low-grade oligoastrocytomas has been more difficult to ascertain.

Although the median survival of patients with low-grade oligoastrocytomas has been demonstrated to be less than that of those with pure low-grade oligodendroglioma, it is still prolonged enough that there is concern about delayed radiation induced neurotoxicity. Many of these patients have very indolent disease and extensive tumor which require large radiotherapy ports. Modern-era radiation therapy may be less likely to cause these delayed effects with their cognitive sequelae, but this remains to be seen. The same indolent nature that causes concern regarding delayed neurotoxicity also raises the question of possible adverse delayed effects of chemotherapy.

At best, the evaluation of the literature on the chemotherapeutic response of the low-grade mixed tumors is challenging. As in the studies of the anaplastic tumors, patients with low-grade oligoastrocytomas comprise a minor percentage of the total subjects in the majority of the literature evaluating chemotherapy in low-grade oligodendroglial tumors. The risk of under-grading a tumor with stereotactic biopsy is present as it is for all the tumors. The genetic characteristics of the low-grade oligoastrocytoma are even less clear than for the high-grade mixed tumors and whether these characteristics can help to determine clinical responsiveness to therapy has yet to be determined. In addition to the difficulties with histology, there is also an absence of agreement on the significance of contrast enhancement. Many studies on recurrent anaplastic tumors use contrast enhancement as an entry criteria, however many studies of low-grade gliomas also include tumors that exhibit contrast enhancement. Contrast enhancement is generally thought to be strongly suggestive of the presence of anaplasia. The validity of considering tumors with contrast enhancement as low-grade tumors is questionable.

Additional difficulties exist in the assessment of response to chemotherapy in all low-grade gliomas and consequently in the interpretation of the results of these studies. These include the inability of investigators to use the standard method of measuring the change in enhancement to assess radiographic response to chemotherapy and the delay in appreciable radiographic response that can occur in these slow-growing tumors. Subjects must be followed for several years with repeat neuroimaging studies before a determination of the effect of the chemotherapy on time to tumor progression or survival can be made. Improvement in clinical measures, such as improved seizure control, may be as important a marker of response as radiographic change in studies of low-grade oligodendroglial tumors.

PCV

Olson *et al.* retrospectively reviewed the outcome of 106 patients with low-grade oligodendroglial tumors [26]. Thirteen patients had contrast enhancement on initial neuroimaging studies. Eighteen patients had histology obtained at the time of tumor progression, not at the time of diagnosis. A presumptive diagnosis of low-grade tumor was made if 12 months had passed between the patient's initial imaging and their progression. This very heterogeneous group of patients had been treated in a variety of ways with a variety of methods. Thirty-eight patients were treated at the time of diagnosis; twelve received chemotherapy, twenty received radiotherapy, and six received adjuvant chemotherapy in addition to radiation. Treatment was deferred in 68 patients, who were observed. Fourteen of the patients treated initially with chemotherapy received PCV. Ultimately 76 patients received chemotherapy. Sixty-eight per cent of the patients recurred by the median follow-up of six years. The median TTP for the entire group was five years. Twenty-nine of the 106 patients had oligoastrocytomas. There was no statistical difference in median TTP between the mixed oligoastrocytoma group and the pure oligodendroglioma group. The median TTP had not been reached after a median follow-up of 5.5 years in the group treated initially with chemotherapy alone, deferring RT. Median TTP of the group treated initially with both chemotherapy and radiation was 8.6 years. Median overall survival was 16.7 years. Median survival of the oligoastrocytoma group had not been reached after a median follow-up of 7.1 years. Median survival for the pure oligodendroglioma group was 12.1 years. Forty-six per cent of the 76 patients treated eventually with chemotherapy had significant myelosuppression. This was defined as myelosuppression that required dose reduction or delay, colony stimulating factor, transfusion, or treatment of neutropenic fever. One-third of the patients treated with radiotherapy experienced delayed neurotoxicities. Fifteen per cent of these patients were diagnosed with radiation necrosis and twenty-one per cent had delayed cognitive impairment. The median dose of radiation in these patients was 5940 cGy (range 4800–6500 cGy). Necrosis occurred a median of 52 months following radiation therapy (range 3–108 months). No prognostic factors had a significant impact on progression or survival in this study, including the order in which treatment was given (at the time of diagnosis or deferral until progression) and treatment modality.

In a phase II study, Soffietti *et al.* assessed the efficacy of PCV in the treatment of low-grade recurrent oligodendroglial tumors [63]. Twenty-six chemotherapy naïve patients were treated with up to six cycles of PCV at the time of recurrence after surgery alone or after initial therapy with surgery and radiation. Nine patients had oligoastrocytomas. Nineteen patients demonstrated enhancing lesions on neuroimaging studies; in ten of these patients minimal enhancement was present at the time of initial diagnosis and in nine patients it developed at the time of progression. The overall response rate was 61.5 per cent (3 CR, 13 PR). All symptomatic responders improved clinically including improved seizure control and improved hemiparesis. Stable disease was seen in 31 per cent (8 patients). Two of these patients had cessation of seizures and one patient had improved hemiparesis and aphasia. The response rate of the oligoastrocytoma group was 55.5 per cent, though none of these patients had complete responses, and it was not statistically different from the response rate of the oligodendroglioma group (65 per cent). Patients with nonenhancing lesions had a significantly lower response rate than those with enhancement (74 *versus* 29 per cent). The presence of enhancement by histologic subtype was not reported. The median TTP after the start of PCV was 24 months for the entire study population. Oligoastrocytoma histology was associated with a significantly shorter time to progression (12 *versus* 32 months). Grade 3 hematologic toxicity (leukopenia or thrombocytopenia) occurred in 4.5 per cent of all cycles.

Temozolomide

The ease of administration, superior safety profile compared to PCV, and demonstrated efficacy of temozolomide have all contributed to its prevalent use in the management of patients with anaplastic oligodendrogliomas and oligoastrocytomas. For these reasons, this agent is now being used with increased frequency in the treatment of patients with low-grade oligodendroglial tumors. The response rate, optimal length of treatment in this population, durability of response, and the risks (if any) of prolonged use have not yet been established. A number of studies have been recently published examining the use of temozolomide in the treatment of low-grade oligodendroglial tumors. Unfortunately, though patients with oligoastrocytomas were included in many of these studies, their results were not usually reported separately from those of patients with oligodendrogliomas. Also many patients in some studies had tumors which exhibited contrast enhancement on neuroimaging studies, once again raising the question if they should even be considered low-grade tumors.

Hoang-Xuan *et al.* investigated the response of low-grade oligodendroglial tumors to temozolomide and correlated this response with chromosome 1p status [64]. Sixty patients were enrolled in the study, 49 with oligodendroglioma and 11 with oligoastrocytoma. Eleven per cent of the patients had contrast

enhancement on neuroimaging studies. Response rates for histologic subtypes (oligoastrocytoma *versus* oligodendroglioma) were not reported separately. The objective response rate of the 59 evaluable patients was 31 per cent. Ten patients (17 per cent) had a partial response, 8 patients (14 per cent) had a minimal or minor response (between 25 and 50 per cent reduction in T2 signal), and 36 patients had stable disease. Responses occurred slowly over time with the median time to maximum radiographic response being 12 months (range 5–20 months), with some responses not becoming apparent until after up to 10 cycles had been received. Fifty-one per cent of the patients had clinical improvement, especially in regards to seizure control. This includes 33 per cent of the patients who were radiologically stable. A median of 11 cycles of temozolomide was given. In the 26 patients assessed there was a significant association between loss of 1p and response to temozolomide. Histological subtype did not correlate with response to chemotherapy. Grade 3/4 hematologic toxicity occurred in 8 per cent of the patients. One patient developed grade 4 thrombocytopenia after one cycle and was removed from the study. This patient was not included in the final analysis. All other patients who experienced toxicity were able to continue treatment with a 25 per cent dose reduction.

Quinn *et al.* treated 46 patients with progressive low-grade glioma with temozolomide [65]. Five of these 46 patients had oligoastrocytomas (11 per cent) and 20 had oligodendrogliomas (43 per cent). Twenty-nine patients (63 per cent) had received prior therapy with at least one treatment modality. Seventy per cent of the patients had contrast enhancement on their baseline neuroimaging studies, suggesting their tumors had progressed to a higher grade. The overall objective response rate was 61 per cent (11 CR and 17 PR). Thirty-five per cent (16) of the patients had stable disease. Two of the five oligoastrocytoma patients responded (1 CR, 1 PR) and two had stable disease. The percentage of any response (CR + PR + SD) was lower in the oligoastrocytoma group (80 *versus* 96 per cent overall), but this difference was not significant. The enhancement status of the oligoastrocytoma patients was not reported separately. The overall median PFS was 22 months. Patients with oligoastrocytoma had a PFS of 14 months, a 6-month PFS of 0.80 and a 12-month PFS of 0.80. These values are in comparison to an overall 6-month PFS of 0.98 and a 12-month PFS of 0.76. The differences between histologic groups were not significant. Grade 3/4 hematologic toxicity occurred in 13 per cent of the patients, with one death in a patient with grade 4 neutropenia who did not seek early medical attention. The response rate reported in this study is considerably greater than that reported by Hoang-Xuan *et al.*, however the patient populations under investigation differ greatly in their baseline characteristics [64,65]. A greater percentage of patients demonstrated contrast enhancement in the Quinn *et al.* study than in the Hoang-Xuan *et al.* study (70 *versus* 11 per cent) [64,65]. Additionally in the Hoang-Xuan *et al.* study temozolomide was being investigated as a first-line agent [64]. The majority of patients (63 per cent) in the Quinn *et al.* study had received previous therapy, some with more than one modality [65]. The interpretation of these and other recent investigations of temozolomide for the treatment of low-grade oligoastrocytomas is restricted by the significant differences in the patient populations under study and the absence of stratification of results by histologic diagnosis. However, it is still probable that some low-grade oligoastrocytomas respond to treatment with PCV or temozolomide. Temozolomide has a safety profile and ease of administration that are superior to that of PCV. It also seems clear that clinical neurologic improvement can occur, especially in seizure control, without an associated radiographic response. In one study, the time to response was not appreciated for multiple cycles of temozolomide and then continued for several months, suggesting extended treatment lengths may be necessary for the optimal treatment of these patients [64]. This of course may then raise the risk of delayed complications of chemotherapy, such as myelodysplastic syndromes and myelogenous malignancies.

The limitations of the current literature on the treatment of low-grade oligoastrocytomas preclude the formation of specific recommendations for the management of these patients. For now the answers to the questions of if, when, and for how long in regards to the use of chemotherapy for the treatment of low-grade oligoastrocytomas will continue to be answered on an individual basis. This will change only when phase III trials with an adequate number of low-grade oligoastrocytomas that compare radiation therapy to chemotherapy, that correlate genetic alterations to response, and that determine the safety of long-term chemotherapy are complete.

Molecular analysis and stratification in clinical trials by genetic characteristics may ultimately allow us to identify which oligoastrocytomas are likely to respond to chemotherapy and which are not. For tumors with a mixed histologic appearance, their genetic characteristics may then become even more crucial to their optimal management than in the pure tumors.

References

1. Bailey, P., and Cushing, H. (1926). *A Classification of the Tumors of the Gliomas Group on a Histologic Basis with a Correlated Study of Prognosis*, pp. 53. Lippincott Co, Philadelphia.
2. Shaw, E., Scheithauer, B. W., O'Fallon, J. R. *et al.* (1994). Mixed oligoastrocytomas: A survival and prognostic factor analysis. *Neurosurgery* **34**(4), 577–582.
3. Hart, M. N., Petito, C. K., and Earle, K. M. (1974). Mixed Gliomas. *Cancer* **33**, 134–140.
4. Gray, F., De Girolami, U., and Poirier, J. (2004). *Escourolle and Poirier Manual of Basic Neuropathology*, 4th ed. pp. 31. Butterworth-Heinemann, Philadelphia.
5. Kleihues, P., and Cavenee, W. K. (Eds.) (2000). *World Health Organization Classification of Tumours: Pathology and Genetics of Tumours of the Nervous System*, pp. 65–67. IARC Press, Lyon.
6. Okamoto, Y., Di Patre, P.-L., Burkhard, C. *et al.* (2004). Population-based study on incidence, survival rates, and genetic alterations of low-grade diffuse astrocytomas and oligodendrogliomas. *Acta Neuropathol* **108**, 49–56.
7. Kim, L., Hochberg, F. H., Thornton, A. F. *et al.* (1996). Procarbazine, lomustine, and vincristine (PCV) chemotherapy for grade III and grade IV oligoastrocytomas. *J Neurosurg* **85**, 602–607.
8. Krouwer, H. G. J., van Duinen, S. G., Kamphorst, W. *et al.* (1997). Oligoastrocytomas: a clinicopathological study of 52 cases. *J Neurooncol* **33**, 223–238.
9. Beckmann, M. J., and Prayson, R. A. (1997). A clinicopathologic study of 30 cases of oligoastrocytoma including p53 immunohistochemistry. *Pathology* **29**, 159–164.
10. van den Bent, M. J. (2004). Advances in the biology and treatment of oligodendrogliomas. *Curr Opin Neurol* **17**(6), 675–680.
11. Maintz, D., Fiedler, K., Koopmann, J. *et al.* (1997). Molecular genetic evidence for subtypes of oligoastrocytomas. *J Neuropathol Exp Neurol* **56**(10), 1098–1104.
12. Burger, P. C., Yuriko Minn, A., Smith, J. S. *et al.* (2001). Losses of chromosomal arms 1p and 19q in the diagnosis of oligodendroglioma. *Mod Pathol* **14**(9), 842–853.
13. Sasaki, H., Zlatescu, M. C., Betensky, R. A. *et al.* (2002). Histopathological-molecular genetic correlations in referral pathologist-diagnosed low-grade "oligodendroglioma". *J Neuropathol Exp Neurol* **61**(1), 58–63.
14. Cairncross, J. G., and Macdonald, D. R. (1988). Successful chemotherapy for recurrent malignant oligodendroglioma. *Ann Neurol* **23**, 360–364.
15. Macdonald, D. R., Gaspar, L. E., and Cairncross, J. G. (1990). Successful chemotherapy for newly diagnosed aggressive oligodendroglioma. *Ann Neurol* **27**, 573–574.
16. Glass, J., Hochberg, F. H., Gruber, M. L. Louis, D. N., Smith, D., Rattner, B. (1992). The treatment of oligodendrogliomas and mixed oligodendroglioma-astrocytomas with PCV chemotherapy. *J Neurosurg* **76**, 741–745.
17. Reifenberger, G., and Louis, D. N. (2003). Oligodendrogliomas: Toward molecular definitions in diagnostic neuro-oncology. *J Neuropathol Exp Neurol* **62**(2), 111–126.
18. Bigner, S. H., Matthews, M. R., Rasheed, B. K. A. *et al.* (1999). Molecular genetic aspects of oligodendrogliomas including analysis by comparative genomic hybridization. *Am J Pathol* **155**(2), 375–386.
19. Bissola, L., Eoli, M., Pollo, B. *et al.* (2002). Association of chromosome 10 losses and negative prognosis in oligoastrocytomas. *Ann Neurol* **52**, 842–845.

20. Smith, J. S., Perry, A., Borell, T. J. *et al.* (2000). Alterations of chromosome arms 1p and 19q as predictors of survival in oligodendrogliomas, astrocytomas, and mixed oligoastrocytomas. *J Clin Oncol* **18**(3), 636–645.

21. Müller, W., Hartmann, C., Hoffmann, A. *et al.* (2002). Genetic signature of oligoastrocytomas correlates with tumor location and denotes distinct molecular subsets. *Am J Pathol* **161**(1), 313–319.

22. Watanabe, T., Nakamura, M., Kros, J. M. *et al.* (2002). Phenotype versus genotype correlation in oligodendrogliomas and low-grade diffuse astroctyomas. *Acta Neuropathol* **103**, 267–275.

23. Shaw, E., Scheithauer, B., and O'Fallon, J. (1992). Astrocytomas, oligoastrocytomas, and oligodendrogliomas: A comparative survival study. *Neurology* **42**(S3), 693P.

24. Celli, P., Nofrone, I., Palma, L., Cantore, G., Fortuna, A. (1994). Cerebral oligodendroglioma: Prognostic factors and life history. *Neurosurgery* **35**(6), 1018–1035.

25. Smith, M. T., Ludwig, C., Godfrey, A. D., Armbrustmacher, V. W. (1983). Grading of oligodendrogliomas. *Cancer* **52**, 2107–2114.

26. Olson, J. D., Riedel, E., and DeAngelis, L. M. (2000). Long-term outcome of low-grade oligodendroglioma and mixed glioma. *Neurology* **54**(7), 1442–1448.

27. Giannini, C., Scheithauer, B.W., Weaver, A. L. *et al.* (2001). Oligodendrogliomas: Reproducibility and prognostic value of histologic diagnosis and grading. *J Neuropathol Exp Neurol* **60**(3), 248–262.

28. Daumas-Duport, C., Varlet, P., Tucker, M. L., Beuvon, F., Cervera, P., Chodkiewicz, JP. (1997). Oligodendrogliomas. Part I: Patterns of growth, histological diagnosis, clinical and imaging correlations: A study of 153 cases. *J Neurooncol* **34**, 37–59.

29. Cillekens, J. M., Belien, J. A., van der Valk, P. *et al.* (2000). A histopathological contribution to supratentorial glioma grading, definition of mixed gliomas and recognition of low grade glioma with Rosenthal fibers. *J Neurooncol* **46**(1), 23–43.

30. Ino, Y., Betensky, R. A., Zlatescu, M. C. *et al.* (2001). Molecular subtypes of anaplastic oligodendroglioma: Implications for patient management at diagnosis. *Clin Cancer Res* **7**, 839–845.

31. van den Bent, M. J., Looijenga, L. H., Langenberg, K. *et al.* (2003). Chromosomal anomalies in oligodendroglial tumors are correlated with clinical features. *Cancer* **97**, 1276–1284.

32. Bauman, G. S., Ino, Y., Ueki, K. *et al.* (2000). Allelic loss of chromosome 1p and radiotherapy plus chemotherapy in patients with oligodendrogliomas. *Int J Radiat Oncol Biol Phys* **48**(3), 825–830.

33. Cairncross, J. G., Ueki, K., Zlatescu, M. C. *et al.* (1998). Specific genetic predictors of chemotherapeutic response and survival in patients with anaplastic oligodendrogliomas. *J Natl Cancer Inst* **90**(19), 1473–1479.

34. Ständer, M., Peraud, A., Leroch, B., Kreth, F. (2004). Prognostic impact of TP53 mutation status for adult patients with supratentorial world health organization grade II astrocytoma or oligoastrocytoma. *Cancer* **101**(5), 1028–1035.

35. Cairncross, G., Seiferheld, W., Shaw, E. *et al.* (2004). An intergroup randomized controlled clinical trial of chemotherapy plus radiation (RT) versus RT alone for pure and mixed anaplastic oligodendrogliomas: Initial report of RTOG 94-02 (Abstract). *Neuro-oncol* **6**(4), 371.

36. Nutt, C. I., Mani, D. R., Betensky, R. A. *et al.* (2003). Gene expression-based classification of malignant gliomas correlates better with survival than histological classification. *Cancer Res* **63**, 1602–1607.

37. Levin, V. A., Edwards, M. S., Wright, D. C. *et al.* (1980). Modified procarbazine, CCNU, and vincristine (PCV 3) combination chemotherapy in the treatment of malignant brain tumors. *Cancer Treat Rep* **64**, 237–244.

38. Kyritsis, A. P., Yung, W. K. A., Bruner, J., Gleason, M. J., Levin, V. A. (1993). The treatment of anaplastic oligodendrogliomas and mixed gliomas. *Neurosurgery* **32**(3), 365–371.

39. Streffer, J., Schabet, M., Bamberg, M. *et al.* (2000). A role for preirradiation PCV chemotherapy for oligodendroglial brain tumors. *J Neurol* **247**, 297–302.

40. Abrey, L. E., Childs, B., Paleologos, N. *et al.* (2003). High-dose chemotherapy with stem cell rescue as initial therapy for anaplastic oligodendroglioma. *J Neurooncol* **65**(2), 127–134.

41. Jeremic, B., Shibamoto, Y., Grujicic, D. *et al.* (1999). Combined treatment modality for anaplastic oligodendroglioma: A phase II study. *J Neurooncol* **43**, 179–185.

42. Jeremic, B., Milicic, B., Grujicic, D. *et al.* (2004). Combined treatment modality for anaplastic oligodendroglioma and oligoastrocytoma: A 10-year update of a phase II study. *Int J Radiat Oncol Biol Phys* **59**(2), 509–514.

43. van den Bent, M. J., Kros, J. M., Heimans, J. J. *et al.* (1998). Response rate and prognostic factors of recurrent oligodendroglioma treated with procarbazine, CCNU, and vincristine chemotherapy. *Neurology* **51**(4), 1140–1145.

44. Brandes, A. A., Tosoni, A., Vastola, F. *et al.* (2004). Efficacy and feasibility of standard procarbazine, lomustine, and vincristine chemotherapy in anaplastic oligodendroglioma and oligoastrocytoma recurrent after radiotherapy. *Cancer* **101**, 2079–2085.

45. Yung, W. K. A., Prados, M. D., Yaya-Tur, R. *et al.* (1999). Multicenter phase II trial of temozolomide in patients with anaplastic astrocytoma or anaplastic oligoastrocytoma at first relapse. *J Clin Oncol* **17**(9), 2762–2771.

46. Levin, V. A., Yung, A., Prados, M. *et al.* (1997). Phase II study of Temodal (temozolomide) at first relapse in anaplastic astrocytoma (AA) patients (meeting abstract). *Proc Am Soc Clin Oncol* 1370.

47. Marzolini, C., Decosterd, L. A., Shen, F. *et al.* (1998). Pharmacokinetics of temozolomide in association with fotemustine in malignant melanoma and malignant glioma patients: Comparison of oral, intravenous, and hepatic intra-arterial administration. *Cancer Chemother Pharmacol* **42**(6), 433–440.

48. Stupp, R., Ostermann, S., Leyvraz, S., Csajka, C., Buclin, T., Decostered, L. (2001). Cerebrospinal fluid levels of temozolomide as a surrogate marker for brain penetration (meeting abstract). *Proc Am Soc Clin Oncol* **20**, 59.

49. Newlands, E. S., O'Reilly, S. M., Glaser, M. G. *et al.* (1996). The Charing Cross Hospital experience with temozolomide in patients with gliomas. *Eur J Cancer* **32A**(13), 2236–2241.

50. Chinot, O.-L., Honore, S., Dufour, H. *et al.* (2001). Safety and efficacy of temozolomide in patients with recurrent anaplastic oligodendrogliomas after standard radiotherapy and chemotherapy. *J Clin Oncol* **19**(9), 2449–2455.

51. van den Bent, M. J., Keime-Guibert, F., Brandes, A. A. *et al.* (2001). Temozolomide chemotherapy in recurrent oligodendroglioma. *Neurology* **57**(2), 340–342.

52. van den Bent, M. J., Chinot, O., Boogerd, W. *et al.* (2003). Second-line chemotherapy with temozolomide in recurrent oligodendroglioma after PCV (procarbazine, lomustine, and vincristine) chemotherapy: EORTC brain tumor group phase II study 26972. *Ann Oncol* **14**(4), 599–602.

53. van den Bent, M. J., Taphoorn, M. J. B., Brandes, A. A. *et al.* (2003). Phase II study of first-line chemotherapy with temozolomide in recurrent oligodendroglial tumors: The European organization for research and treatment of cancer brain tumor group study 26971. *J Clin Oncol* **21**(13), 2525–2528.

54. Costanza, A., Borgognone, M., Nobile, M., Ruda, R., Mutani, R., Soffietti, R. (2001). Temozolomide in recurrent oligodendroglial tumors: A phase II study. *Neuro-oncol* **3**(Suppl 1), 66.

55. Chahlavi, A., Kanner, A., Peereboom, D., Staugaitis, S. M., Elson, P., Barnett, G. (2003). Impact of chromosome 1p status in response of oligodendroglioma to temozolomide: Preliminary results. *J Neurooncol* **61**, 267–273.

56. Mikkelson, T., Doyle, T., Margolis, J. *et al.* (2004). Neoadjuvant temozolomide single-agent chemotherapy for newly diagnosed anaplastic oligodendroglioma and anaplastic oligoastrocytoma with or without radiation therapy. *Neuro-oncol* **64**(4), 378.

57. Stern, J. I., Apisarnthanarax, S., Paleologos, N. A., Vick, N. A. (2003). Temozolomide as long-term maintenance treatment for gliomas. *Ann Neurol* **54**(Suppl 7), S31.

58. Grewal, J., Nestor, V., and Fink, K. (2005). Long term use of monthly temozolomide in patients with oligodendroglioma: Feasibility and tolerability of two years of therapy. *Neurology* **64**(Suppl 1), A25.

59. Boiardi, A., Eoli, M., Salmaggi, A. *et al.* (2000). Cisplatin and BCNU chemotherapy for anaplastic oligoastrocytomas. *J Neurooncol* **49**, 71–75.

60. Soffietti, R., Nobile, M., Rudá, R. *et al.* (2004). Second-line treatment with carboplatin for recurrent or progressive oligodendroglial tumors after PCV (procarbazine, lomustine, and vincristine) chemotherapy. *Cancer* **100**(4), 807–813.

61. Brandes, A. A., Basso, U., Vastola, F. *et al.* (2003). Carboplatin and teniposide as third-line chemotherapy in patients with recurrent oligodendroglioma or oligoastrocytoma: a phase II study. *Ann Oncol* **14**(12), 1727–1731.

62. Belanger, K., MacDonald, D., Cairncross, G. *et al.* (2003). A phase II study of topetecan in patients with anaplastic oligodendroglioma or anaplastic mixed oligoastrocytoma. *Invest New Drugs* **21**(4), 473–480.

63. Soffietti, R., Ruda, R., Bradac, G. B., Schiffer, D. (1998). PCV chemotherapy for recurrent oligodendrogliomas and oligoastrocytomas. *Neurosurgery* **43**(5), 1066–1073.

64. Hoang-Xuan, K., Capelle, L., Kujas, M. *et al.* (2004). Temozolomide as initial treatment for adults with low-grade oligodendrogliomas or oligoastrocytomas and correlation with chromosome 1p deletions. *J Clin Oncol* **22**(15), 3133–3138.

65. Quinn, J. A., Reardon, D. A., Friedman, A. H. *et al.* (2003). Phase II trial of temozolomide in patients with progressive low-grade glioma. *J Clin Oncol* **21**(4), 646–651.

28

Chemotherapy for Primary Central Nervous System Lymphoma

Francois G. El Kamar and Lauren E. Abrey

ABSTRACT: Methotrexate-based chemotherapy has significantly improved the treatment of primary CNS lymphoma. At the present time, several effective single-agent and multiagent methotrexate-based regimens have been reported as well as attempts to enhance delivery using blood–brain barrier disruption, maintenance chemotherapy, and myeloablative regimens requiring stem cell support. As no one approach is clearly superior to the others, the relative risks and benefits as well as newer agents and approaches to therapy will be discussed.

OVERVIEW

Primary central nervous system lymphoma (PCNSL) is a high or intermediate grade lymphoma that is confined to the brain, cerebrospinal fluid (CSF), eyes, or spinal cord [1,2]. More than 90 per cent of cases are B-cell; CD 19+ CD 20+, large cell or immunoblastic in nature and histologically indistinguishable from other extranodal lymphomas [1,3,4]. Only 3 per cent or less are T-cell lymphomas [5,6].

PCNSL accounts for approximately 2 to 3 per cent of central nervous system (CNS) malignancies and less than 2 per cent of all lymphomas. Despite a three-fold increase in incidence in the immunocompetent population during the past three decades, the overall incidence of PCNSL remains low and is estimated to be 0.3/100 000/year. It may have a higher incidence in the immunosuppressed patient, whether immunosuppression is congenital or acquired (AIDS, organ transplant patients or drug-induced). In this subpopulation, PCNSL is usually associated with the Epstein Barr Virus (EBV) infection. PCNSL has a

male to female ratio of 1:5 and a peak incidence at age 60, as compared with 37 years in the AIDS population [7–10].

Older age, ECOG performance status >2, multifocal or meningeal disease, and uncleaved histology have been associated with shorter survival [11]. The International Extranodal Lymphoma Study Group proposed a scoring scale based on age, performance status, serum LDH level, CSF protein concentration, and involvement of the deep structures of the brain. Each variable was assigned a value of 0 if favorable and 1 if not favorable. The two-year overall survival was 80 ± 8 per cent for patients with a score of 0–1, 48 ± 7 per cent for a score of 2–3 and 15 ± 7 per cent for a score of 4–5 [11,12].

Magnetic resonance imaging (MRI) typically shows one or multiple contiguous lesions that are deep seated, intimately related to the ventricular system and involving the thalamus, basal ganglia or corpus callosum. Hypointense, with indistinct borders on the T1-weighted images, these lesions appear dense and enhance homogeneously after gadolinium injection. On the T2-weighted sequences they are hyperintense with little surrounding edema [13,14]. When MRI is not available or contraindicated, a computed tomography (CT) scan can be obtained and shows similar intense, homogeneous enhancement with intravenous (IV) contrast [14]. A fluorodeoxyglucose (FDG) positron emission scan will demonstrate hypermetabolism but cannot differentiate PCNSL from other brain tumors [15]. Single-photon emission computerized tomography (SPECT) scanning using Thallium 201 can distinguish between CNS toxoplasmosis in the AIDS patient and PCNSL with sensitivity and specificity approaching 95 per cent [16,17]. The

TABLE 28.1   Recommended Baseline Evaluation

Contrast Enhanced MRI of the Brain

Clinical examination including a baseline performance status

Serum LDH levels

CSF total protein and cytology

Ophthalmologic evaluation including a slit lamp examination

CT scan contrast-enhanced of the chest, abdomen and pelvis

Bone marrow biopsy

MRI of the spine contrast enhanced if symptomatic

recommended baseline evaluation for patients diagnosed with PCNSL is detailed in Table 28.1.

MANAGEMENT

PCNSL is a highly treatable disease. It is exquisitely chemosensitive and radiosensitive, similar to systemic non-Hodgkin's lymphoma (NHL); however, lower control and higher recurrence rates result in a poor overall prognosis with a five-year survival rate of 25 per cent [18,19] (Fig. 28.1).

Corticosteroids

Steroid hormones activate the endogenous steroid receptor, which triggers the apoptosis cascade and an oncolytic response in malignant lymphocytes. Corticosteroids may lead to complete remission in 15 per cent or partial remissions in 25 per cent of PCNSL patients. Relapse typically occurs shortly after steroids are stopped but responses are occasionally durable; therefore, steroids should be withheld until a biopsy is obtained. Dexamethasone and prednisone are the most commonly used steroids and directly induce apoptosis in lymphoma cells. However, resistance to steroids may develop after prolonged exposure to corticosteroids or during withdrawal of therapy. Long-term steroid therapy should be avoided in an effort to minimize secondary complications [20–22].

Radiotherapy

Radiotherapy for PCNSL using whole brain radiotherapy (WBRT) 40–50 Gy is the fastest means of control, offering a response rate of 60 per cent, a control rate of 39 per cent, and a recurrence rate of 61 per cent [23,24]. In patients with ocular involvement, the eyes are included in the radiation port and should receive 35 to 40 Gy; bilateral treatment is mandatory even if disease is monocular [25]. Tumor recurrence

and severe neurotoxicity are the rule when WBRT is used as a single modality treatment. The median survival is only about 11.5 months and the five-year survival rate is less than 5 per cent. When compared to the five-year survival rate of 50 per cent or more in equivalent stage IE or IIE extranodal NHL treated with RT, WBRT is an unacceptable single modality treatment for PCNSL [23,26,27].

Chemotherapy

The addition of chemotherapy to WBRT has resulted in improved response rates and survival. Many chemotherapeutic agents have shown some activity as single agents in PCNSL. In systemic lymphoma, combination chemotherapy has proven to be better than single drug therapy and avoids the emergence of a resistant tumor [28]. However, standard NHL regimens such as CHOP, MACOP or MACOP-B have failed to produce any survival advantage over WBRT alone in PCNSL [29–33]. It may be that the blood–brain barrier (BBB), initially disrupted by the tumoral process, regains its integrity with the initial response to therapy, thereby offering a shelter for residual microscopic disease permitting tumor recurrence [34]. Therefore, the use of agents that can cross the blood–brain barrier and achieve high concentrations in the brain parenchyma and the CSF, such as methotrexate and ARA-C, are critical to the development of standard PCNSL regimens [35,36].

METHOTREXATE

Methotrexate (MTX) is the most effective and commonly used single agent for the treatment of PCNSL. It has shown excellent responses and control rates with relatively few side effects at intermediate or high doses. (Fig. 28.2) MTX is a water-soluble agent; however, when administered rapidly at doses greater than 1 g/m^2, it is possible to attain tumoricidal levels in brain parenchyma and CSF. MTX was recognized by Loeffler *et al.* as an active agent for PNCSL with the observation that NHL treated with MTX had fewer CNS relapses, and patients who received intravenous or intrathecal MTX followed by WBRT had prolonged median survivals of 44 months [34,37,38].

High dose (HD) -MTX produces response rates in excess of 50 per cent as a single agent and up to 94 per cent when given in combination with other drugs; these chemotherapeutic approaches followed by WBRT are associated with a two-year overall

TABLE 28.2 Single-agent MTX Studies

Reference	N	Dose MTX (g/m^2)	IT CHT	RT dose	Response rate (%)	Median survival (months)
[42]	13	3.5	–	30–44 Gy	92	
[43]	25	3.5	–	30–44 Gy	88	33
[44]	31	1	MTX	40+14.4 Gy boost	64	41
[39]	31	8	–	–	100	30+
[45]	25	8	–	–	74	22.8+

MTX – methotrexate; IT – intra-thecal.

survival of 58 to 72 per cent and 43 to 73 per cent, respectively [39,40]. Various schedules and doses ranging from 1 to 8 g/m^2 of single-agent MTX have been used with comparable responses and survival improvement (Table 28.2) [40,41]. MTX is relatively contraindicated in patients with effusions, and baseline creatinine clearance (CL$_{crea}$) below 50 ml/min. HD-MTX infusions are followed by leucovorin rescue, hydration, and alkalinization to minimize the risks of toxicities, and serum MTX levels should be followed closely until they are below toxic levels $< 1 \times 10^{-6}$ M at 48 h, or $<1 \times 10^{-8}$ at 72 h.

A wide range of high-dose MTX-based regimens are reported in the literature with various dose levels; however, the optimal MTX dose is yet to be identified. CSF MTX concentration is strictly related to the dose administered and to the infusion schedule. A three-hour infusion has been associated with higher CSF levels compared with other schedules of infusions and may obviate the need for intrathecal therapy. Ideally, an initial rapid administration followed by a more prolonged infusion may be optimal to overcome the distribution phase as MTX clearance from plasma is triphasic. Retrospective studies suggest that prolonged CL$_{crea}$ and high MTX area under the curve (AUC$_{MTX}$), although associated with more toxicity, may be independently associated with better outcome in PCNSL patients. The best timing of MTX administration remains undefined, and there seems to be no significant difference in efficacy or toxicity when MTX at 3.5 g/m^2 is administered every three weeks *versus* every 10 days. The total number of cycles and the need for maintenance or consolidation therapy remain to be determined [41,43,46,47].

INTRA-ARTERIAL MTX

Intra-arterial HD-MTX with blood–brain barrier disruption has been used with response rates of

70 per cent comparable to the results of IV HD-MTX-based therapy [48]. Furthermore, blood–brain barrier disruption (BBBD) and intra-arterial HD-MTX without WBRT, demonstrated improved survival outcome and neurocognitive preservation [49]. However, this is a technically demanding procedure that must be performed under general anesthesia limiting widespread use. For more information regarding this approach to chemotherapy for PCNSL, see Chapter 18 by Doolittle and Neuwelt.

MAINTENANCE THERAPY

Maintenance therapy with monthly MTX after initial remission has also been studied. The NABTT (New Approaches to Brain Tumor Therapy) used a dose of 8 g/m^2 of MTX at two week intervals for the first two cycles followed by 28-day intervals for a total of eight cycles or until remission. This was followed by a maintenance dose of 3.5 g/m^2 every month up to a year without radiotherapy. This yielded a response rate of 100 per cent and a median overall survival in excess of 30 months with minimal neurological toxicity; however, the median PFS was 16.7 months [50].

HD-MTX BASED COMBINATION CHEMOTHERAPY FOR PCNSL (TABLE 28.4)

Methotrexate, Procarbazine, Vincristine (MPV)

MPV, a regimen combining HD-MTX, procarbazine, and vincristine has been reported by several groups, including a large multicenter Radiation Therapy Oncology Group (RTOG) trial as having excellent response and survival when used in combination with WBRT [51–53].

Pre-RT MPV chemotherapy consists of MTX 2.5–3.5 g/m^2 infused over two hours given every other week for a total of five doses. Vincristine 1.4 mg/m^2 (maximum dose, 2.8 mg) is given concomitantly with each cycle of systemic MTX. Procarbazine 100 mg/m^2/day for seven days is given with the first, third, and fifth cycle of MTX. Intra-Ommaya MTX (12 mg) is given weekly on alternate weeks after administration of systemic MTX, if initial CSF cytology is positive. Leucovorin rescue, aggressive hydration, and alkalinization are started 24 h following the MTX infusion. Chemotherapy is followed by 45 Gy of whole-brain RT. All patients with evidence of ocular lymphoma should receive 30–40 Gy of ocular RT.

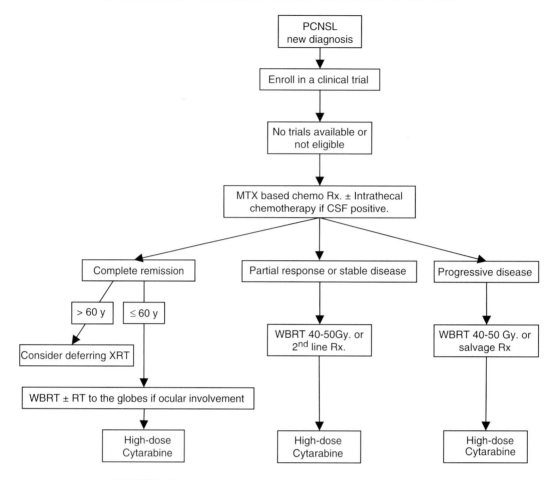

FIGURE 28.1　Treatment Algorithm for Newly Diagnosed PCNSL.

Three weeks after the completion of WBRT, patients receive two courses of high-dose cytarabine; each course consists of two doses of 3 g/m^2 infused over three hours separated by 24 h [51].

The same regimen was used with different MTX doses in two trials. The RTOG trial used 2.5 g/m^2 of MTX per cycle and a European confirmatory trial of MPV used 3 g/m^2 with similar response rates but shorter median overall survival (Table 28.3) [52,53]. The optimal number of cycles of MPV or other HD-MTX-based regimens is not known, but a recent pilot study suggested that additional cycles of MPV may convert a partial remission into a complete remission (CR) [54].

Drug	Dose	Route	Day
MTX	3.5 g/m^2	IV	1
Leucovorin rescue	10 mg qid × 12	IV/PO	2
VCR	1.4 mg/m^2	IV	1
Procarbazine	100 mg/m^2/day	PO	2–8 (cycles 1, 3, 5)
MTX	12 mg	IO	1

Methotrexate, Carmustine (BCNU), Teniposide, Prednisone (MBVP)

MBVP is a regimen developed by the French Groupe Ouest Est d'Etude des Leucemies et Autres Maladies du Sang (GOELAMS) and also studied in

TABLE 28.3　MPV Regimens

Reference	N	MTX dose	WBRT dose	Response to CHT	Medial survival (months)
[51]	52	3.5 g/m^2	45 Gy	90% ORR 56% CR 33% PR	60
[53]	13	3 g/m^2	39.6 Gy	92% ORR 77% CR 15% PR	25+
[52]	102	2.5 g/m^2	45 Gy	94% ORR 58% CR 36% PR	36+

a confirmatory phase II study by the European Organization for Research and Treatment of Cancer Lymphoma Group (EORTC) [55]. MBVP consisted of MTX 3 g/m^2 days 1 and 15, teniposide 100 mg/m^2 days 2 and 3, carmustine 100 mg/m^2 day 4, methylprednisolone 60 mg/m^2 days 1 to 5, and two intrathecal injections of MTX 15 mg, cytarabine 40 mg, and hydrocortisone 25 mg followed by 40 Gy of RT. The overall response rate was 81 per cent and two and three year survival rates were 69 and 58 per cent, respectively. The CR rates after chemotherapy alone were 19 and 33 per cent after one and two cycles, respectively [56].

Drug	Dose	Route	Day
Methylprednisolone	60 mg/m^2	PO	1–5
MTX	3 g/m^2	IV	1, 15
Teniposide	100 mg/m^2	IV	2, 3
Carmustine	100 mg/m^2	IV	4
Repeated every 4 weeks × 2 cycles			
MTX	15 mg	Intrathecally	1, 15
ARA-C	40 mg	Intrathecally	1, 15
Hydrocortisone	25 mg	Intrathecally	1, 15
Weekly × 4			

Carmustine [BCNU], Vincristine [Oncovir], Etoposide, Methotrexate, Methylprednisolone [Solumedrol] (BOMES)

BOMES is a systemic chemotherapy regimen that resembles the MBVP regimen and includes HD-MTX, carmustine, vincristine, etoposide, and methylprednisolone as follows: Carmustine 65 mg/m^2/day IV on days 1–2; Vincristine 2 mg/day IV on days 1 and 8; Methotrexate 1.5 g/m^2 IV on day 15 followed by leucovorin rescue; Etoposide 50 mg/m^2/day IV on days 1–5; and Methylprednisolone 200 mg/day IV on days 1–7; repeated every 4 weeks. Four doses of intrathecal methotrexate were given to patients who had involvement in the cerebrospinal fluid [57].

A total of 19 patients (13 men) were enrolled. Nine patients were previously treated by radiotherapy (four patients), chemotherapy (three patients), or both (two patients). The average number of cycles/patient was 3.3. Radiotherapy was added only at progression or recurrence for patients who had not received RT prior to enrolling in the BOMES protocol.

There were 11 CR (57.9 per cent) and five partial remissions (26.3 per cent), with a total remission rate of 84.2 per cent. The median time to progression of the responders was six months. At last follow-up, four patients were alive without lymphoma at 10, 47, 64, and 66 months, respectively [57].

Methotrexate, Carmustine, Vincristine, Cytarabine (BVAM)

BVAM or CHOD/BVAM is another high-dose methotrexate-based combination chemoradiation regimen that was described initially in 1991 and for which long-term follow-up data was recently published [58]. It is also informative regarding the WBRT dose that was compared at two different doses. This regimen is complex and outlined in the table below.

Drug	Dose	Route	Day
CHOD:			
Cyclophosphamide	750 mg/m^2	IV	1
Doxorubicin	50 mg/m^2	IV	1
VCR	1.4 mg/m^2 [up to 2 mg]	IV	1
Dexamethasone	4 mg qds	PO	1–7
BVAM:			
Carmustine	100 mg/m^2	IV	8 and 50
VCR	1.4 mg/m^2 [up to 2 mg]	IV	15, 29, 43, 57, 71, and 85
MTX	1.5 g/m^2	IV	15, 29, 43, 57, 71, and 85
Leucovorin rescue	15 mg qds	IV/PO	For 3 days
ARA-C	3 g/m^2	IV	16, 30, 44, 58, 72, and 86

This regimen yielded a 62 per cent CR rate after chemotherapy, up to 73 per cent after additional radiotherapy and an overall survival rate of 32 per cent at 10 years for patients younger than 60 years [58,59]. When the same regimen was conducted with a reduced dose radiotherapy to 30.6 Gy, the overall survival at three years in patients younger than 60 years who had achieved a CR after chemotherapy

800 mg/m^2 have been safely infused in order to attain a higher CSF concentration. The measured CSF concentration after six and seven infusions were 1.76 per cent (10 µG/ml) and 1.04 per cent (8.5 µG/ml) of the corresponding serum levels, respectively [70]. Immune and allergic reactions may develop; at higher doses as used in PCNSL, prolonged or delayed neutropenia or other marrow lineage aplasias have been described [71–73].

A recent pilot study combined rituximab with MPV in an attempt to determine feasibility and improve disease control. The rituximab was dosed at 500 mg/m^2 with each cycle of chemotherapy. An increase in the grade 3–4 bone marrow toxicity when compared to the historical results of MPV was observed requiring routine GCSF prophylaxis [54]. Preliminary results show a 93 per cent overall response rate (ORR) and 73 per cent CR rate.

Other Agents

Many agents have shown some activity in PCNSL, but aside from MTX, none has adequate single-agent activity in this disease. Most are used as part of combination regimens and may have an additive activity. Table 28.5 summarizes these agents.

NON-METHOTREXATE COMBINATION REGIMENS

PCV is a combination of procarbazine, lomustine (CCNU), and vincristine that was initially designed for use in gliomas, and was studied as a part of a multimodality approach in newly diagnosed PCNSL, as well as for salvage treatment in refractory or relapsed disease [96].

Chamberlain and colleagues studied post radiotherapy adjuvant PCV. The therapeutic schema consisted of lydroxyurea along with WBRT (55–62 Gy), followed by adjuvant PCV (CCNU 110 mg/m^2 day 1, procarbazine 60 mg/m^2/d days 8–21, and vincristine 1.4 mg/m^2 day 8 and day 29) every 6–8 weeks for one year. The overall response rate was 75 per cent and the median overall survival, 41 months [81].

In the salvage setting, this same regimen was investigated by Herrlinger and colleagues in patients who had relapsed after attaining a CR following HD-MTX and WBRT, and a small subset that had refractory disease and did not receive WBRT. The ORR was 86 per cent, (57 per cent CR and 28 per cent PR), the neurological toxicities were worse in patients pre-treated with WBRT [97]. This regimen may be

TABLE 28.5   Chemotherapeutic agents with known activity in PCNSL

Agent	Properties and use in PCNSL
Methotrexate	Antimetabolite, intravenous high dose single agent or in combination. Intrathecal, intravitreous or intra-arterial administration possible. Requires aggressive hydration and alkalinization with leucovorin rescue [48,49,74–76].
Cytarabine	Antimetabolite, prolonged intra-CSF activity, intravenous and intrathecal administration as a single agent or in combination [51,64,77].
Corticosteroids	Steroid hormones, oral or intravenous, single agent or in combination trigger rapid oncolytic response but have numerous side effects [22,78,79]
Procarbazine	Oral methyl hydrazine derivative used in conjunction with HD-MTX [80–83]
Vincristine	Mitotic spindle inhibitor, intravenous administration in combination regimen, has cumulative neurotoxicity [84]
Thiotepa	Alkylating agent, IV in combination regimen or HD intensification prior to stem cell transplant, also intrathecal and intravitreal administration [85–89].
Temozolomide	Oral imidazotetrazine agent used alone or in combination regimen [90].
Rituximab	Human/murine chimeric anti CD-20 monoclonal antibody, IV, intrathecal or ventricular administration in combination regimen, may prolong post chemotherapy granulocytic recovery [63,69,91,92].
Topotecan	Oral or IV topoisomerase I inhibitor, few reports of activity in PCNSL with dose limiting myelosuppression [93–95].
Lomustine	Oral nitrosourea used in combination regimen, has cumulative pulmonary and renal toxicities [57,81].

useful for salvage or even as first line in patients with severe renal insufficiency or an absolute contraindication to methotrexate.

INTRATHECAL CHEMOTHERAPY

Although the experience from systemic lymphoma and leukemia clearly supports the use of intrathecal chemotherapy to prevent or treat leptomeningeal disease, its role in the treatment of PCNSL remains controversial. Most therapeutic regimens for PCNSL rely on high-dose MTX and high-dose ARA-C, both of which, when administered as a rapid infusion, can achieve tumoricidal CSF levels, thus obviating the need for IT chemotherapy [41,46]. Furthermore, at

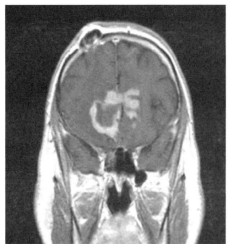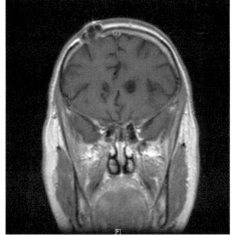

FIGURE 28.2 Post Gadolinium injection T1 weighted Coronal views of PCNSL, prior to and after 5 cycles of HD-MTX based Chemotherapy.

least two retrospective studies have failed to demonstrate any advantage in terms of disease control or survival for patients treated with intrathecal chemotherapy [41,46,47].

SALVAGE THERAPY

Despite high initial CR rates observed after first-line treatment for PCNSL, 10–35 per cent will have primary chemo-refractory disease, and more than half of the patients attaining a remission will relapse. Florid leptomeningeal recurrences are common and 10 per cent of recurrent disease is observed outside the CNS. At relapse, all patients should be completely restaged and treatment must be individualized and based primarily on site of recurrence, prior therapy and the patient's overall clinical and neurological condition.

Patients who deferred initial radiotherapy may derive palliative benefit at relapse. Re-irradiation combined with chemotherapy is also feasible in relapsed previously irradiated patients and prolonged the time to death from 10 to 16.5 months; toxicities were not described.

Most recurrent PCNSL remains chemosensitive with responses reported using re-induction with MTX, temozolomide, topotecan, rituximab, thiotepa, and other single- or multi-agent regimens [90,92,94,98].

Re-induction with HD-MTX in patients who are failing and whose initial therapy included WBRT are at increased risk of leukoencephalopathy. Cytarabine has been the most extensively used cytostatic in this subset of patients at recurrence [99]. Temozolomide for salvage has been tried at 150 mg/m^2/day for five days on a 28-day cycle basis, which yielded a 26 per cent ORR [98]. Addition of intravenous rituximab 500–750 mg/m^2 of RTX weekly × four weeks on a more intensive temozolomide schedule using 150 mg/m^2 daily from day 1–7 and 15–21 every four weeks has resulted in a 53 per cent ORR with a disease-free survival and median survival times of 7.7 and 14 months respectively [100].

VIA is a regimen that was studied in recurrent PCNSL, and consisted of etoposide 100 mg/m^2/day on days 1–3, ifosfamide 1000 mg/m^2/day on days 1–5, and cytarabine 2000 mg/m^2/12 h day 1. The therapy was repeated every 28 days for a total of six cycles. The median number of cycles received was four, for a 37 per cent CR rate, with a 41 per cent overall survival at 12 months [101].

CYVE followed by autologous stem cell transplant has been studied in recurrent/refractory PCNSL in the cases of documented chemosensitivity. It included high-dose cytarabine (2 g/m^2/day on days 2 through 5 in a 3-h infusion; 50 mg/m^2/day on days 1 through 5 in a 12-h infusion) and etoposide (VP-16; 200 mg/m^2/day on days 2 through 5 in a 2-h infusion) (CY). The intensive chemotherapy consisted of high-dose thiotepa (250 mg/m^2/d days 29 through 27) plus busulfan (total dose 10 mg/kg days 26 through 24) and cyclophosphamide (60 mg/kg/d days 23 and 22). This was followed by the autologous stem cell infusion. Dose reduction was allowed for patients > 60 years. Although the treatment-related toxicities and mortality were high, the overall probability of survival at three years was 63.7 per cent. After intensive chemotherapy, that probability was 60 per cent and the three-year probability of event-free survival was 53 per cent [102].

PRIMARY INTRAOCULAR LYMPHOMA

Primary intraocular lymphoma is a subset of PCNSL involving the vitreous, retina and choroids and should be approached in an identical manner. The risk of subsequent recurrence elsewhere in the CNS is as high as 80 per cent [103]. Diagnosis is made with a slit lamp examination, but vitreal aspirates or vitrectomies are sometimes required. While radiotherapy to the globes is almost always successful, relapse is common [104]. Treatment with high-dose MTX or cytarabine in combination with ocular radiotherapy may result in improved disease control. Alternatively, intra-vitreal injections of MTX or thiotepa have been successfully used in some patients [25,105,106].

HIV-RELATED PCNSL

Since the advent of HAART therapy, the incidence of HIV-related PCNSL has substantially declined and therapeutic options for affected patients have improved. A low CD4 count (< 50 cell/μl.) and elevated peripheral viral load are the most important risk factors. The patient population tends to be younger (median 37 years) and predominantly male. Most HIV-related PCNSL is of B-cell phenotype. However, in distinction to PCNSL in the immuno-competent population, the majority of HIV-related PCNSL has associated EBV DNA. Single-photon emission CT (SPECT) scanning using Thallium 201 may be used to distinguish between CNS toxoplas-mosis in the AIDS patient and PCNSL with both sensitivity and a specificity approaching 95 per cent [16,17]. In combination with positive CSF EBV titers, sensitivity and specificity of SPECT to diagnose PCNSL is 100 per cent [107].

Many reports of improved survival with HAART therapy alone were explained by the immune recon-stitution in the AIDS patient, and several small institutional studies have published median survivals approaching or exceeding one year in patients receiving HAART with or without other specific anti-tumor therapy [108,109].

Therefore, treatment of HIV+ patients with newly diagnosed PCNSL should include initiation or opti-mization of HAART and appropriate prophylaxis for opportunistic infections, in addition to MTX-based chemotherapy given the fact that WBRT may exacer-bate or accelerate the risk of HIV related dementia [110,111].

In a series of 15 AIDS patients diagnosed with PCNSL (mean CD4 of 30 cell/μl and a KPS of 50) Jacomet and colleagues used HD-MTX 3 g/m^2 every 14 days for six cycles. This approach yielded a CR rate of 50 per cent and a median survival of 10 months [111].

In 29 patients with a median CD4 count of 133 cell/μl, Tosi and colleagues evaluated oral zidovudine (2, 4, and 6 mg/m^2) and IV MTX at 1 gm/m^2 (moderate-dose) plus leucovorin rescue weekly for three to six cycles. 46 per cent had a complete response and the median survival was 12 months [110].

Given the fact that nearly 100 per cent of AIDS-PCNSL is EBV positive, strategies that directly target EBV or causative viruses such as HHV-8 may be effective and are being explored [112]. A small series of patients treated with parenteral zidovudine (1.6 g twice daily), ganciclovir (5 mg/kg twice daily), and interleukin 2 (2 million units twice daily) produced an excellent response in four of five patients [113].

References

1. Paulus, W. (1999) Classification, pathogenesis and molecular pathology of primary CNS lymphomas. *J Neurooncol* **43**, 203–8.
2. El Kamar, F. G., and Abrey, L. E. (2004) Management of Primary Central Nervous System Lymphoma. *J Natl Cancer Comprehensive Network* **2**, 341–349.
3. Taylor, C. R., Russell, R., Lukes, R. J., and Davis, R. L. (1978) An immunohistological study of immunoglobulin content of primary central nervous system lymphomas. *Cancer* **41**, 2197–205.
4. Tuaillon, N., and Chan, C. C. (2001) Molecular analysis of primary central nervous system and primary intraocular lymphomas. *Curr Mol Med* **1**, 259–72.
5. Gijtenbeek, J. M., Rosenblum, M. K., and DeAngelis, L. M. (2001) Primary central nervous system T-cell lymphoma. *Neurology* **57**, 716–8.
6. Camilleri-Broet, S., Martin, A., Moreau, A. *et al.* (1998) Primary central nervous system lymphomas in 72 immunocompetent patients: pathologic findings and clinical correlations. Groupe Ouest Est d'etude des Leucenies et Autres Maladies du Sang (GOELAMS). *Am J Clin Pathol* **110**, 607–12
7. Schabet, M. (1999) Epidemiology of primary CNS lymphoma. *J Neurooncol* **43**, 199–201
8. Olson, J. E., Janney, C. A., Rao, R. D. *et al.* (2002) The continuing increase in the incidence of primary central nervous system non-Hodgkin lymphoma: a surveillance, epidemiology, and end results analysis. *Cancer* **95**, 1504–10.
9. Cote, T. R., Manns, A., Hardy, C. R., Yellin, F. J., and Hartge, P. (1996) Epidemiology of brain lymphoma among people with or without acquired immunodeficiency syndrome. AIDS/Cancer Study Group. *J Natl Cancer Inst* **88**, 675–9.
10. CBTRUS (2002) Primary Brain Tumors in the United States, 1995–1999. Central Brain Tumor Registry of the United States.

11. Schaller, C., and Kelly, P. J. (1996) Primary central nervous system non-Hodgkin's lymphoma (PCNSL): does age and histology at presentation affect outcome? *Zentralbl Neurochir* **57**, 156–62.

12. Ferreri, A. J., Blay, J. Y., Reni, M. *et al.* (2003) Prognostic scoring system for primary CNS lymphomas: the International Extra-nodal Lymphoma Study Group experience. *J Clin Oncol* **21**, 266–72.

13. Buhring, U., Herrlinger, U., Krings, T., Thiex, R., Weller, M., and Kuker, W. (2001) MRI features of primary central nervous system lymphomas at presentation. *Neurology* **57**, 393–6.

14. Jenkins, C. N., and Colquhoun, I. R. (1998) Characterization of primary intracranial lymphoma by computed tomography: an analysis of 36 cases and a review of the literature with particular reference to calcification haemorrhage and cyst formation. *Clin Radiol* **53**, 428–34.

15. Roelcke, U., and Leenders, K. L. (1999) Positron emission tomography in patients with primary CNS lymphomas. *J Neurooncol* **43**, 231–6.

16. Kessler, L. S., Ruiz, A., Donovan Post, M. J., Ganz, W. I., Brandon, A. H., and Foss, J. N. (1998) Thallium-201 brain SPECT of lymphoma in AIDS patients: pitfalls and technique optimization. *AJNR Am J Neuroradiol* **19**, 1105–9.

17. Lorberboym, M., Wallach, F., Estok, L. *et al.* (1998) Thallium-201 retention in focal intracranial lesions for differential diagnosis of primary lymphoma and nonmalignant lesions in AIDS patients. *J Nucl Med* **39**, 1366–9.

18. Deangelis, L. M., Yahalom, J., Rosenblum, M., and Posner, J. B. (1987) Primary CNS lymphoma: managing patients with spontaneous and AIDS-related disease. *Oncology (Huntingt)* **1**, 52–62.

19. Gurney, J. G., and Kadan-Lottick, N. (2001) Brain and other central nervous system tumors: rates, trends, and epidemiology. *Curr Opin Oncol* **13**, 160–6.

20. DeAngelis, L. M., and Yahalom, J. (2001) Primary Central Nervous lymphoma. *In Cancer: Principles and Practice of Oncology* (V. DeVita, S. Hellman, and S. Rosenberg, Eds.), pp. 2330–2339. Lippincott Williams and Wilkins, Philadelphia, PA.

21. Todd, F. D., 2nd, Miller, C. A., Yates, A. J., and Mervis, L. J. (1986) Steroid-induced remission in primary malignant lymphoma of the central nervous system. *Surg Neurol* **26**, 79–84.

22. Weller, M. (1999) Glucocorticoid treatment of primary CNS lymphoma. *J. Neurooncol* **43**, 237–9.

23. Nelson, D. F., Martz, K. L., Bonner, H. *et al.* (1992) Non-Hodgkin's lymphoma of the brain: can high dose, large volume radiation therapy improve survival? Report on a prospective trial by the Radiation Therapy Oncology Group (RTOG): RTOG 8315. *Int J Radiat Oncol Biol Phys* **23**, 9–17.

24. Berry, M. P., and Simpson, W. J. (1981) Radiation therapy in the management of primary malignant lymphoma of the brain. *Int J Radiat Oncol Biol Phys* **7**, 55–9.

25. Hormigo, A., and DeAngelis, L. M. (2003) Primary ocular lymphoma: clinical features, diagnosis, and treatment. *Clin Lymphoma* **4**, 22–9.

26. Hochberg, F. H., Loeffler, J. S., and Prados, M. (1991) The therapy of primary brain lymphoma. *J Neurooncol* **10**, 191–201.

27. Marks, J. E., Baglan, R. J., Prassad, S. C., and Blank, W. F. (1981) Cerebral radionecrosis: incidence and risk in relation to dose, time, fractionation and volume. *Int J Radiat Oncol Biol Phys* **7**, 243–52.

28. Ferreri, A. J., Reni, M., and Villa, E. (2000) Therapeutic management of primary central nervous system lymphoma: lessons from prospective trials. *Ann Oncol* **11**, 927–37.

29. Stewart, D. J., Russell, N., Atack, E. A., Quarrington, A., and Stolbach, L. (1983) Cyclophosphamide, doxorubicin, vincristine, and dexamethasone in primary lymphoma of the brain: a case report. *Cancer Treat Rep* **67**, 287–91.

30. Shibamoto, Y., Tsutsui, K., Dodo, Y., Yamabe, H., Shima, N., and Abe, M. (1990) Improved survival rate in primary intracranial lymphoma treated by high-dose radiation and systemic vincristine-doxorubicin-cyclophosphamide-prednisolone chemotherapy. *Cancer* **65**, 1907–12.

31. Lachance, D. H., Brizel, D. M., Gockerman, J. P. *et al.* (1994) Cyclophosphamide, doxorubicin, vincristine, and prednisone for primary central nervous system lymphoma: short-duration response and multifocal intracerebral recurrence preceding radiotherapy. *Neurology* **44**, 1721–7.

32. Mead, G. M., Bleehen, N. M., Gregor, A. *et al.* (2000) A medical research council randomized trial in patients with primary cerebral non-Hodgkin lymphoma: cerebral radiotherapy with and without cyclophosphamide, doxorubicin, vincristine, and prednisone chemotherapy. *Cancer* **89**, 1359–70.

33. Schultz, C., Scott, C., Sherman, W. *et al.* (1996) Preirradiation chemotherapy with cyclophosphamide, doxorubicin, vincristine, and dexamethasone for primary CNS lymphomas: initial report of radiation therapy oncology group protocol 88–06. *J Clin Oncol* **14**, 556–64.

34. Reni, M., Ferreri, A. J., Garancini, M. P. *et al.* (1997) Therapeutic management of primary central nervous system lymphoma in immunocompetent patients: results of a critical review of the literature. *Ann Oncol* **8**, 227–34.

35. Ott, R. J., Brada, M., Flower, M. A., Babich, J. W., Cherry, S. R., and Deehan, B. J. (1991) Measurements of blood-brain barrier permeability in patients undergoing radiotherapy and chemotherapy for primary cerebral lymphoma. *Eur J Cancer* **27**, 1356–61.

36. Molnar, P. P., O'Neill, B. P., Scheithauer, B. W., and Groothuis, D. R. (1999) The blood-brain barrier in primary CNS lymphomas: ultrastructural evidence of endothelial cell death. *Neuro-oncol* **1**, 89–100.

37. Blay, J. Y., Conroy, T., Chevreau, C. *et al.* (1998) High-dose methotrexate for the treatment of primary cerebral lymphomas: analysis of survival and late neurologic toxicity in a retrospective series. *J Clin Oncol* **16**, 864–71.

38. Loeffler, J. S., Ervin, T. J., Mauch, P. *et al.* (1985) Primary lymphomas of the central nervous system: patterns of failure and factors that influence survival. *J Clin Oncol* **3**, 490–4.

39. Guha-Thakurta, N., Damek, D., Pollack, C., and Hochberg, F. H. (1999) Intravenous methotrexate as initial treatment for primary central nervous system lymphoma: response to therapy and quality of life of patients. *J Neurooncol* **43**, 259–68.

40. Ferreri, A. J., Abrey, L. E., Blay, J. Y. *et al.* (2003) Summary statement on primary central nervous system lymphomas from the Eighth International Conference on Malignant Lymphoma, Lugano, Switzerland, June 12 to 15, 2002. *J Clin Oncol* **21**, 2407–14.

41. Hiraga, S., Arita, N., Ohnishi, T. *et al.* (1999) Rapid infusion of high-dose methotrexate resulting in enhanced penetration into cerebrospinal fluid and intensified tumor response in primary central nervous system lymphomas. *J Neurosurg* **91**, 221–30.

42. Gabbai, A. A., Hochberg, F. H., Linggood, R. M., Bashir, R., and Hotleman, K. (1989) High-dose methotrexate for non-AIDS primary central nervous system lymphoma. Report of 13 cases. *J Neurosurg* **70**, 190–4.

43. Glass, J., Gruber, M. L., Cher, L., and Hochberg, F. H. (1994) Preirradiation methotrexate chemotherapy of primary central

nervous system lymphoma: long-term outcome. *J Neurosurg* **81**, 188–95.

44. DeAngelis, L. M., Yahalom, J., Thaler, H. T., and Kher, U. (1992) Combined modality therapy for primary CNS lymphoma. *J Clin Oncol* **10**, 635–43.

45. Batchelor, T., Carson, K., O'Neill, A. *et al.* (2003) Treatment of primary CNS lymphoma with methotrexate and deferred radiotherapy: a report of NABTT 96–07. *J Clin Oncol* **21**, 1044–9.

46. Khan, R. B., Shi, W., Thaler, H. T., DeAngelis, L. M., and Abrey, L. E. (2002) Is intrathecal methotrexate necessary in the treatment of primary CNS lymphoma? *J Neurooncol* **58**, 175–8.

47. Ferreri, A. J., Guerra, E., Regazzi, M. *et al.* (2004) Area under the curve of methotrexate and creatinine clearance are outcome-determining factors in primary CNS lymphomas. *Br J Cancer* **90**, 353–8.

48. Doolittle, N. D., Miner, M. E., Hall, W. A. *et al.* (2000) Safety and efficacy of a multicenter study using intraarterial chemotherapy in conjunction with osmotic opening of the blood-brain barrier for the treatment of patients with malignant brain tumors. *Cancer* **88**, 637–47.

49. Neuwelt, E. A., Goldman, D. L., Dahlborg, S. A. *et al.* (1991) Primary CNS lymphoma treated with osmotic blood-brain barrier disruption: prolonged survival and preservation of cognitive function. *J Clin Oncol* **9**, 1580–90.

50. Hochberg, F. H., and Tabatabai, G. (2001) Therapy of PCNSL at the Massachusetts General Hospital with high dose methotrexate and deferred radiotherapy. *Ann Hematol* **80** (Suppl 3), B111–2.

51. Abrey, L. E., Yahalom, J., and DeAngelis, L. M. (2000) Treatment for primary CNS lymphoma: the next step. *J Clin Oncol* **18**, 3144–50.

52. DeAngelis, L. M., Seiferheld, W., Schold, S. C., Fisher, B., and Schultz, C. J. (2002) Combination chemotherapy and radiotherapy for primary central nervous system lymphoma: Radiation Therapy Oncology Group Study 93–10. *J Clin Oncol* **20**, 4643–8.

53. Ferreri, A. J., Reni, M., Dell'Oro, S. *et al.* (2001) Combined treatment with high-dose methotrexate, vincristine and procarbazine, without intrathecal chemotherapy, followed by consolidation radiotherapy for primary central nervous system lymphoma in immunocompetent patients. *Oncology* **60**, 134–40.

54. El Kamar, F. G., DeAngelis, L. M., Yahalom, J. *et al.* (2004). Combined immunochemotherapy with reduced dose whole brain radiation therapy for newly diagnosed patients with Primary Central Nervous System Lymphoma. *Proc Am Soc Clin Oncol* **23**, 111-A 1518.

55. Desablens, B., Francois, S., Sensebe, L. *et al.* (1996). Primary CNS lymphomas in non-HIV patients: Five-year results on 97 cases treated by the POF LCP 88 trial. *Ann Oncol* (suppl 3) **7**, 216.

56. Poortmans, P. M., Kluin-Nelemans, H. C., Haaxma-Reiche, H. *et al.* (2003) High-dose methotrexate-based chemotherapy followed by consolidating radiotherapy in non-AIDS-related primary central nervous system lymphoma: European Organization for Research and Treatment of Cancer Lymphoma Group Phase II Trial 20962. *J Clin Oncol* **21**, 4483–8.

57. Cheng, A. L., Yeh, K. H., Uen, W. C., Hung, R. L., Liu, M. Y., and Wang, C. H. (1998) Systemic chemotherapy alone for patients with non-acquired immunodeficiency syndrome-related central nervous system lymphoma: a pilot study of the BOMES protocol. *Cancer* **82**, 1946–51.

58. Bessell, E. M., Graus, F., Lopez-Guillermo, A., Lewis, S. A., Villa, S., Verger, E., and Petist, J. (2004) Primary non-Hodgkin's lymphoma of the CNS treated with CHOD/BVAM or BVAM

59. Bessell, E. M., Graus, F., Lopez-Guillermo, A. *et al.* (2001) CHOD/BVAM regimen plus radiotherapy in patients with primary CNS non-Hodgkin's lymphoma. *Int. J Radi Oncol Biol Phy* **50**, 457–464.

60. Bessell, E. M., Lopez-Guillermo, A., Villa, S. *et al.* (2002) Importance of Radiotherapy in the Outcome of Patients With Primary CNS Lymphoma: An Analysis of the CHOD/BVAM Regimen Followed by Two Different Radiotherapy Treatments. *J Clin Oncol* **20**, 231–236.

61. Sandor, V., Stark-Vancs, V., Pearson, D. *et al.* (1998) Phase II trial of chemotherapy alone for primary CNS and intraocular lymphoma. *J Clin Oncol* **16**, 3000–6.

62. McAllister, L. D., Doolittle, N. D., Guastadisegni, P. E. *et al.* (2000) Cognitive outcomes and long-term follow-up results after enhanced chemotherapy delivery for primary central nervous system lymphoma. *Neurosurgery* **46**, 51–60.

63. Pels, H., Schmidt-Wolf, I. G., Glasmacher, A. *et al.* (2003) Primary central nervous system lymphoma: results of a pilot and phase II study of systemic and intraventricular chemotherapy with deferred radiotherapy. *J Clin Oncol* **21**, 4489–95.

64. Damon, L. E., Plunkett, W., and Linker, C. A. (1991) Plasma and cerebrospinal fluid pharmacokinetics of 1-beta-D-arabinofuranosylcytosine and 1-beta-D-arabinofuranosyluracil following the repeated intravenous administration of high- and intermediate-dose 1-beta-D-arabinofuranosylcytosine. *Cancer Res.* **51**, 4141–5.

65. DeAngelis, L. M. (1994) Primary central nervous system lymphoma. *Recent Results Cancer Res* **135**, 155–69.

66. Abrey, L. E, Yahalom, J., and Deangelis, L. M. (1997) Relapse and late neurotoxicity in primary central nervous system lymphoma. *Proceedings of the American Academy Neurology* Neurology, pp. **48**, A18.

67. Ferreri, A. J., Reni, M., Pasini, F. *et al.* (2002) A multicenter study of treatment of primary CNS lymphoma. *Neurology* **58**, 1513–20.

68. Ruhstaller, T. W., Amsler, U., and Cerny, T. (2000) Rituximab: active treatment of central nervous system involvement by non-Hodgkin's lymphoma? *Ann Oncol* **11**, 374–5.

69. Grillo-Lopez, A. J., White, C. A., Dallaire, B. K. *et al.* (2000) Rituximab: the first monoclonal antibody approved for the treatment of lymphoma. *Curr Pharm Biotechnol* **1**, 1–9.

70. Rubenstein, J. L., Combs, D., Rosenberg, J. *et al.* (2003) Rituximab therapy for CNS lymphomas: targeting the leptomeningeal compartment. *Blood* **101**, 466–8.

71. Chaiwatanatorn, K., Lee, N., Grigg, A., Filshie, R., and Firkin, F. (2003) Delayed-onset neutropenia associated with rituximab therapy. *Br J Haematol* **121**, 913–8.

72. Voog, E., Morschhauser, F., and Solal-Celigny, P. (2003) Neutropenia in patients treated with rituximab. *N Engl J Med* **348**, 2691–4 discussion 2691–4.

73. Papadaki, T., Stamatopoulos, K., Kosmas, C. *et al.* (2002) Clonal T-large granular lymphocyte proliferations associated with clonal B cell lymphoproliferative disorders: report of eight cases. *Leukemia* **16**, 2167–9.

74. Ramu, A., and Fusner, J. E. (1979) A pharmacokinetic model for predicting the concentration of methotrexate in plasma subsequent to intrathecal injection. *Isr J Med Sci* **15**, 494–9.

75. Chu, E., Mota, A. C., and M, F. (2001). *Pharmacology of cancer chemotherapy antimetabolites.* Lippincott Williams and Wilkins, Philadelphia, PA.

76. Velez, G., Boldt, H. C., Whitcup, S. M., Nussenblatt, R. B., and Robinson, M. R. (2002) Local methotrexate and dexamethasone

phosphate for the treatment of recurrent primary intraocular lymphoma. *Ophthalmic. Surg. Lasers* **33**, 329–33.

77. Chabner, B. (1996) *Cytidine Analogues.* Lippincott-Raven, Philadelphia, PA.

78. O'Neill, B. P., Habermann, T. M., Witzig, T. E., and Rodriguez, M. (1999) Prevention of recurrence and prolonged survival in primary central nervous system lymphoma (PCNSL) patients treated with adjuvant high-dose methylprednisolone. *Med Oncol* **16**, 211–5.

79. Posner, J. (1995) *Supportive care agents and their complications.* F. A. Davis, Philadelphia.

80. Zeller, P., Gutmann, H., Hegedus, B., Kaiser, A., Langemann, A., and Muller, M. (1963). Methylhydrazine derivatives, a new class of cytotoxic agents. *Experientia.* **19**, 129.

81. Chamberlain, M. C., and Levin, V. A. (1992) Primary central nervous system lymphoma: a role for adjuvant chemotherapy. *J Neurooncol* **14**, 271–5.

82. Spivack, S. D. (1974) Drugs 5 years later: procarbazine. *Ann Intern Med* **81**, 795–800.

83. Posner, J. (1995) *Side effects of chemotherapy.* F. A. Davis, Philadelphia.

84. Rosenthal, S., and Kaufman, S. (1974) Vincristine neurotoxicity. *Ann Intern Med* **80**, 733–7.

85. Tew, K., Colvin, M., and B, C. (1996) *Alkylating Agents.* Lippincott-Raven, Philadelphia.

86. Gutin, P. H., Levi, J. A., Wiernik, P. H., and Walker, M. D. (1977) Treatment of malignant meningeal disease with intrathecal thioTEPA: a phase II study. *Cancer Treat Rep* **61**, 885–7.

87. de Smet, M. D., Vancs, V. S., Kohler, D., Solomon, D., and Chan, C. C. (1999) Intravitreal chemotherapy for the treatment of recurrent intraocular lymphoma. *Br J Ophthalmol* **83**, 448–51.

88. Soussain, C., Suzan, F., Hoang-Xuan, K. *et al.* (2001) Results of intensive chemotherapy followed by hematopoietic stem-cell rescue in 22 patients with refractory or recurrent primary CNS lymphoma or intraocular lymphoma. *J Clin Oncol* **19**, 742–9.

89. Freilich, R. J., Delattre, J. Y., Monjour, A., and DeAngelis, L. M. (1996) Chemotherapy without radiation therapy as initial treatment for primary CNS lymphoma in older patients. *Neurology* **46**, 435–9.

90. Reni, M., Ferreri, A. J., Landoni, C., and Villa, E. (2000) Salvage therapy with temozolomide in an immunocompetent patient with primary brain lymphoma. *J Natl Cancer Inst* **92**, 575–6.

91. Maloney, D. G., Grillo-Lopez, A. J., White, C. A. *et al.* (1997) IDEC-C2B8 (Rituximab) anti-CD20 monoclonal antibody therapy in patients with relapsed low-grade non-Hodgkin's lymphoma. *Blood* **90** 2188–95.

92. Pels, H., Schulz, H., Schlegel, U., and Engert, A. (2003) Treatment of CNS lymphoma with the anti-CD20 antibody rituximab: experience with two cases and review of the literature. *Onkologie* **26**, 351–4.

93. Takimoto, C. H., and Arbuck, S. G. (1997) Clinical status and optimal use of topotecan. *Oncology (Huntingt)*, **11**, 1635–46; discussion 1649–51, 1655–7.

94. Ciordia, R., Hochberg, F., and Batchelor, T. (2000) Topotecan as salvage therapy for refractory or relapsed primary central nervous system lymphoma., *Proc Am Soc Clin Oncol* Abstract No: 639.

95. Yeh, K. H., Cheng, A. L., and Tien, H. F. (1995) Primary T cell leptomeningeal lymphoma–successful treatment with systemic chemotherapy. *Oncology* **52**, 501–4.

96. Chamberlain, M. C., and Levin, V. A. (1990) Adjuvant chemotherapy for primary lymphoma of the central nervous system. *Arch Neurol* **47**, 1113–6.

97. Herrlinger, U., Brugger, W., Bamberg, M., Kuker, W., Dichgans, J., and Weller, M. (2000) PCV salvage chemotherapy for recurrent primary CNS lymphoma. *Neurology* **54**, 1707–8.

98. Reni, M., Mason, W., Zaja, F. *et al.* (2004) Salvage chemotherapy with temozolomide in primary CNS lymphomas: preliminary results of a phase II trial. Therapeutic management of refractory or relapsed primary central nervous system lymphomas. *Eur J Cancer* **40**, 1682–8.

99. Reni, M., Ferreri, A. J., and Villa, E. (1999) Second-line treatment for primary central nervous system lymphoma. *Br J Cancer* **79**, 530–4.

100. Enting, R., Demopoulos, A., DeAngelis, L., and Abrey, L. (2004) Salvage therapy for primary CNS lymphoma with a combination of rituximab and temozolomide. *Neurology* **63**, 901–903, 2004.

101. Arellano-Rodrigo, E., Lopez-Guillermo, A., Bessell, E. M., Nomdedeu, B., Montserrat, E., and Graus, F. (2003) Salvage treatment with etoposide (VP-16), ifosfamide and cytarabine (Ara-C) for patients with recurrent primary central nervous system lymphoma. *Eur J Haematol* **70**, 219–24.

102. Soussain, C., Suzan, F., Hoang-Xuan, K. *et al.* (2001) Results of Intensive Chemotherapy Followed by Hematopoietic Stem-Cell Rescue in 22 Patients With Refractory or Recurrent Primary CNS Lymphoma or Intraocular Lymphoma. *J Clin Oncol* **19**, 742–749.

103. Peterson, K., Gordon, K. B., Heinemann, M. H., and DeAngelis, L. M. (1993). The clinical spectrum of ocular lymphoma. *Cancer* **72**, 843–9.

104. Ferreri, A. J. M., Blay, J. -Y., Reni, M. *et al.* (2002) Relevance of intraocular involvement in the management of primary central nervous system lymphomas. *Ann Oncol* **13**, 531–538.

105. Whitcup, S. M., Stark-Vancs, V., Wittes, R. E. *et al.* (1997) Association of interleukin 10 in the vitreous and cerebrospinal fluid and primary central nervous system lymphoma. *Arch Ophthalmol* **115**, 1157–60.

106. Chan, C. C., Buggage, R. R., and Nussenblatt, R. B. (2002) Intraocular lymphoma. *Curr Opin Ophthalmol* **13**, 411–8.

107. Antinori, A., De Rossi, G., Ammassari, A. *et al.* (1999) Value of combined approach with thallium-201 single-photon emission computed tomography and Epstein-Barr virus DNA polymerase chain reaction in CSF for the diagnosis of AIDS-related primary CNS lymphoma. *J Clin Oncol* **17**, 554–60.

108. McGowan, J. P., and Shah, S. (1998) Long-term remission of AIDS-related primary central nervous system lymphoma associated with highly active antiretroviral therapy. *AIDS* **12**, 952–4.

109. Skiest, D. J. and Crosby, C. (2003) Survival is prolonged by highly active antiretroviral therapy in AIDS patients with primary central nervous system lymphoma. *AIDS* **17**, 1787–93.

110. Tosi, P., Gherlinzoni, F., Visani, G. *et al.* (1998) AZT plus methotrexate in HIV-related non-Hodgkin's lymphomas. *Leuk, Lymphoma* **30**, 175–9.

111. Jacomet, C., Girard, P. M., Lebrette, M. G., Farese, V. L., Monfort, L., and Rozenbaum, W. (1997) Intravenous methotrexate for primary central nervous system non-Hodgkin's lymphoma in AIDS. *AIDS* **11**, 1725–30.

112. Roychowdhury, S., Peng, R., Baiocchi, R. A. *et al.* (2003) Experimental Treatment of Epstein-Barr Virus-associated Primary Central Nervous System Lymphoma. *Cancer Res.* **63**, 965–971.

113. Raez, L. E., Cassileth, P. A., Schlesselman, J. J. *et al.* (2004) Allogeneic vaccination with a B7.1 HLA-A gene-modified adenocarcinoma cell line in patients with advanced non-small-cell lung cancer. *J Clin Oncol* **22**, 2800–7.

Chemotherapy of Medulloblastoma

Herbert B. Newton

ABSTRACT: Medulloblastoma is the most common primary brain tumor in children and accounts for 25 per cent of newly diagnosed cases. Recent advances in treatment have extended 5-year survival rates from 3 to over 70 per cent during the past 50 years. These improvements in survival have resulted from a multi-modality approach that includes surgical resection, posterior fossa and craniospinal irradiation (RT), and chemotherapy for selected, high-risk patients. The literature regarding chemotherapy of adult and pediatric patients is reviewed in depth. The most active agents include cisplatin, CCNU, cyclophosphamide, vincristine, and carboplatin. Although patients are living longer with their disease, neuro-cognitive function and quality of life are often impaired after RT to the developing brain. To safely allow reductions in the dose of RT, the specificity and efficacy of chemotherapy must be improved. Recent advances in the molecular genetics of medulloblastoma transformation (e.g., myc, *PTCH*) are also reviewed and discussed. A thorough understanding of these pathways will be critical for the development of more specific, "targeted" therapeutic agents. Further clinical trials will be needed to evaluate the activity of these new drugs and determine their role in the treatment of patients with medulloblastoma.

INTRODUCTION

Medulloblastoma, also known as primitive neuroectodermal tumor (PNET) of the posterior fossa, is the most common primary brain tumor of childhood and accounts for 25 per cent of all newly diagnosed cases [1–5]. In adults, it is much less common and comprises only 1 per cent of new tumors [6,7]. The median age at diagnosis of medulloblastoma is 7 to 9 years; 80 per cent of all cases occur before the age of 20 years. In the adult sub-group, the median age at diagnosis ranges from 26 to 30 years. The overall annual incidence is approximately 5 new cases per million persons in the United States, which corresponds to roughly 350 new patients per year. There is a predilection for males to develop medulloblastoma, with ratios varying from 1.4:1 to 2.3:1 in most studies [2,3].

Medulloblastoma develops sporadically in the vast majority of adult and pediatric patients [2–5,8]. Although there have only been rare reports of familial cases not related to some form of hereditary condition, some studies do suggest an increased incidence of cancer among relatives of patients with medulloblastoma. When genetic factors do play a more direct role, it is usually in the context of a genetic syndrome or inherited disorder. The most commonly associated heritable disorders include Turcot's syndrome (adenomatous polyposis coli [APC] gene), Li-Fraumeni syndrome (p53 gene), ataxia-telangiectasia, and Gorlin's syndrome (*PTCH* gene) [9].

The prognosis and overall survival for patients with medulloblastoma has improved dramatically over the past seven decades. In the original series of 61 pediatric patients reported by Cushing in 1930, the 3-year survival rate was only 1.6 per cent [10]. Through the 1950s and 1960s, there was only modest improvement, with a 5-year survival rate of 3.0 per cent [11]. Since the 1970s, overall survival of patients with medulloblastoma has steadily improved, with recent 5-year survival rates approaching 65 to 80 per cent in many large series [3–5,7]. These extended survival rates have resulted from many factors, including refinements in neurosurgical technique (i.e., use of microsurgery), advances in neuro-imaging of posterior fossa tumors with computed tomography (CT) and magnetic resonance imaging (MRI), use of mega-voltage radiation therapy (RT) equipment and inclusion of the spinal neuraxis within the treatment field, and the addition of chemotherapy in selected patients [12–16].

NEURO-IMAGING AND INITIAL TREATMENT

The most sensitive technique to diagnose medulloblastoma is neuro-imaging with contrast-enhanced MRI [1,3–5,13,17]. MRI is more sensitive than CT for tumors within the posterior fossa and has the added benefit of midsagittal images. Hydrocephalus (75 to 85 per cent of cases) is clearly demonstrated with either modality. On T2 images, the tumor is a heterogeneous high signal lesion within the cerebellum (midline or hemispheric). Regions of calcification, cyst, necrosis, or hemorrhage are not common. On T1 images, the mass appears hypointense or isointense compared to surrounding brain. After administration of gadolinium, there is heterogeneous tumor enhancement in at least 90 per cent of the cases [13,17]. Peritumoral edema is often present, contributing to compression or displacement of the fourth ventricle. It is important to carefully review the brain MRI for leptomeningeal enhancement and infiltration into the brainstem, since this information is important for tumor staging and risk classification (i.e., low *versus* high). Between 30 and 40 per cent of patients will have leptomeningeal dissemination at diagnosis, often with spread into the spinal neuraxis. An enhanced MRI of the entire spine is necessary at diagnosis or several weeks after surgery, in addition to a lumbar puncture and CSF cytology, to evaluate for this possibility (i.e., staging evaluation) [18]. Patients in whom extraneural metastases are suspected require at least a skeletal survey and nuclear medicine screening evaluation.

Initial management involves symptom control and reduction of intracranial pressure. For most patients, dexamethasone (4–6 mg q6h) provides rapid relief of symptoms and allows for semi-elective surgical resection within 24 to 48 h [3–6,14,16,19,20]. In patients with hydrocephalus, a temporary external ventriculostomy (draining only 3–5 ml/h) should be considered. After surgical resection of the tumor, only 35–40 per cent of patients with hydrocephalus will require a permanent shunt. The goals of surgical treatment are to provide a histological diagnosis, re-establish CSF flow, normalize intracranial pressure, and reduce tumor burden. Although the tumor is vascular, it can be easily removed using suction or an ultrasonic aspirator, with minimal risk of hemorrhage

and blood loss. A gross radical resection should be attempted in all patients with non-disseminated disease (i.e., M0), if possible, without inducing neurological injury. The foramina of Luschka must be inspected for residual tumor. A recent Children's Cancer Group study suggests that a 90 per cent or greater resection can be achieved in at least 80 per cent of patients with medulloblastoma [21]. This study also demonstrated that complete removal or extensive subtotal resection of tumor was associated with extended survival. For all patients with M0 disease and less than 1.5 cm^2 of residual tumor, there was a 20 per cent improvement in progression-free survival at 5 years compared to patients with greater than 1.5 cm^2 of residual tumor ($p = 0.065$). However, other studies are contradictory and do not show a survival advantage for patients with gross total resection [3,5,22]. A contrast enhanced CT or MRI should be performed within 24 to 72 h after surgery to screen for residual tumor within the cerebellum or brainstem.

Before and after surgical resection, all patients are evaluated for the amount of residual disease at the resection site, prognostic markers, extent of disease outside the resection site, and then given a Chang staging class designation (i.e., T1–T4, M0–M4) [3]. Once all of this information is available, the patient can be assigned to a "low risk" or "high risk" group. Age is an important risk factor; children younger than 3 years of age are more likely to progress within 5 years of diagnosis than older children or adults. This is mostly due to the withholding of RT, although there may also be intrinsic biological differences in tumors of very young patients. The extent of residual tumor is also critical, with measurements of greater than or equal to 1.5 cm^2 associated with a poor clinical outcome. The presence of metastatic medulloblastoma within the cerebrospinal fluid or distant neuraxis is also a poor prognostic indicator. Patients with low or average risk disease are older than 3 years of age, have had a total or near-total resection (i.e., < 1.5 cm^2 on postoperative enhanced MRI), and do not have evidence for metastases. Patients are considered high risk if they are younger than 3 years of age, have greater than 1.5 cm^2 of residual tumor, or have any evidence for metastatic spread of disease.

Radiation therapy is responsible for most of the improvement in survival that has occurred over the past 50 years in patients with medulloblastoma [3–5,12,14,19,20,23]. The use of megavoltage equipment and the addition of spinal neuraxis irradiation has significantly extended 5-year survival rates (i.e., 50–70 per cent 5-year event-free survival with RT alone). Standard radiotherapy always involves concurrent treatment of the posterior fossa, brain, and spinal neuraxis [12,14,23]. The recommended dose for the posterior fossa is 5000–5500 cGy over 6–7 weeks, in daily fractions of 180–200 cGy. Doses of less than 5000 cGy result in a significantly increased local failure rate. Although the dose for the posterior fossa is generally agreed upon, there is controversy regarding the dose for the whole brain and spinal neuraxis. The traditional recommended doses are 4000 cGy to the brain and 3600 cGy to the spine for patients with low-risk disease [3–5,23]. Doses can be slightly increased for high-risk patients with leptomeningeal spread (i.e., 4500 cGy). Because of the potential damage that RT can cause to the pediatric nervous system, some investigators recommended using lower doses (2500 to 3000 cGy) to the brain and spine in low-risk patients [23,24]. However, recent reports from Deutsch and colleagues suggest that lower doses of neuraxis RT are associated with a higher risk of recurrence [25,26]. In a prospective, randomized trial of 126 patients with low-risk medulloblastoma (combined CCG-923/POG-8631 study), patients receiving the lower dose of neuraxis RT (2340 cGy) were more likely to develop early relapse, have early isolated neuraxis relapse, and have a reduced 5-year event-free survival when compared to patients receiving standard dose RT (3600 cGy). In further efforts to reduce the sequelae of irradiation in low-risk patients, some investigators have combined lower dose RT with multi-agent chemotherapy (cisplatin, lomustine, vincristine) [12,14,27]. Packer and co-workers used this approach in a cohort of 65 patients and demonstrated an overall survival (86 per cent at 3 years, 79 per cent at 5 years) that was comparable to that achieved with more traditional higher dose regimens. Other novel approaches designed to reduce the sequelae of RT and potentially improve efficacy include hyperfractionation techniques [28]. Prados and colleagues used hyperfractionated craniospinal RT (total 7200 cGy; 100 cGy twice per day) in a series of 39 patients. In the low-risk patient cohort, the overall 3- and 5-year survival rates were similar to previous reports of single-fraction RT. However, the failure rate outside the posterior fossa (43.7 per cent) was excessive. This data does not demonstrate any advantage for hyperfractionated irradiation over more traditional approaches and may, in fact, allow more frequent recurrences at distant sites.

CHEMOTHERAPY OF INFANTS AND VERY YOUNG CHILDREN

The developing nervous system of infants and very young children (newborn to 3 years) is highly sensitive to the effects of irradiation [12,14,29–31]. More than 60 per cent of surviving brain tumor patients will develop significant cognitive impairment after exposure to cranial RT. In this group of medulloblastoma patients (roughly 20 per cent of all pediatric cases, which also have the poorest prognosis for survival), chemotherapy has a very prominent role, because it is often applied "up front" to delay the use of RT until the patient matures beyond 2.5 to 3 years of age (see Tables 29.1 and 29.2) [3–5]. Initial reports by van Eys and colleagues have suggested that MOPP chemotherapy, using nitrogen mustard (6 mg/m^2), vincristine (1.4 mg/m^2), procarbazine (100 mg/m^2), and prednisone (40 mg/m^2) was effective against pediatric brain tumors, including medulloblastoma in infants [32–34]. They were able to demonstrate durable responses in 30–40 per cent of their infant cohort. In a subsequent report from the same group, follow-up data was available for 12 infants with medulloblastoma who had received MOPP as "up front" treatment [35]. Three patients remained free of enhancing tumor after complete surgical resection, while 6 patients had a complete response (CR) of residual enhancing disease. Seven patients eventually received radiotherapy for recurrent disease within the posterior fossa and/or CSF. The overall median survival for the cohort was 10.6 years, with a 5-year survival rate of 67 per cent. Serial intelligence testing showed a stable IQ in the non-irradiated cohort and a slow continuous decline of IQ for the irradiated group. In a study of high-risk medulloblastoma patients by Kretschmar and co-workers, six infants received pre-irradiation chemotherapy consisting of a 9 week course of cisplatin (100 mg/m^2) and vincristine (1.5 mg/m^2), followed by MOPP [36]. There were 5 responses (1 PR, 2 MR, 2 SD) to cisplatin and vincristine and 4 responses (2 CR, 2 SD) to MOPP. Although follow-up time was limited, survival ranged from 6 to 57 + months.

In a multicenter study by the Pediatric Oncology Group (POG), Duffner and colleagues reported the use of postoperative chemotherapy and delayed irradiation in 62 infants with medulloblastoma [37]. The regimen consisted of two 28-day cycles of cyclophosphamide (65 mg/kg) and vincristine (0.065 mg/kg), followed by one 28-day cycle of cisplatin (4 mg/kg) and etoposide (6.5 mg/kg). In the group of 27 patients with residual enhancing disease, there were 4 CR, 9 PR, and 9 SD (39 per cent objective response rate). The 2-year progression-free survival for the entire cohort was 34 per cent, with an overall 2-year survival of 46 per cent. In many of the patients, chemotherapy was effective enough to delay RT for one year or more. The extent of surgical resection

TABLE 28.4   Combination Methotrexate-Based Regimens

Ref.	N	Regimen	IT CHT	WBRT dose (Gy)	Response (%)	Median survival (months)
[58]	77	BVAM/CHOD (MTX 1.5 g/m^2)	–	45	73 CR	38
[57]	19	BOMES (MTX 1.5 g/m^2)	MTX	–	84 ORR	PFS = 6 months
[51]	52	MPV (MTX 3.5 g/m^2) Ara-C	MTX	45	90 ORR	60
[62]	74	IA MTX (2.5 g) + Cyclophosphamide and Etoposide	–	–	65 CR	40.7
[53]	13	MPV (MTX 3 g/m^2)	–	39.6	92 ORR	25+
[52]	102	MPV (MTX 2.5 g/m^2) Ara-C	MTX	45	94	30+
[56]	52	MBVP (MTX 3 g/m^2)	MTX ARA-C Hydrocortisone	30 + 10 Gy boost	81 ORR	46
[63]	65	MTX 5 g/m^2 ARA-C 3 g/m^2 Ifosfamide VCR Cyclophosphamide Dexamethasone	MTX ARA-C Prednisolone	–	71 ORR	>60 years = 34 <60 years = 60+

MTX – methotrexate; IA – intra-arterial; MTV – methotrexate, thiotepa, vincristine; IT – intra-thecal; Ara-C – cytarabine; BOMES – BCNU, vincristine, etoposide, methylprednisolone; MPV – methotrexate, procarbazine, vincristine; MTV – methotrexate, thiotepa, vincristine; MBVP – methotrexate, teniposide, carmustine, prednisolone; ORR – overall response rate.

was only 60 per cent, compared to 92 per cent in an identical population of the historical group using 45 Gy [60].

Methotrexate, Thiotepa, Vincristine (MTV)

MTV is a regimen that used an HD-MTX-based combination chemotherapy as a single modality. It was reported in a small phase II trial group of 14 patients. It incorporated a 24-h infusion of 8.4 g/m^2 methotrexate on day 2 with leucovorin rescue starting 12 h after the MTX infusion at 15 mg every six hours until MTX clearance was documented; 35 mg/m^2 of thiotepa on day 1; 1.4 mg/m^2 of vincristine on day 1; and dexamethasone 6 mg 4 times daily days 1–5 with intrathecal cytarabine (50 mg) and methotrexate (12 mg) alternating 5 times, administered in 21-day cycles. Six cycles were intended per patient, the mean number of cycles given was 4.8/patient.

The response rate was 100 per cent with 11 (79 per cent) complete responses and three (21 per cent) partial responses. Cumulative survival and progression-free survival rates at more than 4.5 years were 68.8 and 34.3 per cent, respectively. Median survival had not been reached when the study was presented, and median progression-free survival was 16.5 months [61].

Cytarabine

Cytarabine is the other agent commonly used for the treatment of PCNSL that has excellent penetration into the CSF and the intraocular compartments [64]. Although it may only have a cytostatic effect in this disease, it is often used as consolidation after radiation in the combined modality treatment regimens as outlined above, and is the most widely used drug alone or in combination for salvage therapy at recurrence [51,65]. The use of cytarabine as part of the therapeutic regimen in PCNSL has also been reported to contribute independently to improved survival [66,67].

Rituximab

Rituximab is a human/murine chimeric anti-CD-20 monoclonal antibody. Although it became part of almost all systemic B-cell lymphoma therapy, only recently has its potential role in the treatment of PCNSL been investigated. Single-agent rituximab has reported clinical and radiographic activity in PCNSL, and is being added to investigational protocols. Some authors are also studying its role when administered intrathecally [68–70].

High doses of rituximab are required intravenously to cross the blood–brain barrier and doses up to (i.e., complete resection) was the most significant favorable prognostic factor.

A similar multicenter trial by the Children's Cancer Group (CCG) was reported by Geyer and co-workers [38]. They used the 8 drugs in 1 day regimen (8-in-1; vincristine 1.5 mg/m^2, carmustine 75 mg/m^2, procarbazine 75 mg/m^2, methylprednisolone 300 mg/m^2, hydroxyurea 1500 mg/m^2, cisplatin 60 mg/m^2, cytosine arabinoside 300 mg/m^2, cyclophosphamide 300 mg/m^2) to treat a cohort of 46 infants less than 18 months of age with medulloblastoma; RT was either delayed or withheld. Among the 14 patients with residual enhancing disease, there was 1 CR, 5 PR, and 6 SD after two cycles of chemotherapy (42.8 per cent objective response rate). The median time to progression was 6 months, with a 2-year progression-free survival of 22 per cent. A sub-group of patients remained disease free without exposure to irradiation. The regimen was generally well tolerated, with mostly hematological toxicity.

Other regimens have been used for infants with medulloblastoma. Razzouk and colleagues reported treating 7 evaluable infants with thiotepa (3.25 mg/kg), followed by cisplatin, etoposide, cyclophosphamide, and vincristine in doses similar to the POG study [37,39]. Only 2 patients had brief objective responses. The median time to progression was 7 months, with an overall 3-year survival rate of 45 per cent. Hematological toxicity was severe in all patients. Thiotepa, in combination with the POG regimen, did not improve the progression-free and overall survival documented in the original POG study. In a report from the French Society of Pediatric Oncology, Dupuis-Girod and co-workers treated 20 evaluable infants who had relapsed after more traditional pre-irradiation chemotherapy, using a combination of high-dose thiotepa (900 mg/m^2), busulfan (600 mg/m^2), and autologous bone marrow transplantation [40]. After recovery from transplantation, each patient received involved-field radiotherapy. Among 15 patients with evaluable disease, there were 6 CR, 6 PR, and 3 SD (75 per cent response rate). The event-free survival at 31 months follow-up was 50 per cent. The authors conjectured that high-dose thiotepa and busulfan, autologous transplantation, and involved-field RT could replace craniospinal radiation as prophylaxis against CNS metastases. More recent protocols using high-dose chemotherapy and autologous stem cell rescue have been reported. In a pilot study of five patients aged 14–47 months with newly diagnosed disease, Pérez-Martínez and colleagues used pre-RT chemotherapy consisting of busulfan (16 mg/kg) and melphalan (140 mg/m^2), with or without thiotepa (250 mg/m^2) or topotecan

(2 mg/m^2) [41]. After high-dose chemotherapy, all patients received an autologous peripheral stem cell rescue. Four patients had a CR, while one patient had rapid progression of disease, with a 2-year event-free survival of 71 per cent. Overall, the regimen was well tolerated, without any treatment-related mortality. Using a different approach, Tornesello and colleagues administered a combination of high-dose carboplatin (600 mg/m^2/day), cyclophosphamide (2 g/m^2/day), and vincristine to seven infants after surgical resection [42]. There were 3 PR after the initial two courses of carboplatin and 1 PR after two courses of cyclophosphamide. Overall, although carboplatin demonstrated activity, there was an unacceptable rate of progressive disease during the cyclophosphamide portions of the regimen.

New randomized protocols using more intense chemotherapy regimens have been designed and implemented by several cooperative groups. A recent report by Packer and colleagues (COG A9961) describes a prospective, phase III, randomized trial of reduced dose craniospinal irradiation (2340 cGy) in combination with two forms of chemotherapy for patients with "average-risk" medulloblastoma [43]. All patients received craniospinal irradiation and were then randomized to receive chemotherapy with either CCNU, cisplatin, and vincristine or cytoxan, cisplatin, and vincristine. Of the 421 patients enrolled, 67 (17 per cent) were less than 4 years of age. With a mean follow-up of 44 months, the percentage of patients with disease dissemination was equal between chemotherapy groups and was similar overall to other phase III trials using higher doses of craniospinal RT. The progression-free survival at 3 years was approximately 85 per cent for both chemotherapy groups. Infectious complications were more frequent (18 *versus* 29 per cent; $p < 0.005$) for patients in the cytoxan-based chemotherapy regimen.

Another protocol attempting to reduce craniospinal RT in combination with multiagent chemotherapy has recently been reported by Jacacki and co-workers [44]. After maximal resection, seven patients ranging from 20 to 64 months of age, received multiagent chemotherapy for four months, followed by 1800 cGy RT to the craniospinal axis and 5400 cGy to the posterior fossa. The chemotherapy consisted of cisplatin (100 mg/m^2), cyclophosphamide (1.5 g/m^2), etoposide (75 mg/m^2), and vincristine (Regimen A; 2 patients) or Regimen A alternating with carboplatinum (500 mg/m^2) and etoposide (Regimen B; 5 patients). Two patients with residual disease had PR and long-term progression-free survival of 10.8+ and 7.1+ years. Three patients had progressive disease outside the posterior fossa at 2, 32, and 36 months

TABLE 29.1 Phase III Chemotherapy Trials for Medulloblastoma

Ref. #	N	Regimen	Objective responses	PFS	Overall survival
[43]	421	CCNU, cisplatin, VCR or CTX, cisplatin, VCR; plus 2340 cGy neuraxis RT	NA	85% at 3 yrs in both arms	NA
[48]	179	RT vs RT+CCNU, VCR, prednisone	NA	increased in high-risk disease 46% vs. 0%	65% for both gps; 57% vs 19% for high-risk gp
[49]	286	RT vs RT+CCNU, VCR	NA	48% at 5 yrs; ↑ in chemo gp	53% at 5 yrs; ↑ in chemo gp
[31–33,50]	71	RT vs RT+MOPP	NA	57% vs 68% at 5 yrs	56% vs 74% at 5 yrs
[51]	364	RT vs RT+PCB, VCR, MTX; plus CCNU/VCR in high-risk pts	NA	59% in both gps at 5 yrs	NA
[38,48,52]	203	RT+CCNU, VCR, prednisone vs RT+8 in 1; high-risk pts	NA	64% vs 47% for CVP at 5 yrs	56% at 7 yrs

Abbreviations: Ref. — reference; PFS — progression-free survival; VCR — vincristine; CTX — cyclophosphamide; RT — radiation therapy; NA — not available; gps — groups; yrs — years; PCB — procarbazine; MTX — methotrexate; pts — patients; CVP — CCNU, vincristine, cisplatin.

after treatment. The regimen induced significant neuro-cognitive and neuroendocrine toxicity in long-term survivors.

A recent study from Rutkowski and colleagues attempted "up-front" treatment of a cohort of 62 patients, three years of age or less, with aggressive chemotherapy after surgical resection [45]. Radiotherapy was withheld until evidence of local or distant metastasis. The regimen contained carboplatinum (200 mg/m^2), cyclophosphamide (800 mg/m^2), methotrexate (5 g/m^2), vincristine, etoposide (150 mg/m^2), and intraventricular methotrexate. The 5-year progression-free and overall survival were 55.5 per cent and 63.3 per cent, respectively. Fourteen of 18 patients without residual tumor or metastases remained disease free without RT, with a 5-year progression-free survival of 77.8 per cent. For patients with residual tumor or metastases, the 5-year progression-free survivals were much lower and ranged from 30.7 to 50.0 per cent.

CHEMOTHERAPY OF CHILDREN AND ADOLESCENTS

The application of chemotherapy to medulloblastoma patients in this age range is an extensive topic; several excellent review articles have been published [3,4,14–16,19,20,46,47]. Several important randomized, multicenter, clinical trials have been conducted by CCG, POG, and other groups in newly diagnosed patients to determine the efficacy of chemotherapy after craniospinal RT (see Tables 29.1 and 29.2). In 1990, the CCG-942 trial was reported, in which 179 newly diagnosed patients were randomized to receive

either RT alone (3500–4000 cGy to craniospinal axis, 5000 to 5500 cGy to posterior fossa) or RT plus adjuvant chemotherapy (CCNU 100 mg/m^2; vincristine 1.5 mg/m^2, during RT and with each cycle; prednisone 40 mg/m^2; in 6-week cycles for 1 year) [48]. The 5-year survival rate was 65 per cent in both groups. Chemotherapy was not beneficial for patients with low-risk disease. However, in patients with high-risk disease (i.e., large residual tumors after resection, brainstem infiltration, extraneural metastases, or leptomeningeal dissemination), there was a significant improvement in event-free survival (46 *versus* 0 per cent; $p = 0.006$) and 5-year survival (57 *versus* 19 per cent). In 1990, a similar trial was reported by the International Society of Pediatric Oncology (SIOP), using the same adjuvant chemotherapy regimen, without prednisone [49]. In this study, 286 patients were randomized into RT or RT plus chemotherapy groups. The 5-year overall survival and disease-free survival rates for the entire cohort were 53 and 48 per cent, respectively. At the time of protocol closure, disease-free survival in each group differed significantly ($p = 0.005$). After subsequent long-term follow-up, the overall difference between groups has become less significant. However, significant benefit for disease-free survival was still present for the subgroups that had brainstem involvement, CSF spread, and residual tumor after resection. Another phase III trial was reported by POG in 1991 (POG 7909), in which 71 patients were randomized to receive RT or RT plus MOPP chemotherapy (dosing similar to the van Eys studies, except nitrogen mustard was reduced: 3 mg/m^2/dose) [31–33,50]. There was a significant improvement in 5-year overall survival for the MOPP cohort (74 *versus* 56 per cent; $p = 0.06$, adjusted for race and gender). Although the 5-year progression-free survival rate was not significantly different for the entire cohort (68 *versus* 57 per cent; $p = 0.18$), significance was reached for the subgroup of children 5 years and older ($p = 0.05$). A trend for benefit of MOPP was noted in patients with both high risk and low risk disease. A subsequent phase III trial by SIOP enrolled 364 newly diagnosed patients, stratified into low-and high-risk groups, into either RT alone or RT plus chemotherapy [51]. The chemotherapy was administered as one cycle prior to RT and consisted of procarbazine (100 mg/m^2/day), vincristine, and methotrexate (2 g/m^2). High-risk patients also received six cycles of post-RT chemotherapy, using CCNU and vincristine as in the CCG-942 study. The estimated 5-year progression-free survival rate for the entire cohort was 59 per cent (median follow-up of 76 months). No benefit was noted for neoadjuvanat chemotherapy in this trial, regardless of risk status.

The CCG-921 study was a randomized phase III trial of 203 high risk patients who received either RT plus vincristine, CCNU, and prednisone (VCP; similar to CCG-942) or RT plus 8-in-1 (similar to Geyer and colleagues) [38,48]. A recent 7-year follow-up of patients from this study confirms the efficacy of the CCG-942 regimen [52]. The 7-year overall survival rate and progression-free survival rate for the entire cohort were 56 and 55 per cent, respectively. There was a statistically significant advantage for VCP over 8-in-1 in terms of 5-year progression-free survival (64 *versus* 47 per cent; $p = 0.032$). The presence of CSF dissemination was a powerful negative prognostic factor. Even patients with small residual tumors typically experienced earlier relapses.

The CG-9892 study evaluated 65 newly diagnosed patients with low risk medulloblastoma, to assess the feasibility of reduced-dose craniospinal irradiation in combination with adjuvant chemotherapy [27]. The RT regimen consisted of 2340 cGy to the craniospinal axis and 5580 cGy to the posterior fossa. The chemotherapy protocol was administered during and after RT on 6 week cycles; and included vincristine (1.5 mg/m^2 weekly during RT and each cycle), CCNU (75 mg/m^2), and cisplatin (75 mg/m^2). The protocol was derived from phase II studies by Packer and colleagues, in which newly diagnosed patients had improved survival after treatment with cisplatin, CCNU, and vincristine [53,54]. The progression-free survival rate was 86 per cent at 3 years and 79 per cent at 5 years. Relapse occurred in 14 patients; 2 at the tumor site alone, 9 at the tumor site and CSF, and 3 at non-primary sites. The overall survival rates in this study compare favorably with other large studies using full-dose RT in low-risk patients. However, the results must be interpreted with caution, since only 2 patients had extensive residual disease after surgery and only 17 had brainstem involvement. A recent follow-up study to evaluate the neuropsychological outcome of the CCG-9892 patient cohort was reported by Ris and co-workers [55]. Even though reduced-dose RT was used, intellectual function declined 4.3 IQ-points per year. The effect was more pronounced in young patients and females. As mentioned above, a recent COG phase III study has been reported (A9961) that extends the initial observations of the CG-9892 reduced-dose feasibility protocol [43]. There were 326 patients with ''average-risk'' medulloblastoma between the ages of 5 and 19 who received 2340 cGy craniospinal RT in combination with chemotherapy (CCNU, cisplatinum, vincristine *versus* cyclophosphamide, cisplatinum, vincristine). Similar to the experience in young patients, the older cohort had a 3-year progression-free survival of 85 per cent and the same rate of disease dissemination when compared to protocols with more conventional, higher-dose craniospinal irradiation. The institutional experience of Douglas and colleagues, using a similar chemotherapy regimen (i.e., cisplatin, vincristine, CCNU or cyclophosphamide) in combination with low-dose craniospinal RT (2340 cGy) and a conformal tumor bed boost (3240 cGy), also supports the feasibility and safety of this approach [56]. They reported a 5-year disease-free survival rate of 86 per cent and a low incidence of distant failure.

The POG-8695 study was designed to assess the activity of pre-radiation chemotherapy in newly diagnosed, high-risk medulloblastoma patients [57]. Thirty evaluable patients received a 9 week course of pre-radiation chemotherapy consisting of cisplatin (100 mg/m^2), vincristine, and cyclophosphamide (1200 mg/m^2). On day 64 all patients began a course of RT: 3600 cGy to the craniospinal axis and 5400 cGy to the posterior fossa. The objective response rate was 47 per cent in the posterior fossa, 54 per cent in the spine, and 43 per cent overall. However, there was a poor correlation between response to chemotherapy and length of survival. The 2-year survival rate was 61 per cent, with a 2-year progression-free survival rate of only 40 per cent. A more recent European protocol utilizing pre-RT chemotherapy randomized nonmeta-static patients to receive RT alone or RT in combination with four cycles of up-front vincristine, etoposide (100 mg/m^2), and carboplatinum (500 mg/m^2) alternating with vincristine, etoposide, and cyclophosphamide (1.5 g/m^2) [58]. A total of 179 patients were evaluable, with a median age 7.67 years. The 3-year and 5-year event-free survivals were significantly longer for the RT plus chemotherapy cohort in comparison to RT alone (78.5 and 74.2 per cent *versus* 64.8 and 59.8 per cent, respectively; $p = 0.0366$). The 3-year and 5-year overall survivals were similar between groups. The multivariate analysis identified chemotherapy as having a significant effect ($p = 0.0248$) on event-free survival.

Stewart and co-workers have reported on the use of neoadjuvant topotecan for patients with high-risk medulloblastoma [59]. The topotecan was administered as an intravenous infusion under pharmacokinetic guidance to a cohort of 37 patients, attempting to achieve a topotecan lactone AUC of 120–160 ng/ml·h. This AUC was derived from *in vitro* experiments that defined the CSF exposure duration threshold as a concentration of 1 ng/ml for 8 h. The initial topotecan dosages ranged from 2 to 5.5 mg/m^2/day, with subsequent doses modified as needed to achieve the target AUC. The objective response rate was 28 per cent (CR 4, PR 6), with another 17 patients (47.2 per cent) achieving stable disease.

Another approach to treatment in newly diagnosed patients was reported by Allen and co-workers, who used hyperfractionated RT plus adjuvant chemotherapy [60]. Nineteen patients with large tumors

(i.e., T3b/T4) or CSF dissemination received hyperfractionated RT and multi-agent chemotherapy. Three separate regimens were given as monthly courses, which constituted one complete cycle; a total of three cycles were administered over 9 months after RT. The regimens consisted of cisplatin (20 mg/m^2/day) plus etoposide (50 mg/m^2/day), vincristine plus cyclophosphamide (900 mg/m^2/day), and carboplatin (150 mg/m^2) plus vincristine. Of the 15 patients with large tumors, 14 remained in continuous remission after a median follow-up of 78 months. Two of the 3 evaluable patients with CSF dissemination also remained in remission (79 +, 46 + months).

Numerous phase II trials have evaluated the efficacy of various drugs, either as single agents or in combination, for treatment of patients with recurrent medulloblastoma [3,4,19,20,46,47]. The most effective phase II trials have been summarized and listed in Table 29.2. Active single agents include cisplatin, carboplatin, CCNU, thiotepa, etoposide, and cyclophosphamide. Many of these studies demonstrated objective responses, ranging from 20 to 50 per cent of the evaluable cohort. However, the responses were usually not durable, with progression after 4 to 8 months in most patients. Overall efficacy and the

TABLE 29.2 Phase II Chemotherapy Trials for Medulloblastoma

Ref.#	N	Regimen	Objective responses	PFS	Overall survival
[27]	65	RT (2340 cGy) plus CCNU, cisplatin, VCR; low-risk	NA	86% at 3 yrs 79% at 5 yrs	NA
[35]	12	MOPP	CR – 6	NA	median 10.6 yrs
[36]	6	cisplatin & VCR, followed by MOPP	CR – 2, PR – 1, MR – 2	NA	NA
[37]	27	CTX/VCR × 2, then cisplatin/etoposide	CR – 4, PR – 9	34% at 2 yrs	46% at 2 yrs
[38]	14	8 drugs in 1 day	CR – 1, PR – 5	22% at 2 yrs	NA
[40]	15	thiotepa, busulfan, ABMT	CR – 6, PR – 6	50% at 31 mo.	NA
[45]	62	carbo, CTX, MTX, VCR, VP-16, IT MTX	NA	55.5% at 5 yrs	63.3% at 5 yrs
[57]	30	pre-RT cisplatin, VCR, CTX, then RT (3600 cGy)	47% overall in PF	40% at 2 yrs	61% at 2 yrs
[58]	179	pre-RT carbo, VCR, CTX alt. with CTX, VP-16, VCR	NA	74.2% vs 59.8% at 5 yrs for RT + chemo gp	NA
[61]	16	PCB, CCNU, VCR	NA	median 45 wks	NA
[62]	12	CTX, VCR	CR – 1, PR – 7	NA	NA
[63]	7	CCNU, cisplatin, VCR	CR – 4	18.5 mo overall	NA
[3]	12	Ifos, VP-16, MESNA	CR – 3, PR – 5	NA	NA
[64]	25	carbo, VP-16	CR – 8, PR – 10	median 5 mo	NA
[72]	36	MOPP early, then cisplatin, VP-16, CTX; in high-risk pts	NA	61% at 5 yrs in high-risk pts	NA
[73]	17	cisplatin, CCNU, VCR or cisplatin, VP-16 alt. CTX, VCR	NA	median 48 mo	median 56 mo

Abbreviations: Ref.—references; PFS—progression-free survival; RT—radiation therapy; VCR—vincristine; NA—not available; CR—complete response; yrs—years; PR—partial response; MR—minor response; CTX—cyclophosphamide; ABMT—autologous bone marrow transplant; mo—months; MTX—methotrexate; IT—intrathecal; PF—posterior fossa; alt.—alternating; PCB—procarbazine; Ifos—ifosfamide; carbo—carboplatinum.

durability of responses are improved with combination chemotherapy. The most effective regimens have been MOPP, 8-in-1, CCNU and vincristine, cyclophosphamide and vincristine, and platinum-based combinations [3,32,46,47]. One of the first reports was from Crafts and associates, who used procarbazine (100 mg/m^2/day), CCNU (75 mg/m^2), and vincristine in 16 patients with recurrent disease [61]. There were 10 responders, with a median response duration of 45 weeks. Friedman and colleagues treated 12 patients using cyclophosphamide (1 gm/m^2/day) and vincristine (2 mg/m^2/d), and noted 1 CR and 7 PR [62]. Several of the responses were durable (8,16+, 21+months). Lefkowitz and co-workers used CCNU (100 mg/m^2), cisplatin (90 mg/m^2), and vincristine to treat 7 patients with recurrent tumors [63]. There were 4 CR, with an overall disease-free survival of 18.5 months. Another approach consisted of ifosfamide (1.8 gm/m^2/d), MESNA, and etoposide (100 mg/m^2/d) [3]. Twelve patients received the regimen; there were 3 CR and 5 PR. Gentet and associates treated 25 patients with carboplatin (160 mg/m^2/day × 5 days) and etoposide (100 mg/m^2/day × 5 days) on a 3–4 week schedule [64]. There were 8 CR and 10 PR, with a median response duration of 5 months.

CHEMOTHERAPY OF ADULT PATIENTS

Medulloblastoma is uncommon in adults, occurring in less than 20 per cent of all patients diagnosed after 20 years of age [3,6,7,15]. Survival in adults, like that of pediatric patients, has improved with the advent of microsurgical techniques, megavoltage RT, and the use of craniospinal radiotherapy. Recent estimates of event-free and overall survival at 5-years are approaching 60 and 75 per cent, respectively, in most series [3,6,7,65–69]. The application of chemotherapy to adult patients has been derived from the pediatric experience. Thus far, no randomized, comparative, multi-center trials have been reported in adults. Many of the above-mentioned reports of survival trends in adults have included patients treated with chemotherapy, either up-front or at relapse (see Table 29.2) [6,7,15,65–71]. The efficacy of chemotherapy was variable, with some studies noting a survival benefit, while others were unable to demonstrate any effect on survival. A recent prospective phase II trial by Brandes and colleagues applied multi-agent chemotherapy to a series of 36 adults after classification with the Chang system [72]. Low-risk patients (i.e., T1, T2, T3a, M0, and no residual disease after surgery) received radiotherapy to the

craniospinal axis (3600 cGy) and posterior fossa (5480 cGy), without the addition of chemotherapy. Patients with high-risk disease (i.e., T3b-T4, any M+, or residual disease after surgery) received neoadjuvant and maintenance chemotherapy, as well as the same radiotherapy scheme as the low-risk patients. The chemotherapy regimen was similar to MOPP for the earlier patients, but was modified after 1995 to contain cisplatinum (25 mg/m^2), etoposide (40 mg/m^2), and cyclophosphamide (1 g/m^2). The 5-year progression-free survival was 76 per cent in low-risk patients and 61 per cent in high-risk patients. There was a significant difference in 5-year progression-free survival between M− and M+patients (75 *versus* 45 per cent; $p = 0.01$).

In the report by Bloom and Bessell, 20 of 47 adult patients received chemotherapy with CCNU and vincristine, similar to the CCG-942 trial [6,48]. They noted a trend for improved survival in the chemotherapy cohort. Among 28 patients with a complete or sub-total resection, the 5-year survival rate was 85 per cent for the group receiving chemotherapy, compared to only 60 per cent in those that received radiotherapy alone. Carrie and colleagues report a multi-center retrospective analysis of 156 adults with medulloblastoma [65]. In this series, 75 patients received chemotherapy with either 8-in-1, CCNU and vincristine, or ifosfamide, cisplatin, and vincristine. There was an insignificant trend towards improved 5-year survival for the chemotherapy cohort compared to RT alone. No significant differences were noted in survival between chemotherapy regimens. The chemotherapy seemed more toxic in adults than in the pediatric population. Greenberg and associates treated 17 adult patients using either cisplatin, vincristine, and CCNU (similar to CCG-9892) or cisplatin and etoposide alternating with cyclophosphamide and vincristine (similar to POG study with Duffner *et al.*) [27,37,73]. The median overall survival for the cohort was 56 months, with a median relapse-free survival of 48 months. The 10 patients treated on the CCG protocol arm of the study had median overall and relapse-free survivals of 36 months and 26 months, respectively. The 7 patients on the POG protocol arm of the study had median overall and relapse-free survivals of 57 and 48 months, respectively. Although there was a trend for improved survival on the POG arm, the results were not significantly different. There was no difference in overall or relapse-free survival between high-risk and low-risk patient groups. Both arms of the protocol appeared to be more toxic in adult patients than in the pediatric clinical trials, especially the CCG regimen.

CHEMOTHERAPY WITH BONE MARROW TRANSPLANTATION

In an attempt to improve dose intensity, several investigators have used myeloablative, high-dose chemotherapy followed by bone marrow transplant or rescue (BMT) [3,40,48,74–80]. Preliminary results suggest that durable responses can be achieved in a sub-group of selected patients with high risk or recurrent disease. Chemotherapeutic agents appropriate for high-dose protocols require bone marrow suppression as the dose-limiting toxicity. The most commonly used drugs are busulfan, thiotepa, cyclophosphamide, etoposide, melphalan, and carboplatin. The previously mentioned study by Dupuis-Girod used this approach with a regimen of busulfan and thiotepa in infants and young children [40]. They noted 6 CR, 6 PR, and 3 SD in 15 evaluable patients with progressive disease. The event-free survival at 31 months follow-up was 50 per cent. A similar study of young patients with recurrent brain tumors by Gururangan and co-workers included five with medulloblastoma [77]. High-dose chemotherapy consisted of several combinations of carboplatin, thiotepa, etoposide, and BCNU. There were 3 responders (high risk) with durable survival (10 +, 13 +, 34 + months). In older children and adolescents, several initial studies evaluated the efficacy of high-dose chemotherapy and BMT in patients with recurrent or progressive malignant brain tumors [74–76]. Mahoney and colleagues used high-dose melphalan and cyclophosphamide in a cohort that included 8 patients with medulloblastoma [74]. There was 1 CR, 2 PR, and 1 MR; several of the responses were durable (24 +, 25, 25 + months). In a study that included 19 patients with medulloblastoma, Graham and associates used high-dose cyclophosphamide and melphalan, carboplatin and etoposide, or busulfan and melphalan [76]. There were 4 CR; all of the responses were durable (27 +, 42 +, 47 +, 49 + months). Similar results were described by Finlay and the CCG in a study of high-dose thiotepa and etoposide for recurrent CNS tumors [75]. In a series of 23 patients with recurrent medulloblastoma, Dunkel and co-workers used high-dose carboplatin, thiotepa, and etoposide, followed by autologous stem-cell rescue [77]. Seven patients (30 per cent) had prolonged event-free survival with a median duration of 54 months (range 24 to 78 months). Overall survival estimates at 36 months post-BMT were 46 per cent. Strother and associates applied high-dose cyclophosphamide, cisplatin, and vincristine plus stem-cell rescue to 53 newly diagnosed patients with medulloblastoma and cerebral PNET [80]. The chemotherapy was given in 4 week intervals for four consecutive cycles; each cycle included stem cell rescue. All patients received craniospinal radiotherapy. The 2-year progression-free survival for low-risk and high-risk patients was 93.6 and 73.7 per cent, respectively. A more recent attempt at autologous stem cell rescue in combination with high-dose chemotherapy was focused on adult patients with recurrent medulloblastoma [81]. Six patients were treated at recurrence with two cycles of carboplatinum (600 mg/m^2), etoposide (600 mg/m^2), and cyclophosphamide (50 mg/kg) followed by autologous stem cell rescue. There were 3 CR and 3 PR, with a median time to progression of 13.5 months. Median overall survival for the cohort was 21.5 months.

High-dose chemotherapy and BMT is associated with considerable toxicity and complications (see Chapter 21) [40,74–81]. Potential problems include prolonged neutropenia and sepsis; hemorrhage (e.g., brain); gastrointestinal toxicity such as moderate to severe emesis, mucositis, and diarrhea; hepatotoxicity; nephrotoxicity; pneumonitis; cardiotoxicity and congestive heart failure; and CNS toxicity such as seizures, coma, and psychosis. Although uncommon in most of the studies, toxic deaths were also described that occurred secondary to infection or organ failure.

MOLECULAR GENETICS AND "TARGETED" THERAPEUTICS

The molecular biological basis of medulloblastoma continues to be elucidated [82–85]. Current molecular genetic data indicates that the pattern of oncogene expression and tumor suppressor gene dysfunction in medulloblastoma differs from that in the more common gliomas of adulthood. In adult high-grade gliomas, there is a predilection for amplification of platelet derived growth factor receptors (PDGFR) and epidermal growth factor receptors (EGFR), excessive or constitutive activity of internal signal transduction systems (e.g., ras), and loss or reduced expression of tumor suppressor genes (e.g., p53, Rb, PTEN) [82,86]. In medulloblastoma, different molecular abnormalities are associated with cellular transformation. For example, although there is an elevated frequency of isochromosome 17q and loss of alleles on chromosome 17p, mutations and deletions of the p53 gene are infrequent [3,82,87]. It is theorized that a separate tumor suppressor gene must exist on 17p, distal to the site of p53 [88]. It has also been shown that amplification and over expression of EGF, PDGF, and their associated receptors occurs infrequently in medulloblastoma cell lines and tumor resection specimens. In contrast, amplification and over expression has been demonstrated for members of the *myc* family of oncogenes [89–94]. Amplification of C-*myc* has been reported by several groups and ranges from 5 to 20 per cent of tested specimens. Abnormalities of C-*myc* expression are more frequent and range from 42 to 90 per cent of biopsy specimens and cell lines. In addition, recent laboratory experiments have demonstrated that overexpression of C-*myc* into medulloblastoma cell lines resulted in a more aggressive phenotype [95]. When these C-*myc* cell lines were grown as xenografts in SCID mice, they grew 75 per cent larger over 8 weeks than non-transfected cell lines and histologically resembled large cell/anaplastic medulloblastoma. Although amplification and over expression of N-*myc* have been described, they are much less common.

A recent analysis of the insulin-like growth factor I receptor (IGF-IR) has determined that a constitutively active form of the receptor is frequently present in surgical specimens and cell lines [96]. In addition, when compared to normal cerebellum, higher concentrations of the major substrate for the receptor (IRS-1) were also noted in the majority of cells, suggesting the possibility of an autocrine and paracrine stimulatory mechanism. Several reports by Gilbertson and co-workers have described the expression patterns of HER (ErbB) receptors in medulloblastoma cell lines, tumor specimens, and normal cerebellum [85,97,98]. The HER2 receptor is undetectable in the developing and mature cerebellum. However, in a series of 70 medulloblastoma patients, HER2 was present in 86 per cent of the tumors and was co-expressed with HER4 in 54 per cent of tumors. The co-expression of HER2 and HER4 imparted a poor prognosis for survival ($p = 0.006$), independent of age and tumor stage. Expression of the NRG1-β ligand (binding strongly to the HER2/HER4 heterodimer) was also noted in 87.5 per cent of tumor samples, supporting the presence of an autocrine loop. Co-expression of HER2, HER4, and NRG1-β was significantly correlated ($p < 0.05$) with the presence of central nervous system metastases at diagnosis.

Another important pathway involves the β-catenin gene product, which interacts with other proteins (i.e., glycogen synthase kinase 3, APC) to influence the rate of cell proliferation (see Fig. 29.1) [85,99,100]. In normal cells, β-catenin levels are kept low through constitutive interaction with several multiprotein complexes. The most well-characterized complex contains axin, casein kinase 1α, APC, and glycogen

synthase kinase 3. Casein kinase 1α phosphorylates β-catenin on serine 45 (S45), which primes β-catenin for further phosphorylation by glycogen synthase kinase 3 on S41, S37, and S33. These residues provide binding sites for the β-transducin repeat-containing protein, which promotes polyubiquitination of β-catenin and proteasome degradation. The other multiprotein complex contains protein kinase A, presenilin, and glycogen synthase kinase 3 and functions in a similar manner. Oncogenic mutations of β-catenin eliminate serine phosphorylation sites (i.e., S33, S37) for glycogen synthase kinase 3β. Hypophosphorylation of β-catenin reduces its ubiquitin-mediated degradation, leading to higher cytoplasmic concentrations and more frequent binding to T-cell factor (Tcf-1). The complex of β-catenin and Tcf-1 functions as a transcriptional transactivator and allows constitutive expression of genes that promote cell proliferation (e.g., cyclin D1, C-myc). Several groups have noted oncogenic mutations in the β-catenin pathway in approximately 15 per cent of sporadic medulloblastomas [101,102]. These mutations lead to increased transcription of several genes, including cyclin D1, C-*myc*, and Tcf-1.

Neuropeptides and associated receptors have also been implicated in the cellular transformation of medulloblastoma. Somatostatin (SS-14) and vasoactive intestinal polypeptide (VIP) are both neuromodulators and growth regulators of the developing nervous system [103,104]. Fruhwald and colleagues found that somatostatin and the sst$_2$ receptor subtype were highly expressed in medulloblastoma tumor specimens and cell lines. They postulated that somatostatin was involved in the proliferation and differentiation of these tumors. The same group has reported similar results for VIP and its receptors (VIPR1 and VIPR2). There was high expression of VIP, VIPR1, and VIPR2 in resected specimens, while in cell lines, only the receptors were highly expressed. Application of VIP to the cell lines resulted in significant growth inhibition. The neurotropins are another important class of trophic factors that influence the nervous system. Grotzer and co-workers analyzed the expression of TrkC, the specific receptor for Neurotropin-3, in a series of 81 patients with medulloblastoma [105]. TrkC expression was common in their cohort, and the level of expression was highly prognostic ($p < 0.00005$) for extended survival in univariate and multivariate analyses.

An exciting new area of research involves the *PTCH* gene and the Sonic Hedgehog (SHH) signaling pathway. This pathway was originally described in *Drosophila* and was discovered during a screen for genes involved in embryonic patterning defects

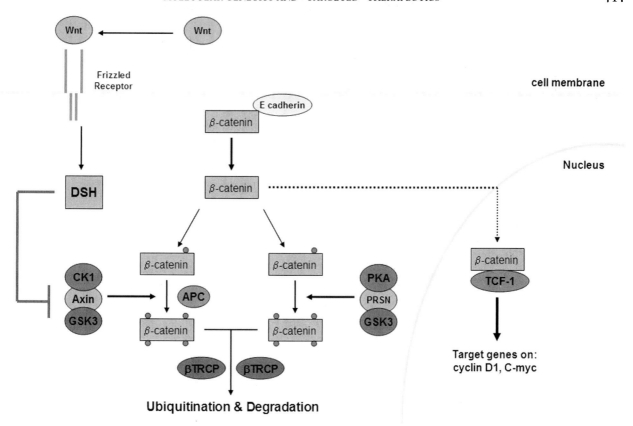

FIGURE 29.1 The Wnt/β-catenin signaling pathway. In normal cells, there is limited binding of WNT to the Frizzled receptor. Without stimulation by WNT, Frizzled does not activate Disheveled (DSH), thereby allowing the multiprotein complex of axin, casein kinase 1α (CK1), adenomatous polyposis coli (APC), and glycogen synthase kinase 3 (GSK3) to phosphorylate β-catenin, which provides binding sites for β-transducin repeat-containing protein (βTRCP). A similar phosphorylation and ubiquitination pathway is mediated by protein kinase A (PKA), presenilin (PRSN), and GSK3. βTRCP mediates poly-ubiquitination of β-catenin, with subsequent proteasome-dependent degradation. With WNT binding, Frizzled phosphorylates and activates DSH, which inhibits the multiprotein complex, leading to hypophosphorylation of β-catenin and an increase in its cytosolic concentrations. β-catenin then shuttles to the nucleus and interacts with T-cell factor 1 (Tcf-1), turning on target genes such as cyclin D1 and C-myc. See Plate 29.1 in Color Plate Section.

[85,106–108]. These genes have been found to be highly conserved and are also important for normal development and morphogenesis in vertebrates. SHH is a secreted protein that undergoes auto-catalytic cleavage into a 19 kDa N-terminal active portion (SHH-N). *PTCH*, located on chromosome 9q22.3, codes for a transmembrane protein that acts as the receptor for SHH-N (see Fig. 29.2). It now appears that *PTCH* functions as a tumor suppressor gene through its interactions with another transmembrane protein, smoothened (Smo), which transduces the SHH-N signal within the cell. When *PTCH* is not bound by SHH-N, it inhibits signal transduction by Smo, thereby down-regulating transcriptional activity. Once SHH-N binds to *PTCH*, the activity of Smo is no longer inhibited and the signal is transduced through activation of the Gli genes (Gli1, Gli2, Gli3). Modulators of Gli activation include fused, which has

a positive influence, and protein kinase A, suppressor of fused, and costal-2, which are all inhibitory. The Gli genes act synergistically as transcription factors to induce downstream expression of several genes, including *PTCH*, *WNT-1*, and members of the bone morphogenetic proteins (BMPs).

It is now clear that dysfunction of the *PTCH* gene and the SHH pathway can cause abnormalities of development and induce a predisposition to several different forms of cancer [85,106–109]. Patients with familial and sporadic Gorlin's syndrome (i.e., nevoid basal cell carcinoma syndrome) have germline mutations of *PTCH*. Affected patients display multiple basal cell carcinomas, jaw cysts, dyskeratotic palmar and plantar pits, skeletal abnormalities, and various other tumors. Approximately 3 per cent of patients with Gorlin's syndrome develop medulloblastoma, often of the desmoplastic variant [109].

Genetic analysis of these tumors reveals frequent mutations or loss of heterozygosity of the *PTCH* gene. In transgenic mouse models, the homozygous animals (*PTCH* −/−) die early *in utero*, while hemizygous animals (*PTCH* +/−) survive, with 14–19 per cent developing medulloblastoma [110–114]. Analysis of the remaining *PTCH* allele showed that mutations were rare and that expression of *PTCH* was not deficient. Therefore, in this model, *PTCH* was not behaving as a classical tumor suppressor gene, which requires the inactivation of both alleles before transformation occurs. However, levels of Gli1 mRNA and protein were increased in the tumors, suggesting that the SHH pathway could be activated despite the persistence of *PTCH* expression. If *PTCH* (+/−) mice were crossed with mice deficient for p53 (−/−), the tumor rate increased to 95 per cent and all mice died before 12 weeks, implying that loss of p53 may enhance genomic instability and the rate of secondary mutations [115]. Microarray analysis of tumor cells from *PTCH* (+/−) mice suggests that SHH activation promotes overexpression of cyclin D1 and N-myc, and that these factors are important mediators of SHH-induced proliferation and tumorigenesis [116]. A more recent transgenic mouse model activated the SHH pathway by expressing a constitutively active form of Smo in cerebellar granule neuron precursors (i.e., ND2:SmoA1 mice), resulting in tumors in 48 per cent of high-expressing mice by a median 6 months of age [117]. Gene expression analysis of tumors demonstrated an increase in Gli1 and N-myc, as well as Notch2 and the Notch target gene HESS. The ND2:SmoA1 model suggests that aberrant Notch signaling contributes to medulloblastoma cell proliferation and survival. Pharmacological inhibition of both the SHH and Notch signaling pathways was shown to induce widespread apoptosis in tumor cells and was more effective than inhibition of either pathway alone.

Mutations of *PTCH* have also been described in 10–15 per cent of sporadic human medulloblastomas [85,118–122]. Although the mutations seem to be more frequent in the desmoplastic variant, they can occur in the classic type as well. Missense, nonsense, and frameshift deletions/insertions have all been described throughout the *PTCH* gene; no hot spots have been identified. The most common mutations predict the generation of a truncated protein. *PTCH2* may also play a role in the transformation of some medulloblastomas, since a truncating mutation was recently described in a sporadic tumor [123]. Oncogenic mutations of other genes involved in the SHH signaling pathway, such as SHH, Smo, GLI, and PKA have also been studied, but appear to be uncommon

events [85,121,124,125]. Mutations in SuFu, which has been mapped to chromosome 10q24.3, appear to be more common and were identified in approximately 9 per cent of tumor samples in a series of 46 medulloblastoma patients [126]. In most cases, truncating mutations were noted, along with a few of the missense variety. All of the truncating SuFu mutations were present in tumors of the desmoplastic subtype. Although mutations of Smo are relatively uncommon, many tumors may demonstrate increased expression of Smo mRNA when compared to normal skin or brain tissue [127]. In a similar expression analysis of *PTCH* and Smo, levels of mRNA were found to have an inverse correlation with grade of malignancy in astrocytic tumors [128]. These results implicated the expression of PTCH and Smo in the transformation process of astrocytic tumors. Expression of the Gli genes did not correlate with astrocytic malignancy.

The SHH/*PTCH* signaling pathway is frequently abnormal and overactive in Gorlin's syndrome-associated and sporadic medulloblastoma and, in addition, may be involved in the transformation of a subset of astrocytic tumors. Due to the relative simplicity of the pathway, it is now under evaluation for "targeted" therapy [85,129]. Small molecule approaches to "targeted" therapy of medulloblastoma are under investigation [85,130–134]. Cyclopamine is a teratogenic steroidal alkaloid derived from the *Veratrum californicum* lily, that can inhibit SHH/*PTCH* signaling through a mechanism that involves binding to Smo and an alteration of its conformation (see Fig. 29.2). The conformational change is induced when cyclopamine binds to the heptahelical domain of Smo, thereby reducing its internal signal transduction capacity. This alteration of conformation and function is similar to the inhibitory action of *PTCH*. *In vitro*, cyclopamine can inhibit proliferation and induce apoptosis of murine and human medulloblastoma cells. Cyclopamine can reduce the growth of tumor allografts in nude mouse models and is also able to inhibit growth and reduce cell viability of freshly cultured human medulloblastoma cells [133]. Because cyclopamine has a relatively low affinity for binding to Smo, second generation inhibitors are under investigation. High-throughput cell-based screening assays for inhibitors of the pathway have identified Hh-Antag691, a benzimidazole derivative with a greater than 10-fold higher binding affinity for Smo than cyclopamine [135]. Hh-Antag691 is able to penetrate the blood–brain barrier after oral delivery, making it an excellent candidate for brain tumor treatment. Using a medulloblastoma mouse model (*PTCH*+/− p53−/−), Romer and colleagues demonstrated that Hh-Antag691 was able to

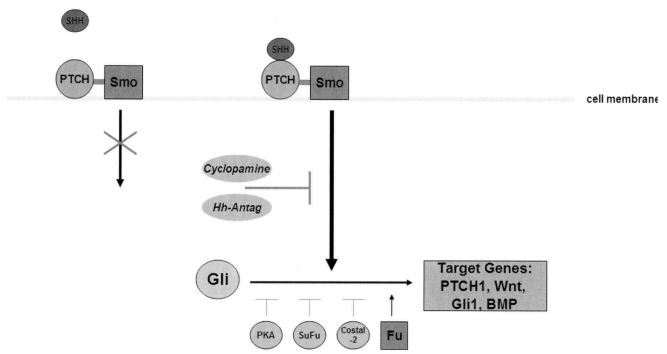

FIGURE 29.2 The SHH/*PTCH* signaling pathway. In the resting state, without binding of SHH, *PTCH* inhibits Smo and does not allow internal signaling. Once SHH binds to *PTCH*, repression of Smo activity is released and the intracellular cascade proceeds, culminating in the expression of the Gli genes. Smo expression of Gli is facilitated by fused (fu) and inhibited by protein kinase A (PKA), costal-2, and suppressor of fused (sup-fu). The Gli genes induce transcription of several target genes, including *PTCH*, WNT-1, and BMP's. New molecular therapeutics such as cyclopamine and Hh-Antag can specifically inhibit the activity of Smo. Adapted from reference [151]. Used with permission from Ashley Publications. See Plate 29.2 in Colour Plate Section.

reduce Gli1 expression and inhibit tumor growth in a dose-dependent manner. Treatment of tumor-bearing mice (20 or 100 mg/kg, twice daily for four days, by oral gavage) resulted in inhibition of cellular proliferation and increased apoptosis. If treatment was extended over two weeks, the 20 mg/kg dose resulted in reduction of tumor volume, while the 100 mg/kg dose induced complete or near-complete eradication of tumors ($p = 0.0159$). In addition, long-term exposure of tumor-bearing mice to 100 mg/kg per day resulted in a significant extension in tumor-free survival (Log-rank test $= 0.0001$). Further pre-clinical testing of Hh-Antag691 is ongoing and phase I clinical trials are under development.

OVERVIEW AND FUTURE CONSIDERATIONS

Now that many medulloblastoma patients are long-term survivors, and a few are even cured of their disease, numerous investigators have begun to focus on the neuropsychological sequelae and quality of life of this cohort after craniospinal RT

[29–31,55,136–146]. In general, these reports demonstrate that irradiated patients have a decrement in overall IQ of 10–20 points compared to non-irradiated controls. All aspects of neuropsychological function are affected, including full-scale IQ, performance IQ, the attention index, and reading and writing achievement. Although these post-treatment deficits affect children at all levels of development, they are most severe in patients less than 7 years of age. A quantitative loss of normal-appearing white matter, as measured by MRI, correlates with the decrement in IQ of these patients [147,148]. In addition to age, the dose of RT also correlates with loss of IQ in a predictable fashion [30,31,55,141]. Using a multiple linear regression model, patients receiving a dose of 3600 cGy can be predicted to have IQ scores 8.2 points less than similar patients receiving only 2400 cGy [31].

In an attempt to minimize the deleterious effects of craniospinal RT, investigators designed clinical trials (i.e., CCG-923/POG-8631) using lower doses to the neuraxis [25,26]. Patients in the reduced-dose cohort (2340 cGy) did have a 10–15 IQ-point advantage compared to patients receiving standard dose RT

(3600 cGy) [144]. However, there was an increase in the rate of early relapse and early isolated neuraxis relapse, and a lower 5-year event-free survival rate, in the reduced-dose cohort. These results prompted several studies combining reduced-dose neuraxis RT with multi-agent chemotherapy in order to preserve neurocognitive function and reduce early relapses in low risk patients [27,43]. The results of these trials have been promising, with the 2- and 3-year progression-free survival and neuraxis relapse rates being comparable to protocols using standard-dose RT. Overall, these results suggest that the combination of reduced-dose RT plus adjuvant chemotherapy is a feasible approach that might eventually lead to prolonged progression-free survival with less severe cognitive dysfunction. For most low-risk patients, reduced-dose neuraxis RT plus multiagent chemotherapy has become the preferred approach [14,85].

In order to expand on these early results and attempt further reductions in the dose of neuraxis RT (e.g., 18 Gy will be used in a new COG study), significant improvements must be made in the efficacy of chemotherapy against medulloblastoma. The mechanisms of action and pharmacology of the commonly used agents are well known [149]. They are simple, non-specific drugs that function as alkylating agents (e.g., CCNU, cisplatin, carboplatin, temozolomide), spindle poisons (e.g., vincristine), and topoisomerase inhibitors (e.g., etoposide). These drugs are crude tools to use against a tumor that is now known to have a very complex transformation process and molecular phenotype. It is imperative that basic research continue to elucidate the molecular abnormalities involved in the transformation, invasion, and angiogenesis of medulloblastoma [150]. These new molecular abnormalities can be used to design drugs that attack more specific, "downstream", therapeutic targets that are intimately involved in maintenance of the malignant phenotype [151,152].

There are three main molecular pathways by which cellular transformation can result in medulloblastoma: the SHH/PTCH pathway, amplification of one of the myc genes (i.e., C- or N-myc), and oncogenic mutations of β-catenin [85,152]. Significant progress has been made in the discovery of "targeted" drugs that can impact on the SHH/PCTH pathway through the inhibition of Smo activity, such as cyclopamine and Hh-Antag691. Further drug discovery of more potent and specific Smo antagonists will be important. Compounds that inhibit the ability of Gli1/Gli2 to induce expression of SHH target genes (i.e., WNT-1, BMPs) could also reduce activity of the pathway. Other potentially useful agents might be agonists to SuFu, protein kinase A, and costal-2, as well

as inhibitors of fused. It remains less clear how drugs could be developed to target the amplified myc gene, oncogenic β-catenin pathways, and HER receptors. Myc functions as a transcriptional regulator in a network with other proteins (i.e., myc/Max/Mad network) [153]. The activity of myc appears to be antagonized by Mad proteins. It may be possible to develop small molecule agonists of Mad, which could abrogate myc activation of cellular proliferation. Alternatively, inhibitors could be developed that reduce the ability of myc to interact with cell cycle proteins and induce expression of target genes. For the β-catenin pathway, potential targets for drug development include inhibitors of the interaction between β-catenin and Tcf-1, as well as compounds that could interfere with the activity of the formed β-catenin/Tcf-1 complex. In addition, agonists of the glycogen synthase kinase 3 and presenilin multiprotein complexes might be developed, with the ability to mediate phosphorylation and ubiquitination of β-catenin. Overactivity of the HER2/HER4 receptors can lead to activation of downstream signal effectors involved in cell survival, such as PI3K, Akt, and mTOR [152]. Tyrosine kinase inhibitors have been developed with activity against HER receptors and are now being considered for application to patients with medulloblastoma [85]. Other drugs that target the activated downstream effectors (e.g., PI3K, Akt) may also be of benefit in selected patients.

Several recent reports would suggest that the current process of risk stratification, which is solely based on clinical information, could be significantly improved if supplemented by molecular data derived from resected tumor specimens [85,154–156]. Using several different methods of molecular analysis (e.g., microarray expression profiles, northern and western blotting, fluorescent in situ hybridization, real-time polymerase chain reaction), the various authors consistently noted that molecular data added to the ability to predict a poor outcome. For example, in the study by Fernandez-Teijeiro and co-workers, the microarray profiles of patients with long-term survival and those with rapid progression were very different, and remained significantly predictive during multivariate analysis [154]. Highly expressed genes that were markers of treatment failure included ribosomal protein S18, V-myb, and ribosomal protein S10. In the other reports, the presence of ERBB2 expression, amplification of myc, and large-cell anaplastic histology were negative predictors [155,156]. In addition to improving the risk stratification process and the ability to prognosticate for medulloblastoma patients, molecular analysis may eventually guide therapeutic decisions. Microarray or "genechip" technology can be used to screen resected tumor tissue for thousands of genes [157,158]. The profile derived from the genechip would outline genes involved in the transformation process of a given tumor, such as *PTCH*, Smo, C-*myc*, Gli1/Gli2, β-catenin, ERBB2, and others. This information could then be used to guide treatment in a prospective fashion, using molecular agents designed to inhibit the activated pathways. Well-designed, multicenter clinical trials will be necessary to test these new molecular drugs as single agents and in combination with cytotoxic drugs. More than likely, molecular chemotherapy agents will be more effective in combination with conventional cytotoxic drugs (e.g., cisplatin, CCNU, temozolomide) than as single agents.

Development of new molecular agents, as well as continued refinements of more conventional chemotherapy, will allow us to move forward in the treatment of medulloblastoma. More experience will be needed in the use of chemotherapy in combination with reduced-dose RT, including conventional dosing regimens and high-dose protocols followed by BMT or stem cell rescue. It is hoped that further progress can be made to reduce the dosage of craniospinal RT, while minimizing the incidence of neuraxis recurrence and continuing to extend the 5-year survival rates. This will most likely occur in the context of multiagent chemotherapy protocols that combine molecular and cytotoxic drugs with different mechanisms of action against the malignant phenotype.

ACKNOWLEDGMENTS

The author would like to thank Ryan Smith for research assistance. Dr. Newton was supported in part by National Cancer Institute grant, CA 16058 and the Dardinger Neuro-Oncology Center Endowment Fund.

References

1. Newton, H. B. (1994). Primary brain tumors: Review of etiology, diagnosis, and treatment. *Am Fam Phys* **49**, 787–797.
2. Roberts, R., Lynch, C., Jones, M., Hart, M. (1991). Medulloblastoma: A population-based study of 532 cases. *J Neuropathol Exp Neurol* **50**, 134–144.
3. Siffert, J., and Allen, J. C. (1997). Medulloblastoma. *In Handbook of Clinical Neurology, Vol. 24 (68): Neuro-Oncology Part II.* (C. J. Vecht Ed.), **8**, 181–209. Elsevier Science, Amsterdam.
4. Tomlinson, F. H., Scheithauer, B. W., Meyer, F. B. *et al.* (1992). Medulloblastoma: I. Clinical, diagnostic, and therapeutic overview. *J Child Neurol* **7**, 142–155.
5. Berger, M. S., Margrassi, L., and Geyer, R. (1995). Medulloblastoma and primitive neuroectodermal tumors. *In Brain Tumors.* *An Encyclopedic Approach.* (A. H. Kaye, and E. R. Laws Eds.), **30**, 561–574. Churchill Livingstone, Edinburgh.
6. Bloom, H. J. G., and Bessel, E. M. (1990). Medulloblastoma in adults: A review of 47 patients treated between 1952 and 1981. *Int J Rad Oncol Biol Phys* **18**, 763–772.
7. Kunchner, L. J., Kuttesch, J., Hess, K., and Yung, W. K. A. (2001). Survival and recurrence factors in adult medulloblastoma: The M. D. Anderson Cancer Center experience from 1978 to 1998. *Neurooncol* **3**, 167–173.
8. Hung, K. L., Wu, C. M., Huang, J. S., and How, S. W. (1990). Familial medulloblastoma in siblings: Report in one family and review of the literature. *Surg Neurol* **33**, 341–346.
9. Bondy, M., Wiencke, J., Wrensch, M., and Kyritsis, A. P. (1994). Genetics of primary brain tumors: a review. *J Neurooncol* **18**, 69–81.
10. Cushing, H. (1930). Experiences with cerebellar medulloblastoma: a critical review. *Acta Pathol Microbiol Scand* **7**, 1–86.
11. Mcintosh, N. (1979). Medulloblastoma: a changing prognosis. *Arch Dis Child* **54**, 200–203.
12. Freeman, C. R., Taylor, R. E., Kortmann, R. D., and Carrie, C. (2002). Radiotherapy for medulloblastoma in children: A perspective on current international clinical research efforts. *Med Pediatr Oncol* **39**, 99–108.
13. Koeller, K. K., and Rushing, E. J. (2003). Medulloblastoma: A comprehensive review with radiologic-pathologic correlation. *Radiographics* **23**, 1613–1637.
14. Rood, B. R., MacDonald, T. J., and Packer, R. J. (2004). Current treatment of medulloblastoma: Recent advances and future challenges. *Semin Oncol* **31**, 666–675.
15. Newton, H. B. (2001). Review of the molecular genetics and chemotherapeutic treatment of adult and paediatric medulloblastoma. *Expert Opin Investig Drugs* **10**, 2089–2104.
16. Mazzola, C. A., and Pollack, I. F. (2003). Medulloblastoma. *Curr Treat Options Neurol* **5**, 189–198.
17. Luh, G. Y., and Bird, C. R. (1999). Imaging of brain tumors in the pediatric population. *Neuroimag Clin N Am* **9**, 691–716.
18. Fouladi, M., Gajjar, A., Boyett, J. M. *et al* (1999). Comparison of CSF cytology and spinal magnetic resonance imaging in the detection of leptomeningeal disease in pediatric medulloblastoma or primitive neuroectodermal tumor. *J Clin Oncol* **17**, 3234–3237.
19. Packer, R. J., and Finlay, J. L. (1988). Medulloblastoma: Presentation, diagnosis, and management. *Oncol* **2**, 35–44.
20. Whelan, H. T., Krouwer, H. G., Schmidt, M. H., Reichert, K. W., and Kovnar, E. H. (1998). Current therapy and new perspectives in the treatment of medulloblastoma. *Pediatr Neurol* **18**, 103–115.
21. Albright, A. L., Wisoff, J. M., Zeltzer, P. M. *et al.* (1996). Effects of medulloblastoma resections on outcome in children: A report from the Children's Cancer Group. *Neurosurg* **38**, 265–271.
22. Sutton, L. N., Phillips, P. C., and Molloy, P. T. (1996). Surgical management of medulloblastoma. *J Neurooncol* **29**, 9–21.
23. Paulino, A. C. (1997). Radiotherapeutic management of medulloblastoma. *Oncol* **11**, 813–923.
24. Brand, W. N., Schneider, P. A., and Tokars, R. P. (1987). Long term results of a pilot study of low dose cranial-spinal irradiation for cerebellar medulloblastoma. *Int J Radiat Oncol Biol Phys* **13**, 1641–1645.
25. Deutsch, M., Thomas, P. R. M., Krischer, J. *et al.* (1996). Results of a prospective randomized trial comparing standard dose neuraxis irradiation (3600 cGy/20) with reduced neuraxis irradiation (2340 cGy/13) in patients with low-stage medulloblastoma. *Pediatr Neurosurg* **24**, 167–177.

26. Thomas, P. R. M., Deutsch, M., Kepner, J. L. *et al.* (2000). Low-stage medulloblastoma: Final analysis of trial comparing standard-dose with reduced-dose neuraxis irradiation. *J Clin oncol* **18**, 3004–3011.

27. Packer, R. J., Goldwein, J., Nicholson, H. S. *et al.* (1999). Treatment of children with medulloblastomas with reduced-dose craniospinal radiation therapy and adjuvant chemotherapy: A Children's Cancer Group study. *J Clin Oncol* **17**, 2127–2136.

28. Prados, M. D., Edwards, M. S. B., Chang, S. M. *et al.* (1999). Hyperfractionated craniospinal radiation therapy for primitive neuroectodermal tumors: Results of a phase II study. *Int J Radiat Oncol Biol Phys* **43**, 279–285.

29. Packer, R. J., Sutton, L. N., Atkins, T. E. *et al.* (1989). A prospective study of cognitive function in children receiving whole-brain radiotherapy and chemotherapy: 2-year results. *J Neurosurg* **70**, 707–713.

30. Moore, B. D., Ater, J. L., and Copeland, D. R. (1992). Improved neuropsychological outcome in children with brain tumors diagnosed during infancy and treated without cranial irradiation. *J Child Neurol* **7**, 281–290.

31. Silber, J. H., Radcliffe, J., Peckam, V. *et al.* (1992). Whole-brain irradiation and decline in intelligence: The influence of dose and age on IQ score. *J Clin Oncol* **10**, 1390–1396.

32. Cangir, A., Van Eys, J., Berry, D. H., Hvizdala, E., and Morgan, S. K. (1978). Combination chemotherapy with MOPP in children with recurrent brain tumors. *Med Pediatr Oncol* **4**, 253–261.

33. Van Eys, J., Cangir, A., Coody, D., and Smith, B. (1985). MOPP regimen as primary chemotherapy for brain tumors in infants. *J Neurooncol* **3**, 237–243.

34. Baram, T. Z., Van Eys, J., Dowell, R. E. *et al.* (1987). Survival and neurologic outcome of infants with medulloblastoma treated with surgery and MOPP chemotherapy. A preliminary report. *Cancer* **60**, 173–177.

35. Ater, J. L., Van Eys, J., Woo, S. Y. *et al.* (1997). MOPP chemotherapy without irradiation as primary postsurgical therapy for brain tumors in infants and young children. *J Neurooncol* **32**, 243–252.

36. Kretschmar, C. S., Tarbell, N. J., Kupsky, W. *et al.* (1989). Pre-irradiation chemotherapy for infants and children with medulloblastoma: a preliminary report. *J Neurosurg* **71**, 820–825.

37. Duffner, P. K., Horowitz, M. E., Krischer, J. P. *et al.* (1993). Postoperative chemotherapy and delayed radiation in children less than three years of age with malignant brain tumors. *New Engl J Med* 328:172501731.

38. Geyer, J. R., Zeltzer, P. M., Boyett, J. M. *et al.* (1994). Survival of infants with primitive neuroectodermal tumors or malignant ependymomas of the CNS treated with eight drugs in 1 day: a report from the Children's Cancer Group. *J Clin Oncol* **12**, 1607–1615.

39. Razzouk, B. I., Heideman, R. L., Friedman, H. S. *et al.* (1995). A phase II evaluation of thiotepa followed by other multiagent chemotherapy regimens in infants and young children with malignant brain tumors. *Cancer* **75**, 2762–2767.

40. Dupois-Girod, S., Hartmann, O., Benhamou, E. *et al.* (1996). Will high dose chemotherapy followed by autologous bone marrow transplantation supplant cranio-spinal irradiation in young children treated for medulloblastoma? *J Neurooncol* **27**, 87–98.

41. Pérez-Martínez, A, Quintero, V., Vicent, M. G., Sevilla, J., Díaz, M. A., and Madero, L. (2004). High-dose chemotherapy with autologous stem cell rescue as first line of treatment in young children with medulloblastoma and supratentorial neuroectodermal tumors. *J Neurooncol* **67**, 101–106.

42. Tornesello, A., Mastrangelo, S., Bembo, D. *et al.* (1999). Progressive disease in children with medulloblastoma/PNET during preradiation chemotherapy. *J Neurooncol* **45**, 135–140.

43. Packer, R. J., Gajjar, A., Vezina, G. *et al.* (2004). 2340 cGy of craniospinal radiotherapy (CSRT) plus chemotherapy for children with "average-risk" medulloblastoma (MB): A prospective randomized Children's Oncology Group study (A9961). *Neuro-Oncol* 6:387.

44. Jakacki, R. I., Feldman, H., Jamison, C., Boaz, J. C., Luerssen, T. G., and Timmerman, R. (2004). A pilot study of preirradiation chemotherapy and 1800 cGy craniospinal irradiation in young children with medulloblastoma. *Int J Rad Oncol Biol Phys* **60**, 531–536.

45. Rutkowski, S., Bode, U., Deinlein, F. *et al.* (2004). Cure of children less than three years of age with medulloblastoma (M0/M1-stage) by postoperative chemotherapy only: Final results of the HIT-SKK'92 study. *Neuro-Oncol* **6**, 469.

46. Packer, R. J. (1990). Chemotherapy for medulloblastoma/primitive neuroectodermal tumors of the posterior fossa. *Ann Neurol* **28**, 823–828.

47. Gajjar, A., Kuhl, J., Epelman, S., Bailey, C., and Allen, J. (1999). Chemotherapy of medulloblastoma. *Child's Nerv Syst* **15**, 554–562.

48. Evans, A. E., Jenkin, D. T., Sposto, R. *et al.* (1990). The treatment of medulloblastoma. Results of a prospective randomized trial of radiation therapy with and without CCNU, vincristine, and prednisone. *J Neurosurg* **72**, 572–582.

49. Tait, D. M., Thornton-Jones, H., Bloom, H. J. G., Lemerle, J., and Morris-Jones, P. (1990). Adjuvant chemotherapy for medulloblastoma: the first multi-centre trial of the International Society of Pediatric Oncology (SIOP I). *Eur J Cancer* **26**, 464–469.

50. Krischer, J. P., Ragab, A. H., Kun, L. *et al.* (1991). Nitrogen mustard, vincristine, procarbazine, and prednisone as adjuvant chemotherapy in the treatment of medulloblastoma. A Pediatric Oncology Group study. *J Neurosurg* **74**, 905–909.

51. Bailey, C. A., Gnekow, S., Wellik, M. *et al* (1995). Prospective randomized trial of chemotherapy given before radiotherapy in childhood medulloblastoma. International Society of Pediatric Oncology (SIOP) and the (German) Society of Pediatric Oncology (GPO): SIOP II. *J Pediatr Hematol Oncol* **25**, 166–178.

52. Zeltzer, P. M., Boyett, J. M., Finlay, J. L. *et al.* (1999). Metastasis stage, adjuvant treatment, and residual tumor are prognostic factors for medulloblastoma in children: Conclusions from the Children's Cancer Group 921 randomized phase III study. *J Clin Oncol* **17**, 832–845.

53. Packer, R. J., Sutton, L. N., Goldwein, J. W. *et al.* (1991). Improved survival with the use of adjuvant chemotherapy in the treatment of medulloblastoma. *J Neurosurg* **74**, 433–440.

54. Packer, R. J., Sutton, L. N., Elterman, R. *et al.* (1994). Outcome for children with medulloblastoma treated with radiation and cisplatin, CCNU, and vincristine chemotherapy. *J Neurosurg* **81**, 690–698.

55. Ris, M. D., Packer, R., Goldwein, J., Jones-Wallace, D., and Boyett, R. M. (2001). Intellectual outcome after reduced-dose radiation therapy plus adjuvant chemotherapy for medulloblastoma: A Children's Cancer Group study. *J Clin Oncol* **19**, 3470–3476.

56. Douglas, J. G., Barker, J. L., Ellenbogen, R. G., and Geyer, J. R. (2004). Concurrent chemotherapy and reduced-dose cranial spinal irradiation followed by conformal posterior fossa tumor bed boost for average-risk medulloblastoma: Efficacy and patterns of failure. *Int J Rad Oncol Biol Phys* **58**, 1161–1164.

57. Mosijczuk, A. D., Nigro, M. A., Thomas, P. R. M. *et al.* (1993). Preradiation chemotherapy in advanced medulloblastoma. A Pediatric Oncology Group pilot study. *Cancer* **72**, 2755–2762.

58. Taylor, R. E., Bailey, C. C., Robinson, K. *et al.* (2003). Results of a randomized study of preradiation chemotherapy versus radiotherapy alone for nonmetastatic medulloblastoma: The International Society of Paediatric Oncology/United Kingdom Children's Cancer Study Group PNET-3 study. *J Clin Oncol* **21**, 1581–1591.

59. Stewart, C. F., Iacono, L. C., Chintagumpala, M. *et al.* (2004). Results of a phase II upfront window of pharmacokinetically guided topotecan in high-risk medulloblastoma and supratentorial primitive neuroectodermal tumor. *J Clin Oncol* **22**, 3357–3365.

60. Allen, J. C., Donahue, B., Darossa, R., and Nirengerg, A. (1996). Hyperfractionated craniospinal radiotherapy and adjuvant chemotherapy for children with newly diagnosed medulloblastoma and other primitive neuroectodermal tumors. *Int J Rad Oncol Biol Phys* **36**, 1155–1161.

61. Crafts, D. C., Levin, V. A., Edwards, M. S., Pischer, T. L., and Wilson, C. B. (1978). Chemotherapy of recurrent medulloblastoma with combined procarbazine, CCNU, and vincristine. *J Neurosurg* **49**, 589–592.

62. Friedman, H. S., Mahaley, M. S., Schold, S. C. *et al.* (1986). Efficacy of vincristine and cyclophosphamide in the therapy of recurrent medulloblastoma. *Neurosurg* **18**, 335–340.

63. Lefkowitz, I. B., Packer, R. J., Siegel, K. R. *et al.* (1990). Results of treatment of children with recurrent medulloblastoma/primitive neuroectodermal tumors with lomustine, cisplatin, and vincristine. *Cancer* **65**, 412–417.

64. Gentet, J. F., Doz, E., Bouffet, D. *et al.* (1994). Carboplatin and VP-16 in medulloblastoma: a phase II study of the French Society of Pediatric Oncology (SFOP). *Med Pediatr Oncol* **23**, 422–427.

65. Carrie, C., Lasset, C., Alapetite, C. *et al.* (1994). Multivariate analysis of prognostic factors in adult patients with medulloblastoma. Retrospective study of 156 patients. *Cancer* **74**, 2352–2360.

66. Le, Q. T., Weil, M. D., Wara, W. M. *et al.* (1997). Adult medulloblastoma: An analysis of survival and prognostic factors. *Cancer J Sci Am* **3**, 238–2.45.

67. Brandes, A. A., Palmisano, V., and Monfardini, S. (1999). Medulloblastoma in adults: clinical characteristics and treatment. *Cancer Treat Rev* **25**, 3–12.

68. Chan, A. W., Tarbell, N. J., Black, P. M. *et al.* (2000). Adult medulloblastoma: Prognostic factors and patterns of relapse. *Neurosurg* **47**, 623–632.

69. Malheiros, S. M. F., Franco, C. M. R., Stávale, J. N. *et al.* (2002). Medulloblastoma in adults: a series from Brazil. *J Neurooncol* **60**, 247–253.

70. Kortmann, R. D., and Brandes, A. A. (2003). Current and future strategies in the management of medulloblastoma in adults. *FORUM Trends Exp Clin Med* **13**, 99–110.

71. Eisenstat, D. D. (2004). Clinical management of medulloblastoma in adults. *Expert Rev Anticancer Ther* **4**, 795–802.

72. Brandes, A. A., Ermani, M., Amista, P. *et al.* (2003). The treatment of adults with medulloblastoma: A prospective study. *Int J Rad Oncol Biol Phys* **57**, 755–761.

73. Greenberg, H. S., Chamberlain, M. C., Glantz, M. J., and Wang, S (2001). Adult medulloblastoma: Multiagent chemotherapy. *Neuro-Oncol* **3**, 29–34.

74. Mahoney, D. H., Strother, D., Camitta, B. *et al.* (1996). High-dose melphalan and cyclophosphamide with autologous bone marrow rescue for recurrent/progressive malignant brain tumors in children: A pilot Pediatric Oncology Group study. *J Clin Oncol* **14**, 382–388.

75. Finlay, J. L., Goldman, S., Wong, M. C. *et al.* (1996). Pilot study of high-dose thiotepa and etoposide with autologous bone marrow rescue in children and young adults with recurrent CNS tumors. *J Clin Oncol* **14**, 2495–2503.

76. Graham, M. L., Herndon, J. E., Casey, J. R. *et al.* (1997). High-dose chemotherapy with autologous stem-cell rescue in patients with recurrent and high-risk pediatric brain tumors. *J Clin Oncol* **15**, 1814–1823.

77. Dunkel, I. J., Boyett, J. M., Yates, A. *et al.* (1998). High-dose carboplatin, thiotepa, and etoposide with autologous stem-cell rescue for patients with recurrent medulloblastoma. *J Clin Oncol* **16**, 222–228.

78. Gururangan, S., Dunkel, I. J., Goldman, S. *et al.* (1998). Myeloablative chemotherapy with autologous bone marrow rescue in young children with recurrent malignant brain tumors. *J Clin Oncol* **16**, 2486–2493.

79. Millot, F., Delval, O., Giraud, C. *et al.* (1999). High-dose chemotherapy with hematopoietic stem cell transplantation in adults with bone marrow relapse of medulloblastoma: report of two cases. *Bone Marrow Transplan* **24**, 1347–1349.

80. Strother, D., Ashley, D., Kellie, S. J. *et al.* (2001). Feasibility of four consecutive high-dose chemotherapy cycles with stem-cell rescue for patients with newly diagnosed medulloblastoma or supratentorial primitive neuroectodermal tumor after craniospinal radiotherapy: Results of a collaborative study. *J Clin Oncol* (2001) **19**, 2696–2704.

81. Zia, M. I., Forsyth, P., Chaudhry, A., Russell, J., and Stewart, D. A. (2002). Possible benefits of high-dose chemotherapy and autologous stem cell transplantation for adults with recurrent medulloblastoma. *Bone Marrow Transplan* **30**, 565–569.

82. Mclenden, R. E., Enterline, D. S., Tien, R. D., Thorstad, W. L., and Bruner, J. M. (1998). Tumors of central neuroepithelial origin. *In Russell & Rubinstein's Pathology of Tumors of the Nervous System* (D. D. Bigner, R. E. McLendon, and J. M. Bruner, Eds.) 6th ed., pp. 456–479, Arnold Press, London.

83. Friedman, H. S., Oakes, W. J., Bigner, S. H., Wikstrand, C. J., and Bigner, D. D. (1991). Medulloblastoma: tumor biological and clinical perspectives. *J Neurooncol* **11**, 1–15.

84. Tomlinson, F. H., Scheithauer, B. W., and Jenkins, R. B. (1992). Medulloblastoma: II. A pathobiologic overview. *J Child Neurol* **7**, 240–252.

85. Gilbertson, R. J. (2004). Medulloblastoma: signaling a change in treatment. *Lancet Oncol* **5**, 209–218.

86. Shapiro, J. R., and Coons, S. W. (1998). Genetics of adult malignant gliomas. *BNI Quarterly* **14**, 27–42.

87. Saylors, R. L., Sidransky, D., Friedman, H. S. *et al.* (1991). Infrequent p53 mutations in medulloblastomas. *Cancer Res* **51**, 4721–4723.

88. Biegel, J. A., Burk, C. D., Barr, F. G., and Emanuel, B. S. (1992). Evidence for a 17p tumor related locus distinct from p53 in pediatric primitive neuroectodermal tumors of the central nervous system. *Cancer Res* **52**, 3391–3395.

89. Garson, J. A., Pemberton, L. F., Sheppard, P. W. *et al.* (1989). N-*myc* gene expression and oncoprotein characterisation in medulloblastoma. *Br J Cancer* **59**, 889–894.

90. Macgregor, D. N., and Ziff, E. B. (1990). Elevated c-*myc* expression in childhood medulloblastomas. *Pediatr Res* **28**, 63–68.

91. Bigner, S. H., Friedman, H. S., Vogelstein, B., Oakes, W. J., and Bigner, D. D. (1990). Amplification of the c-*myc* gene in human medulloblastoma cell lines and xenografts. *Cancer Res* **50**, 2347–2350.

92. Tomlinson, F. H., Jenkins, R. B., Scheithauer, B. W. *et al.* (1994). Aggressive medulloblastoma with high-level N-*myc* amplification. *Mayo Clin Proc* **69**, 359–365.

93. Bruggers, C. S., Tai, K. F., Murdock, T. *et al.* (1998). Expression of the C-*myc* protein in childhood medulloblastoma. *J Pediatr Hematol Oncol* **20**, 18–25.

94. Herms, J., Neidt, I., Luscher, B. *et al.* (2000). C-*myc* expression in medulloblastoma and its prognostic value. *Int J Cancer* **89**, 395–402.

95. Stearns, D., Chaudhry, A., Burger, P. C., and Eberhart, C. G. (2004). C-myc expression recapitulates the large cell/anaplastic medulloblastoma phenotype in xenografts. *Neuro-Oncol* **6**, 414.

96. Wang, J. Y., Valle, L. D., Gordon, J. *et al.* (2001). Activation of the IGF-IR system contributes to malignant growth of human and mouse medulloblastomas. *Oncogene* **20**, 3857–3868.

97. Gilbertson, R. J., Perry, R. H., Kelly, P. J., Pearson, A. D. J., and Lunec, J. (1997). Prognostic significance of HER2 and HER4 coexpression in childhood medulloblastoma. *Cancer Res* **57**, 3272–3280.

98. Gilbertson, R. J., Clifford, S. C., Macmeekin, W. *et al.* (1998). Expression of the ErbB-Neuregulin signaling network during human cerebellar development: Implications for the biology of medulloblastoma. *Cancer Res* **58**, 3932–3941.

99. Polakis, P. (2000). Wnt signaling and cancer. *Genes Dev* **14**, 1837–1851.

100. Henderson, B. R., and Fagotto, F. (2002). The ins and outs of APC and beta-catenin nuclear transport. *EMBO Rep* **3**, 834–839.

101. Zurawel, R. H., Chiappa, S. A., Allen, C., and Raffel, C. (1998). Sporadic medulloblastomas contain oncogenic β–catenin mutations. *Cancer Res* **58**, 896–899.

102. Koch, A., Waha, A., Tonn, J. C. *et al.* (2001). Somatic mutations of WNT/wingless signaling pathway components in primitive neuroectodermal tumors. *Int J Cancer* **93**, 445–449.

103. Fruhwald, M. C., O'dorisio, M. S., Pietsch, T., and Reubi, J. C. (1999). High expression of somatostatin receptor subtype 2 (sst2) in medulloblastoma: Implications for diagnosis and therapy. *Pediatr Res* **45**, 697–708.

104. Fruhwald, M. C., O'dorisio, M. S., Fleitz, J., Pietsch, T., and Reubi, J. C. (1999). Vasoactive intestinal peptide (VIP) and VIP receptors: Gene expression and growth modulation in medulloblastoma and other central primitive neuroectodermal tumors of childhood. *Int J Cancer* **81**, 165–173.

105. Grotzer, M. A., Janss, A. J., Fung, K. M. *et al.* (2000). TrkC expression predicts good clinical outcome in primitive neuroectodermal brain tumors. *J Clin Oncol* **18**, 1027–1035.

106. Ming, J. E., Roesseler, E., and Muenke, M. (1998). Human developmental disorders and the sonic hedgehog pathway. *Mol Med Today* Aug: 343–349.

107. Ingham, P. W. (1998). The patched gene in development and cancer. *Curr Opin Gen Devel* **8**, 88–94.

108. Wicking, C., Smyth, I., and Bale, A. (1999). The hedgehog signaling pathway in tumorigenesis and development. *Oncogene* **18**, 7844–7851.

109. Saldanha, G. (2001). The hedgehog signaling pathway and cancer. *J Pathol* **193**, 427–432.

110. Vortmeyer, A. O., Stavrou, T., Selby, D. *et al.* (1999). Deletion analysis of the adenomatous polyposis coli and PTCH gene loci in patients with sporadic and nevoid basal cell carcinoma syndrome-associated medulloblastoma. *Cancer* **85**, 2662–2667.

111. Wetmore, C., Eberhart, D. E., and Curran, T. (2000). The normal patched allele is expressed in medulloblastomas from mice with heterozygous germ-line mutation of patched. *Cancer Res* **60**, 2239–2246.

112. Zurawel, R. H., Allen, C., Wechsler-Reya, R., Scott, M. P., and Raffel, C. (2000). Evidence that haploinsufficiency of PTCH leads to medulloblastoma in mice. *Genes Chromosomes Cancer* **28**, 77–81.

113. Corcoran, R. B., and Scott, M. P. (2001). A mouse model for medulloblastoma and basal cell nevus syndrome. *J Neuro-oncol* **53**, 307–318.

114. Hasselager, G., and Holland, E. C. (2003). Using mice to decipher the molecular genetics of brain tumors. *Neurosurg* **53**, 685–695.

115. Wetmore, C., Eberhart, D. E., and Curran, T. (2001). Loss of p53 but not ARF accelerates medulloblastoma in mice heterozygous for patched. *Cancer Res* **61**, 513–516.

116. Wechsler-Reya, R. J., Oliver, T. G., Grasfeder, L. L. *et al.* (2003). Sonic hedgehog-induced proliferation of neuronal precursors is mediated by N-myc and cyclin D1. *Proc Am Assoc Cancer Res* **44**, *1193.*

117. Hallahan, A. R., Pritchard, J. I., Hansen, S. *et al.* (2004). The SmoA1 mouse model reveals that notch signaling is critical for the growth and survival of sonic hedgehog-induced medulloblastomas. *Cancer Res* **64**, 7794–7800.

118. Wolter, M., Reifenberger, J., Sommer, C., Ruzicka, T., and Reifenberger, G. (1997). Mutations in the human homologue of the Drosophila segment polarity gene patched (PTCH) in sporadic basal cell carcinomas of the skin and primitive neuroectodermal tumors of the central nervous system. *Cancer Res* **57**, 2581–2585.

119. Pietsch, T., Waha, A., Koch, A. *et al.* (1997). Medulloblastomas of the desmoplastic variant carry mutations of the human homologue of drosophila patched. *Cancer Res* **57**, 2085–2088.

120. Raffel, C., Jenkins, R. B., Frederick, L. *et al.* (1997). Sporadic medulloblastomas contain PTCH mutations. *Cancer Res* **57**, 842–845.

121. Zurawel, R. H., Allen, C., Chiappa, S. *et al.* (2000). Analysis of PTCH/SMO/SHH pathway genes in medulloblastoma. *Genes Chromosomes Cancer* **27**, 44–51.

122. Dong, J., Gailani, M. R., Pomeroy, S. L., Reardon, D., and Bale, A. E. (2000). Identification of patched mutations in medulloblastomas by direct sequencing. *Human Mut* (Online) **339**, 1–7.

123. Smyth, I., Narang, M. A., Evans, T. *et al.* (1999). Isolation and characterization of human patched 2 (*PTCH2*), a putative tumour suppressor gene in basal cell carcinoma and medulloblastoma on chromosome 1p32. *Human Mol Genet* **8**, 291–297.

124. Reifenberger, J., Wolter, M., Weber, R. G. *et al.* (1998). Missense mutations in SMOH in sporadic basal cell carcinomas of the skin and primitive neuroectodermal tumors of the central nervous system. *Cancer Res* **58**, 1798–1803.

125. Erez, A., Ilan, T., Amariglio, N. *et al.* (2002). Gli3 is not mutated commonly in sporadic medulloblastomas. *Cancer* **95**, 28–31.

126. Taylor, M. D., Liu, L., Raffel, C. *et al.* (2002). Mutations in SUFU predispose to medulloblastoma. *Nature Genet* **31**, 306–310.

127. Kinzler, K. W., Bigner, S. H., Bigner, D. D. *et al.* (1987). Identification of an amplified, highly expressed gene in a human glioma. *Science* **236**, 70–73.

128. Katayama, M., Yoshida, K., Ishimori, H. *et al.* (2002). Patched and smoothened mRNA expression in human astrocytic tumors inversely correlates with histological malignancy. *J Neurooncol* **59**, 107–115.

129. Brandes, A. A., Paris, M. K., and Basso, U. (2003). Medulloblastomas: do molecular and biologic markers indicate different prognoses and treatments? *Expert Rev Anticancer Ther* **3**, 615–620.

130. Incardona, J. P., Gaffield, W., Kapur, R. P., and Roelink, H. (1998). The teratogenic Veratrum alkaloid cyclopamine inhibits sonic hedgehog signal transduction. *Develop* **125**, 3553–3562.

131. Taipale J, Chen, J. K., Cooper, M. K. *et al.* (2000). Effects of oncogenic mutations in smoothened and patched can be reversed by cyclopamine. *Nature* **406**, 1005–1009.

132. Chen, J. K., Taipale J, Cooper, M. K. *et al.* (2002). Inhibition of hedgehog signaling by direct binding of cyclopamine to smoothened. *Genes Devel* **16**, 2743–2748.

133. Berman, D. M., Karhadkar, S. S., Hallahan, A. R. *et al.* (2002). Medulloblastoma growth inhibition by hedgehog pathway blockage. *Science* **297**, 1559–1561.

134. Chen, J. K., Taipale, J., Young, K. E., Maiti, T., and Beachy, P. A. (2002). Small molecule modulation of smoothened activity. *Proc Natl Acad Sci* **99**, 14071–14076.

135. Romer, J. T., Kimura, H., Magdaleno, S. *et al.* (2004). Suppression of the Shh pathway using a small molecule inhibitor eliminates medulloblastoma in Ptc1 + /- p53-/- mice. *Cancer Cell* **6**, 229–240.

136. Duffner, P. D., Cohen, M. E., and Thomas, P. (1983). Late effects of treatment on the intelligence of children with posterior fossa tumors. *Cancer* **51**, 233–237.

137. Duffner, P. K., Cohen, M. E., and Parker, M. S. (1988). Prospective intellectual testing in children with brain tumors. *Ann Neurol* **23**, 575–579.

138. Duffner, P. K., and Cohen, M. E. (1991). The long-term effects of central nervous system therapy on children with brain tumors. *Neurol Clin* **9**, 479–495.

139. Mulhern, R. K., Ochs, J., and Kun, L. E. (1991). Changes in intellect associated with cranial radiaton therapy. *In Radiation Injury to the Nervous System* (P. H. Gutin, S. A. Leibel, and G. E. Sheline, Eds.), pp. 325–340, Ravenpress, New York.

140. Mulhern, R. K., Hancock, J., Fairclough, D., and Kun, L. (1992). Neuropsychological status of children treated for brain tumors: A critical review and integrative analysis. *Med Pediatr Oncol* **20**, 181–191.

141. Silverman, C. L., Palkes, H., Talent, B. *et al.* (1984). Late effects of radiotherapy on patients with cerebellar medulloblastoma. *Cancer* **54**, 825–829.

142. Johnson, D. L., Mccabe, M. A., Nicholson, S. H. *et al.* (1994). Quality of long-term survival in young children with medulloblastoma. *J Neurosurg* **80**, 1004–1010.

143. Dennis, M., Spiegler, B. J., Hetherington, C. R., and Greenberg, M. L. (1996). Neuropsychological sequelae of the treatment of children with medulloblastoma. *J Neurooncol* **29**, 91–101.

144. Mulhern, R. K., Kepner, J. L., Thomas, P. R. *et al.* (1998). Neuropsychologic functioning of survivors of childhood medulloblastoma randomized to receive conventional or reduced-dose craniospinal irradiation: A Pediatric Oncology Group study. *J Clin Oncol* **16**, 1723–1728.

145. Walter, A. W., Mulhern, R. K., Gajjar, A. *et al.* (1999). Survival and neurodevelopmental outcome of young children with medulloblastoma at St. Jude Children's Research Hospital. *J Clin Oncol* **17**, 3720–3728.

146. Palmer, S. L., Goloubeva, O., Reddick, W. E. *et al.* (2001). Patterns of intellectual development among survivors of pediatric medulloblastoma: A longitudinal analysis. *J Clin Oncol* **19**, 2302–2308.

147. Mulhern, R. K., Reddick, Palmer, S. L. *et al.* (1999). Neurocognitive deficits in medulloblastoma survivors and white matter loss. *Ann Neurol* **46**, 834–841.

148. Mulhern, R. K., Palmer, S. L., Reddick, W. E. *et al.* (2001). Risks of young age for selected neurocognitive deficits in medulloblastoma are associated with white matter loss. *J Clin Oncol* **19**, 472–479.

149. Newton, H. B., Turowski, R. C., Stroup, T. J., and McCoy, L. K. (1999). Clinical presentation, diagnosis, and pharmacotherapy of patients with primary brain tumors. *Ann Pharmacother* **33**, 816–832.

150. Newton, H. B. (2000). Novel chemotherapeutic agents for the treatment of brain cancer. *Expert Opin Invest Drugs* **9**, 2815–2829.

151. Garrett, M. D., and Workman, P. (1999). Discovering novel chemotherapeutic drugs for the third millenium. *Eur J Cancer* **35**, 2010–2030.

152. Newton, H. B. (2004). Molecular neuro-oncology and the development of "targeted" therapeutic strategies for brain tumors. Part 2 – PI3K/Akt/PTEN, mTOR, SHH/PTCH, and angiogenesis. *Expert Rev Anticancer Ther* **4**, 105–128.

153. Henrikkson, M., and Luscher, B. (1996). Proteins of the myc network: essential regulator of cell growth and differentiation. *Adv Cancer Res* **68**, 109–182.

154. Fernandez-Teijeiro, A., Betensky, R. A., Sturla, L. M., Kim, J. Y. H., Tamayo, P., and Pomeroy, S. L. (2004). Combining gene expression profiles and clinical parameters for risk stratification in medulloblastomas. *J Clin Oncol* **22**, 994–998.

155. Gajjar, A., Hernan, R., Kocak, M. *et al.* (2004). Clinical, histopathologic, and molecular markers of prognosis: Toward a new disease risk stratification system for medulloblastoma. *J Clin Oncol* **22**, 984–993.

156. Lamont, J. M., McManamy, C. S., Pearson, A. D., Clifford, S. C., and Ellison, D. W. (2004). Combined histopathological and molecular cytogenetic stratification of medulloblastoma patients. *Clin Cancer Res* **10**, 5482–5493.

157. Greenberg, S. A. (2001). DNA microarray gene expression analysis technology and its application to neurological disorders. *Neurol* **57**, 755–761.

158. Gilbertson, R., Wickramasinghe, C., Hernan, R. *et al.* (2001). Clinical and molecular stratification of disease risk in medulloblastoma. *Br J Cancer* **85**, 705–712.

30

Chemotherapy of Ependymoma

Mark G. Malkin

ABSTRACT: Ependymomas are rare neoplasms of the central nervous system (CNS) seen most commonly in children. The most effective treatment modality is surgical resection, and it is widely held that gross total removal of these tumors confers a better prognosis. Radiation therapy (RT) is commonly used to treat residual, recurrent, or progressive disease. The experience with chemotherapy has been a long and frustrating one, both up front and at relapse. Certainly, significant radiographic and clinical responses have been observed with a number of agents, particularly with the platinum compounds, but these successes have been uncommon and unpredictable, have rarely produced prolonged disease-free or overall survival, and have almost never been associated with cure. Alone or in combination, in conventional doses or in the high doses used with bone marrow/stem cell rescue, chemotherapy has been a great disappointment in the management of this tumor. One can only hope with the advent of newer biological agents, and with a better understanding of the molecular events underlying neoplastic transformation of ependymal cells into ependymomas, that the trying experience with drug therapy of ependymoma will become a thing of the past.

BACKGROUND

Ependymal tumors are a group of neoplasms that are derived from the ependymal lining of the ventricles and from the central canal of the spinal cord. The World Health Organization (WHO) pathologic classification includes ependymomas, anaplastic ependymomas, myxopapillary ependymomas, and subependymomas. Ependymomas are the most common of these tumors and may develop at any age. Uncommon, ependymomas constitute 8–10 per cent of pediatric brain tumors, and 1–3 per cent of brain tumors in adults. Sixty per cent of ependymomas occur in children younger than 16 and 25 per cent occur in children younger than 4. Intracranial ependymomas most often present with signs and symptoms of raised intracranial pressure either due to the mass effect of the tumor or obstructive hydrocephalus. Cerebrospinal fluid (CSF) dissemination occurs in 3–12 per cent of all intracranial ependymomas. Studies of prognostic variables in ependymomas are hampered by the rarity of this tumor. Controversy exists regarding the relationship between tumor grade and outcome, partly because of discordance among neuropathologists regarding the criteria to distin-guish low-grade tumors from high-grade ones. Survival does appear to be somewhat improved in patients who have had gross total resection. Optimal management of these tumors includes surgical resection and evaluation of the extent of CNS involvement using both CSF cytology and craniospinal contrast-enhanced magnetic resonance imaging (MRI). This staging permits stratification of patients into those with (M1) or without metastasis (M0), and patients with or without residual disease following surgery, the two most important clinical parameters affecting outcome. Surgery remains the mainstay of therapy for ependymomas. Re-operation should be considered, if feasible, to remove residual disease. Re-operation at the time of local recurrence should also be contemplated if there is no evidence of disseminated disease. For completely resected low-grade ependymomas, local field RT to a tumor dose of 54–55.8 Gy *versus* observation only is recommended. For incompletely resected low-grade ependymomas, local field RT to a tumor dose of 54–55.8 Gy is recommended. For completely or incompletely resected high-grade ependymomas local field RT to a tumor dose of 59.4 Gy is recommended. There is no clear role for craniospinal RT in patients with localized disease. Most recurrences are within the original tumor bed. Recently published comprehensive reviews concerning the salient issues in the management of ependymomas are cited in references [1–9] below.

TRIALS DEDICATED TO EPENDYMOMA

Ten trials devoted solely to the treatment of patients with ependymoma have been reported over the years, and are presented here roughly in chronological order of their publication.

Fifteen children younger than 18 months old with ependymoma were treated on a Children's Cancer Group (CCG) protocol with an eight-drug-in-one-day chemotherapeutic regimen (vincristine 0.05 mg/kg, carmustine 7.25 mg/kg, procarbazine 2.25 mg/kg, hydroxyurea 45 mg/kg, cisplatin 1.8 mg/kg, cytarabine 9 mg/kg, methylprednisolone 10 mg/kg, and cyclophosphamide 9 mg/kg) following surgery and postoperative staging [10]. Chemotherapy began within six weeks from the date of surgery. A second course of chemotherapy was given two weeks after the first course. Subsequent courses of chemotherapy were administered monthly for a total of eight courses. Delayed or reduced-volume RT was to be administered to all patients but, in fact, was omitted in most cases. On central neuropathologic review, six ependymomas were considered anaplastic; nine were not. The three-year progression-free survival (PFS) rate for ependymoma was 26 per cent. Of five patients with ependymoma assessable for response following two cycles of chemotherapy, there were no complete (CR) or partial (PR) responses. The most frequent toxicity was bone marrow suppression. Since disease progression occurred early after diagnosis, in most cases while chemotherapy was still being administered, the authors recommended new treatment strategies, including intensification of initial chemotherapy and second-look surgical resection.

Grill et al. entered 16 children with refractory or relapsed ependymoma into a phase II study of high-dose chemotherapy followed by autologous bone marrow transplantation (ABMT) [11]. The conditioning regimen consisted of busulfan 150 mg/m^2/day for four days followed by thiotepa 300 mg/m^2/day, for three days. Busulfan and thiotepa were chosen because they cross the blood–brain barrier readily. Moreover, previous studies had demonstrated the value of thiotepa in CNS tumors. Three tumors were

low-grade and 13 high-grade according to the revised WHO classification. All patients had previously been treated by surgery and conventional chemotherapy. Eight of them had also received RT at tumor doses ranging from 45 to 55 Gy. At the time of transplantation, the median patient age was 5 years. Fifteen patients were evaluable for response. No CRs or PRs were observed. Toxicity was severe, mainly profound myelosuppression, vomiting, diarrhea, mucositis, and diffuse perineal rash, and one toxicity-related death occurred presenting as acute renal failure, hyponatremia, coma, and intractable seizures. From this experience, the authors concluded that ependymomas did not appear to be sensitive to this combination therapy and advocated new therapeutic approaches. Mason et al. reported the CCG experience using intensive chemotherapy followed by bone marrow reconstitution for children with recurrent intracranial ependymoma [12]. All children had undergone maximum surgical debulking. Prior RT had been administered in 13 of 15 children. Prior chemotherapy had been administered to 14 of the 15 patients. Thiotepa 300 mg/m^2/day (total 900 mg/m^2) and etoposide 250–500 mg/m^2/day (total 750–1500 mg/m^2) were administered for three consecutive days with or without the addition of carboplatin 500 mg/m^2/day (total 1500 mg/m^2) for an additional three consecutive days, and autologous bone marrow was reinfused 72 h following chemotherapy. The children ranged in age from 5 months to 12 years (median 22 months). Five patients died of treatment-related toxicities within 62 days of marrow reinfusion. This toxic mortality rate of 33 per cent was unexpected and unacceptable. No CRs or PRs were observed. All patients relapsed locally. This regimen is not an effective strategy for retrieving heavily pre-treated children with recurrent ependymoma.

The only randomized study of adjuvant chemotherapy for the treatment of ependymoma was published by the CCG in 1996 [13]. Patients aged 2–16 with newly diagnosed, previously untreated, infratentorial ependymomas were registered within three weeks of diagnostic surgery. Eligible patients were randomly assigned to receive either craniospinal RT alone or with adjuvant chemotherapy. Adjuvant chemotherapy consisted of vincristine 1.5 mg/m^2 given by intravenous injection weekly for eight weeks during RT. Following a four-week rest period, maintainance chemotherapy was to begin with cycles of six-weeks duration, consisting of CCNU (100 mg/m^2 orally on day 1), vincristine (1.5 mg/m^2 intravenously on days 1, 8, and 15) and prednisone (40 mg/m^2/day orally, days 1 through 14, inclusive). Thirty-six patients were evaluable. Median age at diagnosis was 5 years, with a median follow-up of 10 years. There were two toxic deaths, both on the chemotherapy arm. There were no significant differences between the two treatment arms regarding clinical characteristics possibly affecting outcome such as age, sex, stage, brain stem involvement, or extent of resection. There were no statistically significant differences in outcome between the two regimens. Ten-year survival for the chemotherapy plus RT arm was 40 per cent, and 35 per cent for the RT-only arm. Median PFS was 17.1 months in the chemotherapy plus RT arm, and 20.6 months in the RT-only arm.

In a pilot study published by Needle et al., 19 children age 3–14 (median 7.5 years) were treated with postoperative RT and chemotherapy [14]. In this pilot, physicians were left to choose the RT field and fractionation schedule. Chemotherapy consisted of carboplatin 560 mg/m^2 with vincristine 1.5 mg/m^2, weekly for three weeks, alternating at four-week intervals with ifosfamide 1.8 g/m^2 and etoposide 100 mg/m^2, for five consecutive days for a total of four cycles. The estimated five-year PFS rate was 74 per cent. All treatment failures occurred within the primary site without evidence of metastatic spread. Toxicity, limited predominantly to myelosuppression, was manageable. The PFS for children with postoperative residual ependymoma treated with RT and chemotherapy in this study was higher than published survival results for RT alone. Although these results suggest a role for multialkylator chemotherapy in incompletely resected intracranial ependymoma, and provide the rationale for a randomized trial comparing this strategy with conventional postoperative RT, such a trial has not been done.

Fouladi et al. reviewed the overall survival of 11 consecutive patients with anaplastic ependymoma treated over a ten-year period with surgery and ICE chemotherapy, with or without RT, at the Hospital for Sick Children in Toronto [15]. The chemotherapy regimen consisted of ifosfamide 3 g/m^2 on day 1 and day 2; etoposide 150 mg/m^2 on days 1 and 2; carboplatin 500 mg/m^2 on day 3, for two cycles prior to RT and for eight cycles after RT. In children younger than 2, the regimen consisted of ten cycles of ICE only. The median age of the children was 3.4 years, with a range of 1.2–11 years. Of the six patients with subtotal resection, two had a CR and three had a PR. All recurrences were at the site of the original tumor. Overall survival was 39 per cent at 7 years.

Chamberlain reported the results of a single institution phase II study of oral etoposide in 12 children with recurrent intracranial ependymoma [16]. All were refractory to surgery, RT and

chemotherapy with carboplatin or PCV. Their median age was 8 years. The dose of oral etoposide was 50 mg/m^2/day for 21 consecutive days followed by a 14-day break and then an additional 21 days. The most serious complication was reversible grade 3 and 4 myelopsuppression but there were no treatment-related deaths. There were two PRs lasting 8 and 10 months, respectively. No improvement was seen in either neurologic status or Lansky performance score.

Timmerman et al. summarized the results of two German prospective trials (HIT 88/89 and HIT 91) of combined postoperative RT and chemotherapy for anaplastic ependymomas in childhood [17]. Gross total resection was accomplished in 28 of 55 patients. All patients received chemotherapy before ($n = 40$) or after RT ($n = 15$). Pre-RT chemotherapy consisted of ifosfamide, etoposide, methotrexate, and cisplatin. Vincristine was given during RT. After RT, chemotherapy consisted of cisplatin, CCNU, and vincristine. Median follow-up was 38 months, and the overall survival rate at three years was 76 per cent. Disease progression occurred in 25 children, with local progression occurring in 20. The mode of chemotherapy did not affect survival.

Between 1990 and 1998, the French Society of Pediatric Oncology (SFOP) evaluated a strategy that avoids RT in first-line treatment of children under 5 with ependymoma by exclusively administering 16 months of adjuvant multi-agent chemotherapy after surgery [18]. Seventy-three children were enrolled onto this multicenter trial. Patients received adjuvant conventional chemotherapy after surgery consisting of seven cycles of three courses alternating two drugs at each course (procarbazine and carboplatin, etoposide and cisplatin, and vincristine and cyclophosphamide). No CRs or PRs were observed. With a median follow up of 4.7 years, the overall survival rate in this series was 59 per cent. In the multivariate analysis, gross total resection was associated with favorable outcome ($p = 0.0009$). These results raise the question of the need for adjuvant therapy in completely resected ependymomas.

Finally, the first review of data on the role of chemotherapy in recurrent ependymoma in adults was just published by Brandes et al. [19]. A retrospective review was made of the charts of 28 adults (≥ 18 years) with progressive or recurrent ependymal tumors after surgery and RT, who received chemotherapy between 1993 and 2003. Thirteen patients received cisplatin-based chemotherapy and 15 received regimens without cisplatin. The mean age at diagnosis was 43.9 years. The median KPS was 90. The median follow-up period was 19.1 months. Platinum-based chemotherapy yielded 2 CRs and

2 PRs. There were no CRs and only 2 PRs in non-platinum-based regimens. However, cisplatin-based chemotherapy did not prolong PFS (9.9 months *versus* 10.9 months, NS) or overall survival (31 months *versus* 40.7 months, NS) compared to non-cisplatin-based regimens.

TRIALS INCLUDING EPENDYMOMA

Nineteen trials of various agents, primarily for children with recurrent or progressive primary brain tumors of various histologies, including ependymoma, have been published over the years. The results of those studies are presented here roughly in chronological order of their publication in the peer-reviewed medical literature.

The Pediatric Oncology Group (POG) treated 46 children with recurrent primary brain tumors using cisplatin, 60 mg/m^2/day for two days every three to four weeks [20]. Three of 15 patients with ependymoma had a CR. Dose limiting major toxicities were renal and auditory. The authors recommended that the new analogues of cisplatin with less toxicity (e.g., carboplatin, iproplatin) be used to study these tumors. Between October 1985 and March 1988, the CCG entered 95 patients with recurrent brain tumors into a phase II trial of carboplatin 560 mg/m^2 every four weeks [21]. Two PRs were observed in 14 children with ependymomas. Moderate to severe neutropenia, thrombocytopenia, ototoxicity, and nausea and vomiting were observed. The authors recommended further investigation of this agent in children with recurrent primary brain tumors, although it appears that the drug had marginal effectiveness using this dosing schedule. POG conducted a randomized phase II study to evaluate the activity of carboplatin and iproplatin in children with progressive or recurrent brain tumors [22]. Treatment consisted of carboplatin 560 mg/m^2 at four-week intervals or iproplatin 270 mg/m^2 at three-week intervals. Neither drug demonstrated appreciable activity in the treatment of ependymoma (two responses in 17 patients given carboplatin; no responses in seven patients given iproplatin). The major toxicity observed was myelosuppression, particularly thrombocytopenia, for both agents.

POG treated 17 children with recurrent medulloblastoma and ependymoma using piperidine-2, 6-dione chloroethyl nitrosourea (PCNU), 100 mg/m^2 every six weeks [23]. There were no CRs or PRs in ten patients with ependymoma. Administration of PCNU was associated with significant thrombocytopenia.

The CCG treated 75 children with recurrent, progressive, or metastatic primary brain tumors with aziridinylbenzoquinone (AZQ) at 9 mg/m^2/day by 30-minute intravenous infusion for five days every three weeks [24]. A single CR lasting for 35+ months occurred in one of 12 patients with ependymoma. Profound and prolonged myelosuppression was the significant toxicity observed. As administered in this study, AZQ had marginal activity and severe toxicity.

The SFOP treated 42 children with a variety of recurrent primary brain tumors in a phase II trial of ifosfamide 3 g/m^2/day for two days every two weeks [25]. A PR was demonstrated in one of eight patients with ependymoma. Toxicity was primarily neurologic. The authors recommended further trials with other dose schedules to assess the activity of this drug. Subsequently, POG conducted a phase II trial of ifosfamide 3 g/m^2 every other day for three doses in 87 recurrent pediatric brain tumors [26]. A PR was observed in one of 12 patients with ependymoma. Neurotoxicity was less common than in studies of daily ifosfamide. The authors concluded that ifosfamide monotherapy possessed little meaningful clinical activity in brain tumors.

Heideman *et al.* treated 60 children with recurrent primary brain tumors on a multi-institutional phase II study of intravenous thiotepa at a dose of 65 mg/m^2 administered every three weeks [27]. There were no responses in nine patients with ependymoma. Myelopsuppression was the principal toxic effect encountered.

Needle *et al.* enrolled 28 children with recurrent brain tumors in a phase II study of oral etoposide [28]. The dose was 50 mg/m^2/day for 21 consecutive days. Courses were repeated every 28 days pending bone marrow recovery. Two of five patients with ependymoma responded, one with a CR and one with a PR. Toxicity was manageable and consisted primarily of myelosuppression. The authors recommended daily oral etoposide in combination with DNA cross-linking agents in future phase III trials. The Australian and New Zealand Children's Cancer Study Group (ANZCCSG) enrolled 42 patients <4 years old with brain tumors into a two-phase chemotherapy protocol following surgery and before RT [29]. The initial phase consisted of four courses of the three-drug regimen vincristine, etoposide, and intensive cyclophosphamide. The continuation phase was comprised of two-drug courses: cyclophosphamide plus vincristine, cisplatin plus etoposide, and carboplatin plus etoposide. Six of seven patients with ependymoma had a CR or PR. The authors recommended further dose escalation.

Arndt *et al.* reported a phase II trial of idarubicin in children with relapsed brain tumors [30]. Patients received idarubicin at a dose of 5 mg/m^2/day for three days by intravenous bolus, followed by granulocyte-colony stimulating factor (G-CSF) at a dose of 5 µg/kg/day, starting on Day 7 of each cycle and continuing for at least seven days, until the absolute neutrophil count was $\geq 10\,000$/mm^3. There were no CRs or PRs in 13 patients with ependymoma.

Topotecan was studied as a 72-h infusion given every three weeks for the treatment of recurrent or progressive CNS tumors in a POG phase II study [31]. Treatment began at a dose of 1.0 mg/m^2/day and was escalated to 1.25 mg/m^2/day. There were no CRs or PRs in 17 patients with ependymoma. In a phase I study of topotecan administered as a 21-day continuous infusion in children with recurrent solid tumors, two of three patients with ependymoma had PRs [32]. Myelosuppression was the dose-limiting toxicity.

On behalf of POG, Mulne *et al.* reported a pilot study of low-dose oral methotrexate in children with recurrent or progressive CNS tumors [33]. Eight doses of methotrexate 7.5 mg/m^2 every six hours were administered on a weekly schedule for as long as 18 months. There was one PR in seven patients with ependymoma. The authors did not recommend this regimen for front-line therapy.

Hurwitz *et al.* attempted to assess the efficacy and define the toxicity of paclitaxel given at a dosage of 350 mg/m^2 every three weeks as a 24-h continuous infusion to 75 children with recurrent or progressive primary brain tumors on the POG 9330 protocol [34]. There were no CRs or PRs in 13 patients with ependymoma. Median time to progression was 2.1 months in the ependymoma patients.

A phase II study of irinotecan (CPT-11) was conducted at Duke University Medical Center to evaluate the activity of this agent in children with high-risk malignant brain tumors [35]. A total of 22 patients were enrolled in this study, including five with ependymomas. Each course of CPT-11 consisted of 125 mg/m^2 per week given intravenously for four weeks followed by a two-week rest period. Patients with both newly diagnosed and recurrent tumors were included in this trial. One PR was observed in the five patients with recurrent ependymoma. Toxicity was mainly myelosuppression. Ongoing studies will demonstrate if the activity of CPT-11 can be enhanced when combined with alkylating agents, including carmustine and temozolomide.

Hargrave *et al.* enrolled patients younger than 21 with recurrent/resistant brain tumors in a conventional phase I study of fotemustine [36]. The dose was increased from 100 to 175 mg/m^2 every

three weeks. No CRs or PRs were observed. Delayed and cumulative myelopsuppression was noted.

Rojas-Marcos *et al.* reported a CR lasting eight months in a patient with recurrent anaplastic ependymoma given the combination of tamoxifen (200 mg/day) and isoretinoin (80 mg/m^2/day for three weeks, followed by one week of rest) [37].

Finally, Wolff and Finlay summarized the experience with high-dose chemotherapy in childhood brain tumors in a review article in 2004 [38]. Thiotepa-based regimens had a high toxic death rate (5 of 15 patients), but no response in ependymomas, regardless if BCNU, VP-16 or busulfan were added. They concluded that high-dose technology has not yet changed the disappointing results of conventional chemotherapy in ependymomas.

References

1. Grill, J., Pascal, C., and Chantal, K. (2003). Childhood ependymoma. A systematic review of treatment options and strategies. *Pediatr Drugs* **5**(8), 533–543.
2. Moynihan, T. J. (2003). Ependymal tumors. *Curr Treat Options Oncol* **4**, 517–523.
3. Chamberlain, M. C. (2003). Ependymomas. *Curr Neurol Neurosci Rep* **3**, 193–199.
4. Merchant, T. E. (2002). Current management of childhood ependymoma. *Oncology* **16**(5), 629–644.
5. Bouffet, E., and Foreman, N. (1999). Chemotherapy for intracranial ependymomas. *Child's Nerv Syst* **15**, 563–570.
6. Siffert, J., and Allen, J. C. (1998). Chemotherapy in recurrent ependymoma. *Pediatr Neurosurg* **28**, 314–319.
7. Souweidane, M. M., Bouffet, E., and Finlay, J. (1998). The role of chemotherapy in newly diagnosed ependymoma of childhood. *Pediatr Neurosurg* **28**, 273–278.
8. Robertson, P. L., Zeltzer, P. M., Boyett, J. M. *et al.* (1998). Survival and prognostic factors following radiation therapy and chemotherapy for ependymomas in children: a report of the Children's Cancer Group. *J Neurosurg* **88**, 695–703.
9. Bouffet, E., Perilongo, G., Canete, A., and Massimino, M. (1998). Intracranial ependymomas in children: a critical review of prognostic factors and a plea for cooperation. *Med Pediatr Oncol* **30**, 319–331.
10. Geyer, J. R., Zeltzer, P. M., Boyett, J. M. *et al.* (1994). Survival of infants with primitive neuroectodermal tumors or malignant ependymomas of the CNS treated with eight drugs in 1 day: a report from the Children's Cancer Group. *J Clin Oncol* **12**(8), 1607–1615.
11. Grill, J., Kalifa, C., Doz, F. *et al.* (1996). A high-dose busulfan-thiotepa combination followed by autologous bone marrow transplantation in childhood recurrent ependymoma. *Pediatr Neurosurg* **25**, 7–12.
12. Mason, W. P., Goldman, S., Yates, A. J., Boyett, J., Li, H., and Finlay, J. L. (1998). Survival following intensive chemotherapy with bone marrow reconstitution for children with recurrent intracranial ependymoma. A report of the Children's Cancer Group. *J Neurooncol* **37**, 135–143.
13. Evans, A. E., Anderson, J. R., Lefkowitz-Boudreaux, I. B. and Finlay, J. L. (1996). Adjuvant chemotherapy of childhood posterior fossa ependymoma: cranio-spinal irradiation with or

without adjuvant CCNU, vincristine, and prednisone: a Children's Cancer Group Study. *Med Pediatr Oncol* **27**, 8–14.

14. Needle, M. N., Goldwein, J. W., Grass, J. *et al.* (1997). Adjuvant chemotherapy for the treatment of intracranial ependymoma of childhood. *Cancer* **80**(2), 341–347.

15. Fouladi, M., Baruchel, S., Chan, H., *et al.* (1998). Use of adjuvant ICE chemotherapy in the treatment of anaplastic ependymoma. *Childs Nerv Syst* **14**, 590–595.

16. Chamberlain, M. C. (2001). Recurrent intracranial ependymoma in children: salvage therapy with oral etoposide. *Pediatr Neurol* **24**(2), 117–121.

17. Timmerman, B., Kortmann, R.-D., Kühl, J., *et al.* (2000). Combined postoperative irradiation and chemotherapy for anaplastic ependymomas in childhood: results of the German prospective trials HIT 88/89 and HIT 91. *Int J Radiat Oncol Biol Phys* **46**(2), 287–295.

18. Grill, J., Le Deley, M.-C., Gambarelli, D., *et al.* (2001). Postoperative chemotherapy without irradiation for ependymoma in children under 5 years of age: a multicenter trial of the French Society of Pediatric Oncology. *J Clin Oncol* **19**(5), 1288–1296.

19. Brandes, A. A., Cavallo, G., Reni, M. *et al.* (2005). A multicenter retrospective study of chemotherapy for recurrent intracranial ependymal tumors in adults by the Gruppo Italiano Cooperativo di Neuro-Oncologia. *Cancer* **104**(1), 143–148.

20. Sexauer, C. L., Khan, A., Burger, P. C., *et al.* (1985). Cisplatin in recurrent pediatric brain tumors. A POG phase II study. A Pediatric Oncology Group study. *Cancer* **56**(7), 1497–1501.

21. Gaynon, P. S., Ettinger, L. J., Baum, E. S., Siegel, S. E., Krailo, M. D., and Hammond, G. D. (1990). Carboplatin in childhood brain tumors. A Children's Cancer Study Group phase II trial. *Cancer* **66**(12), 2465–2469.

22. Friedman, H. S., Krischer, J. P., Burger, P. *et al.* (1992). Treatment of children with progressive or recurrent brain tumors with carboplatin or iproplatin: A Pediatric Oncology Group randomized phase II study. *J Clin Oncol* **10**(2), 249–256.

23. Ragab, A. A. H., Burger, P., Badnitsky, S., Krischer, J., and Van Eys, J. (1986). PCNU in the treatment of recurrent medulloblastoma and ependymoma. *J Neurooncol* **3**, 341–342.

24. Ettinger, L. J., Ru, N., Krailo, M., Ruccione, K. S., Krivit, W., and Hammond, G. D. (1990). A phase II study of diaziquone in children with recurrent or progressive primary brain tumors: A report from the Children's Cancer Study Group. *J Neurooncol* **9**, 69–76.

25. Chastagner, P., Sommelet-Olive, D., Kalifa, C. *et al.* (1993). Phase II study of ifosfamide in childhood brain tumors: A report by the French Society of Pediatric Oncology (SFOP). *Med Pediatr Oncol* **21**, 49–53.

26. Heideman, R. L., Douglass, E. C., Langston, J. A. *et al.* (1995). A phase II study of every other day high-dose ifosfamide in pediatric brain tumors: A Pediatric Oncology Group study. *J Neurooncol* **25**, 77–84

27. Heideman, R. L., Packer, R. J., Reaman, G. H. *et al.* (1993). A phase II evaluation of thiotepa in pediatric central nervous system malignancies. *Cancer* **72**(1), 271–275.

28. Needle, M. N., Molloy, P. T., Geyer, J. R. *et al.* (1997). Phase II study of daily oral etoposide in children with recurrent brain tumors and other solid tumors. *Med Pediatr Oncol* **29**, 28–32.

29. White, L., Kellie, S., Gray, E. *et al.* (1998). Postoperative chemotherapy in children less than 4 years of age with malignant brain tumors: promising initial response to a VETOPEC-based regimen. *J Pediatr Hematol/Oncol* **20**(2), 125–130.

30. Arndt, C. A. S., Krailo, M. D., Steinherz, L., Scheithauer, B., Liu-Mares, W., and Reaman, G. H. (1998). A phase II clinical trial of idarubicin administered to children with relapsed brain tumors. *Cancer* **83**(4), 813–816.

31. Kadota, R. P., Stewart, C. F., Horn, M. *et al.* (1999). Topotecan for the treatment of recurrent or progressive central nervous system tumors – a Pediatric Oncology Group phase II study. *J Neurooncol* **43**, 43–47.

32. Frangoul, H., Ames, M. M., Mosher, R. B. *et al.* (1999). Phase I study of topotecan administered as a 21-day continuous infusion in children with recurrent solid tumors: a report from the Children's Cancer Group. *Clin Cancer Res* **5**, 3956–3962.

33. Mulne, A. F., Ducore, J. M., Elterman, R. D., *et al.* (2000). Oral methotrexate for recurrent brain tumors in children: a Pediatric Oncology Group study. *J Pediatr Hematol/Oncol* **22**(1), 41–44.

34. Hurwitz, C. A., Strauss, L. C., Kepner, J. *et al.* (2001). Paclitaxel for the treatment of progressive or recurrent childhood brain tumors: A Pediatric Oncology phase II study. *J Pediatr Hematol/Oncol* **23**(5), 277–281.

35. Turner, C. D., Gururangan, S., Eastwood, J. *et al.* (2002). Phase II study of irinotecan (CPT-11) in children with high-risk malignant brain tumors: the Duke experience. *Neuro-Oncol* **4**, 102–108.

36. Hargrave, D. R., Bouffet, E., Gammon, J., Tariq, N., Grant, R. M., and Baruchel, S. (2002). Phase I study of fotemustine in pediatric patients with refractory brain tumors. *Cancer* **95**(6), 1294–1301.

37. Rojas-Marcos, I., Calvet, D., Janoray, P., and Delattre, J. Y. (2003). Response of recurrent anaplastic ependymoma to a combination of tamoxifen and isoretinoin. *Neurology* **61**, 1019–1020.

38. Wolff, J. E. A., and Finlay, J. L. (2004). High-dose chemotherapy in childhood brain tumors. *Onkologie* **27**, 239–245.

Chemotherapy for Glioneuronal Tumors

Nimish Mohile and Jeffrey J. Raizer

ABSTRACT: Glioneuronal tumors are rare primary brain tumors. They are composed of both glial and neuronal elements. In most cases, they are benign tumors that are often cured with surgery. Radiation therapy is sometimes used to treat residual or aggressive tumors. The role of chemotherapy for these tumors is not well defined or not indicated. This chapter reviews the available literature on the use of chemotherapy to treat these tumors and when it may be indicated. A framework for which agents may be of benefit should assist physicians when chemotherapy is needed.

INTRODUCTION

Neuronal and mixed neuronal–glial tumors are a rare but important group of primary brain tumors. With the exception of ganglioglioma and dysplastic gangliocytoma of the cerebellum, the remaining tumors have been described in the past 30 years [1–7]. These tumors are usually benign and often curable with surgical resection. For this reason, histologic diagnosis is important in order to avoid unnecessary treatment with radiation therapy (RT) or chemotherapy (CTX). In the cases where there has been incomplete resection or malignant degeneration occurs, RT or CTX may have a role but the "true" benefits, if any, often remain unknown. This chapter reviews the available literature using CTX to treat glioneuronal tumors. Reasons to use CTX include recurrent tumors after radiation, in place of RT to avoid the long-term sequelae in young patients and for patients who refuse RT. Definitive recommendations cannot be made as most reports are limited to case reports or small numbers of patients and the types of CTX are often different for an individual

tumor type. We hope that this review will provide a framework to assist physicians in choosing CTX, knowing that much remains unknown and new drugs are rapidly becoming available.

CENTRAL NEUROCYTOMA

Central neurocytoma was first described in 1982 and further characterized in 1993 [4,8]. They make up 0.25–0.5 per cent of intracranial tumors and there is no predilection for gender [8]. They usually present in the second and third decades of life with symptoms of increased intracranial pressure, due to their intraventricular location within the lateral and third ventricles [8,9]. Reports of extraventricular neurocytomas have been described [10,11]. CT scan appearance is an isodense to hyperdense tumor with cystic changes and sometimes calcifications; they enhance with contrast. MRI appearance is high signal on T1 and T2 with variable contrast enhancement [4,8]. The histopathology of central neurocytomas is similar to that of oligodendrogliomas. They consist of small uniform cells with little cytoplasm and rounded nuclei. Mitoses are rare and vascular proliferation and necrosis are not commonly observed. On immunohistochemistry they stain for neuron-specific enolase (NSE) and synaptophysin positive. Glial fibrillary acidic protein (GFAP) staining may be seen and the labeling index (LI) is usually low. Atypical neurocytomas have a MIB-1 > 5 per cent, focal necrosis, vascular proliferation, and increased mitotic activity [10]. Atypical lesions may have a more aggressive course. Their intraventricular location lends neurocytomas to disseminate within the neuro-axis and even intra-peritoneally *via* VP shunt [12–16]. The primary treatment of these tumors is surgical irrespective of location [9–11,17,18]. Gross total resection (GTR) has resulted in 5-year local control rates of 95–100 per cent and 5-year survival rates of 90–99 per cent [17]. For subtotal resection (STR) without any adjuvant therapy, these rates were 46–70 per cent and 77–86 per cent, respectively. Radiation therapy to the primary tumor only has been advocated for lesions that are subtotally resected to increase local control and survival [17,19]. However, the exact role of radiation is unclear as some patients with a STR do not recur and have similar survivals to those with GTR. Recently gamma knife radiosurgery has been used in some cases of STR tumors with good local control [20,21].

The use of CTX in these tumors is limited. Louis *et al.* [22] reported on two patients with subtotally resected tumors treated with a combination of cyclophosphamide and cisplatin prior to craniospinal radiation. Both the patients were alive at 11 and 14 months; which modality had the greatest impact is unclear. Eng *et al.* [14] reported on two patients treated with a regimen of etoposide (60 mg/m^2 × 3 days), cisplatin (30 mg/m^2 × 3 days) and cyclophosphamide (1000 mg on day 3) but outcomes and responses are not mentioned. Schild *et al.* [17] reviewed 32 patients retrospectively, of which 4 were treated with CTX after GTR (2) or STR (2) as follows: lomustine, cisplatin + lomustine, lomustine + carmustine, vincristine, lomustine, and prednisone. All four received the CTX after RT and maintained local control based on follow-up neuroimaging. However, the impact of CTX is unclear as it followed RT. Dodds *et al.* [23] treated a 15-year-old patient with four cycles of carboplatin (500 mg/m^2 on week 1 day 1 and 2), etoposide (100 mg/m^2 days 1–3 of week 1 and 3), and ifosfamide (3 g/m^2 days 1–3 of week 3) due to inoperability of his tumor. The patient had regression of the solid component of the tumor by CT scan but a specific percentage of shrinkage was not reported. This was followed by 22 months of stabilization. Sgouros *et al.* [18] reported on a 19-year-old woman who temporarily responded to carboplatin after a second recurrence (previously treated with surgery and RT). Three patients with progressive disease, despite surgery and RT, were treated with a regimen of etoposide (40 mg/m^2 per day × 4 days), cisplatin (25 mg/m^2 per day × 4 days), and cyclophosphamide (1000 mg on day 4) [24]. One patient had a partial response which peaked at 8 months, one had stable disease, and one had a complete response in the spinal cord and stable disease in the third ventricle. Responses lasted 15, 18, and 36 months, respectively. von Koch *et al.* [25] treated a 20-year-old female with procarbazine, CCNU, and vincristine (PCV) for tumor recurrence (2 cycles PCV, 3 cycles CCNU and vincristine, and 1 cycle CCNU) after

several subtotal resections. She was noted to have tumor reduction after the second cycle with continued reduction until stabilization after the sixth cycle, which was present sixteen months later.

GANGLIOGLIOMA AND GANGLIOCYTOMA

Perkins named this entity in 1926 [1]. Gangliogliomas (WHO grade I or II, some grade III and IV lesions) make up 1.3 per cent of all brain tumors and contain both neuronal and glial elements, while gangliocytomas (WHO grade I) are composed predominantly of neoplastic neurons [26]. They typically present with seizures because of their predilection for the temporal lobe. They usually present before the age of 30 with a male:female ratio of 1–2:1[27]. Imaging on CT scan with contrast shows a solid mass or cyst with a mural nodule, both well circumscribed. They are usually isodense or hypodense and enhance with contrast; calcifications may be seen. MRI appearance on T1 is hypointense and T2 is hyperintense; the mass is well-circumscribed. Enhancement is variable and may be solid, rim, or nodular [26,27].

Gangliocytomas are composed of irregular groups of large multipolar neurons often with dysplastic features; non-neoplastic glial elements and reticulin fibers make up the stroma [26]. Gangliogliomas have a neoplastic glial component, usually astrocytes [26]. By immunohistochemistry, these tumors stain for synaptophysin, GFAP, and less so for chromogranin. Labeling indices less than 3 per cent but increased growth fraction and p53 may be associated with tumor recurrence. Despite these tumors being benign, reports of malignant degeneration, anaplastic gangliogliomas, drop metastases, and dissemination *via* a VP shunt are published [27–35]. Malignant change is usually of the glial elements leading to features similar to a glioblastoma. Features of anaplasia appear to be a negative prognostic factor [36–39].

Many large retrospective case series highlight the important role of surgery in these tumors [40–42]. Long-term survival rates with surgery alone are excellent [27,40]. Several series have found no increased survival with adjuvant RT [36,37,41,43,44], but for anaplastic lesions this may be of some benefit especially when not completely resected [38,45,46]. Radiation may predispose to malignant degeneration [31,33].

There are a handful of case reports of patients being treated with various CTX regimens for gangliogliomas, some of which where anaplastic or malignant [31,35,47–50]. In most of the case reports, patients who have been treated with CTX (often not noted as to type) have also been treated with RT, so the true benefit is difficult to assess; in several series either there was no benefit, patients did poorly, or there was no specific mention of outcome [37–40,44,47,48,51].

Prados et al. [52] treated one pediatric patient with a ganglioglioma with 6-thioguanine (30 mg/m^2 q6h days 1–3), Procarbazine (50 mg/m^2 q6h starting day 3), Dibromodulcitol (400 mg/m^2 × 1 dose on day 3), CCNU (110 mg/m^2 × 1 dose on day 4), and Vincristine (1.4 mg/m^2 capped at 2 mg × 1 dose on days 14 and 28) but outcome was not noted. In a series of 34 patients with gangliogliomas, one patient was treated with a nitrosourea and one with accutane after STR and irradiation, with stable disease at 4 and 39 months, respectively [41]. Sasaki et al. [50] treated one patient with a recurrent ganglioglioma that became progressively more anaplastic with each recurrence with local methotrexate without any tumor activity. Kaba et al.[53] treated a patient with a ganglioglioma that underwent malignant transformation with Cis-Retinoic acid with almost complete resolution of the lesion. Johnson et al. [48] treated a 17-year-old woman with escalated doses of topotecan in combination with fixed doses of cisplatin, cyclophosphamide, and vincristine for 4 cycles, but then required craniospinal RT for progressives disease; she remained stable 28 months after surgery. Dash et al. [49] treated a 6-year-old girl with a mixed glioblastoma–ganglioglioma with 4 cycles of etoposide, carboplatin, and vincristine after surgery, and then RT with weekly vincristine without tumor stabilization. She was then treated with I^{131} monoclonal antibody after a second resection 6 months from diagnosis. This was followed by autologous stem cell transplantation with carboplatin, etoposide, and thiotepa she presented with leptomeningeal dissemination several months later and eventually died. Hassell et al. [54] treated one patient with recurrent spinal cord ganglioglioma with carboplatin (560 mg/m^2 per month) for 11 cycles; the patient had a radiographic and clinical stabilization for 17 months. Another report describes a patient with leptomeningeal dissemination who had no response to post-radiation ACNU [34].

DESMOPLASTIC INFANTILE GANGLIOGLIOMA (DIG) AND DESMOPLASTIC INFANTILE ASTROCYTOMA (DIA)

DIA, first described by Taratuto in 1982 [7] and DIG, described by VandenBerg in 1987 [5] are rare WHO grade I tumors. They are characterized by a mix of neuronal and glial elements surrounded by a desmoplastic reaction [5,7]. Almost all cases occur in children <2 years with a male:female ratio of 1.7:1 [7,55]. Children commonly present with increase in head circumference, bulging fontanelles, and downward ocular deviation ("sunset sign"); other symptoms include seizures, paresis, increased muscle tone, and reflexes [7,55]. These are large hypodense cystic masses on CT scan with a solid isodense or slightly hyperdense superficial portion which shows contrast enhancement. MRI on T1 shows a hypointense cystic mass with an isointense peripheral solid component that enhances; on T2 the cyst is hyperintense and the solid portion is heterogeneous [7,55,56]. They are composed of neoplastic astrocytes and a prominent reticulin rich, desmoplastic stroma (DIA) in some cases with a variable extent of neuronal differentiation (DIG); they are often adherent to the dura and involve multiple lobes. Immunohistochemical stains are positive for GFAP, vimentin, and synaptophysin and the LI is low [7].

Although the tumors appear histologically malignant, their course is usually benign [57–59]. At least one report of CNS dissemination exists [60]. The primary mode of treatment is surgical with the aim of gross total resection [55,57–59,61]. In fact, some reports have demonstrated stable follow-up and even spontaneous regression after partial resection [61,62]. One author even recommends repeated surgery before resorting to CT [63].

There are only a few patients who have received CTX for DIG/DIA. Duffner et al. [64] reported on four patients with DIG treated with vincristine (0.065 mg/kg; max 1.5 mg days 1–8), cyclophosphamide (65 mg/kg on day 1) for two 28 day cycles and then cisplatin (4 mg/kg on day 1) and etoposide (6.5 mg/kg on days 3 and 4) for one 28 day cycle; this sequence was repeated for up to 2 years and RT was not used. Case 1 was a 5-month-old boy disease free more than 36 months after diagnosis from 24 months of treatment. Case 2 was an 8-month-old boy treated for 12 months whose tumor remained stable 42 months after diagnosis. Case 3 was a 4-month-old boy treated for 24 months; he had a complete response and was well 5 years after diagnoses. Case 4 was a 7-month-old boy treated for 18 months who had a partial response and was well 4 years after diagnosis. VandenBerg et al. [5] reported on 11 patients, 3 of whom had surgery, RT, and then CTX. Of these three, one was treated with vincristine and dactinomycin while the CTX used in the other two was not specified. Ultimately, two of the three patients died. De Munnynck et al. [65] reported a case of a 2-year-old girl with a malignant DIG based on histology, radiographic appearance, and clinical aggressiveness. MRI one month after surgery revealed tumor progression and she was initially started on a combination of vincristine (0.15 mg/m^2) and carboplatin (550 mg/m^2) for one cycle with a decrease in the tumor. Three cycles of carboplatin (200 mg/m^2 for 24 h days 1–4) and etoposide (100 mg/m^2 for 24 h days 1–4) was then given followed by one cycle of high-dose carboplatinum (500 mg/m^2 for 24 h days 1–4) and etoposide (200 mg/m^2 for 24 h days 1–4) with autologous stem cell rescue. She had a further reduction in tumor volume. She remained stable on maintenance vincristine (1.5 mg/m^2) and cisplatinum (80 mg/m^2) for four weeks × 3 cycles but then progressed 11 months after surgery. She was subsequently placed on Tamoxifen (120 mg/m^2) without response and died soon after. Fan *et al.* [66] reported on a 6-month-old boy treated with carboplatin and vincristine after surgery but outcome is not reported. Bock *et al.* [67] treated a 4-month-old boy with DIA and central nervous system dissemination with vincristine and carboplatin without benefit; treatment was changed to vincristine, cyclophosphamide, methotrexate, carboplatin, and etoposide, again without benefit. One patient with a partially resected DI tumor was treated with RT and CTX (type not noted) but died nonetheless [56].

DYSEMBRYOPLASTIC NEUROEPITHELIAL TUMOR

Dysembryoplastic neuroepithelial tumors (DNET) are low-grade, cortically based tumors that were described as a distinct entity by Daumas-Duport *et al.* in 1988 [3]. DNET are located primarily in the temporal lobes and often lead to drug-resistant epilepsy. They occur most frequently in the second and third decades with a male predominance. On CT scan they may be hypodense with a pseudocystic appearance, isodense, or hyperdense and may contain calcifications. On T1 MRI they are hypointense and on T2 hyperintense. About one-third enhance on CT or MRI and often have ring-like features [3,68]. These are nodular appearing lesions with glioneuronal elements that have some GFAP positivity and the "floating neurons" are synaptophysin positive. The LI is low and cortical dysplasia may be seen. These are usually benign tumors but there are two reports of malignant transformation; in one case it may have been due to prior RT and CTX [69,70]. Surgical resection is the main treatment with excellent long-term outcomes

even for subtotally resected tumors [3,68,71,72]. There does not appear to be a role for RT for these tumors. There are no reports of CTX in the treatment of DNET, except in the cases where the diagnosis was incorrect, but there does not appear to be any indication for its use.

DYSPLASTIC GANGLIOCYTOMA OF THE CEREBELLUM

Dysplastic gangliocytoma of the cerebellum (WHO grade I), also known as Lhermitte-Duclos disease, is a rare hamartomatous or non-neoplastic tumor of the cerebellar cortex that most commonly presents with increased intracranial pressure and hydrocephalus, headache, cerebellar symptoms, and cranial nerve palsies [73,74]. It usually presents in the third or fourth decade with no gender predominance [73,74]. Recently this condition has been noted to occur in association with Cowden disease, one of the phakomoteses that is associated with multiple hamartomas and neoplasms of the breast, thyroid, endometrium, and genitourinary tract [73,74]. MRI imaging shows a hypointense lesion on T1 with little or no enhancement and hyperintense on T2 [73,74]. On histopathology, the lesion is characterized by large neuronal cells that expand the granular and molecular layers [73,74]. There have been no reports of malignant transformation of this tumor and its clinical course is benign. The optimal treatment is surgical excision and decompression to relieve symptoms of increased intracranial pressure [73,74]. There are no reports of CTX or RT for this lesion and there does not appear to be any indication to use these, even when incompletely resected.

CEREBELLAR LIPONEUROCYTOMA

First recognized by Bechtel *et al.* in 1978 [6], this is a well-differentiated neurocytic tumor (WHO grade I or II) that occurs in the fifth and six decades [75]. They occur in the cerebellum and present with symptoms referable to the posterior fossa, similar to dysplastic gangliogliocytoma of the cerebellum. They contain round neoplastic cells with a consistent neuronal and focal lipomatous differentiation and a low rate of proliferation. There is focal GFAP expression and also NSE and synaptophysin staining; labeling index (LI) is low [75]. The lipomatous component may be hyperintense to a variable degree on T1 MRI and enhancement is minimal [75,76].

One report for this tumor acting more aggressively exists; this tumor recurred despite radiation therapy (RT); the patient underwent a second resection and was treated with chemotherapy (CTX) per medulloblastoma protocol but outcome or response was not noted [77]. The treatment is surgical but for STR lesions, radiation may prevent relapse but the limited number of cases makes any definite recommendations difficult [78]. There are no reports of CTX for these tumors so its role is unknown.

References

1. Perkins, O. C. (1926). Ganglioglioma. *Arch Pathol Lab Med* **2**, 11–17.

2. Lhermitte, J., and Duclos, P. (1920). Sur un ganglioneurome diffus du coertex du cervelet. *Bull Assoc Fran Etude Cancer* **9**, 99–107.

3. Daumas-Duport, C., Scheithauer, B. W., Chodkiewicz, J. P., Laws E. R., Jr., Vedrenne, C. (1988). Dysembryoplastic neuroepithelial tumor: a surgically curable tumor of young patients with intractable partial seizures. Report of thirty-nine cases. *Neurosurgery* **23**, 545–556.

4. Hassoun, J., Gambarelli, D., Grisoli, F. *et al.* (1982). Central neurocytoma. An electron-microscopic study of two cases. *Acta Neuropathol (Berl)* **56**, 151–156.

5. VandenBerg, S. R., May, E. E., Rubinstein, L. J. *et al.* (1987). Desmoplastic supratentorial neuroepithelial tumors of infancy with divergent differentiation potential ("desmoplastic infantile gangliogliomas"). Report on 11 cases of a distinctive embryonal tumor with favorable prognosis. *J Neurosurg* **66**, 58–71.

6. Bechtel, J. T., Patton, J. M., and Takei, Y. (1978). Mixed mesenchymal and neuroectodermal tumor of the cerebellum. *Acta Neuropathol (Berl)* **41**, 261–263.

7. Taratuto, A. L., Monges, J., Lylyk, P. *et al.* (1982). Meningo-cerebral astrocytoma attached to the dura with "desmoplastic" reaction. Proceedings of the IX International Congress of Neuropathology (Vienna), 5–10.

8. Hassoun, J., Soylemezoglu, F., Gambarelli, D. *et al.* (1993). Central neurocytoma: a synopsis of clinical and histological features. *Brain Pathol* **3**, 297–306.

9. Schmidt, M. H., Gottfried, O. N., von Koch, C. S., Chang, S. M., McDermott, M. W. (2004). Central neurocytoma: a review. *J Neurooncol* **66**, 377–384.

10. Brat, D. J., Scheithauer, B. W., Eberhart, C. G., Burger, P. C. (2001). Extraventricular neurocytomas: pathologic features and clinical outcome. *Am J Surg Pathol* **25**, 1252–1260.

11. Giangaspero, F., Cenacchi, G., Losi, L. *et al.* (1997). Extra-ventricular neoplasms with neurocytoma features. A clinico-pathological study of 11 cases. *Am J Surg Pathol* **21**, 206–212.

12. Soylemezoglu, F., Scheithauer, B. W., Esteve, J., Scheithauer, B. W., Esteve, J., Kleihues, P. (1997). Atypical central neurocytoma. *J Neuropathol Exp Neurol* **56**, 551–556.

13. Rades, D., Fehlauer, F., and Schild, S. E. (2004). Treatment of atypical neurocytomas. *Cancer* **100**, 814–817.

14. Eng, D. Y., DeMonte, F., Ginsberg, L., Fuller, G. N., and Jaeckle, K. (1997). Craniospinal dissemination of central neurocytoma. Report of two cases. *J Neurosurg* **86**, 547–552.

15. Elek, G., Slowik, F., Eross, L. *et al.* (1999). Central neurocytoma with malignant course. Neuronal and glial differentiation and craniospinal dissemination. *Pathol Oncol Res* **5**, 155–159.

16. Coelho, N. M., Ramina, R., de Meneses, M. S. Arruda, W. O., Milano, J. B. (2003). Peritoneal dissemination from central neurocytoma: case report. *Arq Neuropsiquiatr* **61**, 1030–1034.

17. Schild, S. E., Scheithauer, B. W., Haddock, M. G. *et al.* (1997). Central neurocytomas. *Cancer* **79**, 790–795.

18. Sgouros, S., Carey, M., Aluwihare, N., Barber, P., Jackowski, A. (1998). Central neurocytoma: a correlative clinicopathologic and radiologic analysis. *Surg Neurol* **49**, 197–204.

19. Rades, D., and Fehlauer, F. (2002). Treatment options for central neurocytoma. *Neurology* **59**, 1268–1270.

20. Anderson, R. C., Elder, J. B., Parsa, A. T., Issacson, S. R., Sisti, M. B. (2001). Radiosurgery for the treatment of recurrent central neuro-cytomas. *Neurosurgery* **48**, 1231–1237.

21. Tyler-Kabara, E., Kondziolka, D., Flickinger, J. C. *et al.* (2001). Stereotactic radiosurgery for residual neurocytoma. Report of four cases. *J Neurosurg* **95**, 879–882.

22. Louis, D. N., Swearingen, B., Linggood, R. M. *et al.* (1990). Central nervous system neurocytoma and neuroblastoma in adults–report of eight cases. *J Neurooncol* **9**, 231–238.

23. Dodds, D., Nonis, J., Mehta, M., Rampling, R. (1997). Central neurocytoma: a clinical study of response to chemotherapy. *J Neurooncol* **34**, 279–283.

24. Brandes, A. A. (2000). mist inverted question mP, Gardiman M. *et al.* Chemotherapy in patients with recurrent and progressive central neurocytoma. *Cancer* **88**, 169–174.

25. von Koch, C. S., Schmidt, M. H., Uyehara-Lock, J. H. Berger, M. S., Chang, S. M. (2003). The role of PCV chemotherapy in the treatment of central neurocytoma: illustration of a case and review of the literature. *Surg Neurol* **60**, 560–565.

26. Nelson, J. S., Bruner, J. M., Wiestler, O. D., VandenBerg, S. R. (2000). Ganglioglioma and gangliocytoma. In *Pathology & Genetics: Tumours of the Nervous System*. (P. Kleihues, and W. K. Cavenee, Eds.), pp. 96–98.

27. Zentner, J., Wolf, H. K., Ostertun, B. *et al.* (1994). Gangliogliomas: clinical, radiological, and histopathological findings in 51 patients. *J Neurol Neurosurg Psychiatry* **57**, 1497–1502.

28. Wacker, M. R., Cogen, P. H., Etzell, J. E. *et al.* (1992). Diffuse leptomeningeal involvement by a ganglioglioma in a child. *Case report J Neurosurg* **77**, 302–306.

29. Hukin, J., Siffert, J., Velasquez, L., Zagzag D., Allen J. (2002). Leptomeningeal dissemination in children with progressive low-grade neuro-epithelial tumors. *Neuro-oncol* **4**, 253–260.

30. Araki, M., Fan, J., Haraoka, S. *et al.* (1999). Extracranial metastasis of anaplastic ganglioglioma through a ventriculo-peritoneal shunt: a case report. *Pathol Int* **49**, 258–263.

31. Jay, V., Squire, J., Becker, L. E., Humphreys, R. (1994). Malignant transformation in a ganglioglioma with anaplastic neuronal and astrocytic components. Report of a case with flow cytometric and cytogenetic analysis. *Cancer* **73**, 2862–2868.

32. Jay, V., Squire, J., Blaser, S., Hoffman, H. J., Hwang, P. (1997). Intracranial and spinal metastases from a ganglioglioma with unusual cytogenetic abnormalities in a patient with complex partial seizures. *Childs Nerv Syst* **13**, 550–555.

33. Rumana, C. S., and Valadka, A. B. (1998). Radiation therapy and malignant degeneration of benign supratentorial gangliogliomas. *Neurosurgery* **42**, 1038–1043.

34. Nakajima, M., Kidooka, M., and Nakasu, S. (1998). Anaplastic ganglioglioma with dissemination to the spinal cord: a case report. *Surg Neurol* **49**, 445–448.

35. Kurian, N. I., Nair, S., and Radhakrishnan, V. V. (1998). Anaplastic ganglioglioma: case report and review of the literature. *Br J Neurosurg* **12**, 277–280.

36. Rumana, C. S., Valadka, A. B., and Contant, C. F. (1999). Prognostic factors in supratentorial ganglioglioma. *Acta Neurochir (Wien)* **141**, 63–68.

37. Lang, F. F., Epstein, F. J., Ransohoff, J. *et al.* (1993). Central nervous system gangliogliomas. Part 2: Clinical outcome. *J Neurosurg* **79**, 867–873.

38. Krouwer, H. G., Davis, R. L., McDermott, M. W., Hoshino, T., Prados, M. D. (1993). Gangliogliomas: a clinicopathological study of 25 cases and review of the literature. *J Neurooncol* **17**, 139–154.

39. Hirose, T., Scheithauer, B. W., Lopes, M. B. *et al.* (1997). Ganglioglioma: an ultrastructural and immunohistochemical study. *Cancer* **79**, 989–1003.

40. Luyken, C., Blumcke, I., Fimmers, R. *et al.* (2004). Supratentorial gangliogliomas: histopathologic grading and tumor recurr–ence in 184 patients with a median follow-up of 8 years. *Cancer* **101**, 146–155.

41. Selch, M. T., Goy, B. W., Lee, S. P. *et al.* (1998). Gangliogliomas: experience with 34 patients and review of the literature. *Am J Clin Oncol* **21**, 557–564.

42. Im, S. H., Chung, C. K., Cho, B. K. *et al.* (2002). Intracranial ganglioglioma: preoperative characteristics and oncologic outcome after surgery. *J Neurooncol* **59**, 173–183.

43. Celli, P., Scarpinati, M., Nardacci, B., Cervoni, L., Cantore, G. P. (1993). Gangliogliomas of the cerebral hemispheres. Report of 14 cases with long-term follow-up and review of the literature. *Acta Neurochir(Wien)* **125**, 52–57.

44. Silver, J. M., Rawlings, C. E., III, Rossitch, E., Jr., Zeidman, S. M. Friedman, A. H. (1991). Ganglioglioma: a clinical study with long-term follow-up. *Surg Neurol* **35**, 261–266.

45. Haddad, S. F., Moore, S. A., Menezes, A. H., VanGilder, J. C. (1992). Ganglioglioma: 13 years of experience. *Neurosurgery* **31**, 171–178.

46. Johannsson, J. H., Rekate, H. L., and Roessmann U. (1981). Gangliogliomas: pathological and clinical correlation. *J Neurosurg* **54**, 58–63.

47. Johnson, J. H., Jr., Hariharan, S., Berman, J. *et al.* (1997). Clinical outcome of pediatric gangliogliomas: ninety-nine cases over 20 years. *Pediatr Neurosurg* **27**, 203–207.

48. Johnson, M. D., Jennings, M. T., and Toms, S. T. (2001). Oligodendroglial ganglioglioma with anaplastic features arising from the thalamus. *Pediatr Neurosurg* **34**, 301–305.

49. Dash, R. C., Provenzale, J. M., McComb, R. D. *et al.* (1999). Malignant supratentorial ganglioglioma (ganglion cell-giant cell glioblastoma): a case report and review of the literature. *Arch Pathol Lab Med* **123**, 342–345.

50. Sasaki, A., Hirato, J., Nakazato, Y., Tamura, M., Kadowaki, H. (1996). Recurrent anaplastic ganglioglioma: pathological characterization of tumor cells. Case report. *J Neurosurg* **84**, 1055 1059.

51. Hakim, R., Loeffler, J. S., Anthony, D. C., Black, P. M. (1997). Gangliogliomas in adults. *Cancer* **79**, 127–131.

52. Prados, M. D., Edwards, M. S., Rabbitt, J. *et al.* (1997). Treatment of pediatric low-grade gliomas with a nitrosourea-based multiagent chemotherapy regimen. *J Neurooncol* **32**, 235–241.

53. Kaba, S. E., Langford, L. A., Yung, W. K., Kyritsis A. P. (1996). Resolution of recurrent malignant ganglioglioma after treatment with cis-retinoic acid. *J Neurooncol* **30**, 55–60.

54. Hassall, T. E., Mitchell, A. E., and Ashley, D. M. (2001). Carboplatin chemotherapy for progressive intramedullary spinal cord low-grade gliomas in children: three case studies and a review of the literature. *Neuro-oncol* **3**, 251–257.

55. Tamburrini, G., Colosimo, C., Jr., Giangaspero, F., Riccardi R., Di R. C. (2003). Desmoplastic infantile ganglioglioma. *Childs Nerv Syst* **19**, 292–297.

56. Trehan, G., Bruge, H., Vinchon, M. *et al.* (2004). MR imaging in the diagnosis of desmoplastic infantile tumor: retrospective study of six cases. *AJNR Am J Neuroradiol* **25**, 1028–1033.

57. Sugiyama, K., Arita, K., Shima, T. *et al.* (2002). Good clinical course in infants with desmoplastic cerebral neuroepithelial tumor treated by surgery alone. *J Neurooncol* **59**, 63–69.

58. VandenBerg, S. R. (1993). Desmoplastic infantile ganglioglioma and desmoplastic cerebral astrocytoma of infancy. *Brain Pathol* **3**, 275–281.

59. Sperner, J., Gottschalk, J., Neumann, K. *et al.* (1994). Clinical, radiological and histological findings in desmoplastic infantile ganglioglioma. *Childs Nerv Syst* **10**, 458–462.

60. Setty, S. N., Miller, D. C., Camras, L., Charbel F., Schmidt, M. L. (1997). Desmoplastic infantile astrocytoma with metastases at presentation. *Mod Pathol* **10**, 945–951.

61. Mallucci, C., Lellouch-Tubiana, A., Salazar, C. *et al.* (2000). The management of desmoplastic neuroepithelial tumours in childhood. *Childs Nerv Syst* **16**, 8–14.

62. Takeshima, H., Kawahara, Y., Hirano, H. *et al.* (2003). Postoperative regression of desmoplastic infantile ganglioglio-mas: report of two cases. *Neurosurgery* **53**, 979–983.

63. Bachli, H., Avoledo, P., Gratzl, O., Tolnay, M. (2003). Ther-apeutic strategies and management of desmoplastic infantile ganglioglioma: two case reports and literature overview. *Childs Nerv Syst* **19**, 359–366.

64. Duffner, P. K., Burger, P. C., Cohen, M. E. *et al.* (1994). Desmoplastic infantile gangliogliomas: an approach to therapy. *Neurosurgery* **34**, 583–589.

65. De, M. K., Van, G. S., Van, C. F. *et al.* (2002). Desmoplastic infantile ganglioglioma: a potentially malignant tumor? *Am J Surg Pathol* **26**, 1515–1522.

66. Fan, X., Larson, T. C., Jennings, M. T. *et al.* (2001). December 2000: 6 month old boy with 2 week history of progressive lethargy. *Brain Pathol* **11**, 265–266.

67. Bock, D., Rummele, P., Friedrich, M., Wolff, J. E. (2002). Multifocal desmoplastic astrocytoma, frontal lobe dysplasia, and simian crease. *J Pediatr* **141**, 445.

68. Daumas-Duport, C., Pietsch, T., and Lantos, P. L. (2000). Dysembryoplastic neuroepithelial tumour. *Pathology & Genetics: Tumours of the Nervous System* 103–106.

69. Hammond, R. R., Duggal, N., Woulfe, J. M., Girvin, J. P. (2000). Malignant transformation of a dysembryoplastic neuro-epithelial tumor. *Case report J Neurosurg* **92**, 722–725

70. Rushing, E. J., Thompson, L. D., and Mena H. (2003). Malignant transformation of a dysembryoplastic neuro-epithelial tumor after radiation and chemotherapy. *Ann Diagn Pathol* **7**, 240–244.

71. Daumas-Duport, C., Varlet, P., Bacha, S. *et al.* (1999). Dysembryoplastic neuroepithelial tumors: nonspecific histo-logical forms – a study of 40 cases. *J Neurooncol* **41**, 267–280.

72. Raymond, A. A., Halpin, S. F., Alsanjari, N. *et al.* (1994). Dysembryoplastic neuroepithelial tumor. Features in 16 patients. *Brain* **117** (Pt 3), 461–475.

73. Nowak, D. A., and Trost, H. A. (2002). Lhermitte-Duclos disease (dysplastic cerebellar gangliocytoma): a malformation, hamartoma or neoplasm? *Acta Neurol Scand* **105**, 137–145.

74. Wiestler, O. D., Padberg, G. W., and Steck, P. A. (2000). Cowden disease and dysplastic gangliocytoma of the cerebellum/

Lhermitte-Duclos disease. *In Pathology & Genetics: Tumours of the Nervous System* (P. Kleihues, and W. K. Cavenee, Eds.), pp. 235–237.

75. Kleihues, P., Chimelli, L., and Giangaspero, F. (2000). Cerebellar liponeurocytoma. *Pathology & Genetics: Tumours of the Nervous System.* 110–111.

76. Cacciola, F., Conti, R., Taddei, G. L., Buccoliero, A. M. Di L. N. (2002). Cerebellar liponeurocytoma. Case report with considerations on prognosis and management. *Acta Neurochir (Wien)* **144**, 829–833.

77. Jenkinson, M. D., Bosma, J. J., Du, P. D. *et al.* (2003). Cerebellar liponeurocytoma with an unusually aggressive clinical course: case report. *Neurosurgery* **53**, 1425–1427.

78. Jackson, T. R., Regine, W. F., Wilson, D., Davis D. G. (2001). Cerebellar liponeurocytoma. Case report and review of the literature. *J Neurosurg* **95**, 700–703.

CHAPTER

32

Chemotherapy of Pineal Parenchymal Tumors

Herbert B. Newton

ABSTRACT: Pineal region tumors encompass a large group of neoplasms, including germ cell tumors, meningiomas, neuroectodermal tumors, and pineal parenchymal tumors (pineocytoma, pineoblastoma). They have an incidence of 0.3–0.7 per cent and usually affect patients between the ages of 20 and 35 years. Common presenting signs and symptoms include headache, Parinaud's syndrome, hydrocephalus, papilledema, and ataxia. Pineal tumors and hydrocephalus are clearly delineated on enhanced magnetic resonance imaging. The main form of treatment for pineocytoma is surgical resection, which may be curative in some cases. For patients with pineoblastoma, surgical resection and radiation therapy are usually required. The role of chemotherapy continues to be defined, but should be considered for all the patients with malignant tumors, as well as for patients with lower grade tumors that recur or become disseminated. Platinum-based regimens appear to be most effective, usually in combination with other drugs, such as etoposide, vincristine, or lomustine.

PINEAL ANATOMY AND PHYSIOLOGY

The pineal gland has been a topic of intense interest since antiquity, and yet its function remains somewhat obscure [1,2]. The average gland is 7.4 mm in length, 6.9 mm in width, and 2.5 mm thick [3]. It is surrounded by a capsule and is composed of lobules separated by connective tissue septae. The gland forms during the seventh week of embryonic development, when the roof plate of the primitive diencephalon thickens medially and then evaginates posteriorly. In the adult, the pineal gland is located under the splenium of the corpus callosum and over the superior colliculus. The stalk of the gland forms part of the posterior superior wall of the third ventricle and lies in between the posterior commissure and the more dorsal habenular commissure. Structurally, the stalk is composed of two laminae, a cranial or superior lamina and a caudal or inferior lamina. The pineal recess projects posteriorly into the pineal body between the two laminae. The ventricular side of the stalk of the pineal gland is lined by ependymal cells and encloses the pineal recess of the third ventricle. The vascular supply to the pineal gland is *via* the pineal artery, which is one of the branches of the medial posterior choroidal artery.

The human adult pineal gland contains two main types of cells: parenchymal cells called pinealocytes and fibrillary astrocytes [1–3]. Pinealocytes stain positively with neuron–specific enolase (i.e., a marker for neurons and central neuroendocrine cells) and have slender argyrophilic cytoplasmic processes of varying length. The pinealocytes contain numerous organelles, including dense-core, clear, granular, and agranular vesicles, which suggest active secretory processes. The fibrillary astrocytes stain for glial fibrillary acidic protein (GFAP) and have numerous processes that form a "stroma", surrounding each of the pinealocytes and producing a barrier between the perivascular spaces and the pineal parenchyma [1]. In addition, the astrocytic processes form a limiting lamina at the periphery of the gland. Admixed with the cells and stromal elements are plexuses of nerve fibers which are associated with synapses, gap junctions, and tight junctions. The nerve fibers are thought to be of sympathetic origin, with cell bodies arising from the superior cervical ganglion (SCG). Although the nerve terminals come in close contact with the pinealocytes, they do not form true synapses.

Photosensory information reaches the mammalian pineal gland *via* a complex polyneuronal pathway that culminates in the nerve fibers of the SCG [2]. In order to reach the SCG, ambient environmental light must first be converted into electrical impulses by the retinal photoreceptors, and then transmitted *via* the retinohypothalamic tract (RHT; within the optic nerve and tract) bilaterally to the suprachiasmatic nuclei of the hypothalamus (SCN), which function as the neural pacemakers for circadian rhythms [4,5]. Each SCN receives a bilateral retinal projection from the RHTs, with a predominantly contralateral innervation from fibers that cross over at the optic chiasm. Integrated photoreceptor data is then transmitted from the SCN to other synapses within the lateral hypothalamus, followed by descending projections into the spinal cord, which may involve the medial forebrain bundle. In the spinal cord, the descending pathways synapse with the intermediolateral cell column, from which preganglionic sympathetic fibers project to the SCG. Postganglionic noradrenergic fibers from the SCG reach the pineal gland *via* the nervi conarii, which pass through the tentorium cerebelli.

As the pineal gland is able to convert electrical photoreceptor signals into hormonal signals, it is considered to function as a "neuroendocrine transducer" [1,2,5]. Sympathetic adrenergic stimulation of pinealocytes increases the cellular levels of cAMP, thereby increasing the levels and activity of *N*-acetyltransferase, the most critical enzyme in the synthesis of the hormone melatonin, the major secretory product of the pineal gland. The synthesis of melatonin starts with tryptophan and requires several steps and intermediate compounds, including serotonin. Melatonin synthesis is dependent on the level of ambient environmental light; it is stimulated during darkness and inhibited during light [2,6]. Once synthesized by the pinealocytes, melatonin is released and enters the blood and CSF *via* simple diffusion. Melatonin levels in humans are higher in serum than in the lumbar or ventricular CSF, suggesting that the majority of melatonin is secreted into the bloodstream. In general, plasma levels of melatonin are low during the day and then increase by 10- to 50-fold at night. Circulating melatonin is taken up by all organs and is not hindered by the blood–brain barrier. Melatonin is active in many sites within the nervous system, including the hypothalamus, midbrain, cerebral cortex, pituitary gland, pineal gland, peripheral

nerves, and SCG. Sites of activity outside the nervous system include the thyroid glands, adrenal glands, gonads, eyes, skin, and Harderian glands. Further discussion of the biochemistry and physiology of melatonin is beyond the scope of this chapter, but are reviewed in detail in several of the references [1,2,6].

EPIDEMIOLOGY

Pineal region tumors encompass a large group of neoplasms, with quite variable histology and malignant potential (see Table 32.1) [7–12]. The most common tumors are of germ cell origin (60 per cent) and include germinoma, teratoma, choriocarcinoma, embryonal carcinoma, and yolk sac tumors. Primary neuroectodermal tumors comprise 15 per cent of pineal region malignancies and include astrocytoma, oligodendroglioma, ependymoma, and other less common histologies. Pineal parenchymal tumors account for 15 per cent of neoplasms in this region, while primary non-neuroectodermal and miscellaneous tumors (e.g., lipoma, meningioma, and metastases) comprise the remaining 10 per cent of the cases.

Pineal region neoplasms have an overall incidence of 0.3–0.7 per cent in the USA [7–9,11,12]. For patients less than 20 years of age, the incidence increases to approximately 1.8 per cent. Incidence rates are similar in Europe, but may be much higher in Japan, where the rates approach 4–6 per cent [8,10,13]. However, other reports by Japanese investigators have demonstrated incidence rates comparable to, but still higher than, the USA and European data (0.07 per 100 000 person years) [14]. Pure pineal parenchymal tumors represent 15 per cent of all pineal region neoplasms and only 0.1 per cent of all intracranial malignancies. This corresponds to an estimated incidence of 0.01 per 100 000 persons per year in the USA [15]. The remaining sections of this chapter will focus on pineal parenchymal tumors. Pineal region germ cell tumors are covered in Chapter 33 (Chemotherapy of CNS Germ Cell Tumors). The other pineal region tumor types are discussed in their specific chapters.

The mean age at presentation for all types of pineal parenchymal tumors is in the third and fourth decades (i.e., 20–35 years), with a range from early childhood to the eighth decade [7–9,15]. Pineoblastomas tend to occur in younger patients, with a mean age of onset at approximately 18–20 years. Pineocytomas arise in a slightly older age group, with a mean age of onset between 30 and 35 years. In most of the studies, there is no preference for gender [7–9,15]. However, in those studies with a preference, it was in favor of male incidence and was very small.

TABLE 32.1 Differential Diagnosis of Pineal Region Tumors

Germ Cell Tumors (60 per cent)

  Germinoma

  Embryonal carcinoma

  Yolk sac tumor

  Choriocarcinoma

  Teratoma

  Mixed histology

Pineal Parenchymal Tumors (15 per cent)

  Pineocytoma

    with neuronal differentiation

    with astrocytic differentiation

    with neuronal and astrocytic differentiation

  Pineoblastoma

    with pineocytic differentiation

    with retinoblastomatous differentiation

Primary Neuroectodermal Tumors (15 per cent)

  Astrocytoma

  Oligodendroglioma

  Ependymoma

  Medulloepithelioma

  Paraganglioma

  Ganglioneuroma

  Melanoma

Primary Non-Neuroectodermal Tumors (5 per cent)

  Lipoma

  Hemangioma

  Meningioma

  Hemangiopericytoma

  Craniopharyngioma

  Choroids plexus papilloma

Miscellaneous (5 per cent)

  Pineal cyst

  Arachnoid cyst

  Epidermoid cyst

  Dermoid cyst

  Metastatic tumors

The vast majority of pineal parenchymal tumors are sporadic and do not have any familial association [8,9,15]. On rare occasions, pineal tumors can be part of a familial tumor syndrome, such as "trilateral retinoblastoma", or occur during the course of an unexplained familial cluster [16,17]. In trilateral retinoblastoma, patients have germline mutations or deletions of the retinoblastoma gene (Rb) on chromosome 13, with subsequent development of spontaneous bilateral retinoblastomas and intracranial masses. The intracranial tumors are similar histologically to the retinoblastomas (i.e., primitive neuroectodermal morphology) and are noted in the pineal gland or suprasellar region. All pineal tumors in this syndrome are malignant (i.e., pineoblastomas) and are thought to arise from transformed cells originating within the vestigial photoreceptors of the pineal gland.

BIOLOGY AND MOLECULAR GENETICS

As pineal parenchymal tumors are relatively uncommon, the biology and molecular genetics of these neoplasms remains incomplete [18]. Cytogenetic information has been published, although it has mainly been limited to small institutional series of patients. Several pineocytomas have been reported with variable loss of chromosomes 22, 11, and 1, suggesting the presence of tumor suppressor genes in these regions [19,20]. Pineoblastomas have also been described with deletions of chromosome 11 and complex rearrangements including monosomy for chromosomes 20 and 22, and trisomy for chromosome 14 [21,22]. It is important to note that the loss of material on chromosome 13q (i.e., the site of the retinoblastoma gene) has been infrequent in cytogenetic studies of sporadic or familial pineoblastomas. Two cell lines derived from a pineoblastoma have been reported to show an isochromosome for 17q, as well as elevated expression of N-myc [23]. However, there was no amplification of N-myc at the genomic level. Another group has reported overexpression of C-myc in a pineocytoma in comparison to normal pineal parenchymal tissue [24].

As mentioned above, although germline mutations and deletions of the retinoblastoma gene are present in pineoblastomas from patients with trilateral retinoblastoma, the gene is not affected in sporadic tumors or those associated with other genetic syndromes [18]. However, recent data would suggest that pineoblastomas that do harbor a mutant form of Rb are more aggressive and resistant to treatment than sporadic tumors that have intact Rb function [25]. In a comparison of patients with retinoblastoma-related and sporadic pineoblastomas, those with tumors that had mutant Rb were more likely to progress rapidly through radiotherapy and chemotherapy, and had significantly lower 5-year overall survival rates (12 *versus* 89 per cent; $p = 0.002$). The mechanism of treatment resistance mediated by mutant Rb remains unclear and is under investigation.

In contrast to astrocytic tumors, the tumor suppressor genes p53 and p21 are infrequently mutated in sporadic pineal parenchymal tumors [26,27]. Other oncogenes and tumor suppressor genes that have been implicated in the transformation of gliomas (e.g., epidermal growth factor and receptor, platelet-derived growth factor and receptor, Ras, PTEN, mTOR) have thus far not been associated with pineal parenchymal tumors [28,29].

CLINICAL PRESENTATION

Pineal parenchymal tumors can become symptomatic through several different mechanisms, including direct compression of local neural structures (i.e., brainstem, cerebellum), obstruction of cerebrospinal fluid (CSF) pathways and hydrocephalus, and endocrine dysfunction [8,9,15,30]. Direct compression of the dorsal midbrain can cause Parinaud's Syndrome, which is characterized by paralysis of upgaze, retraction nystagmus, impaired convergence, and light-near pupillary dissociation [31]. Further compression or infiltration of the dorsal midbrain and peri-aqueductal region may cause paralysis of downgaze, ptosis, and lid retraction. Ataxia and dysmetria can result from compression of the cerebellum or superior cerebellar peduncles. In addition, hemiparesis and long-tract signs can develop with infiltration or compression of the brainstem. Hydrocephalus is very common in patients with pineal tumors and is caused by obstruction of third ventricular CSF outflow through the aqueduct of Sylvius. Endocrine dysfunction is an infrequent occurrence in patients with pineal parenchymal tumors, and usually suggests direct involvement of tumor in the hypothalamus. Rarely, endocrine disturbances can result from the secondary effects of hydrocephalus.

The most common symptom at presentation is headache, which is noted in 70–75 per cent of the patients (see Table 32.2) [8,9,15,30]. The headache occurs secondary to increased intracranial pressure from hydrocephalus and regional mass affect. Impaired vision (e.g., diplopia, blurring) is noted in 35–45 per cent of the patients and can be quite variable. Nausea and emesis are usually related to the severity of hydrocephalus and occur in approximately 40 per cent of the patients. Abnormalities of gait are noted in 30–35 per cent of the patients, while impaired memory and dizziness are each present in about 20–25 per cent of the patients. Other less common symptoms include fatigue, alterations of speech, tinnitus, and decreased level of alertness.

The most common neurological signs at presentation are papilledema (40–60 per cent) and Parinaud's syndrome (30–55 per cent) (see Table 32.2) [8,9,15,30]. Ataxia is noted in 25–50 per cent of the

TABLE 32.2   Presenting Signs and Symptoms of Pineal Parenchymal Tumors

Signs and Symptoms	% of patients
Signs	
Papilledema	40–60
Parinaud's syndrome	30–55
Ataxia	25–50
Tremor	15–25
Hemiparesis	15–25
Abnormal papillary reflexes	10–15
Hyperactive deep tendon reflexes	10–15
Abducens nerve palsy	5–8
Oculomotor nerve palsy	<5
Seizure	<2
Sensory loss	<2
Symptoms	
Headache	70–75
Impaired vision	35–45
Nausea and emesis	35–40
Gait abnormalities	30–35
Impaired memory	20–25
Dizziness	20–25
Fatigue	15–20
Speech alterations	10–15
Tinnitus	10–15
Decreased alertness	10–15

Adapted from [8,9,15]

patients, while tremor and hemiparesis are each noted in 15–25 per cent of the cases. Abnormal pupillary reflexes and hyperactive deep tendon reflexes are each present in 10–15 per cent of the patients. Other less common neurological signs include palsies of the abducens and oculomotor nerves, seizures, and sensory loss.

NEURO-IMAGING AND LABORATORY DIAGNOSIS

Although neuro-imaging of pineal region masses has significantly improved with the advent of magnetic resonance technology (i.e., MRI), accurate prediction of tumor histology based on radiographic features remains unreliable [8,9,32–34]. MRI has replaced computed tomography (CT) as the imaging modality of choice for pineal region masses. Enhanced MRI has the ability to clearly visualize the tumor in relation to surrounding neural structures and demonstrate its size, vascularity, and enhancement pattern.

In addition, the margination pattern and irregularities of the tumor border are clearly delineated, which may provide information about tumor grade and degree of invasiveness. Hydrocephalus is well visualized, along with the relationship of the tumor to the third ventricle and deep venous structures. Pineocytomas and pineoblastomas tend to be hypo- or isointense on T1-weighted imaging and hyperintense on T2 and Flair images. Both the tumor types are usually homogeneous and have dense, uniform enhancement after the administration of gadolinium. Pineoblastomas are more likely to have regions of hemorrhage and necrosis, while pineocytomas more often demonstrate cysts. Pineocytomas are not encapsulated, but have well-defined borders in comparison to pineoblastomas. Calcification can be present within the tumor and is more likely with pineocytomas. In some cases, pineoblastomas can be distinguished by their irregular shape and large size (i.e., more than 4 cm).

CT scans can sometimes be helpful for initial screening of masses in the pineal region [35,35]. On unenhanced CT, pineocytomas and pineoblastomas usually appear as masses of slightly increased density with significant amounts of calcification. The tumors strongly enhance after the administration of contrast.

The laboratory diagnosis of pineal region tumors involves the measurement of tumor markers in the serum and CSF [8,9]. Tumor markers are most helpful in the differential diagnosis of pineal region germ cell tumors and include α-fetoprotein (AFP) and β-human chorionic gonadotropin (HCG). AFP is a glycoprotein normally produced by fetal yolk sac elements that is markedly elevated with endodermal sinus tumors. Embryonal cell carcinomas and immature teratomas also have elevated levels of AFP, but to a lesser degree than endodermal sinus tumors. HCG is a glycoprotein normally secreted by placental trophoblastic tissue that is markedly elevated with choriocarcinomas. Further discussion of the utility of tumor markers in the diagnostic work-up of CNS germ cell tumors can be found in Chapter 33. Pineal parenchymal tumors do not cause an elevation of AFP or HCG. At this time there are no reliable markers for pineocytoma or pineoblastoma. Melatonin and S antigen have been evaluated as markers for these tumors, but the results remain inconsistent.

INITIAL EVALUATION AND TREATMENT

All the patients suspected of having a pineal region tumor should undergo further evaluation with an MRI scan of the brain, with and without the administration of gadolinium [8,9,13,37]. This should include those patients that appear to have a pineal mass by CT scan. The MRI should be screened thoroughly for any evidence of CSF dissemination. Pituitary function should be screened in those patients suspected of having endocrine abnormalities. If the tumor has any extension into the suprasellar region, a formal visual field examination is required. Tumor markers should be measured in the serum and CSF. In addition, the CSF should undergo a thorough cytological examination. Most of the authors recommend waiting until after shunt placement and tumor resection (i.e., 2–3 weeks) before obtaining CSF for analysis. This would also be the appropriate time for a complete spinal MRI, with and without enhancement, for those patients with malignant tumors that require screening for distant metastases.

Surgical management of pineal parenchymal tumors involves treatment of hydrocephalus and biopsy or resection of the lesion for histological diagnosis [8,9,13,37–39]. Hydrocephalus can be treated by the placement of an external ventricular drain, or with a permanent ventriculoperitoneal shunt. In some cases, extensive resection of the tumor can result in normalization of CSF flow, obviating the need for a diversionary shunt. An open tumor resection procedure, with significant debulking of the mass, is the preferred surgical option for most of the patients. A stereotactic biopsy would be reserved for those patients with severe medical co-morbidities, widely disseminated CNS disease, and malignant, locally invasive tumors. Open surgical resection has been shown to benefit patients postoperatively, by removing most or all of the tumor, thereby reducing intracranial pressure and regional mass affect. For patients with benign pineocytomas, a gross total resection may be curative. Although pineoblastomas cannot be completely resected, patients often benefit from debulking and regional decompression. In addition, reducing the tumor burden may improve the effectiveness of subsequent treatment modalities (i.e., radiotherapy, chemotherapy). Recent experience with surgical treatment of pineal region tumors suggests that gross total removal is possible in approximately 90 per cent of benign tumors and 25 per cent of malignant tumors [13,37–39]. Minimally invasive approaches to surgery for pineal region tumors are now under investigation [40,41]. Several groups are analyzing the utility of endoscopic management of these tumors, including hydrocephalus. Preliminary results suggest the technique is well tolerated and effective in carefully selected patients.

External beam radiotherapy is necessary for all the patients with pineoblastoma and might be appropriate for those rare patients with progressive pineocytoma that are not amenable to further surgical therapy [8,9,15,37]. Irradiation is not recommended for pineocytomas that have undergone complete tumor removal. These patients can be followed closely and monitored with serial MRI scanning. For patients that require irradiation, the recommended dose is 4000 cGy to the whole-brain (or a large conformal field) followed by 1500 cGy to the tumor bed. The dosage should be given in 180 cGy daily fractions over 5–6 weeks. When patients receive doses of less than 5000 cGy, there appears to be a significant risk for local treatment failure [15]. Prophylactic spinal irradiation is not recommended for pineal parenchymal tumors. The risk of spinal seeding is in the range of 10–20 per cent, with the majority of affected patients having pineoblastoma. Spinal axis radiotherapy should be administered only to patients with documented seeding by spinal MRI or CSF cytology. The currently recommended dose is 3500 cGy to the spinal axis.

Stereotactic radiosurgery is a form of focused, volumetric irradiation that is under study for application to pineal region tumors [8,9]. Several recent reports would suggest that radiosurgery can be considered an alternative to microsurgical resection in patients with pineocytoma, with excellent local tumor control rates [42,43]. In contrast, pineoblastomas are less likely to have objective responses and tend to progress rapidly. For either tumor type, radiosurgery can also be considered for boosting into the tumor bed after incomplete surgical resection and fractionated irradiation. Patients should not receive radiosurgery if they have large, diffusely infiltrative tumors or widespread disseminated disease.

CHEMOTHERAPY OF PINEAL PARENCHYMAL TUMORS

The use of chemotherapy for the treatment of pineal region neoplasms has been applied mainly to CNS germ cell tumors, which is extensively reviewed in Chapter 33. The experience with pineal parenchymal tumors is more limited and mainly consists of small institutional series of patients [8,9, 15,37]. In the majority of cases, chemotherapy has been applied to patients with recurrent or progressive pineoblastoma, with less frequent application to those with recurrent pineocytoma. Objective responses have been limited, using a variety of chemotherapy agents including vincristine, lomustine, cisplatin, etoposide, carboplatin, cyclophosphamide, actinomycin D, and methotrexate. Thus far, no single drug or combination regimen has been shown to be most effective.

Initial experience with chemotherapy of pineal parenchymal tumors was gained in pediatric patients. Friedman and colleagues reported a small series of young children with poor-prognosis medulloblastoma ($N = 14$) and pineoblastoma ($N = 2$) that were treated with intravenous melphalan (45 mg/m^2 every 4 weeks) [44]. Both the patients with pineoblastoma had extensive partial responses (PR) that remained durable for 2–4 months, after which they received neuraxis irradiation. Sakoda and co-workers then reported a 6-year-old female with a pineocytoma that was treated with intravenous ACNU (25 mg) [45]. She received four cycles of chemotherapy and had a complete response (CR), which was maintained during 18 months of observation.

In a series of three children (3–7 years of age) with newly diagnosed pineoblastoma, Ghim and associates reported the use of neoadjuvant chemotherapy followed by radiotherapy [46]. The chemotherapy regimen consisted of intravenous cisplatin (100 mg/m^2 on day 1), etoposide (100 mg/m^2 on days 1–3), and vincristine (1.5 mg/m^2 on day 1) every four weeks, for a total of four cycles. Each child was then treated with radiotherapy to the primary site (5040–5440 cGy) and the craniospinal axis (2520–3060 cGy). One patient had a CR that lasted 5+ years, while another had a near-CR that was maintained for more than two years. The last patient had a PR that lasted for five months before disease progression.

In a series of 30 patients with pineal parenchymal tumors (pineoblastoma 15, pineocytoma 9, mixed or intermediate tumors 6) reported by Schild and colleagues, six received chemotherapy as part of their initial treatment [15]. The regimens consisted of various combinations of vincristine, lomustine, prednisone, cisplatin, procarbazine, etoposide, and cyclophosphamide. Among the five patients receiving chemotherapy that were assessable, four developed tumor shrinkage consistent with PR.

In a Children's Cancer Group (CCG) evaluation of chemotherapy for infants and young children with pineoblastoma, patients were nonrandomly assigned to receive eight-drugs-in-one-day without irradiation (8 patients less than 18 months of age) or were randomized to receive craniospinal radiotherapy plus lomustine, vincristine, and prednisone or eight-drugs-in-one-day (17 patients) [47]. All the infants that received "up-front" chemotherapy with eight-drugs-in-one-day had progression of disease after a median of four months. Of the older group of patients receiving irradiation and chemotherapy, the 3-year progression-free survival (PFS) was 61 per cent. Twelve of the seventeen patients (70.6 per cent) had at least some residual pineal mass, which remained stable or resolved during five years of follow-up observation. There were only four patients that had subsequent progression of disease. The efficacy was similar between the eight-drugs-in-one-day and lomustine, vincristine, and prednisone regimens.

Another CCG trial treated a larger cohort of 55 patients (aged 1.5–19.3 years) with supratentorial primitive neuroectodermal tumors (PNET, including pineoblastoma) with craniospinal radiotherapy plus eight-drugs-in-one-day or lomustine, vincristine, and prednisone [48]. For the sub-group of patients with pineoblastoma, the overall survival and PFS rates were 73 and 61 per cent, respectively. The PFS rates were equivalent between the two chemotherapy arms of the study. Patients with pineoblastoma appeared to be more sensitive to chemotherapy than patients with other supratentorial PNET, demonstrating significantly ($p < 0.03$) better PFS rates.

In a series of eight patients with pineoblastoma (2 recurrent, 6 newly diagnosed), Ashley and co-workers reported on the use of single-agent high-dose cyclophosphamide (2 gm/m^2 per day × 2 days per month × 4 cycles) [49]. Among the newly diagnosed patients, there were three with PR and three with stable disease, during the treatment period. All the six patients remained alive during further observation. Of the patients receiving treatment at recurrence, one had a brief period of disease stabilization before progression. High-dose cyclophosphamide was associated with hematologic and pulmonary toxicity.

Kurisaka and associates reported the results of chemotherapy for a small series of patients with pineal parenchymal tumors (pineoblastoma 1, pineocytoma 1, and mixed tumors 2), some of which had distant spread of disease to the spinal axis [50]. The patients received low-dose irradiation (2500–3000 cGy) with various combinations of several intravenous drugs, including cisplatin (20–100 mg/m^2 on days 1–5), vinblastin (1.0–1.5 mg/m^2 on days 1 and 2), bleomycin (15 mg/m^2 on days 1 and 15), ifosfamide (900 mg/m^2 on days 1–5), as well as intra-arterial carboplatin (300 mg/m^2 on day 1). All four patients have responded with some degree of tumor shrinkage, with responses ranging from 25 to 145 months.

Several groups have reported the results of treatment in adult patients with pineoblastoma [51,52]. The first report by Chang and co-workers reviewed the responses of seven patients (median age 36 years, range 17–59 years) with biopsy proven

pineoblastoma to irradiation (5400–7200 cGy) plus chemotherapy [51]. The chemotherapy regimens consisted of lomustine, procarbazine, vincristine, dibromodulcitol, and 6-thioguanine or lomustine, cisplatin, and vincristine. There were five patients with PR and two that had progression during chemotherapy. Several patients that only received radiotherapy were also noted to have CR and PR. In a more recent multi-institutional, retrospective review of treatment in 101 adult patients with malignant pineal parenchymal tumors, chemotherapy was attempted in 34 cases [52]. It was mainly used for patients with pineoblastoma or mixed tumors with disseminated disease. Numerous drug regimens were used (e.g., cisplatin, lomustine, cyclophosphamide, vincristine, and PCV) at various times during the treatment, including neoadjuvantly, concomitantly with irradiation, after irradiation, or only after progression of disease. The use of chemotherapy was not a significant factor for overall survival or PFS during univariate and multivariate analyses.

Chemotherapy is not required for patients with pineocytoma after a gross total resection. However, a recent report suggests it may be of benefit for patients with recurrent or progressive disease [53]. Jackson and Plowman describe three adult patients with progressive pineocytoma that responded well to a regimen of carboplatin (200 mg/m^2 × 2–3 days), etoposide (100 mg/m^2 × 2–3 days), and vincristine (2 mg × 1 day). Case 1 received three cycles of chemotherapy before irradiation and was noted to have slight shrinkage on follow-up MRI. Case 2 developed leptomeningeal dissemination after radiotherapy, with enhancing nodules along the spine. After five cycles of chemotherapy, an extensive PR was noted of all enhancing tumor along the spinal axis. Case 3 also had disseminated disease with minor progression after radiation therapy. Following two cycles of chemotherapy, there was disease stabilization that has lasted for over 5 years.

CONCLUSIONS

The role for chemotherapy in the treatment of pineal parenchymal tumors remains undefined, due to the low incidence of these neoplasms and the lack of clinical trial results. There is consensus that patients in whom a complete resection has been performed for a pineocytoma should not receive chemotherapy, and can be followed closely with serial MRI scans. Chemotherapy may be of benefit for these patients after they have developed locally recurrent tumor or disseminated disease. Patients with pineoblastoma or mixed tumors with significant regional infiltration or dissemination should be strongly considered for some form of chemotherapy. This is especially important for young children (less than 3 years of age), as a means to delay the use of radiotherapy on the developing brain. There is no consensus on which chemotherapy agents to use for patients with recurrent or malignant pineal tumors. The most compelling evidence supports the use of a platinum-based regimen, using either cisplatin or carboplatin, in combination with one or two other drugs, such as etoposide, vincristine, lomustine, or ifosfamide. The most commonly reported combination was cisplatin, etoposide, and vincristine, similar to what has been reported for medulloblastoma (see Chapter 29). Other potential options include high-dose cyclophosphamide or melphalan.

Further clinical research will be necessary to determine if newer chemotherapy agents, such as temozolomide, might have activity against pineal parenchymal tumors [54]. In addition, as the molecular mechanisms for transformation of pineal parenchymal cells are further clarified, it will be important to evaluate the activity of molecular-based therapeutics, similar to the trend for malignant gliomas (e.g., imatinib, geftinib, erlotinib, and CCI-779) [28,29].

ACKNOWLEDGMENTS

The author would like to thank Ryan Smith for research assistance. Dr. Newton was supported in part by National Cancer Institute grant, CA 16058 and the Dardinger Neuro-Oncology Center Endowment Fund.

References

1. Erlich, S. S., and Apuzzo, M. L. J. (1985). The pineal gland: anatomy, physiology, and clinical significance. *J Neurosurg* **63**, 321–341.
2. Macchi, M. M., and Bruce, J. N. (2004). Human pineal physiology and functional significance of melatonin. *Front Neuroendocrinol* **25**, 177–195.
3. Yamamoto, I., and Kageyama, N. (1980). Microsurgical anatomy of the pineal region. *J Neurosurg* **53**, 205–221.
4. Moore, R. Y. (1973). Retinohypothalamic projection in mammals: a comparative study. *Brain Res* **49**, 403–409.
5. Moore, R. Y. (1992). The fourth C. U. Ariëns Kappers lecture. The organization of the human circadian timing system. *Prog Brain Res* **93**, 99–115.
6. Reiter, R. J. (1991). Melatonin: The chemical expression of darkness. *Mol Cell Endocrinol* **79**, C153–C158.
7. Zimmerman, R. A., and Bilaniuk, L. T. (1982). Age-related incidence of pineal calcification detected by computed tomography. *Neuroradiol* **142**, 659–662.

8. Bruce, J. N., Connolly, E. S., Stein, and B. M. (1995). Pineal cell and germ cell tumors. *In Brain Tumors. An Encyclopedic Approach* (A. H. Kaye, and E. R. Laws, Eds.), pp. 725–755. Churchill Livingstone, Edinburgh.
9. Parsa, A. T., Pincus, D. W., Feldstein, N. A., Balmaceda, C. M., Fetell, M. R., and Bruce, J. N. (2001). Pineal region tumors. *In Tumors of the Pediatric Nervous System* (R. F. Keating, J. T. Goodrich, and R. J. Packer, Eds.), pp. 308–325. Thieme, New York.
10. Barker, D., Weller, R., and Garfield, J. (1976). Epidemiology of primary tumors of the brain and spinal cord: a regional survey in southern England. *J Neurol Neurosurg Psych* **39**, 290–296.
11. Schoenberg, B., Christine, B., and Whisnant, J. (1976). The descriptive epidemiology of primary intracranial neoplasms: the Connecticut experience. *Am J Epidemiol* **104**, 499–510.
12. Central Brain Tumor Registry of the United States 2002–2003 (1995–1999 Years Collected Data).
13. Sano, K. (1987). Pineal region and posterior third ventricular tumors: a surgical overview. *In Surgery of the Third Ventricle* (Apuzzo, Ed.), pp 663–683. Williams & Wilkins, Baltimore.
14. Ojeda, V. J., Ohama, E., and English, D. R. (1987). Pineal neoplasms and third-ventricular teratomas in Niigata (Japan) and western Australia. A comparative study of their incidence and clinicopathological features. *Med J Austral* **146**, 357–359.
15. Schild, S. E., Scheithauer, B. W., Schomberg, P. J. *et al.* (1993). Pineal parenchymal tumors. Clinical, pathologic, and therapeutic aspects. *Cancer* **72**, 870–880.
16. Kivelä, T. (1999). Trilateral retinoblastoma: A meta-analysis of hereditary retinoblastoma associated with primary ectopic intracranial retinoblastoma. *J Clin Oncol* **17**, 1829–1837.
17. Fleitz, J., Donson, A., Manchester, D., and Winston, K. (2000). Pineoblastoma: report of two familial cases in half-siblings. *Neurooncol* **2**, S78.
18. Taylor, M. D., Mainprize, T. G., Squire, J. A., and Rutka, J. T. (2001). Molecular genetics of pineal region neoplasms. *J Neurooncol* **54**, 219–238.
19. Rainho, C. A., Rogatto, S. R., de Moraes, L. C., and Barbieri-Neto, J. (1992). Cytogenetic study of a pineocytoma. *Cancer Genet Cytogenet* **64**, 127–132.
20. Bello, M. J., Rey, J. A., de Campos, J. M., and Kusak, M. E. (1993). Chromosomal abnormalities in a pineocytoma. *Cancer Genet Cytogenet* **64**, 185–186.
21. Sreekantaiah, C., Jockin, H., Brecher, M. L., and Sandberg, A. A. (1989). Interstitial deletion of chromosome 11q in a pineoblastoma. *Cancer Genet Cytogenet* **39**, 25–131.
22. Bigner, S. H., McLendon, R. E., Fuchs, H., McKeever, P. E., and Friedman, H. S. (1997). Chromosomal characteristics of childhood brain tumors. *Cancer Genet Cytogenet* **97**, 125–134.
23. Kees, U. R., Biegel, J. A., Ford, J. *et al.* (1994). Enhanced, MYCN expression and isochromosome 17q in pineoblastoma cell lines. *Genes Chromsomes Cancer* **9**, 129–135.
24. Fevre-Montange, M., Jouvet, A., Privat, K. *et al.* (1998). Immunohistochemical, ultrastructural, biochemical and in vitro studies of a pineocytoma. *Acta Neuropathol (Berl)* **95**, 532–539.
25. Plowman, P. N., Pizer, B., and Kingston, J. E. (2004). Pineal parenchymal tumours: II. On the aggressive behaviour of pineoblastoma in patients with an inherited mutation of the RB1 gene. *Clin Oncol* **16**, 244–247.
26. Tsumanuma, I., Sato, M., Okazaki, H. *et al.* (1995). The analysis of p53 tumor suppressor gene in pineal parenchymal tumors. *Noshuyo Byori* **12**, 39–43.

27. Tsumanuma, I., Tanaka, R., Abe, S., Kawasaki, T., Washiyama, K., and Kumanishi, T. (1997). Infrequent mutation of Waf1/p21 gene, a, CDK inhibitor, in brain tumors. *Neurol Med Chir (Tokyo)* **37**, 150–156.

28. Newton, H. B. (2003). Molecular neuro-oncology and the development of "targeted" therapeutic strategies for brain tumors. Part 1 – growth factor and ras signaling pathways. *Expert Rev Anticancer Ther* **3**, 595–614.

29. Newton, H. B. (2004). Molecular neuro-oncology and the development of "targeted" therapeutic strategies for brain tumors. Part 2 – PI3K/Akt/PTEN, mTOR, SHH/PTCH, and angiogenesis. *Expert Rev Anticancer Ther* **4**, 105–128.

30. Edwards, M. S. B., Hudgins, R. J., Wilson, C. B., Levin, V. A., and Wara, W. M. (1988). Pineal region tumors in children. *J Neurosurg* **68**, 689–697.

31. Parinaud, H. (1886). Paralysis of the movement of convergence of the eyes. *Brain* **9**, 330–341.

32. Smirniotopoulos, J. G., Rushing, E. J., and Mena, H. (1992). Pineal region masses: differential diagnosis. *Radiographics* **12**, 577–596.

33. Tien, R. D., Barkovich, A. J., and Edwards, M. S. B. (1990). MR imaging of pineal tumors. *Am J Neuroradial* **11**, 557–565.

34. Stringaris, A. K., Limperopoulos, K., and Samara, C. (2002). Pineal tumors. *In Imaging of Brain Tumors with Histological Correlations* (A. Drevelegas, Ed.), pp. 137–146. Springer, Berlin.

35. Ganti, S. R., and Hilal, S. K. (1986). CT of pineal region tumors. *Am J Roentgenal* **146**, 451–458.

36. Chang, T., Teng, M. M. H., Guo, W. Y., and Sheng, W. C. (1989). CT of pineal tumors and intracranial germ-cell tumors. *Am J Neuroradiol* **10**, 1039–1044.

37. Konovalov, A. N., and Pitskhelauri, D. I. (2003). Principles of treatment of the pineal region tumors. *Surg Neurol* **59**, 250–268.

38. Little, K. M., Friedman, A. H., and Fukushima T (2001). Surgical approaches to pineal region tumors. *J Neurooncol* **54**, 287–299.

39. Bruce, J. N., and Ogden, A. T. (2004). Surgical strategies for treating patients with pineal region tumors. *J Neurooncol* **69**, 221–236.

40. Oi, S., Shibata, M., Tominaga, J. *et al.* (2000). Efficacy of neuroendoscopic procedures in minimally invasive preferential management of pineal region tumors: a prospective study. *J Neurosurg* **93**, 245–253.

41. Yamini, B., Refai, D., Rubin, C. M., and Frim, D. M. (2004). Initial endoscopic management of pineal region tumors and associated hydrocephalus: clinical series and literature review. *J Neurosurg (Ped)* **100**, 437–441.

42. Kobayashi, T., Kida, Y., and Mori, Y. (2001). Stereotactic gamma radiosurgery for pineal and related tumors. *J Neurooncol* **54**, 301–309.

43. Hasegawa, T., Kondziolka, D., Hadjipanayis, C. G., Flickinger, J. C., and Lunsford, L. D. (2002). The role of radiosurgery for the treatment of pineal parenchymal tumors. *Neurosurg* **51**, 880–889.

44. Friedman, H. S., Schold, S. C., Mahaley, M. S. Jr. *et al.* (1989). Phase II treatment of medulloblastoma and pineoblastoma with melphalan: clinical therapy based on experimental models of human medulloblastoma. *J Clin Oncol* **7**, 904–911.

45. Sakoda, K., Uozumi, T., Kawamoto, K., Fujoika, Y., Hasada, J., Hatayama, T., and Nakahara, T. (1989). Responses of pineocytoma to radiation therapy and chemotherapy – report of two cases. *Neurol Med Chir (Tokyo)* **29**, 825–829.

46. Ghim, T. T., Davis, P., Seo, J. J., Crocker, I., O'Brien, M., and Krawiecki, N. (1993). Response to neoadjuvant chemotherapy in children with pineoblastoma. *Cancer* **72**, 1795–1800.

47. Jakacki, R. I., Zeltzer, P. M., Boyett, J. M. *et al.* (1995). Survival and prognostic factors following radiation and/or chemotherapy for primitive neuroectodermal tumors of the pineal region in infants and children: a report of the Children's Cancer Group. *J Clin Oncol* **13**, 1377–1383.

48. Cohen, B. H., Zeltzer, P. M., Boyett, J. M. *et al.* (1995). Prognostic factors and treatment results for supratentorial primitive neuroectodermal tumors in children using radiation and chemotherapy: a Children's Cancer Group randomized trial. *J Clin Oncol* **13**, 1687–1696.

49. Ashley, D. M., Longee, D., and Tien, R. *et al.* (1996). Treatment of patients with pineoblastoma with high dose cyclophosphamide. *Med Pediatr Oncol* **26**, 387–392.

50. Kurisaka, M., Arisawa, M., Mori, T. *et al.* (1998). Combination chemotherapy (cisplatin, vinblastin) and low-dose irradiation in the treatment of pineal parenchymal cell tumors. *Child's Nerv Syst* **14**, 564–569.

51. Chang, S. M., Lillis-Hearne, P. K., Larson, D. A., Wara, W. M., Bollen, A. W., and Prados, M. D. (1995). Pineoblastoma in adults. *Neurosurg* **37**, 383–391.

52. Lutterbach, J., Fauchon, F., Schild, S. E. *et al.* (2002). Malignant pineal parenchymal tumors in adult patients: Patterns of care and prognostic factors. *Neurosurg* **51**, 44–56.

53. Jackson, A. S. N., and Plowman, P. N. (2004). Pineal parenchymal tumours: I. Pineocytoma: A tumour responsive to platinum-based chemotherapy. *Clin Oncol* **16**, 238–243.

54. Newton, H. B. (2000). Novel chemotherapeutic agents for the treatment of brain cancer. *Expert Opin Investig Drugs* **12**, 2815–2829.

33

Current Therapeutic Management Strategies for Primary Intracranial Germ Cell Tumors

Mark T. Jennings

ABSTRACT: Two factors appear dominant in determining the prognosis of children and adolescents with primary intracranial germ cell tumors (GCT): histological diagnosis and the extent of disease dissemination. Germinoma is associated with significantly longer survival than the nongerminatous germ cell tumors (NG-GCT), which often fail conventional treatment. Consensus exists for managing germinomas, without evident dissemination, with involved field external beam radiotherapy (EBRT). The long-term efficacy of chemotherapy alone remains disputed. Establishing the histopathologic diagnosis of NG-GCT represents *a priori* justification for aggressive surgical resection, intensive induction, and/or adjunctive chemotherapy as well as craniospinal irradiation. It cannot be stated with certainty that the presence of elevated biomarkers, beta-human chorionic gonadotropin (β-HCG) and/or alpha-fetoprotein (α-FP), in a patient with a histologically *"pure"* germinoma, without dissemination, warrants intensification of therapy. Patients with elevated biomarkers will probably continue to be considered as NG-GCT and so treated. An ominous prognosis is also associated with neoplastic involvement of the hypothalamus, III ventricle, and spinal cord. The direction of therapy for germinomas is evolving towards balancing the significant, but often transient, efficacy of induction chemotherapy against the long-term neurotoxicity of EBRT. Another controversy will be the role of total surgical resection *versus* pharmacologic cytoreduction in establishing long-term control with acceptable risk among those patients with a favorable prognosis. Patients with

NG-GCT will continue to be treated with increasingly aggressive combination chemotherapy regimens, including myeloablative chemotherapy with autologous bone marrow transplantation.

INTRODUCTION

Tumors of germ-cell derivation are comprised of five interrelated neoplasms which demonstrate a hierarchical order of increasing malignant behavior: germinoma, teratoma (including mature, immature, and malignant subtypes), embryonal carcinoma, endodermal sinus tumor, and choriocarcinoma [1,2]. Each represents the malignant correlate of a normal stage of embryonal development: the primordial germ cell (germinoma), the embryonic differentiated derivative (teratoma) of the pluripotential stem cell of the embryo proper (embryonal carcinoma), as well as the extraembryonic differentiated derivatives which form the yolk sac endoderm (endodermal sinus tumor) and trophoblast (choriocarcinoma) [3,4]. Cancers of germinal cell origin arise in specific midline sites: the gonads, sacrococcygeum, retroperitoneum, mediastinum, diencephalon, and rarely the orbit or nasopharynx [5,6]. Irrespective of the site of origin, the histological features of GCTs are "identical" by light [7,8] and electron microscopy [9], as well as by enzyme and fluorescent histochemical examination [10].

Germ-cell tumors are present clinically in three principal age periods. Congenital GCT are typically benign teratomas. During infancy (one month to three years), endodermal sinus tumors and/or malignant teratomas predominate. Adolescence and young adulthood (ten to thirty years) are usually associated with endodermal sinus tumors and embryonal carcinomas, often with teratomatous elements. In contrast to the prevalence of the NG-GCT among the young, the majority (60 per cent) of adult GCT are *"pure"* germinomas of gonadal origin [11,12].

EPIDEMIOLOGY

In Japan, Taiwan, and South Korea, GCTs comprise 2.1–11.1 per cent of primary intracranial neoplasms [13–18]. This is consistently higher than the 0.4–3.4 per cent reported in Western series [19]. In North America, origin within the brain is more common among males (2.3/100 0000 per annum) than females (0.9/100 0000 per annum); these account for as many as 14 per cent of all GCT occurring in persons less than 20 years of age. One Turkish referral center observed that the incidence of primary GCT of the central nervous system (CNS) to be only 1.1 per cent of all primary intracranial malignant tumors and 2.2 per cent of all GCT [20].

CLINICAL PRESENTATION

The natural history of primary GCTs of the CNS has been established through a retrospective analysis of the literature, which selected 389 appropriate cases. Their histopathologic spectrum ranges from germinoma (65 per cent), teratoma (18 per cent), embryonal carcinoma (5 per cent), endodermal sinus tumor (7 per cent), to the most malevolent, the choriocarcinoma (5 per cent) [11,12].

Tumor Origin

Ninety-five per cent of primary intracranial GCTs originated adjacent to the III ventricle, along an axis from the suprasellar cistern (37 per cent) to the pineal gland (48 per cent). Involvement of both the sites, either sequentially or simultaneously, occurred rarely (6 per cent), as did origin within the III ventricle (3 per cent), basal ganglia-thalamus (3 per cent), or other ventricular sites (3 per cent). Germinomas preferentially involved the suprasellar region (57 per cent, including patients with multicentric involvement), while 68 per cent of NG-GCTs arose in the pineal recess ($p < 0.0001$). The GCTs arising within the basal ganglia-thalamus were all germinomas, whereas those in the lateral ventricular-cerebral region, of IV

ventricular-cerebellar origin, or those that appeared holocranial were NG-GCTs [12].

Correlation of Gender and Age at Diagnosis

This series studied 269 males and 120 females, for a ratio of 2.24:1. This ratio was increased for NG-GCTs (3.25:1) relative to germinomas (1.88:1) ($p = 0.01$). Germ-cell tumors were found in the suprasellar region in 75 per cent of the female patients; among males, pineal involvement was more frequent (67 per cent) ($p = 0.0001$). The age distribution peaked for both the sexes between 10 and 12 years of age, with 68 per cent of patients diagnosed between 10 and 21 years of age. Onset during early puberty was especially common among the germinoma patients. Nongerminomatous GCTs (24 per cent) were more frequently diagnosed between birth and 9 years than were germinomas (11 per cent) ($p < 0.0001$). Presentation during infancy and early childhood was more common among teratomas (31 per cent) and choriocarcinomas (36 per cent) [12].

Duration of Symptomatic Interval Prior to Diagnosis

Among patients with germinomas, 35 per cent were reported to be symptomatic for six months or longer, with half of these in excess of twenty-four months. The prodrome was much shorter among patients with NG-GCT ($p = 0.0007$). Typically, the diagnosis of an intracranial teratoma, embryonal carcinoma, endodermal sinus tumor, and choriocarcinoma was reached within six months of the onset of clinical complaints. Prediagnosis symptomatic intervals were longer among GCTs with suprasellar origin ($p = 0.001$) and female patients ($p = 0.02$) [12].

Presenting Signs and Symptoms

Patients with germinomas, which were commonly suprasellar, presented with chiasmal visual field defects (33 per cent), diabetes insipidus (41 per cent), and hypothalamic-pituitary dysfunction (33 per cent). The neuroendocrine deficits included delay or regression of sexual development (16 per cent), hypopituitarism (16 per cent), and growth failure (9 per cent). Less frequent neurological deficits at presentation were hydrocephalus (21 per cent), obtundation (15 per cent), Parinaud's sign (14 per cent), pyramidal tract signs (11 per cent), diplopia (10 per cent), ataxia (9 per cent), and seizures (3 per cent). Precocious puberty was rare (5 per cent) and occurred in three patients with tumors originating in the suprasellar region, in five with diencephalic tumors, and in four with pineal tumors.

Presenting symptoms and signs among the NG-GCTs typically localized the lesion to the pineal recess, including hydrocephalus (47 per cent), Parinaud's sign (34 per cent), obtundation (26 per cent), pyramidal tract findings (21 per cent), and ataxia (19 per cent). Less common symptoms included hypothalamic-pituitary failure (19 per cent) and diabetes insipidus (18 per cent). Choriocarcinomas were often (55 per cent) associated with sexual precocity and elevated β-HCG and/or luteinizing hormone in the serum and/or cerebrospinal fluid (CSF) [12].

Diagnostic Utility of Ectopic Hormonal Secretion

Considerable effort has been made to use serological biomarkers, such as β-HCG and α-FP for diagnostic, monitoring and prognostic purposes. Among primary intracranial GCT, pathologically increased levels of β-HCG have been reported among patients with choriocarcinomas, germinomas with syncytiotrophoblastic elements, embryonal carcinomas, teratomas, and endodermal sinus tumors. Alpha-fetoprotein levels have been reported to be elevated among children with intracranial germinoma, teratoma, embryonal carcinoma, endodermal sinus tumor, and choriocarcinoma. Therefore, there appears to be a relative rather than an absolute correlation between elevated CSF levels of β-HCG and/or α-FP with the diagnosis of NG-GCT [12].

The identification and localization of the site of β-HCG, luteinizing hormone, and α-FP secretion may be made through comparison of CSF and serum levels. Cerebrospinal fluid β-HCG levels are considered positive if they are more than 2 per cent of the serum levels. However, the CSF concentration of α-FP may not be a dependable marker of CNS involvement [21]. Serial CSF sampling of β-HCG is recommended as more reliable for the detection of occult disease recurrence than comparative quantitation against serum levels. One series of seven patients demonstrated that the CSF biomarker levels became elevated prior to any increase, or even evidence of their presence in the serum [22].

Routes of GCT Dissemination

Germ-cell tumors metastasized both by infiltration into the adjacent hypothalamus (11 per cent) as well

as *via* the ventricular and subarachnoid pathways. Involvement of the III ventricle was especially common with the more aggressive endodermal sinus tumor (42 per cent) and with choriocarcinoma (42 per cent). Spinal cord metastases were more prevalent among the patients with germinomas (11 per cent) and endodermal sinus tumor (23 per cent). Systemic dissemination, especially to lung and bone, occurred in 3 per cent of GCTs (especially those patients with choriocarcinoma, embryonal carcinoma, and germinoma). Abdominal and pelvic metastases developed in approximately 10 per cent of the 106 patients who were known to have required ventriculoperitoneal shunting [12].

PROGNOSTIC VARIABLES

Histopathologic Grading

In retrospective analysis, histopathologic diagnosis exerted the greatest prognostic impact upon overall survival (OS) among primary intracranial GCT. Germinomas were associated with longer survival than the NG-GCT ($p < 0.0001$). Conversely, choriocarcinoma exhibited a singularly dismal prognosis ($p = 0.009$). Survival curves for patients with teratomas, embryonal carcinomas, and endodermal sinus tumors showed no significant difference; one-half of the patients died within the first year [12].

Analysis of the 153 cases treated between 1963 and 1994 at the University of Tokyo Hospital confirmed pathologic diagnosis to be the dominant prognostic determinant. The 5-, 10-, and 20- year overall survival rates (5Y-, 10Y-, 20Y-OS) for germinomas were 96, 93, and 81 per cent, respectively. The 10Y-OS for mature and immature teratoma were 93 and 86 per cent, respectively. Patients, whose teratoma which underwent malignant transformation, demonstrated a 3Y-OS of 50 per cent. The diagnoses of embryonal carcinoma and endodermal sinus tumor had a 3Y-OS of 27 per cent. All the children with choriocarcinoma died within a year of diagnosis [23,24]. The Korean experience has been similar. A series of 107 patients with primary intracranial GCT treated at Yonsei University demonstrated that the 5Y-OS ranged from 91 per cent among germinoma patients and 80 per cent for mature teratoma to 49 per cent among those with other NG-GCT [17].

Extent of Disease Staging

Neoplastic involvement of the hypothalamus ($p = 0.0002$), III ventricle ($p = 0.02$), and spinal cord

$(p = 0.01)$ demonstrated ominous significance [12]. The most sensitive indicator of hypothalamic infiltration is thought to be neuroendocrinological evaluation of the hypothalamic-pituitary axis [25].

Cytologic examination of the CSF in a series of newly diagnosed 42 patients with germinoma was positive in 52 per cent. Neuroradiographic confirmation of metastatic disease into the ventricles or spinal subarachnoid space was documented in 36 per cent of these patients and none of those with a negative cytology. All the patients received EBRT as part of their treatment. Following therapy, CSF dissemination was identified in 18 per cent of those with an initially positive cytology as well as 5 per cent of the originally negative subjects. Remarkably, despite some differences in management, the 5Y-OS was 93 per cent for the cytology positive patients and 94 per cent for those who had no evidence of CSF involvement [26].

Biomarker Expression

It remains controversial whether β-HCG production by a primary CNS germinoma adversely alters the patient's prognosis [27]. Several recent studies are germane to the difficulty in distinguishing between the poorer prognosis of GCT with nongerminomatous elements *versus* a possible selective advantage conferred by gonadotropin secretion. The *First International Germ Cell Tumor Study* observed that elevated β-HCG was associated with increased risk of disease progression $(p = 0.06)$ following chemotherapy alone, but did not affect OS in a series of 71 patients [28].

Retrospective analysis of a group of 44 patients found that 45 per cent had an elevated β-HCG titer in the blood and/or CSF. However, histologic diagnosis was available in only 43 per cent, with the diagnosis being made on clinical criteria in the remainder. The characteristics of those patients with *"pure"* germinoma and those with presumptive syncytiotrophoblastic elements were otherwise similar, including extent of disease evaluation. The former group was treated with 46.4 Gray (Gy, one Gy equals 100 rad) to the primary tumor and the latter received 47.5 Gy. The 10-year event-free survival (10Y-EFS) and 10Y-OS was 100 per cent for the germinoma patients with elevated β-HCG levels and 89 per cent for those with *"pure"* germinoma [29].

The opposite result was found in another trial comprised of 33 patients with GCT (16 germinomas, 11 β-HCG expressing germinomas, 3 mixed teratoma-germinoma, and 3 NG-GCT) treated with preradiation chemotherapy. Patients with *"pure"* germinomas demonstrated an 86 per cent 5Y-EFS and 100 per cent 5Y-OS. In contrast, those germinoma patients with measurable β-HCG expression exhibited only a 44 per cent 5Y-EFS and 75 per cent 5Y-OS [30].

In a third study, a case report and review of the literature examined the prognostic factors among primary intracranial choriocarcinoma/germ-cell tumors expressing high levels of β-HCG. The patient population consisted of primary intracranial choriocarcinoma (35 patients), mixed GCT with choriocarcinomatous differentiation (23 subjects), and GCT with high levels of β-HCG but without pathologic confirmation of syncytiotrophoblastic elements. The adverse prognostic markers demonstrated by univariate analysis were suprasellar origin and tumor hemorrhage. In fact, the latter may be a surrogate marker for choriocarcinoma. Multivariate analysis revealed that extent of surgery, EBRT, and chemotherapy were the independent variables predicting the outcome [31]. Unfortunately, this study did not isolate the prognostic effect of β-HCG on outcome against a control group. Perhaps a more pertinent way to phrase the first portion of the question is: Are GCT to be graded by the most prevalent cell population or the worst histopathologic element identified in order to assign a specific histopathologic diagnosis? Within defined subgroups, the prognostic effect of biomarker expression might be more scientifically dissected.

THERAPEUTIC EFFECTIVENESS AND CONSEQUENT PROGNOSIS

The Norton–Simon hypothesis predicts that a cancer's regression rate is a direct function of its pretreatment growth rate and the intensity of therapy [32]. The Goldie–Coldman hypothesis states that resistance to therapy develops due to spontaneous mutations within cancer cells. Consequently, the absolute number of resistant cells increases as a function of progressive tumor proliferation [33,34]. The treatment intensity concept may help to dissect the mechanism(s) of failure when the strategic plan is critically analysed by the relative degree of cytoreduction achieved at each sequential stage of multimodality intervention.

Surgical Intervention

The emergence of effective chemotherapy for testicular GCT has changed the management of intracranial GCT; a review of the evolution of therapy is necessary to understand current trends. Prior to the 1970s, the attempted extirpation of a posterior III ventricular or pineal tumor carried a 25–70 per cent operative mortality risk. This naturally emphasized EBRT as the treatment of choice [21]. Takeuchi *et al.* [35] advocated exploiting the radiosensitivity of pineal germinomas as a diagnostic as well as therapeutic procedure. In Japan, it became common practice to first irradiate a pineal tumor thought to be a germinoma with 20 Gy ("*the radiation test*"), and then, if the tumor regressed, to continue EBRT. If there was no reduction in tumor size, surgical excision was then to be considered [35,36].

Chapman and Linggood [37] showed that early radiation response was not in itself diagnostic of tumor histology or curability. The postirradiation recurrence rate of large germinomas led Sano and Matsutani [38] to advocate direct surgery and postoperative EBRT for all such patients, with the exception of those with small or multiple germinomas. Over time, advances in radiographic imaging, neuroanesthesia, and microsurgical technique produced an operative mortality rate of less than 5 per cent among the patients with a pineal tumor. In fact, the extent of surgical resection has been proven to be an important determinant of survival in the single largest series of 153 patients with primary intracranial GCT [23]. Several recent retrospective studies have emphasized the importance of macroscopic total resection at the time of diagnosis, in addition to EBRT and chemotherapy, in order to achieve durable remissions among the NG-GCT patients with the poorest prognosis [39–41]. Another compelling justification for establishing a histological diagnosis was the identification of patients who could benefit from more intensive therapy specifically directed against the NG-GCT (*vide infra*).

The pendulum has swung back again in Asia. Due to the increasing effectiveness of nonsurgical therapies, Sawamura *et al.* [42] found that the attempted radical resection of primary intracranial germinomas offered no further survival advantage. Among the Japanese and Korean adolescents, the incidence of GCT among pineal region tumors has been so high that tissue diagnosis is no longer felt to be obligatory if the neuroradiographic findings and biomarker serologic studies favor this diagnosis. Japanese Society for Pediatric Neurosurgery and the Korean Society for Pediatric Neurosurgery have emphasized the radio- and chemo-sensitivity of pineal region GCT. Jointly these groups have published a position paper, which advocates minimally invasive surgical procedures, such as stereotactic biopsy, to be followed by platinum-based chemotherapy and/or targeted EBRT [36]. It is only among the cases demonstrating a poor response to "*trial therapy*" that surgical intervention is to be considered to clarify the pathologic diagnosis and/or to debulk the tumor in preparation for further treatment [17].

Neuroendoscopic procedures are increasingly explored as a means to select cases requiring a microsurgical approach, such as medium to large size NG-GCT, from cases better treated with EBRT and/or chemotherapy. Hydrocephalus is more often being controlled through the use of III ventricular fenestration to avoid the complications of ventriculoperitoneal shunting [43,44].

External Beam Radiotherapy

In the older literature, conventional EBRT achieved 5Y-OS rates of 60 per cent (range 25–88 per cent) among a heterogeneous group of pineal tumors, which were being empirically treated [45–47]. Higher radiation dosages (50–55 Gy) have reduced the local recurrence rate from 47 to 10 per cent [48]. Among the patients with known germinomas, combined surgical resection and radiotherapeutic intervention improved 10Y-OS rates from 69 to 93 per cent [23,38]. Radiotherapy prescriptions of a primary tumor dose of 50–55 Gy, with or without 18–36 Gy to the neuraxis, have been proven to be reliably effective in sustaining complete remissions; EBRT remains the standard of care for germinoma patients in a number of Western series [49–52]. Some investigators find that germinomas are not only curable with irradiation but advocate that craniospinal doses of ≤ 25.5 Gy to the whole brain and < 22 Gy to the spine are comparable to higher doses of combined EBRT–chemotherapy [53].

Subsequent controversy regarding the treatment of "*pure*" germinomas, without known metastatic disease, has centered on the issues of dosimetry, treatment portals (involved field *versus* ventricular) and the indication for craniospinal irradiation. As experience accrued to prove that germinomas of the CNS were as radiosensitive as those of the testis, there has been increasing consensus that those patients without evident CSF dissemination could be controlled with involved-field EBRT alone [23,50,54–56]. Investigators at Kyoto University have prospectively studied the relationship between postoperative residual tumor volume and radiotherapy prescription among 35 patients with germinomas. Subjects with no evident disease were irradiated to 36 Gy, for those with less than 2.5 cm diameter residual disease the dose was 40 Gy. Among patients with a tumor diameter of 2.5–4 cm, doses of 45 Gy were administered, and therapy to ≥ 50 Gy was reserved for dimensions larger than 4 cm. The decision regarding involved-field *versus* craniospinal treatment (20–24 Gy) was determined by the staging evaluation at diagnosis. The 10- year relapse-free survival has been 95 per cent with a 10Y-OS of 91 per cent. Two suffered meningeal dissemination but there were no local recurrences [57].

Stereotactic radiotherapy has been suggested as a means of precisely administering the tumor boost following involved field, whole brain or craniospinal EBRT. Median doses of 25.2 Gy (range 15–36) to the whole brain and 21.6 Gy (range of 21–26) to the neuraxis were supplemented by a median boost of 26 Gy (range 21.6–36) delivered to the 95 per cent isodose line. At a median follow-up time of 40 months, there were no local or marginal recurrences among 13 germinoma patients. Two of the five subjects with mixed GCT experienced either relapse or inadequate control despite the addition of chemotherapy. Stereotactic radiotherapy was well tolerated in this group of children and is thought to have reduced long-term neurotoxicity [58].

There is no debate regarding the necessity of treating NG-GCT with EBRT to the tumor (50–55 Gy) and craniospinal axis (36–40 Gy) due to their significantly higher rate of metastasis and recurrence [12,17,54,59–61]. Unfortunately, the NG-GCTs, as a group, have shown limited sensitivity to radiotherapy doses of more than 50 Gy [62].

Multimodality Therapy including Chemotherapy

Among localized and metastatic testicular GCTs, investigational protocols have combined synergistically cytotoxic agents, including vinblastine, actinomycin D, bleomycin, adriamycin, cyclophosphamide, and/or cisplatin (the "VAB" regimens), following surgery and irradiation to produce durable remission rates of 60–90 per cent. Such agents can be delivered systemically in sufficient concentration to cross the blood–brain barrier and significantly lengthen survival time in cases of GCT metastases to the brain [63]. This experience initially encouraged the use of chemotherapy among primary intracranial GCTs at relapse, and then adjunctively during the 1980s [64,65].

Several principles have emerged in the pharmacologic treatment of gonadal GCT:

(1) Cisplatin-based combination chemotherapy has significantly enhanced the survival rates among patients with bulky or metastatic disease,

(2) The dose–response relationship of cisplatin among GCT is such that intensive induction chemotherapy may obviate the need for maintenance treatment among patients who demonstrate a complete response (CR),

(3) Carboplatin has comparable efficacy to cisplatin, with less oto- and nephro-toxicity and may not elicit cross-resistance among patients previously treated with cisplatin, and

(4) Vinblastine has been largely replaced by etoposide for synergism with the platinator, while the necessity for bleomycin remains uncertain.

To enhance the response to EBRT, investigators at Memorial Sloan-Kettering [66] used three cycles of high-dose cyclophosphamide or vinblastin–bleomycin–cyclophosphamide–cisplatin as induction therapy among 11 children with primary intracranial GCT. Patients with disease restricted to the primary site were treated with 30 Gy to the involved field. Children with disseminated disease received 30 Gy to the primary site and 20 Gy to the neuraxis. Of the original group, 10 remained disease-free for four years from diagnosis (Table 33.1) [66].

A second phase II trial by this group administered two courses of carboplatinum to 11 patients with primary CNS germinomas without known metastatic disease. Children achieving a CR were treated with reduced dose EBRT (30 Gy to the involved field with 21 Gy to the neuraxis). Those patients with lesser responses went on to two additional courses of chemotherapy followed by full-dose irradiation (50 Gy to the involved field with 36 Gy to the neuraxis). Five subjects (45 per cent) demonstrated a CR after two courses, an additional two (64 per cent) after the fourth course. Two patients experienced a partial response (PR; ≥ 50 per cent tumor reduction) after two courses and one after a total of four cycles of chemotherapy. Ninety-one per cent of the participants maintained complete remission for a median of twenty-five months. There was one relapse and death in a child who had responded to the two courses of induction chemotherapy. His serum α-FP and β-HCG were found to be elevated at that time suggesting the emergence of resistant NG-GCT elements [67].

Not all studies have segregated the germinoma and NG-GCT patients. A Japanese study prescribed two to three courses of cisplatin–etoposide prior to irradiation. Twelve patients were treated—five at initial diagnosis and seven at recurrence. Of the five patients with elevated biomarkers felt to indicate the diagnosis of a NG-GCT, all demonstrated normalization of these levels prior to EBRT. A CR was noted in seven, while the remaining five patients demonstrated a PR. One patient with recurrent disease progressed and died despite further treatment; there was one death from sepsis. The remaining ten patients were free of disease at a mean of thirteen months from diagnosis [68].

The University of Eppendorf group has brought the attention to the risk of iatrogenic dissemination following surgical biopsy and/or resection among biomarker-positive GCT of the pineal recess. These investigators have proposed using one course of bleomycin–etoposide–cisplatin to induce significant regression and better delineation of the tumor's margins. An infratentorial, supracerebellar surgical approach is then used for attempted gross total resection. This is to be followed by vinblastine–ifosfamide–cisplatin and craniospinal EBRT (50 Gy tumor boost with 30 Gy to the neuraxis). This protocol was used for the treatment of three boys, two with elevated β-HCG and one with α-FP expression. Each of the children received surgical resection *in toto*. No one required a permanent ventriculoperitoneal shunt. Following induction therapy, pathologic analysis revealed benign, mature teratoma with derivatives from all three germinal layers. There was also hemorrhagic necrosis with granulomatous and lymphocytic inflammation. In one case, there were small admixtures of immature teratoma with mitoses. None demonstrated malignant extraembryonal differentiation. The boys remained disease-free at 66, 71, and 78 months following surgery [69,70].

A series of 18 patients with known (14 cases) and presumed NG-GCT (four with elevated biomarkers) were treated with three to four courses of cisplatin–etoposide followed by EBRT. Of the twelve patients with evaluable disease, five demonstrated a CR and four patients exhibited PR, two patients had stable disease and one progressed. Patients were irradiated either to the involved field (eleven), craniospinal axis (four), or whole brain (two). Twelve subjects received an additional four cycles of postradiation chemotherapy with vinblastine–bleomycin–etoposide–carboplatin. The six patients with no evident postoperative disease remained in remission following additional chemotherapy. Four patients had died by the time of reporting, three were disease related. The 4Y-EFS and 4Y-OS rates were 67 and 74 per cent, respectively [71].

The French Society of Pediatric Oncology reported 29 patients with biopsy proven, localized germinomas. Induction chemotherapy with carboplatin–etoposide–ifosfamide was administered for two courses prior to irradiation of the initial tumor volume. Of the 26 evaluable patients, 58 per cent

achieved a CR. Twenty-eight patients remained in their first full remission over a median follow-up period of 32 months (range 7–68 months). The 4Y-EFS was 93 per cent without any fatalities [72].

The *Cooperative German/Italian Study* accepted the diagnostic entry criteria to be the appropriate neuroradiographic findings and elevated biomarkers (β-HCG and/or α-FP). Nineteen patients (16 males and 3 females) were placed into the study. The therapeutic design consisted of two induction courses of cisplatinum–etoposide–ifosfamide (PEI); patients responding to chemotherapy were to receive an additional two courses. Patients who did not respond and those with progressive disease were to be advanced to surgical resection, if feasible, prior to craniospinal EBRT (30 Gy with a tumor boost of 24 Gy). This cohort demonstrated an elevated α-FP and/or β-HCG level in 16/19 patients at the time of diagnosis. Thirteen of the sixteen had normalization of biomarker levels following the second course of PEI induction chemotherapy, which paralleled objective neuroradiographic evidence of a cytoreductive effect in 10/13. Three children demonstrated a CR after two courses of chemotherapy. Eight demonstrated no evident disease following the fourth course, with two children showing lesser responses. Three patients with teratomas demonstrated progressive disease for which tumor resection was attempted; one child died postoperatively due to intratumoral hemorrhage. Seventeen of the patients survived; 81 per cent have remained in remission over a median follow-up of 11 months (range 7–39 months). The toxicity of therapy was reported to be tolerable [73].

The practice at the University of Tokyo has been to rank GCT patients by relative risk into three subgroups:

(1) mixed germinoma and teratoma,
(2) mixed GCT with predominance of germinoma or teratoma with some *"pure malignant tumor"* (embryonal carcinomas, endodermal sinus tumors. and choriocarcinomas), and
(3) mixed tumors with predominance of *"pure malignant tumor"*.

Surgery and EBRT produced a 10Y-OS rate of 91.7 per cent among the germinoma patients. Combination chemotherapy (cisplatin–vinblastine–bleomycin, cisplatin–etoposide, or carboplatin–etoposide) and radiotherapy was shown to significantly reduce the risk of disease recurrence in the intermediate prognostic group when compared to irradiation alone ($p = 0.049$). Forty per cent of these patients experienced a CR following induction chemotherapy as well as a much lower recurrence rate (21.4 *versus* 45 per cent). The high-risk patients did better with chemotherapy (3Y-OS rate of 27.3 per cent) than with EBRT alone (3Y-OS of 10.2 per cent) although the difference did not reach statistical significance [23].

The Japanese Pediatric Brain Tumor Study Group has expanded upon these observations. Germinoma patients were treated with carboplatin–etoposide or cisplatin–etoposide for three courses and followed with EBRT. The NG-GCT subjects received ifosfamide–cisplatin–etoposide for three courses after which they underwent irradiation. Among those with germinomas, 84 per cent achieved a CR with induction chemotherapy. Over a median follow-up period of 2.9 years, recurrent disease developed in 12 per cent. In the majority of these, relapse occurred outside the radiation portal. Among the 10 patients with β-HCG secreting germinomas, 78 per cent achieved a CR with chemotherapy. Following EBRT, 90 per cent were disease-free. Over a monitoring period of 3.4 years, none have recurred. The intermediate prognosis group consisted of 18 patients with malignant teratoma and mixed tumors. There were no CRs noted with chemotherapy. However, following the completion of radiotherapy, 56 per cent were tumor-free. Two patients relapsed during a median observation period of 3.7 years. The poor prognosis group was made up of nine patients, of whom two were still alive without recurrence more than two years after the treatment [74].

The Societe Francaise d'Oncologie Pediatrique (SFOP) initiated a study of induction chemotherapy with either limited-field irradiation (40 Gy) for germinoma patients with localized disease (51 children) or "*low dose*" craniospinal EBRT for those with dissemination (six patients). Four alternating courses of etoposide–carboplatin and etoposide–ifosfamide were administered postoperatively. With a median follow-up of 42 months, the 3Y-EFS was 96.4 per cent and the 3Y-OS was 98 per cent. Of the four patients who relapsed, three achieved a second CR with chemotherapy with or without irradiation [75].

The investigators at Kumamoto University have proposed treating patients with NG-GCT with neoadjuvant chemotherapy and EBRT to sufficiently cytoreduce the neoplasm as to allow gross total resection. In a population of 11 such patients (five with yolk sac tumor, one with embryonal carcinoma, one immature teratoma, and four subjects with mixed malignant GCT), induction therapy produced two CR and six PR. Two patients demonstrated stable disease and one progressed. Nine patients underwent surgical removal of residual disease. Ten of the

eleven are alive at a mean of 96 months (range 30–177) following the diagnosis [76].

The Hokkaido University protocol is to combine induction chemotherapy with "*low-dose*" involved-field EBRT for newly diagnosed GCT patients. Solitary, "*pure*" germinomas are initially treated with etoposide–cisplatin. Ifosfamide–cisplatin–etoposide (ICE) is used for the induction of those patients with NG-GCT, β-HCG secreting germinomas and for those with disseminated disease. Patients with pineal germinomas are treated with attempted gross total resection while others were biopsied or partially removed. The clinical treatment volume for EBRT includes the tumor site for germinomas and immature teratomas. Multifocal germinomas received irradiation (24 Gy) to the "*whole ventricle*", which includes the III and lateral ventricles, for pineal region tumors the IV ventricle is also treated. The β-HCG secreting germinoma patients are administered an additional 6 Gy to the neurohypophysis and 10 Gy to the pineal region. Highly malignant and disseminated tumors are treated with craniospinal EBRT with a tumor boost (50–54 Gy). This group reported their results with 16 germinoma patients, 11 with β-HCG secreting germinomas, three with immature teratoma/germinoma, and three patients with either embryonal carcinoma or malignant teratoma. Eight patients had multifocal origin and three suffered from metastatic disease. Every germinoma patient achieved a CR by the third course of chemotherapy; the presence of β-HCG did not affect the response to therapy. For the entire group of 33 patients, 5Y-OS was 93 per cent. Relapse-free survival rates at 5 years were 69 per cent for the whole population, 90 per cent for the germinoma patients, and 44 per cent for those with β-HCG secreting germinomas (the difference between the last two being $p = 0.025$). Patients with relapsed disease did respond to further therapy. All the six patients with NG-GCT were alive at a median of 65 months (range of 23–92 months). There were no treatment or disease-related deaths. Furthermore, no cognitive deterioration was observed using Wechsler Intelligence Scale testing for children and adults, with the exception of one patient who died from hypothalamic dysfunction (Table 33.1) [30].

Chemotherapy Alone

A Taiwanese study evaluated 11 newly diagnosed cases of primary intracranial germinoma who were treated only with six courses of vinblastine–bleomycin–cisplatin–etoposide (VBPE). Every patient achieved a CR, however 55 per cent relapsed at a

TABLE 33.1   Complete Response Rates to Induction Chemotherapy among Newly Diagnosed Germinomas and NonGerminomatous Germ Cell Tumor Patients

Germinoma	Course 2 (%)	Course 4 (%)	Course 6 (%)	EBRT	2Y-EFS (%)	3Y-EFS (%)	4Y-EFS (%)	5Y-EFS (%)	Reference
HD-CPM or VBL–BLM–CPM–cDDP (11 pts)	91			Yes			91		[66]
CBDCA (11 pts)	45	64		Yes	91				[67]
1st IGCTS									[28]
CBDCA–VP16–BLM (45 pts)			84	No	84				
SFOP									[72]
VP16–CBDCA → VP16–IFOS (29 pts)		58		Yes	93				
SFOP									[75]
VP16–CBDCA → VP16–IFOS (57 pts)				Yes		96			
Hokkaido									[30]
VP16–cDDP (16 pts)		100		Yes				90	
βHCG (+) germinoma (17 pts)		100		Yes				44	
VBL–BLM–VP16–cDDP (11 pts)			100	No	45				[77]
JPBTSG									[74]
CBDCA–VP16 or cDDP–VP16 (75 pts)			84	Yes	88				
β-HCG(+) germinoma (10 pts)			78	Yes					
2nd IGCST									[80]
cDDP–VP16–CPM–BLM; CBDCA–VP16–BLM (19 pts)			100	No				47	
NG-GCT									
VP16-BLM–CPM–cDDP (11 pts)			0	Yes					[66]
1st IGCST									[28]
CBDCA–VP16–BLM (26 pts)			78	No	62				
BLM–VP16–cDDP (preop) VBL–IFOS–cDDP (postop) cDDP–VP16 (18 pts)				Yes				100	[70]
VPB–BLM–CBDCA (post-RT, 12 pts)			42	83			67		[71]
CG/IS									[73]
cDDP-VP16-IFOS (19 pts)	16	42		Yes					
SFOP									[78]
VBL–BLM–CBDCA; VP16–CBDCA/IFOS–VP16 (18 pts)			100	13				8	
JPBTSG									[74]
IFOS–cDDP–VP16 (27 pts)			0	Yes					
2nd IGCST									[79]
cDDP-VP16-CPM-BLM; CBDCA–VP16–BLM (20 pts)			62–65	18				36	

Abbreviations: external beam radiotherapy (EBRT), two-, four- and five- year event free survival (2Y-, 4Y-, 5Y-EFS), high dose cyclophosphamide (HD-CPM), carboplatin (CBDCA), cisplatin (cDDP), etoposide (VP16), bleomycin (BLM), cyclophosphamide (CPM), ifosfamide (IFOS), vinblastine (VBL), patients (pts), nongerminatous germ cell tumors (NG-GCT), First International Germ Cell Tumor Study (1st IGCTS), Societe Francaise d'Oncologie Pediatrique (SFOP), Japanese Pediatric Brain Tumor Study Group (JPBTSG), beta-human chorionic gonadotropin (β-HCG), Cooperative German/Italian Study (CG/IS), preoperative administration (preop), postoperative administration (postop).

mean of 16.8 months. These patients were all retrieved with focal EBRT. The authors felt that the strategy was beneficial in delaying or eliminating the need for irradiation and its complications (Table 33.1) [77].

The *SFOP* group treated 18 patients with GCT secreting either α-FP or β-HCG with six cycles of chemotherapy (combinations of vinblastine, bleomycin, carboplatin, etoposide, and/or ifosfamide) with surgical resection of residual tumor. Focal EBRT was reserved for cases with viable residual tumor. Thirteen patients (72 per cent) received only chemotherapy and two were additionally treated with radiation. Twelve of the cases (67 per cent), who did not receive irradiation, developed recurrent disease. Of the original 18, 12 patients survived, however 11 required EBRT (61 per cent) at some point. These investigators felt that α-FP or β-HCG secreting GCTs were not curable with conventional chemotherapy and recommended that focal irradiation should be part of the treatment schema [78].

The *First International Germ Cell Tumor Study* proposed a chemotherapy only regimen of carboplatin - etoposide - bleomycin. This study accessioned 45 patients with germinoma and 26 with NG-GCT, of whom 68 were considered evaluable. The protocol for germinoma patients, who achieved a CR after four induction courses, prescribed two additional cycles. Those subjects with less than a CR were treated with a chemotherapy regimen fortified by cyclophosphamide followed by EBRT. A CR was achieved in 57 per cent of the patients after four induction courses of chemotherapy, and additional 24 per cent were left with no evident disease after intensified chemotherapy or *"second-look"* surgery. Thus 55 of the 71 patients (78 per cent) were rendered disease-free without irradiation (84 per cent of germinomas and 78 per cent of NG-GCT patients). Thirty-nine per cent of the patients achieved a durable remission over a median follow-up period of 31 months. The incidence of progressive disease on therapy was 10 per cent with a recurrence rate of 39 per cent, which occurred at a median of 13 months. Ninety-three per cent of the patients with recurrent disease responded to *"salvage"* therapy. Ten per cent of the patients died of chemotherapy related complications. Consequently, 41 per cent of the survivors and 50 per cent of all the patients were treated successfully with only chemotherapy and had not required EBRT at the time of reporting. The 2Y-OS was 84 per cent for germinoma and 62 per cent for NG-GCT patients [28].

The *Second International CNS Germ Cell Study Group* employed two courses of cisplatin–etoposide–cyclo-phosphamide–bleomycin to assess chemosensitivity in a group of 20 patients with NG-GCT. Pathologic diagnosis was available for 14, 11 of whom demonstrated mixed GCT. The study design was that patients achieving a CR would receive two additional courses of carboplatin–etoposide–bleomycin, as well as an additional cycle of the original treatment regimen. Those not achieving or sustaining a CR underwent *"second look"* surgery and/or EBRT. At least a 50 per cent tumor reduction was achieved in 94 per cent of evaluable patients [16,17]. The median EFS for patients experiencing a CR was 62 months, compared with those individuals with a PR who demonstrated a 23 month mean EFS. However, 69 per cent of the evaluable patients developed progressive disease during or following chemotherapy. The 5Y-EFS and OS for all the patients have been 36 and 75 per cent, respectively. This trial did not observe a relationship between biomarker expression prior to the treatment and the outcome [79].

A second report from these same investigators analyzed the results among 19 patients with germinomas, who were treated with this protocol. Of the 11 patients with postoperative residual disease, everyone achieved a CR. However at a mean of 6.5 years, only 42 per cent remained in remission. Three patients died of treatment-related toxicity and one from an uncharacterized leukoencephalopathy, constituting a *"toxic-death"* rate of 19 per cent. The 5Y-EFS was 47 per cent and 5Y-OS was 68 per cent. The presence of diabetes insipidus was an adverse prognostic marker. These investigators acknowledged the effectiveness of intensive chemotherapy to induce remission but were critical of its toxicity and inadequacy for maintaining long-term disease control even among germinoma patients (Table 33.1) [80].

Retrospective review of 126 patients enrolled in both these studies found 10 who underwent delayed surgical resection because of residual radiographic abnormalities despite improvement or resolution of elevated serum/CSF biomarkers following induction chemotherapy. There were two with *"pure"* germinoma, without elevated α-FP or β-HCG, and eight with NG-GCT at the time of initial diagnosis. Surgical reexploration revealed mature teratoma (three patients), immature teratoma (two) or necrotic/scar tissue (five subjects). Of the four patients with NG-GCT and persistent elevation of the biomarkers, three required EBRT due to disease progression. This was necessary despite a tissue diagnosis of teratoma or necrotic/scar tissue. With an average follow-up period of 37 months (range 3–96), 70 per cent have maintained durable remissions [81].

Dose Intensification with Autologous Hematopoietic Support

High-dose chemotherapy with autologous stem cell rescue has been studied at Shinshu University among six patients with NG-GCT. The patients were treated when in CR following resection, EBRT and four to seven courses of chemotherapy. The myeloablative regimen consisted of cisplatin–etoposide–ANCU. All six have survived from one to seven years with good performance status and have not required further therapy [82].

Patients experiencing progressive disease on therapy and those with relapse following multimodality therapy suffered a dismal prognosis. High-dose chemotherapy utilizing thiotepa and autologous stem-cell rescue has been studied among a group of 21 such patients. The 4Y-EFS and 4Y-OS were 52 and 57 per cent, respectively. Seven of nine (78 per cent) germinoma subjects remained disease free with a median survival of 48 months. The survival among the NG-GCT patients was much worse, as only 33 per cent were alive at a median of 35 months. The majority succumbed to progressive disease with a post-treatment median survival of four months. The difference between the germinoma and NG-GCT patients was significant in terms of EFS ($p = 0.014$) and OS ($p = 0.016$). Subjects achieving a CR did much better than those with lesser responses ($p < 0.001$ for EFS and OS). There were no *"toxic deaths"* in this study, leading the authors to recommend it for recurrent germinoma patients and for those NG-GCT with minimal residual tumor burden [83].

LONG-TERM COMPLICATIONS OF THE DISEASE AND ITS THERAPY

Patterns of Relapse

Four patterns of disease recurrence following EBRT have been recognized. *Type I* is local recurrence outside the original field of therapy. It is recommended that these patients are to be treated with craniospinal irradiation. The second pattern is characterized by progression of residual tumor which is *"benign"* or mature teratoma. This appears best managed by reoperation. *Type III* consists of emergence of non-germinomatous elements, often secreting α-FP and/or β-HCG, which required aggressive intervention, such as chemotherapy, radiosurgery, etc., due to their high mortality rate. The fourth pattern of recurrence is that of extraneural dissemination without evidence of CNS involvement [84]. In contrast to other CNS neoplasms following chemotherapy, recurrent GCT do not appear to acquire cross-resistance to EBRT and remain responsive to doses of 40–47 Gy [85].

Functional Impairment—Cognitive and Endocrinologic

There is relatively little data available regarding the long-term toxicity of therapy among patients with primary intracranial GCT. Due to the peculiarities of age at onset, site of origin, radio- and chemosensitivity, the iatrogenic risk for these children and young adults cannot be readily extrapolated from the experience of those treated for malignant gliomas or primitive neuroectodermal tumors [86]. For example, children with GCT involving the neurohypophyseal region demonstrate lower intelligence quotient scores *at the time of diagnosis* than those with pineal origin [87].

Retrospective review of 54 cases of suprasellar germinoma at Kitasato University found 12 patients who had received EBRT as their only treatment modality. Diabetes insipidus and growth hormone deficiency were found in 75 and 42 per cent, respectively, before irradiation. Over a mean follow-up period of 161 months (range 63–262 months), the survival rate was 100 per cent. Fifty per cent were shown to have developed *"remarkably low mental function"*. Furthermore, 92 per cent required hormone replacement for the above-noted neuroendocrine deficits, adrenal insufficiency and/or hypothyroidism [88]. Other workers have reported that induction chemotherapy appears more likely to reverse or obviate the presenting hormonal inadequacy than EBRT [89].

Quality of life assessment was determined among 43 patients treated in the *First International CNS Germ Cell Tumor Study,* at a median of 6.1 years following diagnosis (range 4.5–8.8 years). Patients entered on study at 19 years of age or older retained normal psychosocial and physical functioning. In contrast, the younger patients were rated by their parents to become low average or borderline in performance. Patients with germinomas significantly outperformed those with NG-GCT on all neuropsychological measures administered. The need for EBRT adversely affected overall physical functioning. In this study, the histopathologic diagnosis (germinoma *versus* NG-GCT), with its attendant consequences regarding therapy, appeared to be the salient risk factor for cognitive outcome [90].

Recently, Lutterbach *et al.* [91] reviewed the literature regarding the long-term sequelae of treatment for malignant GCT, which had metastasized to the brain from a systemic primary tumor. Even so, there still remains considerable controversy regarding the cognitive and functional consequences of therapy, among these predominantly adult patients being treated for a potentially curable malignancy.

CONCLUSIONS

Two factors appear dominant in determining the prognosis of primary intracranial GCT; histopathologic diagnosis and the extent of disease dissemination. Patients with germinoma demonstrate a significantly longer survival ($p < 0.0001$) than those with NG-GCT, which often fail to respond adequately to conventional treatment. There is considerable confidence in managing primary intracranial germinomas, without evident dissemination, with involved-field EBRT. The long-term efficacy, and appropriateness, of chemotherapy alone continues to be disputable. Establishing the pathologic diagnosis of a NG-GCT represents *a priori* justification for aggressive surgical resection, intensive induction and/or adjunctive chemotherapy as well as craniospinal radiation therapy. An ominous prognosis has been associated with neoplastic dissemination to the hypothalamus ($p = 0.0002$), III ventricle ($p = 0.02$), and spinal cord ($p = 0.01$), however the significance of a positive CSF cytology alone is uncertain. While caution must be exercised in the interpretation of these latter associations, as the assessment of extent of disease was made retrospectively by ante- and post-mortem description, patients with metastatic disease deserve intensive multimodality therapy.

Our original review suggested that the neuroendocrinological events of puberty were an *"activating"* influence in the expression of malignant behavior among intracranial GCTs [11]. The observation has been made that gonadotropin-secreting GCTs are associated with a worse prognosis, even for tumors within the same pathologic grade. It cannot be stated with certainty that the presence of elevated biomarker levels, in and of itself, warrants intensification of therapy in a patient with a histologically *"pure"* germinoma, in the absence of dissemination. In practical terms though, patients with elevated biomarkers will probably continue to be considered as NG-GCT and so treated.

The direction of therapy for germinomas appears to be evolving towards balancing the significant, but transient efficacy of induction chemotherapy against the longer term neurotoxicity of surgery and EBRT. Another controversial issue, which may be determined in future trials, is the role of attempted gross total surgical resection *versus* pharmacologic cytoreduction in establishing long-term disease control with acceptable risk for those patients with a favorable prognosis. Among the NG-GCT, currently available maximal therapy is considered largely inadequate. These patients will continue to be treated with increasingly aggressive combination chemotherapy regimens, including myeloablative chemotherapy with autologous bone marrow transplantation [92].

References

1. Hajdu, S. I. (1979). Pathology of germ cell tumors of the testis. *Semin Oncol* **6**, 14–25.
2. Mostofi, F. K. (1980). Pathology of germ cell tumors of the testis. A progress report. *Cancer* **45**, 1735–1754.
3. Pierce, G. B., and Abell, M. R. (1970). Embryonal carcinoma of the testis. *Pathol Annu* **5**, 27–60.
4. Takei, Y., and Pearl, G. S. (1981). Ultrastructural study of intracranial yolk sac tumor: with special reference to the oncologic phylogeny of germ cell tumors. *Cancer* **48**, 2038–2046.
5. Brodeur, G. M., Howarth, C. B., Pratt, C. B., Caces, J., and Hustu, H. O. (1981). Malignant germ cell tumors in 57 children and adolescents. *Cancer* **48**, 1890–1898.
6. Gonzalez-Crussi, F. (1982). "Extragonadal Teratomas. Atlas of Tumor Pathology", Series 2, Fascicle 18. Armed Forces Institute of Pathology, Washington D.C.
7. Friedman, N. B. (1947). Germinoma of the pineal. Its identity with germinoma ("seminoma") of the testis. *Cancer Res* **7**, 363–368.
8. Friedman, N. B. (1951). The comparative morphogenesis of extragenital and gonadal teratoid tumors. *Cancer* **4**, 265–276.
9. Markesbery, W. R., Brooks, W. H., Milsow, L., and Mortara, R. H. (1976). Ultrastructural study of the pineal germinoma *in vivo* and *in vitro*. *Cancer* **37**, 327–337.
10. Beeley, J. M., Daly, J. J., Timperley, W. R., and Warner, J. (1973). Ectopic pinealoma: an unusual clinical presentation and a histochemical comparison with seminoma of the testis. *J Neurol Neurosurg Psychiatry* **36**, 864–873.
11. Jennings, M. T., Gelman, R., and Hochberg, F. (1984). Intracranial germ cell tumors: natural history and pathogenesis. *In Diagnosis and Treatment of Pineal Region Tumors*, pp. 116–138. Williams & Wilkins, Baltimore.
12. Jennings, M. T., Gelman, R., and Hochberg, F. (1985). Intracranial germ cell tumors: natural history and pathogenesis. *J. Neurosurg* **63**, 155–167.
13. Araki, C., and Matsumoto, S. (1969). Statistical reevalution of pinealoma and related tumors in Japan. *J Neurosurg* **30**, 146–149.
14. Shih, C. J. (1977). Intracranial tumors in Taiwan. A cooperative study of 1,200 cases with special reference to the intracranial tumors in children. *J Formosan Med Assoc* **76**, 515–528.
15. Koide, O., Watanabe, Y., and Sato, I. (1980). A pathologic survey of intracranial germinoma and pinealoma in Japan. *Cancer* **45**, 2119–2130.
16. Lin, I. J., Shu, S. G., Chu, H. Y., and Chi, C. S., (1997). Primary intracranial germ cell tumor in children. *Chinese Med J* **60**, 259–264.
17. Choi, J.-U., Kim, D.-S., Chung, S.-S., and Kim, T. S., (1998). Treatment of germ cell tumors in the pineal region. *Child's Nerv Syst* **14**, 41–48.

18. Nomura, K. (2001). Epidemiology of germ cell tumors in Asia of pineal region tumor. *J Neuro-Oncol* **54**, 211–217.

19. Jellinger, K. (1973). Primary intracranial germ cell tumours. *Acta Neuropathol* **25**, 291–306.

20. Akyuz, C., Koseoglu, V., Bertan, V., Spylemezoglu, F., Kutluk, M. T., and Buyukpamukcu, M. (1999). Primary intracranial germ cell tumors in children: a report of eight cases and review of the literature. *Turkish J Pediatr* **41**, 161–172.

21. Schmidek, H. H., (1977). Surgical management of pineal region tumors. *In Pineal Tumors* (H. H. Schmidek, Ed.), pp. 99–113. Masson, New York.

22. Fujimaki, T., Mishima, K., Asai, A., *et al.* (2000). Levels of beta-human chorionic gonadotropin in cerebrospinal fluid of patients with malignant germ cell tumor can be used to detect early recurrence and monitor the response to treatment. *Jap J Clin Oncol* **30**, 291–294.

23. Matsutani, M., Sano, K., Takakura, K. *et al.* (1997). Primary intracranial germ cell tumors: a clinical analysis of 153 histologically verified cases. *J Neurosurg* **86**, 446–455.

24. Sano, K. (1999). Pathogenesis of intracranial germ cell tumours reconsidered. *J Neurosurg* **90**, 258–264.

25. Grote, E., Lorenz, R., and Vuia, O. (1980). Clinical and endocrinological findings in ectopic pinealoma and spongioblastoma of the hypothalamus. *Acta Neurochir* **53**, 87–98.

26. Shibamoto, Y., Oda, Y., Yamashita, J., Takahashi, M., Kikuchi, H., and Abe, M. (1994). The role of cerebrospinal fluid cytology in radiotherapy planning for intracranial germinoma. *Int J Radiat Oncol Biol Phys* **29**, 1089–1094.

27. Arita, N., Ushio, Y., Hayakawa, T. *et al.* (1980). Serum levels of alpha fetoprotein, human chorionic gonadotropin and carcinoembryonic antigen in patients with primary intracranial germ cell tumors. *Oncodev Biol Med* **1**, 235–240.

28. Balmaceda, C., Heller, G., Rosenblum, M., Diez, B., Villablanca, J. G., Kellie, S. *et al.* (1996) Chemotherapy without irradiation – a novel approach for newly diagnosed CNS germ cell tumors: results of an international cooperative trial. *J Clin Oncol* **14**, 2908–2915.

29. Shibamoto, Y., Takahashi, M., and Sasai, K. (1997). Prognosis of intracranial germinoma with syncytiotrophoblastic giant cells treated by radiation therapy. *Int J Radiat Oncol Biol Phys* **37**, 505–510.

30. Aoyama, H., Shirato, H., Ikeda, J., Fujieda, K., Miyasaka, K., and Sawamura, Y. (2002). Induction chemotherapy followed by low-dose involved-field radiotherapy for intracranial germ cell tumors. *J Clin Oncol* **20**, 857–865.

31. Shinoda, J., Sakai, N., Yano, H., Hattori, T., Ohkuma, A., and Sakaguchi, H. (2004). Prognostic factors and therapeutic problems of primary intracranial choriocarcinoma/germ cell tumors with high levels of HCG. *J Neuro-Oncol* **66**, 225–240.

32. Norton, L., and Simon, R. (1977). Tumor size, sensitivity to chemotherapy and the design of treatment schedules. *Cancer Treat Rep* **61**, 1307–1317.

33. Goldie, H. S., and Coldman, A. J. (1979). A mathematical model for relating drug sensitivity of tumors to their spontaneous mutation rate. *Cancer Treat Rep* **63**, 1727–1733.

34. Goldie, J. H. (1993). Neoadjuvant combined modality therapy. *In Chemoradiation: an Integrated Approach to Cancer Treatment* (M. J. John, M. S. Flam, S. S. Legha, and T. L. Phillips, Eds.), pp. 18–26. Lea & Febiger, Philadelphia.

35. Takeuchi, J., Handa, H., and Nagata, I. (1978). Suprasellar germinoma. *J Neurosurg* **49**, 41–48.

36. Oi, S., Matsuzawa, K., Choi, J. -U., Kim, D. S., Kang, J. K., and Cho, B. K. (1998). Identical characteristics of the patient populations with pineal region tumors in Japan and in Korea and therapeutic modalities. *Child's Nerv Syst* **14**, 36–40.

37. Chapman, P. H., and Linggood, R. M. (1980). The management of pineal area tumors: a recent reappraisal. *Cancer* **46**, 1253–1257.

38. Sano. K., and Matsutani, M. (1981). Pinealoma (germinoma) treated by direct surgery and postoperative irradiation. A long term follow-up. *Child's Brain* **8**, 81–97.

39. Schild, S. E., Haddock, M. G., Scheithauer, B. W., Marks, L. B., Norman, M. G., Burger, P. C. *et al.* (1996). Nongerminomatous germ cell tumors of the brain. *Int J Radiat Oncol Biol Phys* **36**, 557–563.

40. Nishizaki, T., Kajiwara, K., Adachi, N., Tsuba, M., Nakayama, H., Ohshita, N. *et al.* (2001). Detection of craniospinal dissemination of intracranial germ cell tumours based on serum and cerebrospinal fluid levels of tumour markers. *J Clinic Neurosci* **8**, 27–30.

41. Ogawa, K., Toita, T., Nakamura, K., Uno, T., Onishi, H., Itami, J. *et al.* (2003). Treatment and prognosis of patients with intracranial nongerminomatous malignant germ cell tumors: a multiinstitutional retrospective analysis of 41 patients. *Cancer* **98**, 369–376.

42. Sawamura, Y., de Tribolet, N., Ishii, N., and Abe, H. (1997). Management of primary intracranial germinomas: diagnostic surgery or radical resection? *J Neurosurg* **87**, 262–266.

43. Oi, S., Shibata, M., Tominaga, J., Honda, Y., Shinoda, M., Takei, F. *et al.* (2000). Efficacy of neuroendoscopic procedures in minimally invasive preferential management of pineal region tumors: a prospective study. *J Neurosurg* **93**, 245–253.

44. Gangemi, M., Maiuri, F., Colella, G., and Buonamassa, S. (2001). Endoscopic surgery for pineal region tumors. *Minimally Invasive Neurosurg* **44**, 70–73.

45. Jenkin, R. D. T., Simpson, W. J. K., and Keen, C. W. (1978). Pineal and suprasellar germinomas. Results of radiation treatment. *J Neurosurg* **48**, 99–107.

46. Abay, E. O., III, Laws, E. R., Jr., Grado, G. L., Bruckman, J. E., Forbes, G. S., Gomez, M. R. *et al.* (1981). Pineal tumors in children and adolescents. Treatment by CSF shunting and radiotherapy. *J Neurosurg* **55**, 889–895.

47. Rao, Y. T. R., Medini, E., Haselow, R. E., Jones, T. K., Jr, and Levitt, S. H. (1981). Pineal and ectopic pineal tumors: the role of radiation therapy. *Cancer* **48**, 708–713.

48. Sung, D. I., Hariasidis, L., and Chang, C. H. (1978). Midline pineal tumors and suprasellar germinomas: highly curable by irradiation. *Radiology* **128**, 745–751.

49. Ledigo, A., Packer, R. J., Sutton, L. N., D'Angio, G., Rorke, L. B., Bruce, D. E. *et al.* (1989). Suprasellar germinomas in childhood: a reappraisal. *Cancer* **63**, 340–344.

50. Wolden, S. L., Wara, W. M., Larson, D. A., Prados, M. D., Edwards, M. S., and Sneed P. K. (1995). Radiation therapy for primary intracranial germ-cell tumors. *Int J Radiat Oncol Biol Phys* **32**, 943–949.

51. Merchant, T. E., Sherwood, S. H., Mulhern, R. K., Rose, S. R., Thompson, S. J., Sanford, R. A. *et al.* (2000). CNS germinoma: disease control and long term functional outcome for 12 children treated with craniospinal irradiation. *Int J Radiat Oncol Biol Phys* **46**, 1171–1176.

52. Maity, A., Shu, H. K., Janss, A., Belasco, J. B., Rorke, L., Phillips, P. C. *et al.* (2004). Craniospinal radiation in the treatment of biopsy-proven intracranial germinomas: twenty-five years experience in a single center. *Int J Radiat Oncol Biol Phys* **58**, 1165–1170.

53. Hardenbergh, P. H., Golden, J., Billet, A., Scott, R. M., Shrieve, D. C., Silver, B. *et al.* (1997). Intracranial germinoma: the case for lower dose radiation therapy. *Int J Radiat Oncol Biol Phys* **39**, 419–426.

54. Shibamoto, Y., Abe, M., Yamashita, J., Takahashi, M., Hiraoka, M., Ono, K. *et al.* (1988). Treatment results of intracranial germinoma as a function of the irradiated volume. *Int J Radiat Oncol Biol Phys* **15**, 285–290.

55. Lindstadt, D., Wara, W. M., Edwards, M. S. B., Hudgins, R. J., and Sheline, G. E. (1988). Radiotherapy of primary intracranial germinomas. The case against routine craniospinal irradiation. *Int J Radiat Oncol Biol Phys* **15**, 291–297.

56. Dattoli, M. F., and Newall, J. (1990). Radiation therapy for intracranial germinoma: the case for limited volume treatment. *Int J Radiat Oncol Biol Phys* **19**, 429–433.

57. Shibamoto, Y., Sasai, K., Oya, N., and Hiraoka, M. (2001). Intracranial geminoma: radiation therapy with tumor volume-based dose selection. *Radiology* **218**, 452–456.

58. Zissiadis, Y., Dutton, S., Kieran, M., Goumnerova, J., Scott, R. M., Kooy, H. M. *et al.* (2001). Stereotactic radiotherapy for pediatric intracranial germ cell tumors. *Int J Radiat Oncol Biol Phys* **51**, 108–112.

59. Smith, D. B., Newlands, E. S., Begent, R. H., Rustin, G. J., and Bagshawe, K. D. (1991). Optimum management of pineal germ cell tumours. *Clin Oncol* **3**, 96–99.

60. de Goede, E. C., Vandertop, W. P., Struikmans, H., and ter Bruggen, J. P. (1991). The value of tumor markers in germ cell tumors. *Tijdrschr Kindergeneeskd* **59**, 85–87.

61. Nakagawa, K., Aoki, Y., Akanuma, A., Sakata, K., Karasawa, K., Terahara, A. *et al.* (1992). Radiation therapy of intracranial germ cell tumors with radiosensivity assessment. *Radiat Medic* **10**, 55–61.

62. Calaminus, G., Bamberg, M., Baranzelli, M. C., Benoit, Y., di Montezemoli, L. C., Fossati-Bellani, F. *et al.* (1994). Intracranial germ-cell tumors: a comprehensive update of the European data. *Neuropediatrics* **25**, 26–32.

63. Logothetis, C. J., Samuels, M. L., and Trindale, A. (1982). The management of brain metastases in germ cell tumors. *Cancer* **49**, 12–18.

64. Kirshner, J. J., Ginsberg, S. J., Fitzpatrick, A. V., and Comis, R. L. (1981). Treatment of primary intracranial germ cell tumor with systemic chemotherapy. *Med. Ped. Oncol* **9**, 361–365.

65. Allen, J. C., Bosl, G., and Walker, R. (1985). Chemotherapy trials in recurrent primary intracranial germ cell tumors. *J Neuro-Oncol* **3**, 147–152.

66. Allen, J. C., Kim, J. H., and Packer, R. J. (1987) Neoadjuvant chemotherapy for newly diagnosed germ cell tumors of the CNS. *J Neurosurg* **67**, 65–70.

67. Allen, J. C., DaRosso, R. C., Donahue, B., and Nirenberg, A. (1994). A phase II trial of preirradiation carboplatin in newly diagnosed germinoma of the central nervous system. *Cancer* **74**, 940–944.

68. Kobayashi, T., Yoshida, J., Ishiyama, J., Noda, S., Kito, A., and Kida, Y. (1989). Combination chemotherapy with cis-platin and etoposide for malignant intracranial germ cell tumors. An experimental and clinical study. *J Neurosurg* **70**, 678–681.

69. Herrmann, H.-D., Westphal, M., Winkler, K., Laas, R., and Schulte, F.-J. (1994). Treatment of nongerminomatous germ-cell tumors of the pineal region. *Neurosurgery* **34**, 524–529.

70. Knappe, U. J., Bentele, K., Horstmann, M., and Herrmann, H. D. (1998) Treatment and long-term outcome of pineal nongerminomatous germ cell tumors. *Pediatr. Neurosurg* **28**, 241–245.

71. Robertson, P. L., DaRosso, R. C., and Allen, J. C. (1997). Improved prognosis of malignant intracranial non-germinoma germ cell tumors with multimodality therapy. *J Neuro-Oncol* **32**, 71–80.

72. Baranzelli, M. C., Patte, C., Bouffet, E. *et al.* (1997) Nonmetastatic intracranial germinoma: the experience of the French Society of Pediatric Oncology. *Cancer* **80**, 1792–1979

73. Calaminus, G., Andreussi, L., Garrè, M.-L., Kortmann, R. D., Schober, R., and Gobel, U. (1997). Secreting germ cell tumors of the central nervous system (CNS). First results of the cooperative German/Italian pilot study (CNS sGCT). *Klin Pädiatr* **209**, 222–227.

74. Matsutani, M., and The Japanese Pediatric Brain Tumor Study Group (2001). Combined chemotherapy and radiation therapy for CNS germ cell tumors – the Japanese experience. *J Neuro-Oncol* **54**, 311–316.

75. Bouffet, E., Baranzelli, M. C., Patte, C., Portas, M., Edan, C., Chastagner, P. *et al.* (1999). Combined treatment modality for intracranial germinomas: results of a multicenter SFOR experience. Societe Francais d'Oncologie Pediatrique. *Brit J Cancer* **79**, 1199–1204.

76. Kochi, M., Itoyama, Y., Shiraishi, S., Kitamura, I., Marubayashi, T., and Ushio, Y. (2003). Successful treatment of intracranial nongerminomatous malignant germ cell tumors by administering neoadjuvant chemotherapy and radiotherapy before excision of residual tumors. *J Neurosurg* **99**, 106–114.

77. Farng, K. T., Chang, K. P., Wong, T. T., Guo, W. Y., Ho, D. M., and Hu, W. L. (1999). Pediatric intracranial germinoma treated with chemotherapy alone. *Chinese Med Journal* **62**, 859–866.

78. Baranzelli, M. C., Patte, C., Bouffet, E., Portas, M., Mechinaud-Lacroix, F., Sariban, E. *et al.* (1998). An attempt to treated pediatric intracranial alpha-FP and beta-HCG secreting germ cell tumors with chemotherapy alone. SFOP experience with 18 cases. Societe Francais d'Oncologie Pediatrique. *J Neuro-Oncol* **37**, 229–239.

79. Kellie, S. J., Boyce, H., Dunkel, I. J., Diez, B., Rosenblum, M., Brualdi, L. *et al.* (2004a). Primary chemotherapy for intracranial nongerminomatous germ cell tumors: results of the Second International CNS Germ Cell Study Group protocol. *J Clin Oncol* **22**, 846–853.

80. Kellie, S. J., Boyce, H., Dunkel, I. J., Diez, B., Rosenblum, M., Brualdi, L. *et al.* (2004b). Intensive chemotherapy and cyclo-phosphamide-based chemotherapy without radiotherapy for intracranial germinomas: failure of a primary chemotherapy approach. *Pediatr. Blood & Cancer* **43**, 126–133.

81. Weiner, H. L., Lichtenbaum, R. A., Wisoff, J. H., Snow, R. B., Souweidane, M. M., Bruce, J. N. *et al.* (2002). Delayed surgical resection of central nervous system germ cell tumors. *Neurosurgery* **50**, 727–733.

82. Tada, T., Takizawa, T., Nakazato, F., Kobayashi, K., Koike, K., Oguchi, M. *et al.* (1999). Treatment of intracranial nongerminomatous germ-cell tumor by high dose chemo-therapy and autologous stem-cell rescue. *J Neuro-Oncol* **44**, 71–76.

83. Modak, S., Gardner, S., Dunkel, I. J., Balmaceda, C., Rosenblum, M. K., Miller, D. C. *et al.* (2004). Thiotepa-based high-dose chemotherapy with autologous stem-cell rescue in patients with recurrent or progressive CNS germ cell tumors. *J Clin Oncol* **22**, 1934–1943.

84. Ono, N., Isobe, I., Uki, J., Kurihara, H., Shimuzu, T., and Kohno, K. (1994). Recurrence of primary intracranial germinomas after complete response with radiotherapy: recurrence patterns and therapy. *Neurosurgery* **35**, 615–620.
85. Shibamoto, Y., Sasai, K., Kokubo, M., and Hiraoka, M. (1999). Salvage radiation therapy for intracranial germinoma recurring after primary chemotherapy. *J Neuro-Oncol* **44**, 181–185.
86. Jennings, M. T. (1995). Neurological Complications of Radiation Therapy. *In Neurological Complications of Cancer* (R. Wiley, Ed.), pp. 219–240. Marcel Dekker Inc., New York.
87. Kitamura, K., Shirato, H., Sawamura, Y., Suzuki, K., Ikeda, J., and Miyasaka, K. (1999). Preirradiation evaluation and technical assessment of involved-field radiotherapy using computed tomographic (CT) simulation and neoadjuvant chemotherapy for intracranial germinoma. *Int J Radiat. Oncol Biol Phys* **43**, 783–788.
88. Oka, H., Kawano, N., Tanaka, T., Utsuki, S., Kobayashi, I., Maezawa, H. *et al.* (1998). Long-term functional outcome of suprasellar germinomas: usefulness and limitations of radiotherapy. *J Neuro-Oncol* **40**, 185–190.
89. Kumanogoha A., Kasayama, S., Kouhara, H., Koga, M., Arita, N., Hayakawa, T. *et al.* (1994) Effects of therapy on anterior pituitary functions in patients with primary intracranial germ cell tumors. *Endocrine J* **41**, 287–292.
90. Sands, S. A., Kellie, S. J., Davidow, A. L., Diez, B., Villablanca, J., Weiner, H. L. *et al.* (2001). Long-term quality of life and neuropsychologic functioning for patients with CNS germ cell tumors: from the First International CNS Germ-Cell Tumor Study. *Neuro-Oncology* **3**, 174–183.
91. Lutterbach, J., Spetzger, U., Bartelt, S., and Pagenstecher, A. (2002). Malignant germ cell tumors metastatic to the brain: a model for a curable neoplasm? The Freiburg experience and a review of the literature. *J Neuro-Oncol* **58**, 147–156.
92. Balmaceda, C., Modak, S., and Finlay, J. (1998). Central nervous system germ cell tumors. *Semin. Oncol* **25**, 243–250.

Chemotherapy of Meningiomas

Herbert B. Newton

ABSTRACT: Meningiomas are slow growing, extra-axial tumors that derive from the arachnoidal cap cells of the meninges. They comprise 18–20 per cent of intracranial tumors and have an overall incidence of 2.6 per 100 000. Common presenting signs and symptoms include headache, hemiparesis, visual loss, seizures, gait difficulty, and confusion. Meningiomas are clearly delineated with enhanced computed tomography and magnetic resonance imaging scans. The main form of treatment for meningiomas is surgical resection, which can be curative in some cases. External beam radiotherapy and radiosurgery can be of benefit in selected patients with recurrent or progressive disease. The role of chemotherapy continues to be defined, but should be considered for patients with inoperable or frequently recurrent meningiomas. Although numerous agents have been studied, the majority have had minimal efficacy. The most active agents appear to be interferon-alpha, the combination regimen of cyclophosphamide, adriamycin, and vincristine, and single-agent hydroxyurea.

INTRODUCTION AND EPIDEMIOLOGY

Meningiomas are usually slow growing, benign tumors of extra-axial origin, that arise from the arachnoidal cap cells associated with the arachnoidal villi at the dural venous sinuses, cranial nerve foramina, cribiform plate, and medial middle fossa [1–3]. Meningiomas are classified by their site of origin within the nervous system, which is most commonly the intracranial cavity. In adults, 85–90 per cent of tumors occur supratentorially, with 30–40 per cent arising along the base of the anterior and middle fossae (see Table 34.1) [4]. The most common sites are the parasagittal or falcine region (25 per cent), convexity (19 per cent), sphenoid ridge (17 per cent), and suprasellar area (10 per cent). In children, meningiomas occur more commonly within the posterior fossa and ventricular system.

Meningiomas account for 18–20 per cent of all intracranial tumors in most series [2,5–7]. The frequency appears to be similar between series in North America, Europe, and Japan. However, studies of African populations suggest an elevated frequency (mean 30.1 per cent; range 24–38 per cent) [2,8]. The incidence rates vary across different studies from 0.08 to 13.72 per 100 000, with an overall incidence of 2.6 per 100 000 [6]. Incidence rates may be higher in African Americans in comparison to Caucasian Americans (3.1 per 100 000 *versus* 2.3 per 100 000). Most studies support an increased incidence in females, with male to female ratios ranging from 1:1.4 to 1:2.8. However, male to female ratios may be more equal in African and African–American populations. The incidence of intracranial meningioma increases with increasing age, peaking in the seventh decade for males (6.0 per 100 000) and the eighth decade for females (9.5 per 100 000) [4].

For the majority of patients in whom a meningioma develops, the etiology remains unclear [1–3,6,7]. Potential etiologic factors that have been investigated include radiation exposure, cranial trauma, viruses, hormonal stimulation, and molecular genetic events. The most consistently documented etiological factor appears to be prior cranial radiation exposure. Meningiomas have been shown to be induced by low- and high-dose irradiation [6,7,9,10]. To meet the criteria for a radiation-induced tumor, the mass must occur within the irradiated field, develop after a period of latency following irradiation (e.g., 20–30 years), and be histologically different from any pre-existing neoplasm in the region. Children treated with

TABLE 34.1 Common Sites and Incidence of
Intracranial Meningiomas

Site	Incidence (%)
Parasagittal	25
Convexity	19
Sphenoid ridge	17
Suprasellar	10
Posterior fossa	8
Olfactory groove	8
Middle fossa/Meckel's cave	4–5
Tentorial	3
Pertorcular	3
Lateral ventricle	2
Foramen magnum	1–2
Orbit/optic nerve sheath	1–2

Adapted from [1–4].

low-dose irradiation (i.e., 250–1000 cGy) for tinea capitus had their risk for developing a meningioma increased by a factor of 9.5. Meningiomas were the most common form of brain tumor noted in these irradiated children. Similar results have been noted in patients after high-dose radiotherapy (i.e., 5500–7500 cGy) for primary head and neck cancer. In general, the higher the radiation dose, the shorter the interval of latency before the tumor develops. For low-dose irradiation, the interval is typically 30 years or more, while for high-dose irradiation, it is usually around 20 years.

Other etiological factors, such as cranial trauma, viruses, and hormonal stimulation are not as strongly associated with the development of meningiomas [1–3,6,7]. Initial studies by Cushing and others in the 1930s suggested an association between cranial trauma and meningiomas. More recent studies do not support this contention, including a prospective review of 2953 patients with head injuries by Annegers and colleagues [11]. They did not find an increased incidence of meningiomas in this cohort in comparison to the general population. In addition, of those head injury patients that subsequently developed a brain tumor, the location of the tumor did not correlate with the area of cranial trauma. Several types of DNA and RNA viruses (e.g., simian virus 40; SV40) are capable of inducing transformation and brain tumor growth in cultured cells and animal models [1–3,12,13]. However, no virus is capable of de novo formation of a typical dural-based meningioma in rodent models. Earlier studies using DNA hybridization techniques demonstrated BK, SV40, and

adenoviral DNA within meningiomas, but the viral DNA material was not always integrated into the tumor DNA. A more recent analysis by Weggen and co-workers, using polymerase chain reaction assays, only found evidence for SV40 DNA sequences in 1 of 131 meningiomas, suggesting that DNA viruses do not play an important role in the pathogenesis of these tumors [14]. The preponderance of meningiomas in women and the tendency for these tumors to enlarge during pregnancy have led to speculation that meningiomas are stimulated by hormonal influences [1–3,6,7]. Studies using modern molecular techniques have verified that the majority of sporadic meningiomas (85–90 per cent) express high concentrations of progesterone receptors, with minimal expression of estrogen receptors [15–18]. The high expression of progesterone receptors has been documented in meningiomas from men and children, and not just in sexually mature women. Progesterone has been shown to stimulate some meningioma cell lines, which can then be reversed by the use of progesterone receptor antagonists. The expression of progesterone receptors is inversely proportional to the degree of tumor proliferation and histological grade, so that the highest concentrations are found in lower-grade, more benign tumors. Further studies of meningioma patients and age-matched controls have not found any correlation between the risk for tumor development and age at the time of first delivery or parity status [19].

BIOLOGY AND MOLECULAR GENETICS

Significant progress has been made in the understanding of the biology and molecular genetics of meningiomas [1–3,20–22]. The tumors are thought to arise from malignant transformation of arachnoidal cap cells, which are meningothelial cells with both mesenchymal (e.g., spindle cell morphology, collagenous stroma) and epithelial (e.g., numerous intercellular junctions, expression of epithelial membrane antigen) features. Arachnoidal cap cells form the outer layer of the arachnoid mater and arachnoid villi, and have a diverse group of proposed functions. The most important functions include the formation of anatomic barriers and ensheathing of other cells and vessels, formation of conduits for CSF drainage into venous sinuses and veins, production of collagen and stromal proteins, secretion of CSF proteins and glioneuronal differentiation and proliferation factors, participation in foreign-body reactions, trophic support for glial and neuronal cells, and participation in reactive and reparative processes within the meninges [20,22]. In addition, arachnoidal cap cells perform other important duties such as intercellular communications *via* desmosomes, emperipolesis (i.e., lymphoplasmacytic engulfment), and HLA-DR expression.

Due to reports of familial clustering of meningiomas and the increased incidence of meningiomas in patients with neurofibromatosis type 2 (NF2), many groups have begun to investigate the cytogenetic and molecular biological background of these tumors [1–3,20–22]. Recent reports would suggest that non-NF2-associated familial clustering of meningioma is rare, but does occur [23,24]. Several histological varieties of tumor have now been described that lack any association with NF2 or other genetic diseases. For patients with NF2, meningiomas are the second most common tumor, noted in approximately 50 per cent of the cases. Although children with meningiomas are generally uncommon, almost 40 per cent will have NF2. NF2-associated meningiomas differ from sporadic tumors in several ways, including a tendency to arise several decades earlier in life, to be multiple in most patients, and to belong to the fibroblastic variant. Meningiomas are also known to occur in other genetically mediated diseases, including Turcot's syndrome, Cowden's syndrome, Li-Fraumeni syndrome, Gorlin's nevoid basal cell syndrome, and von Hippel-Lindau disease.

Early cytogenetic studies of meningiomas documented numerous genetic alterations, including monosomy of chromosome 22 in up to 70 per cent of cases, as well as deletions of chromosomes 1p, 6q, 9p, 10q, 14q, and 18q [20–22,25,26]. Subsequent molecular studies have noted loss of heterozygosity on chromosome 22q as the most common genetic alteration, present in 40–70 per cent of all meningiomas. Further research has suggested that the NF2 gene is the usual target within 22q, with bi-allelic mutation or deletion of the gene occurring as an early event in the transformation process in 50–60 per cent of sporadic meningiomas and all NF2-associated meningiomas (see Fig. 34.1). The NF2 gene encodes the protein, merlin (also called schwannomin), which has an open reading frame of 595 amino acids, and is a member of the Protein 4.1 family. The 4.1 family are a group of structural proteins (including the ERM proteins ezrin, radixin, and moesin) that link the cytoskeleton to proteins of the cytoplasmic membrane. Merlin appears to play a role in the regulation of cell growth and motility through interaction with numerous proteins, including paxillin, β1-integrin, CD44, hepatocyte growth factor-regulated tyrosine kinase substrate, βII-spectrin, schwannomin interacting protein-1, and other ERM proteins [22]. Fibroblasts and keratinocytes that are deficient in NF2 exhibit increased proliferation and accelerated cellular movement *in vitro*. Genetically engineered mice (i.e., NF2 +/−) with reduced expression of NF2 have increased cell growth and develop a variety of invasive and highly metastatic tumors, including fibrosarcoma, adenocarcinoma, hepatocellular carcinoma, and osteosarcoma [27]. Inactivation of NF2 in leptomeningeal cells can lead to meningioma formation in mice [28]. If wild type merlin is re-expressed into

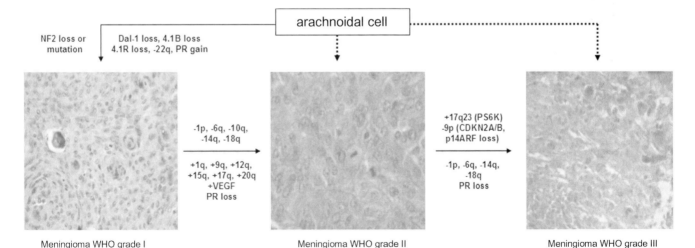

FIGURE 34.1 Overview of the current molecular model of the stepwise pathogenesis from a normal arachnoidal cell to meningioma grades I, II, and III (solid lines and arrows). In some cases, there may be a more direct pathway from a precursor cell to a grade II or III tumor (dotted lines and arrows). Genetic alterations that may be involved in each step are listed and explained in the text.
Abbreviations: PR—progesterone receptors, WHO—World Health Organization, VEGF—vascular endothelial growth factor. Adapted from references [1,2,20–22]. See Plate 34.1 in Color Plate Section.

tumor cell lines, the cells have reduced growth and motility *in vitro* and *in vivo* [29].

The majority of meningiomas have absent or reduced immunoreactivity to merlin, which correlates strongly with loss of heterozygosity of 22q [20–22, 30–32]. This is consistent with the mutational spectrum of NF2 in meningiomas, since most mutations cause a truncation of the gene, with reduced or absent merlin expression. The frequency of NF2 mutations is variable among different histological varieties of meningioma [31]. It is most common in the fibroblastic and transitional forms, noted in 70–80 per cent of cases. In contrast, the mutation is only present in roughly 25 per cent of the cases with the meningothelial form. Immunohistochemical results are consistent with the mutational data, showing a lack of staining for merlin in fibroblastic and transitional meningiomas, but rarely in meningothelial tumors. There are some tumors with reduced or absent expression of NF2 that do not harbor mutations within the gene [32]. In these cases, it is possible that other mechanisms may be involved, such as homozygous deletions, methylation, or undetected NF2 mutations. The paucity of NF2 mutations in the meningothelial variant suggests that other transformation pathways are more important. NF2 mutations are present in sporadic patients with multiple meningiomas, but in tumors from familial clusters that have multiple tumors [33].

In addition to the NF2 gene, other genes have been implicated on chromosome 22, as well as on several other chromosomes [20–22]. Several candidate tumor suppressor genes located on 22q (e.g., ADTB1, RRP22, hSNF5/InI1, CLH-22) have been evaluated and found to have reduced expression in some cases. However, mutations within these genes have been infrequent in tumors with abnormal expression. Further research is needed to determine the role of these genes in the transformation of meningiomas. DAL-1 is a protein belonging to the 4.1 Family of membrane-associated proteins that has significant sequence homology with merlin [18]. It is a putative tumor suppressor gene and maps to chromosome 18p11.3. DAL-1 expression is reduced or absent in 76 per cent of sporadic meningiomas, which is similar to merlin. Loss of DAL-1 expression also appears to be an early event in the pathogenesis of meningiomas, since the expression pattern was only slightly different between benign and atypical tumors (70–76 per cent) *versus* anaplastic tumors (87 per cent). Loss of heterozygosity of chromosomes 1, 14, and 10 are common in meningiomas and tend to correlate with severity of grade [20–22,34]. Several putative tumor suppressor genes have been evaluated on chromosomes 1 (e.g., CDKN2C, RAD54L, ALPL, TP73) and 10 (e.g., PTEN,

DMBT1) for evidence of mutation, abnormal expression, or methylation. Thus far, none of these genes appears to play a significant role as a tumor suppressor in the pathogenesis of meningioma. In addition, no specific meningioma suppressor genes have been identified on chromosome 14. Chromosome 9 has been studied intensely due to the presence of several genes that are important for the regulation of the cell cycle and p53 (i.e., CDKN2A, p14ARF, CDKN2B) [35]. Recent reports suggest that the majority of malignant meningiomas (55–70 per cent) have mutations or expression abnormalities related to all three genes; these findings are significantly less frequent in benign meningiomas.

The expression of growth factors and their receptors have also been studied in meningiomas [20–22]. Several groups have provided evidence for paracrine and autocrine loops related to the activity of epidermal growth factor (EGF), platelet-derived growth factor (PDGF), and their receptors (EGFR, PDGFR) (see Fig. 34.1) [36,37]. Another growth factor related to angiogenesis and vascular re-modeling, vascular endothelial growth factor (VEGF), is also highly expressed in meningiomas [20–22, 38]. The expression levels of VEGF appear to be related to the degree of edema associated with the tumor and, to a lesser extent, to tumor grade. However, VEGF expression levels do not correlate well with the vascular density of meningiomas.

CLINICAL PRESENTATION

The spectrum of signs and symptoms in meningioma patients are quite variable and will depend on the location of the tumor within the intracranial cavity, and the proximity to other neural structures [2,4]. Symptoms referable to chronic increased intracranial pressure (e.g., papilledema, sixth nerve palsy) can occur, but only with the largest tumors. In general, meningiomas are slow growing, so there is a gradual onset and worsening of symptoms. Overall, the most common symptoms are headache and hemiparesis, which occur in 36 and 30 per cent of patients, respectively [4]. Other frequent symptoms include seizures, gait difficulty, visual abnormalities, confusion, memory deficits, and personality changes. However, it is the combination of symptoms and specific deficits on the neurological examination that allow for localization of the tumor.

Meningiomas of the olfactory groove present with symptoms once they have reached a moderately large size [1,2,4]. The most common symptoms include alterations of mental status and personality that may appear similar to depression, impaired insight and judgment, lack of motivation, and headache. Patients rarely complain of olfactory dysfunction. Tumors of the planum sphenoidale usually present with headaches, seizure activity, alterations of mental status, and visual impairment if the optic chiasm is involved. If the tumor arises from the tuberculum sellae, the main symptom is progressive asymmetric visual loss that may resemble a pituitary adenoma (i.e., bitemporal defect). Meningiomas of the sphenoid wing can develop with a medial or lateral orientation. Medial sphenoid tumors present with slowly progressive unilateral visual loss, headaches, and seizure activity. More laterally placed tumors are also characterized by headaches and seizures, with the addition of proptosis and fullness in the temporal region. Cavernous sinus meningiomas develop double vision, facial numbness, and headaches. Seizures are uncommon unless the tumor is large enough to extend into the middle fossa and compress the temporal lobe. Tumors of the parasagittal region typically present with headaches and focal motor or sensory seizures. Focal motor deficits and reflex asymmetries are also possible, depending on tumor size and the amount of compression of underlying motor cortex. Meningiomas of the convexity are often asymptomatic as they grow over the frontal, temporal, parietal, and occipital lobes. If the tumor becomes large enough, patients can develop headaches, seizures, and focal neurological deficits (e.g., hemiparesis, dysphasia, visual loss). Tentorial meningiomas present with headache, limb and/or gait ataxia, nausea and emesis, and visual disturbances; seizures are uncommon. The neurological examination often reveals papilledema, dysfunction of cranial nerves IV, V, and VI, and extremity ataxia. Tumors of the petroclival region slowly compress the anterior pons and brainstem, and present with headache, gait disturbance, diplopia, vertigo, and reduced hearing. Findings on neurological examination include palsy of cranial nerves IV, V, VI, and VIII, as well as papilledema, ataxia, and dysmetria in larger tumors. Meningiomas of the foramen magnum describe slowly progressive suboccipital and neck pain that is exacerbated with flexion and Valsalva's maneuver. Motor and sensory deficits can develop, including full-blown myelopathy, in patients with large tumors that compress the spinal cord.

NEURO-IMAGING

Plain radiographs and angiographic evaluation of meningiomas have mainly been supplanted by computerized axial tomography (CT), magnetic resonance imaging (MRI), and MR angiography (MRA) [1,2,39]. Plain radiographs can show calcified tumors and regions of hyperostotic bone. Angiography may be helpful for the evaluation of the safety and feasibility of preoperative embolization of highly vascular meningiomas [40]. Angiography is able to demonstrate the main arterial blood supply and also clearly delineates the extent of collateral branches. CT is capable of detecting the majority of meninigomas, especially at wide window and level settings that optimally visualize bone involvement [39,41]. Hyperostosis and lytic regions of bone are well delineated by CT. On nonenhanced CT, most meningiomas appear isodense to slightly hyperdense compared to brain. The tumors are usually homogeneous in density and have a variable amount of calcification (i.e., punctate or confluent). Tumors with significant calcification are more likely to remain stable at the time of subsequent follow-up. After the administration of contrast, the majority of meningiomas demonstrate intense enhancement with sharply delineated margins. Peritumoral edema appears as a region of hypodensity around the enhancing tumor mass. Aggressive tumors may be distinguished by the presence of indistinct or irregular margins, mushroom-like projections from the main tumor mass, and invasion of the underlying brain. In general, tumors with extensive peritumoral edema and brain invasion are more likely to recur after treatment.

MRI is a very sensitive technique for the detection of intracranial meningiomas, especially with the application of its multiplanar capabilities [1,2,39]. In general, it is considered to be superior to CT in this capacity, except for the evaluation of tumor calcification and involvement of bone. On T1 images, 60–90 per cent of meningiomas are isointense in comparison to gray matter, while 10–30 per cent are mildly hypointense. On T2 images, 30–45 per cent of meningiomas will have increased signal intensity, while approximately 50 per cent will be isointense, in comparison to brain. Tumor margins and vascularity, as well as vascular distortion and encasement, are clearly delineated. In some cases, MRI may be able to differentiate between the histological sub-types of meningioma [42]. On T2 images, meningiothelial and angioblastic variants are sometimes noted to have persistently higher signal intensity than fibroblastic and transitional tumors. In addition, cerebral edema is typically more prominent in surrounding brain with the meningothelial and angioblastic sub-types. Contrast-enhanced MRI is the most sensitive method for detecting a meningioma. Virtually, all tumors enhance densely and homogeneously, often with the presence of a small "dural tail". Postoperative enhanced MRI is also very sensitive and specific for detection of residual or recurrent tumor.

INITIAL EVALUATION AND TREATMENT

Once a meningioma has been discovered by CT or MRI, its location must be correlated with the patient's symptoms to determine if the tumor is incidental or causally related. If the symptoms do not match the tumor location, and the tumor appears to be benign by imaging criteria, then a period of observation (i.e., serial MRI every 6–12 months) is appropriate [1,2,43,44]. This strategy is reasonable, since it is well known that a certain percentage of meningiomas (approximately 35–60 per cent) will spontaneously stop growing and remain dormant for various lengths of time [45,46]. Once tumor growth has been demonstrated by CT or MRI, more definitive intervention can be instituted. Large vascular tumors should be screened for endovascular embolization, which can reduce tumor vascularity and surgical bleeding, and may soften tumors, thereby improving the ease of surgical removal [40,44,47].

Surgical resection is the mainstay of treatment for intracranial meningiomas [1,2,43,44]. Complete surgical resection of the tumor, associated dural margins, and any involved bone is the goal for all patients. However, this is not possible for many tumors due to location (e.g., cavernous sinus, medial sphenoid wing) or involvement with delicate neurovascular structures (e.g., carotid artery). In addition, it is usually difficult to perform a complete resection of atypical and malignant meningiomas, due to extensive infiltration along the dura and invasion of underlying cortex. The completeness of resection is important because it impacts directly on the risk of tumor recurrence. For patients with a gross total resection, the 5-, 10-, and 15-year local control rates were 93, 80, and 76 per cent, respectively [48]. In contrast, after a subtotal resection, patients were much more likely to develop recurrent tumor, with 5-, 10-, and 15-year local control rates of 53, 40, and 30 per cent, respectively. Survival is also impacted by the degree of tumor resection. In a review of 581 patients with meningioma that had undergone surgery at the Mayo Clinic, the 5- and 10-year progression-free survival rates following a gross-total resection were 88 and 75 per cent, respectively [49]. For patients, who had received a subtotal resection, the 5- and 10-year

progression-free survival rates were only 61 and 39 per cent, respectively.

External beam radiotherapy should be considered for selected patients with meningioma [1,2,43,44]. It is a viable option for patients who cannot undergo surgery because of tumor location and medical complications, as well as for those that refuse the procedure. Irradiation is not indicated after gross total resection of histologically benign meningiomas, but may be of benefit for subtotally resected tumors or those with atypical or malignant features [49,50]. In addition, radiotherapy may be active against tumors at the time of recurrence or progression. The use of irradiation has been found to be a significant factor associated with improved survival in an extensive review of over 9500 patients by McCarthy and co-workers [50]. Longer survival was noted for patients with benign ($p < 0.0001$) and malignant tumors ($p < 0.001$). In a separate study by Goldsmith and colleagues, the use of irradiation for benign meningiomas after subtotal resection resulted in 5- and 10-year progression-free survival rates of 89 and 77 per cent, respectively [51]. Effective doses are in the range of 4500–6000 cGy (pre-surgical tumor volume plus 1–2 cm margin) for benign tumors, administered in 180–200 cGy daily fractions, over five to six weeks. For malignant meningiomas, most authors recommend a more aggressive treatment plan, with doses ranging from 6000 to 6500 cGy with a 3–4 cm margin [50].

Stereotactic radiosurgery is a newer radiation modality that has been applied to meningiomas in recent years [1,2,43,44]. Radiosurgery can be administered by linear accelerator or gamma knife, into a well-defined intracranial volume (typically less than 3.5 cm in diameter) that contains the tumor, with minimal exposure to normal brain. The median effective dose in most series is 15 Gy, with a range of 13–20 Gy. Radiosurgery has been applied most often to tumors that cannot be resected due to patient-related clinical factors (e.g., elderly, medical complications), tumor location (e.g., cavernous sinus, petroclival), or proximity to eloquent regions of brain (e.g., parasagittal, convexity). Some authors have also recommended radiosurgery after incomplete resection of benign and malignant meningiomas or, in selected cases, as an alternative to surgery [52–54]. In a recent review of 190 consecutive meningioma patients treated with radiosurgery, Stafford and colleagues noted an overall 5-year local control rate of 89 per cent [52]. The local control rate varied significantly ($p < 0.0001$) by histology, with benign, atypical, and malignant tumors having rates of 93, 68, and 0 per cent, respectively. The overall 5-year survival rate for the cohort was 82 per cent, with cause-specific rates for benign, atypical, and malignant meningiomas of 100, 76, and 0 per cent, respectively.

CHEMOTHERAPY OF MENINGIOMAS

The role of adjuvant chemotherapy in meningioma patients remains unclear and continues to evolve [1,2,43,44,55]. It has mainly been applied to inoperable patients, especially in the setting of progression or recurrence after some form of radiotherapy. Numerous approaches have been taken, including traditional cytotoxic drugs, molecular agents, immunomodulators, and hormone manipulating drugs. Although none of these drugs have been particularly effective, some have been able to demonstrate modest activity in sub-groups of patients.

Several investigators have reported the results of cytotoxic and immune modulating chemotherapy approaches in meningioma patients [1,2,43,44,55]. Initial reviews of treatment responses in older sets of patients did not suggest any benefit from chemotherapy [56]. In a review of 25 patients with malignant meningioma treated at MD Anderson from 1944 through 1992, Younis and associates noted 10 that had received cytotoxic chemotherapy. The regimens consisted of intra-arterial or intravenous cisplatin, intravenous dacarbazine, and intravenous doxorubicin. None of the patients had an objective response or clinical improvement that could be attributed to chemotherapy. These results were in contrast to earlier descriptions of malignant meningioma patients that had had modest responses to adriamycin [57]. Bernstein and colleagues reported a patient with invasive rectal cancer who was receiving chemotherapy with 5-fluorouracil (425 mg/m^2/day × 5 days), folinic acid (20 mg/m^2/day × 5 days), and levamisole (50 mg q8h × 3 days) and was noted to have a large falcine meningioma [58]. After one cycle of chemotherapy the patient developed headache, somnolence, and left-sided weakness. Follow-up MRI revealed a new low intensity region within the tumor. After tumor resection, pathological examination revealed a large area of necrosis that matched the low signal portion of the MRI. Groves and co-workers reported minimal activity using cisplatin, dacarbazine, doxorubicin, and interferon-alpha (IFN-α), either alone or in combination [59]. Only one patient treated with IFN-α had an objective response. This is consistent with a case report by Wober-Bingol and colleagues, which also described a meningioma patient that had had a response to IFN-α [60]. In contrast to the experience of

Groves, other authors have reported activity with cisplatin-based regimens [61]. In two patients with recurrent, inoperable malignant meningiomas, intra-arterial cisplatin (60 mg/m^2) and intravenous doxorubicin (75 mg/m^2) was administered. One patient experienced a partial response (PR) that was durable for five years. The other patient had stabilization of disease that was maintained for almost two years. Chamberlain used multiagent chemotherapy for a series of fourteen patients with malignant meningioma [62]. The regimen consisted of three to six cycles of intravenous cyclophosphamide (500 mg/m^2/day × 3 days), adriamycin (15 mg/m^2/day × 3 days), and vincristine (1.4 mg/m^2 × 1 day). There were 3 patients with PR and 11 with stable disease (SD). The median time to tumor progression was 4.6 years, with a median survival of 5.3 years. The regimen was relatively toxic; four patients required dosage reductions and two had neutropenic fever. Based on the two responses with IFN-α noted above, a more extensive phase II trial was performed in six patients with recurrent unresectable and malignant meningiomas [63]. The regimen consisted of IFN-α-2B administered subcutaneously at a dosage of 4 mU/m^2/day × 5 days per week. Five of the six patients responded to treatment, with 1 minor response (MR) and four SD. The mean time to progression was 8.3+ months, with a range of 6+ to 14+ months. Interferon-α-2B was generally well tolerated, with frequent mild flu-like symptoms and occasional leukopenia.

Reports in the literature have described an increased incidence of meningiomas in patients with acromegaly, suggesting an association between growth hormone, insulin-like growth factor-1 (IGF-1), and meningioma development [64–66]. Initial studies demonstrated the presence of IGF-1 receptors in roughly 75 per cent of meningioma specimens, while more recent data suggests that growth hormone receptors are almost universally expressed in meningiomas of all grades [65,66]. *In vitro* studies have shown that meningioma cells can be stimulated to increase DNA synthesis and replicate after exposure to IGF-1 and growth hormone. Treatment of cultures with B2036 (pegvisomant), a competitive growth hormone receptor antagonist, resulted in a 20 per cent reduction in growth rates [65]. Further studies by McCutcheon and co-workers have tested the activity of pegvisomant in an *in vivo* model of meningioma [66]. The model consisted of primary explant cultures from resected meningioma specimens xenografted into the flank of athymic mice. Animals received pegvisomant (315 mg/kg/week) or saline control for an 8-week treatment period. At the onset of treatment, the tumor volumes were equivalent between the experimental and control animals (291 *versus* 284 mm^3). After the treatment interval, there was a significant difference in tumor volume between the pegvisomant-treated and control animals (198.3 *versus* 350.1 mm^3; $p < 0.001$). Pegvisomant has not yet undergone human testing and is under consideration for Phase I clinical trials.

As mentioned above, estrogen and progesterone receptor expression and surface immunoreactivity have been demonstrated in meningiomas, and may mediate the mitogenic effect of circulating hormones [1,2,15–18,20–22]. Based on these observations, hormone receptor antagonists and modulators have been developed as therapeutic approaches for these tumors. Several groups have investigated the activity of tamoxifen as an anti-estrogenic treatment for inoperable patients. Markwalder and colleagues treated six patients with tamoxifen over an 8- to 12-month period [67]. The 6-year progression-free survival and overall survival rates were 43 and 72 per cent, which are similar to the rates for resection alone. In a larger study by the Southwest Oncology Group, Goodwin and associates treated 19 evaluable patients with nonresectable, refractory meningiomas with tamoxifen (40 mg/m^2 twice daily $\times$ 4 days, then 10 mg twice daily) [68]. One patient achieved a PR that was maintained for 38+ months, while two other patients had brief MR (4 and 20 months). Six patients had SD that lasted for 31+ months and 10 patients had progression. The overall median time to progression was 15.1 months. A recent case report describes a patient with a presumed meningioma (MRI criteria only) that developed an objective response to mepitiostane, an antiestrogenic agent similar to tamoxifen [69]. After two years of chronic treatment with mepitiostane, follow-up CT demonstrated a robust PR, with approximately 73 per cent shrinkage of the tumor. Other authors have evaluated the activity of medroxyrogesterone acetate (MPA) against meningiomas [70–72]. MPA is a semisynthetic progestin that acts as a competitive agonist for progesterone receptors. It has been shown to reduce progesterone receptor activity in meningioma cytosolic preparations. Further *in vitro* testing in a large series of meningioma monolayer tissue cultures demonstrated that MPA was able to inhibit or delay growth in 35 per cent of the cultures [71]. The progesterone receptor content of the parent tumor did not correlate well with the response to MPA in culture. Jaaskelainen and co-workers administered MPA (1000 mg intramuscular weekly) to five postmenopausal female patients with meningiomas [70]. Four patients with low-grade tumors stabilized with treatment, while a patient with an anaplastic meningioma continued to grow rapidly. In another small study by Grunberg and Weiss, nine patients with unresectable meningiomas received a different progestational agent, megestrol acetate (Megace; 40–80 mg qid) [72]. No objective responses or durable stabilizations were noted. The drug was not well tolerated, with three patients noting visual deterioration within 3 months of onset of treatment and several other patients having significant weight gain.

An alternative approach has been to use antiprogestins, in an attempt to block potential mitogenic stimulation mediated *via* progesterone receptors [1,2,15–18,20–22]. The most well-studied drug in this class is RU 486 (mifepristone), an 11 beta substituted derivative of the progestin norethindrone, which has a high affinity for progesterone and glucocorticoid receptors [73]. Several *in vitro* and animal studies suggest that RU 486 can inhibit the growth of meningioma cells. Olson and colleagues tested three meningiomas in a cell culture assay and demonstrated 18–36 per cent growth inhibition after treatment with RU 486 [74]. The same group then tested RU 486 in a meningioma nude mouse model, and noted objective shrinkage of implanted tumor in two of three mice with implanted tumors [75]. In another *in vitro* study, Blankenstein and co-workers used RU 486 to treat explanted cultures from 13 resected meningioma specimens [76]. They were able to demonstrate a significant decrease in thymidine labeling index with increasing concentrations of RU 486. Based on these pre-clinical results, Grunberg and associates initiated a small study of RU 486 in patients with unresectable meningiomas [77]. Fourteen patients were placed on long-term therapy with RU 486 (200 mg/day), ranging from 2 to 31+ months. There were five patients with MR, including one with a sphenoid wing tumor that had improvement in extraocular muscle function. Another patient had a stable tumor but also noted improvement in extraocular muscle function. Of the remaining patients, five had SD that ranged from 12 to 27+ months, including one with an improvement in headache, while three had tumor progression. A similar study by Lamberts and colleagues reported the results of ten patients with recurrent or inoperable meningiomas who were each treated with RU 486 (200 mg/day) over a 12-month period [78]. Three patients had transient MR, while another three had stabilization of disease. Because the results of these early studies suggested that RU 486 was active against meningioma, a double-blind, randomized placebo-controlled trial was organized by Grunberg in cooperation with the Southwest Oncology Group and the Eastern Cooperative Oncology Group [79]. A total of 160 evaluable patients were enrolled and received RU 486 (200 mg/day) or placebo over a two year observation period. The median time to progression between the RU 486 and placebo arms of the study were similar (10 *versus* 12 months, respectively; $p = 0.44$). These results suggested that RU 486 had minimal activity against meningiomas when administered at this dosage on a daily basis.

A more recent approach to meningioma therapy was initially espoused by Schrell and colleagues, and involved the use of hydroxyurea [80–82]. Hydroxyurea is known to inhibit ribonucleotide reductase and is commonly used against chronic myelogenous leukemia [83]. Schrell established early-passage explant cell lines from 20 surgically removed meningiomas and then treated the cultures with 5×10^{-4} to 5×10^{-3} M hydroxyurea over 5 to 9 days [80,81]. In comparison to control cultures, there was a significant reduction in cell proliferation and growth. DNA flow cytometry revealed cell cycle block in the S phase of many of the inhibited cells. In addition, electron micrographic and light microscopic analysis of inhibited cells revealed DNA fragmentation and the "DNA ladder effect", consistent with the induction of apoptosis. Apoptosis was also noted using an *in vivo* model of meningioma fragments transplanted into nude mice, followed by treatment with hydroxyurea (0.5 mg/g) over 15 days. These results suggested that hydroxyurea was a potent inhibitor of meningioma cell growth, mainly through blockade of the cell cycle in S phase and the induction of apoptosis. Hydroxyurea (20 mg/kg/day) was then administered to four patients with unresectable and recurrent meningiomas [82]. Two patients with grade I meningiomas had extensive PR (i.e., 70–75 per cent shrinkage over a 10–13 month follow-up), while a third patient with a low-grade tumor had minimal regression. The fourth patient with a grade III tumor remained free of recurrent disease while on hydroxyurea, following a sixth surgical resection. Based on these early promising results, Newton and associates reported the use of hydroxyurea (20 mg/kg/day) in a larger cohort of 16 evaluable patients with unresectable or residual meningioma [84]. Eleven patients had actively growing tumors or neurological progression at the onset of treatment. Fourteen of the sixteen patients (88 per cent) had stabilization of disease that was maintained for a median of 80 weeks. Three of the initial responders progressed after 20, 36, and 56 weeks of treatment. In contrast to the Schrell data, there were no PR or MR in this cohort. Toxicity was mainly hematological, with leukopenia being the most common manifestation. Nine patients (53 per cent) required minor dosage reductions

(250–500 mg/day) secondary to hematological side effects. A similar study by Mason and co-workers evaluated the activity of hydroxyurea (20 mg/kg/day over a two-year treatment period) on a group of twenty patients with recurrent or unresectable meningiomas, including four with atypical or malignant tumors [85]. Tumor enlargement was documented in all patients before the onset of chemotherapy. In the sub-group of patients with low-grade meningiomas, there was 1 MR, 12 SD (median length of treatment 122 weeks), and 3 with disease progression (after 41, 55, and 66 weeks of treatment). The 1-year freedom from progression rate was 93 per cent for patients with benign tumors. In the sub-group with atypical and malignant meningiomas, disease progression was relatively rapid. In a more recent report, a patient was described with painless, right-sided visual loss who was noted to have an enlarging meningioma of the optic nerve [86]. Hydroxyurea was started (20 mg/kg/day) as the initial form of therapy, before irradiation or resection. After seven months of treatment, the patient reported subjective improvement in vision. In addition, there was also improvement in formal visual field testing and normalization of visual-evoked potentials. The tumor remained stable by MRI and the patient was clinically improved during 18 months of follow-up evaluation. A subsequent report by Newton and associates described an expanded cohort and longer follow-up interval on their original set of patients [87]. Eighteen of twenty evaluable patients (90 per cent) stabilized after treatment with hydroxyurea, with a median time to progression of 176 weeks (range 20–328+ weeks). Five of the stabilized patients eventually progressed after 20, 56, 36, 216, and 56 weeks of treatment. Several of the patients remained stable by MRI even after hydroxyurea was discontinued.

In contrast to the studies outlined above, two recent reports are less suggestive of activity of hydroxyurea against meningiomas [88,89]. Loven and colleagues describe twelve patients with non-resectable, slow-growing tumors that received hyroxyurea at the usual dosing regimen over a 24-month observation period [88]. One patient had a MR that lasted for the 24-month period of treatment, while nine patients had progressive disease with a median time to progression of 13 months (range 4–24 months). The two remaining patients were withdrawn from the study due to severe hematological toxicity. In the study by Fuentes and co-workers, 43 patients with meningioma were treated with hydroxyurea (20 mg/kg/day), 36 of whom had had documentation of progression by MRI and/or clinical criteria at the onset of chemotherapy [89]. Of the evaluable patients, there were 2 with objective tumor shrinkage by MRI, 13 with disease stabilization, and 21 with tumor progression.

CONCLUSIONS

Meningiomas are slow growing extra-axial tumors, typically of benign histology, that are generally amenable to surgical resection and irradiation. However, it is important that effective medical treatment options be developed for these tumors since many patients cannot undergo surgery or, in some cases, recur after surgical resection and radiotherapy. Of the many chemotherapy options that have been reviewed in this chapter, none has been particularly effective. The agents or regimens with the most convincing data suggestive of minimal or modest activity include IFN-α, the multiagent regimen of cyclophosphamide, adriamycin, and vincristine, and single agent hydroxyurea. Well-designed clinical trials with larger patient cohorts will be required to prove or disprove the efficacy of these agents. In addition, it will be important to continue to elucidate the signal transduction pathways and molecular phenotype that contribute to the growth of meningiomas (e.g., NF2, PDGF, EGF, VEGF). This information will be critical in the development of molecular-based therapeutics, similar to what has occurred in high-grade astrocytomas and medulloblastomas [90,91].

ACKNOWLEDGMENTS

The author would like to thank Dr. Ray Chaudhury for the meningioma pathology slides and Ryan Smith for research assistance. Dr. Newton was supported in part by National Cancer Institute grant, CA 16058 and the Dardinger Neuro-Oncology Center Endowment Fund.

References

1. McDermott, M. W., Quinones-Hinjosa, A., Bollen, F. W., Larson, D. A., and Prados, M. (2002). Meningiomas. *In Brain Cancer* (M. Prados, Ed.), pp. 333–364. BC Decker Inc., Hamilton.

2. DeMonte, F., and Al-Mefty, O. (1995). Meningiomas. *In Brain Tumors. An Encyclopedic Approach* (A. H. Kaye, and E. R., Laws Eds.), Vol. 35, pp. 675–704. Churchill Livingstone, Edinburgh.

3. Bruner, J. M., Tien, R. D., and Enterline, D. S. (1998). Tumors of the meninges and related tissues. *In Russel & Rubinstein's Pathology of Tumors of the Nervous System.* (D. D. Bigner, R. E. McLendon, and J. M. Bruner Eds.), Vol. 11, 6th ed., pp. 67–139. Arnold, London.

4. Rohinger, M., Sutherland, G. R., Louw, D. F. *et al.* (1989). Incidence and clinicopathological features of meningioma. *J Neurosurg* **71**, 665–672.

5. Longstreth, S. T., Dennis, L. K., McGuire, V. M., Drangsholt, M. T., and Koepsell, T. D. (1993). Epidemiology of intracranial meningioma. *Cancer* **72**, 639–648.

6. Preston-Martin, S. (1996). Epidemiology of primary CNS neoplasms. *Neurol Clin* **14**, 273–290.

7. Wrensch, M., Minn, Y., Chew, T., Bondy, M., and Berger, M. S. (2002). Epidemiology of primary brain tumors: Current concepts and review of the literature. *Neuro-Oncol* **4**, 278–299.

8. Manfredonia, M. (1973). Tumors of the nervous system in the African in Eritrea (Ethiopia). *African J Med Sci* **4**, 383–387.

9. Ron, E., Modan, B., and Boice, J. D. (1988). Mortality after radiotherapy for ringworm of the scalp. *Am J Epidemiol* **127**, 713–725.

10. Strojan, P., Popovic, M., and Jereb, B. (2000). Secondary intracranial meningiomas after high-dose cranial irradiation: report of five cases and review of the literature. *Int J Radiat Oncol Biol Phys* **48**, 65–73.

11. Annegers, J. F., Laws, E. R., Kurland, L. T. *et al.* (1979). Head trauma and subsequent brain tumors. *Neurosurg* **4**, 203–206.

12. Bouton, A. H., and Parsons, J. T. (1993). Retroviruses and cancer: Models for cancer in animals and humans. *Cancer Investig* **11**, 70–79.

13. Butel, J. S., and Lednicky, J. A. (1999). Cell and molecular biology of simian virus 40: Implications for human infections and disease. *J Natl Cancer Inst* **91**, 119–134.

14. Weggen, S., Bayer, T. A., von Deimlin, A. *et al.* (2000). Low frequency of SV40, JC and BK polyomavirus sequences in human medulloblastomas, meningiomas and ependymomas. *Brain Pathol* **10**, 85–92.

15. Schrell, U. M. H., Adams, E. F., Fahlbusch, R. *et al.* (1990). Hormonal dependency of cerebral meningiomas. *J Neurosurg* **73**, 743–749.

16. Black, P. M. (1997). Hormones, radiosurgery and virtual reality: new aspects of meningioma management. *Can J Neurol Sci* **24**, 302–306.

17. Hsu, D. W., Efird, J. T., and Hedley-White, E. T. (1997). Progesterone and estrogen receptors in meningiomas: prognostic considerations. *J Neurosurg* **86**, 113–120.

18. Perry, A., Cai, D. X., Scheithauer, B. W. *et al.* (2000). Merlin, DAL-1, and progesterone receptor expression in clinicopathologic subsets of meningiomas: a correlative immunohistochemical study of 175 cases. *J Neuropathol Exp Neurol* **59**, 872–879.

19. Lambe, M., Coogan, P., and Baron, J. (1997). Reproductive factors and the risk of brain tumors: a population-based study in Sweden. *Int J Cancer* **72**, 389–393.

20. Sanson, M., and Cornu, P. (2000). Biology of meningiomas. *Acta Neurochir (Wien)* **142**, 493–505.

21. Lamszus, K. (2004). Meningioma pathology, genetics, and biology. *J Neuropathol Exp Neurol* **63**, 275–286.

22. Perry, A., Gutmann, D. H., and Reifenberger, G. (2004). Molecular pathogenesis of meningiomas. *J Neurooncol* **70**, 183–202.

23. Maxwell, M., Shih, S. D., Galanopoulos, T., Hedley-White, E. T., and Cosgrove, G. R. (1998). Familial meningioma: Analysis of expression of neurofibromatosis 2 protein Merlin. Report of two cases. *J Neurosurg* **88**, 562–569.

24. Heth, J. A., Kirby, P., and Menezes, A. H. (2000). Intraspinal familial clear cell meningioma in a mother and child. Case report. *J Neurosurg* **93**, 317–321.

25. Collins, V. P., Nordenskjold, M., and Dumanski, J. P. (1990). The molecular genetics of meningiomas. *Brain Pathol* **1**, 19–24.

26. Zang, K. D. (2001). Meningioma: A cytogenetic model of a complex benign human tumor, including data on 394 karyotyped cases. *Cytogenet Cell Genet* **93**, 207–220.

27. McClatchey, A. I., Saotome, I., Mercer, K. *et al.* (1998). Mice heterozygous for a mutation at the NF2 tumor suppressor locus develop a range of highly metastatic tumors. *Genes Dev* **12**, 1121–1133.

28. Kalamarides, M., Niwa-Kawakita, M., Leblois, H. *et al.* (2002). NF2 gene inactivation in arachnoidal cells is rate-limiting for meningioma development in the mouse. *Genes Dev* **16**, 1060–1065.

29. Gutmann, D. H., Sherman, L., Seftor, L., Haipek, C., Lu, K. H., and Hendrix, M. (1999). Increased expression of the NF2 suppressor gene product, merlin, impairs cell motility, adhesion and spreading. *Hum Mol Genet* **8**, 267–276.

30. Huynh, D. P., Mautner, V., Baser, M. E., Stavrou, D., and Pulst, S. M. (1997). Immunohistochemical detection of schwannomin and neurofibromin in vestibular schwannomas, ependymomas and meningiomas. *J Neuropathol Exp Neurol* **56**, 382–390.

31. Wellenreuther, R., Kraus, J. A., Lenartz, D. *et al.* (1995). Analysis of the neurofibromatosis 2 gene reveals molecular variants of meningioma. *Am J Pathol* **146,** 827–832.

32. Ueki, K., Wen-Bin, C., Narita, Y., Asai, A., and Kirino, T. (1999). Tight association of loss of merlin expression with loss of heterozygosity at chromosome 22q in sporadic meningiomas. *Cancer Res* **59**, 5995–5998.

33. Heinrich, B., Hartmann, C., Stemmer-Rachamimov, A. O., Louis, D. N., and MacCollin, M. (2003). Multiple meningiomas: Investigating the molecular basis of sporadic and familial forms. *Int J Cancer* **103**, 483–488.

34. Simon, M., von Deimling, A., Larson, J. J. *et al.* (1995). Allelic losses on chromosomes 14, 10, and 1 in atypical and malignant meningiomas: a genetic model of meningioma progression. *Cancer Res* **55**, 4696–4701.

35. Bostrom, J., Meyer-Puttlitz, B., Wolter, M. *et al.* (2001). Alterations of the tumor suppressor genes CDKN2A (p16[INK4a]), p14(ARF), CDKN2B (p15[INK4b]), and CDKN2C (p18[INK4c]) in atypical and anaplastic meningiomas. *Am J Pathol* **159**, 661–669.

36. Carroll, R. S., Black, P. M., Zhang, J. *et al.* (1997). Expression and activation of epidermal growth factor receptors in meningiomas. *J Neurosurg* **87**, 315–323.

37. Yang, S. Y., and Xu, G. M. (2000). Expression of PDGF and its receptor as well as their relationship to proliferating activity and apoptosis of meningiomas in human meningiomas. *J Clin Neurosci* **8**, 49–53.

38. Bitzer, M., Opitz, H., Popp, J. *et al.* (1998). Angiogenesis and brain edema in intracranial meningiomas: Influence of vascular endothelial growth factor. *Acta Neurochir (Wien)* **140**, 333–340.

39. Drevelagas, A., Karkavelas, G., Chourmouzi, D., Boulogianni, G., Petridis, A., and Dimitriadis, A. (2002). Meningeal tumors. *In Imaging of Brain Tumors with Histological Correlations* (A. Drevelegas, Ed.), pp. 177–214. Springer, Berlin.

40. Nelson, P. K., Setton, A., Choi, I. S., Ransohoff, J., and Berenstein, A. (1994). Current status of interventional neuroradiology in the management of meningiomas. *Neurosurg Clin N Am* **5**, 235–259.

41. Latchaw, R. E., and Hirsch, W. L. (1991). Computerized tomography of intracranial meningiomas. *In Meningiomas* (O. Al-Mefty, Ed.), pp. 195–207. Raven Press, New York.

42. Kaplan, R. D., Coons, S., Drayer, B. P. *et al.* (1992). MR characteristics of meningioma subtypes at 1.5 Tesla. *J Comp Assist Tomogr* **16**, 366–371.

43. De Monte, F. (1995). Current management of meningiomas. *Oncology* **9**, 83–96.

44. Akeyson, E. W., and McCutcheon, I. E. (1996). Management of benign and aggressive intracranial meningiomas. *Oncology* **10**, 747–756.

45. Olivero, S. C., Lister, J. R., and Elwood, P. W. (1995). The natural history and growth rate of asymptomatic meningiomas: a review of 60 patients. *J Neurosurg* **83**, 222–224.

46. Kuratsu, J. I., Kochi, M., and Ushio, Y. (2000). Incidence and clinical features of asymptomatic meningiomas. *J Neurosurg* **92**, 766–770.

47. Dean, B. L., Flom, R. A., Wallace, R. C. *et al.* (1994). Efficacy of endovascular treatment of meningiomas: evaluation with matched samples. *Am J Neruoradiol* **15**, 1675–1680.

48. Condra, K. S., Buatti, J. M., Mendenhall, W. M. *et al.* (1997). Benign meningiomas: primary treatment selection affects survival. *Int J Radiat Oncol Biol Phys* **39**, 427–436.

49. Miralbell, R., Linggood, R. M., de la Monte, S. *et al.* (1992). The role of radiotherapy in the treatment of subtotally resected benign meningiomas. *J Neurooncol* **13**, 157–164.

50. McCarthy, B. J., Davis, F. G., Freels, S. *et al.* (1998). Factors associated with survival in patients with meningioma. *J Neurosurg* **88**, 831–839.

51. Goldsmith, B. J., Wara, W. M., Wilson, C. B. *et al.* (1994). Postoperative irradiation for subtotally resected meningiomas. A retrospective analysis of 140 patients treated from 1967 to 1990. *J Neurosurg* **80**, 195–201.

52. Stafford, S. L., Pollock, B. E., Foote, R. L. *et al.* (2001). Meningioma radiosurgery: Tumor control, outcomes, and complications among 190 consecutive patients. *Neurosurg* **49**, 1029–1038.

53. Ojemann, S. G., Sneed, P. K., Larson, D. A., *et al.* (2000). Radiosurgery for malignant meningioma: results in 22 patients. *J Neurosurg* **93**(Suppl 3), 62–67.

54. Pollock, B. E., Stafford, S. L., Utter, A., Giannini, C., and Schreiner, S. A. (2003). Stereotactic radiosurgery provides equivalent tumor control to simpson grade 1 resection for paitents with small- to medium-size meningiiomas. *Int J Radiat Oncol Biol Phys* **55**, 1000–1005.

55. Kyritsis, A. P. (1996). Chemotherapy for meningiomas. *J Neurooncol* **29**, 269–272.

56. Younis, G. A., Sawaya, R., DeMonte, F., Hess, K. R., Albrecht, S., and Bruner, J. M. (1995). Aggressive meningeal tumors. *J Neurosurg* **82**, 17–27.

57. Steward, D. J., Maroun, J. A., Peterson, E. *et al.* (1986). Adriamycin in the treatment of malignant meningiomas. *In Biology of Brain Tumours* (M. D. Walker and D. G. T. Thomas Eds.), pp. 453–455. Martinus Nijhoff, Boston.

58. Bernstein, M., Villamil, A., Davidson, G., and Erlichman, C. (1994). Necrosis in a meningioma following systemic chemotherapy. Case report. *J Neurosurg* **81**, 284–287.

59. Groves, M. D., DeMonte, F., and Yung, W. K. A. (1995). Chemotherapeutic treatment of aggressive meningeal tumors. *Skull Base Surgery* **5**(Suppl 1), 2.

60. Wober-Bingel, C., Wober, C., Marosi, C., and Prayer, D. (1995). Interferon-alpha-2b for meningioma. *Lancet* **345**, 331.

61. Stewart, D. J., Dahrouge, S., Wee, M., Aitken, S., and Hugenholtz, H. (1995). Intraarterial cisplatin plus intravenous doxorubicin for inoperable recurrent meningiomas. *J Neurooncol* **24**, 189–194.

62. Chamberlain, M. C. (1996). Adjuvant combined modality therapy for malignant meningiomas. *J Neurosurg* **84**, 733–736.

63. Kaba, S. E., DeMonte, F., Bruner, J. M. *et al.* (1997). The treatment of recurrent unresectable and malignant meningiomas with interferon alpha-2B. *Neurosurg* **40**, 271–275.

64. Bunick, E. M., Mills, L. C., and Rose, L. I. (1978). Association of acromegaly and meningiomas. *J Am Med Assoc* **240,** 1267–1268.

65. Friend, K. E., Radinsky, R., and McCutcheon, I. E. (1999). Growth hormone receptor expression and function in meningiomas: effect of a specific receptor antagonist. *J Neurosurg* **91,** 93–99.

66. McCutcheon, I. E., Flyvbjerg, A., Hill, H. *et al.* (2001). Antitumor activity of the growth hormone receptor antagonist pegvisomant against human meningiomas in nude mice. *J Neurosurg* **94,** 487–492.

67. Markwalder, T. M., Seiler, R. W., and Zava, D. T. (1985). Antiestrogenic therapy of meningiomas – a pilot study. *Surg Neurol* **24,** 245–249.

68. Goodwin, J. W., Crowley, J., Eyre, H. J., Stafford, B., Jaeckle, K. A., and Townsend, J. J. (1990). A phase II evaluation of tamoxifen in unresectable or refractory meningiomas: a southwest oncology group study. *J Neurooncol* **15,** 75–77.

69. Oura, S., Sakurai, T., Yoshimura, G. *et al.* (2000). Regression of a presumec meningioma with the antiestrogen agent mepitiostane. Case report. *J Neurosurg* **93,** 132–135.

70. Jaaskelainen, J., Laasonen, E., Karkkainen, J., Haltia, M., and Troupp, H. (1986). Hormone treatment of meningiomas: lack of response to medroxyprogesterone acetate (MPA). A pilot study of five cases. *Acta Neurochir (Wien)* **80,** 35–41.

71. Waelti, E. R., and Markwalder, T. M. (1989). Endocrine manipulation of meningiomas with medroxyprogesterone acetate. Effect of MPA on growth of primary meningioma cells in monolayer tissue culture. *Surg Neurol* **31,** 96–100.

72. Grunberg, S. M., and Weiss, M. H. (1990). Lack of efficacy of megestrol acetate in the treatment of unresectable meningioma. *J Neurooncol* **8,** 61–65.

73. Spitz, I. M., and Barding, D. W. (1993). Mifepristone (RU 486) – A modulator of progestin and glucocordticoid action. *New Engl J Med* **329,** 404–412.

74. Olson, J. J., Beck, D. W., Schlechte, J. *et al.* (1986). Hormonal manipulation of meningiomas in vitro. *J Neurosurg* **65,** 99–107.

75. Olson, J. J., Beck, D. W., Schlechte, J. A. *et al.* (1987). Effect of the antiprogesterone RU-38486 on meningioma implanted into nude mice. *J Neurosurg* **66,** 584–587.

76. Blankenstein, M. A., van't Verlaat, J. W., and Croughs, R. J. M. (1989). Hormone dependency of meningiomas. *Lancet* **1,** 1381.

77. Grunberg, S. M., Weiss, M. H., Spitz, I. M. *et al.* (1991). Treatment of unresectable meningiomas with the antiprogesterone agent mifepristone. *J Neurosurg* **74,** 861–866.

78. Lamberts, S. W. J., Tanghe, H. L. J., Avezaat, C. J. J. *et al.* (1992). Mifepristone (RU 486) treatment of meningiomas. *J Neurol Neurosurg Psych* **55,** 486–490.

79. Grunberg, S. M., Rankin, C., Townsend, J. *et al.* (2001). Phase III double-blind randomized placebo-controlled study of mifepristone (RU) for the treatment of unresectable meningioma. *Proc ASCO* **20,** 56a.

80. Schrell, U. M. H., Rittig, M. G., Koch, U., Marschalek, R., and Anders, M. (1996). Hydroxyurea for treatment of unresectable meningiomas. *Lancet* **348,** 888–889.

81. Schrell, U. M. H., Rittig, M. G., Anders, M. *et al.* (1997). Hydroxyurea for treatment of unresectable and recurrent meningiomas. I. Inhibition of primary human meningioma cells in culture and in meningioma transplants by induction of the apoptotic pathway. *J Neurosurg* **86,** 845–852.

82. Schrell, U. M. H., Rittig, M. G., Anders, M. *et al.* (1997). Hydroxyurea for treatment of unresectable and recurrent meningiomas. II. Decrease in the size of meningiomas in patients treated with hydroxyurea. *J Neurosurg* **86,** 840–844.

83. Krakoff, I. H., Brown, N. C., and Reichard, P. (1968). Inhibition of ribonucleoside diphosphate reductase by hydroxyurea. *Cancer Res* **28,** 1559–1565.

84. Newton, H. B., Slivka, M. A., and Stevens, C. (2000). Hydroxyrea chemotherapy for unresectable or residual meningioma. *J Neurooncol* **49,** 165–170.

85. Mason, W. P., Gentill, F., Macdonald, D. R., Hariharan, S., Cruz, C. R., and Abrey, L. E. (2002). Stabilization of disease progression by hydroxyurea in patients with recurrent or unresectable meningioma. *J Neurosurg* **97,** 341–346.

86. Paus, S., Klockgether, T., Schlegel, U., and Urbach, H. (2003). Meningioma of the optic nerve sheath: treatment with hydroxyurea. *J Neurol Neurosurg Psych* **74,** 1348–1353.

87. Newton, H. B., Scott, S. R., and Volpi, C. (2004). Hydroxyurea chemotherapy for meningiomas: enlarged cohort with extended follow-up. *Br J Neurosurg* **18,** 495–499.

88. Loven, D., Hardoff, R., Sever, Z. B. *et al.* (2004). Non-resectable slow-growing meningiomas treated by hydroxyurea. *J Neurooncol* **67,** 221–226.

89. Fuentes, S., Chinot, O., Dufour, H. *et al.* (2004). Hydroxyurea treatment for unresectable meningioma. *Neurochir* **50,** 461–467.

90. Newton, H. B. (2003). Molecular neuro-oncology and the development of "targeted" therapeutic strategies for brain tumors. Part 1 – growth factor and ras signaling pathways. *Expert Rev Anticancer Ther* **3,** 595–614.

91. Newton, H. B. (2004). Molecular neuro-oncology and the development of "targeted" therapeutic strategies for brain tumors. Part 2 – PI3K/Akt/PTEN, mTOR, SHH/PTCH, and angiogenesis. *Expert Rev Anticancer Ther* **4,** 105–128.

CHAPTER

35

Chemotherapy for Brain Metastases

Herbert B. Newton

ABSTRACT: Metastatic brain tumors (MBT) are the most common complication of systemic cancer and affect 20–40 per cent of all adult cancer patients. Whole-brain radiotherapy and surgical resection of accessible, solitary lesions have been the mainstay of treatment. Recently, chemotherapy has become a more viable treatment option for MBT. Many different drugs and administrative approaches have been shown to be clinically active. Traditional chemotherapy, given before or during irradiation, can be effective with agents, such as cyclophosphamide, cisplatin, and etoposide. Non-traditional approaches, such as temozolomide and intra-arterial administration of carboplatin, have demonstrated activity against recurrent metastatic disease. In early clinical trials of interstitial chemotherapy, biodegradable polymers have shown some clinical efficacy and have been well tolerated. Molecular approaches are also under investigation in response to new information regarding the metastatic phenotype. Potential targets include growth factor receptors and other protein tyrosine kinases, internal signal transduction pathways, *ras* activation, and matrix metalloproteinase activity. New clinical trials will be needed to investigate these new molecular-based therapeutics, alone and in combination with currently available treatment options, to determine the optimal application of chemotherapy to MBT.

INTRODUCTION

Brain metastases (MBT) are the most common complication of systemic cancer [1]. They develop in 20–40 per cent of all adult cancer patients, which corresponds to an estimated 100 000–150 000 new cases per year. Therefore, MBT occur far more frequently than gliomas and other primary brain tumors, which have an annual incidence of 15 000–17 500 new cases per year [2]. Brain metastases most often arise from primary tumors of the lung (50–60 per cent), breast (15–20 per cent), melanoma (5–10 per cent), and gastrointestinal tract (4–6 per cent) [1,3–7]. However, they can develop from virtually any systemic malignancy, including primary tumors of the prostate, ovary and female reproductive system, kidney, esophagus, soft tissue sarcoma, bladder, and thyroid [8–17]. Post-mortem autopsy studies in adults would suggest that melanoma, renal carcinoma, and testicular carcinoma have the greatest propensity for spread to the brain [3]. In children and young adults, MBT arise most often from sarcomas (e.g., osteogenic, Ewing's), germ cell tumors, and neuroblastoma [3–5]. In 65–75 per cent of the patients, two or more metastatic tumors will develop simultaneously and be present at the time of diagnosis. Single brain metastasis are less common, and are most often noted in patients with breast, colon, and renal cell carcinoma. Patients with malignant melanoma and lung carcinoma are more likely to have multiple metastatic lesions.

Radiation therapy (i.e., external beam) has been the mainstay of treatment for MBT during the past 40 years, and continues to be the most frequently used modality [1,3–5,18]. Median survival after conventional irradiation is quite variable, but ranges from 12 to 20 weeks in most of the studies. Radiosurgical techniques (LINAC, gamma knife) have also been shown to be effective and have demonstrated excellent rates of local growth control. Recent reports would suggest that in selected, high-performance patients, radiosurgery without whole brain irradiation should be considered [19–21]. For carefully selected patients with solitary and multiple lesions, surgical resection can also be of benefit [1,3–5,22,23].

In recent years, chemotherapy has become a more viable option for the treatment of brain metastases, especially for recurrent disease [24–30]. The prior reluctance to use chemotherapy stemmed from concerns about the ability of chemotherapy drugs to cross the blood–brain barrier (BBB) and penetrate tumor cells, intrinsic chemoresistance of metastatic disease, and the high probability of early death from systemic progression. However, recent animal data suggests that metastatic tumors that strongly enhance on CT or MRI have an impaired blood–brain barrier and will allow entry of chemotherapeutic drugs [5,24,26]. In addition, systemic resistance to a given drug does not always preclude sensitivity of the metastasis within the brain [24]. Several types of metastatic brain tumors are relatively chemosensitive and may respond, including breast cancer, small-cell lung cancer, non-small-cell lung cancer, germ cell tumors, and ovarian carcinoma. Although the role of chemotherapy in the treatment of MBT has not been precisely defined, the available literature will be reviewed in detail and recommendations will be made about the use of this modality. In addition, potential trends will be explored regarding clinical trials and therapy with newly discovered drugs that exploit molecular targets and internal signal transducton pathways.

BIOLOGY AND MOLECULAR GENETICS

Systemic tumor cells usually travel to the brain by hematogenous spread through the arterial circulation, often after genetic alterations that produce a more motile and aggressive phenotype [1,3–5,31–35]. The metastasis most often originates from the lung, either from a primary lung tumor or from a pulmonary metastasis. Occasionally, cells reach the brain through the Batson's venous plexus or by direct extension from adjacent structures (e.g., sinuses, skull). The distribution of brain metastases follows the relative volume of blood flow to each area, so that 80 per cent of the tumors arise in the cerebral hemispheres, 15 per cent in the cerebellum, and 5 per cent in the brain stem. Tumor cells typically lodge in small vessels at the gray–white junction and then spread into the brain parenchyma, where they proliferate and induce their own blood supply by neoplastic angiogenesis [34]. Expansion of the MBT disrupts the function of adjacent neural tissue through several mechanisms, including direct displacement of brain structures, perilesional edema, irritation of overlying gray matter, and compression of arterial and venous vasculature.

The metastatic phenotype is the result of a complex alteration of gene expression that affects tumor cell adhesion, motility, protease activity, and internal signaling pathways [33,34]. Initial changes involve down regulation of surface adhesion molecules, such as integrins and cadherins, which reduces cell-to-cell interactions and allows easier mobility through the surrounding extracellular matrix (ECM). Cell motility is also accelerated in response to specific ligands, such as scatter factor and autocrine motility factor [32–34]. Several oncogenes and signal transduction pathways are also commonly activated in these aggressive cells, including members of the *Ras* family, *Src*, *Met*, and downstream molecules such as Raf, MAPK 1/2, Rac/ Rho, PI3-kinase, and focal adhesion kinase. Cellular invasive capacity is augmented in the metastatic phenotype by increased tumor cell secretion of matrix metalloproteinases (e.g., collagenases, gelatinases, elastases) and other enzymes that degrade the ECM [34]. In addition, metastatic cells often have down-regulated secretion of tissue inhibitors of metalloproteinases (i.e., TIMP-1, TIMP-2), which further enhances their invasive potential and access to the vasculature. Loss of certain metastasis-suppressor genes have also been implicated in the metastatic phenotype, including *nm23*, *KA11*, *KiSS1*, *PTEN*, *Maspin*, and others [34,35]. Reduced expression of these genes removes inhibitory control over the formation of macroscopic metastases. A recent case control study of non-small-cell lung cancer patients with and without MBT, attempted to correlate the expression of EGFR, cyclooxygenase-2, and Bax with the risk for developing brain metastases [36]. It was found that expression of the biomarkers was similar for patients with and without MBT, and could not be used to predict the potential for developing a MBT. In addition, expression levels of EGFR, cyclooxygenase-2, and Bax did not correlate with patient survival in multivariate analysis.

Once the metastatic bolus of cells has traveled to the nervous system and lodged within the brain, neoplastic angiogenesis is required for the tumor to grow to a clinically relevant size [34,37,38]. The angiogenic phenotype requires up-regulation of angiogenic promoters such as vascular endothelial growth factor (VEGF), fibroblast growth factors (basic FGF, acidic FGF), angiopoietins (Ang-1, Ang-2), platelet-derived growth factor (PDGF), epidermal growth factor (EGF), transforming growth factors (TGFα and TGFβ), interleukins (IL-6, IL-8), and the various growth factor receptors (e.g., VEGFR, PDGFR, EGFR) [37,38]. During the "angiogenic switch" to the metastatic phenotype, tumor cells also reduce secretion of angiogenesis inhibitors, such as thrombospondin-1, platelet factor-4, and interferons α and β [37]. This reduced concentration of inhibitory factors further "tips the balance" in the local environment to permit angiogenic activity within and around the tumor mass.

CLINICAL PRESENTATION AND PROGNOSTIC FACTORS

The majority of patients with metastatic brain tumors develop one or more progressive symptoms that are caused by enlargement of the mass within the brain. Although virtually any symptom is possible, depending on the location of the MBT, the most frequent complaints include headache (25–40 per cent), alterations of thinking and memory (20–25 per cent), focal weakness (20–30 per cent), and seizure activity (15–20 per cent) [1,3–5]. Less common symptoms consist of gait difficulty (15 per cent), visual loss (8 per cent), speech abnormalities (5–8 per cent), and sensory loss (5 per cent). At the time of diagnosis, the neurological examination will disclose hemiparesis and impaired cognitive function in 55–60 per cent of all patients. Other common neurological findings include papilledema (20 per cent), dysphasia (15–20 per cent), gait disturbance (10–20 per cent), hemianopsia (5 per cent), and hemisensory loss (5 per cent).

Metastatic brain tumors should be suspected in any cancer patient with new, progressive neurological symptoms. The diagnosis can be confirmed by enhanced computed tomographic scanning or magnetic resonance imaging (MRI)[1,39]. On either type of scan, metastatic tumors will present as rounded, well-circumscribed, noninfiltrative masses surrounded by a large amount of edema. With contrast administration, metastatic tumors usually demonstrate dense, homogeneous enhancement. Although CT is a good screening tool, MRI is significantly more sensitive for detecting multifocal tumors, small tumors, and lesions within the posterior fossa (i.e., cerebellum and brain stem) [1,39].

The overall prognosis for patients with MBT is quite poor and is dependent on the histological tumor type, number, and size of the metastatic lesions, neurological status, and degree of systemic involvement. The natural history is such that, left untreated, patients with MBT will usually die of neurological deterioration within 4 weeks. The addition of steroids will typically extend survival to 8 weeks. External beam radiotherapy, the most common modality of treatment, can further extend survival from 12 to 20 weeks in many patients [1,3–5,18]. Several studies have assessed how various prognostic factors relate to metastatic brain tumor patients at the time of diagnosis. A recent recursive partitioning analysis (RPA) of three RTOG radiation therapy brain metastases trials evaluated a wide range of prognostic factors and their impact on patient survival [40]. The most important factors were age (older or younger than 65 years; $p < 0.0001$), Karnofsky Performance Status (KPS) score (greater than or less than 70; $p < 0.0001$), and extent of systemic disease ($p < 0.0001$). Using these criteria, patients could be grouped into three distinct classes. Class 1 included patients who were less than 65 years of age, had KPS scores greater than 70, and had well-controlled systemic disease; Class 3 consisted of all patients with KPS scores less than 70; while Class 2 included all other patients who did not fit into Class 1 or Class 3. The median overall survival varied significantly between groups: 28.4 weeks for patients in Class 1, 16.8 weeks for those in Class 2, and 9.2 weeks for Class 3 patients. In addition, patients with multiple MBT had a significantly reduced survival in comparison to those with solitary lesions during univariate analyses ($p = 0.021$).

In a similar study by Nussbaum and colleagues, the number of metastatic lesions present at diagnosis was found to correlate with overall survival [41]. They noted a significant difference ($p = 0.0001$) in median survival between patients with solitary brain metastases and those with multifocal disease: 5 *versus* 3 months, respectively.

TRADITIONAL CHEMOTHERAPEUTIC APPROACHES

The most common approach to chemotherapy for brain metastases is to administer it "up front", before or during conventional RT or radiosurgery [42–51]. Several authors have demonstrated that combination regimens given intravenously can be active in this context (see Table 35.1). The most frequently used agents included cisplatin, etoposide, and cyclophosphamide (CTX). In a series of 19 patients with small-cell lung cancer and brain metastases, Twelves and co-workers used intravenous (IV) CTX, vincristine, and etoposide (days 1–3) every three weeks before any form of irradiation [42]. Ten of the 19 patients (53 per cent) had a radiological or clinical response. In 9 patients, there was CT evidence of tumor shrinkage (1 CR, 8 PR); while in 1 patient there was neurological improvement, without neuro-imaging follow-up. The mean time to progression (TTP) was 22 weeks, with a median overall survival of 28 weeks.

TABLE 35.1 Summary of Traditional Chemotherapy Approaches for Treatment of Metastatic Brain Tumors

Author [ref]	MBT type	N	Design	Chemo Tx	Route	ORR(%)	Overall TTP	Overall Survival
Twelves [42]	SCLCA	19	Phase II, upfront	CTX 1 gm/m^2	IV	47	Mean 22 weeks	Median 28 weeks
				VCR	IV			
				VP-16 100 mg/m$^{2/d}$	IV			
Cocconi [43]	Breast	22	Phase II, upfront	DDP 100 mg/m^2	IV	55	Median 25	Median 58
				VP-16 100 mg/m^2	IV			
Franciosi [44]	Mix	107	Phase II, upfront	Same as above	IV	30	Median 15	Median 27
Bernardo [45]	NSCLCA	22	Phase II, upfront	VNR 25 mg/m^2/d	IV	45	Median 25	Median 33
				GMC 1000 mg/m^2/d	IV			
				carbo AUC 5	IV			
Rosner [46]	Breast	87	Phase II, upfront	CTX 150 mg/m^2	IV	>50	N/A	N/A
				5FU 300 mg/m^2/d	IV			
				prednisone; or	Po			
				MTX 25 mg	IV			
				VCR	IV			
				5FU 500 mg	IV			
				CTX 100 mg	Po			
Boogerd [47]	Breast	11	Phase II, upfront	CTX 100 mg/m^2	Po	73	N/A	Mean 24
				MTX 40 mg/m^2	IV			
				5FU 600 mg/m^2	IV			
				or				
				CTX 500 mg/m^2	IV			
				Dox 50 mg/m^2	IV			
				5FU 50 mg/m^2	IV			
Robinet [48]	NSCLCA	176	Phase III, with RT or delayed RT	DDP 100 mg/m^2	IV	27 vs. 33	Median 13 vs. 11	Median 24 vs. 21
				VNR 30 mg/m^2	IV			
Postmus [49]	NSCLCA	120	Phase III, 50% patients receive RT	TNP 120 mg/m^2	IV	22 vs. 57	Median 7 vs. 12	Median 3.2 vs. 3.5 months
Ushio [50]	Lung	100	Phase III, 100% patients receive RT; 50% receive chemo	methyl-CCNU 100–120 mg/m^2	IV	36 vs. 72	N/A	Median 27 vs. 30
				or				
				ACNU 80–100 mg/m^2	IV			
				plus				
				tegafur 300 mg/m^2	IV			
Guerrieri [51]	NSCLC	42	Phase III. 100% patients receive RT; 50% receive chemo	Carboplatin 70 mg/m^2/day × five days with RT	IV	10 vs. 29	N/A	Median 4.4 vs. 3.7 months
Oberhoff [55]	Breast	16	Phase II, up-front	Topotecan 1.5 mg/m^2/day × 5 days, every 3 weeks	IV	38	N/A	6.25 months

Abbreviations: ref — references; MBT — metastatic brain tumor; Tx — treatment; ORR — objective response rate; TTP — time to progession; SCLCA — small cell lung carcinoma; RT — radiation therapy; NSCLCA — non-small cell lung carcinoma; CTX — cyclophosphamide; VCR — vincristine; VP-16 — etoposide; DDP — cisplatin; VNR — vinorelbine; GMC — gemcitobine; carbo — carboplatin; 5FU — 5-fluorouracil; MTX — methotrexate; Dox — doxurubicin; TNP — tenoposide; CCNU — lomustine; ACNU - N/A — not available; IV — intravenous; po-oral

Cocconi and colleagues used up front IV cisplatin (day 1) and etoposide (days 4, 6, 8) every three weeks for 22 evaluable patients with brain metastases from breast carcinoma [43]. There were 5 CR and 7 PR, for an overall objective response rate of 55 per cent. The median TTP was 25 weeks overall and 40 weeks in the objective response cohort. Overall median survival was 58 weeks. The same authors have expanded their series to include patients with brain metastases from breast (56), non-small-lung carcinoma (43), and malignant melanoma (8) [44]. Objective responses were noted in the breast (7 CR, 14 PR) and lung (3 CR, 10 PR) cohorts. None of the patients with melanoma had objective responses. The overall objective response rate was 30 per cent (34/89 patients). Median TTP was 15 weeks, with a median survival for the cohort of 27 weeks.

In a study of 22 patients with MBT from non-small-cell lung cancer, Bernardo and colleagues administered vinorelbine, gemcitabine, and carboplatin (for 3 days) IV every three weeks before radiotherapy [45]. The objective response rate was 45 per cent (3 CR, 6 PR), with a median TTP of 25 weeks. The median survival was 33 weeks overall and 48 weeks in responders.

Several authors have used up front CTX-based regimens for treatment of patients with MBT from breast carcinoma [46,47]. An early study by Rosner and associates used several different regimens prior to irradiation [46]. Fifty-two patients received IV CTX, 5-fluorouracil (for 5 days), and prednisone every five weeks. Another 35 patients received methotrexate (MTX), vincristine, and 5-fluorouracil IV qweek, plus daily oral CTX and prednisone. Objective responses were noted by radionuclide brain scan or CT in over 50 per cent of the cohort (46/87 patients). For patients with CR, the median TTP and survival were 40 weeks and 39.5 months, respectively; while for those with PR, it was 28 weeks and 10.5 months, respectively. Boogerd and colleagues treated 11 evaluable patients with either CTX (days 1–14), MTX (days 1,8), and 5-fluorouracil (days 1,8) or CTX, doxorubicin, and 5-fluorouracil IV every 3 weeks [47]. Objective responses were noted in 8 patients (73 per cent; 2 CR, 6 PR). The mean survival of the cohort was 24 weeks.

There have been several reports of phase III trials of chemotherapy for MBT patients, usually in combination with radiotherapy [48–50]. Robinet and associates treated 176 patients with non-small-cell lung carcinoma with cisplatin and vinorelbine (days 1, 8, 15, 22) every four weeks [48]. Whole-brain irradiation (3000 cGy/10 fractions) was administered to all patients. However, in 50 per cent of the patients it was concomitant with chemotherapy, while for the other half of the cohort it was delayed by two weeks. The objective response rate (27 versus 33 per cent), median TTP (13 versus 11 weeks), and median survival (24 versus 21 weeks) were similar for both groups, demonstrating that the timing of irradiation had no significant impact on any of the study endpoints. In a similar study of 120 patients with non-small-cell lung carcinoma, Postmus and colleagues administered teniposide (days 1, 3, 5) every 3 weeks to all patients [49]. One half of the cohort also received whole-brain radiotherapy (3000 cGy/10 fractions) concomitantly with chemotherapy. The TTP (7 versus 12 weeks; $p = 0.005$) and objective response rate (22 versus 57 per cent) were significantly improved by the addition of irradiation. However, irradiation did not improve survival, which was poor in both groups (3.2 versus 3.5 months). Equivalent results were reported by Ushio and associates after the treatment of 100 patients with lung carcinoma [50]. They used whole-brain radiotherapy (3500–3900 cGy) alone or in combination with methyl-CCNU or ACNU, and tegafur. The objective response rate was improved with the addition of chemotherapy (36 versus 72 per cent). However, overall median survival was poor and was comparable between groups (27–30 weeks).

A recent report by Guerrieri and colleagues described a randomized phase III trial of 42 patients with non-small cell lung carcinoma and brain metastases that compared whole brain RT (2000 cGy × 5 fractions) with RT plus carboplatin [51]. The carboplatin regimen consisted of 70 mg/m^2/day administered intravenously for five days, concomitant with RT. The median survival was 4.4 months for RT alone and 3.7 months for RT plus carboplatin ($p = 0.64$). The objective response rates suggested a trend for the RT plus carboplatin arm (29 versus 10 per cent), but failed to reach significance ($p = 0.24$).

Topotecan is a semisynthetic camptothecan derivative that selectively inhibits topoisomerase I in the S phase of the cell cycle, thereby causing dysfunction of replication and transcription within tumor cells [52]. It demonstrates excellent penetration of the blood–brain barrier in primate animal models and humans, possibly due to low protein binding in the serum in comparison to other camptothecins (20 versus > 95 per cent). Because of these unique characteristics, topotecan has been evaluated for activity against MBT [53–56]. Several European groups have evaluated single agent topotecan (1.5 mg/m^2/day, days 1–5 every 3 weeks) in patients with MBT from small-cell lung carcinoma [53,54,56]. Summating the data of more than 60 patients, the objective response rates have been encouraging, with 30 to 60 per cent of patients demonstrating a CR or PR [56]. In a report focusing on breast cancer patients with MBT, Oberhoff and co-workers treated 16 patients with topotecan (1.5 mg/m^2/day for 5 days, every 3 weeks) [55]. They noted six patients with objective responses, 1 CR and 5 PR, for a response rate of 38 per cent. Topotecan is also being investigated in combination with radiotherapy and other cytotoxic chemotherapy agents, such as temozolomide. A recent phase I trial has evaluated the tolerability of temozolomide (50–200 mg/m^2) and topotecan (1–1.5 mg/m^2), given daily for five days every 28 days [57]. Twenty-five patients with systemic solid tumors (e.g., melanoma, non-small-cell lung, breast) were treated. Toxicity was mainly hematological, with frequent neutropenia and thrombocytopenia. Three patients were noted to have PR.

TEMOZOLOMIDE

Temozolomide (Temodar) is an imidazotetrazine derivative of the alkylating agent dacarbazine with activity against systemic and CNS malignancies [25,58–60]. The drug undergoes chemical conversion at physiological pH to the active species 5-(3-methyl-1-triazeno)imidazole-4-carboxamide (MTIC). Temozolomide exhibits schedule dependent antineoplastic activity by interfering with DNA replication through the process of methylation. The methylation of DNA is dependent upon formation of a reactive methyldiazonium cation, which interacts with DNA at the following sites: N^7-guanine (70 per cent), N^3-adenine (9.2 per cent), and O^6-guanine (5 per cent). The cytotoxicity of temozolomide can be modulated by the activity of DNA-mismatch repair enzymes and O^6-alkylguanine-DNA alkyltransferase [58,59].

The antitumor activity of temozolomide is schedule dependent as shown by *in vitro* and *in vivo* experiments [25,58–60]. A 5-day administration schedule is superior to single day dosing for numerous malignancies, including lymphoma, leukemia, and CNS tumor xenografts. Several ongoing studies are evaluating whether dose intensified schedules (e.g., 7 days on, 7 days off; 6 weeks on, 2 weeks off) will improve efficacy [60]. Because this drug is stable at acid pH, it can be taken orally in capsules. Oral bioavailability is approximately 100 per cent, with rapid absorption of the drug. Temozolomide has excellent penetration of the BBB and brain tumor tissue.

Temozolomide is currently FDA approved for patients with recurrent malignant gliomas, but is also being studied for use against brain metastases

(see Table 35.2) [25,60–64]. Abrey and colleagues have used single-agent temozolomide (for 5 days, every 28 days) in a phase II study of 41 patients with MBT (24 lung, 10 breast, 3 melanoma) that had progressed through irradiation [61]. In 34 patients evaluated by neuro-imaging, there were 2 PR (both non-small-cell lung) and 9 SD, with an overall median survival of 26.4 weeks. Temozolomide was well tolerated, with mostly hematological toxicity. A similar phase II study has been reported by Christodoulou and associates, in which 28 patients with recurrent or progressive MBT (17 lung, 4 breast) were treated with temozolomide (for 5 days, every 28 days) [62]. In the evaluable cohort of 24 patients, there was 1 PR, 4 SD, and 19 PD. The median TTP and survival were 12 and 18 weeks, respectively. Using a dose intensified schedule of temozolomide (7 days on, 7 days off), Friedman and co-workers treated 24 patients with refractory MBT (12 breast, 9 lung) [63]. Of the 19 evaluable patients, there were 0 PR, 5 SD, and 14 PD. Survival and TTP data were not yet available. In a more recent report, 2 patients with MBT from lung cancer were treated with temozolomide plus oral etoposide or gemcitabine [64]. Both patients had substantial PR that were quite durable, with improvement in neurologic symptoms. Temozolomide has also been used for a rare patient with intracranial spread of esthesioneuroblastoma [65]. The patient had been heavily pre-treated, but stabilized with treatment for almost two years. However, not all reports demonstrate activity from temozolomide. In a series of 12 patients with non-small-cell lung carcinoma, Dziadziuszko and colleagues used single-agent temozolomide (200 mg/m^2/d × 5 days every 28 days) [66]. No objective responses or significant tumor stabilizations were noted.

Other authors feel temozolomide may have a role as a radiation sensitizing agent [67,68]. In a phase II study of 20 patients with MBT (11 lung, 3 breast, 3 rectal), temozolomide was administered during and after (for 5 days, every 28 days) radiation therapy [67]. After the completion of irradiation and six cycles of temozolomide, there were 3 CR, 8 PR, and 8 SD. Follow-up at 8 months revealed that 14 of 20 patients were still alive. Antonadou and co-workers describe the results of a randomized phase II study, in which 52 newly diagnosed MBT patients (lung and breast) were treated with either irradiation (4000 cGy) alone; or irradiation plus temozolomide (dosing as above) [68]. The addition of temozolomide improved the objective response rate when compared to radiotherapy alone (CR 38 per cent, PR 58 per cent *versus* CR 33 per cent, PR 33 per cent). In addition, neurologic improvement during treatment was more

TABLE 35.2   Summary of Non-Traditional Chemotherapy Approaches for Treatment of Metastatic Brain Tumors

Author [ref]	MBT type	N	Design	Chemo Tx	Route	ORR (%)	Overall TTP	Overall survival
Abrey [61]	Mix	41	Phase II, after RT failure	Temozolomide 150–200 mg/m^2 × 5 days	Po	6	N/A	Median 26.4 weeks
Christodoulou [62]	Mix	28	Phase II, after RT failure	Same as above	Po	0.4	Median 12 weeks	Median 18
Friedman [63]	Mix	24	Phase II, after RT failure	Temozolomide 150 mg/m^2, 7 days on, 7 days off	Po	0	N/A	N/A
Dardoufas [67]	Mix	20	Phase II, during and after RT	Temozolomide 60 mg/m^2 during RT; 200 mg/m^2 × 5 days after RT	Po	55	N/A	N/A
Antonadou [68]	Mix	45	Phase II, randomized, 100% pts receive RT, 50% receive chemo	Same as above	Po	96 vs. 66	N/A	N/A
Yamada [76]	Lung	9	Phase II, after RT	BCNU 100 mg/m^2	IA	44	Range 16 to 24	N/A
Madajewicz [77]	Mix	35	Phase II, after RT	Same as above	IA	34.3	N/A	Median 16
Cascino [78]	Mix	31	Phase II, after RT	Same as above	IA	16.25	N/A	Median 17
Madajewicz [79]	Mix	28	Phase II, pre-RT	DDP 40 mg/m^2 VP-16 20 mg/m^2	IA IA	42.8	N/A	Median 28
Newton [85]	Mix	24	Phase II, after RT	Carbo 200 mg/m^2 VP-16 100 mg/m^2	IA IV	54	Median 16	Median 20
Doolittle [88]	Mix	13	Phase II, after RT	Carbo 200 mg/m^2 CTX 330 mg/m^2 VP-16 200 mg/m^2	IA IV IV	38.4	N/A	N/A
Ewend [96]	Mix	25	Phase I/II, with RT	BCNU-implanted polymer	IS	N/A	N/A	Median 15.5 months
Brem [97]	Mix	42	Phase II, during or after RT	BCNU-implanted polymer	IS	N/A	N/A	Mean 16.8 months
Newton [99]	Mix	17	Phase III, during RT	BCNU-implanted Polymer	IS	N/A	N/A	Mean 40.6+ vs. 39.2+ weeks
Ceresoli [125]	NSCLC	41	Phase II, before or after RT	gefitinib, 250 mg/day	po	9.8	Median 3.0 months	Median 5.0 months

Abbreviations: ref — references; MBT — metastatic brain tumor; Tx — treatment; ORR — objective response rate; TTP — time to progression; SCLCA — small cell lung carcinoma; RT — radiation therapy; po — oral; IA — intra-arterial; N/A — not available; BCNU — carmustine; DDP — cisplatin; VP-16 — etoposide; carbo — carboplatin; IV — intravenous; CTX — cyclo- Phosphamide; IS — interstitial.

pronounced in the cohort of patients receiving temozolomide.

Several investigators have reported that patients with brain metastases from melanoma are responsive to temozolomide, either alone or in combination with other chemotherapeutic agents [69–72]. Biasco and colleagues reported a 57-year-old male with multifocal MBT who received standard temozolomide (200 mg/m^2/day × 5 days every 28 days) and achieved a CR after six cycles of treatment [69]. In a phase II study of 151 melanoma patients that had not received any form of brain irradiation, treatment consisted of standard-dose single-agent temozolomide (150–200 mg/m^2, depending on prior chemotherapy exposure) [70]. Overall, objective responses

were noted in nine patients (6 per cent; CR 1, PR 8), with another 40 patients (26.5 per cent) experiencing stabilization of disease. For the entire cohort, the median survival ranged from 2.2 to 3.5 months. In a series of 25 patients from Bafaloukos and co-workers, temozolomide was used alone (6) or in combination with docetaxel (10), or cisplatin (9) [71]. There were objective responses in 6 patients (24 per cent; all PR), with another 5 patients (20 per cent) exhibiting stabilization of disease. The overall median TTP was only 2.0 months, with an overall median survival of 4.7 months. Thalidomide, a moderately potent angiogenesis inhibitor, has been used in combination with temozolomide by several investigators [73]. Hwu and associates initially reported a 43-year-old woman with multifocal MBT who received a combination of temozolomide (75 mg/m^2/d × 6 weeks, 2 weeks off every 8 weeks) and thalidomide (200 mg/day, titrated to 400 mg/day) [72]. After nine cycles of treatment, the patient developed a CR that was quite durable. In a randomized phase II trial, 181 patients with stage IV melanoma (21 with MBT) were treated with temozolomide (given every 8 h or once daily) in addition to interferon alfa-2b or thalidomide [74]. Of the cohort with MBT, there was one patient that had received temozolomide and thalidomide in whom a PR was documented; however, there was fairly rapid progression in the other 20 patients.

INTRA-ARTERIAL TREATMENT APPROACHES

In an effort to improve dose intensity, some authors have given some or all of the chemotherapy drugs by the intra-arterial (IA) route [25,75–79]. There are several advantages to administering chemotherapy IA instead of by the conventional IV route, including augmentation of the peak concentration of drug in the region of the tumor and an increase in the local area under the concentration–time curve [75]. Pathologically, metastatic brain tumors are excellent candidates for IA approaches, because they tend to be well circumscribed and non-infiltrative [1]. This is in contrast to most high-grade primary brain tumors, which are very infiltrative and extend beyond the region of contrast enhancement [2]. In addition, MBT universally enhance on CT and MRI imaging, indicating excellent arterial vascularization and impairment of the blood–tumor barrier. Therefore, with metastatic tumors the vast majority of tumor cells receive significant exposure to agents when administered *via* the localized arterial vasculature. Pharmacologic studies using animal models of IA and IV drug infusion have shown that the IA route can increase the intra-tumoral concentration of a given agent by at least a factor of 3 × to 5.5 × [80,81]. For chemosensitive tumors, improving the intra-tumoral concentrations of drug should augment tumor cell kill and the ability to achieve objective responses [75].

An important consideration in the application of IA chemotherapy is choosing the proper drug or drugs to be infused. Not all drugs have the appropriate metabolism and pharmacokinetic profile for IA usage. It is important that the drug have a rapid total body clearance, as defined by the concept of R_a, the *Regional Advantage* (see Chapter 17 for a more detailed explanation) [82]. The R_a is the pharmacological advantage a drug has (or may not have) when

administered IA *versus* the IV route; it is maximized by a rapid (i.e., large) total body clearance. The drugs with the most appropriate R_a for IA chemotherapy of MBT (ranked in descending order) include BCNU > cisplatin >> carboplatin > etoposide. The two drugs with the highest R_a (i.e., BCNU, cisplatin) are also known to have the most significant neuro-toxicity.

Intra-arterial chemotherapy has been attempted with MBT patients in the past (see Table 35.2), beginning with the 1979 report of Yamada and colleagues [76]. They used IA BCNU (every 4 weeks; in addition to systemic chemotherapy) for nine patients with MBT from lung carcinoma. There were 4 responders, with TTP ranging from 16 to 28 weeks. In a more extensive phase II study, the same investigators used IA BCNU for 35 patients with lung carcinoma and malignant melanoma [77]. Twelve patients had objective responses (34.3 per cent; 6 CR, 6 PR; lung only), with a median survival of 16 weeks. Toxicity included infusional pain, seizures, and confusional episodes. Cascino and co-workers also used IA BCNU (every 5–6 weeks) for 31 patients with MBT that had progressed through radiation therapy [78]. They noted 5 PR (16.2 per cent; 3 lung, 1 breast, 1 melanoma), with a median survival of 17 weeks, and similar toxicity to the Yamada studies. In a more recent report by Madajewicz and associates, a regimen of IA cisplatin and etoposide was administered every three weeks to twenty-eight patients with MBT [79]. There were 6 CR and 6 PR, with a median survival of 28 weeks. Significant neuro-toxicity was noted (e.g., seizures, infusional pain, confusion) in several patients.

At the Ohio State University Medical Center and James Cancer Hospital, Newton and colleagues have recently developed a carboplatin-based IA regimen for treatment of MBT [25,83–85]. The regimen consists of IA carboplatin and intravenous etoposide for two days, every 4 weeks. They report a series of 24 evaluable patients with MBT (11 lung, 9 breast, 2 colon) that have received IA treatment [25,85]. Most of the patients had multifocal metastases and had failed whole-brain radiotherapy. There were 6 CR, 6 PR, and 1 minor response, with a median TTP of 16 weeks overall and 30 weeks in responders. The overall median survival was 20 weeks. The regimen has been well tolerated, with mainly hematological toxicity. Procedural complications and neuro-toxicity during IA administration of carboplatin have been very rare (< 1 per cent). Based on these initial encouraging results with IA carboplatin and the previously mentioned activity of temozolomide, a new phase I/II multi-center trial has been opened for accrual [Newton *et al.* unpublished data]. The regimen consists of IA carboplatin (200 mg/m^2 × 2 days) and temozolomide (150 mg/m^2 × 5 days) every 4 weeks, for patients with recurrent or symptomatic residual brain metastases. The phase I portion of the protocol has been completed. The regimen has been well tolerated, except for hematological toxicity in heavily pre-treated patients. Several patients have demonstrated PR and SD during MRI follow-up examinations.

Neuwelt and associates have attempted to further dose intensify IA chemotherapy by using it in combination with IA mannitol and osmotic BBB disruption [25,81,86–88]. The rationale for this approach is to augment the general advantages of regular IA treatment by further intentional disruption of the blood–brain and blood–tumor barriers of metastatic tumors (see Chapter 18 for more details). Disruption of the barrier significantly increases intratumoral concentrations of drug and may improve response rates [79,81,86]. Initial experience was gained using a regimen of IA methotrexate (1–5 g) and mannitol, IV CTX (15–30 mg/kg), and oral procarbazine (100 mg/day × 14 days) every 4–6 weeks [87]. Several objective responses were noted on CT follow-up in a series of 7 patients with MBT (breast, lung, testicular). In a more recent report of 13 MBT patients by Doolittle and co-workers, the regimen consisted of IA carboplatin and mannitol, IV CTX, and IV etoposide for two days every 4 weeks [88]. Results included 3 CR, 2 PR, and 7 SD; TTP and survival data were not available. The risk of procedural complications and neuro-toxicity is greater with IA treatment in combination with BBB disruption, in comparison to regular IA approaches (e.g., brain swelling, herniation, arrhythmia).

INTERSTITIAL CHEMOTHERAPY

Interstitial chemotherapy involves the placement of a drug or drugs directly into the brain tumor or resection cavity of the brain parenchyma [89]. With this approach, the tumor and surrounding brain are exposed to high concentrations of drug, which is not limited by the BBB. In addition, there is minimal exposure of the drug to the systemic circulation and organs. Interstitial chemotherapy delivery systems were originally developed for application to malignant gliomas, due to their high local recurrence rate and limited potential for systemic metastases [89–92]. Brem and colleagues were able to devise a biodegradable polymer wafer, "loaded" *via* anhydride bonds to BCNU (3.85 per cent; 7.7 mg). After placement of the wafers into the tumor resection cavity, water in the

interstitial fluid breaks the anhydride bonds and releases the BCNU in a controlled fashion. The wafer slowly degrades in layers through "surface erosion", like a bar of soap. All of the BCNU is released from the polymer by three weeks, and the polymer degrades in the brain after 4–6 months. BCNU can be measured within the CSF and serum; however, the concentrations are negligible. Systemic toxicity that is typical for IV BCNU (e.g., leukopenia, thrombocytopenia, interstitial pulmonary fibrosis) does not occur with the BCNU-loaded polymer.

Recent research by Ewend and co-workers has investigated the use of chemotherapy-loaded wafers as a treatment method for MBT [93–95]. In a mouse model, 20 per cent BCNU-loaded wafers were implanted after surgical resection of melanoma MBT and compared to animals that received only external beam irradiation [93]. Survival was significantly longer in the chemotherapy wafer group ($p = 0.032$), with 57 per cent of the animals still alive at 150 days follow-up. For the radiation alone group, only 50 per cent of animals were still alive during follow-up after 25 days. In a more extensive study using a mouse model, several different MBT (melanoma, renal cell, colon, lung) and chemotherapy wafer (BCNU, carboplatin, camptothecin) types were studied [94]. Each type of wafer was used against each type of MBT, with or without radiotherapy. The results demonstrated a significant survival benefit for the BCNU-loaded wafer over the carboplatin and camptothecin wafers. In addition, there appeared to be a synergistic effect between the BCNU-polymer and irradiation, for all tumor types, when compared to irradiation alone ($p = 0.0005$). A similar study, used a BALB/c mouse model, focused on chemotherapy wafer treatment of breast metastases [95]. Mice with intracranial EMT-6 breast tumors were treated with implanted wafers loaded with BCNU, carboplatin, or camptothecin. For each wafer type a control was used. Radiation therapy was administered to half of the control and wafer animals, as well as to each tumor type as sole treatment. The BCNU wafer alone ($p < 0.0001$) or in combination with irradiation ($p = 0.02$), significantly extended survival time compared to controls or animals implanted with carboplatin or camptothecin.

Based on the successful results of the mouse studies, Ewend and co-workers initiated a multicenter phase I/II study of 25 patients with solitary brain metastases (13 lung, 4 melanoma, 3 renal, 2 breast) [96]. After surgical resection of the metastatic lesion, each patient was implanted with 3.85 per cent BCNU-loaded biodegradable wafers and irradiated. Interstitial placement of the wafers was generally well tolerated, similar to the experience with gliomas, with adverse events in only 7 patients (e.g., seizures, eye pain, re-operation for cerebral edema). No local recurrences were noted in the cohort; there have been 4 remote CNS failures. The median survival in follow-up of 16 evaluable patients was 15.5 months. The authors concluded that the BCNU wafer was well tolerated and active against MBT; and should be tested in a phase III trial. In a recent report of 42 consecutive patients with MBT, Brem and colleagues described their experience with the implantable BCNU wafers [97]. Thirty-four patients were newly diagnosed, while 8 were treated at recurrence. Radio-therapy (3000–4400 cGy) was used, in addition to surgical resection and wafer placement, for all newly diagnosed patients. The majority of the cohort had MBT from non-small-cell lung cancer (20), melanoma (11), renal carcinoma (4), and breast (3). There were no localized recurrences in newly diagnosed patients; 3 patients developed distant CNS recurrences. The overall mean survival was 16.8 months; 17.8 months in newly diagnosed patients and 12.9 months for those with recurrent disease. Similar results have been reported by the PROLONG Study Group in a cohort of 36 patients, with minimal evidence for localized recurrence of implanted MBT [98].

A multi-center, randomized, phase III trial has been initiated at the Ohio State University Medical Center and several other sites [99]. The trial is attempting to compare surgical resection, BCNU wafer placement, and irradiation *versus* surgical resection and irradiation alone for patients with solitary, resectable MBT. Seventeen evaluable patients (15 with MBT from lung) have been enrolled into the study, with 10 of the cohort receiving the 3.85 per cent BCNU-loaded biodegradable polymers. The mean age of the patients was 62.5 years, with a range of 39–78 years. Placement of the wafers has been well tolerated, with infrequent seizures, cerebral edema, or wound infections. The mean survival has been similar between groups thus far (wafers–40.6+ weeks *versus* no wafers–39.2+ weeks), although follow-up has been incomplete.

MOLECULAR TREATMENT APPROACHES

The recent explosion of knowledge regarding the molecular biology of neoplasia and the metastatic phenotype has led to intense development of therapeutic strategies designed to exploit this new information [100]. Several targets of therapeutic intervention have been developed, including growth factor receptors and their tyrosine kinase activity, disruption of aberrant internal signal transduction pathways, inhibition of excessive matrix metalloproteinase activity, down-regulation of cell cycle pathways, and manipulation of the apoptosis pathways. The most promising approach thus far has been the development of small-molecule drugs or monoclonal antibodies to the major growth factor receptors (e.g., PDGFR, EGFR, Her2, CD20) [101–105]. Monoclonal antibody agents such as rituximab (i.e., Rituxan) and trastuzumab (i.e., Herceptin) have proven to be clinically active against non-Hodgkin's lymphoma and breast cancer, respectively. Several small-molecule inhibitors of the tyrosine kinase activity of the EGFR (e.g., ZD 18339, OSI-774) are under clinical evaluation in phase I trials of patients with solid tumors [105–107]. Similar efforts are underway to develop agents that target the tyrosine kinase activity of PDGFR [86,105]. Other investigators are developing agents that can inhibit the *ras* molecule and its signaling pathway [104,105,110–112]. The majority of drugs are being designed to target the *ras* farnesyltransferase enzyme, which is critical for activating *ras* and allowing it to attach to the inner surface of the plasma membrane. Other agents under development will target downstream effectors, such as Raf, MAPK, and Rac/Rho. Another strategy under investigation is the development of drugs that inhibit matrix metalloproteinase activity, which is up-regulated in the metastatic phenotype [113,114]. Several agents, including marimastat and batimistat, are currently being tested in clinical phase I, II, and III trials.

A recent pre-clinical study by Grossi and associates investigated the activity of trastuzumab against breast MBT in an animal model [115]. Metastatic breast tumor cells transfected to overexpress Her2 were implanted intracranially into athymic rats. Trastuzumab (2 mg/kg) was administered to the rats by intracerebral microinfusion or intraperitoneal injection, along with saline and isotype monoclonal antibody controls. Animals treated with intraperitoneal trastuzumab had a median survival of 26.5 days, while those receiving the drug by intracerebral microinfusion had a significantly longer median survival of 52 days ($p = 0.009$). The authors concluded that trastuzumab was active against MBT from breast carcinomas that overexpress Her2, as long as the agent can bypass the BBB with a technique such as intracerebral microinjection.

Another "molecular-based" drug, designed as a protein tyrosine kinase inhibitor, is imatinib mesylate (i.e., Gleevec; formerly known as STI571) [116,117]. Imatinib blocks the ATP binding site of tyrosine kinase receptors, inhibiting this activity and interfering with the transduction of receptor-mediated signals to internal effectors. It has demonstrated significant *in vitro* activity against BCR-ABL, C-KIT, and PDGFR [116–118]. Imatinib has also shown impressive clinical activity against chronic myelogenous leukemia and gastrointestinal stromal tumors (GIST). It is administered orally and is well tolerated in phase I studies. The drug is metabolized by the cytochrome P450 system, and so may be affected by drugs that alter the activity of this enzyme system, such as anticonvulsants (e.g., phenytoin, carbamazepine). A recent case report by Brooks and colleagues suggests that imatinib may have activity against MBT that express C-KIT, such as GIST [119]. They report a 75-year-old male with a C-KIT positive GIST, that developed neurological deterioration and gait difficulty. An MRI demonstrated leptomeningeal disease with brain infiltration and edema. After treatment with imatinib mesylate (400 mg bid) for 2 months, his neurological function and gait improved. A follow-up MRI scan revealed complete resolution of the meningeal and intra-parenchymal abnormalities.

Gefitinib (ZD 1839, Iressa) is another protein tyrosine kinase inhibitor with selective activity against EGFR [105–107]. It has been clinically effective against several different types of solid tumors, including non-small-cell lung cancer and brain tumors. Several authors have recently described case reports of the use of gefitinib in patients with MBT from non-small-cell lung cancer [120–124]. Several of these initial patients had objective responses, including a few CR, that were quite durable. These early reports lead Ceresoli and colleagues to perform a prospective phase II trial of gefitinib in patients with non-small-cell lung cancer [125]. Forty-one consecutive patients were treated with gefitinib (250 mg/day); 37 had received prior chemotherapy and 18 had undergone brain irradiation. There were four patients with a PR and seven with SD. The overall progression-free survival was 3 months. However, the median duration of responses in the patients with a PR was 13.5 months.

OVERVIEW AND FUTURE CONSIDERATIONS

Metastatic brain tumors are the most common form of brain cancer and may become even more prevalent with further advancements in the treatment of primary systemic malignancies. It is imperative that all modes of treatment for MBT continue to be

explored and investigated. Although chemotherapy has been intermittently evaluated over the past few decades as a potential treatment modality for MBT, it has not been studied as intensively as other options, such as external beam radiotherapy and radiosurgery (i.e., gamma knife, LINAC). It is apparent from this review of the available literature that MBT are responsive to chemotherapy, in both the up front setting and at recurrence. In addition, both the traditional methods of drug delivery, and "non-traditional" approaches (e.g., temozolomide, IA, interstitial) have significant activity. Efficacy of chemotherapy is probably maximized when it is administered up front, before or concomitant with some form of radiotherapy. In this context, the tumors are chemotherapy naïve and do not have their vascular perfusion impaired by radiation effects, which could reduce regional chemotherapy drug concentrations and dose intensity. If the drug is given during irradiation, a synergistic effect is possible. The most active drugs administered by traditional methods appear to be cisplatin, CTX, and etoposide. However, as shown above, several different combinations of IV drugs have been able to induce objective responses. The phase III study data would suggest that the combination of irradiation and chemotherapy together are more efficacious than either modality alone [42–44].

For recurrent disease, both temozolomide (alone and in combination with other drugs) and the carboplatin-based IA regimens have demonstrated activity. It is important, though, with this group of patients, to screen carefully before initiating treatment. The benefits of chemotherapy and control of CNS disease may be abrogated if there is uncontrolled systemic tumor burden and significant damage to critical organs (e.g., lungs, liver). In addition, based on the experience noted above with traditional chemotherapy, it is possible that temozolomide and IA carboplatin may have more activity if used before any form of irradiation has been administered. However, new clinical trials will be necessary before this issue can be clarified.

More phase I, II, and III clinical trials will be needed over the next five years to discern the optimal application of chemotherapy to patients with MBT, especially with the advent of new, "molecular" therapeutic agents. It is important that these clinical trials be stratified by RPA class. This will improve the ability to compare results, between studies, of patients of the same class. In addition, it will be important during phase I trials to monitor for any possible interactions between cytochrome P450 enzyme-inducing anticonvulsants and the new chemotherapy drugs. Separate treatment arms may be necessary for these studies, so that patients can be stratified according to anticonvulsant status. Some of the newly developed drugs that should be tested in clinical trials against MBT include angiogenesis inhibitors, such as thalidomide, angiostatin, and endostatin. These agents will probably be most effective in combination with traditional chemotherapeutics (e.g., alkylating agents), similar to the experience in gliomas [25]. The biology of the common primary malignancies that cause MBT suggests that receptor tyrosine kinase inhibitors and farnesyltransferase inhibitors may also be effective, alone or in combination with other agents. A similar strategy would be to combine monoclonal antibodies directed against growth-factor receptors with more traditional drugs or non-traditional approaches (e.g., temozolomide, IA). Modulators of apoptosis, inhibitors of the cell cycle, and matrix metalloproteinase inhibitors target other aspects of the metastatic phenotype, and should also be considered as therapeutic options for clinical trials.

Consideration should also be given to apply some of the already mentioned agents and approaches in different ways. For instance, temozolomide could be used in novel combinations with other agents (e.g., traditional, IA, molecular) and in more intensive dosing schedules (e.g., 6 weeks on, 2 weeks off). Similarly, IA chemotherapy (with or without BBB disruption) could be attempted with new agents that might have activity, such as melphalan.

During the design and implementation of these new therapeutic trials, clinical parameters should be carefully monitored, as should the traditional endpoints of objective responses on neuro-imaging, TTP, and survival. Neurological function (and KPS) in response to treatment, as well as neurological TTP, will need to be followed closely. Quality of life during chemotherapy, and in response to treatment, should be carefully evaluated. The studies should also be careful to limit enrollment of patients with extensive systemic disease. Only patients with relatively well-controlled systemic malignancies will allow an accurate determination of the effectiveness of these novel therapeutic approaches against MBT.

References

1. Newton, H. B. (1999). Neurological complications of systemic cancer. *Am Fam Phys* **59**, 878–886.
2. Newton, H. B. (1994). Primary brain tumors: review of etiology, diagnosis, and treatment. *Am Fam Phys* **49**, 787–797.
3. Oneill, B. P., Buckner, J. C., Coffey, R. J. *et al.* (1994). Brain metastatic lesions. *Mayo Clin Proc* **69**, 1062–1068.
4. Patchell, R. A. (1996). The treatment of brain metastases. *Cancer Investig* **14**, 169–177.
5. Wen, P. Y., and Loeffler, J. S. (1999). Management of brain metastases. *Oncol* **13**, 941–961.
6. Farnell, G. F., Buckner, J. C., Cascino, T. L. *et al.* (1996). Brain metastases from colorectal carcinoma. The long term survivors. *Cancer* **78**, 711–716.
7. Sampson, J. H., Carter, J. H., Friedman, A. H., and Seigler, H. F. (1998). Demographics, prognosis, and therapy in 702 patients with brain metastases from malignant melanoma. *J Neurosurg* **88**, 11–20.
8. Leroux, P. D., Berger, M. S., Elliott, J. P., and Tamimi, H. K. (1991). Cerebral metastases from ovarian carcinoma. *Cancer* **67**, 2194–2199.
9. Martinez-Manas, R. M., Brell, M., Rumia, J., and Ferrer, E. (1998). Case report. Brain metastases in endometrial carcinoma. *Gyn Oncol* **70**, 282–284.
10. Mccutcheon, I. E., Eng, D. Y., and Logothetis, C. J. (1999). Brain metastasis from prostate carcinoma. Antemortem recognition and outcome after treatment. *Cancer* **86**, 2301–2311.
11. Lowis, S. P., Foot, A., Gerrard, M. P. *et al.* (1998). Central nervous system metastasis in Wilms' tumor. A review of three consecutive United Kingdom trials. *Cancer* **83**, 2023–2029.
12. Culine, S., Bekradda, M., Kramar, A. *et al.* (1998). Prognostic factors for survival in patients with brain metastases from renal cell carcinoma. *Cancer* **83**, 2548–2553.
13. Qasho, R., Tommaso, V., Rocchi, G. *et al.* (1999). Choroid plexus metastasis from carcinoma of the bladder: case report and review of the literature. *J Neurooncol* **45**, 237–240.
14. Salvati, M., Frati, A., Rocchi, G. *et al.* (2001). Single brain metastasis from thyroid cancer: Report of twelve cases and review of the literature. *J Neurooncol* **51**, 33–40.
15. Ogawa, K., Toita, T., Sueyama, H. *et al.* (2002). Brain metastases from esophageal carcinoma. Natural history, prognostic factors, and outcome. *Cancer* **94**, 759–764.
16. Espat, N. J., Bilsky, M., Lewis, J. J. *et al.* (2002). Soft tissue sarcoma brain metastases. Prevalence in a cohort of 33829 patients. *Cancer* **94**, 2706–2711.
17. Schouten, L. J., Rutten, J., Huveneers, H. A. M., and Twijnstra, A. (2002). Incidence of brain metastases in a cohort of patients with carcinoma of the breast, colon, kidney, and lung and melanoma. *Cancer* **94**, 2698–2705.
18. Berk, L. (1995). An overview of radiotherapy trials for the treatment of brain metastases. *Oncol* **9**, 1205–1212.
19. Chitapanarux, I., Goss, B., Vongtama, R. *et al.* (2003). Prospective study of stereotactic radiosurgery without whole brain radiotherapy in patients with four or less brain metastases: incidence of intracranial progression and salvage radiotherapy. *J Neurooncol* **61**, 143–149.
20. Lutterbach, J., Cyron, D., Henne, K., and Ostertag, C. B. (2003). Radiosurgery followed by planned observation in patients with one to three brain metastases. *Neurosurg* **52**, 1066–1074.
21. Hawegawa, T., Kondziolka, D., Flickinger, J. C., Germanwala, A., and Lunsford, L. D. (2003). Brain metastases treated with radiosurgery alone: An alternative to whole brain radiotherapy? *Neurosurg* **52**, 1318–1326.
22. Lang, F. F., and Sawaya, R. (1996). Surgical management of cerebral metastases. *Neurosurg Clin N Am* **7**, 459–484.
23. Tan, T. C., and Black, P. M. (2003). Image-guided craniotomy for cerebral metastases: Techniques and outcomes. *Neurosurg* **53**, 82–90.
24. Lesser, G. J. (1996). Chemotherapy of cerebral metastases from solid tumors. *Neurosurg Clin N Am* **7**, 527–536.

25. Newton, H. B. (2000). Novel chemotherapeutic agents for the treatment of brain cancer. *Expert Opin Investig Drugs* **12**, 2815–2829.

26. Newton, H. B. (2002). Chemotherapy for the treatment of metastatic brain tumors. *Expert Rev Anticancer Ther* **2**, 495–506.

27. Tosoni, A., Lumachi, F., and Brandes, A. A. (2004). Treatment of brain metastases in uncommon tumors. *Expert Rev Anticancer Ther* **4**, 783–793.

28. van den Bent, M. J. (2003). The role of chemotherapy in brain metastases. *Eur J Cancer* **39**, 2114–2120.

29. Schuette, W. (2004). Treatment of brain metastases from lung cancer: chemotherapy. *Lung Cancer* **45**(suppl 2), S253–S257.

30. Bafaloukos, D., and Gogas, H. (2004). The treatment of brain metastases in melanoma patients. *Cancer Treatment Rev* **30**, 515–520.

31. Liotta, L. A., Stetler-Stevenson, W. G., and Steeg, P. S. (1991). Cancer invasion and metastasis: Positive and negative regulatory elements. *Cancer Investig* **9**, 543–551.

32. Reilly, J. A., and Fidler, I. J. (1993). The biology of metastatis. *Contemp Oncol November,* 32–46.

33. Aznavoorian, S., Murphy, A. N., Stetler-Stevenson, W. G., and Liotta, L. A. (1993). Molecular aspects of tumor cell invasion and metastasis. *Cancer* **71**, 1368–1383.

34. Webb, C. P., and Vande Woude, G. F. (2000). Genes that regulate metastasis and angiogenesis. *J Neurooncol* **50**, 71–87.

35. Yoshida, B. A., Sokoloff, M. M., Welch, D. R., and Rinker-Schaeffer, C. W. (2000). Metastasis-suppressor genes: a review and perspective on an emerging field. *J Natl Cancer Inst* **92**, 1717–1730.

36. Milas, I., Komaki, R., Hachiya, T. *et al.* (2003). Epidermal growth factor receptor, cyclooxygenase-2, and BAX expression in the primary non-small cell lung cancer and brain metastases. *Clin Cancer Res* **9**, 1070–1076.

37. Hanahan, D., and Folkman, J. (1996). Patterns of emerging mechanisms of the angiogenic switch during tumorigenesis. *Cell* **86**, 353–364.

38. Beckner, M. E. (1999). Factors promoting tumor angiogenesis. *Cancer Investig* **17**, 594–623.

39. Schellingerf, P. D., Meinick, H. M., and Thron, A. (1999). Diagnostic accuracy of MRI compared to CCT in patients with brain metastases. *J Neurooncol* **44**, 275–281.

40. Gaspar, L., Scott, C., Rotman, M. *et al.* (1997). Recursive partitioning analysis (RPA) of prognostic factors in three radiation therapy oncology group (RTOG) brain metastases trials. *Int J Rad Oncol Biol Phys* **37**, 745–751.

41. Nussbaum, E. S., Djalilian, H. R., Cho, K. H., and Hall, W. A. (1996). Brain metastases: Histology, multiplicity, surgery, and survival. *Cancer* **78**, 1781–1788.

42. Twelves, C. J., Souhami, R. L., Harper, P. G. *et al.* (1990). The response of cerebral metastases in small cell lung cancer to systemic chemotherapy. *Br J Cancer* **61**, 147–150.

43. Cocconi, G., Lottici, R., Bisagni, G. *et al.* (1990). Combination therapy with platinum and etoposide of brain metastases from breast carcinoma. *Cancer Investig* **8**, 327–334.

44. Franciosi, V., Cocconi, G., Michiarava, M. *et al.* (1999). Front-line chemotherapy with cisplatin and etoposide for patients with brain metastases from breast carcinoma, nonsmall lung carcinoma, or melignant melanoma. A prospective study. *Cancer* **85**, 1599–1605.

45. Bernardo, G., Cuzzoni, Q., Strada, M. R. *et al.* (2002). First-line chemotherapy with vinorelbine, gemcitabine, and carboplatin in the treatment of brain metastases from non-small-cell lung cancer: A phase II study. *Cancer Investig* **20**, 293–302.

46. Rosner, D., Nemoto, T., and Lane, W. W. (1986). Chemotherapy induces regression of brain metastases in breast carcinoma. *Cancer* **58**, 832–839.

47. Boogerd, W., Dalesio, O., Bais, E. M., and Van Der Sande, J. J. (1992). Response of brain metastases from breast cancer to systemic chemotherapy. *Cancer* **69**, 972–980.

48. Robinet, G., Thomas, P., Breton, J. L. *et al.* (2001). Results of a phase III study of early *versus* delayed whole brain radiotherapy with concurrent cisplatin and vinorelbine combination in inoperable brain metastasis of non-small-cell lung cancer: Groupe Français de Pneumo-Cancérologie (GFPC) protocol 95–1. *Ann Oncol* **12**, 59–67.

49. Postmus, P. E., Haaxma-Reiche, H., Smit, E. F. *et al.* (2000). Treatment of brain metastases of small-cell lung cancer: Comparing teniposide and teniposide with whole-brain radiotherapy – A phase III study of the European Organization for the Research and Treatment of Cancer Lung Cancer Cooperative Group. *J Clin Oncol* **18**, 3400–3408.

50. Ushio, Y., Arita, N., Hayakawa, T. *et al.* (1991). Chemotherapy of brain metastases from lung carcinoma: A controlled randomized study. *Neurosurg* **28**, 201–205.

51. Guerrieri, M., Wong, K., Ryan, G., Millward, M., Quong, G., and Ball, D. L. (2004). A randomized phase III study of palliative radiation with concomitant carboplatin for brain metastases from non-small cell carcinoma of the lung. *Lung Cancer* **46**, 107–111.

52. Slichenmyer, W. J., Rowinsky, E. K., Donehower, R. C., and Kaufmann, S. H. (1993). The current status of camptothecin analogues as antitumor agents. *J Natl Cancer Inst* **85**, 271–291.

53. Ardizzoni, A., Hansen, H., Dombernowsy, P. *et al.* (1997). Topotecan, a new active drug in the second-line treatment of small-cell lung cancer: A phase II study in patients with refractory and sensitive disease. *J Clin Oncol* **15**, 2090–2096.

54. Korfel, A., Oehm, C., von Pawel, J. *et al.* (2002). Response to topotecan of symptomatic brain metastases of small-cell lung cancer also after whole-brain irradiation: a multicentre phase II study. *Eur J Cancer* **38**, 1724–1729.

55. Oberhoff, C., Kieback, D. G., Würstlein, R. *et al.* (2001). Topotecan chemotherapy in patients with breast and brain metastases: Results of a pilot study. *Onkologie* **24**, 256–260.

56. Wong, E. T., and Berkenblit, A. (2004). The role of topotecan in the treatment of brain metastases. *The Oncologist* **9**, 68–79.

57. Eckardt, J. R., Martin, K. A., Schmidt, A. M., White, L. A., Greco, A. O., and Needles, B. M. (2002). A phase I trial of IV topotecan in combination with temozolomide daily time 5 every 28 days. *Proc ASCO* **21**, 83b.

58. Newlands, E. S., Stevens, M. F. G., Wedge, S. R. *et al.* (1997). Temozolomide: a review of its discovery, chemical properties, pre-clinical development and clinical trials. *Cancer Treat Rev* **23**, 35–61.

59. Hvisdos, K. M., and Goa, K. L. (1999). Temozolomide. *CNS Drugs* **12**, 237–243.

60. Stupp, R. K., Gander, M., Leyvraz, S., and Newlands, E. (2001). Current and future developments in the use of temozolomide for the treatment of brain tumors. *Lancet Oncol* **2**, 552–560.

61. Abrey, L. E., Olson, J. D., Raizer, J. J. *et al.* (2001). A phase II trial of temozolomide for patients with recurrent or progressive brain metastases. *J Neuro-Onc* **53**, 259–265.

62. Christodoulou, C., Bafaloukos, D., Kosmidos, P. *et al.* (2001). Phase II study of temozolomide in heavily pretreated cancer patients with brain metastases. *Ann Oncol* **12**, 249–254.

63. Freidman, H., Quinn, J., Reardon, D. *et al.* (2001). Phase II treatment of adults with brain metastases with Temodar. *Neuro-Oncol* **3**, 358.

64. Ebert, B. L., Niemierko, E., Shaffer, K., and Salgia, R. (2003). Use of temozolomide with other cytotoxic chemotherapy in the treatment of patients with recurrent brain metastases from lung cancer. *The Oncologist* **8**, 69–75.

65. Wick, W., Wick, A., Küker, W., Dichgans, J., and Weller, M. (2004). Intracranial metastatic esthesioneuroblastoma responsive to temozolomide. *J Neurooncol* **70**, 73–75.

66. Dziadziuszko, R., Ardizzoni, A., Postmus, P. E. *et al.* (2003). Temozolomide in patients with advanced non-small cell lung cancer with and without brain metastases: a phase II study of the EORTC Lung Cancer Group (08965). *Eur J Cancer* **39**, 1271–1276.

67. Dardoufas, C., Miliadou, A., Skarleas, C. *et al.* (2001). Concomitant temozolomide (TMZ) and radiotherapy (RT) followed by adjuvant treatment with temozolomide in patients with brain metastases from solid tumours. *Proceedings ASCO* **20**, 75b.

68. Antonadou, D., Paraskaveidis, M., Sarris, N. *et al.* (2002). Phase II randomized trial of temozolomide and concurrent radiotherapy in patients with brain metastases. *J Clin Oncol* **20**, 3644–3650.

69. Biasco, G., Pantaleo, M. A., and Casadei, S. (2001). Treatment of brain metastases of malignant melanoma with temozolomide. *New Engl J Med* **345**, 621–622.

70. Agarwala, S. S., Kirdwood, J. M., Gore, M. *et al.* (2004). Temozolomide for the treatment of brain metastases associated with metastatic melanoma: A phase II study. *J Clin Oncol* **22**, 2101–2107.

71. Bafaloukos, D., Tsoutsos, D., Fountzilas, G. *et al.* (2004). The effect of temozolomide-based chemotherapy in patients with cerebral metastases from melanoma. *Melanoma Res* **14**, 289–294.

72. Hwu, W. J., Raizer, J. J., Panageas, K. S., and Lis, E. (2001). Treatment of metastatic melanoma in the brain with temozolomide and thalidomide. *Lancet Oncol* **2**, 634–635.

73. Kumar, S., Witzig, T. E., and Rajkumar, S. V. (2004). Thalidomid: Current role in the treatment of non-plasma cell malignancies. *J Clin Oncol* **22**, 2477–2488.

74. Danson, S., Lorigan, P., Arance, A. *et al.* (2003). Randomized phase II study of temozolomide given every 8 hours or daily with either interferon alpha-2b or thalidomide in metastatic malignant melanoma. *J Clin Oncol* **21**, 2551–2557.

75. Stewart, D. J. (1989). Pros and cons of intra-arterial chemotherapy. *Oncol* **3**, 20–26.

76. Yamada, K., Bremer, A. M., West, C. R. *et al.* (1979). Intra-arterial BCNU therapy in the treatment of metastatic brain tumor from lung carcinoma. A preliminary report. *Cancer* **44**, 2000–2007.

77. Madajewicz, S., West, C. R., Park, H. C. *et al.* (1981). Phase II study–Intra-arterial BCNU therapy for metastatic brain tumors. *Cancer* **47**, 653–657.

78. Cascino, T. L., Byrne, T. N., Deck, M. D. F., and Posner, J. B. (1983). Intra-arterial BCNU in the treatment of metastatic tumors. *J Neurooncol* **1**, 211–218.

79. Madajewicz, S., Chowhan, N., Iliya, A. *et al.* (1991). Intracarotid chemotherapy with etoposide and cisplatin for malignant brain tumors. *Cancer* **67**, 2844–2849.

80. Barth, R. F., Yang, W., Rotaru, J. H. *et al.* (1997). Boron neutron capture therapy of brain tumors: Enhanced survival following intracarotid injection of either sodium borocaptate or boronophenylalanine with or without blood–brain barrier disruption. *Cancer Res* **57**, 1129–1136.

81. Kroll, R. A., and Neuwelt, E. A. (1998). Outwitting the blood–brain barrier for therapeutic purposes: Osmotic opening and other means. *Neurosurg* **42**, 1083–1100.

82. Eckman, W. W., Pattack, C. S., and Fenstermacher, J. D. (1974). A critical evaluation of the principles governing the advantages of IA infusion. *J Pharmacokinet Biopharmacokinet* **2**, 257–285.

83. Gelman, M., Chakares, D., and Newton, H. B. (1999). Brain tumors: Complications of cerebral angiography accompanied by intra-arterial chemotherapy. *Radiol* **213**, 135–140.

84. Newton, H. B., Stevens, C., and Santi, M. (2001). Brain metastases from fallopian tube carcinoma responsive to intra-arterial carboplatin and intravenous etoposide: A case report. *J Neurooncol* **55**, 179–184.

85. Newton, H. B., Snyder, M. A., Stevens, C. *et al.* (2003). Intra-arterial carboplatin and intravenous etoposide for the treatment of brain metastases. *J Neurooncol* **61**, 35–44.

86. Kroll, R. A., Pagel, M. A., Muldoon, L. L. *et al.* (1998). Improving drug delivery to intracerebral tumor and surrounding brain in a rodent model: A comparison of osmotic versus bradykinin modification of the blood–brain and/or blood–tumor barriers. *Neurosurg* **43**, 879–889.

87. Neuwelt, E. A., and Dahlborg, S. A. (1987). Chemotherapy administered in conjunction with osmotic blood–brain barrier modification in patients with brain metastases. *J Neurooncol* **4**, 195–207

88. Doolittle, N. D., Miner, M. E., Hall, W. A. *et al.* (2000). Safety and efficacy of a multicenter study using intra-arterial chemotherapy in conjunction with osmotic opening of the blood–brain barrier for the treatment of patients with malignant brain tumors. *Cancer* **88**, 637–647.

89. Tomita, T. (1991). Interstitial chemotherapy for brain tumors: review. *J Neurooncol* **10**, 57–74.

90. Brem, H., Mahaley, M. S., Vick, N. A. *et al.* (1991). Interstitial chemotherapy with drug polymer implants for the treatment of recurrent gliomas. *J Neurosurg* **74**, 441–446.

91. Tamargo, R. J., Myseros, J. S., Epstein, J. I. *et al.* (1993). Interstitial chemotherapy of the 9L gliosarcoma: Controlled release polymers for drug delivery in the brain. *Cancer Res* **53**, 329–333.

92. Brem, H., and Langer, R. (1996). Polymer-based drug delivery to the brain. *Science & Med* **3**, 2–11.

93. Ewend, M. G., Anderson, R. C., Tabassi, K. *et al.* (1996). Local delivery of BCNU from biodegradable polymer is superior to radiation therapy in treating intracranial melanoma metastases. *Surg Forum* **XLVII**, 564–566.

94. Ewend, M. G., Williams, J. A., Tabassi, K. *et al.* (1996). Local delivery of chemotherapy and concurrent external beam radiotherapy prolongs survival in metastatic brain tumor models. *Cancer Res* **56**, 5217–5223.

95. Ewend, M. G., Sampath, P., Williams, J. A. *et al.* (1998). Local delivery of chemotherapy prolongs survival in experimental brain metastases from breast carcinoma. *Neurosurg* **43**, 1185–1193.

96. Ewend, M. G., Brem, S., Gilbert, M. *et al.* (2001). The treatment of single brain metastasis with surgery, BCNU-polymer wafers, and radiation therapy: *Results of a phase I-II trial. Proceed ASCO* **20**, 57a.

97. Brem, S., Staller, A., Wotoczek-Obadia, M. *et al.* (2004). Interstitial chemotherapy for local control of CNS metastasis. *Neuro-Oncol* **6**, 370–371.

98. Golden, G. A., Meldorf, M. *et al.* (2004). Patients with metastatic brain cancer undergoing resection and gliadel implantation

experienced low local recurrence rates in the PROLONG registry. *Neuro-Oncol* **6**, 375–376.

99. Newton, H. B., Scott, S. R., Volpi, C., Miner, M. E., McGregor, J., and Ryken, T. (2002). Phase III randomized multicenter trial of surgical resection, Gliadel wafer, and irradiation *versus* surgical resection and irradiation alone for treatment of operable, solitary brain metastases from systemic cancer. *Ann Oncol* **13**, 166–167.

100. Garrett, M. D., and Workman, P. (1999). Discovering novel chemotherapeutic drugs for the third millennium. *Eur J Cancer* **35**, 2010–2030.

101. Livitzki, A., and Gazit, A. (1995). Tyrosine kinase inhibition: An approach to drug development. *Science* **267**, 1782–1788.

102. Gibbs, J. B. (2000). Anticancer drug targets: growth factors and growth factor signaling. *J Clin Investig* **105**, 9–13.

103. Dillman, R. O. (2001). Monoclonal antibodies in the treatment of malignancy: Basic concepts and recent developments. *Cancer Investig* **19**, 833–841.

104. Hao, D., and Rowinsky, E. K. (2002). Inhibiting signal transduction: Recent advances in the development of receptor tyrosine kinase and ras inhibitors. *Cancer Investig* **20**, 387–404.

105. Newton, H. B. (2003). Molecular neuro-oncology and the development of "targeted" therapeutic strategies for brain tumors. Part 1 – growth factor and ras signaling pathways. *Expert Rev Anticancer Ther* **3**, 595–614.

106. Woodburn, J. R. (1999). The epidermal growth factor receptor and its inhibition in cancer therapy. *Pharmacol Ther* **82**, 241–250.

107. Arteaga, C. L. (2001). The epidermal growth factor receptor: From mutant oncogene in honhuman cancers to therapeutic target in human neoplasia. *J Clin Oncol* **19**, 32s–40s.

108. Hidalgo, M., Siu, L. L., Nemunaitis, J. *et al.* (2001). Phase I and pharmacologic study of OSI-774, an epidermal growth factor receptor tyrosine kinase inhibitor, in patients with advanced solid malignancies. *J Clin Oncol* **19**, 3267–3279.

109. Giese, N. A., Lokker, N. A., Yu, J. C. *et al.* (2001). Development of antagonists of the platelet-derived growth factor receptor family. *Oncol Spectrums* **2**, 550–556.

110. Rowinsky, E. K., Windle, J. J., and Von Hoff, D. D. (1999). Ras protein farnesyltransferase: A strategic target for anticancer therapeutic development. *J Clin Oncol* **17**, 3631–3652.

111. Adjei, A. A. (2001). Blocking oncogenic ras signaling for cancer therapy. *J Natl Cancer Inst* **93**, 1062–1074.

112. Crul, M., De Klerk, G. J., Swart, M. *et al.* (2002). Phase I clinical and pharmacologic study of chronic oral administration of the farnesyl protein transferase inhibitor R115777 in advanced cancer. *J Clin Oncol* **20**, 2726–2735.

113. Yip, D., Ahmad, A., Karapetis, C. S. *et al.* (1999). Matrix metalloproteinase inhibitors: applications in oncology. *Investig New Drugs* **17**, 387–399.

114. Hidalgo, M., and Edkhardt, S. G. (2001). Development of matrix metalloproteinase inhibitors in cancer therapy. *J Natl Cancer Inst* **93**, 178–193.

115. Grossi, P. M., Ochiai, H., Archer, G. E. *et al.* (2003). Efficacy of intracerebral microinfusion of trastuzumab in an athymic rat model of intracerebral metastatic breast cancer. *Clin Cancer Res* **9**, 5514–5520.

116. Mauro, M. J., O'Dwyer, M., Heinrich, M. C., and Druker, B. J. (2001). STI571, A paradigm of new agents for cancer therapeutics. *J Clin Oncol* **20**, 325–334.

117. Savage, D. G., and Antman, D. H. (2002). Imatinib mesylate – a new oral targeted therapy. *New Engl J Med* **346**, 683–693.

118. Heinrich, M. C., Blanke, C. D., Druker, B. J., and Corless, C. L. (2002). Inhibition of KIT tyrosine kinase activity; A novel molecular approach to the treatment of KIT-positive malignancies. *J Clin Oncol* **20**, 1692–1703.

119. Brooks, B. J., Bani, J. C., Fletcher, C. D. M., and Demeteri, G. D. (2002). Response of metastatic gastrointestinal stromal tumor including CNS involvement to imatinib mesylate (STI-571). *J Clin Oncol* **20**, 870–872.

120. Cappuzzo, F., Ardizzoni, A., Soto-Parra, H. *et al.* (2003). Epidermal growth factor receptor targeted therapy by ZD 1839 (Iressa) in patients with brain metastases from non-small cell lung cancer (NSCLC). *Lung Cancer* **41**, 227–231.

121. Cappuzzo, F., Calandri, C., Bartolini, S., and Crinò, L. (2003). ZD 1839 in patients with brain metastases from non-small cell lung cancer (NSCLC): report of four cases. *Br J Cancer* **89**, 246–247.

122. Poon, A. N. Y., Ho, S. S. M., Yeo, W., and Mok, T. S. K. (2004). Brain metastases responding to gefitinib alone. *Oncology* **67**, 174–178.

123. Ishida, A., Kanoh, K., Nishisaka, T. *et al.* (2004). Gefitinib as a first line of therapy in non-small cell lung cancer with brain metastases. *Intern Med* **43**, 718–720.

124. Katz, A., and Zalewski, P. (2003). Quality-of-life benefits and evidence of antitumor activity for patients with brain metastases treated with gefitinib. *Br J Cancer* **89**, S15–S18.

125. Ceresoli, G. L., Cappuzzo, F., Gregorc, V., Bartolini, S., Crinò, L., and Villa, E. (2004). Gefitinib in patients with brain metastases from non-small cell lung cancer: a prospective trial. *Ann Oncol* **15**, 1042–1047.

36

The Role of Chemotherapy in Pediatric Gliomas

Lorna K. Fitzpatrick and Patricia K. Duffner

ABSTRACT: Pediatric gliomas are a heterogeneous group of brain tumors that include low-grade and high-grade malignancies. Historically, treatment for these tumors has been a combination of surgery and radiation therapy. In certain situations, addition of chemotherapy to treatment regimens has proven beneficial in the treatment of pediatric gliomas. This chapter will address the uses of chemotherapy in the treatment of pediatric gliomas, including low-grade gliomas, optic pathway/hypothalamic gliomas, high-grade gliomas, and diffuse intrinsic pontine gliomas.

INTRODUCTION

Brain tumors are the most common solid tumor in children with an annual incidence of 4 per 100 000 person years. Of these, astrocytomas represent the largest histopathologic grouping. In the SEER data of children with brain tumors (0–15 years), astrocytomas represented 57 per cent of the tumors while in the Central Brain Tumor Registry of the United States (CBTRUS) Registry (0–19 years) astrocytomas accounted for 20.5 per cent [1]. Treatment of these tumors is diverse and depends upon the pathologic diagnosis, surgical options, and age of the child. Although surgery and radiation therapy are often considered to be standard of care in patients with intracranial gliomas, the use of chemotherapy is expanding.

Historically, delivery of chemotherapy to the central nervous system (CNS) has been difficult because of a variety of factors. Primarily, the tight endothelial junctions of the blood–brain barrier (BBB) can restrict the delivery of chemotherapy into the CNS. Penetration of the BBB is enhanced in compounds that are nonionizable at a physiologic pH and have a low molecular weight, high lipid solubility, and lack of plasma protein binding. A better understanding of these properties has led to the development of newer chemotherapeutic agents that can penetrate the central nervous system. Attempts have also been made to interrupt the BBB in order to improve the drug delivery (see Chapter 18 on Blood–Brain Barrier Disruption Chemotherapy). Many of these techniques, however, are limited by toxicity and complexity [2–4]. Additionally, many of the gliomas have been relatively resistant to chemotherapy. This resistance can be complicated by the use of anti-epileptic drugs that induce hepatic P450, significantly altering the blood levels and hence the efficacy of chemotherapeutic agents. Development of newer chemotherapy agents that can cross the BBB as well as the use of newer anti-epileptic agents that are not metabolized in the liver, such as gabapentin and levetiracetam has helped to minimize some of these obstacles.

The role of chemotherapy in pediatric gliomas is now expanding. Two main reasons exist for the addition of chemotherapy to treatment plans for children with these neoplasms. First, the historically grave prognosis with high-grade gliomas and diffuse intrinsic pontine gliomas has prompted the addition of chemotherapy agents to standard radiation and surgery in an attempt to improve survival. Second, the adverse effects of radiation on children are

TABLE 36.1   Mechanism of Action and Side Effects of Chemotherapies Frequently
Used in Pediatric CNS Gliomas

Chemotherapy	Mechanism of action	Side effects
Carboplatin	Heavy-metal/Alkylating-like agent	Anaphylaxis, thrombocytopenia,rash, abnormal liver function, rarely: hepatic veno-occlusive disease, peripheral neuropathy, renal abnormalities, hearing loss, optic neuritis
Temozolamide	Alkylating agent	Pancytopenia, mouth sores, fatigue, and abnormal liver function, rarely: seizures, dizziness, prolonged thrombocytopenia
Cisplatin	Heavy-metal/ Alkylating-like agent	Pancytopenia, nephrotoxicity, abnormal liver function, peripheral neuropathy, ototoxicity, hypomagnesemia, Rarely: cardiovascular toxicity
Cyclophosphamide	Alkylating agent	Neutropenia, anemia, abnormal liver enzymes, headaches, hemorrhagic cystitis, sterility, rarely: interstitial pulmonary fibrosis, cardiac necrosis, acute pericarditis
Vincristine	Vinca alkyloid/tubulin inhibitor	Constipation, abdominal pain, jaw/leg pain, elevated liver enzymes, peripheral neuropathy, ptosis, rarely: SIADH, bronchospasm, pancreatitis fatal if given intrathecally
Etoposide	Topoisomerase II inhibitor	Pancytopenia, stomatitis, anaphylactic reactions, rarely: peripheral neuropathy, somnolence, second malignancy (acute myelogenous leukemia)
Carmustine/BCNU	Nitrosurea	Pancytopenia, elevated liver enzymes, pulmonary fibrosis, hypotension, azotemia, rarely: second malignancies
Lomustine/CCNU	Nitrosurea	Pancytopenia, rarely: liver/renal dysfunction, second malignancy, pulmonary dysfunction/fibrosis, ataxia, dysarthria

well established. Neuropsychological, developmental, growth, and endocrine disorders have been described in young patients who have received radiation therapy to the brain and neuroaxis. These effects are potentially devastating in very young children [5]. Chemotherapy, however, has proven to be less toxic in these patients [6–8]. Due to this, chemotherapy is now being used for some patients with low-grade gliomas in order to postpone radiation therapy, thereby minimizing the toxic effects.

This chapter will address the use of chemotherapy in pediatric gliomas, including low-grade astrocytomas, optic pathway gliomas, high-grade gliomas, and diffuse intrinsic pontine gliomas (Table 36.1.).

LOW-GRADE GLIOMAS

Low-grade gliomas are a heterogeneous group of brain tumors that have varied pathology and location. Included in this group are WHO grade 1 and 2 tumors, pilocytic astrocytomas, fibrillary, gemistocytic, and mixed variants. Less common are the pleomorphic xanthroastrocytomas and subependymal giant cell astrocytomas [9]. Optic pathway gliomas are also considered to be low-grade gliomas and are discussed separately in the next section.

Pilocytic astrocytomas (WHO grade I) are the most common histopathologic grouping of the low-grade gliomas. They are characterized by compact and spongy regions consisting of elongated cells surrounding spongy loose areas in which stellate astrocytes are observed. Necrosis, hemorrhage, and Rosenthal fibers may be found. They are often microcystic or may have a single large cyst with a tumor nodule. The diffuse astrocytoma (fibrillary pattern) has meningeal fibrils which impact a firm consistency to the tumor. Mitoses are absent and there is no endothelial proliferation. Gemistocytic astrocytomas have homogeneous, lightly eosinophilic cytoplasm. They have a tendency to dedifferentiate into glioblastoma multiforme [10].

Surgery is the treatment of choice for these tumors. For patients with a total excision, prognosis is excellent. For some patients, tumor location or size can preclude complete surgical excision. In these instances, additional treatment may be necessary for progressive or symptomatic patients. Treatment options for these patients include radiation therapy or chemotherapy. Side effects from radiation therapy can be detrimental to young children, and can negatively affect intellectual functioning, growth, and endocrinologic function [11]. The role of chemotherapy in the treatment of progressive or symptomatic low-grade tumors in young children has, therefore, been expanded in an effort to spare them the effects of radiation therapy. In a Phase II Children's Oncology Group randomized study, patients with progressive

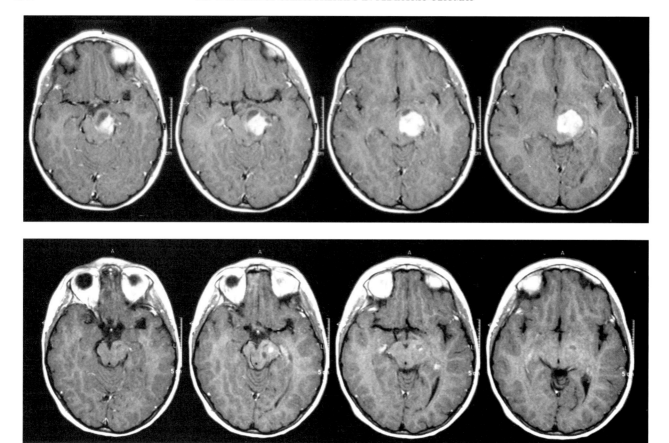

FIGURE 36.1   MRI scan of a seven-year-old patient with low-grade astrocytoma. Top: MRI done at
diagnosis. Bottom: MRI done after treatment with vincristine/carboplatin for one year.

or recurrent brain tumors were randomized to receive
either carboplatin or iproplatin. Patients were given
carboplatin 560 mg/m^2 at four week intervals or
iproplatin 270 mg/m^2 at three week intervals.
Twenty-four patients on the study had low-grade
astrocytic neoplasms. Overall, 71 per cent of the
patients with low-grade astrocytomas had response
to these agents. It was noted that patients treated
with carboplatin had sustained stable disease (median
40+ months) as compared with iproplatin (median
7 months). Major toxicities of therapy included
myelosuppression, particularly thrombocytopenia,
and ototoxitity [12]. Results from this study confirmed
the potential use of carboplatin in future chemotherapy regimens for low-grade gliomas.

Vincristine and carboplatin have also been used
together for the treatment of low-grade gliomas
(Fig. 36.1). Twenty-three children with recurrent and
thirty-seven children with newly diagnosed low-
grade gliomas were treated with a ten week induction
of weekly carboplatin (175 mg/m^2) and vincristine
(1.5 mg/m^2) followed by maintenance treatment with
the same drugs. Fifty-two percent of the children with

recurrent disease had an objective response
to treatment and 62 per cent of newly diagnosed
patients had an objective response. Hematologic
toxicity, specifically thrombocytopenia, was the most
common therapy induced toxicity, resulting in one
patient being removed from the study. Six children
had allergic reactions, presumably to the carboplatin [13]. Follow-up studies confirmed a 3-year
progression-free survival rate of 74 per cent in newly
diagnosed children less than five years old. Older
patients demonstrated only a 39 per cent progression-
free survival rate. There was no difference in response
in those patients with Neurofibromatosis Type I.
A current study in the Children's Oncology Group
randomizes patients to receive either vincristine/
carboplatin or thioguanine/procarbazine/CCNU/
vincristine for progressive or symptomatic tumors in
children less than ten years old. Results from this
study are still pending. More recently, temozolomide
has shown promising results in adult patients
with gliomas. Although more commonly used in
high-grade gliomas, the efficacy of temozolomide is
being evaluated in the low-grade gliomas.

OPTIC PATHWAY/ HYPOTHALAMIC GLIOMAS

Optic pathway gliomas/hypothalamic gliomas (OPHG) represent about five percent of intracranial neoplasms in young children. Children with these tumors tend to have an excellent long-term prognosis, and long-term survival is estimated at about 85 per cent [1,14]. Patients with optic pathway tumors are generally diagnosed on neuroimaging after presenting with decreased vision, optic atrophy, proptosis, or nystagmus. Optic gliomas are also the most common tumor associated with Neurofibromatosis Type I (NF-1), and patients with this disorder are often identified on routine screening done while they are asymptomatic [15].

The distinction between optic pathway and hypothalamic gliomas is difficult, as anterior hypothalamic tumors often blend imperceptibly into the area of the optic chiasm or tracts. Children with hypothalamic tumors may present with the diencephalic syndrome, characterized by failure to thrive, emaciation, and loss of subcutaneous fat in the presence of accelerated long bone growth [16]. Other presenting symptoms associated with hypothalamic tumors include precocious puberty, obesity, diabetes insipidus, hypogonadism, and lethargy. If the tumor extends into the region of the foramen of Monro, obstructive hydrocephalus with signs and symptoms of increased intracranial pressure will ensue.

Most OPHGs are pilocytic astrocytomas. Leptomeningeal seeding may occur with these tumors, particularly in those with the diencephalic syndrome.

Treatment of OPHGs requires a multi-disciplinary approach. Careful consideration needs to be given to the child's symptoms, visual acuity, and progression of symptoms. Complete surgical excision is often not possible, and biopsy of the tumors may compromise vision. Spontaneous regression is well documented, and it is generally recommended that treatment be reserved for patients with progressive symptoms [15]. Patients with NF-1 have a higher rate of regression than the general population [16,17].

The decision to start therapy should take into consideration multiple factors including extent of disease, rate of progression, and potential for useful vision [19]. Watchful waiting is often the approach to children with optic pathway tumors, particularly those identified on screening neuroimaging for NF-1. Treatment is initiated upon signs of progressive visual loss, proptosis, signs of increased intracranial pressure, or progression on neuroimaging. In children with intraorbital optic gliomas who have no useful vision, surgical resection of the nerve to the level of the chiasm is warranted. Otherwise the role of surgery in OPT is limited. In contrast, debulking surgery for hypothalamic tumors has gained in popularity. Radiation therapy in doses of 45–60 Gy has been used with good success, and generally can cause regression or stable disease. Risks of using this treatment in young children, however, is significant, with 69 per cent of children less than ten years old developing hypothalamic-pituitary deficiency. Additionally, use of radiation therapy in these patients has been associated with detrimental intellectual sequelae and the development of moya moya disease, especially in children less than 5 years with NF-1 [18].

Due to this, chemotherapeutic options have been explored. Combination chemotherapy with actinomycin D and vincristine was used to treat twenty-four patients with progressive OPHGs [22]. Patients had a median age of 1.6 years at the time of diagnosis and were followed for a median of 4.3 years. Patients were treated with six 8-week cycles of vincristine and actinomycin-D. Of these patients, 62.5 per cent had stable or responsive disease and 37.5 per cent had clinical progression. Median time to progression was 3 years. Mean Intelligence Quotient (IQ) testing on patients who received only chemotherapy was 103. Seven patients had sequential IQ testing before and after therapy, and of these, none demonstrated a decrease in IQ scores. Based on this study, it was concluded that chemotherapy could significantly delay the need for radiation therapy in some children while avoiding severe neurotoxicity.

The use of other chemotherapy agents was then investigated. Patients with progressive or recurrent OPHGs were included in the Pediatric Oncology Group study discussed above that used carboplatin or iproplatin in progressive brain tumors [12]. Carboplatin was then investigated as a single agent in patients younger than five years old with progressive optic pathway tumors. Carboplatin was given at a dose of 560 mg/m^2 per week. It was concluded that carboplatin was active and well tolerated in children with progressive optic pathway tumors [21]. Carboplatin was then used in conjunction with vincristine in patients with OPHGs. Other studies have confirmed that the use of carboplatin when used in conjunction with vincristine can defer or prevent the need for radiation therapy in patients with OPHGs [20,21]. A current Children's Oncology Group study is evaluating the efficacy of thioguanine, procarbazine, CCNU, and vincristine (TPCV). In this study, patients are randomized to carboplatin and vincristine or TPCV. Patients with Neurofibromatosis Type I are nonrandomly assigned to receive carboplatin and vincristine. Results from this study are pending.

As the long-term survival from these tumors is excellent, therapy is generally reserved for patients with progressive or symptomatic disease. Current treatment options aim to maximize benefits and minimize long-term side effects in this group of patients.

HIGH-GRADE GLIOMAS

High-grade gliomas include tumors, such as anaplastic astrocytoma and glioblastoma multiforme. Anaplastic astrocytomas are characterized on histopathology by increased cellularity, cellular pleomorphism and mitoses, while glioblastoma multiforme has pleomorphism, endothelial proliferation, necrosis, multinucleated giant cells, mitoses (70 per cent) and hemorrhage (60 per cent) [10,23,24]. Children with these diagnoses tend to have a very poor prognosis despite the use of combinations of surgery, chemotherapy, and radiation therapy. Surgery alone is not considered to be curative for these tumors and the addition of radiation therapy has been considered to be standard treatment. Despite the use of radiation therapy, long-term survival remains poor, prompting the addition of chemotherapy in attempts to improve survival.

The effectiveness of chemotherapy in high-grade gliomas was addressed in a prospective randomized trial of the Children's Cancer Study Group [25]. Patients with high-grade gliomas were treated with surgery and radiation therapy. Patients were randomized to receive either chemotherapy or no chemotherapy. Chemotherapy consisted of CCNU, vincristine, and prednisone (PCV). Addition of chemotherapy appeared to prolong survival and event-free survival, with five-year event-free survival reaching 46 per cent for those patients who received radiation and chemotherapy as opposed to 18 per cent for those who received radiation alone. Patients with recurrent disease did not survive despite chemotherapy. Attempts to improve chemotherapy outcomes resulted in later studies in which an "eight-drugs-in-one-day" chemotherapy regimen was compared in a randomization with the previous PCV regimen [26]. The eight drugs were chosen to minimize bone marrow toxicity, maximize possible penetration of the blood–brain barrier and overcome resistance. There was no difference in outcomes as compared with the PCV regimen, and toxicities were greater, limiting the usefulness of the "eight-drugs-in-one-day" regimen [26]. Some studies

have suggested that cisplatin has a limited role in the treatment of high-grade gliomas, and it is not clear if the addition of cisplatin to a nitrosurea provides any survival advantage. Other studies have suggested a limited role of cisplatin in recurrent astrocytic tumors [24].

High-grade gliomas in infants are rare. There is some evidence, however, that these patients may fare better than adult patients with high-grade gliomas. In the Pediatric Oncology Group experience, eighteen children less than three years of age were treated on a regimen of prolonged post-operative chemotherapy with vincristine /cyclophosphamide alternating with etoposide/cisplatin. Response rates to two cycles of cyclophosphamide and vincristine were 60 per cent. Progression-free survival at one year was 54 per cent, and progression-free survival at three and five years were 43 per cent. This suggests that malignant gliomas in children may be more responsive to chemotherapy than in adult patients and that gliomas in infants may have a better prognosis [28].

Temozolomide is an oral alkylating agent that has proven efficacy in the treatment of gliomas. This drug has been studied in pediatric Phase I and II studies as detailed above. Temozolomide has been studied in adult populations and has become a first-line agent for the treatment of malignant gliomas [29]. In pediatrics, temozolomide has been used as a treatment for recurrent high-grade gliomas. In a study of recurrent high-grade gliomas, 20 per cent of patients had a measurable response and the chemotherapy was well tolerated [30]. The good response rate combined with the low toxicity profile have led to trials in pediatrics with temozolomide being used as a first-line agent in conjunction with radiation after surgical resection. Results from this trial are still pending. Temozolomide has also been used in combination with nitrosoureas in adult trials of patients with high-grade gliomas, and the therapy has been well tolerated. Combination of these two agents in pediatric clinical trials is forthcoming.

DIFFUSE INTRINSIC PONTINE GLIOMAS

Children with brainstem gliomas typically present with the clinical triad of long tract signs, cranial neuropathies, and ataxia. While those with focal or exophytic tumors, cystic lesions, a cervical medullary location, and/or pilocytic histology may fare better, those with diffuse intrinsic location in the pons, a duration of symptoms less than six months, two or three brainstem signs and tumor that engulfs the basilar artery have an extremely poor prognosis with survivals less than 10 per cent [31]. Survival rates for those patients with diffuse intrinsic pontine gliomas have changed little over the last several decades despite multi-modality treatment with radiation therapy and chemotherapy. Outlook for these patients remains dismal.

Surgical resection is not possible in these patients, and radiation therapy is the standard of care. Several studies of hyperfractionated radiation were conducted by the Pediatric Oncology Group ranging in dose from 6400 cGy to as high as 7800 cGy. Despite these very high doses, no improvement in survival was identified [32,33]. Therefore, in an attempt to improve long-term survival, various chemotherapeutic agents have been investigated. A Children's Cancer Group study evaluated the use of pre-irradiation chemotherapy in newly diagnosed brainstem gliomas. Patients were randomly assigned to receive carboplatin, etoposide, and vincristine or to cisplatin, etoposide, cyclophosphamide, and vincristine. Neither chemotherapy regimen improved overall survival, with event-free survival of 6 per cent at two years [34]. Use of chemotherapy with radiation therapy has not improved survival in published reports [35]. Due to the grim prognosis, high-dose chemotherapy has been used, including the use of autologous stem cell rescue (see Chapter 29 on Bone Marrow Transplant Chemotherapy). Again, there has been no improved survival [36]. Several Phase I trials are available for these patients.

HIGH-DOSE CHEMOTHERAPY AND AUTOLOGOUS STEM CELL RESCUE IN PEDIATRIC GLIOMAS

The dismal prognosis for patients with high-grade gliomas and diffuse intrinsic brainstem gliomas has led to the investigation of high-dose chemotherapy followed by autologous stem cell rescue. The rationale for this treatment is multi-factorial. Primarily, conventional treatments have failed to produce adequate survival results. Radiation therapy alone has not provided adequate outcomes, and the addition of standard-dose chemotherapy to radiation has not significantly improved survival [37]. The failure of chemotherapy to improve survival is likely due to the restrictions of the BBB as well as to intrinsic drug resistance of these tumors [38,39]. The theoretical advantage to high-dose chemotherapy is the ability to overcome these barriers.

High-dose chemotherapy with autologous stem cell rescue was first tried in adult patient with high-grade gliomas in the early 1980s. Results from these studies are mixed, with some demonstrating unacceptable toxicities and some showing modest improvement in survival. Toxicities included severe CNS and pulmonary toxicity, failure of engraftment, veno-occlusive disease of the liver and severe infections. No consistent improvement in survival has been documented [40–44].

Dismal prognosis for pediatric patients with high-risk gliomas, including glioblastoma multiforme and diffuse intrinsic brainstem glioma, has prompted the evaluation of autologous stem cell rescue after high-dose chemotherapy. Preparative regimens have included melphalan/cytoxan, thiotepa/etoposide/BCNU, Thiotepa/etoposide and single agent BCNU [41,45–51]. In comparison with adult patients, there is an overall trend of less toxic deaths associated with such regimens. These intensive regimens have failed to significantly improve long-term survival in these patients.

In summary, pediatric gliomas are a heterogeneous group of tumors with very varied prognoses. Treatment of choice for these tumors is generally surgical resection and radiation therapy. The use of chemotherapy is proving to be promising both as a way to delay radiation in young children and as an adjuvant to radiation therapy in patients with poor prognoses.

References

1. Central Brain Tumor Registry of the United States www.cbtrus.org.
2. Lashford, L. S., Thiesse, P., Jouvet, T. *et al.* (2002). Temozolomide in malignant gliomas of childhood: a United Kingdom Children's Cancer Study Group and French Society for Pediatric Oncology Intergroup Study. *J Clin Oncol* **20**, 4684–4691.
3. Grossman, S. A., and Batara, J. F. (2004). Current management of glioblastoma multiforme. *Semin Oncol* **31**, 635–644.
4. Kemper, E. M., Boogerd, W., Thuis, I. *et al.* (2004). Modulation of the blood-brain barrier in oncology: therapeutic opportunities for the treatment of brain tumours? *Cancer Treat Rev* **30**, 415–423.
5. Duffner, P. K., and Cohen, M. E. (1991). The long-term effects of central nervous system therapy on children with brain tumors. *Neurol Clin* **9**, 479–495.
6. Duffner, P. K. (2004). Long-term effects of radiation therapy on cognitive and endocrine function in children with leukemia and brain tumors. *Neurologist* **10**, 293–310.
7. Duffner, P. K., Marc, E. Horowitz, Jeffrey, P. *et al.* (1993). Postoperative chemotherapy and delayed radiation in children less than three years of age with malignant brain tumors. *N Engl J Med* **328**, 1725–1731.
8. Lacaze, E., Kieffer, V., Streri, A. *et al.* (2003). Neuropsychological outcome in children with optic pathway tumours when first-line treatment is chemotherapy. *Br J Cancer* **89**, 2038–2044.

9. Stupp, R., Janzer, R. C., Hegi, M. E. et al.(2003). Prognostic factors for low-grade gliomas. *Semin Oncol* **30(6 Suppl 19)** 23–28.

10. Russell, D. S., and Rubinstein, L. F. (1989). *Pathology of tumors of the nervous system*, 4th Ed, Edward Arnold, London.

11. Kortmann, R. D., Timmermann, B., Taylor, R. E. et al. (2003). Current and future strategies in radiotherapy of childhood low-grade glioma of the brain. Part I: Treatment modalities of radiation therapy. *Strahlenther Onkol* **179(8)**, 509–520.

12. Friedman, H.S., Krischer, J. P., Burger, P. et al. (1992). Treatment of children with progressive or recurrent brain tumors with carboplatin or iproplatin: a Pediatric Oncology Group ·randomized phase II study. *J Clin Oncol* **10(2)**, 249–256.

13. Packer, R.J., Ater, J., Allen, J. et al. (1997). Carboplatin and vincristine chemotherapy for children with newly diagnosed progressive low-grade gliomas. *J Neurosurg* **86(5)**, 747–754.

14. Gnekow, A.K., Kortmann, R. D. Pietsch, T., and Emser, A. (2004). Low grade chiasmatic-hypothalamic glioma-carboplatin and vincristine chemotherapy effectively defers radiotherapy within a comprehensive treatment strategy – report from the multicenter treatment study for children and adolescents with a low grade glioma – HIT-LGG 1996 – of the Society of Pediatric Oncology and Hematology (GPOH). *Klin Padiatr* **216(6)**, 331–342.

15. Grabenbauer, G. G., Schuchardt, U., Buchfelder, M. (2000). Radiation therapy of optico-hypothalamic gliomas (OHG)–radiographic response, vision and late toxicity. *Radiother Oncol* **54(3)**, 239–245.

16. White, P. T., and Ross, A. T. (1963). Inanition syndrome in infants with anterior hypothalamic neoplasms. *Neurology* **13**, 974.

17. Parsa, C. F., Hoyt, C. S., Lesser, R. L. et al. (2001). Spontaneous regression of optic gliomas: thirteen cases documented by serial neuroimaging. *Arch Ophthalmol* **119(4)**, 516–529.

18. Korf, B. R. (2000). Malignancy in neurofibromatosis type 1. *Oncologist* **5(6)**, 477–485.

19. Allen, J. C. (2000). Initial management of children with hypothalamic and thalamic tumors and the modifying role of neurofibromatosis-1. *Pediatr Neurosurg* **32(3)**, 154–162.

20. Packer, R. J., Lange, B., Ater, J. et al. (1993). Carboplatin and vincristine for recurrent and newly diagnosed low-grade gliomas of childhood. *J Clin Oncol* **11(5)**, 850–856.

21. Mahoney, D. H., Jr. Cohen, M. E., Friedman, H. S. et al. (2000). Carboplatin is effective therapy for young children with progressive optic pathway tumors: a Pediatric Oncology Group phase II study. *Neuro-oncol* **2(4)**, 213–220.

22. Packer, R. J., Sutton, L. N., Bilaniuk, L. T. et al. (1988). Treatment of chiasmatic/hypothalamic gliomas of childhood with chemotherapy: an update. *Ann Neurol* **23(1)**, 79–85.

23. Frankel, S. A., and German, W. J. (1958). Glioblastoma multiforme:review of 219 cases with regard to natural history, pathology, diagnostic methods, and treatment. *J Neurosurg* **15**, 489.

24. Rubinstein, L. J. (1972). *Tumors of the central nervous system*. Armed Forces Institute of Pathology, Washington, DC.

25. Sposto, R., Ertel, I. J., Jenkin, R. D. et al. (1989). The effectiveness of chemotherapy for treatment of high grade astrocytoma in children: results of a randomized trial. A report from the Childrens Cancer Study Group. *J Neurooncol* **7(2)**, 165–177.

26. Pendergrass, T. W., Milstein, J. M., Geyer, J. R. et al. (1987). Eight drugs in one day chemotherapy for brain tumors: experience in 107 children and rationale for preradiation chemotherapy. *J Clin Oncol* **5(8)**, 1221–1231.

27. Sexauer, C. L., Khan, A., Burger, P. C. et al. (1985). Cisplatin in recurrent pediatric brain tumors. A POG Phase II study. A Pediatric Oncology Group Study. *Cancer* **56(7)**, 1497–1501.

28. Duffner, P. K., Krischer, J. P., Burger, P. C. et al. (1996). Treatment of infants with malignant gliomas: the Pediatric Oncology Group experience. *J Neurooncol* **28(2–3)**, 245–256.

29. Chibbaro, S., Benvenuti, L., Caprio, A. et al. (2004). Temozolomide as first-line agent in treating high-grade gliomas: phase II study. *J Neurooncol* **67(1–2)**, 77–81.

30. Verschuur, A. C., Grill, J., Lelouch-Tubiana, A. et al. (2004). Temozolomide in paediatric high-grade glioma: a key for combination therapy? *Br J Cancer* **91(3)**, 425–429.

31. Fisher, P. G., Breiter, S. N., Carson, B. S. et al. (2000). A clinicopathologic reappraisal of brainstem tumor classification. Identification of a pilocystic astrocytoma and fibrillary astrocytoma as distinct entities. *Cancer* **89(7)**, 1569.

32. Mandell, L. R., Kadota, R., and Freeman, C. (1999). There is no role for hyperfractionated radiotherapy in the management of children with newly diagnosed diffuse intrinsic brainstem tumors: results of a Pediatric Oncology Group Phase III trial comparing conventional vs. hyperfractionated radiotherapy. *Int J Radiat Oncol Biol Phys* **43(5)**, 959.

33. Freeman, C. R., Krischer, J., Sanford, R. A. et al (1991). Hyperfractionated radiotherapy in brain tumors: results of treatment at the 7020 cGy dose level: Pediatric Oncology Group Study #8495. *Cancer* **68**, 474.

34. Jennings, M. T., Sposto, R., Boyett, J. M. et al. (2002). Preradiation chemotherapy in primary high-risk brainstem tumors: phase II study CCG-9941 of the Children's Cancer Group. *J Clin Oncol* **20(16)**, 3431–3437.

35. Wolff, J. E. et al. (2002). Treatment of paediatric pontine glioma with oral trophosphamide and etoposide. *Br J Cancer* **87(9)**, 945–949.

36. Wolff, J. E., and Finlay, J. L. (2004). High-dose chemotherapy in childhood brain tumors. *Onkologie* **27(3)**, 239–245.

37. Joohnson, D. B., Thompson, J. M, Corwin, J. A. et al. (1987). Prolongation of survival for high-grade malignant gliomas with adjuvant high-dose BCNU and autologous bone marrow transplantation. *J Clin Oncol* **5(5)**, 783–789.

38. Fine, H. A., and Antman, K. H. (1992). High-dose chemotherapy with autologous bone marrow transplantation in the treatment of high grade astrocytomas in adults: therapeutic rationale and clinical experience. *Bone Marrow Transplant* **10(4)**, 315–321.

39. Petersdorf, S. H., and Livingston, R. B. (1994). High dose chemotherapy for the treatment of malignant brian tumors, *J Neurooncol* **20(2)**, 155–163.

40. Wolff, S. N., Phillips, G. L., and Herzig, G. P. (1987). High-dose carmustine with autologous bone marrow transplantation for the adjuvant treatment of high-grade glioma of the central nervous system. *Cancer Treat Rep* **71(2)**, 183–185.

41. Papadakis, V., Dunkel, I. J., Cramer, L. D. et al. (2000). High-dose carmustine, thiotepa and etoposide followed by autologous bone marrow rescue for the treatment of high risk central nervous system tumors. *Bone Marrow Transplant* **26(2)**, 153–160.

42. Mbidde, E. K., Selby, P. J., and Perren, T. J. (1998). High dose BCNU chemotherapy with autologous bone marrow transplantation and full dose radiotherapy for grade IV astrocytoma. *Br J Cancer* **58(6)**, 779–782.

43. Mortimer, J. E., Hewlett, J. S., Bay, J. et al. (1983). High dose BCNU with autologous bone marrow rescue in the treatment of recurrent malignant gliomas. **1(3)**, 269–273.

44. Nomura, K., Watanabe, T., Nakamura, O. et al. (1984). Intensive chemotherapy with autologous bone marrow rescue for recurrent malignant gliomas **7(1)**, 13–22.

45. Finlay, J.L., August, C., Packer, R. *et al.* (1990). High dose multi-agent chemotherapy followed by bone marrow 'rescue' for malignant astrocytomas of childhood and adolescence, *J Neurooncol* **9**(3), 239–248.

46. Hedieman, R. L., Douglass, E. C., Krance, R. A. *et al* (1993) High-dose chemotherapy and autologous bone marrow rescue followed by interstitial and external -beam radiotherapy in newly diagnosed pediatric malignant gilomas. *J Clin Oncol* **11**(8), 1458–1465.

47. Kedar, A., Maria, B. L., Graham-Pole, J. *et al.* (1994). High-dose chemotherapy with marrow reinfusion and hyperfractionated irradiation for children with high-risk brain tumors. *Med Pediatr Oncol* **23**, 428–436.

48. Mahoney, D. H., Jr., Strother, D., Camitta, B. *et al.* (1996). High-dose melphalan and cyclophosphamide with autologous bone marrow rescue for recurrent/progressive malignant brain tumors in children: a pilot pediatric oncology group study. *J Clin Oncol* **14**(2), 328–338.

49. Finlay, J. L., Goldman, S., Wong, M. C. *et al.* (1996). Pilot study of thiotepa and etoposide with autologous bone marrow rescue in children and youg adults with recurrent CNS tumors. The Children's Cancer Group. *J Clin Oncol* **14**(9), 2495–2503.

50. Bouffet, E., Khelfaoui, F., Philip, I. *et al.* (1997). High-dose carmustine for high-grade gliomas in childhood. *Cancer Chemother Pharmacol* **39**(4), 376–379.

51. Bouffet, E., Mottolese, C., Jouvet, A. *et al.* (1997). Etoposide and thiotepa followed by ABMT (autologous bone marrow transplantation) in children and young adults with high grade gliomas. *Eur J Cancer* **33**(1), 91–95.

Index

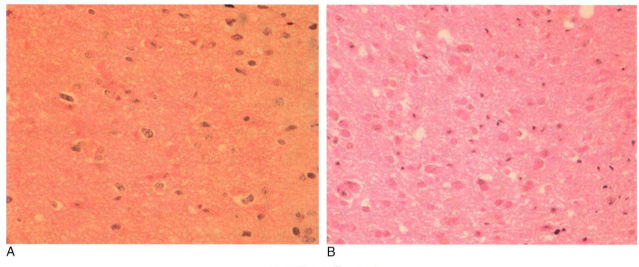

A   B

PLATE 1.1 (Fig. 1.1)

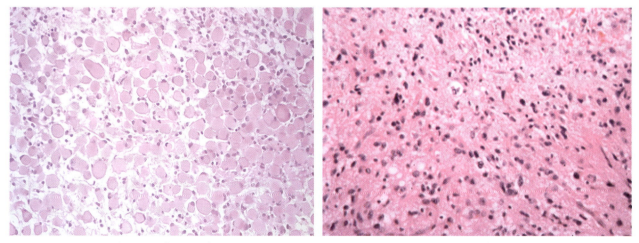

PLATE 1.2 (Fig. 1.2)   PLATE 1.3 (Fig. 1.3)

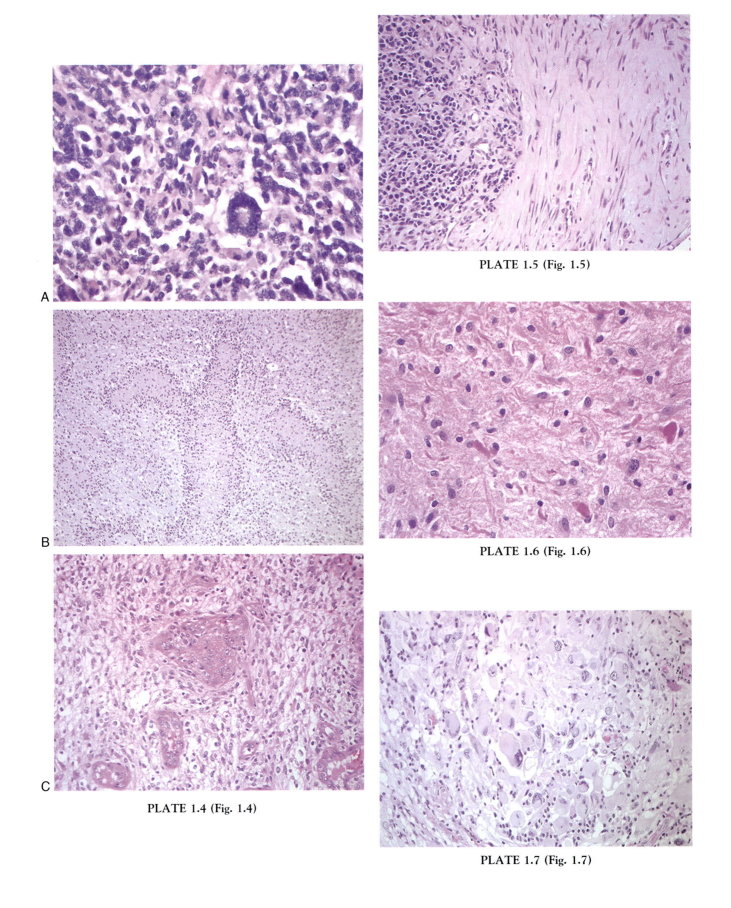

A

B

PLATE 1.4 (Fig. 1.4)

C

PLATE 1.5 (Fig. 1.5)

PLATE 1.6 (Fig. 1.6)

PLATE 1.7 (Fig. 1.7)

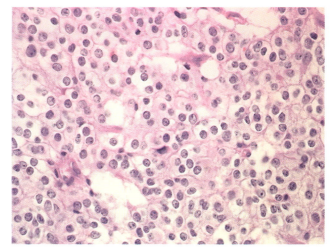

PLATE 1.8 (Fig. 1.8)

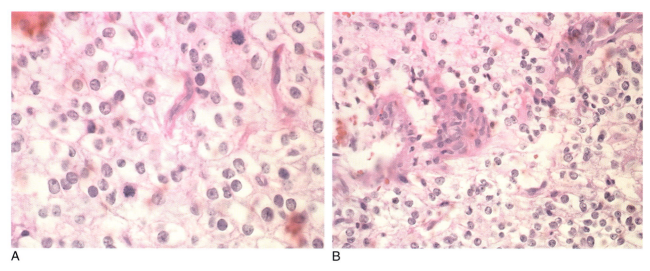

A   B

PLATE 1.9 (Fig. 1.9)

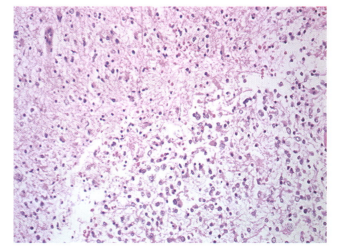

PLATE 1.10 (Fig. 1.10)

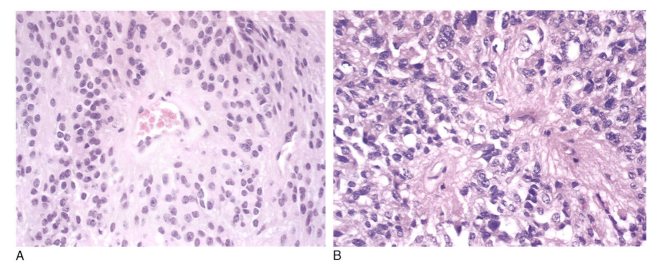

A   B

PLATE 1.11 (Fig. 1.11)

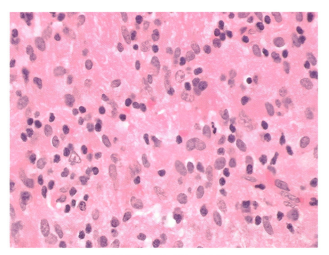

PLATE 1.12 (Fig. 1.12)

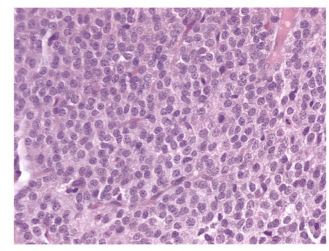

PLATE 1.13 (Fig. 1.13)

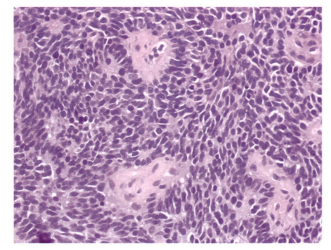

PLATE 1.14 (Fig. 1.14)

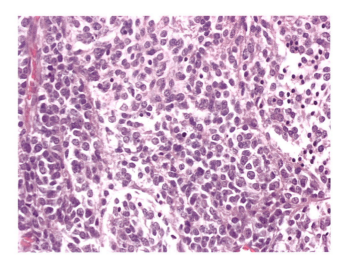

PLATE 1.15 (Fig. 1.15)

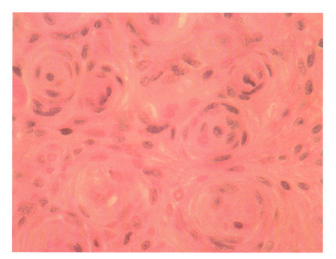

PLATE 1.16 (Fig. 1.16)

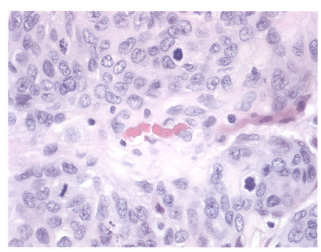

PLATE 1.17 (Fig. 1.17)

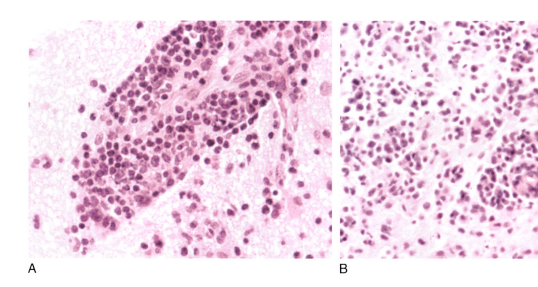

A

B

PLATE 1.18 (Fig. 1.18)

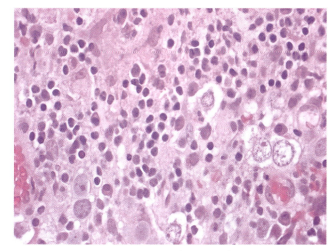

PLATE 1.19 (Fig. 1.19)

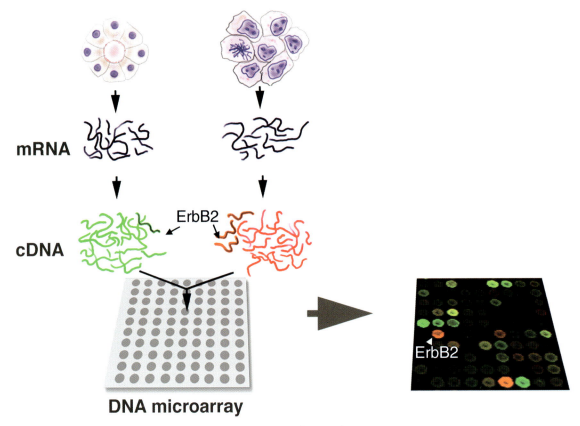

mRNA

cDNA

ErbB2

DNA microarray

ErbB2

PLATE 5.1 (Fig. 5.1)

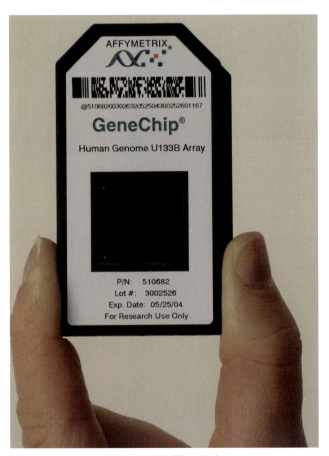

PLATE 5.2 (Fig. 5.2)

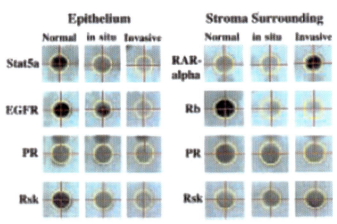

PLATE 5.3 (Fig. 5.3)

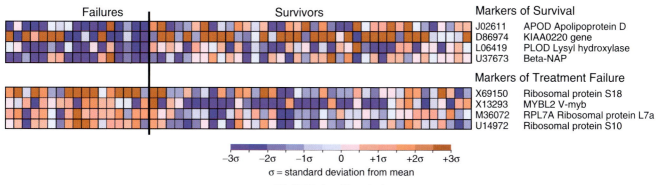

	Markers of Survival	
J02611	APOD	Apolipoprotein D
D86974	KIAA0220 gene	
L06419	PLOD	Lysyl hydroxylase
U37673	Beta-NAP	

	Markers of Treatment Failure	
X69150	Ribosomal protein S18	
X13293	MYBL2	V-myb
M36072	RPL7A	Ribosomal protein L7a
U14972	Ribosomal protein S10	

Failures   Survivors

−3σ   −2σ   −1σ   0   +1σ   +2σ   +3σ

σ = standard deviation from mean

PLATE 5.4 (Fig. 5.4)

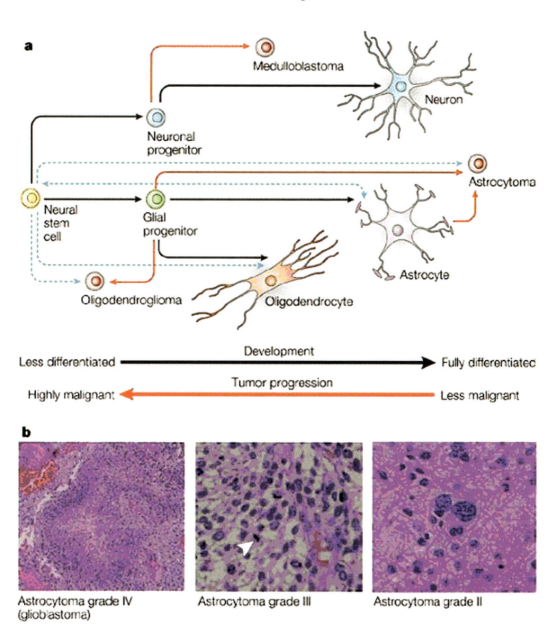

PLATE 5.5 (Fig. 5.5)

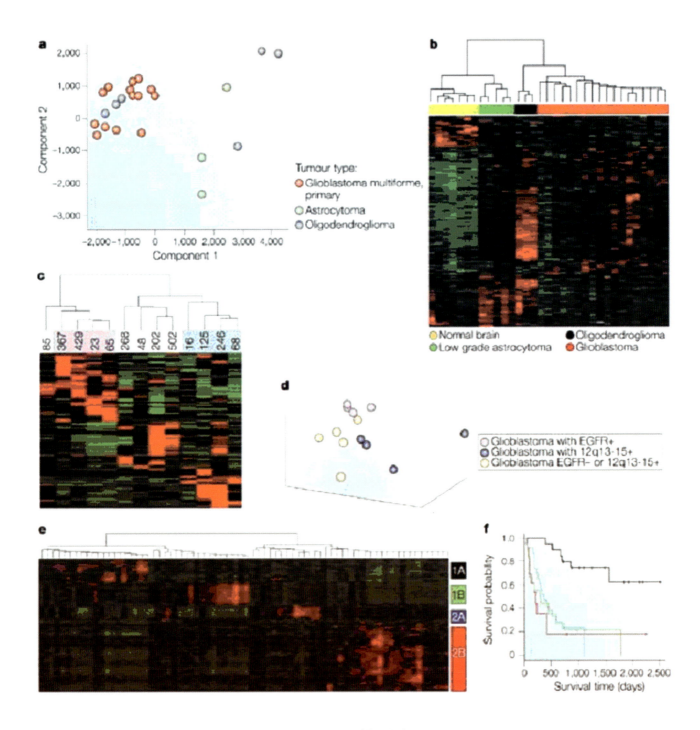

PLATE 5.6 (Fig. 5.6)

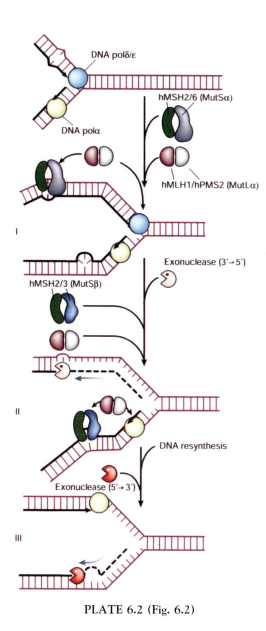

PLATE 6.2 (Fig. 6.2)

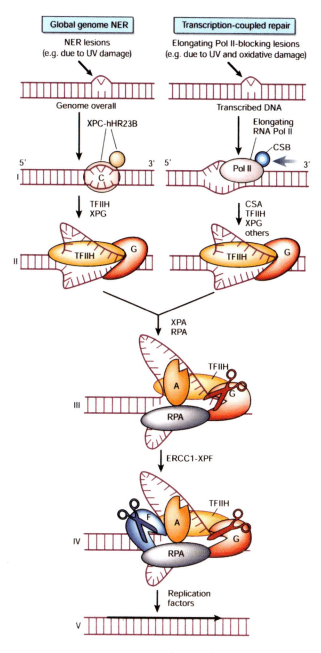

PLATE 6.3 (Fig. 6.3)

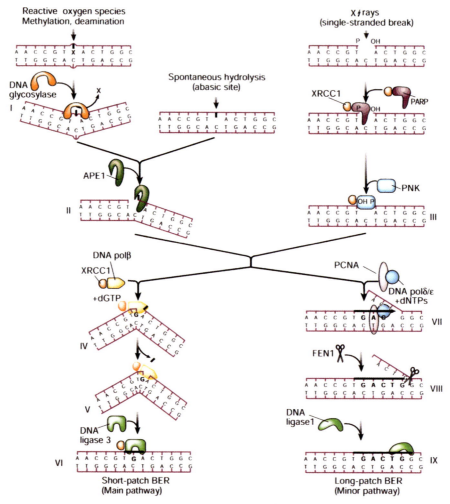

PLATE 6.4 (Fig. 6.4)

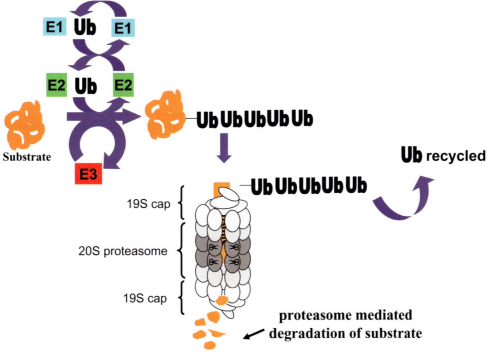

Substrate

19S cap

20S proteasome

19S cap

Ub recycled

proteasome mediated
degradation of substrate

PLATE 9.2 (Fig. 9.2)

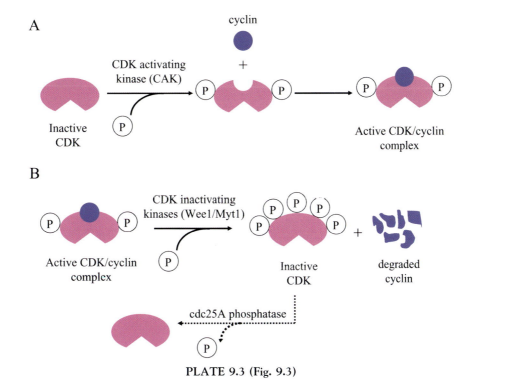

A

cyclin

+

CDK activating
kinase (CAK)

Inactive
CDK

Active CDK/cyclin
complex

B

CDK inactivating
kinases (Wee1/Myt1)

Active CDK/cyclin
complex

Inactive
CDK

+

degraded
cyclin

cdc25A phosphatase

PLATE 9.3 (Fig. 9.3)

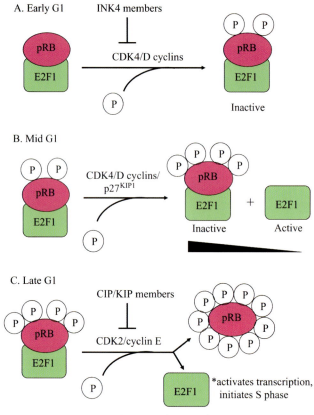

PLATE 9.4 (Fig. 9.4)

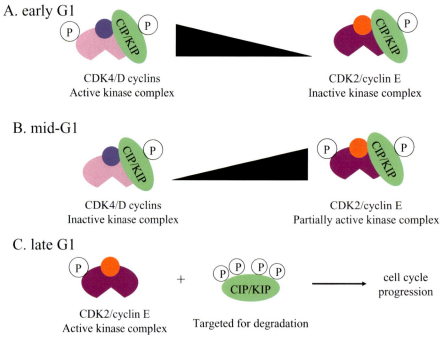

PLATE 9.5 (Fig. 9.5)

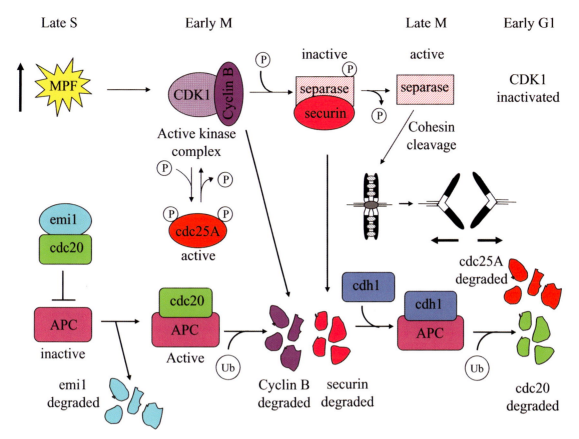

PLATE 9.6 (Fig. 9.6)

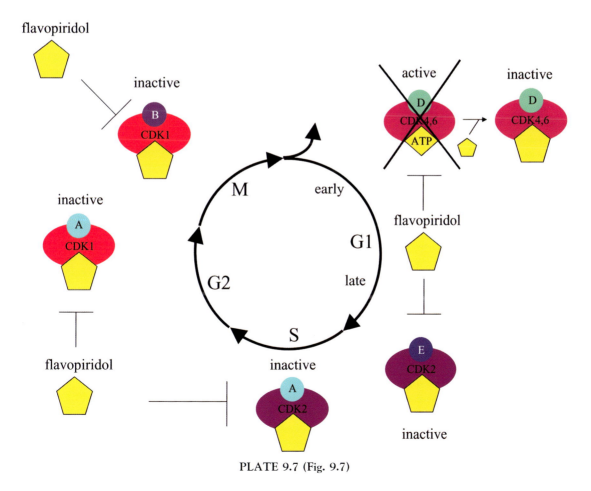

PLATE 9.7 (Fig. 9.7)

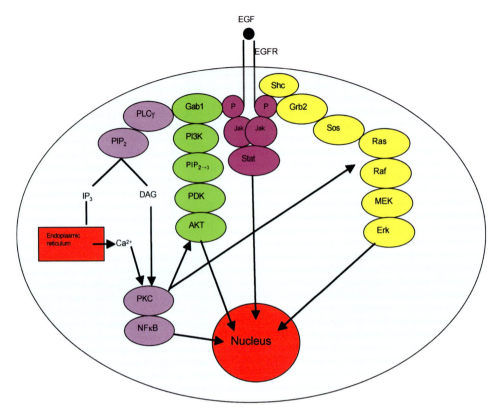

PLATE 11.1 (Fig. 11.1)

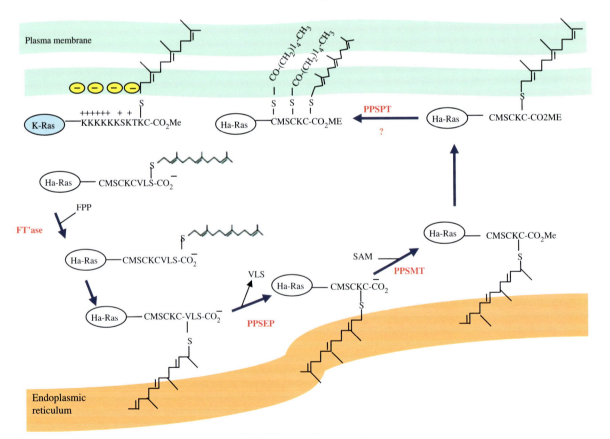

PLATE 12.1 (Fig. 12.1)

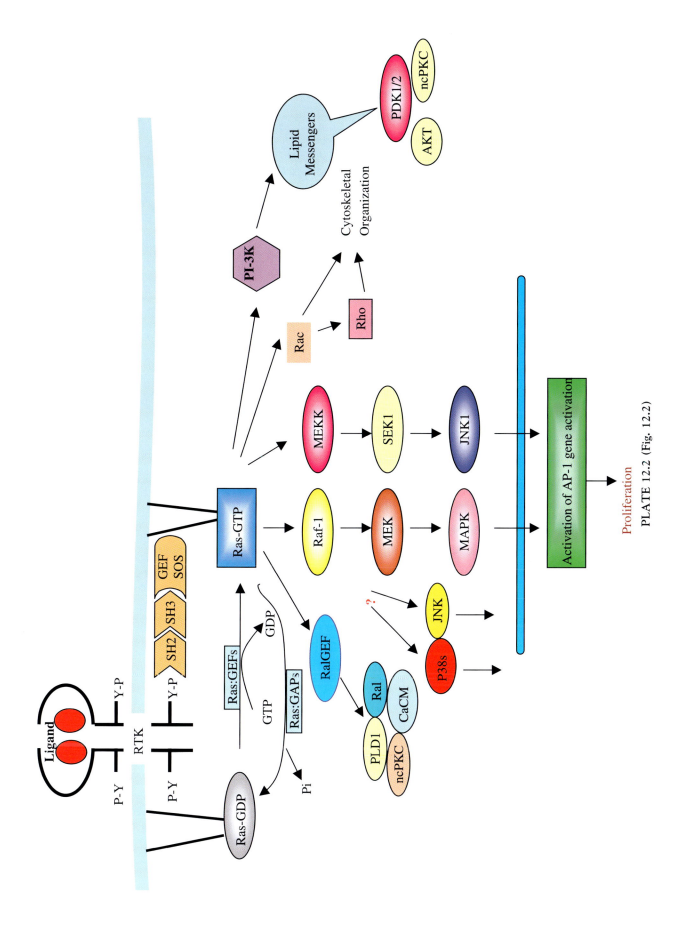

PLATE 12.2 (Fig. 12.2)

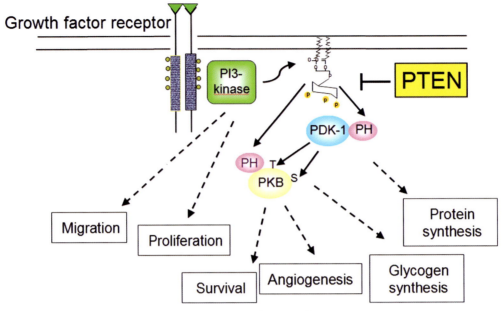

Growth factor receptor

Migration
Proliferation
Survival
Angiogenesis
Glycogen synthesis
Protein synthesis

PI3-kinase
PTEN
PDK-1 PH
PH T
PKB S

PLATE 13.1 (Fig. 13.1)

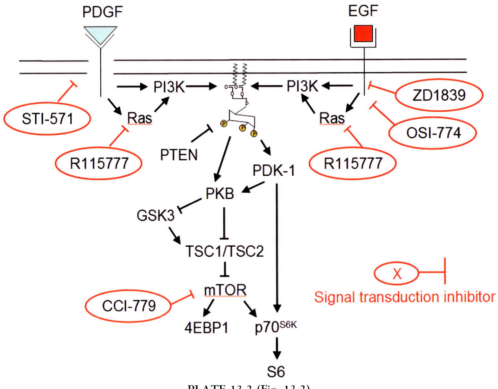

PDGF
EGF

PI3K
PI3K
Ras
Ras
STI-571
ZD1839
OSI-774
R115777
PTEN
PDK-1
R115777
PKB
GSK3
TSC1/TSC2
mTOR
CCI-779
4EBP1
p70^{S6K}
S6

X
Signal transduction inhibitor

PLATE 13.2 (Fig. 13.2)

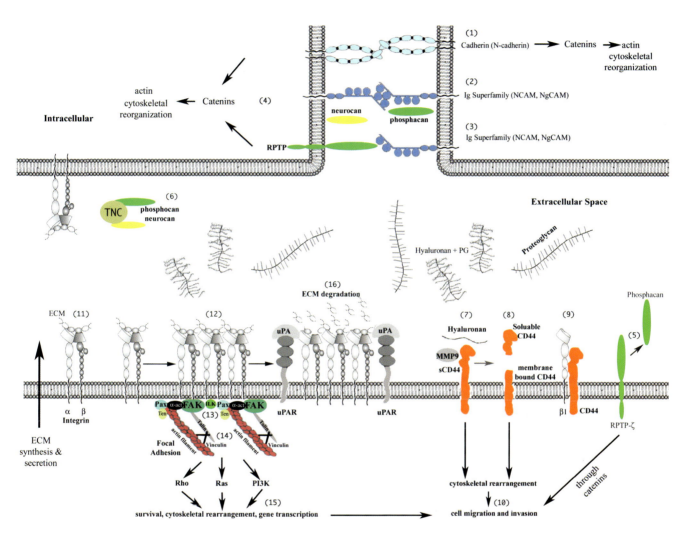

PLATE 14.1 (Fig. 14.1)

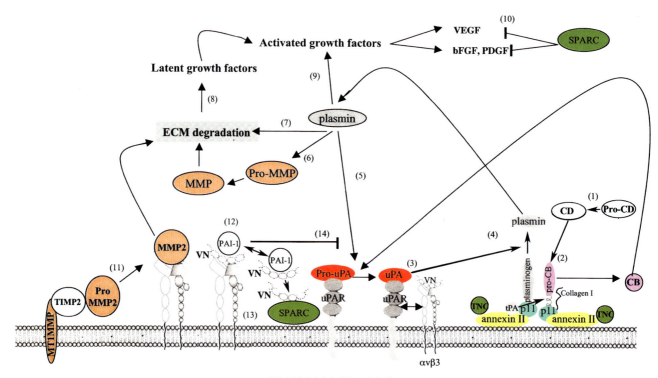

PLATE 14.2 (Fig. 14.2)

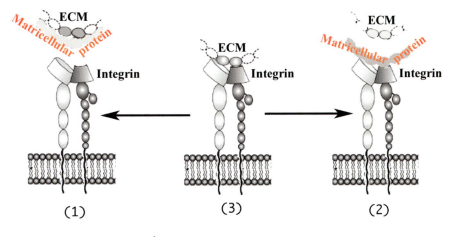

Matricellular Proteins / ECMs / Integrins / Matricellular Proteins

(4)   (5)   (6)   (4)

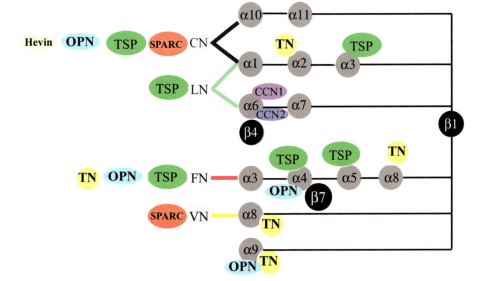

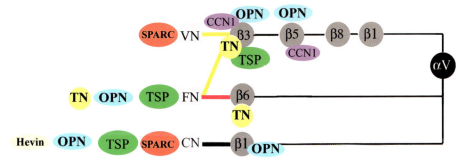

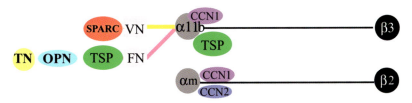

PLATE 14.3 (Fig. 14.3)

Hypoxic Tumor Tissue

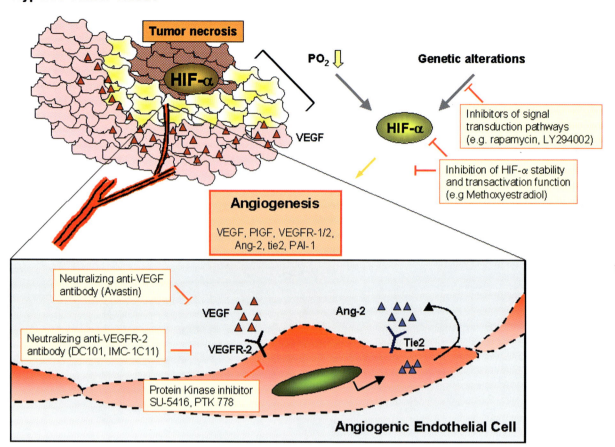

PLATE 15.1 (Fig. 15.1)

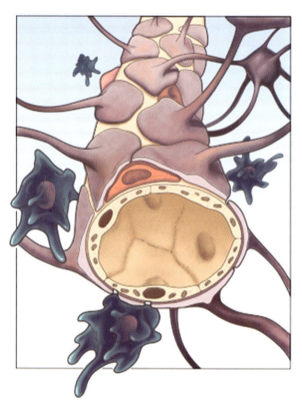

PLATE 16.1 (Fig. 16.1)

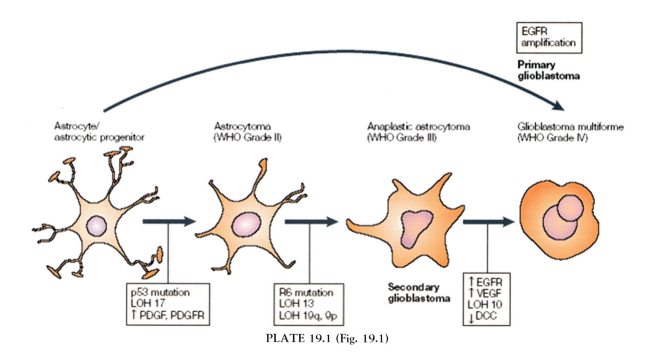

EGFR
amplification

**Primary
glioblastoma**

Astrocyte/
astrocytic progenitor

Astrocytoma
(WHO Grade II)

Anaplastic astrocytoma
(WHO Grade III)

Glioblastoma multiforme
(WHO Grade IV)

p53 mutation
LOH 17
↑ PDGF, PDGFR

R6 mutation
LOH 13
LOH 19q, 9p

**Secondary
glioblastoma**

↑ EGFR
↑ VEGF
LOH 10
↓ DCC

PLATE 19.1 (Fig. 19.1)

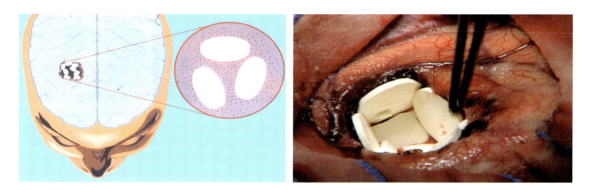

PLATE 19.4 (Fig. 19.4)

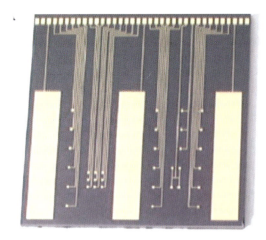
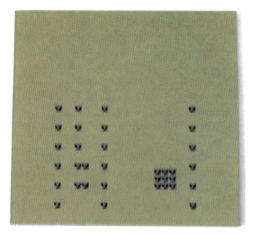

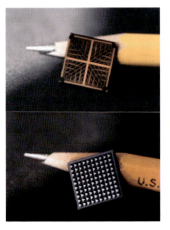

A

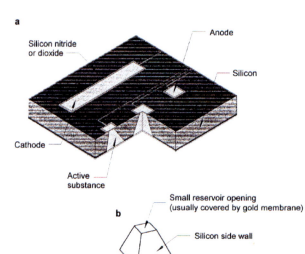

B

PLATE 19.7 (Fig. 19.7)

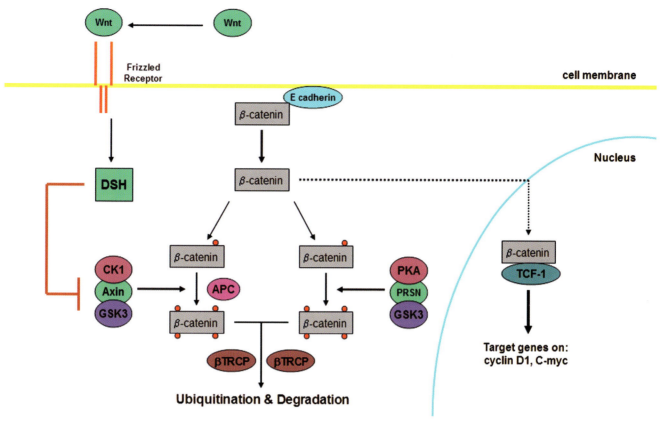

PLATE 29.1 (Fig. 29.1)

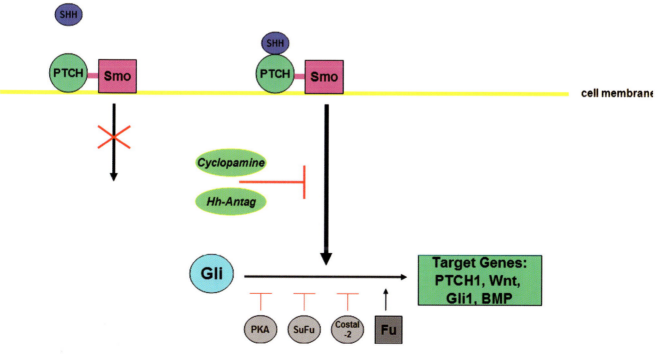

PLATE 29.2 (Fig. 29.2)

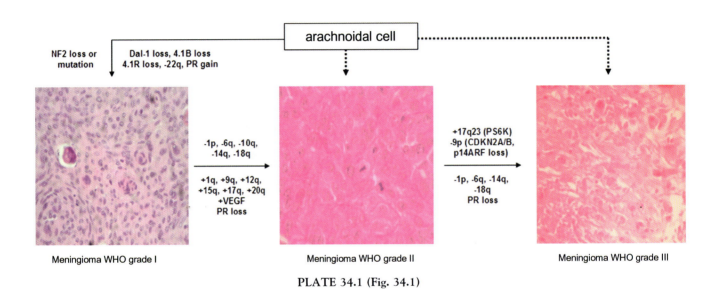

arachnoidal cell

NF2 loss or
mutation

Dal-1 loss, 4.1B loss
4.1R loss, -22q, PR gain

-1p, -6q, -10q,
-14q, -18q

+1q, +9q, +12q,
+15q, +17q, +20q
+VEGF
PR loss

+17q23 (PS6K)
-9p (CDKN2A/B,
p14ARF loss)

-1p, -6q, -14q,
-18q
PR loss

Meningioma WHO grade I

Meningioma WHO grade II

Meningioma WHO grade III

PLATE 34.1 (Fig. 34.1)